Handbook on Experimental Mechanics

Second Revised Edition

Handbook on Experimental Mechanics

Second Revised Edition

Edited by

Albert S. Kobayashi

Editor: Albert S. Kobayashi
Department of Mechanical Engineering
University of Washington
Seattle, WA 98195

This book is printed on acid-free paper.

Library of Congress Cataloging-in-Publication Data

Handbook on experimental mechanics / Society for Experimental
  Mechanics, Inc.; edited by Albert S. Kobayashi.—2nd rev. ed.
      p.   cm.
    Includes biographical references and index.
    ISBN 1-56081-640-6
      1. Mechanics, Applied—Handbooks, manuals, etc.   2. Materials—
  Testing—Handbooks, manuals, etc.   I. Kobayashi, Albert S., 1924–
    II. Society for Experimental Mechanics.
  TA350.H24   1993
  620.1—dc20   93-4523
                                                                   CIP

No responsibility is assumed by the publisher for any injury and/or damage to persons or property as a matter of products liability, negligence or otherwise, or from any use or operation of any methods, products, instructions or ideas contained in the material herein.

Special regulations for readers in the USA

This publication has been registered with the Copyright Clearance Center Inc. (CCC), Salem, Massachusetts. Information can be obtained from the CCC about conditions under which photocopies of parts of this publication may be made in the USA. All other copyright questions, including photocopying outside the USA, should be referred to the publisher.

All rights reserved. No part of this publication may be reproduced, stored in a retrieval system, or transmitted in any form or by any means, electronic, mechanical, photocopying, recording, or otherwise, without the prior written permission of the publisher.

© 1993 by Society for Experimental Mechanics
7 School Street, Bethel, CT 06801 U.S.A.

This work is subject to copyright.
All rights reserved, whether the whole or part of the material is concerned, specifically those of translation, reprinting, re-use of illustrations, broadcasting, reproduction by photocopying machine or similar means, and storage data banks.
Registered names, trademarks, etc. used in this book, even when not specifically marked as such, are not to be considered unprotected by law.

Printed in the United States of America

To Dr. M. Hetényi

Preface

Since the publication of the first edition, considerable progress has been made in automated image processing, greatly reducing the heretofore laborious task of evaluating photoelastic and moiré fringe patterns. It is therefore appropriate to add Chapter 21: "Digital Image Processing" before the final chapter, "Statistical Analysis of Experimental Data." Apart from the new chapter, this second edition is essentially same as the first edition with minor corrections and updating. Exceptions to this are the addition of a section on optical fiber sensors in Chapter 2: "Strain Gages," and extensive additions to Chapter 14, which is retitled "Thermal Stress Analysis," and to Chapter 16: "Experimental Modal Analysis."

To reiterate, the purpose of this handbook is to document the principles involved in experimental mechanics rather than the procedures and hardware, which evolve over time. To that extent, we, the twenty-seven authors, judging from the many appreciative comments which were received upon the publication of the first edition, have succeeded.

Albert S. Kobayashi
April 1993

Preface to the First Edition

The *Handbook on Experimental Stress Analysis*, which was published under the aegis of the Society for Experimental Stress Analysis in 1950, has been the comprehensive and authoritative reference in our field for more than thirty years. Under the able editorship of the late M. Hetényi, 31 authors contributed without compensation 18 chapters and 3 appendices to this handbook. It received international acclaim and brought considerable income to the Society for Experimental Mechanics.

Since 1950, new experimental techniques, such as holography, laser speckle interferometry, geometric moiré, moiré interferometry, optical heterodyning, and modal analysis, have emerged as practical tools in the broader field of experimental mechanics. The emergence of new materials and new disciplines, such as composite materials and fracture mechanics, resulted in the evolution of traditional experimental techniques to new fields such as orthotropic photoelasticity and experimental fracture mechanics. These new developments, together with the explosive uses of on/off-line computers for rapid data processing and the combined use of experimental and numerical techniques, have expanded the capabilities of experimental mechanics far beyond those of the 1950's.

Sensing the need to update the handbook, H. F. Brinson initiated the lengthy process of revising the handbook during his 1978-79 presidency of the Society. Since M. Hetényi could not undertake the contemplated revision at that time, the decision was made to publish a new handbook under a new editor. Opinions ranging from topical coverage to potential contributors were solicited from various SEM members, and after a short respite I was chosen as editor by the ad hoc Handbook Committee chaired by J. B. Ligon. Despite the enormous responsibility, our task was made easier by inheriting the legacy of the Hetényi *Handbook* and the numerous suggestions that were collected by H. Brinson.

The new handbook, appropriately entitled *Handbook on Experimental Mechanics*, is dedicated to Dr. Hetényi. Twenty-five authors have contributed 21 chapters that include, among others, the new disciplines and developments that are mentioned above. The handbook emphasizes the principles of the experimental techniques and de-emphasizes the procedures that evolve with time. I am grateful to the contributors, who devoted many late afterhours in order to meet the manuscript deadlines and to J. B. Ligon who readily provided welcomed assistance during the trying times associated with this editorship.

Albert S. Kobayashi
1987

Contents

1 Mechanical Responses of Materials   1
Satya N. Atluri and Albert S. Kobayashi

   1-0 Introduction and Notation,   1
   1-1 Elementary Theories of Material Responses,   2
   1-2 Boundary and Initial Value Problems in Elasticity,   14
   1-3 Numerical Methods,   20
   1-4 Mechanical Properties of Materials,   28
   1-5 Summary,   33
         References,   34

2 Strain Gages   39
James W. Dally, William F. Riley, and James S. Sirkis

   2-0 Introduction,   39
   2-1 Strain Sensitivity,   40
   2-2 Bonded Resistance Strain Gages,   43
   2-3 Strain-Gage Calibration,   48
   2-4 Behavior of Foil-Type Strain Gages,   51
   2-5 Resistance Strain Gages at Low and Moderate Temperatures,   61
   2-6 Semiconductor Strain Gages,   64
   2-7 Optical Fiber Strain Sensors,   67
   2-8 Summary and Future Trends,   75
         References,   76

3 Strain-Gage Instrumentation and Data Analysis 79
Kenneth G. McConnell and William F. Riley

3-0 Introduction, 79
3-1 Signal-Conditioning Devices, 79
3-2 Recording Instruments for Strain Gages, 93
3-3 Calibration, 103
3-4 Problem Areas, 105
3-5 Data Reduction, 114
3-6 Summary, 117
References, 117

4 Force–Pressure–Motion Measuring Transducers 119
Kenneth G. McConnell and William F. Riley

4-0 Introduction, 119
4-1 Main Elements of a Measuring System, 120
4-2 Common Electronic Circuits Found in Dynamic Instruments, 121
4-3 Displacement and Velocity Measurement: Fixed Reference, 126
4-4 Seismic Instruments: Mechanical Characteristics, 133
4-5 Piezoelectric Sensor Circuits, 140
4-6 Transient Signal Measurements, 146
4-7 Calibration, 152
4-8 Summary and Future Trends, 161
References, 162

5 Photoelasticity 165
Christian P. Burger

5-0 Introduction, 165
5-1 Theory of Photoelasticity, 165
5-2 Two-Dimensional Photoelasticity, 219
5-3 Three-Dimensional Photoelasticity, 233
5-4 Scattered-Light Methods, 239
5-5 Birefringent Coatings, 245
5-6 Computer-Aided Photoelasticity, 248
5-7 Dynamic Photoelasticity, 252
5-8 Photothermoelasticity, 256
5-9 Photoplasticity, 258
5-10 Flow Birefringence, 258
5-11 Orthotropic Photoelasticity, 260
5-12 Major New Developments and Applications of Photoelastic Methods, 261
References, 261

6 Geometric Moiré   267
Vincent J. Parks

  6-0 Introduction,   267
  6-1 In-Plane Moiré,   267
  6-2 Out-of-Plane Moiré,   287
  6-3 Summary,   292
       List of Symbols,   294
       References,   294

7 Moiré Interferometry   297
Daniel Post

  7-0 Introduction,   297
  7-1 Fundamentals,   298
  7-2 Interference of Light,   301
  7-3 The Camera,   308
  7-4 Diffraction Gratings and Diffraction,   309
  7-5 Moiré Interferometry,   313
  7-6 Specimen Gratings,   320
  7-7 Theoretical Limit,   323
  7-8 Optical Systems,   323
  7-9 Quantitative Analysis,   328
  7-10 Bridges,   330
  7-11 Anomalies and Subtraction of Uniform Gradients,   330
  7-12 Alternative Optical Systems,   333
  7-13 Achromatic Systems,   335
  7-14 Vibration Control,   338
  7-15 Sensitivity to Out-of-Plane Motion,   338
  7-16 Coping with the Initial Field,   339
  7-17 Mechanical Differentiation for Normal and Shear Strains,   341
  7-18 Vector Treatment,   347
  7-19 ±45° Gratings,   355
  7-20 Determination of u_z,   355
  7-21 Variations of Moiré Interferometry,   358
  7-22 Data Processing,   359
  7-23 Prospectus,   359
       List of Symbols,   360
       References,   361

8 Holographic and Laser Speckle Interferometry   365
W.F. Ranson, M.A. Sutton, and W.H. Peters

  8-0 Introduction,   365
  8-1 Basic Theory of Fringe Formation,   367
  8-2 Holographic Interferometry Applications,   378
  8-3 Speckle Metrology Applications,   386
  8-4 Image-Processing Applications in Speckle Metrology,   392
  8-5 Summary and Future Trends,   404
       References,   404

9 Shadow Optical Method of Caustics   407
Jörg F. Kalthoff

   9-0 Introduction,   407
   9-1 Physical Principle,   407
   9-2 Quantitative Description of Shadow Optical Images,   412
   9-3 Crack-Tip Caustics,   424
   9-4 Experimental Techniques,   444
   9-5 Applications,   455
   9-6 Summary,   471
         List of Symbols,   473
         References,   474

10 Optical Heterodyning   477
Karl A. Stetson

   10-0 Introduction,   477
   10-1 Basic Technology,   478
   10-2 Holographic Applications,   479
   10-3 Speckle Photogrammetry,   482
   10-4 Concomitant Strain-Measuring Systems,   484
   10-5 Summary and Future Possibilities   488
         References,   489

11 Brittle Coating   491
A. J. Durelli, John Hall, and Ferdi Stern

   Part I Stresses and Strains in Brittle Coatings,   491
         A. J. Durelli
    11-0 Introduction,   491
    11-1 State of Stress in Coating,   491
    11-2 Law of Failure of Brittle Coatings,   494
    11-3 Determination of Ultimate Strength of Coating,   494
    11-4 Analysis of Brittle-Coating Data Using Maximum Tensile Stress Law of Failure,   497
    11-5 Failure Chart of the Coating when σ_1^c Reaches Ultimate Tensile Strength,   500
    11-6 Formula for the Calculation of "Apparent" Stress,   503
    11-7 Cracking of Coating by Direct Load Followed by Perpendicular Cracking by Higher Direct Load,   503
    11-8 Direct Load Cracking Followed by Perpendicular Cracking by Relaxation Load,   506
    11-9 Cracking by Direct Load Followed by Perpendicular Cracking by Highest Direct Load and Highest Relaxation Load at Different Location,   507
    11-10 Determination of Second Principal Stress by Graphical Integration,   509
    11-11 Effect of Refrigeration,   509
    11-12 Applications to the Analysis of Stress Fields,   512

CONTENTS

Part II Application of Brittle Coating, 516
John Hall and Ferdi Stern
11-13 Test Procedures, 516
11-14 Applications, 521
11-15 Summary, 521
References, 525

12 Nondestructive Evaluation 527
James E. Doherty

12-0 Introduction, 527
12-1 Penetrant Inspection, 527
12-2 Magnetic Inspection, 531
12-3 Eddy-Current Inspection, 533
12-4 Radiographic Inspection, 539
12-5 Ultrasonic Inspection, 544
12-6 Summary, 553
References, 554

13 High-Temperature Strain/Displacement Measurement 557
William N. Sharpe, Jr.

13-0 Introduction, 557
13-1 Resistance Strain Gages, 559
13-2 Capacitance Strain Gages, 565
13-3 Electromechanical Extensometers, 570
13-4 Electro-Optical Extensometers, 572
13-5 Optical Techniques, 573
13-6 Summary, 575
References, 576

14 Thermoelastic Stress Analysis 581
Brian J. Rauch and Robert E. Rowlands

14-0 Introduction, 581
14-1 History and Theoretical Foundation, 581
14-2 Equipment, 583
14-3 Test Materials and Methods, 586
14-4 Calibration and Experimental Considerations, 588
14-5 Composite Materials, 590
14-6 Applications, 591
14-7 Summary, 596
List of Symbols 596
References, 597

15 Similitude, Modeling, and Dimensional Analysis   601
Donald F. Young

15-0 Introduction   601
15-1 Dimensions, Units, and Equations,   601
15-2 Theory of Dimensional Analysis,   604
15-3 Some Applications of Dimensional Analysis,   606
15-4 Theory of Models,   611
15-5 Structural Models,   614
15-6 Fluid-Flow Models,   622
15-7 Thermal Models,   625
15-8 True, Adequate, and Distorted Models,   627
15-9 Similarity Laws from Differential Equations,   629
15-10 Summary,   632
List of Symbols,   632
References,   634

16 Experimental Modal Analysis   635
Randall J. Allemang and David L. Brown

16-0 Introduction,   635
16-1 Modal Analysis Theory,   638
16-2 Experimental Modal Analysis Methods,   659
16-3 Modal Data Acquisition,   672
16-4 Modal Parameter Estimation,   711
16-5 Summary,   743
List of Symbols   743
References,   745

17 Hybrid Experimental-Numerical Stress Analysis   751
Albert S. Kobayashi

17-0 Introduction,   751
17-1 Hybrid Experimental-Numerical Stress Analysis,   753
17-2 Elastic Analysis of Structural Components,   754
17-3 Elastic-Plastic Fracture Mechanics,   763
17-4 Dynamic Fracture,   766
17-5 Summary,   781
References,   781

18 Residual Stresses   785
Robert E. Rowlands

18-0 Introduction,   785
18-1 Hole-Drilling Methods,   786
18-2 Hole Boring—Rock Mechanics,   806
18-3 Ultrasonic (Acoustical) Techniques,   809
18-4 Acoustoelasticity,   812

CONTENTS   xvii

    18-5 X-Ray, 815
    18-6 Photomechanical Techniques, 819
    18-7 Numerical Analyses, 821
    18-8 Relieving Residual Stresses, 822
    18-9 Summary, Discussion, and Future, 822
         List of Symbols, 822
         References, 823

19 Composite Materials   829
Isaac M. Daniel

    19-0 Introduction 829
    19-1 Mechanics of Composites, 830
    19-2 Experimental Methods for Composite Materials, 850
    19-3 Composite Material Characterization, 855
    19-4 Biaxial Testing, 865
    19-5 Effects of Stress Concentration, 871
    19-6 Nondestructive Evaluation, 885
    19-7 Summary and Future Trends, 897
         References, 899

20 Experimental Fracture Mechanics   905
C. W. Smith and Albert S. Kobayashi

    20-0 Introduction, 905
    20-1 Theoretical Background, 906
    20-2 Experimental Methods, 919
    20-3 Summary, 960
         List of Symbols, 960
         References, 961

21 Digital Image Processing   969
Yoshiharu Morimoto

    21-0 Introduction, 969
    21-1 Analog and Digital Images, 970
    21-2 Image-Processing Systems (Hardware), 975
    21-3 Image-Processing Programs (Software), 979
    21-4 Algorithms for Image Processing, 981
    21-5 Examples of Image Processing, 997
    21-6 Fringe Pattern Analysis, 997
    21-7 Fourier Transform Moiré and Grid Method, 1008
    21-8 Special Image-Processing Systems (Hardware), 1017
    21-9 Summary and Future Trends, 1025
         References, 1026

22 Statistical Analysis of Experimental Data 1031
James W. Dally

    22-0   Introduction,   1031
    22-1   Characterizing Statistical Distributions,   1032
    22-2   Statistical Distribution Functions,   1035
    22-3   Confidence Intervals for Predictions,   1040
    22-4   Comparison of Means,   1042
    22-5   Statistical Safety Factor,   1043
    22-6   Statistical Conditioning of Data,   1044
    22-7   Regression Analysis,   1044
    22-8   Field Applications of Least-Squares Method,   1048
    22-9   Chi-Square Analysis,   1051
    22-10 Summary,   1053
           Readings on Statistics,   1053

Author Index 1055

Subject Index 1070

CHAPTER

1

Mechanical Responses of Materials

Satya N. Atluri

Georgia Institute of Technology Atlanta, Georgia

Albert S. Kobayashi

University of Washington Seattle, Washington

1-0 Introduction and Notation

In this chapter we consider certain useful fundamental topics from the vast panorama of the analytical mechanics of solids, which, by itself, has been the subject of several handbooks [1–3]. The specific topics that are briefly summarized include (1) elementary theories of material response such as elasticity, dynamic elasticity, viscoelasticity, plasticity, viscoplasticity, and creep; (2) some useful analytical results for boundary/initial value problems in elasticity; (3) numerical methods such as the finite-difference, finite-element, boundary element, and general weighted residuals; (4) a compendium of mechanical properties of materials; and (5) a discussion of possible future trends.

Herein, we employ Cartesian coordinates exclusively. We use a fixed Cartesian system with base vectors $\mathbf{e}_i (i = 1, 2, 3)$. The coordinates of a material particle before and after deformation are x_i and y_i, respectively. The deformation gradient, denoted as F_{ij}, is defined to be

$$\frac{\partial y_i}{\partial x_j} \equiv y_{i,j} \tag{1-1}$$

The displacement components will be denoted by $u_i (= y_i - x_i)$, such that

$$F_{ij} = \delta_{ij} + u_{i,j} \tag{1-2}$$

where δ_{ij} is a Kronecker delta. The Green–Lagrange strain tensor, ε_{ij}, is given by

$$\varepsilon_{ij} = \tfrac{1}{2}[F_{ki}F_{kj} - \delta_{ij}] \equiv \tfrac{1}{2}[u_{i,j} + u_{j,i} + u_{k,i}u_{k,j}] \tag{1-3}$$

When displacements and their gradients are infinitesimal, Eq. (1-3) may be approximated as

$$\varepsilon_{ij} = \tfrac{1}{2}[u_{i,j} + u_{j,i}] \equiv u_{(i,j)} \tag{1-4}$$

A wide variety of other strain measures may be derived [4,5].

Let (da) be a differential area in the *deformed* body, and let n_i be direction cosines of a unit outward normal to (da). If the differential force acting on this area is df_i, the "true stress" or "Cauchy stress," τ_{ij}, is defined from the relation

$$df_i = (da)n_j\tau_{ji} \qquad (1\text{-}5)$$

Thus τ_{ij} is stress per unit area in the deformed body. The "nominal stress" (or the transpose of the so-called first Piola–Kirchhoff stress) t_{ij} and the "second Piola–Kirchhoff stress" S_{ij} are defined through the relations

$$df_i = (dA)N_j t_{ji} \qquad (1\text{-}6)$$

$$= (dA)N_j S_{jk} y_{i,k} \qquad (1\text{-}7)$$

where $(dA)N_j$ is the image in the undeformed configuration, of the oriented area $(da)n_j$ in the deformed configuration. Note that both t_{ji} and S_{ji} are stresses per unit area in the undeformed configuration, and t_{ji} is unsymmetric, while S_{ji} is, by definition, symmetric [4,5]. It should also be noted that a wide variety of other "stress measures" may be derived [4,5].

From the geometric theory of deformation [6], it follows that

$$(da)n_j = (J)(dA)N_k\left(\frac{\partial x_i}{\partial y_j}\right) \qquad (1\text{-}8)$$

where

$$J = \frac{dv}{dV} = \frac{\rho_0}{\rho} \qquad (1\text{-}9)$$

In the above dv is a differential volume in the deformed body, and dV is its image in the undeformed body. From Eqs. (1-5) through (1-9), it follows that

$$t_{ij} = J\frac{\partial x_i}{\partial y_k}\tau_{kj} \quad \text{and} \quad S_{ij} = J\frac{\partial x_i}{\partial y_m}\tau_{mn}\frac{\partial x_j}{\partial y_n} \qquad (1\text{-}10)$$

Another useful stress tensor is the so-called Kirchhoff stress tensor, denoted σ_{ij}, and defined as

$$\sigma_{ij} = J\tau_{ij} \qquad (1\text{-}11)$$

When displacements and their gradients are infinitesimal, $J \simeq 1$, $\partial x_i/\partial y_k \simeq \delta_{ik}$, and so on, and thus the distinction between all the stress measures largely disappears. Hence, in an infinitesimal deformation theory, one may speak of *the* stress tensor σ_{ij}.

1-1 Elementary Theories of Material Responses

The mathematical characterization of behavior of solids is one of the most baffling aspects of solid mechanics. Most of the time, the general behavior of a material defies our mathematical ability to characterize it. The theories discussed below must be viewed simply as *idealizations* of regimes of material response under specific types of loading and/or environmental conditions.

Elasticity

In this idealization, the underlying assumption is that stress is a single-valued function of strain and is independent of the *history* of straining. Also, for such materials, one may define a "potential" for stress in terms of strain, in the form of a "strain-energy density function," denoted here as W.

MECHANICAL RESPONSES OF MATERIALS

It is customary [4] to measure W per unit of *undeformed* volume. In the general case of finite deformations, different stress measures are related to the derivative of W with respect to specific strain measures, labeled as conjugate strain measures. Thus it may be shown [4] that

$$t_{ij} = \frac{\partial W}{\partial F_{ji}} \qquad S_{ij} = \frac{\partial W}{\partial \varepsilon_{ij}} \tag{1-12}$$

Note that for finite deformations, the Cauchy stress does not have a simple conjugate strain measure. When W does not depend on the location of the material particle (in the undeformed conjugation), the material is said to be *homogeneous*. A material is said to be isotropic if W depends on ε_{ij} only through the basic *invariants* of ε_{ij}. These invariants may be defined as

$$I_1 = 3 + 2\varepsilon_{kk} \qquad I_2 = 3 + 4\varepsilon_{kk} + 2(\varepsilon_{kk}\varepsilon_{mm} - \varepsilon_{km}\varepsilon_{km})$$

and

$$I_3 = \det |\delta_{mn} + 2\varepsilon_{mn}| \equiv 1 + 2\varepsilon_{kk} + 2(\varepsilon_{kk}\varepsilon_{mm} - \varepsilon_{km}\varepsilon_{km}) + \tfrac{4}{3}e_{ijk}e_{rst}\varepsilon_{ir}\varepsilon_{js}\varepsilon_{kt} \tag{1-13}$$

where e_{ijk} is equal to $(+1)$ if (ijk) take on values (123) in a cyclic order, is equal to (-1) if in anticyclic order, and is zero if two of the indices take on identical values. Sometimes, invariants J_1, J_2, and J_3, defined as

$$J_1 = (I_1 - 3); \qquad J_2 = (I_2 - 2I_1 + 3); \qquad J_3 = (I_3 - I_2 + I_1 - 1) \tag{1-14}$$

are also used. When the material is isotropic, the Kirchhoff stress tensor may be shown to be the derivative of W with respect to a certain logarithmic strain measure [4]. Also, by decomposing the deformation gradient F_{ij} into pure stretch and rigid rotation [4,6], one may derive certain other useful stress measures, such as the Biot–Luré stress, Jaumann stress, and so on [4].

An isotropic nonlinear elastic material may be characterized, in its behavior at finite deformations, by

$$W = \sum_{r,s,t=0}^{\infty} C_{rst}(I_1 - 3)^r(I_2 - 3)^s(I_3 - 1)^t \qquad C_{000} = 0 \tag{1-15}$$

The ratio of volume change due to deformation, dv/dV, is given, for finite deformations, by I_3. Thus, for incompressible materials, $I_3 = 1$. For incompressible materials, stress is determined from strain only to within a scalar function called the hydrostatic pressure. For such materials, one may define a "modified" strain-energy function, say $\overline{W}$, in which the incompressibility condition, $I_3 = 1$, is introduced as a constraint through the Lagrange multiplier, p. Thus

$$\overline{W} = W(\varepsilon_{ij}) + p(I_3 - 1) \tag{1-16}$$

that is,

$$t_{ij} = \frac{\partial W}{\partial F_{ji}} + p\frac{\partial I_3}{\partial F_{ji}} \qquad S_{ij} = \frac{\partial W}{\partial \varepsilon_{ij}} + p\frac{\partial I_3}{\partial \varepsilon_{ij}} \tag{1-17}$$

For isotropic incompressible elastic materials,

$$W(\varepsilon_{ij}) = W(I_1, I_2) \tag{1-18}$$

Thus Eqs. (1-17) and (1-18) yield, for instance,

$$S_{ij} = 2\frac{\partial W}{\partial I_1}\delta_{ij} + 4[\delta_{ij}(1 + \varepsilon_{mm}) - \delta_{im}\delta_{jn}\varepsilon_{mn}]\frac{\partial W}{\partial I_2}$$
$$+ p[\delta_{ij}(1 + 2\varepsilon_{mm}) - 2\delta_{im}\delta_{jn}\varepsilon_{mn} + 2e_{imn}e_{jrs}\varepsilon_{mr}\varepsilon_{ns}] \tag{1-19}$$

A well-known representation of Eq. (1-18) is due to Mooney [7], where

$$W(I_1, I_2) = C_1(I_1 - 3) + C_2(I_2 - 3) \tag{1-20}$$

So far, we have discussed isotropic materials. In general, for a homogeneous solid, one may write

$$W = E_{ij}\varepsilon_{ij} + \tfrac{1}{2}E_{ijmn}\varepsilon_{ij}\varepsilon_{mn} + \tfrac{1}{3}E_{ijmnrs}\varepsilon_{ij}\varepsilon_{mn}\varepsilon_{rs} + \cdots \quad (1\text{-}21)$$

We use, for convenience, S_{ij} and ε_{ij} as conjugate measures of stress and strain. Since S_{ij} and ε_{ij} are both symmetric, one must have

$$\begin{array}{l} E_{ij} = E_{ji} \quad E_{ijmn} = E_{jinm} = E_{ijnm} = E_{mnij} \\ E_{ijmnrs} = E_{jimnrs} = E_{ijnmrs} = E_{ijmnsr} = \cdots = E_{rsijmn} = \cdots \end{array} \quad (1\text{-}22)$$

Thus

$$S_{ij} = E_{ij} + E_{ijmn}\varepsilon_{mn} + E_{ijmnrs}\varepsilon_{mn}\varepsilon_{rs} + \cdots \quad (1\text{-}23)$$

Henceforth, we will consider the case when deformations are *infinitesimal*. Thus $\varepsilon_{ij} \approx (1/2)[u_{i,j} + u_{j,i}]$. Further, the differences in the definitions of various stress measures disappear, and one may speak of *the* stress σ_{ij}. Thus Eq. (1-23) may be rewritten as

$$\sigma_{ij} = E_{ij} + E_{ijmn}\varepsilon_{mn} + E_{ijmnrs}\varepsilon_{mn}\varepsilon_{rs} + \cdots \quad (1\text{-}24)$$

A material is said to be linearly elastic if a linear approximation of Eq. (1-24) is valid for the magnitude of strains under consideration. For such a material,

$$\sigma_{ij} = E_{ij} + E_{ijmn}\varepsilon_{mn} \quad (1\text{-}25)$$

The stress at zero strain (i.e., E_{ij}) most commonly is due to temperature variation from a reference state. The simplest assumption in thermal problems is to set

$$E_{ij} = -\beta_{ij}\Delta T$$

where $\Delta T\ [\ = (T - T_0)]$ is the temperature increment from the reference value T_0.

For an anisotropic linear elastic solid, in view of the symmetries in Eq. (1-23), one has 21 independent elastic constants E_{ijkl} and six constants β_{ij}. In the case of isotropic linear elastic materials, an examination of Eqs. (1-13) through (1-15) reveals that the number of independent elastic constants E_{ijkl} is reduced to two, and the number of independent β's to one. Thus, for an isotropic elastic material,

$$\sigma_{ij} = \lambda\varepsilon_{kk}\delta_{ij} + 2\mu\varepsilon_{ij} - \beta\,\Delta T\delta_{ij} \quad (1\text{-}26a)$$

where λ and μ are Lamé parameters, which are related to the Young's modulus E and Poisson's ratio ν, through

$$\lambda = \frac{E\nu}{(1 + \nu)(1 - 2\nu)} \qquad \mu = \frac{E}{2(1 + \nu)}$$

The bulk modulus K is defined as

$$K = \frac{3\lambda + 2\mu}{3}$$

The inverse of Eq. (1-26a) is

$$\varepsilon_{ij} = -\frac{\nu}{E}\sigma_{mm}\delta_{ij} + \frac{1 + \nu}{E}\sigma_{ij} + \alpha\,\Delta T\delta_{ij} \quad (1\text{-}26b)$$

where β and α are related through

$$\beta = \frac{E\alpha}{1 - 2\nu} \quad (1\text{-}26c)$$

and α is the linear coefficient of thermal expansion.

The state of "plane strain" is characterized by the conditions of validity that $u_\alpha = u_\alpha(X_\beta)$, α, $\beta = 1, 2$, and $u_3 = 0$. Thus $\varepsilon_{3i} = 0$, $i = 1, 2, 3$. In plane strain,

$$\varepsilon_{11} = \frac{1-v^2}{E}\left(\sigma_{11} - \frac{v}{1-v}\sigma_{22}\right) + \alpha(1+v)\Delta T \tag{1-27a}$$

$$\varepsilon_{22} = \frac{1-v^2}{E}\left(\sigma_{22} - \frac{v}{1-v}\sigma_{11}\right) + \alpha(1+v)\Delta T \tag{1-27b}$$

$$\varepsilon_{12} = \frac{1+v}{E}\sigma_{12} \tag{1-27c}$$

$$\sigma_{33} = v(\sigma_{11} + \sigma_{22}) - \alpha E \Delta T \tag{1-27d}$$

The state of "plane stress" is characterized by the conditions of validity that $\sigma_{3k} = 0$, $k = 1, 2, 3$. Here one has

$$\varepsilon_{11} = \frac{1}{E}(\sigma_{11} - v\sigma_{22}) + \alpha \Delta T \tag{1-28a}$$

$$\varepsilon_{22} = \frac{1}{E}(\sigma_{22} - v\sigma_{11}) + \alpha \Delta T \tag{1-28b}$$

$$\varepsilon_{12} = \frac{1+v}{E}\sigma_{12} \tag{1-28c}$$

$$\varepsilon_{33} = -\frac{v}{E}(\sigma_{11} + \sigma_{22}) + \alpha \Delta T \tag{1-28d}$$

Note that in Eqs. (1-27c) and (1-28c), ε_{12} is the tensor component of strain. Sometimes it is customary to use the engineering strain component $\gamma_{12} = 2\varepsilon_{12}$. Note also that in the case of a linear elastic material, the strain-energy density W is given by

$$W = \tfrac{1}{2}\sigma_{ij}\varepsilon_{ij} = \tfrac{1}{2}(\sigma_{11}\varepsilon_{11} + \sigma_{22}\varepsilon_{22} + \sigma_{33}\varepsilon_{33} + 2\varepsilon_{12}\sigma_{12} + 2\varepsilon_{13}\sigma_{13} + 2\varepsilon_{23}\sigma_{23}) \tag{1-29}$$
$$\equiv \tfrac{1}{2}(\sigma_{11}\varepsilon_{11} + \sigma_{22}\varepsilon_{22} + \sigma_{33}\varepsilon_{33} + \gamma_{12}\sigma_{12} + \gamma_{23}\sigma_{23} + \gamma_{13}\sigma_{13})$$

From Eq. (1-26b) it is seen that for linear elastic isotropic materials,

$$\varepsilon_{kk} = \frac{1-2v}{E}\sigma_{mm} + 3\alpha \Delta T \equiv \frac{\sigma_{mm}}{3k} + 3\alpha \Delta T \tag{1-30}$$

When the bulk modulus $k \to \infty$ (or $v \to \tfrac{1}{2}$), it is seen that $\varepsilon_{kk} \to 3\alpha \Delta T$ and is independent of mean stress. Note also from Eq. (1-26c) that $\beta \to \infty$ as $v \to \tfrac{1}{2}$. For such materials, the mean stress is indeterminate from deformation alone. In this case, the relation Eq. (1-26a) is replaced by

$$\sigma_{ij} = -p\delta_{ij} + 2\mu\varepsilon'_{ij} \tag{1-31a}$$

with the constraint

$$\varepsilon_{kk} = 3\alpha \Delta T \tag{1-31b}$$

where p is the "hydrostatic pressure" and ε'_{ij} the deviator of the strain. Note that the strain-energy density of a linear elastic incompressible material is

$$W = \mu\varepsilon'_{ij}\varepsilon'_{ij} - p(\varepsilon_{kk} - 3\alpha \Delta T) \tag{1-32}$$

wherein p acts as a Lagrange multiplier to enforce Eq. (1-31b).

Viscoelasticity

A linear elastic solid, by definition, is one that has the memory of only its unstrained state. Viscoelastic materials are those for which the current deformation is a function of the entire history of loading, and conversely, the current stress is a function of the entire history of straining. Linear viscoelastic materials are those for which the hereditary relations above are expressed in terms of linear superposition integrals, which, for infinitesimal strains, take the forms

$$\sigma_{ij}(t) = \varepsilon_{kl}(0^+)E_{ijkl}(t) + \int_0^t E_{ijkl}(t - \tau)\frac{\partial \varepsilon_{kl}}{\partial \tau} d\tau \qquad (1\text{-}33a)$$

$$\equiv E_{ijkl}(0^+)\varepsilon_{kl}(t) + \int_0^t \varepsilon_{kl}(t - \tau)\frac{\partial E_{ijkl}}{\partial \tau} d\tau \qquad (1\text{-}33b)$$

In the above it has been assumed that $\sigma_{kl} = \varepsilon_{kl} = 0$ for $t < 0$ and that $\varepsilon_{ij}(t)$ and $E_{ijkl}(t)$ are piecewise continuous. $E_{ijkl}(t)$ is called the relaxation tensor. Conversely, one may write

$$\varepsilon_{ij}(t) = \sigma_{kl}(0^+)C_{ijkl}(t) + \int_0^t C_{ijkl}(t - \tau)\frac{\partial \sigma_{kl}}{\partial \tau} d\tau \qquad (1\text{-}34)$$

where $C_{ijkl}(t)$ is called the creep compliance tensor.

For isotropic linear viscoelastic materials,

$$E_{ijkl}(t) = \mu(t)[\delta_{ik}\delta_{jl} + \delta_{il}\delta_{jk}] + \lambda(t)\delta_{ij}\delta_{kl} \qquad (1\text{-}35)$$

where $\mu(t)$ is the shear-relaxation modulus and $B(t) \equiv [3\lambda(t) + 2\mu(t)]/3$ is the bulk relaxation modulus. It is often assumed that $B(t)$ is independent of time; thus the material is assumed to have purely elastic volumetric change. The definition of a Poisson's ratio is somewhat ambiguous in viscoelasticity. For instance, in a uniaxial tension test, let the stress be σ_{11}, the longitudinal strain ε_{11}, and the lateral strain ε_{22}. For creep at constant stress, the ratio of lateral contraction, denoted as $v_c(t)$, is $v_c(t) \equiv -\varepsilon_{22}(t)/\varepsilon_{11}(t)$. For relaxation at constant strain, $v_R(t) \equiv -\varepsilon_{22}(t)/\varepsilon_{11}$. One may write

$$B(t) = \frac{2\mu(t)[1 + v_R(t)]}{1 - 2v_R(t)}$$

It is often convenient to assume that v_R = constant, which renders $B(t)$ similar to $\mu(t)$. In this case, the material is said to have a constant Poisson's ratio.

The Laplace transforms of Eqs. (1-33) and (1-34) may be written as

$$\bar{\sigma}_{ij}(p) = p\bar{E}_{ijkl}(p)\bar{\varepsilon}_{kl}(p) \qquad (1\text{-}36a)$$

and

$$\bar{\varepsilon}_{ij}(p) = p\bar{C}_{ijkl}(p)\bar{\sigma}_{kl}(p) \qquad (1\text{-}36b)$$

where $(\bar{\cdot})$ is the Laplace transform of $(\cdot)$ and p is the Laplace variable. From Eqs. (1-36a) and (1-36b), it follows that

$$p^2\bar{E}_{ijkl}\bar{C}_{klmn} = \delta_{im}\delta_{nj} \qquad (1\text{-}37)$$

It is also customary to represent the relaxation moduli, $\mu(t)$ and $B(t)$, in series form, as

$$\mu(t) = \mu_0 + \sum_{m=1}^{M} \mu_m \exp(-\alpha_m t)$$

$$B(t) = B_0 + \sum_{m=1}^{M} B_m \exp(-\beta_m t)$$

Plasticity

Most structural metals behave elastically for only very small values of strain, after which the materials yield. During yielding, the apparent instantaneous "tangent" moduli of the material are reduced from those in the prior elastic state. Removal of load causes the material to unload elastically with the initial elastic moduli. Such materials are usually labeled as "elastic–plastic." Observed phenomena in the behavior of such materials include the so-called Bauschinger effect (a specimen initially loaded in tension often yields at a much reduced stress when reloaded in compression), cyclic hardening,[1] and so on. Various levels of sophistication of elastic–plastic constitutive theories are necessary to incorporate some or all of these observed phenomena. Here we give a rather cursory review of this still burgeoning literature.

In most theories of metal plasticity, it is assumed that plastic deformations are entirely distortional in nature, and that volumetric strain is purely elastic. The elastic limit of the material is assumed to be specified by a "yield function," which is a function of stress (or of strain, but most commonly of stress). Since plastic deformation is assumed to be insensitive to hydrostatic pressure, the yield function is assumed, in general, to depend on the stress deviator, σ'_{ij} ($\sigma'_{ij} = \sigma_{ij} - \frac{1}{3}\sigma_{mm}\delta_{ij}$). The commonly used yield functions are:

von Mises:

$$f(\sigma_{ij}) = J_2 - k^2 = 0 \qquad J_2 = \tfrac{1}{2}\sigma'_{ij}\sigma'_{ij} \tag{1-38}$$

Tresca:

$$f(\sigma_{ij}) = [(\sigma_1 - \sigma_2)^2 - 4k^2][(\sigma_2 - \sigma_3)^2 - 4k^2] \times [(\sigma_1 - \sigma_3)^2 - 4k^2] = 0 \tag{1-39}$$

In Eqs. (1-38) and (1-39), k may be a function of plastic strain. Both Eqs. (1-38) and (1-39) imply the equality of the tensile and compressive yield stresses at all times—the so-called isotropic hardening. Thus the yield surface expands while its center remains fixed in the stress space. The relation of k to test data follows: (1) in Eq. (1-38), $k = \bar{\sigma}/\sqrt{3}$, where $\bar{\sigma}$ is the yield stress in uniaxial tension, which may be a function of plastic strain for strain-hardening materials, or $k = \bar{\tau}/\sqrt{2}$, where $\bar{\tau}$ is the yield stress in pure shear; (2) Eq. (1-39), $k = \bar{\sigma}/2$ or $\bar{\tau}$. Experimental data appear to favor the use of the Mises condition [8,9].

To account for Bauschinger effect, one may use the representation of the yield surface:

$$f(\sigma_{ij} - \alpha_{ij}) = 0 = \tfrac{1}{2}(\sigma'_{ij} - \alpha'_{ij})(\sigma'_{ij} - \alpha'_{ij}) - \tfrac{1}{3}\bar{\sigma}^2 = 0 \tag{1-40}$$

where α'_{ij} represents the center of the yield surface in the deviatoric stress space. The evolution equations suggested for α_{ij} by Prager [10] and Ziegler [11], respectively, are:

$$d\alpha'_{ij} = c\, d\varepsilon^p_{ij} \tag{1-41}$$

and

---

[1] When a specimen is subjected to cyclic straining of amplitude $-\varepsilon$ to $+\varepsilon$, the stress for the same value of tensile strain ε, prior to unloading, increases monotonically with the number of cycles and eventually saturates.

$$d\alpha_{ij} = d\mu(\sigma_{ij} - \alpha_{ij}) \tag{1-42}$$

In the above, $(\cdot)'$ denotes the deviatoric part of the second-order tensor $(\cdot)$ and an additive decomposition of differential strain into elastic and plastic parts (i.e., $d\varepsilon_{ij} = d\varepsilon^e_{ij} + d\varepsilon^p_{ij}$) has been used.

Elastic processes (with no increase in plastic strain) and plastic processes (with increase in plastic strain) are defined [9] as follows:

Elastic process:

$$f < 0 \quad \text{or} \quad f = 0 \quad \text{and} \quad \frac{\partial f}{\partial \sigma_{ij}} d\sigma_{ij} \leq 0 \tag{1-43}$$

Plastic process:

$$f = 0 \quad \text{and} \quad \frac{\partial f}{\partial \sigma_{ij}} d\sigma_{ij} > 0 \tag{1-44}$$

The flow rule for strain-hardening materials, arising out of consideration of stress working in a cyclic process and stability of the process—often referred to as Drucker's [12] postulates—is given by

$$d\varepsilon^p_{ij} = d\lambda \frac{\partial f}{\partial \sigma_{ij}} \tag{1-45}$$

The scalar $d\lambda$ is determined from the fact that $df = 0$ during a plastic process—the so-called consistency condition. Using the isotropic-hardening J_2 flow theory for which f is given in Eq. (1-38), this consistency condition then leads to

$$d\varepsilon^p_{ij} = \frac{9}{4} \sigma'_{ij} \frac{\sigma'_{mn} d\sigma_{mn}}{H'\bar{\sigma}^2} \tag{1-46}$$

where H' is the slope of the stress versus plastic strain curve in uniaxial tension (or more correctly, the slope of the true-stress versus logarithmic-strain curve in pure tension). On the other hand, for the linear kinematic hardening rules of Prager, given in Eqs. (1-40) and (1-41), the consistency condition leads to

$$d\varepsilon^p_{ij} = \frac{3}{2c\bar{\sigma}^2} [(\sigma'_{mn} - \alpha'_{mn})d\sigma_{mn}](\sigma'_{ij} - \alpha'_{ij}) \tag{1-47}$$

For pressure-insensitive plasticity, the stress–strain laws may be written as

$$d\sigma_{mn} = (3\lambda + 2\mu) d\varepsilon_{mm} \tag{1-48a}$$

$$d\sigma'_{ij} = 2\mu(d\varepsilon'_{ij} - d\varepsilon^p_{ij}) \tag{1-48b}$$

Choosing a parameter α such that $\alpha = 1$ when $d\varepsilon^p_{ij} \neq 0$ and $\alpha = 0$ when $d\varepsilon^p_{ij} = 0$, we have

$$d\sigma'_{ij} = 2\mu\left(d\varepsilon'_{ij} - \frac{9}{4}\sigma'_{ij}\frac{\sigma'_{mn}d\sigma_{mn}}{H'\bar{\sigma}^2}\alpha\right) \tag{1-49}$$

for isotropic hardening, and

$$d\sigma'_{ij} = 2\mu\left[d\varepsilon'_{ij} - \frac{3}{2c\bar{\sigma}^2}(\sigma'_{mn} - \alpha'_{mn})\,d\sigma_{mn}(\sigma'_{ij} - \alpha'_{ij})\alpha\right] \qquad (1\text{-}50)$$

for Prager's linear kinematic hardening. Taking the tensor product of both sides of Eq. (1-49) with σ'_{ij} (and noting that $\sigma'_{mn}d\sigma_{mn} \equiv \sigma'_{mn}d\sigma'_{mn}$ by definition), we have

$$d\sigma'_{ij}\sigma'_{ij} = 2\mu\left(d\varepsilon'_{ij}\sigma'_{ij} - \frac{3\alpha}{2H'}\sigma'_{mn}d\sigma'_{mn}\right) \qquad (1\text{-}51a)$$

or

$$d\sigma'_{ij}\sigma'_{ij} = \frac{2\mu H'}{H' + 3\mu}\,d\varepsilon'_{ij}\sigma'_{ij} \qquad \text{when } \alpha = 1 \qquad (1\text{-}51b)$$

Use of Eq. (1-51b) in (1-49) results in

$$d\sigma'_{ij} = 2\mu\left[d\varepsilon'_{ij} - 2\mu\frac{9\alpha}{4(H' + 3\mu)\bar{\sigma}^2}\sigma'_{ij}\sigma'_{mn}d\varepsilon'_{mn}\right] \qquad (1\text{-}52)$$

Combining Eqs. (1-48a) and (1-52), one may write the *isotropic-hardening* elastic–plastic constitutive law in differential form as

$$d\sigma_{ij} = \left[2\mu\delta_{im}\delta_{jn} + \lambda\delta_{ij}\delta_{mn} - 2\mu\frac{9\alpha\mu}{(2H' + 6\mu)\bar{\sigma}^2}\sigma'_{ij}\sigma'_{mn}\right]d\varepsilon_{mn} \qquad (1\text{-}53)$$

wherein $\sigma'_{mn}d\varepsilon'_{mn} \equiv \sigma'_{mn}d\varepsilon_{mn}$ has been noted. Similarly, by taking the tensor product of Eq. (1-50) with $(\sigma'_{ij} - \alpha'_{ij})$ and repeating steps analogous to those in Eqs. (1-51) through (1-53), one may write the kinematic-hardening elastoplastic constitutive law as

$$d\sigma_{ij} = \left[2\mu\delta_{im}\delta_{jn} + \lambda\delta_{ij}\delta_{mn} - 2\mu\frac{3\mu}{(c + 2\mu)\bar{\sigma}^2}(\sigma'_{ij} - \alpha'_{ij})(\sigma'_{mn} - \alpha'_{mn})\right]d\varepsilon_{mn} \qquad (1\text{-}54)$$

Note that all the developments above are restricted to the infinitesimal strain and small deformation case. Discussion of finite deformation[2] plasticity is beyond the scope of this work. Here the objectivity of stress–strain relation plays an important role. We refer the reader to Refs. 5, 13, and 14.

We now briefly examine the elastic–plastic stress–strain relations, in the isotropic-hardening case, for plane strain and plane stress, leaving it to the reader to derive similar relations for kinematic hardening. In the plane-strain case, $d\varepsilon_{3n} = 0$, $n = 1, 2, 3$. Using this in Eq. (1-53), we have

$$d\sigma_{\alpha\beta} = \left[2\mu\delta_{\alpha\theta}\delta_{\beta\nu} + \lambda\delta_{\alpha\beta}\delta_{\theta\nu} - 2\mu\frac{9\alpha\mu}{(2H' + 6\mu)\bar{\sigma}^2}\sigma'_{\alpha\beta}\sigma'_{\theta\nu}\right]d\varepsilon_{\theta\nu} \qquad (1\text{-}55)$$

and

$$d\sigma_{33} = \lambda d\varepsilon_{\theta\theta} - 2\mu\frac{9\alpha\mu}{(2H' + 6\mu)\bar{\sigma}^2}\sigma'_{33}\sigma'_{\theta\nu}\,d\varepsilon_{\theta\nu} \qquad \alpha, \beta, \theta, \nu = 1, 2) \qquad (1\text{-}56)$$

---

[2] Even in small deformation plasticity, if the current tangent moduli of the stress–strain relation are of the same order of magnitude as the current stress, one must use an objective stress rate, instead of the material rate $d\sigma_{ij}$, in Eq. (1-54).

Note that in the plane-strain case, σ_{33}, as integrated from Eq. (1-56), enters the yield condition. In the plane-stress case, the stress–strain relation is somewhat tedious.

Noting that in the plane-stress case, $d\varepsilon_{\alpha\beta} = d\varepsilon^e_{\alpha\beta} + d\varepsilon^p_{\alpha\beta}$, one may, using the elastic strain–stress relations as given in Eq. (1-33), write

$$d\varepsilon_{\alpha\beta} = d\varepsilon^e_{\alpha\beta} + d\varepsilon^p_{\alpha\beta} = \frac{1}{2\mu}\left(d\sigma_{\alpha\beta} - \frac{\nu}{1+\nu}d\sigma_{\theta\theta}\delta_{\alpha\beta}\right) + d\varepsilon^p_{\alpha\beta} \quad (1\text{-}57)$$

$$= \frac{1}{2\mu}\left(d\sigma_{\alpha\beta} - \frac{\nu}{1+\nu}d\sigma_{\theta\theta}\delta_{\alpha\beta}\right) + \sigma'_{\alpha\beta}(\sigma'_{\theta\nu}d\sigma_{\theta\nu})\frac{9}{4H'\bar\sigma^2} \quad (1\text{-}58)$$

wherein Eq. (4-16) has been used. Equation (1-58) may be inverted to obtain $d\sigma_{\alpha\beta}$ in terms of $d\varepsilon_{\theta\nu}$. This 3×3 matrix inversion may be carried out, leading to the result [15]

$$\frac{Q}{E}d\sigma_{11} = [(\sigma'_{22})^2 + 2P]\,d\varepsilon_{11} + (-\sigma'_{11}\sigma'_{22} + 2\nu P)\,d\varepsilon_{22} - \frac{\sigma'_{11} + \nu\sigma'_{22}}{1+\nu}\sigma_{12}2\,d\varepsilon_{12}$$

$$\frac{Q}{E}d\sigma_{22} = (-\sigma'_{11}\sigma'_{22} + 2\nu P)\,d\varepsilon_{11} + [(\sigma'_{11})^2 + 2P]\,d\varepsilon_{22} - \frac{\sigma'_{22} + \nu\sigma'_{11}}{1+\nu}\sigma_{12}2\,d\varepsilon_{12}$$

$$\frac{Q}{E}d\sigma_{12} = -\left(\frac{\sigma'_{11} + \nu\sigma'_{22}}{1+\nu}\sigma_{12}\right)d\varepsilon_{11} - \frac{\sigma'_{22} + \nu\sigma'_{11}}{1+\nu}\sigma_{12}\,d\varepsilon_{22}$$

$$+ \left[\frac{R}{2(1+\nu)} + \frac{2H'}{9E}(1-\nu)\bar\sigma^2\right]d\varepsilon_{12}$$

where

$$P = \frac{2H'}{9H}\bar\sigma^2 + \frac{\sigma^2_{12}}{1+\nu} \qquad Q = R + 2(1-\nu^2)P \quad (1\text{-}59)$$

and

$$R = \sigma'^2_1 + 2\nu\sigma'_{11}\sigma'_{22} + \sigma'^2_{22} \quad (1\text{-}60)$$

As noted earlier, the classical plasticity theory has several limitations. Intense research is under way to improve constitutive modeling in cyclic plasticity, and so on—some notable avenues of current research being multisurface plasticity model, endochronic theories, and related internal variable theories (see, e.g., Refs. 16–18).

Viscoplasticity and Creep

A viscoplastic solid is similar to a viscous fluid, except that the former can resist shear stress even in a rest configuration; but when the stresses reach critical values as specified by a yield function, the material flows. Consider, for instance, the loading case of simple shear with the only applied stress being σ_{12}. Restricting ourselves to infinitesimal deformations and strains, let the shear-strain rate be $\dot\varepsilon_{12}[=(d\varepsilon_{12}/dt)]$. Then

$$\dot\varepsilon_{12} = 0$$

until the magnitude of σ_{12} reaches a value k, called the yield stress. When $|\sigma_{12}| > k$, $\dot\varepsilon_{12}$, by definition for a simple viscoplastic material, is proportional to $|\sigma_{12}| - k$ and has the same sign as σ_{12}. Thus, defining F^1 for this one-dimensional problem as

$$F^1 = \frac{|\sigma_{12}|}{k} - 1 \tag{1-61}$$

the property may be characterized by the equation

$$2\eta\dot{\varepsilon}_{12} = k \langle F^1 \rangle \sigma_{12} \tag{1-62}$$

where F^1 is a specific function, defined as

$$\langle F^1 \rangle = 0 \quad \text{if } F^1 < 0 \tag{1-63}$$
$$ F^1 \quad \text{if } F^1 \geq 0$$

and where η is the coefficient of viscosity.

The relation for simple shear above is due to Bingham [19]. Recognizing that $J = \sigma_{12}^2$ for simple shear, a generalization of the above for three-dimensional case was given by Hohenemser and Prager [20] as

$$2\eta\dot{\varepsilon}_{ij}^{vp} = 2k \langle F \rangle \frac{\partial F}{\partial \sigma_{ij}} \tag{1-64a}$$

where

$$F = \frac{J_2^{1/2}}{k} - 1 \equiv \frac{(\sigma'_{ij}\sigma'_{ij}/2)^{1/2}}{k} - 1 \tag{1-64b}$$

and the specific function $\langle F \rangle$ is defined similar to $\langle F^1 \rangle$.

For an elasto-viscoplastic solid undergoing infinitesimal straining, one may use the additive decomposition

$$\dot{\varepsilon}_{ij} = \dot{\varepsilon}_{ij}^e + \dot{\varepsilon}_{ij}^{vp} \tag{1-65}$$

and the stress–strain rate relation

$$\dot{\sigma}_{ij} = E_{ijkl}(\dot{\varepsilon}_{kl} - \dot{\varepsilon}_{kl}^{vp}) \tag{1-66}$$

where E_{ijkl} are the instantaneous *elastic* moduli. Note that the viscoplastic strains in Eq. (1-64a) are purely deviatoric, since $\partial F/\partial \sigma_{ij} = \sigma'_{ij}/2$ is deviatoric. Thus, for an isotropic solid, Eq. (1-66) may be written as

$$\dot{\sigma}_{mm} = (3\lambda + 2\mu)\dot{\varepsilon}_{mm} \tag{1-67a}$$

and

$$\dot{\sigma}'_{ij} = 2\mu(\dot{\varepsilon}'_{ij} - \dot{\varepsilon}_{ij}^{vp}) \tag{1-67b}$$

On the other hand, for metals operating at elevated temperatures, the strain in uniaxial tension is known to be a function of time, for a constant stress of magnitude even below the conventional elastic limit. Most often, based on extensive experimental data [21], the creep strain under constant stress, in uniaxial tests, is expressed as

$$\varepsilon_c = A\sigma^n t^m \tag{1-68}$$

where σ is the uniaxial stress and t is the time. The creep rate may be written as

$$\dot{\varepsilon}^c = f(\sigma, t) \tag{1-69a}$$

or, equivalently,

$$\dot{\varepsilon}^c = g(\sigma, \varepsilon_c) \tag{1-69b}$$

The expression of Eq. (1-69a) is often referred to as *time hardening* and Eq. (1-69b) as *strain hardening*. Inasmuch as Eqs. (1-68), (1-69a), and (1-69b), are valid for constant stress, Eqs. (1-69a) and (1-69b), when integrated for variable stress histories, do not necessarily give the same results. Usually, strain hardening leads to better agreement with experimental findings for variable stresses.

In the study of creep at long times, called steady-state creep, the creep strain rate in uniaxial loading is usually expressed as

$$\dot{\varepsilon}^c = f(\sigma, T) \tag{1-70}$$

where T is the temperature. Assuming that the effect of σ or T are separable, the relation

$$\dot{\varepsilon}^c = f_1(\sigma) \times f_2(T) \tag{1-71a}$$

$$= A\sigma^n \times f_2(T) = B\sigma^n \tag{1-71b}$$

where B is thus a function of temperature, is usually employed.

The steady-state creep strains are associated largely with plastic deformations and are usually observed to involve no volume change. Thus, in the multiaxial case, $\dot{\varepsilon}^c_{ij}$ is a deviatoric tensor. The relation Eq. (1-71) may be generalized to the multiaxial case as

$$\dot{\varepsilon}^c_{eq} = B\sigma^n_{eq} \tag{1-72}$$

where the subscript "eq" denotes and "equivalent" quantity, defined, analogous to the case of plasticity, as

$$\sigma_{eq} = \left(\frac{3}{2}\sigma'_{ij}\sigma'_{ij}\right)^{1/2} \quad \text{and} \quad \dot{\varepsilon}^c_{eq} = \left(\frac{2}{3}\dot{\varepsilon}_{ij}\dot{\varepsilon}_{ij}\right)^{1/2} \tag{1-73}$$

such that $\sigma_{eq}\dot{\varepsilon}^c_{eq} = \sigma'_{ij}\dot{\varepsilon}^c_{ij}$. Thus Eq. (1-72) implies that

$$\dot{\varepsilon}^c_{ij} = \tfrac{3}{2}B(\sigma_{eq})^{n-1}\sigma'_{ij} \tag{1-74}$$

For the elastic-creeping solid, one may again write

$$\dot{\varepsilon}_{ij} = \dot{\varepsilon}^e_{ij} + \dot{\varepsilon}^c_{ij} \tag{1-75}$$

and once again, the stress–strain rate relation may be written as

$$\dot{\sigma}_{ij} = E_{ijkl}(\dot{\varepsilon}_{ij} - \dot{\varepsilon}^c_{ij}) \tag{1-76}$$

In the above, the applied stress level has been assumed to be such that the material remains within the elastic limit. If the applied loads are of such a magnitude as to cause the material to exceed its yield limit, one must account for plastic or viscoplastic strains.

An interesting unified viscoplastic/plastic/creep constitutive law has been proposed by Perzyna [22]. Under multiaxial conditions, the relation for inelastic strain rate suggested in Ref. 22 is

$$\dot{\varepsilon}^a_{ij} = A\langle\psi(f)\rangle\frac{\partial q}{\partial \sigma_{ij}} \tag{1-77}$$

where A is the fluidity parameter, the superscript a denotes an elastic strain rate, and f is a loading function, expressed, analogous to the plasticity case, as

$$f(\sigma_{ij}, k) = \phi(\sigma_{ij}) - k = 0 \tag{1-78}$$

q is a viscoplastic potential defined as

$$q = q(\sigma_{ij}) \tag{1-79}$$

and $\langle\psi(f)\rangle$ is a specific function such that

$$\langle\psi(f)\rangle = \begin{cases} 0 & \text{if } f < 0 \\ \psi(f) & \text{if } f \geq 0 \end{cases} \tag{1-80}$$

If $q \equiv f$, one has the so-called associative law, and if $q \neq f$, one has a "nonassociative" law. Perzyna [22] suggests a fairly general form for ψ as

$$\psi(f) = f^n \tag{1-81a}$$

and

$$f = (\tfrac{3}{2}\sigma'_{ij}\sigma'_{ij})^{1/2} - \bar{\sigma} = \sigma_{eq} - \bar{\sigma} \tag{1-81b}$$

By letting $\bar{\sigma} = 0$ and $q = f$, one may easily verify that $\dot{\varepsilon}^a_{ij}$ of Eq. (1-77) tends to the creep strain rate $\dot{\varepsilon}^c_{ij}$ of Eq. (1-74).

Letting $\bar{\sigma}$ be a specified value and $q = f$, we obtain, using Eq. (1-81b) in Eq. (1-77), that

$$\dot{\varepsilon}^a_{ij} = \frac{3}{2}A(\sigma_{eq} - \bar{\sigma})^n \frac{\sigma'_{ij}}{\sigma_{eq}} \quad \text{for } f > 0 \quad (\text{i.e., } \sigma_{eq} > \bar{\sigma}) \tag{1-82}$$

The equivalent inelastic strain may be written as

$$\dot{\varepsilon}^a_{eq} = \left(\frac{2}{3}\dot{\varepsilon}^a_{ij}\dot{\varepsilon}^a_{ij}\right)^{1/2} = A(\sigma_{eq} - \bar{\sigma})^n \tag{1-83a}$$

or

$$\sigma_{eq} - \bar{\sigma} = \left(\frac{1}{A}\dot{\varepsilon}^a_{eq}\right)^{1/n} \tag{1-83b}$$

Thus, if a stationary solution of the present inelastic model (i.e., when $\dot{\varepsilon}^a_{eq} \to 0$) is obtained, it is seen that $\sigma_{eq} \to \bar{\sigma}$. Thus a classical inviscid plasticity solution is obtained. This fact has been utilized in obtaining classical rate-independent plasticity solutions from the general model of Eq. (1-77), by Zienkiewicz and Cormeau [23]. An alternative way of obtaining an inviscid plastic solution from Perzyna's model is to let $A \to \infty$. This concept has been implemented numerically by Argyris and Kleiber [24].

Also, as seen from Eq. (1-83b), σ_{eq}, or equivalently the size of the yield surface, is governed by (1) isotropic work-hardening effects as characterized by the dependence of $\bar{\sigma}$ on viscoplastic work, and (2) the strain-rate effect as characterized by the term $(\dot{\varepsilon}^a_{eq})^{1/n}$. Thus the rate-sensitive plastic problems may also be treated by Perzyna's model [22].

Thus, by appropriate modifications, the general relation Eq. (1-77) may be used to model creep, rate-sensitive plasticity, and rate-insensitive plasticity. By a linear combination of strain rates resulting from these individual types of behavior, combined creep, plasticity, and viscoplasticity may be modeled. However, such a model is more or less formalistic and does not lead to any physical insights into the problem of *interactive effects* between creep, plasticity, and viscoplasticity. Modeling of such interactions is the subject of a large number of current research studies.

1-2 Boundary and Initial Value Problems in Elasticity

Basic Field Equations

When deformations are finite and the material is nonlinear, the field equations governing the motion of a solid may become quite complicated. When the constitutive equation is of a differential form, such as in plasticity, viscoplasticity, and so on, it is often convenient to express the field equations in rate form as well. On the other hand, when stress is a single-valued function of strain as in nonlinear elasticity, the field equations may be written in a total form. In general, when numerical procedures are employed to solve boundary/initial value problems for arbitrary-shaped bodies, it is often convenient to write the field equations in rate form, for arbitrary deformations and general constitutive laws. A wide variety of equivalent but alternative forms of these equations is possible, since one may use (1) a wide variety of stress and strain measures, (2) a wide variety of rates of stress and strain, and (3) a variety of coordinate systems, such as those in the initial undeformed configuration (total Lagrangian), the currently deformed configuration (updated Lagrangian), or any other known intermediate configuration. For a detailed discussion, see Refs. 4 and 5. Each of the alternative forms may offer advantages in specific applications.

It is beyond the scope of this chapter to discuss the foregoing alternative forms. Here we state, for a finitely deformed nonlinear elastic solid, the relevant field equations governing stress, strain, and deformation when the solid undergoes dynamic motion. For this purpose, let x_i be Cartesian coordinates of a material particle in the undeformed solid. Let $u_k(x_i)$ be the arbitrary displacement of a material particle from the undeformed to the deformed configuration. Let S_{ij} be the second Piola–Kirchhoff stress in the finitely deformed solid. Note that S_{ij} is measured per unit area in the undeformed configuration. Let the Green–Lagrange strain tensor, which is work-conjugate to S_{ij} [4], be ε_{ij}. Let ρ_o be the mass density in the undeformed solid; b_i be body forces per unit mass; $\bar{t}_i$ be tractions measured per unit area in the undeformed solid, prescribed at surface S_t of the undeformed solid; and let $\bar{u}_i$ be prescribed displacements at S_u. The field equations are [4]:

Linear momentum balance:

$$[S_{ik}(\delta_{jk} + u_{j,k})]_{,i} + \rho_o b_j = \rho_o \ddot{u}_j \tag{1-84}$$

Angular momentum balance:

$$S_{ij} = S_{ji} \tag{1-85}$$

Strain displacement relation:

$$\varepsilon_{ij} = \tfrac{1}{2}[u_{i,j} + u_{j,i} + u_{k,i} u_{k,j}] \tag{1-86}$$

Constitutive law:

$$S_{ij} = \frac{\partial W}{\partial \varepsilon_{ij}} \tag{1-87}$$

Traction boundary condition:

$$n_i S_{ik}(\delta_{jk} + u_{j,k}) \equiv \bar{t}_j \quad \text{at} \quad S_t \tag{1-88}$$

Displacement boundary condition:

$$u_j = \bar{u}_i \quad \text{at} \quad S_u \tag{1-89}$$

In the above, $(\cdot)_{,k}$ denotes $\partial(\cdot)/\partial x_k$; $(\ddot{\cdot})$ denotes $\partial^2(\cdot)/\partial t^2$; n_i are components of a unit normal to S_t; and W is the strain-energy density, measured per unit volume in the undeformed body.

When the deformations and strains are infinitesimal, the differences in the alternate stress and

MECHANICAL RESPONSES OF MATERIALS

strain measures disappear. Further, considering only a linear elastic solid, the equations above simplify as

$$\sigma_{ij,i} + \rho_0 b_j = \rho_0 \ddot{u}_j \tag{1-90}$$

$$\sigma_{ij} = \sigma_{ji} \tag{1-91}$$

$$\varepsilon_{ij} = \tfrac{1}{2}[u_{i,j} + u_{j,i}] \tag{1-92}$$

$$\sigma_{ij} = E_{ijkl}\varepsilon_{kl} \tag{1-93}$$

$$n_i \sigma_{ij} = \bar{t}_j \quad \text{at} \quad S_t \tag{1-94}$$

$$u_i = \bar{u}_i \quad \text{at} \quad S_u \tag{1-95}$$

and the initial condition,

$$u_i(x_k, 0) = u_i^*(x_k) \quad \dot{u}_i(x_k, 0) = \dot{u}_i^*(x_k) \quad \text{at} \quad t = 0 \tag{1-96}$$

Some Useful Analytical Results

Here we list some classical analytical solutions to the linear boundary/initial value problem in Eqs. (1-90) through (1-96) for specific cases that are often of interest.

Solutions for Point Loads

Consider a three-dimensional, linear elastic, homogeneous, and isotropic *infinite* solid that is subject to a point load in the jth coordinate axis (i.e., along the unit vector e_j) at the location ξ_k. Then the solution for the displacement in the ith direction at x_m, denoted by $u_{ij}(x_m)$, is given by [25]

$$u_{ij}(x_m, \xi_m) = \frac{1}{16\pi\mu(1-\nu)} \frac{1}{r}\left[(3 - 4\nu)\delta_{ij} + \frac{(x_i - \xi_i)(x_j - \xi_j)}{r^2}\right] \tag{1-97a}$$

where

$$r^2 = (x_m - \xi_m)(x_m - \xi_m) \tag{1-97b}$$

Similarly, the traction component t_i on an oriented surface at x_m, with a unit normal n_k, due to a unit load in the jth direction at the location ξ_m is given by

$$t_{ji}(x_m, \xi_m) = \frac{-1}{8\pi(1-\nu)} \frac{1}{r^2} \left\{ \frac{1 - 2\nu}{r}[n_j(x_i - \xi_i) - n_i(x_j - \xi_j)] \right.$$
$$\left. + \frac{x_k - \xi_k}{r} n_k \left[\frac{3(x_i - \xi_i)(x_j - \xi_j)}{r^2} + (1 - 2\nu)\delta_{ij}\right]\right\} \tag{1-98}$$

Note that the solutions at x_m above are due to a unit load at ξ_m in the jth direction, denoted mathematically as a "singular" body force,

$$\bar{F}_j(x_m) = \delta(x_m - \xi_m)e_j \tag{1-99}$$

Thus e_j is simply used to indicate the direction of the point force.

Similarly, in a plane-strain problem, with an *infinite* domain, the corresponding two-dimensional solution $u_\alpha(x_\beta)$ due to a point load in the μth direction at ξ_β is given by:

Load:

$$\bar{F}_\mu(x_\beta) = \delta(x_\beta - \xi_\beta)e_\mu \tag{1-100}$$

$$u_\alpha(x_\beta) = -\frac{1}{8\pi\mu(1-\nu)}\left\{(3-4\nu)\delta_{\alpha\mu}\ln r - \frac{(x_\alpha - \xi_\alpha)(x_\mu - \xi_\mu)}{r^2}\right\} + \text{constant} \tag{1-101}$$

where

$$r^2 = (x_\beta - \xi_\beta)(x_\beta - \xi_\beta)$$

The corresponding traction on an oriented surface, with a normal n_σ at x_β, is

$$t_\alpha(x_\beta) = \frac{1}{4\pi(1-\nu)r^2}\left\{(1-2\nu)[n_\mu(x_\alpha - \xi_\alpha) - n_\alpha(x_\mu - \xi_\mu)] + (x_\sigma - \xi_\sigma)n_\sigma\left[(1-2\nu)\delta_{\alpha\mu} + \frac{2(x_\alpha - \xi_\alpha)(x_\mu - \xi_\mu)}{r^2}\right]\right\} \tag{1-102}$$

The corresponding plane-stress solution is obtained by replacing ν by $\bar{\nu} = \nu/(1+\nu)$.

Another interesting problem is that of a point normal load acting at the origin of the coordinate system on the horizontal plane of a semi-infinite solid, the solution of which is due to Boussinesq [26]:

$$\bar{F}_3 = \delta(x_m)e_3 \tag{1-103}$$

$$\sigma_{rr} = \frac{1}{2\pi R^2}\left[-\frac{3r^2 x_3}{R^2} + \frac{(1-2\nu)R}{R + x_3}\right] \tag{1-104}$$

$$\sigma_{\theta\theta} = \frac{1-2\nu}{2\pi R^2}\left(\frac{x_3}{R} - \frac{R}{R + x_3}\right) \tag{1-105}$$

$$\sigma_{33} = -\frac{3x_3^3}{2\pi R^5} \qquad \sigma_{r3} = -\frac{3rx_3^2}{2\pi R^5} \tag{1-106}$$

where $r^2 = (x_1^2 + x_2^2)$ and $R^2 = r^2 + x_3^2$. The solution for the corresponding problem for a point tangential load on a half-space is due to Cerutti, per Love [27].

Solutions Near Holes, Voids, and Inclusions

Consider a plane problem of an infinite linear elastic isotropic body, with a hole of radius a. Let the body be subjected to uniaxial tension, say, along the x_1 axis. Let the Cartesian system be centered at the center of the hole and let r, θ be the corresponding polar coordinates, with θ being the angle measured from the x_1 axis. The stress state near the hole is given [28] by

$$\sigma_{rr} = \frac{\sigma_{11}^\infty}{2}\left(1 - \frac{a^2}{r^2}\right) + \frac{\sigma_{11}^\infty}{2}\left(1 - 4\frac{a^2}{r^2} + 3\frac{a^4}{r^4}\right)\cos 2\theta \tag{1-107}$$

$$\sigma_{\theta\theta} = \frac{\sigma_{11}^\infty}{2}\left(1 + \frac{a^2}{r^2}\right) - \frac{\sigma_{11}^\infty}{2}\left(1 + \frac{3a^4}{r^4}\right)\cos 2\theta \tag{1-108}$$

$$\sigma_{r\theta} = -\frac{\sigma_{11}^\infty}{2}\left(1 + \frac{2a^2}{r^2} - \frac{3a^4}{r^4}\right)\sin 2\theta \tag{1-109}$$

The solutions for a biaxial stress state may be obtained by superposition. For a compendium of solutions for holes in anisotropic bodies, and for shapes of holes other than circular, such as elliptical, and so on, see Savin [29].

All the solutions considered above pertain to homogeneous media. A basic problem that underlies most of the theories of heterogeneous media (composites, etc.) is that of inclusions. In a plane problem of isotropic elasticity, consider a rigid inclusion of radius a and assume that there is perfect bonding (for other interface conditions, see Mushkelishvili [28]) between the elastic domain and the rigid inclusion. The solution for stresses near the inclusion due to far-field uniaxial tension are [28]

$$\sigma_{rr} = \frac{\sigma_{11}^\infty}{2}\left[1 - \nu\frac{a^2}{r^2} + \left(1 - \frac{2\beta a^2}{r^2} - \frac{3\delta a^4}{r^4}\right)\cos 2\theta\right] \tag{1-110}$$

$$\sigma_{\theta\theta} = \frac{\sigma_{11}^\infty}{2}\left(1 + \nu\frac{a^2}{r^2} - \left[1 - \frac{3\delta a^4}{r^4}\right]\cos 2\theta\right) \tag{1-111}$$

$$\sigma_{r\theta} = -\frac{\sigma_{11}^\infty}{2}\left(1 + \beta\frac{a^2}{r^2} + \frac{3\delta a^4}{r^4}\right) \tag{1-112}$$

$$\beta = -\frac{2(\lambda + \mu)}{\lambda + 3\mu} \qquad \nu = -\frac{\mu}{\lambda + \mu} \qquad \delta = \frac{\lambda + \mu}{\lambda + 3\mu} \tag{1-113}$$

where λ and μ are Lamé constants.

An *important* three-dimensional problem of an ellipsoidal inclusion has been solved by Eshelby [30], which is too detailed to be discussed here cursorily.

Cracks

Consider a plane linear elastic isotropic body with a straight-line crack along the x_1 axis. Let r, θ be polar coordinates centered at the crack tip. Let the body be subjected to a general far-field loading in the $(x_1 - x_2)$ plane. The solutions for stresses and displacements are [31]

$$\begin{Bmatrix}\sigma_{11}\\ \sigma_{22}\\ \sigma_{12}\end{Bmatrix} = \sum_{n=1}^{\infty}\frac{n}{2}A_{In}r^{(n-2)/2}\begin{Bmatrix}\left\{2+(-1)^n+\frac{n}{2}\right\}\cos\left(\frac{n}{2}-1\right)\theta - \left(\frac{n}{2}-1\right)\cos\left(\frac{n}{2}-3\right)\theta\\ \left\{2-(-1)^n-\frac{n}{2}\right\}\cos\left(\frac{n}{2}-1\right)\theta + \left(\frac{n}{2}-1\right)\cos\left(\frac{n}{2}-3\right)\theta\\ -\left\{(-1)^n+\frac{n}{2}\right\}\sin\left(\frac{n}{2}-1\right)\theta + \left(\frac{n}{2}-1\right)\sin\left(\frac{n}{2}-3\right)\theta\end{Bmatrix}$$

$$-\sum_{n=1}^{\infty}\frac{n}{2}A_{IIn}r^{(n-2)/2}\begin{Bmatrix}\left\{2-(-1)^n+\frac{n}{2}\right\}\sin\left(\frac{n}{2}-1\right)\theta - \left(\frac{n}{2}-1\right)\sin\left(\frac{n}{2}-3\right)\theta\\ \left\{2+(-1)^n-\frac{n}{2}\right\}\sin\left(\frac{n}{2}-1\right)\theta + \left(\frac{n}{2}-1\right)\cos\left(\frac{n}{2}-3\right)\theta\\ -\left\{(-1)^n-\frac{n}{2}\right\}\cos\left(\frac{n}{2}-1\right)\theta - \left(\frac{n}{2}-1\right)\cos\left(\frac{n}{2}-3\right)\theta\end{Bmatrix}$$

(1-114)

As $r \to 0$, the value $n = 1$ leads to singularities in stresses. The corresponding coefficients, A_{I1}

and $A_{\text{II}2}$, respectively, are defined as the strengths of the singularities. Most commonly, one defines the so-called stress-intensity factors K_I and K_II through the relations $A_{\text{II}} = K_\text{I}\sqrt{2\pi}$ and $A_{\text{III}} = -K_\text{II}\sqrt{2\pi}$. The corresponding displacements, for plane strain, are

$$\begin{Bmatrix} u \\ v \end{Bmatrix} = \sum_{n=1}^{\infty} \frac{1}{2\mu} A_{\text{I}n} r^{n/2} \begin{Bmatrix} (3-4\nu)\cos\frac{n}{2}\theta - \frac{n}{2}\cos\left(\frac{n}{2}-2\right)\theta + \left\{\frac{n}{2}+(-1)^n\right\}\cos\frac{n}{2}\theta \\ (3-4\nu)\sin\frac{n}{2}\theta + \frac{n}{2}\sin\left(\frac{n}{2}-2\right)\theta - \left\{\frac{n}{2}+(-1)^n\right\}\sin\frac{n}{2}\theta \end{Bmatrix}$$

$$- \sum_{n=1}^{\infty} \frac{1}{2\mu} A_{\text{II}n} r^{n/2} \begin{Bmatrix} (3-4\nu)\sin\frac{n}{2}\theta - \frac{n}{2}\sin\left(\frac{n}{2}-2\right)\theta + \left\{\frac{n}{2}+(-1)^n\right\}\sin\frac{n}{2}\theta \\ (3-4\nu)\cos\frac{n}{2}\theta - \frac{n}{2}\cos\left(\frac{n}{2}-2\right)\theta + \left\{\frac{n}{2}+(-1)^n\right\}\cos\frac{n}{2}\theta \end{Bmatrix}$$

(1-115)

The corresponding displacements for the plane-stress case are obtained by replacing ν in Eq. (1-115) by $\nu/(1+\nu)$. For cracks in plane anisotropic media, and so on, see Ref. 32.

Now, consider a crack propagating at a constant velocity v along the x_1 axes in a plane elastic isotropic body, subject to in-plane, far-field loading. The general solution for the stresses near the propagating crack tip is given by [33]

$$\sigma_{xn} = \frac{K_n^0 B_\text{I}(C)}{\sqrt{2\pi}} \frac{n(n+1)}{2}$$
$$\left\{(1 + 2\beta_1^2 - \beta_2^2) r_1^{n/2-1} \cos\left(\frac{n}{2}-1\right)\theta_1 - 2h(n) r_2^{n/2-1} \cos\left(\frac{n}{2}-1\right)\theta_2\right\}$$
$$+ \frac{K_n^* B_\text{II}(C)}{\sqrt{2\pi}} \frac{n(n+1)}{2} \left\{(1 + 2\beta_1^2 - \beta_2^2) r_1^{n/2-1} \sin\left(\frac{n}{2}-1\right)\theta_1 \right.$$
$$\left. - 2h(\bar{n}) r_2^{n/2-1} \sin\left(\frac{n}{2}-1\right)\theta_2\right\}$$

$$\sigma_{yn} = \frac{K_n^0 B_\text{I}(C)}{\sqrt{2\pi}} \frac{n(n+1)}{2}$$
$$\left\{-(1 + \beta_2^2) r_1^{n/2-1} \cos\left(\frac{n}{2}-1\right)\theta_1 + 2h(n) r_2^{n/2-1} \cos\left(\frac{n}{2}-1\right)\theta_2\right\}$$
$$+ \frac{K_n^* B_\text{II}(C)}{\sqrt{2\pi}} \frac{n(n+1)}{2} \left\{-(1 + \beta_2^2) r_1^{n/2-1} \sin\left(\frac{n}{2}-1\right)\theta_1\right.$$
$$\left. + 2h(\bar{n}) r_2^{n/2-1} \sin\left(\frac{n}{2}-1\right)\theta_2\right\}$$

(1-116)

MECHANICAL RESPONSES OF MATERIALS

$$\sigma_{xyn} = \frac{K_n^0 B_I(C)}{\sqrt{2\pi}} \frac{n(n+1)}{2}$$
$$\left\{ -2\beta_1 r_1^{n/2-1} \sin\left(\frac{n}{2}-1\right)\theta_1 + \frac{1+\beta_2^2}{\beta_2} h(n) r_2^{n/2-1} \sin\left(\frac{n}{2}-1\right)\theta_2 \right\}$$
$$+ \frac{K_n^* B_{II}(C)}{\sqrt{2\pi}} \frac{n(n+1)}{2} \left\{ +2\beta_1 r_1^{n/2-1} \cos\left(\frac{n}{2}-1\right)\theta_1 \right.$$
$$\left. -\frac{1+\beta_2^2}{\beta_2} h(\bar{n}) r_2^{n/2-1} \cos\left(\frac{n}{2}-1\right)\theta_2 \right\}$$

$$u_n = \frac{K_n^0 B_I(C)}{2\mu} \sqrt{\frac{2}{\pi}} (n+1) \left\{ r_1^{n/2} \cos\frac{n}{2}\theta_1 - h(n) r_2^{n/2} \cos\frac{n}{2}\theta_2 \right\}$$
$$+ \frac{K_n^* B_{II}(C)}{2\mu} \sqrt{\frac{2}{\pi}} (n+1) \left\{ r_1^{n/2} \sin\frac{n}{2}\theta_1 - h(\bar{n}) r_2^{n/2} \sin\frac{n}{2}\theta_2 \right\}$$

$$v_n = \frac{K_n^0 B_I(C)}{2\mu} \sqrt{\frac{2}{\pi}} (n+1) \left\{ -\beta_1 r_1^{n/2} \sin\frac{n}{2}\theta_1 + \frac{h(n)}{\beta_2} r_2^{n/2} \sin\frac{n}{2}\theta_2 \right\}$$

where n is an integer $1 \leq n \leq \alpha$; $\bar{n} = n+1$; and

$$\beta_1^2 = 1 - \left(\frac{v}{c_d}\right)^2 \qquad \beta_2^2 = 1 - \left(\frac{v}{c_s}\right)^2$$

$$c_d^2 = \frac{k+1}{k-1} \frac{\mu}{\rho} \qquad c_s^2 = \frac{\mu}{\rho}$$

$$k = \frac{3-\nu}{1+\nu} \qquad \text{plane stress}$$

$$= 3 - 4\nu \qquad \text{plane strain}$$

$r_j e^{i\theta_j} = x + i\beta_j y \qquad x, y$ coordinates centered at the propagating crack tip

$$h(n) = \frac{2\beta_1\beta_2}{1+\beta_2^2} \qquad n \text{ odd} \qquad (1\text{-}117)$$

$$= \frac{1+\beta_2^2}{2} \qquad n \text{ even}$$

$$B_1(v) = \frac{1+\beta_2^2}{D(v)} \qquad B_2(v) = \frac{2\beta_2}{D(v)}$$

$$D(v) = 4\beta_1\beta_2(1+\beta_2)^2$$

K_1^0, K_1^* related to dynamic stress intensity factors

For details of stress solutions for dynamic crack propagation in mode III, see Ref. 33.

We close this section by considering certain three-dimensional problems of cracks in linear elastic isotropic solids. Consider an arbitrary crack (surface of discontinuity) in the plane $x_3 = 0$. Let the crack border be parametrized by the parameter s (say, length along the crack front).

Consider an orthogonal coordinate system: x_3 perpendicular to the crack plane t locally tangential to the crack border, and n locally normal to the crack plane (thus n and z are in the plane $x_3 = 0$). Then the stress field asymptotically close to the crack border is given [32] by

$$\sigma_{nm} = \frac{K_I}{4(2r)^{1/2}}\left(3\cos\frac{\theta}{2} + \cos\frac{5\theta}{2}\right) - \frac{K_{II}}{4(2r)^{1/2}}\left(7\sin\frac{\theta}{2} + \sin\frac{5\theta}{2}\right) + O(r)$$

$$\sigma_{tt} = \frac{K_I}{(2r)^{1/2}}2\nu\cos\frac{\theta}{2} - \frac{K_{II}}{(2r)^{1/2}}2\nu\sin\frac{\theta}{2} + O(r)$$

$$\sigma_{33} = \frac{K_I}{4(2r)^{1/2}}\left(5\cos\frac{\theta}{2} - \cos\frac{5\theta}{2}\right) - \frac{K_{II}}{4(2r)^{1/2}}\left(\sin\frac{\theta}{2} - \sin\frac{5\theta}{2}\right) \quad (1\text{-}118)$$

$$\sigma_{t3} = \frac{K_{III}}{(2r)^{1/2}}\cos\frac{\theta}{2} + O(r)$$

$$\sigma_{n3} = -\frac{K_I}{4(2r)^{1/2}}\left(\sin\frac{\theta}{2} - \sin\frac{5\theta}{2}\right) + \frac{K_{II}}{4(2r)^{1/2}}\left(3\cos\frac{\theta}{2} + \cos\frac{5\theta}{2}\right)$$

$$\sigma_{nt} = \frac{-K_{III}}{(2r)^{1/2}}\sin\frac{\theta}{2} + O(r)$$

A general solution for an infinite solid containing an elliptical crack, the faces of which are subjected to arbitrary tractions (normal as well as shear) has been found [34]. The details are too complex to be summarized here. However, the stress-intensity factor for constant normal pressure is [34]

$$K_I = \frac{\sqrt{\pi}}{E(\beta)}\sqrt{\frac{a_2}{a_1}}\,p_0(a_1^2\sin^2\theta + a_2^2\cos^2\theta)^{1/4} \quad (1\text{-}119a)$$

In the above, the ellipse, in the $x_3 = 0$ plane, is described by $(x_1/a_1)^2 + (x_2/a_2)^2 = 1$; $\beta^2 = 1 - (a_2/a_1)^2$, $E(\beta)$ is an elliptic integral of the second kind, and θ is the elliptic angle. For a crack-face pressure of $\sigma_{33} = p_0(x_1/a_2)$, the solution is

$$K_I = \frac{p_0\sqrt{\pi a_2}\sin\theta\,\beta^2}{(1+\beta^2)E(\beta) - (a_2/a_1)^2 F(\beta)}\left(\sin^2\theta + \frac{a_2}{a_1}\cos^2\theta\right)^{1/4} \quad (1\text{-}119b)$$

where $F(\beta)$ is an elliptical integral of the first kind.

1-3 Numerical Methods

The basic field equations and boundary conditions for a boundary/initial value problem of linear elasticity were given in Eqs. (1-90) through (1-96). Some analytical ("exact") solutions were presented in the preceding section for certain "idealized" problems in which, for instance, (1) the domain was considered to be finite, (2) the material was considered to be homogeneous and isotropic, and (3) the boundary conditions were rather simple. In practice, however, it is common to encounter problems wherein (1) the domain is finite and of an arbitrary and, usually, a complex shape; (2) the material is neither homogeneous nor isotropic; and (3) the boundary conditions are themselves complex—for instance, when certain components of displacement (say, μ_1) and complementary components of traction (say, t_2 and t_3) are specified at each point on the boundary. To cope with these practical problems of common engineering occurrence, the use of numerical (and "large-computer" -oriented) methods is mandatory. In the following, we discuss briefly (1) the finite-difference method, (2) the finite-element method, and (3) the boundary element method. The scope and space limitations of this chapter, however, permit only a discussion of *linear* problems; for general nonlinear problems, the reader is referred to Refs. 35–37.

Finite-Difference Method

Consider, for instance, the elastostatic problem (inertia being considered to be negligible) for an isotropic, homogeneous, linear elastic solid. For such a case, Eqs. (1–90) through (1–96) may be reduced to the form of the well-known Navier equations [27]

$$\mu \Delta^2 u_i + (\lambda + \mu)\theta_{,i} + F_i = 0 \tag{1-120a}$$

where

$$\theta = u_{j,j} \tag{1-120b}$$

and the boundary conditions

$$u_i = \bar{u}_i \quad \text{at} \quad S_u \tag{1-121}$$

and

$$[\lambda u_{k,k}\delta_{ij} + \mu(u_{i,j} + u_{j,i})]n_j = \bar{t} \quad \text{at} \quad S_t \tag{1-122}$$

The displacement components u_i ($i = 1, 2, 3$) are the independent variables in the three-dimensional boundary value problem in Eqs. (1-120) through (1-122), where $\nabla^2 = (\partial/\partial x_1^2) + (\partial^2/\partial x_2^2) + (\partial^2/\partial x_3^2)$. In plane stress, the Navier equations are

$$\mu\left(\frac{\partial^2 u_1}{\partial x_1^2} + \frac{\partial^2 u_1}{\partial x_2^2}\right) + \mu\left(\frac{1+\nu}{1-\nu}\right)\frac{\partial}{\partial x_1}\left(\frac{\partial u_1}{\partial x_1} + \frac{\partial u_2}{\partial x_2}\right) + F_1 = 0 \quad \text{in} \quad \Omega \tag{1-123}$$

and

$$\mu\left(\frac{\partial^2 u_2}{\partial x_1^2} + \frac{\partial^2 u_2}{\partial x_2^2}\right) + \mu\left(\frac{1+\nu}{1-\nu}\right)\frac{\partial}{\partial x_2}\left(\frac{\partial u_1}{\partial x_1} + \frac{\partial u_2}{\partial x_2}\right) + F_2 = 0 \quad \text{in} \quad \Omega \tag{1-124}$$

Similar equations for plane strain may be obtained by simply replacing ν in Eqs. (1-123) and (1-124) by $(\nu/1+\nu)$.

We shall consider, for instance, the numerical solution of Eqs. (1-123) and (1-124) with boundary conditions of the type of Eqs. (1-121) and (1-122). In the finite-difference method, the derivatives occurring in Eqs. (1-123) and (1-124) are simply replaced with respective finite-difference approximations. Let Ω be an arbitrary-shaped plane domain. On this domain Ω, we place a square mesh of points with a mesh spacing h. Note that unless Ω is a regular domain such as a square, rectangle, or a right triangle, the boundary of Ω will *not*, in general, intersect the square mesh above only at mesh points. Consider the mesh point (j, k) in the interior of Ω. The finite-difference approximations to the relevant derivatives at (j, k) are

$$(\nabla^2 u_\alpha)_{j,k} \equiv \left(\frac{\partial^2 u_\alpha}{\partial x_1^2} + \frac{\partial^2 u_\alpha}{\partial x_2^2}\right)_{j,k} \quad (\alpha = 1, 2)$$

$$= \frac{1}{h^2}\{(u_\alpha)_{j+1,k} + (u_\alpha)_{j-1,k} + (u_\alpha)_{j,k+1}$$

$$+ (u_\alpha)_{j,k+1} - 4(u_\alpha)_{j,k}\} + O(h^2) \tag{1-125a}$$

$$\left(\frac{\partial^2 u_1}{\partial x_1^2}\right)_{j,k} = \frac{1}{h^2}\{(u_1)_{j+1,k} + (u_1)_{j-1,k} - 2(u_1)_{j,k}\} + O(h^2) \tag{1-125b}$$

$$\frac{\partial^2 u_1}{\partial x_1 \partial x_2} = \frac{1}{4h^2}\{(u_1)_{j-1,k+1} + (u_1)_{j-1,k+1}$$

$$+ (u_1)_{j+1,k+1} + (u_1)_{j+1,k-1}\} + O(h^2) \tag{1-125c}$$

and so on. When equations of the type of Eqs. (1-125a) through (1-125c) are substituted into Eqs. (1-123) and (1-124), there results a system of simultaneous equations for the values of u_1 and u_2 at the mesh points.

For further details of finite-difference methods, see Ref. 38. The generally recognized drawbacks of the finite-difference methods are (1) the difficulties in the treatment of irregular boundaries, (2) the difficulties in the treatment of higher-order boundary conditions such as the traction boundary condition expressed in terms of displacements [i.e., Eq. (1-122)], (3) the difficulties in the treatment of nonhomogeneous domains wherein the differential equation must be written in each homogeneous subdomain and jump and/or continuity conditions must be enforced at the interfaces, and (4) the difficulties in solving problems of structural assemblies, such as stiffened plates and shells.

Attempts at overcoming the foregoing difficulties have resulted in the birth of the finite-element methods, which, under a broad philosophical classification, are analogous to the finite-difference methods but possess certain fundamentally novel features that make it possible to solve problems of complicated geometry, higher-order boundary conditions, nonhomogeneous domains, and so on. These are briefly discussed below.

Finite-Element Method

Here we consider the general system of Eqs. (1-90) through (1-96) as such, without reducing them to the Navier form as in Eq. (1-120) involving u_i alone as independent variables. We restrict ourselves, for simplicity, to problems wherein inertia plays an insignificant role.

We consider the arbitrary-shaped domain Ω, with boundary $\partial\Omega$, to be subdivided into finite elements $\Omega_m (m = 1 \cdots M)$, each with a boundary $\partial\Omega_m$. In general, $\partial\Omega_m = \rho_m + S_{um} + S_{tm}$, where ρ_m is an interelement boundary (i.e., adjoining other elements), and S_{tm} (or S_{um}) is that portion of $\partial\Omega_m$ falling on the external boundary S_t (or S_u). Note that for a majority of the elements in the interior of the solid, $\partial\Omega_m \equiv \rho_m$. Suppose that σ_{ij}, ε_{ij}, and u_i are independently assumed trial functions for solution in each finite element. Thus σ_{ij}, ε_{ij}, and u_i in each Ω_m are ultimately expected to obey Eqs. (1-90) through (1-95), respectively. Further, at the interelement boundary ρ_m, the trial solutions in neighboring elements should obey:

Displacement compatibility:

$$u_i^+ = u_i^- \quad \text{at} \quad \rho_m \tag{1-126}$$

Traction reciprocity:

$$(\sigma_{ij}n_j)^+ + (\sigma_{ij}n_j)^- = 0 \quad \text{at} \quad \rho_m \tag{1-127}$$

Note that the (+) and (−) refer, arbitrarily, to either side of ρ_m. Note also that even though $n_j^+ = -n_j^-$, not *all* the stress components σ_{ij} are required to be continuous at ρ_m.

Henceforth, we assume that Eq. (1-126) is always satisfied (i.e., the trial functions for displacement are continuous at ρ_m). The situation when Eq. (1-126) is not satisfied is beyond the scope of this chapter; for a discussion of nonstandard finite-element methods, such as hybrid methods, when Eq. (1-126) is not satisfied a priori, see Refs. 39 and 40.

In the remainder of this chapter, trial functions σ_{ij} and ε_{ij} are arbitrary but symmetric ($\sigma_{ij} =$

MECHANICAL RESPONSES OF MATERIALS

σ_{ji}, $\varepsilon_{ij} = \varepsilon_{ji}$), while u_i satisfy Eq. (1-126). We now consider test functions (or weight functions) τ_{ij} and u_{ij} that are also arbitrary but symmetric, and test functions v_i such that $v_i^+ = v_i^-$ at ρ_m. The weighted residual form of Eqs. (1-90) through (1-95) and Eq. (1-127) may be written as

$$(\sigma_{ij,j} + \rho_0 b_i)v_i \, d\Omega = 0 = \int_{\Omega_m} \{-\sigma_{ij}v_{(i,j)} + \rho_0 b_i v_i\} d\Omega \tag{1-128}$$

$$+ \int_{\partial\Omega_m} n_j \sigma_{ij} v \, ds \tag{1-129}$$

$$\int_{\Omega_m} [\varepsilon_{ij} - u_{(i,j)}]\tau_{ij} \, d\Omega = 0 \tag{1-130}$$

$$\int_{\Omega_m} (\sigma_{ij} - E_{ijkl}\varepsilon_{kl})u_{ij} \, d\Omega = 0 \tag{1-131}$$

$$\int_{S_{tm}} (n_i\sigma_{ij} - \bar{t}_j)v_j \, ds = 0 \tag{1-132}$$

$$\int_{S_{um}} (u_i - \bar{u}_i)\tau_{ij}n_j \, ds = 0 \tag{1-133}$$

$$\int_{\rho_m} \{(\sigma_{ij}n_j)^+ + (\sigma_{ij}n_j)^-\}v_i \, ds = 0$$

Equations (1-128) through (1-133) are valid for each element. We shall express them as domain equations or global equations, by noting the connectivity conditions $u_i^+ = u_i^-$ and $v_i^+ = v_i^-$ at ρ_m. Prior to doing so, we note a notational detail: ρ_m is in fact a common boundary for *two* neighboring elements. Thus, when Eq. (1-133) is summed for all elements, one obtains the term

$$\sum_m \int_{\rho_m} (\sigma_{ij}n_j v_i) ds \tag{1-134}$$

(see Refs. 39 and 40 for further explanation). In writing the global-level equation, we observe from Eqs. (1-128) through (1-134) that certain terms are repeated. Based on this observation, we "add" Eqs. (1-128) through (1-133) and take a sum over all elements, as follows:

$$0 = -\sum_m \{(1\text{-}128) + (1\text{-}129) + (1\text{-}130) - (1\text{-}131) - (1\text{-}132) - (1\text{-}133)\}$$

$$= \sum_m \left[\int_{\Omega_m} \{E_{ijkl}\varepsilon_{kl}\mu_{ij} - (\sigma_{ij}\mu_{ij} + \tau_{ij}\varepsilon_{ij}) + (\sigma_{ij}v_{(i,j)} + \tau_{ij}\mu_{(i,j)}) \right.$$

$$- \rho_0 b_i v_i\} d\Omega - \int_{S_{tm}} \bar{t}_j v_i \, ds - \int_{S_{um}} (u_i - \bar{u}_i)n_j \tau_{ji} \, ds$$

$$\left. - \int_{S_{um}} v_i n_j \sigma_{ji} \right] ds \tag{1-135}$$

Now in the sense of calculus of variations, one may formally identify

$$\mu_{ij} \equiv \delta\varepsilon_{ij} \qquad \tau_{ij} \equiv \delta\sigma_{ij} \qquad v_i \equiv \delta u_i \tag{1-136}$$

Thus Eq. (1-135) may be written as

$$0 = \delta\left[\sum_m \int_{\Omega_m} \{A(\varepsilon_{ij}) - \sigma_{ij}\varepsilon_{ij} + \sigma_{ij}u_{(i,j)} - \rho_0 b_i u_i\} d\Omega \right.$$

$$\left. - \int_{S_{tm}} \bar{t}_i u_i \, ds - \int_{S_{um}} (u_i - \bar{u}_i)n_j\sigma_{ji} \, ds \right] \tag{1-137}$$

Equation (1-137) thus forms the basis of the most general mixed finite-element method of linear elasticity wherein σ_{ij}, ε_{ij}, and u_i are all independently approximated, with the only a priori restrictions, $\sigma_{ij} = \sigma_{ji}$, $\varepsilon_{ij} = \varepsilon_{ji}$, and $u_i^+ = u_i^-$ at ρ_m. Equation (1-137) is a variational equation. When finite-dimensional approximations are used, and the functional in Eq. (1-137) is varied with respect to the undetermined parameters in σ_{ij}, ε_{ij}, and u_i, one obtains the finite-element algebraic equations (see Refs. 39 and 40).

We now consider what is usually meant when one refers to "the finite-element method" in solid mechanics. Here u_i alone is treated as an independent variable; ε_{ij} is derived from u_i using Eq. (1-92) and σ_{ij} is derived from $\varepsilon_{ij}(u_k)$ using Eq. (1-93). Further, in the "usual finite-element method," u_i is restricted to satisfy $u_i^+ = u_i^-$ at ρ_m a priori, as well as the boundary conditions at S_u [i.e., Eq. (1-95)]. Thus one has to satisfy only Eqs. (1-90), (1-94), and (1-127) through a weighted residual technique. The weighted residual forms of Eqs. (1-90), (1-94), and (1-127) are, respectively, Eqs. (1-128), (1-131), and (1-133). In these equations it should be understood that now $\sigma_{ij} = \sigma_{ij}(u)$ (i.e., σ_{ij} is derived from the assumed u_k). By adding Eqs. (1-128), (1-130), and (1-133), we obtain the formal basis of the common "compatible displacement finite-element method," as follows:

$$0 = \sum_m \left\{ \int_{\Omega_m} [\sigma_{ij}(u_k) v_{(i,j)} - \rho_0 b_i v_i] dv - \int_{S_{tm}} \bar{t}_i v_i \, ds \right\} \tag{1-138}$$

$$= \delta \sum_m \int_{\Omega_m} [\tfrac{1}{2} E_{ijkl} u_{(i,j)} u_{(k,l)} - \rho_0 b_i u_i] dv - \int_{S_{tm}} \bar{t}_i u_i \, ds \tag{1-139}$$

In each element one may introduce an approximation

$$\{u\} = \{u_i\} = \underset{3 \times N}{[A]} \underset{N \times 1}{\{q\}} \tag{1-140}$$

where $\{q\}$ is an $N \times 1$ vector of nodal displacement for each element. The strains can be written as

$$\{\varepsilon\} = \begin{Bmatrix} \varepsilon_{11} \\ \cdot \\ \cdot \\ \cdot \\ \varepsilon_{12} \end{Bmatrix} = [B]\{q\}$$

and represent E_{ijkl} as a (6×6) matrix $[D]$ for the general three-dimensional case (or a 3×3 matrix for plane stress/strain). Thus Eq. (1-139) becomes

$$0 = \delta \left\{ \sum_m \int_{\Omega_m} \langle \tfrac{1}{2} [q][B]^T[D][B]\{q\} - \rho_0[b][A]\{q\} \rangle dv \right.$$

$$\left. - \int_{S_{tm}} [\bar{t}][A]\{q\} ds \right\} \tag{1-141}$$

$$= \delta \left\{ \sum_m \tfrac{1}{2} [q][k]\{q\} - [Q]\{q\} \right\} \tag{1-142}$$

$$= \delta \{\tfrac{1}{2}[q^*][K]\{q^*\} - [Q]\{q^*\}\} \tag{1-143}$$

where $\{q^*\}$ is a global vector of nodal displacements, $[K]$ is the global "stiffness" matrix, and

[Q] is the global vector of nodal forces. For admissible arbitrary variations (such that $\delta q^* = 0$ at S_u), one gets the system of equations

$$[\mathbf{K}_{11}]\{\mathbf{q}_1^*\} = \{\mathbf{Q}_1\} - [\mathbf{K}_{12}]\{\mathbf{q}_2^*\} \tag{1-144}$$

where $[\mathbf{q}_1^*]$ are the unknown displacements, $\{\mathbf{Q}_1\}$ are the known conjugate forces, $\{\mathbf{q}_2^*\}$ are prescribed displacements at S_u, and $[\mathbf{K}_{11}]$ and $[\mathbf{K}_{12}]$ are appropriately partitioned stiffness matrices.

From Eqs. (1-138) to (1-144), the following advantages of the finite-element method may be noted: (1) for the linear elastic problem on hand, the trial functions u_k as well as the test functions v_k need only be continuous and once differentiable; (2) the higher-order boundary conditions (i.e., those on tractions) are included in the weighted residual form Eq. (1-138) and pose no particular difficulties (even though they are satisfied only in an average sense); and (3) complicated geometries and/or nonhomogeneous material properties pose no special problems. It may also be seen that each of the nodal equations in Eq. (1-144) represents in a sense a balance between the summed weighted residual errors in the equilibrium equations in elements meeting at the node, the summed weighted residual error in the traction reciprocity conditions at interelement boundaries meeting at the node, and the weighted residual error in the traction boundary conditions at S_{tm} for any elements meeting at the node. Thus it may be seen that when the number of elements is finite, none of the equations of equilibrium, traction reciprocity, and traction boundary condition is satisfied exactly.

From a computer-implementation point of view, the algebraic equation (1-144), for the linear elastic problem, can be seen to have the properties: (1) the matrix $\mathbf{K}_{11}$ is symmetric and positive definite, and (2) the matrix $\mathbf{K}_{11}$ is banded and sparsely populated. This "bandedness" comes about because each "node" in the system is coupled only to the nodes of all the elements meeting at the node in question. This "local support" for the trial and test functions corresponding to each node is one of the fundamentally novel features of the finite-element method.

It is fair to say that the finite-element method has become one of the most indispensable tools of analysis of problems in varied engineering disciplines. It has been extended to general nonlinear solid mechanics problems of large strain and inelastic material behavior, fluid flows, heat transfer, and other sundry field problems. However, as with any "tool," the finite-element method should be used with caution; it also has, unfortunately, brought in an era of indiscriminate "number crunching" and a broad euphoria that all problems are "solvable" if only a large enough computer is available for a long enough time. Such is not the case, in general; a fundamental understanding of the underlying mechanics of the problem is still often necessary prior to embarking on "computations."

Boundary Element Method

It has been shown above that in the *usual finite-element method* (1) the interior of the domain is discretized into a finite number of elements; (2) neither the differential equation, nor the interelement traction reciprocity condition, nor the traction boundary condition is satisfied exactly; and (3) for linear elasticity, both the trial functions u_k and test functions v_k are required only to be continuous and once-differentiable.

Suppose that it is possible to assume trial and test functions that satisfy the differential equation exactly. Thus the "error" in the interior of the domain is precisely zero. Then one needs only satisfy the boundary conditions in some weighted residual sense. Thus, in principle, in certain problems, one needs to discretize only the boundary of the domain. Such methods are known as "boundary element" methods. For *linear* elasticity problems, there are two such methods that have been studied extensively: (1) the "integral equation" method [41,42], and (2) the "edge-function" method [43].

Here we briefly discuss the fundamental concepts of boundary element methods in linear elasticity, based on an integral equation approach. Consider a *global* trial function u_k (i.e., over the whole domain Ω) and a global test function v_k. Let the compatibility equation (1-92) and stress-strain relation (1-93) be satisfied a priori, that is,

$$\sigma_{ij}(u_k) = E_{ijkl}u_{(k,l)} \tag{1-145}$$

and

$$\sigma_{ij}(v_k) = E_{ijkl}v_{(k,l)} \tag{1-146}$$

In view of the symmetry properties of E_{ijkl} as in Eq. (1-22), σ_{ij} as defined in Eqs. (1-145) and (1-146) is symmetric. Thus the only equations that remain to be satisfied are Eqs. (1-87), (1-90), and (1-95). Note that since no interior elements are contemplated, the interelement conditions (1-126) and (1-127) have no bearing on the present approach. The weighted residual forms of Eqs. (1-90), (1-94), and (1-95) are, respectively,

$$\int_\Omega [\sigma_{ij}(u_k)_{,i} + \rho_0 b_j]v_j \, d\Omega = 0 \tag{1-147a}$$

$$\int_{S_t} [\bar{t}_i - n_j\sigma_{ij}(u_k)]v_i \, ds = 0 \tag{1-147b}$$

$$\int_{S_u} (u_i - \bar{u}_i)t_i(v_k)d\Omega = 0 \tag{1-147c}$$

Using the divergence theorem in Eq. (1-147a), one can "add" Eqs. (1-148a) to (1-147c) to obtain the combined equation [analogous to that in Eq. (1-139)]

$$\int_\Omega [\sigma_{ij}(u_k)v_{(i,j)} - \rho_0 b_j v_j]d\Omega - \int_{S_t} \bar{t}_i v_i \, ds - \int_{S_u} (u_i - \bar{u}_i)n_j\sigma_{ji}(v_k)ds$$
$$- \int_{S_u} \sigma_{ij}(u_k)n_j v_i \, ds = 0 \tag{1-148}$$

Since the material is linear, we have

$$\sigma_{ij}(u_k)v_{(i,j)} = E_{ijkl}u_{(k,l)}v_{(i,j)} \equiv \sigma_{kl}(v_i)u_{(k,l)} \tag{1-149}$$

Using Eq. (1-149) in (1-148) and using the divergence theorem on the volume integral, it is easy to obtain

$$\int_\Omega [\sigma_{ij}(v_k)_{,j}u_i + \rho_0 b_i v_i]d\Omega + \int_{S_t} \bar{t}_i v_i \, ds + \int_{S_u} t_i(u_k)v_i \, ds$$
$$- \int_{S_t} t_i(v_k)u_i \, ds - \int_{S_u} t_i(v_k)\bar{u}_i \, ds = 0 \tag{1-150a}$$

or, more concisely,

$$\int_\Omega [\sigma_{ij}(v_k)_{,j}u_i + \rho_0 b_i v_i]d\Omega + \int_{\partial\Omega} t_i(u_k)v_i \, ds - \int_{\partial\Omega} t_i(v_k)u_i \, ds = 0 \tag{1-150b}$$

where, by definition, $\partial\Omega = S_t + S_u$. It is implied in Eq. (1-150b) that the values of $t_i(u_k)$ and u_i (wherein u_i is the sought-after trial solution) may be prescribed at appropriate locations on $\partial\Omega$.

The key step in the integral equation formulation based on Eq. (1-150b) is in making a specific choice for the test function v_k, that is, the singular solution for a point load in an infinite space, as discussed in Eqs. (1-97) to (1-99) for the three-dimensional case. Let the point load be in the e_1 direction at the location $x_m = \xi_m$. Thus it is seen that v_k satisfies the equation

$$[\sigma_{ij}(v_k)]_{,j} + \delta(x_m - \xi_m)\delta_{il}e_i = 0 \qquad \text{(for any } l = 1, 2, 3) \tag{1-151}$$

where $\delta(x_m - \xi_m)$ is the Dirac function, δ_{il} is the Kronecker delta, and e_l denotes simply the direction of the point load. Using Eq. (1-151) and the property of a Dirac function, it is seen that

$$\int_\Omega [\sigma_{ij}(v_k)]_{,j} u_i(x_m) \, d\Omega = -u_l(\xi_m) \tag{1-152}$$

Note that the displacement in the ith direction at x_m due to a unit load in the lth direction at ξ_m is given by $v_{il}(x_m, \xi_m)$ as in Eq. (1-97), while the traction in the ith direction on an oriented surface with normal cosines n_k at x_m, due to a unit load in the lth direction at ξ_m, is given by $t_{li}(x_m, \xi_m)$ as in Eq. (1-98). Thus, from Eqs. (1-150) and (1-152), one obtains

$$\mu_l(\xi_m) = \int_\Omega \rho_0 b_i(x_m) v_{il}(x_m, \xi_m) \, dv + \int_{\partial\Omega} t_i(u_k, x_m) v_{il}(x_m, \xi_m) \, ds$$

$$- \int_{\partial\Omega} t_{li}(x_m, \xi_m) u_i(x_m) \, ds \tag{1-153}$$

In the above, ξ_m is an interior point, while x_m is the "dummy" variable in Ω or on $\partial\Omega$ as indicated. Note that v_{il} and t_{li} are the Kelvin solutions, which are singular as $\xi_m \to x_m$. When ξ_m is taken in the limit to approach a boundary point, one must consider the Cauchy principal values of the integrals in Eq. (1-153); and thus one obtains [41,42], the so-called boundary-integral equation,[3]

$$\tfrac{1}{2}u_l(\xi_m^b) = \int_\Omega \rho_0 b_i(x_m) v_{il}(x_m, \xi_m^b) \, dv + \int_{\partial\Omega} t_i(x_m^b) v_{il}(x_m^b, \xi_m^b) \, dv$$

$$- \int_{\partial\Omega} t_{li}(x_m^b, \xi_m^b) u_i(x_m^b) \, ds \tag{1-154}$$

wherein a superscript b denotes a "boundary variable."

The so-called boundary element method seeks to satisfy the boundary-integral equation above in a weighted residual sense. Note first that although a volume integral does appear in Eq. (1-154), it does *not* involve the sought-after solution u_i. Thus, although interior discretization may have to be used to evaluate the volume integral, there are no interior elements in the finite-element sense. Over each boundary element, $u_i(\xi_m^b)$ and $t_i(x_m^b)$ may be interpolated. Note that some nodal values $t_i(x_m^b)$ are prescribed at S_t, while some nodal values $u_i(x_m^b)$ are prescribed at S_u. It can be shown [41,42] that the boundary element method, for the present linear elastic problem, leads to equations of the type

$$[A]\{q^*\} = [B]\{Q^*\} \tag{1-155}$$

where $\{q^*\}$ is the global vector of boundary nodal displacements (some of which are specified) and $\{Q^*\}$ is the global vector of boundary nodal tractions (some of which are specified), and $[A]$ and $[B]$ are in general unsymmetric and fully populated.

The boundary element methods are most effective in linear elastic isotropic material problems, especially when the volume/surface ratio is quite high. For anisotropic, nonhomogeneous problems, the singular solution (for a point load in infinite space) is generally not available. The boundary element method has natural advantages for problems involving unbounded domains. As for solution

---

[3]Here it is assumed that the boundary is smooth.

strategies, note that the solution of Eq. (1-155) yields only boundary data, which must be reused in equations of the type Eq. (1-153) to compute the needed interior data at *each interior point*.

It is generally recognized now that the boundary element and the finite-element methods should not be viewed as competing numerical methodologies, but should be used in conjunction with each other to capitalize on the specific advantages each one offers. There exist several ways of rationally coupling these methodologies [42,44].

1-4 Mechanical Properties of Materials

The boundary and initial value problems in Section 1-1 were solved by using the constitutive relation for an isotropic, linearly elastic material as represented by Eqs. (1-26). Since most engineering structures are designed to sustain only elastic loads and are fabricated from isotropic and homogeneous materials, the literature is abundant with data on material constants used in Eqs. (1-26). Analysis of structural components fabricated with composite materials, on the other hand, requires the use of anisotropic elasticity theory, which is based on the generalized Hooke's law of Eq. (1-25). Analysis of problems in metal-forming and ductile fracture are based on the inelastic and plastic responses of materials, particularly those under large deformation. Lesser known are the relaxation and creep moduli and the yield function quoted in the latter part of Section 1-1. Although these and other selected mechanical properties are discussed in the following, the traditional list of various engineering materials is not included in this chapter. For updated listings of mechanical properties, readers are referred to current editions of handbooks, such as those of Refs. 45–47.

Elastic Response

The most common test procedure for determining the elastic constants in Eqs. (1-26) is the uniaxial tension test of a round bar specimen, which is shown in the legend of Fig. 1-1. Figure 1-1 shows a typical test result of a common engineering material (i.e., mild steel) in terms of the Kirchhoff stress σ and infinitesimal strain ε. The former is usually approximated by $\sigma = P/A_0$ and latter by $\varepsilon = \Delta L/L_0$, where P, A_0, ΔL, and L_0 are the applied load, original cross-sectional area, elongation, and original gage length, respectively. Under a "sufficiently small" stress level, the loading and unloading curve in Fig. 1-1 coincides. This portion of the load–unload curve corresponds to the elastic response of the material, and its straight-line portion represents a linear elastic response. The slope of this straight line is the elastic modulus E of the material.

The ratio of transverse strain, which can be determined by the reduction in the specimen diameter, to ε during a linear elastic loading is Poisson's ratio ν. Other elastic constants, such as the Lamé parameters λ and μ, and the bulk modulus K, can be computed once E and ν are known. Conversely, any two of the five elastic constants above will determine the remaining three constants. Thus other test procedures, such as the torsion test, which measures one of the Lamé parameters (i.e., the shear modulus μ) can be used in conjunction with the simple tension test to characterize completely the elastic constants of an isotropic homogenous material.

Plastic Response

The maximum stress, referred to as the yield stress, beyond which the loading and unloading curves differ, is the elastic limit of the material. Beyond the elastic limit, say point B in Fig.

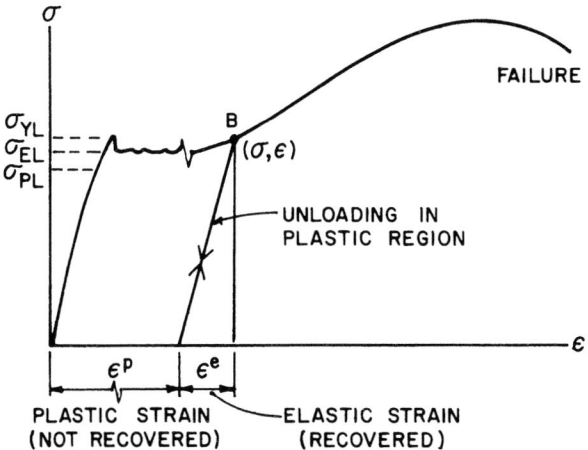

Figure 1-1 Stress-strain curve for mild steel.

1-1, the unloading curve is usually parallel to the elastic loading curve and results in a permanent set or a plastic strain, ε^p, at complete unloading. The reloading curve after a permanent set retraces the unloading curve elastically but reverts to plastic straining at point B. The difference in yield stresses between the virgin and plastically deformed material is caused by its strain-hardening response. The isotropic strain hardening, which is inferred by Eqs. (1-38) and (1-39), in a simple tension specimen is illustrated schematically in Fig. 1-2. The Bauschinger and the kinematic-hardening effects, both of which are inferred by Eq. (1-40), are illustrated schematically in Fig. 1-3 for a simple tension test.

When the material is subjected to multiaxial states of stress or strain, the elastic limit is generally specified by a functional of the stresses or strains [e.g., the yield functions of Eqs. (1-38) through

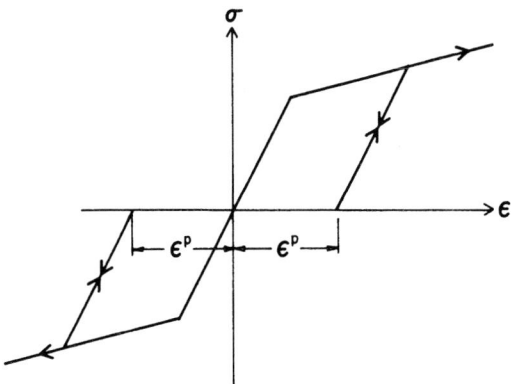

Figure 1-2 Isotropic strain hardening in simple tension.

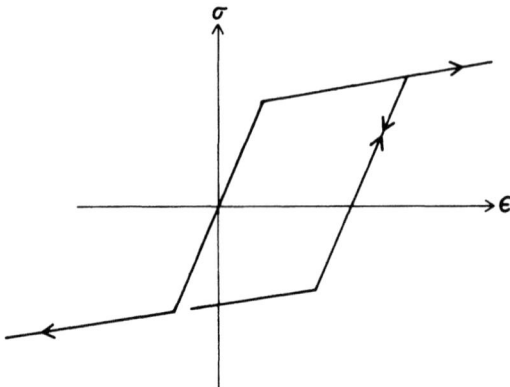

Figure 1-3 Kinematic strain hardening in simple tension.

(1-40)]. The yield functions, which are based on plastic incompressibility and the volumetric strains, are purely elastic, as noted in Section 1-1. The von Mises yield criterion of Eqs. (1-38) through (1-40) can then be represented by the surface of a circular cylinder in the stress space, as shown by Fig. 1–4. Another popular yield criterion is the maximum shear stress or Trescan yield criterion, which is represented by the hexagonal cylinder inscribed to the circular cylinder in Fig. 1-4. In terms of principal stress component, Tresca's yield criterion can be satisfied by any of the following six equations:

$$\sigma_1 - \sigma_2 = \pm \sigma_0$$
$$\sigma_2 - \sigma_3 = \pm \sigma_0 \quad (1\text{-}156)$$
$$\sigma_3 - \sigma_2 = \pm \sigma_0$$

Experimental efforts to verify von Mises's and Tresca's as well as other yield criteria have been summarized in various classical textbooks on plasticity [9,48,49]. Also discussed are the limited experimental verifications of the strain–stress relations of Eqs. (1-45) through (1-47). Although none of these yield criteria or strain–stress relations can universally and accurately match the experimental results, the small deviations in experimental data do not cause large differences in the final structural responses, particularly in the numerical results generated by finite-element and finite-difference methods.

Creep Response

The mechanical properties of all solids are affected to varying degrees by the temperature and rate of deformation. Although such effects are not measurable at "low" temperature, they become noticeable at "high" temperatures relative to the glassy transition temperature for polymers or the melting temperature for metals. Above the glassy transition temperature, many amorphous polymers flow like a Newtonian fluid and are referred to as linear viscoelastic materials. The stress–strain relation of such a material can be represented as a convolution integral:

$$\sigma_{ij}(t) = G_{ijkl}(t)\varepsilon_{kl}(0) + \int_0^\tau G_{ijkl}(t - \tau) \frac{d\varepsilon_{kl}(\tau)}{d\tau} d\tau \quad (1\text{-}157)$$

MECHANICAL RESPONSES OF MATERIALS

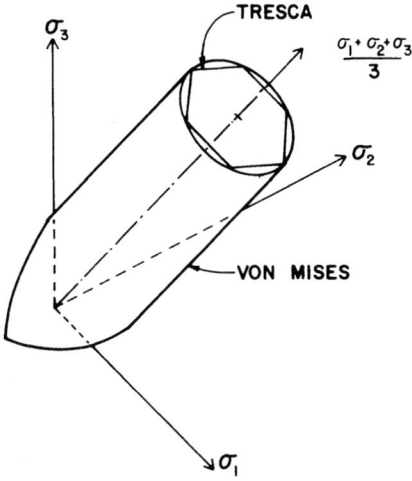

Figure 1-4 Yield surfaces.

where the integrating function $G_{ijkl}(t)$, which is shown in Fig. 1-5, is called the relaxation function. The strain–stress relation, on the other hand, can be represented as

$$\varepsilon_{ij}(t) = \int_{-\infty}^{\tau} J_{ijkl}(t - \tau) \frac{d\sigma_{kl}(\tau)}{d\tau} d\tau \tag{1-158}$$

where the integrating function $J_{ijkl}(t)$, which is shown in Fig. 1-6, called the creep function. These relaxation and creep functions are often modeled by various combinations of linear spring and dashpots, which provide visual representation of the material responses [50,51].

Both relaxation and creep functions change with temperature as shown in Fig. 1-7. For a large class of polymers, these changes can be modeled as a simple shift in the time scale on which the

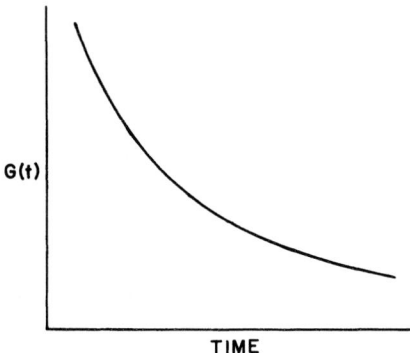

Figure 1-5 Relaxation function.

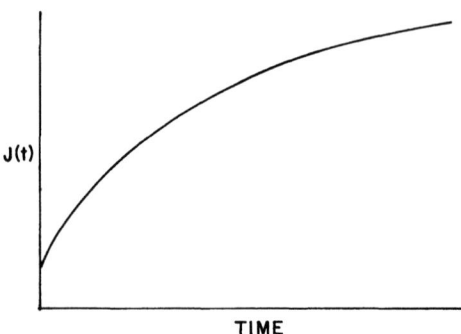

Figure 1-6 Creep function.

relaxation functions all coincide. Such materials are termed thermorheologically simple materials based on the time–temperature superposition principle, which provides a solution procedure for analyzing linear viscoelastic response in a varying temperature field.

The nonlinear viscoelastic or creep response, on the other hand, must be represented by the empirical relation of Eqs. (1-69) through (1-71). For the short-time, uniaxial creep extension of 6% Al–4% V titanium alloy at 650°C as shown in Fig. 1-8, the empirical coefficients of Eqs. (1-69a) and (1-69b) are [52]

$A = 200.10^{-8}$ m/m

$m = 1.66$

$n = 0.605$

When this material is subjected to a three-step loading as shown in Fig. 1-9, the strain-hardening

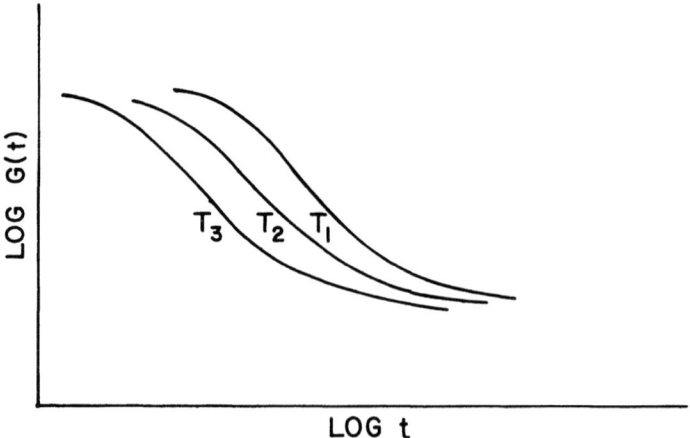

Figure 1-7 Temperature dependence of relaxation functions.

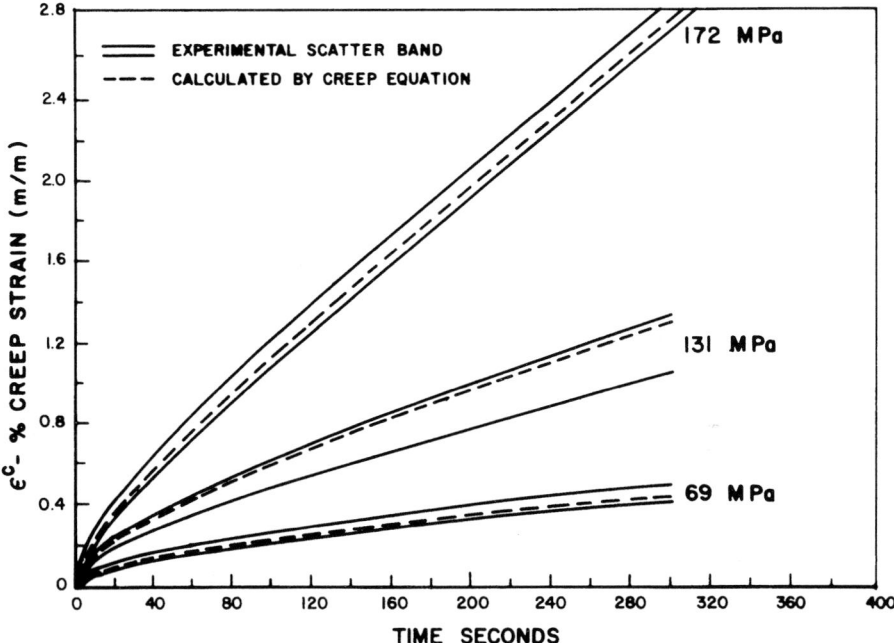

Figure 1-8 Creep strain versus time.

theory of Eq. (1-69b) appears to match the experimental results better than does the time-hardening theory of Eq. (1-69a). Obviously, such an isolated example should not be construed as a general conclusion for the nonlinear and often problem-dependent creep response of structural materials.

1-5 Summary

It is obviously impossible in this extremely brief review even to mention all of the important subjects and recent developments in the theories of elasticity, plasticity, visoelasticity, viscoplasticity, as well as associated numerical techniques. For further details readers are referred to the many excellent books and survey articles, many of which are referenced in the succeeding chapters, in each of the disciplines.

Acknowledgements

The first author thanks Georgia Tech, AFOSR, ONR, NSF, NASA, and other agencies that enabled him to learn and think about the various topics discussed herein. He also thanks Ms. J. Webb for her assistance in the preparation of the manuscript.

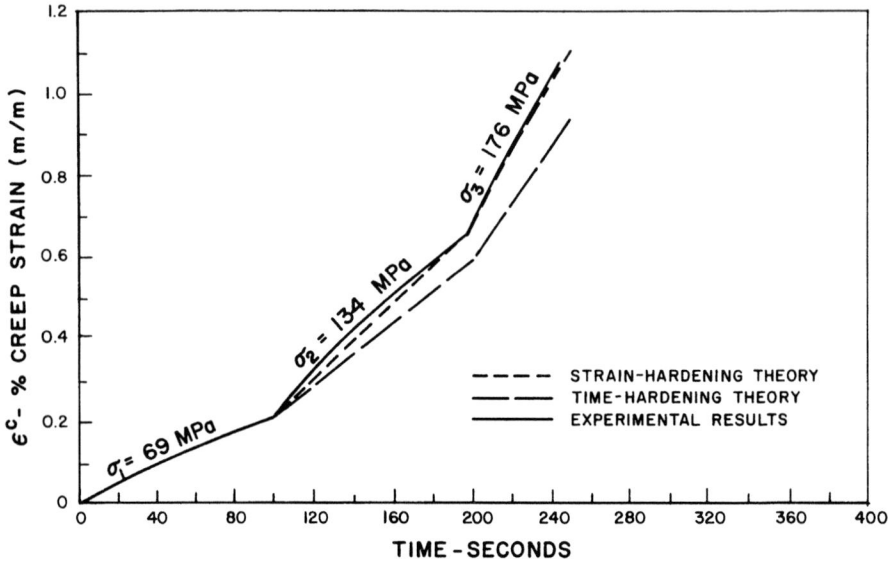

Figure 1-9 Cumulative creep strain versus time.

References

1. H. Kardestuncer, F. Brezzi, S. N. Atluri, D. Norrie, and W. Pilkey (Eds.), *Handbook of Finite Elements*, McGraw-Hill, New York, 1984.

2. C. Truesdell (Ed.), *Encyclopedia of Physics*, Vol. VIa/2, *Mechanics of Solids II*, Springer-Verlag, New York, 1972.

3. S. Flugge (Ed.), *Encyclopedia of Physics*, Vol. III/3, *The Nonlinear Field Theories of Mechanics* (by C. Truesdell and W. Noll), Springer-Verlag, New York, 1965.

4. S. N. Atluri, Alternate Stress and Conjugate Strain Measures, and Mixed Variational Formulations Involving Rigid Rotations, for Computational Analyses of Finitely Deformed Solids, with Application to Plates and Shells—I. Theory, *Comput. Struct.,18*, no. 1 (1984), 93–116.

5. S. N. Atluri, On Some New General and Complementary Energy Theorems for the Rate Problems in Finite Strain, Classical Elastoplasticity, *J. Struct. Mech.*, 8, no. 1 (1980), 61–92.

6. A. C. Eringen, *Nonlinear Theory of Continuous Media*, McGraw-Hill, New York, 1962.

7. M. Mooney, A Theory of Large Elastic Deformation, *J. Appl. Phys.*, 11, (1940), 582–592.

8. G. I. Taylor and H. Quinney, The Plastic Deformation of Metals, *Phil. Trans. R. Soc. London, A* 230 (1931) 323–362.

9. R. Hill, *The Mathematical Theory of Plasticity*, Oxford University Press, New York, 1950.

10. W. Prager, A New Method of Analyzing Stresses and Strains in Work-Hardening Plastic Solids, *J. Appl. Mech.*, 23 (1956), 493–496.

11. H. Ziegler, A Modification of Prager's Hardening Rule, *Q. Appl. Math.*, 17 (1959), 55–65.

12. D. C. Drucker, A More Fundamental Approach to Plastic Stress–Strain Relations, *Proc. 1st U.S. Nat. Cong. Appl. Mech.*, 1951, 487–491.

13. S. Nemat-Nasser, Continuum Bases for Consistent Numerical Formulations of Finite Strains in Elastic and Inelastic Structures, in *Finite Element Analysis of Transient Nonlinear Structural Behavior* (T. Belytschko et al., Eds.) AMD Vol. 14, ASME, New York, 1975, 85–98.

14. S. N. Atluri, On Constitutive Relations at Finite Strain: Hypoelasticity and Elastoplasticity with Isotropic or Kinematic Hardening, *Comput. Methods Appl. Mech. Eng.*, 43 (1984), 137–171.

15. Y. Yamada, N. Yoshimura, and T. Sakurai, Plastic Stress–Strain Matrix and Its Application to the Solution of Elastic–Plastic Problems by the Finite Element Method, *Int. J. Mech. Sci.*, 10 (1968), 343–354.

16. K. C. Valanis, Fundamental Consequences of a New Intrinsic Tune Measure: Plasticity as a Limit of the Endochronic Theory, *Arch. Mech.*, 32, no. 2 (1980), 171–191.

17. Z. Mroz, An Attempt to Describe the Behavior of Metals Under Cyclic Loads Using a More General Workhardening Model, *Acta Mech.*, 7 (1969), 199–212.

18. O. Watanabe and S.N. Atluri, Constitutive Modeling of Cyclic Plasticity and Creep Using an Internal Time Concept, *Int. J. Plasticity*, 2, no. 2 (1986), 107–134.

19. E. C. Bingham, *Fluidity and Plasticity*, McGraw-Hill, New York, 1922.

20. K. Hohenemser and W. Prager, Über die Ansatze der Mechanik isotroper Kontinua, *Z. Angew. Math. Mech.*, 12 (1932), 216–226.

21. I. Finnie and W. R. Heller, *Creep of Engineering Materials*, McGraw-Hill, New York, 1959.

22. P. Perzyna, The Constitutive Equations for Rate Sensitive Plastic Materials, *Q. J. Appl. Math. Mech.*, XX, no. 4 (1963), 321–332.

23. O. C. Zienkiewicz and I. C. Cormeau, Visco-plasticity, Plasticity, and Creep in Elastic Solids—A Unified Numberial Solution Approach, *Int. J. Num. Methods Eng.*, 8 (1974), 821–845.

24. J. H. Argyris and M. Kleiber, Incremental Formulation in Nonlinear Mechanics and Large Strain Elasto-Plasticity—Natural Approach—Part I, *Comput. Methods Appl. Mech. Eng.*, 11 (1977), 215–247.

25. W. Thomson (Lord Kelvin), On the Equations of Equilibrium of an Elastic Solid, *Cambridge Dublin Math. J.*, 3 (1848), 87–89.

26. J. Boussinesq, Equilibre d'élasticité d'un solide istorope sans pesanteur, supportant différents poids, *C.R. Acad. Sci. Paris*, 86 (1878), 1260–1263.

27. A. E. H. Love, *A Treatis on Mathematical Theory of Elasticity*, Dover, New York, 1963.

28. N. I. Mushkelishvili, *Some Basic Problems of the Mathematical Theory of Elasticity*, Noordhoff, Groningen, The Netherlands, 1953.

29. G. N. Savin, *Stress Concentration Around Holes*, Pergamon Press, Elmsford, NY, 1961.

30. J. D. Eshelby, The Determination of the Elastic Field of an Ellipsoidal-Inclusion and Related Problems, *Proc. R. Soc. London*, A241 (1957), 376–396.

31. A. S. Kobayashi, Linear Elastic Fracture Mechanics, in *Computational Methods in the Mechanics of Fracture* (S.N. Atluri, Ed.), North-Holland, Amsterdam, 1985, 21–53.

32. G. C. Sih and H. Liebowitz, Mathematical Theories of Brittle Fracture, in *Fracture, Vol. II* (H. Liebowitz, Ed.), Academic Press, New York, 1968.

33. T. Nishioka and S. N. Atluri, Path-Independent Integrals, Energy Release Rates, and General Solutions of Near-Tip Fields in Mixed Mode Dynamic Fracture Mechanics, *Eng. Fract. Mech.*, 18, no. 1 (1983), 1–22.

34. K. Vijayakumar and S. N. Atluri, An Embedded Elliptical Flaw in an Infinite Solid Subjected to Arbitrary Crack Face Tractions, *ASME J. Appl. Mech.*, 48 (1981), 88–96.

35. S. N. Atluri and H. Murakawa, in *On Hybrid Finite Element Models in Nonlinear Solid Mechanics* (P.G. Bergan et al., Eds.), Tapir Press, Trondheim, Norway, 1978, 3–40.
36. K. W. Reed and S. N. Atluri, Analyses of Large Quasistatic Deformations of Inelastic Bodies by a New Hybrid-Stress Finite Element Algorithm, *Comput. Methods Appl Mech. Eng.*, 39 (1983), 245–295.
37. K. J. Bathe, *Finite Element Procedures in Engineering Analysis*, Prentice-Hall, Englewood Cliffs, NJ, 1975.
38. L. Collatz, *Numerical Treatment of Differential Equations* (translated by P. G. Williams), Springer-Verlag, Berlin, 1960.
39. S. N. Atluri, On Hybrid Finite Element Methods in Solid Mechanics, in *Advances in Computer Methods of Partial Differential Equations* (S. Vishnevetsky, Ed.), AICA, Rutgers University, 1975, 346–356.
40. S. N. Atluri, R. H. Gallagher, and O. C. Zienkiewicz (Eds.), *Hybrid and Mixed Finite Element Methods*, Wiley, New York, 1983.
41. T. A. Cruse, *Mathematical Foundations of the Boundary Integral Equation Method in Solid Mechanics*, AFOSR TR 77-1002, U.S. Air Force, 1977.
42. S. N. Atluri and J. J. Grannell, *Finite Elements, Boundary Elements, and Combined FEM, BEM*, Tech. Rep., Center for the Advancement of Computational Mechanics, Georgia Institute of Technology, 1978.
43. P. M. Quinlan, J. J. Grannell, S. N. Atluri, and J. E. Fitzgerald, The Edge Function Method for Three-Dimensional Stress Analysis, Including Embedded Elliptical Cracks and Surface Flaws, in *Boundary Element Methods in Engineering* (C. A. Brebbia, Ed.), Springer-Verlag, New York, 1982, 457–471.
44. S. N. Atluri and T. Nishioka, Hybrid Methods of Analysis, in *Unification of Finite Element Methods* (H. Kardestuncer, Ed.), North-Holland, Amsterdam, 1984.
45. *Metals Handbook*, American Society for Metals, Metals Park, OH; use latest edition.
46. *Index to Standards*, American Society for Testing and Materials, Philadelphia; use latest edition.
47. *Data Book*, Metal Progress; use latest edition.
48. W. Prager and P. G. Hodges, *Theory of Perfectly Plastic Solids*, Wiley, New York, 1951.
49. W. Johnson and P. B. Mellor, *Plasticity for Engineers*, D. Van Nostrand, Princeton, NJ, 1962.
50. D. R. Bland, *The Theory of Linear Viscoelasticity*, Pergamon Press, Elmsford, NY, 1960.
51. R. M. Christensen, *Theory of Viscoelasticity, An Introduction*, Academic Press, New York, 1971.
52. A. D. Russell and A. S. Kobayashi, Short-Time Creep Response of 6% Al–4% V Titanium Alloy Subjected To Variable Uniaxial Tensile Loading, *Proc. Jt. Int. Conf. Creep*, Institute of Mechanical Engineers, London, 1963, 4-33-4-37.

General References. The following additional references, although not specifically cited in the text, provide useful information concerning the topics discussed herein.

53. A. P. Boresi and P. P. Lynn, *Elasticity in Engineering Mechanics*, Prentice-Hall, Englewood Cliffs, NJ, 1964.
54. Y. C. Fung, *Foundations of Solid Mechanics*, Prentice-Hall, Englewood Cliffs, NJ, 1965.
55. Y. C. Fung, *A First Course in Continuum Mechanics*, Prentice-Hall, Englewood Cliffs, NJ, 1969.
56. A. E. Green and W. Zerna, *Theoretical Elasticity*, Oxford University Press, New York, 1954.
57. A. E. Green and J. E. Adkins, *Large Elastic Deformations and Nonlinear Continuum Mechanics*, Oxford University Press, New York, 1960.
58. M. A. Biot, *Mechanics of Incremental Deformations*, Wiley, New York, 1965.
59. A. C. Eringen, *Nonlinear Theory of Continuous Media*, McGraw-Hill, New York, 1962.
60. L. E. Malvern, *Introduction to the Mechanics of a Continuous Medium*, Prentice-Hall, Englewood Cliffs, NJ, 1969.
61. W. Prager, *Introduction to Mechanics of Continua*, Ginn and Company, Lexington, MA, 1961.
62. K. Washizu, *Variational Methods in Elasticity & Plasticity*, 3rd ed., Pergamon Press, Elmsford, NY, 1982.

63. E. W. Billington and A. Tate, *The Physics of Deformation and Flow*, McGraw-Hill, New York, 1981.
64. W. Prager, *An Introduction to Plasticity*, Addison-Wesley, Reading, MA, 1959.
65. A. Mendelson, *Plasticity: Theory and Application*, Macmillan, New York, 1959.
67. T. Y. Thomas, *Plastic Flow and Fracture in Solids*, Academic Press, New York, 1961.
68. A. K. Noor and W. D. Pilkey (Eds.), *State-of-the-Art Surveys on Finite Element Technology*, American Society of Mechanical Engineers, New York, 1983.
69. L. I. Sedov, *Introduction to the Mechanics of a Continuous Medium*, Addison-Wesley, Reading, MA, 1965.
70. R. D. Cook, *Concepts and Applications of Finite Element Analysis*, Wiley, New York, 1981.
71. O. C. Zienkiewicz, *The Finite Element Method in Engineering Science*, 3rd ed., McGraw-Hill, New York, 1977.
72. R. H. Gallagher, *Finite Element Analysis: Fundamentals,*, Prentice-Hall, Englewood Cliffs, NJ, 1975.
73. K. J. Bathe, *Finite Element Procedures in Engineering Analysis*, Prentice-Hall, Englewood Cliffs, NJ, 1982.
74. S. S. Rao, *The Finite Element Method in Engineering*, Pergamon Press, Elmsford, NY, 1982.
75. B. M. Irons and S. Ahmad, *Techniques of Finite Elements*, (Ellis Horwood Series), Wiley, New York, 1981.
76. P. K. Banerjee, *The Boundary Element Method in Engineering Science*, McGraw-Hill, New York, 1981.

CHAPTER

2

Strain Gages

James W. Dally
University of Maryland, College Park, Maryland

William F. Riley
Iowa State University, Ames, Iowa

James S. Sirkis
University of Maryland, College Park, Maryland

2-0 Introduction

Stress analysis is often performed by measuring strains on the surfaces of prototype or production components. Of the many experimental methods available for measuring strain (mechanical, acoustical, electrical, optical, moiré, grids, etc.), the electrical-resistance strain gage is the most common choice of the great majority of experimentalists. This choice is based on the fact that the electrical-resistance strain gage is usually the most cost-effective method of measurement because the gage exhibits most of the optimum characteristics commonly used to judge the adequacy of a strain-gage system. These characteristics are:

1. The calibration constant for the gage should be stable with respect to both temperature and time.
2. The gage should be capable of measuring strains with an accuracy of ± 1 μm/m (μin./in.) over a range of $\pm 5\%$ strain ($\pm 50,000$ μm/m).
3. The gage length and width should be small so that the measurement approximates strain at a point.
4. The inertia of the gage should be minimal to permit the recording of high-frequency dynamic strains.
5. The response or output of the gage should be linear over the entire strain range of the gage.
6. The gage and its associated electronics should be economical.
7. Installation and readout of the gage should require minimal skills and understanding.

Perry [1] estimated that the electrical-resistance strain gage was used for 80% or more of the industrial experimental stress analysis performed in the United States in the 1980s. Also, electrical-resistance strain gages are widely used as sensors in transducers that have been developed to measure load, torque, and pressure.

The electrical-resistance strain gage is based on principles established by Lord Kelvin [2] in a classic series of experiments conducted in 1856. In these experiments, Kelvin proved that:

1. Wires subjected to loads and/or strains change resistance.
2. Wires fabricated from different materials exhibit different strain sensitivities.
3. Resistance changes in the wires resulting from the imposed strains are small, but they can be accurately measured with a Wheatstone bridge.

Early development of the electrical-resistance strain gage in the United States occurred in the 1930s, and gages were independently fabricated by Simmons [3] and Ruge. These laboratory gages eventually led to the SR-4 type of strain gage that was marketed exclusively as a wire-grid/paper-carrier gage for nearly two decades.

Metal-foil strain gages were first developed by Sanders and Roe in England in 1952. The photographic production of a master image of the grid configuration followed by a foil-etching process used to produce this type of electrical-resistance strain gage permits miniaturization of the grids, versatility of the grid configuration, economies of production, and enhanced control of quality of the gage. After the metal-foil strain gage was introduced in the United States in the late 1950s, it rapidly replaced the wire-grid SR-4 strain gage except for a very few special applications.

Semiconductor strain gages were developed as a by-product of research at Bell Telephone Laboratories with semiconducting materials and junctions that eventually led to the development of the transistor. The characterization of semiconducting silicon and germanium by Smith [4] and the initial development of laboratory transducers by Mason and Thurston [5] provided the foundations for the commercial development of piezoresistive strain gages that became available in the 1960s.

The metal alloys used in the production of metal-foil electrical-resistance strain gages have maximum strain limits of approximately $\pm 5\%$. For very-large-strain applications where specimen elongations of 100% may be encountered, liquid-metal strain gages have been developed. The liquid-metal strain gage is simply a rubber tube filled with mercury or a gallium–indium–tin alloy. The tube is formed into a grid configuration and bonded to the specimen. When subjected to strain, the volume of the tube cavity remains constant, since Poisson's ratio ν of rubber is approximately 0.50. The resistance of the liquid-metal strain gage increases with a tensile strain, however, since the length of the tube increases and the diameter of the tube decreases to maintain the constant volume. Liquid-metal strain gages require care in application because they exhibit a very small resistance due to the relatively large diameter of the tube and because they exhibit a nonlinear response to increasing strain.

The three basic types of electrical-resistance strain gage, including metal foil, semiconductor, and liquid metal, will be reviewed in this chapter with the primary emphasis on the metal-foil gage, which has been the standard for the past 20 years.

2-1 Strain Sensitivity

Kelvin [2] first observed that a change in the strain imposed on a wire is accompanied by a change in resistance ΔR of the wire. The relationship between resistance change ΔR and strain ε can be derived by considering a uniform conductor of length L, cross-sectional area A, and specific resistance ρ. The resistance R of such a conductor is given by the expression

$$R = \frac{\rho L}{A} \qquad (2\text{-}1)$$

STRAIN GAGES

Differentiation of Eq. (2-1) and division of the result by R gives

$$\frac{dR}{R} = \frac{d\rho}{\rho} + \frac{dL}{L} - \frac{dA}{A} \tag{2-2}$$

However,

$$\frac{dA}{A} = -2\nu \frac{dL}{L} + \nu^2 \left(\frac{dL}{L}\right)^2 \tag{2-2a}$$

provided the deformation of the conductor is elastic. The strain sensitivity S_A of the metal used to fabricate the conductor is defined as

$$S_A = \frac{dR/R}{dL/L} \tag{2-3}$$

Substituting Eqs. (2-2) and (2-2a) into Eq. (2-3) yields

$$S_A = \frac{d\rho/\rho}{dL/L} + 1 + 2\nu - \nu^2 \left(\frac{dL}{L}\right) \tag{2-4}$$

The last three terms in Eq. (2-4) are due to the dimensional changes in the conductor (dL and dA). The first term is due to the change in specific resistance with strain. The last term in Eq. (2-4) is usually neglected for elastic strains in metals, since it is small ($<0.1\varepsilon$) compared to the two middle terms (≈ 1.60). For large strains, the approach to the derivation of Eq. (2-4) must be modified, since the conductor undergoes plastic deformation.

Note that the volume V of the conductor is $V = AL$, which can be substituted into Eq. (2-1) to give

$$R = \frac{\rho L^2}{V} \tag{2-5}$$

Differentiating and dividing by R gives

$$\frac{dR}{R} = \frac{d\rho}{\rho} + 2\frac{dL}{L} - \frac{dV}{V} \tag{2-6}$$

Under plastic deformation $d\rho/\rho \to 0$ and $dV = 0$; therefore, Eq. (2-6) reduces to

$$\frac{dR}{R} = 2\frac{dL}{L} \tag{2-7}$$

If the strains are plastic but small, the alloy sensitivity obtained by substituting Eq. (2-7) into Eq. (2-3) is $S_A = 2$. If the plastic strains are not small, the change in resistance ΔR must be utilized in place of the derivative dR because second-order terms cannot be neglected. The change in resistance ΔR can be determined by integrating Eq. (2-7) as follows:

$$\int_{R_0}^{R} \frac{dR}{R} = 2 \int_{L_0}^{L} \frac{dL}{L}$$

$$\ln \frac{R}{R_0} = 2 \ln \frac{L}{L_0} \quad \text{or} \quad \ln\left(1 + \frac{\Delta R}{R_0}\right) = \ln\left(1 + \frac{\Delta L}{L}\right)^2$$

Table 2-1 Strain Sensitivity S_A and Composition for Common Strain-Gage Alloys

Material	Composition (%)	S_A
Advance or constantan	45 Ni, 55 Cu	2.1
Karma	74 Ni, 20 Cr, 3 Al, 3 Fe	2.0
Isoelastic	36 Ni, 8 Cr, 0.5 Mo, 55.5 Fe	3.6
Nichrome V	80 Ni, 20 Cr	2.1
Platinum–tungsten	92 Pt, 8 W	4.0
Armour D	70 Fe, 20 Cr, 10 Al	2.0

Thus

$$\frac{\Delta R}{R_0} = 2\frac{\Delta L}{L_0} + \left(\frac{\Delta L}{L_0}\right)^2 \tag{2-8}$$

If $S_A = (\Delta R/R_0)/(\Delta L/L_0)$, then from Eq. (2-8) it is evident that

$$S_A = 2 + \varepsilon \tag{2-9}$$

where $\varepsilon = \Delta L/L_0$ is the engineering strain.

In this instance, where the plastic strains are large, the alloy sensitivity S_A is not a constant, but instead varies linearly with engineering strain ε. Although Eq. (2-9) is well known and widely used in practice, the assumptions of $d\rho/\rho \to 0$ and $dV/V = 0$ have yet to be verified for the metal-foil gage.

A list of the metallic alloys commonly used in the manufacture of commercially available strain gages is presented in Table 2-1 together with typical values of their strain sensitivity S for elastic deformation. Actual values of S_A for a specific lot of material will depend on trace impurities in the alloy, the degree of cold working used in producing the thin foil, and the range of strain used to determine S_A. Note that S_A varies from 2.0 to 4.0 for these common alloys. This fact implies that the specific resistance term $(d\rho/\rho)/(dL/L)$ is a significant contributor to strain sensitivity S_A since the term $(1 + 2\nu)$, which is approximately equal to 1.6, does not totally account for the magnitude of S_A. Indeed, for semiconductor gages, the specific resistance term $(d\rho/\rho)/(dL/L)$ is dominant and the contribution from the term $1 + 2\nu$ is usually neglected.

Advance is the alloy used for most strain gages, since its strain sensitivity S_A is constant over a very wide range of strain and the magnitude of S_A does not appear to change significantly as the alloy undergoes the transition from elastic to plastic deformation. The Advance alloy also has a high specific resistance ($\rho = 0.49\ \mu\Omega \cdot$ m) and excellent thermal stability. The response diagram ($\Delta R/R$ as a function of $\Delta L/L$) shown in Fig. 2-1 indicates that S_A is essentially constant for the Advance alloy over the strain range from 0 to 8%. In an annealed state, the Advance alloy can be utilized to measure strains up to 20%; however, in the annealed condition, the Advance alloy exhibits zero shift, indicating a permanent resistance change with each cycle of loading. Another advantage of the Advance alloy is the ability to temperature compensate the strain gage by matching the temperature coefficient of expansion (TCE) of the gage with that of many different engineering materials (within the range 0 to 50 ppm/°F).

Karma alloy is also widely used, and it has three advantages over the Advance alloy. First, it can be temperature compensated over a wider range of temperatures than Advance alloy. Second, the nickel–chromium content of Karma alloy gives it a high strength, and the performance of the Karma gages in fatigue exceeds that of Advance gages. Third, the alloy exhibits excellent stability with time; therefore, Karma is the preferred alloy when measuring static strains over periods ranging from months to years. The primary disadvantage of Karma is the difficulty encountered in soldering lead wires to the grid tabs.

STRAIN GAGES

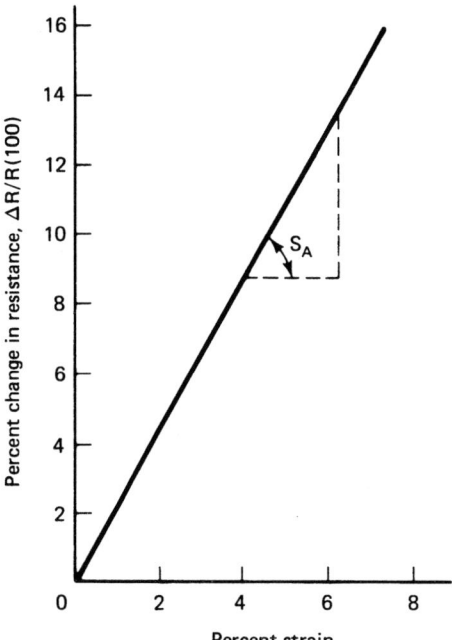

Figure 2-1 Percent change in resistance as a function of percent strain for Advance alloy.

Isoelastic alloy has the advantage of high sensitivity ($S_A = 3.6$) and the highest fatigue strength of the strain-gage alloys. Since the alloy is extremely sensitive to temperature, it cannot be temperature compensated; therefore, its use is limited to dynamic or cyclic measurements where temperature-induced instabilities are not of importance.

Nichrome V, platinum–tungsten, and Armour D alloys are used in very special applications where high temperatures are involved and resistance to oxidation becomes important.

2-2 Bonded Resistance Strain Gages

Early designs of the electrical-resistance strain gage (the unbonded electrical-resistance strain gage) were fabricated with a wire filament as the sensor and a special frame as a carrier that was attached to the specimen through knife edges and a clamping arrangement designed for the particular application. These strain transducers were reusable but they were large and bulky, exhibited limited dynamic response, and were awkward to mount. The primary advantage of the "new" gage developed by Simmons and Ruge was the feature that permitted the gage to be bonded to the specimen with an adhesive.

The bonded resistance strain gage in its present form consists of a strain-sensing element, a thin film that serves as an insulator and a carrier for the strain-sensing element, and tabs (or terminals) for lead-wire connections.

The strain-sensing element (illustrated in Fig. 2-2) consists of a grid that is either stamped or

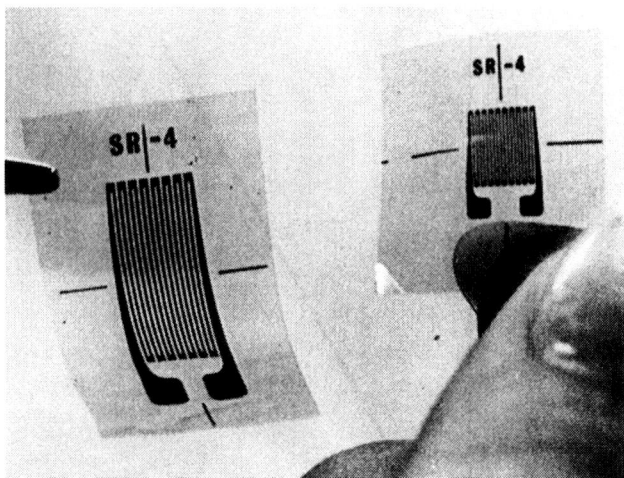

Figure 2-2 Electrical-resistance strain gages. (Courtesy of BLH Electronics.)

photoetched from a very thin sheet of metallic foil. The grid configuration is used to develop the required resistance ($\approx 100\ \Omega$) of the sensing element while keeping the length l_0 and the width w_0 of the gage small. The size of the grid is adjusted by the manufacturer to suit a wide variety of applications. Gage lengths are available throughout the range from 0.178 mm (0.007 in.) to 150 mm (6 in.). The size of the sensing element is selected by the experimentalist to minimize error due to nonlinear strain gradients, and the alloy is selected to accommodate test conditions, including temperature, duration of the measurement, and dynamic or cyclic loading.

Multiple-element (chain) gages are available with up to 10 sensing elements arranged in a row. These chain gages are usually used at a structural discontinuity, such as a fillet, where a high strain gradient exists and the location of the peak strain is not known. Two-element rectangular rosettes are available in either a stacked or a planar configuration for applications where the principal directions are known. Three-element rosettes in either delta (60°) or rectangular (45°) orientations are available for the general two-dimensional problem where both the principal strains and the principal directions are unknown. Special-purpose gages such as the diaphragm pressure gage, the stress gage, and the shear-strain gage are also available. The primary point to be made here is that the manufacturing of strain gages is a mature industry; the processes are versatile and under excellent control. The result is a wide selection of standard grid configurations and many special-purpose grid configurations that extend the range of applications of the gage. Some of the numerous different grid designs are displayed in Fig. 2-3.

The strain-sensing foils are extremely thin [approximately 0.0001 in. (0.0025 mm)] and very fragile. Therefore, they must be supported by a thin film that is bonded to the foil prior to photoetching. The film serves as a carrier during production and testing before packaging and for handling the sensing element prior to installation. The carrier also serves as an insulator after installation.

The different materials employed as carriers in gage construction include acrylic, polyimide, polyimide–epoxy, epoxy, epoxy–phenolic, epoxy–glass, phenolic, phenolic–glass, and paper. The carrier material employed with the wire gages of the 1940s was paper; however, in most instances, paper has been replaced with thin-film polymers. Polyimide, a carrier material that is flexible,

STRAIN GAGES

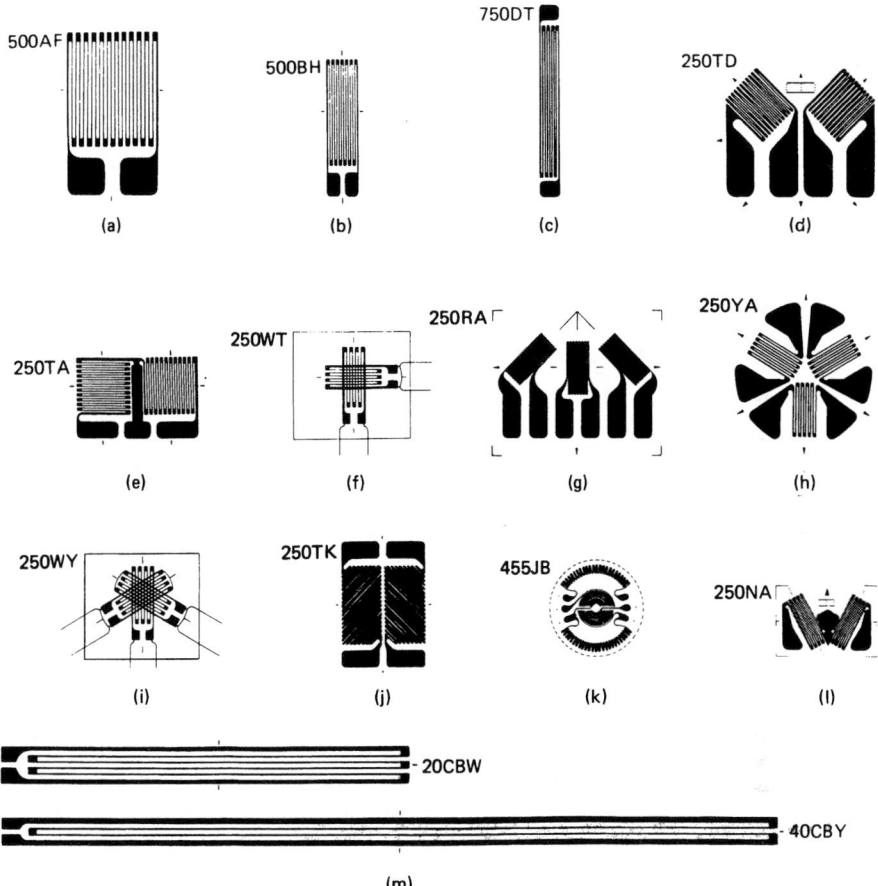

Figure 2-3 Selection of metal-foil electrical-resistance strain gages: (a) single-element gage; (b) single-element gage; (c) single-element gage; (d) two-element rosette; (e) two-element rosette; (f) two-element stacked rosette; (g) three-element rosette; (h) three-element rosette; (i) three-element stacked rosette; (j) torque gage; (k) diaphragm gage; (l) stress gage; (m) single-element gages for use on concrete. (Courtesy of Measurements Group, Inc.)

tough, and compatible with most adhesive systems, is used for most general-purpose gages. A very thin epoxy film is used for gages designed for transducer applications. The linear elastic behavior of this carrier material helps the designer to meet the tight linearity and hysteresis specifications normally imposed on transducer performance. Glass-fiber-reinforced polymers are used for gages that will be exposed to large-magnitude cyclic strains. The reinforcement provided by the carrier and the overlayer (encapsulated gages) enhances the fatigue life of the gage. Glass-fiber-reinforced epoxy and phenolic carriers are used for gages that will be exposed to elevated

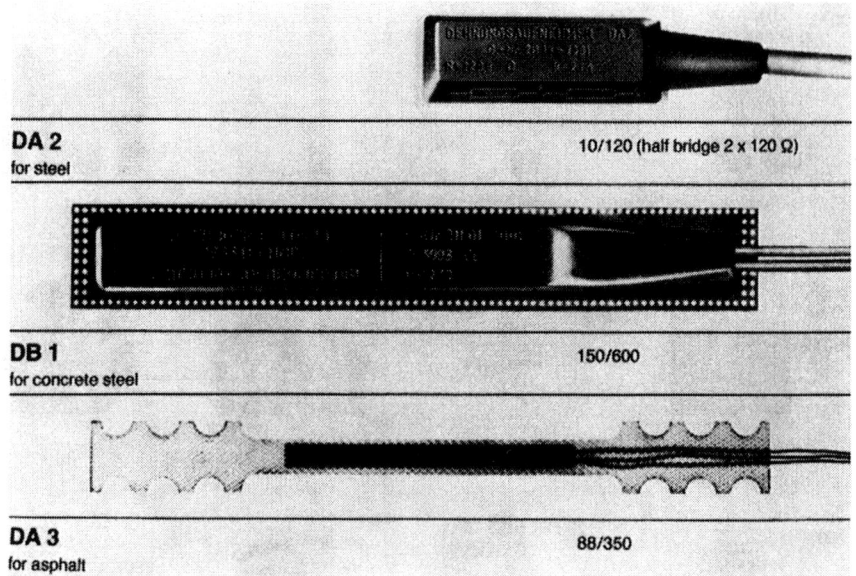

Figure 2-4 Encapsulated electrical-resistance strain gages. (Courtesy of Hottinger Baldwin Measurements, Inc.)

temperatures. The glass fibers extend the operating temperature range of these carrier materials by inhibiting viscoelastic deformation in the polymer film. For very-high-temperature applications, the carrier is removed and the gage grid is bonded and insulated with a ceramic adhesive.

In certain field applications it is advantageous to employ encapsulated strain gages where environmental conditions are very severe. These encapsulated gages, illustrated in Fig. 2-4, are protected from weather and abuse. During fabrication, the gages are mounted on a thin sheet of stainless steel or brass and then covered with sheet metal, rubber, or polycarbonate.

Another form of gage construction for field application and for high-temperature measurement of strain is the weldable gage illustrated in Fig. 2-5. In this gage, the strain-sensing element is in the form of a thin Ni-Cr wire that is swaged into a stainless steel tube. The swaging operation compacts MgO powder about the strain-sensing wire to provide support and high-temperature insulation. The stainless steel tube is brazed to a thin stainless steel shim, which in turn is spot welded to the structure under investigation.

Strain gages with flexible carriers are usually mounted to the specimen by using the tape-transfer technique [6]. This technique simplifies handling of the gage and provides a means for precisely locating and orienting the gage on the specimen during the bonding process.

Gages should be bonded to the specimen with an adhesive that is strong, linearly elastic, and stable over extended periods of time. Many adhesives are available that meet these requirements; however, the hardware-store variety of adhesive should not be used, since it has not been formulated specifically for use with strain gages and it may not be sufficiently stable to ensure accurate strain measurements. Strain-gage manufacturers provide several different adhesives which are formulated specifically for different strain applications. The combination of gage, carrier, and adhesive forms the strain-measuring transducer, and care should be exercised

STRAIN GAGES

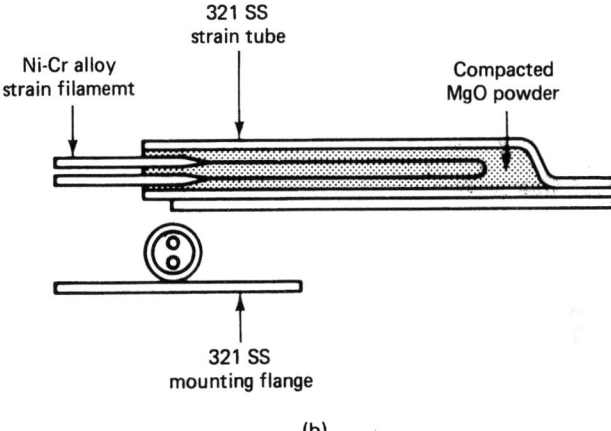

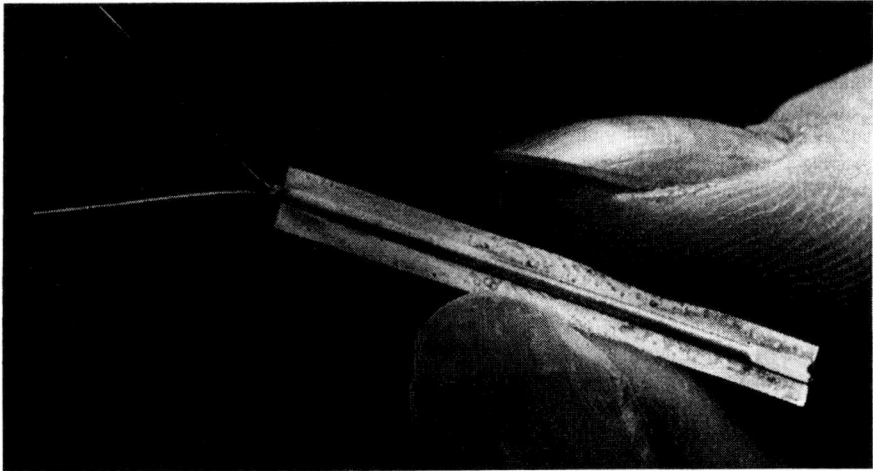

Figure 2-5 Weldable type of electrical-resistance strain gage; (a) external appearance; (b) construction details.

to ensure the integrity of the system by using certified adhesives and the proper application and curing procedures. The most commonly employed adhesives are methyl-2-cyanoacrylate, epoxy, polyimide, and several ceramics.

Cyanoacrylate adhesives do not require either heat or a hardening agent to induce polymerization. When this adhesive is spread in a thin film between the strain gage and the specimen and a slight pressure is applied, the small traces of water and/or weak bases present on the surfaces are sufficient to induce polymerization. A catalyst can be applied to one of the surfaces to enhance the polymerization process, but it is not essential. The very rapid polymerization of the cyanoacrylate adhesive makes it ideal for general-purpose strain-gage applications. One minute of thumb pressure followed by a 2-minute pause before tape removal and the polymerization that occurs during lead wire attachment is adequate. The cyanoacrylate can be used over the temperature range from -32 to $65°C$ (-25 to $+150°F$). It is capable of transmitting strains up to 6%. Since the strength of this adhesive is degraded by time and moisture absorption, the installation must be waterproofed if the gage is to be maintained over any extended period of time.

An epoxy adhesive consists of a monomer and a hardening agent that reacts with the monomer to produce polymerization. In some instances, a solvent is added to the monomer to reduce its viscosity. The solvent-thinned epoxies (or epoxy–phenolics) are preferred because they produce extremely thin, high-strength, and void-free glue lines with low creep and hysteresis. A pressure of 70 to 210 kPa (10 to 30 psi) must be applied to the gage during the curing process to ensure a thin, void-free bond. Epoxy adhesives are usually cured at elevated temperatures for several hours to ensure complete polymerization. Epoxy adhesives, formulated for specific strain-gage applications, are commercially available in kit form with the components preweighed to facilitate preparation of the adhesive. The operating temperature range of the epoxy adhesives depends on the formulation. The epoxy–phenolic is probably the best, with a range of -269 to $260°C$ (-452 to $500°F$). The elongation capabilities also depend on formulation and are usually in the neighborhood of 3 to 10%.

Polyimide is a single-component polymer that can be used over a very wide temperature range of -269 to $399°C$ (-452 to $750°F$). The polyimide is cured under pressure 275 kPa (40 psi) at a temperature of $260°C$ ($500°F$). This adhesive is useful for strain measurements at elevated temperatures up to $315°C$ ($600°F$). The polyimide is capable of strains in excess of 2% at room temperature.

Ceramic adhesives used with strain gages usually consist of a mixture of aluminum phosphate and silica. They are usually applied in a two-step process. First, a thin film is applied and dried to serve as the insulation barrier. The gage with a strippable carrier is then applied with a second coat. The cement cures without pressure at a temperature of $315°C$ ($600°F$). The operating temperature range is -269 to $650°C$ (-452 to $1200°F$) and a leakage to ground resistance of at least $10^6 \, \Omega$ is maintained at $555°C$ ($1031°F$).

After the adhesive has cured (with the exception of the ceramics), the strain-gage installation should be coated with a sealant such as wax, rubber, or polyurethane. These coatings protect the grid, carrier, and adhesive from the detrimental effects of absorbed moisture and extend the useful life of the strain transducer.

The strain-gage grid is designed with tabs that facilitate the connection of lead wires. Large tabs help simplify the attachment procedure and desensitize the portion of the gage that is removed from the strain-sensing area. A large number of tab geometries and lead attachment arrangements are available. In cases where the tabs are small (very small gages), the lead wires should be attached to the gages by the manufacturer prior to installation.

2-3 Strain-Gage Calibration

The strain sensitivity S_A of a uniform conductor of length L, cross-sectional area A, and resistance R was defined by Eq. (2-3) as

$$S_A = \frac{dR/R}{dL/L} \tag{2-3}$$

STRAIN GAGES

When the conductor is formed into a grid to reduce its length, the strain sensitivity changes and the gage exhibits sensitivity to both axial and transverse strain. The response of a surface-mounted gage that is subjected to an axial strain ε_a, a transverse strain ε_t, and a shearing strain γ_{at} can be expressed as

$$\frac{dR}{R} = S_a \varepsilon_a + S_t \varepsilon_t + S_s \gamma_{at} \qquad (2\text{-}10)$$

where S_a = sensitivity of the gage to axial strain
S_t = sensitivity of the gage to transverse strain
S_s = sensitivity of the gage to shearing strain

In general, the gage sensitivity to shearing strain is small and can be neglected. The response of the gage can then be expressed as

$$\frac{dR}{R} = S_a(\varepsilon_a + K_t \varepsilon_t) \qquad (2\text{-}11)$$

where $K_t = S_t/S_a$ is defined as the transverse sensitivity factor for the gage.

Strain-gage manufacturers provide the transverse sensitivity factor K_t and a calibration constant known as the gage factor S_g for each gage. The gage factor represents the calibration constant for a batch or lot of gages and is determined by testing sample gages drawn from a lot of gages produced in a given production run. Resistance change dR experienced by a gage is related to the gage factor S_g and the axial strain ε_a in the calibration beam by the expression

$$\frac{dR}{R} = S_g \varepsilon_a \qquad (2\text{-}12)$$

The stress field is uniaxial in the calibration beam used for the determination of S_g; therefore, the gage is subjected to a biaxial strain field with

$$\varepsilon_t = -\nu_0 \varepsilon_a \qquad (2\text{-}13)$$

where $\nu_0 = 0.285$ is Poisson's ratio for the calibration beam material. Substituting Eq. (2-13) into Eq. (2-11) yields

$$\frac{dR}{R} = S_a \varepsilon_a (1 - \nu_0 K_t) \qquad (2\text{-}14)$$

A comparison of Eqs. (2-12) and (2-14) indicates that

$$S_g = S_a (1 - \nu_0 K_t) \qquad (2\text{-}15)$$

Equation (2-15) indicates that the gage factor S_g relates resistance change dR to axial strain ε_a experienced by the gage only if $K_t = 0$ or if the gage is used in a uniaxial stress field on a material with Poisson's ratio $\nu = 0.285$. In all other situations, some error will be associated with the measurement due to the limitation in determining S_g.

The magnitude of the error can be determined by considering the response of a gage in a general biaxial field with strains ε_a and ε_t. Substituting Eq. (2-15) into Eq. (2-11) yields

$$\frac{dR}{R} = \frac{S_g \varepsilon_a}{1 - \nu_0 K_t} \left(1 + K_t \frac{\varepsilon_t}{\varepsilon_a}\right) \qquad (2\text{-}16)$$

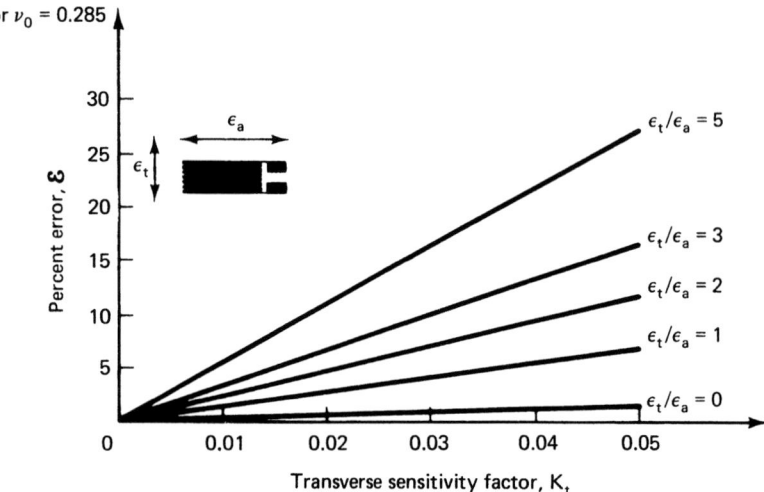

Figure 2-6 Percent error as a function of transverse sensitivity factor for different ratios of transverse to axial strain.

Solving Eq. (2-16) for the true axial strain yields

$$\varepsilon_a = \frac{dR/R}{S_g} \frac{1 - \nu_0 K_t}{1 + K_t(\varepsilon_t/\varepsilon_a)} \qquad (2\text{-}17)$$

The apparent strain ε_a' obtained from Eq. (2-12) is

$$\varepsilon_a' = \frac{dR/R}{S_g} \qquad (2\text{-}18)$$

Substituting Eq. (2-18) into Eq. (2-17) yields

$$\varepsilon_a = \varepsilon_a' \frac{1 - \nu_0 K_t}{1 + K_t(\varepsilon_t/\varepsilon_a)} \qquad (2\text{-}19)$$

The present error $\mathscr{E}$ involved in neglecting the transverse sensitivity of the gage can be expressed as

$$\mathscr{E} = \frac{\varepsilon_a - \varepsilon_a'}{\varepsilon_a}(100) \qquad (2\text{-}20)$$

Substituting Eq. (2-19) into Eq. (2-20) yields

$$\mathscr{E} = \frac{K_t[(\varepsilon_t/\varepsilon_a) + \nu_0]}{1 - \nu_0 K_t} \qquad (2\text{-}21)$$

The results of Eq. (2-21), shown graphically in Fig. 2-6, indicate that the error is a function of K_t and the strain biaxiality ratio $\varepsilon_t/\varepsilon_a$. Since the errors can be significant when both K_t and $\varepsilon_t/\varepsilon_a$

STRAIN GAGES 51

are large, it is important that corrections be made to account for the transverse sensitivity of the gage.

Corrections for the cross-sensitivity effect, when the strain field is unknown, require the experimental determination of strain in two perpendicular directions. If ε_a' and ε_t' are apparent strains in the axial and transverse directions, respectively, then from Eq. (2-19) it is evident that

$$\varepsilon_a' = \frac{1}{1 - \nu_0 K_t} (\varepsilon_a + K_t \varepsilon_t)$$
$$\varepsilon_t' = \frac{1}{1 - \nu_0 K_t} (\varepsilon_t + K_t \varepsilon_a)$$
(2-22)

where the unprimed quantities are the true strains ε_a and ε_t. Solving Eq. (2-17) for the true strains yields

$$\varepsilon_a = \frac{1 - \nu_0 K_t}{1 - K_t^2} (\varepsilon_a' + K_t \varepsilon_t')$$
$$\varepsilon_t = \frac{1 - \nu_0 K_t}{1 - K_t^2} (\varepsilon_t' - K_t \varepsilon_a')$$
(2-23)

Correction equations for transverse strains in two- and three-element rosettes can be found in Ref. 7.

2-4 Behavior of Foil-Type Strain Gages

Foil strain gages are small precision resistors mounted on a flexible carrier that can be bonded to a component part in a typical application. The gage resistance is accurate to $\pm 0.4\%$, and the gage factor S_g, based on lot calibration, is certified to $\pm 1.5\%$. These specifications indicate that foil-type strain gages provide a means for making precise measurements of strain. The results actually obtained, however, are a function of the installation procedures, the state of strain being measured, and environmental conditions during the test. Usually, the results obtained are much less accurate than the specifications on the gage factor and the resistance indicate.

Strain Cycling

One measure of the performance of a strain-gage system ("system" here implies gage, adhesive, and instrumentation) involves considerations of linearity, hysteresis, and zero shift. If gage output, in terms of measured strain, is plotted as a function of applied strain as the load on the component is cycled, results similar to those shown in Fig. 2-7 will be obtained. A slight deviation from linearity is typically observed, and the unloading curve normally falls below the loading curve to form a hysteresis loop. Also, when the applied strain is reduced to zero, the gage output indicates a small negative strain, termed zero shift. The magnitude of the deviation from linearity, hysteresis, and zero shift depends on the strain level, the adequacy of the bond, the degree of cold working of the foil material, and the carrier material.

For properly installed gages, deviations from linearity should be approximately 0.1% of the maximum strain for polyimide carriers and 0.05% for epoxy carriers. First-cycle hysteresis and zero shift are more difficult to predict; however, zero shifts of 1% of the maximum strain are frequently observed in typical applications. If possible, strain cycling to 125% of the maximum

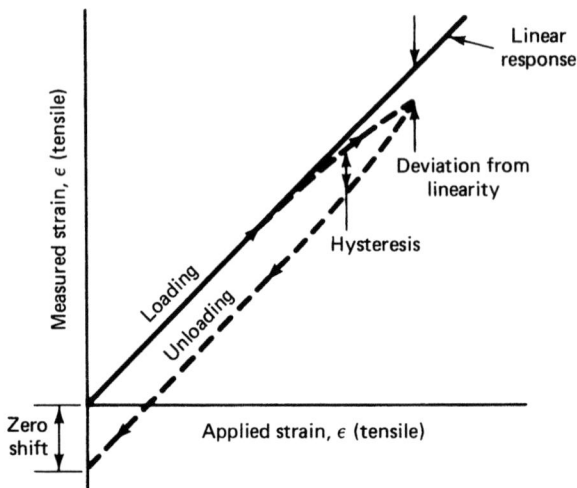

Figure 2-7 Typical strain cycle showing nonlinearity, hysteresis, and zero shift (scale exaggerated).

test strain is recommended, since the amount of hysteresis and zero shift will decrease to less than 0.2% of the maximum strain after four or five cycles.

Temperature

In many test programs, the strain-gage installation is subjected to temperature changes during the test period, and careful consideration must be given to determining whether the change in resistance observed is due to applied strain or temperature change. When the ambient temperature changes, four effects occur that may alter the performance characteristics of the gage:

1. The strain sensitivity S_A of the metal alloy changes.
2. The gage grid elongates or contracts ($\Delta L/L = \alpha \Delta T$).
3. The specimen elongates or contracts ($\Delta L/L = \beta \Delta T$).
4. The resistance of the gage changes ($\Delta R/R = \gamma \Delta T$).

The strain sensitivities S_A of the two most commonly used alloys (Advance and Karma) are linear functions of temperature as shown in Fig. 2-8. These plots indicate that $\Delta S_A/\Delta T$ equals 0.00735 and -0.00975% per degree Celsius for Advance and Karma alloys, respectively. Since these changes are small (less than 1% for $\Delta T = 100°C$), variations in S_A with temperature are usually neglected in routine stress analysis work; however, in thermal-stress studies where temperature variations of several hundred degrees are common, changes in S_A become significant and must be considered.

Effects 2, 3, and 4 are much more significant and combine to produce a change in resistance of the age with temperature $(\Delta R/R)_{\Delta T}$, which can be expressed as

$$\left(\frac{\Delta R}{R}\right)_{\Delta T} = (\beta - \alpha)S_g \Delta T + \gamma \Delta T \tag{2-24}$$

STRAIN GAGES

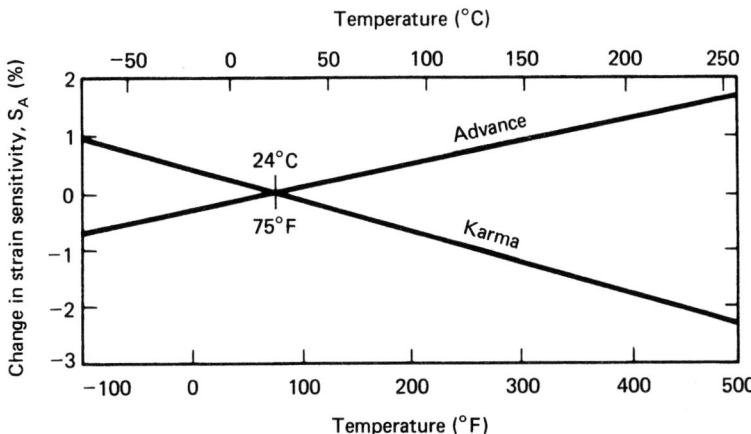

Figure 2-8 Alloy sensitivity as a function of temperature for Advance and Karma alloys.

where α = thermal coefficient of expansion of the gage alloy
β = thermal coefficient of expansion of the specimen
S_g = gage factor for the gage
γ = temperature coefficient of resistivity of the gage alloy

A differential expansion between the gage grid and the specimen due to a change in temperature subjects the gage grid to a thermally induced mechanical strain $\varepsilon_T = (\beta - \alpha)\Delta T$, which does not occur in the specimen. The gage responds to the strain ε_T in the same way that it responds to a load-induced strain ε in the specimen. Unfortunately, it is impossible to separate the component of the response due to temperature change from the response due to the load.

If the thermal coefficients of the gage and specimen match ($\alpha = \beta$), the first term in Eq. (2-24) will not produce a response; however, the second term will produce a response (an apparent strain) that does not exist in the specimen. A temperature-compensated gage is obtained only if both terms in Eq. (2-24) are zero or if they cancel.

The values of α and γ are quite sensitive to the composition of the alloy and to the degree of cold working imparted to the alloy during the rolling of the foil. It is common practice for strain-gage manufacturers to measure the thermal response characteristics of a few gages from each roll of foil that they use in manufacturing the gages. Because of variations in α and γ between melts and rolls of foil, it is possible to select gages fabricated from Advance or Karma alloys for many different specimen materials. These gages are known as selected-melt or temperature-compensated gages.

Unfortunately, selected-melt gages are not perfectly compensated over a wide range of temperatures because of the nonlinear terms that were omitted in Eq. (2-24). A typical selected-melt strain gage exhibits an apparent strain with temperature as shown in Fig. 2-9. The apparent strain produced by a temperature change of a few degrees in the neighborhood of 24°C (75°F) is small (less than 0.5 μm/m°C); however, when the temperature change is large, the apparent strain becomes large and corrections to account for it must be made. These corrections require the measurement of temperature at the gage location with a temperature sensor and the use of an apparent strain versus temperature curve similar to that shown in Fig. 2-9.

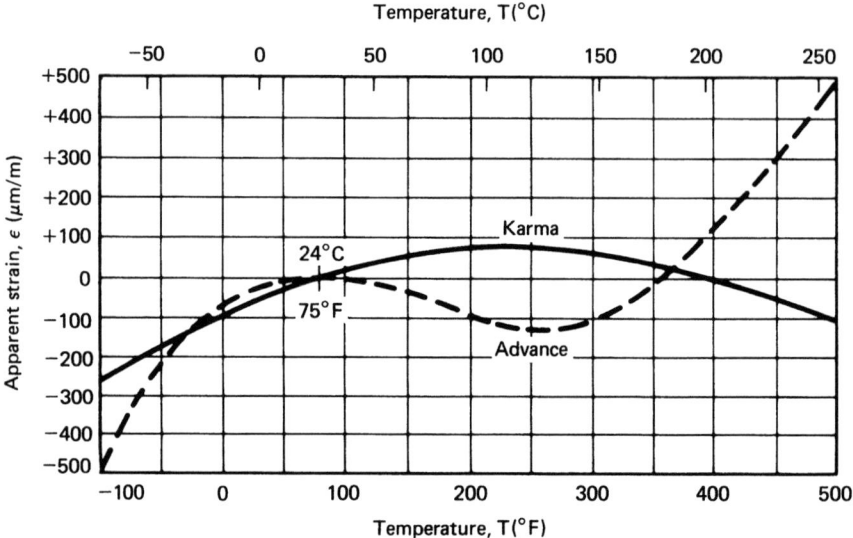

Figure 2-9 Apparent strain as a function of temperature of Advance and Karma alloys.

Maximum Strain

The maximum strain that can be measured with a foil strain gage depends on the gage length, the foil alloy, the carrier material, and the adhesive. The Advance and Karma alloys with polyimide carriers, used for general-purpose strain gages, can be employed to strain limits of ±5 and ±1.5% strain, respectively. This strain range is adequate for elastic analyses on metallic and ceramic components, where yield or fracture strains rarely exceed 1%; however, these limits can easily be exceeded in plastic analyses, where strains in the postyield range can become very large. In these instances, a special postyield gage is normally employed. Postyield gages are fabricated by using a double-annealed Advance foil grid with a high-elongation polyimide carrier. Urethane-modified epoxy adhesives are used to bond the gage to the structure. If proper care is exercised in preparing the surface of the specimen, roughening the back of the gage, formulating a high-elongation plasticized adhesive system, and attaching the lead wires without significant stress raisers, it is possible to approach strain levels of 20% before cracks begin to occur in the solder tabs or at the ends of the grid loops. Attention to the gage factor is necessary as the strains become larger, as indicated by Eq. (2-9). The gage factor becomes a function of the strain (i.e., $S_g = 2 + \varepsilon$) and it is necessary to correct the results for this nonlinear effect.

For very large strains, where foil gages cannot be employed, it is possible to employ liquid-metal strain gages. This type of gage is fabricated by filling a rubber capillary tube with a gallium–indium–tin alloy, with lead wires serving to close the ends of the tube. This gage is bonded to the specimen with a high-elongation rubber cement. As the specimen is strained, the length of the tube increases and its diameter decreases so that the volume of the cavity remains constant. The resistance change $\Delta R/R_0$ increases with strain in accordance with Eq. (2-8), and as a result, the gage factor is given by Eq. (2-9).

The initial resistance of the gage is very small (less than 1 Ω); therefore, a Wheatstone bridge

STRAIN GAGES

Table 2-2 Allowable Power Densities

Specimen Conditions	Power Density, P_D	
	W/mm^2	W/in.^{2}
Heavy aluminum or copper sections	0.008–0.016	5–10
Heavy steel sections	0.003–0.008	2–5
Thin steel sections	0.0015–0.003	1–2
Fiberglass, glass, ceramics	0.0003–0.0008	0.2–0.5
Unfilled plastics	0.00003–0.00008	0.02–0.05

designed with an appropriate power supply and resistance balance method must be used to accommodate the low resistance of the strain gage.

Power Dissipation

It is well recognized that temperature variations can significantly influence the output of strain gages, particularly those that are not temperature compensated. The temperature of the gage is influenced by ambient-temperature variations and by the power dissipated by the gage when it is connected into a measurement circuit. The power P is dissipated in the form of heat; therefore, the temperature of the gage must increase above the ambient temperature to accomplish the heat transfer. The exact temperature increase required is very difficult to specify, since many factors influence the heat balance for the gage. Factors that govern the heat dissipation include:

1. Gage size, w_0 and L_0
2. Gage configuration, spacing, and size of the conducting elements
3. Carrier, type of polymer, and thickness
4. Adhesive, type of polymer, and thickness
5. Specimen material, thermal diffusivity
6. Specimen volume in the local area of the gage
7. Type and thickness of overcoat used to protect the grid
8. Velocity of air flowing over the gage installation

A parameter often used to characterize the heat-dissipation characteristics of a strain-gage installation is the power density $P_D = P/A$, where P is the power that must be dissipated by the gage and A is the area of the grid of the gage. Power densities that can be tolerated by a gage are strongly related to the specimen that serves as the heat sink. Recommended values of P_D for different materials and conditions are listed in Table 2-2.

Allowable bridge voltages for specified grid areas and power densities are shown for 120-Ω gages in Fig. 2-10: typical grid configurations are identified along the abscissa to illustrate the effect of gage size on allowable bridge excitation. It should be noted that small gages mounted on a poor heat sink result in lower allowable bridge voltages than those employed in most commercial strain indicators (3 to 5 V). In these instances it is necessary to use a higher-resistance gage, a gage with a larger grid area, or a fixed resistor in series with the gage to reduce the voltage on the gage in both the active and compensating arms of the bridge.

Fatigue

Strain gages are frequently mounted on components subjected to fatigue or cyclic loading, and the life of the component during which the gage must be monitored can exceed several million

56   HANDBOOK ON EXPERIMENTAL MECHANICS

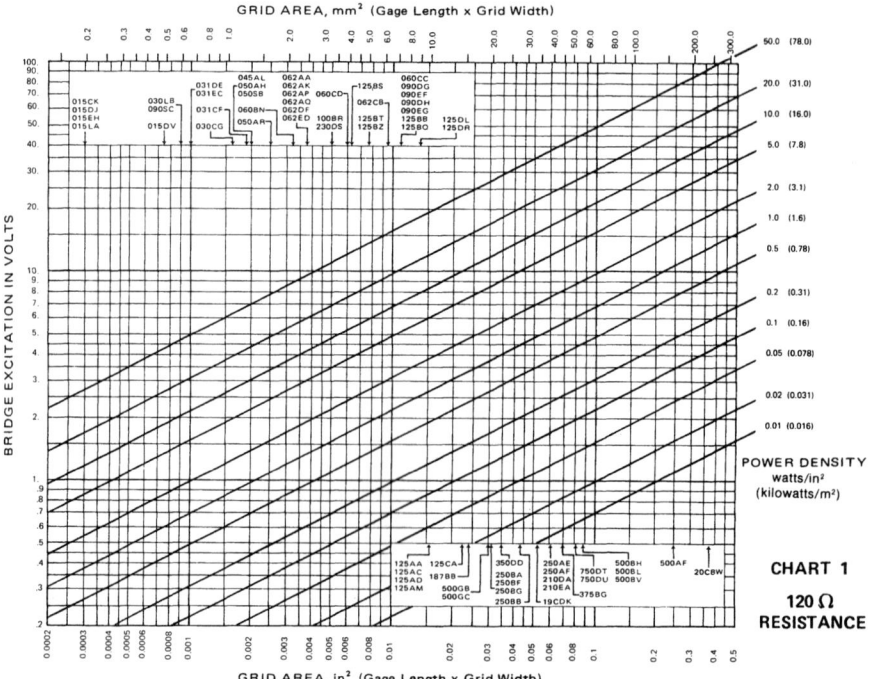

Figure 2-10 Allowable bridge excitation voltage as a function of grid area for different power densities. (Courtesy of Measurements Group, Inc.)

cycles. Three factors that must be considered for this particular type of application are zero shift, change in gage factor, and failure of the gage in fatigue.

As the strain gage is subjected to repeated cyclic strain, the gage gridwork hardens and its specific resistance changes. The specific-resistance change produces a zero shift. The amount of the zero shift depends on the magnitude of the strain, the number of cycles, the grid alloy, the original state of cold work of the grid alloy, and the type of carrier employed in the gage construction. Typical examples of zero shift as a function of number of strain cycles are shown in Fig. 2-11 for four different gages. The poorest gage for this type of application is the open-faced Advance grid (EA) gage, which begins to exhibit noticeable zero shift at 10^3 cycles. Encapsulating the grid in a glass-reinforced epoxy–phenolic resin (WA type) improves the life, and exposure of 10^5 cycles at 2100 μm/m occurs before zero shift becomes apparent. Further improvements can be achieved by using Karma (K) or isoelastic (D) alloys fully encapsulated in the glass-reinforced epoxy–phenolic carrier material with factory-installed lead wires. The changes in gage factor are quite small due to strain cycling and in general can be neglected if the zero shift is less than a few hundred micrometers per meter.

STRAIN GAGES

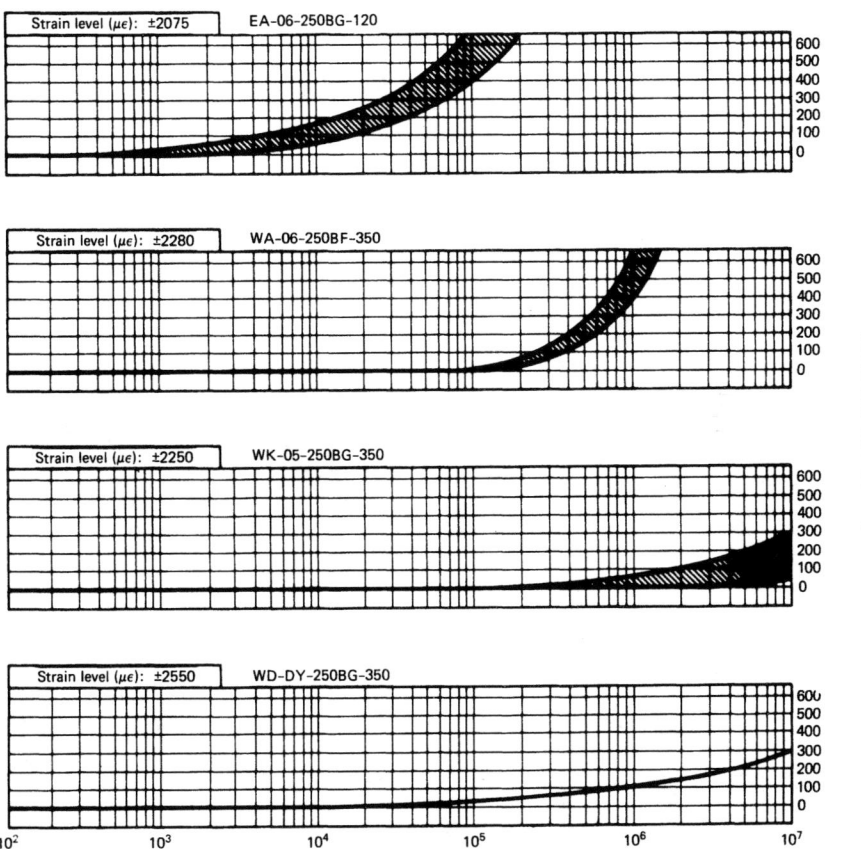

Figure 2-11 Zero shift as a function of fatigue exposure for several types of strain gage. (Courtesy of Measurements Group, Inc.)

A very important factor to consider when using strain gages under fatigue conditions is the increase in sensitivity that occurs when a fatigue crack develops in the grid of the gage. The crack will usually develop in the grid near the lead wire on the tab. The presence of the crack produces a small increase in the resistance of the gage, which is monitored as an apparent strain. Thus the apparent gage factor has increased with the development and subsequent propagation of the fatigue crack.

Time

In certain strain-gage applications it is necessary to record strains over a period of months or years without having the opportunity to unload the specimen and recheck the zero reading. The duration

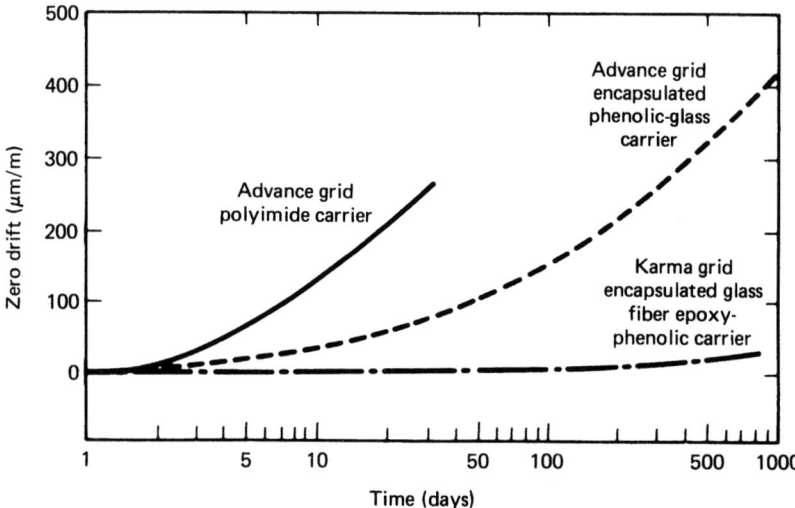

Figure 2-12 Zero shift as a function of time for several types of strain gage at 167°F (75°C). (Data from Ref. 8.)

of the readout period is important and makes this application of strain gages one of the most difficult. All the factors that can influence the behavior of the gage have an opportunity to contribute; moreover, there is sufficient time for the individual contribution to the error from each of the factors to become significant.

Drift in the zero reading from an electrical-resistance strain-gage installation is due to the effects of moisture or humidity variations on the carrier and the adhesive, the effects of long-term stress relaxation of the adhesive, the carrier, and the strain-gage alloy, instabilities in the resistors in the inactive arms of the Wheatstone bridge, and drift of the power supply with time.

Results from a series of stability tests by Freynik [8] are presented in Fig. 2-12. A typical general-purpose strain gage with an Advance alloy grid and a polyimide carrier exhibited a zero shift of 270 μm/m after 30 days. Since this installation was carefully waterproofed, the large drift was attributed to stress relaxation in the polyimide carrier over the period of observation. An Advance alloy gage with a glass-fiber-reinforced phenolic carrier exhibited a zero drift of approximately 100 μm/m after 50 days. The presence of the glass fibers essentially eliminated drift due to stress relaxation in the carrier material; the drift that occurred was attributed to instabilities in the Advance alloy grid. The most satisfactory results were obtained with a Karma grid and an encapsulating glass-reinforced epoxy–phenolic carrier. The zero shift was only 30 μm/m after an observation period of 900 days.

These results show that electrical-resistance strain gages can be used for long-term measurements provided Karma grids with glass-fiber-reinforced epoxy–phenolic carriers are employed with a well-cured epoxy adhesive system. The gage installation must be waterproofed to minimize the effects of humidity. Hermetically sealed bridge completion resistors should also be specified to ensure stability of the bridge over the long observation period.

STRAIN GAGES

Strain Gradients

Normal strains, which are measured with electrical-resistance strain gages, are defined as derivatives of the displacement field u, v, w as

$$\varepsilon_{xx} = \frac{\delta u}{\delta x} \qquad \varepsilon_{yy} = \frac{\delta v}{\delta y} \qquad \varepsilon_{zz} = \frac{\delta w}{\delta z} \qquad (2\text{-}25)$$

Strains are defined at a prescribed point $P(x, y, z)$; however, strains measured with a strain gage are the integral effect of the strains acting over the area of the gage. The integral effect produces the output dR/R.

Electrical-resistance strain gages are calibrated on a cantilever beam where the gradient of the strain in the longitudinal direction (say, x) is linear in x and the gradient in the transverse direction (say, y) is zero. In the calibration, the integrating effect is accounted for by noting that $dR/R = S_g x$ at the midpoint of the gage. However, when strain gages are utilized in strain fields that exhibit nonlinear gradients, errors are produced that can be significant.

To illustrate the magnitude of the error, consider a strain distribution with a nonlinear gradient in the x direction and no gradient in the y direction. Thus

$$\varepsilon_{xx} = k_1 x^2 + k_2 x + k_3 \qquad (2\text{-}26)$$

The average strain, measured over the gage length L_0, can be obtained from Eq. (2-26). Thus

$$\varepsilon_{xx}|_{avg} = \frac{1}{L_0} \int_0^{L_0} (k_1 x^2 + k_2 x + k_3)\, dx = \frac{k_1 L_0^2}{3} + \frac{k_2 L_0}{2} + k_3 \qquad (2\text{-}27)$$

The strain at the midpoint of the gage (where $x = L_0/2$) is the true strain. Thus

$$\varepsilon_{xx}|_{x=L_0/2} = \frac{k_1 L_0^2}{4} + \frac{k_2 L_0}{2} + k_3 \qquad (2\text{-}28)$$

The absolute error $\mathscr{E}$ is then

$$\mathscr{E} = -\frac{k_1 L_0^2}{12} \qquad (2\text{-}29)$$

Equation (2-29) indicates that the measured strain is less than the true strain. The magnitude of the error depends upon the strain gradient (which is indicated by k_1) and the square of the gage length L_0.

Perry [1] has demonstrated the size of errors produced by strain gradients by considering the problem of a hole of radius R_0 in an infinite plate subjected to a uniaxial stress σ_0. The maximum strain at the hole is

$$\varepsilon|_{max} = \frac{\sigma_0}{4E} \left[4 + \frac{2}{r^2}(\nu - 3)\cos 2\theta - \frac{2}{r^2}(1 + \nu)\left(2 - \frac{3}{r^2}\right)\cos 4\theta \right] \qquad (2\text{-}30)$$

To illustrate the percent error associated with a specific measurement, Perry considered a 13-mm (0.5-in.)-diameter hole and a strain gage with a square grid and an active grid length of 1.6 mm (0.0625 in.) mounted on the surface of the plate near the edge of the hole. The key geometric ratio in this example is $(L_0/R_0) = 0.25$, which produces a measurement error of -24%. The error can be reduced by decreasing the active length and width of the gage as indicated by Eq. (2-30); however, accuracies of 5% or less cannot be achieved with commercially available strain gages when the hole size is 13 mm (0.5 in.) or smaller. It is reasonable to assume that similar errors result in peak strain measurement in other stress concentration regions. It is also important to note that the strain-gage measurement underestimates the peak strain.

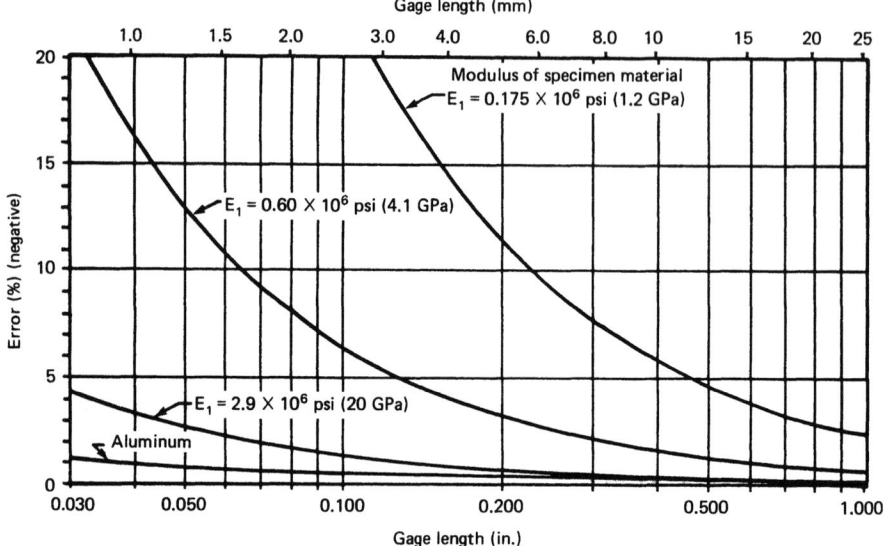

Figure 2-13 Error due to local reinforcement by a strain gage. (Data from Ref. 10.)

Reinforcement Effects

When strain gages are employed on metallic structures, the effect of the strain-gage installation (the adhesive, carrier grid, and overcoat) on the rigidity of the structure is negligible if the thickness of the section is large relative to the thickness of the strain-gage installation. In cases where strain gages are employed on low-modulus materials such as paper or plastics or on thin cross sections of any material, the gage installation tends to produce a reinforcement that alters the strain in the local area around the gage. The effective modulus of elasticity of a strain-gage installation ranges from 7 to 20 GPa (1 to 3 × 10^6 psi) depending on the gage selected and the adhesive used.

Studies [9–12] of reinforcing effects are usually divided into two categories: local reinforcement, where the strain is affected only in the immediate region of the gage installation, and global reinforcement, where the strain throughout the body is affected. Local reinforcement occurs when a strain gage is mounted on the edge of a thin plate. The strain under the gage is not significantly lower than the far-field strain; however, there may be measurement errors due to strain transmission, which depends on gage length and results from shear lag at the gage–plate interface. Results of a qualitative analysis of the strain transmission error are shown in Fig. 2-13. It is evident from these results that local reinforcement on deep sections fabricated from metallic materials is negligible, but the errors can be very large for the lower-modulus polymeric materials when $L_0 < 13$ mm (0.5 in.). To accommodate the local reinforcing effect for a particular gage installation, it is possible to recalibrate the gage by using a calibration procedure that simulates the local reinforcing effect. An effective gage factor S_{ge} is determined which accounts for the shear lag present with the low-modulus materials.

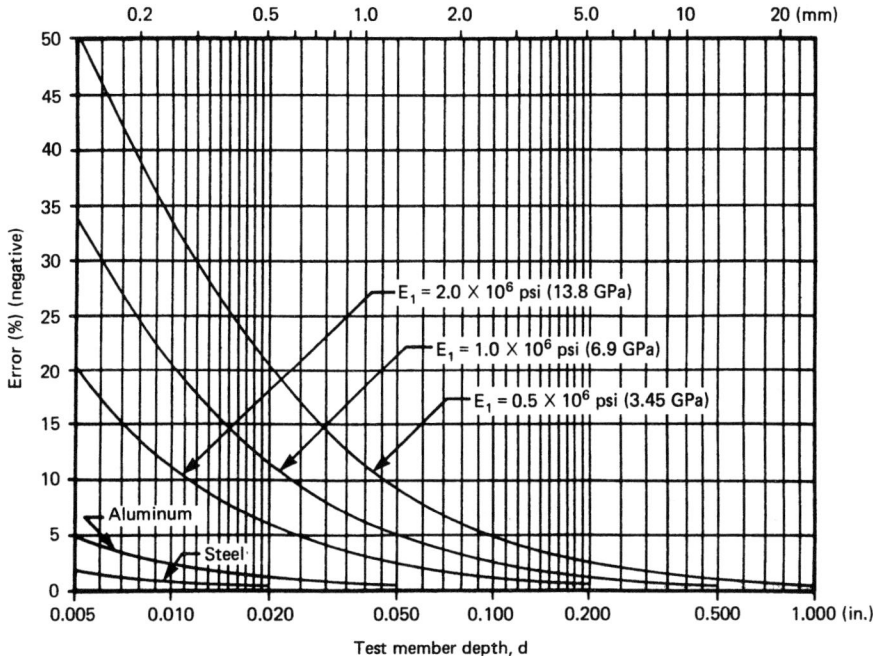

Figure 2-14 Error due to global reinforcement by a strain gage. (Data from Ref. 1.)

When the reinforcing effect is global and the presence of the strain gage affects the strain distribution over the cross section of the specimen, the problem is more difficult. Global reinforcement occurs when the section is thin (equal to gage width) and the depth is small relative to the gage length. Errors resulting from global reinforcement are shown in Fig. 2-14 as a function of test member depth and modulus, assuming perfect strain transmission to the gage. These results overestimate the error that would occur in a wider specimen, since a gage of reasonable width would not reinforce the entire width of the section. The results underestimate error for depth $d >$ 2.5 mm (0.1 in.), since the gage length L_0 is not long relative to the specimen depth and perfect strain transmission to the grid does not occur.

2-5 Resistance Strain Gages at Low and Moderate Temperatures

Strain Measurement Methods at Moderate Temperatures

Resistance-type strain gages can be employed at elevated temperatures for both static and dynamic stress analyses; however, the measurements require many special precautions that depend primarily on the temperature and the time of observation. At elevated temperatures, the resistance R of a strain gage must be considered to be a function of temperature T and time t in addition to strain ε. Thus

$$\frac{\Delta R}{R} = S_g \Delta \varepsilon + S_T \Delta T + S_t \Delta t \qquad (2\text{-}31)$$

where S_g = gage sensitivity to strain (gage factor)
S_T = gage sensitivity to temperature
S_t = gage sensitivity to time

In the discussion of performance characteristics of foil strain gages it was shown that sensitivity of the gages to temperature and time could be minimized at normal operating temperatures (24°C or 75°F) by proper selection of the strain-gage alloy and carrier material. As the test temperature increases, it is more difficult to make accurate strain measurements because the sensitivities S_T and S_t become more significant.

The effects of temperature on the behavior of Advance and Karma alloys between -73 and $+260°C$ (-100 and $+500°F$) were indicated in Figs. 2-8 and 2-9. For large temperature variations, the gage factor changes linearly with temperature. Corrections to S_g are relatively small and easy to make if the temperature of the strain-gage installation is known. The effect $S_T \Delta T$ produces an apparent strain (see Fig. 2-9) that must be subtracted from the strain measurement to obtain accurate results. Temperature measurements at or near the gage locations are necessary to determine the apparent strains from Fig. 2-9.

The results shown in Fig. 2-9 indicate that the Karma alloy is more suitable for measurements involving large temperature variations than Advance. The Karma gages can be employed over a temperature range from -18 to $260°C$ (0 to $500°F$) with temperature-induced apparent strains held to within ± 100 μm/m.

The stability of a strain-gage installation is also affected by temperature; and strain-gage drift becomes a more serious problem as the temperature and the time of observation are increased. Stability is affected by stress relaxation in the adhesive bond and in the carrier material and by metallurgical changes (phase transformations and annealing) in the strain-gage alloy. The upper temperature limit on commercially available Karma gages is controlled by the carrier material. Glass-reinforced epoxy–phenolic carriers are rated at 288°C (550°F); however, Karma gages with this type of carrier exhibit zero drift with time. If the time of loading and observation is long, corrections must be made for this zero drift. Drift rates will depend on both the strain level and the temperature. For high-temperature strain analyses, a series of strain–time calibration curves that cover the range of strains and temperatures to be encountered should be developed. Zero-drift corrections can then be taken from the appropriate curve.

The problem of gage stability and apparent strain due to temperature changes is greatly reduced if the period of observation is short. For analyses where relatively short strain–time records are required (times less than 1 second), any temperature change is small and as a result temperature effects are insignificant. Thus short-time strain-gage analyses can be made at very high temperatures with good precision and with relative ease.

Resistance strain-gage measurements at temperatures higher than 288°C (550°F) require special gages and special techniques for mounting and monitoring the strain-gage signal. At these higher temperatures, polymeric materials can no longer be used for the carrier or the adhesive. The subject of strain measurement at very high temperatures (temperatures higher than 288°C or 550°F) is covered in Chapter 12.

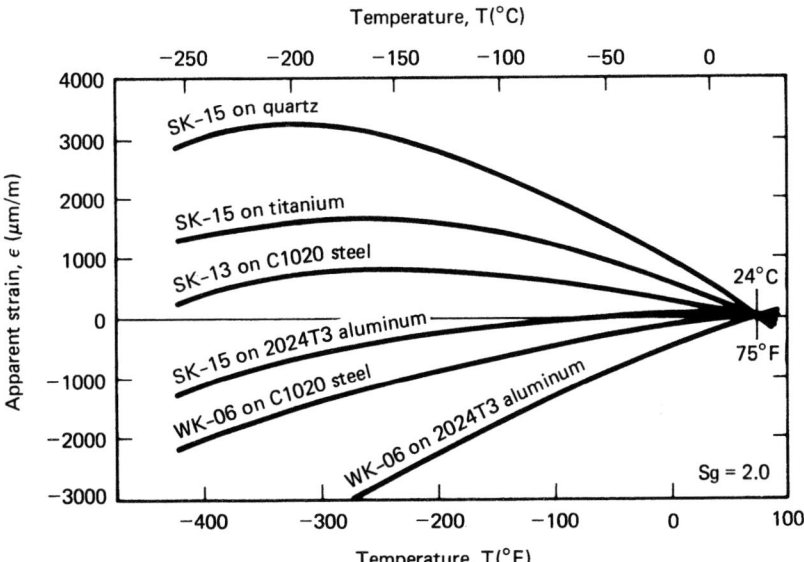

Figure 2-15 Apparent strain as a function of temperature for several selected-melt Karma gages mounted on different materials. [Data from J. C. Telinde, Strain Gages in Cryogenic Environment, *Exp. Mech.*, 10, no. 9 (1970).]

Strain Measurement Methods at Cryogenic Temperatures

Strain measurements at cryogenic temperatures are possible if corrections similar to those noted in the preceding section are made. The change of gage factor with temperature remains linear (0.0043%/°F for Advance and −0.010%/°F for Karma). Again the correction can be made and accurate results achieved if the temperature is known.

A second correction for temperature-induced apparent strain must also be made, since very large apparent strains occur with small changes in test temperature (see Fig. 2-9). The curves shown in Fig. 2-9 indicate that Karma alloy is superior to Advance alloy in this application; however, both alloys exhibit large sensitivity to temperature in the cryogenic region. The apparent strain is shown in Fig. 2-15 as a function of temperature for a number of different Karma-type gages. It is apparent from these results that it is possible to minimize the effects of temperature variations at cryogenic temperatures by selecting compensating gage types that are mismatched relative to the component material. For example, the SK-13 gage is compensated for use on a material with $\alpha = 13$ ppm/°F; however, it exhibits better performance on C1020 steel with $\alpha = 6$ ppm/°F than the WK-06 gage, which is intended for use on steels.

Corrections must also be made to account for changes in the modulus of elasticity of the specimen material with temperature. The effect of cryogenic temperatures is to increase the elastic modulus from 5 to 20% as illustrated in Fig. 2-16. The higher values of the elastic modulus must be used in the stress–strain relations required for data reduction and analysis.

Cryogenic temperatures are usually obtained in tests by using nitrogen, hydrogen, or helium in the liquid state. All three are insulating materials in the liquid state, and for this reason it is not necessary to electrically isolate the gage grid or the lead wires from the fluids. However, it is advisable to use insulated wire and to coat the exposed grid and leads with a silicone grease to

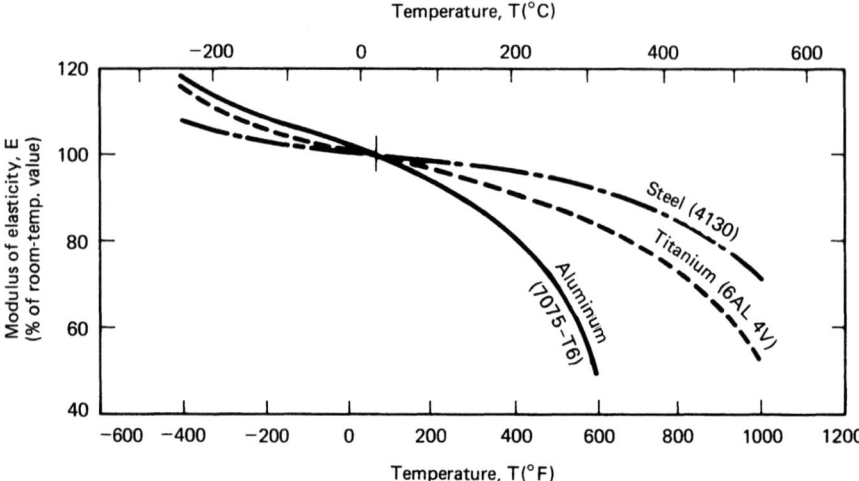

Figure 2-16 Change in elastic modulus as a function of temperature for several different materials. (Data from Telinde: see Fig. 2-15.)

provide a thermal barrier and to eliminate boiling of the liquid in the local neighborhood of the gage.

2-6 Semiconductor Strain Gages

The development of semiconductor strain gages was a by-product of research at the Bell Telephone Laboratories that led to the development of the transistor in the early 1950s. The piezoresistive properties of semiconducting silicon and germanium were determined by Smith [4] in 1954. Further development of semiconductor transducers by Mason and Thurston [5] in 1957 eventually led to the commercial marketing of piezoresistive strain gages in 1960.

Basically, the semiconductor strain gage consists of a small, ultrathin rectangular filament of a single crystal of silicon that is mounted on a carrier to simplify handling. Since the resistivity of silicon is approximately 1000 times greater than the resistivity of the Advance alloy used for foil gages, it is not necessary to utilize grid configurations with semiconductor strain gages to obtain an initial gage resistance in the range 10^2 to 10^3 Ω.

When a state of stress is imposed on the crystal, it responds by exhibiting a piezoresistive effect that can be described by the expression.

$$\rho_{ij} = \delta_{ij}\rho + \pi_{ijkl}\tau_{kl} \tag{2-32}$$

where ρ_{ij} = resistivity tensor
π_{ijkl} = piezoresistivity tensor
τ_{kl} = stress tensor

A complete discussion of the results obtained from Eq. (2-32), which is beyond the scope of this treatment of semiconductor strain gages, can be found in the original paper by Mason and Thurston

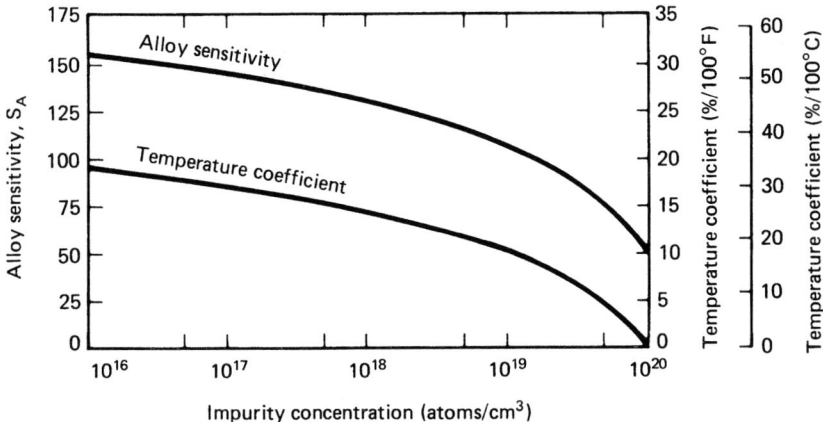

Figure 2-17 Alloy sensitivity and temperature coefficient of sensitivity as a function of impurity concentration for P-type silicon.

[5]. The results indicate that the sensitivity of the gage to stress can be varied by changing the orientation of the strain-gage element relative to the crystal axes or by changing the type and the level of doping of the silicon. Boron is used as the trace impurity in producing the P-type (positive age factor) material, and arsenic is used to produce the N-type (negative gage factor) material.

The strain sensitivity of a conductor as defined in Eq. (2-4) is approximately:

$$S_A = \frac{d\rho/\rho}{\varepsilon} + (1 + 2\nu)$$

For the common strain-gage alloys, the term $1 + 2\nu$ is approximately equal to 1.6; thus the term $(d\rho/\rho)/\varepsilon$ ranges from 0.4 to 2.0. For semiconducting materials, the term $(d\rho/\rho)/\varepsilon$ can be varied between -125 and $+175$ by selecting the type and concentration of the doping material. This very high sensitivity, together with the very high resistivity of semiconductor strain gages, has led to their application in measuring extremely small strains, as sensing elements in miniaturized transducers, and in transducers with very high signal output.

Temperature Effects with Semiconductor Strain Gages

The strain sensitivity S_A of lightly doped (10^{19} atoms/cm^3 or less) semiconducting materials depends on both temperature and strain. The variation of S_A as a function of doping level for P-type silicon is shown in Fig. 2-17. As the impurity concentration is increased from 10^{16} atoms/cm^3 to 10^{20} atoms/cm^3, the strain sensitivity S_A decreases from 155 to 50. Two significant advantages related to the temperature effect compensate for this loss of sensitivity:

1. The effect of temperature on the strain sensitivity S_A is greatly diminished (since the temperature coefficient of sensitivity approaches zero) when the impurity concentration approaches 10^{20} atoms/cm^3.
2. The temperature coefficient of resistance decreases from 0.009/°C (0.005/°F) for impurity concentrations of 10^{16} atoms/cm^3 to 0.00036/°C (0.0002/°F) for impurity concentrations of 10^{19} atoms/cm^3.

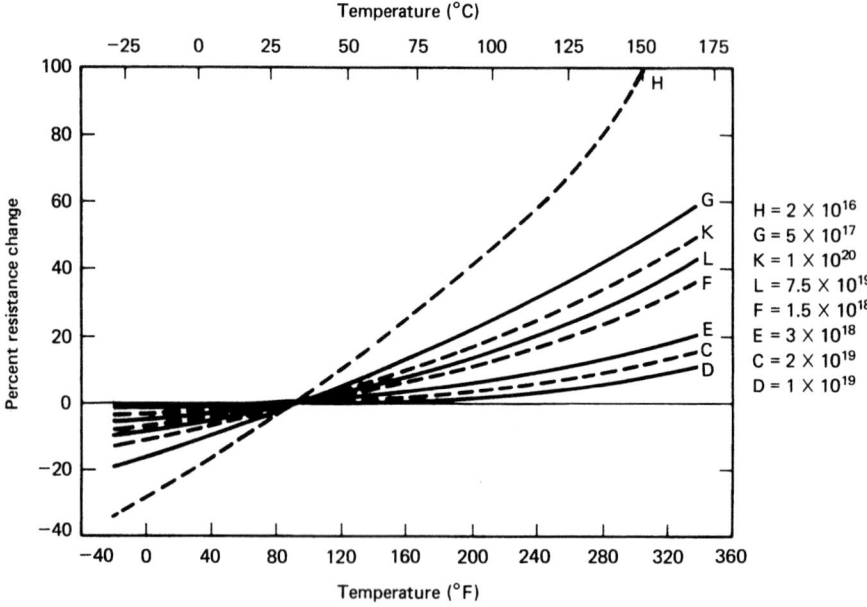

Figure 2-18 Resistance change as a function of temperature for several P-type silicon semiconductors with different impurity concentrations, relative to 27°C (81°F). (Courtesy of Kulite Semiconductor Products, Inc.)

The effect of impurity concentration on resistance change as a function of temperature for P-type silicon is shown in Fig. 2-18. These results indicate that the effects of temperature are minimized with impurity concentrations of 10^{19} atoms/cm^3. This concentration yields a strain sensitivity S_A of 105. Temperature compensation of single-element P-type semiconductor strain gages is not possible; therefore, gage response resulting in apparent strain occurs when the temperature changes during the test period. One method for achieving temperature compensation requires two gages (one of P-type and the other of N-type material) that have the same temperature coefficient of resistance. The two gages are connected into adjacent arms of a Wheatstone bridge, where the output due to temperature is canceled while the output due to strain is added due to the positive and negative sensitivities of the two gages. Dual-element gages are commercially available which are compensated for apparent strain to within ±0.5 µm/m°C over the temperature range 10 to 65°C (50 to 150°F). Such gages exhibit strain sensitivities in the range 220 to 265. Temperature compensation can also be accomplished by fabricating a gage from an N-type material with a positive temperature coefficient of resistance. The positive coefficient is selected (by controlling the impurity concentration) to cancel the negative change in resistance produced by differential expansion between the gage and specimen materials. Single-element N-type gages are available from commercial suppliers for use on materials with coefficients of expansion of 4, 6, 9, and 12 ($\times$ 10^{-6})/°F.

Linearity and Maximum Strain of Semiconductor Strain Gages

Piezoelectric materials with a low impurity concentration exhibit significant nonlinear behavior, as shown in Fig. 2-19. Increasing the impurity level to 10^{19} to 10^{20} atoms/cm^3 markedly improves

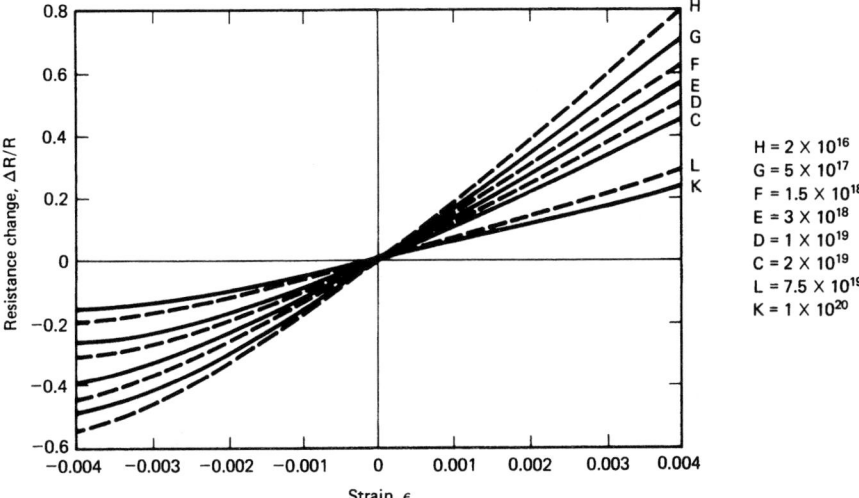

Figure 2-19 Resistance change as a function of strain for P-type semiconductor gages with different impurity concentrations. (Courtesy of Kulite Semiconductor Products, Inc.)

the linearity, especially if P-type gages are used in tensile stress fields and N-type gages are used in compressive fields. Linearity characteristics of dual-element gages (one P-type and one N-type) are usually satisfactory since the nonlinear components tend to cancel in the Wheatstone bridge circuit.

The silicon material is glasslike; therefore, it must be treated with extreme care during mounting. On flat surfaces, installation usually does not present a serious problem, and the gages can be subjected to a maximum strain of approximately 3000 μm/m before they begin to fail by brittle rupture. When semiconductor gages are used on curved surfaces, the strains induced during installation reduce the strain range available for load-induced strain.

Other Characteristics of Semiconductor Strain Gages

For semiconductor gages, the power that can be dissipated per unit length P/L_0 ranges from 4 to 8 W/m (0.1 to 0.2 W/in.). When the specimen on which the gage is mounted has poor thermal characteristics, the lower ratio should be used. For specimens with good heat-sink characteristics, the higher ratio is usually satisfactory. Thermal characteristics of different types of specimen are listed in Table 2-2.

The fatigue life of semiconductor gages is rated at 10^7 cycles for cyclic strains of ± 500 μm/m. This limit is considerably less than the limit for metallic-foil-type strain gages; however, the strain sensitivity of semiconductor gages is high, and they are usually employed in low-magnitude strain applications.

2-7 Optical Fiber Strain Sensors

Optical fiber strain sensors are a relatively new technology that was made possible by the development of low-loss, single-mode optical fibers in the early 1970s. Still, it was not until the late

1970s that Bucaro et al. [13] and Butter and Hocker [14] provided laboratory demonstrations of sensors based on optical fibers. Since then extensive research with optical fibers has led to the development of a wide variety of optical fiber sensors. These include sensors for acoustic pressure, temperature, magnetic fields, electric current, displacement, rotation, acceleration, force, flow velocity, and of course strain [15–19]. The primary advantages offered by optical fiber sensors are high sensitivity, immunity to electromagnetic interference, relative insensitivity to harsh environments, large bandwidth and high data rates, geometric versatility, compatibility with optical fiber telemetry, and no Joule heating. All these attributes have contributed to the rapid development of optical fiber sensors and have significantly enhanced the design of optical fiber sensors of all types. Because of the diversity of the optical fiber sensor technology, we will not attempt to describe all the types of optical fiber sensors; rather, the discussion is limited to developments of the optical fiber strain sensors. However, many of the features described are valid for optical fiber sensors in general.

Optical fiber strain sensors usually use one or both of the basic principles of light propagation in their operation: (1) a change in amplitude (intensity) of light or (2) a change in phase of light as it propagates through an optical fiber waveguide. Both amplitude and phase can change as a function of applied strain and sensor configuration. Sensors are classified as either *extrinsic* or *intrinsic*. Extrinsic sensors incorporate a transducer element in addition to the optical fiber and are not discussed here. Instead, this section treats the intrinsic strain sensors that have demonstrated the most potential for both laboratory and commercial applications. These sensors are classified as Mach–Zehnder, Fabry–Perot, Michelson, polarimetric, modal-domain, twin-core, and Bragg grating types. All except the Mach–Zehnder sensor can be designed to be insensitive to optical effects in the leads. Accordingly, short gage length measurements are possible. Intrinsic optical fiber strain sensors usually employ a *sensing fiber* and a *reference fiber*, either or both of which can be exposed to the strain field. However, the geometric or optical properties of the two fibers must be different if both the reference and sensing fibers are exposed to strain [20]. Each fiber is composed of a *lead-in fiber*, which carries light from an optical source to the strain-sensitive *active fiber*, followed by a *lead-out fiber*, which carries the light from the active fiber to an optical detector. The lead-in and lead-out fibers are also sensitive to the strain, so methods of designing *lead-insensitive* optical fiber sensors are of practical importance and will be described.

The feature common to all intrinsic optical fiber sensors is that they are constructed using optical fiber components only: fiber couplers replace beam splitters, aluminum-coated fiber ends replace mirrors, optical fiber retarders replace wave plates, and so on. The only non-fiber components are the light source and detection electronics, and these are usually *pig-tailed* (fiber attached directly to the electronic component) into the optical fiber sensor, effectively forming a fully guided optical system. The sensors will become completely closed optical systems with the incorporation of optical fiber lasers and all-optical (passive) phase detection schemes. While optical fiber sensors can be produced from suitable combinations of fiber and bulk optics, these hybrid designs are generally more cumbersome to use in operation because the bulk optics add undesirable instabilities. The electronic detection schemes employed with optical fiber sensors are as varied as the sensors and are beyond the scope of this review. However, appropriate references will be provided for the interested reader.

Mach–Zehnder, Michelson, and Fabry–Perot Strain Sensors

The Mach–Zehnder, Michelson, and Fabry–Perot strain sensors use all-fiber versions of their namesake classical interferometers. The Mach–Zehnder optical fiber strain sensor, illustrated in Fig. 2-20, is perhaps the best known, since it was the first to be developed [13,14]. Much of the early strain sensor research involved the Mach–Zehnder fiber interferometer; and because it is

STRAIN GAGES

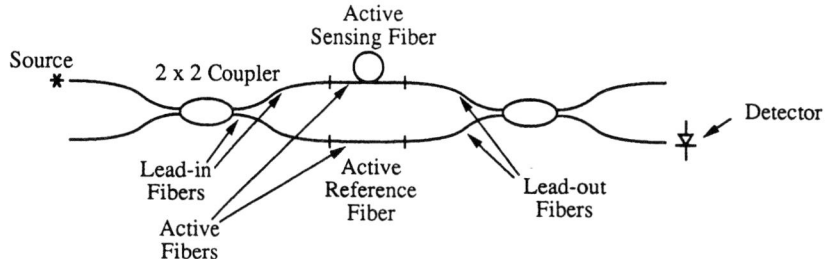

Figure 2-20 The Mach–Zehnder optical fiber strain sensor.

simple to employ, the Mach–Zehnder strain sensor is still used as a test-bed for optical fiber sensor concepts. This interferometer acts in the classic sense, since the light propagating in the reference arm is optically interfered with the light propagating in the sensing arm. This resulting coherent interference occurs in the second 2 × 2 coupler. The intensity is modulated by a strain-induced optical path length change, which results in a strain-induced optical phase shift. The change in optical path length is due to a change in the active length caused by the applied strain, and the birefringent effect. The birefringence is a result of the dependence of the refractive index on strain. Both these contributions to the change in optical path length are integrated over the active fiber length [21,22]. The phase shift is measured by directly counting the fringes, or by using active or passive homodyne or heterodyne techniques [15,23,24].

While it is possible to actually form interferometric fringes [14,23] in space that can be employed to give the strain-induced phase change, this approach is generally not recommended, since it adds considerable instability to the optical arrangement. Instead, the second 2 × 2 coupler is added to provide a stable interference location. The photodetector records the intensity, which varies as a sinusoid with increasing or decreasing strain as shown in Fig. 2-21. The results in Fig. 2-21 also serve to illustrate the difficulties encountered when attempting to count the fringes directly with an interferometer. Except for the point where the slope of the strain–time trace changes sign, the intensity–time trace is identical. This example illustrates that unless the strain history is monotonic, it is necessary to employ the more advanced detection schemes discussed in Refs. 15 and 24 to determine when the strain changes sign. This restriction applies to all fiber interferometers that undergo a phase shift exceeding 2π, since the intensity is a periodic function of phase. The active homodyne is the simplest of these detection schemes to implement in a laboratory environment [15].

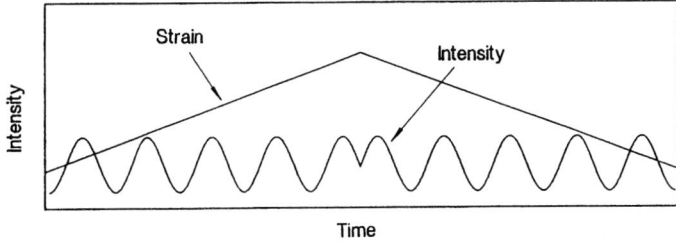

Figure 2-21 Sinusoidal interferometer time trace.

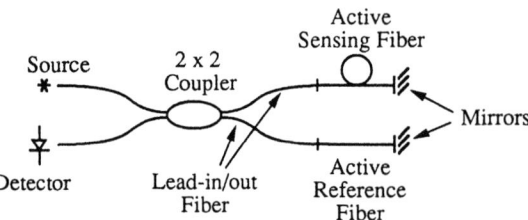

Figure 2-22 The Michelson optical fiber strain sensor.

Various strain gages, stress gages, and strain rosettes have been developed by using the Mach–Zehnder fiber interferometer by simply making appropriate alterations to the geometry of either the active reference or the sensing fiber [20,25]. The greatest disadvantage of the Mach–Zehnder strain sensor is the difficulty in isolating the leads from external strain fields.

The Michelson optical fiber strain sensors are Mach–Zehnder strain sensors operating in reflection as shown in Fig. 2-22. The optical fibers are cleaved after the first 2 × 2 coupler, and mirrors are formed on the optical fibers by depositing aluminum on the cleaved fiber ends. As with Mach–Zehnder strain sensors, lead-insensitive Michelson sensors are difficult to design. The most common approach is to subject the lead-in/out fibers to nearly identical strain fields, thus canceling strain-induced effects [26,27]. This approach to localized measurements works well only when little bending is present in the plane containing the lead-in and lead-out fibers. The sensitivity of the Michelson strain sensor is relatively high, so that gage lengths on the order of millimeters are practical. References 28 and 29 describe detection schemes commonly used with Michelson optical fiber strain sensors.

As the application for structurally embedded optical fiber sensors developed, it became important to reduce the number of optical leads required. Lee et al. [30] introduced the Fabry–Perot optical fiber strain sensor to produce a system in which a single optical fiber acts in both the lead-in and the lead-out capacities. The all-fiber Fabry–Perot strain sensor arranged in the reflection mode is shown in Fig. 2-23. An in-line Fabry–Perot cavity, which results in multiple interference of the light entering and leaving the cavity, is formed by a three-step process:

1. A short segment of the optical fiber is removed through carefully controlled cleaving techniques [30].
2. A partially reflecting material (e.g., TiO_2) is deposited on the cleaved face between the lead fiber and the cavity, and a fully reflective coating (e.g., Al) on the other cleaved end.

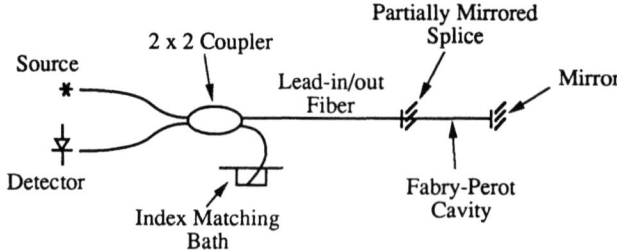

Figure 2-23 The all-fiber Fabry–Perot strain sensor: reflection mode.

3. The segment is replaced in its original position by using a slow, low-arc current fusion splicing technique [30].

A transmission Fabry–Perot strain sensor may be produced by following a very similar procedure if the mirror is replaced by a partial mirror and spliced to a lead-out fiber. Retardation measurements based on changes in the cavity length with the applied strain produce an optical phase shift, as described in the generalized theory presented in Ref. 22. The Fabry–Perot strain sensor combines the advantages of high sensitivity of the Mach–Zehnder and Michelson strain sensors with the advantage of a single optical lead. However, the semireflective fusion splice is difficult to fabricate and the partial mirror at the interface of the splice degrades the strength of the splice by as much as 50%. Depositing the semireflective coating on only the core of the optical fiber has been suggested as a means of improving the strength of the splice [31]. This approach shows promise, since the core accounts for less than 1% of the area of a single mode fiber.

Polarimetric Strain Sensor

The polarimetric strain sensor, first suggested by Rashleigh [32] in 1983, has become one of the most popular methods for measuring strain with an optical fiber. This sensor is essentially an integrated all-fiber polariscope with a high-birefringence (HiBi) optical fiber acting as the active sensing element. HiBi fibers have two preferred polarization states, which are due to residual stresses introduced in the fiber manufacturing process. Light propagating in these fibers is decomposed into two orthogonal components that travel with different velocities. These two components, labeled the fast and slow components (following photoelasticity nomenclature), interfere to produce fluctuation in intensity that are related to the strain field. Both circular and plane polariscope versions of a polarimetric strain sensor exist.

Polarimetric sensors measure localized strains, require only a single lead-in/out fiber, and can be self-temperature compensated. While polarimetric sensors are less sensitive than most optical fiber sensors, their sensitivity is adequate for structural strain measurements. A plane polariscope optical fiber polarimetric strain sensor operating in the reflection mode [33] is illustrated in Fig. 2-24A. The light exiting a plane-polarized source is launched into a HiBi lead-in/out optical fiber coincident with either the fast or slow axis to preserve the state of polarization (see Fig. 2-24B). HiBi fibers used in this manner are known as polarization-maintaining (PM) fibers. This lead-in/out fiber is spliced at a 45° angle with respect to the fast and slow axes of the sensing fiber, resolving the incident radiation equally onto the two polarization axes of the active fiber. An aluminum film deposited on the cleaved face of the sensing fiber acts as a mirror, which redirects the light back through the sensing and lead fibers. The load-induced phase shift in the active segment of the polarimetric sensor is related to the load parameters by the general theory presented by Ref. 22.

Polarimetric strain sensor rosettes [34], self-temperature-compensated polarimetric strains sensors [35], and several effective demodulation schemes have been demonstrated [15,35]. Since the sensitivity of this optical fiber sensor is relatively low, the sensor may not exhibit a phase shift greater than 2π. The strain in these situations is directly proportional to the intensity of the output fiber, and a demodulation scheme is not necessary. This linear relationship is valid only when an appropriate phase bias is applied, so that the sensor zero strain point is nearly centered on the zero of the sensor's cosine-squared output intensity function [15]. Because demodulation is not necessary, the low sensitivity is considered to be an advantage. Another attraction of the polarimetric sensor is that long gage lengths can be used without exceeding the coherence length of the optical source, since the optical path length difference always remains small. The principal disadvantage in implementing systems with polarimetric sensors is that HiBi components (fiber, couplers, all-fiber polarizers) are required.

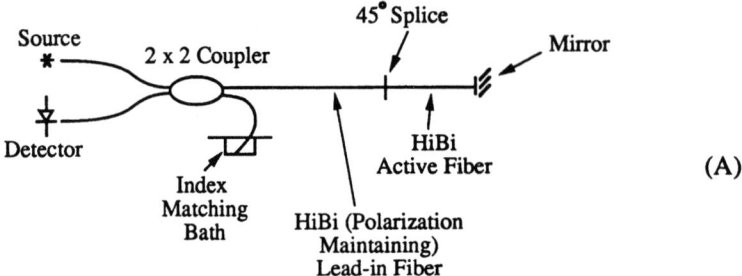

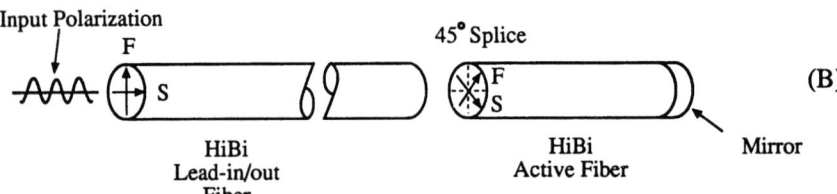

Figure 2-24 (a) A plane polariscope optical fiber polarimetric strain sensor: reflection mode; (b) passage of light from plane-polarized source to HiBi lead-in/out fiber.

Modal-Domain Sensor

Modal-domain sensors, introduced by Layton and Bucaro [36] in 1979, act similarly to the polarimetric sensor because two components of light propagating in a single optical fiber are made to interfere with each other. The launch conditions and the wavelength of the source are selected to excite two transverse modes that propagate in a nominally single-mode optical fiber. A typical single-mode optical fiber is specified with a cutoff wavelength, and this fiber supports only a single propagating mode if the light launched into the fiber has a wavelength greater than the cutoff [37]. Any number of propagating modes can be supported if the wavelength of the source is less than the cutoff. One can tailor the launch conditions and/or use high-birefringence optical fibers to select the precise modes the optical fiber will support. The most popular selection, called the dual-mode sensor, supports the LP_{01} and LP_{11}^{even} modes [38]. These two modes propagate over the same legnth of the optical fiber, but with different velocities. The two modes produce a strain-dependent beating effect. If one excites only the LP_{01} and LP_{11}^{even} modes and projects the light output from the fiber on to a screen, a two-lobed spatial intensity pattern is observed, as depicted in Fig. 2-25a. As a monotonic strain is applied, one lobe gains power at the expense of the next, until a maximum is reached (Fig. 2-25b through 2-25d); then with continued application of the strain, the lobe pattern reverses as indicated in Fig. 2-25e. The cyclic behavior of the lobe intensity is a sinusoidal function of strain, and monitoring the power of a single lobe through a spatial filter provides a measure of the strain.

STRAIN GAGES

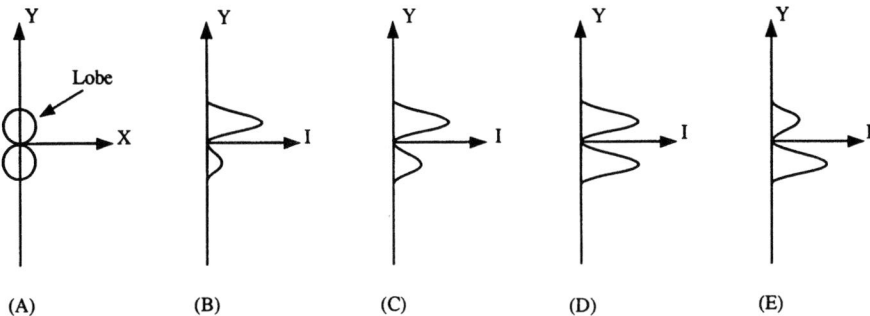

Figure 2-25 Cyclic intensity of a modal-domain strain sensor.

Lead-insensitive, modal-domain strain sensors can be designed by using fibers of two different types for lead-in and lead-out. The lead-in fiber is a HiBi fiber acting as a polarization maintaining fiber. This fiber is actively spliced [38] to a HiBi active sensing fiber with a higher cutoff wavelength so that dual-mode operation occurs. The active sensing fiber is then spliced with a diametral offset to a standard single-mode fiber so that the light from only one lobe enters the lead-out fiber [38]. In principle, it is also possible to design a single-ended, localized, modal-domain sensor. The sensitivities of modal-domain and polarimetric sensors are about the same because, in both cases, the signal output is due to differential interference as opposed to the division of wavefront interference typical of the Mach–Zehnder, Michelson, and Fabry–Perot sensors.

Two of the demodulation schemes used with modal-domain sensors are the passive spitter–recombiner approach [38] and heterodyne detection [39]. In many instances the relatively low sensitivity of the modal-domain sensor enables operation without demodulation schemes (provided a phase bias is applied). However both the modal-domain and the polarimetric sensors require specialized splicing techniques that are considered to constitute a disadvantage.

Twin-Core Sensors

The twin-core optical fiber sensor operates on the principle of evanescent coupling of optical power between two closely spaced waveguides [40,41]. A special optical fiber is produced which contains two cores. The electromagnetic field of each core extends beyond its confines and overlaps the other core. As a result, the electromagnetic energy is transferred back and forth between the two waveguides within the same fiber (see Fig. 2-26). This transfer of energy is called evanescent coupling or "cross talk." The amount of cross talk is a function of the core diameters, core separation, and applied strain [42]. It is possible to measure the strain by monitoring the intensity output from one of the cores. The output intensity of twin-core sensors behaves similarly to the modal-domain sensors (see Fig. 2-25), with the two cores defining odd and even spatial modes. To obtain the form of cross talk appropriate for sensing applications, the wavelength of the illuminating source should be less than the single-mode cutoff, but greater than that required for dual-mode operation. Offset splices, as described for the modal-domain sensors, are used to produce lead-insensitive sensors. The most commonly cited disadvantages of the twin-core sensor include the difficulty in producing offset splices and the expense of the twin-core optical fiber. Nevertheless, the twin-core sensor has proven effective in applications such as monitoring internal fatigue damage in bolted composite joints and high-temperature strain measurements [43].

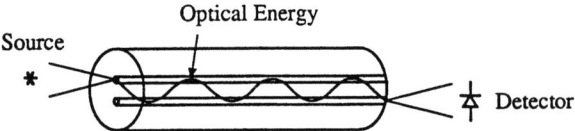

Figure 2-26 Cross talk in a twin-core optical fiber sensor.

Bragg Grating Sensors

The Bragg grating sensor differs markedly from the interferometric sensors just described. Qualitatively this optical fiber sensor is an optical grating whose pitch, and therefore wavelength tuning, is a function of the strain transferred to the optical fiber [44,45]. A fine pitch grating is impressed in a germanium-doped optical fiber using high-intensity ultraviolet dual-beam interference that spatially modulates the index of refraction of the fiber core [45]. The pitch of the grating is controlled by adjusting the angle between the two coherent beams forming the interference. The Bragg grating sensor can use either a narrow- or broadband light source. The narrowband sources offer a high temporal bandwidth signal, while the broadband sources provide the opportunity to multiplex sensors using only a single fiber. Multiplexing is accomplished by producing a fiber with a sequence of spatially separated gratings, each with a different pitch. The output of the multiplexed sensor is processed in an optical spectrum analyzer, and the optical signal associated with a given pitch p is centered at the $1/p$ location in the spectral domain. A strain-induced change in grating pitch is seen as a perturbation of the resonant frequency of the unstrained grating pitch and provides a direct measure of the strain. The advantages of this sensor include lead insensitivity, gage lengths on the order of millimeters, and the capability for simple multiplexing. The disadvantages include the high-power laser source required to produce the Bragg grating in the fiber core, and the high expense and limited frequency response of the spectrum analyzer. The frequency response of Bragg grating sensors can be improved by using a narrowband source, but this practice requires more complex multiplexing procedures [46].

Other Optical Fiber Strain Sensors

Our discussion of different optical fiber strain sensors provides an insight to the wide array sensor configurations that have been developed; however, this list of sensors is by no means complete. Other sensors that merit review are the extrinsic Fabry–Perot [47], microbend [48], optical time-domain reflectometry (OTDR) [49], and Brillouin optical time-domain analysis (BOTDA) types [50]. The extrinsic Fabry–Perot device has many of the qualities of the intrinsic Fabry–Perot sensor, but there is no need for a partially reflecting splice. The advantages gained by removing the partially mirrored splice are offset by the addition of other implementational difficulties [47]. The microbend sensor is one of the simplest optical fiber strain sensors from both theoretical and implementational points of view, yet sensors of this type require in-situ calibration. The OTDR techniques have the advantage of using off-the-shelf optical fiber diagnostics equipment, but they suffer from limited spatial resolution (on the order of meters) and very low frequency response (on the order of 100 Hz). While reentrant loops have significantly advanced the capabilities of OTDR sensors, they are still limited to large structural applications [49]. BOTDA sensors suffer from many of the same problems as the OTDR devices; however, they have the unique ability to measure the continuous and absolute (relative to zero strain) distribution of strain along the entire fiber length. The BOTDA strain sensors are the only optical fiber strain sensors developed to date that have this feature [50].

The choice of optical fiber strain sensor is dependent on the application and level of expertise of the user, and obvious tradeoffs exist between performance and difficulty in implementation. While optical fiber strain sensors have not reached the level of commercialization of resistance strain gages, the twin-core, Bragg grating, intrinsic Fabry–Perot, OTDR, and micro-Michelson, as well as a host of other sensors, are commercially available in small quantities. One may expect that many of the difficulties encountered in implementing these measurements will be resolved in the next few years as the methods and hardware are optimized. The intrinsic costs of the sensors and instruments required for signal processing are already comparable to resistance strain-gage technology.

The optical fiber sensor technology is finding applications in niche areas not well suited for resistance gages. This is particularly true in electromagnetic noise environments, explosive environments (where electrical sparks are intolerable), and applications requiring a large degree of design flexibility or passive sensor-multiplexing capabilities [49].

2-8 Summary and Future Trends

Since the mid-1950s and the introduction of the foil-grid construction and selected-melt gages there has been no major development in metallic alloy electrical-resistance strain gages. Continued research and development by the user community has led to an improved awareness of the range of application of the gages and an appreciation of the degree of precision that can be obtained. Continued development and investment by strain gage manufacturers has led to a broad product line with very tightly controlled specifications that ensure a high-quality gate that will perform in a reliable and stable manner. As the electrical-resistance strain gage begins its sixth decade of development, it has become part of a mature technology, where advances are made in very small increments through the contributions of many users and suppliers.

Development also continues in instrumentation as improvements in semiconductor technology have led to high-gain stable amplifiers, more highly regulated power suppliers, precision digital voltmeters, multichannel data acquisition systems, and on-line computer processing of strain gage data in real time. These developments have reduced both the time and the costs and improved the accuracies achieved in tests where strain gages are used in large numbers to characterize completely the stress field in complex structures.

The improvements in high-gain amplifiers have inhibited the application of semiconductor strain gages for general-purpose stress analysis. The cost of the semiconductor gages is justified only in those cases where measurements of extremely low strains with great precision are required. Piezoresistive technology is still important today in developing transducers to measure load and pressure. In these transducers, silicon is often used as the mechanical element and the gaging for a complete Wheatstone bridge is incorporated directly into the silicon by utilizing integrated-circuit manufacturing procedures. The result is a very high output voltage with excellent linearity and essentially no hysteresis.

New developments are occurring in the strain measurement field, but in an area where several new technologies, such as fiber optics, laser diodes, and pin diodes, are being combined to develop optical fiber sensors. Laboratory work by several investigators shows promise of providing a new and novel method for measuring strain that will extend the measurement capabilities of the experimentalist.

References

1. C. C. Perry, The Resistance Strain Gage Revisited, W. Murray Lecture, 5th Int. Cong. Exp. Mech., Montreal, Quebec, Canada, June 10–15, 1984.
2. W. Thompson (Lord Kelvin), On the Electrodynamic Qualities of Metals, *Proc. R. Soc. London, 146* (1856), 649–751.
3. E. E. Simmons, Jr., Material Testing Apparatus, U.S. patent 2,292,549, Feb. 23, 1940.
4. C. S. Smith, Piezoresistive Effect in Germanium and Silicon, *Phys. Rev., 94* (1954), 42–49.
5. W. P. Mason and R. N. Thurston, Piezoresistive Materials in Measuring Displacement, Force, and Torque, *J. Acoust. Soc. Am. 29*, no. 10 (1957), 1096–1101.
6. J. W. Dally and W. F. Riley, *Experimental Stress Analysis*, 2nd ed., McGraw-Hill, New York, 1978, 162–164.
7. Measurements Group, Inc., Errors Due to Transverse Sensitivity in Strain Gages, *Tech. Note 509*, 1982, 1–8.
8. H. S. Freynik and G. R. Dittbenner, Strain Gage Stability Measurements for a Year at 75°C in Air, *Univ. Calif. Rad. Lab. Rep. 76039*, 1975.
9. P. Stehlin, Strain Distribution in and Around Strain Gauges, *J. Strain Anal., 7*, no. 3 (1972), 228–235.
10. M. F. Beatty and S. W. Chewning, Numerical Analysis of the Reinforcement Effect of a Strain Gage Applied to a Soft Material, *Int. J. Eng. Sci., 17* (1979), 907–915.
11. L. F. McCalvey, Strain Measurements on Low Modulus Materials, presented at the Br. Soc. Strain Conf., Univ. Surrey, United Kingdom, September 1982.
12. R. N. White, Model Study of the Failure of a Steel Bin Structure, presented at the ASCE/SESA Exchange Session on Physical Modeling of Shell and Space Structures, ASCE Annu. Conv., New Orleans, October 1982.
13. J. A. Bucaro, H. D. Dardy, and E. Carome, Fiber Optic Hydrophone, *J. Acoust. Soc. Am., 62*, (1977), 1302–1304.
14. C. D. Butter and G. B. Hocker, Fiber Optics Strain Gage, *Appl. Opt., 17*, no. 18 (1978), 2867–2869.
15. T. G. Giallorenzi, J. A. Bucaro, A. Dandridge, G. H. Sigel, J. H. Cole, S. C. Rashleigh, and R. G. Priest, Optical Fiber Sensor Technology, *IEEE J. Quant. Electron., QE-18*, no. 4 (1982), 626–665.
16. R. De Paula, Fiber Optic Sensor Overview, *Proc. SPIE, 566*, (1985), 2–15.
17. B. Culshaw, and J. Dankin, *Optical Fiber Sensor: Systems and Applications*, Vol. 1, Artech House, Norwood, MA, 1988.
18. B. Culshaw and J. Dankin, *Optical Fiber Sensor: Systems and Applications*, Vol. 2, Artech House, Norwood, MA, 1989.
19. R. D. Turner, T. Valis, W. Hogg, and R. Measures, Fiber-Optic Sensors of Smart Structures. *J. Intell. Mater. Syst. Struct., 1*, no. 1 (1990), 26–49.
20. H. W. Haslach, and J. S. Sirkis, Surface-Mounted Optical Fiber Strain Sensor Design, *Appl. Opt. 30*, no. 25 (1991).
21. J. S. Sirkis and H. W. Haslach, Complete Phase–Strain Model for Structurally Embedded Interferometric Optical Fiber Sensors, *J. Intell. Mat. Syst. Struct., 2*, no. 1 (1991), 3–24.
22. J. S. Sirkis, Phase–Strain–Temperature Models for Smart Struct Interferometric Optical Fiber Sensors: Parts I and II. *Optical Engr., 32*, no. 4 (1993), 752–772.
23. J. S. Sirkis and C. E. Taylor, Interferometric Fiber Optic Strain Sensor, *Exp. Mech., 28*, no. 2 (1988), 170–176.
24. R. S. Medlock, A Review of Modulating Techniques for Fibre Optical Sensors, *J. Opt. Sensors*, (1986), 43–68.
25. J. S. Sirkis and H. W. Haslach, Strain Measurement with Surface Mounted Single Mode Optical Fibers, in *Developments in Theoretical and Applied Mechanics*, Vol. XV (S. V. Hanagud, M. P. Kamat, and C. E. Ueng, Eds.), 1990, 544–551.
26. K. A. Murphy, M. Gunther, and A. Vensarkar, R. O. Claus, Miniaturized Fiber Optical Michelson Type Interferometric Sensor, *Appl. Opt., 30*, no. 26 (1991). Vol. 30, No. 34, pp 5063–5067, (1991).
27. T. Valis, E. Tapanes, K. Liu, and R. M. Measures, Passive-Quadrature Demodulation Localized-Michelson Fiber-Optical Strain Sensor Embedded in Composite Materials, *J. Lightwave Technol., 9*, no. 4, (1991), 535–544.

28. K. P. Koo, A. B. Tventen, and A. Dandridge, Passive Stabilization Scheme for Fiber Interferometers Using (3 × 3) Fiber Directional Couplers, *Appl. Phys. Lett., 41*, no. 7 (1982), 616–618.

29. D. A. Jackson, Monomode Optic Fibre Interferometers for Precision Measurement, *J. Phys. E: Sci. Instrum., 18,* (1985), 981–1001.

30. C. E. Lee, H. F. Taylor, A. M. Markus, and E. Udd, Optical-Fiber Fabry–Perot Embedded Sensors, *Opt. Lett., 13,* no. 21 (1989), 1225–1227.

31. T. Valis, D. Hogg, and R. M. Measures, Composite Material Embedded Fabry–Perot Fiber Optic Strain Rosette, *Proc. SPIE, 1370,* (1990), 145–161.

32. S. C. Rashleigh, Origins and Control of Polarization in Single Mode Optical Fibers, *J. Lightwave Technol., 1,* no. 2 (1983), 312–331.

33. M. Corke, A. D. Kersey, K. Lui, and D. A. Jackson, Remote Temperature Sensing Using Polarization-Preserving Fibre, *Electron. Lett., 20,* no. 2 (1984), 67–69.

34. R. M. Measures, D. Hogg, R. D. Turner, T. Vallis, and M. J. Giliberto, Structurally Integrated Fiber Optic Strain Rosette, *Proc. SPIE, 986,* (1988), 32–42.

35. J. P. Dankin and C. A. Wade, Compensated Polarimetric Sensor Using Polarisation-Maintaining Fibre in a Differential Configuration, *Electron. Lett., 20,* (1984), 51–53.

36. M. R. Layton and J. A. Bucaro, Optical Fiber Acoustic Sensor Utilizing Mode–Mode Interference, *Appl. Opt., 18,* no. 5 (1979), 666–670.

37. C. Yeh, *Handbook on Fiber Optics: Theory and Applications,* Academic Press, San Diego, CA, 1990.

38. K. A. Murphy, M. S. Miller, A. M. Vensarkar, and R. O. Claus, Elliptical-Core Two-Mode Optical-Fiber Sensor Implementation Methods, *J. Lightwave Technol., 8,* no. 11, (1990), 1688–1696.

39. J. N. Blake, Q. Li, and B. Y. Kim, Elliptical Core Two-Mode Fiber Strain Gauge With Heterodyne Detection, *OFS'88,* New Orleans, 1988, 416–419.

40. G. Meltz, J. R. Dunphy, W. W. Morey, and E. Snitzer, Fiber Optic Temperature and Strain Sensors, *Proc. SPIE, 798,* (1987), 103–114.

41. G. Meltz, J. R. Dunphy, W. W. Morey, and E. Snitzer, Cross-Talk Fiber-Optic Sensor, *Appl. Opt., 22,* no. 3 (1983), 464–477.

42. E. Snitzer, J. R. Dunphy, and G. Meltz, Fiber Optic Strain Sensor, in *Fiber-Optic Rotation Sensors and Related Technologies* (S. Ezekiel and H. J. Arditty, Eds.), Springer Verlag, New York, 1982, 406–412.

43. R. Mack, Fiber Sensors Provide Key for Monitoring Stresses in Composite Materials, *Laser Focus,* May 1987.

44. J. R. Dunphy, G. Meltz, and W. W. Morey, Multi-Function, Distributed Optical Fiber Sensor for Composite Cure and Response Monitoring, *Proc. SPIE, 1370* (1990), 116–118.

45. G. R. Meltz, W. W. Morey, and W. H. Glen, Formation of Bragg Gratings in Optical Fibers by a Transverse Holographic Method, *Opt. Lett., 14,* (1989), 823–825.

46. A. D. Kersey and A. Dandridge, Distributed and Multiplexed Fiber-Optic Sensors, *OFS'88,* New Orleans, 1988, 60–71.

47. K. A. Murphy, M. Gunther, A. Vensarkar, and R. O. Claus, Quadrature Phase-Shifted Extrinsic Fabry–Perot Optical Fiber Sensors, *Opt. Lett., 16* (1991), 783.

48. E. Udd, J. P. Theriault, A. Markus, and Y. Bar-Cohen, Microbending Fiber Optic Sensors for Smart Structures, *Proc. SPIE, 1170* (1989), 478–482.

49. B. D. Zimmerman, R. O. Claus, D. A. Kapp, and K. A. Murphy, Fiber Optic Sensors Using High Resolution Optical Time Domain Instrumentation Systems, *J. Lightwave Technol., 8,* (1990), 1273–1277.

50. T. Kurashima, T. Horiguchi, N. Yoshizawa, H. Tada, and M. Tateda, Measurement of Distributed Strain Due to Laying and Recovery of Submarine Optical Fiber Cable, *Appl. Opt., 30,* no. 3 (1991), 334–337.

CHAPTER
3
Strain-Gage Instrumentation and Data Analysis

Kenneth G. McConnell and William F. Riley

Iowa State University
Ames, Iowa

3-0 Introduction

A large number of devices have been developed to measure strains on the surfaces of machine components and structural members. The electrical-resistance strain gage has proven to be the most versatile of many devices, since it can be fabricated in a wide variety of configurations to meet many different measurement requirements. The electrical-resistance strain gage is also used as the sensing element in the design of many force, pressure, and acceleration transducers. The dynamic characteristics of these transducers are discussed in detail in Chapter 4. In this chapter, basic characteristics of electronic devices such as amplifiers, filters, converters, power supplies, and switches used with constant-voltage and constant-current potentiometers and Wheatstone bridge circuits are considered. Calibration procedures and problems associated with lead wires, switches, data transmission, and electrical noise are discussed. Finally, the equations required to reduce measured strain data to stress and strain at a point are presented.

3-1 Signal-Conditioning Devices

The resistance change that occurs when a strain gage is deformed is very small. As a result, the Wheatstone bridge unbalance and voltage change associated with the resistance change also are very small. This small voltage change must be detected, transmitted, and recorded. Several useful electronic devices [amplifiers, filters, analog-to-digital (A/D) converters, etc.] that are common to many modern measurement systems are discussed first in this chapter.

Amplifiers

An amplifier increases the low-level transducer voltage to a level sufficient for use by a voltage-recording device. An amplifier has the following characteristics:

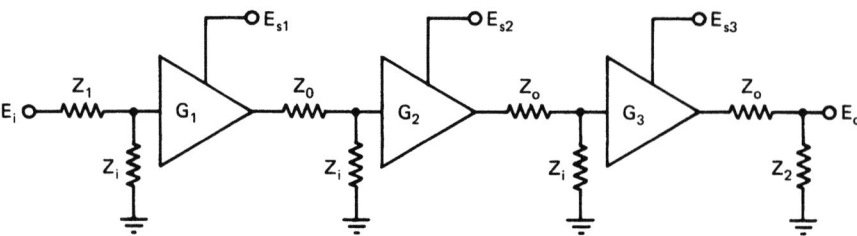

Figure 3-1 High-gain amplifier system.

1. Gain G is the ratio of the output voltage E_o to the input voltage E_i in the linear range of the amplifier. Thus

$$E_o = GE_i \tag{3-1}$$

All amplifiers have a limited range where the performance is linear.
2. Input impedance Z_i controls the amount of current that the amplifier draws from the input voltage source. Since an amplifier with a low input impedance can draw excessive current from the transducer, a high input impedance is desirable.
3. Output impedance Z_o controls the amount of voltage drop that occurs inside the amplifier for a given amount of output current. A low output impedance is desirable.
4. Frequency response $H(\omega)$ (also called transfer function) is a measure of gain change with frequency of the input voltage signal.
5. Time constant τ is a measure of amplifier response to a step input.

The interrelationship between the impedances and gains of the high-gain amplifier system shown in Fig. 3-1 can be expressed as

$$E_o = G_1 G_2 G_3 \left(\frac{Z_i}{Z_i + Z_1}\right)\left(\frac{Z_i}{Z_i + Z_o}\right)^2 \left(\frac{Z_2}{Z_o + Z_2}\right) E_i \tag{3-2}$$

where Z_i = amplifier input impedance
Z_o = amplifier output impedance
Z_1 = internal impedance of the transducer
Z_2 = input impedance of the voltage recorder

In properly designed voltage amplifiers, Z_i is much larger than either Z_1 or Z_o. Therefore, Eq. (3-2) reduces to

$$E_o = G_1 G_2 G_3 \left(\frac{Z_2}{Z_o + Z_2}\right) E_i \tag{3-3}$$

where the term $Z_2/(Z_o + Z_2)$ represents a voltage attenuation due to the current required to drive the voltage recorder. By using a recorder with a high input impedance relative to the amplifier output impedance, the overall gain of the cascaded amplifier system becomes the product of the gains of the individual amplifiers. Thus

$$E_o = G_1 G_2 G_3 E_i \tag{3-4}$$

The frequency response of the amplifier in any instrumentation system requires careful con-

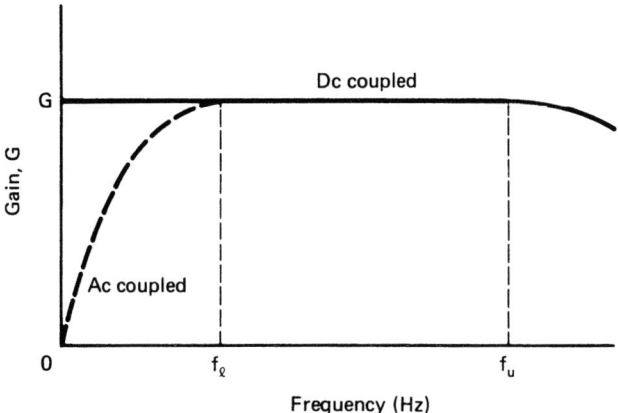

Figure 3-2 Frequency response of an amplifier-recorder system.

sideration. As shown graphically in Fig. 3-2, the gain is flat *between* the lower and upper frequency limits f_l and f_u. In a *dc-coupled* amplifier the gain is constant down to a frequency of zero Hertz. In an *ac-coupled* amplifier, the gain drops to zero as the frequency of the input signal goes to zero. Above the upper limiting frequency the gain drops to zero as the frequency increases. Thus both low- and high-frequency data can be seriously altered both in magnitude and phase by the amplifier's transfer function.

A second way to measure the frequency response of an amplifier is to observe its response to a step-input voltage, as shown in Fig. 3-3. This response is often closely approximated by the exponential function.

$$E_o = G(1 - e^{-t/\tau})E_i \qquad (3\text{-}5)$$

where τ is the amplifier time constant.

Operational Amplifiers

An operational amplifier (commonly called an op-amp) is a complete high-gain ($G > 10^5$ is typical) amplifier circuit that is easily used with many different external passive resistive and capacitive components. The high gain, coupled with a high input impedance ($Z_i > 10$ MΩ) and a low output impedance ($Z_o < 100$ Ω), makes the op-amp an ideal element to use in voltage signal-conditioning circuits. The commonly used schematic for an op-amp is shown in Fig. 3-4. The two input terminals are known as the *inverting* ($-$) terminal and the *noninverting* ($+$) terminal. The output voltage E_o of an op-amp is given by the expression

$$E_o = G(E_{i2} - E_{i1}) \qquad (3\text{-}6)$$

which shows that the op-amp is a *differential amplifier*, since the output voltage depends on the difference between the input voltages. Op-amps can be used in a number of different ways. Typical examples are as voltage followers, inverting amplifiers, summing amplifiers, integrating amplifiers, and differentiating amplifiers.

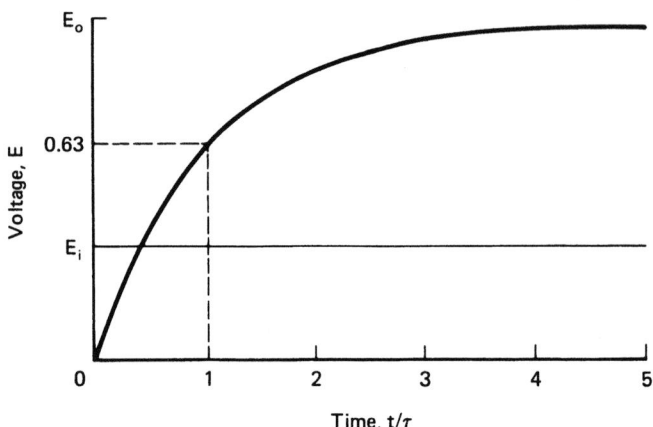

Figure 3-3 Response of an amplifier to a step-input voltage.

Voltage Followers

The voltage-follower circuit shown schematically in Fig. 3-5 uses an op-amp to provide a very large input impedance to the transducer so that little current (power) is required from the transducer, and to provide a low output impedance to the voltage recorder so that the recorder load effects on the transducer's circuit are minimized. The voltage follower serves as a buffer between the transducer and the recorder. The gain of the voltage follower is

$$G_c = \frac{E_o}{E_i} = \frac{G}{1 + G} \sim 1 \tag{3-7}$$

The effective input impedance is

$$R_{ci} = \frac{E_i R_i}{E_i - E_o} = (1 + G)R_i \tag{3-8}$$

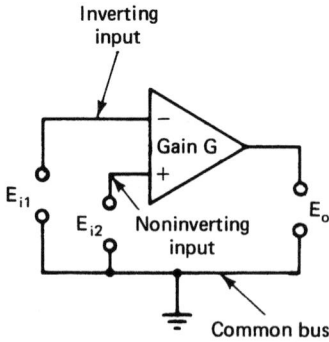

Figure 3-4 Op-amplifier circuit.

STRAIN-GAGE INSTRUMENTATION AND DATA ANALYSIS

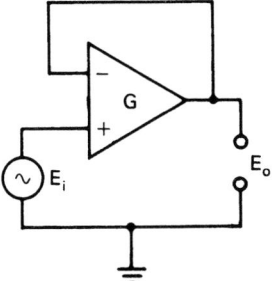

Figure 3-5 Voltage-follower circuit.

where R_i is the amplifier input impedance. The effective output impedance is

$$R_{co} = \frac{R_o}{(1 + G)} \tag{3-9}$$

where R_o is the amplifier output impedance.

Inverting Amplifiers

The inverting amplifier circuit constructed with an op-amp and input and feedback resistors R_1 and R_f is shown in Fig. 3-6a. The amplifier gain G_c for this circuit is given by the expression

$$G_c = \frac{E_o}{E_i} = -\frac{R_f}{R_1} \frac{1}{1 + (1/G)(1 + R_f/R_1 + R_f/R_a)} \tag{3-10}$$

Equation (3-10) reduces to

$$G_c \sim -\frac{R_f}{R_1} \tag{3-11}$$

when the op-amp gain G is large so that the term $(1 + R_f/R_1 + R_f/R_a)/G$ is small compared to unity.

Noninverting Amplifiers

The noninverting amplifier circuit is shown in Fig. 3-6b. The amplifier gain G_c for this circuit is given by the expression

$$G_c = \frac{E_o}{E_i} = \frac{G}{1 + GR_1/(R_1 + R_f)} \tag{3-12}$$

which reduces to

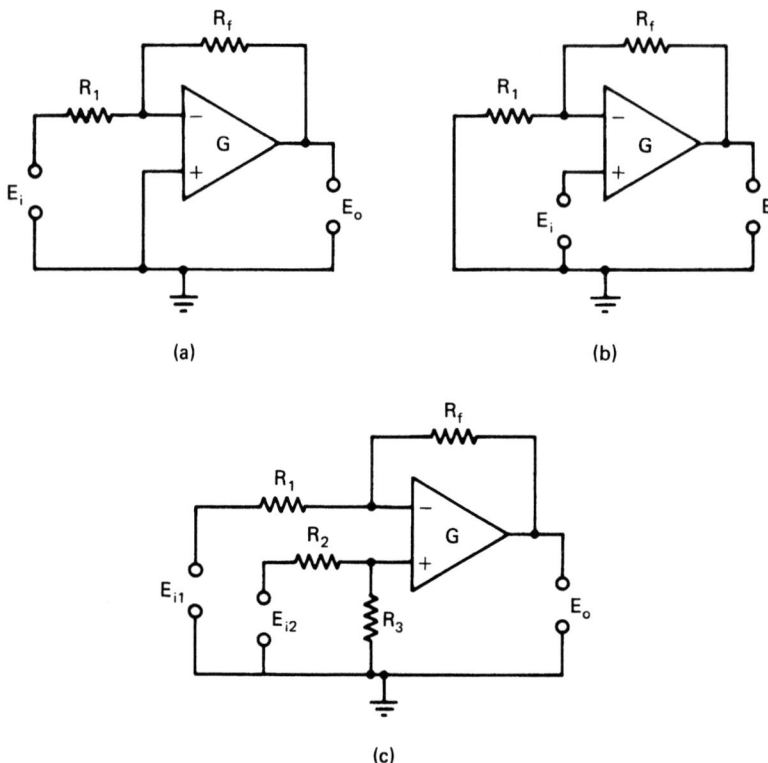

Figure 3-6 Simple instrument amplifiers constructed with operational amplifiers: (a) inverting; (b) noninverting; (c) differential.

$$G_c \sim 1 + \frac{R_f}{R_1} \tag{3-13}$$

when the gain G of the op-amp is large.

Differential Amplifiers

The differential amplifier circuit is shown in Fig. 3-6c. For this circuit,

$$E_o \sim \frac{R_3}{R_2}\left(\frac{1 + R_f/R_1}{1 + R_3/R_2}\right)E_{i2} - \frac{R_f}{R_1}E_{i1} \tag{3-14}$$

which can be expressed as

$$G_c \sim \frac{E_o}{E_{i2} - E_{i1}} \sim \frac{R_f}{R_1} \tag{3-15}$$

when $R_f/R_1 = R_3/R_2$.

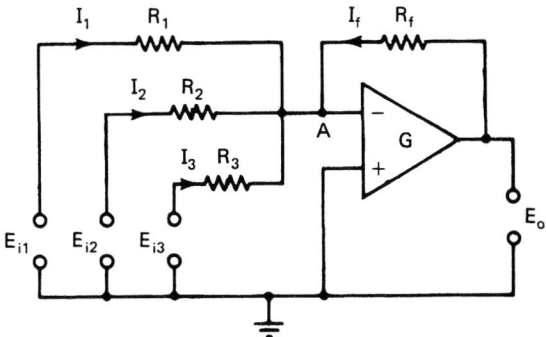

Figure 3-7 Summing amplifier circuit.

Summing Amplifiers

A summing amplifier circuit is shown in Fig. 3-7. For this circuit,

$$E_o = -\frac{E_{i1}/R_1 + E_{i2}/R_2 + E_{i3}/R_3}{1/R_f + (1/G)(1/R_1 + 1/R_2 + 1/R_3 + 1/R_f + 1/R_i)} \tag{3-16}$$

which reduces to

$$E_o = -R_f \sum_{i=1}^{n} \frac{E_i}{R_i} \tag{3-17}$$

when the op-amp gain G is large with respect to the resistance ratio sum. This circuit can be used for adding and subtracting voltages if both the positive and the negative inputs of the op-amp are used.

Integrating Amplifiers

An integrating amplifier uses a feedback capacitor C_f in place of a feedback resistor R_f as shown in Fig. 3-8. The voltage input–output relationship for this circuit is given by the expression

$$E_o = -\frac{1}{R_1 C_f} \int_0^t E_i \, dt \tag{3-18}$$

if the gain G of the op-amp is large so that the input voltage at point A is small compared to E_i. It is clear from Eq. (3-18) that output voltage is the integral of the input voltage times $-1/R_1 C_f$, where $R_1 C_f$ is the integration time constant.

Differentiating Amplifiers

A differentiating amplifier is produced when the resistor and the capacitor in the circuit shown in Fig. 3-8 are interchanged. The output voltage of the differentiating amplifier is given by the expression

$$E_o = -(R_f C_1) \frac{dE_i}{dt} \tag{3-19}$$

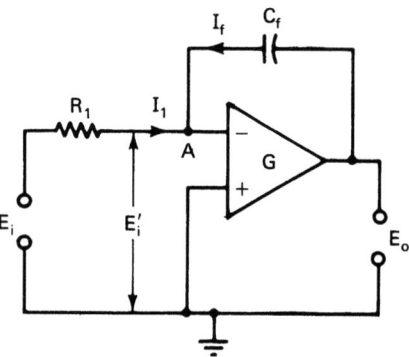

Figure 3-8 Integrating amplifier circuit.

where $R_f C_1$ is the differentiation time constant. Extreme care must be exercised when this circuit is used, since a significant amount of high-frequency noise is generated in addition to the desired signal.

The foregoing sample of op-amp circuits shows that the device is very useful in measurement systems. The gain of the op-amp must be large for these applications, but no specific value is required provided the gain exceeds a minimum amount. Obviously, voltage offset, thermal drift, and other factors beyond the scope of this handbook need to be considered. For more details, consult any of the excellent reference texts on the subject.

Filters

Often, the transducer signal is contaminated with noise or other undesirable voltage components. If there is a significant difference between the frequency content of the desired transducer signals and the undesirable noise signals, filters can be used to eliminate or significantly reduce the unwanted signals. Some of the common filters employed in signal conditioning are the high-pass *RC* filter, the low-pass *RC* filter, and the second-order *RC* filter. These filters types can be constructed from either passive or active (with op-amps) circuits. Schematic diagrams for each filter type are shown in Fig. 3-9. Good filter designs use op-amps in a voltage-follower configuration to buffer the passive filter elements from both the voltage source (transducer) and the voltage recording devices. In the following sections, equations are derived by assuming that the voltage source has zero internal impedance and the voltage recording device has nearly infinite impedance, thus drawing no current from the filter. Both of these impedance requirements are satisfied by using op-amps on the input and output sides of the passive circuits so that nearly ideal filter performance is achieved.

High-Pass RC *Filter*

This simple, yet effective, first-order filter is used to attenuate low-frequency signals and pass high-frequency signals. A schematic diagram for a simple high-pass *RC* filter is shown in Fig. 3-9a. The transfer function for the filter can be obtained by assuming a sinusoidal input voltage. The resulting output voltage is observed to be attenuated and phase shifted relative to the input

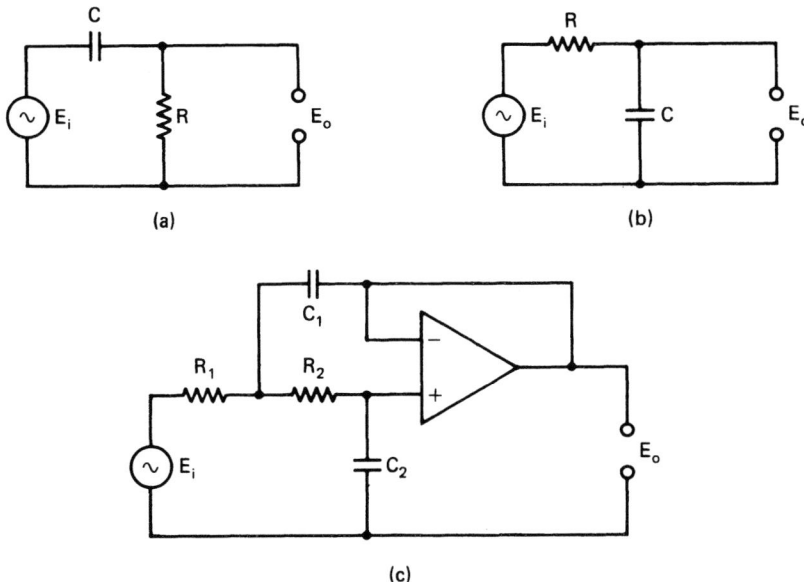

Figure 3-9 Common filters employed for signal conditioning: (a) high-pass RC; (b) low-pass RC; (c) second-order low-pass.

voltage by the filter. This magnitude distortion and phase distortion depends on the frequency of the input signal. The resulting transfer function for this type of filter is

$$\frac{E_o}{E_i} = \frac{jRC\omega}{1 + jRC\omega} = \frac{RC\omega \exp[j(\pi/2 - \beta)]}{\sqrt{1 + (RC\omega)^2}} \quad (3\text{-}20)$$

and

$$\tan \beta = RC\omega \quad (3\text{-}21)$$

where $j = \sqrt{-1}$
ω = frequency of input signal, rad/s

Equation (3-20) indicates that zero-frequency signals vanish so that the filter effectively blocks dc signals. At frequencies somewhat higher than the cutoff frequency $f_c = 1/(2\pi RC)$, the voltage ratio goes to unity as shown in Fig. 3-10a. The signal is attenuated 3 dB (0.707) at the critical frequency ($RC\omega = 1$) and there is a phase shift of $\pi/4$ rad (45°) between the input and output signal. One of the primary uses of this filter is to isolate the desired higher-frequency transducer signals from a high dc voltage.

Low-Pass RC Filter

A schematic diagram for a simple low-pass RC filter is shown in Fig. 3-9b. The transfer function for this type of filter is

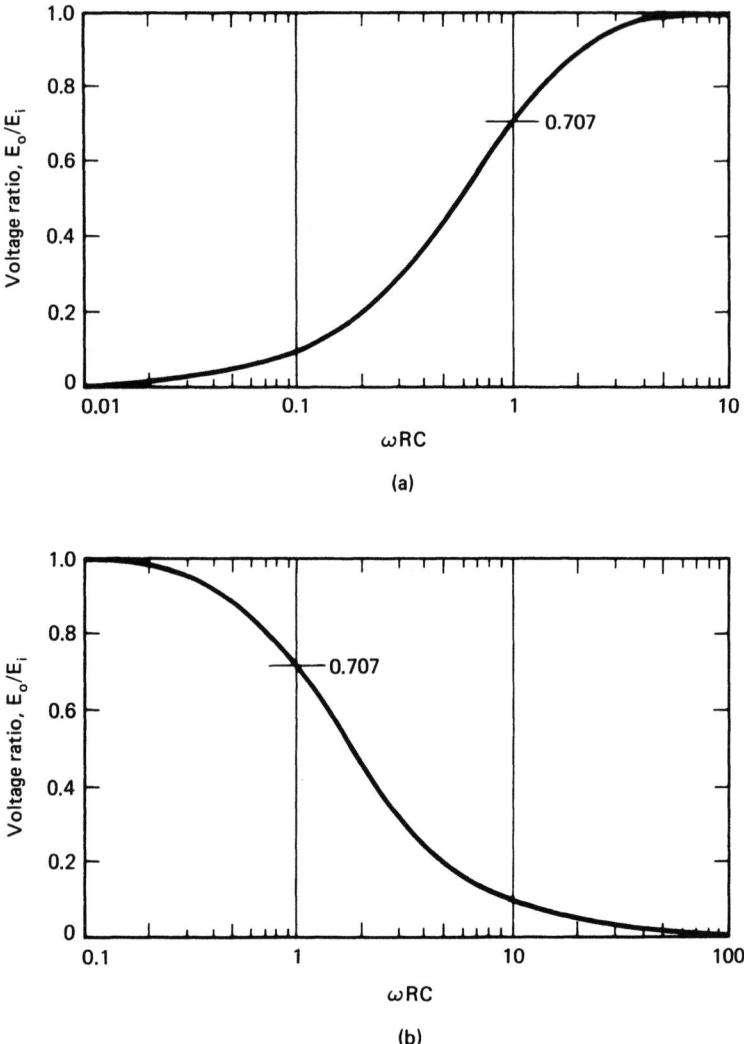

Figure 3-10 Response curves for (a) high-pass and (b) low-pass RC filters.

$$\frac{E_o}{E_i} = \frac{1}{1 + jRC\omega} = \frac{\exp(-j\beta)}{\sqrt{1 + (RC\omega)^2}} \tag{3-22}$$

where the lagging phase angle of the output signal relative to the input signal is given by Eq. (3-21). The magnitude of the transfer function is shown in Fig. 3-10b. Here it is seen that signals above the critical frequency

$$f_c = \frac{1}{2\pi RC} \quad \text{Hz} \tag{3-23}$$

decrease with increasing frequency. An attenuation of 3 dB occurs at the critical frequency f_c. A 2% attenuation occurs when $RC\omega = 0.2$.

Second-Order Filter

A schematic diagram for a simple second-order filter is shown in Fig. 3-9c. This type of filter is constructed with two resistors (R_1 and R_2), two capacitors (C_1 and C_2), and one op-amp. The transfer function for this type of filter is

$$\frac{E_o}{E_i} = \frac{1}{1 - r^2 + j2\,rd} = \frac{\exp(-j\beta)}{\sqrt{(1 - r^2)^2 + (2\,rd)^2}} \tag{3-24}$$

where r = a dimensionless frequency ratio = f/f_c
f_c = break frequency = $1/(2\pi\sqrt{R_2C_2R_1C_1})$
d = damping ratio = $(R_1 + R_2)C_2/2\sqrt{R_1C_1R_2C_2}$

and

$$\tan \beta = \frac{2\,rd}{1 - r^2} \tag{3-25}$$

Optimum behavior of this filter occurs when the damping d is in the range 0.64 to 0.707. Within this range of damping, the phase shift is nearly linear with frequency and the attenuation is a minimum up to the cutoff frequency. The following procedure can be used to design a filter for a particular application. First, let $R_2 = AR_1$ and $C_2 = BC_1$, so that $R_2C_2 = (AB)R_1C_1$ and $f_c = 1/(2\pi R_1 C_1 \sqrt{AB})$. The damping equation then yields the design equation

$$(AB)^2 + (2B - 4d^2)(AB) + B^2 = 0 \tag{3-26}$$

Once the damping value d and the capacitor ratio B are selected, the value of AB is obtained from Eq. (3-26). Once AB is known and the desired cutoff frequency is selected, the remaining filter parameters are easily obtained.

The principal advantage of the second-order filter is that the attenuation is a function of $1/\omega^2$ (12 dB/octave). The attenuation of the first-order filter is a function of $1/\omega$ (6 dB/octave). The difference in performance is dramatically illustrated by noting that the frequency ratio required to obtain a 99% attenuation is 100 for the first-order system and 10 for the second-order system.

Analog-to-Digital Converters

Often it is desirable to digitize the amplified transducer voltage for the purposes of storing, transmitting, processing, and displaying data. The two most popular analog-to-digital (A/D) converter types are the shift-programmed successive approximation A/D converter for use in high-speed applications and the dual-slope integrating A/D converter for use in low-cost, low-speed applications.

Successive Approximation Converter

During operation, this A/D converter compares the input analog voltage to a reference voltage obtained from a programmed digital-to-analog converter (D/A). The reference voltage E_i corresponding to the $i$the digital bit position is given by the equation.

$$E_i = \left(\frac{2^{i-1}}{2^N}\right)E_r \tag{3-27}$$

where E_r = reference voltage for the A/D converter
N = most significant bit number

The process starts with the most significant digital bit ($i = N$) being set true in the reference voltage-generating digital-to-analog converter. If the input voltage is greater than this reference voltage, the binary bit remains true and the process continues to the next-most-significant bit. If the input voltage is less than the reference voltage, this binary bit is set to false and the process continues to the next-most-significant bit in the binary word. This process continues through each of the binary bits down to the least significant bit ($i = 1$). The data are set in the output digital latch and the converter sends an end-of-conversation (EOC) signal when finished so that the digital data system knows that the data are available for its use. For a 10-bit A/D converter, this process takes place 10 times and the resulting voltage is obtained to an accuracy of 1 part in 2^{10} (1024 parts). The accuracy is limited to $\pm$ 1 least significant bit (LSB) under the best of circumstances. The problem of signal variation during the conversion process is overcome by using a *sample-and-hold* (S/H) circuit.

Dual-Slope Integrating Converter

This A/D converter operates on the basis of integrating both an unknown and a known voltage. First, the unknown voltage E_i is integrated for a fixed period of time t_i by charging a capacitor. Second, an internal known voltage E_r is applied to the integrator, causing the charged capacitor to discharge toward zero. A counter measures the time t_r required to discharge the capacitor to zero. The unknown voltage is calculated by using the equation

$$E_i = \left(\frac{E_r}{t_i}\right)t_r \tag{3-28}$$

since E_r and t_i are known constants of the converter and t_r is measured.

The integrating converter is preferred for most low-speed applications because of its high resolution (17 bits give 1 part in 131,072), true averaging of the signal during conversion (60-Hz noise is averaged out when the integration period is 1/60s), autozeroing, and low cost.

Amplitude Modulation

Amplitude modulation occurs whenever a strain-gage bridge is excited by an ac voltage. The carrier frequency is the ac voltage frequency ω_c, and the modulation (transducer) frequency is the frequency of the measurand ω_t. The resulting output voltage from the bridge can be represented by the expression

$$\begin{aligned} E_o &= E_b \sin(\omega_c t)\sin(\omega_t t) \\ &= \frac{E_b[\cos(\omega_c - \omega_t)t - \cos(\omega_c + \omega_t)t]}{2} \end{aligned} \tag{3-29}$$

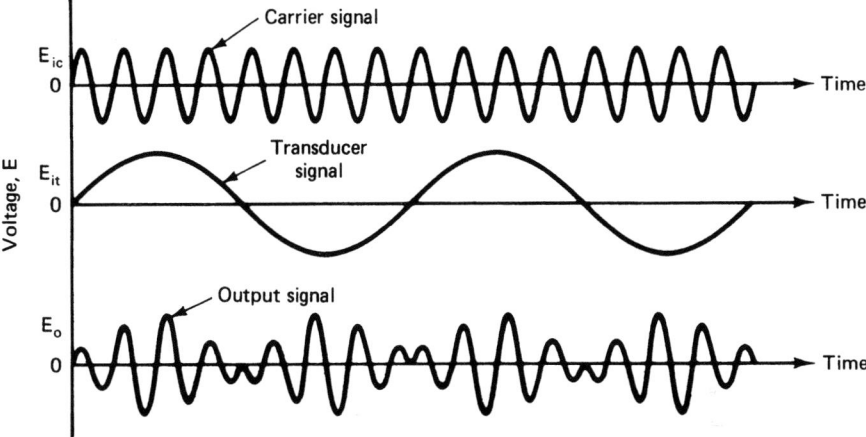

Figure 3-11 Carrier, transducer, and amplitude-modulated output signals.

where E_b is the amplitude of the bridge voltage, which is proportional to the bridge excitation voltage and the amount of bridge unbalance.

Equation (3-29) indicates that the signal is composed of two equal-amplitude sinusoidal waveforms with two distinct frequencies ($\omega_c - \omega_t$) and ($\omega_c + \omega_t$). One frequency is smaller than the carrier frequency and the other is large than the carrier frequency by an amount equal to the transducer frequency. These two frequencies are referred to as sidebands. For practical applications, the carrier frequency should be at least 10 times the maximum anticipated transducer frequency. One of the main advantages of a carrier system is that a low-frequency transducer signal is converted to two high-frequency signals that can pass through a high-pass filter to eliminate low-frequency noise, such as 60-Hz line-frequency noise, which often occurs when signals are transmitted over long distances. A second advantage is that ac-coupled amplifiers can be used to overcome thermal drift problems. The combination of two signals to obtain amplitude modulation is illustrated in Fig. 3-11.

The output signal from amplitude modulation is not suitable for display and interpretation until the transducer signal has been separated from the carrier signal. This separation process is called *demodulation*. The amplitude-modulated signal is processed by a full-wave, phase-sensitive rectifier, as shown in Fig. 3-12. The rectifier output consists of a series of half-sine waves, the envelope of which follows the transducer waveform as shown. The half-sine waves have a frequency spectrum consisting of the transducer frequency plus two and four times the carrier frequency with sidebands. These carrier and sideband frequencies are easily filtered to give the transducer signal when the carrier frequency is at least 10 times the maximum transducer signal frequency.

Constant-Voltage Source

Many instruments require a constant-voltage power source. Power obtained directly from rectified line power often varies significantly with time as other power demands change on the power line. Therefore, other means must be used to ensure that a constant voltage is provided to the instrument. A simple single-sided, constant-voltage circuit is shown in Fig. 3-13. This circuit employs a

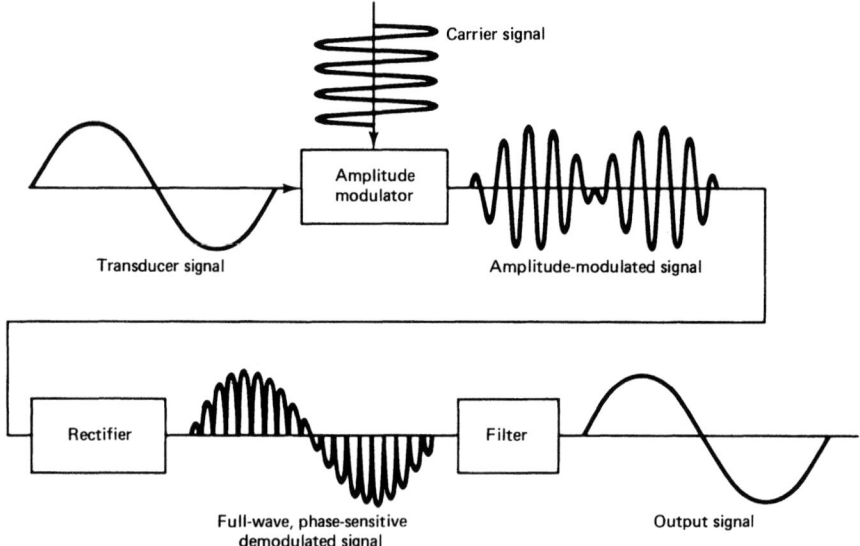

Figure 3-12 Amplitude modulation and demodulation.

temperature-stable voltage-reference diode D_1, an op-amp with input bias current that is temperature compensated, a feedback potentiometer R_1 and R_2, and a transistor for power gain. The load resistor R_L represents the instrument that is to be driven at constant voltage. The output voltage E_o for the circuit is given by the expression

$$E_o = \left(1 + \frac{R_2}{R_1}\right) V_0 = \frac{V_0}{a} \tag{3-30}$$

where V_0 is the reference voltage and a is the position of the potentiometer wiper. It is clear from Eq. (3-30) that the long-term stability of the voltage supply depends on the thermal stability of the reference voltage source and the op-amp. The resistance ratio R_2/R_1 is extremely stable when a potentiometer is used, since the thermal resistance coefficient is the same for both the R_1 and the R_2 parts. When two different resistors are used (even from the same batch), the thermal coefficient can be not only different for each resistor but also of opposite sign. This situation can cause a large thermal sensitivity. Note that neither the thermal offset of the op-amp nor the characteristics of the transistor play a direct role in this circuit as long as the op-amp can provide the control required by the feedback signal. The op-amp simply controls the circuit to produce a constant voltage across the load resistance.

Constant-Current Source

A simple constant-current circuit is shown in Fig. 3-14a. In this circuit, the reference voltage V_0 is applied to the positive op-amp terminal. The output voltage E_o increases until the voltage at point A is equal to V_0. This causes a constant current I to pass through resistor R_1 regardless of the changes that occur in resistor R_2 (provided these changes are within the control limit of the

STRAIN-GAGE INSTRUMENTATION AND DATA ANALYSIS

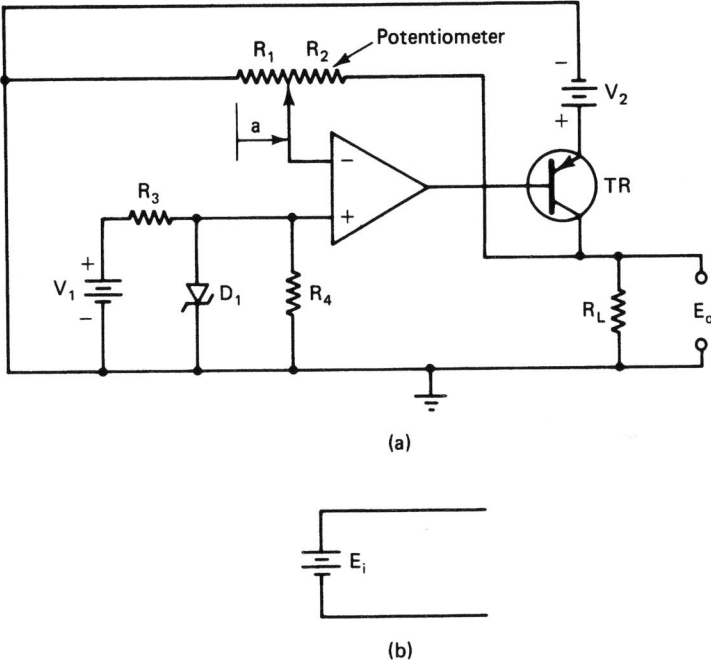

Figure 3-13 Constant-voltage transducer-excitation source with power transistor (TR) and load resistance R_L: (a) circuit diagram; (b) equivalent circuit.

op-amp). A similar circuit, shown in Fig. 3-14b, uses a power transistor TR to produce a larger current I. It should be noted that the stability of this constant-current circuit depends on the thermal stability of the reference voltage V_0, the thermal stability of the reference resistor R_1, and the thermal stability of the op-amp.

Both constant-voltage and constant-current power supplies are used with strain-gage circuit.

3-2 Recording Instruments For Strain Gages

Many different recording instruments can be used to monitor the change in resistance that occurs in a strain gage. The instrument chosen usually depends on the type of strain measurement (static or dynamic) and the number of gages and/or channels employed.

Short-term static readings are usually the easiest and least expensive to make. Long-term static measurements can present extreme difficulties due to strain-gage and recording-instrument drift, especially when it is impossible to establish a zero benchmark reference because the load cannot be removed from the structure. Dynamic strain-recording problems are usually associated with frequency response and noise.

Strain-gage resistance changes are very small, approximately 0.000240 Ω per microstrain for a 120-Ω strain gage. Thus a strain of 1000 μm/m will produce a resistance change of 0.240 Ω

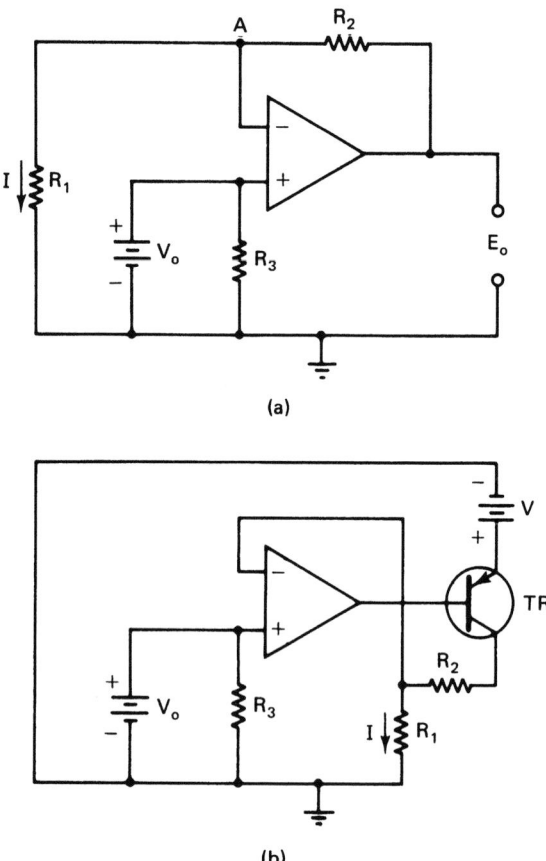

Figure 3-14 Constant-current sources for transducers and signal conditioning circuits: (a) simple constant-current source: (b) constant-current source with a power transistor.

in a 120-Ω gage. Potentiometer and Wheatstone bridge circuits are widely used with strain gages to convert these small changes in resistance to output voltage signals that can be observed and recorded. Both circuits are used with either constant-voltage or constant-current excitation.

Potentiometer Circuits

Constant-Voltage Excitation

The constant-voltage potentiometer circuit has three distinctive parts, as shown in Fig. 3-15. These parts are the constant-voltage supply, the potentiometer circuit itself, and the output recording interface. The potentiometer circuit consists of the ballast resistor R_b and the strain-gage resistor

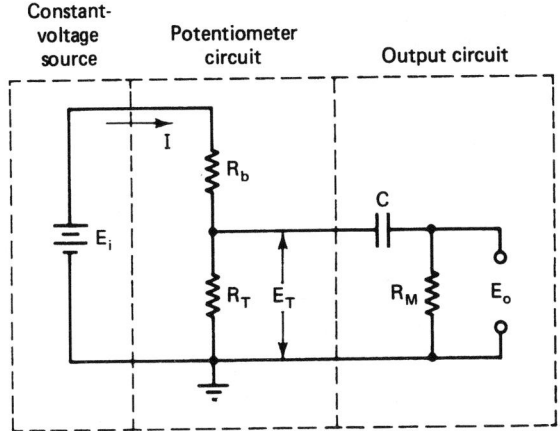

Figure 3-15 Constant-voltage potentiometer circuit.

R_T. The ideal open-circuit output voltage E_T is obtained when the recorder resistance R_M is assumed to be infinite. Thus

$$E_T = \frac{R_T}{R_T + R_b} E_i = \frac{1}{1 + r} E_i \tag{3-31}$$

where r is the resistance ratio R_b/R_T. The change in output voltage due to resistance changes ΔR_b and ΔR_T can be obtained from Eq. (3-31) as

$$\Delta E_T = \frac{r}{(1 + r)^2} \left(\frac{\Delta R_T}{R_T} - \frac{\Delta R_b}{R_b} \right)(1 - \eta) E_i \tag{3-32}$$

where

$$\eta = 1 - \left[1 + \frac{1}{1 + r} \left(\frac{\Delta R_T}{R_T} - r \frac{\Delta R_b}{R_b} \right) \right]^{-1} \tag{3-33}$$

The term η contains the nonlinear terms associated with the output of the circuit. If the ballast resistor R_b is also a strain gage, it is clear from Eq. (3-32) that temperature compensation is achieved if both gages are mounted on the same material and are exposed to the same temperature change. However, the nonlinear term in Eq. (3-33) is largest when $r = 1$.

The power P_T that can be dissipated by the transducer (strain gage) limits the maximum voltage that can be applied to the circuit. This power is related to the excitation voltage E_i by the expression

$$P_T = I E_T = \frac{E_i^2}{R_T (1 + r)^2}$$

The circuit sensitivity can be expressed in terms of r, R_T, and P_T as

$$S_{cv} = \frac{\Delta E_T}{\Delta R_T/R_T} = \frac{r}{1 + r} \sqrt{P_T R_T} \tag{3-34}$$

The presence of a large ballast resistor tends to provide a circuit with nearly constant-current behavior. Also, if r is large, the nonlinear term is minimum for a given $\Delta R_T/R_T$. The value of r is limited by the voltage sources normally available for use with the circuit. Since values of r above $r = 9$ do not significantly increase the sensitivity, larger values are seldom used. For $r = 9.0$ and $\Delta R_b = 0$, the error term can be approximated as

$$\eta = 0.1\left(\frac{\Delta R_T}{R_T}\right) - \left(\frac{0.1\,\Delta R_T}{R_T}\right)^2 + \cdots \qquad (3\text{-}35)$$

from which it is clear that a 1% error in η ($\eta = 0.01$) corresponds to $\Delta R_T/R_T = 0.1$. This change in resistance ratio corresponds to a strain of 50,000 μm/m when the strain gage has a gage factor of 2.0. A resistance ratio change of 0.1 is very large for most applications using metal-foil gages but is not unreasonable for semiconductor strain gages, which have gage factors greater than 100.

A load resistance R_M can cause the output voltage E_o to be attenuated due to the current that is drawn by this resistor. This attenuation can be expressed as

$$\Delta E_o = \Delta E_T(1 - \eta_1) \qquad (3\text{-}36)$$

where the loss factor η_1 is given by the expression

$$\eta_1 = 2r\frac{R_T}{R_M}\frac{1 + r[1 + 0.5(R_T/R_M)]}{1 + r(1 + R_T/R_M)} \qquad (3\text{-}37)$$

It is clear from Eq. (3-37) that loading errors of less than 1% occur when $R_M > 400R_T$ for $r = 1$ and when $R_M > 200R_T$ for $r = 9$. The resistance ratio is easily obtained in most instances. Furthermore, it should be noted that proper calibration will negate this recorder load effect, since it will be present in the calibration signal and thus automatically accounted for. The major point to be emphasized here is that small values of R_M will result in an undesirable reduction in an already small signal, thus requiring additional amplification for proper recording.

The capacitive coupling shown in Fig. 3-15 is used to remove the high-level dc voltage component present in the signal E_T from the low-level voltage changes produced by the strain. The frequency response of this RC circuit is that of a high-pass filter with the time constant controlled by $R_M C$. Clearly, this circuit is suited for use only with dynamic strains with frequency components above the low-frequency cutoff $[f_c = 1/(2\pi R_M C)]$.

Constant-Current Excitation

The constant-current potentiometer circuit is shown in Fig. 3-16. This circuit consists of a constant-current supply, the potentiometer elements R_b and R_T, and the output circuit. In this case, the presence of the ballast resistor R_b is superfluous since the circuit is designed to operate at constant current and $R_b = 0$ does not change circuit performance.

The transducer voltage E_T is given by the simple expression

$$E_T = IR_T \qquad (3\text{-}38)$$

when the load resistance R_M is very large. The corresponding change in voltage becomes

$$\Delta E_T = IR_T\frac{\Delta R_T}{R_T} = E_T\frac{\Delta R_T}{R_T} \qquad (3\text{-}39)$$

Note that the change in voltage ΔE_T is a linear function of the resistance change ΔR_T irrespective

STRAIN-GAGE INSTRUMENTATION AND DATA ANALYSIS

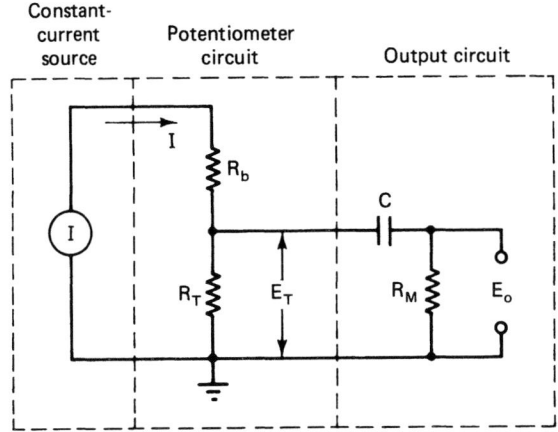

Figure 3-16 Constant-current potentiometer circuit.

of the magnitude of the change. This linear behavior extends the useful range for many applications. For the constant-current potentiometer, the circuit sensitivity is given by the expression

$$S_{cc} = \frac{\Delta E_T}{\Delta R_T/R_T} = IR_T = \sqrt{P_T R_T} \tag{3-40}$$

A comparison of this equation with Eq. (3-34) shows that the constant-current circuit is more efficient by a factor of $(1 + r)/r$, which is 2 for $r = 1$ and 1.111 for $r = 9$.

The load resistance R_M can affect the sensitivity of the constant-current circuit by desensitizing the change in resistance. For this case, the change in output voltage due to a change in resistance ΔR_T becomes

$$\Delta E_o = I \, \Delta R_T (1 - \eta_1) \tag{3-41}$$

where the loss factor η_1 is given by the expression

$$\eta_1 = \frac{(R_T/R_M)[2 + R_T/R_M + (\Delta R_T/R_T)(1 + R_T/R_M)]}{(1 + R_T/R_M)[1 + (R_T/R_M)(1 + \Delta R_T/R_T)]} \tag{3-42}$$

Thus the loss factor η_1 depends on both R_T/R_M and $\Delta R_T/R_T$. For errors of less than 1%, $R_M > 210 R_T$ when $\Delta R_T/R_T = 0.1$ and $R_M > 300 R_T$ when $\Delta R_T/R_T = 1$. Thus, if $R_M > 300 R_T$, the loss factor is less than 1% for most applications.

Capacitive coupling is used with this circuit. This capacitive coupling serves the same purpose as in the constant-voltage potentiometer circuit.

Wheatstone Bridge

Constant-Voltage Excitation

The constant-voltage Wheatstone bridge consists of three parts, as shown in Fig. 3-17. These are the constant-voltage source E_i, *four resistors* R_1, R_2, R_3, and R_4 arranged in the bridge configuration;

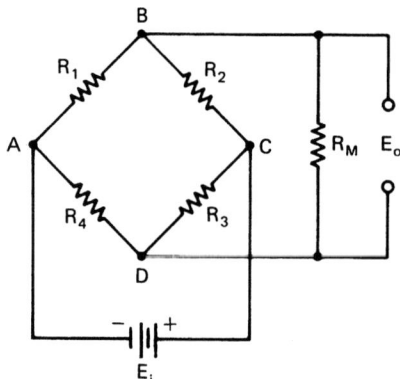

Figure 3-17 Constant-voltage Wheatstone bridge circuit.

and the readout circuit, consisting of a load resistance R_M. In the equations that follow, the value of R_M is assumed to be infinite so that no current is drawn from the bridge. The effect of using a finite-load resistance is discussed at the end of the section.

For the Wheatstone bridge, the output voltage E_o (the difference between voltages E_B and E_D) is given by

$$E_o = \frac{R_1 R_3 - R_2 R_4}{(R_1 + R_2)(R_3 + R_4)} E_i \tag{3-43}$$

Equation (3-43) indicates that the initial output voltage will vanish ($E_o = 0$) if

$$R_1 R_3 = R_2 R_4 \tag{3-44}$$

When Eq. (3-44) is satisfied, the bridge is said to be *balanced*. This means that the small unbalanced voltage caused by a change in resistance is measured from a zero or near-zero condition. This small signal can easily be amplified to significant levels for recording.

An output voltage ΔE_o develops when resistances R_1, R_2, R_3, and R_4 change by amounts ΔR_1, ΔR_2, ΔR_3, and ΔR_4, respectively. Such changes in resistance in strain gages can be due to strain or temperature. From Eq. (3-43), the change in output voltage ΔE_o resulting from these small resistance changes is

$$\Delta E_o = \frac{r}{(1 + r)^2} \left(\frac{\Delta R_1}{R_1} - \frac{\Delta R_2}{R_2} + \frac{\Delta R_3}{R_3} - \frac{\Delta R_4}{R_4} \right)(1 - \eta) E_i \tag{3-45}$$

where the error term η is given by

$$\eta = \left[1 + \frac{1 + r}{\frac{\Delta R_1}{R_1} + \frac{\Delta R_4}{R_4} + r\left(\frac{\Delta R_2}{R_2} + \frac{\Delta R_3}{R_3}\right)} \right]^{-1} \tag{3-46}$$

This error term reduces to

$$\eta = \frac{\sum_{i=1}^{4} \frac{\Delta R_i}{R_i}}{\sum_{i=1}^{4} \frac{\Delta R_i}{R_i} + 2} \quad (3\text{-}47)$$

for the common case where all four resistances are equal (i.e., $r = 1.0$). It is evident from both Eqs. (3-46) and (3-47) that the error term will be zero when ΔR_1 and ΔR_4 are equal and of opposite sign with $\Delta R_2 = \Delta R_3 = 0$ as well as when ΔR_2 and ΔR_3 are equal and of opposite sign with $\Delta R_1 = \Delta R_4 = 0$. This result is important, since nonlinear bridge behavior is nonexistent when two active gages are used in combination in either arms R_1 and R_4 or arms R_2 and R_3. The nonlinear effect does not exist when four active gages of the same nominal resistance are used in the bridge. For a single active gage, the nonlinear effect is limited to a 1% error when $\Delta R_1/R_1$ is less than 0.02 (a value that normally corresponds to a strain of 10,000 μm/m).

The sensitivity S_{cv} of a single active strain gage in a constant-voltage Wheatstone bridge circuit is given by the expression

$$S_{cv} = \frac{\Delta E_o}{\Delta R_1} = \frac{r}{(1+r)^2} E_i = \frac{r}{1+r} \sqrt{P_g R_g} \quad (3\text{-}48)$$

where P_g = power dissipated by the strain gage
R_g = gage resistance

The power dissipation characteristics of the strain gage limit the maximum sensitivity that can be achieved, since the circuit efficiency $r/(1 + r)$ is limited to approximately 90% ($r = 9$) by practical upper limits on the supply voltage E_i. Strain gages that can dissipate high power should be selected. However, the base material and its heat transfer characteristics also limit the power dissipation of a given gage in a given application.

The four different ways that strain gages are commonly connected in the Wheatstone bridge are shown in Fig. 3-18. The advantages and limitations of each of these cases are discussed relative to Eqs. (3-45) and (3-46), which control the output-voltage sensitivity and the nonlinear bridge behavior.

CASE 1. A single active strain gage is used in location R_1 as shown in Fig. 3-18a. This arrangement is used for both static and dynamic measurements when temperature compensation is not an important consideration. The resistances R_2, R_3, and R_4 are usually adjusted to give an initially balanced bridge and maximum circuit sensitivity. This is the most nonlinear bridge-operating case, and $\Delta R_g/R_g$ must be limited to 0.02 if errors are to be less than 1%.

The sensitivity of the strain-gage Wheatstone bridge system is defined in terms of the change in voltage ΔE_o per unit strain, which is also the product of the gage sensitivity S_g and the bridge circuit sensitivity S_{cv} given in Eq. (3-48). Thus

$$S_s = \frac{\Delta E_o}{\varepsilon} = S_g S_{cv} = \frac{\Delta R_g/R_g}{\varepsilon} \frac{\Delta E_o}{\Delta R_g/R_g} = \frac{r}{1+r} S_g \sqrt{P_g R_g} \quad (3\text{-}49)$$

It is clear from Eq. (3-49) that there are four factors that control the system sensitivity. The first is the circuit efficiency $r/(1 + r)$, which is limited to the practical range of [50% ($r = 1$), 80% ($r = 4$), and 90% ($r = 9$)] due to voltage limitations. The second factor is the strain gage sensitivity S_g, which is usually between 2 and 4 for metal gages. The third factor is the power dissipated by the gage P_g, which is limited by the material on which the gage is mounted and the strain-gage size required for the given application. The fourth factor is the gage resistance R_g,

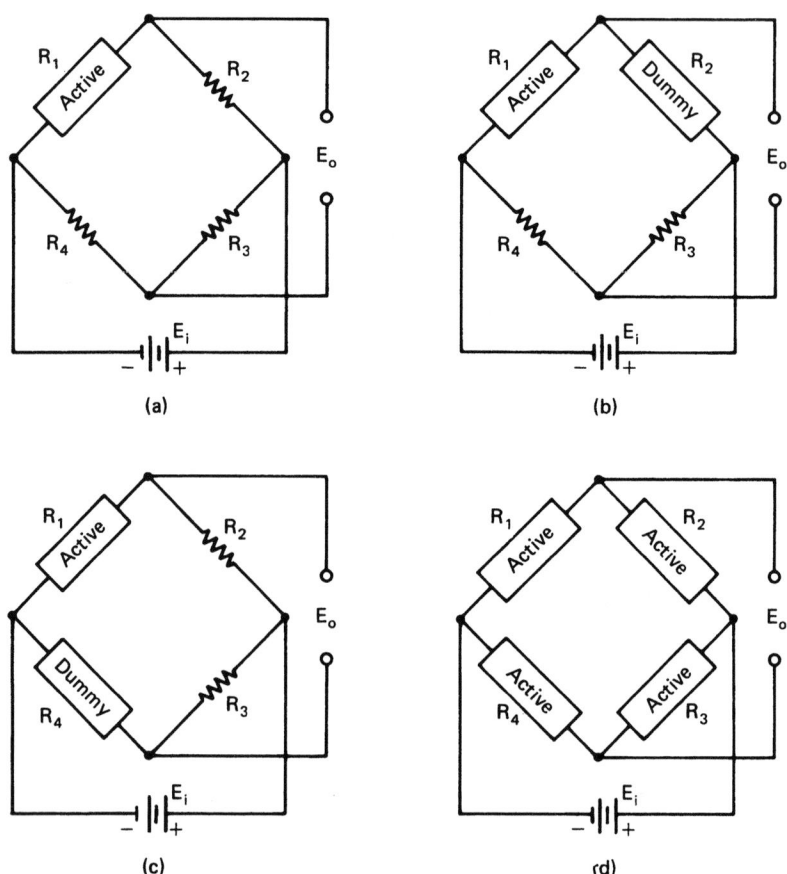

Figure 3-18 Four common strain-gage arrangements in a Wheatstone bridge.

which ranges from 120 to 1000 Ω for metal gages. The 120-, 350-, and 500-Ω gages are most common, so the choice here is often limited by commercial availability.

CASE 2. This bridge arrangement uses two active arms, R_1 R_2, with R_3 and R_4 as fixed resistors, as shown in Fig. 3-18b. There are two ways in which this bridge arrangement is used.

The first provides temperature compensation for the single active gage located in position 1 by using a dummy gage in position 2 that is subjected to the same temperature as the active gage. It is clear from Eq. (3-45) that the portion of ΔR_1 due to temperature would be canceled if ΔR_2 changes the same amount due to temperature. This requires that identical gages be mounted on the same material and be subjected to the same temperature changes. The nonlinear effect given by Eq. (3-46) is the same for this case as it is for the single active gage in Case 1. The only advantage over Case 1 is reduction of temperature sensitivity of the circuit.

The second use of this strain-gage circuit is to increase the circuit sensitivity while maintaining temperature compensation and removing or reducing the nonlinear behavior of the circuit. Suppose that gage 1 is mounted on the top and gage 2 is mounted on the bottom of a beam at the same cross section. In this case, both gages are subjected to essentially the same temperature (assuming that both are shielded from direct sunlight or other heat sources that could cause strong temperature differences to occur with time) so that temperature-induced resistance changes essentially cancel according to Eq. (3-45). The strain experienced by the two gages due to loads on the beam should be of opposite sign (and will be equal in magnitude for symmetrical beams). The resulting voltage output from Eq. (3-45) doubles when the strains are equal and of opposite sign and the nonlinear term is zero, as seen from Eq. (3-46) when $\Delta R_1/R_1 = -\Delta R_2/R_2$. Thus, using two gages can increase sensitivity, reduce the nonlinearity, and remove temperature sensitivity if the two gages are carefully mounted to assure similar temperature changes. The major disadvantage of this circuit arrangement is that $r = 1$, so that circuit efficiency is only 50%.

CASE 3. Two active gages in arms 1 and 4 are used as shown in Fig. 3-18c. As seen from Eq. (3-45), arm 4 can be used like arm 2 in Case 2 to achieve both temperature compensation and increased sensitivity, since both $\Delta R_2/R_2$ and $\Delta R_4/R_4$ have the same sign relative to $\Delta R_1/R_1$. The principal advantage of Case 3 over Case 2 is that the circuit efficiency $r/(1 + r)$ can be increased from 50% for $r = 1$ to near 80 to 90% for $r = 4$ to 9, since r does not have to be unity to balance the bridge initially. Also, the nonlinearity is zero for any value of r as long as the $\Delta R/R$ ratio is the same magnitude and of opposite sign. For the case where arm 1 is the active arm and arm 4 is the temperature-compensating dummy arm, the nonlinear term is that of a single active gage. The sensitivity approaches that of a four-arm bridge, described next.

CASE 4. Four active strain gages are used in this case, as shown in Fig. 3-18d, so that r must be unity. There is complete temperature compensation in this circuit if identical gages are mounted on surfaces of the same material that experience the same changes in temperature with time. In addition, Eq. (3-45) shows that the output voltage is four times that of a single active gage, that for Case 2 with $r = 1$, and slightly greater than that for Case 3 when $r = 9$, provided the gages are mounted so that

$$\Delta R_1 = \Delta R_3 = -\Delta R_2 = -\Delta R_4 \tag{3-50}$$

The main advantages of the four-arm bridge are temperature compensation, linear behavior (if all ΔR's are the same magnitude), and ability to design measuring elements that eliminate undesired bending or axial load sensitivity.

A close examination of system sensitivity for each case shows that the sensitivity can range from $0.5S_g\sqrt{P_gR_g}$ for Case 2 to $2.0S_g\sqrt{P_gR_g}$ for Case 4, with values near $S_g\sqrt{P_gR_g}$ for Case 1 ($r = 9$). Generally, high-gain differential amplifiers can be used more economically than multiple gages to increase circuit activity.

The effect of load resistance R_M on the ideal output voltage E_o can be obtained from an analysis of the load resistance in series with the bridge. The result can be expressed as

$$E_{o1} = \frac{R_M}{R_M + R_B}E_o = \frac{1}{1 + R_B/R_M}E_o \tag{3-51}$$

where E_{o1} is the actual output voltage and

$$R_B = \frac{R_1R_2}{R_1 + R_2} + \frac{R_3R_4}{R_3 + R_4} \tag{3-52}$$

is the effective bridge resistance. It is evident from Eq. (3-52) that $R_B = R_g$ when all arms of the

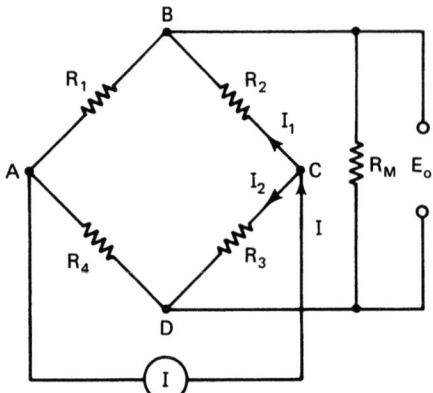

Figure 3-19 Constant-current Wheatstone bridge circuit.

bridge are of equal resistance R_g. Thus an error of less than 1% results when $R_M > 100R_g$ for an equal-arm bridge. This condition is easily achieved with most commercial instruments, since R_M is usually greater than $1000R_g$. In addition, this load effect can be accounted for with proper calibration procedures when R_M is the same for measurements and calibration.

Constant-Current Excitation

A constant-current power supply can be used with the Wheatstone bridge as shown in Fig. 3-19. The supply current I is divided into two currents I_1 and I_2 at point C of the bridge. Thus the current relationship is

$$I = I_1 + I_2 \tag{3-53}$$

The corresponding output voltage E_o is given by the expression

$$E_o = I_1 R_1 - I_2 R_4 = \frac{I}{R_1 + R_2 + R_3 + R_4}(R_1 R_3 - R_2 R_4) \tag{3-54}$$

which has the same balance condition as given by Eq. (3-44). The change in output voltage ΔE_o resulting from a resistance change in each arm of the bridge is

$$\Delta E_o = \frac{I R_1 R_3}{\Sigma R + \Sigma \Delta R}\left(\frac{\Delta R_1}{R_1} - \frac{\Delta R_2}{R_2} + \frac{\Delta R_3}{R_3} - \frac{\Delta R_4}{R_4} + \frac{\Delta R_1 \Delta R_3}{R_1 R_3} - \frac{\Delta R_2 \Delta R_4}{R_2 R_4}\right) \tag{3-55}$$

where $\Sigma R = R_1 + R_2 + R_3 + R_4$
$\Sigma \Delta R = \Delta R_1 + \Delta R_2 + \Delta R_3 + \Delta R_4$

There are two nonlinear terms in this equation. The first occurs in the ΔR collection of terms and the second occurs in the second-order product terms in parentheses. The effect of these terms can most easily be illustrated by considering an example. Let R_1 be the active gage, with the remaining gages fixed in value so that $R_4 = R_1 = R_g$, $R_2 = R_3 = rR_g$, and $\Delta R_2 = \Delta R_3 = \Delta R_4 = 0$. Note that the nonlinear terms in parentheses drop out when $\Delta R_3 = \Delta R_4 = 0$. The corresponding output voltage under these conditions becomes

$$\Delta E_o = \frac{I R_g r}{2(1 + r) + (\Delta R_g/R_g)} \frac{\Delta R_g}{R_g} \tag{3-56}$$

$$= \frac{I R_g r}{2(1 + r)} (1 - \eta) \frac{\Delta R_g}{R_g} \tag{3-57}$$

where the nonlinear term η is given by the expression

$$\eta = \frac{\Delta R_g/R_g}{2(1 + r) + \Delta R_g/R_g} \tag{3-58}$$

Equation (3-58) indicates that errors are only about half as large as the errors associated with the constant-voltage bridge [see Eq. (3-46)]. Thus the constant-current bridge extends the useful range of the Wheatstone bridge.

The circuit sensitivity of the constant-current Wheatstone bridge can be expressed as

$$S_{cc} = \frac{\Delta E_o}{\Delta R_g/R_g} = \frac{I R_g r}{2(1 + r)} \tag{3-59}$$

which reduces to

$$S_{cc} = \frac{r}{1 + r} \sqrt{P_g R_g} \tag{3-60}$$

for a symmetric bridge where $I_g = I/2$. This sensitivity is the same as that for the constant-voltage bridge [see Eq. (3-48)].

3-3 Calibration

The strain-measurement system shown in Fig. 3-20 usually contains one or more strain gages, one or more power supplies, bridge completion resistors, bridge balance resistors, an amplifier, and a recording instrument. Each of these elements contributes to the overall system sensitivity, and an overall calibration scheme is preferred over the costly process of calibrating each component separately.

The recorder output displacement d_g is related to the strain in each bridge arm by the expression

$$d_R = \frac{G r E_i (1 - \eta) S_g}{(1 + r)^2} (\varepsilon_1 - \varepsilon_2 + \varepsilon_3 - \varepsilon_4) = S_R \varepsilon_B \tag{3-61}$$

where G = amplifier and recorder gain
r = resistance ratio R_2/R_1 or R_3/R_4
E_i = bridge excitation voltage
η = nonlinear term [see Eq. 3-46)]
S_g = gage factor of the strain gage
ε_i = strain in the ith bridge arm
S_R = overall system sensitivity
ε_B = effective strain for the bridge

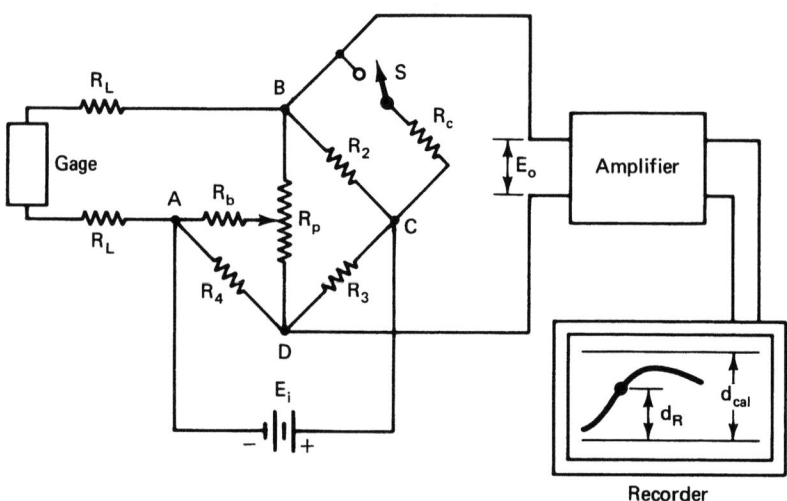

Figure 3-20 Schematic diagram of a strain-measurement system.

The effective strain ε_B can be expressed as

$$\varepsilon_B = \varepsilon_1 - \varepsilon_2 + \varepsilon_3 - \varepsilon_4 = N\varepsilon \tag{3-62}$$

where $N = 1$ for a single active gage, 2 for two active gages, and 4 for four active gages when the *magnitude* of the strain in each arm is *equal*. The system sensitivity from Eq. (3-61) is

$$d_{cal} = S_R \varepsilon_{cal}$$

so that the effective strain is

$$\varepsilon_B = \frac{\varepsilon_{cal}}{d_{cal}} d_R \tag{3-63}$$

There are two basic methods of determining system sensitivity, mechanical and electrical.

In the mechanical calibration method, a strain gage (with the same gage factor S_g as those to be used) is mounted on a cantilevered calibration bar and connected in arm R_1 of the Wheatstone bridge. The recorder deflection is observed for known amounts of strain applied to the gage. For a cantilever beam, the average calibration strain is

$$\varepsilon_{cal} = \frac{3ha}{2l^3}\delta \tag{3-64}$$

where h = beam thickness
a = distance from the load to the center of the gage
l = length of the beam
δ = beam deflection under the load

The measurement system output is observed before, during, and after application of the calibration load, and these observations are used to determine the calibration deflection, d_{cal}. This loading procedure is repeated for several different loads in order to establish system repeatability and linearity. The mechanical calibration method is good for establishing the validity of the measurement system but is inconvenient and costly for regular use.

Electrical calibration uses a calibration resistor R_c that is shunted across one of the arms of the Wheatstone bridge as shown in Fig. 3-20. With the calibration switch S closed, the effective resistance for arm R_2 becomes

$$R_{2e} = \frac{R_2 R_c}{R_2 + R_c} \tag{3-65}$$

Therefore, the apparent change in resistance and the corresponding calibration strain become

$$\frac{\Delta R_2}{R_2} = \frac{R_{2e} - R_2}{R_2} = -\frac{R_2}{R_2 + R_c} = -S_g \varepsilon_{\text{cal}} \tag{3-66}$$

By using different calibration resistors, it is possible to quickly simulate standard strains (for $S_g = 2.0$) in a 100-, 200-, 500-, 1000-μm/m sequence.

The effective output from the bridge due to a strain in the gage shown in Fig. 3-20 is

$$\frac{\Delta R_g}{R_1} = \frac{R_g}{R_g + 2R_L} S_g \varepsilon_1 \tag{3-67}$$

Equation (3-67) indicates that the lead-wire resistance R_L desensitizes the bridge. In this case it is best to calibrate by shunting a calibration resistor across the gage itself (not across the resistor R_2 as shown in Fig. 3-20) in order for the effects of the lead-wire resistance to be fully accounted for. Standard resistors are available for this type of electrical calibration. Once the calibration from the gage has been established, calibration across R_2 can be used on a daily basis as long as the lead wires remain unchanged.

3-4 Problem Areas

Strain-gage circuits are designed to measure the extremely small resistance changes that occur in the gages. Any disturbance that causes a change in resistance of any component in the bridge is extremely important, since this affects the output voltage. The main circuit components are the strain gages, lead wires, connecting circuit joints (including both solder and mechanical binding posts), and in certain circumstances switches and slip rings. While solder joints, terminals, and binding posts can contribute significant errors, these can be overcome by avoiding cold solder joints and using tight mechanical joints. It should be mentioned that the joints should be at the same temperature to avoid thermal–electrical effects. In this section the more complex problems of lead wires, switches, slip rings, telemetry, and electric noise are addressed.

Lead Wires

Single Gage

A *single gage* with significant lead-wire resistance is shown in Fig. 3-20. Each lead wire is assumed to have the same resistance R_L. The lead wires can cause two problems, signal attenuation and temperature sensitivity.

Signal Attenuation

Signal attenuation is due to the change in resistance ΔR_g being divided by the arm resistance $R_g + 2R_L$ instead of the gage resistance R_g so that

$$\frac{\Delta R_g}{R_1} = \frac{R_g}{R_g + 2R_L} \frac{\Delta R_g}{R_g} = (1 - \eta) \frac{\Delta R_g}{R_g} \tag{3-68}$$

where the loss factor η is given by the expression

$$\eta = \frac{2R_L/R_g}{1 + 2R_L/R_g} \tag{3-69}$$

Equation (3-69) shows that R_L/R_g must be less than 0.005 if the attenuation is to be less than 1%. For a 120-Ω gage, R_L must be less than 0.60 Ω, which is equivalent to 30.5 m (100 ft) of number 18 solid copper wire. Clearly, long, small-diameter lead wires should be avoided or carefully calibrated as described in Section 3-3.

Temperature

Temperature compensation can be lost if long lead wires are used with a single gage and the connection is made as shown in Fig. 3-20, where gage 1 has long lead wires and gage 4 has short lead wires. Suppose that the gages and the lead wires are subjected to *temperature change ΔT* during a time when strain is being measured by gage 1. The bridge output voltage can be expressed as

$$E_o = \frac{rE_i}{(1+r)^2} \left(\frac{\Delta R_{g\epsilon}}{R_1} + \frac{\Delta R_{g\Delta T}}{R_1} + \frac{2\Delta R_{L\Delta T}}{R_1} - \frac{\Delta R_{4\Delta T}}{R_4} \right) \tag{3-70}$$

where $R_1 = R_g + 2R_L$ is the resistance of arm 1 and $R_4 = R_g$ is the dummy gage resistance with negligible lead-wire resistance. The terms considered are:

1. Active gage resistance change due to strain being measured
2. Active gage resistance change due to temperature ΔT
3. Lead-wire resistance change in arm 1 due to temperature ΔT
4. Dummy gage resistance change due to temperature ΔT

It is apparent that temperature compensation is not achieved, since the second and fourth terms do not cancel exactly and the third (lead wire) term is not canceled. Such a circuit can produce significant errors due to temperature changes.

Three-Lead-Wire Connection

The *three-lead-wire* bridge connection shown in Fig. 3-21 is often employed to reduce the detrimental effects of long lead wires. Here the active and the dummy gages are connected with same-length lead wires so that both bridge arms (1 and 4) contain lead-wire resistance R_L (assumed to be the same in each case). Both arms of the bridge have one long lead wire and one very short lead wire. The third lead wire effectively moves point A in the conventional bridge to point A' and is not considered to be a lead wire.

STRAIN-GAGE INSTRUMENTATION AND DATA ANALYSIS

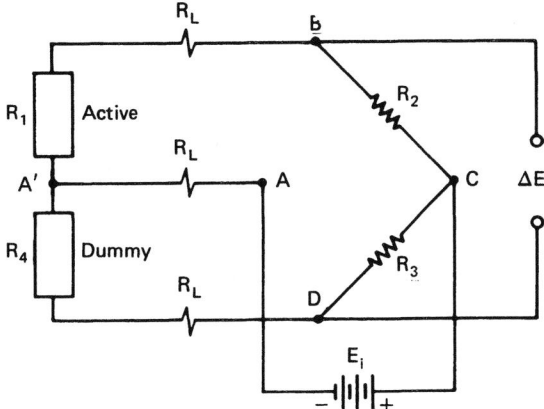

Figure 3-21 Gage connections to a bridge with a three-lead-wire system.

Signal Attenuation

The signal loss factor associated with the three-lead-wire bridge connection is the same for either a single active gage or two active gages and is given by the expression

$$\eta = \frac{R_L/R_g}{1 + R_L/R_g} \tag{3-71}$$

A comparison of Eqs. (3-69) and (3-71) shows that the signal attenuation is reduced by a factor of approximately 2 when the three-lead-wire system is used.

Temperature

Temperature compensation is vastly improved when the three-lead-wire system is used. In this case, Eq. (3-70) becomes

$$E_o = \frac{rE_i}{(1+r)^2}\left(\frac{\Delta R_{g\varepsilon 1}}{R_1} + \frac{\Delta R_{g\Delta T1}}{R_1} + \frac{\Delta R_{L\Delta T1}}{R_1} - \frac{\Delta R_{g\varepsilon 4}}{R_4} - \frac{\Delta R_{g\Delta T4}}{R_4} - \frac{\Delta R_{L\Delta T4}}{R_4}\right) \tag{3-72}$$

Temperature compensation in the gages occurs if $(\Delta R_{g\Delta T1}/R_1) = (\Delta R_{g\Delta T4}/R_4)$, which requires that both gages be subjected to the same temperature changes. Temperature compensation in the lead wires occurs if $(\Delta R_{L\Delta T1}/R_1) = (\Delta R_{L\Delta T4}/R_4)$. This requirement appears to be easily satisfied. However, experience with twisted cables lying in the sun shows that significant thermal drift occurs, since thermal heating of the two wires is often different. Extreme caution must be exercised under such conditions.

Switches

Often a large number of strain gages are employed to evaluate a structure or component and each gage is read numerous times during the test. The cost of a separate readout system for each gage would be prohibitive; therefore, a single recording device is used with a switching arrangement that allows each gage to be read in a predetermined sequence. Two different switching arrangements are employed in multiple-gage installations.

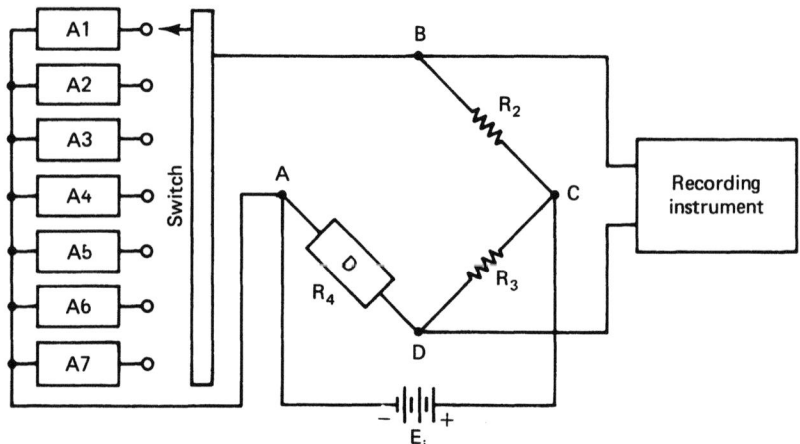

Figure 3-22 Switching a large number of gages into a Wheatstone bridge with a single-pole switch.

Single-Pole Switching

Single-pole switching is the least costly arrangement and involves switching one side of each gage, in turn, into the R_1 position of the bridge as shown in Fig. 3-22. The other side of each gage is permanently wired into the bridge. When this switching arrangement is used, the switch contact resistance R_s is in series with the gage. Since the strain-gage resistance change is very small, the switch resistance must be reproducible within a few hundred micro-ohms in order to control strain errors (240 $\mu\Omega$ corresponds to one microstrain for a 120-Ω gage with S_g = 2.0) and to maintain consistent signal attenuation. Most switching units use silver- or gold-tipped contacts and two or more sets of contacts per switch. The quality of the switch should be checked by taking repeated readings, since very poor contacts will give erratic data. The switches must be cleaned periodically to remove dirt and oxidation materials from the contact surfaces. The apparent strain resulting from switch resistance during single-pole switching is

$$\varepsilon_s = \frac{\Delta R_s / R_g}{S_g} \qquad (3\text{-}73)$$

Three-Pole Switching

Three-pole switching is a more costly arrangement in which the entire bridge circuit is switched into the recording instrument and power supply, as shown in Fig. 3-23. Usually, one side of the power supply C is connected to each gage through a common connection so that only the power connection at A and the two signal connections at B and D must be switched for each bridge. It is evident that switch resistance does not play a significant role in this arrangement, since the switch resistances are in series with the high input impedance of the recording system. An additional advantage of this arrangement is that the bridge can initially be balanced so that the readout is referenced to the balanced condition. The major disadvantage of the arrangement is the additional cost of the resistors required to complete the individual bridges. This is a one-time cost, however, since the set of three quarter-bridges can be used in future testing.

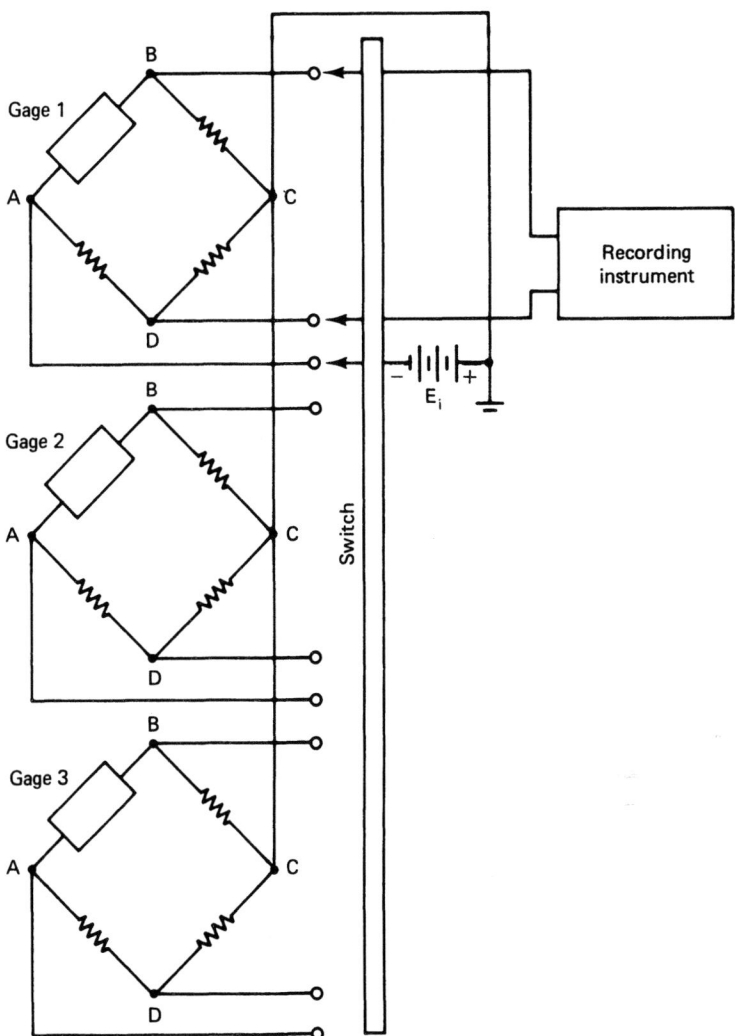

Figure 3-23 Switching several complete bridges into the power supply and recording instrument with a three-pole switch.

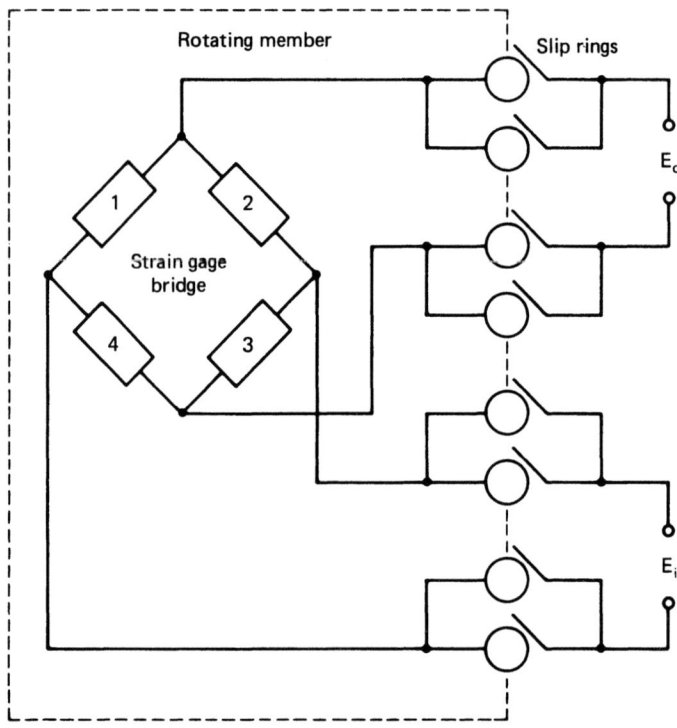

Figure 3-24 Schematic illustration of a slip-ring connection between a rotating member and a fixed instrumentation station.

Both switch systems suffer from thermal drift, since the power is applied to the strain gage only during the period when it is switched into the bridge. The single-pole system suffers the greater disadvantage since only the gage must warm up, while the other components are warm and do not change. In the three-pole arrangement, all components must warm up at the same time, and chances are better that they will do this in a more uniform way, thus reducing the time before the strain reading can be recorded. Thermal drift should be carefully watched in both cases, and data should be recorded only after the reading has stabilized.

Data Transmission

Often strains on rotating structural members must be measured, which requires transmission of the strain data from the Wheatstone bridge on the rotating member to a stationary data-recording system. Signal transmission under these conditions is usually done with slip rings or telemetry.

Slip Rings

Slip rings are used to connect the Wheatstone bridge on the rotating structure to the power supply E_i and the voltage recording instrument E_o as shown in Fig. 3-24. The slip-ring assembly contains a series of insulated rings mounted on the rotating shaft and a series of insulated brushes mounted on the stationary outside case. Bearings are used to keep the brushes and the rings aligned and concentric so that brush bounce is minimized. Two or more rings can be used for each connection to reduce brush noise. Even though the circuit is relatively independent of brush contact resistance, noise is always a problem and requires that brush assemblies be kept clean and free of serious wear. The rings are usually made from Monel metal (copper–nickel alloy) and the brushes from a silver–graphite mixture. Brush pressure should be maintained in the range 50 to 100 psi. Slip-ring assemblies with speed ratings in excess of 6000 rpm are available. One disadvantage of using slip-ring assemblies is that they must be mounted at the end of the rotating shaft.

Telemetry

Telemetry has several advantages over slip rings: the unit can be placed anywhere on the rotating shaft or structure; there are no slip-ring noise problems; and the recording instrument can be located at a greater distance if the appropriate radio frequencies are used. A simple system consists of the Wheatstone bridge, the power supply (some manufacturers power the moving system through inductive coils), and the radio transmitter on the moving part. The radio receiver and the recording instrument are located nearby (which may be a few feet or hundreds of feet) so that low-power radio transmitters can be used. A *multiplexing process* is used to handle multiple-channel data for transmission by a single transmitter-and-receiver combination. The two basic methods of processing multichannel data for transmission and eventual recording are *frequency-division multiplexing* and *time-division multiplexing*.

Frequency-division multiplexing of three channels of data is shown schematically in Fig. 3-25. Each channel uses a VCO (voltage-controlled oscillator) with a different center frequency (400, 560 and 730 Hz in Fig. 3-25). The analog strain signals cause each VCO to change frequency [giving a frequency-modulated (FM) signal] within a range of ± 7.5% of the center frequency. There is always a significant dead space between the individual channel frequencies, so that there is no channel overlap. These variable-frequency signals are mixed to form one composite signal. This composite signal is then broadcast by the transmitter at the transmitter's frequency (2200 MHz in this example). At the receiving station, the signal passes through selective filters, which pass only that portion of the signal belonging to a given channel. This signal is then demodulated and filtered before being recorded. Advances in solid-state electronics have made these systems extremely useful, compact, low in power consumption, and highly reliable. An operating license is required for use over long distances. Two crowded frequency ranges are available for transmissions in the United States, 1435 to 1535 MHz and 2200 and 2300 MHz.

Time-division multiplexing uses one transmitting channel, but each strain gage uses this channel for only a small portion of the time in a repeated sample sequence. For two channels of data transmission, the sample frequency must be at least 10 times the largest frequency present in either strain signal (five samples of each signal). For an n-channel system, the required sample rate is $10n$ times the highest frequency in any strain signal, since each channel is sampled only $1/n$ times for each sequence.

Electrical Noise

Electrical noise induced into the bridge and signal lines by magnetic fields from adjacent power lines, as shown in Fig. 3-26, can be a serious problem. The magnitude of the voltage induced in the signal lines by the current flowing in the power line is proportional to the area enclosed by the signal lines and inversely proportional to the distances (d_1 and d_2) between lines. Three precautions can be taken to minimize this noise.

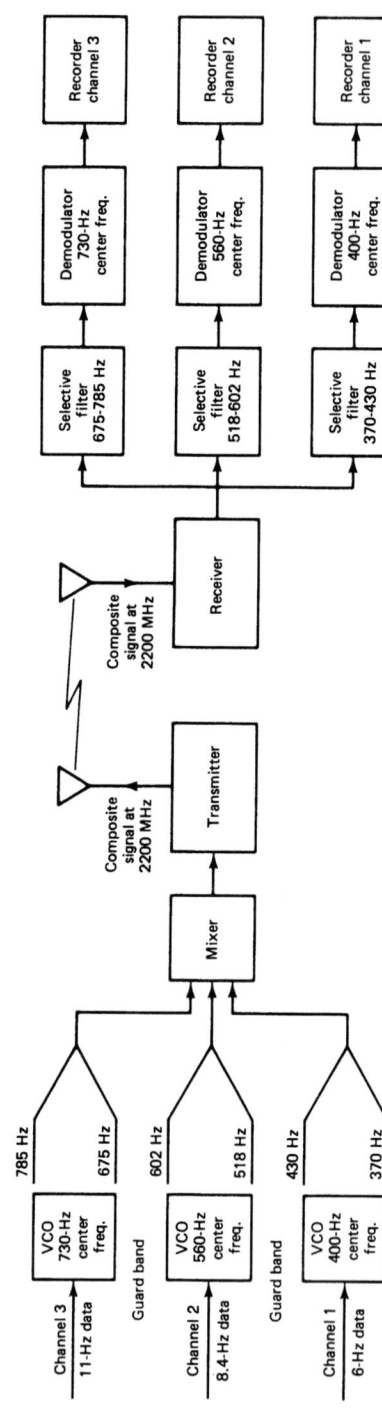

Figure 3-25 Schematic diagram of a data transmission system that utilizes frequency-division multiplexing.

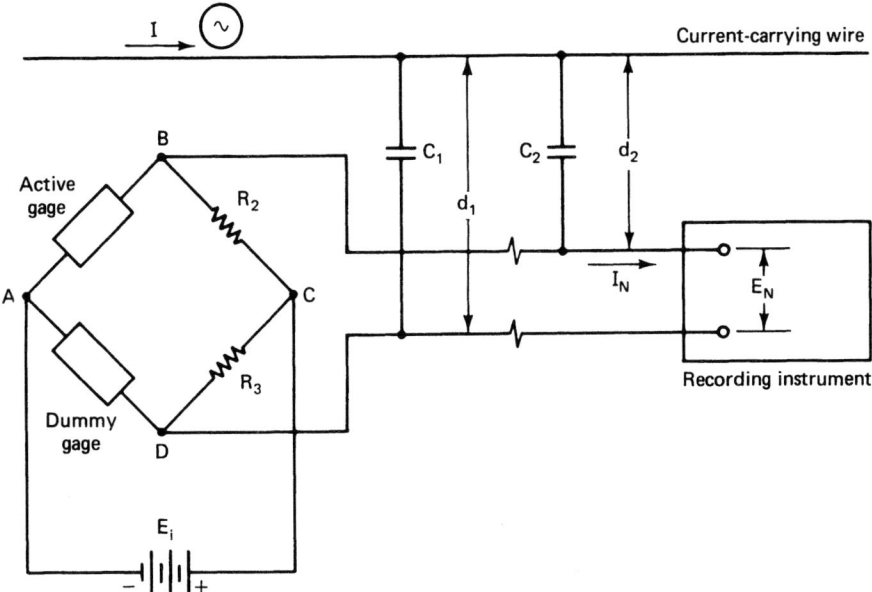

Figure 3-26 Schematic diagram showing generation of electrical noise.

First, only twisted lines should be used to minimize the area between the wires. The length of the wires should be kept at a minimum. To avoid any transformer effect, extra-long wires should not be coiled.

Second, use only shielded cables. The shield should be grounded to the negative terminal of the power supply. In this way, no ground currents can be generated in the shield and the shield is held to a near-zero voltage potential. The power supply should float relative to the system ground (the third wire in the power cord) so as to avoid any ground loops in the power supply.

Third, use only differential amplifiers with high common-mode noise reduction. In twisted lead wires, nearly equal noise is induced into each wire and the common-mode rejection will reduce this noise significantly.

Solution to Long Leads and Noise

One way to eliminate noise problems associated with long lead wires is to place a differential amplifier close to the strain-gage bridge, as shown in Fig. 3-27. With the short leads from the bridge, only a minimum of noise is generated at the amplifier input. For both amplifiers 1 and 2, the differential signal has a gain of R_5/R, while the noise is common and has a gain of unity. This means that the signal-to-noise ratio is improved before the common-mode rejection of amplifier 3 is utilized. The overall voltage gain for the amplifier system is

$$G = \left(1 + \frac{2R_5}{R}\right)\frac{R_7}{R_6} \tag{3-74}$$

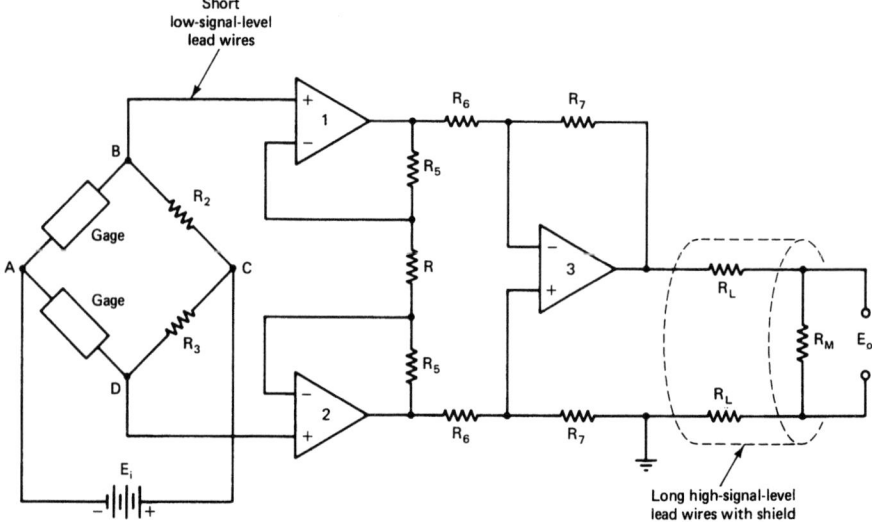

Figure 3-27 Schematic diagram of a low-noise circuit for use with a strain-gage bridge.

where the resistance ratio R_5/R is often 250 to 500 and the resistance ratio R_7/R_6 is 2 to 10, giving an overall gain of 500 to 5000. It is desirable to have the R_5/R ratio much larger than the R_7/R_6 ratio in order to minimize amplifier noise problems and enhance the common-mode rejection capabilities of the third op-amp. The output amplifier is used to drive the long twisted shielded lines, which are terminated with a low-resistance recording instrument of several hundred ohms. The magnetically induced signals in the long lead wires are usually of low power. Therefore, they cannot generate a significant voltage signal across the small resistance R_M in comparison to the high-level signal provided by amplifier 3. The cable noise (associated with conventional circuits where long lead wires operate at low signal levels) is significantly reduced by the three op-amp differential amplifiers and further reduced by the low-resistance termination.

3-5 Data Reduction

Strain gages are normally used to determine the state of stress at a point on the free surface ($\sigma_{zz} = \tau_{zx} = \tau_{zy} = 0$) of a specimen subjected to a system of loads. The conversion of strain data to stresses on planes normal to the free surface at a point of interest requires knowledge of the elastic constants E and ν and the strains in three directions on the free surface. *It is assumed here that any necessary corrections for cross-axis sensitivity of the strain gages are made before using any of these equations.*

In the *two-dimensional state of stress* σ_{xx}, σ_{yy}, and τ_{xy} are nonzero and the principal stress directions are not known. Therefore, there are three unknowns that must be determined in order to specify completely the state of stress at the point. These unknowns are the principal stresses σ_1 and σ_2 and the principal angle β_1.

The required strain data are obtained from a three-element strain rosette mounted on the free

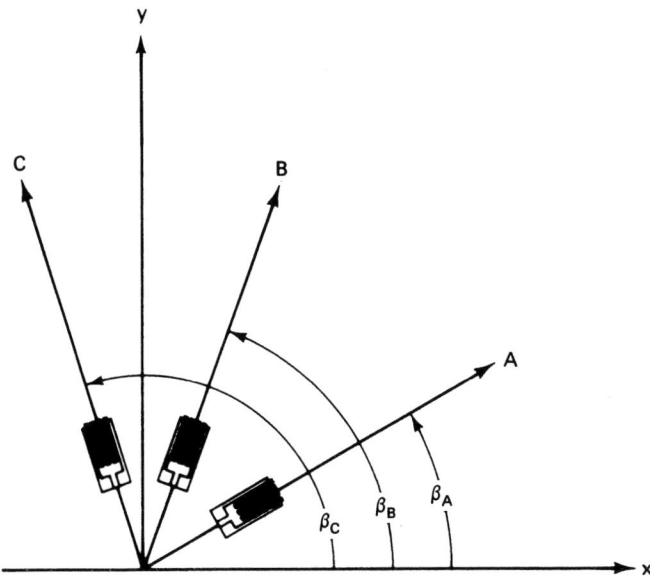

Figure 3-28 Three strain gages oriented at arbitrary angles with repect to the x axis.

surface of the specimen. Consider three gages aligned along axes A, B, and C, as shown in Fig. 3-28. The equations of strain transformation give

$$\varepsilon_A = \varepsilon_{xx} \cos^2\beta_A + \varepsilon_{yy} \sin^2\beta_A + \gamma_{xy} \sin\beta_A \cos\beta_A$$
$$\varepsilon_B = \varepsilon_{xx} \cos^2\beta_B + \varepsilon_{yy} \sin^2\beta_B + \gamma_{xy} \sin\beta_B \cos\beta_B \quad (3\text{-}75)$$
$$\varepsilon_C = \varepsilon_{xx} \cos^2\beta_C + \varepsilon_{yy} \sin^2\beta_C + \gamma_{xy} \sin\beta_C \cos\beta_C$$

For a given set of angels β_A, β_B, and β_C, the Cartesian components of strain (ε_{xx}, ε_{yy}, and γ_{xy}) can be obtained by solving Eqs. (3-75) when the normal strains ε_A, ε_B, and ε_C are known. The principal strains ε_1 and ε_2 and the principal direction are

$$\varepsilon_1 = \tfrac{1}{2}(\varepsilon_{xx} + \varepsilon_{yy}) + \tfrac{1}{2}\sqrt{(\varepsilon_{xx} - \varepsilon_{yy})^2 + \gamma_{xy}^2}$$
$$\varepsilon_2 = \tfrac{1}{2}(\varepsilon_{xx} + \varepsilon_{yy}) - \tfrac{1}{2}\sqrt{(\varepsilon_{xx} - \varepsilon_{yy})^2 + \gamma_{xy}^2} \quad (3\text{-}76)$$
$$2\beta = \tan^{-1}\frac{\gamma_{xy}}{\varepsilon_{xx} - \varepsilon_{yy}}$$

where β is the principal angle between ε_1 and the x axis.

The two most commonly used rosettes are the three-element rectangular rosette and the delta rosette. The three-element rectangular rosette is constructed with angles

$$\beta_A = 0 \qquad \beta_B = 45° \qquad \beta_C = 90°$$

With these fixed angles Eqs. (3-75) become

$$\varepsilon_A = \varepsilon_{xx}$$

$$\varepsilon_B = \frac{\varepsilon_{xx} + \varepsilon_{yy} + \gamma_{xy}}{2} \qquad (3\text{-}77)$$

$$\varepsilon_C = \varepsilon_{yy}$$

The shear strain γ_{xy}, from Eq. (3-77) is

$$\gamma_{xy} = 2\varepsilon_B - \varepsilon_A - \varepsilon_C \qquad (3\text{-}78)$$

The principal strains ε_1 and ε_2 and the principal angle β can be expressed in terms of ε_A, ε_B, and ε_C by combining Eqs. (3-76), (3-77), and (3-78) to give

$$\begin{aligned}
\varepsilon_1 &= \tfrac{1}{2}(\varepsilon_A + \varepsilon_C) + \tfrac{1}{2}\sqrt{(\varepsilon_A - \varepsilon_C)^2 + (2\varepsilon_B - \varepsilon_A - \varepsilon_C)^2} \\
\varepsilon_2 &= \tfrac{1}{2}(\varepsilon_A + \varepsilon_C) - \tfrac{1}{2}\sqrt{(\varepsilon_A - \varepsilon_C)^2 + (2\varepsilon_B - \varepsilon_A - \varepsilon_C)^2} \qquad (3\text{-}79) \\
\beta &= \tfrac{1}{2}\tan^{-1}\frac{2\varepsilon_B - \varepsilon_A - \varepsilon_C}{\varepsilon_A - \varepsilon_C}
\end{aligned}$$

There are two values of β that satisfy Eq. (3-79). One value is the angle β_1 between ε_1 and the x axis, while the other value is the angle β_2 between ε_2 and the x axis. The range of values for β_1 can be obtained from the following scheme:

$$0° < \beta_1 < 90° \qquad \text{when } 2\varepsilon_B > (\varepsilon_A + \varepsilon_C)$$
$$-90° < \beta_1 < 0° \qquad \text{when } 2\varepsilon_B < (\varepsilon_A + \varepsilon_C)$$
$$\beta_1 = 0° \qquad \text{when } \varepsilon_A > \varepsilon_C \text{ and } \varepsilon_A = \varepsilon_1$$
$$\beta_1 = \pm 90° \qquad \text{when } \varepsilon_A < \varepsilon_C \text{ and } \varepsilon_A = \varepsilon_2$$

The corresponding principal stresses can be expressed in terms of the measured strains. The expressions are

$$\begin{aligned}
\sigma_1 &= E\left[\frac{\varepsilon_A + \varepsilon_C}{2(1-\nu)} + \frac{1}{2(1+\nu)}\sqrt{(\varepsilon_A - \varepsilon_C)^2 + (2\varepsilon_B - \varepsilon_A - \varepsilon_C)^2}\right] \\
\sigma_2 &= E\left[\frac{\varepsilon_A + \varepsilon_C}{2(1-\nu)} - \frac{1}{2(1+\nu)}\sqrt{(\varepsilon_A - \varepsilon_C)^2 + (2\varepsilon_B - \varepsilon_A - \varepsilon_C)^2}\right]
\end{aligned} \qquad (3\text{-}80)$$

The *delta rosette* has angles of 0°, 60°, and 120° for direction A, B, and C, respectively. The corresponding principal strains ε_1 and ε_2 and the principal direction β are given by the expression

$$\begin{aligned}
\varepsilon_1 &= \frac{\varepsilon_A + \varepsilon_B + \varepsilon_C}{3} + \sqrt{\left(\varepsilon_A - \frac{\varepsilon_A + \varepsilon_B + \varepsilon_C}{3}\right)^2 + \left(\frac{\varepsilon_B - \varepsilon_C}{\sqrt{3}}\right)^2} \\
\varepsilon_2 &= \frac{\varepsilon_A + \varepsilon_B + \varepsilon_C}{3} - \sqrt{\left(\varepsilon_A - \frac{\varepsilon_A + \varepsilon_B + \varepsilon_C}{3}\right)^2 + \left(\frac{\varepsilon_B - \varepsilon_C}{\sqrt{3}}\right)^2} \qquad (3\text{-}81) \\
\beta &= \tfrac{1}{2}\tan^{-1}\frac{(\varepsilon_B - \varepsilon_C)/\sqrt{3}}{\varepsilon_A - (\varepsilon_A + \varepsilon_B + \varepsilon_C)/3}
\end{aligned}$$

STRAIN-GAGE INSTRUMENTATION AND DATA ANALYSIS 117

The corresponding principal stresses are

$$\sigma_1 = E\left[\frac{\varepsilon_A + \varepsilon_B + \varepsilon_C}{3(1 - \nu)} + \frac{1}{1 + \nu}\sqrt{\left(\varepsilon_A - \frac{\varepsilon_A + \varepsilon_B + \varepsilon_C}{3}\right)^2 + \left(\frac{\varepsilon_B - \varepsilon_C}{\sqrt{3}}\right)^2}\right]$$

$$\sigma_2 = E\left[\frac{\varepsilon_A + \varepsilon_B + \varepsilon_C}{3(1 - \nu)} - \frac{1}{1 + \nu}\sqrt{\left(\varepsilon_A - \frac{\varepsilon_A + \varepsilon_B + \varepsilon_C}{3}\right)^2 + \left(\frac{\varepsilon_B - \varepsilon_C}{\sqrt{3}}\right)^2}\right]$$

(3-82)

3-6 Summary

Signal-conditioning circuits commonly used with electrical-resistance strain gages were considered in this chapter. The characteristics of electronic components such as amplifiers, filters, analog-to-digital converters, and power supplies that can be incorporated into potentiometer and Wheatstone bridge arrangements to improve the accuracy of strain measurements were discussed. It was shown that nonlinear effects can be reduced to less than 1% for strains less than 50,000 μm/m when a strain gage with a gage factor of 2.0 is used in a potentiometer circuit. With the Wheatstone bridge circuit, the nonlinear effects are largest when only a single active gage is employed in the circuit. In this case, a 1% error occurs at a strain of 10,000 μm/m if the strain gage has a gage factor of 2.0. The strain-gage system should be calibrated by shunting a known calibration resistor across one of the active gages. In this way, the signal attenuation effects of long lead wires are completely accounted for. The three-wire, single-gage configuration minimizes long-lead-wire effects but does not completely eliminate all temperature effects, as is often claimed. Strain gages are often used in a strain rosette configuration when principal strains and stresses must be obtained. Equations commonly used to convert the measured strains to principal strains and stresses were provided at the end of the the chapter.

References

1. J. J. Brophy, *Basic Electronics for Scientists*, 3rd ed., McGraw-Hill, New York, 1977.
2. Bruel and Kjaer Instruments, *Application of B and K Equipment to Strain Measurements*, available from Bruel and Kjaer Instruments, Inc., Marlborough, MA.
3. J. W. Dally and W. F. Riley, *Experimental Stress Analysis*, 2nd ed., McGraw-Hill, New York, 1978, 153–336.
4. J. W. Dally, W. F. Riley, and K. G. McConnell, *Instrumentation for Engineering Measurements*, 2nd ed., Wiley, New York, 1993.
5. J. D. Lenk, *Manual for Operational Amplifier Users*, Reston, Reston, VA, 1976.
6. Edward B. Magrab and Donald S. Blomquist, *The Measurement of Time-Varying Phenomena*, Wiley-Interscience, New York, 1971.
7. Measurements Group, Inc., Noise Control in Strain Gage Measurements, *Tech. Note 501*, 1980, 1–5.
8. Measurements Group, Inc., Optimizing Strain Gage Excitation Levels, *Tech. Note 502*, 1979, 1–5.
9. Measurements Group, Inc., Temperature-Induced Apparent Strain and Gage Factor Variation in Strain Gages, *Tech. Note 504*, 1976, 1–9.
10. Measurements Group, Inc., Errors Due to Wheatstone Bridge Nonlinearity, *Tech. Note 139*, 1974, 1–4.
11. C. C. Perry and H. R. Lissner, *The Strain Gage Primer*, 2nd ed., McGraw-Hill, New York, 1962, 200–217.

CHAPTER

4

Force–Pressure–Motion Measuring Transducers

Kenneth G. McConnell and William F. Riley
Iowa State University
Ames, Iowa

4-0 Introduction

Many methods have been developed to measure force, pressure, and motion. In the following sections, the common elements that are involved in these measurements are explored. Both linear and angular motions are often measured. Either type of motion can be measured relative to a *fixed* (absolute) or a moving reference. For many engineering problems, absolute motion is essentially the same as motion relative to the earth, even though the earth is moving around the sun. For other problems, the earth's motion is significant and cannot be neglected.

The three basic motion parameters for rectilinear and angular motion are *displacement* (s and β), *velocity* (v and ω), and *acceleration* (a and α), as shown in Fig. 4-1. The relationships between displacement, velocity, and acceleration can be expressed mathematically as

$$v = \frac{ds}{dt}$$
$$\omega = \frac{d\beta}{dt}$$
(4-1)

for velocities and

$$a = \frac{dv}{dt} = \frac{d^2s}{dt^2}$$
$$\alpha = \frac{d\omega}{dt} = \frac{d^2\beta}{dt^2}$$
(4-2)

for accelerations. These simple relationships suggest that any of the quantities can be measured and the others obtained by simply integrating or differentiating the recorded signal. Since the integration process is a smoothing process, it is preferred over the differentiation process, which is an error-amplifying process. The two most commonly measured motion quantities are accelerations and displacements. Acceleration measurements are usually made in vibration and shock

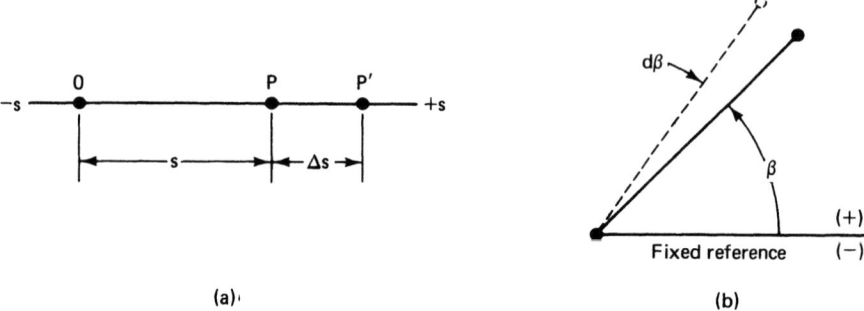

Figure 4-1 Rectilinear and angular coordinate definition: (a) rectilinear motion of a particle; (b) angular motion of a line.

monitoring situations, while displacements are most frequently used in manufacturing and process control applications.

4-1 Main Elements of a Measuring System

The main elements of a dynamic measuring system are shown schematically in Fig. 4-2. These elements are:

1. *Measurand.* The measurand is the quantity being measured and is often called the input.
2. *Transducer.* The transducer converts the change in the measurand (the measured mechanical or thermal quantity) into an electrical quantity that can be more easily amplified and conditioned before it is recorded.
3. *Power supply.* The power supply provides the energy required to drive the transducer. This power may be provided by either *dc* (direct current) or *ac* (alternating current) voltage sources, depending on the nature of the transducer. The power supply is also used to drive the other elements of the recording system. A serious concern should always be the amount of power (energy) removed from the physical system being measured. Either externally powered or high-impedance, signal-conditioning circuits should be employed.

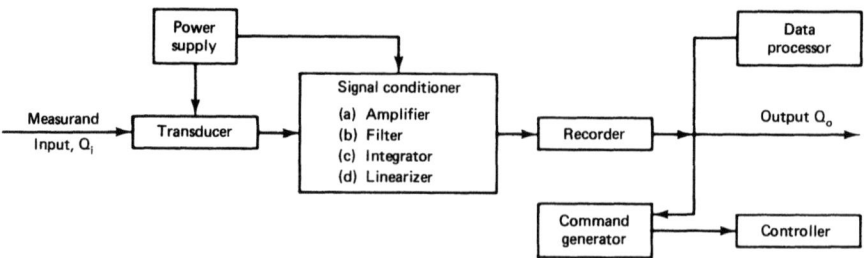

Figure 4-2 Components of a measuring system.

FORCE–PRESSURE–MOTION MEASURING TRANSDUCERS

4. *Signal conditioner.* The signal-conditioning part of the measurement system usually consists of electronic circuits that convert, compensate, amplify, filter (in the frequency domain), or manipulate the transducer output into a more usable form. Common elements are:
   a. Amplifiers, which provide voltage gain and impedance matching between the transducer and the other parts of the recording system.
   b. Filters, which remove unwanted frequency components or enhance other frequency components of the signal.
   c. Integrators, which integrate the signal in the time domain.
   d. Linearizers, which convert a nonlinear transducer characteristic into a linear input–output relationship. The need for linearizers is being greatly reduced by the use of digital techniques where the nonlinear characteristics of the transducer are corrected in the digital domain rather than electronically.
5. *Recorders.* Recorders are devices that display the measurement in a form that can be read, interpreted, and stored as a permanent record. Digital voltmeters are used for static signals, while oscillographs and oscilloscopes are used for dynamic signals. The digital oscilloscope/data processor, the newest of the recording devices, is having a great impact on how data are recorded and analyzed.
6. *Data processor.* Data processors convert the output signal into digital form for data reduction and/or process control.
7. *Process controller.* Process controllers monitor and adjust mechanical or thermal quantities in accordance with a command signal generated by a command generator. Usually, there is direct control which involves a closed-loop situation. The advent of digital systems allows for open-loop adaptive control systems, which adjust themselves to the operating conditions obtained from multiple system measurements.

In the sections that follow the various aspects of dynamic instrumentation will be explored. This means looking at the mechanical and electrical characteristics of commonly used instruments.

4-2 Common Electronic Circuits Found in Dynamic Instruments

There are three widely used electronic circuits in dynamic measurements that can significantly affect the measurement results. These circuits are discussed here, since they will be used frequently throughout the remainder of this chapter.

RC Integrators

The *RC* integrator circuit is shown in Fig. 4-3a. In the analysis that follows, it is assumed that the input voltage source has zero resistance and the output voltage is measured with an instrument with infinite input impedance. Under these conditions, the equation describing the circuit's behavior is

$$RC\dot{E}_o + E_o = E_i \tag{4-3}$$

where E_i = input voltage
R = circuit resistance
C = circuit capacitance
E_o = output voltage
$\dot{E}_o$ = time derivative of the output voltage

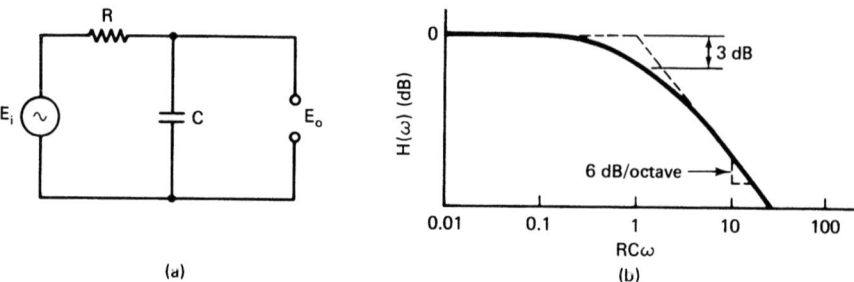

Figure 4-3 (a) *RC* integrator circuit; (b) transfer function.

For a sinusoidal excitation voltage, the input phasor is assumed to be $E_i = E_i e^{j\omega t}$ and the response phasor is assumed to be given by $E_o = A_o e^{j\omega t}$. Substituting E_i and E_o into Eq. (4-3) gives the transfer function $H(\omega)$. Thus

$$H(\omega) = \frac{E_o}{E_i} = \frac{1}{1 + jRC\omega} = \frac{e^{-j\beta}}{\sqrt{1 + (RC\omega)^2}} \qquad (4\text{-}4)$$

where $j = \sqrt{-1}$ and the phase angle β is given by

$$\tan \beta = RC\omega \qquad (4\text{-}5)$$

This transfer function is often displayed as a Bode plot, as shown in Fig. 4-3b, where the amplitude is given in *decibels* (dB) and the $RC\omega$ term is plotted on a logarithmic scale. The phase angle β is initially 0, increases to $\pi/4$ (45°) when $RC\omega = 1.0$, and approaches $\pi/2$ (90°) when $RC\omega > 10$.

The *decibel* is defined as $10 \log(P/P_r)$, where P is a powerlike quantity and P_r is a reference-power-like quantity. For voltages, power is proportional to the square of the voltage. Therefore, for a fixed resistance,

$$dB = 20 \log \frac{E_o}{E_i} = -10 \log[1 + (RC\omega)^2] \qquad (4\text{-}6)$$

It is evident from Fig 4-3b that the output amplitude is down 3 dB (output = 0.707 of the input amplitude) when $RC\omega = 1.0$ and the phase angle β is 45°. As the frequency ω increases, the output decreases linearly on the Bode plot, with a slope of 6 dB/octave (20 dB/decade). When $RC\omega > 3$, the circuit begins to perform like an integrator for sinusoidal signals, since the integral of a sinusoid is equivalent to dividing by $j\omega$. The accuracy limits for integration are: $RC\omega = 3.0$ gives 5% magnitude and 18.2° phase errors, $RC\omega = 5$ gives 2% magnitude and 11.5° phase errors, and $RC\omega = 7.0$ gives 1% magnitude and 8.0° phase errors. Thus, for practical purposes, this circuit performs like an integrator for values of $RC\omega > 5$, where it filters unwanted higher-frequency signals and passes the useful lower-frequency signals (those below $RC\omega < 1$). In either case, it is a very useful circuit for signal conditioning.

RC Differentiators

The *RC* differentiator circuit is shown in Fig. 4-4a. For the analysis that follows, the source is assumed to have zero impedance and the readout instrument is assumed to have infinite impedance. Under these conditions, the differential equation describing the circuit is

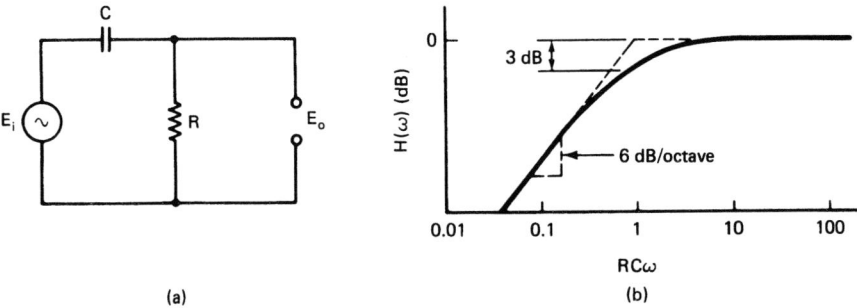

Figure 4-4 (a) *RC* differentiating circuit; (b) transfer function.

$$RC\dot{E}_o + E_o = RC\dot{E}_i \tag{4-7}$$

The corresponding transfer function $H(\omega)$ between the output and input voltages for sinusoidal signals is given by

$$H(\omega) = \frac{E_o}{E_i} = \frac{jRC\omega}{1 + jRC\omega} = \frac{RC\omega e^{j(\pi/2 - \beta)}}{\sqrt{1 + (jRC\omega)^2}} \tag{4-8}$$

where the phase angle β is given by Eq. (4-5). The Bode plot of Eq. (4-8) is shown in Fig. 4-4b. The magnitude increases at the rate of 6 dB/octave (20 dB/decade) for values of $RC\omega \ll 1.0$. In this region, the output is proportional to $j\omega$. Therefore, the circuit effectively differentiates the input signal. Since any dc signal is completely attenuated or blocked from passing through the system, this circuit is frequently referred to as a *blocking circuit*. From Eq. (4-8) it is seen that the phase angle is $\pi/2$ when $RC\omega \ll 1.0$; that is, the output leads the input by 90°. As ω increases, the phase angle decreases to $\pi/4$ (45°) when $RC\omega = 1.0$ and eventually approaches zero when $RC\omega \gg 5$. Thus the output is essentially equal to and in phase with the input when $RC\omega > 5$. The errors are 30% and 45° at $RC\omega = 1.0$, 3% and 14.0° at $RC\omega = 4.0$, 2% and 11.3° at $RC\omega = 5.0$, and 1% and 8.1° at $RC\omega = 7.0$. Similar results occur for the differentiation line when $RC\omega < 0.2$.

Second-Order Filters

Two common second-order filter circuits (12 dB/octave or 40 dB/decade) can easily be constructed using inexpensive op-amps.

Three-Op-Amp Model

In the three-op-amp model, the positive ($+$) input is connected to ground and the circuit components are connected to the negative ($-$) input, as shown in Fig. 4-5. The corresponding transfer function $H(\omega)$ is

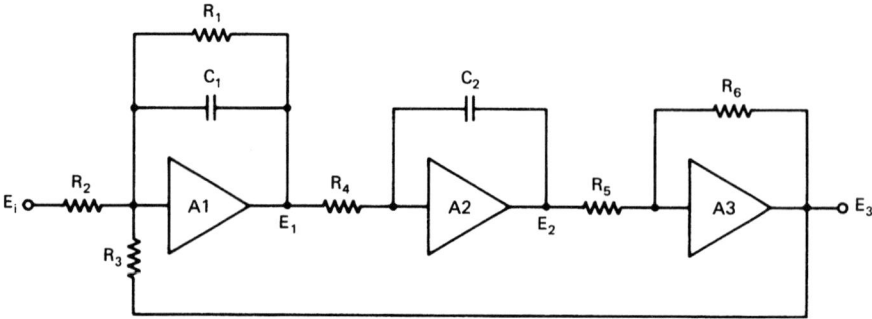

Figure 4-5 Schematic diagram of a second-order low-pass filter with three op-amps.

$$H(\omega) = \frac{E_o}{E_i} = \frac{-K}{1 - (\omega/\omega_n)^2 + j2\,d(\omega/\omega_n)} = \frac{-Ke^{-j\beta}}{\sqrt{(1 - r^2)^2 + (2\,rd)^2}} \quad (4\text{-}9)$$

Equation (4-9) represents the response of a standard second-order equation. The variables are defined in terms of the circuit parameters by the following equations:

$$\omega_n^2 = \frac{R_6}{R_5}\frac{1}{\tau_2}\frac{1}{\tau_3} \qquad \text{natural frequency}$$

$$\tau_2 = R_4 C_2 \qquad \text{time constant op-amp 2}$$

$$\tau_3 = R_3 C_1 \qquad \text{time constant op-amp 1} \qquad (4\text{-}10)$$

$$K = \frac{R_3}{R_2} \qquad \text{system gain constant}$$

$$d = \frac{\omega_n}{2}\frac{R_3}{R_1}\frac{R_5}{R_6}\tau_2 \qquad \text{for system damping}$$

$$r = \frac{\omega}{\omega_n} \qquad \text{dimensionless frequency ratio}$$

Design of the filter and selection of circuit components usually starts with a consideration of natural frequency. Usually, τ_2 and τ_3 are chosen to be equal. Then the value of the resistance ratio is adjusted to give the desired frequency. The system gain constant is then controlled by R_2, while the damping ratio d is controlled by the selection of resistance R_1.

A typical Bode plot for the transfer function given by Eq. (4-9) is shown in Fig. 4-6. This plot indicates that low-frequency signals ($r < 1$) are not attenuated, while high-frequency signals ($r \gg 1$) are attenuated at the rate of 12 dB/octave (40 dB/decade). The signal attenuation is 3 dB (0.707 K) and the phase shift is 90° when $r = 1.0$.

The optimum range for damping is between 0.6 and 0.707. In this range, magnitude errors are small and phase-angle shift with frequency is nearly linear, as shown in Fig. 4-7. The nearly linear phase shift with frequency is required when complex periodic signals are processed so that each frequency component is time shifted the proper amount to maintain signal wave shape.

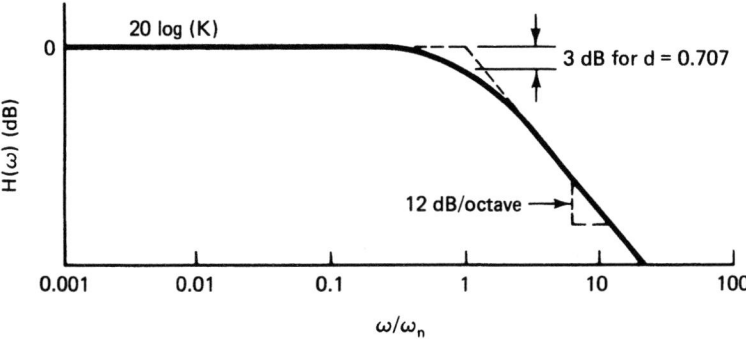

Figure 4-6 Bode plot for a second-order filter.

Single-Op-Amp, Second-Order Filter

The circuit for a single-op-amp second-order filter is shown in Fig. 4-8. The transfer function for this arrangement is given by Eq. (4-9) when $K = 1.0$. The equation variables and the circuit parameters are related by the following expression:

Natural frequency:

$$\omega_n^2 = \frac{1}{(R_1 C_1)(R_2 C_2)}$$

Damping ratio: (4-11)

$$d = \frac{\omega_n}{2}(R_1 + R_2)C_2$$

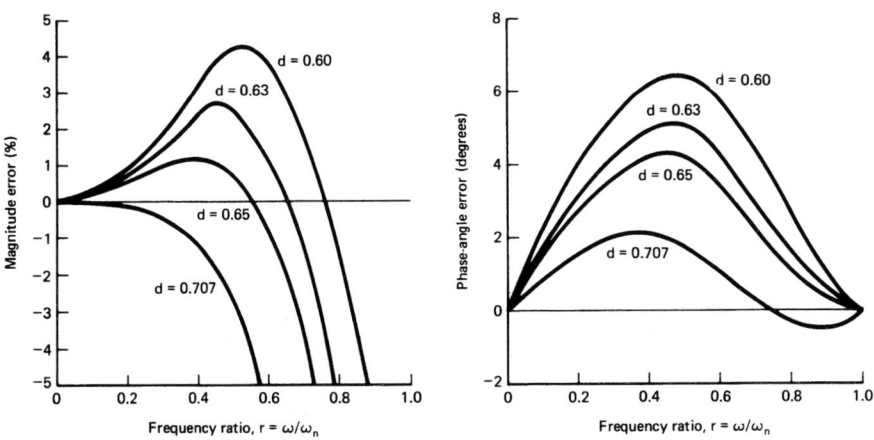

Figure 4-7 Magnitude and phase errors as a function of frequency for different degrees of damping.

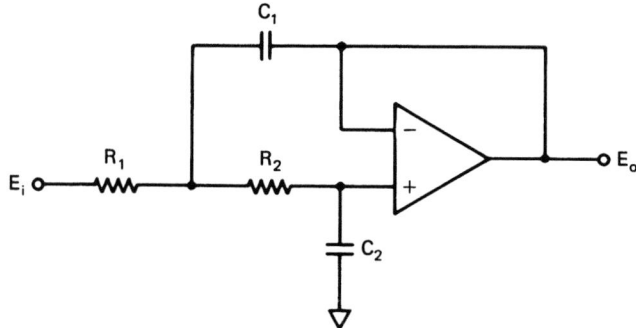

Figure 4-8 Schematic diagram of a second-order filter with a single op-amp.

The two resistors R_1 and R_2 and the two capacitors C_1 and C_2 are selected to provide the desired natural frequency and damping ratio. It is obvious from Eqs. (4-11) that circuit design freedom is more limited for this circuit than for the more elaborate three-op-amp circuit.

4-3 Displacement and Velocity Measurement: Fixed Reference

Displacement and velocity are usually measured relative to a fixed reference in production-type applications.

Displacement Sensors

There are a number of displacement sensors. Five common types are presented here.

Variable-Resistance Potentiometers

Potentiometer devices have been developed to measure both linear and angular positions. The resolution of wire-wound devices is limited by the wire size used in their manufacture. Potentiometers manufactured by using resistance films do not exhibit this limitation. The sensitivity of a potentiometer is limited by the amount of power that the resistance element can dissipate without overheating.

A schematic of the potentiometer circuit is shown in Fig. 4-9. Here R_p is the potentiometer resistance, R_M is the recording instrument resistance, E_i is the power supply voltage, and C is the output capacitance. The capacitor is often used with wire-wound units to smooth voltage fluctuations associated with wiper movement. The load resistance R_M can cause serious measurement errors. An analysis of the circuit of Fig. 4-9 (neglecting C) gives the output voltage as

$$E_o = \frac{R_M/R_p}{(R_M/R_p) + (R/R_p) - (R/R_p)^2} \frac{R}{R_p} E_i \qquad (4\text{-}12)$$

where $R = (x/l)R_p$ is the resistance proportional to position x of the wiper on the potentiometer coil of length l. The nonlinear behavior of the output voltage E_o as a function of resistance R (also position x) can be expressed in terms of a nonlinear factor η so that Eq. (4-12) becomes

$$E_o = (1 - \eta) \frac{R}{R_p} E_i \qquad (4\text{-}13)$$

FORCE–PRESSURE–MOTION MEASURING TRANSDUCERS

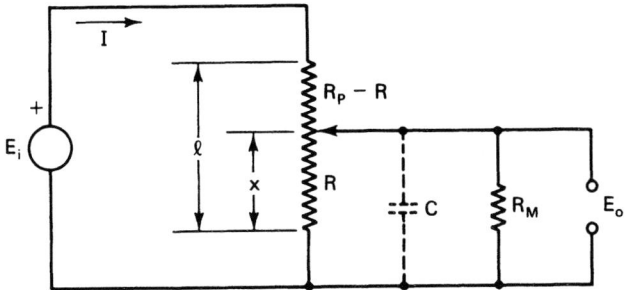

Figure 4-9 Potentiometer-type displacement sensor.

Values of η as a function of R/R_p (wiper position x) for different values of R/R_M are shown in Fig. 4-10. It is evident that the maximum error occurs when $R/R_p = 0.5$ for all values of R_M/R_p. An expression for the error can be obtained from Eq. (4-12). Thus

$$\mathscr{E} = \frac{1}{1 + 4(R_M/R_p)} \tag{4-14}$$

The maximum error due to nonlinear effects is limited to 2.44% when $R_M/R_p = 10.0$ and 0.25% when $R_M/R_p = 100$.

The nonlinear behavior problem can be overcome by introducing a voltage follower between the potentiometer and the recording instrument as shown in Fig. 4-11. The voltage follower converts the high-impedance potentiometer into a low-impedance voltage source that is capable of driving

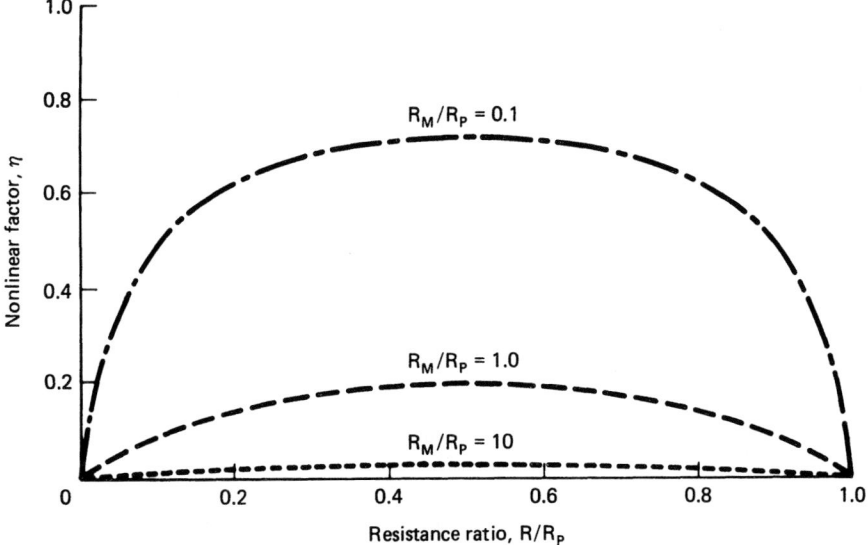

Figure 4-10 Nonlinear factor η as a function of resistance ratio R/R_p (position) for different values of R_m/R_p.

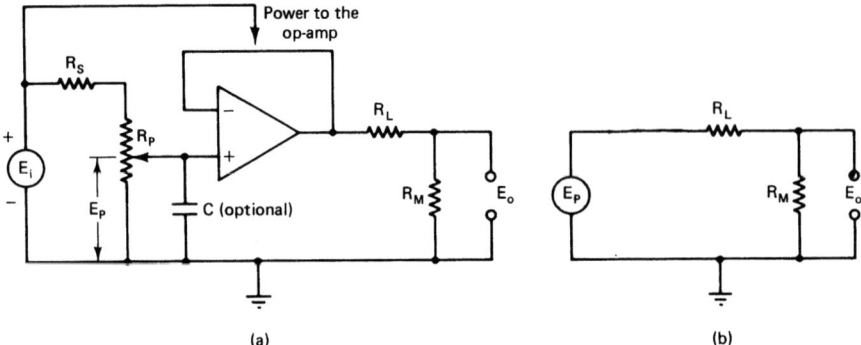

Figure 4-11 (a) Displacement-measuring circuit with a potentiometer as a sensor and a voltage-follower interface between the sensor and the indicator; (b) equivalent circuit.

long lead wires and a low-resistance recording instrument. Long lead wires are often susceptible to power-line (60-Hz) noise. The use of recording instruments with low resistance (R_M), twisted-wire pairs, and shielded wires will reduce this type of noise.

Multiple-Resistor Devices

Displacement or position can also be measured by using an array of resistors connected in parallel as shown in Fig. 4-12a. The output voltage E_o is related to the input voltage E_i and the array resistance by the expression

$$E_o = \frac{R_o}{R_o + R_e} E_i \qquad (4\text{-}15)$$

where

$$\frac{1}{R_e} = \frac{1}{R_1} + \frac{1}{R_2} + \frac{1}{R_3} + \cdots + \frac{1}{R_n}$$

The moving object either breaks the connecting wire or opens a switch to alter the effective resistance of the parallel resistor array. The result is a decreasing stair-step output voltage as shown in Fig. 4-12b.

Variable-Inductance Displacement Transducers

Both linear and angular motion can be measured with linear variable-differential transformers (LVDTs). This type of displacement transducer consists of three wire coils wound on an insulated circular shell which contains a movable magnetic core as shown in Fig. 4-13. An ac voltage is used to drive the center coil. The core magnetically couples the center primary coil to the two adjacent secondary coils, through mutual inductance, to produce a zero-output voltage when the

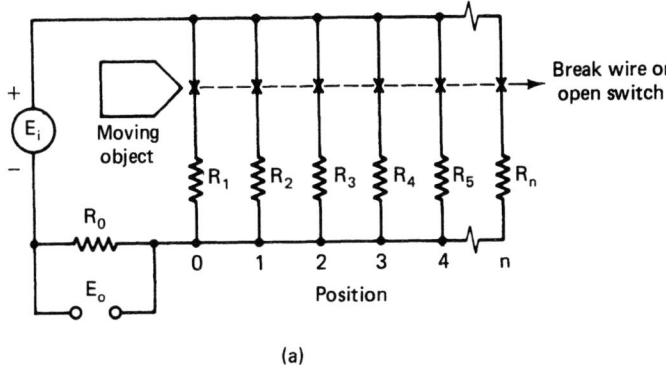

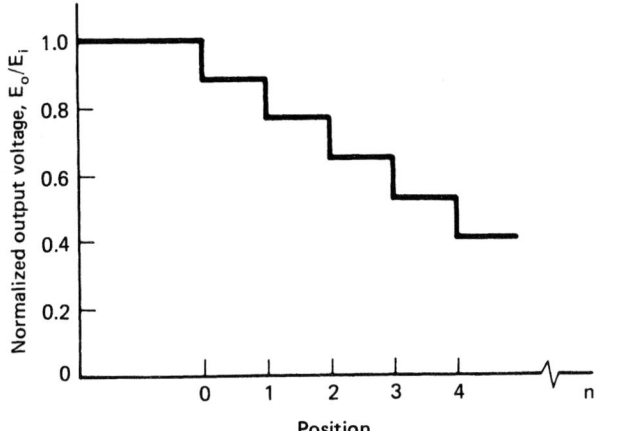

Figure 4-12 (a) Multiple-resistor displacement-measuring circuit; (b) output voltage E_o from the circuit.

core is at its magnetic center. Displacement of the core from this central position causes a linear voltage over the linear range of motion of the transducer, as shown in Fig. 4-14. The direction of motion can be obtained through phase demodulation of the output voltage with respect to the excitation voltage.

The frequency of the excitation voltage can range from 50 Hz to over 25 kHz. The greatest sensitivity is obtained in the 1 to 5 kHz frequency range. The excitation frequency should be at least 10 times the maximum frequency being measured, since it takes about 10 carrier cycles per measured cycle to describe the measured cycle adequately. The range of motion for commercially available LVDTs can be as small as ± 2 mm or as large as ± 0.5 m.

The LVDT is a passive sensor, since outside power must be supplied to the instrument. The instrument removes no energy from the object being measured if there is no friction between the core and the shell of the transducer. Careful core alignment will provide this ideal behavior.

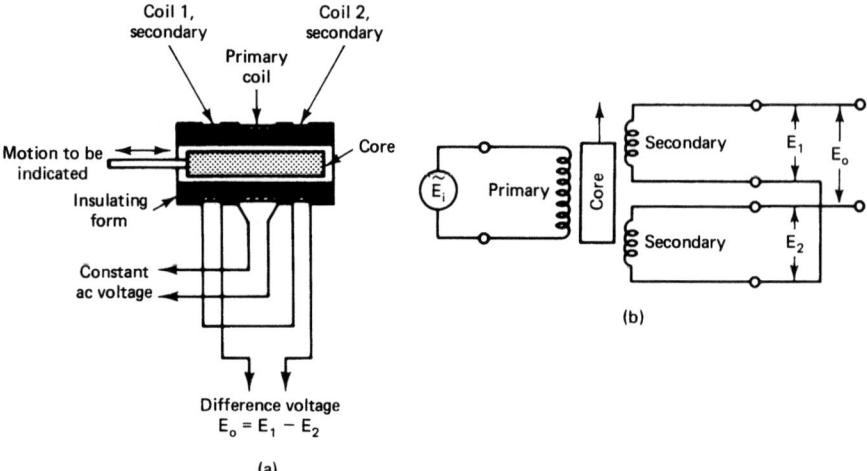

Figure 4-13 (a) Sectional view of a linear variable-differential transformer (LVDT); (b) schematic diagram of an LVDT circuit.

Photovoltaic Sensors

An opaque object, a light source, and a flat photovoltaic cell can be used to construct a noncontacting displacement-measuring system. The photovoltaic cell acts like a current generator in parallel with a capacitor, as shown in Fig. 4-15a. The output voltage E_o depends on load resistance R_M, and the relationship between voltage and illumination is usually nonlinear. Dynamic response and linearity can be improved by using the *current amplifier* circuit shown in Fig. 4-15b. The effect of any capacitance is eliminated since no significant voltage builds up on the capacitor C ($E = E_o/G$ due to the large open-loop gain G of the amplifier). The feedback resistor R can be adjusted to give the best linearity and output-voltage range. Common error sources for this type of measuring system are light-source variations, light reflections from other sources, and nonparallel light source.

Microswitch Sensors

A simple device for measuring position and counting passage of objects is shown schematically in Fig. 4-16a. The device consists of a voltage source E_i, a series resistor R_s to protect the power supply from circuit shorts, a microswitch with positions of NO (normally open) and NC (normally closed), and a measuring instrument with output resistance R_M. Contact bounce, illustrated in Fig. 4-16b, is a common problem encountered with the closure of all mechanical switches. The effect of contact bounce in counting applications is an error in the count. The contact-bounce problem can be completely eliminated by using dual-TTL NAND gate logic elements connected as shown in Fig. 4-17. The output states of both gates switch according to the truth table shown in the figure. When one contact is made, even if only momentarily (<0.1 μs), the logic state changes and remains the same no matter how many times the switch makes contact on that side without contacting the opposite side. The output logic state will change only when the opposite side is touched, if only for a moment. In this way the output signal is completely debounced, and reliable position and/or counting information can be obtained.

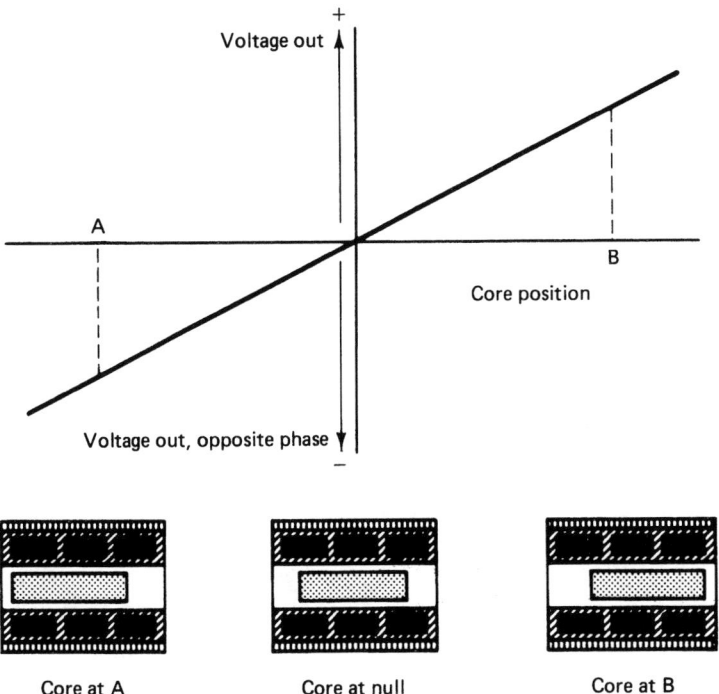

Figure 4-14 Phase-referenced output voltage as a function of LVDT core position.

Velocity Sensors

Nearly all direct-reading linear- and angular-velocity sensors are based on the principle of electromagnetic induction, where either a conductor moves through a magnetic field or the magnetic field moves past the conductor. In either case, the voltage generated E_T is given by the equation

$$E_T = Blv \tag{4-16}$$

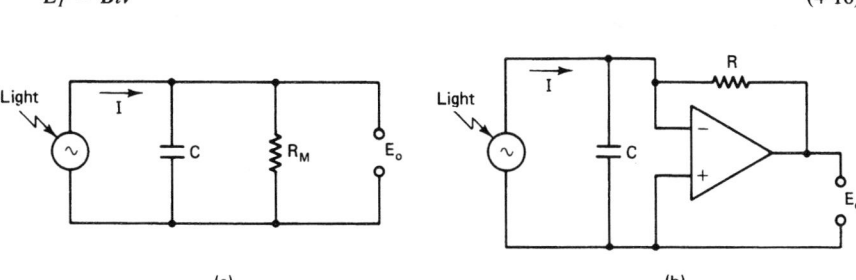

Figure 4-15 (a) Displacement-measuring circuit with a photovoltaic sensor; (b) displacement-measuring circuit with a photovoltaic sensor and an op-amp as a current amplifier.

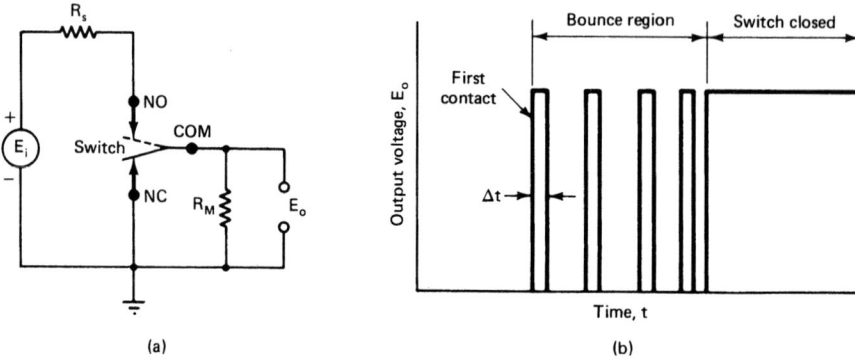

Figure 4-16 (a) Simple microswitch circuit for counting applications; (b) typical voltage output from a switch with mechanical contacts.

where B = flux density of the magnetic field normal to velocity
l = length of conductor
v = velocity of the magnetic field over the wire

Ac generators, which generate an even number of voltage cycles per revolution, are often used to measure average angular velocities by recording the frequency of the voltage with a frequency counter.

A common linear velocity transducer, constructed as shown in Fig. 4-18a, has the equivalent electrical circuit shown in Fig. 4-18b. For a sinusoidal input velocity, the output voltage can be expressed as

$$E_o = \left[\frac{R_M}{(R_M + R_T) + jL_T\omega}\right] S_v V_i e^{j\omega t} \qquad (4\text{-}17)$$

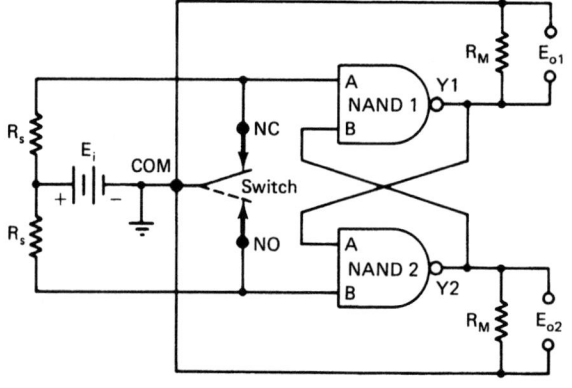

NAND-gate truth table

A	B	Y
L	L	H
L	H	H
H	L	H
H	H	L

L, low logic state less than 0.8 volts for 5-V TTL logic chips

H, high logic state greater than 2.4 V for 5-V TTL logic chips

Figure 4-17 Dual-NAND-gate, switch-debouncer circuit.

FORCE–PRESSURE–MOTION MEASURING TRANSDUCERS

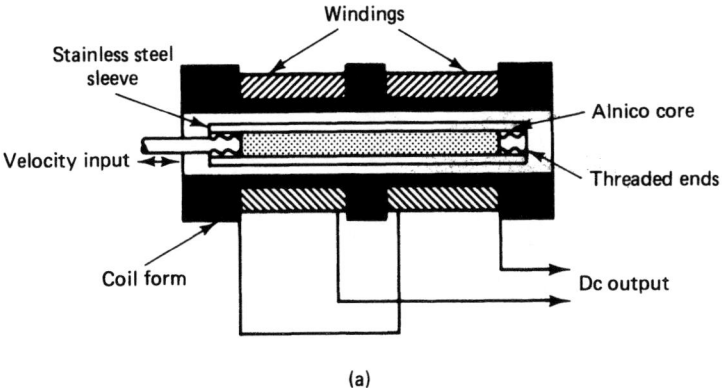

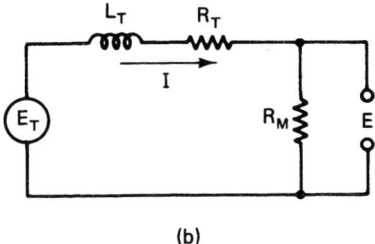

Figure 4-18 Linear-velocity transducer and circuit; (a) cross section; (b) circuit.

where L_T = inductance of the windings
R_T = resistance of the windings
R_M = resistance of the readout instrument
S_v = voltage sensitivity of the instrument
V_i = input velocity
ω = frequency of the motion

From previous discussions it is evident that the terms in brackets can cause high-frequency roll-off similar to that encountered with the RC circuit at frequencies above the cutoff frequency. Note also that R_M should be at least 100 times as large as R_T to minimize the load effects. The voltage sensitivity of these instruments usually ranges from 10 to 100 mV/in./s. Therefore voltage amplification is not required (a definite advantage). The major disadvantage of these instruments is their inherent sensitivity to magnetic fields. Use near power lines carrying large currents is not recommended.

4-4 Seismic Instruments: Mechanical Characteristics

In many situations it is impossible to use a fixed reference when making displacement, velocity, and acceleration measurements. Seismic instruments make these measurements by measuring the

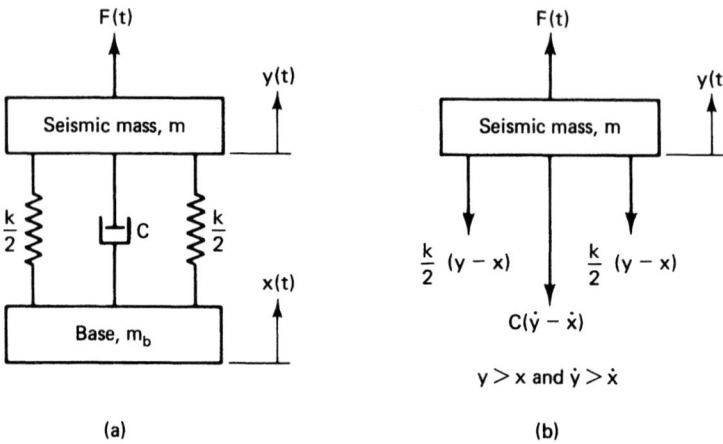

Figure 4-19 Single-degree-of-freedom model of a seismic instrument; (a) schematic of a seismic transducer; (b) free-body diagram of the seismic mass.

relative motion between the object and a seismic mass. Pressure and force transducers can also exhibit the same mechanical characteristics when the transducer is used in an accelerating environment. Thus a general theory of mechanical characteristics is developed and specialized for each type of instrument.

General Mechanical Theory

A single-degree-of-freedom mechanical model with viscous damping, as shown in Fig. 4-19a, can be used to describe the mechanical behavior of instruments that measure force, pressure, and motion. From the free-body diagram shown in Fig. 4-19b and Newton's second law of motion, the equation of motion for the seismic mass is obtained. Thus

$$m\ddot{y} + C(\dot{y} - \dot{x}) + k(y - x) = F(t) \qquad (4\text{-}18)$$

where m = mass of the seismic body
k = linear spring constant
C = viscous damping constant
$x, \dot{x}, \ddot{x}$ = base displacement, velocity, and acceleration
$y, \dot{y}, \ddot{y}$ = seismic mass displacement, velocity, and acceleration
$F(t)$ = time-dependent external force or pressure applied to the transducer

The force required to attach the base of the transducer to the structure, but not measured directly by the transducer, is given by

$$F_b = F(t) - m\ddot{y} - m_b\ddot{x} \qquad (4\text{-}19)$$

The relative motion between the base and the seismic mass, which is measured by the sensing elements, can be expressed as

$$\begin{aligned} z &= y - x \\ \dot{z} &= \dot{y} - \dot{x} \\ \ddot{z} &= \ddot{y} - \ddot{x} \end{aligned} \qquad (4\text{-}20)$$

Equation (4-18) can then be written in terms of the relative motion as

$$m\ddot{z} + C\dot{z} + kz = F(t) - m\ddot{x} = R(t) \quad (4\text{-}21)$$

where the external forcing function

$$R(t) = F(t) - m\ddot{x} \quad (4\text{-}22)$$

depends on both the base motion and any external forces [including pressure since $F(t) = Ap(t)$]. Equation (4-21) shows that the relative motion of the sensor depends on the external forcing function $R(t)$, the instrument damping C, and the frequency of the forcing function relative to the transducer's natural frequency ($\omega_n^2 = k/m$). The steady-state response of Eq. (4-21) to a sinusoidal excitation

$$R(t) = R_o e^{j\omega t} \quad (4\text{-}23)$$

is

$$z = \frac{R_o e^{j\omega t}}{k - m\omega^2 + jC\omega} = \frac{R_o e^{j(\omega t - \beta)}}{k\sqrt{(1 - r^2)^2 + (2rd)^2}} \quad (4\text{-}24)$$

where $r = \omega/\omega_n = f/f_n$ = dimensionless frequency ratio
$d = C/2\sqrt{km}$ = dimensionless damping ratio

and

$$\tan \beta = \frac{C\omega}{k - m\omega^2} = \frac{2rd}{1 - r^2} \quad (4\text{-}25)$$

where β is the phase angle of the relative output motion z with respect to the input excitation $R(t)$.

Seismic Displacement Transducers

Seismic displacement transducers are designed to measure the displacement of the base relative to the seismic mass. To illustrate this behavior, the base motion is assumed to be sinusoidal with amplitude X_0, so that

$$x(t) = X_0 e^{j\omega t} \quad (4\text{-}26)$$

The corresponding excitation function $R(t)$ becomes

$$R(t) = m\omega^2 X_0 e^{j\omega t} \quad (4\text{-}27)$$

since $F(t)$ is zero for this type of instrument. The corresponding sinusoidal solution from Eq. (4-24) becomes

$$z = \frac{m\omega^2 X_0 e^{j\omega t}}{k - m\omega^2 + jC\omega} = \frac{r^2 X_0 e^{j(\omega t - \beta)}}{\sqrt{(1 - r^2)^2 + (2rd)^2}} \quad (4\text{-}28)$$

Equation (4-28) represents the transfer function of the instrument. The magnitude of the ratio z/X_0 and the phase angle β are plotted as functions of the frequency ratio r for different values of the

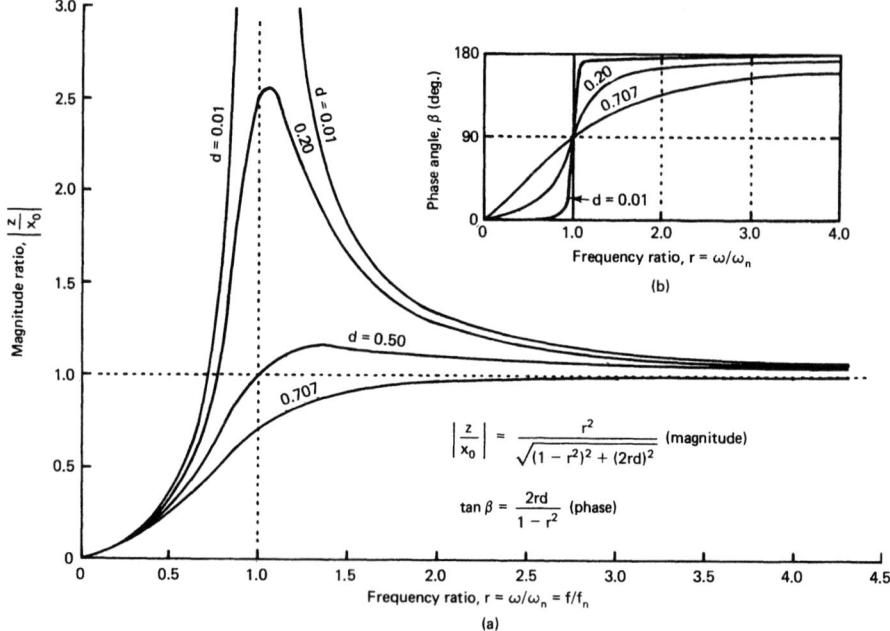

Figure 4-20 Transfer function [(a) magnitude and (b) phase] for a displacement transducer.

damping ratio d in Fig. 4-20. The magnitude ratio approaches a value of unity and the phase angle β approaches a value of 180° for values of $r > 4$ regardless of the amount of damping. There are no peaks for damping values in excess of 0.707. These results suggest that the natural frequency should be as low as possible so that the value of r will be as large as possible under most measurement situations. For $r \gg 1$, Eq. (4-28) shows that $z = -x$. This means that the seismic mass remains stationary and the base vibrates relative to it. Thus soft springs and a large rattle space are required to accommodate the large relative motion that takes place in the instrument. Common sensing elements are LVDTs and strain gages on flexible elastic members.

Seismic Velocity Transducers

Seismic velocity transducers are obtained from displacement transducers by employing velocity-dependent magnetic sensing elements to measure the relative velocity between the base and the seismic mass. The transfer function for a seismic velocity transducer is obtained from Eq. (4-28) by differentiating with respect to time to obtain

$$\dot{z} = \frac{r^2 j\omega X_0}{\sqrt{(1 - r^2)^2 + (2rd)^2}} e^{j(\omega t - \beta)} \tag{4-29}$$

which reduces to $\dot{z} = -\dot{x}$ and $\beta = \pi$ when $r \gg 1$. This result shows that the transfer function is identical to that shown for the displacement transducer in Fig. 4-20. The major disadvantages of this type of instrument are size, weight, and sensitivity to stray magnetic fields. Major disadvantages are the high sensitivity and the self-generating characteristics of the sensor.

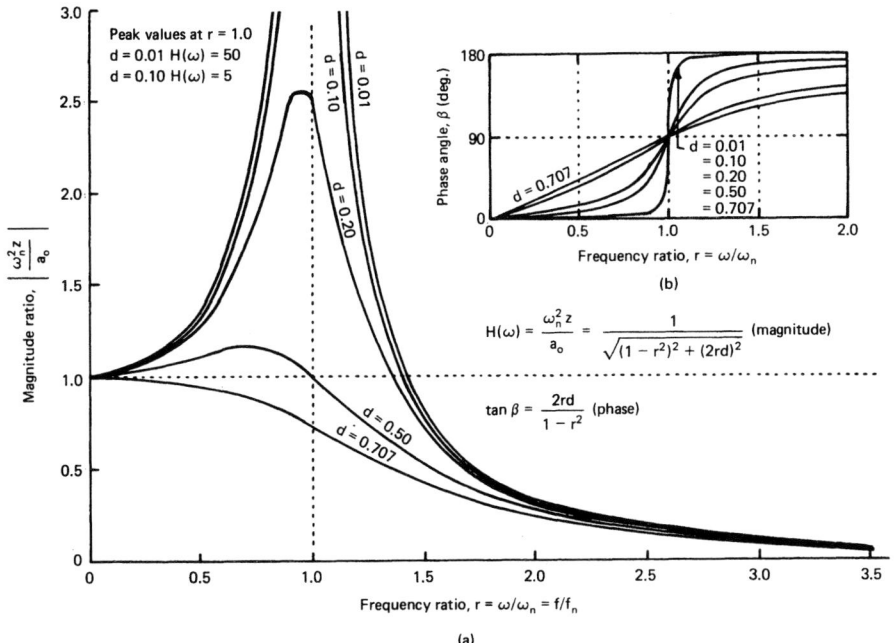

Figure 4-21 Transfer function [(a) magnitude and (b) phase] for acceleration, force, and pressure transducers.

Seismic Acceleration Transducers

Seismic acceleration transducers are obtained by using displacement sensors with a high natural frequency (stiff spring k). Then, for frequencies far below the natural frequency of the transducer, it is evident from Eq. (4-28) that the relative motion is proportional to the base acceleration ($a_0 = -\omega^2 X_0$). Thus

$$z = \frac{-ma_0 e^{j\omega t}}{k - m\omega^2 + jC\omega} = \frac{-a_0 e^{j(\omega t - \beta)}}{\omega_n^2 \sqrt{(1 - r^2)^2 + (2rd)^2}} \tag{4-30}$$

where the phase angle β is given by Eq. (4-25).

The transfer function (magnitude and phase) as described by Eq. (4-30) is shown in Fig. 4-21. Here it is seen that the output is near unity and the phase angle approaches zero when the frequency ratio r is near zero, regardless of the amount of damping. For $d = 0$, the magnitude error is 5% when $r = 0.2$. The peak response occurs around $r = 1.0$ for small values of damping and decreases for increasing values of damping in the range between 0 and 0.707. There are no magnitude peaks for damping ratios greater than 0.707. The usable frequency range for this type of instrument is limited to frequencies below 20% of the natural frequency since $d < 0.01$.

When $r \to 0$, Eq. (4-30) reduces to

$$z = -\frac{1}{\omega_n^2} a_0 = -\frac{ma_0}{k} \tag{4-31}$$

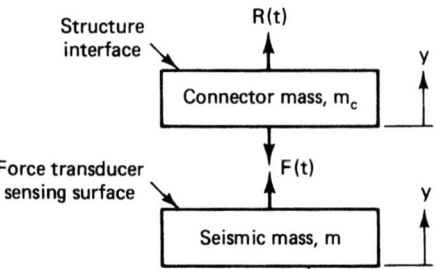

Figure 4-22 Schematic representation of a force transducer with a connector mass.

Thus it is clearly seen that the basic sensing mechanism consists of an inertia force ma_0 being resisted by the spring force kz. It is also clearly seen that the requirement for high natural frequencies is detrimental in obtaining large sensitivities, since the relative motion is inversely proportional to the square of the natural frequency. Fortunately, an accelerometer with a piezoelectric sensing element can simultaneously provide the required stiffness and sensitivity.

Dynamic Force and Pressure Transducers

Dynamic force and pressure transducers require stiff structures and high natural frequencies. Therefore, they respond to base acceleration and are considered to be seismic instruments. The transfer function for these transducers is given by Eq. (4-24). Hence they have the same magnitude and phase curves as those shown in Fig. 4-21. Some unique characteristics of force and pressure transducers are considered here.

The mounting employed with these transducers can seriously alter the results obtained. Consider the situation shown in Fig. 4-22, where a connection mass m_c (bolt and washer) is used to attach the sensing seismic mass to the structure; the force to be measured is $R(t)$ [or area times pressure $Ap(t)$] and the force sensed by the seismic mass is $F(t)$. For this situation, Newton's second law of motion applied to the connector mass gives:

Force transducer:

$$F(t) = R(t) - m_c \ddot{y} \qquad (4\text{-}32)$$

Pressure transducer:

$$F(t) = Ap(t) - m_c \ddot{y}$$

Substituting of Eq. (4-32) into Eq. (4-21) gives

$$(m + m_c)\ddot{z} + C\dot{z} + kz = R(t) - (m + m_c)\ddot{x} \qquad (4\text{-}33)$$

from which it is evident that the effective mass is increased by an amount $m(1 + m_c/m)$. This means that the natural frequency is decreased in proportion to the square root of $(1 + m_c/m)$ and the sensitivity of the transducer to base motion is increased by the same amount. Transducer sensitivity to base motion can be eliminated by electronically subtracting a signal that is proportional to the base motion. Several pressure transducers have been designed to compensate automatically for the base motion sensitivity by having a built-in accelerometer as part of its construction. This is possible with a pressure transducer since the connector mass m_c is nearly constant for a given fluid, while that for a force transducer is highly variable and dependent on the user.

FORCE–PRESSURE–MOTION MEASURING TRANSDUCERS

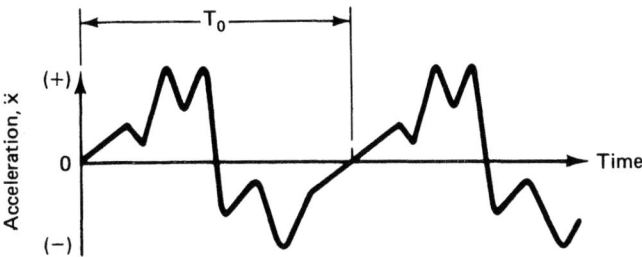

Figure 4-23 Periodic acceleration input.

When a force transducer is being used to drive a structure at or near its resonant frequency, the base motion sensitivity problem is most acute. Consider a structure described in its first mode of vibration by a single-degree-of-freedom model whose governing equation is

$$m_s \ddot{x} + C_s \dot{x} + k_s x = R(t) = R_0 e^{j\omega t} \tag{4-34}$$

where m_s = structure mass
C_s = structure damping
k_s = structure stiffness

The resulting force sensed by the transducer, when the excitation frequency becomes equal to the natural frequency of the structure $\omega_{ns}^2 = k_s/m_s$), is

$$F(t) = R_0 \left[1 - \frac{jm(1 + m_c/m)}{2d_s m_s} \right] \tag{4-35}$$

where d_s is the structural damping ratio $C_s/2\sqrt{k_s m_s}$. The bracketed term can be considerably greater than unity when the connector mass m_c is large compared to the seismic mass m and the structural damping C_s is small. For R_o to be measured within 2% when the structural damping is 5% and the connector mass is large compared to the seismic mass, the connector mass m_c must be less than 2% of the structural mass m_s. When the damping decreases to 1%, the connector mass decreases to 0.4% of the structural mass. Thus great care must be exercised in making such dynamic measurements if serious errors are to be avoided.

Periodic Signals

Periodic signals contain a multitude of frequency components. The requirements for faithful reproduction of a periodic signal can be obtained by considering a signal such as the one shown in Fig. 4-23, which can be expressed analytically in a Fourier series as

$$x = a_1 \sin \omega_0 t + a_2 \sin 2\omega_0 t + \cdots + a_q \sin q\omega_0 t \tag{4-36}$$

where $\omega_0 = 2\pi/T_0$ = fundamental frequency
T_0 = fundamental period of the signal

The magnitude and the phase of each of these frequency components are modified by the transfer function of the transducer to give an output signal that can be expressed as

$$z = b_1 \sin(\omega_0 t - \beta_1) + b_2 \sin(2\omega_0 t - \beta_2) + \cdots + b_q \sin(q\omega_0 t - \beta_q) \tag{4-37}$$

Equations (4-36) and (4-37) have the same form if:

1. The b_q coefficients are constant multiples of the a_q coefficients.
2. The phase angle is either zero or a linear function of frequency.

The magnitude and phase errors shown in Fig. 4-7 were calculated as a function of frequency for different amounts of damping from Eq. (4-30). It is evident that the amplitude is modified by less than 5% for the lightly damped case when the frequencies are limited to 20% of the natural frequency. There is little or no phase shift for the lightly damped case. The useful range of these transducers can be increased to 85% of the natural frequency for $\pm 5\%$ amplitude errors when the damping is 0.59. The phase angles are nearly linear functions of frequency when the damping d is between 0.59 and 0.707. Within this range of damping, the maximum phase error is about 6°. When the phase shift is linear, the argument for the qth term becomes

$$q\omega_0 t - \beta_q = q\omega_0(t - a) \qquad (4\text{-}38)$$

This equation shows that a linear phase shift produces an output signal that is time shifted relative to the input signal by a seconds. The signal shape is not altered. Most seismic instruments are designed to have a high natural frequency and low damping, since most damping schemes add considerable mass to the transducer and can be highly temperature dependent.

4-5 Piezoelectric Sensor Circuits

Piezoelectric sensing elements are commonly used with accelerometers, force gages, and pressure transducers, since the elements are stiff, provide light damping, provide good sensitivity, and allow for small and lightweight transducer design. These sensors are charge generators and require high-input-impedance, signal-conditioning units such as voltage followers, charge amplifiers, and built-in amplifiers. The performance and operating characteristics of these units are discussed in this section.

Voltage Followers

The unity-gain voltage-follower circuit consists of the piezoelectric sensor, cable, and voltage follower, as shown in Fig. 4-24a. The sensor generates a charge q that is proportional to the quantity a (acceleration, force, or pressure) being measured. Thus

$$q = S_q a \qquad (4\text{-}39)$$

where S_q is the charge sensitivity of the transducer (pC/g, etc.)

There are five circuit components that must be considered in an analysis. These components are the transducer capacitance C_t, the cable capacitance C_c, the standardizing capacitance C_s on the amplifier input, the blocking capacitance C_b that is used in some designs to protect the amplifier from large voltages, and the resistance R (combined effect of amplifier input resistance and a resistance placed in parallel with the amplifier). These circuit elements can be idealized as shown in Fig. 4-24b, where the three capacitances in parallel are combined into a single capacitance C given by

$$C = C_t + C_c + C_s \qquad (4\text{-}40)$$

Analysis of the circuit shows that its governing differential equation is

$$\dot{E} + \frac{E}{RC_{eq}} = \frac{S_q}{C}\dot{a} \qquad (4\text{-}41)$$

FORCE–PRESSURE–MOTION MEASURING TRANSDUCERS

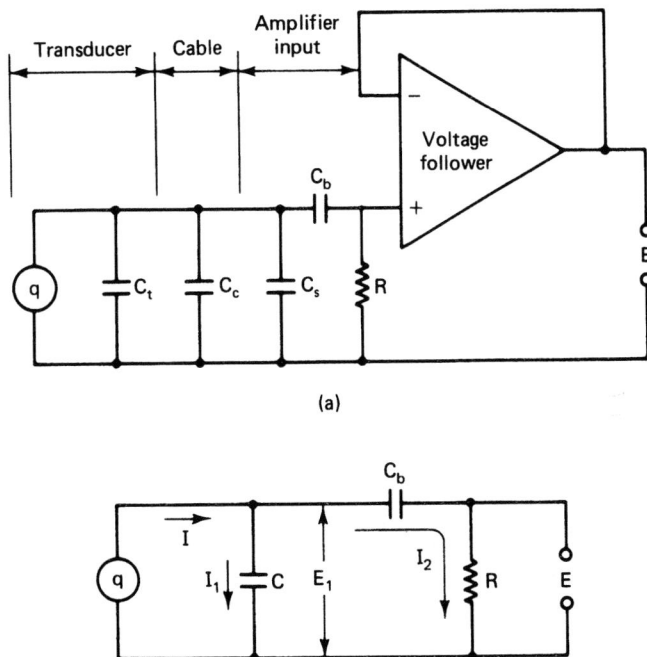

Figure 4-24 (a) Measurement circuit with a piezoelectric sensor and a voltage follower; (b) equivalent circuit for analysis.

where the equivalent capacitance C_{eq} is

$$\frac{1}{C_{eq}} = \frac{1}{C} + \frac{1}{C_b} \tag{4-42}$$

The value of C_b is usually of the order of 100,000 pF, while C usually ranges from 300 to 10,000 pF. Thus the equivalent capacitance C_{eq} is usually not significantly different from that of the combined capacitance C in Eq. (4-40). The form of Eq. (4-41) is precisely that of Eq. (4-7), so that the corresponding response to a sinusoidal input becomes

$$\frac{E_o}{a_o} = \left(\frac{jRC_{eq}\omega}{1 + jRC_{eq}\omega}\right)\frac{S_q}{C} \tag{4-43}$$

where the voltage sensitivity of the transducer is given by

$$S_v = \frac{S_q}{C}$$

The voltage sensitivity of this circuit is directly affected by the value of C, which depends on the capacitances of the cable, transducer, and amplifier input. Usually, changing cables will change system sensitivity. The low-frequency response of the circuit (given by the terms in parentheses)

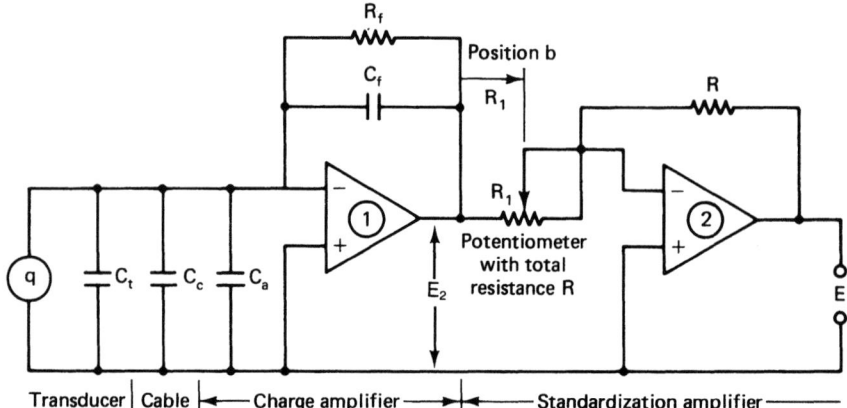

Figure 4-25 Measurement circuit with a piezoelectric sensor and a charge amplifier.

was shown in Fig. 4-4 for values of $RC_{eq}\omega < 100$. It is apparent that significant low-frequency magnitude and phase distortion can take place when the value of $RC_{eq}\omega < 3$. Typical values of R range from 100 to 1000 MΩ, so that good low-frequency response is obtained with values of $C_{eq} > 1000$ pF. However, dirt and moisture (including fingerprints, etc.) on cable connectors can significantly reduce the value of R. This type of situation can cause serious low-frequency measurement errors and must be carefully watched when using high-input-impedance devices like the piezoelectric sensor.

Charge Amplifiers

Two operational amplifiers, as shown in Fig. 4-25, can be used to describe the behavior of a piezoelectric sensor and charge amplifier. The first amplifier, with both resistive R_f and capacitive C_f feedback, is the charge amplifier that converts charge q into an output voltage E_2. The second amplifier is used to standardize the sensitivity of the output voltage by changing the gain with the resistive input R_1. An equivalent circuit can be constructed as shown in Fig. 4-26, where the transducer, cable, and amplifier input capacitance are combined into a single equivalent capacitance C according to Eq. (4-40). The differential equation that describes the behavior of this circuit is

$$\dot{E} + \frac{E}{R_f C_{eq}} = \frac{G_2 S_q}{C_{eq}} \dot{a} \qquad (4\text{-}45)$$

with an equivalent capacitance of

$$C_{eq} = \frac{C}{G_1} + C_f = C_f\left(1 + \frac{C}{C_f G_1}\right) \qquad (4\text{-}46)$$

where G_1 = open-loop gain of the first op-amp
   $G_2 \, (=1/b)$ = circuit gain of the second op-amp

Equation (4-46) shows that the input capacitance C has little effect on the measurement system, since the term $C/C_f G_1$ will be small compared to unity ($G_1 > 1 \times 10^4$). Charge amplifier performance is often specified in terms of the maximum input capacitance C that can be tolerated

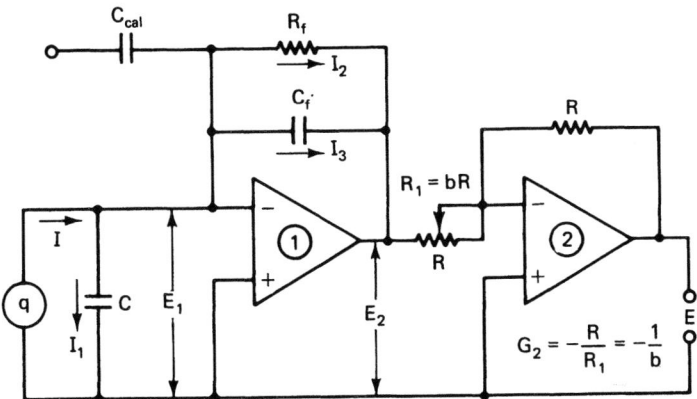

Figure 4-26 Equivalent circuit for analysis.

in order to limit the measurement error to specified bounds (usually 2%) for each built-in feedback capacitance C_f. When the input capacitance limitation is satisfied, the governing differential equation becomes

$$\dot{E} + \frac{E}{R_f C_f} = \frac{S_q}{bC_f} \dot{a} \qquad (4\text{-}47)$$

which is identical in form to Eq. (4-41) for the voltage-follower circuit, hence has identical performance characteristics.

The corresponding voltage sensitivity of the charge amplifier circuit is given by

$$S_v = \frac{S_q}{bC_f} = \frac{S_q^*}{C_f} \qquad (4\text{-}48)$$

which shows that this sensitivity can be controlled by the two parameters (b and C_f) for a given transducer. A standard charge sensitivity S_q^* ($= S_q/b$) can be obtained by adjusting the potentiometer position b. The instrument range is selected by changing the feedback capacitor C_f to give convenient voltage sensitivities in a 1-2-5-10 sequence with units of (volts/unit of a). The circuit time constant T ($= R_f C_f$) is controlled by the charge amplifier feedback resistance and capacitance rather than that of the input circuit as long as the input capacitance is less than its maximum value for a given feedback capacitance. This makes the system performance independent of the external circuit parameters and dependent only on the charge amplifier parameters.

The charge amplifier has four distinct advantages. These are:

1. The time constant is controlled by the charge amplifier feedback elements rather than the input circuit elements.
2. The system performance is independent of the input circuit as long as the input capacitance is within the specified bounds.
3. The charge sensitivity can be easily standardized by adjusting the gain of the second amplifier using position b control.
4. The voltage sensitivity of the instrument is easily changed by changing the feedback capacitance in a 1-2-5-10 sequence.

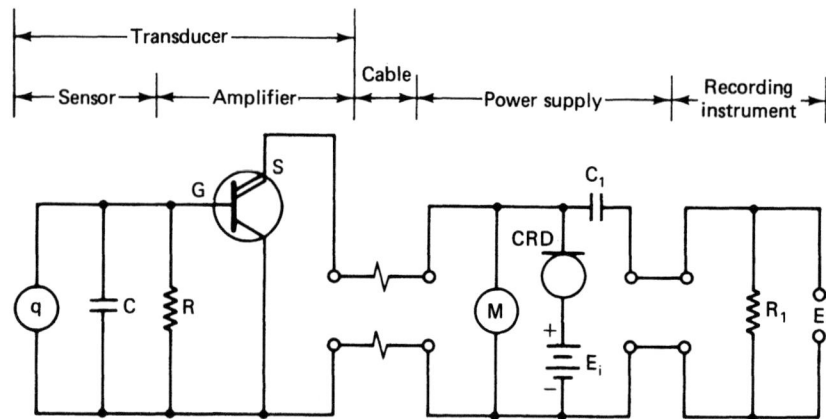

Figure 4-27 Schematic diagram of a measurement circuit that utilizes a piezoelectric transducer with a built-in amplifier.

Built-In Amplifiers

Solid-state electronics has made it possible to build a miniature integrated-circuit, unity-gain amplifier inside the transducer housing using MOSFET (metal-oxide-semiconductor field-effect transistor) technology. A typical circuit is shown in Fig. 4-27, where the amplifier is positioned between the sensing element and the power supply and output circuitry.

The cable capacitance C_c is now on the low-impedance-output side of the unit, so its value has no effect on measurement accuracy. The transducer capacitance C and amplifier input resistance R control the transducer time constant T, which is fixed when the instrument is manufactured and is not influenced by the environmental conditions other than temperature, since the elements are internal. The power supply provides a nominal voltage (usually $+11$ V) to the transistor source S when there is no transducer input by using a CRD (current-regulating diode) to regulate the supply voltage E_i. The blocking capacitor C_1 shields the recording instrument from the dc voltage.

The equivalent circuit shown in Fig. 4-28 is useful in circuit analysis and gives two different

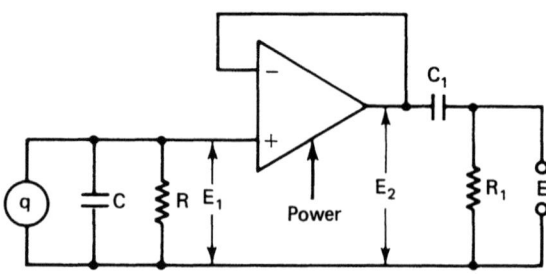

Figure 4-28 Equivalent circuit for analysis.

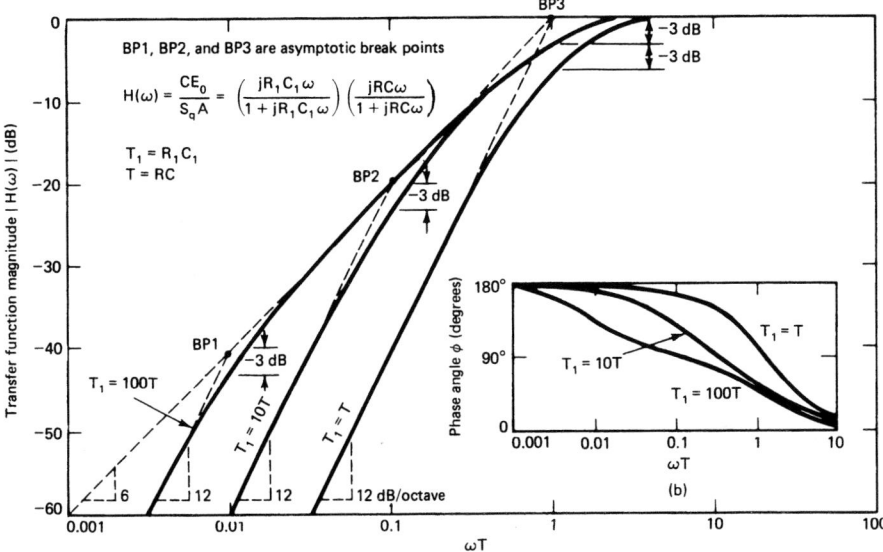

Figure 4-29 Transfer function (magnitude and phase) for a piezoelectric transducer with a built-in amplifier.

equations, one for the transducer voltage E_1 and one for the output voltage sensed by the recorder. These equations are

$$\dot{E}_1 + \frac{E_1}{RC} = \frac{S_q}{C} \dot{a} \tag{4-49}$$

and

$$\dot{E} + \frac{E}{R_1 C_1} = \dot{E}_2 = \dot{E}_1 \tag{4-50}$$

These equations give a transfer function that can be expressed as

$$H(\omega) = \frac{CE_o}{S_q a_0} = \frac{jR_1 C_1 \omega}{1 + jR_1 C_1 \omega} \frac{jRC\omega}{1 + jRC\omega} \tag{4-51}$$

from which two circuit time constants $T_1 = R_1 C_1$ and $T = RC$ are evident. The internal time constant T is fixed at time of manufacture and should be large, typically 0.5 to 2000 s. The value of R is limited by amplifier current requirements, while the voltage sensitivity ($S_v = S_q/C$) limits the value of C. The external time constant T_1 is controlled by the blocking capacitor C_1 supplied by the power supply manufacturer and instrument input impedance R_1 (usually in the range 0.1 to 1.0 MΩ), which is supplied by the device user. Care must be exercised to avoid making T_1 too small for a given application.

The magnitude and phase shift associated with this transfer function are shown in Fig. 4-29 for various ratios of T/T_1. Note that the break point is affected by the time-constant ratio. For $T_1 > 100T$, the low-frequency behavior is nearly identical to that for a single time constant circuit,

Table 4-1 Values of ωT for 2 and 5% Errors

T/T_1	2%	5%
1	7.00	4.96
10	4.95	3.06
100	4.93	3.04

with T as the time constant. This is also true when $T > 100T_1$. Then data are plotted against ωT_1 instead of ωT, since T and T_1 play identical roles in Eq. (4-51). The values of ωT for 2 and 5% amplitude errors are dependent on the T/T_1 ratio, as shown in Table 4-1, where the values for $T/T_1 = 100$ are the same as those for a single-time-constant system.

The advantages of this type of device are:

1. Voltage sensitivity is fixed by manufacturer.
2. Cable capacitance has little effect on performance.
3. The circuit operates directly with most readout instruments.
4. Inexpensive power supplies can be used.

4-6 Transient Signal Measurements

In a transient input signal, rapid changes occur over a short period of time that is preceded and followed by a constant input for long periods. These transients cause different behavior in the mechanical and the electrical parts of the measurement system.

Mechanical Response

Consider a force or pressure transducer that is *rigidly mounted* so that the differential equation of motion becomes

$$m\ddot{z} + C\dot{z} + kz = F(t) \tag{4-52}$$

and consider the ramp–hold force–time history shown in Fig. 4-30a. The corresponding *undamped* ($C = 0$) mechanical response is given by

$$\frac{z}{z_0} = \frac{t}{t_0} - \frac{\sin \omega_n t}{\omega_n t_0} \quad \text{for } t < t_0 \tag{4-53}$$

where
ω_n = transducer natural frequency
t_0 = ramp time
$z_0 \,(= F_0/k)$ = static deflection

and

$$\frac{z}{z_0} = 1 - \frac{D \cos(\omega_n t - \beta)}{\omega_n t_0} \quad \text{for } t > t_0 \tag{4-54}$$

where $D = \sqrt{2(1 - \cos \omega_n t_0)}$
$\beta = \arctan[(1 - \cos(\omega_n t_0))/\sin(\omega_n t_0)]$

The second term in Eq. (4-53) shows an oscillation about the ramp function with an amplitude of $1/\omega_n t_0$ (when $\omega_n t_0 > 2\pi$), as shown in Fig. 4-30b. The maximum error that can occur during the ramp (for $\omega_n t_0 > 2\pi$) is the amplitude of the oscillation. The amplitude of the oscillation

FORCE–PRESSURE–MOTION MEASURING TRANSDUCERS

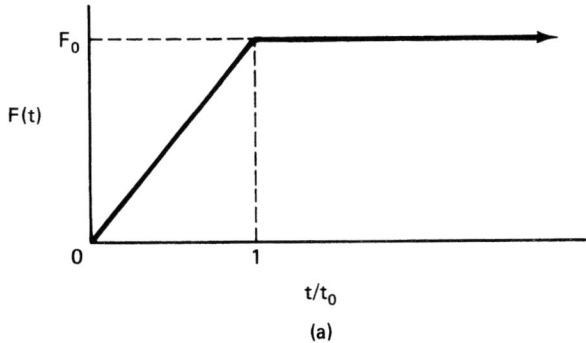

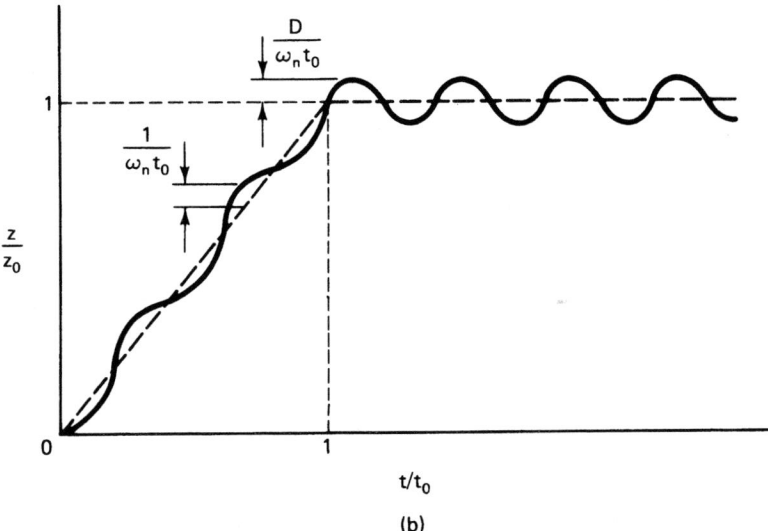

Figure 4-30 (a) Terminated-ramp type of input; (b) response of a load cell.

during the hold portion [see Eq. (4-54)] is given by $D/\omega_n t_0$, which ranges from zero to a maximum value of $2/\omega_n t_0$. The adequacy of a transducer to measure a given ramp type of transient with a specified error can be estimated from:

For the ramp portion:

$$\omega_n > \frac{1}{(\text{error})t_0} \qquad (4\text{-}55)$$

For the hold portion:

$$\omega_n > \frac{2}{(\text{error})t_0} \qquad (4\text{-}56)$$

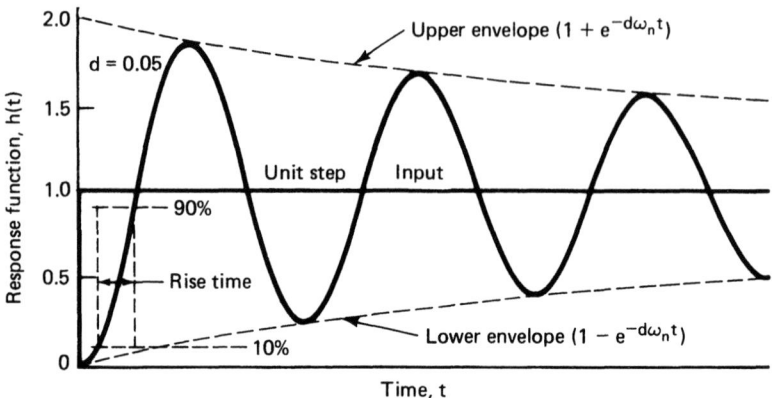

Figure 4-31 Response of a lightly damped transducer to a unit-step input.

When the value of $\omega_n t_0 \ll \pi/2$, the ramp-function response of Eq. (4-53) approaches that of a quadratic function while the hold portion of Eq. (4-54) approaches $(1 - \cos \omega_n t)$. When this happens, the transducer is totally incapable of following the input function and exhibits the response to a step input, a response that is often called *transducer ringing*. Whenever the natural frequency of the transducer is strongly evident in the output signal, there is a good chance that the measurement is in error, since the transducer rings only when the rate of input signal change exceeds the transducer's ability to respond correctly.

The damped transducer response to a step input of F_0 at time $t = 0$ (while not physically possible, it gives the limiting case of transducer performance when $\omega_n t_0 \ll \pi/5$) is given by the expression

$$\frac{zk}{F_0} = 1 - \frac{1}{\sqrt{1 - d^2}} e^{-d\omega_n t} \cos(\omega_d t - \beta) \qquad (4\text{-}57)$$

where $\tan \beta = d/\sqrt{1 - d^2}$ (phase)
$\omega_d = \sqrt{1 - d^2}\, \omega_n$ (damped frequency)

A typical damped response to the step input is shown in Fig. 4-31 together with the upper and lower decay envelopes. Instrument damping can be estimated from the decay envelopes by using the log decrement technique. It is evident from Fig. 4-31 that magnitude errors approaching 100% are associated with step inputs. Such large errors are clearly unacceptable for most measurements. A common measure of transducer response is the rise time, which is defined as the time required for the signal to rise from 10% to 90% of the static or final value as shown in Fig. 4-31. The rise time for a lightly damped instrument can be estimated from the approximate $(1 - \cos \omega_n t)$ response function, which indicates that the rise time is approximately 0.1623 times the natural period.

The undamped mechanical responses to a half-sine and to a triangular transient are shown in Fig. 4-32. It is evident that the transducer should experience a minimum of five cycles during the pulse duration if any reasonable signal output is to be measured with the mechanical device.

Electrical Response

The typical low-frequency, roll-off characteristic of a piezoelectric sensor can also cause transient measurement errors. The rectangular input pulse shown in Fig. 4-33a is physically impossible to

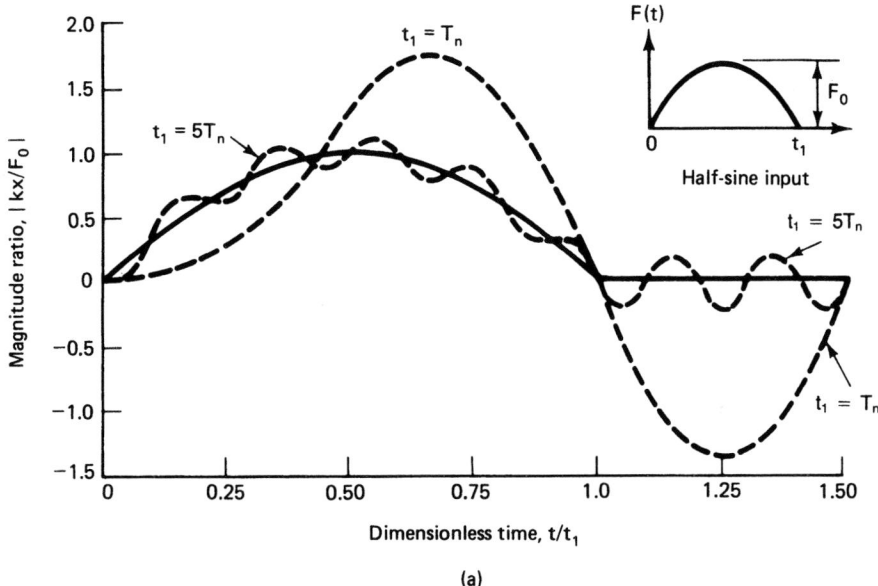
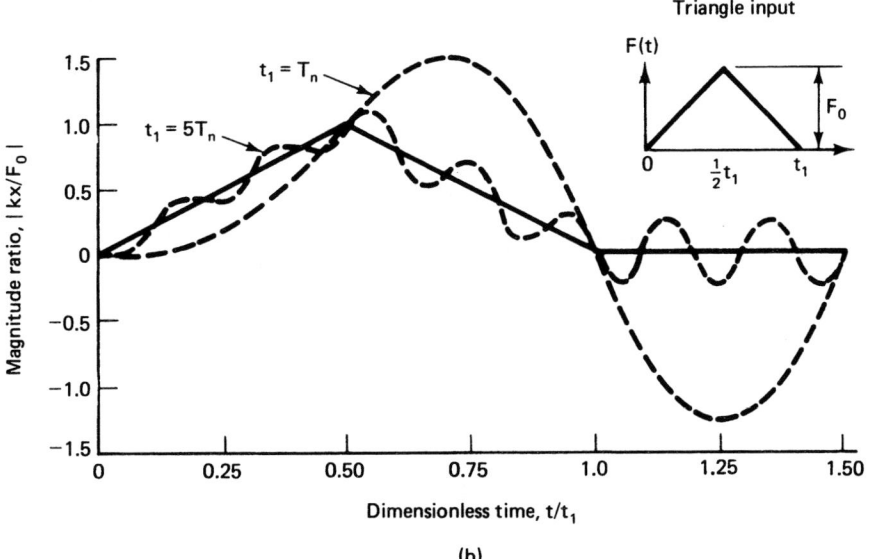

Figure 4-32 Mechanical response of a transducer to half-sine (a) and triangular (b) inputs.

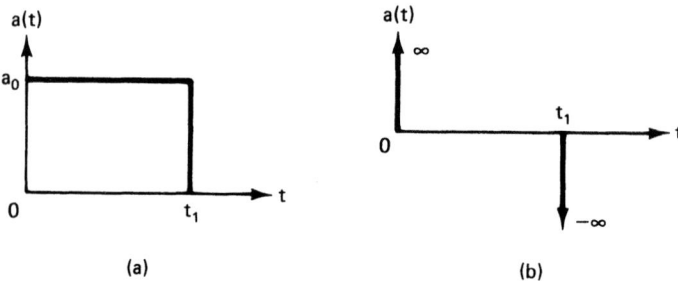

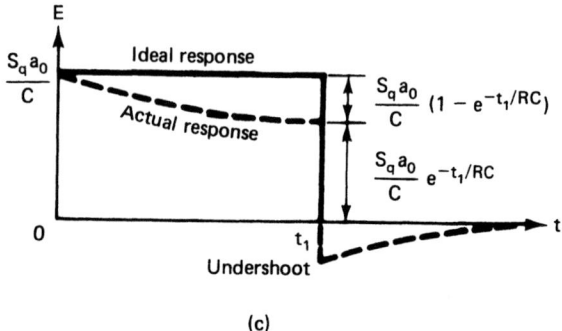

Figure 4-33 Electrical response of an RC circuit with a piezoelectric sensor to a rectangular pulse (transient) type of input; (a) rectangular input pulse; (b) time derivative of the input pulse; (c) output voltage.

generate with mechanical systems but it provides a valuable limiting case for judging the adequacy of the measuring system. The time derivative of the rectangular pulse can be expressed as

$$a = a_0 \delta(t) - a_0 \delta(t - t_1) \tag{4-58}$$

where $\delta(t)$ is the Dirac delta function (see Fig. 4-33b), which is zero except when its argument is zero, has infinite amplitude when the argument t is zero and has an area of unity under the curve. Integration of the differential equation that describes circuit behavior [see Eq. (4-41)] yields

$$E = \frac{S_q a_0}{C} [e^{-t/RC} - u(t - t_1) e^{-(t-t_1)/RC}] \tag{4-59}$$

where $u(t - t_1)$ is the unit-step function. The response described by Eq. (4-59) (see Fig. 4-33c) shows two important characteristics of the low-frequency response: the exponential decay $(1 - e^{-t/RC})$, which generates errors as time progresses, and the classic undershoot, which occurs at the end of the pulse. This undershoot is a clear indication that the time constant RC of the measuring system is inadequate for the measurement being made. The maximum error that occurs at the end of the pulse duration can be estimated from a series expansion of the exponential decay function. Thus

$$\eta_{max} = \frac{t_1}{RC} \quad \text{or} \quad T = RC = \frac{t_1}{\eta_{max}} \tag{4-60}$$

FORCE–PRESSURE–MOTION MEASURING TRANSDUCERS

Table 4-2 Time Constant Requirements for Various Error Levels

Pulse shape	2% Error	5% Error	10% Error
Rectangular pulse	$50t_1$	$20t_1$	$10t_1$
Triangular pulse	$25t_1$	$10t_1$	$5t_1$
Half-sine pulse	A: $16t_1$ B: $31t_1$	$6t_1$ $12t_1$	$3t_1$ $6t_1$

For $t_1/RC < 0.10$, Eq. (4-60) indicates that the error will be less than 10%. Similar equations can be developed for both the triangular and rectangular transient pulses. The values shown in Table 4-2 were derived from these equations and provide a guide to the time constants required to maintain measurements within a specified error. The greatest demand on the time-constant requirement comes from the rectangular pulse. Fortunately, the rectangular pulse is easy to generate electronically, so that it can be used to check the fidelity of a measuring system.

The built-in amplifier transducer presents special problems since it has two time constants. The response of this system to a step input can be expressed as

$$E = \frac{S_v a_0}{T - T_1} (Te^{-t/T_1} - T_1 e^{-t/T}) \tag{4-61}$$

where S_v is the voltage sensitivity of the transducer and a_0 is the magnitude of the step input. This response is highly dependent on the value of the time-constant ratio. The effect of the two time constants can be combined into a single equivalent time constant for use with Table 4-2 in estimating the errors that may occur during a transient measurement. The equivalent time constant is given by the expression

$$T_e = \frac{TT_1}{T + T_1} \tag{4-62}$$

When $T/T_1 > 100$, the response is effectively controlled by the time constant T_1.

A typical transfer function is shown in Fig. 4-34 for a mechanical transducer with a piezoelectric sensor. It is apparent that a working frequency range from ω_1 to ω_2 exists where there is no significant magnitude distortion. For frequencies below ω_1, there are amplitude attenuation and phase distortion due to the RC time constant of the circuit for sinusoidal inputs and RC decay, usually indicated by undershoot, for transient inputs. For frequencies above ω_2, there are amplitude magnification and phase distortion due to mechanical resonance of the transducer for sinusoidal inputs and mechanical transducer ringing for transient inputs when the input rise time is less than five natural periods of the transducer. It is important to keep all of these factors in mind when selecting an instrument for making a given measurement.

4-7 Calibration

Transducer charge and voltage sensitivities are important quantities supplied by the manufacturer at the time of purchase. The user must recalibrate the instrument from time to time to ensure that the original calibration has not changed. While the National Institute of Standards and Technology (NIST) and manufacturers often use absolute techniques, users usually employ comparison methods that utilize constant, sinusoidal, or transient inputs. The user is also faced with the problem of checking the overall system calibration on a regular basis under field conditions, where it is often impossible to test the transducer except to establish whether it is working or not working.

Transducer Calibration

Accelerometer with Constant Input

A simple means to check the nominal operational status of an accelerometer is to rotate the transducer's sensitive axis in the earth's gravitational field, where it will experience a $-1g$ to $+1g$ change. The major disadvantages of this method are limited range, near-zero check on frequency-response characteristics, and dependence on the local value of gravity. The range limitation can be solved by using a centrifuge system, where the transducer can be exposed to acceleration levels from 0 to $60,000g$'s.

Accelerometer with Sinusoidal Input

Sinusoidal motion can easily be generated at various frequencies and amplitudes by using a vibration generator. This simple sinusoidal motion is used for both comparison and absolute motion calibration.

In the comparison method, two accelerometers are usually mounted back to back on the moving head of the vibration generator so that they experience the same input acceleration. One accelerometer is a standard that is used only for calibration (its calibration being traceable to the NIST), while the other is the accelerometer being calibrated. The standard accelerometer is usually supplied with a charge amplifier that gives it a standard sensitivity of 10.0 mV/g over a broad range of frequencies. The linearity of the test accelerometer is obtained from a plot of test accelerometer output voltage versus standard accelerometer output voltage over a range of input amplitudes at a given frequency. The sensitivity as a function of frequency is often obtained by keeping the input constant and measuring the output at different frequencies. This type of calibration is often

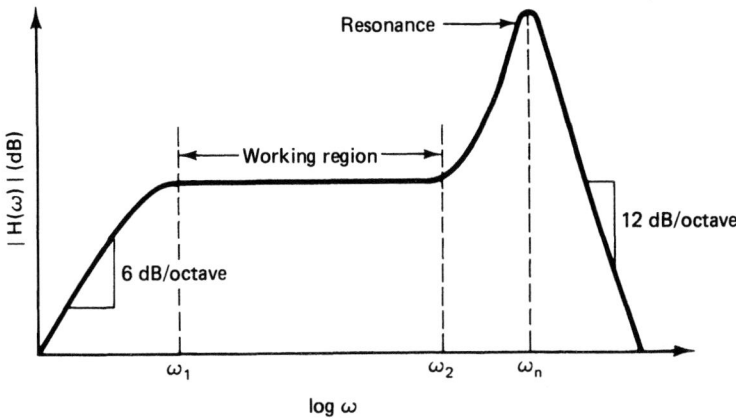

Figure 4-34 Typical transfer function $H(\omega)$ for a piezoelectric transducer.

performed with automated vibration test equipment. The comparison method gives accuracies within ±2%. Portable accelerometer calibrators are available for field use.

In the absolute calibration method, precise frequency (ω) and amplitude (x) measurements are required since the maximum acceleration a_{max} is given by the equation

$$a_{max} = \omega^2 x_{max} \qquad (4\text{-}63)$$

As an example of the magnitudes of the quantities involved in such measurements, an amplitude of 0.001 in. at a frequency of 100 Hz gives an acceleration of $1.02g$. Peak-to-peak measurements are usually used to improve the resolution when a traveling microscope is used for displacement measurements below 100 Hz. For frequencies above 100 Hz, proximity gages or interferometry methods are used to improve the displacement-measurement accuracy. Accuracies of ±0.5% have been obtained using the interferometer method over the frequency range 100 to 2000 Hz.

Accelerometer with Transient Input

The calibration technique shown schematically in Fig. 4-35 is known as the gravimetric method and is based on Newton's second law of motion, which can be written as

$$F = ma = mg\frac{a}{g} \qquad (4\text{-}64)$$

The measuring system consists of a force transducer with an impact cushion mounted on a rigid base, a test mass in the form of a steel cylinder on which the test accelerometer is mounted, a plastic guide tube for the steel cylinder, preamplifiers, and an oscilloscope (digital is preferred). This technique requires two distinct steps during which three voltages are measured.

First, the test mass is resting on the impact cushion and force transducer, and the voltage E_{mg} is measured as the test mass is quickly removed. This voltage is a measure of the weight of the test mass.

Second, the test mass is dropped from a convenient height onto the impact cushion and force

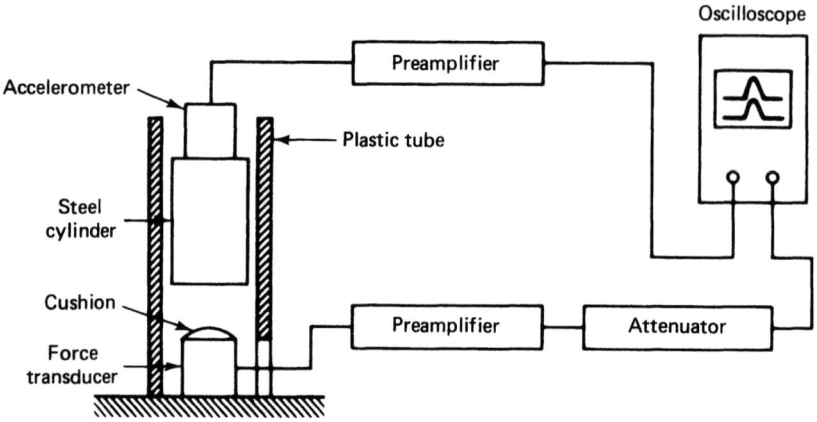

Figure 4-35 Schematic diagram of a gravimetric calibration system.

transducer while the force (E_f) and acceleration (E_a) voltages are recorded simultaneously. The resulting forces, accelerations, and voltages are related by the expression

$$F_{mg} = mg = \frac{E_{mg}}{S_f}$$

$$F = \frac{E_f}{S_f} \qquad (4\text{-}65)$$

$$\frac{a}{g} = \frac{E_a}{S_a}$$

where S_f = voltage sensitivity of the force transducer
S_a = voltage sensitivity of the accelerometer

The voltage sensitivity of the accelerometer is then obtained by using Eqs. (4-64) and (4-65). Thus

$$S_a = \frac{E_{mg} E_a}{E_f} \qquad (4\text{-}66)$$

Equation (4-66) indicates that accelerometer sensitivity S_a is directly proportional to the acceleration voltage signal E_a and inversely proportional to the force transducer voltage ratio E_f/E_{mg}. The force transducer sensitivity for a linear transducer is eliminated in this calibration method. However, the calibration is dependent on the local acceleration of gravity since weight is measured with voltage E_{mg}.

The impact pulse duration is controlled by the cushion material, while the amplitude of the impulse is controlled by changing the drop height. Calibration is performed by using different combinations of cushion material and drop heights to cover a range of frequencies and amplitudes. This calibration method is field portable and can give calibration values within ±1%.

Force with Constant Input

Static force calibration can be performed with calibrated loading machines, calibrated weight standards, or by comparison with calibrated load cells traceable to the NIST. Piezoelectric sensors

FORCE–PRESSURE–MOTION MEASURING TRANSDUCERS

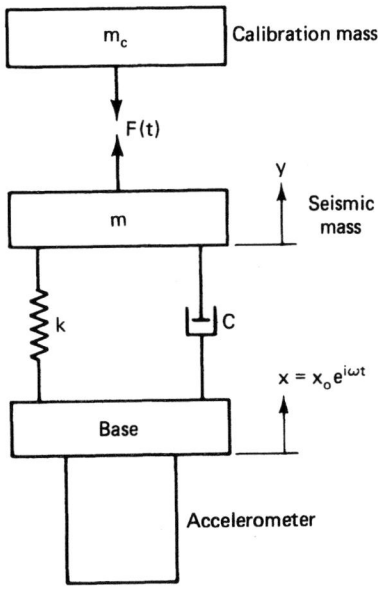

Figure 4-36 Calibration of a force transducer with a vibration generator.

are not capable of measuring static loads, so the loads must be quickly applied and removed to prevent serious exponential signal decay during calibration.

Force with Sinusoidal Input

The vibration generator can be used with a reference accelerometer (to measure the base motion) and a set of known masses to calibrate a force gage under dynamic (sinusoidal) loading. The method is illustrated schematically in Fig. 4-36. The external force applied to the transducer is the inertia force $F(t) = -m_c\ddot{y}$. The resulting differential equation of motion for the transducer calibration is Eq. (4-33) with $R(t) = 0$. Under these circumstances, the natural frequency of the transducer is lowered by addition of the calibration mass, and care must be exercised to avoid errors due to transducer resonance effects. From Eq. (4-33) it is seen that the effective force during calibration is the seismic mass inertia force $m\ddot{x}$ and the calibration mass inertia force $m_c\ddot{x}$. The corresponding output voltages from the force and acceleration transducers are

$$E_f = S_f(m + m_c)\ddot{x} \tag{4-67}$$

$$E_a = S_a \frac{\ddot{x}}{g} \tag{4-68}$$

The voltage ratio obtained from Eqs. (4-67) and (4-68) is

$$\frac{E_f}{E_a} = \frac{S_f}{S_a}(m + m_c)g = \frac{S_f}{S_a}(W + W_c) \tag{4-69}$$

where W = weight of the seismic mass
W_c = weight of the calibration mass

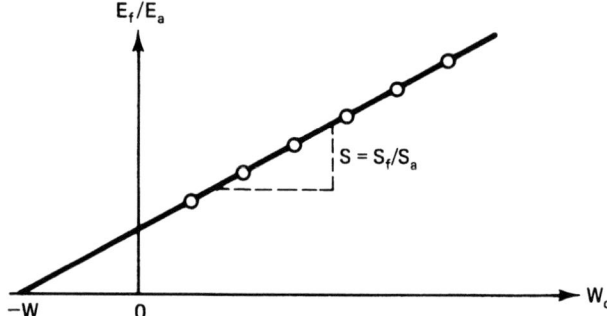

Figure 4-37 Voltage ratio E_f/E_a versus calibration weight W_c from a sinusoidal calibration of a force transducer.

Equation (4-69) indicates that the sensitivity ratio $S = S_f/S_a$ can be obtained by plotting the voltage ratio E_f/E_a as a function of the calibration weight W_c as shown in Fig. 4-37. The slope of the line is the sensitivity ratio $S = S_f/S_a$. The horizontal intercept provides a check of the seismic weight W. A wide range of vibration amplitudes (to check for linearity) and frequencies (to check for resonance effects) should be used to accurately establish the slope of the line. The force sensitivity is then obtained by using the slope S and the voltage sensitivity S_a of the reference accelerometer. Thus

$$S_f = SS_a \tag{4-70}$$

Accuracies of $\pm 2\%$ can be obtained using this method of calibration.

Force with Impulse Input

Impact hammers are used extensively in tests to determine the dynamic response characteristics of a structure. A force transducer is mounted on the head of a hammer to measure the impact force. The duration of impact depends on the cushion material used between the hammer and the structure as well as the dynamic response characteristics of the structure being tested.

Calibration of the force transducer used for these applications can be performed by using the pendulum system shown in Fig. 4-38. The differential equation of motion for the force transducer is

$$m_h \ddot{z} + C\dot{z} + k_f z = F(t) - m_h \ddot{x}_b \tag{4-71}$$

where m_h = hammer head and transducer seismic mass
m_b = mass of the hammer body and transducer case
k_f = spring constant of the transducer
z = relative motion of the transducer

Since the spring force $k_f z$ is the dominant force when the impact frequencies are well below the natural frequency of the transducer and the relative motion z is small compared to the base and head motion, the transducer behavior can be described by the approximate expression

$$F(t) = m_b \ddot{x}_b + m_h \ddot{x}_h = (m_b + m_h)\ddot{x}_b$$

or

$$z = \frac{m_b}{m_h + m_b} \frac{F(t)}{k_f}$$

FORCE–PRESSURE–MOTION MEASURING TRANSDUCERS

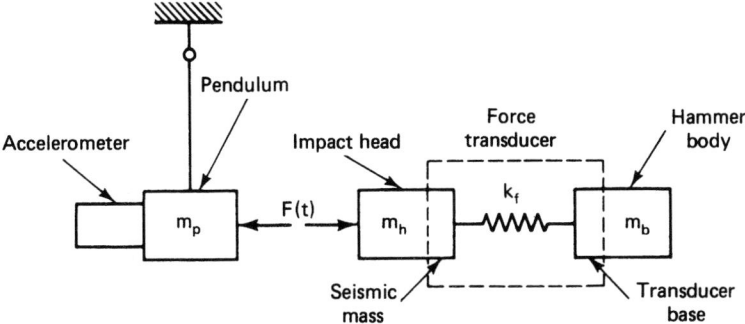

Figure 4-38 Calibration of an impact hammer with a pendulum system.

The output voltage E_f is then given by the following expression

$$E_f = S_z z = \left(\frac{m_b}{m_b + m_h} \frac{S_z}{k_f}\right) F(t) = S_f F(t) \tag{4-72}$$

where S_z is the voltage sensitivity of the transducer to relative motion z. Both S_z and k_f are sensor properties that remain constant, while the mass ratio depends on changes in both m_b (hammer mass) and m_h (impact head mass). These masses are easily changed by the user so that the voltage sensitivity of the transducer S_f can be varied by the user.

The force $F(t)$ applied to the pendulum by the impacting hammer can be expressed as

$$F(t) = m_p a = m_p g \frac{a}{g} = W_p \frac{a}{g} \tag{4-73}$$

where m_p = pendulum and accelerometer mass
W_p = pendulum and accelerometer weight
a = pendulum acceleration during impact

The corresponding accelerometer output voltage E_a is $E_a = S_a(a/g)$, so that the voltage ratio becomes

$$\frac{E_f}{E_a} = \frac{S_f}{S_a} W_p = S W_p \tag{4-74}$$

Equation (4-74) is the same form as eq. (4-69), so the same technique of plotting voltage ratio versus the pendulum weight can be used to obtain the slope of the line. The voltage sensitivity of the force gage can be obtained from Eq. (4-70). Accuracies of $\pm 2\%$ can be achieved by using this technique.

System Calibration

When a measurement system is composed of a number of components, separate calibration of each of the elements is a costly and time-consuming process. A more efficient and accurate method involves calibration of the entire system so that the interaction of each component is fully accounted for. In the ideal situation, the transducer is excited with a known input and the recorder output is observed and compared with the known input. This ideal procedure is nearly impossible to use

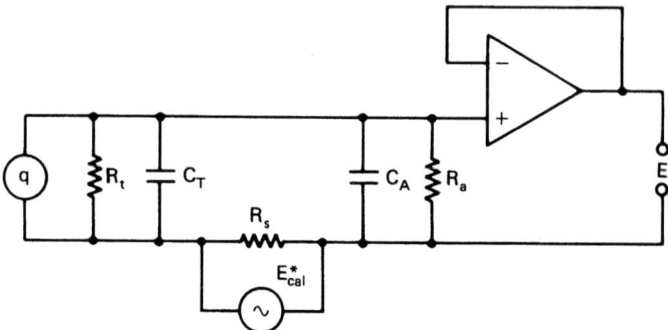

Figure 4-39 Voltage-insertion method of system calibration.

once the transducer has been installed in the field. As a result, voltage substitution methods have been developed that allow the overall system fidelity, which includes voltage sensitivity, rise time, overshoot, undershoot, and time-constant characteristics of the system to be checked.

Charge Amplifier Systems

Charge amplifiers are usually supplied with a calibration capacitor as shown in Fig. 4-26. A calibration voltage can be applied through this capacitor to simulate a charge. The required calibration voltage E_{cal}^* is obtained from the expression

$$E_{cal}^* = \frac{q_{cal}}{C_{cal}} = \frac{S_q a_{cal}}{C_{cal}} \qquad (4\text{-}75)$$

where q_{cal} = calibration charge to be simulated
a_{cal} = calibration input variable (g's, etc.)
C_{cal} = calibration capacitor in series with the voltage source as shown in Fig. 4-26

When $R_f C_f \omega \gg 1$, the corresponding output voltage E_{cal} applied to the additional components of the measuring system is

$$E_{cal} = S_v a_{cal} = \frac{S_q a_{cal}}{bC_f} = \frac{S_{cal} E_{cal}^*}{bC_f} \qquad (4\text{-}76)$$

where S_q is the charge sensitivity of the transducer. The feedback capacitor C_f is used to set the instrument's range. The potentiometer (position b) is used to establish a standard voltage sensitivity.

Square-wave and step-voltage changes, generated by precision hand-held calibrators, are commonly used calibration waveforms. Such instruments provide voltages adjustable from 0 to 10.0 V with a resolution of 0.02% full scale. Since this system calibration method does not test the transducer, it must be calibrated separately by using one of the methods outlined previously.

High-Input-Impedance Amplifier Systems

A special adapter is commercially available for use with this type of system to provide the voltage-insertion type of system calibration shown schematically in Fig. 4-39. The calibration voltage E_{cal}^* is applied across a series resistance R_s that is small (10 to 100 Ω) in comparison to the

FORCE–PRESSURE–MOTION MEASURING TRANSDUCERS

amplifier input resistance R_a (100 MΩ or greater). The capacitance C_T includes all capacitance on the transducer side of the adapter, while capacitance C_A includes all capacitance on the amplifier side of the adapter. The total system capacitance equals $C = C_T + C_A$. The effective resistance of the system is the amplifier resistance R_a since the transducer resistance $R_t >>> R_a$. The normal sinusoidal input-output relationship for the circuit shown in Fig. 4-39, when no calibration voltage is present, is given by the expression

$$E = \frac{jR_aC\omega}{1 + jR_aC\omega} \frac{GS_q}{C} a_0 \tag{4-77}$$

where G is the amplifier gain and S_q is the transducer charge sensitivity. When the calibration voltage E^*_{cal} is applied with an inactive transducer connected to the circuit, the corresponding output voltage E_{cal} becomes

$$E_{\text{cal}} = \frac{jR_aC\omega}{1 + jR_aC\omega} \frac{GC_T}{C} E^*_{\text{cal}} \tag{4-78}$$

A comparison of Eqs. (4-77) and (4-78) shows that

$$E^*_{\text{cal}} = \frac{S_q}{C_T} a_{\text{cal}} \tag{4-79}$$

and for $R_aC\omega >> 1$,

$$E_{\text{cal}} = \frac{GC_T}{C} E^*_{\text{cal}} = S_a a_{\text{cal}} \tag{4-80}$$

where S_a is the effective voltage sensitivity of the system.

The advantage of this method of calibration is that all components of the system are present during calibration. Therefore, a step-voltage input will establish not only the sensitivity of the system but also the effective time constant for the system. The output voltage E resulting from a step input voltage E_s is

$$E = \frac{GC_T}{C} E_s e^{-t/RC} = S_a a_{\text{cal}} e^{-t/RC} \tag{4-81}$$

A typical system response to a step input is shown in Fig. 4-40. The rise time, overshoot, exponential decay, and recorder natural frequency can be established from the output signal. Each of these quantities may limit system usefulness in a given application. Unfortunately, the transducer's natural frequency is not provided by this test.

Environmental Considerations

The calibration of a transducer or an overall system can be easily negated by environmental factors. Eight factors that need to be carefully considered at all times are:

1. *Temperature.* Since both the charge sensitivity and capacitance of a piezoelectric sensor are temperature dependent, significant sensitivity changes can occur as a result of temperature changes. Some transducers also generate an output when subjected to temperature transients. This can be a serious problem when the signal levels are low. In high-temperature applications, a heat sink and cooling air jet may be required. Consult the manufacturer for information on temperature problems.
2. *Mounting.* Mechanical mounting of many transducers is accomplished with a small (often 10-

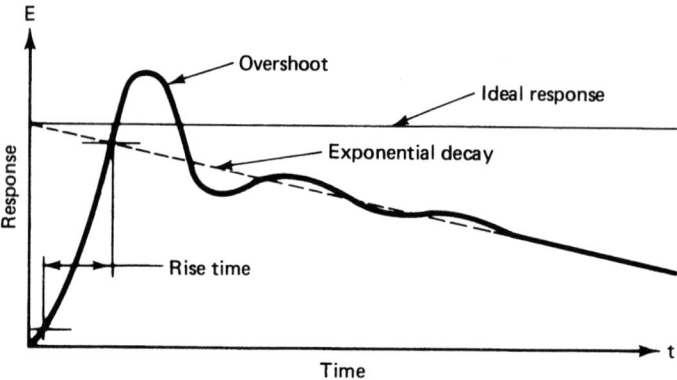

Figure 4-40 Typical response of a measurement system to a step input.

32 stud) bolt. At high frequencies, the flexibility of this type of attachment may permit the transducer to resonate against the mounting surface. This problem can be overcome by using either wax or high-wax-content grease between the mounting surfaces. The highly viscous nature of the grease essentially glues the transducer to the surface at high frequencies.

3. *Dirt and humidity.* For piezoelectric circuits, it is normally assumed that the transducer has a very large resistance with respect to the nominal 100-MΩ amplifier input impedance. The presence of fingerprints, grease, dirt and/or humidity on the cable connection surfaces can significantly reduce this high resistance and thus alter the *RC* time constant and the corresponding low-frequency response of the circuit. Cable connection surfaces must be clean and dry and kept that way during the tests. Silicone grease can be used on the transducer socket and cable plug for short-term exposure to moisture, while acid-free RTV silicon rubber compound is recommended for complete-immersion or high-humidity applications, along with Teflon-insulated cables.
4. *Cable noise.* Noise signals in systems with high-input-impedance amplifiers and cables may result from ground loops, tribo-electric effects, and electromagnetic fields.
    a. *Ground loops.* Ground loops result when the ground of the measuring system and the ground of the surface being measured are at different potentials, which causes a current to flow through the cable shield. Such a ground loop can often be broken by electronically isolating the transducer from the surface being measured. The presence of the isolation washer can reduce the natural frequency of the mounted transducer, however, and must be used with caution.
    b. *Tribo-electric effects.* Tribo-electric effects are generated when a transducer cable is damaged and a charge is generated by friction between the conductor and the insulation material when the cable vibrates. This problem is usually avoided by using graphite in the cables at time of manufacture and limiting cable motion. Cable condition can be checked at the time of installation by shaking the cable and observing the output signal. Any cable that generates a significant signal should be replaced.
    c. *Electromagnetic fields.* Electromagnetic fields can generate noise problems when the transducer cable is close to power lines or operating electrical machinery that generates significant time-variable magnetic fields. Double-shielded cables and special differential amplifiers can be used to control this problem.
5. *Base strains.* Surface deformation of the object on which the transducer is mounted can cause

unwanted signals to be generated if deformation of the transducer base is significant. New designs that protect the piezoelectric sensor have significantly reduced this problem.
6. *Nuclear radiation.* Nuclear radiation can change the operating characteristics of piezoelectric sensing materials. Consult with the manufacturer for a given application in order to avoid undesirable transducer change.
7. *Corrosive substances.* Most modern transducers are manufactured with stainless steel, titanium, and beryllium case materials and have a high resistance to corrosive environments.
8. *Operator error.* Improper training of personnel who install and use the measuring system can easily lead to the largest error.

4-8 Summary and Future Trends

The main elements of systems used for the dynamic measurement of force, pressure, displacement, velocity, and acceleration were examined in this chapter. The characteristics of first- and second-order filters and an *RC* differentiating circuit were considered. The measurement of displacement relative to a fixed reference was explored by considering commonly used techniques such as a potentiometer and multiple-resistor-type circuits, as well as variable-inductance, photovoltaic, and microswitch sensors. Magnetic sensors for measuring velocity were described. The mechanical characteristics of seismic instruments for measuring displacement, velocity, acceleration, force, and pressure were developed. It was shown that all force and pressure transducers are inherently accelerometers. Therefore, this characteristic must be carefully considered and accounted for when force and pressure measurements are made.

Characteristics of piezoelectric circuits were explored in depth, since piezoelectric sensors are commonly used in many dynamic measuring instruments. Three commonly used signal-conditioning arrangements that employ either a voltage follower, a charge amplifier, or a built-in amplifier were described in detail. It was shown that all three of these signal-conditioning units have the same frequency-response characteristics. The mechanical and electrical response characteristics of piezoelectric transducers to periodic and transient input signals were considered. It was shown that the mechanical limits are a result of the natural frequency of the transducer, while the electrical limits are a result of the low-frequency *RC* time constant. In a discussion of calibration procedures for motion and force measuring systems, consideration was given to both sinusoidal and transient events. Finally, questions of overall system fidelity and considerations of temperature, transducer mounting, dirt and humidity, cable noise, base strains, nuclear radiation, corrosive substances, and operator error were explored.

Recent research into the effects of accelerometer cross-axis sensitivity in experimental modal analysis by Han and McConnell [1-5] has shown that apparently small amounts of cross-axis sensitivity can create the appearance of ghost resonant frequencies and mode shapes. These contaminated results can lead to considerable confusion when dealing with an unknown structure and can even contaminate finite-element analysis that is being modified to match the "correct" experimental results. These cross-axis sensitivity effects can be corrected for in postprocessing of the measured frequency-response functions according to the method developed by McConnell and Han [5].

Recent research by McConnell [6,7] into modeling a force transducer shows that three classical application configurations occur. These configurations are fixed base, impulse hammer, and attached between a vibration exciter and structure under test. The fixed-base and impulse-hammer models give behavior identical to the traditional single degree of freedom resonating instrument as presented here. When the force transducer is attached between the vibration exciter and the structure under test, however, it has a completely different mechanical characteristic that does not have the traditional resonance response. This does not mean that the transducer does not resonate; it means that this resonance has no effect on the force being measured so that the measurement is correct.

Hu and McConnell and Hu [8,9] are conducting research on the interaction of the structure under test, force, transducers, stingers, vibration exciters, and the mass compensation technique recommended by Ewins [10] during vibration testing. This research is giving new insights into how the test system parameters effect the measured frequency-response function. It is found that the force transducer behaves differently when employed in this classical test configuration [7] and that significant measurement errors (on the order of 1000%) can occur at the structure's resonant frequency under certain rather innocent-looking conditions [8]. These errors are largest at a structure's resonances, the very frequency regions where the best data are required for experimental model analysis. Consequently, this literature should be followed during the coming years as these measurement issues are discussed and the author's papers are published in the *International Modal Analysis Conference Proceedings* and in the *International Journal for Analytical and Experimental Modal Analysis*.

References

1. S. Han and K. G. McConnell, Analysis of Frequency Response Functions Affected by the Coupled Modes of the Structure, *Intl J. Anal. Exp. Modal Analy.*, 6, no. 3 (July 1991), 147–160.

2. K. G. McConnell and S. Han, A Theoretical Basis for Cross-Axis Corrections in Tri-Axial Accelerometers, *Proc. 9th Int. Modal Analysis Conf.*, Florence, Italy, Vol. I, April 1991, 171–175.

3. K. G. McConnell and S. Han, Measuring True Acceleration Vectors with Tri-Axial Accelerometers, *Proc. 1989 Spring SEM Conf.*, Cambridge, MA, May 1989, 106–111.

4. K. G. McConnell and S. Han, The Effects of Cross-Axis Sensitivity in Modal Analysis, *Proc. 7th Int. Modal Analysis Conf.*, Las Vegas, Vol. I, January 1989, 505–511.

5. S. Han, The Effects of Transducer Cross-Axis Sensitivity in Modal Analysis, Ph.D. thesis, Iowa State University, 1988.

6. K. G. McConnell, Errors in Using Force Transducers, *Proc. 8th Int. Modal Analysis Conf.*, Kissimmee, FL, Vol. II, February 1990, 884–890.

7. K. G. McConnell, The Interaction of Force Transducers with Their Test Environment, *Int'l Journal for Analytical and Experimental Modal Analysis*, 8, No. 2, April 1993, 137–149.

8. X. Hu and K. G. McConnell, Mass Cancellation and Instrument Phase Shift Cause Large FRF Measurement Errors Near Resonance, *Proc. 10th Int. Modal Analysis Conf.*, San Diego, CA, February 1992.

9. X. Hu, A Study of Experimental Modal Analysis Topics Related to Stinger Axial Dynamics and Mass Compensation, Ph.D. thesis, Iowa State University, December 1991.

10. D. J. Ewins, *Modal Analysis: Theory and Practice*, Research Studies Press Ltd, Letchworth, Hertfordshire, England. (Also available from Bruel and Kjaer Instruments, Marlborough, Ma.)

11. Jens Trampe Broch, *Mechanical Vibration and Shock Measurements*, available from Bruel and Kjaer Instruments, Inc., Marlborough, MA.

12. Bruel and Kjaer Instruments, *Piezoelectric Accelerometer and Vibration Preamplifier Handbook*, available from Bruel and Kjaer Instruments, Inc., Marlborough, Mass.

13. Bruel and Kjaer Instruments, *Technical Review*, a quarterly publication available from Bruel and Kjaer Instruments, Inc., Marlborough, MA
    (a) Vibration Testing of Components, no. 2, 1958.
    (b) Measurement and Description of Shock, no. 3, 1969.
    (c) Vibration Testing, no. 3, 1967.
    (d) Vibration Measurement by Laser Interferometer, no. 1, 1971.

(e) A Portable Calibrator for Accelerometers, no. 1, 1971.
(f) High Frequency Response of Force Transducers, no. 3, 1972.
(g) On the Measurement of Frequency Response Functions, no. 4, 1975.

14. N. D. Change, *General Guide to ICP Instrumentation*, available from PBC Piezotronics, Inc., P.O. Box 33, Buffalo, NY.
15. F. L. Crosswy and H. T. Kalb, *Dynamic Force Measurement Techniques-I. Dynamic Compensation, Instrum. Control Syst.*, February 1970, 81–83.
16. F. L. Crosswy and H. T. Kalb, *Dynamic Force Measurement Techniques-II. Experimental Verification, Instrum. Control Syst.*, February 1970, 117–121.
17. J. W. Dally, W. F. Riley, and K. G. McConnell, *Instrumentation for Engineering Measurements*, Wiley, New York, 1984.
18. E. Jones, S. Edelman, and K. S. Sizemore, *Calibration of Vibration Pickups at Large Amplitudes, J. Acoust. Soc. Am., 33*, no. 11 (1961), 1462–1466.
19. E. Jones, S. Edelman, E. R. Smith, and E. T. Pierce, *Modulated Photoelectric Measurement of Vibration, J. Acoust. Soc. Am., 34*, no. 4 (1966), 455–458.
20. W. P. Kistler, *Precision Calibration of Accelerometers for Shock and Vibration, Test Engl.*, May 1966, 16.
21. R. W. Lally, *Gravimetric Calibration of Accelerometers*, available from PCB Piezotronics, Inc., P.O. Box 33, Buffalo, NY.
22. E. B. Magrab and D. S. Blomquist, *The Measurement of Time-Varying Phenomena*, Wiley-Interscience, New York, 1971.
23. K. G. McConnell and Y. S. Park, *Electronic Compensation of a Force Transducer for Measuring Fluid Forces Acting on an Accelerating Cylinder, Exp. Mech., 21*, no. 4 (1981), 169–172.
24. D. Pennington, *In-place Calibration of Piezoelectric Crystal Accelerometer Amplifier Systems, ISA National Flight Test Instrument Symposium*, San Diego, CA, May 1960. See Also Endevco Corp. Tech. Paper nos. 211 and 216.
25. D. Pennington, *Piezoelectric Accelerometer Manual*, Endevco Corporation, Pasadena, CA, 1965.
26. K. G. McConnell and S. E. Han, Effect of Mass on Force Transducer Sensitivity, *Exp. Tech., 10*, no. 7 (1986), 23–27.

CHAPTER
5

Photoelasticity

Christian P. Burger

Texas A & M University
College Station, Texas

5-0 Introduction

Photoelasticity is an experiment method for analyzing stress or strain fields in mechanics. It deals with the effects on light of stress and/or strain. As light traverses certain transparent materials, photoelasticity infers the stresses from their optical effect.

While the traditional areas of application of photoelasticity have been largely taken over by modern numerical techniques, the same computers that made this possible are now being used to extend the range of photoelasticity into new and exciting directions. Developments in lasers, fiber optics, and digital procedures for pattern recognition, image analysis, data acquisition, and data analysis continue to extend the range of problems for which photoelasticity is the most feasible means of investigation. The result is that instead of diminishing, the importance of the photoelastic method in pure and applied mechanics is increasing.

The aim of this chapter is to present photoelasticity in its present state. The reader is encouraged to consult the current literature to stay abreast of developments as the "new photoelasticity" using electronic imaging and new materials technologies unfolds.

5-1 Theory of Photoelasticity

Polarization of Light [1–5]

The photoelastic effect can be adequately described by the electromagnetic theory of light. According to this view, light propagates as transverse electromagnetic waves. The electric and magnetic fields are perpendicular to the direction of propagation and to each other, as shown in Fig. 5-1. Accordingly, the disturbance producing light is described as a wave motion where the instantaneous magnitude E of the electric vector, as observed at a fixed point along the direction of propagation is

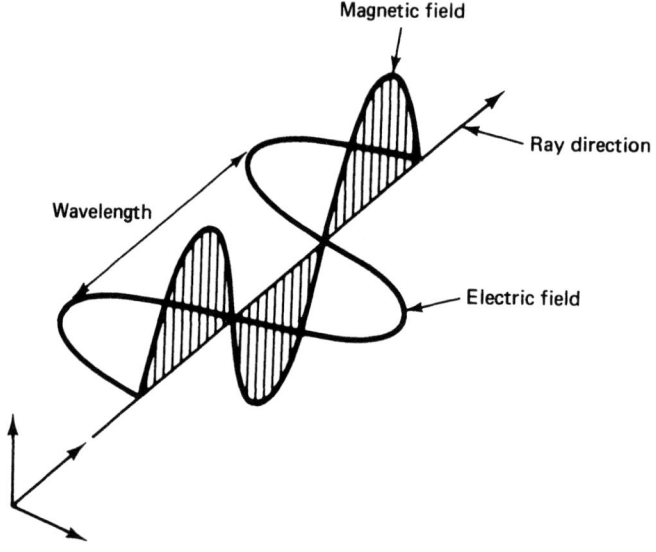

Figure 5-1 Instantaneous field pattern of a linearly polarized ray.

$$E = a \sin \frac{2\pi}{\lambda} ct$$

$$= a \sin 2\pi ft = a \sin \omega t$$

(5-1)

where t = time
         c = velocity of propagation (3×10^8 m/s in vacuum)
         λ = wavelength
         a = amplitude
         f = frequency of the light
         $\omega = 2\pi f$ = circular frequency of the light

Defining the direction of a ray and its electric field specifies the three vector directions: propagation, and electric and magnetic fields. Most light sources consist of a large number of randomly oriented atomic or molecular emitters. The rays emitted in any direction from such a source will have electric fields that have no preferred orientation. These rays make an unpolarized light beam.

If a light beam is made up of rays with their electric fields oriented in the same direction, the beam is said to be *linearly polarized* or *plane polarized*. The field at any point varies in magnitude and sign; its direction of vibration is used to describe the polarization. If the field vector is vertical, the light is said to be vertically polarized, and the *plane of polarization* is the vertical plane. This is shown diagramatically in Fig. 5-2.

Any ray with linear polarization can be resolved into its components polarized along any pair of arbitrary orthogonal axes by the usual vector sum rules as in Fig. 5-3.

Linear polarization is a special case of *elliptical polarization* that describes any kind of light that is not randomly polarized. Any polarized beam can be defined by the ellipse traced by the electric-field vector. For full characterization, three factors must be specified: the degree of ellipticity, the orientation of the major axis of the ellipse, and the handedness (right- or left-handed).

PHOTOELASTICITY

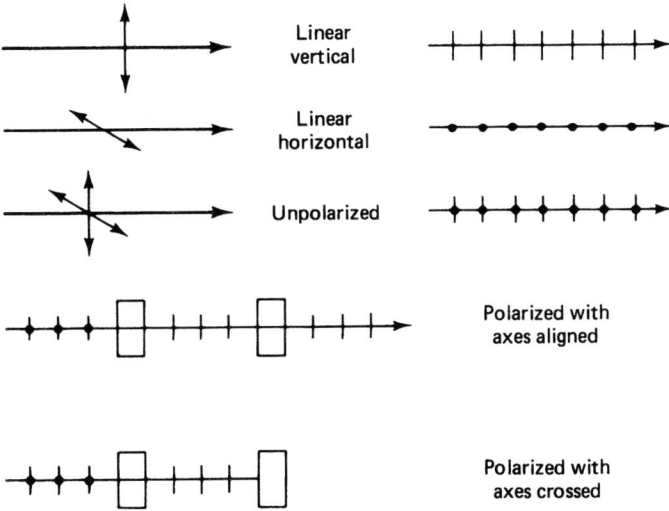

Figure 5-2 Representation of polarized light.

Another special case of elliptical polarization is *circular polarization*. In this case the electric field at any point rotates around the direction of propagation with constant magnitude and only the handedness—that is, the direction of rotation—need be specified. The three forms of polarization are depicted in Fig. 5-4. Thus plane-polarized light occurs when all components of the light vector lie in a single plane known as the plane of polarization. Circularly polarized light is obtained when the tip of the light vector describes a circular helix as the light propagates along the z axis. Elliptically polarized light is obtained when the tip of the light vector describes an elliptical helix as the light propagates along the z axis.

There are several ways to produce polarized light from a natural (randomly polarized) source:

1. Polarization by reflection
2. Polarization by scattering
3. Polarization by fine grids
4. Polarization with Polaroid sheets
5. Polarization through double refraction or birefringence

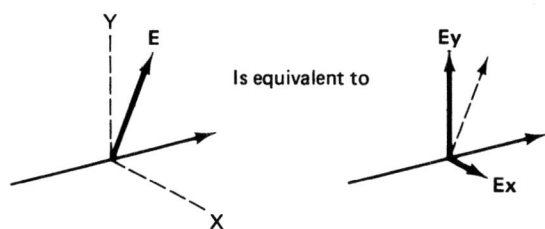

Figure 5-3 Resolving linearly polarized light into two orthogonal components.

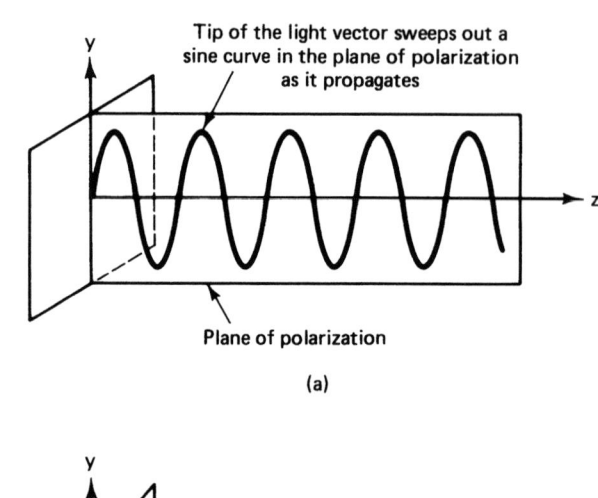
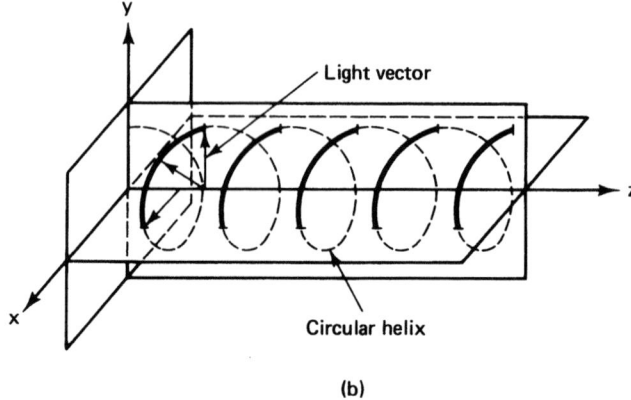
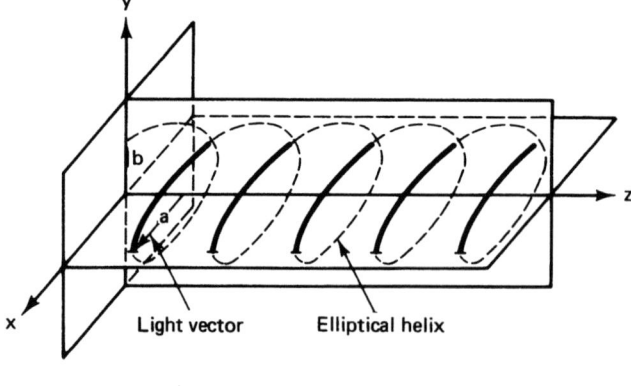

Figure 5-4 Motion of the light vector for (a) plane; (b) circularly; (c) elliptically polarized light.

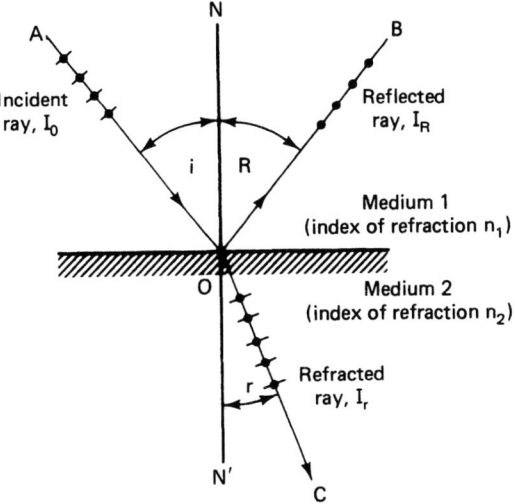

Figure 5-5 Reflection and refraction at a plane surface, $n_2 > n_1$.

Polarization by Reflection

When incidence occurs at an angle other than normal to an interface between two dielectric media of different refractive indices, such as air and glass, there will be some degree of polarization of the reflected as well as the transmitted beams.

At off-normal incidence, the direction of the transmitted beam does not coincide with that of the incident beam and is said to be refracted. The angles between the incident, reflected, and refracted beams and the normal to the surface at the point of incidence are known as the angles of incidence, reflection, and refraction (Fig. 5-5) and are denoted by i, R, and r, respectively, where $i = R$.

The ratio of the sine of the angle of incidence to the sine of the angle of refraction is a constant, depending on the nature of the media and the wavelength of the light. This constant is called the relative refractive index for light traveling from medium 1 into medium 2 and is denoted by n_{12}. Thus

$$\frac{\sin i}{\sin r} = n_{12} \tag{5-2}$$

This is known as Snell's law. If the incident medium is a vacuum, this ratio is called the absolute refractive index of medium 2 and is denoted simply by n_2.

Refraction is explained in terms of the wave theory of light by assuming that the velocity of light depends on the medium through which it passes and is a maximum in a vacuum. If v_1 and v_2 are the velocities of light in the two media, the relative refractive index $n_{12} = v_1/v_2$. If c is the velocity of light in a vacuum (about 3×10^8 m/s), the absolute refractive indices are

$$n_1 = \frac{c}{v_1} \quad \text{and} \quad n_2 = \frac{c}{v_2}$$

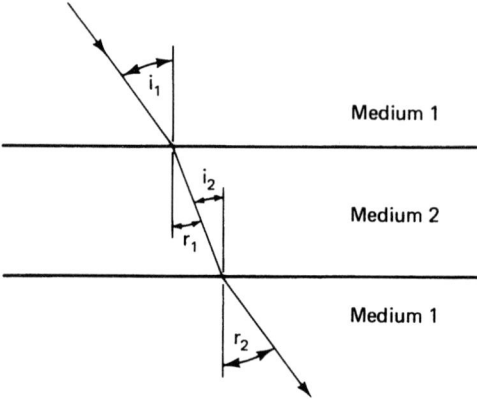

Figure 5-6 Refraction through a parallel plate.

so that

$$n_{12} = \frac{n_2}{n_1}$$

The refractive index for air at standard temperature and pressure is 1.000278. So, for most practical purposes, the absolute refraction index of a solid material and its relative refractive index from air are regarded as the same.

When light travels from one isotropic medium to another, the number of waves passing in any given time interval must be the same at every point in both media; that is, the frequency is unchanged. Therefore, the wavelength must vary according to the medium. Since the frequency is expressed by

$$f = \frac{v}{\lambda}$$

we have, for the two media, that

$$\frac{\lambda_1}{\lambda_2} = \frac{v_1}{v_2} = n_{12} \tag{5-3}$$

When light passes through a parallel plate, the angles of incidence and refraction at the upper and lower surfaces, respectively, are as shown in Fig. 5-6. Since the surface are parallel, $r_1 = i_2$, $i_1 = r_2$, and the emergent rays are parallel to the incident rays but have a lateral displacement.

If a beam of natural light is incident on the surface of a transparent medium at an angle i equal to approximately 57°, it is found that the reflected beam is plane polarized such that the plane of vibration is perpendicular to the plane of incidence and parallel to the reflecting surface, as indicated in Fig. 5-5.

The refracted ray is also partially plane polarized but with the vibrations lying in the plane of incidence. The exact angle of incidence giving the greatest degree of polarization depends on the refractive index of the material and is called the polarizing angle.

Fresnel has shown that if the incident ray is plane polarized in the plane of incidence (i.e., the plane of vibration is in the plane of the diagram), the relative amplitudes a, b, and c of the incident, reflected, and refracted rays are related as follows:

$$\frac{b}{a} = \frac{\tan(i - r)}{\tan(i + r)}$$

$$\frac{c}{a} = \frac{2 \cos i \sin r}{\sin(i + r) \cos(i - r)}$$

(5-4)

If the incident wave is polarized perpendicular to the plane of incidence (i.e., the vibration is perpendicular to the paper), the amplitude ratios are

$$\frac{b}{a} = \frac{-\sin(i - r)}{\sin(i + r)}$$

$$\frac{c}{a} = \frac{2 \cos i \sin r}{\sin(i + r)}$$

(5-5)

Since the intensity of light is the square of the amplitude,

$$I = (\text{amplitude})^2 \tag{5-6}$$

$(b/a)^2$ will be the relative intensity of the reflected and incident light. This ratio is known as the coefficient of reflection of a surface.

Unpolarized light can be assumed to consist of equal amounts of two components that are plane polarized in mutually perpendicular planes. Hence the intensity I_R of natural light reflected at the surface is given by

$$\frac{I_R}{I_0} = \frac{\sin^2(i - r)}{2 \sin^2(i + r)} + \frac{\tan^2(i - r)}{2 \tan^2(i + r)} \tag{5-7}$$

where I_0 is the intensity of the incident light.

The intensity I_r of the refracted wave is given by

$$\frac{I_r}{I_0} = \frac{\cos^2 i \sin^2 r}{\sin^2(i + r)} + \frac{\cos^2 i \sin^2 r}{\sin^2(i + r) \cos^2(i - r)} \tag{5-8}$$

When the angle of incidence is small, such as for near-normal incidence, Eqs. (5-4) and (5-5) simplify and from Eqs. (5-2), (5-7), and (5-8) we have that

$$\frac{I_R}{I_0} = \left(\frac{n - 1}{n + 1}\right)^2$$

$$\frac{I_r}{I_0} = \left(\frac{2}{n + 1}\right)^2$$

(5-9)

For a glass plate in air we can assume a refractive index from air to glass of $n = 1.5$ and from glass to air of $1/n = 0.67$. Then Eq. (5-9) shows that 4% of the incident light is reflected at each surface.

When $i + r = 90°$, $\sin r = \cos i$, and from Snell's law

$$\tan i = n$$

(5-10)

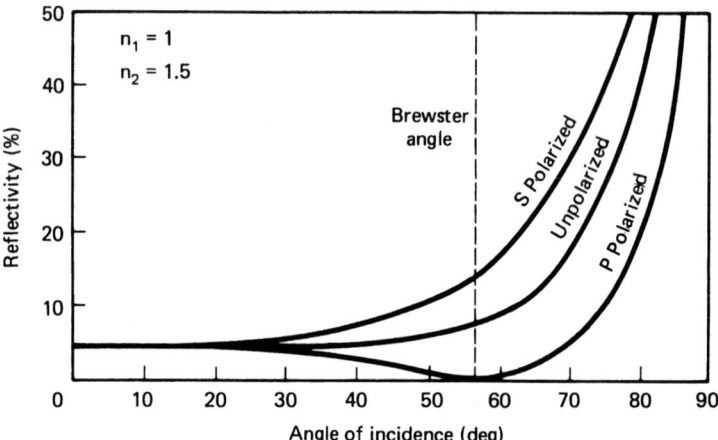

Figure 5-7 Reflectivity against angle of incidence.

The second term on the right of Eq. (5-7) becomes zero, and the reflected light is completely plane polarized with the vibrations parallel to the surface of the plate. If the incident light is plane polarized with the plane of vibration coinciding with the plane of incidence, no light is reflected. The angle of incidence so defined is the polarizing angle or Brewster's angle and has a value of about 57° for air and glass. (Equation (5-10) expresses Brewster's law, which states that the tangent of the polarizing angle equals the refractive index (for air to medium).

Figure 5-7 shows reflectance against angle of incidence for light polarized in the plane of incidence and light polarized perpendicular to the plane of incidence. If the angle of incidence is the Brewster angle, the reflected beam is entirely polarized perpendicular to the plane of incidence. Light polarized perpendicularly to the plane of incidence is said to have "S polarization," while light polarized parallel to the plane of incidence is said to have "P polarization." S and P polarizations, which are meaningful only for light striking a surface, are frequency used in the specification of dielectric coatings. A plane-polarized beam with proper polarization passes through a Brewster window with no reflection loss. Figure 5-8 shows the reflections and transmissions for incidence at the polarizing angle.

The metal surfaces of front-faced mirrors have large reflection coefficients, as shown in Fig. 5-9. At oblique incidence a P beam has a lower reflectivity than an S beam. A phase change also occurs with metallic reflection. This phase change varies with both the angle of incidence and the direction of polarization, so that plane-polarized light may be changed by oblique reflection to elliptically polarized light.

When the refracted beam is considered, the second term on the right of Eq. (5-8) is seen to be larger than the first. This means that some of the refracted light is plane polarized, with the vibration coinciding with the plane of incidence (except in the limiting case of normal incidence). The maximum degree of polarization occurs when the angle of incidence is equal to the polarizing angle, but polarization is only partial. The degree of polarization can be increased by employing a pile of parallel glass plates arranged so that the light is incident on them at the polarizing angle. Light is progressively polarized as it passes through successive plates until after seven or eight plates a high degree of polarization is obtained.

It is important to note that waves with amplitudes of opposite sign differ in phase by π. When

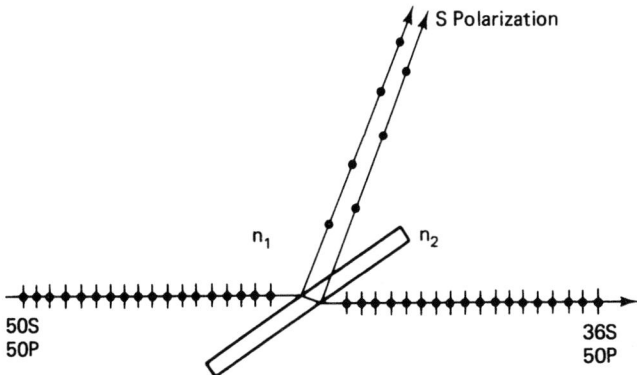

Figure 5-8 Incidence at Brewster's angle.

reflection occurs in the rarer medium so that $i > r$, the ratio b/a in Eq. (5-5) is negative. For an S-polarized beam, a change of phase equal to π is introduced. This is important when constructing reflection polariscopes. On the other hand, if reflection occurs internally, such as in the medium of greater density, so that $i < r$, b/a is positive and there is no change in phase.

If the incident wave is P polarized, the ratio b/a of Eq. (5-4) is positive with external reflection for small values of i, but it diminishes to zero and changes sign at the polarizing angle. Thus there is a phase change of π when the angle of incidence exceeds the polarizing angle but none for smaller angles of incidence. This is reversed when reflection occurs internally.

Polarization by Scattering

Light scattering by small charged particles is sensitive to wavelength and angle of polarization. In a perfectly clear atmosphere, the eye would see a ray of light only when the line of vision was

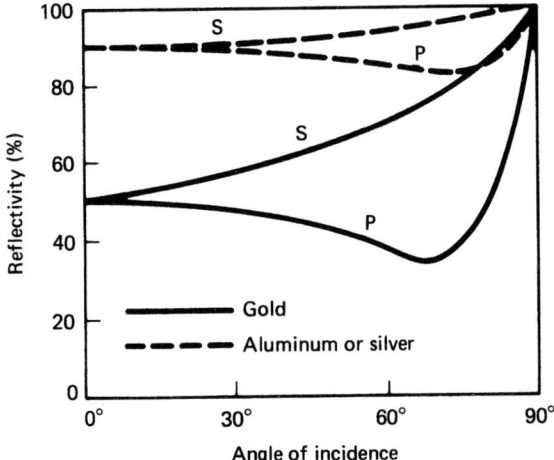

Figure 5-9 Reflection coefficients for several air–metal interfaces.

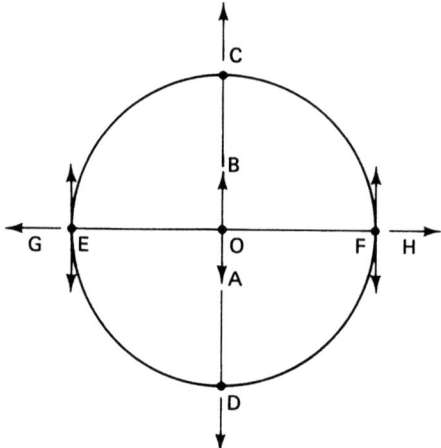

Figure 5-10 Dynamic model for light scattering.

along the ray and directed toward the source. In space, the sky does indeed appear black. However, owing to the presence at low altitudes of small particles (or even comparatively large particles such as the drops of water in a cloud), light is scattered, causing illumination of objects not directly in the path of the rays from the sun (e.g., in the shadow of a building). On a cloudless day about one-fifth of the total illumination on a horizontal surface is caused by scattered light from the sky. Small particles scatter short waves more than long waves; therefore, in the light reaching the eye there is a predominance of the shorter waves, giving the sensation of blue—hence the blue color of the sky.

The mechanism of this scattering of light may be illustrated by an elastic–vibration analogy. Consider a small particle at O (Fig. 5-10) and a wave of plane-polarized elastic vibrations traveling normal to the plane of the paper, incident on the particle. The vibrations of this incident wave are in the plane of the paper along AB, for example. If the particle is free to move, it will be set vibrating also along the line AOB. In general, there will be a phase difference between the vibrations of the particle and those of the incident wave, giving rise to a relative motion between the particle and the elastic medium. This will produce a secondary wave radiating from O, the vibrations of which will be parallel to AB. At any instant this secondary wave surface will be a sphere of center O. If the medium can transmit only transverse vibrations, there will be no waves propagated in directions OC and OD. In all other directions there will be waves whose amplitudes increase from zero at C and D to a maximum at points in the horizontal plane through O. All the rays of *this secondary wave will be polarized*, their vibrations at any point being in the vertical plane, that is, the plane of AB.

This effect can be demonstrated by passing a beam of plane-polarized white light through a glass tube containing a cloud of finely suspended particles. If the plane of vibration of the incident light is vertical, no light will be seen if the eye is vertically above or below the tube, but maximum intensity of light (which will appear blue) will be observed when the tube is viewed horizontally from either side.

Polarization by Fine Grids

The most familiar linear polarizers, H-type Polaroid sheets, operate in a similar way. These sheets are molecular polarizers operating by dichroism or selective absorption. A dichroic material is a birefringent material in which one of the two orthogonal polarizations (either the ordinary or the extraordinary) is strongly absorbed while the other is not. The word *dichroic* literally means two-color and has its origin in the fact that certain crystalline materials, when subjected to white linearly polarized illumination, characteristically exhibit two distinct colors. The early sheet polarizers were made with tiny dichroic crystals, all oriented in a special way and embedded in a plastic film. Modern H-type sheets are made from a sheet of polyvinyl alcohol in which the long-chain molecules are oriented by stretching, thereby achieving birefringence. Pigment molecules of iodine are absorbed by these long chains by selective attachment. The selective absorption of one polarization occurs because of the orientation of the chemical bonds at the attachment sites. In this way the absorbed iodine forms effective thin iodine wires that absorb light polarized parallel to the "wires" but transmit orthogonally polarized light (Fig. 5-11a). The transmission of the sheet and the polarization/extinction ratio are controlled by the amount of iodine absorbed by the sheet. Because dichroic materials function by selective absorption, they cannot be used in high-power laser beams.

The transmittance T of a single-sheet polarizer in a beam of linearly polarized incident light is given by

$$T = k_1 \cos^2\theta + k_2 \sin^2\theta \tag{5-11}$$

where θ = angle between the plane of polarization of the incident beam
(more accurately, the plane of the electric-field vector
of the incident beam) and the plane of preferred
transmission of the polarizer.

k_1, k_2 = principal transmittances of the polarizer,
and both are functions of wavelength; ideally,
$k_1 = 1$, and $k_2 = 0$, in reality, k_1 is always somewhat
less than unity, and k_2 always has some small but nonzero value

If the incident beam is unpolarized, and the angle θ is redefined to be the angle between the planes of preferred transmission (planes of polarization) of two sheet polarizers in near contact, the transmittance of the pair is given by

$$T_{\text{pair}} = k_1 k_2 \sin^2\theta + \tfrac{1}{2}(k_1^2 + k_2^2) \cos^2\theta$$

If we define H_{90} as the transmittance when $\theta = 90°$, then

$$H_{90} = T_{\text{pair}}(90°) = k_1 k_2$$
$$H_0 = T_{\text{pair}}(0°) = \tfrac{1}{2}(k_1^2 + k_2^2)$$

The formula above then simplifies to

$$\begin{aligned} T_{\text{pair}} &= H_{90} \sin^2\theta + H_0 \cos^2\theta \\ &= H_{90} + (H_0 - H_{90}) \cos^2\theta \end{aligned} \tag{5-12}$$

The quantity H_{90} is called the closed transmittance or extinction ratio, while the quantity H_0 is called the open transmittance. Both quantities are wavelength dependent; their effect is shown in Fig. 5-11b.

The transmission spectra for a typical Polaroid sheet are given in Fig. 5-11c. Because of the

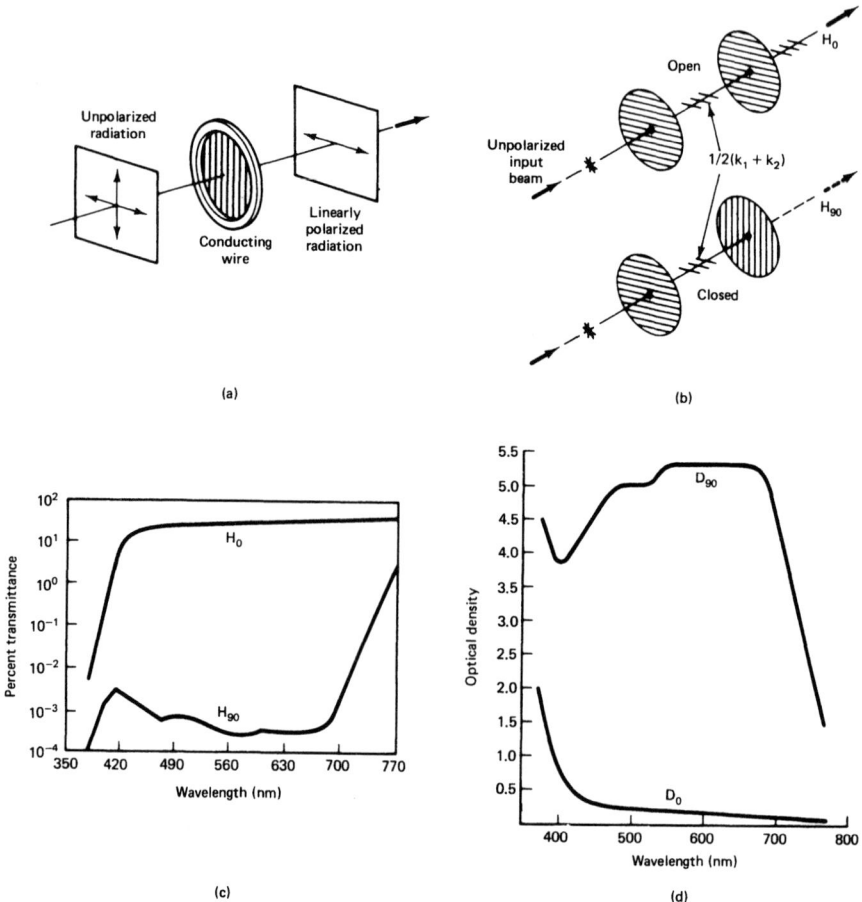

Figure 5-11 (a) Wire-grid polarizer; (b) parallel and crossed dichroic polarizers; (c) transmittance spectra for parallel and crossed sheets of Polaroid HN38S; (d) spectral curves for polaroid HN38S.

large ranges of open and closed transmission, the optical densities are sometimes graphed, rather than the transmissions themselves. The open and closed optical densities are defined as follows

$$D_{90} = \log_{10} \frac{1}{H_{90}}$$

and

$$D_0 = \log_{10} \frac{1}{H_0} \tag{5-13}$$

Figure 5-11d shows typical values of optical densities for the same sheet of materials shown in Fig. 5-11c.

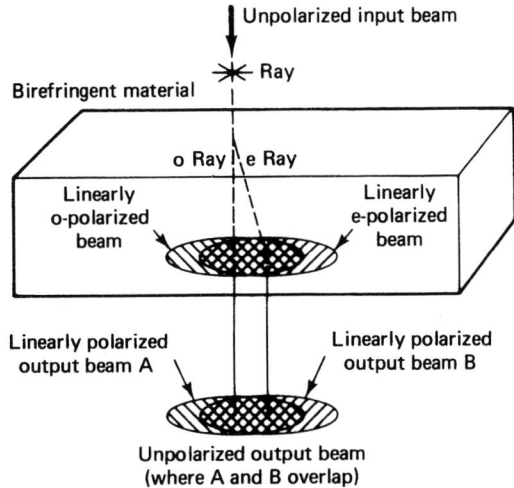

Figure 5-12 Double refraction in a birefringent crystal.

A major advantage of sheet polarizers is the wide useful field angle, that is, the angle of incidence with relation to the normal, at which the sheet is an effective polarizer. For typical cases this angle is almost 90° or grazing incidence.

Polarization Through Double Refraction or Birefringence

Many crystalline materials do not have a single value for their refractive index, because of structural anisotropy. Fused silica, being isotropic, has one refractive index at any wavelength. Crystal quartz, on the other hand, has a refractive index that depends on the polarization of the propagating light wave and the direction of propagation through the crystal. Any material that can be characterized with two different indices of refraction is said to be *birefringent*.

A birefringent crystal will divide an entering ray of monochromatic radiation into two rays having orthogonal (perpendicular) polarizations and usually having two somewhat different directions of propagation through the crystal. The two rays usually propagate with different velocities. Depending on whether the crystal is uniaxial or biaxial, there are only one or two directions in the crystal, known as optic axis directions, along which the orthogonally polarized rays remain exactly collinear and propagate with equal speeds. Similarly, a beam (an extended family of rays) of unpolarized monochromatic radiation is resolved within the medium into two orthogonally polarized beams. If the medium is in the form of a plane-parallel plate, these beams recombine at emergence to form an unpolarized beam, where they overlap. In Fig. 5-12 the linear polarizations of beams A and B are orthogonal. Separation of beams A and B has been exaggerated to clarify the distinction between ray and beam behavior.

The two new rays within the birefringent medium are easily distinguished by more than just polarization and velocity. One of the rays is extraordinary in the sense that it manages to violate, under suitable circumstances, both Snell's law and the law of reflection. This ray need not be confined to the plane of incidence. Its velocity, furthermore, changes in a continuous way with direction, that is, with angle of incidence. Thus the refractive index for this ray, known as the

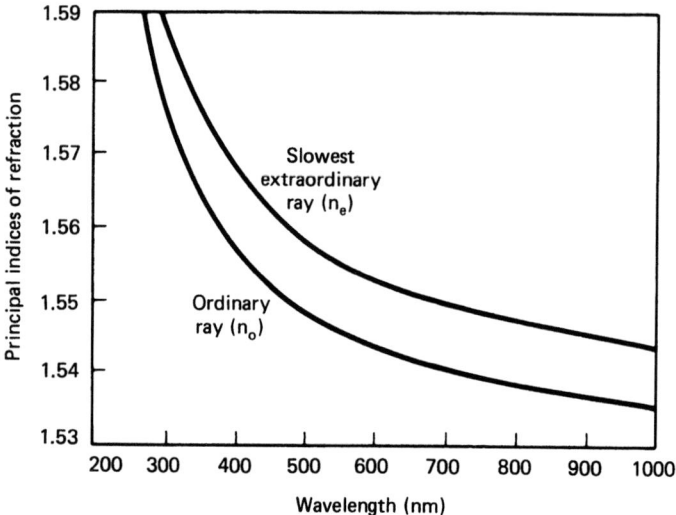

Figure 5-13 Refractive indices of crystalline quartz (at 18°C) versus wavelength.

extraordinary ray, e, is also a continuous function of direction. The refractive index for the other ray, known as the ordinary ray, o, is a constant independent of direction. Ray o refracts according to Snell's law. The two indices of refraction for rays e and o are equal only in the direction of an optic axis.

A graph of refractive index versus wavelength for quartz is shown in Fig. 5-13. It is known as a dispersion curve. On such graphs the behavior of the ordinary ray is described by a single unique curve, while the behavior of the extraordinary ray is described by a family of curves—different curves for different directions. The extraordinary index may have any value between n_o and n_e (the slowest extraordinary ray), depending on the direction of propagation in relation to the direction of the crystalline optic axis.

Except for a special polarization state, or unless the crystalline surface is perpendicular to an optic axis, a ray incident normal to the surface of a birefringent material will divide into two components, as shown in Fig. 5-14.

The ordinary ray o travels without deviation through the crystal and obeys the laws of refraction.

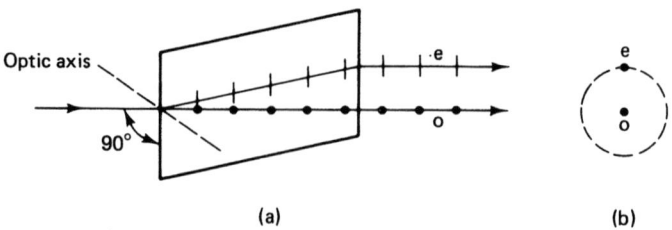

Figure 5-14 (a) Transmission of light through a principal section of a calcite rhomb; (b) rotation of emergent e ray when crystal is rotated about.

PHOTOELASTICITY

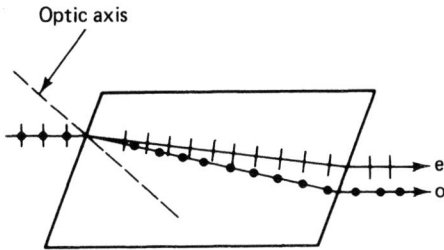

Figure 5-15 Polarization splitting in calcite.

The extraordinary ray e is deviated in the crystal and thus suffers a lateral displacement on emergence. It does not obey Snell's law.

If the crystal (often called a rhomb) is rotated about ray o as axis, ray e describes a circle around o, showing that the direction of propagation of ray e is dependent, while o is independent of the orientation of the crystal. The emergent rays are both plane polarized, with the extraordinary ray vibrating in the plane containing the optic axis and the light ray (called the principal section) and the ordinary ray in a plane perpendicular to this. If the incident ray is plane polarized parallel to the plane of the section, the ordinary ray is extinguished and the extraordinary ray has maximum brightness. This will be reversed when the plane of vibration is perpendicular to the principal section.

When the incident ray falls obliquely on a face of the rhomb, generally both the ordinary and extraordinary rays are refracted but by different amounts. The two components are transmitted with the same velocity and in the same direction for only one direction of transmission. This direction is known as the optic axis, and in calcite this coincides with the principal crystallographic axis. Crystals of this type, having only one optic axis, are called uniaxial crystals. Many crystals have two optic axes and as such are known as biaxial crystals.

Figure 5-15 shows a typical, unpolarized ray obliquely incident onto the face of a cleaved crystal of calcite. The unpolarized ray is split into its o and e components. These separate at the entrance face and exit the crystal at different points. The exiting rays are orthogonally polarized. The ordinary ray with fixed index n_o, and the extraordinary ray in the direction where its index n_e differs most (i.e., at maximum or minimum values, depending on the type of birefringent crystal) from the value for the ordinary ray are together known as the principal indices of refraction. Similarly, a beam of linearly polarized radiation is resolved into two components, which are orthogonally polarized within the medium according to the principal section of the crystal. If the traversed medium is a parallel plate, these beams have a relative phase shift and recombine vectorially at emergence to form a beam which is, in general, elliptically polarized.

Crystals of calcite (also called Iceland spar) are widely used for their birefringent properties. For light with a wavelength of 550 nm, calcite has a refractive index of 1.66 for the o ray, which is independent of direction of travel. For the e rays, however, the index varies from 1.66 for light propagating parallel to the optic axis to 1.49 for light traveling perpendicular to the optic axis. To follow the propagation of light rays through calcite crystals, the rays are resolved into components polarized parallel to the optic axis (e rays) and orthogonal to the optic axis (o rays). There are three conditions:

1. *Rays parallel to the optic axis.* These travel as in glass; the o rays and e rays have the same refractive index.

Table 5-1 Indices of Refraction at 589 nm

	n_o	n_e	$n_e - n_o$
Calcite	1.658	1.486	−0.172
Crystalline quartz	1.544	1.553	0.009
Mica	1.598	1.593	−0.005
Sapphire (Al_2O_3)	1.768	1.760	−0.008

2. *Rays perpendicular to the optic axis.* The e ray will travel faster than the o ray because of its lower refractive index, but it will travel in the same direction.
3. *Rays traveling at some angle between 0 and 90° from the optic axis.* The o ray will travel undeviated and will be refracted according to Snell's law; the e ray will deviate from the o ray because of the different refractive indices with direction. The direction of deviation will be away from the optic axis and generally out of the *plane of incidence*.

The optical behavior of any particular crystal is determined by that crystal's symmetry. Sodium chloride and other cubic crystals are optically isotropic. Crystals such as calcite, quartz, sapphire, and magnesium fluoride have one optic axis and are thus known as *uniaxial*. Other crystal types have two optic axes and are called *biaxial*. These are characterized by three indices of refraction. Mica, frequency used for wave plates, is biaxial. Calcite is widely used because of the large difference in its two refractive indices and its good transmission; these properties simplify separation of the two different polarizations. Crystal quartz has a much smaller refractive index difference that renders it less suitable for use in making polarizing prisms; however, other properties of quartz, such as its transmission and suitability for optical finishing, have led to its use in wave plates and other polarization-dependent devices.

Tables 5-1 and 5-2 give the indices of refraction for a few widely used birefringent materials.

Polarizing prisms are designed to separate light into two orthogonal polarizations. They are also constructed to pass one polarization undeviated and to have an acceptance angle of several degrees. The designs rely on the difference in refractive indices for the o and e rays. The original Nicol prism, now little used, and the Glan Thompson and Glan Foucault prisms operate by separation of the two polarizations at an angled interface that is designed to exceed the critical angle for one of the rays. At the critical angle of incidence, a refracted ray will propagate at grazing incidence along the interfacial surface of the second material; that is, the angle of refraction is 90°. At larger angles of incidence there will be no transmitted (refracted) wave but rather total internal reflection

Table 5-2 Variation of $n_e - n_o$ with Wavelength

Wavelength (μm)	Calcite	Crystalline Quartz	Mica	Sapphire
0.2	−0.326	0.0130		−0.0117
0.3	−0.206	0.0103		−0.0091
0.4	−0.184	0.0091		−0.0085
0.5	−0.176	0.0093	−0.0047	−0.0082
0.6	−0.172	0.0091	−0.0048	−0.0081
0.7	−0.169	0.0090	−0.0048	−0.0080
0.8	−0.167	0.0089		−0.0079
0.9	−0.165	0.0088		−0.0079
1.0	−0.164	0.0088		

PHOTOELASTICITY

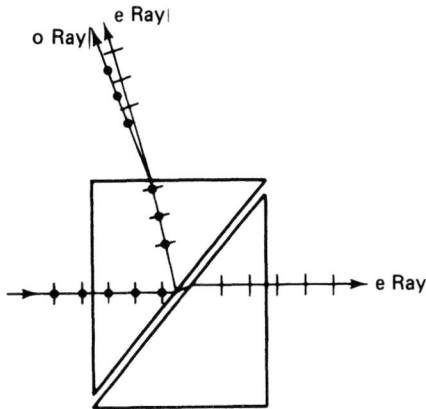

Figure 5-16 Oriel Glan Taylor prism.

in the first medium. This is the property exploited by the Glan Taylor prism as shown in Fig. 5-16. The o ray is totally internally reflected at the interface and emerges through the side of the prism or is absorbed there. The e ray passes through the interface with some reflection loss and exits the crystal, slightly laterally displaced and parallel to the input beam. Note that if the input beam satisfies the acceptance angle condition, the transmitted beam is highly polarized.

Extinction ratios of 10^{-5} to 10^{-6} can be achieved by these prisms. The beam that exits the side of the polarizer is mainly o ray, but it is accompanied by the orthogonally polarized beam reflected at the interface. This spurious beam is not exactly parallel to the o ray; thus it can be spatially separated.

We have seen that the rays along which the oppositely polarized waves are transmitted deviate within the crystal. There are two exceptions, however; when the direction of transmission is perpendicular to the principal axis in a uniaxial crystal or to two principal axes in a biaxial crystal. The deviation is extremely small in crystals of low birefringence and in most photoelastic models. As such, it is usually disregarded but should not be forgotten, because in some special cases (such as in regions of high stress levels in photoelasticity) the deviation is not small and can be used to measure stress levels.

The foregoing description of birefringence applies to nonactive crystals. The optic axis has been defined as the direction in which the o and e rays have equal refractive indices, that is, equal wave velocities. In an active uniaxial crystal such as quartz, there is no direction in which the indices are equal, and the optic axis direction must instead be defined as that direction in which the ordinary (o) and extraordinary (e) indicates are most nearly equal. In the inactive uniaxial crystal the ordinary (o) and extraordinary (e) rays are in orthogonal states of linear polarization, meaning that electric-field vectors are perpendicular. In an active crystal such as quartz, o and e are orthogonal in a more general mathematical sense, and both are in general elliptically polarized. In directions perpendicular to the optic axis, which are the only directions of interest in connection with linear retarders, the ellipse degenerates into a straight line, both the o and e rays are linearly polarized, and behavior is the same as for an inactive uniaxial crystal. In directions *almost* perpendicular to the optic axis, the polarizations are *almost* linear (elliptic and highly eccentric). As the direction of an incident ray departs further from normal to the optic axis, the polarization ellipse becomes less eccentric. In the optic axis direction, the ordinary and extraordinary rays are circularly polarized in opposite (orthogonal) senses and propagate at different speeds.

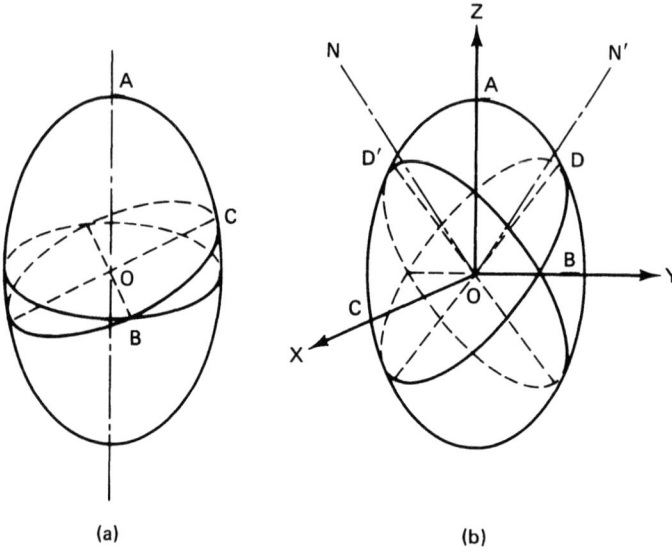

Figure 5-17 Index ellipsoid for (a) a positive uniaxial crystal and (b) a biaxial crystal.

If a quartz plate is cut so that the optic axis is normal to its surfaces, and a ray of linearly polarized light is incident parallel to the optic axis, the difference in speeds of the o and e circular polarizations has the effect of causing the resultant plane of polarization to rotate about the optic axis as the ray penetrates into the plate. The amount of rotation is directly proportional to the depth of penetration, and ultimately, to plate thickness. The result of superpositions of the counterrotating circular polarizations is always to produce linear polarization, and there are no intermediate, elliptical polarization states. This is the distinction between optical activity or circular retardation and the more commonly known linear retardation. For crystalline quartz and yellow sodium light ($\lambda = 589.3$ nm), the difference in refractive indices for opposite circular polarizations propagating along the optic axis is only about 1 part in 10,000. Correspondingly, the plane of resultant linear polarization rotates at a rate of about 21.7° per millimeter travel. Unfortunately, the polarization of the o and e rays in quartz changes rapidly from circular to elliptical when the incident angle departs from the direction of the optic axis. For sodium light at an angle of only 5% from the axis, the polarization ellipse has a major axis-to-minor axis ratio of 2.37. For the ellipse to be even approximately circular, much smaller angles are required. For this reason, devices that depend on circular retardation, such as the Cornu pseudodepolarizer, are effective only in highly collimated light that is propagating quite accurately in a direction parallel to the optic axis.

Crystalline quartz exists in two distinct forms. Natural crystals of these forms have shapes that are mirror images of each other. The plane of resultant linear polarization rotates in opposite senses in these forms (known as left-handed and right-handed quartz).

The effects produced when a beam of light falls on a crystal at varying angles of incidence can readily be visualized by a geometrical representation known as the index ellipsoid or Fresnel's ellipsoid. This is shown in Fig. 5-17, where a straight line from a central point o, drawn in any

PHOTOELASTICITY

direction, represents the reciprocal of the velocity, that is, the refractive index for a ray vibrating in the direction of that line. The end points of all such lines lie on the surface of an ellipsoid.

Figure 5-17a depicts a positive uniaxial crystal. The principal axis, OA, corresponds to the optic axis of the crystal. The section through O normal to OA is circular, so that when a ray passes through the crystal in the direction OA the component rays have equal velocities; that is, the refractive index is the same for both. For any other direction of transmission, the normal section through O is an ellipse. The directions of polarization coincide with the directions of the principal semiaxes OB and OC of this ellipse. The length of one of these semiaxes (i.e., OB is obviously always equal to the radius of the circular section, showing that the velocity of the ordinary ray is independent of the direction of transmission. The velocity of the extraordinary ray can vary between $1/OA$ and $1/OB$, depending on the direction of transmission. The index of refraction for the extraordinary ray represented by OA is known as the principal index of refraction for this ray and is denoted by n_e. For any random direction of vibration, that will be denoted by n'_e. The index of refraction n_o for the ordinary ray is represented by the radius of the circular section.

Uniaxial crystals are classified as positive or negative according to whether the velocity v_e of the extraordinary ray is respectively smaller or greater than that of the ordinary ray, that is, on whether n_e is greater than or less than n_o. For a positive uniaxial crystal the ellipsoid is a prolate shperoid of revolution as shown, while for a negative crystal it is oblate.

In a biaxial crystal, the principal semiaxes of the ellipsoid all have different lengths. We denote the principal refractive indices by n_x, n_y, and n_z and assume $n_x < n_y < n_z$, so that in Fig. 5-17b, $OC < OB < OA$. Then there must exist two central sections of the ellipsoid normal to the xz plane for which the semiaxes OD and OD' are equal to OB. These sections are circular and are equally inclined on either side of OA. Light passing in the directions normal to these sections behaves as if the crystal were isotropic. These directions (i.e., ON and ON') are the optic axes of the crystal. The lengths of both principal axes of the central section vary, however, with the direction of transmission, so that no ordinary ray exists in a biaxial crystal.

Retardation Plates and Wave Plates

The polarization of light can be changed, rather than selected, by using the properties of birefringence as described previously.

Since the phase difference between the oppositely polarized waves emerging from a crystal depends on the length of path within the crystal, it is evident that any desired phase difference can be obtained by using a crystal of the appropriate thickness. Birefringent optics of this type are called retarders, phase shifters, or wave plates.

When a light travels a distance d in a medium, it would travel a distance nd and in a vacuum in the same time. The product nd is called the optical path length.

The absolute retardation, R_0, for a ray passing through a thickness d of a material of refractive index n, is then

$$R_0 = (nd - d) = (n - 1)d \tag{5-14}$$

If the two rays of light, o and e, originating from a monochromatic source, propagate through a retardation plate, they will have the same frequency but different velocities, v_o and v_e. Their respective optical paths for a plate thickness d will be $n_o d$ and $n_e d$, and they will accumulate a relative retardation

$$R_d = (n_o - n_e)d \tag{5-16}$$

This is the basis for the design of optical retarders. The simplest retardation plate is a slice cut out of a uniaxial crystal such that the plane of the slice contains the crystalline optic axis. The velocity difference between the ordinary (o) and extraordinary (e) beams within the plate, which results from an unpolarized beam of light that is normally incident upon it, is therefore maximized. As the o and e beams traverse the plate, a phase difference accumulates between these beams; that difference is proportional to the distance traveled within the plate. At emergence the o and e beams recombine to form a second unpolarized beam.

If, on the other hand, the input beam is plane polarized at 45° to the optic axis, the beam will be resolved into two equal components with orthogonal polarizations, one parallel to and one normal to the optic axis. The vector sum of these two beams is, of course, equal to the original, and the action of the plate is to delay one beam with respect to the other.

If the thickness of the plate is such that the phase difference (retardation of the slow ray by comparison with the fast ray at emergence) if $\frac{1}{4}$ wavelength, or $\pi/2$ radians, the plate is called a first-order, quarter-wave plate. If the phase difference at emergence is $\frac{1}{2}$ wavelength (π radians), the plate is called a first-order, half-wave plate. If the phase difference at emergence is some multiple of $\frac{1}{4}$ or $\frac{1}{2}$ wavelength, the plate is called a multiorder or higher-order plate. Notice that it is the phase difference, not the physical thickness of the plate, to which these names refer. Mica and quartz are frequently used to make retardation plates, and the actual thickness of even a first-order, quarter-wave plate is a large number of wavelengths. The usual form is a flat, thin disk sandwiched between layers of glass for added strength. Mica happens to be biaxial, and quartz exhibits optical activity (rotation of the plane of linear polarization as a function of distance through the medium), but those effects do not significantly affect the end results.

Retarders may also be manufactured from noncrystalline materials such as glass, cellophane, or any of a large number of polymers that display artificial anisotropy as a result of residual stresses introduced during manufacture.

Both the o- and e-ray refractive indices of most materials are strongly dependent on wavelength; the thickness required to accumulate a specific retardation is therefore also wavelength dependent. If a particular value of retardation is desired, this value can be exactly achieved for normal incidence at only one wavelength, and the wavelength itself must also be specified. Mica is exceptional in that its principal indices of refraction vary slowly across the visible spectrum. Thus a retarder made for a wavelength of 550 nm and normal incidence will produce approximately the same retardation at other visible wavelengths.

In crystal quartz the indices at 546 nm are 1.5553 and 1.5462. The o ray is retarded by 0.006 wave for every wavelength traveled. To obtain a retardance of one quarter-wave, we need a quartz plate with a thickness of only 15 μm. Since it is not practical to make so thin a plate, either of two ways is used to make a quartz quarter-wave plate: (1) choosing a thickness so the retardance is an integer plus one-quarter ($n + \frac{1}{4}$) waves or (2) combining two plates, where the second is designed to compensate all the retardance of the first except one-quarter. For 1-in.-diameter plates, the order n (or sometimes $n + 1$) is usually around 30. The single plate is called a multiple-order retardation plate and the two-plate combination is referred to as a zero (or sometimes first-order) plate. These zero-order plates are equivalent to a thin, true-zero-order plate and have advantages in their insensitivity of the retardance to temperature and wavelength. Table 5-3 compares different quality plates. Note that tilting the plates alters the retardance, so the plates can be tuned to nearby wavelengths by tilting.

Wave plates are also made of mica for less critical applications. Zero-order, cleaved mica plates are around 0.35 mm thick and are cemented with an index-matching cement between glass plates. Even the best-quality mica has impurities and contains zones in which the optic axes are slightly misaligned. Crystal quartz plates should thus be selected for critical applications.

Table 5-3 Comparison of Properties for 632.8-nm Quarter-Wave Quartz Plates

	Good Zero Order	Good Multiorder	Standard Multiorder
Temperature sensitivity: degrees change in retardance per °C	0.009	0.4	1.1
Angle sensitivity to tilt of fast axis			
Phase change (degs) for 10-mrad tilt	0.0075	0.1	0.22
Phase change (degs) for 100-mrad tilt	0.75	10.1	21.8
Bandwidth (nm): waveband undergoing 89–91° retardance	14	0.265	0.124

Achromatic retarders with large apertures consist of three or more layers of birefringent polymer plates cemented between windows that have been optically polished to $\frac{1}{10}$ wave and antireflection coated. These retarders are usually available with center wavelengths of 420, 540, 620, and 850 nm for the quarter-wave retarders and 400, 540, and 800 nm for the half-wave retarders. The achromaticity for these retarders is typically within 0.10λ in the range from 0.85 to 1.2 times the center wavelength. Figure 5-18 represents retardance variations and fast axis directions as functions of wavelength for quarter- and half-wave plates of this type.

Large-aperture achromatic retarders have a larger field of view than crystal rhombs but have much lower power-handling capabilities. Rhombs are much more achromatic than polymer plates but require accurate mounting because of their small acceptable range of angle of incidence.

For larger-diameter beams, such as large diffuse-field polariscopes, film-type retardation sheets are needed. These retarders, produced by Polaroid Corporation, consist of an oriented, polyvinyl alcohol film that is laminated to a cellulose–acetate–butyrate substrate. They have fast and slow axes, which resolve light into two polarized components and offer an economical method for providing an approximate retardation. Retardation film is available with linear retardations of 140, 200, 280, and 520 nm, all within ±20 nm. Two or three of these wave plates, used in combination, can be tuned by angle shifting to provide the exact retardance desired. Within the plane of a retarder, the optic axis may be "fast" or "slow," depending on whether the uniaxial crystal behavior is positive or negative. Rotation around the optic axis increases the effective thickness of the plate but does not affect the velocity difference between the o and e rays, thus increasing the accumulated retardation. Rotation around the other axis both increases the effective thickness of the plate and reduces the velocity difference between the o and e rays. The latter effect dominates for small rotations, reducing the accumulated retardation. Thus, by correct tilting, a narrowband retarder (or combination of retarders) may be tuned over a limited range of retardations at fixed wavelengths, or over a limited range of wavelengths at fixed retardation.

The effect of a retarder on a beam of polarized light depends on three things: the initial state of polarization, the orientation of the retarder about an axis perpendicular to the disk (measured with respect to the axes of polarization of the incident beam), and the net value of the retardation. The principal plane of the retarder is the plane passing through both the symmetry axis of the disk and the crystalline optic axis within the disk. In the examples of retarder applications that follow, we assume that the monochromatic incident light is collimated and normally incident upon the disk.

The most familiar retarders are "wave plates" designed to change linearly polarized light of one wavelength into circularly polarized light or to rotate the plane of linear polarization. These plates are frequently disks for which the optic axis of the crystal lies parallel to the entrance and

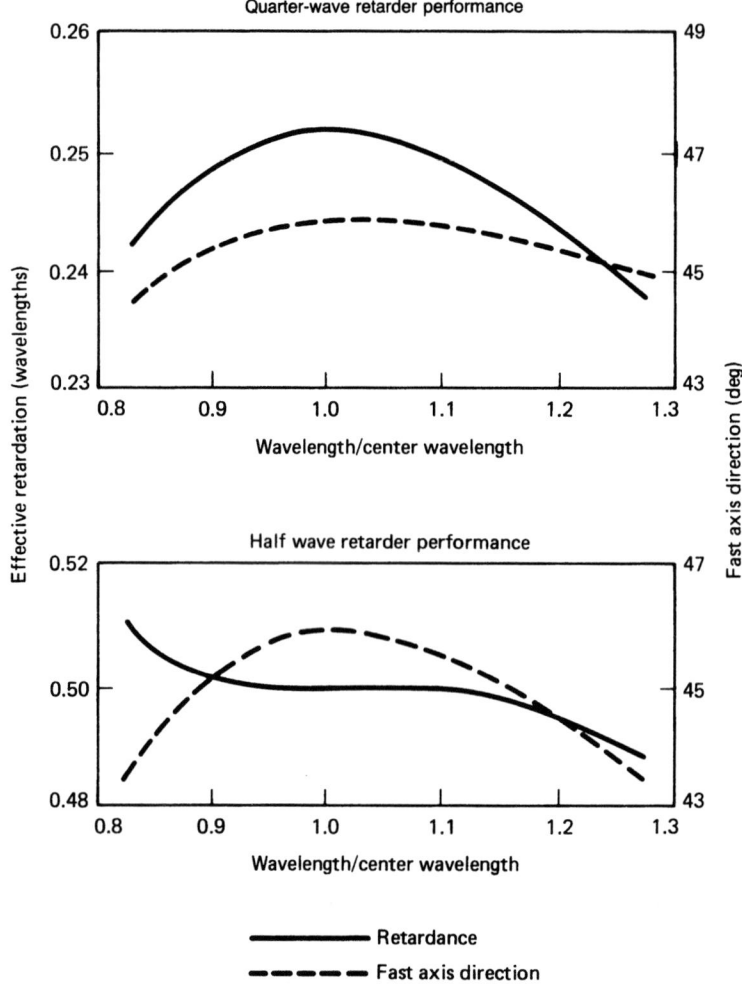

Figure 5-18 Retardance curves for typical birefringent polymer, large-aperture, achromatic retarders.

exit faces. Any ray incident normal to the plate can be thought of as two rays—one polarized parallel to the optic axis (e ray) and the other perpendicular (o ray). The relative intensity of these rays will, of course, depend on the polarization of the incident light. The plates are designed to exploit the phase differences between the two rays to produce a new polarization state or to leave the polarization unchanged for only one wavelength. Figures 5-19 and 5-20 illustrate these effects.

If the retarder is a half-wave plate (either first- or multiple-order) and the angle between the electric-field vector of a plane-polarized incident beam and the retarder's principal plane is θ (assumed to be acute), the electric-field vector in the emergent beam (also plane or linearly polarized) will make a negative acute angle $-\theta$ with the retarder's principal plane. Thus the effect

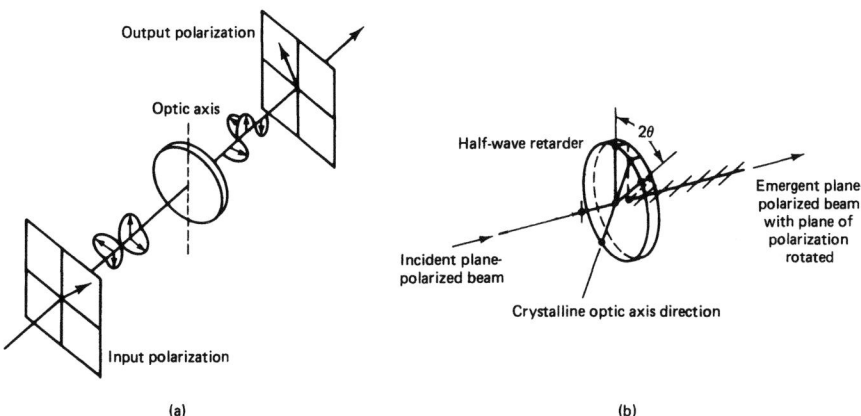

Figure 5-19 (a) Rotation of polarization by a half-wave retarder; (b) half-wave plate effect on linearly polarized laser beam.

of the half-wave plate is to rotate the plane of polarization through an angle 2θ. Since the angle θ is continuously adjustable, the plane of polarization of the emergent beam is continuously adjustable. In this way half-wave plates are used as "laser-line rotators." They allow the plane of polarization of a laser beam, as determined by its Brewster's angle windows, to be rotated without having to physically rotate the laser itself. The same result can be accomplished by placing a pair of quarter-wave plates, identically oriented, in series. A half-wave plate will convert left-circularly polarized light into right-circularly polarized light, or vice versa, by reversing the direction of propagation. In the same way it will convert left-elliptically polarized light into right-elliptically polarized light, or vice versa, effectively reflecting the major axis of the ellipse in the retarder's principal plane.

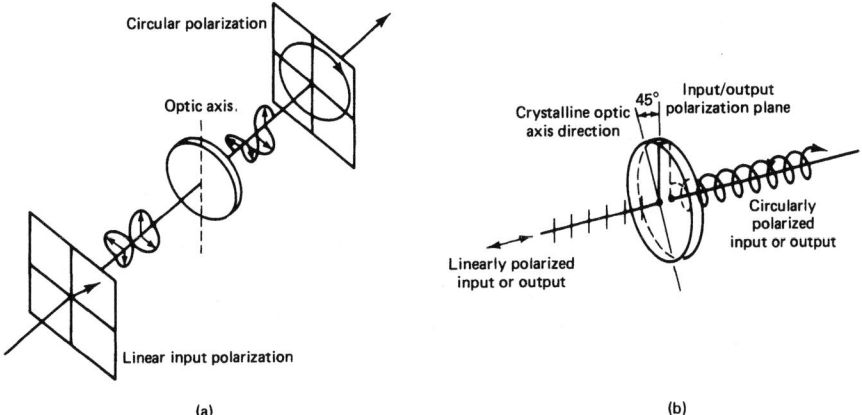

Figure 5-20 (a) Quarter-wave retarder; (b) quarter-wave retardation plate.

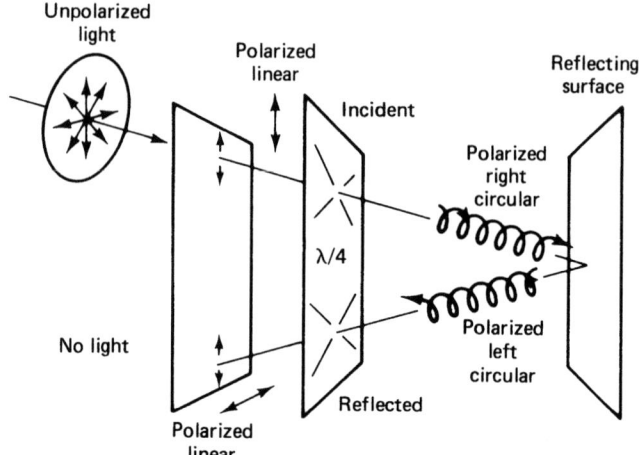

Figure 5-21 Linear polarized layer + quarter-wave retarder = circular polarizer.

If the retarder is a quarter-wave plate and the angle θ (between the electric-field vector of the incident-linearly polarized beam and the retarder principal plane) is +45°, the emergent beam is circularly polarized. Reversing θ to −45° reverses the sense of the emergent circular polarization. If the direction of propagation is reversed, a quarter-wave plate will transform circularly polarized light into linearly polarized light.

Especially important is the case in which θ is ±45° and the direction of propagation of the circularly polarized beam is reversed by reflection in a mirror or other surface at normal incidence. In this case the sense of the circular polarization is reversed at the mirror, before the second transformation into linear polarization. The result is that the ultimate linear polarization of the reflected beam, as it finally emerges from its second trip through the retarder, is orthogonal to the incident linear polarization with which we began. This is shown in Fig. 5-21. If the polarization of the incident beam had been determined by a polarizer, this polarizer would extinguish the reflected beam. The combination of a polarizer and a quarter-wave plate, with θ equal to either ±45 or −45°, traversed in the order stated, is opaque to specularly reflected radiation. Such a combination is called an isolator.

Isolators are used to prevent feedback into lasers, with the resulting troublesome mode pulling and detuning of the lasers. It is imperative that both faces of the polarizing element be multilayer, antireflection, and "V"-coated for the laser wavelength, or inclined so that the laser beam is not normally incident upon them. The side of the retarder facing the laser should be similarly treated. In photoelasticity the arrangement of Fig. 5-21 is used to produce a reflection polariscope with dark-field, circular polarization.

If θ has some value other than ±45°, a quarter-wave plate transforms linearly polarized light into (right or left) elliptically polarized light. Again the transformation is reversed by reversing the direction of propagation.

Polaroid Corporation produces circular polarizers consisting of a sheet-type linear polarizer in combination with a quarter-wave (λ/4) sheet retarder, whose slow and fast axes are at 45° to the axis of the polarizer. When a beam of unpolarized light passes through the linear portion of the sheet, the beam becomes polarized at 45° to the axis of the retarder that constitutes the second portion of the sheet. On passing through the retarder, the beam is converted from plane to elliptical

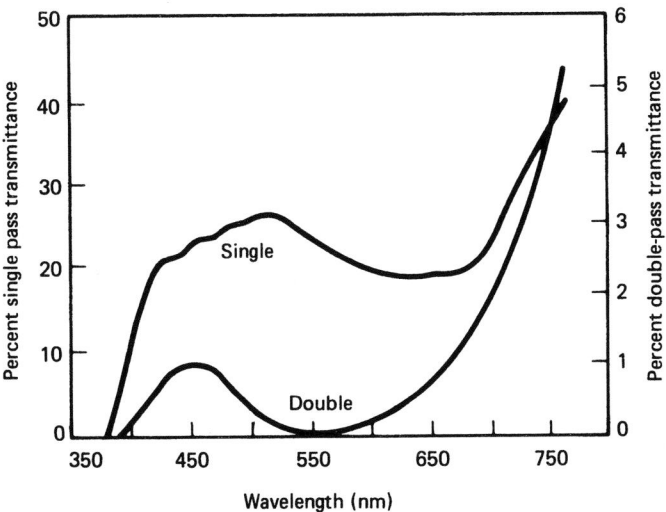

Figure 5-22 Spectral curves for polarized HNCP22 circular polarizer.

polarization. Light of the correct wavelength will be circularly polarized. Figure 5-22 shows the spectral transmittance curves for unpolarized light passing through a single sheet of a broadband, circular Polaroid sheet. The transmittance for the double pass is for the arrangement of Fig. 5-21. Such a sheet is a "correct" circular polarizer only at the minimum point for the double-pass transmittance.

Full-wave plates are useful as spectral filter and tuning elements in tunable dye lasers. A full-wave plate for wavelength λ has no effect on the polarization of a beam of wavelength λ. Adjacent wavelengths do not undergo exact full-wave retardation; thus a polarizer, full-wave plate, and analyzer (second polarizer) constitute the spectral filter.

The multipass nature of lasers and the polarizing surfaces in laser resonators make wave plates useful laser tuners. In practice, combinations of plates are tilted or rotated in the beam. The tuner design depends on laser parameters, particularly the amplitude and shape of the gain curve.

A combination of tapered retardation plates can be designed to act as a continuously adjustable wave plate. The Soleil–Babinet compensator is the most useful arrangement for this purpose. This arrangement is shown in Fig. 5-23. Consider two crystalline-quartz plates of equal thickness, each cut so that the crystalline optic axes are parallel to the surfaces of the plates. If these are stacked on top of one another so that the crystalline optic axes of the plates are perpendicular to each other, the roles of the ordinary and extraordinary rays are interchanged as light passes from one plate into the other. Because of this interchange and because the plates are of equal thickness, any phase difference or retardation that may have accumulated in the first plate is exactly canceled out in passing through the second plate. By replacing one of the fixed-thickness plates with an effective plate of variable thickness (a pair of wedges that move relative to one another), any desired net retardation can be obtained, and a Soleil–Babinet compensator results.

One wedge is optically contacted to the lower plate, while the other wedge, separated from the first by a small air space, is moved by a micrometer screw. The retardation exhibited by the emergent beam is proportional to the difference in thickness between the fixed and effective plates. Unknown retardations can be measured by adjusting the compensator to produce an exactly

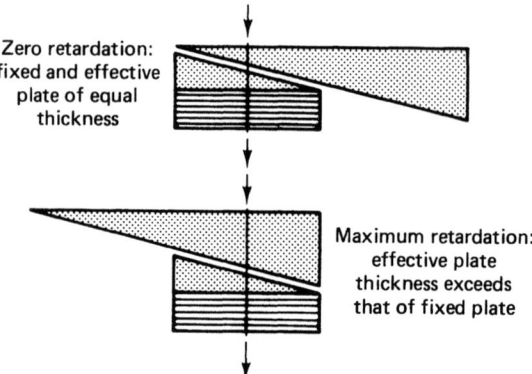

Figure 5-23 Soleil–Babinet compensator.

canceling retardation. With the compensator, one can accurately determine the planes of polarization, phase difference, and intensity ratio of the linear polarization states of which any elliptically polarized beam is composed. Compensators also permit extremely accurate measurement of strain- or stress-induced birefringence.

The Stress-Optic Law

Certain noncrystalline transparent materials, notably some polymeric plastics, are optically isotropic under normal conditions but become doubly refractive or birefringent when stressed. Like crystals, they then also have the ability to resolve an impinging light vector into orthogonal components and to transmit each with a different velocity. This effect normally persists while the loads are maintained but vanishes almost instantaneously or after some interval of time depending on the material and the conditions of loading, when the loads are removed. This phenomenon of temporary or artificial birefringence was first observed by Brewster in 1816 and is the physical characteristic on which photoelasticity is based.

In a general, three-dimensional stress system there are three principal stresses and three principal stress directions at any point in the stressed body. In a body displaying temporary birefringence, the directions of the principal stresses coincide with the temporary principal crystallographic axes, and the principal planes of stress correspond to the principal sections of the equivalent crystal represented by an element of the material at that point. There is, therefore, a direct relationship between the ellipsoid of stress and the index ellipsoid. The optical properties of the material at any point can thus be represented by an index ellipsoid in which the principal axes coincide with the principal axes of stress.

The relations between stresses and the indices of refraction for temporary birefringent materials were formulated by Maxwell in 1852 as

$$\begin{aligned} n_1 - n &= C_1\sigma_1 + C_2(\sigma_2 + \sigma_3) \\ n_2 - n &= C_1\sigma_2 + C_2(\sigma_3 + \sigma_1) \\ n_3 - n &= C_1\sigma_3 + C_2(\sigma_1 + \sigma_2) \end{aligned} \quad (5\text{-}16)$$

where n = index of refraction of the unstressed material in its optically isotropic state
n_1, n_2, n_3 = principal indices of refraction for waves vibrating parallel to the principal stresses
$\sigma_1, \sigma_2, \sigma_3$ = principal stresses
C_1, C_2 = varying constants (depending on the material), called the "stress-optic coefficients" for the specific material

Subtracting to eliminate n gives the stress-optic law for a material under a general triaxial stress system as

$$n_1 - n_2 = (C_1 - C_2)(\sigma_1 - \sigma_2) \qquad (5\text{-}17\text{a})$$
$$= C(\sigma_1 - \sigma_2)$$

where $C = C_1 - C_2$
= relative stress-optic coefficient

Similarly,

$$n_1 - n_3 = C(\sigma_1 - \sigma_3) \qquad (5\text{-}17\text{b})$$
$$n_2 - n_3 = C(\sigma_2 - \sigma_3) \qquad (5\text{-}17\text{c})$$

If the model is a plate and the stress normal to the plate (i.e., in the direction of the propagation of the light) is either zero or a constant, and the in-plane principal stresses do not vary through the thickness of the plate, the stressed plate acts as a temporary retardation plate.

Light will propagate through the plate polarized in the principal planes, and the two orthogonal rays will accumulate a stress-induced, relative linear retardation

$$\delta = hC(\sigma_1 - \sigma_2) \qquad (5\text{-}18\text{a})$$

where h is the thickness of the plate, or more specifically, the distance traversed by the light in the stressed plate.

It is generally more convenient to express the relative retardation in radians. Thus the relative angular retardation R is

$$R = \frac{2\pi h}{\lambda} C(\sigma_1 - \sigma_2) \qquad (5\text{-}18\text{b})$$

where R = angular phase shift between the two component vectors after passing through the stressed plate
 = $(2\pi/\lambda)\delta$
λ = wavelength of the monochromatic light traversing the plate

Equation (5-18b) expressed the stress-optic law, which forms the foundation of photoelasticity. The stress-optic law states that the relative angular retardation R is linearly proportional to the difference in the in-plane principal stresses $\sigma_1 - \sigma_2$ and to the thickness of the plate; the relative retardation, R is inversely related to the wavelength of the light.

The stress-optic law is more commonly written as

$$\sigma_1 - \sigma_2 = \frac{N f_\sigma}{h} \qquad (5\text{-}19)$$

where $N = R/2\pi = \delta/\lambda$ = relative retardation in terms of complete cycles of the radiation used; N is also called the "fringe order" and is wavelength dependent

$f_\sigma = \lambda/C$ = material fringe value or fringe-stress coefficient with typical units of N/m fringe or lb/in. fringe; it represents the principal stress difference necessary to produce unit change in the fringe order in a model of unit thickness

Therefore, the principal stress difference can be determined if the material fringe value of the model can be established by a calibration procedure and if N can be measured at each point.

The linear proportionality and constancy with time under load of Eq. (5-18b) are not in full agreement with the actual behavior of light in most photoelastic materials. Second-order effects, such as time-dependent changes in C and the failure of ordinary and extraordinary rays to traverse exactly the same path through the material, are usually ignored. When stresses are high and/or stress gradients are steep, these effects can cause serious errors.

Polariscopes [5-8]

A polariscope is an instrument that measures the relative retardations or phase differences that occur when polarized light passes through a stressed photoelastic model. The simplest arrangement consists of a model placed between two linear polarizers, called the polarizer and analyzer, respectively, with their axes crossed as shown in Fig. 5-24. This is the arrangement for a plane, dark-field polariscope.

The behavior of the light upon passage through the stressed model is similar to that in a retardation plate producing elliptically polarized light from a plane-polarized input. The retardive retardations at different points in a plate with a varying stress field is given by Eqs. (5-18), where σ_1 and σ_2 are the principal stresses at the point.

An alternative arrangement is the circular polariscope shown in Fig. 5-25. This arrangement differs from that in Fig. 5-24 by having two quarter-wave plates ($\lambda/4$ plates) placed between the polarizer and analyzer. The fast and slow axes of these retarders may be crossed or parallel but are at 45° with the polarizer to convert the light from plane to circularly polarized and back to plane polarized before entering the analyzer. A stressed model placed between the wave plates will introduce an additional relative retardation. This changes the beam between the two retarders from circularly polarized to elliptically polarized.

There are five useful arrangements of the polariscope elements. These are listed in Table 5-4.

In analyzing the propagation of light through the polariscope, it is convenient to assume that the electric vector $\mathbf{E}$, incident on the model, consists of two components. Thus, in a completely general approach, the light incident on the stressed model consists of two orthogonally polarized components, $\mathbf{E}_x$ and $\mathbf{E}_y$, of Fig. 5-26a

$$E_x = k_x \cos(\omega t + \Delta_x)$$
$$E_y = k_y \cos(\omega t + \Delta_y)$$
(5-20)

where E_x, E_y = component waves vibrating in the x and y directions, respectively
ω = circular frequency
t = time
Δ_x, Δ_y = phase of the light components polarized in the x and y directions
k_x, k_y = amplitudes of the orthogonally polarized components

PHOTOELASTICITY

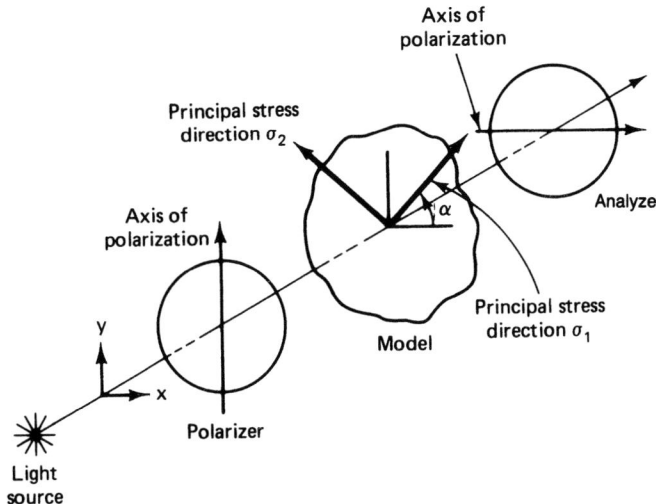

Figure 5-24 Stressed photoelastic model in a plane polariscope.

If these two component rays are in phase, the incident light is plane polarized. If they are out of phase, the incident light is elliptically polarized. In the special case in which the phase difference is a quarter-wavelength ($\pi/2$ radians), and the two components are of equal magnitude, the light is circularly polarized.

With this notation, the total incident light vector **E** is

$$\mathbf{E} = E_x \hat{x} + D_y \hat{y}$$

where $\hat{x}$ = unit vector in the x direction
$\hat{y}$ = unit vector in the y direction

To simplify our equations, we change to rectangular phasor notation.

Since the frequency ω is known and remains unchanged upon passage through the polariscope, it can be ignored, and the transformation simplifies to

$$E_x = k_x e^{j\Delta_x} \qquad (5\text{-}21)$$
$$E_y = k_y e^{j\Delta_y}$$

and its magnitude is

$$E = \sqrt{E_x^2 + E_y^2}$$

In passing through the model, the two components suffer retardations Δ_1 and Δ_2, respectively. The relative retardation $R = \Delta_1 - \Delta_2$.

At exit from the model, the condition of the light is

$$E_1' = E_1 e^{j\Delta_1} \qquad (5\text{-}22)$$
$$E_2' = E_2 e^{j\Delta_2}$$

or in terms of their x and y components,

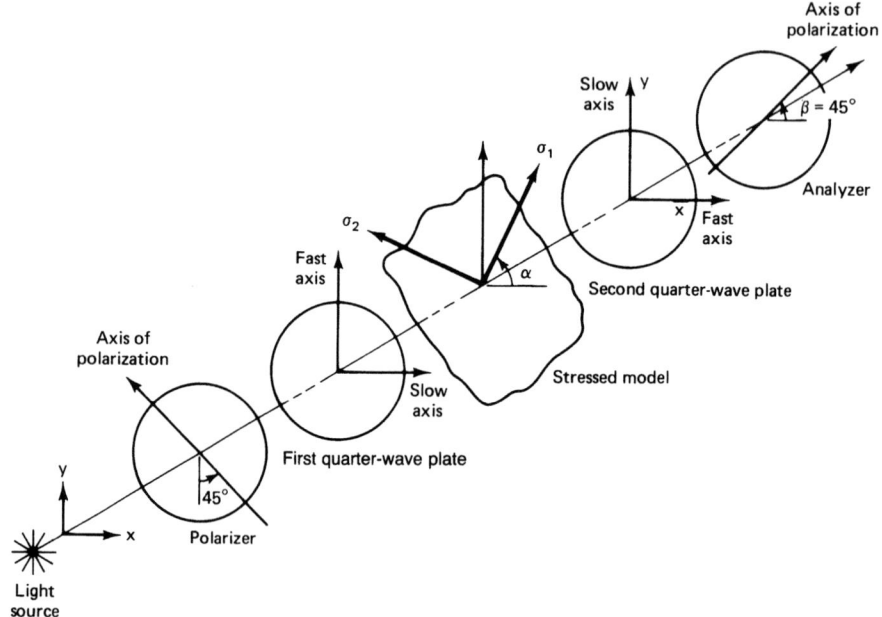

Figure 5-25 Stressed photoelastic model in a dark-field circular polariscope (crossed polarizer and analyzer and crossed quarter-wave plates $\beta = 45°$).

$$E'_x = E'_1 \cos \alpha - E'_2 \sin \alpha$$

$$E'_y = E'_2 \cos \alpha + E'_1 \sin \alpha$$

Then, substitute for E'_1 and E'_2, factor out $e^{j\Delta_2}$, and express the last equations in terms of the relative retardation, $R = \Delta_1 - \Delta_2$. Finally,

$$E'_x = \tfrac{1}{2}e^{j\Delta_2}[E_x(e^{jR} + 1) + (e^{jR} - 1)(E_x \cos 2\alpha + E_y \sin 2\alpha)]$$
$$E'_y = \tfrac{1}{2}e^{j\Delta_2}[E_y(e^{jR} + 1) - (e^{jR} - 1)(E_y \cos 2\alpha - E_x \sin 2\alpha)]$$ (5-23)

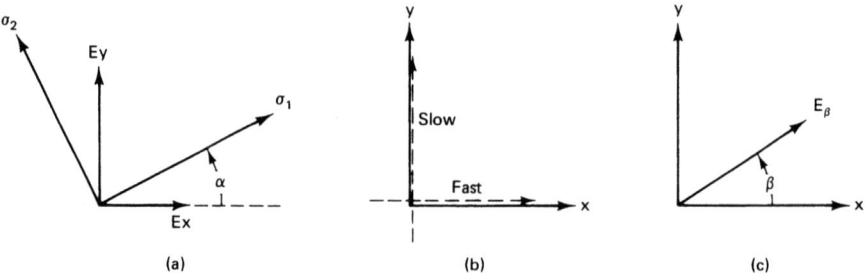

Figure 5-26 Light vectors at different stages in a general polariscope: (a) at entrance into a stressed model with principal stress direction at angle α with x and y; (b) axes of second retardation plate aligned with x and y; (c) axis of analyzer at angle β with x direction.

Table 5-4 Polariscope Setups[a]

Setup (Refer to Figs. 5-24 and 5-25)	Polarizer and Analyzer	Quarter-wave plates	Polariscope Field[b]
Plane polariscope	×	None	Dark
Circular polariscopes			
Crossed/crossed	×	×	Dark
Crossed/parallel	×	‖	Bright
Parallel/crossed	‖	×	Bright
Parallel/parallel	‖	‖	Dark

[a] ×, Crossed axis; ‖, parallel axis.
[b] When model is unstressed or when there is no model in the polariscope, the background field as viewed from beyond the analyzer is dark or bright as indicated.

These are the general expressions for the light emerging from the stressed model.

The addition of a retardation plate after the model, aligned on the x-y axes (as shown in Fig. 5-26b) and having absolute retardations for each of its axes as Q_x and Q_y, once more changes the condition of the light. If the fast axis of the retarder is aligned with the x axis, the light emerging from this retarder is

$$E_x'' = \tfrac{1}{2} e^{jQ_x} e^{j\Delta_2} [E_x(e^{jR} + 1) + (e^{jR} - 1)(E_x \cos 2\alpha + E_y \sin 2\alpha)] \quad (5\text{-}24)$$
$$E_y'' = \tfrac{1}{2} e^{jQ_y} e^{j\Delta_2} [E_y(e^{jR} + 1) - (e^{jR} - 1)(E_y \cos 2\alpha - E_x \sin 2\alpha)]$$

Finally, adding a linear polarizer (called an "analyzer") with angle β between its plane of polarization and the x direction (Fig. 5-26c) yields an emerging ray.

$$E_\beta = E_x'' \cos \beta + E_y'' \sin \beta$$

Factoring out e^{jQ_y} gives the rather long equation

$$E_\beta = \tfrac{1}{2} e^{j\Delta_2} e^{jQ_y} [e^{jQ}(E_x(e^{jR} + 1) + (e^{jR} - 1)(E_x \cos 2\alpha + E_y \sin 2\alpha)) \cos \beta \quad (5\text{-}25)$$
$$+ (E_y(e^{jR} + 1) - (e^{jR} - 1)(E_y \cos 2\alpha - E_x \sin 2\alpha)) \sin \beta]$$

where Q = relative retardation in the second quarter-wave plate
  R = relative retardation (stress-induced) in the stressed model
  α = angle between the x direction and the direction of the first principal stress in the model
  β = angle between the x direction and the plane of polarization of the analyzer

The x direction coincides with the fast axis of the analyzing quarter-wave plate.

Equation (5-25) is the general expression for the output from any polariscope. Consider now how it is simplified when certain special conditions apply:

CASE 1: LINEAR ANALYZER. For this case,

  $\beta = 0$ (analyzer axis coincides with reference direction)

Then

$$E_\beta = \tfrac{1}{2} e^{j\Delta_2} e^{jQ_x} [E_x(e^{jR} + 1) + (e^{jR} - 1)(E_x \cos 2\alpha + E_y \sin 2\alpha)] \quad (5\text{-}26)$$

There are two special forms of illumination onto the stressed model, linear and circular. The expressions and significance of these are as follows:

Case 1a: Linear Analyzer with Linear Illumination. The light incident upon the stressed model is linearly polarized. Then, if

θ = angle between the planes of polarization of the incident light (polarizer axis) and the analyzing polarizer (reference axis)

$E_x = E \cos \theta$ and $E_y = E \sin \theta$

Since the intensity of a ray of light is equal to the product of the light vector and its conjugate, the intensity of the light emerging from the polariscope at any point in the field is

$$I = E_\beta E_\beta^* \qquad (5\text{-}27)$$

Substitute from Eq. (5-26) and simplify to yield

$$I = E_x^2 \cos^2 \frac{R}{2} + (E_x \cos 2\alpha + E_y \sin 2\alpha)^2 \sin^2 \frac{R}{2} \qquad (5\text{-}28)$$

Note that when, additionally, $Q = 0$ (i.e., when there is no added differential retardation after the model), Eq. (5-28) is unchanged. It is then the general equation for the emerging light when a stressed model is placed between two linear polarizers with their axes at an angle, θ, with respect to each other: that is, as if there are no quarter-wave plates in the polariscope. It shows that when the two crossed quarter-wave plates in a general polariscope are rotated so that the axes align with the analyzer axis, they are "optically" removed from the polariscope. The output from the polariscope is as if the wave plates were absent.

The linear plariscope can be specially arranged so that the polarizer and analyzer have their axes crossed. This yields the basic plane, or linear, dark-field polariscope. For crossed polarizers, $\theta = 90°$, so that

$E_x = 0 \qquad$ and $\qquad E_y = E$

From Eq. (5-28),

$$I = E^2 \sin^2 2\alpha \sin^2 \frac{R}{2} \qquad (5\text{-}29)$$

When a stressed model is viewed through such a polariscope, it appears as in Fig. 5-27. The background of the model is dark. Also, the model when unstressed is dark, because there will be no stress-induced relative retardation R.

According to Eq. (5-29), the intensity of the emerging light from a plane, dark-field polariscope is a function of the principal stress directions α and the stress-induced relative retardation R in the model.

There are two conditions for $I = 0$, expressed by Eqs. (5-30) and (5-31).

$$\alpha = 0, \frac{\pi}{2}, \pi, \frac{3\pi}{2}, \ldots \text{radians} \qquad (5\text{-}30)$$

With all four of these values for α, the polarizing axes of the polariscope coincide with the two in-plane, principal stress directions in the model. The loci of points of equal stress direction will therefore be dark bands called isoclinics. The angle of the stress directions along such loci is called the isoclinic angle.

$$\frac{R}{2} = 0, \pi, \ldots \text{radians} \qquad \text{or} \qquad R = 0, 2\pi, \ldots, N(2\pi) \qquad (5\text{-}31)$$

PHOTOELASTICITY

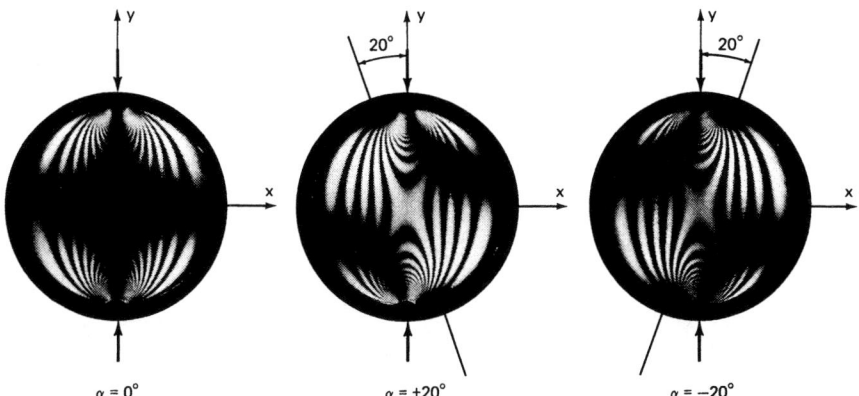

Figure 5-27 Dark-field images of a stress-frozen disk under diametral compression: α is the angle between the polariscope axis and the direction of loading.

where N is an integer, with sign unknown, from 0 to ∞. N equals the number of full wavelengths or full cycles of stress-induced relative retardation in the model. Loci of equal values of N are also black lines and are known as the isochromatic fringes or just "fringes."

Equation (5-19) has shown that

$$N = \frac{h(\sigma_1 - \sigma_2)}{f_\sigma} \tag{5-19}$$

so isochromatics are lines of equal principal stress difference. Also, since the maximum in-plane shear stress is

$$\tau^1\text{max} = \tfrac{1}{2}(\sigma_1 - \sigma_2) \tag{5-32}$$
$$= \frac{1}{2}\frac{Nf_\sigma}{h}$$

Isochromatics are also lines of equal in-plane maximum shear stress.

On a free boundary, the principal stress normal to the boundary, σ_1 or σ_2, is zero. Then, from Eq. (5-19), the fringe order N at the boundary is proportional to the tangential stress at the boundary.

Since in Eq. (5-29) the relative retardation term is squared, the isochromatics at the boundary carry no information about the sign (tension or compression) of the boundary stresses. Therefore,

$$N_\text{boundary} \propto |\sigma_\text{boundary}| \tag{5-33}$$

The two sets of fringes, isoclinics and isochromatics, occur simultaneously when viewed in a linear polariscope. They interfere where they coincide in the plane of the model. This is illustrated in the three photographs of Fig. 5-27, where a disk in diametral compression is shown for three positions α of the polariscope axes. In the first photo, $\alpha = 0$, the axes coincide with the direction of loading on the disk. In the next two photographs the polariscope axes are rotated 20° counterclockwise and clockwise, respectively. Since the tangential stress at the boundary is a principal stress, the isoclinic should intersect the boundary at points where the slope of the boundary is equal to α.

Notice that there are two sets of black lines. One set, the isoclinics, moves when the polariscope is rotated with respect to the model. The second set, the isochromatics, remains unchanged.

The isoclinics are eliminated when the circular polariscope setup shown in Fig. 5-25 is used. This is discussed under Case 2.

Case 1b: Linear Analyzer with Circular Illumination. The light incident on the model is circularly polarized. This case is known as a half-circular polariscope. Now

$$E_y = jE_x$$

From Eqs. (5-26) and (5-27),

$$I = E_x^2(1 + \sin R \sin 2\alpha) \tag{5-34}$$

This equation is sometimes used as an intermediate stage in automated photoelasticity.

CASE 2: CIRCULAR ANALYZER. There is a quarter-wave plate between the model and the analyzing Polaroid. For this condition to be met, $\beta = 45°$, so that from Eq. (5-25)

$$E_\beta = \frac{\sqrt{2}}{4} e^{j\Delta_2} e^{jQ_y} [(e^{jR} + 1)(E_y + jE_x)$$
$$+ (e^{jR} - 1)(jE_x - E_y)(\cos 2\alpha - j \sin 2\alpha)] \tag{5-35}$$

When this condition is used in conjunction with circular illumination, we get a circular polariscope. Table 5-4 shows that there are two useful arrangements: polarizers crossed and polarizers aligned.

CASE 2a: POLARIZER AND ANALYZER AXES CROSSED. This is a dark-field circular polariscope with

$$E_y = -jE_x$$

From Eq. (5-35), we break down

$$E_\beta = \frac{\sqrt{2}}{4} e^{j\Delta_2} e^{jQ_y} (e^{jR} - 1)(-2E_y)(\cos 2\alpha - j \sin 2\alpha)$$

into

$$I = E_\beta E_\beta^* \tag{5-27}$$

$$I = E_y^2(1 - \cos R)$$

$$= 2E_y^2 \sin^2 \frac{R}{2}$$

$$= K \sin^2 \frac{R}{2} \tag{5-36}$$

where K is a constant for any particular polariscope. This equation is similar to Eq. (5-29), except that the term containing α, the isoclinic parameter, has disappeared. Only one condition for extinction remains:

$$I = 0 \text{ for } \frac{R}{2} = 0, \pi, \ldots \text{ radians} \quad \text{or} \quad R = 0, 2\pi, \ldots, N(2\pi)$$

where N = integer, sign undefined, from 0 to ∞
      = isochromatic fringe order (full-order fringes)

PHOTOELASTICITY

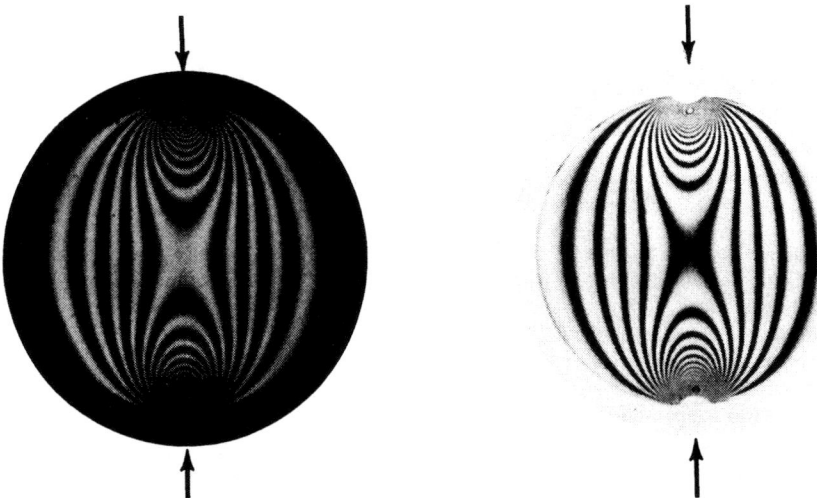

Figure 5-28 Dark-field and light-field isochromatic images of a disk in diametral compression.

Since extinction is independent of the orientation α of the model in the polariscope, the isochromatic fringe pattern (Fig. 5-28, left) remains unchanged when the model is rotated in the polariscope. This is a good test for correct alignment of a circular polariscope.

Points in the stressed model at which the stress-induced relative retardation is an integer number of full cycles ($N = 1, 2, \ldots$) will be dark, and Eq. (5-19) or (5-32) can be used to find the magnitude of the in-plane maximum shear stresses.

The intensity I will be zero when there is no model in the field and at points where $R = 0$. The background against which the model is viewed is therefore dark, and an unstressed model will also be dark. Under these conditions it is sometimes difficult to identify the boundaries of the model.

CASE 2b: POLARIZER AND ANALYZER AXES ALIGNED   This is a bright-field circular polariscope with

$$E_y = jE_x$$

From Eqs. (5-35) and (5-27),

$$\begin{aligned} I &= E_y^2(1 + \cos R) \\ &= 2E_y^2 \cos^2 \frac{R}{2} \\ &= K \cos^2 \frac{R}{2} \end{aligned} \quad (5\text{-}37)$$

where K is a constant. The conditions for extinction of the light are now

$$I = 0 \quad \text{for} \quad \frac{R}{2} = \frac{\pi}{2}, \frac{3\pi}{2}, \ldots, \text{radians}$$

$$\text{or} \quad R = \pi, 3\pi, 5\pi, \ldots$$

$$= \tfrac{1}{2}(2\pi), 1\tfrac{1}{2}(2\pi), \ldots, (N + \tfrac{1}{2})2\pi$$

where N = integer, sign undefined, from 0 to ∞
= half-order isochromatic fringe count

Rotating the analyzer through 90° from its position in the dark-field circular polariscope (Case 2a) shifted the isochromatics by 0.5 fringe order to coincide with the lines of maximum intensity in the dark-field setup.

When there is no model in the field or when $R = 0$, then $I = I_{max} = 2E_y^2$. The background against when the model is viewed is bright and at maximum intensity. Since there are always some reflections or shadows along the edges of the model, the boundaries will be clearly outlined.

When the two setups—bright-field and dark-field circular—are used sequentially, the effective resolution of the method is $N/2$.

Polariscope Errors

There are three common sources of error in a correctly set up polariscope. These are:

1. Quarter-wave plates that are not matched to the frequency of the light source
2. Averaging through the thickness of a stressed model in a diffuse-field polariscope
3. Nonuniform light intensity across the field of a polariscope caused by variations in transmittance of individual elements

Equations (5-18) and (5-19) show that the relative retardation, that is, the fringe order, is inversely proportional to the wavelength of the radiation used to illuminate the stressed model in a polariscope. This means that if the quarter-wave plates are not correctly matched to the light source, they will not produce a circularly polarized field. If the mismatch is such that the relative retardation through the wave plates is $R' = (\pi/2 + \varepsilon)$ instead of $\pi/2$, the intensity of the light emerging from a dark-field circular polariscope (crossed polarizers) is

$$I' = I_0(1 - \sin^2 2\alpha \sin^2\varepsilon) \sin^2 \frac{R}{2} \tag{5-38}$$

with the same notations as in Eqs. (5-25) and (5-29). Extinction still occurs when $\sin^2 R/2 = 0$, but the overall intensity of the isochromatic field will be reduced because of the second term in the parentheses. This reduction is zero only when $\sin^2 2\alpha = 0$, that is, when the principal stress directions coincide with the axes of the quarter-wave plates ($\alpha = 0$). This occurs only along the 45° isoclinic. For all other isoclinic angles, the intensity is reduced, with the largest effect occurring at $\alpha = 45°$ (i.e., when the principal stresses coincide with the polarizer and analyzer axes).

The error due to mismatch between the wavelength of the light source and the quarter-wave plates can be corrected by combining two identical wave plates with their optic axes at appropriate angles relative to each other and to the plane of polarization of the incident plane-polarized light. If two such sets of adjustable quarter-wave plates or "achromatic retarders" are used in a polariscope, they can be adjusted so that a single polariscope can be used with different sources of monochromatic light [1,6,9,10].

PHOTOELASTICITY

Table 5-5 Setup Angles for Composite Quarter-Wave Plates Produced from Two Sheets of Polaroid Retarders with Relative Retardation of 140 nm ($\lambda_0 = 560$ nm)

Source	Source Wavelength, λ (nm)	Single Plate R for Source Wavelength (deg)	ϕ (deg)	i (deg)	Composite R (deg)
Sodium: yellow	589.3	85.5	44.8	2.2	90
Mercury					
Yellow	577.0	87.3	44.9	1.3	90
Green	546.1	92.3	45.0	−1.1	90
Blue	435.0	115.9	38.0	−14.5	90

If

R = relative angular retardation desired

    = $2\pi\lambda_0/\lambda$ radians or $360\lambda_0/\lambda$ degrees

λ_0 = wavelength for which a single plate is correct

λ = wavelength of source to which plates must be matched

ψ = angle between fast axes of two plates in a composite assembly

i = angle between plane of polarization (plane of vibration of entering plane-polarized light) and the fast axis of the first plate in an assembly

then

$$\cos 2\psi = \cot^2 R$$
$$\sin 2i = \cos R$$

for
$$\begin{cases} \dfrac{\pi}{4} \leq R \leq \dfrac{3}{4}\pi \\ 45° \leq R \leq 135° \end{cases}$$

Table 5-5 gives the values of these variables when correct quarter-wave plates are produced from two plates of Polaroid retarders with $\lambda_0 = 140$ nm.

Polariscope Construction

A basic polariscope consists of a light source with rotational polarizing and retardation filters arranged in sequence as in Fig. 5-25. The last element, not shown in this figure, is usually a photographic or video camera for recording the isoclinic and isochromatic fringe data. The most convenient and cheapest arrangement has a light source that consists of several lamps mounted behind a diffusing screen in such a way that the output from the diffuser is a relatively uniform light field. Figure 5-29a shows a typical arrangement for such a "diffuse-light polariscope."

The isochromatic stress-optic law is strictly true only for a single ray of monochromatic light passing through a model in an instantaneous sense and with normal incidence. With a diffuse light source, the birefringence recorded at a point on the focal plane of the camera does not come from a single ray through the corresponding point in the stressed model. It is rather an integrated value for all the rays that lie within a cone of material subtended by the aperture of the camera lens as shown in Fig. 5-29b. The volume of material in this case depends on four factors:

1. The distance from the camera lens to the model

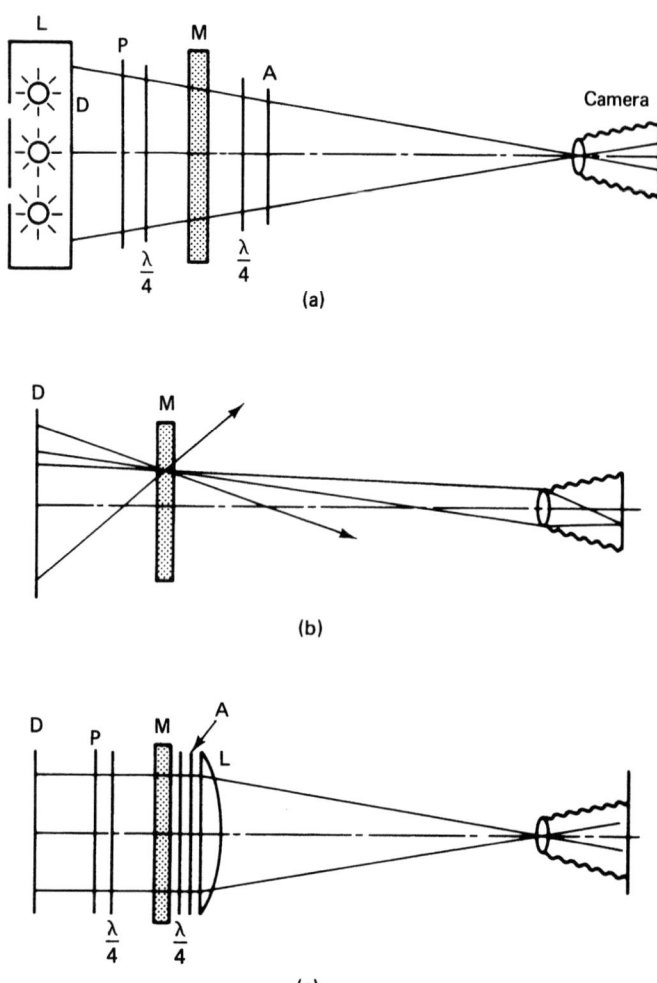

Figure 5-29 Diffuse-light polariscope arrangements: (a) standard diffuse-light setup (L, light source; D, diffuser plate; P, polarizer; A, analyzer; λ/4, quarter-wave plates; M, stressed model); (b) errors from interrogation of a cone of material in the model rather than at a point; (c) diffuse-light setup with collecting, or field, lens reduces the error that appears in (b). (L, collecting lens).

2. The aperture of the camera lens
3. The thickness of the model
4. The focus of the camera, that is, how well it is focused onto the model and the plane within the model onto which it is focused

These effects are illustrated in Fig. 5-30.

PHOTOELASTICITY

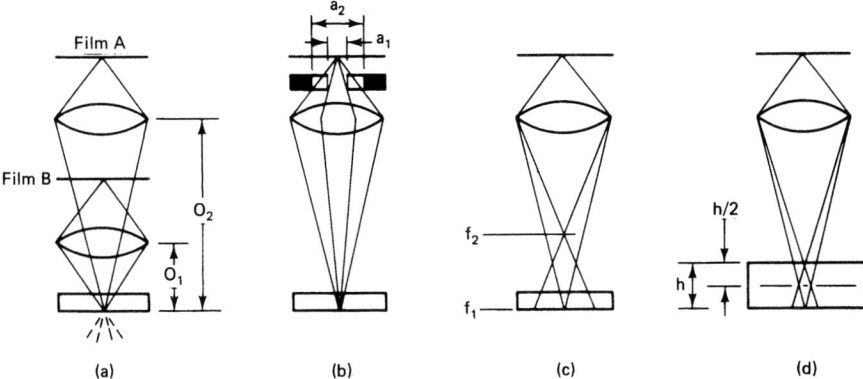

Figure 5-30 Sources of error due to averaging over a cone of material in the model: (a) effect of distance; (b) effect of aperture; (c) effect of focus; (d) effect of thickness and plane of focus. Each case illustrates a large error and a means for reducing the magnitude of that error.

Averaging can be reduced by:

1. Increasing the distance between the camera lens and the model. For the same magnification this will require a lens with a longer focal length.
2. Reducing the aperture of the camera lens. This also increases the depth of field at the model and makes focusing easier.
3. Decreasing the thickness of the model.
4. Focusing onto the midplane of the model rather than onto the front or back surfaces.

Averaging from a diffuse-light polariscope can be reduced by adding a large collecting lens to the assembly as shown in Fig. 5-29c. This lens must be large enough to cover the desired area of view on the model and should be as close to the model as possible so that the image is formed essentially by rays that are parallel in the model.

The most convenient light sources are high-pressure arc lamps. These are compact sources, which need to be expanded so that the field of illumination on the model will be of the required size. To do this, condensing and collimating lenses are used to produce a parallel light beam of sufficient diameter incident onto the polarizer. We then have a simple collimated "parallel beam" or "lens polariscope," as shown in Fig. 5-31.

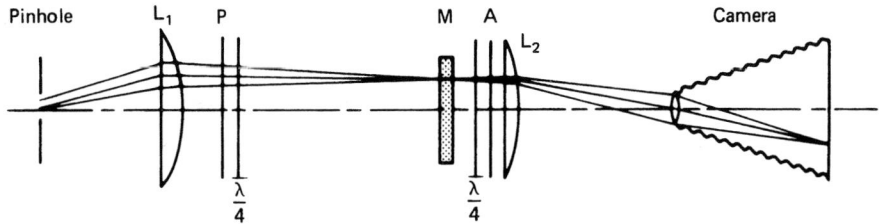

Figure 5-31 Standard collimated parallel-beam polariscope setup.

Table 5-6 Spectral and Laser Line Wavelengths for Light Sources in the Visible Range

Laser Type	Wavelength (nm)	Analytic Line	Wavelength (nm)
Argon	488	Hg Blue	453.8
	514.5	Green	546.1
Nd-YAG	½(1064)	Yellow	577.0
	= 532	Na Yellow	589.3
He-Ne	632.8		
Kr	647.1		
Ruby	694.3		

Light Sources

Photoelasticity requires incoherent light sources. The speckle associated with coherent sources, such as lasers, is generally undesirable. When lasers are used it is often necessary to spoil the coherence of the light that they produce. However, laser light is highly monochromatic and is usually polarized. Lasers are therefore attractive light sources that will find increased use in studies of photoelasticity. When used in conjunction with a spatial filter and beam expander, lasers provide high-quality illumination for polariscopes.

Spectral lamps or arc lamps are the most widely used light source. Sodium electric discharge lamps emit a strong yellow light. Mercury vapor sources produce a spectrum that has several bright spectral lines that can be isolated with narrow band filters. Table 5-6 lists the characteristic frequencies for the most popular light sources.

Equations (5-18) and (5-19) show that the phase shift, or relative retardation R, is inversely proportional to the wavelength of the radiation used to illuminate the model. The fringe values and material fringe coefficient are, therefore, wavelength dependent, and from Eq. (5-19)

$$N = \frac{\delta}{\lambda}$$

and

$$f_c = \frac{\lambda}{C}$$

For two different light sources with wavelengths λ_1 and λ_2, the recorded fringe values and the effective fringe stress coefficients will be related by

$$N_2 = \frac{\lambda_2}{\lambda_1} N_1$$
$$f_{\sigma 2} = \frac{\lambda_1}{\lambda_2} f_{\sigma 1}$$
(5-39)

Multiplication of Isochromatic Fringes [11]

In regions of low fringe order or low fringe gradient, the ordinary isochromatic pattern may be inadequate to allow an accurate evaluation of the stresses. This is not only because of the sparsity of the fringes, but also because the fringes in such regions consist of broad bands within which the lines of maximum darkness or brightness are difficult to locate. A procedure known as fringe

PHOTOELASTICITY

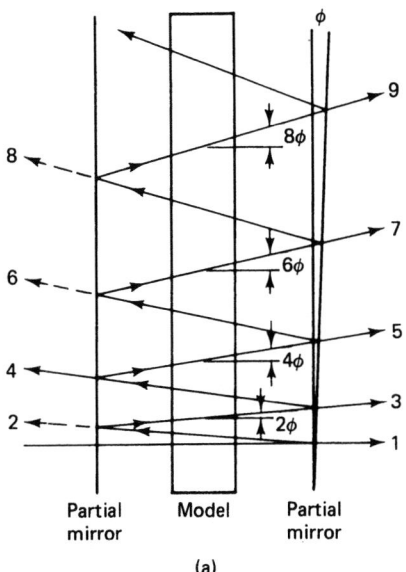

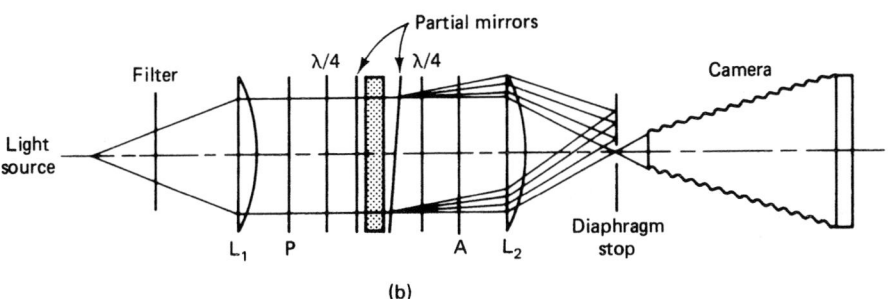

Figure 5-32 Fringe multiplication: (a) mechanism of fringe multiplication; (b) polariscope setup for fringe multiplication.

multiplication increases the fringe order at every point and produces an isochromatic pattern containing several times as many fringes as the ordinary pattern.

Fringe multiplication is achieved by causing the light to pass through the model several times through multiple reflections between two partial mirrors inserted, one on either side of the model, in a parallel field polariscope. One of the mirrors is inclined slightly, as shown in Fig. 5-32a. In practice, the inclination is small (about 0.005 rad). The effect of the inclined partial mirror on the light passing back and forth through the model is that rays that have traveled through the model different numbers of times emerge in different directions. All rays that have passed through the model the same number of times emerge in the same direction to form parallel beams of light. Figure 5-32b shows how the forward-transmitted beams are converged by the second lens to form a row of discrete images of the source in its focal plane, where they can be separated with a diaphragm stop. Rays 1, 3, 5, and 7, which have traversed the model the same number of times as their ray number, emerge at angles 0, 2ϕ, 4ϕ, and 6ϕ. Although the rays do not pass through the same point, the inclination angle ϕ in Fig. 5-32a is greatly exaggerated. The length of the line over which the photoelastic effect is averaged depends on the angle of inclination ϕ, the ray number, and the separation distance between the mirrors. In practice, multiplication by factors of 5 to 7 can be achieved without introducing objectionable errors due to the averaging process that is inherent in this method.

The intensity relationship for the mth ray, where $m = 1, 2, 3, \ldots$, can be established by modifying Eq. (5-36) to account for the loss in intensity and the added thickness effects as the light traverses the model m times. Thus, from Eq. (5-36),

$$I = K \sin^2 \frac{\Delta}{2}$$

where $\Delta = R/m$ = relative retardation for a single pass through the model. The intensity of each ray can then be written as

$$I_1 = K(1 - r)^2 \sin^2 \frac{\Delta}{2} = KT^2 \sin^2 \frac{\Delta}{2}$$

where r and T are the reflection and transmission coefficients of the two identical partial mirrors. Ray 3 has undergone two reflections and two transmissions, hence the T and r terms squared. Also, the light has passed through the model three times, and the argument of the sine function is multiplied by 3 to account for this fact. Then

$$I_3 = KT^2 r^2 \sin^2 \frac{3\Delta}{2}$$

$$I_m = KT^2 r^{m-1} \sin^2 \frac{m\Delta}{2}$$

(5-40)

Thus fringe multiplication, by Post's partial-mirror method, is accompanied by a considerable loss of light intensity. The intensity of the multiplied fringe pattern compared with the ordinary fringe pattern is decreased by the term $T^2 r^{m-1}$, which is always much less than 1. The loss in intensity for a particular ray can be minimized by properly selecting the mirror coefficient r and T. Assuming that the mirrors are perfect,

$$T + r = 1$$

The reflection coefficient for the mirrors, which minimize the intensity lost, is

$$r = \frac{m - 1}{m + 1}$$

Since r is a function of m (the number of light traverses through the model), it is not possible to optimize the mirrors for all rays simultaneously. The optimum coefficients of reflection and transmission for each value of m are presented in Table 5-7.

First Invariant of a Polariscope

Take the elements of a circular polariscope in any general orientation with respect to each other and the expression for the light emerging from the analyzer will be as in Eq. (5-25).

Table 5-7 Optimum Mirror Properties for Fringe Multiplication

Multiplication Factor, m	r	T	Intensity Coefficient, $T^2 r^{m-1}$
1	0	1	1
3	0.500	0.500	0.0625
5	0.667	0.333	0.0219
7	0.750	0.250	0.0111
9	0.800	0.200	0.0067

Rotating the analyzer through $-90°$ changes Eq. (5-25) to

$$E'_\beta = \tfrac{1}{2}e^{j\Delta_2}e^{jQ_y}[e^{jQ}(E_x(e^{jR} + 1) + (e^{jR} - 1)(E_x \cos 2\alpha + E_y \sin 2\alpha)) \sin \beta \\ - (E_y(e^{jR} + 1) - (e^{jR} - 1)(E_y \cos 2\alpha - E_x \sin 2\alpha)) \cos \beta]$$

Define

$$A = \tfrac{1}{2}e^{jQ}[E_x(e^{jR} + 1) + (e^{jR} - 1)(E_x \cos 2\alpha + E_y \sin 2\alpha)]$$

$$B = \tfrac{1}{2}[E_y(e^{jR} + 1) - (e^{jR} - 1)(E_y \cos 2\alpha - E_x \sin 2\alpha)]$$

so that

$$E_\beta = e^{j\Delta_2}e^{jQ_y} (A \cos \beta + B \sin \beta) \quad (5\text{-}25)$$

$$E'_\beta = e^{j\Delta_2}e^{jQ_y} (A \sin \beta - B \cos \beta) \quad (5\text{-}41)$$

The sum of the two intensities for analyzer at reference angle and reference angle $-90°$ is

$$I_0 = I + I'$$
$$= E_\beta E_\beta^* + E'_\beta E_\beta'^*$$
$$= AA^* + BB^*$$

So I_0 is independent of β, the angle between the fast axis of the analyzing quarter-wave plate and the analyzing polarizer ("analyzer").

Substitution yields

$$I_0 = E_x E_x^* + E_y E_y^*$$

For a plane polariscope (linear polarization)

$$E_x = E \cos \theta \quad \text{and} \quad E_y = E \sin \theta$$

and

$$I_{O_p} = E^2(\cos^2 \theta + \sin^2 \theta)$$
$$= E^2$$

which is independent of θ, the original angle between the polarizer and analyzer axes.

Similarly, for a circular setup

$$E_y = jE_x \quad \text{and} \quad I_{O_c} = 2E_x^2$$

which is also independent of θ.

For any polariscope setup, I_0 is equal to the maximum intensity (brightest point) attainable in the photoelastic field for that polariscope. I_0 is also the sum of the light intensities for any two orthogonal settings of the analyzer. It is a constant known as the first invariant of a polariscope. The first polariscope invariant,

$$I_0 = I_{\beta°} + I_{(\beta+90)°} = \text{constant} \tag{5-42}$$

is useful for calibrating automatic polariscopes.

Tardy Compensation [5,7]

In many cases of stress exploration in a plate, isochromatic fringes will be distributed so closely over the whole plate that the stress difference at any point can be determined with sufficient accuracy by a graphical interpolation. Where this is not the case, it becomes necessary to measure relative retardations involving fractions of a wavelength. This is most often done with a method known as Tardy compensation. The method is a point-by-point one that is best understood by considering the effects of a stressed model in a circular polariscope with the local stress direction aligned to the polarizer. The analyzer position is arbitrary. The easiest procedure is to start with a correctly set plane polariscope. Align the polariscope so that the isoclinic falls directly across the desired point. Then insert a set of quarter-wave plates at 45° to the polarizer. This arrangement results in the following conditions:

1. Principal stress direction is 45° from fast axis: that is, $\alpha = \pi/4$
2. Circular illumination: that is, $E_y = -jE_x$
3. Quarter-wave analyzer: that is, $E^{jQ} = j$

Then Eq. (5-25) becomes

$$E_\beta = \tfrac{1}{2} e^{j\Delta} 2 e^{jQ} y E_x [(e^{jR} + je^{jR} - 1 + j1) \cos \beta \\ + (e^{jR} - je^{jR} - 1 - j1) \sin \beta]$$

and

$$I = E_\beta E_\beta^*$$

becomes

$$I = E_x^2 [1 + \sin(R - 2\beta)] \tag{5-43}$$

At the dark-field circular polariscope setting, $\beta_0 = 45°$. If a fringe does not pass through the desired point for this position of the analyzer, the partial fringe value can be found by rotating the analyzer an additional angle $\gamma = (\beta - 45°)$ until one of the adjacent fringes has moved to the desired point. Extinction then exists and

$$R - 2\beta = (2n - \tfrac{1}{2})\pi$$

or

$$R - 2\gamma = 2n\pi$$

where n is an integer.

In terms of isochromatic fringe value, the relative retardation $R = N(2\pi)$ and extinction occurs when

$$N = n \pm \frac{\gamma}{\pi} \quad (\gamma \text{ in radians})$$
$$n \pm \frac{\gamma}{180°} \quad (\gamma \text{ in degrees}) \tag{5-44}$$

PHOTOELASTICITY

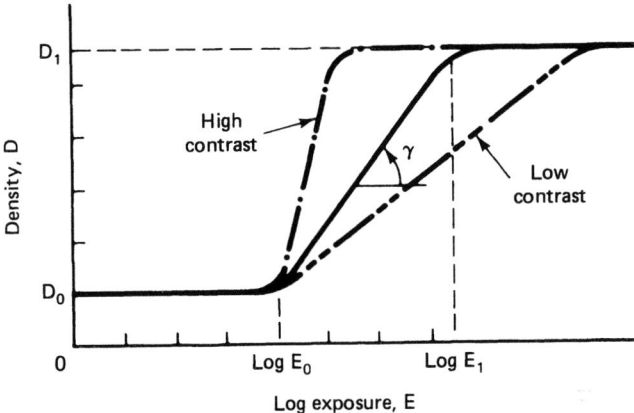

Figure 5-33 Response curves for photographic film with different levels of contrast.

The γ term is additive when the next lower fringe is moved to pass through the point to be measured. It is subtractive when the next higher fringe is brought into coincidence with the point.

Photograph [12]

When exposed to light and developed, a photographic negative has a density that is a function of the exposure (light intensity × time) presented to the film. The density is a measure of the opaqueness of the negative. The relationship between density and exposure is shown in Fig. 5-33. The log-linear portion of the curve has the equation

$$D = D_0 + \gamma(\log E - \log E_0) \qquad \text{for } E_0 \leq E \leq E_1 \qquad (5\text{-}45)$$

where D_0 = minimum density or "fog"
D_1 = maximum density
$D = \log(I_i/I_e)$
I_i = intensity of light incident upon the developed negative
I_e = intensity of light emerging from the developed negative
$E = It$
I = intensity of light incident upon film at time of exposure
t = time of exposure
γ = slope of the log-linear portion of the density curve

High-contrast films have a steep slope γ, as shown by the dashed-dotted line in Fig. 5-33. A small change in exposure beyond E_0 changes the negative to the saturation density D_1. The negative records images as "black" (D_1) or "white" (D_0). When printed onto photographic paper, the positive image will be white negative areas D_1 and black for negative areas D_0. These are generally desirable characteristics for recording photoelastic fringe patterns.

A low-contrast negative has a continuous range of gray tones spread out over a longer exposure bracket. These films generally have a finer grain structure, which gives them better spatial resolution. They are preferred in video image analysis systems, where low fringe gradients are resolved by recording the intermediate density levels for low-level, partial fringe analysis.

The initial value E_0 produces the fringe width on a negative where, in theory, the fringe should be a line. This can be overcome by preexposing the negative to E_0 before loading it into the polariscope camera. Any small amount of additional exposure will then cause the density D to increase along the slope γ. Such preexposure is useful in recording dynamic events, where the exposure time is short and the source intensity cannot be adjusted. All the energy in the short-duration light pulse is then used to increase the density in the resulting photograph.

Model-to-Prototype Relations [5]

The equations of equilibrium and compatibility show us that the distribution of stress in the elastic state is independent of the magnitude of the loads and the scale of the model. In two-dimensional problems, it is also independent of the elastic constants if the body forces are zero or uniform. For the investigation of such problems, the material and the scale of the model and the magnitudes of the loads may therefore be chosen as convenient.

If the model and prototype are geometrically similar and the distribution of the loading is the same for both, the model results can be related to the prototype, with certain exceptions, following the laws of similarity. In a plate that is in a state of plane stress, the stress distribution is constant through the thickness; the scale of the thickness need not, therefore, be the same as that of the dimensions in the plane of the plate.

In certain two-dimensional problems, the distribution of stress depends on the value of Poisson's ratio for the material. This occurs when the body forces vary over the field of the plate. An example of this type of problem is the distribution of centrifugal stresses in a rotating disk. The stress distribution also depends on Poisson's ratio in multiply connected bodies (i.e., bodies with holes) in thermal stress problems and if a resultant force acts at the boundary of a hole. For such cases, the distribution of stress in the model will differ to some extent from that in the prototype unless Poisson's ratio is the same for both materials. In three-dimensional problems, the distribution of stress depends on the value of Poisson's ratio.

The laws of similarity may also not apply near the loading points, where the stresses may not be truly two-dimensional and where the area of contact may alter considerably under load. It follows from St.-Venant's principle, however, that the effect on the stress distribution will be negligible except in the immediate vicinity of such points.

The difference in the values of Poisson's ratio for the usual hard model plastics and the prototype material in room-temperature investigations is often small, and its effect can usually be neglected. Even when Poisson's ratio for the model material is approximately 0.5, errors arising from this effect are likely to be comparatively small.

In problems of plane stress and in three-dimensional problems, the stresses in the prototype are related to those in the model through

$$\sigma_p = \sigma_m \frac{F_p}{F_m} \left(\frac{L_m}{L_p}\right)^2 \tag{5-46}$$

where it is assumed that $\nu_m = \nu_p$ and

ν = Poisson's ratio
F = force
L = characteristic dimensions
σ = stress
m,p = subscripts referring to model and prototype, respectively

For exact similarity, the model and prototype should be geometrically similar when deformed by their respective loads. If they are geometrically similar before loading, this means that cor-

PHOTOELASTICITY

responding strains in the model and prototype should be equal, that is, $\varepsilon_m = \varepsilon_p$. Thus, if $\nu_m = \nu_p$,

$$\frac{\varepsilon_m}{\varepsilon_p} = 1 = \frac{\sigma_m}{\sigma_p}\frac{E_p}{E_m} = \frac{F_m}{F_p}\left(\frac{L_p}{L_m}\right)^2\frac{E_p}{E_m}$$

from which

$$F_m = F_p\left(\frac{L_m}{L_p}\right)^2\frac{E_m}{E_p} \tag{5-47}$$

The force scale thus depends on the scale of the model and the ratio of the elastic moduli of the materials. For true similarity, the loads applied to the model should conform with Eq. (5-47). However, if the absolute deformations are small, the influence of different relative deformations usually can be ignored.

Similarly, for loading by pressure p, we obtain

$$\sigma_p = \sigma_m\frac{p_p}{p_m}$$

and

$$p_m = p_p\frac{E_m}{E_p}$$

for plane-stress problems, where the thickness ratio between model and prototype, $d_m/d_p \neq L_m/L_p$ [Eq. (5-46)], is replaced by

$$\sigma_p = \sigma_m\frac{F_p}{F_m}\frac{L_m}{L_p}\frac{d_m}{d_p}$$

For stresses due to rotation with $\nu_p = \nu_m$,

$$\sigma_p = \sigma_m\left(\frac{L_p}{L_m}\right)^2\frac{\rho_p}{\rho_m}\left(\frac{\omega_p}{\omega_m}\right)^2$$

where ρ is the density and ω the angular velocity.

For gravitational stresses in dams and the like,

$$\sigma_p = \sigma_m\frac{\rho_p}{\rho_m}\frac{L_p}{L_m}$$

Similar relations for other forms of loading can be written down from elementary considerations.

Use of Compensators [5,7,8]

If an additional birefringent material is introduced into the polariscope immediately before or after the model, the two retardation fields, compensator and model, will add tensorially. If the birefringence of the compensator is adjustable, as in the double wedge of the Babinet–Soleil compresator of Fig. 5-34a, it can be adjusted to give an integral fringe order at the point under observation. The correct procedure is to align the retardation axes of the compensator with those at the point in the model in such a way that the two birefringences, compensator and model, subtract. Then adjust the compensator until a zero-order fringe passes through the point. If the signs of the birefringence in the material of the model and the compensator are the same, the

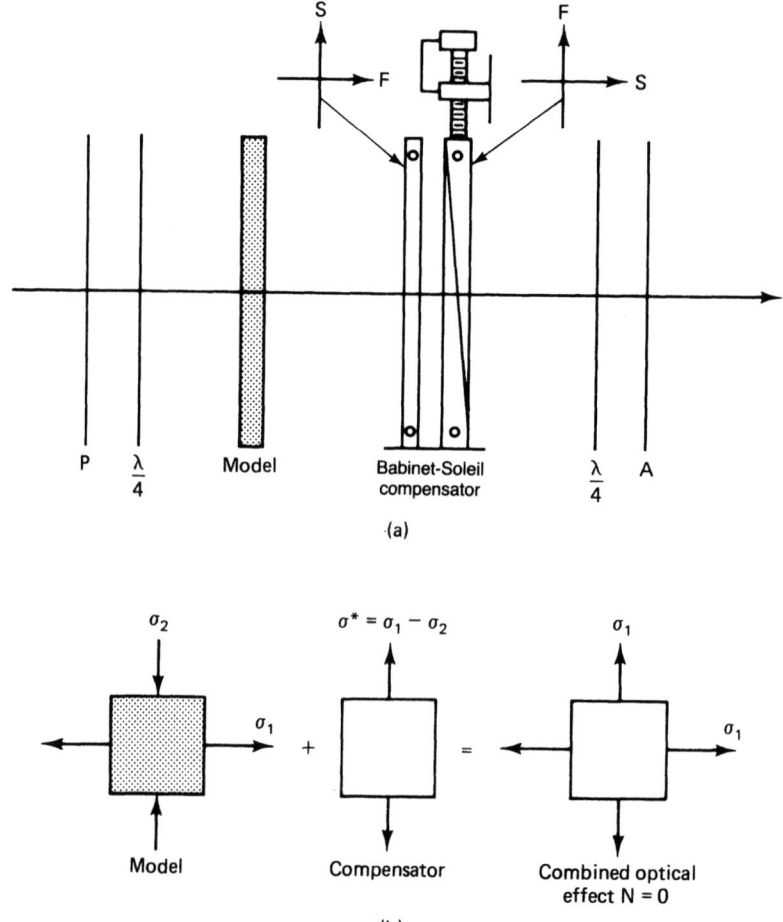

Figure 5-34 Compensator operation: (a) light passing through the model and a Babinet–Soleil compensator; (b) superposition of retardation exhibited by model and compensator.

setting for the compensator will give the retardation at the point in the model. This is shown in Fig. 5-34b.

Relative Directions of Principal Stresses

The directions of principal stresses at a point may be determined by rotating the crossed polarizer and analyzer in a plane polariscope until an isoclinic is coincident with the point under observation. Coincidence will be obtained at two angles, θ and $\theta + 90°$, but there is no way of telling which of these directions is that of σ_1 and which is that of σ_2. This information may be found by observing the fringe movement produced by a given rotation of the analyzer in a Tardy compensation setup.

Table 5-8 Color Matching in White Light for Dark-Field Polariscope

Approximate Relative Retardation (nm)	Color Extinguished	Approximate order of extinction	Color observed[a]	
0	Retardation less than		Black	
50	wavelength of		Gray	
200	visible light		White	
400	Violet		Yellow	
450	Blue	First-order	Orange	
500	Green	colors	Red	
590	Yellow	($i = 1$)	Tint of passage	
650	Orange		Blue	
700	Red		Green	
800	Deep red ($i = 1$), violet ($i = 2$)		Yellow	
900	Blue ($i = 2$)	Second order	Orange	
1000	Green ($i = 2$)	($i = 2$)	Red	
1180	Yellow ($i = 2$) violet ($i = 3$)		Tint of passage (approximate)	
1300	Orange ($i = 2$), blue ($i = 3$)		Green	
1400	Red ($i = 2$), blue ($i = 3$)	Third order	Yellow	
1550	Deep red ($i = 2$), green ($i = 3$), violet ($i = 4$)	($i = 3$)	Pink-red	
1770			Tint of passage (approximate)	
1800	Yellow ($i = 3$), deep blue ($i = 4$), deep violet ($i = 5$)	Fourth order ($i = 4$)	Pale green	
2100			Pink	
2360			Tint of passage	four or more extinctions in visible spectrum
2500			Pale green	
⋮			⋮	
			White	

[a] Progression of observed colors for increasing retardation is yellow → red → green. This sequence can be identified for three cycles.

First, calibrate the polariscope with reference to a simply loaded specimen of the same material as the model, in which the direction of σ_1 is known. A tensile specimen or a beam in bending would be good. Then set the transmission axis of the analyzer parallel to σ_1 in the model and rotate the analyzer clockwise. Note whether the fringe order at a point increases or decreases. In a model in which several fringes are in the field of view, this can be done by noting whether the higher or lower fringe order moves toward the point. In some instances, such as a tensile specimen, however, there is a uniform stress field over the test portion, and one must determine the nature of the fringe-order change by observing the color sequence in white light. For example, if the applied load gives a first-order orange clock in a dark-field circular polariscope, then, on rotating the analyzer clockwise, the color sequence will either be that of decreasing fringe order (orange, yellow, white, . . .) or of increasing fringe order (orange, red, tint of passage) as shown in Table

5-8. If a clockwise rotation produces a decrease in fringe order, we know that at any point in a photoelastic medium the direction of σ_1 will be given by the same criterion.

Now, to find the direction of σ_1 at any point in a photoelastic specimen, first rotate the crossed polarizer and analyzer in a plane polariscope to bring an isoclinic to the point. This will occur for two angles, θ or $\theta + 90°$, and the transmission axis of the analyzer will be parallel to either σ_1 or σ_2. Insert the quarter-wave plates to produce a dark-field circular polariscope and rotate the analyzer clockwise. Then, if the fringe order at the point decreases (lower-order fringe moves toward the point), the transmission axis of the analyzer is aligned with σ_1. If the fringe order increases (higher-order fringe moves toward the point), it is aligned with σ_2. Thus the relative directions of σ_1 and σ_2 may be quickly determined.

If the sign of the stress-induced birefringence for the model material is known, the relative directions as well as the sign of the principal stresses can be found by using a compensator. A calibrated tensile specimen or a Babinet–Soleil compensator may be used.

Sign of Boundary Stresses

The procedure described above can also be used to determine whether the tangential boundary stresses in a model are tensile or compressive. Assuming the same criteria as before, we align the analyzer transmission axis in a dark-field circular polariscope, with the tangent to the boundary at the point at which readings are to be taken. Then, if a clockwise rotation produces a reduction in fringe order, we know that the tangential stress σ_t is the greater principal stress σ_1, since at a free boundary the stress normal to the boundary, σ_n, is zero. By definition, σ_1 is algebraically greater than σ_2, and σ_t must be tensile. Conversely, if the fringe order increases at the point on the boundary, $\sigma_t = \sigma_2$, $\sigma_n = \sigma_1 = 0$, and consequently σ_t must be compressive.

The sign of the edge stresses can also be determined by means of a simple compensator aligned either parallel or perpendicular to the edge. If a precision instrument such as the Babinet–Soleil compensator is used, both the magnitude and the sign of the edge stress can be determined.

In many cases, consideration of the loading will indicate regions at the edges of a model where the sign of the stress is known with certainty. The stresses must vary continuously along the edge and can change sign only at a point where the isochromatic order is zero. It sometimes happens, however, that the sign cannot be determined with certainty from inspection of the fringe pattern alone. In such cases, the "nail test" provides a simple method of determining the sign. Local compression is applied normal to the edge at the point where the sign is to be determined; compression is applied using a fingernail or other sharp edge. The sign is found by observing the resulting displacement of the isochromatics in the region of the superimposed load. If σ_t denotes the stress tangential to the edge and σ_n the compressive stress applied normal to it, then clearly (if σ_t is tensile).

$$\sigma_t - \sigma_n > \sigma_1$$

The fringe order will thus increase and higher-order isochromatics will appear at the edge. On the other hand, if σ_t is compressive, then

$$\sigma_t - \sigma_n < \sigma_1$$

The fringe order at the edge will now be reduced, and the isochromatics of the lower order will be displaced toward the edge.

In cases where local variations of fringe order are too small for the foregoing effects to be observed, it is possible to recognize an increase or a decrease in the fringe order by observing the color sequence using white light.

Use of White Light [5,8]

When a white-light source is used in a polariscope, the shortest wavelengths (violet and blue) will be the first to be extinguished with increasing stress. The color for the complementary spectrum, which happens to be yellow-orange, will be observed. Further increases in stress will produce larger phase shifts due to birefringence, and the longer wavelengths will be extinguished in sequence. The progression of extinguished and the complementary observed colors is shown in Table 5-8.

At a relative retardation R of approximately 800 nm, the second extinction for violet light coincides with the first extinction of deep red. With increasing retardation, overlapping between extinctions occurs more frequently, and the intensity of the observed colors begins to fade. With $R > 3200$ nm, no colors (i.e., no fringes) will be observed. Thus at high fringe orders a model viewed in white light will appear to be unstressed. When the white-light source is filtered to provide a narrowband monochromatic illumination, the fringes will appear. This explains why spectral lamps with narrowband filters are used in high-quality polariscopes. Broadband monochromatic sources cannot be used to resolve high-order fringes ($N > 11$ for green light).

Black is observed only when $R = 0$. White light is therefore useful for identifying zones of zero fringe order.

With increasing retardation the observed colors appear in sequence as yellow, red, and green. Knowledge of this series is often used to find whether fringe orders are increasing or decreasing in any particular direction.

Photoelastic Material [5–7]

The properties desirable in a photoelastic material differ according to the nature of the data required and the procedure to be employed. For the production of an isochromatic pattern, which is the most common photoelastic operation, the stress-optic coefficient should be as high as possible. Furthermore, to ensure that the shape of the model (when deformed under load) does not differ appreciably from that of the prototype, the elastic modulus should be as high as possible. By combining these two requirements, the relative merits of photoelastic materials are compared using a quantity known as the figure of merit Q, which is the ratio of the elastic modulus to the material fringe value.

$$Q = \frac{E}{f_\sigma}$$

The figure of merit should be as high as possible and should remain constant with load and time under load and over the temperature range encountered during a photoelastic investigation.

Additionally, the materials should have linear proportionality between stress and strain, and freedom from creep. For photoplastic investigations, the material should have creep and yield properties that simulate the behavior of the material of the prototype.

Most model materials exhibit initial stresses resulting from the manufcturing process. These can be eliminated in some cases by annealing: that is, heating a model to a temperature above the softening point of the material and cooled at a very low rate to avoid freezing in thermal stresses. For the production of castings, the materials and casting procedure should be chosen to avoid the development of initial stresses during gelation and cure.

Plastics often suffer from "time-edge effects." Such phenomena most commonly are due to the evaporation or absorption of water, which causes changes of volume and results in tensile or compressive stresses in a narrow zone bordering the edges of the model. The optical effects produced are superimposed on those caused by the applied loading and distort the true fringe

Table 5-9 Approximate Properties for a Few Photoelastic Model Materials

Material	Generic Types and Generic Names	Stress Fringe Value, f_σ (green light, $\lambda = 546$ nm)		Room-temperature Properties					Figure of Merit, Q
				Young's Modulus, E		Proportional Limit		Poisson's Ratio	
		kN/m Fringe	lb/in. Fringe	MPa	ksi	MPa	ksi		E/f_σ
Glass		−300 +400	−1700 +2200	70,000	10^4	60	8.7	0.25	5600
Plexiglas	Polymethyl methacrylate: Lucite Perspex	−130	−700	2,800	400			0.38	0.570
Celluloid	Cellulose nitrate	30–300	170–1700	2,200	300	35	5	0.33	1700–200
Homolite 100	Polyester	24	140	3,900	560	48	7	0.35	4000
Homolite 911 (CR-39)	Allyl diglycol carbonate	16	90	1,700	250	21	3	0.4	2800
Epoxy	Araldite, Epon, Bakelite	11	60	3,300	430	55	8	0.37	8000
Polycarbonate	Makrolon PSM-1	7	40	2,600	360	3.5	5	0.28	9000
Polyurethane	Hysol	0.2	1	3	0.5	0.14	0.02	0.46	500
Gelatin		0.09	0.5	0.3	0.04	—	—	0.5	80

pattern, especially near the edges. For all normal procedures, freedom from initial (or residual) stress and time-edge effect is desired. However, the development of such stresses is sometimes used to model the development of residual compressive stresses from various diffusion processes designed to improve hardness, strength, and fatigue life.

Typical properties at room temperature for some photoelastic model materials are given in Table 5-9. Since considerable variation in these values is possible across different samples of the same material, the values to be applied in an investigation should be determined from calibration tests of the actual material used.

Epoxy resins, the most widely used photoelastic materials today, are suitable for the manufacture of models for both two- and three-dimensional investigations and for use as birefringent coatings. They possess good optical and mechanical properties, are only slightly susceptible to edge effect, and are available at comparatively low cost. At room temperature these materials have a slight but unimportant tendency to creep. Epoxies have good casting properties and parts may be cemented together at room temperature, allowing the fabrication of complicated models. The finished resin is formed by the chemical reaction that takes place between a base resin and a hardener when the two are mixed together at a suitable temperature. The resin is available in both solid and liquid forms, but the type used as little effect on the properties of the finished product. The properties depend greatly, however, on the proportion and type of hardener used. Optimum properties are obtained using phthalic anhydride as the dominant curing agent. The relative merits of various epoxies and hardeners available in the United States have been studied in detail by Leven [13] and by Cernosek [14].

Polycarbonate has a very high optical sensitivity combined with a reasonably high modulus of elasticity, which gives it an exceptionally high figure of merit. Polycarbonate is practically free from time-edge effects and shows little creep at room temperature. Some initial stresses resulting from the manufacturing process are usually present and can be eliminated only by extremely careful heat treatment. Commercially annealed sheets of polycarbonate are popular for two-dimensional model studies, but they are expensive.

Columbia resin CR39 is an allyl diglycol resin. It is highly transparent, colorless, and has good optical sensitivity. It is supplied by the Homolite Corporation as H911 in the form of flat plates of uniform thickness with highly polished surfaces; consequently, it is very popular for general two-dimensional work. CR39 is only slightly susceptible to time-edge effect but suffers to some extent from creep. Over a considerable range, the isochromatic order as measured after a given time at load is linearly related to the stress. Plates of CR39 possess marked initial stresses through the thickness of the plates because they are manufactured under pressure. This disadvantage does not interfere with the use of this material in two-dimensional work, but care must be taken that the plate is kept truly normal to the incident light. Models of CR39 should not be used in oblique-incidence analyses. The material is rather brittle; thus special care is required when shaping to avoid chipping the model at the edges. This is especially likely to happen during drilling. Sheets of CR39 can be readily machined by high-speed routing.

Homolite 100 is a polyester resin available in sheets of good optical quality. It exhibits very little creep and, since moisture absorption is extremely slow, time-edge effects develop only slowly.

Glass was the first material employed for the determination of stresses by photoelastic methods. It has many advantages as a photoelastic material. It can be obtained in large plates or thick blocks; it is homogeneous, perfectly transparent, and stress-free. It also has a high Young's modulus and a large linear loading range. Unlike the plastics, its stress-optic properties are not subject to any appreciable change with time or with ordinary variations of temperature or atmospheric conditions. It can be perfectly annealed, but the temperature required is high, and the annealing process slow. The disadvantages of glass are that it is difficult to shape, its brittleness is likely to result in chipping at load points, and its stress-optic coefficient is small. However, its figure of merit is high.

Glass is undergoing a resurgence of popularity as a photoelastic material for some special types of problem. In one such application it serves in long-life optical transducers for use in mining and civil engineering structures. The advent of digital gray-level analyzers for studies at low levels of stress-induced birefringence is permitting investigators to exploit the desirable properties of glass despite its low stress-optic response.

The stress-optic properties of glasses of different compositions differ widely. Coker and Filon [15] carried out an extensive investigation of glasses and found that the stress-optic coefficient varied from about 45 brewsters for a glass containing no lead oxide to -2 brewster for one containing 80% lead oxide. This means that for the most optically sensitive glass, the fringe value is about 140 kN/m fringe or 80 lb/in. fringe.

Polymethyl methacrylate, marketed under the trade names Plexiglas, Lucite, and Perspex, can be purchased at moderate cost in the form of flat sheets, round bars, and tubes, all with highly polished surfaces and in a wide variety of sizes. The material is highly transparent, colorless, and free from initial stress. It exhibits practically no creep or edge effect but has a very high stress-fringe value. For this reason, it is sometimes used as part of a built-up model or for making loading tackle in cases when the polarized light must pass through parts without being affected by the stresses in them.

The new developments in digital, partial, fringe-order photoelasticity can take advantage of the low cost and other desirable features of the methacrylates. These materials will be used more extensively as the new technology develops.

Polyurethane rubber is a rubberlike plastic with low elastic modulus and high optical sensitivity. It is a transparent amber-colored material and is practically free from time-edge effects. Optical and mechanical creep are negligible at room temperature after the load has been applied for a few seconds, and the relation between stress and strain is linear over a wide range [16].

Urethane rubber is a popular material for the investigation of problems of dynamic stresses. The velocity of propagation of stress waves in urethane rubber is only about one-thirtieth of that in the conventional hard plastics. The problem of recording transient fringe patterns is thus greatly simplified. Urethane rubber is also suitable for the investigation of problems caused by self-weight. Since the material is available in grades covering a considerable range of elastic moduli and over which Poisson's ratio is approximately constant, it may be used for the manufacture of composite models free from the complicating effects of the difference of Poisson's ratio.

Gelatin has an optical sensitivity greatly exceeding that of any other photoelastic material. It is so high, in fact, that isochromatics are produced by the stresses arising from the weight of the substance itself. For this reason, gelatin is suitable for the investigation of stress problems in which the influence of self-weight is important—for example, in determining the stress distribution around tunnels. Gelatin has been used to study problems in soil mechanics and the stresses in structures such as dams.

Celluloid was one of the earliest photoelastic materials but has been superseded for conventional work by modern materials having greater uniformity and higher optical sensitivity. Celluloid is susceptible to creep. Its yield behavior, as represented by the stress–strain diagram, resembles that of aluminum and carbon steel according to the grade of celluloid, the rate of loading, the time at load, and to some extent, the temperature and humidity of the environment. The yield behavior has led to a revival of interest in celluloid as a photoplastic material for the study of elastoplastic states of stress and strain.

Polyesters are a family of materials that can be cut into thick sections. All polyesters exhibit viscoelastic behavior at room temperature. Their yield behavior can be adjusted to match that of some structural materials by mixing different base resins. For this reason, polyester has been used in photoplasticity studies of large deformation processes.

Calibration of Model Materials [7,8]

Since the stress-fringe values of model materials vary with time and also from batch to batch, it is necessary to calibrate each sheet or casting at the time of the test. Calibration is performed on simple specimens with a known stress state at a specific point. The load is first increased and then decreased incrementally. The equivalent stress corresponding to the load is plotted against the exact fringe order at the point. Tardy compensation is usually used to read the fringe order to the desired accuracy. A best-fit straight line is then drawn through the points and the slope of this line is the stress-optic response in MPa/fringe, N/fringe meter, or lb/fringe inch. In this form f_σ is normalized to unit thickness. It must be corrected for the actual thickness of the model.

Most laboratories use any one of three geometries for calibration.

1. *Circular disk in diametral compression.* This specimen is compact, easy to machine, and easily loaded. At the center point of the disk

    $$f_\sigma = \frac{8P}{\pi DN}$$

    where P = load
    D = diameter of the disk
    N = observed fringe value at the center

The ratio of diameter to thickness of the disk should be chosen so that the disk does not buckle under the load.
2. *Tensile specimen.* This specimen has a simple state of stress, but it is difficult to load such a model without introducing some bending. Anywhere in the uniform section of the specimen

$$f_\sigma = \frac{P}{wN}$$

where w is the width of model.
3. *Beam in pure bending.* This, too, is a simple model, but it is hard to load a beam in pure bending without introducing axial stresses. The ratio of depth to length should be chosen to avoid out-of-plane buckling at higher loads
For any specific point on the beam,

$$f_\sigma = \frac{My}{IN}$$

where M = constant bending moment
I = second moment of area of the beam with respect to the neutral axis
y = distance of the calibration point from the neutral axis
N = fringe value at the point

If N is measured at various distances y from the neutral axis, a calibration can be completed with only one level of loading.

5-2 Two-Dimensional Photoelasticity

In conventional two-dimensional photoelasticity, a geometrically correct scaled model is machined from a flat, optically isotropic plate of a suitable transparent material. The model is placed in a polariscope with its plane normal to the axis of the polariscope; it is loaded in directions in the plane of the model plane. The two sets of fringes, isoclinics and isochromatics, are then recorded and interpreted. When the stresses perpendicular to the plate (z direction) are zero ($\sigma_z = 0$), the following results can be obtained:
1. The difference between the principal stresses
2. The directions of the principal stresses

In the case of result 1, the difference between the in-plane principal stresses at every point is found from the stress-optic law [Eq. (5-19)] by observing the isochromatic fringes in a circular polariscope.

If the in-plane principal stresses have opposite signs ($\sigma_1 > 0 > \sigma_2$), the isochromatics yield directly the maximum shear stress at every point in the model as

$$\tau_{max} = \frac{\sigma_1 - \sigma_2}{2} = \frac{1}{2}\frac{Nf_\sigma}{h}$$

If the two in-plane principal stresses have the same sign, the maximum shear stress does not occur in the plane of the plate, and Eq. (5-19) gives only the secondary maximum shear stress at each point (maximum shear stress in the plane of the plate).

At the free boundaries, or edges, of the plate, the in-plane principal stress normal to the boundary is zero, and the isochromatic fringe order is proportional to the magnitude of the tangential stress along the boundary.

If $\sigma_2 = 0$: $\sigma_1 = \dfrac{Nf_\sigma}{h}$

If $\sigma_1 = 0$: $\sigma_2 = \dfrac{-Nf_\sigma}{h}$

The sign of the tangential stress is found by any of the methods described in the preceding section.

When $\sigma_1 = 0$, the directions of the in-plane principal stresses at any point in the model are found from the isoclinic parameter for that point by viewing the model in a plane polariscope adjusted so that an isoclinic fringe passes through the point.

An isoclinic fringe is a line along which the direction of the principal stresses is constant and coincides with the axes of the polarizer and analyzer of the crossed-plane polariscope. It is usual to record the positive acute angle between the axis of polarization of the analyzer and a convenient reference axis on the model. Compensation methods can be used to find which of the two orthogonal directions associated with an isoclinic is σ_1 or σ_2.

Iosclinic data are correct only if the directions of the principal stresses do not rotate through the thickness of the plate. While, under normal incidence, isochromatic data are insensitive to residual stresses normal to the plane of the model ($\sigma_z \neq 0$), the isoclinics frequently are affected by such stresses, especially if they are accompanied by variations in the state of stress through the thickness of the plate. Residual stresses frequently cause rotation of the stress directions through the thickness. These rotations may make the isoclinics appear vague.

Singular Points [5,6,8,17]

When $\sigma_1 = \sigma_2$, the material is in a plane, hydrostatic state of stress. All directions in the plane of the plate are principal, and points where these occur are called *isotropic points*, because the material behaves as if it were optically isotropic in the plane of the model at those points. At isotropic points, $\sigma_1 - \sigma_2 = 0$ and $N = 0$. If the stress in every direction is zero ($\sigma_1 = \sigma_2 = \sigma_z = 0$), the isotropic point is called a singular point. Isotropic points are easily identified, because as the plane polariscope is rotated, all isoclinics will pass through an isotropic point.

An isotropic point situated at a free boundary must also be a singular point. Since the stress is zero in the direction normal to the boundary, it must also be zero in all directions. Such points usually indicate a change of sign of the boundary stress.

Two different forms of isotropic points can be distinguished. As the crossed polaroids are rotated, the isoclinic lines rotate about an isotropic point. In some cases the lines rotate in the same sense as the polaroids, which in others they rotate in the opposite sense. Such points are referred to as positive and negative isotropic points, respectively, and are shown in Fig. 5-35a and b.

A singular point at a load-free edge of a plate can only be of the negative type. A free corner of a plate is always a negative singular point.

When the isoclinics pass through several isotropic points in sequence, the points must be alternatively positive and negative, because the order of succession of the isoclinics at one point will be reversed at the next (Fig. 5-35c). If an isoclinic passing through an isotropic point on a free boundary also passes through another isotropic point, the latter must be of the positive type. Furthermore, if an isoclinic passes through two singular points on the free edges of a plate, it must also pass through a positive isotropic point in the interior.

PHOTOELASTICITY

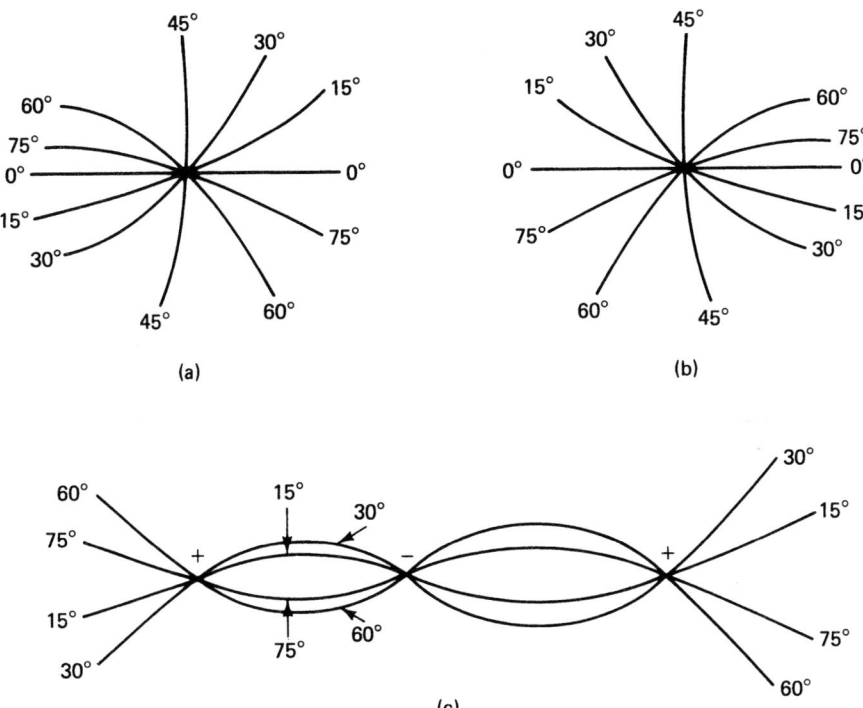

Figure 5-35 Isoclinic patterns about (a) positive and (b) negative isotropic points. (c) Alternation of signs of successive isotropic points.

Isoclinics of all parameters also pass through a point on the boundary of a plate at which a concentrated load acts. Such a point is not isotropic in the sense of our previous definition, since the difference of principal stresses is not zero.

The following additional rules help in plotting isoclinics:

1. Isoclinic lines do not intersect one another except at an isotropic point.
2. An isoclinic intersects a free boundary or one subjected to normal forces only at a point where the inclination of the tangent to the boundary is identical with the parameter of the isoclinic, the reference direction being the same for each. It follows that a straight shear-free edge is an isoclinic.
3. One isoclinic always coincides with an axis of symmetry. In plates that are symmetrical in form and loading, the isoclinic pattern is also symmetrical. An isoclinic of parameter θ on one side of a section of symmetry corresponds, however, with the isoclinic of parameter $90° - \theta$ on the other side.
4. At a point on a shear-free boundary where the stress parallel to the boundary has a maximum or a minimum value, the isoclinic intersects the boundary at right angles.

Stress Trajectories [5,6,8,17]

Isoclinics may be used to obtain stress trajectories that cover the field of the plate.

A stress trajectory, line of principal stress, or *isostatic,* is a line such that its direction at any point coincides with that of one of the principal stresses at the point. Since the two principal stresses at any point of a two-dimensional stress system are mutually perpendicular, it follows that a system of stress trajectories will consist of two orthogonal families of curves. One of these families indicates the directions of the σ_1 (algebraically greater) principal stresses and the other those of the σ_2 stresses. Near a positive isotropic point, the system consists of two families of curves of parabolic appearance that partly enclose the isotropic point, forming what is known as an interlocking system. In the vicinity of a negative isotropic point, the system consists of families of hyperbolic-type curves separated by asymptotes; this is known as a noninterlocking system. Each asymptote separates two families of the same class. These two patterns are shown in Fig. 5-36a and b.

The following useful properties of stress trajectories can be identified:

1. At points on a load-free boundary or at a point subjected to normal forces only, the directions of the principal stresses are normal and tangential to the boundary. A stress trajectory of one family will therefore coincide with such a boundary, while those of the other family will intersect it orthogonally.
2. The distance between a load-free boundary and a neighboring stress trajectory of the same family varies inversely as the tangential stress at the boundary; that is, $\sigma_1 \propto 1/\Delta s_2$ (Fig. 5-36c).
3. Since a section of symmetry is shear-free, it coincides with a stress trajectory. Such a section also coincides with an isoclinic.
4. In doubly or multiply connected plates (such as rings) or in plates where loads or thermal stresses act within the field of the plate as well as at the edges, the stress trajectories may form closed loops but not spirals.
5. The sign of $(\sigma_1 - \sigma_2)$ is constant along each stress trajectory. Since it is possible to move from any one point of a plate to any other by following one or more stress trajectories that do not pass through an isotropic point, the sign of $(\sigma_1 - \sigma_2)$ is constant throughout the plate. An apparent change in the sign of $(\sigma_1 - \sigma_2)$ occurs when a stress trajectory passes through an isotropic point; however, the two parts of such stress trajectories lying on opposite sides of an isotropic point (e.g., AO and BO in Fig. 5-36a) are in fact parts of two different trajectories belonging to opposite families.
6. The principal stress associated with one stress trajectory has a maximum or a minimum value at a point where the orthogonal stress trajectory is straight or has a point of inflection. This can be shown if the equations of equilibrium are written for a small element of dimensions ds_1 and ds_2 in a two-dimensional state of stress bounded by two stress trajectories, as in Fig. 5-36d [5]. If ρ_1 and ρ_2 are the radii of the curvature of the stress trajectories in the directions σ_1 and σ_2, respectively, we obtain the Lamé–Maxwell equations of equilibrium, according to which

$$\frac{\delta \sigma_1}{ds_1} + \frac{\sigma_1 - \sigma_2}{\rho_2} = 0 \quad \text{and} \quad \frac{\delta \sigma_2}{ds_2} + \frac{\sigma_1 - \sigma_2}{\rho_1} = 0$$

In the first equation the necessary condition for a maximum or a minimum value of $\sigma_1 (\partial \sigma_1/\partial s_1 = 0)$ is satisfied when $\rho_2 = \infty$, that is, when s_2 is straight or has a point of inflection. Then, since $\partial \phi/\partial s_1 = 1/\rho_1$ and $\partial \phi/\partial s_2 = 1/\rho_2$,

$$\frac{1}{\rho_2} = \frac{d\phi}{ds_2} = 0$$

PHOTOELASTICITY

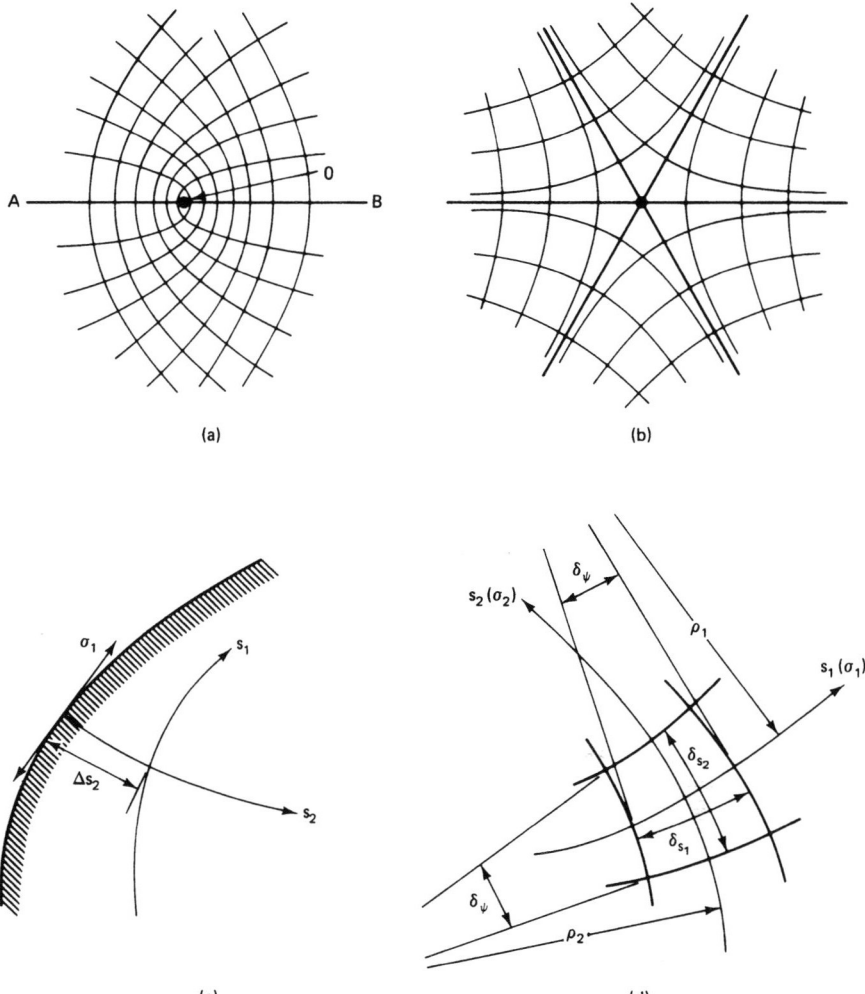

Figure 5-36 Characteristic stress trajectory patterns: (a) interlocking system at a positive isotropic point; (b) noninterlocking system at a negative isotropic point; (c) stress trajectory adjacent to a load-free boundary; (d) element bounded by stress trajectories.

which means that the direction of s_2 coincides with that of the isoclinic passing through the point. The isoclinic is therefore perpendicular to the line s_1.

Since a shear-free edge is a stress trajectory, it follows from the discussion above that the stress tangential to the edge is a maximum or a minimum at a point where an isoclinic intersects the edge at right angles. Similarly, since a section of symmetry coincides with both an isoclinic and a stress trajectory, the stress normal to the section has everywhere a maximum or a minimum value.

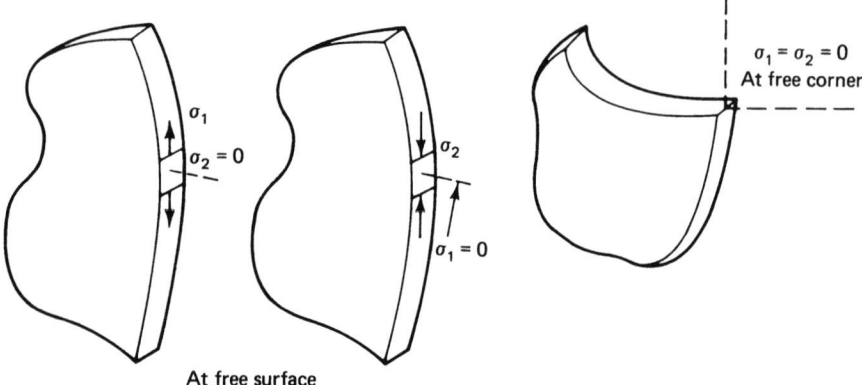

Figure 5-37 Free boundary stresses. At a free boundary, one of the principal stresses is zero. At a corner, both principal stresses are zero.

Separation of Principal Stresses [5,7,16]

On a free boundary, one of the principal stresses is zero. The remaining principal stress is determined uniquely by the isochromatic fringe order. Its sign, compression or tension, is found by any one of the compensation techniques. At a free external corner, the absence of external forces must be accompanied by an absence of internal resistive forces or stresses. Thus $\sigma_1 = \sigma_2 = 0$ and the isochromatic fringe order is always zero. This also holds true for dynamic stress systems. Figure 5-37 illustrates the determination of the principal stresses at free boundaries.

At interior points the complete state of plane stress is determined by three quantities, involving σ_1, σ_2, and their directions θ. However, the photoelastic method provides only two quantities $(\sigma_1 - \sigma_2)$ and θ. Thus the isochromatic pattern and the isoclinic curves are not sufficient by themselves to provide the complete stress solution for interior points. Supplementary information must be provided before the values of the individual principal stresses can be found.

Several methods have been developed to separate σ_1 and σ_2, some by calculation and others by subsidiary experiments. These are discussed below.

Oblique-Incidence Method [5,7,18]

The frequently used oblique-incidence method employs a second or additional measurement of fringe order where the light ray traverses the model obliquely with respect to the model surface. Thus a supplementary relationship between stresses and measured fringe order is obtained, which, together with knowledge of $\sigma_1 - \sigma_2$ and θ, permits calculation of the individual principal stresses, σ_1 and σ_2.

The relative retardation, or interference order, for light incident normal to the stressed model is given by Eq. (5-19) as

$$N = \frac{h(\sigma_1 - \sigma_2)}{f_\sigma}$$

By considering Eqs. (5-17) and (5-18), the fringe order corresponding to each of the separate principal stresses may be expressed as

$$N_1 = \frac{\sigma_1 h}{f_\sigma} \quad \text{and} \quad N_2 = \frac{\sigma_2 h}{f_\sigma} \quad \text{where} \quad N = N_1 - N_2$$

PHOTOELASTICITY

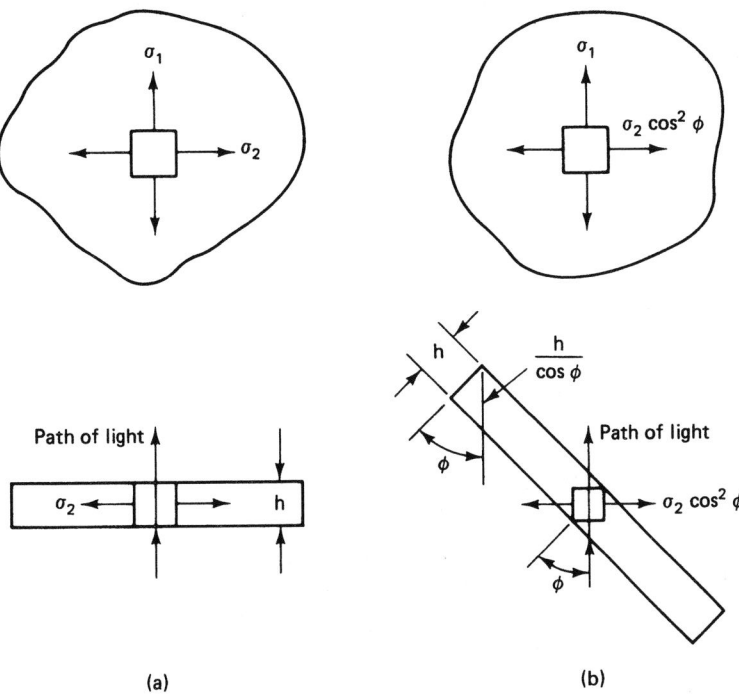

Figure 5-38 Stresses producing retardation for (a) normal incidence and (b) oblique incidence.

With the principal stress directions known, rotate the model about the σ_1 axis through an angle ϕ_1. Accordingly, the stresses normal to the light path are as shown in Fig. 5-38, and the effective model thickness is $h/\cos\phi$. Then, for a rotation ϕ_1, the fringe order for oblique incidence, N_{01}, is

$$N_{01} = \frac{\sigma_1 - \sigma_2 \cos^2\phi_1}{f_\sigma} \frac{h}{\cos\phi_1} = \frac{N_1 - N_2 \cos^2\phi_1}{\cos\phi_1}$$

Combining oblique incidence with normal incidence then gives

$$N_1 = \frac{\cos\phi_1 (N_{01} - N\cos\phi_1)}{\sin^2\phi_1} \qquad N_2 \frac{N_{01}\cos\phi_1 - N}{\sin^2\phi_1}$$

from which the individual stresses can be calculated.

This method is useful for separating stresses along an axis of symmetry where one rotation of the model about the axis provides sufficient data to separate the stresses along the entire line. If the directions of the principal stresses are for some reason not known, the model may be rotated about arbitrary axes Oy and Ox. Three fringe patterns are then obtained, normal incidence and two oblique incidence values.

Since for normal incidence

$$N = \frac{h}{f_\sigma}[(\sigma_x - \sigma_y)^2 + 4\tau_{xy_2}]^{1/2}$$

we find three equations for the three unknowns, σ_x, σ_y, and τ_{xy}. If the two rotations $\phi_x = \phi_y = \phi$ are equal, we have

$$\sigma_x = \frac{f_\sigma}{h} \cot \phi \left[\frac{N_{\phi x}^2 + N_{\phi y}^2 \cos^2\phi - N^2(1 + \cos^2\phi)}{1 - \cos^4\phi} \right]^{1/2}$$

$$\sigma_y = \frac{f_\sigma}{h} \cot \phi \left[\frac{N_{\phi y}^2 + N_{\phi x}^2 \cos^2\phi - N^2(1 + \cos^2\phi)}{1 - \cos^4\phi} \right]^{1/2}$$

Together with the first invariant of stress

$$I_1 = \sigma_1 + \sigma_2 = \sigma_x + \sigma_y$$

the state of stress can be solved with data for the quantities $(\sigma_1 - \sigma_2)$, $(\sigma_1 + \sigma_2)$, σ_x and σ_y or σ_x, σ_y, and τ_{xy}, where

$$\tau_{xy} = \frac{1}{2} \left[\left(\frac{Nf_\sigma}{h} \right)^2 - (\sigma_x - \sigma_y)^2 \right]^{1/2}$$

Methods Based on Equilibrium Equations [5,7,8]

Two methods are used, both of which are based on graphical integration of the equations of equilibrium and are therefore independent of the elastic constants of the photoelastic model material.

Like all graphical integration techniques, these methods suffer from the limitation that errors are accumulated as the integration proceeds. Thus extreme care must be exercised to ensure a high degree of accuracy in the original experimental data (isochromatics and isoclinics).

The *shear difference method* uses the equations of equilibrium (body forces are zero):

$$\frac{\partial \sigma_x}{\partial x} + \frac{\partial \tau_{yx}}{\partial y} = 0 \qquad \frac{\partial \sigma_y}{\partial y} + \frac{\partial \tau_{yx}}{\partial x} = 0$$

where σ_x, σ_y, and τ_{xy} are the normal and shear components of stress at any arbitrary point in the plane of the model. Solutions of these equations can be in the form

$$\sigma_x = (\sigma_x)_0 - \int \frac{\partial \tau_{yx}}{\partial y} dx \qquad \sigma_y = (\sigma_y)_0 - \int \frac{\partial \tau_{xy}}{\partial x} dy$$

which can be approximated by the finite-difference expressions

$$\sigma_x = (\sigma_x)_0 - \sum \frac{\Delta \tau_{yx}}{\Delta y} \Delta x \qquad \sigma_y = (\sigma_y)_0 - \sum \frac{\Delta \tau_{xy}}{\Delta x} \Delta y$$

The terms $(\sigma_x)_0$ and $(\sigma_y)_0$ represent known stresses at points that have been selected as starting points for the integration process. Usually, these points are selected on free boundaries where the nonzero stress can be computed directly from the isochromatic data. At any interior point of the model, where θ_1 is the angle between the x axis and the direction of σ_1 as given by the isoclinic parameter, the shear stress is

$$\tau_{xy} = \tfrac{1}{2}(\sigma_1 - \sigma_2) \sin 2\theta_1 = \frac{Nf_\sigma (\sin 2\theta_1)}{2h}$$

It is important to maintain the correct algebraic sign for τ_{xy}. The expression as presented gives the sign of the shear stress in accordance with the theory-of-elasticity sign convention, according

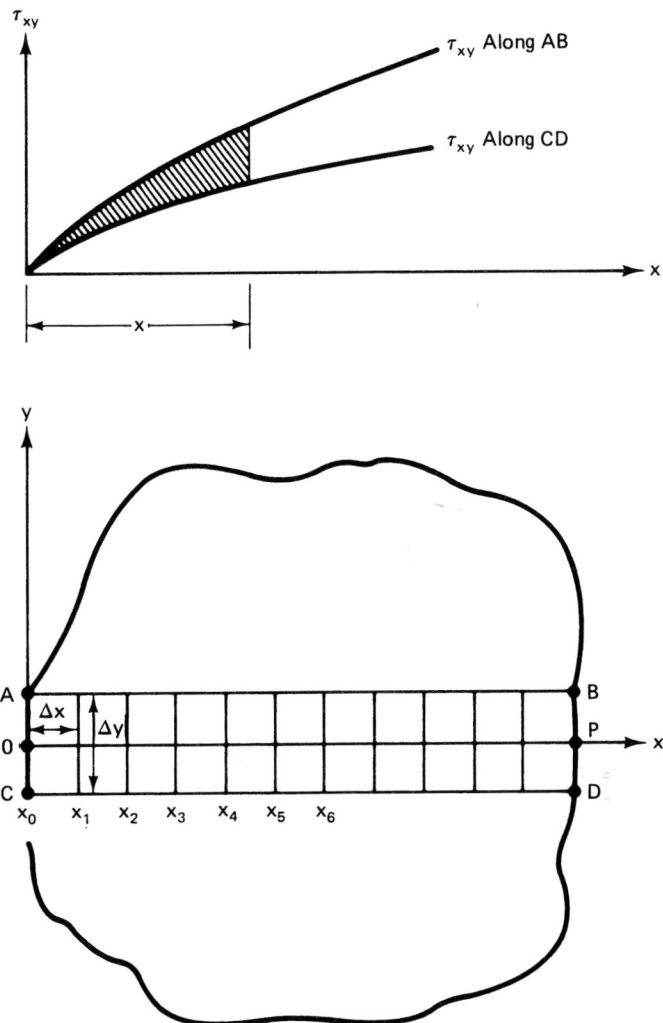

Figure 5-39 Grid system often employed in the application of the shear-difference method.

to which τ_{xy} is positive when it acts in the positive y direction on a plane with an outward normal in the direction of increasing x.

The term $\Delta\tau_{xy}$ is determined from a plot of the shear-stress distributions along two auxiliary lines (parallel to the line of interest and symmetrically located with respect to it), as illustrated in Fig. 5-39. Note that the hatched area between the τ_{xy} curves represents the quantity $\Sigma\, \Delta\tau_{xy}\Delta x$. Thus the accumulated area between the origin and a point at a distance x from the origin along the line of interest can be used to compute the difference between $(\sigma_x)_0$ and $(\sigma_x)_i$, simply by dividing by the distance Δy between the two auxiliary lines. Once σ_x is known at a given point, the value for σ_y can be computed from the expression

$$\sigma_y = \sigma_x - (\sigma_1 - \sigma_2)\cos 2\theta_1$$

and from the stress invariants and the stress-optic law

$$\sigma_1 = \tfrac{1}{2}(\sigma_x + \sigma_y) + \tfrac{1}{2}(\sigma_1 - \sigma_2) = \tfrac{1}{2}(\sigma_x + \sigma_y) + \frac{Nf_\sigma}{2h}$$

$$\sigma_2 = \tfrac{1}{2}(\sigma_x + \sigma_y) - \tfrac{1}{2}(\sigma_1 - \sigma_2) = \tfrac{1}{2}(\sigma_x + \sigma_y) - \frac{Nf_\sigma}{2h}$$

Filon's method [17] uses the Lamé–Maxwell equation for a small curvilinear rectangle bounded by four isostatics (stress trajectories), as shown in Fig. 5-36d:

$$\frac{\partial \sigma_1}{\partial s_1} = -\frac{\sigma_1 - \sigma_2}{\rho_2} \qquad \frac{\partial \sigma_2}{\partial s_2} = -\frac{\sigma_1 - \sigma_2}{\rho_1}$$

where s_1 and s_2 are orthogonal curvilinear coordinates measured along the σ_1 and σ_2 isostatics, respectively, and ρ_1 and ρ_2 are the respective radii of curvature of these isostatics.
Integrating along one of the isostatics yields

$$\sigma_1 = (\sigma_1)_0 - \int \frac{\sigma_1 + \sigma_2}{\rho_2} ds_1 \qquad \sigma_2 = (\sigma_2)_0 - \int \frac{\sigma_1 - \sigma_2}{\rho_1} ds_2$$

Integration along an isostatic line is performed by starting at a point of known stress, usually on a free boundary, and evaluating the integral from a plot of $(\sigma_1 - \sigma_2)/\rho_2$ versus s_1 or $(\sigma_1 - \sigma_2)/\rho_2$ versus s_2. The change in principal stress between the origin and the point of interest along the isostatic is the accumulated area under the curve.

When the isostatics are accurately known, this form of the equilibrium equations is easier to use than the Cartesian form. This method is therefore the most convenient when the problem of interest is axisymmetric. In this case the isostatics are concentric circles and radial lines, and the isochromatics alone are sufficient for the integration. For an integration along a radial line, the radius of curvature needed for the integration is simply the distance from the axis of symmetry to the point under consideration. The appropriate plot for evaluating the integral would be $(\sigma_1 - \sigma_2)/r$ versus r.

In the more general case of integration along an isostatic in an arbitrary stress field, the radius-of-curvature data are difficult to obtain with sufficient accuracy from photoelastic observations of the isoclinics.

Methods Based on Compatibility Equations [7,8,19]

It is shown in textbooks on the theory of elasticity that in the absence of body forces, the sum of the principal stresses satisfies Laplace's equation,

$$\frac{\partial^2}{\partial x^2}(\sigma_1 + \sigma_2) + \frac{\partial^2}{\partial y^2}(\sigma_1 + \sigma_2) = 0$$

If the boundary values of $(\sigma_1 + \sigma_2)$ are known, it is possible to solve this equation. Thus the value of this function can be uniquely determined at all interior points if the boundary values are known. Since the photoelastic isochromatics provide an accurate means for determining both the principal difference $(\sigma_1 - \sigma_2)$ at all interior points of a two-dimensional model and, in many instances, complete boundary-stress information, knowledge of the principal-stress sum $(\sigma_1 +$

σ_2) throughout the interior provides an effective means for evaluating the individual principal stresses.

Rigorous mathematical solution of Laplace's equation is possible only when the boundary is relatively simple. Approximate solutions can, however, be obtained by numerical and experimental means. Also, since the Laplace equation serves as the governing equation in many other fields of engineering, including electrostatic fields, steady-state temperature distributions, and shapes of uniformly stretched films or membranes, the behavior of these different physical systems can be expressed in the same mathematical form, permitting the use of analogy methods.

The most important limitation of the method is the requirement for complete knowledge of the boundary-stress distribution. Three methods are in general use for solving the Laplace equation: an analytic method, a numerical method, and an analogy method.

An analytic solution of Laplace's equation by the method of separation of variables [19] yields a sequence of harmonic functions that can be added together in a linear combination to give a series representation H of the first stress invariant I. Solutions for H referred to several coordinate systems are presented in Ref. 7.

If a particular region of a model conforms to a regular coordinate system, the sequence of functions when evaluated on the boundaries of the region reduces to a Fourier series. The unknown coefficients are the Fourier coefficients and can be obtained by integrating the prescribed boundary values.

If the chosen region of a model does not conform to a particular coordinate system, Fourier analysis cannot be employed. Instead, the method of least squares is used to determine the coefficients. A finite number of harmonic functions is first selected to appear in the series solution. Coefficients are then chosen such that the mean-square difference between the prescribed boundary values and the evaluation of the series along the boundary is minimized. If N harmonic functions $F_1, F_2, \ldots, F_N$ are selected, and the associated unknown coefficients are denoted as $C_1, C_2, \ldots, C_N$, the series solution for the first stress invariant is

$$H = \sum_{n=1}^{N} C_n F_n \tag{5-48}$$

If $I(s)$ is used to represent the distribution of the first stress invariant along a boundary of total length L, then H must be selected such that

$$\int_0^L \left[I(s) - \sum_{n=1}^{N} C_n F_n \right]^2 ds = \text{minimum} \tag{5-49}$$

and Ref. 19 shows that

$$\sum_{N=1}^{N} C_n \int_0^L F_n F_k \, ds = \int_0^L I(s) F_k \, ds \qquad k = 1, 2, \ldots, N$$

This equation yields N simultaneous equations in terms of the N unknown coefficients. Solution of this set of equations gives the coefficients that provide the best match of boundary values possible with the initial selection of N harmonic functions. By increasing the number of functions in the series for H, the fit can be made as accurate as the original photoelastic determination of $I(s)$.

Numerical methods can be used very effectively to solve Laplace's equation. A popular method uses an iteration procedure by which estimated values of the harmonic function at points of a network are systematically improved by making use of the fact that the value of the function at an point depends on the values of the function at neighborhood points [7,8]. The basic relationship

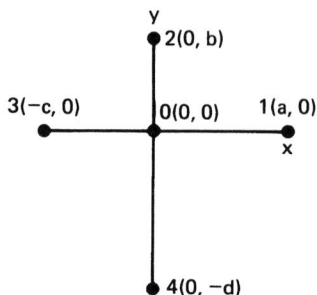

Figure 5-40 Network points for the finite-difference solution of Laplace's equation.

between values of the function at different points can be expressed by the four-point influence equation as

$$\Phi_0 = C_1\Phi_1 + C_2\Phi_2 + C_3\Phi_3 + C_4\Phi_4$$

where neighborhood points in the most general case are located as shown in Fig. 5-40, and the constants, C_1, C_2, C_3, and C_4 have the following values:

$$C_1 = \frac{bcd}{(bd + ac)(a + c)} \qquad C_2 = \frac{acd}{(bd + ac)(b + d)}$$

$$C_3 = \frac{abd}{(bd + ac)(c + a)} \qquad C_4 = \frac{abc}{(bd + ac)(d + b)}$$

In regions where a square array or network of points can be used, the computations are simplified, since $C_1 = C_2 = C_3 = C_4 = \frac{1}{4}$.

Known boundary values are assigned to all points where the network intersects the boundary. Estimated values of zero can be assigned to interior points. The value of each interior point is then improved by traversing the network in a definite sequence and using the influence equation. Each time the network is traversed, the values are improved. The process is continued until further corrections do not alter the values more than a predetermined amount.

The method suffers from the limitation that complete boundary-stress data must be available. Also, the method cannot be used to evaluate stress distributions along selected lines of interest without performing the evaluation for the complete model. Only symmetry considerations can be used to reduce the number of network points to be evaluated. The method has two advantages: (1) isochromatic data alone are sufficient for the determinations, and (2) errors made during the iteration process do not influence the final results. Errors merely increase the number of iterations required to reach the necessary level of convergence.

Electrical analogies [7,20] use the knowledge that electrical fields in isotropic conducting media obey Laplace's equation in the same manner as the elastic field for the sum of the principal stresses. Baths of uniformly conducting liquid or paper with a coating of uniformly conducting material generally is used.

An electrical model of the same geometry as the photoelastic model is prepared from the uniformly conducting medium. Teledeltos paper, which consists of a uniform layer of graphite particles over a thin paper carrier, is a very suitable material from which to fabricate the electrical model for two-dimensional problems.

Since on a free boundary either σ_1 or $\sigma_2 = 0$, the sum of the principal stresses is proportional

PHOTOELASTICITY

to the isochromatic fringe order. When voltages proportional to the value N determined on the boundary of the photoelastic model are applied to the boundary of the electrical model, the voltage at any internal point on the conducting sheet will be similarly proportional to the sum of the principal stresses at the corresponding point in the photoelastic model.

The voltages applied to the boundary of the electrical model have to be either positive or negative, so that care should be exercised in applying the correct sign to the magnitude of the values for N.

With $(\sigma_1 - \sigma_2)$ known from the photoelastic model and $(\sigma_1 + \sigma_2)$ derived from the electrical model, the individual stresses can be computed.

Methods Based on Hooke's Law

From Hooke's law for plane stress, the strain in the photoelastic model in the direction of light propagation, that is, in the thickness of the plate, is

$$\varepsilon_z = \frac{\Delta h}{h} = -\frac{\nu}{E}(\sigma_1 + \sigma_2)$$

where Δh is the change in thickness and ν is Poisson's ratio for the material.

Thus any procedure that can measure accurately enough the change in thickness Δh can be used to obtain the sum of the principal stresses. Together with the values for $(\sigma_1 - \sigma_2)$ from the photoelastic analysis at corresponding points, the two principal stresses can be separated. Methods that have been used include accurate lateral extensometers and various interferometric methods.

Separation of Principal Stresses

Since stress is a tensor, the isochromatic fringe order for two superposed states of stress does not add up numerically. The stresses have to be added as tensors; that is, the directions as well as the magnitudes of the two states of stress have to be considered. Fringe orders can be added or subtracted numerically only if the two sets of principal stresses have the same directions, such as along a line of symmetry and at a free boundary.

Superpositioning of two states of stress occurs whenever a model contains regions with residual (or initial) birefringence. The photoelastic response of the model to experimental loads can then be found correctly by tensorial subtraction of the initial birefringence from the resultant birefringence after loading.

For any convenient coordinate system (x, y), the difference of the principal stress is

$$\sigma_1 - \sigma_2 = \sqrt{(\sigma_x - \sigma_y)^2 + 4\tau_{xy}^2}$$

$$\tan 2\phi = \frac{2\tau_{xy}}{\sigma_x - \sigma_y}$$

where ϕ is the direction or σ_1 with respect to x.

If two states of stress represented by the components $(\sigma_{x1}, \sigma_{y1}, \tau_{xy1})$ and $(\sigma_{x2}, \sigma_{y2}, \tau_{xy2})$ are superimposed, the resultant values of $(\sigma_x - \sigma_y)$ and τ_{xy} are

$$\sigma_x - \sigma_y = (\sigma_{x1} - \sigma_{y1}) + (\sigma_{x2} - \sigma_{y2}) = (\sigma_x - \sigma_y) + (\sigma_x - \sigma_y)_2$$

$$\tau_{xy} = \tau_{xy1} + \tau_{xy2}$$

Since the stress-optic effect is proportional only to the maximum in-plane shear stress, the hydrostatic component $\frac{1}{2}(\sigma_1 + \sigma_2)$ of the stress field can be neglected, and the center of the Mohr's circles for two states of stress coincide as shown in Fig. 5-41a, where OA represents the photoe-

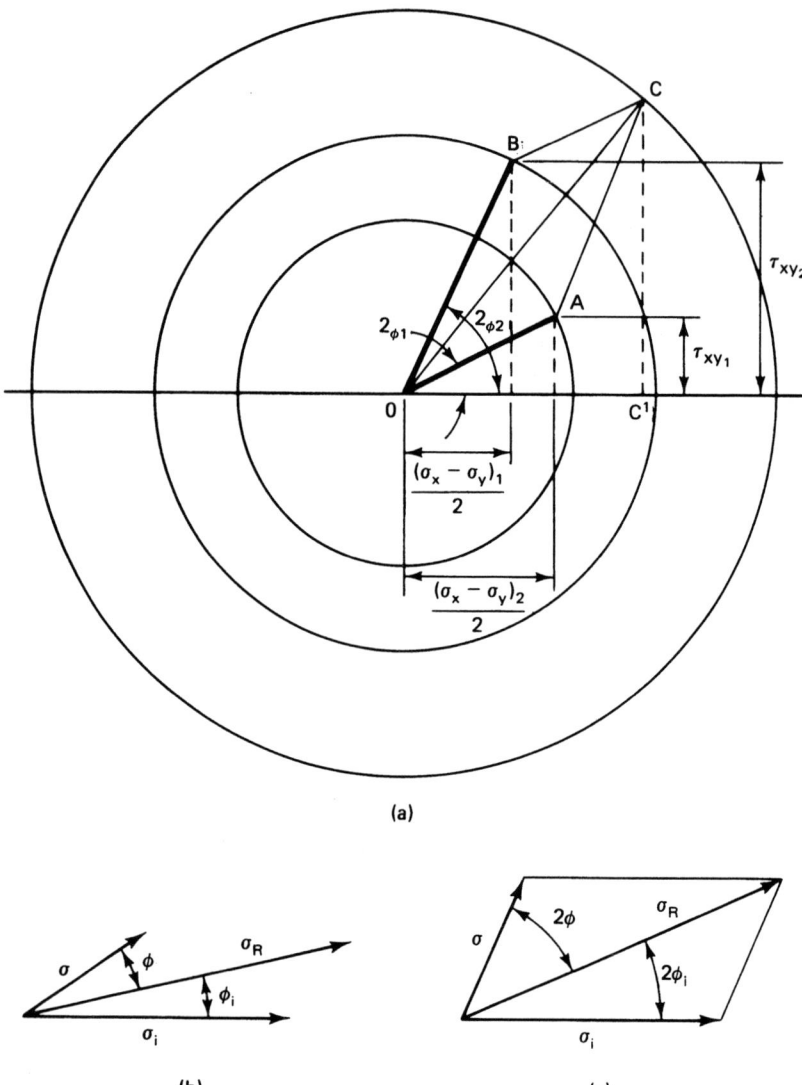

Figure 5-41 (a) Mohr's circle representation for the effective components causing birefringence for two superimposed states of stress; (b) two superposed stress states in their true directions; (c) two superposed stress states with relative angles doubled as required by the Mohr's circle representation.

lastically effective component of the first state of stress, OB the second state, and OC the resultant state of stress

$$OC = [(OC')^2 + (CC')^2]^{1/2}$$
$$\text{where } OC' = \tfrac{1}{2}[(\sigma_x - \sigma_y)_1 + (\sigma_x - \sigma_y)_2]$$
$$CC' = \tau_{xy1} + \tau_{xy2}$$

and

$$<COC' = 2\phi_R$$

indicates the direction of the resultant state of stress.

The resultant double refraction (i.e, the resultant state of stress) can then be determined in the following way. The difference of the principal stresses (or the maximum shear stress) for each state of stress is represented by a vector. The direction of each vector is parallel to the algebraically greater principal stress, and its length is proportional to the difference of principal stresses. The angle between these two vectors is doubled, and the vectors are added. The length of the resulting vector is proportional to the difference of principal stresses, that is, to the birefringence of the resultant state of stress. The direction of the algebraically greater resultant principal stress is found by bisecting the angel between the resulting vector and the component vector that was originally drawn in its true direction.

This procedure gives only the resultant shear stress or double refraction. If the resultant state of stress is to be determined completely, the hydrostatic parts of the component states of stress must be added to the results as scalars.

Consider now a model with initial birefringence, σ_i, which modifies the birefringence caused by the applied loads, σ to yield a resultant birefringence σ_R. The vectors σ_i, σ, and σ_R correspond to virtual stress. These stresses are shown in their true directions in Fig. 5-41b. In Fig. 5-41c the angels between the virtual stresses have been doubled; σ_R is drawn in its correct direction; and the initial component σ_i is indicated at double the angle between the two states as found from their respective isoclinic parameters. The construction yields the birefringence caused by the applied stress. The actual angle is then used in Fig. 5-41b to show the resultant effect, σ.

5-3 Three-Dimensional Photoelasticity [5,7,14]

Three-dimensional photoelasticity by the stress-freezing method is possibly the most powerful method of experimental stress analysis. Stress-freezing phenomena in glass were described by Filon in 1931 (see Ref. 15), and the foundations of our present techniques were laid by Opel, Kuske, Hetényi, and Frocht. The use of stress freezing in industry declined in the 1980s, because it was neither cost- nor time-effective when compared to numerical methods. Cernosek [14], however, developed improved materials and procedures, which once again made three-dimensional photoelasticity by stress freezing quick to perform and cost-effective.

Materials used for three-dimensional photoelasticity by stress freezing are thermosetting polymers that experience a sudden change in mechanical and optical properties above their glass-transition temperatures T_G, as their thermodynamic conformance changes from glasslike to rubberlike. This behavior can be explained in terms of a multiphase theory according to which the polymer is viewed as consisting of an elastic phase and some plastic or viscous phases analogous to a sponge (elastic phase) impregnated with the viscous phase. The viscous phases soften at high temperatures, so that the sponge deforms elastically when loaded. When the composites cooled while under load, the soft phases harden and the deformation of the sponge is locked in. At room temperature the material can be sliced without removing the locked-in deformations. Deformation of the elastic phase is governed by Hooke's law, and its modulus of elasticity is practically independent of temperature. The viscous phase is assumed to consist of a multitude of components differing in viscosity and temperature dependence. The elastic deformation of the primary or elastic component is permanently locked into the model by the re-formed secondary phases. This interaction occurs at the molecular level, so that the deformation and accompanying birefringence are maintained in any small section cut from the original model. The cutting or slicing process may relieve some stresses on each face of a slice, but this layer is so thin compared to the thickness of a slice that the effect is not detectable.

The elastic component is characterized by a low magnitude of modulus of elasticity (2000 to 5000 psi). Poisson's ratio changes from 0.33 to 0.37 in the glasslike state at room temperature to about 0.5 in the rubberlike states above the transition temperature. The material's fringe constant (the measure of optical response of the material to applied mechanical stress) exhibits a similar change if the material is heated above a certain temperature T_K, which is close but not necessarily identical to the glass-transition temperature. Changes in mechanical and optical properties occur within the temperature range ($T_{GE1} \rightarrow T_{GE2}$, $T_{GK1} \rightarrow T_{GK2}$) (see Fig. 5-42) in which the material exhibits pronounced mechanical and optical creep. It is therefore possible to construct a rheological model of the material in which a spring is connected to a number of Maxwell's models. The viscosity η of dashpots of Maxwell's models is temperature dependent and drops to zero when the transition temperature, T_{GE2} or T_{GK2}, is reached. Thus the spring represents the instantaneous response of the material at temperatures higher than the glass-transition temperatures. This is shown in Fig. 5-43.

The process of three-dimensional photoelasticity is quite straightforward. A model made from optically sensitive polymers is heated to a temperature above the stress-freezing temperature. The stress-freezing temperature is the higher of T_{GE2} or T_{GK2}. It is a parameter that is specific for various materials and depends on the chemical composition and curing process of the material. After reaching this temperature, the model is thermally soaked for some time to ensure uniformity of temperature throughout. It is then loaded to levels commensurate with the lower modulus (larger deformations) associated with the softening at or near T_G, slowly cooled to room temperature, and the loads are removed. The optical response (optical anisotropy) of the material, which can be related to the mechanical stress, remains fixed in the model. The optical response is not disturbed even when the model is cut into thin slices. A polariscope can be used to analyze these slices by utilizing essentially the same technique as that developed for two-dimensional photoelasticity. Constitutive equations can be used to determine stress distribution at any surface point directly from optical measurements. If stress at an interior point is of concern, a supplementary mathematical procedure (e.g., numerical integration of equations of equilibrium) has to be employed.

Materials

Serious difficulties arose with phenolic resins, especially because of the edge effect. Condensation polymerization of phenolics involves evolution of water molecules. Evaporation of water from the surface causes the surface layer to shrink. The volume change of the surface layer is accompanied by a parasitic, time-dependent birefringence that causes difficulties in evaluating the fringe pattern induced by a load.

Epoxy resins, introduced as a photoelastic material by Leven [13], gave great impetus to three-dimensional photoelastic stress analysis because they come closest to satisfying the requirements of an ideal photoelastic material. Epoxies are castable in large sizes—no trapped air bubbles, no residual stresses, no optical nonhomogeneities or mottles; they also have high optical sensitivity, high modulus of elasticity, good machinability, and linearity of stress–strain, and stress-fringe

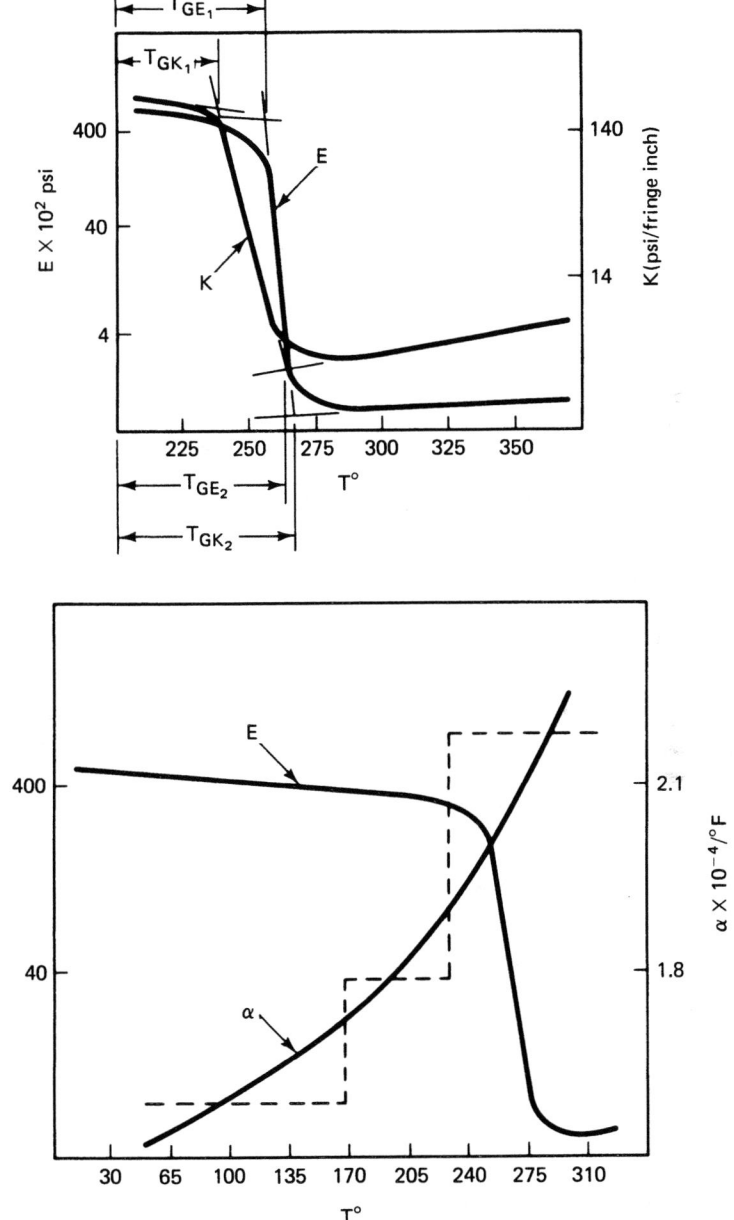

Figure 5-42 Dependence of modulus of elasticity, material fringe constant, and coefficient of thermal expansion on temperature. Material: Epoxy cured with a mixture of maleic and phthalic anhydrides. (From Ref. 14.)

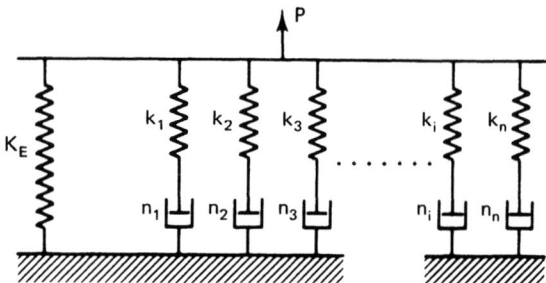

Figure 5-43 Rheological model of photoelastic material.

relations. Epoxies are castable "on shape" in the mold, have a short curing time, a low stress-freezing temperature, virtually no relaxation of frozen-fringe patterns, and only small or negligible parasitic birefringence on the surface due to moisture absorption and time effects.

Epoxy resins are condensation products of epichlorohydrin and polyhydric phenol. They can be cured by numerous curing agents, but only two types of hardener, acid anhydrides and amines, are actually used to cure photoelastic model material. In general, amine hardeners are more reactive. The curing reaction is more exothermic, and curing usually can be accomplished at room temperature. Curing with an hydrides requires elevated temperatures.

Long, linear molecular chains are formed in the initial curing phase of epoxies cured by amine hardeners. These chains are then crosslinked to form a three-dimensional structure during the second curing phase. If the molecular chains are crosslinked after having been subjected to deformation, the deformation is permanently cured into the material. This deformation is accompanied by residual birefringence, which cannot be annealed by heating the material above the glass-transition temperature. This phenomenon, although useful at times [9], precludes the use of amine hardeners for curing epoxies when models are cast "on shape" or in larger volumes. The shrinkage of the material during polymerization and its restriction by the rigid mold causes the residual birefringence to be permanently cured into the material.

A three-dimensional, crosslinked molecular structure is formed immediately in the initial phase of curing when acid anhydrides are used. This seems to prevent the formation of permanent residual birefringence in the cured material, even if the epoxy is cast around rigid enclosures.

The most commonly used curing agents are phthalic, hexahydrophthalic, and maleic anhydrides, or their mixtures. The mechanical properties (modulus of elasticity) and optical properties (material-fringe constant) depend on the base epoxy resin and the mixture of the hardeners. These materials exhibit various degrees of viscoelasticity at room temperature, but their behavior is perfectly elastic if loaded at or above stress-freezing temperature. A widely used mixture and casting procedure was proposed by Leven [13], who also tested many mixture ratios for their particular behaviors.

A common disadvantage for these materials is that the curing process requires a long time (frequently several weeks). Also, the machinability is very poor, and the products exhibit a so-called rind effect (parasitic birefringence), which renders the load-induced fringe pattern impossible to evaluate as an at-cast surface. Cernosek developed a photoelastic material that can be cast into the mold without resulting in the rind effect. The material is based on a commercially available epoxy (Epon 828, Shell Chemical) and is cured by a mixture of phthalic anhydride (50% of the resin weight) and a hardener CA-1 (1% of total resin and phthalic anhydride weight), which is a mixture of aromatic amines. This material cures overnight into a hard material exhibiting excellent machinability and negligible rind effect. The preparation of this material is as follows.

PHOTOELASTICITY

Hardener CA-1 is added to the epoxy resin at room temperature and stirred until uniform mixture color is achieved. If the hardener CA-1 is in a solid form, it can be liquefied by heating to 150°F. The hardener CA-1/epoxy mixture is heated to 220°F and then solid phthalic anhydride hardener is added. This cools the mixture, which must be reheated to 220°F while stirring to dissolve the phthalic anhydride. After the anhydride is dissolved, the mixture is stirred for 10 to 15 minutes at 220°F and then cooled to 206°F. The mixture is placed into an oven preheated to this temperature. After 2 or 3 hours, the mixture will begin to thicken. From time to time, small samples are removed from the mixture, cooled to room temperature, and pulled into a long string. If the string snaps back after breaking, the material has been advanced sufficiently and is ready to be remixed and then poured. Remixing the material is necessary to break up mottles (polymerization centers) that form frequently during polymerization [14]; the mottles do not form again before the mixture gels.

The mixture is poured into the mold, held for 2 hours at 206°F, then heated at 15°F/h to 240°F, held for 4 hours, and then cooled at 15°F/h to 200°F. Cooling below this temperature can be done naturally. This completes the preliminary curing of the material, which is, at this stage, dark brown-green. Some birefringence may be present but is due solely to "frozen" thermal stresses, which will be relieved subsequently in the stress freezing or during the final curing cycle. A unique property of this material is its lack of brittleness. It can be machined easily using standard production tools, machines, and fixtures.

The modulus of elasticity for this material after initial curing is approximately 1500 to 1800 psi at the stress-freezing temperature. If a higher modulus of elasticity is required, the model can be postcured in the loading fixture. The postcuring procedure is to heat the model to 255°F at 20°F/h, hold at this temperature for 4 to 6 hours, and heat to the critical temperature (285°F) at 15°F/h. At this temperature, the material does not exhibit optical creep.

The model is then mechanically loaded after 1 or 2 hours of thermal soaking at 285°F and cooled to 200°F at a rate of 5°F/h or less. After the 200°F temperature has been reached, the cooling rate can be increased to 10 to 20°F/h. The fully cured material is light yellow-brown. At the stress-freezing temperature the modulus of elasticity is 2800 to 3000 psi and the tensile strength exceeds 125 psi. The material's fringe constant is 1.96 to 2.15 (lb/fringe in.). The material exhibits no relaxation of frozen-fringe pattern and small or negligible time-edge effect due to moisture absorption; it can easily be cut into slices.

The material's fringe constant depends, for a particular material, on the degree of polymerization. It can vary even for materials of the same chemical composition that were cured as identically as possible. It is therefore highly desirable that a calibration specimen accompany the model and undergo the same thermal treatment (curing, postcuring, stress freezing, etc.). A circular disk loaded in diametral compression [17] is a suitable calibration specimen. The disk is easily machinable from a rod cast with the model in a separate mold, or it can be prepared from a header in the casting. If the modulus of elasticity above the stress-freezing temperature is to be determined, the ring method [18] seems to be the most versatile. Table 5-10 lists the approximate properties for a few epoxy resins.

Models can be machined whole or in parts from cast billets of material and cemented together [7] to buildup composite models. However, casting to size, which is possible by using a technique for mold preparation developed by Cernosek [14], is preferred.

Evaluation of Stress from Three-Dimensional Models

After stress freezing, the three-dimensional model is usually sliced to remove planes of interest; these can then be examined individually to determine the state of stress existing in that particular

Table 5-10 Approximate Properties of Epoxy Resins at Their Critical Temperatures

Material	f		E		Q		T
	N/m fringe $\times 10^{-3}$	lb$_f$/in. fringe	N/m$^2 \times 10^{-7}$	lb$_f$/in.^{2}	fringes/cm (fringes/m $\times 10^{-2}$)	fringes/in.	(°C)
Araldite CT200							
30 pph phthalic	0.23	1.28	1.3	1850	566	1400	140
Araldite 6020[a]							
50 pph phthalic	0.415	2.32	3.59	5100	866	220	162
Bakelite ERL 2774[a]							
50 pph phthalic	0.444	2.48	3.67	5210	828	2100	160
Standard HEX-phthalic[a]							
2 pph phthalic, 220 pph HEX	0.479	2.68	4.54	6450	945	2400	170
PLM-4B[b]	0.393	2.20	1.72	2500	447	1136	120
Epon 828[c]	0.357	2.00	2.04	2900	571	1450	140

[a] After Leven
[b] Available from Photolastic, Inc.
[c] After Ref. 14. Available from Stress-Strain Laboratories.

plane or slice. In studies of this type it must be assumed that the slice is sufficiently thin in relation to the size of the model to ensure that the stresses do not change in either magnitude or direction through the thickness of the slice.

The preferred slicing method consists of cutting the model with a diamond-impregnated wheel rotating at very high speed. The wheel must be cooled with a large amount of coolant to prevent overheating the model, which would result in distortion or annealing of the stress-frozen fringe pattern. After cutting, the slices are hand sanded using 220-grit sandpaper. The slices do not have to possess constant thickness because the evaluation of stress at a point on a free surface is a point-by-point method and requires three measurements of fringe order. The preparation of a constant-thickness slice is economical only when complete evaluation of stress in certain cross sections of the part is required.

The interpretation of the data is discussed in detail in the literature, notably by the authors of Refs. 5, 7, 14, and 16. In general, three stages of slicing and recording of the data have to be executed, as shown in Fig. 5-44.

Two methods are currently used to measure the fringe orders: subslicing as in Fig. 5-44 and the oblique-incidence method [7,21].

Even though simpler relationships are used to calculate the stress when subslicing is used, the oblique incidence is often more practical. To minimize the inherent error due to the usual assumption that the stress is constant within the subslice, it is necessary to keep the sub-subslice as small as possible. The size of sub-subslice determines the averaging effect and has the same effect on the accuracy of stress determination as the grid length of a strain gage. The smallest practical size seems to be 0.1 in. It is difficult to handle such a sub-subslice in a polariscope, and subslice preparation can be time-consuming and therefore expensive.

When using the oblique-incidence method, the slice, which is cut from a high-stress area, can be made quite thin (0.050 in. or less), to minimize the error due to the gradient of stress. This error can be further minimized by measurement of fringe orders in a tangential plane to the free surface of the model.

PHOTOELASTICITY

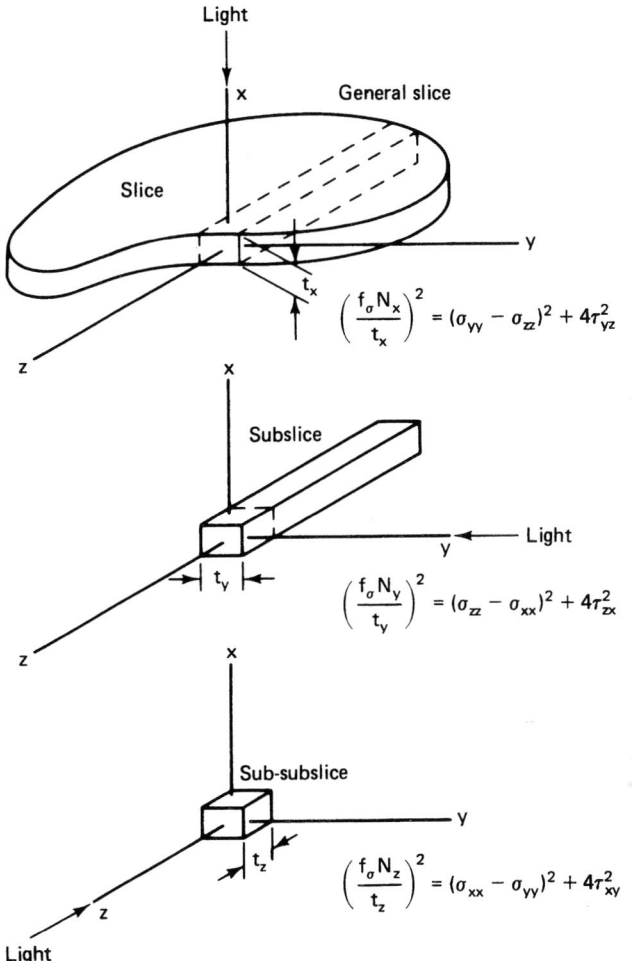

Figure 5-44 Subslicing of stress-frozen model.

5-4 Scattered-Light Methods [5,7,8,22]

A photoelastic model may be considered to have an infinite number of scattering sources uniformly distributed throughout the material. Incident light will be scattered at every point and will produce a secondary source of plane-polarized light that propagates radially outward from the source. The polarization produced by this internal scattering may be used in place of either the polarizer or the analyzer in a polariscope. This is equivalent to locating a polarizer (or an analyzer) in the interior of the model. Since either the polarizer or the analyzer can be optically positioned at arbitrary planes in the photoelastic model, stress information can be obtained without stress freezing

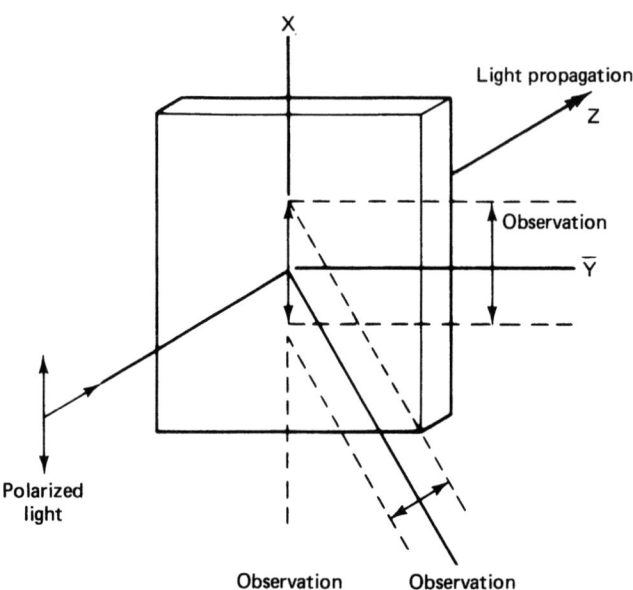

Figure 5-45 Scattered-light intensities from incident plane-polarized beam.

or slicing of the model. Scattered-light methods therefore provide nondestructive means for optical slicing in three-dimensional problems [23–25].

Figure 5-45 shows that the light scattered by a medium is linearly polarized at right angles to the direction of the transmitted light. Components of the transmitted light (optic vector) are scattered in the x-y plane but not in the x-z plane.

If an observer were to look at the "scattered" optic vector, the intensity would be zero when observed along the x axis, and the intensity, which is proportional to the vibration amplitude of the optic vector squared, would be maximum when observed along the y axis.

In scattered-light photoelasticity, the concept of secondary principal stresses is important. Thus, when any axis (such as T in Fig. 5-46) is passed through a stress point, a pair of secondary principal stresses exist that are perpendicular to T. These secondary principal stress values are not necessarily principal stresses, since they are specified relative to the T axis rather than to the stressed point in general. Such secondary principal stresses are identified as σ_1^* and σ_2^*, and if the reference axis T is a principal axis, the secondary principal stresses σ_1^* and σ_2^* are principal stresses.

Consider a monochromatic plane-polarized beam of light entering a stressed birefringent model as in Fig. 5-47. Mathematically, the optic vector of the plane-polarized beam is represented as

$$\overline{E} = \overline{A} \cos \frac{2\pi v_1 t}{\lambda}$$

where $\overline{A}$ = maximum amplitude of the polarized light
v_1 = velocity of light in its respective medium
λ = wavelength of light
t = time

PHOTOELASTICITY

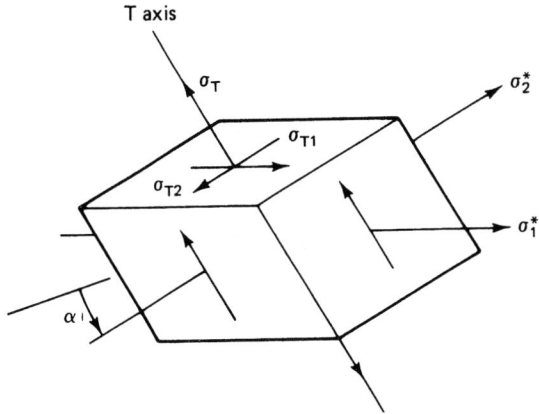

Figure 5-46 Secondary principal axes.

As the optic vector propagates, it intersects a quarter-wave plate at 45° to the principal optic axis, and the light becomes circularly polarized. The light can be represented mathematically by two orthogonal components in any direction (i.e., the secondary principal axes of the model) with one component one quarter of a wave out of phase with the other:

$$\bar{E}_1^* = \frac{\bar{A}}{\sqrt{2}} \cos \omega t$$

$$\bar{E}_2^* = \frac{\bar{A}}{\sqrt{2}} \cos \left(\omega t + \frac{\pi}{2} \right) = -\frac{\bar{A}}{\sqrt{2}} \sin \omega t$$

where $\omega = 2\pi v/\lambda$, the angular frequency of light
E_1^*, E_2^* = optic vector entering the model along the secondary principal axes

As the optic vector E^* and E^* propagates into the stressed birefringent model, one component optically slows down or is retarded relative to the other because of the state of stress in the model; thus

$$\bar{E}_1^* = \frac{\bar{A}}{\sqrt{2}} \cos \omega t$$

$$\bar{E}_2^* = -\frac{\bar{A}}{\sqrt{2}} \sin(\omega t + \delta\phi)$$

where $\delta\phi$ is the phase difference of one component relative to the other. The resultant magnitude R of the scattered light emerging from within the model can be evaluated by algebraically adding projections of E_1^* and E_2^* along a line perpendicular to the line of sight.

$$R = E_1^* \sin \alpha - E_2^* \cos \alpha = \frac{A}{\sqrt{2}} [\cos \omega t \sin \alpha + \sin(\omega t + \delta\phi)\cos \alpha]$$

or

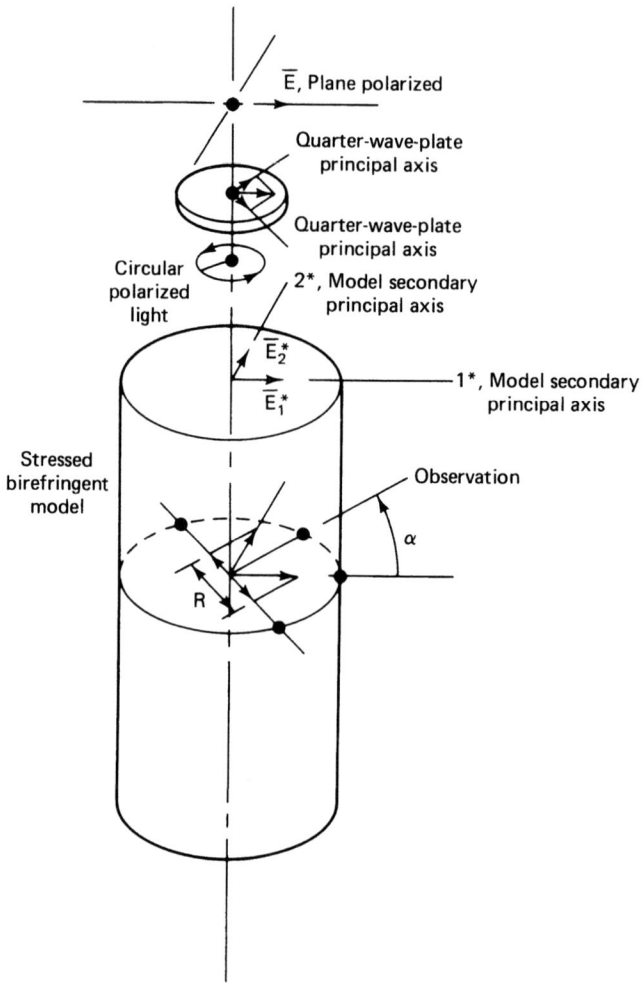

Figure 5-47 Stressed scattered-light model.

$$R = \frac{A}{\sqrt{2}} [(\sin \alpha + \sin \delta\phi \cos \alpha)\cos \omega t + (\cos \delta\phi \cos \alpha)\sin \omega t]$$

At any given point of observation, α and $\delta\phi$ are constant; therefore, the coefficients preceding $\cos \omega t$ and $\sin \omega t$ are constants. The resultant can then be expressed as

$$R = C_1 \cos \gamma \cos \omega t + C_1 \sin \gamma \sin \omega t$$

or in more compact form,

$$R = C_1 \cos(\omega t - \gamma)$$

where

$$C_1 = \frac{A}{\sqrt{2}}[\sin^2\alpha + \cos^2\alpha(\sin^2\delta\phi + \cos^2\delta\phi) + 2\sin\alpha\cos\alpha\sin\delta\phi]^{1/2}$$

or

$$C_1 \frac{A}{\sqrt{2}}(1 + \sin 2\alpha \sin \delta\phi)^{1/2}$$

and

$$\gamma = \arcsin\frac{\cos\delta\phi\cos\alpha}{1 + \sin 2\alpha \sin \delta\phi}$$

The intensity I of the resultant scattered light at a point is porportional to C_1^2. Thus

$$I = \mu \frac{A^2}{2}(1 + \sin 2\alpha \sin \delta\phi) \tag{5-50}$$

where μ is a proportionality constant.

Equation (5-50) is the basic scattered-light equation. If α is set equal to 45° (i.e., if the scattered light is observed from an angle of 45° to the maximum secondary principal axis, 1*), then Eq. (5-50) reduces to

$$I_{45} = \mu \frac{A^2}{2}(1 + \sin \delta\phi) \tag{5-50a}$$

Now, as $\delta\phi$ increases, or as the light propagates into the stressed model, the intensity expressed in Eq. (5-50a) changes from $I_{45} = \mu(A^2/2)$ to μA^2 to $\mu(A_2/2)$ to 0 and repeats while $\delta\phi$ changes from $\delta\phi = 0$ to $\pi/2$ to π to $3/2\pi$ and repeats. If instead of $\alpha = 45°$, the angle of observation is set at $\alpha = 135°$.,

$$I_{135} = \mu \frac{A^2}{2}(1 - \sin \delta\phi)$$

For this case, as $\delta\phi$ increases or as the light propagates into the stressed model, the intensity changes from $I_{135} = \mu(A^2/2)$ to 0 to $\mu(A^2/2)$ to μA^2 and repeats while $\delta\phi$ changes from $\delta\phi = 0$ to $\pi/2$ to π to $3/2\pi$ and repeats. It is noted that zero intensity occurs with $\alpha = 45°$ when

$$\delta\phi = (2n_{11} + \tfrac{3}{2}\pi) \quad \text{for } n_{11} = 0, 1, 2, 3, \ldots$$

where n_{11} is defined as "whole"-order fringes.

Furthermore, if $\alpha = 135°$, zero intensity occurs when

$$\delta\phi = (2n_{11} + \tfrac{1}{2})\pi \quad \text{for } n_{11} = 0, 1, 2, 3, \ldots$$

where n_{11} is defined as "half"-order fringes. The stress-optic law relating $\delta\phi$ to stress is

$$\sigma_1^* = \sigma_2^* = \frac{dn_{11}}{dt}f_\sigma$$

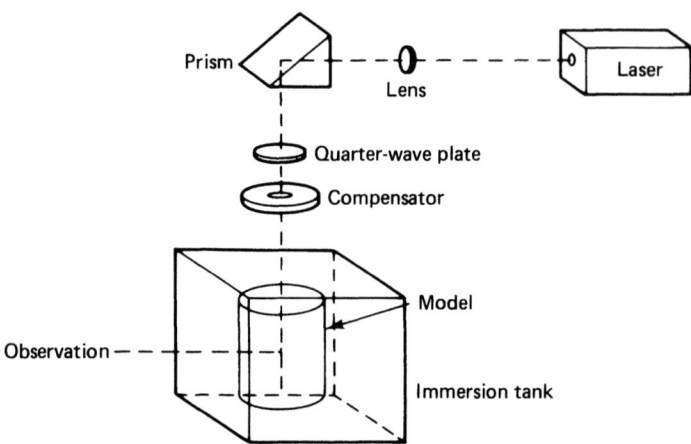

Figure 5-48 Diagrammatic representation of a scattered-light polariscope.

where $\sigma_1^* - \sigma_2^*$ is the difference in secondary principal stresses with reference to the T axis, which is the direction of the propagating light beam, and f_σ is the material-fringe constant related to the photoelastic material and the wavelength of the light source.

The direction of the secondary principal axis can be found by rotating the angle of observation until $\alpha = 0$ or $\pi/2$, at which instant the intensity of Eq. (5-50) becomes equal to $\mu(A^2/2)$. This alignment can be verified, since moving from the position in one direction reveals whole-order fringes and, in the opposite direction, half-order fringes.

The factor dn_{11}/dt of Eq. (5-51) is the change in fringe order per distance that the light travels along its respective coordinate direction. These values are found by plotting fringe order versus distance and taking the slope of the curve at the point of interest. The point of interest will be the model boundary in a surface-stress analysis. The value of f_σ is obtained by taking fringe data from a loaded model for which the stress solution is known and calculating f_σ. A tensile bar is recommended; the model is simple and relatively easy to load, and scattered-light fringe spacing is uniform.

Scattered-Light Polariscopes

The basic scattered-light polariscope is illustrated in Fig. 5-48. The laser should be at least 10 mW in power and continuous. Most continuous lasers have polarized output; if not, the beam must be polarized externally. The plane of polarization should be horizontal; otherwise, the loss incurred in being reflected by the prism is significant. The laser is mounted horizontally to minimize contamination of the laser windows from inside the tube.

A long, focal-point lens can be employed effectively to converge the light to a small size as it enters the model. This eliminates the necessity of several lenses to parallel the light and simplifies alignment. In some investigations, a cylindrical-lens arrangement can be used to produce a sheet of light, and fringe contours can be analyzed. Capability to traverse the model in a plane perpendicular to the light path is needed, as in an immersion tank with index-matching fluid.

Some materials scatter light with such small intensities that fringe patterns cannot be detected. Addition to the material of the casting of a silica-gel material having particle size on the order of micronmeters works very well (about 5 to 10 mL per pound of mix is representative). This produces extremely vivid scattered-light fringes in the polariscope. Too much gel material will cause excessive scattering, and the scatter pattern will be "washed out."

New Developments

Developments in nondestructive three-dimensional photoelasticity are moving this technique toward being the prime method for the experimental investigation of many categories of three-dimensional problems. One of these is the method of isodynes developed by Pinera [23]; another is a series of developments by Lagarde [24,25] for using the coherent properties of laser light toward effective optical slicing in scattered-light investigations for full-field as well as point-by-point analyses of problems in fracture mechanics and dynamic stresses.

5-5 Birefringent Coatings [7,26,27]

The method of birefringent coatings, or photoelastic coatings, extends transmission photoelasticity to the measurement of surface strains on opaque bodies. A thin sheet of photoelastic material is bonded to the surface of the part to be analyzed. The interface between the prototype material and the coating either is naturally reflective or is rendered reflective by including in the cement reflective particles that provide adhesion between the coating and the surface. When the prototype is loaded, the displacements on its surface are transmitted to the coating, and the resultant strain-induced birefringence is observed through a reflection polariscope as shown in Fig. 5-49a. The light traverses the coating twice along slightly different paths. The fringe pattern reveals the strain field over a large area so that it is easy to identify regions where detailed analysis should be performed. The method is nondestructive, and the need for models is eliminated.

Coating Stresses and Strains

If it is assumed that the coating is sufficiently thin, the strains on the surface of the prototype are transmitted to the coating with minimal distortion. Writing superscripts c for "coating" and s for "specimen," it is then reasonable to assume that the third principal stress $\sigma_{zz} = \sigma_3 = 0$ in both coating and specimen, so that

$$\varepsilon_1^c(x, y) = \varepsilon_1^s(x, y)$$

$$\varepsilon_2^c(x, y) = \varepsilon_2^s(x, y)$$

The surface strains are the same in the plastic coating as in the prototype. If ε_1 and ε_2 are principal strains in plastic (and metal) and β is the direction between ε and a selected reference direction, the elastic stresses in the prototype can be established from the strain by Hooke's laws,

$$\sigma_1 = \frac{E}{1 - \nu^2} (\varepsilon_1 + \nu\varepsilon_2)$$

$$\sigma_2 = \frac{E}{1 - \nu^2} (\varepsilon_2 + \nu\varepsilon_1)$$

and

$$\sigma_1 - \sigma_2 = \frac{E}{1 - \nu} (\varepsilon_1 - \varepsilon_2)$$

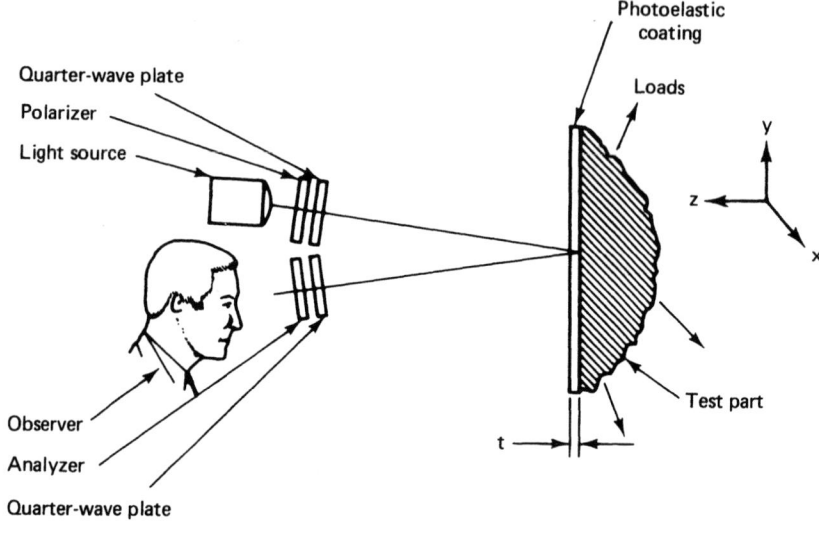

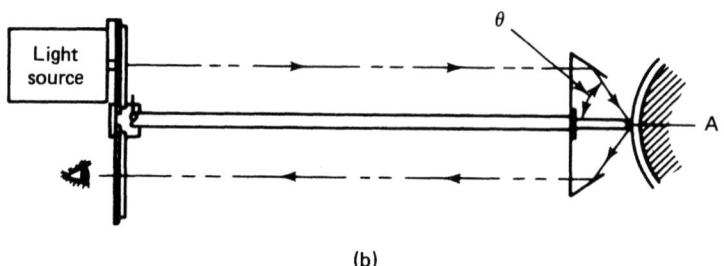

Figure 5-49 Reflection polariscopes for viewing photoelastic coatings: (a) standard setup; (b) oblique-incidence adaptor.

At every point we record the relative retardation δ between two light beams, one polarized along ε_1, the other along ε_2,

$$\delta = N\lambda = 2tK(\varepsilon_1 + \varepsilon_2)$$

where δ = retardation
λ = wavelength
N = "fringe order"

Then

$$\varepsilon_1 - \varepsilon_2 = N\frac{\lambda}{2tK} = Nf$$

PHOTOELASTICITY

where $f = \lambda/2tK =$ "fringe value" or coating sensitivity accounting for the thickness of the coating (in./in. fringe or m/m fringe

$t =$ thickness of coating; since the light passes twice through the coating, the effective path length is $2t$

$K =$ sensitivity of the plastic; the value for K is supplied by the manufacturer and derives from Brewster's law [i.e., $(n_1 - n_2) = K(\varepsilon_1 - \varepsilon_2)$]; it is known as the "strain-optical coefficient"

The difference of principal stresses in the structure is then

$$\sigma_1 - \sigma_2 = (\varepsilon_1 - \varepsilon_2)\frac{E}{1 + \nu} = Nf\frac{E}{1 + \nu}$$

In normal incidence measurements, the quantity determined is the difference of principal strains $\varepsilon_1 - \varepsilon_2$, which facilitates calculation of the difference of principal stresses $\sigma_1 - \sigma_2$. In many practical applications (edges, uniaxial fields, corners, long members) one of the principal stresses is zero or nearly zero and

$$\sigma = N\frac{fE}{1 + \nu}$$

In the case of a biaxial stress field, two measurements are needed to determine the individual principal stresses σ_x and σ_y. This usually is done by taking two oblique-incidence readings or an oblique-incidence reading combined with a normal-incidence reading.

A typical oblique-incidence adapter, shown in Fig. 5-49b, has a fixed mirror angle θ, which provides for simplified data reduction. In normal incidence

$$N_n\lambda = \delta_{\text{normal}} = 2tK(\varepsilon_1 - \varepsilon_2) \tag{5-53}$$

In oblique incidence

$$N_{01}\lambda = \delta_{\text{oblique}} = 2tK(A\varepsilon_1 - B\varepsilon_2) \tag{5-54}$$

with a rotation θ about the direction of ε_1. The coefficients A and B depend on the angle θ and Poisson's ratio for the coating.

$$A = \frac{1 - \nu^c \cos^2\theta}{(1 - \nu^c)\cos\theta} \qquad B = \frac{\cos^2\theta - \nu^c}{(1 - \nu^c)\cos\theta}$$

For a rotation θ about the ε_2 direction

$$N_{02} = 2tK(A\varepsilon_2 - B\varepsilon_1) \tag{5-55}$$

The directions of the principal strains are found from the isoclinics.

Simultaneous solution of Eqs. (5-53) and (5-54) yields the principal strains. Alternatively, two oblique measurements give the values for $\varepsilon_1 + \varepsilon_2$ from Eqs. (5-54) and (5-55).

Selection of a Coating

The sensitivity of the coating, f in Eq. (5-52), represents the difference in principal strains (or maximum in-plane shear strain) required to produce one fringe. The lower this parameter, the more sensitive the coating. From Eq. (5-52) we find

$$f = \frac{\varepsilon_1 - \varepsilon_2}{N} = \frac{\gamma_{\max}}{N} = \frac{\text{expected strain level}}{\text{maximum number fringes desired}}$$

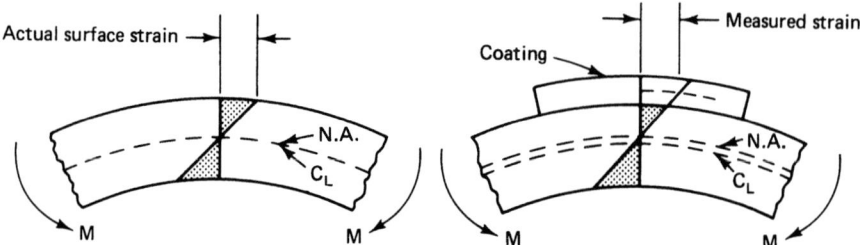

Figure 5-50 Effect of birefringent coating on thin plates in bending.

If the expected strain level corresponds to incipient yielding of the specimen material, the maximum number of fringes that can be observed is $\sigma_{\tau yp}/f$ or $\sigma_{\tau yp}/(\lambda/2tK)$. For elastic analysis, N is generally chosen to be less than 4.

The thickness t of the coating should, in general, be kept small. For curved surfaces it should generally be less than 20% of the radius of curvature. Furthermore, when thin plates are subject to bending (Fig. 5-50), the average strain in the coating is greater than the strain at the surface of the specimen. If the prototype material has a low modulus, the coating will reinforce the specimen. The neutral plane shifts toward the coating and the curvature produced by a prescribed bending moment is smaller than in the prototype. Also, the photoelastic effect is averaged through the coating. The necessary correction factor to obtain the actual strain is given in Ref. 26.

5-6 Computer-Aided Photoelasticity

Over the years, several automated polariscope systems have been developed. Notable among these are proposals by Nurse and Allison [28], Redner [29], and Mueller and Saachel [30].

However, direct exploration of digital computers was introduced by Burger and Voloshin in the early 1980 [31–33]. Their technique is based on the ability of digital image analysis systems to rapidly acquire information on light intensity over a whole field and to transfer this information into a digital storage device. Later, this information can be accessed by computer and processed to improve the quality of the image and to yield stress fields.

The system, called "half-fringe photoelasticity" (HFP) because it operates effectively with less than 0.5 wavelength of relative retardation, is outlined in Fig. 5-51. It is best visualized in terms of a traditional polariscope setup consisting of a radiation source, a polarizer plus quarter-wave plate, a model, a quarter-wave plate plus analyzer, and a means of recording the image. In place of the classical camera or the more recent photodiodes, a wholly new computer-based image analysis system now exists. This system includes:

1. A video scanner, sweeps the choosen image area. The picture is divided into 480 lines and each line is divided into 640 parts. the brightness (Z value) is converted into a video signal. This division represents 307,200 points or picture elements, called "pixels," in the picture. X values range from 0 to 639 and Y values form 0 to 479. Z values range from 0 to 255; that is, the total gray scale from white to black is divided into 256 different gray levels. This represents an 8-bit resolution.
2. A high-resolution digitizer, which digitizes the video signal in 21 seconds and places it in the Z register. The databus provides computer access to the Z register through the interface controller.

PHOTOELASTICITY

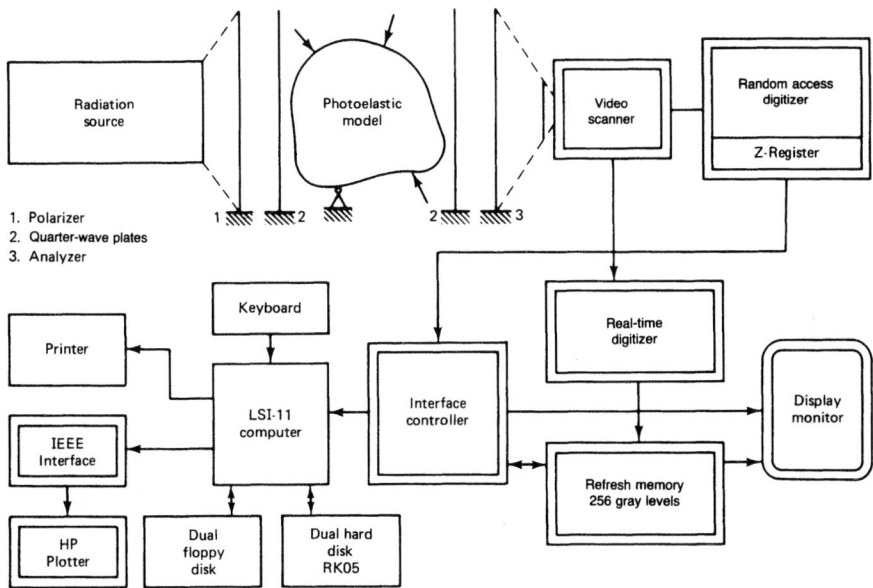

Figure 5-51 Diagram of the HFP system.

3. A real-time digitizer, which can digitize the video signal in 1/30 second. This is too fast for direct transfer to the computer, so a special digitizer data bus transfers the data to the refresh memory, where it can be accessed later by the computer.
4. A display system or monitor, which visualizes the information and acts as a graphics and numeric terminal for data processing, program development and graphical data displays.

As we discovered earlier, the intensity of the transmitted light for a dark-field circular polariscope is

$$I = K \sin^2 \frac{R}{2} \tag{5-36}$$

where K is a constant, R is the phase difference and $I = 0$ when $R/2 = n\pi$ for $n = 0, 1, 2, 3$. For these values $N = R/2\pi = n = 0, 1, 2, \ldots, n$.

In the light-field setup for a circular polariscope

$$I = K \cos^2 \frac{R}{2} \tag{5-37}$$

Extinction occurs when $I = 0$, that is, when $R/2 = 1 + 2n/2\pi$ for $n = 0, 1, 2, \ldots, n$. Then

$$N = \frac{R}{2\pi} = \frac{1}{2} + n$$

$$= \frac{1}{2}, \frac{3}{2}, \frac{5}{2}, \ldots, \frac{2n-1}{2}$$

The light-intensity distribution for Eq. (5-36) is given in Fig. 5-52.

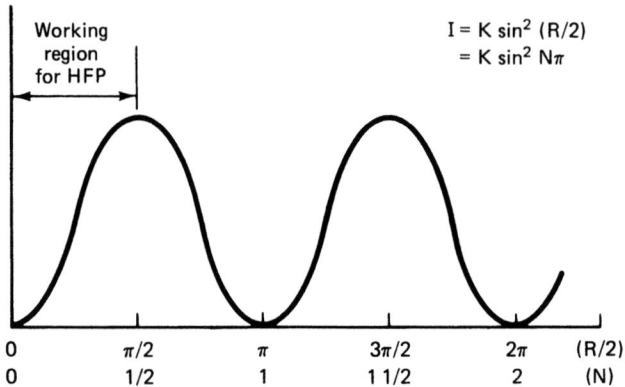

Figure 5-52 Plot of radiation intensity for a dark-field circular polariscope ($I = K \sin^2 R/2$).

In half-fringe photoelasticity the image analysis system with 80-bit resolution distinguishes between 256 gray levels in any interval from 0 to $\pi/2$, $\pi/2$ to π, and so on. in cannot identify the interval; that is, it cannot identify which half-fringe it is reading. The experimental parameters must, therefore, be chosen so that $0 > \Delta/2 > \pi/2$; $0 > N > 1/2$. The digitization of the intensity plot occurs on the I axis so that equal gray-level divisions represent uneven stress increments.

To keep $N = (\sigma_1 - \sigma_2)h/f_\sigma < \frac{1}{2}$, the parameters h, f_σ, and $(\sigma_1 - \sigma_2)$ must be selected accordingly. This requires that one or all of the following be true:

1. Low $(\sigma_1 - \sigma_2)$; that is, low shear stresses in the models. This usually means that the model loads will be low.
2. High f_σ; that is, low stress birefringence in the model material.
3. Small h; that is, thin models.

The first and last requirements are desirable in all photoelastic work. Low loads mean small deformations, reduced nonlinearity, and better modeling of prototype response. Thin models approximate plane stress better and make it easier to meet the requirement that the $(\sigma_1 - \sigma_2)$ field not vary through the thickness of the model or slice. High f_σ (i.e., lower sensitivity to loads) is generally an undesirable characteristic. The drive for low values of f_σ has precluded the use of many model materials. for HFP, a high f_σ is often desirable, since one must guard against having a fringe order that is too high.

Now consider Eq. (5-36) rewritten to incorporate the reference intensity I_0 from Eq. (5.42) for a particular model and polariscope combination.

$$I = I_0 \sin^2 \frac{R}{2}$$

or

$$\frac{R}{2} = \arcsin \sqrt{\frac{I}{I_0}}$$

Then Eq. (5-18) becomes

PHOTOELASTICITY

$$\sigma_1 - \sigma_2 = \frac{\delta}{Ch} = \frac{2\lambda}{\pi Ch} \arcsin \sqrt{\frac{I}{I_0}} \tag{5-56}$$

The digital optical system here described is unequaled in its ability to measure I/I_0 to a very high degree of resolution. The response Z of the video scanner to light is log linear

$$Z = K'I^\gamma$$

where Z = digitized output value for each pixel in the field
I = light intensity of that picture element
γ = log-linear slope of the camera sensitivity
K' = a proportionality constant

If the response to the brightest point I_{max} in the field is Z_{max}, then

$$Z_{max} = K'I^\gamma_{max}$$

Thus the intensity at any point in a particular field can be qualitatively calibrated with respect to the brightest point in that field. This is related to the photoelastic effect, such that

$$\sin^2(N\pi) = \left(\frac{Z}{Z_{max}}\right)^{1/\gamma} \sin^2(N_{max}\pi)$$

The HFP procedure requires that $N_{max} \leq \tfrac{1}{2}$, so that for the highest permissible response

$$\sin^2(N_{max}\pi) = 1$$

or

$$N = \frac{1}{\pi} \sin^{-1}\left(\frac{Z}{Z_{max}}\right)^{1/2\gamma}$$

The exact values for N at any point can be obtained from Tardy compensation by aligning the polariscope with the isoclinics at the point and then rotating the analyzer from the dark-field position until the indicated intensity (Z value) is a minimum at that point. If this procedure is repeated for a number of different points in the field, a calibration curve can be drawn. On a log-log plot, a straight line is a best fit through the data points, so that

$$N = \frac{1}{\pi} \sin^{-1}\left(\frac{Z}{BZ_{max}}\right)^{1/2\gamma}$$

with B close to unity ($B = 1 \pm 0.05$) and γ the slope of the line.

From the equations for an arbitrary analyzer position, that is, for the equations used in Tardy compensation, we have the light intensity emerging from a circular polariscope with the analyzer rotated an arbitrary angle β from the dark-field position as

$$I_\beta = K(1 - \cos 2\beta \cos \Delta - \sin 2\beta 2 \cos 2\alpha \sin \Delta)$$

where β = angle of rotation of the analyzer from the crossed or dark-field position
α = isoclinic angle
Δ = relative retardation

If the polariscope has been first aligned with the directions of the principal stresses at the point under observation (i.e., if it is aligned with the isoclinics), the isoclinic angle α is zero, and

$$I_\beta = K(1 - \cos 2\beta \cos \Delta - \sin 2\beta \sin \Delta)$$

If the analyzer is rorated plus or minus 90° from this position,

$$I_{\beta \pm 90°} = K(1 + \cos 2\beta \cos \Delta + \sin 2\phi \sin \Delta)$$

Adding yields

$$I_\beta + I_{\beta \pm 90°} = 2K = 2I_0$$

Then

$$I_0 = \tfrac{1}{2}(I_\beta + I_{\beta \pm 90°}) \tag{5-57}$$

where I_0 is considered to be a reference intensity, so that

$$I = I_0 \sin_2(N\pi)$$

Note that I_0 or K can be found directly from the equation for I_β without first aligning the polariscope with the isoclinic. Thus the sum of intensities from two mutually perpendicular settings of the analyzer is a constant, $2I_0$. This value does not depend on whether the model is loaded. It is a reference intensity for each point in the field of the model and includes all variations in the field density, whether introduced by the model, the light source, or the optical elements.

If there is any long-term drift in the calibration of the camera, that is, in K' or γ, this Tardy procedure will confirm it. It is therefore a standard procedure to perform a Tardy compensation procedure before and after each test.

This technique has been applied to a wide range of problems, including fracture mechanics under quasi-static [34,35] and transient thermal stresses [36], in two- and three-dimensional problems [37], and in composite materials [38,39].

5-7 Dynamic Photoelasticity [5,16,40]

Dally [41] established the validity of the stress-optic law for the dynamic case. Thus the difference in the principal stresses $\sigma_1 - \sigma_2$ can be determined in the normal manner with

$$\sigma_1 - \sigma_2 = \frac{Nf_\sigma^*}{h}$$

where N = fringe order
h = model thickness
f_γ^* = dynamic material-fringe value, which is usually 10 to 30% higher than the static-fringe value

Separation of the principal stresses to obtain the individual values of the principal stresses is usually much more difficult than in the static problem, because the most common static-separation methods are based on the stress equations of equilibrium or Laplace's equation; these are not valid for the dynamic problem. Separation of σ_1 and σ_2 is possible in certain cases by using only isochromatic data [41].

Free Boundaries

For any wave propagating along a free boundary with the x axis defining the free boundary, $\sigma_1 = \sigma_x$, $\sigma_2 = \sigma_y = 0$, and $\tau_{xy} = 0$. The stress-optic law reduces to

$$\sigma_1 = \frac{Nf_\sigma^*}{h}$$

Rigid Boundaries

For a dilatational wave propagating along a rigid boundary that is coincident with the x axis, it is evident that the rigid boundary constrains the displacement field, so that

$$\frac{\partial u_x}{\partial x} = \frac{\partial u_x}{\partial y} = \frac{\partial u_y}{\partial y} = \frac{\partial u_y}{\partial x} = 0$$

and it follows that the Oxy coordinates are principal and that

$$\sigma_1 = \frac{1}{1 - \nu} \frac{Nf_\sigma^*}{h}$$

$$\sigma_2 = \frac{-\nu}{1 - \nu} \frac{Nf_\sigma^*}{h}$$

Axisymmetric Radial-Wave Propagation [40]

An axisymmetric dilation wave will be produced if the load is a center of dilation at some interior point in the model. The wave propagates radially outward from the center with a displacement field at any instant of time represented by

$$u_r = f(r) \quad \text{and} \quad u_\theta = 0$$

Because of the limited nature of the displacement field, it can be shown that the strain field is given by

$$\varepsilon_{\theta\theta} = -\frac{1 + \nu}{E} \frac{f_\sigma^*}{h} \int_r \frac{N}{r} dr$$

$$\varepsilon_{rr} = -\frac{1 + \nu}{E} \frac{f_\sigma^*}{h} \left(\int_r \frac{N}{r} dr + N \right)$$

and the stresses are obtained from the biaxial form of Hooke's law as

$$\sigma_{rr} = \frac{E}{1 - \nu^2} (\varepsilon_{rr} + \nu \varepsilon_{\theta\theta})$$

$$\sigma_{\theta\theta} = \frac{E}{1 - \nu^2} (\varepsilon_{\theta\theta} + \nu \varepsilon_{rr})$$

Shear-Wave Propagation

Since the shear wave propagates with no volume change, the first invariants of stress I_1 and strain J_1 are zero, and it is evident that

$$\sigma_1 = -\sigma_2$$

$$\sigma_1 = -\sigma_2 = \frac{Nf_\sigma^*}{2h}$$

For the general case of two or more stress waves superimposed at interior regions of the model, separation of the principal stresses cannot be accomplished without additional experimental data. Riley [42] has shown that the moiré method can be applied to determine one component of the dynamic strain field, which can be used in conjunction with the isochromatic data to separate the stresses. Holloway et al. [43] have developed an ingenious optical system where the isochromatic and isopachic patterns are recorded simultaneously to give both $\sigma_1 - \sigma_2$ and $\sigma_1 + \sigma_2$, which permits direct separation.

Dynamic Calibration

Dynamic calibration of photoelastic materials [44] is usually accomplished by impacting the end of a long rod with a square cross section by a projectile fired from an airgun. Strain gages (two-element rosettes) are mounted in the center of the rod on the top and bottom surfaces, and a beam of polarized light is transmitted through the vertical surfaces of the rod at this same location. The intensity of light from the polariscope is monitored with a photodiode to obtain $N(t)$, and the strain gages provide $\varepsilon_a(t)$ and $\varepsilon_t(t)$. The signals are recorded simultaneously on an oscilloscope. Strain gages are also mounted at a second station, so that the speed of propagation of the bar wave can be measured. The optical and mechanical properties of the material are obtained from these data by using the following equations:

$$\text{Poisson's ratio } \nu = \frac{-\varepsilon_t(t_1)}{\varepsilon_a(t_1)}$$

$$C_B = \frac{s}{t_*}$$

where t_* is the transit time for the bar wave to travel the distance s between the two strain-gage stations and C_B is the bar-wave velocity. Also, with the density ρ,

$$E = C_B^2 \rho$$

$$f_e^* = \frac{h(1+\nu)}{N(t)} \varepsilon_a(t)$$

$$f_\sigma^* = \frac{E}{1+\nu} f_e^*$$

Model Laws for Dynamic Investigations

In addition to the usual requirement of geometrical similarity, in dynamic investigations the shape of the pulse should be similar in model and prototype, and it should be to the same scale as the linear dimensions. In problems of impact, this requires that the striking body also be reproduced true to scale and that the densities and elastic moduli of the striking and struck bodies have the same ratios as for the prototype, that is,

$$\frac{\rho_{sp}}{\rho_p} = \frac{\rho_{sm}}{\rho_m} \quad \text{and} \quad \frac{E_{sp}}{E_p} = \frac{E_{sm}}{E_m}$$

PHOTOELASTICITY

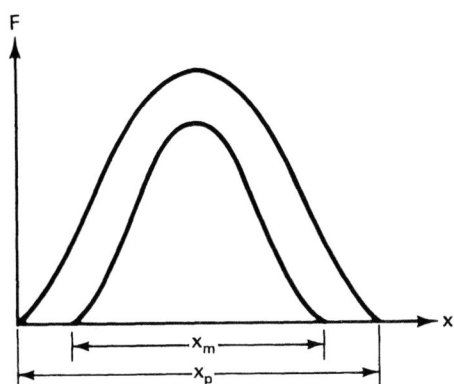

Figure 5-53 Similitude between waves in model and prototype.

where the subscript s refers to the striking body, while p and m indicate model and prototype, respectively.

For equal inertia forces in rods, the equation of motion gives

$$\frac{\partial \sigma_p/\partial x_p}{\partial \sigma_m/\partial x_m} = \frac{\rho_p \alpha_p}{\rho_m \alpha_m} = 1$$

while similarity of the pulse (see Fig. 5-53) requires

$$\frac{\partial F_p}{\partial x_p} = \frac{\partial F_m}{\partial x_m} \quad \text{and} \quad \frac{F_p}{x_p} = \frac{F_m}{x_m}$$

The relation between the velocities of two striking bodies follows:

$$\frac{m_p v_p}{m_m v_m} = \frac{\int F_p dt}{\int F_m dt} = \frac{\dfrac{A_p}{c_p}\int \sigma_p dx}{\dfrac{A_m}{c_m}\int \sigma_m dx}$$

In this equation, $\int \sigma\, dx$ can be replaced by σ_x or σ_L; observing also that $m \doteq \rho L^3$ and $A = L^2$, it yields

$$\frac{v_p}{v_m} = \frac{c_m \rho_m \sigma_p}{c_p \rho_p \sigma_m}$$

Assuming that the requirements for analogous conditions are otherwise satisfied, corresponding stress distributions will exist in the model and in the prototype at different instants during the event when the stress waves have traveled corresponding distances. This occurs when

$$\frac{t_m}{t_p} = \frac{L_m c_p}{L_p c_m}$$

where t is the time elapsed since the beginning of the event and c is the velocity of propagation. In rods and in other problems with $v_m = v_p$,

$$\frac{C_p}{C_m} = \sqrt{\frac{E_p}{E_m} \frac{\rho_m}{\rho_p}}$$

The equation above then becomes

$$\frac{t_m}{t_p} = \frac{L_m}{L_p} \left(\frac{E_p}{E_m}\right)^{1/2} \left(\frac{\rho_m}{\rho_p}\right)^{1/2}$$

Dynamic photoelasticity has been effective in solving many problems in such fields as geophysics [45], fracture [46], and ultrasonic inspection [47].

Recording Methods

Perhaps the most difficult aspect of dynamic photoelasticity is recording the isochromatic-fringe pattern, which represents the transient state of stress. Recording is difficult because the fringes propagate at extremely high velocities, ranging up to 2500 m/s for dilation waves in two-dimensional models fabricated from Homalite 100. Of the many systems for high-speed photography, two have found general application in dynamic photoelasticity.

The Cranz–Schardin spark camera [48] has typical exposure times of between 400 and 500 ns. A modification of this system using multiple pulsed-laser sources rather than arcs has exposure times of less than 20 ns for each frame. These systems are well suited to studies of wave propagation [49,50] and fracture [51].

5-8 Photothermoelasticity [52]

Both transient and steady-state thermal stresses can be determined from photoelastic models because the birefringent response is generated by self-equilibrated thermal stresses, which, in turn, are generated by temperature distributions in the models.

When a body is subjected to a uniform change in temperature but is free to expand, it exhibits a thermal strain. If the coefficient of thermal expansion of the material is constant over the temperature range T_1 to T_2, the thermal strain is $\varepsilon_t = \alpha(T_2 - T_1) = \alpha(\Delta T)$. If the body is not free to expand, thermal stresses are induced. These stresses are proportional to the coefficient of thermal expansion α, the modulus of elasticity E, and the temperature change ΔT. For uniaxial constraint, such as the ease of a long bar that cannot expand axially, the stress $\sigma_t = E\alpha (\Delta T)$. If the constraint is biaxial, as when a plate is constrained along its edges, $\sigma_t = \alpha E(\Delta T)/(1 - \nu)$, where ν is Poisson's ratio for the material. Fortunately, restraints are not always absolutely rigid, and the stresses are reduced by a factor K, known as the coefficient of restraint. Then $\sigma_t = KE\alpha(\Delta T)$ with $K < 1$.

Transient thermal stresses arise when the temperature in one region or on one surface of a part changes. The resulting stresses vary with time as the temperatures adjust to new equilibrium conditions [52,53]. Both the magnitudes and distributions of thermal stresses depend on how "suddenly" the surface temperature changes. In a few rare instances, such as near-surface explosions or pulsation of a high-energy laser beam onto a small surface area, the thermal shock is so sudden that elastic waves are generated in the material so that the thermal stresses are "coupled" with the inertia response of the material. This is true thermal shock.

True thermal shock, which sets up stress waves analogous to impact waves, requires substantial temperature changes in times short compared to the mechanical response time of a structure. Since periods of free vibration are of the order 10^{-4} to 10^{-1} s for most structures, inertia effects will be important only if heating times are 10^{-6} to, perhaps 10^{-2} s.

When the heat flow is distributed by some discontinuity, there occurs an elevation of load temperature that produces rapidly changing stress concentrations or stress intensity factors [54]. Thermoelastic stress intensity factors at cracks maximize during a transient [55].

Similarity Relations [56]

Several of the factors that influence experiments in thermal stresses do not occur in isothermal tests. In addition to the material and heat-transfer parameters discussed above, scaling factors are needed for size, temperature, and rate of heating. Hovanesian and Kowalski [56] worked out suitable similarity relations, which include temperature-scaling factors, size factors, and time-scaling factors. With subscripts m for model and p for prototype, these relations predict that for plane-strain conditions, the following stress ratio from prototype to model:

$$\frac{\sigma_p}{\sigma_m} = \left(\frac{E_p}{E_m} \frac{\alpha_p}{\alpha_m} \frac{1 - \nu_m}{1 - \nu_p}\right) \frac{\Delta T_p}{\Delta T_m}$$

where E is the elastic modulus, α the coefficient of thermal expansion, ν Poissons's ratio, and ΔT the temperature difference. Similarly, the time ratio τ, which relates the time for a certain temperature profile to develop in a real structure to that in a model, is

$$\tau = \frac{t_p}{t_m} = \left(\frac{k_m}{k_p} \frac{C_p}{C_m} \frac{\rho_p}{\rho_m}\right) \left(\frac{l_p}{l_m}\right)^2 = \frac{\beta_m}{\beta_p} \left(\frac{l_p}{l_m}\right)^2$$

where C is the specific heat, ρ the density, and k the thermal conductivity; l_p/l_m is the scale of the model, and $\beta = k/\rho C$ is the thermal diffusivity. Approximate values for β are: steel = 0.54 ft²/h (500 cm²/h), concrete = 0.02 ft²/h (19 cm²/h), and epoxy model materials = 0.006 ft²/h (6 cm²/h). Thus τ for a full-scale plastic model of a steel prototype will be between 0.01 and 0.012; that is, the time for an event to occur in the model may be 100 times longer than in a metal prototype of the same size. Such a slowdown of events in the model is an obvious advantage when studying thermal shock and other rapidly occurring events, but it comes at a price. Since events occur slowly, any induced irregularity in the temperature–time history will persist in the model for a long time, maybe 30 minutes or more, control over surface temperatures must be exact and must be maintained for long periods. This is often difficult and usually expensive to do. In problems concerning thermal stress in concrete, models are usually much smaller than the prototype. For a 1:100 scale model, $\tau \simeq 3000$. Heat development during hydration of concrete in large structures occurs over a period of from 20 to 30 days, so that the resulting thermal stress problem can be modeled in epoxy in 10 to 15 minutes.

Modeling Materials [53,57]

Since the thermal diffusivity, $\beta = k/\rho C_p$, is the governing parameter for heat flow within a solid in transient conditions, the properties of the thermal conductivity k, specific heat C_p, and density ρ, together with other properties, such as the coefficient of thermal expansion α, Young's modulus E, Poissons's ratio ν, and the photoelastic fringe coefficient f_σ, feature strongly when the time of occurrence of certain stress states in a thermal stress model is scaled with respect to a prototype.

It is common practice in photoelasticity to employ a figure of merit Q_t to compare different model materials. For photothermoelasticity, the figure of merit is of the form $Q_t = E\alpha/f_\sigma$. The important heat-transfer parameter (thermal diffusivity) also should be used in conjunction with the figure of merit in the evaluation of model materials for transient thermal stress analysis.

Reference 57 presented the properties for three materials: a room-temperature-cured epoxy, Araldite 502 with hardener 951; a hot-cured epoxy, EPon 828 with phthalic anhydride hardener; and a polyester PSM-1, available from Photoplastic Inc.

5-9 Photoplasticity [58,59]

Optical birefringence in the inelastic range was first observed in glass in the early twentieth century by Filon [15]. It was, however, not until the early 1950s that much additional published work on the subject appeared. Fried observed the optical effects in different materials when they were stressed beyond the elastic limit. The study included polystyrene, Lucite, Plexiglas, nylon, cellulose acetate, silver chloride, and celluloid. He found celluloid to be an acceptable material and used it to determine stress-concentration factors and strain distributions under creep conditions.

Burger et al. [60] showed that polyesters may be used to obtain three-dimensional photoplastic data for rolled billets after unloading. By varying the mixture ratio and the test temperature, the plastic-flow properties of the material could be adjusted to match the flow of hot metals reasonably well. Their preferred materials were two mixtures of a "rigid" and a "flexible" resin [i.e., 60/40 and 70/30 parts by weight of Laminac polyester resin 4116 (rigid) and 4134 (flexible)]. These were deformed at temperature between 40 and 51°C. Tables 5-11 [62–72] and 5-12 [58,60,63–67,70–83] and Refs. 59 and 61 review the methods and applications of photoplasticity.

5-10 Flow Birefringence [84]

The birefringent- or double-refractive fluid-flow method is a technique for visually studying complex fluid-flow problems. The majority of the existing visual techniques yield only qualitative information on the fluid motion. The flow-birefringent method for quantitative studies of two-dimensional laminar-flow problems makes use of the phenomenon of flow birefringence or flow double refraction, which certain liquids and colloidal systems exhibit. The interference fringe patterns relate quantitatively to the deformation rate of the flowing fluid. The method has a definite advantage over many experimental fluid-dynamics measurements in that it requires no probe or measuring device in the fluid; hence no local disturbance is introduced.

The theories of flow birefringence can be divided into two groups: (1) stretching and orientation of long polymerlike chains and (2) deformation or orientation of suspended macromolecules or colloidal particles [85,86].

Theories in the first group have been applied to flow-birefringent studies of solutions of high polymers. In these theories, the stress-optic law, analogous to that used in photoelasticity, relates the orientation of the optical axes to the principal directions of the stress field in the polymer solution, and the relative retardation to the difference of the principal stresses in the plane perpendicular to the polarized-light beam. The second group of theories can be further divided into deformation theory, where the birefringence is assumed to be related to the deformation of ordinarily optically isotropic particles, and the orientation theory, which assumes that flow birefringence is caused by the orientation of geometrically and optically anisotropic-particles. In both theories of

Table 5-11 Methods in Photoplasticity

Reference	Material	Comments
Bidimensional: Loaded		
Frocht and Cheng [62]	Celluloid	Isochromatics and isoclinics are related to principal stress differences and directions; stress concentrations are studied; shear difference method is applied
Monch and Loreck [63]	Celluloid	Dispersion of birefringence is used to determine elastoplastic boundaries
Ito [64], Brill [65], Whitfield and Smith [66]	Polycarbonate	Birefringence is related to strain
Zachary and Riley [67]	Polyester	Checks on equilibrium using four-point bending of beams; birefringence is related to strains
Ohashi [68]	Celluloid and Araldite	Photorheological method; time-dependent rates of creep strain and instantaneous elastoplastic strain are considered
Bidimensional: Unloaded		
Zachary and Riley [67]	Polyester	Birefringence is related to strains; stress concentration around a circular hole is a plate
Three-Dimensional: Unloaded		
Burger et al. [69]	Polyester	Locked-in plastic birefringence is used to determine plastic strains; slicing techniques are used; models for hot metal forming processes (e.g., rolling, upset forging, and extrusion)
Dally and Mulc [70]	Polycarbonate	Unloaded birefringence is used to determine plastic strains
Hunter [71]	Epoxy resin	Epoxy is subjected to a thermal cycle whose maximum temperature is well below the critical temperature; creep and frozen characteristics are studied
Three-Dimensionial: Loaded		
Johnson [72]	Cellulose proprionate	Viscoelastic behavior is used to simulate elastoplastic behavior of metals; method uses scattered-light photomechanical techniques

Source: Ref. 61.

the second group, the birefringence is related to the shear-strain rate. The relationship is linear at low shear-rates and becomes nonlinear at high shear-strain rates [86].

Reference 84 gives a comprehensive review of developments in two-dimensional birefringent flow. In investigations of such phenomena, the birefringence is taken as the integrated value through the light path. The boundary effects of the channel walls perpendicular to the light beam cannot be evaluated. A three-dimensional method is needed to relate the distribution of local strain rate and velocity to the birefringence of the flow. The scattered-light method offers the potential for three-dimensional birefringent-flow problems [87].

Birefringent Fluids

The fluids used for birefringent-flow investigations can be divided into two groups. The first group includes various polymer solutions. Polyisobutylene, polystyrene polymer solutions, and ethylcellulose polymer solutions belong to this group. They have been shown to follow the flow-optic theory of stretching and orientation of long-chain molecules. These fluids have been used to study flows of viscoelastic fluid by flow birefringence. To the second group belong birefringent fluids containing macromolecules or colloidal suspensions. Some of them have been assumed to follow

Table 5-12 Applications of Photoplasticity [61]

Applications	References
Stress concentration	
Bar with notches or grooves	Hunter [71], Frocht and Thomson [73], Javornicky [58]
Bar with hole	Ito [64], Frocht and Thomson [73], Brill [65], Whitfield and Smith [66], Monch and Loreck [63], Javornicky [74], Morris and Riley [75]
Grooved shaft in torsion	Johnson [72]
Metal forming	
Compression of cylinder	Dally and Mulc [70], Freire et al. [76]
Extrusion	Burger and Koenig [77]
Hot rolling	Oyinlola, Burger, et al. [60, 78]
Ring upset	Burger et al. [60], Gomide and Burger [79]
Indentation	Nisida et al. [80]
Four-point bending	Zachary and Riley [67]
Compression of wedge	Nisida et al. [80]
Others	
Brittle fracture of polyester	Ito [81]
Dugdale model (polycarbonate)	Brinson [82], Theocaris and Gdoutos [83]

the birefringent theory of deformation of these ordinarily isotropic particles. Most of the others have been established from experimental results to follow the orientation theory of the originally anisotropic solid particles. Vanadium pentoxide solutions, bentonite solutions, sesame oil, tobacco mosaic virus, pure ethyl cinnamate, and milling yellow colloidal suspensions belong to this group. Most of these fluids are difficult to prepare and exhibit the undesirable properties of extremely high viscosity, low optical sensitivity, and instability on contact with common structural materials. The fluid that does not have these undesirable features is the aqueous colloidal suspension of milling yellow N.G.S., a commercial dye (supplied by Keystone Aniline and Chemical Co.).

The popularity of milling yellow is due to its enormous optical sensitivity compared to other birefringent fluids. Its optical sensitivity as well as viscosity increase for increased dye concentration and also depend on temperature. The optically sensitive liquids have milling yellow concentration between 1.2 and 2.0% by weight in water. Only less than 1% by weight of milling yellow powder can be dissolved in water at room temperature. The basic method for preparing a working fluid is to add about 1% by weight powder to be distilled water at boiling temperature. The mixture is then boiled to obtain the desired concentration of the dye suspension. A complete discussion of milling yellow suspension preparation was given by Hirsh [86]. The solubility properties of milling yellow were discussed by Swanson and Green [87]. A survey of properties of milling yellow solutions was presented by Peebles et al. [88].

Milling yellow suspension exhibits non-Newtonian rheological characteristic in general. For the dye concentrations mentioned above and at a temperature range between 22.4 and 29.8°C, the fluid exhibits quasi-Newtonian behavior at the very low shear rate of $3 \times 10 \text{ s}^{-1}$ and again above 500 s^{-1}. In between these shear rates it is clearly non-Newtonian. For quantitative evaluation, it is desirable to work in the low-shear-rate Newtonian region and consequently with low Reynolds numbers.

5-11 Orthotropic Photoelasticity

The application of transmission photoelasticity to birefringent, orthotropic model materials is known as orthotropic photoelasticity. The model materials are composites, usually glass-fiber-reinforced materials, rendered transparent by careful matching of the refractive indices of the fibers and the matrix [89,90]. Although the material is heterogeneous on the microscopic scale, it is treated as a continuum on the macroscopic scale and, as such, exhibits well-defined photoelastic responses [91].

Several proposed theories of photo-orthotropic elasticity are reviewed in Ref. 92, and an exact strain-optic law was proposed by Agarwal and Chaturvedi in 1982 [93]. Stress separation through oblique incidence is also possible [94], and the inherent variability of the optical properties in the different directions can be overcome by a least-squares approach to the calibration of orthotropic model materials [95]. The application of digital image procedures to orthotropic photoelasticity [96] promises to overcome difficulties in analysis associated with the low levels of birefringence that are characteristic of these materials.

5-12 Major New Developments and Applications of Photoelastic Methods

The outstanding feature of the photoelastic method is the precise analysis of the shear-stress distribution within models. However, for three-dimensional models it can establish only five independent equations. This is insufficient to solve for six unknown stress components.

Parks and Sanford [97] showed that if correct account is taken of material behavior, holophotoelastic methods can be used for rapid and complete stress analysis. Sanford reviews *photoelastic holography* in an excellent paper [98] that describes the equipment and typical applications.

The method of *integrated photoelasticity* was introduced by Aben [99,100]. The technique itself is simpler than other methods of three-dimensional analysis, but the theory is complicated because it has to take account of the laws that govern light propagation in anisotropic, nonhomogenous media. It is very suitable for full stress analysis of transparent items that contain symmetry.

Lagarde has introduced several related methods that utilize the coherence of laser light sources in two- and three-dimensional problems for nondestructive evaluation of stress states [101]. These methods, which include *ellipsometry, punctual point methods,* and three-dimensional *optical slicing,* merit serious attention because they represent a major advance in the concepts of photomechanics.

Photoelasticity is making major contributions in the exciting new area of experimental fracture mechanics [102]. Particularly notable is the work by Smith, which is reviewed in Ref. 102 and refined in a series of papers summarized in Refs. 103 and 104. Strong contributions continue to be made by Dally [105], Sanford [106], Burger et al. [55,107], and Kobayashi and Barker [108].

Many materials have desirable properties for modeling but are opaque to radiation in the visible range. They are, however, transparent and birefringent to near-infrared and ultraviolet light. There is great potential for the use of photoelastic techniques with these materials.

It is impossible to cover all the successful and potentially successful applications of stress or strain birefringence to important problems in technology. Two applications, however, need to be named because they are in fields normally considered to be beyond the interests of experimental stress analysts. Here we mention the applications of photoelasticity in biomechanics practiced by Arcan and Brull [109] and Robert Mark's [110] exciting studies on historical buildings, which produced new insights into the history of the design and construction of monumental buildings.

References

1. W. G. Driscoll and W. Vaughan, *Handbook of Optics* (Optical Society of America), McGraw-Hill, New York, 1978.

2. *Optics and Filters*, Vol. III, Oriel Corporation, Stratford, CT, 1984.

3. *Optics Guide*, Melles Griot, Irvine, CA, 1981.

4. *The Optical Industry and Systems Purchasing Directory*, Book 2, Optical Publishing Company, Pittsfield, MA, 1983.

5. A. Kuske and G. Robertson, *Photoelastic Stress Analysis*, Wiley, New York, 1974.

6. H. T. Jessop and F. C. Harris, *Photoelasticity: Principles and Methods*, Dover, New York, 1949.

7. J. W. Dally and W. F. Riley, *Experimental Stress Analysis*, McGraw-Hill, New York, 1978.

8. A.S. Holister, *Experimental Stress Analysis*, Cambridge University Press, New York, 1967.

9. L.H. Adams and R. M. Wasler, Superimposed Birefractory Plates, *J. Res. Natl. Bur. Stand.*, 69c (1965), 103–114.

10. W. H. J. Childs, Composite Quarter- and Half-Wave Plates, *J. Sci. Instrum.*, 1956, 298–301.

11. D. Post, Photoelastic Fringe Multiplication for Ten Fold Increase in Sensitivity, *Exp. Mech.*, 10 (1970), 305–312.

12. J. H. Flanagan, Photoelastic Photography, *Proc. SESA*, XV (1958), 1–10.

13. M. M. Leven, Epoxy Resins for Photoelastic Use, *Proc. Int. Symp. Photoelasticity* (M. M. Frocht, Ed.) Pergamon Press, Elmsford, NY, 1962, 145–165.

14. J. Cernosek, Three-Dimensional Photoelasticity by Stress Freezing, *Exp. Mech.*, 20 (1980), 417–426.

15. E. G. Coker and L. N. G. Filon, *A Treatise on Photoelasticity*, Cambridge University Press, New York, 1957.

16. A. J. Durelli and W. F. Riley, *Introduction to Photomechanics*, Prentice-Hall, Englewood Cliffs, NJ, 1965.

17. A. J. Durelli, E. A. Phillips, and C. H. Tsao, *Introduction to the Theoretical and Experimental Analysis of Stress and Strain*, McGraw-Hill, New York, 1958.

18. D. C. Drucker, The Method of Oblique Incidence of Photoelasticity, *Proc. SESA*, VIII, no. 1 (1950), 51–66.

19. J. W. Dally and E. R. Erisman, An Analytic Separation Method for Photoelasticity, *Exp. Mech.*, 6 (1966), 493–499.

20. W. F. Stokey and W. F. Hughes, Tests on the Conducting Paper Analogy for Determining Isopachic Lines, *Proc. SESA, XIII (1955)*, 77–82.

21. S. K. Chaturvedi, A Rational Theory of Oblique Incidence and Its Extension to Stress Separation in Birefringent Composites, *Exp. Mech.*, 23 (1983), 36–41.

22. W. F. Swinson, J. L. Turner, and W. F. Ranson, Designing with Scattered Light Photoelasticity, *Exp. Mech.*, 20 (1980), 397–402.

23. J. T. Pindera and S. B. Mazurkiewicz, Studies of Contact Problems Using Photoelastic Isodynes, *Exp. Mech.*, 21 (1981), 448–455.

24. A. Lagarde (Ed.), Optical Methods in Mechanics of Solids, *Proc. IUTAM Symp.*, Poitiers, France, 1979, Sijthoff en Noordhoff, Alphen aan den Rijn, The Netherlands, 1981.

25. A. Lagarde, Modern Nondestructive Methods of Coherent Light Photoelasticity with Applications in Two and Three Dimensional Problem in Statics, Contact Stresses, Fracture Mechanics and Dynamic Impulse, IUTAM, 1985.

26. F. Zandeman, S. Redner, and J. W. Dally, Photoelastic Coatings, *SESA Monograph No. 3*, Iowa State University Press, Ames, 1977.

27. Measurements Group Inc., Education Division, Student Manual on the Photoelastic Coating Technique, *Bulletin 315*, 1984.

28. P. Nurse and I. M. Allison, Automatic Acquisition of Photoelastic Data, *JBCSA Conf.*, 1972.

29. S. Redner, New Automatic Polariscope System, *Exp. Mech.*, 14 no. 12 (1974), 486–491.

30. R. K. Mueller and L. Saachel, Complete Automatic Analysis of Photoelastic Fringes, presented at the 1978 SESA Spring Meeting, Wichita, KS, 1978.

31. C. P. Burger and A. S. Voloshin, A New Instrument for Whole Field Stress Analysis, *ISA Trans.*, 22, no. 2 (1982), 85–95.
32. C. P. Burger and A. S. Voloshin, Half-Fringe Photoelasticity: A New Lease on Life for an Old Method, *Proc. J. Int. Conf. Exp. Mech.*, SESA/JSME, Honolulu, May 23–28, 1982, 972–977.
33. A. S. Voloshin and C. P. Burger, Half-Fringe Photoelasticity: A New Approach to the Whole Field Stress Analysis, *Exp. Mech.*, 23, no. 3 (1983), 304–313.
34. A. S. Voloshin and C. P. Burger, Evaluation of Stress Intensity Factors for Near Surface Cracks Through a Synthesis of Photoelasticity and Image Analysis Procedures, *Eng. Appl. Opt. Meas., Soc. Exp. Stress Anal.*, 1982, 54–58.
35. C. P. Burger and A. S. Voloshin, Stress Intensity Factors in Glass and Polycarbonate by Half-Fringe Photoelasticity, *9th Can. Cong. Appl. Mech. (CANCAM 83)*, Saskatoon, Saskatchewan, Canada, June 1983, 915–916.
36. P. Zhang and C. P. Burger, Determination of Stress Intensity Factors Due to Thermal Stresses Using HFP, *Proc. Int. Conf. Exp. Mech.*, Beijing, China, 1985.
37. C. P. Burger, T. K. Baek, and A. S. Voloshin, Half-Fringe Photoelasticity Determines K-factors for Double Cracks by Stress Freezing, Developments in Theoretical and Applied Mechanics, Vol. XII, *Proc. SECTAM XII*, 1984.
38. A. S. Voloshin and C. P. Burger, Half-Fringe Photoelasticity for Orthotropic Materials, *Fibre Sci. Technol.*, 21 (1984), 341–351.
39. D. Mallik, C. P. Burger, A.S. Voloshin, and E. Matsumoto, Stress Analysis of Adhesive Joints in Composite Structures through HFP, *Compos. Struct.*, 4 (1985), 97–109.
40. J. W. Dally, An Introduction to Dynamic Photoelasticity, *Exp. Mech.*, 20 (1980), 409–416.
41. J. W. Dally, Data Analysis in Dynamic Photoelasticity, *Exp. Mech.*, 7 (1967), 332–338.
42. W. F. Riley and A. J. Durelli, Application of Moiré Methods to the Determination of Transient Stress and Strain Distributions, *J. Appl. Mech.*, 26 (1962), 23–29.
43. D. C. Holloway, W. F. Ranson, and C. E. Taylor, A Neoteric Intrerferometer for Use in Holographic Photoelasticity, *Exp. Mech.*, 3 (1972), 461–465.
44. A. B. J. Clark and R. J. Sanford, A Comparison of Static and Dynamic Properties of Photoelastic Materials, *Exp. Mech.*, 3 (1963), 148–151.
45. C. P. Burger and W. F. Riley, Effects of Impedance Mismatch on the Strength of Waves in Layered Solids, *Exp. Mech.*, 14, no. 4 (1974), 129–137.
46. H. P. Rossmanith, A Hybrid Technique for Improved K Determination from Photoelastic Data, *Exp. Mech.*, 23 (1983), 152–157.
47. C. P. Burger, A. J. Testa, and A. Singh, Dynamic Photoelasticity as an Aid in Developing New Ultrasonic-Test Methods, *Exp. Mech.*, 22, no. 4 (1982), 147–154.
48. W. F. Riley and J. W. Dally, Recording Dynamic Fringe Patterns with a Cranz–Schardin Camera, *Exp. Mech.*, 9 (1969), 27N–33N.
49. W. F. Riley and J. W. Dally, A Photoelastic Analysis of Stress Wave Propagation in a Layered Model, *Geophysics*, 31 (1966), 881–899.
50. C. P. Burger, I Miskioglu, R. G. Hughes, and J. Waskey, Improved Ultrasonic Evaluation of Surface Defects through Photoelastic Visualization of Elastic Waves, in *Technology Advances in Engineering and Their Impact on Detection, Diagnosis and Prognosis Methods* (G. A. Whittaker, T. R. Shives, and G. J. Philips, Eds.), Cambridge University Press, New York, 1983, 39–44.
51. J. W. Dally, Dynamic Photoelastic Studies of Dynamic Fracture, *Exp. Mech.*, 19 (1979), 349–361.
52. C. P. Burger, Thermal Modeling, *Exp. Mech.*, 15 (1975), 430–442.
53. C. P. Burger, Photothermoelastic Study of Stress Concentrations in a Plate with Internal Heating, *Exp. Mech.*, 12 (1972), 483–488.
54. C. P. Burger and I. Miskioglu, Transient Thermal Stresses at Elliptical Holes Near Edge, *J. Therm. Stresses*, 7 (1984), 19–33.

55. P. Zhang and C. P. Burger, Stress Intensity Factors for Edge Cracks Under Transient Thermal Stresses by Photoelasticity, *Proc. 1985 SEM Spring Conf. Exp. Mech.*, Las Vegas, 1985, 907–915.
56. J. D. Hovanesian and H. C. Kowalski, Similarity in Thermoelasticity, *Exp. Mech.*, 7 (1967), 82–84.
57. I. Miskioglu, J. Gryzagoridis, and C. P. Burger, Material Properties in Thermal Stress Analysis, *Exp. Mech.*, 21 (1981), 295–301.
58. J. Javornicky, *Photoplasticity*, Elsevier Scientific Publishing Company, New York, 1974.
59. C. P. Burger, Non-Linear Photomechanics, *Exp. Mech.*, 20 (1980), 381–389.
60. C. P. Burger, A. K. Oyinlola, and T. E. Scott, Full-Field Strain Distribution in Hot Rolled Billets by Simulation, *Trans. ASME, Met. Eng. Q.*, 1976, 26–29.
61. J. L. F. Freire, Studies on the Yield Behavior of Photoplastic Material, Ph.D. dissertation, Iowa State University, Ames, 1979.
62. M. M. Frocht and Y. F. Cheng, An Experimental Study of the Laws of Double Refraction in the Plastic State in Cellulose Nitrate Foundations for Three-Dimensional Photoplasticity, *Proc. Int. Symp. Photoelasticity* (M. M. Frocht, Ed.), Pergamon Press, Elmsford, NY, 1963.
63. E. Monch and R. Loreck, A Study of the Accuracy and Limits of Application of Plane Photoplastic Experiments, *Proc. Int. Symp. Photoelasticity* (M. M. Frocht, Ed.), Pergamon Press, Elmsford, NY, 1963.
64. K. Ito, New Model Material for Photoelasticity and Photoplasticity, *Exp. Mech.*, 2 (1962), 373–376.
65. W. A. Brill, Basic Studies in Photoplasticity, Ph.D. dissertation, Stanford University, Stanford, CA, 1966.
66. J. K. Whitfield and C. W. Smith, Characterization Studies of a Potential Photoelastoplastic Material, *Exp. Mech.*, 12 (1972), 67–74.
67. L. W. Zachary and W. F. Riley, Optical Response and Yield Behavior of a Polyester Model Material, *Exp. Mech.*, 17 (1977), 321–326.
68. Y. Ohashi, Experimental Stress Analysis by Photorheologic Method, *Exp. Mech.*, 13 (1973), 287–293.
69. C. P. Burger, J. N. El-Hout, and H. A. Gomide, Three-Dimensional Strain Field for a Hot Rolled Billet by Photoplastic Simulation, *Proc. 3rd Int. Conf. Mech. Behav. Mater.* (K. J. Miller and R. F. Smith, Eds), University of Cambridge, Cambridge, England, 1979, 557–567.
70. J. W. Dally and A. Mulc, Polycarbonate as a Model Material for Three-Dimensional Photoplasticity, *Trans. ASME, J. Appl. Mech.*, E95 (1973), 600–605.
71. A. R. Hunter, Photoelastoplastic Analysis of Notched-Bar Configuration Subjected to Bending, *Exp. Mech.*, 10 (1970), 281–287.
72. R. L. Johnson, Measurement of Elastic–Plastic Stresses by Scattered-Light Photomechanics, *Exp. Mech.*, 16 (1976), 201–208.
73. M. M. Frocht and R. A. Thomson, Experiments in Mechanical and Optical Coincidence in Photoplasticity, *Exp. Mech.*, 1 (1961), 43–47.
74. J. Javornicky, Evaluation of Stresses and Strains in Two-Dimensional Photoplasticity, *J. Strain Anal.*, 3 (1968), 33–38.
75. D. H. Morris and W. F. Riley, A Photomechanics Material for Elasto-Plastic Stress Analysis, *Exp. Mech.*, 12 (1972), 448–453.
76. J. L. F. Freire, H. A. Gomide, and W. F. Riley, Photoplastic Study of Axially Compressed Cylinders, *Proc. COBEM79*, Vth Congresso Brasileiro de Engenharia Mechanica, Brasil, December 1979.
77. C. P. Burger and L. N. Koenig, Photoplastic Modeling of Strains in the Hot Forming of Aluminum, SESA Pap WR-19-1975, presented at SESA Spring Meeting, 1975.
78. C. P. Burger and H. A. Gomide, Three-Dimensional Strain in Rolled Slabs by Photoplastic Simulation, *Exp. Mech.*, 22 (1982), 441–447.
79. H. A. Gomide and C. P. Burger, Three-Dimensional Strain Distribution in Upset Rings by Photoplastic Simulation, *Exp. Mech.*, 21 (1981), 361–370.

80. M. Nisida, M. Hondo, and T. Hasumuma, Studies of Plastic Deformation by the Photoplastic Method, *Proc. 6th Jpn. Natl. Cong. Appl. Mech.*, 6 (1956), 137–140.
81. K. Ito, Photoelasto-Plastic Studies on Brittle Fracture of High-Polymer Solids, *Exp. Mech.*, 1 (1961), 159–168.
82. H. F. Brinson, An Interpretation of Inelastic Birefringence, *Exp. Mech.*, 11 (1971), 467–471.
83. P. S. Theocaris and E. E. Gdoutos, The Size of Plastic Zones in Cracked Plates Made of Polycarbonate, *Exp. Mech.*, 15 (1975), 169–176.
84. H. Pih, Birefringent-Fluid-Flow Method in Engineering, *Exp. Mech.*, 20 (1980), 437–444.
85. S. P. Sutera and H. Wayland, Quantitative Analysis of Two-Dimensional Flow by Means of Streaming Birefringence, *J. Appl. Phys.*, 32 (1961), 721.
86. A. E. Hirsh, The Flow of a Non-Newtonian Fluid in a Diverging Duct, Experimental, Ph.D. dissertation, University of Tennessee, Knoxville, 1964.
87. W. M. Swanson and R. L. Green, Colloidal Suspension Properties of Milling Yellow Dye, *J. Colloid Interface Sci.*, 29 (1969), 161.
88. F. N. Peebles, J. W. Prados, and E. H. Honeycutt, Birefringent and Rheological Properties of Milling Yellow Suspensions, *J. Polym. Sci.: Part C*, 5 (1964), 37.
89. R. Prabhakaran, Fabrication of Birefringent Anisotropic Model Materials, *Exp. Mech.*, 20 (1980), 320–321.
90. I. M. Daniel, G. M. Koller, and T. Niiro, Development and Characterization of Orthotropic-Birefringent Materials, *Exp. Mech.* 24, (1984), 134–143.
91. J. W. Dally and R. Prabhakaran, Photo-Orthotropic Elasticity, *Exp. Mech.*, 11 (1971), 346–356.
92. B. D. Agarwal and S. K. Chaturvedi, Improved Birefringent Composites and Assessment of Photoelastic Theories, *Fibre Sci. Technol.*, 11 (1978), 399–412.
93. B. D. Agarwal and S. K. Chaturvedi, Exact and Approximate Strain-Optic Laws for Photoelastic Composites, *Polym. Compos.*, 3 (1982), 146–151.
94. S. K. Chaturvedi, A Rational Theory of Oblique Incidence and Its Extension to Stress Separation in Birefringent Composites, *Exp. Mech.*, 24 (1984), 17–21.
95. R. Prabhakaran and R. G. Chermahini, Application of Least-Squares Method to Elastic and Photoelastic Calibration of Orthotropic Composites, *Exp. Mech.*, 24 (1984), 17–21.
96. D. Mallik, C. P. Burger, A. S. Voloshin, and E. Matsumoto, Stress Analysis of Adhesive Joints in Composites Structures through Half Fringe Photoelasticity, *Compos. Struct.*, 4 (1985), 97–109.
97. V. J. Parks and R. J. Sanford, On the Role of Material and Optical Properties in Complete Photoelastic Analysis, *Exp. Mech.*, 16 (1976), 441–447.
98. R. J. Sanford, Photoelastic Holography—A Modern Tool for Stress Analysis, *Exp. Mech.*, 20 (1980), 427–436.
99. H. Aben, *Integrated Photoelasticity*, McGraw-Hill, New York, 1979.
100. H. Aben, Integrated Photoelasticity and Its Applications for Stress Determination in Single Crystals, in *Optical Methods in Mechanics of Solids* (A. Lagarde, Ed.), Sijthoff en Noordhoff, Alphen aan den Rijn, The Netherlands, 1981, 41–54.
101. A. Lagarde, On Some Aspects in the Development of Photoelastic Measurements, in *Optical Methods in Mechanics of Solids* (A. Lagarde, Ed.), Sijthoff en Noordhoff, Alphen aan den Rijn, The Netherlands, 1981, 1–40.
102. A. S. Kobayashi (Ed.), *Experimental Techniques in Fracture Mechanics*, Vols. 1 and 2, Iowa State University Press, Ames, 1973 and 1975.
103. C. W. Smith and O. Olaosebikan, Use of Mixed Mode Stress Intensity Algorithms for Photoelastic Data, *Exp. Mech.*, 24 (1984), 300–307.
104. C. W. Smith, W. H. Peters, and G. C. Kirby, Crack-Tip Measurements in Photoelastic Models, *Exp. Mech.*, 22 (1982), 448–453.
105. J. W. Dally, Dynamic Photoelastic Studies of Fracture, *Exp. Mech.*, 19 (1979), 349–367.

106. R. J. Sanford and J. W. Dally, A General Method for Determining Mixed Mode Stress Intensity Factors from Isochromatic Fringe Patterns, *Eng. Fract. Mech., 11* (1979), 621–633.

107. A. S. Voloshin and C.P. Burger, Evaluation of Stress Intensity Factors for Near Surface Cracks Through a Synthesis of Photoelasticity and Image Analysis Procedures, *Eng. Appl. Opt. Meas.,* 1982, 54–58.

108. M. Ramula, A. S. Kobayashi, B. S. J. Kang, and D. B. Barker, Further Studies in Dynamic Crack Branching, *Exp. Mech., 23* (1983), 431–437.

109. M. Arcan and M. A. Brull, A Fundamental Characteristic of the Human Body and Foot—The Foot Ground Pattern, *J. Biochem., 9* (1976), 453–457.

110. R. Mark, Modeling Architectural Structure: Experimental Mechanics in Historiography and Criticism, 1982 Murray Lecture for Society of Experimental Mechanics, *Exp. Mech., 22* (1982), 361–371.

CHAPTER

6

Geometric Moiré

Vincent J. Parks

The Catholic University of America
Washington, D.C.

6-0 Introduction [1–7]

Hooke determined elongation in wires under load by noting the change in length between two scribed lines on the wire. This is such a simple and direct approach that it hardly warrants designation as a method. Following Hooke, other experimentalists applied grids of many lines, and in crossed directions, to specimens. The method of grids is summarized elsewhere [3]. It has long been recognized [12,32] that interference between two sets of lines can be used to determine the same quantities as those obtained by measuring between the lines. This use of interference was termed moiré because of the similarity of the visual effect to that of watered silk, well known in France as moiré.

Moiré is defined in this chapter as the visual pattern produced by the superposition of two regular motifs that geometrically interfere. This chapter is restricted to geometric interference; optical interference is treated in Chapter 7.

Examples of regular motifs are equispaced parallel lines, rectangular arrays of dots, concentric circles, and radial lines. The motifs may be superposed by reflection, by shadowing, by double-exposure photography, or by direct contact. In most cases, at least one of the superposed motifs must have transparent regions.

Moiré patterns are used to measure variables such as displacements, rotations, curvature, and strain. The moiré patterns essentially magnify the distortions of the motifs to provide a visual picture of the variable throughout the viewed area.

6-1 In-Plane Moiré [8–24]

The most common use of moiré, to determine strains and displacements that act in and parallel to the plane of analysis, is presented in this section. Other uses are presented in the following section. In-plane moiré is typically conducted with gratings of equispaced, parallel lines. One set

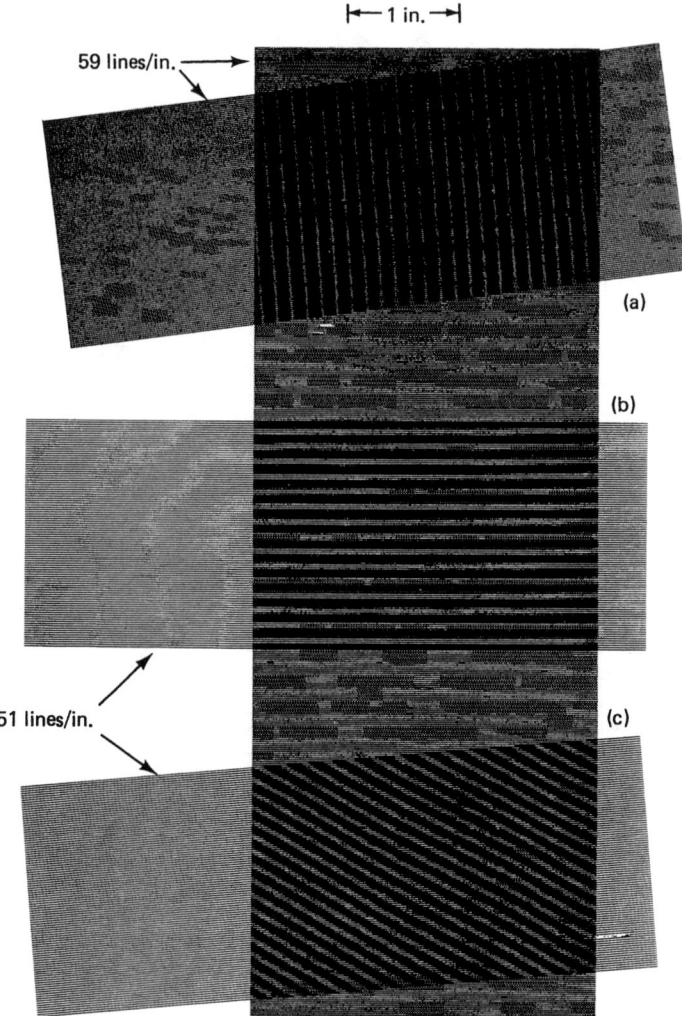

Figure 6-1 Moiré patterns of three combinations of strain and rotation: (a) 6.5° rotation; (b) 16% strain; (c) 16% strain and 4° rotation.

of lines is applied to a flat surface of the specimen to be analyzed, and a second set (called the reference grating) is put in contact with the specimen grating.

When the specimen is loaded, or moved, interference patterns such as that shown in Fig. 6-1 are generated. If the lines of the specimen grating are initially interspaced between the lines of the reference grating, the overall field appears dark. Under load, any region of the specimen that does not move will remain dark. If a region moves half the distance between the grating lines,

GEOMETRIC MOIRÉ

the specimen and grating lines will overlap, leaving a light space between each pair of overlapping lines, and that region of the specimen will appear lighter than it was before loading. If a region moves the whole distance between lines, it will be as dark as an unmoved portion. The distance between grating lines is called the pitch and is symbolized by a lowercase g. The motion causes the light–dark sequence to be repeated in steps of g. As illustrated with Fig. 6-1, the dark and light regions alternate to correspond to the changing motion of the specimen. The dark regions are usually called *fringes*. Sometimes the light regions are referred to as light fringes. Assuming that the specimen and reference grating lines are initially interspaced, the dark fringes in Fig. 6-1 represent motions of . . . , $-3g$, $-2g$, $-g$, 0, $2g$, $3g$, $4g$, ...[1]

The displacement described above, and represented by the moiré fringes, is that component of the displacement that is normal to the reference grating lines. Any component of motion parallel to the reference grating lines does not alter the relative line positions, and so does not contribute to creation of fringes. A second grating on the surface with lines perpendicular to the first is required to obtain complete motion of a plane. With the reference grating lines initially parallel to this second set of lines, the motion is repeated. A new fringe pattern is produced, which represents the components of displacement perpendicular to the first, again in steps of g units. The two sets of fringes give a complete description of motion in the plane. Unless the specimen is symmetrical, such that one set of lines can be used alternately in both directions, an orthogonal (crossed) specimen grating is often used. Either a one-way or a crossed reference grating can be used.

Under load, deformation will change the line spacing of the specimen grating. This changed spacing on the specimen is important in determining strain. Both the specimen line spacing and the specimen rotation can be determined geometrically from the spacing and angle of the moiré fringes. This result, in turn, can be used to determine strain.

Strains can also be determined by using strain–displacement equations. The four derivatives of the fringes in the two reference grating directions are directly related to the four Cartesian displacement derivatives found in the two-dimensional strain–displacement equations of theory of elasticity. Despite the source of these equations, they are strictly geometric equalities and do not require elastic behavior of the specimen.

A number of techniques can be used to improve the accuracy of the analyses described above. These include using a reference grating of a multiple pitch, or one of a slightly different pitch, to that of the specimen pitch. The multiple pitch multiplies the number of fringes. The mismatched pitch increases the number of fringes. Both aim at increasing fringe sensitivity. Shifting one grating over another or over its duplicate can aid the analysis. The concepts above are all discussed in detail in the following sections.

Gratings and Specimen Preparation

Gratings are scribed, etched, ruled, printed, stamped, photographed, or cemented onto specimens. Pitches vary from 10 μm to 1 mm (i.e., 1 to 100 lines per millimeter). In English units the usual range is 0.001 to 0.1 in (e.g., 10 to 1000 lines per inch). The coarse gratings, 1 to 5 lines/mm, can be obtained from graphic arts suppliers, in sheets of 25 by 50 cm. Commercial artists use both cross gratings and one-way gratings, of various line densities, for shading. The cross gratings are often rectangular dot arrays (as found in newspaper and magazine "halftone" photographs). Finer gratings (5 to 10 lines/mm) are available from high-quality lithographers or photoengravers

---

[1] A list of the symbols used in this chapter precedes the References.

and their suppliers. A few suppliers have gratings with frequencies greater than 40 lines/mm. Scientific supply houses carry Ronchi gratings. Optical grating companies supply gratings beyond 40 lines/mm and do have some high-quality gratings in the 5-line/mm range. Sieves, etched from sheet metal to form cross gratings, could have application in moiré.

Suppliers to the stress analysis industry provide 20- to 40-line/mm (500- and 1000-lines/in.) rulings in both one-way and cross gratings, on glass, up to 20 by 25 cm (8 by 10 in.), as well as metallic dot arrays of 20 lines/mm (500 lines/in.). The developing technology of manufacturing integrated circuits and computer chips gives promise of more precise, denser, and less expensive gratings.

Most commercial gratings have a line width of approximately 50% of the pitch. Two gratings with 50% line width are optimum to achieve maximum fringe contrast. If the grating lines are wider (say 60%), the photographic negative (40% line width) is the optimum choice for good fringe contrast. Photography can be used to duplicate gratings and in special cases to vary line width, by varying exposure and developing times.

A grating can be made by exposing a photographic plate to two laser beams from the same source. The angle between the beams determines the pitch. Fine optical gratings are scribed on glass with ruling engines and replicated with plastics. Regardless of how the grating is made, it can be copied again and again photographically. It can be enlarged or reduced. Accordingly, most moiré gratings are photographs on film or glass plate. These procedures are discussed in further detail in Chapter 7.

The grating can be transferred to a metal specimen by lithography using a photosensitive coating such as photoresist or dicromated gelatin on the specimen. The grating image can be printed on the metal by direct contact or through a lens. The image can be darkened for use, used as is, or acid etched to make a grating of the metal itself (photoengraving).

Where a metal specimen is not required, transparent plastic is often used as a model material. The grating can be applied to the plastic, as in lithography but without etching, or a photographic film can be bonded to the plastic (stripping film is designed for this type of application). Plastic sheet with photographic emulsion is available from supplies of photographic materials. For three-dimensional applications, gratings have been printed on interior planes of plastic models before the parts are bonded together. Thus a myriad of methods are available to apply gratings to specimens and models, depending on the application.

Generating and Recording the Pattern

Once the specimen grating has been formed on the specimen, there are two ways for the reference grating to interact with the specimen grating. The first way is the direct approach. If the specimen grating surface is horizontal, the reference grating can simply be laid on top of it (grating to grating). Adding an oil film between the two gratings improves the contrast of the fringe pattern. If the specimen surface is not completely flat (and of course, under load, the specimen surface will usually warp), a reference grating on film may fit better than a reference grating on glass. Some precision is lost, however. The same ideas apply to vertical specimen surfaces, but in addition some arrangement must be made to hold the reference grating on the specimen. Gratings on film can often be held in place by the oil film used to improve contrast. Besides fixing the reference grating in intimate contact with the specimen grating, in principle, the reference grating should remain absolutely still, as the specimen moves due to load. This requirement is not always important in strain analysis, as will be seen later.

The other way to contact the specimen and reference grating is optically. A simple form of optical contact is to photograph the specimen grating before loading, load, and rephotograph the same specimen grating on the same film, producing a double exposure. The unloaded specimen grating serves as the reference grating. When developed, the double-exposed film will be the moiré pattern of the in-plane displacement component of the specimen grating. This form has the disadvantage that a part of the moiré pattern is lost along any portion that moves outward from a boundary during loading. A variation that preserves the boundary is to place a reference grating in the camera directly in front of the focal plane (the ground glass). Here camera magnification should be initially adjusted to match the reference and specimen grating, to give a uniform dark field, with no fringes, over the whole view of the specimen. The specimen can then be loaded and the moiré pattern viewed or photographed. Other variations include photographing the optically formed moiré pattern with a second camera, photographing the deformed specimen grating directly and later selecting a reference grating for analysis, and double exposures of steps of loading. In general, the optical contact approach does not produce as high-contrast a moiré pattern as the direct contact (as discussed below), and usually requires a better lens, since the grating rather than the moiré pattern is being photographed.

With regard to illuminating the grating, it should be noted that there is a decided advantage in the use of transparent specimens so that the grating can be lighted from behind. Backlighting of the gratings greatly increases the contrast of the moiré pattern.

Occasionally, the moiré pattern is recorded with a video camera or other scanning device [15]. The pattern is stored for later visual display or automated analysis. A variation for coarse gratings is to scan the distorted grating directly. In this case the reference grating is sometimes represented by its mathematical equivalent in an automated analysis.

Fringe Intensity: Superposition Versus Double Exposure

As noted above, the moiré fringes can be formed by superposition of two gratings (direct approach), or by a double exposure on film of the two grating images. The fringes will have the same interpretation in either case but will have different intensity distributions.

The intensity distribution due to superposition is easier to understand. The dark lines block the light either individually or where the lines overlap. As seen in Figs. 6-1 and 6-2, two levels of intensity are created. Any region with a single line or two superposed lines appears black. Any region not covered by either set of lines will appear light. Where the two sets of lines are at an angle to each other, the dark fringes appear as herringbone (or zigzag) patterns. The light fringes are rows of light diamonds interspersed between the overlapping dark lines. If the two sets of lines are parallel, the dark fringes form where the two sets of lines fall between each other; the light fringes are the region where the lines overlap. The intensity of the dark fringes is at or near zero; the average intensity of the light fringes, created by alternate light and dark spaces, is about half the total light intensity. When viewed at a distance, such that individual grating lines are not resolved, the pattern appears as black and light fringes. The contrast is intrinsically high.

Double-exposure intensity is somewhat more complicated to explain. The first exposure forms dark lines on the film. If the second set is at some angle to the first, it crosses the dark lines of the first. *Three* levels of intensity are created: (1) regions with no exposure to light (zero intensity), (2) regions with double exposure to light (full intensity), and (3) regions with only one exposure to light (half-intensity).

When the positive image of the double-exposure moiré pattern negative is printed, the light fringes are generated by light diamonds (full intensity) interspaced between dark diamonds (zero intensity). But the dark fringes are now generated by a zigzag pattern of half-intensity. If the line widths are equal to the interspace widths, and if the photography faithfully reproduces both images,

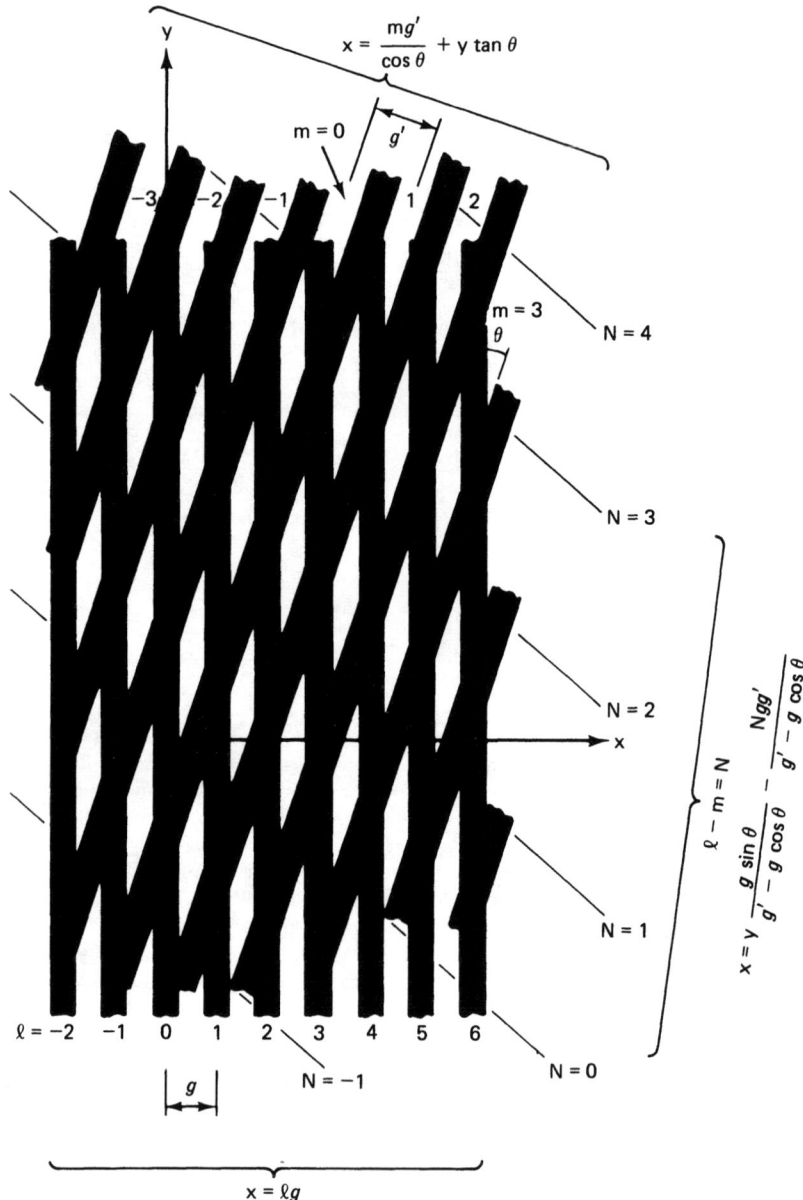

Figure 6-2 Detail of fringe formation between two parallel line gratings set at an angle.

when the positive image is viewed from a distance, such that the individual grating lines are not resolved, both light and dark fringes will have an average intensity of one-half and will be indistinguishable. No pattern is seen.

In practice this seldom happens. Variation in line width and line intensity will create contrast. Nonlinear high-contrast films and papers will cause the continuous half-intensity region to become more opaque, and light transmission is much less than the average transmission in the alternate full intensity–zero intensity regions. The result is light and dark fringes seen at a distance. Still the contrast is less than in superposition. A method of improving the contrast of double-exposure moiré patterns is optical filtering, discussed in Chapter 7.

Analysis

The description of fringe formation at the beginning of this section leads directly to a quantitative displacement–fringe relationship. Each new fringe indicates an increase or decrease of specimen displacement by one grating pitch, so the basic equation relating displacement to fringes is

$$\delta = gN \tag{6-1}$$

where N = fringe order
g = reference grating pitch
δ = component of the displacement perpendicular to the reference grating lines

The first step in the analysis is determination of the fringe order N. Sometimes this is as simple as watching as the specimen moves and counting the fringes, as suggested by Fig. 6-1. Sometimes it is possible to watch a specific point during loading, taking note of the direction of motion, and count from the zero fringe until the full load is reached. In more complicated patterns it is often necessary to consider the continuity, uniqueness, and direction of the displacements.

The displacement and components of displacements within the boundaries of a body are continuous and single valued. Intermediate displacement values must occur between any two points with different displacements. If the displacements were not continuous, gaps (breaks) would be created in the body. Continuity is illustrated by an analogous property that is found in contours representing topological elevations. From physical considerations, it is obvious that between two points with different elevations, the intermediate elevations must occur. The moiré fringes are similar to topological contours that end at the boundary of the map or form closed curves.

The requirement that the moiré pattern be single valued is analogous to requiring that the topological surface have no vertical drop-offs (cliffs) within the field. The analogy may be carried one step further by noting that because the pattern represents a physical phenomenon, there are no infinitely high or infinitely low values.

The properties of continuity and uniqueness help in ordering the fringes. Assigning a displacement value to the first fringe is often arbitrary, since it defines only a datum reference displacement (which may be associated with a rigid-body motion). Consecutive fringes will have a difference in displacement equal to the pitch. Thus if a fringe is assigned a displacement of zero, consecutive fringes will have displacement values of g, $2g$, $3g$, and so on.

From the first fringe value, it is sometimes possible to assign values to the entire pattern by a simple count. This is true if each fringe order appears only once on the pattern. Sometimes considerations of the symmetry and the load aid in the determination of the displacement values.

Beyond continuity and uniqueness, the fringe orders must be assigned a direction. This can be decided by arbitrarily choosing one direction perpendicular to the reference grating as positive, and then using external evidence to decide whether the displacements increase or decrease in that direction. (Thus for a set of horizontal reference lines, if "up" is chosen as positive, and a vertical load is compressing the body, in general, the fringes have lower orders near the top.) Sometimes this is done by choosing the order of fringes and then assigning the positive direction, using the external evidence, to agree with the fringe direction.

Although the positive direction, and therefore the sign of the fringe, is arbitrary, the ratio of the change of fringe order to the change of position (i.e., the fringe derivative) does have a specific sign. If the sign is not known from external evidence, it can be determined simply by shifting the reference grating with respect to the specimen grating. The rule is this: if the reference grating is shifted perpendicular to its grating lines, the derivative is positive if the fringes move in the direction of the shift and negative if they move opposite to the shift. If the only movement is perpendicular to the shift, the derivative is zero. In summary, the sign of the fringe derivative cannot be decided by simply viewing the final pattern; it must be determined either by external evidence (such as the applied load) or by shifting the reference grating.

Once the fringes have been ordered, the components of displacements are obtained directly from Eq. (6-1). And once two sets of displacement components have been obtained, the experimental analysis is complete. The two sets of displacement components completely describe the surface motion. Other quantities, specifically normal strain, shear strain, and rotation, can be determined from geometric analysis of the displacements, without recourse to the fringe patterns. However, because strain analysis does not require knowledge of the absolute displacement values, it is customary to take fringe measurements directly from the fringe pattern to determine strain. A geometric approach uses measurements of fringe spacing and inclination at points of interest to determine strain. The alternative approach uses the slopes of fringe-position curves to determine strain. Both approaches are described below.

Geometric Approach

It has been shown [18] that a moiré pattern can be described analytically, in a parametric manner, if the analytic expressions for the two grating lines that produce the moiré pattern are known. But, in general, in the application of moiré to the determination of motion, only one grating, the reference grating, can be expressed analytically. Both the specimen grating and the moiré pattern exhibit complex shapes associated with the motion and deformation of the specimen. However, if analysis is limited to a small area of homogeneous strain, or to some simple displacement that has an analytic expression, the analytic method of determining the parameters of the resulting curves can be a useful description of the moiré pattern in that small area or in the simple field.

Figure 6-2 is a set of parallel opaque lines alternated with equiwidth transparent spaces, superposed on a second similar set of opaque lines and spaces, still of equal width, but of slightly larger width than the first set. The figure represents the small area mentioned in the paragraph above. The direction of the lines in the first set differs somewhat from the direction of the lines in the second set. If a Cartesian coordinate system is applied with directions perpendicular and parallel to the lines of the first set as shown in Fig. 6-2, the family of lines in the first set will be considered the reference grating and can be described analytically by

$$x = lg \qquad (6\text{-}2)$$

where g is the width between lines (the pitch of the grating) and l is the order number of each line. The grating line passing through the origin has the number zero, and the lines away from the origin are ordered 1, 2, 3 in the positive x direction and -1, -2, -3 in the negative x direction.

The family of lines in the second set (considered as the specimen grating) can be described analytically by

GEOMETRIC MOIRÉ

$$x = \frac{mg'}{\cos \theta} + y \tan \theta \tag{6-3}$$

where g' is the pitch of the specimen grating and θ is the angle between the two sets of grating lines. The Cartesian coordinates have been chosen with the origin on a transparent space of the specimen grating. Then the parameter m will refer to the transparent spaces, with zero order at the origin. $m = 1, 2, 3, \ldots$ to the right and $m = -1, -2, -3, \ldots$ to the left in the figure.

The origin has been chosen at a point of interference and the orders chosen so that at the origin $l = m = 0$. "Interference" here is taken to mean that the opaque line of one set fills a clear space of the other set. Note that the interference at the origin is on the same moiré fringe as the interference occurring at $l = m = 1$, and that the interferences occurring at $l = m = 2$ and at $l = m = 3$ are also on that same fringe. The general relation of l to m for the moiré fringe through the origin is then $l = m$. Next consider the interference just above the origin. Here the interference occurs at $l = 0$, $m = -1$. Again, notice that the fringe continues through the points $l = 1$, $m = 0$; $l = 2$, $m = 1$; and $l = 3$, $m = 2$, at which interference also takes place. The fringe is described in general by the equation $l - m = 1$. Similar considerations show that the moiré fringes in general are described by

$$l - m = N \tag{6-4}$$

where N is an integer and the parameter of the moiré fringes.

Both l and m have been introduced as integers; however, note that the moiré fringe described by N is located by any values of l and m (integer or noninteger) which give an integral value of N. As such, Eq. (6-4) is continuous over the field and represents the moiré fringes as a whole, not just the intersections of opaque and light areas in the field. In fact, the value of N can also be interpreted as a noninteger to give a continuous fringe family of which the integral moiré fringe orders are only selected members. Equation (6-4) is a basic equation of moiré and describes the fringe order of any moiré pattern in terms of the grating line orders.

If the angle θ were to increase greatly, say beyond $\pm 30°$, the fringes shown in Fig. 6-2 would fade out. If θ were to increase to over $\pm 150°$, a new set of fringes would appear, the mirror image, so to speak, of those shown in Fig. 6-2. Mathematically, these new fringes are described by the expression $l + m = N'$. Although these fringes are not normally visible, they can be made visible by using advanced optical techniques, including diffraction, which are beyond the scope of this geometric analysis.

A similar fade-out occurs if either pitch increases beyond about 30% of the other. As with rotation, a new fringe pattern forms as one pitch reaches twice the width of the other. Subsequently, a new fringe pattern forms with each new integer multiple. This phenomenon is called fringe multiplication and its application to fringe analysis is described later in this section.

The ranges above serve as broad guidelines for the formation of the normal moiré pattern. The moiré pattern is limited to a strain of $\pm 30\%$, or a rotation of $\pm 30°$, or lesser amounts for combinations of the two.

By substituting Eqs. (6-2) and (6-3) into (6-4), an expression for the fringes produced by two straight-line gratings with pitches g and g' and at an angle θ is obtained

$$\frac{x}{g} - \frac{x - y \tan \theta}{g'} \cos \theta = N \tag{6-5}$$

or in the form

$$Ax + By + C = 0$$
$$x(g' - g \cos \theta) + yg \sin \theta - Ng'g = 0$$

The space between fringe lines can be computed using analytic geometry. Thus, in Fig. 6-2, the shortest distance from the origin to the fringe of order $N = 1$ is

$$d = \frac{|C|}{\sqrt{A^2 + B^2}} = \frac{gg'}{\sqrt{g^2 \sin^2 \theta + (g' - g \cos \theta)^2}} \qquad (6\text{-}6)$$

This is the spacing of fringes throughout the portion of a pattern with pitch g' and grating angle θ. Equation (6-6) is valid regardless of what orders are assigned to the two sets of grating lines as long as they are continuous, and so applies to any moiré pattern formed by two sets of parallel grating lines. Again using analytic geometry, the slope of fringes measured from the reference grating is given by

$$\tan \phi = -\frac{g \sin \theta}{g' - g \cos \theta} \qquad (6\text{-}7)$$

The pitch and angle of the specimen grating are described completely from the spacing (d), the angle (ϕ) of the moiré fringes, and the pitch of the reference grating (g).

$$g' = \frac{d}{\sqrt{1 + (d/g)^2 + 2(d/g) \cos \phi}} \qquad (6\text{-}8)$$

$$\theta = \arctan \frac{\sin \phi}{d/g + \cos \phi} \qquad (6\text{-}9)$$

These equations have been developed directly by another author [4] without using analytic geometry. Nomographs [2,16] can be used to determine g' and θ.

If the line segment g, which was initially perpendicular to the specimen grating lines, rotates with respect to the specimen lines due to load, its final length will not be g'. It will be slightly larger than g'. This difference is usually neglected in the geometric approach and g' is used as the final length of the line segment. The strain is obtained directly as

$$\varepsilon = \frac{g' - g}{g} \qquad (6\text{-}10)$$

This is an exact expression in the principal directions of strain.

Calculus Approach

In the theory of elasticity, the relation of strain to displacement is often reduced to

$$\begin{aligned}
\varepsilon_{xx} &= \frac{\partial u_x}{\partial x} \\
\varepsilon_{yy} &= = \frac{\partial u_y}{\partial y} \\
\varepsilon_{xy} &= \frac{\partial u_x}{\partial y} + \frac{\partial u_y}{\partial x} \\
\omega_{xy} &= \frac{1}{2}\left(\frac{\partial u_x}{\partial y} - \frac{\partial u_y}{\partial x}\right)
\end{aligned} \qquad (6\text{-}11)$$

GEOMETRIC MOIRÉ

where u_x, u_y = components of displacement in the x and y directions
ε_{xx}, ε_{yy} = strains in the x and y directions
ε_{xy} = shear strain
ω_{xy} = average rotation of the x and y coordinates

Where there is rigid-body rotation due to load, in order to apply the equations above, it often becomes necessary to reorient the Cartesian coordinates. Or the load can be assumed to be small enough that such rotations can be neglected. In moiré analysis these ploys are seldom convenient, and a more general form of the strain displacement relations is used:

$$\varepsilon_{xx} = \sqrt{1 + 2\frac{\partial u_x}{\partial x} + \left(\frac{\partial u_x}{\partial x}\right)^2 + \left(\frac{\partial u_y}{\partial x}\right)^2} - 1$$

$$\varepsilon_{yy} = \sqrt{1 + 2\frac{\partial u_y}{\partial y} + \left(\frac{\partial u_y}{\partial y}\right)^2 + \left(\frac{\partial u_x}{\partial y}\right)^2} - 1 \qquad (6\text{-}12)$$

$$\varepsilon_{xy} = \arcsin \frac{\frac{\partial u_x}{\partial y} + \frac{\partial u_y}{\partial x} + \frac{\partial u_x}{\partial x}\frac{\partial u_x}{\partial y} + \frac{\partial u_y}{\partial x}\frac{\partial u_y}{\partial y}}{(1 + \varepsilon_{xx})(1 + \varepsilon_{yy})}$$

Equations (6-11) are an approximate form of Eqs. (6-12). The symbols all have essentially the same meaning as before. This version applies to large strains, as well as large rotations, and the definitions need to be more specific. The strain ε_{xx} (or ε_{yy}) is defined as the change in length of a unit line segment originally in the x (or y) direction, without restriction of magnitude. The shear strain ε_{xy} is defined as the decrease in radians of a right angle originally located between the positive x and y directions. The dimension z, and the corresponding displacement component u_z, are omitted above since the moiré method is restricted to a plane. Various simplifications of Eq. (6-12) are discussed in Ref. 19.

To obtain the derivatives of u_x and u_y, Eq. (6-1) is rewritten as

$$u_x = gN_x \qquad (6\text{-}13)$$
$$u_y = gN_y$$

and the derivatives are

$$\frac{\partial u_x}{\partial x} = g\frac{\partial N_x}{\partial x} \qquad \frac{\partial u_y}{\partial x} = g\frac{\partial N_y}{\partial x}$$
$$\frac{\partial u_x}{\partial y} = g\frac{\partial N_x}{\partial y} \qquad \frac{\partial u_y}{\partial y} = g\frac{\partial N_y}{\partial y} \qquad (6\text{-}14)$$

where N_x and N_y are the fringe orders in the moiré patterns produced by grating lines initially perpendicular to the x direction and y direction, respectively. Equations (6-14) indicate that knowledge of the absolute fringe orders is unnecessary for strain analysis, as long as the relative orders assigned in counting have the right direction of change.

The displacement derivatives in Eqs. (6-14) can be estimated simply, by measuring the fringe spacing in the x and y directions and dividing those values into g. More precisely, the derivatives are obtained as shown in Fig. 6-3 by (1) recording the fringe orders along lines originally in the x and y directions, (2) plotting fringes versus position, and (3) taking the slope of the plot at each point at which strain is required. A grid of lines in the x and y directions on the specimen grating, at the intervals needed in analysis, greatly aids this procedure (e.g., lines spaced every centimeter on a 900-line/cm grating). Points can be taken every centimeter by estimating the fringe fraction, or at every whole fringe by measuring the fringe's position on the line.

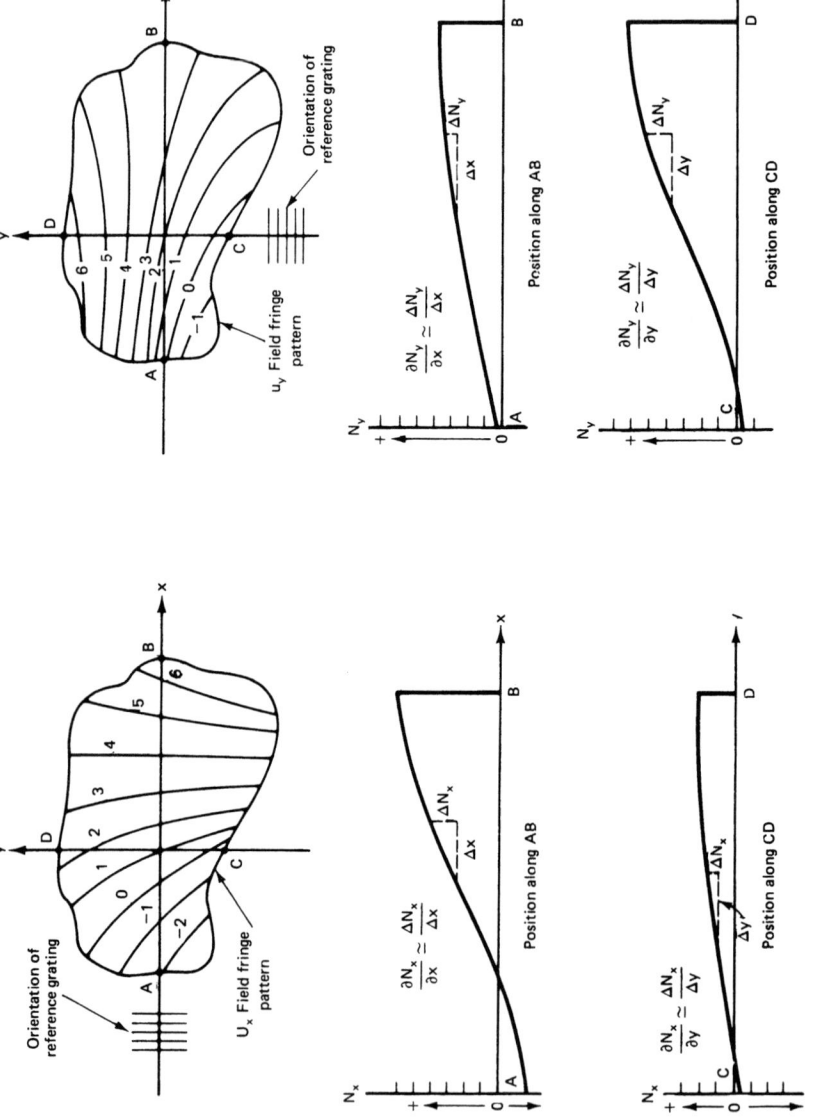

Figure 6-3 Illustration of procedure to obtain the four Cartesian fringe derivatives from the two Cartesian fringe patterns.

GEOMETRIC MOIRÉ

If even greater precision is desired, the fringe readings can be numerically adjusted by curve fitting [24] or fitting to a surface [9]. If such numerical methods are automated, it becomes convenient to combine them with the subsequent strain and stress analyses described below.

Once the fringe derivatives have been obtained, they are multiplied by g to obtain the displacement derivatives in Eqs. (6-14). Strains can then be obtained by substituting these values in either Eq. (6-11) or (6-12).

Note that for each specimen two moiré fringe patterns are required, the u_x displacement field and the u_y displacement field. For each field, derivatives must be obtained in both directions. So, at each point at which the complete two-dimensional strain tensor is required, four derivatives are needed. If the Cartesian strains, once obtained, are small, even though Eqs. (6-12) were used to account for rotations, the principal strains can be obtained from

$$\varepsilon_{1,2} = \tfrac{1}{2}[\varepsilon_{xx} + \varepsilon_{yy} \pm \sqrt{(\varepsilon_{xx} - \varepsilon_{yy})^2 + \varepsilon_{xy}^2}] \tag{6-15}$$

If the material is elastic, stresses can be obtained from Hooke's law.

The calculus approach involves taking derivatives of the fringe patterns. This is usually done as shown in Fig. 6-3. An alternative way is to generate "super" moiré, or moiré-of-moiré, patterns from the initial patterns.

All the moiré patterns discussed previously were composed of fringes representing the difference of pairs of grating lines [i.e., Eq. (6-4)]. Similarly, the fringe derivative (in finite form) is the difference of two fringe orders over a specific distance. If a fringe pattern is duplicated photographically and the negative copy is laid on the original copy, offset by a specific distance, the fringes of the two copies will also interfere (just as the grating lines interfere) to form a moiré-of-moiré pattern. As with grating lines, the new superfringe pattern will represent the difference of the original fringes at every point in the field. If the offset is a linear shift in one direction (without rotation), every point in the combined field will have the images of *two* points, actually separated in the original field by the distance of the shift. The superfringes will be proportional to the fringe difference per unit shift, in the direction of the shift.

Ordering the superfringes is more involved than ordering ordinary fringes. Since the original moiré fringes are visible, it is possible to establish the magnitude of the superfringes by noting any two original fringes that interfere to form the superfringe and counting by how many fringes they differ in the original moiré pattern. The sign of the superfringe may be obvious from the gradient in the original moiré. In any case, assuming that the ordinary fringes are ordered and that one image is shifted in the positive coordinate direction, and the other image an equal amount in the negative direction, the rule for ordering the superfringe is as follows. The original fringe order on the image that was shifted in the negative direction, minus the original fringe order on the image that was shifted in the positive direction, gives the superfringe order and its sign.

The displacement derivatives are obtained using superfringes, with the equations

$$\frac{\partial u_x}{\partial x} = g\frac{N_{xx}}{s_{xx}} \qquad \frac{\partial u_y}{\partial x} = g\frac{N_{yx}}{s_{yx}}$$
$$\frac{\partial u_x}{\partial y} = g\frac{N_{xy}}{s_{xy}} \qquad \frac{\partial u_y}{\partial y} = g\frac{N_{yy}}{s_{yy}} \tag{6-16}$$

where N_{xx}, N_{xy}, N_{yx}, and N_{yy} are the superfringe orders and s_{xx}, s_{xy}, s_{yx}, and s_{yy} are the respective shifts. The first subscript refers to the direction perpendicular to the grating lines and the second to the direction of shift. A complete strain analysis can be conducted as in ordinary moiré by using Eqs. (6-11) or (6-12) and (6-15) as required.

The ordinary fringes must have sufficient density to generate supermoiré fringes. In some cases, wide ordinary fringes confuse the superfringe patterns. It is possible to obtain the superfringes without generating ordinary moiré fringes. This is done by photographing the deformed grating, making a negative copy, and shifting the two deformed grating images, just as the ordinary fringe pattern is shifted on its image. The pattern is the same as the moiré-of-moiré pattern, without the intermediate ordinary fringes. To avoid fractionating the fringe orders, the shift should be an integral number of pitches. The ratio s/g then gives the number of grating lines shifted, and that number divided into the superfringe order gives the derivative. The method is presented in detail in Ref. 20. The same effect can sometimes be achieved by shifting the specimen, the film, or even the lens, and making a double exposure of the deformed grating.

Sensitivity

When the amount of displacement is small (or a coarse grating is used) and the derivatives must be determined over short spans, the two techniques described below can be used to increase fringe response.

Fringe Multiplication

It can be shown [22] that if two gratings are superposed and the frequency of one is an integral multiple n of the other, there will be n times the number of fringes produced by the same load or displacement, as on a pattern where both gratings were of the frequency of the coarser grating. The sensitivity is increased by a factor of n. As an example, suppose that a specimen grating has a nominal frequency of 10 lines/mm and is laid on a reference grating of about the same frequency, and suppose that four fringes are generated. If the same specimen grating is laid on a reference grating with 50 lines/mm, 20 fringes will be produced. Therefore, a fringe pattern produced by a specimen grating, and a reference grating with an integer multiple frequency, can then be analyzed using Eqs. (6-13) and (6-14), with the pitch g of the reference grating.

With special gratings in which the ratio of opaque line width to clear line width conforms to specifications of Ref. 22, excellent contrast is achieved. In addition, the moiré fringes are dramatically sharpened (the fringes are narrower) relative to the ordinary unmultiplied fringes, which permits improved accuracy in locating the center of each fringe. If gratings with arbitrary line widths are used for fringe multiplication, fringe contrast is reduced. Multiplication of up to 5 has been achieved [22].

A variation of the fringe multiplication method is to precisely shift the reference grating over the deformed specimen grating in integral fractions of the pitch [11]. Shifting the reference grating by half the pitch generates a new set of fringes, which fall between those in the original pattern, thus providing twice the number of fringes for analysis. And if shifts of one-third and two-thirds of the pitch are made, two additional patterns are obtained that when combined with the original pattern give three times the fringes. This variant, like fringe multiplication, has limits. The primary limit is the precision with which the reference grating can be shifted. Even if shifting is done with complete accuracy, the precision of the grating line pitch will eventually limit the method. The shifting can be self-calibrated by attaching one or more pieces of undeformed grating to the specimen grating. The pieces should be rotated slightly with respect to the reference grating so as to produce a fringe at each end of the pieces. Calibration is achieved by marking the two fringes and dividing the space between them equally into the number of shift steps required. The desired shifts are obtained by moving the reference grating until the fringe on the piece of grating moves, in turn, to each of the division marks.

Another variant [5] of fringe multiplication is to use a densitometer to calibrate the variation in density of light between the two fringes in a region with a uniform displacement gradient. The density can then be related to the fraction of the pitch displacement between fringes. Measurement of the density of the actual fringe pattern provides the displacement of the body between fringes in terms of fractions of the pitch.

Mismatch [21]

If the specimen grating has an initial pitch 1 or 2% different from the reference grating, then before the specimen is loaded there will be an initial fringe pattern of a frequency 1 or 2% of the reference grating frequency.

If the pitch of the specimen grating is greater than the reference grating, and if the load further increases the specimen pitch, the number of fringes will increase over those in the initial mismatch pattern. If there are initially 10 mismatch fringes and the loading generates two fringes, there will be 12 fringes instead of the two that would be generated from initially matched gratings. Even if the load decreases the pitch, there would still be eight $(10 - 2)$.

In general, at least two reference gratings are used, one with slightly greater pitch and one with slightly less pitch than the specimen grating. With the choice of a proper reference grating, the number of fringes in the pattern can always be increased.

For analysis of a mismatch fringe pattern, the expressions for the derivatives must be modified as follows:

$$\frac{\partial u_x}{\partial x} = g\left(\frac{\partial N_x}{\partial x} - \frac{\partial N_0}{\partial x}\right) \quad \frac{\partial u_y}{\partial x} = g\frac{\partial N_y}{\partial x}$$
$$\frac{\partial u_x}{\partial y} = g\left(\frac{\partial N_x}{\partial y}\right) \quad \frac{\partial u_y}{\partial y} = g\left(\frac{\partial N_y}{\partial y} - \frac{\partial N_0}{\partial y}\right) \quad (6\text{-}17)$$

where g is the reference grating pitch and $\partial N_0/\partial x$ and $\partial N_0/\partial y$ are the fringe derivatives of the mismatch fringes before loading in the u_x and u_y fields, respectively. The remaining four fringe derivatives are taken from the loaded mismatch patterns. Mismatch is presented in detail in Ref. 21.

Analysis of a Rubber Model

Figure 6-4 shows a 152-mm (6-in.)-wide by 19-mm ($\frac{3}{4}$-in.)-thick polyurethane rubber model on which a 12-line/mm (300-line/in.) cross grating is printed. The grating lines are in a 30° counterclockwise direction from the vertical axis and the horizontal direction. The line tangent to the cutout and hole is divided in 6-mm steps for analysis. The model was stretched vertically about 3% overall.

The reference grating is a one-way grating with a mismatch pitch of 312 lpi, thus producing a mismatch before loading of 12 fringes/in. ($\partial N_0/\partial x = 12$), which as indicated in Eq. (6-17) must be subtracted from the fringe gradient shown in the figure. Note the patch to the left of the semicircular cutout. It is a free-mounted piece of 300-line/in. cross grating, lined up with the unloaded model grating to aid in aligning the reference grating and to indicate the amount of mismatch.

Analysis of a Metal Plate

Figure 6-5 shows a 177-mm (7-in.) by 13-mm ($\frac{1}{2}$-in.)-thick stainless steel plate with a 38-mm ($1\frac{1}{2}$-in.)-diameter hole. A 20-line/mm (500-line/in.) cross grating with a 6-mm cross grid was acid

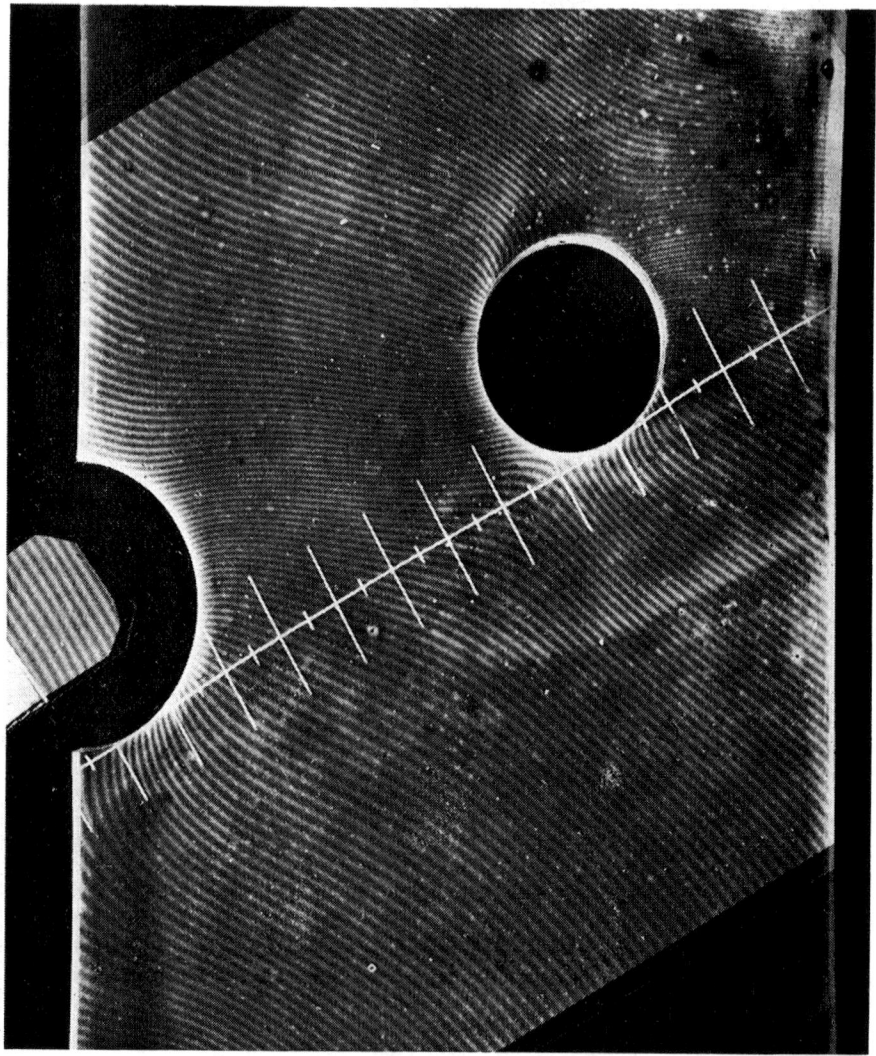

Figure 6-4 Mismatched moiré pattern of displacement components on a rubber specimen ($g = 1/312$ in.).

etched on the surface by a commercial photoengraver. Both negative and positive films were available. It was found that the etched lines (as opposed to the etched dots) provided more contrast. The reference grating is a 40-line/mm (1000-line/in.) one-way grating of horizontal lines. The 40-line/mm grating produces a fringe multiplication of 2, and the grating pitch by which the fringes are multiplied is $g = 0.025$ mm (0.001 in.).

The plate was loaded with a stress of approximately 70 MPa and held at that load and a

GEOMETRIC MOIRÉ 283

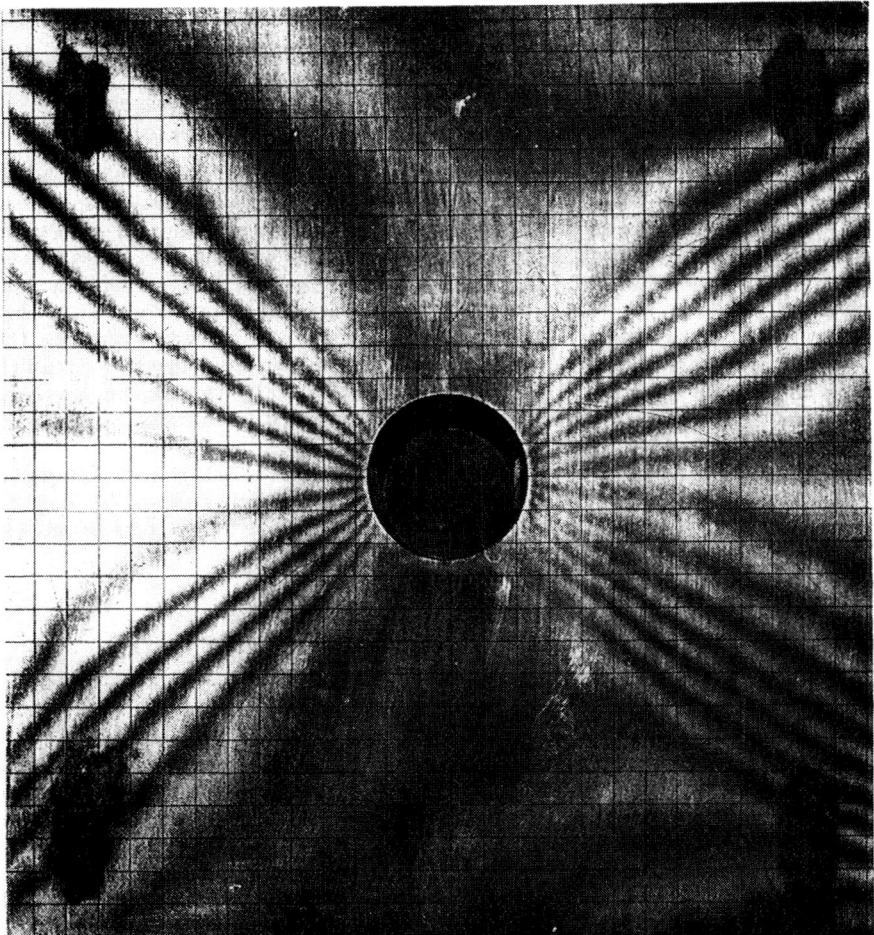

Figure 6-5 Moiré pattern multiplication of vertical displacement components on a stainless steel specimen etched with a 20-line/mm cross grating.

temperature of 600°C for some 40 days. The cross grating was photographed directly in the oven at one-to-one magnification and the photographs analyzed with a reference grating overlay. A disk of the stainless steel, etched with the same grating, rested unloaded in the central hole and allowed determination of the thermal expansion due to temperature. Thermal expansion can be looked at as a self-induced mismatch.

The u_y-field pattern as seen in Fig. 6-5 was not obtained from the high-temperature photographs; rather, it is the residual strain pattern, due to creep, left after unloading and cooling the specimen. A one-way horizontal 40-line/mm grating was laid directly on the plate and the pattern photographed at room temperature. Since the specimen was stretched, if the y direction is assumed to be up, the fringe orders increase from bottom to top. The fringes at the horizontal axis of the hole have a spacing of less than 2.5 mm, indicating a residual strain of over 1%.

Analysis of the Interior Planes of a Stress-Frozen Model [14]

The procedure to freeze stresses and strains in a three-dimensional model, described in Chapter 5, is not repeated here. The moiré patterns in Fig. 6-6 are on two perpendicular slices cut from the interior of a three-dimensional stress-frozen, pressure-loaded model. Gratings of 40 lines/mm were printed on the stress-frozen slices. Advantage was taken of the symmetries in the slices to print what correspond to perpendicular gratings on the same slice, thus giving a complete pair of patterns on each slice. The patterns were generated by placing the two slices with the printed gratings in an oven and heating until the stress-frozen strains were relieved by annealing. On removal from the oven, the fringe patterns produced with the reference grating are the negative of the fringes that would be produced by the displacements due to the pressure load. However, for purposes of analysis the fringes are labeled with orders corresponding to the actual displacements that occurred under load.

Using Eqs. (6-14), (6-15), and (6-11) or (6-12) with the pitch and the model size, the photographs contain all the data necessary to determine all the strains in the two planes shown. The complete three-dimensional strain tensor can be obtained along the line of intersection of the two planes. Reference 14 contains additional details of the analysis.

Supermoiré Analysis of the Meridian Plane of a Sphere [13]

Large strains on the interior central plane of a sphere were analyzed by printing a 20-line/mm grating on the flat side of a transparent rubber hemisphere, and then bonding it to another rubber hemisphere to form a sphere with the grating on its meridian plane. The sphere was placed in a tank with flat parallel glass sides and filled with liquid of approximately the same index of refraction as that of the rubber. The sphere was loaded axially by a compressive force from the top. The grating was photographed at one-to-one magnification with the grating lines horizontal. Advantage was taken of the symmetry in load and geometry and the sphere was turned 90°, to place the grating lines vertical, reloaded, and rephotographed. Contact "positives" of the two negatives were made on film. Each pair was then shifted, negative on positive, in the vertical and horizontal directions, and the four supermoiré patterns shown in Fig. 6-7 formed.

Figure 6-7 contains all the data necessary to compute the strain in the meridian plane using Eq. (6-12). Because both the geometry and load are axisymmetric, the normal strains perpendicular to the meridian plane are principal strains and can be obtained by analyzing the displacement components perpendicular to the axis of symmetry, the u_x displacement field. This third strain is $\varepsilon_{zz} = u_x/x$, where x is the distance of u_x and ε_z from the axis of symmetry. The maximum strain magnitude is $\varepsilon_{yy} = -0.40$ on the sphere's axis. The horizontal strain at the same point is $\varepsilon_{xx} = 0.16$. Reference 13 contains additional details of the analysis.

Dynamic Moiré [23]

Figure 6-8 gives patterns from the left and right halves of a 125-mm-diameter transparent rubber disk impacted by a 30-g weight dropped from a height of 98 cm. The photograph was taken 1.8 ms after impact with a microflash that exposed a 100 mm × 125 mm sheet of film. This is one of a series of patterns reported in Ref. 23.

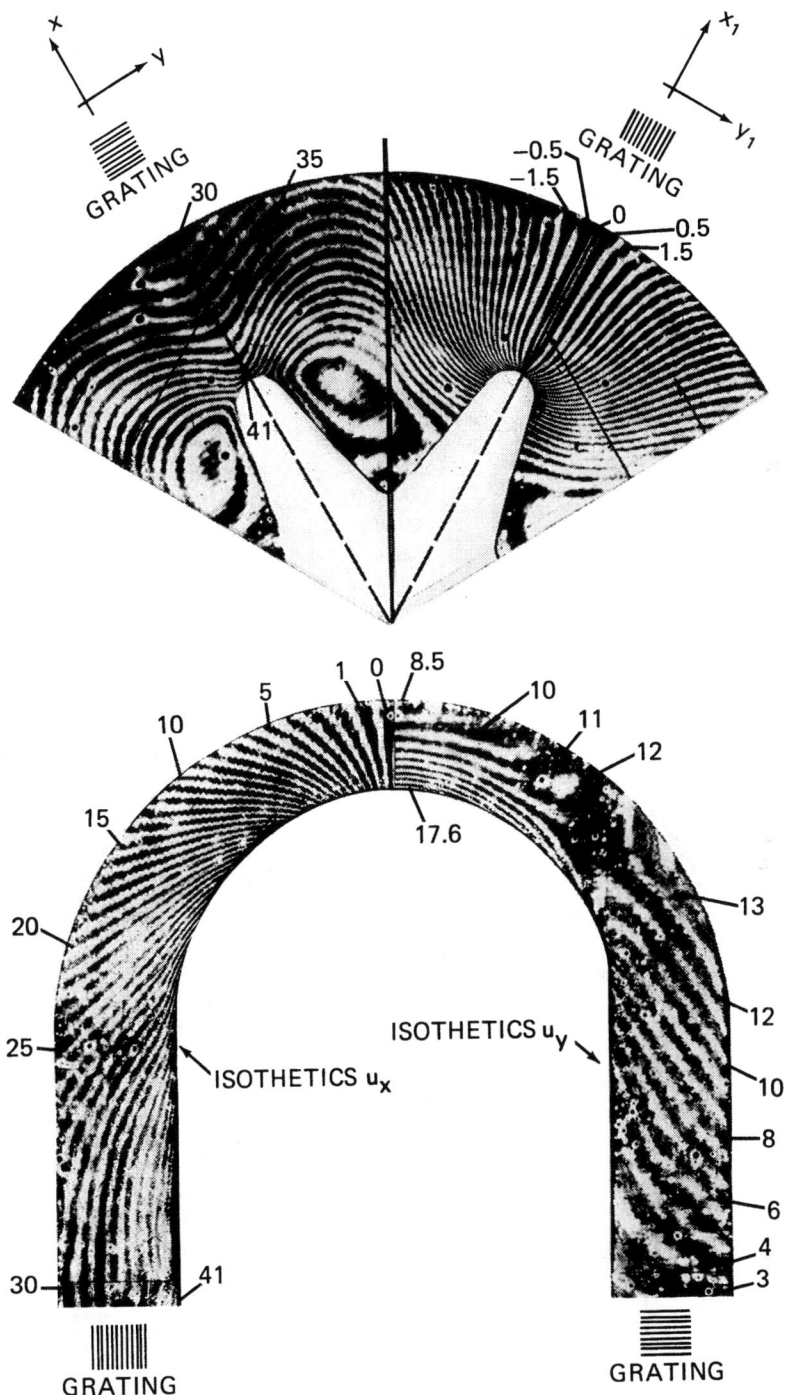

Figure 6-6 Moiré patterns of orthogonal displacement components on (a) vertical and (b) horizontal slices taken from the interior of a stress-frozen model. (From Ref. 14.)

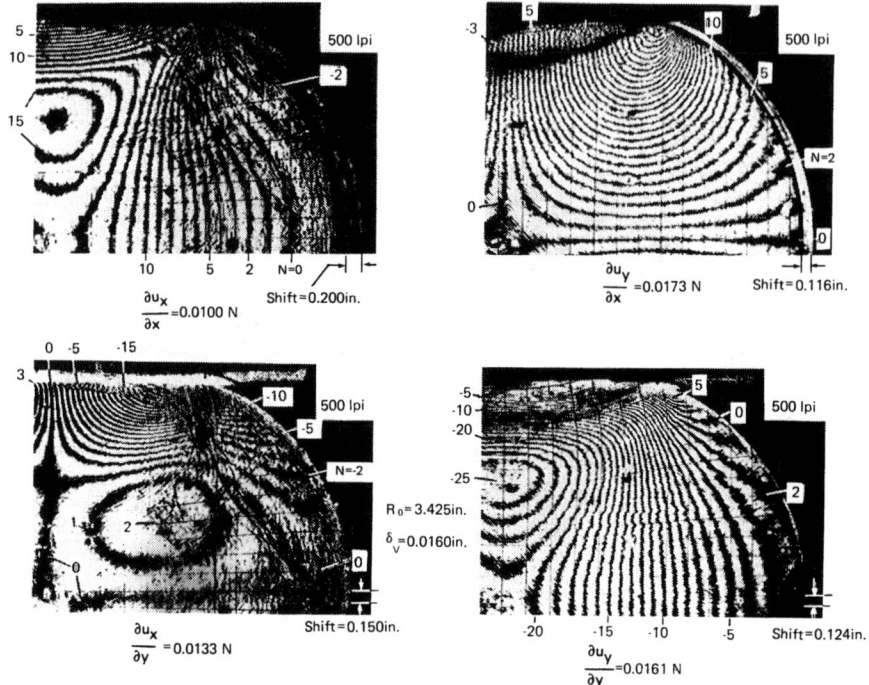

Figure 6-7 Supermoiré fringes corresponding to the four derivatives of displacements in the meridian plane at a sphere subjected to a large compression: lpi = lines/in. (From Ref. 13.)

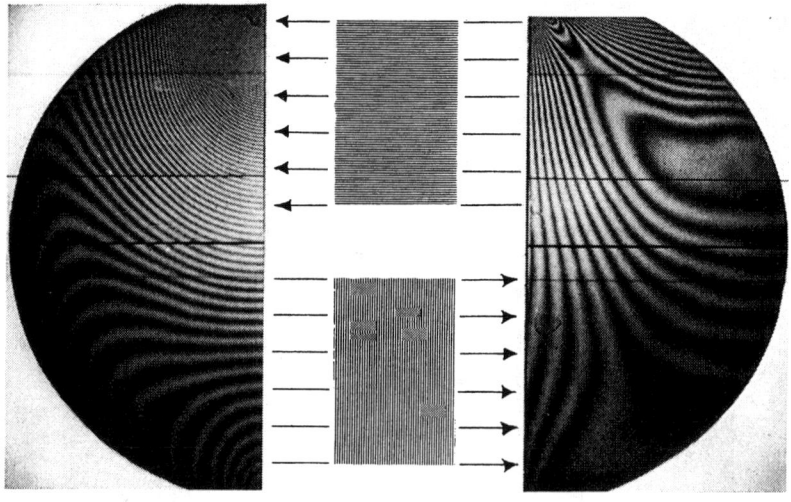

Figure 6-8 Dynamic moiré fringe patterns of vertical and horizontal displacement components on a disk shortly after an impact at the top ($g = 0.025$ mm). (From Ref. 23.)

GEOMETRIC MOIRÉ

6-2 Out-of-Plane Moiré

Analysis of Slope and Curvature Using Reflection Moiré [25–28]

It was mentioned in Section 6-1 that in-plane moiré can be conducted with a single grating, the specimen grating, by making a double-exposure photograph of the grating before and after loading. Out-of-plane moiré is *normally* conducted with one grating, in this case, the reference grating. The reference grating is made to interfere with either its reflection or its shadow.

Ligtenberg [26] proposed a method to analyze the out-of-plane rotation and curvature of a plate using reflection. A simplified version of the method is shown in Fig. 6-9. A reference grating of equispaced parallel lines at least twice the size of the plate is positioned at a suitable distance from the polished plate. The reflected image of the grating is photographed. The plate is loaded, producing out-of-plane rotation of the plate. The component of out-of-plane rotation, perpendicular to the reference grating lines, shifts the image of the grating lines. A second exposure of the film results in a double-exposure photograph which, when developed, is a moiré pattern representing the interference of the two grating images.

As with the in-plane moiré pattern, the fringe orders here represent the difference between the orders of the increasing grating lines, with the same relationship specified by Eq. (6-4). In Fig. 6-9, the lth line and the mth line that produce the Nth fringe are shown, at a distance $e = gN$ apart on the grating. If the distance d is large enough that the out-of-plane motions of the plate are small compared to d, and also that the angle subtended by the plate at the camera lens is small, and finally if the plate rotations are small, the paraxial assumption will allow the angle θ to be approximated as $e/2d$. The angle of rotation in the plane perpendicular to the grating lines will be

$$\theta = \frac{gN}{2d} \tag{6-18}$$

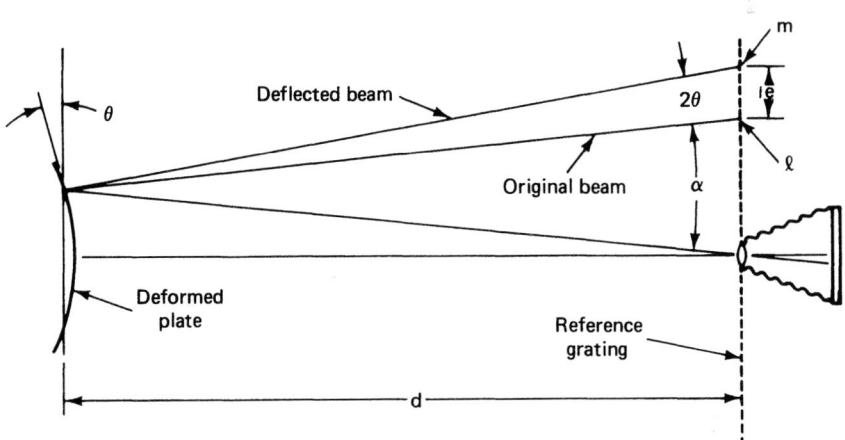

Figure 6-9 Schematic arrangement to produce a reflection moiré pattern representing the slope components of the plate.

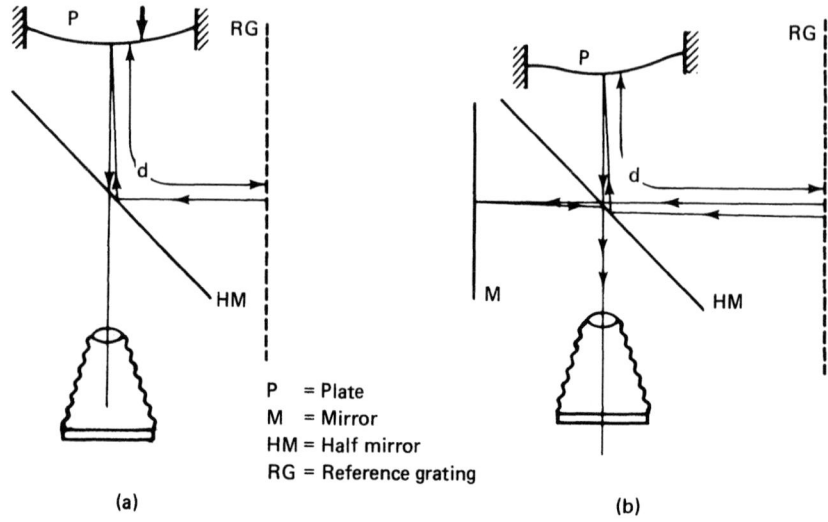

Figure 6-10 Two arrangements for the analysis of slope and curvature components on a reflective plate.

Ligtenberg [26] suggested using a curved grating to improve the accuracy of Eq. (6-18). The usual practice is to use a flat grating and arrange d, α, and θ to minimize error. To allow for these angular adjustments and to avoid having to view the plate through the grating, a half-mirror can be placed at 45° to the camera axis and the flat grating placed off to the side as shown in Fig. 6-10a. As described earlier under "Fringe Intensity" in Section 6-1, the double-exposure method produces poor contrast fringes. These patterns can be improved with the optical filtering methods discussed in Chapter 7. Alternatively, the setup shown in Fig. 6-10b gives a fringe pattern with a single exposure.

The analysis of curvature with out-of-plane moiré is similar to the analysis of strain with in-plane moiré. As in in-plane moiré, two moiré patterns are necessary to describe the full field. One difference from in-plane moiré is that since the grating is not applied to the specimen, it is only necessary to rotate the specimen, or the grating, 90° and rephotograph to obtain the second moiré pattern.

The two out-of-plane Cartesian rotations at each point obtained from Eq. (6-18) can be represented by θ_x and θ_y. If the plate is initially flat and vertical, these are also the Cartesian slopes of the loaded plate. If the plate is flat but has an initial inclination, they are still the slopes with respect to the initial plane. Since most analyses are conducted on flat plates, after noting the qualifications above, θ_x and θ_y will be referred to as slopes.

The maximum slope (or gradient) at a point is the vector sum of the Cartesian slopes, as the maximum displacement at a point is the vector sum of the displacement components. Similarly, in the direction perpendicular to the maximum slope, the slope is zero.

Curvature is the derivative of slope. The two Cartesian curvatures are $\partial\theta_x/\partial x$ and $\partial\theta_y/\partial y$. The derivative of the slope perpendicular to its direction is called warp. The warp terms are $\partial\theta_x/\partial y$ and $\partial\theta_y/\partial x$.

GEOMETRIC MOIRÉ

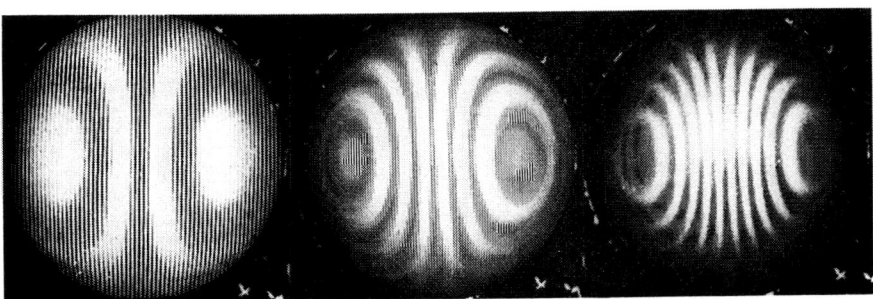

Figure 6-11 Reflection moiré pattern of fringe representing the horizontal components of slope in a circular plate supported on its edge and subjected to uniform pressure. The three levels of density were obtained by varying the distance d, shown in Fig. 6-10. (From Ref. 4.)

Figure 6-11 shows reflection moiré fringe patterns on a circular plate clamped around its edges and subjected to a uniform pressure. The optical setup used was similar to the one shown in Fig. 6-10a. By varying the distance d, the fringe pattern density was varied as shown. The load for all three patterns was the same. The slope θ_x is zero along the vertical axis of each pattern and increases positively to the left and negatively to the right.

Figure 6-12 shows reflection moiré fringes of the θ_x and θ_y patterns of a free-standing rectangle subjected to a dynamic load in its upper right-hand corner. The optical setup was similar to the one in Fig. 6-10a. Both these examples are drawn from Ref. 4.

An extensive series of moiré methods to measure the slopes and curvatures of reflective surfaces is presented in Ref. 25. The methods involve one or two gratings placed at various positions along the optical axes of the system. The analysis is too lengthy to be presented here.

Analysis of Topography or Out-of-Plane Displacement with Shadow Moiré [29–35]

In addition to the in-plane displacements due to load that were discussed in Section 6-1, there are usually out-of-plane displacements normal to the surface plane. In the case of plane stress, these out-of-plane displacements can be analyzed to determine the sum of the in-plane strains ε_{xx} and ε_{yy}, point by point. In the case of membranes and plates subjected to out-of-plane loads, such displacements are also of interest. The topology of three-dimensional surfaces, membranes, and films (both liquid and solid), whether produced by load or not, are useful in the detection of irregularities, dents, and subsurface flaws. Many of these cases have been analyzed with shadow moiré.

The slopes and curvatures analyzed using reflection moiré are but the second and third steps in a description of surface topology. Thus for a surface lying above the x-y plane, the $z(x, y)$ coordinate describes the surface. The slope of the surface is the derivative of z (e.g., $\theta_x = \partial z/\partial x$ and $\theta_y = \partial z/\partial y$). The curvature of the surface is the derivative of the slope (e.g., $\partial \theta_x/\partial x$ and $\partial \theta_y/\partial y$, etc). Thus the fringe pattern representing slopes obtained from reflection moiré could be integrated to obtain the height z above the datum plane of a warped or curved surface. The same information can be obtained directly from shadow moiré.

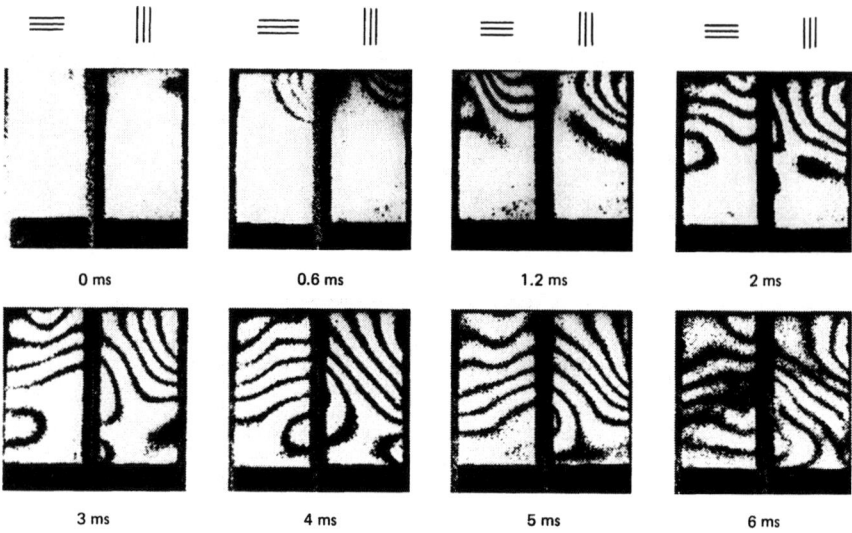

Figure 6-12 Dynamic reflection moiré patterns of vertical and horizontal slope components of a rectangular plate supported at its base at the time shown after being struck by a load at the upper right-hand corner. (From Ref. 4.)

To obtain a shadow moiré pattern, a matte rather than a polished surface is required. The surface is often painted with white tempera or flat paint. A grating is placed directly in front of the matte surface. A typical surface and grating with different arrangements of camera and lighting are shown in Fig. 6-13. With a collimated light source, the camera should be placed with its axis perpendicular to the grating (Fig. 6-13a). With a point source, the camera and light source should be equidistant from the grating (Fig. 6-13b). The reason for these choices is explained below in the development of the fringe equation.

Consider the arrangement in Fig. 6-13a. The grating lines, as seen by the camera, interfere with their shadow on the surface. Referring again to Eq. (6-4), the distance e, between the lth line of the grating and the mth line of the grating that produces the shadow will depend on the angle α of the collimated light and the distance z of the surface from the grating.

$$\tan \alpha = \frac{e}{z} \qquad (6\text{-}19)$$

The fringe order N, which is the difference between the l and m grating lines, is given by $e = Ng$. The distance z from the grating to the surface is then given by

$$z = \frac{gN}{\tan \alpha} \qquad (6\text{-}20)$$

This equation is exact on the axis of the camera. As in the case of reflection moiré, it is a good approximation for small viewing angles. The angle, subtended by the surface, at the camera lens

GEOMETRIC MOIRÉ

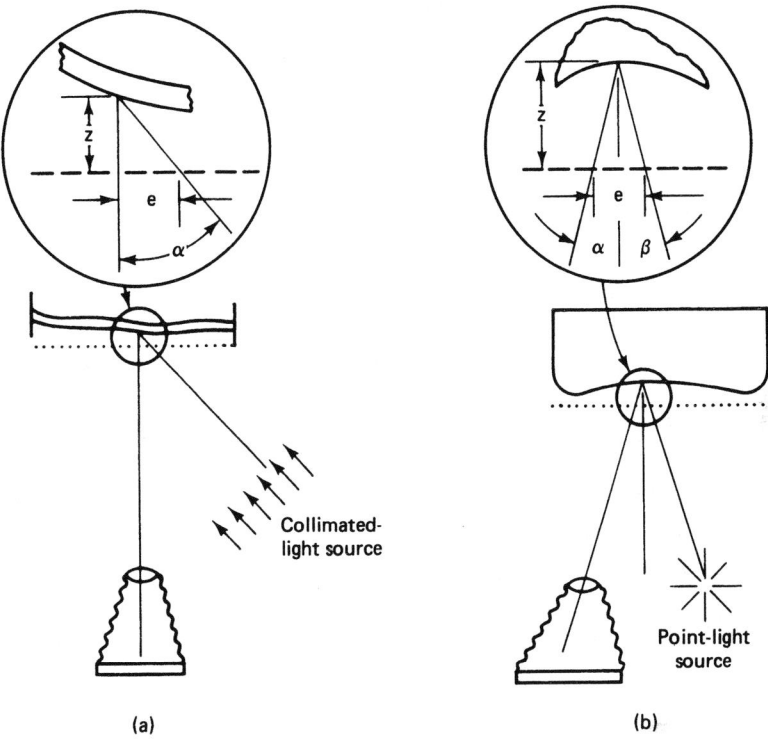

Figure 6-13 Two arrangements for the analysis of out-of-plane displacement, or topography of the surface: (a) with collimated light; (b) with a point source.

should be small compared to α, the angle of the collimated light. As with reflection moiré, the larger the camera lens focal length, the better the approximation.

The point-lighting arrangement shown in Fig. 6-13b puts the same restrictions on the light source as are on the camera. As with the camera, the angle, subtended by the surface, at the light source should be small compared to the angles α and β of camera and light to the grating normal. This approximation is best with the grating and surface placed midway between camera and light, as shown in Fig. 6-13b.

The formula in this case becomes

$$z = \frac{gN}{\tan \alpha + \tan \beta} \tag{6-21}$$

Figure 6-14 shows the shadow moiré pattern of the surface of a thick rubber tube with four holes after it has been turned inside out. A 2-line/mm grating was placed directly in front of the inverted tube and light directed from 45° to the normal as illustrated in Fig. 6-13a. The grating lines and their shadows are just visible in the photograph.

Figure 6-14 Two-line-per-millimeter shadow moiré pattern of a tube with four holes that has been turned inside out.

Figure 6-15 shows shadow moiré patterns on a 50-mm-wide reinforced rubber strip of variable thickness while under test in a fatigue-bending test machine. A 2-line/mm grating was used and the light was at 37.5° to the normal. Because of the machine frame, it was necessary to view and photograph the pattern through a mirror, but this is not significant. Patterns were photographed at about 2 Hz, but to find the total displacement the machine was stopped at the two extreme positions and the patterns shown recorded and analyzed as shown.

A variant of the shadow moiré method is projection moiré. A grating is placed at the light source and its shadow projected with the use of lenses. A companion grating in front of the camera is used as a reference grating. Pirodda [33] describes a number of shadow and projection moiré variants, as well as many static and dynamic applications. Andrews [29] and Halioua et al. [30] show that in shadow moiré, the fringes remains stationary when the grating is moved in its plane, perpendicular to the grating lines. In addition, fringe contrast is improved by such motion during photographing.

6-3 Summary

Geometric moiré consists of a family of optical methods that take advantage of the interference of various motifs, in particular equidistant parallel lines, to determine displacements and rotations, in the plane of view and out from the plane of view. In addition to their own value, these variables provide the data for analysis of in-plane strain and out-of-plane curvatures.

GEOMETRIC MOIRÉ

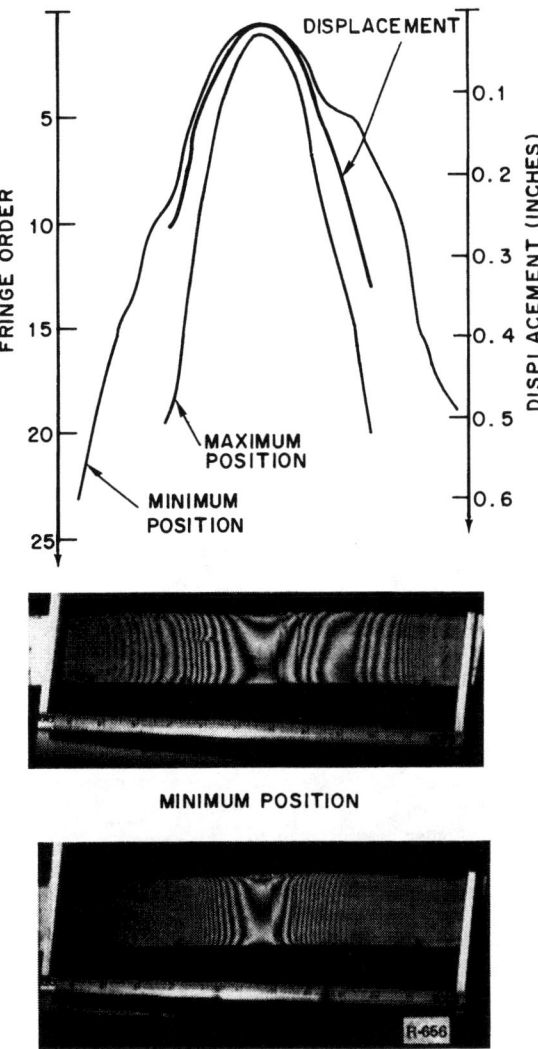

Figure 6-15 Shadow moiré patterns of surface height on a reinforced rubber strip, which were subtracted to determine the out-of-plane displacement between the maximum and minimum positions.

The geometric methods described here are the precursors of the interferometric moiré described in Chapter 7. Interferometric moiré increases sensitivity by more than an order of magnitude. Contemporary moiré analysis, as exemplified by applications in this and the following chapter, should be improved and extended by use of automated pattern-recognition equipment.

Acknowledgments

The support of the Catholic University of America, the U.S. Naval Research Laboratory, and the American Society for Engineering Education in preparing this chapter is gratefully acknowledged. Professor Daniel Post was kind enough to read the chapter and make many valuable suggestions. Madeline Sapienza is given special thanks for her careful preparation and editing of the manuscript.

List of Symbols

A, B, C	equation constants
d	distance between moiré fringes
d	distance from reflecting grating to a plate
e	distance between grating lines of the images that intersect in photograph (out-of-plane moiré)
g, g'	pitches of reference and specimen gratings
l, m	order of reference and specimen grating lines
N	moiré fringe order
N'	moiré fringe order due to extraordinary strain or rotation
N_o	mismatch moiré fringe order
N_x, N_y	moiré fringe orders due to grating lines initially perpendicular to x and y
$N_{xx}, N_{yy}, N_{xy}, N_{yx}$	supermoiré fringe orders
$S_{xx}, S_{yy}, S_{xy}, S_{yx}$	shift of like specimen grating images
u_x, u_y	components of displacements in the x-y plane (see δ)
x, y	Cartesian coordinates along which in-plane moiré gratings are aligned
z	distance from shadow moiré grating to object
α	original angle in reflection moiré
α, β	setup angles in shadow moiré
δ	displacement component perpendicular to grating lines
ε	strain, change in length per unit length
$\varepsilon_1, \varepsilon_2$	principal strains
$\varepsilon_{xx}, \varepsilon_{yy}$	strains in x and y directions
ε_{xy}	shear strain
θ	angle between reference and specimen in-plane grating lines
θ_x, θ_y	out-of-plane rotation in x-z and y-z planes
ϕ	in-plane moiré fringe angle measured from reference grating
ω_{xy}	average rotation of the x and y coordinates

References

Books, Chapters, and Reviews

1. J. W. Dally and W. F. Riley, *Experimental Stress Analysis*, 3rd ed., McGraw-Hill, New York, 1991, Chap. 11.
2. A. J. Durelli and V. J. Parks, *Moiré Analysis of Strain*, Prentice-Hall, Englewood Cliffs, NJ, 1970.

3. V. J. Parks, Strain Measurement Using Grids, *Opt. Eng., 21*, no. 4 (1982), 633–639.
4. A. S. Kobayashi (Ed.), *Manual of Engineering Stress Analysis*, 3rd ed., Prentice-Hall, Englewood Cliffs, NJ, 1982, Chap. 6 (by F. P. Chiang).
5. C. A. Sciammarella, The Moiré Method—A Review, *Exp. Mech. 22*, no. 11 (1982), 418–433; Discussion, *23*, no. 12 (1983), 446–449.
6. P. S. Theocaris, *Moiré Fringes in Strain Analysis*, Pergamon Press, Elmsford, NY, 1969.
7. P. S. Theocaris, Moiré Fringes, a Powerful Measuring Device, *Appl. Mech. Rev., 15*, 333–339 (1962); up-dated in *Applied Mechanics Surveys*, H. N. Abramson, H. Liebowitz, J. M. Crowley, and S. Juhasz, (Eds.), Spartan Books, Washington, DC, 1966, 613–626.

In-Plane Analysis

8. P. M. Boone, A. G. Vinckier, R. M. Denys, W. M. Sys, and E. N. Deleu, Application of Specimen-Grid Moiré Techniques in Large Scale Steel Testing. *Opt. Eng., 21*, no. 4 (1982), 615–625.
9. W. Bossaert, R. Dechaene, and A. Vinckier, Computation of Finite Strains from Moiré Displacement Patterns, *J. Strain Anal., 3*, no. 1 (1968), 65–75.
10. J. M. Burch and C. Forno, High Resolution Moiré Photography, *Opt. Eng., 21* no. 4 (1982), 602–614.
11. F. P. Chiang, V. J. Parks, and A. J. Durelli, Moiré Fringe Interpolation and Multiplication by Fringe Shifting, *Exp. Mech., 8*, no. 12 (1968) 554–560.
12. M. Dantu, Recherches diverses d'extensométrie et de détermination des contraintes, *Conf. GAMAC*, Feb. 22, 1954; Utilization des réseaux pour l'étude des déformations, Laboratoire Central des Ponts et Chaussées, Paris, Publ. 57-6, 1957 (seminal papers).
13. A. J. Durelli and T. L. Chen, Displacement and Finite-Strain Fields in a Sphere Subjected to Large Deformation, *Int. J. Non-Linear Mech., 8* (1973), 17–30.
14. A. J. Durelli, V. J. Parks, and C. J. del Rio, Experimental Determination of Stresses and Displacements in Thick-Wall Cylinders of Complicated Shape, *Exp. Mech. 8*, no. 7 (1968), 319–326.
15. Y. Morimoto and T. Hayashi, Deformation Measurement During Powder Compaction by a Scanning-Moiré Method, *Exp. Mech., 24*, no. 2 (1984), 112–116.
16. S. Morse, A. J. Durelli, and C. A. Sciammarella, Geometry of Moiré Fringes in Strain Analysis, *Trans ASCE, 126-1* (1961).
17. D. W. Oplinger, Application of Moiré Methods to Evaluation of Structural Performance of Composite Materials, *Opt. Eng., 21*, no. 4 (1982), 626–632.
18. G. Oster, M. Wasserman, and C. Zwerling, Theoretical Interpretation of Moiré Patterns, *J. Opt. Soc. Am., 54*, no. 2 (1964), 169–175; see also *55*, no. 10 (1965), 1329–1330.
19. V. J. Parks and A. J. Durelli, Various Forms of the Strain-Displacement Relations Applied to Experimental Stress Analysis, *Exp. Mech., 4*, no. 2 (1964), 37–47.
20. V. J. Parks and A. J. Durelli, Moiré Patterns of Partial Derivatives of Displacement Components, *J. Appl. Mech., E33*, no. 4 (1966), 901–906.
21. D. Post, Moiré Grid-Analyzer Method for Stress Analysis, *Exp. Mech., 5*, no. 11 (1965), 366–377, discussion *6*, no. 5 (1966), 287–288.
22. D. Post, Sharpening and Multiplication of Moiré Fringes, *Exp. Mech., 7*, no. 4 (1967), 154–159.
23. W. F. Riley and A. J. Durelli, Application of Moiré Methods to the Determination of Transient Stress and Strain Distributions, *J. Appl. Mech., 29*, no. 1 (1962).
24. R. E. Rowlands, T. Liber, I. M. Daniel, and P. G. Rose, Higher-Order Numerical Differentiation of Experimental Information, *Exp. Mech., 13*, no. 3 (1973), 105–113.

Reflection Method

25. T. Y. Kao and F. P. Chiang, Family of Grating Techniques of Slope and Curvature Measurements for Static and Dynamic Flexure of Plates, *Opt. Eng., 21*, no. 4 (1982), 721–742.

26. F. K. Ligtenberg, The Moiré Method: A New Experimental Method for the Determination of Moments in Small Slab Models, *Proc. SESA, 12*, no. 2 (1954), 83–98.

27. B. Ranganayakamma and F. P. Chiang, Mismatches Applied to Ligtenberg's Reflective Moiré Methods. *J. Strain Anal., 8*, no. 1 (1973), 24–29.

28. R. Ritter, Reflection Moiré Methods for Plate Bending Studies, *Opt. Eng., 21*, no. 4 (1982), 663–671.

Shadow Moiré

29. D. R. Andrews, Shadow Moiré Contouring of Impact Craters, *Opt. Eng., 21*, no. 4 (1982), 650–654.

30. M. Halioua, R. S. Krishnamurthy, H. Liu, and F. P. Chiang, Projection Moiré with Moving Gratings for Automated 3-D Topography, *Appl. Opt., 22*, no. 6 (1983), 850–855.

31. K. G. Harding and J. S. Harris, Projection Moiré Interferometer for Vibration Analysis, *Appl. Opt., 22*, no. 6 (1983), 856–861.

32. M. Marcel Mulot, Application du moiré à l'étude des déformations du mica, *Rev., d'Opt., 4* (1925), 252–259 (seminal paper).

33. L. Pirodda, Shadow and Projection Moiré Techniques for Absolute or Relative Mapping of Surface Shapes, *Opt. Eng., 21*, no. 4 (1982), 640–649.

34. H. Takasaki, Moiré Topology, *Appl. Opt., 9*, no. 6 (1970), 1457–1472; *12*, no. 4 (1973), 845–850.

35. P. S. Theocaris, Moiré Fringes of Isopachics, *J. Sci. Instrum., 41*, no. 3 (1964), 133–138.

CHAPTER

7

Moiré Interferometry

Daniel Post

Virginia Polytechnic Institute and State University
Blacksburg, Virginia

7-0 Introduction

Moiré interferometry combines the concepts and techniques of geometrical moiré and optical interferometry. In his definitive monograph, Guild [1] shows that all moiré phenomena can be treated as cases of optical interference. All moiré techniques are members of one family. Moiré generated by low-frequency bar-and-space gratings, however, can also be explained on the basis of obstruction or mechanical interference, and that is the subject of Chapter 6. Of the remaining topics under moiré, this chapter concentrates on one aspect—a technique called moiré interferometry—which is explored in depth.

We will study a high-sensitivity moiré interferometry method of whole-field, in-plane displacement measurements. It offers a unique combination of high sensitivity and excellent contrast, range, and spatial resolution. It is reasonably easy to understand and apply. In 1965 Dally and Riley [2] wrote: "If and when higher quality line arrays with line spacings of 10,000 to 30,000 lines/in. become available, the moiré method may see more general usage in a wider range of experimental stress analysis." That stage has been reached. A reference grating frequency of 1200 lines/mm (30,480 lines/in.) was used for one fringe pattern shown in this chapter, 4000 lines/mm (101,600 lines/in.) for another, and 2400 lines/mm (60,960 lines/in.) for all the rest.

Moiré interferometry provides contour maps of in-plane displacement fields u_x and u_y.[1] Traditionally, the displacement fields have been used to determine small strains by the relationships

$$\varepsilon_{xx} = \frac{\partial u_x}{\partial x} \qquad \varepsilon_{yy} = \frac{\partial u_y}{\partial y} \tag{7-1}$$

$$\varepsilon_{xy} = \frac{\partial u_x}{\partial y} + \frac{\partial u_y}{\partial x} \tag{7-2}$$

---

[1] A list of the symbols used in this chapter precedes the References.

Certain aspects of the determination of strains are addressed in this chapter. However, experimental determinations of displacement fields are especially important in their own right.

Moiré interferometry offers an experimental counterpart to the powerful computational methods of solid mechanics, where displacement fields are also primary output. The experimental measurements provide a dependable datum to evaluate computational programs and their results, and the experimental measurements offer data for powerful hybrid experimental/computational methods. Their mutuality is exciting.

This chapter is not a survey of a broad field. With its narrow scope, much of the contents is treated as a self-contained tutorial. It begins with certain fundamentals of optics needed in our studies. This is a useful review, too, for other chapters on photomechanics in this book. Moiré interferometry for in-plane measurements is treated in considerable detail. Related topics of fringe vectors, emergence vectors, and out-of-plane measurements by moiré interferometry and data processing are discussed at the end.

The techniques of moiré interferometry discussed here emerged quickly, but they are part of an evolutionary process. Their foundation is geometrical moiré, classical optics, and innumerable creative contributions of predecessors and colleagues. Reference 3 provides an extensive bibliography; Refs. 4 and 5 are excellent textbooks covering background topics; Refs. 6 through 20 provide historic and technical background more closely aligned with the techniques studied here; and Refs. 21 through 56 cover descriptions and studies that were published during the development or refinement of techniques discussed in this chapter.

7-1 Fundamentals

The wave theory of light is sufficient to explain all the characteristics of moiré interferometry. A parallel beam of light emitted in the z direction is depicted (at a given instant) as a train of regularly spaced disturbances that vary with z as

$$A = a \cos 2\pi \frac{z}{\lambda} \tag{7-3}$$

The symbol A describes the amplitude or strength of the disturbance, which is usually viewed as the strength of an electromagnetic field at a point in space, particularly, the electric-field strength. The coefficient a is a constant. The field strength varies harmonically along z, where the distance between neighboring maxima is λ, called the *wavelength*. Length z is not endless—for that would represent an eternal light—but it is very long compared to λ; in the case of laser light, length z of the wave train may be 1 million wavelengths or more. For classical light sources (not lasers), many short wave trains exist simultaneously in the beam.

However, the wave train is not stationary. It travels or propagates through space with a very high constant velocity C. At any fixed point along the path of the wave train, the disturbance is a periodic variation of field strength. Field strength varies through one full cycle in the brief time interval λ/C (seconds/cycle); its frequency ω is C/λ (cycles/second). During the passage of the wave train through any fixed point $z = z_0$, the light disturbance varies with time t as

$$A = a \cos 2\pi \frac{C}{\lambda} t = a \cos 2\pi \omega t \tag{7-4}$$

The variable $2\pi\omega t$ is called the *phase* of the disturbance. For a parallel beam of light, the phase is constant along any plane perpendicular to the beam; in other words, the field strength is the same at all points in the cross section.

MOIRÉ INTERFEROMETRY

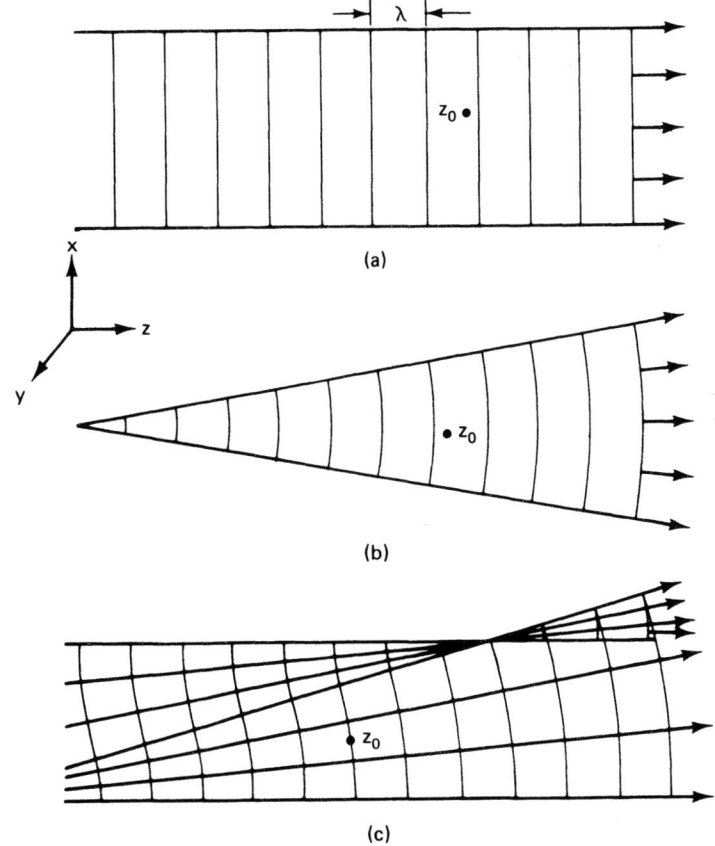

Figure 7-1 Beams of light and their corresponding wavefronts: (a) plane; (b) spherical; (c) irregular or warped.

We will define a *wavefront* as any continuous surface along which the field strength is constant, that is, along which

$$\omega t = k \tag{7-5}$$

where k may be any numerical value. In a parallel beam (also called a *collimated* beam), the wavefronts are always plane cross sections of the beam, as illustrated in Fig. 7-1a. These move along the length of the beam with the speed of light.

In a conical beam (Fig. 7-1b), the wavefronts are spherical and their radii of curvature increase with distance from the origin. The shape of the wavefront changes constantly as it propagates through space, but all wavefronts in the wave train have a fixed shape as they intercept any fixed point z_0 in their path.

Beams may acquire irregular shapes (Fig. 7-1c)—for example, by reflection from an irregular mirror. Again the wavefronts constantly change shape as they propagate, and they may even fold and become double valued, as shown. This is a consequence of the rule that light propagates in the direction of the *wave normal*, that is, in the direction perpendicular to the wavefront; this is always true for light propagation in homogeneous media. As before, every wavefront in the beam has the same unique shape as it propagates through a fixed point z_0. Wavefronts of smooth but irregular shapes are called *warped* wavefronts.

More Facts and Features

Light trains propagate with an astounding velocity of about 3×10^8 m/s or 186,000 miles/s in free space. The visible spectrum encompasses the wavelength range from about 400 nm (i.e., nanometers or 10^{-9} m) for blue light to about 700 nm for red light. The popular helium–neon laser emits visible light of wavelength 632.8 nm (24.91×10^{-6} in.), and the popular argon laser can be adjusted for several different colors, with strongest emissions at 514.5 nm (green) and 488.0 nm (blue-green).

The visible spectrum is only a tiny portion of the electromagnetic spectrum, in which wavelengths extend at least from 10^{-14} m to 10^3 m, a span of 17 decades. All the phenomena discussed here—interference, diffraction, and so on—apply for the entire electromagnetic spectrum.

When a wave train travels through a fixed point in space, its frequency of oscillation, $\omega = C/\lambda$, is about 6×10^{14} Hz for visible light. No instruments can detect individual cycles in this frequency range. Instead, receivers like the eye, photographic film, and photoelectric cells respond to energy.

When a wave train, or a beam of light, is intercepted by a receiver, the *energy* available to be dissipated into the receiver is the *intensity* of the light multiplied by the exposure time. During the passage of the wave train in a volume of space, the light has an intensity everywhere in that space. The energy associated with it is a potential energy. This potential is realized only when the light is intercepted by a receiver and its energy is dissipated into the receiver.

It is shown in electromagnetic theory that the intensity of light of amplitude $A = a \cos 2\pi\omega t$ is

$$I = a^2 \tag{7-6}$$

Thus intensity is a time-averaged quantity and it is *independent of the frequency and phase* of the light beam.

The velocity C' of light propagation in a transparent material is less than its velocity in free space. The relationship is

$$C' = \frac{C}{n}$$

where n has a value for each material; n is a material property called *index of refraction* or *refractive index*. For gases, n is only slightly greater than unity; for liquids, values between 1.3 and 1.5 are common; for solids, values between 1.5 and 1.7 are common.

Sometimes we are interested in *polarization* of the light. This relates to the electromagnetic disturbance described by Eq. (7-4), which shows A as a function of t, but independent of x and y. This is a simplification of the facts, although it models unpolarized light very well. With plane-polarized light, the field strength is unidirectional, say in the y direction, and the disturbance is written

$$A_y = a \cos 2\pi\omega t \tag{7-4a}$$

Then field strength in the x direction is zero, or

$$A_x = 0 \tag{7-4b}$$

which means that there is no light with polarization parallel to x.

MOIRÉ INTERFEROMETRY

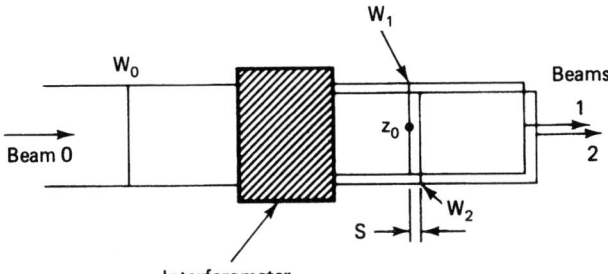

Figure 7-2 Schematic representation of an interferometer. This device divides the input beam into two coherent beams that travel along paths of different lengths inside the interferometer; upon emergence, the total path lengths traveled by the beams differ by S.

7-2 Interference of Light

The superposition of two *coherent* beams of light is called two-beam *interference*. For our purposes, coherent beams are made up of long wave trains in which the wavelength λ is a unique value. Numerous different optical systems are used to produce two-beam interference for different purposes, including systems for moiré interferometry. They are all described schematically by Fig. 7-2. A parallel beam of coherent light with wavefront W_0 enters the optical apparatus (often called an *interferometer*), in which it is divided into two separate beams, 1 and 2, with wavefronts W_1 and W_2. Wavefronts W_1 and W_2 have the same positions in their new respective wave trains as W_0 had at a previous time. They evolved from one wavefront and they show the progress, with time, of subdivisions of that wavefront. As such, the phase of W_1 is identical to the phase of W_2 for all time. The apparatus causes beam 1 to travel a longer path than beam 2, longer by a distance S. When W_1 reaches any point z_0, its mate W_2 has traveled further by S. At any time during the passage of the two wave trains, they pass z_0 with a relative phase (i.e., with a phase difference) of S/λ cycles or $2\pi S/\lambda$ radians.

Pure Two-Beam Interference

Referring to Fig. 7-2, let emergent beams 1 and 2 have equal intensities I_1 and I_2, each of them equal to half the original intensity I_0 of the incoming beam. If at z_0 the amplitude of wave train I_1 is assumed to vary with time as

$$A_1 = a \cos 2\pi\omega t \qquad (7\text{-}4c)$$

then the phase of wave train I_2 must be greater by $2\pi S/\lambda$ when it crosses z_0 and the amplitude of I_2 at z_0 must be

$$A_2 = a \cos 2\pi\left(\omega t + \frac{S}{\lambda}\right) \qquad (7\text{-}4d)$$

Since the phase difference at z_0 is the same for all time during the passage of the wave trains, the resultant amplitude of the emergent light is

$$A = A_1 + A_2 = K \cos 2\pi(\omega t + \phi)$$

where

$$K = \left[2a^2\left(1 + \cos 2\pi\frac{S}{\lambda}\right)\right]^{1/2}$$

and the value of ϕ is unimportant. By Eq. (7-6), the resultant intensity of the recombined wave trains is K^2 or

$$I = 2a^2\left(1 + \cos 2\pi\frac{S}{\lambda}\right) = 4a^2 \cos^2 \pi\frac{S}{\lambda} \tag{7-7}$$

This is a fundamental relationship, called the intensity distribution of pure two-beam interference, Intensity I applies to *every* case of division into two equally intense beams and their subsequent recombination. Equation (7-7) shows that the intensity of the combined beams is persistent, or independent of time. It varies cyclically from maximum to zero to maximum, and so on, as a function of the difference S of path lengths traveled by the two beams.

In the schematic arrangement of Fig. 7-2, S might have any value. If S is an integral number of wavelengths, or $S/\lambda = 0, 1, 2, 3, \ldots$, the intensity at arbitrary point z_0 is maximum and *constructive* interference occurs. If $S = \lambda/2$ plus an integral number of wavelengths, or $S/\lambda = \frac{1}{2}, \frac{3}{2}, \frac{5}{2}, \ldots$, the intensity reaches a minimum value and *destructive* interference occurs. Since W_1 and W_2 are parallel and travel with identical velocities, and since z_0 is an arbitrary point, the intensity—or the condition of constructive or destructive interference—is the same at every point in space where beams 1 and 2 coexist. The case of nonparallel wavefronts will be considered subsequently. The present example is called *pure* two-beam interference, since the interfering beams have equal strengths and destructive interference is complete, producing $I_{min} = 0$.

Interference can occur only if the two beams have common polarizations. If the beams are unpolarized or if their polarizations are not parallel, the amplitude of each beam can be divided into its vector components A_{x1} and A_{y1}, A_{x2} and A_{y2}. Then A_{x1} and A_{x2} interfere according to Eq. (7-8) to give I_x; similarly for I_y. The resultant intensity is their scalar sum, $I = I_x + I_y$.

Impure Two-Beam Interference

Suppose that emergent beams 1 and 2 (Fig. 7-2) have unequal intensities I_1 and I_2, and unequal phases $2\pi\omega t$ and $2\pi(\omega t + S/\lambda)$. The resultant intensity is [23]

$$I = I_1 + I_2 \pm 2\sqrt{I_1 I_2} \cos 2\pi\frac{S}{\lambda} \tag{7-8}$$

This is the general expression for interference, or superposition of any two coherent wave trains. The resultant intensities are plotted in Fig. 7-3 as a function of path difference $N = S/\lambda$. Interference of beams of unequal strengths is *impure* because destructive interference is incomplete, or $I_{min} \neq 0$.

Fringe Patterns and Walls of Interference

The apparatus represented by Fig. 7-4 is similar to that of Fig. 7-2, but here collimated beams 1 and 2 emerge with a relative angle 2θ. As always (in free space and isotropic media) their wavefronts W_1 and W_2 are perpendicular to the beams and they too intersect at angle 2θ.

Wavefronts W_1 and W_2 originated from division of wavefront W_0. Consequently, W_1 and W_2 have identical phase. Where their separation S is zero or an integral number of wavelengths, interference is constructive; midway between neighboring locations of constructive interference, the interference is destructive.

Analysis is aided by Fig. 7-5, which indicates the two wave trains at a given instant of time and depicts wavefronts W_1 and W_2 and neighboring wavefronts. Separation between wavefronts in each train is λ. The harmonic curves with ordinates labeled A_1 and A_2 represent the amplitude

MOIRÉ INTERFEROMETRY

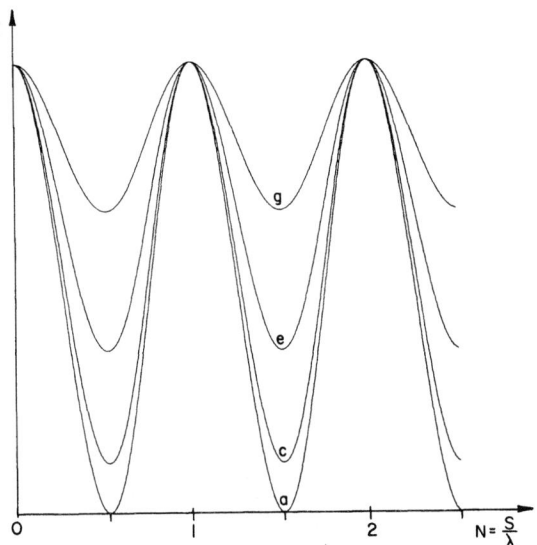

Figure 7-3 Resultant intensity versus path difference S/λ for two-beam interference. Curves correspond to input beam intensity ratios: (a) $I_1/I_2 = 1$; (c) $I_1/I_2 = 4$; (e) $I_1/I_2 = 16$; (g) $I_1/I_2 = 100$.

of the two wave trains in space at the given instant. Phases marked ϕ_1 and ϕ_2 are identical, since they represent wavefronts of common origin. Consequently, when W_1 reaches point a, W_2 already has advanced five wavelengths beyond a, and similarly at b, where the path lengths of the two wave trains again differ by five wavelengths. This is true for every point along a–c; Eq. (7-7) [or Eq. (7-8)] prevails at all these points, where $S/\lambda = 5$ in the equation. The result is constructive interference at every point along a–c.

By the same argument, along d–e the difference S of optical path lengths traveled by the two wave trains is 4.5 wavelengths, or $S/\lambda = 4.5$. Along q–r, $S/\lambda = 4.0$. This explains the formation of constructive and destructive interference at the given instant, but why is the effect persistent, or independent of time?

At a later instant the two wavefronts have advanced, but their velocities are identical and the

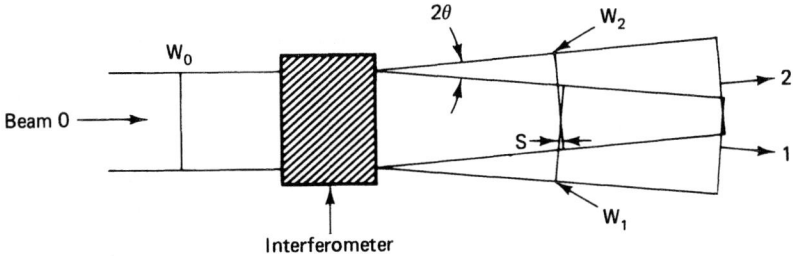

Figure 7-4 In this case, the interferometer causes the two output beams to emerge with a relative angle 2θ.

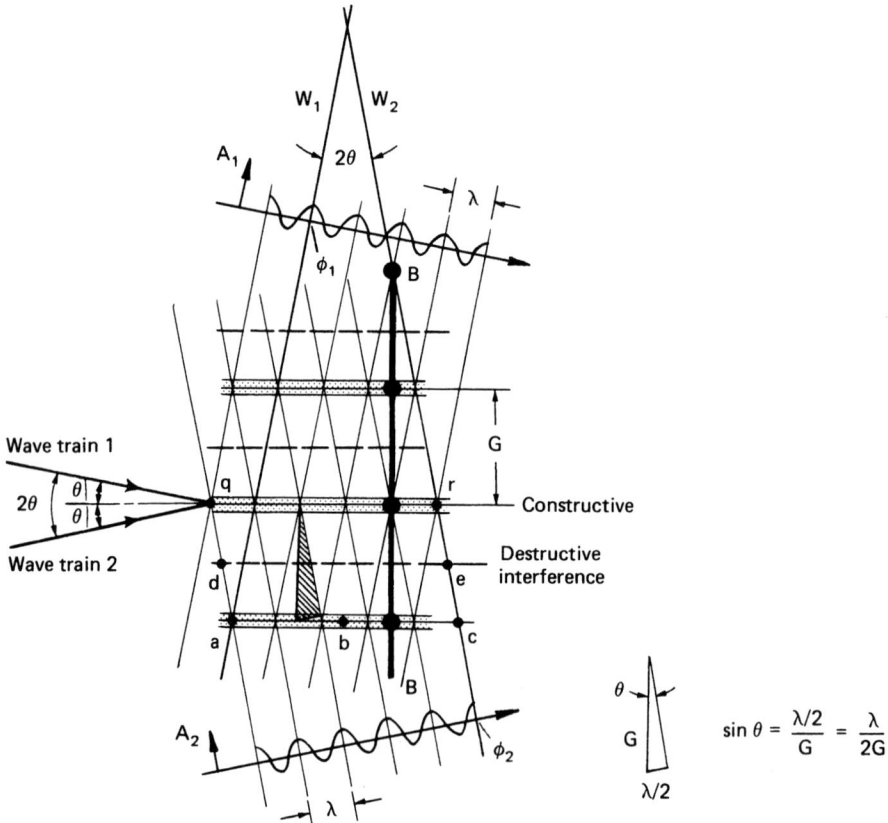

Figure 7-5 Regions of constructive and destructive interference in space where two coherent beams intersect.

distances traveled are identical. Therefore, the phase difference at point a remains five cycles, and the phase difference at all other points remains unchanged. A steady-state condition of constructive and destructive interference is formed! It is clear from Fig. 7-5 that these lines of interference lie parallel to the bisector of the incoming beams.

Furthermore, the wavefronts of Fig. 7-5 are not merely the lines shown, but they are planes that stand perpendicular to the diagram. The optical interference occurs in the three-dimensional space where the two beams intersect. Constructive and destructive interference occurs in a series of parallel planes that stand perpendicular to the diagram and contain the lines labeled constructive and destructive interference.

The planes of constructive interference lie in thin volumes of space where the resultant intensity is relatively high. The planes of destructive interference lie in intermediate thin volumes of space where the resultant intensity is relatively low. One may visualize these thin volumes of space surrounding the planes of constructive interference as *walls* of constructive interference—walls in which light intensity exists. These walls are separated by voids that represent the (relative) absence of light intensity. The mental model is that of three-dimensional space occupied by parallel,

MOIRÉ INTERFEROMETRY

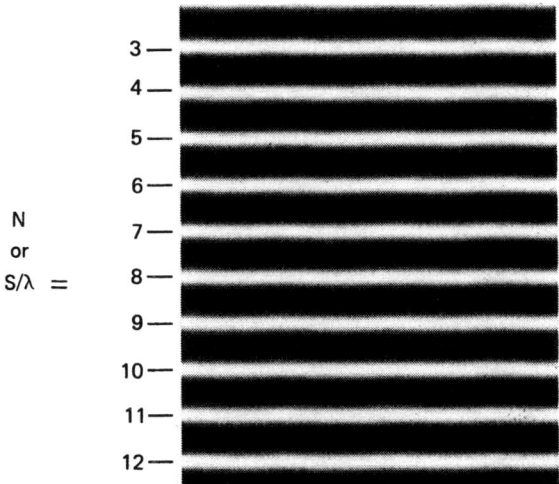

Figure 7-6 Interference fringes recorded by a photographic plate installed in plane B–B of Fig. 7-5.

uniformly spaced *walls of light intensity*. They will be designated by that name and, more simply, by the term *walls of interference*.

Such walls may be visualized wherever two beams of coherent light intersect in space, regardless of the angle of intersection. Of course, the walls do not obstruct anything. Other beams of light can pass through them and emerge uninfluenced by their presence, just as beams of light can cross in space and emerge unimpaired from the intersection zone.

A photographic plate that cuts this space absorbs energy from the walls of light intensity and records their presence. It receives no energy where it cuts through the voids. Thus the photographic plate exhibits dark and light bands such as those in Fig. 7-6. The bands are called *fringes*, and the array of fringes is called an *interference fringe pattern*.

Let the photographic plate be interposed along B–B in Fig. 7-5. A fundamental relationship defines the distance G between adjacent fringes on the plate, or between adjacent walls of interference in space. From the hatched triangle in Fig. 7-5, whose hypotenuse is G and whose short leg is $\lambda/2$, we find

$$\sin \theta = \frac{\lambda/2}{G} \tag{7-9}$$

The frequency of the fringes, or the *fringe gradient* on the plate (e.g., fringes/mm), is F, where

$$F = \frac{1}{G} \tag{7-10}$$

The fringe gradient is determined by

$$F = \frac{2}{\lambda} \sin \theta \tag{7-11}$$

Like Eq. (7-7) or (7-8), the relationship of Eq. (7-11) pertains to every case of optical interference; it will arise repeatedly in our studies. The relationship applies for any value of θ between 0 and

Table 7-1 Two-Beam Interference for Coherent Beams of Unequal Intensities

Case	Two-beam input: $\dfrac{I_1}{I_2}$ or $\dfrac{a_1^2}{a_2^2}$	Output: Result of Coherent Summation	
		$\dfrac{I_{max}}{I_{min}}$	Contrast (%) [Eq. (7-13)]
a	1	∞	100
b	2	34	97
c	4	9	89
d	9	4	75
e	16	2.8	64
f	25	2.2	56
g	100	1.5	33

90° and it applies for any separation S, except that S must be small compared to the length of the wave train or the coherence length of the light source.

If a photographic plate is interposed along B–B, it would record a pattern similar to Fig. 7-6; this is a positive print and shows destructive interference as black, or the absence of light. The bright bands (or bright fringes) represent zones of constructive interference, where

$$\frac{S}{\lambda} = N = 0, 1, 2, 3, \ldots \tag{7-12}$$

just as in the case of Fig. 7-2. Here N is called the *interference fringe order*, or simply the *fringe order*. Equation (7-12) shows that the fringe pattern is a contour map of the separation S between wavefronts W_1 and W_2, where the contour interval is one wavelength, λ. We measure S in the direction of travel of either wave train.

Figure 7-6 illustrates a case of pure two-beam interference. If the two interfering beams had unequal intensities, the resultant intensity at each point (in plane B–B) would be governed by Eq. (7-8) and Fig. 7-3; the dark fringes would be photographed as shades of gray instead of black. Table 7-1 lists parameters relating to the visibility of the fringes for different relative intensities of the input beams. The output is given by the intensity ratio I_{max}/I_{min} for impure interference and also by contrast, where

$$\% \text{ contrast} = \frac{I_{max} - I_{min}}{I_{max}} \times 100\% \tag{7-13}$$

It becomes clear that pure two-beam interference is not required for good fringe visibility. If the intensities of two interfering beams are in the ratio of 4 : 1, the resultant fringe contrast is good, namely 89%. Two beams with a 100 : 1 intensity ratio produce a discernible interference pattern with a contrast of 33%. [This, incidentally, is why extraneous interference patterns or "ghost patterns" are so commonly observed when coherent laser light is used. Small amounts of extraneous light that reflects accidentally from a local zone (e.g., the frame of an optical element) combines with the main beam to produce low-contrast interference fringes.]

Warped Wavefronts

Referring again to Fig. 7-4, let wavefront W_1 emerge as a surface that is not flat, but deviates from flatness in a smooth continuous manner. It is smoothly warped (e.g., in the manner sketched

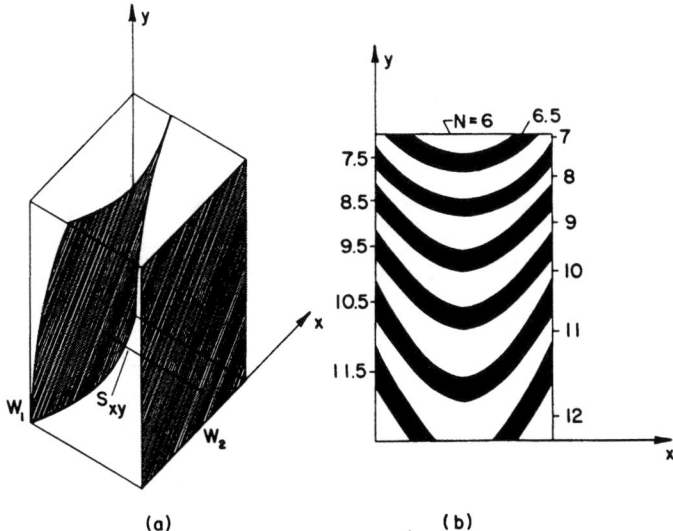

Figure 7-7 When one or both of the wavefronts of the two interfering beams is smoothly warped (a), the interference fringes are continuous curves (b).

in Fig. 7-7a). Wavefront W_2 remains plane in this example. If a photographic plate is installed in the beams, it would record an interference pattern. Interference between the two wave trains would produce the nonlinear fringe pattern of Fig. 7-7b, which is a contour map of separation S_{xy} between the two wavefronts.

It should be evident that Eqs. (7-12) and (7-9) are both operative. The warped wavefront is generated because angle θ varies continuously with x and y. As a result, the wavefront separation S and the distance G between neighboring fringes vary systematically with x and y. Along any continuous bright contour, S_{xy}/λ has the constant value given by fringe order N, where N is an integer. N is constant at all points along any dark contour; N is constant at all points along a continuous contour of *any* intermediate intensity, so N assumes all values: integer, half-order, and intermediate or fractional values.

Since W_1 is smoothly warped, fringe orders must vary in a smooth, continuous manner. Fringes must be smooth contours and the change of fringe order between adjacent fringes must be either 1 or 0.

In moiré interferometry, both beams usually have warped wavefronts. However, the interpretation is the same—the fringe pattern recorded on the photographic plate is a contour map of the separation S_{xy} between wavefronts W_1 and W_2, that is, between wavefronts of equal phase in the two output beams.

Two-beam interference with coherent light is said to be nonlocalized. This means that the predictions of Eqs. (7-7) and (7-8) remain effective at every point in the space where the two beams coexist. The contrast of interference fringes recorded on a photographic plate is the same at every location in this space.

The concept of walls of interference applies in the three-dimensional space where any pair of coherent light beams coexist, including beams with warped wavefronts. The walls are flat, parallel, and uniformly spaced only for the special case of interference between beams with plane wave-

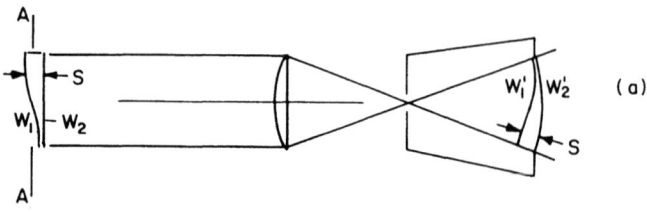

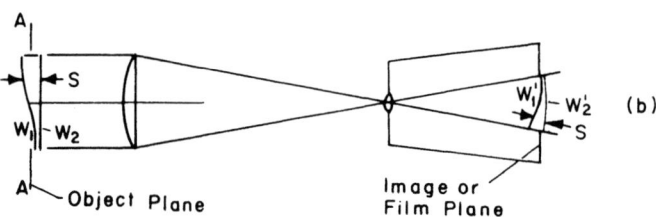

Figure 7-8 A camera reproduces the wavefront separation S in the film plane; S is the same at corresponding object and image points, regardless of image magnification. (a) A single lens is used to collect and focus the light. (b) A compound lens comprised of two widely separated elements is used.

fronts. In the general case of warped wavefronts, the walls have irregular shapes. As a consequence of the wavefront changes depicted in Fig. 7-1c, each wall of interference has a continuous but warped shape, with its warpage systematically different from that of the neighboring wall. A photographic plate that cuts the walls at one location might record the fringe pattern of Fig. 7-7, but at a different location the fringe pattern would be somewhat different.

7-3 The Camera

A predicament is arising. The fringe pattern is not unique—the pattern changes with the location at which it is observed. Where in space must the fringe pattern be recorded?

The fringes are nonlocalized and they exhibit the same fringe contrast everywhere. Consequently, the answer is not some special location at which the fringes are clearest. Instead, the answer is this: In most interferometric techniques for metrological purposes, including moiré interferometry, the fringe pattern must be recorded in the plane where the light exits from the body under test. In Fig. 7-4, the two beams travel different paths inside the box and reunite as they exit. Assume that the body under test also extends to this exit plane. Then the pattern should be photographed where the light emerges from the box. This is the location of the object plane. A photographic plate could be positioned outside the box in Fig. 7-4 to record the pattern generated there. In usual practice, however, this is impossible or inconvenient. Instead, it can be recorded by means of a camera, or the human eye, or any similar optical system.

Camera arrangements are illustrated in Fig. 7-8. A camera reproduces in the image (or film)

MOIRÉ INTERFEROMETRY

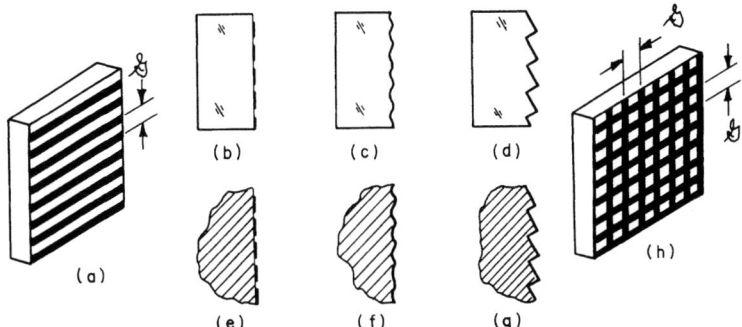

Figure 7-9 Diffraction gratings are comprised of regularly spaced bars (a) or furrows. Cross-sectional views (b), (c), and (d) illustrate transmission gratings, while (e), (f), and (g) illustrate reflection gratings; (b) and (e) represent bar-and-space gratings called amplitude gratings; (c) and (f) represent symmetrical phase gratings; (d) and (g) represent blazed phase gratings; (h) illustrates a crossed-line grating, which can be either amplitude or phase type.

plane the phase relationships of the light that crosses the object plane. Let A–A be the exit plane of the interferometer of Fig. 7-4. A camera focused on that plane would record the same interference pattern as a photographic plate installed in plane A–A.

This is true because the additional wavefront distortions introduced as the wavefront travels from the object plane to the image plane are identical for every wavefront. The difference in warpage between W_1 and W_1' is identical to the difference between W_2 and W_2'. Consequently, the wavefront separation S is equal at every corresponding object and image point. Since it is this separation that governs the locations of constructive and destructive interference, the fringe patterns recorded in the object and image planes are equal.

Lateral magnification of the image, as depicted in Fig. 7-8b by reduced image size, does not alter the result. Distance S remains equal for corresponding object and image points and therefore the interference fringe order N is the same at corresponding points.

In Fig. 7-8a, a single camera lens is used to collect and focus the image, while in Fig. 7-8b a compound-lens system comprised of a large collecting element and a small element performs the same functions. The single lens is usually preferred, except for large fields of view.

7-4 Diffraction Gratings and Diffraction

Moiré interferometry depends on diffraction of light as well as interference, and its explanation requires an understanding of diffraction by gratings. Diffraction gratings are illustrated in Fig. 7-9, where (a) indicates that a grating is a surface with regularly spaced bars or furrows. In the case of transmission gratings, the incident light and the diffracted light appear on opposite sides of the diffracting surface; with reflection gratings, they are on the same side and the substrate may be opaque. *Amplitude gratings* consist of opaque bars and transparent spaces, or else reflective bars and nonreflective spaces. *Phase gratings* have furrowed or corrugated surfaces, with either symmetrical or nonsymmetrical furrow profiles. The period or *pitch* of a grating is the distance $\mathcal{G}$ between corresponding features of adjacent bars or furrows. A *crossed-line* grating has the same repeating arrangement of bars or furrows in two orthogonal directions.

Frequency $\mathcal{F}$ of a grating is the number of bars or furrows per unit length, usually expressed

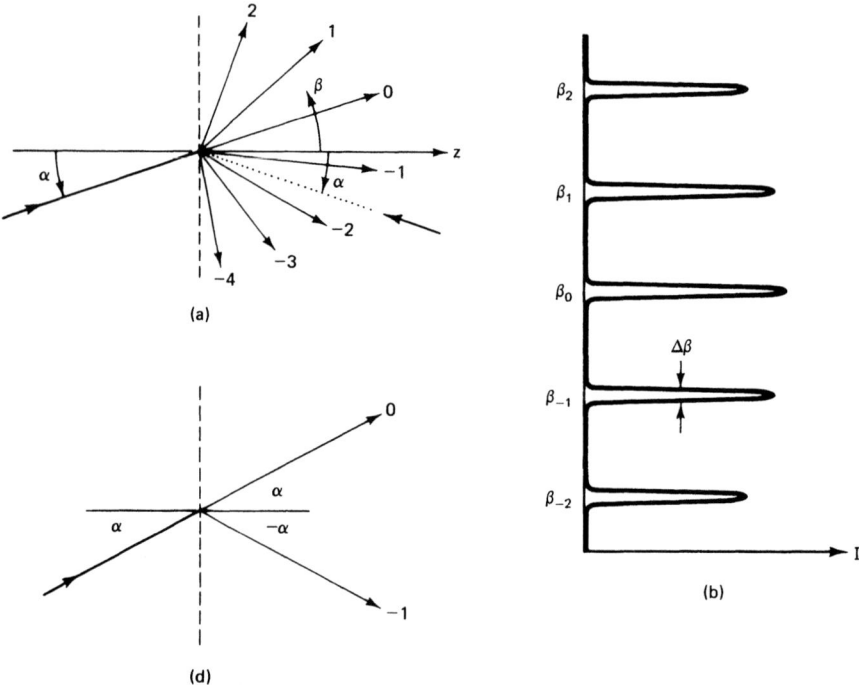

Figure 7-10 (a) A grating divides an incident beam of light into a number of diffracted beams, where the diffracted beams lie in a systematic array of preferred directions. The beams are represented by rays. (b) Diffracted light is confined to very narrow angular zones near the preferred directions. The shape of the envelope connecting intensity peaks depends on furrow shape (phase gratings) or bar width (amplitude gratings). (c) The general case of three-dimensional diffraction is illustrated, where **A** and $\mathbf{D}_m$ are unit vectors that define the directions of incident and diffracted rays (beams), respectively. (d) A special case of two-dimensional diffraction in which the zero and -1 diffraction orders have symmetrical directions.

in *lines* per millimeter, meaning repetitions or cycles per millimeter. Frequency and pitch are related by

$$\mathcal{F} = \frac{1}{\mathcal{G}} \tag{7-14}$$

Our interest lies in gratings of relatively high frequency.

A grating divides every incident wave train into a multiplicity of wave trains of smaller intensities; and it causes these wave trains to emerge in certain preferred directions. This is indicated in Fig. 7-10a, in which beams of light are represented by rays. The vertical dashed line represents either an amplitude or a phase grating. When a parallel beam is incident at angle α from the left, the grating divides it into a series of beams that emerge at preferred angles: . . . , β_{-1}, β_0, β_1, β_2, ... These beams are called *diffraction orders* and are numbered in sequence beginning with the zero order, which is an extension of the incident beam; diffraction orders whose angles are counterclockwise with respect to the zero order are considered to be positive for diffracted beams that propagate in a direction having a positive z component.

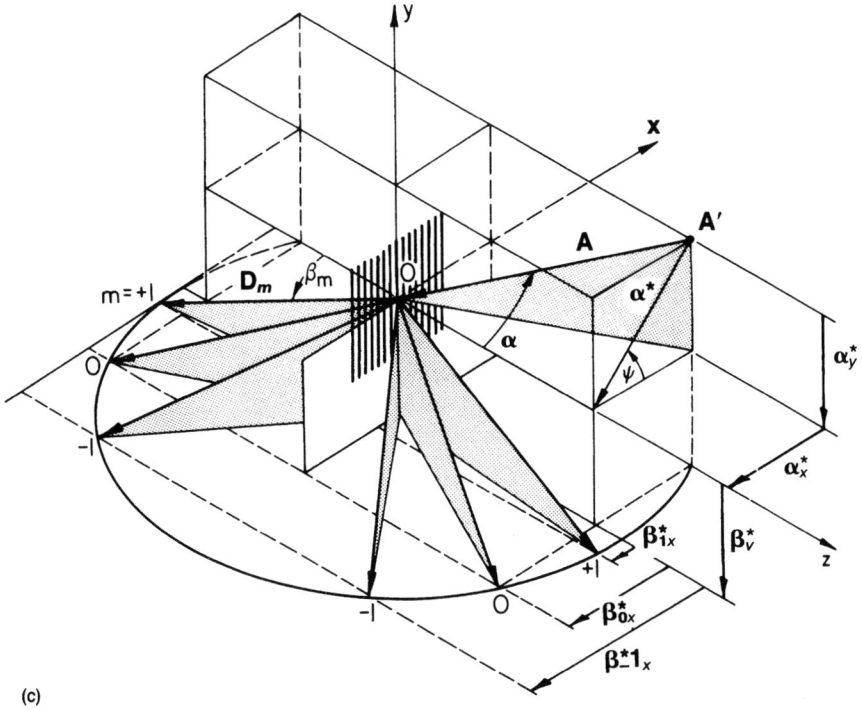

(c)

Figure 7-10 (*continued*)

Reflection gratings diffract light in just the same way if the incident light is the symmetrical image of that for a transmission grating, as shown by the dotted ray in Fig. 7-10a. In reflection, the zeroth-order diffraction is the mirror reflection of the incident beam.

Figure 7-10b is a graph of intensity versus diffraction angle β for the emergent light. Light is diffracted into beams with preferred directions, with very little light emerging in other directions. For high-quality gratings, the angular spread $\Delta\beta$ within a diffraction order is exceedingly narrow and may be assumed to be zero for our purposes.

Figure 7-10a illustrates diffraction in two dimensions, where the plane of incidence (i.e., the plane containing the incident ray and the grating normal) is perpendicular to the grating lines. This special case is violated if the grating is rotated by an angle ψ about the z axis, and three-dimensional diffraction must be considered. The general case is illustrated in Fig. 7-10c, where **A** and **D**$_m$ represent unit vectors that define the directions of incident and diffracted beams, respectively [55,57]. For generality, both the reflected system of diffracted beams (in the $+z$ space) and the transmitted system (in the $-z$ space) are shown here.

The projections of unit vectors **A** and **D**$_m$ that lie parallel to the x, y plane, namely α^* and β_m^* (not shown), are defined as *incidence* and *emergence vectors*, respectively. The x, y components of **A** and **D**$_m$ (i.e., the x, y components of the incidence and emergence vectors) are α_x^*, α_y^*, and β_{mx}^*, β_{my}^*, respectively [55]. These x, y components fully define the directions of incident and diffracted light; they are illustrated in Fig. 7-10c.

The three-dimensional grating equations define the directions of diffracted beams. The relationships in terms of these vector components are very simple, corresponding to the pictorial relationships represented in Fig. 7-10c,

$$\beta_{mx}^* = \alpha_x^* + m\lambda\mathcal{F} \tag{7-15a}$$

$$\beta_{my}^* = \alpha_y^* \tag{7-16a}$$

Frequency $\mathcal{F}$ of the grating is a vector quantity, too, and this will be discussed in Section 7-18.

The equivalent scalar relationships can be expressed if the angle of diffraction β_m is divided into two components, β_{mx} and β_{my}; let β_{mx} be defined as the angle between the unit vector $\mathbf{D}_m$ and the y-z plane and let β_{my} be the angle between $\mathbf{D}_m$ and the x-z plane. Then

$$\sin \beta_{mx} = \sin \alpha \cos \psi + m\lambda\mathcal{F} \tag{7-15b}$$

$$\sin \beta_{my} = \sin \alpha \sin \psi \tag{7-16b}$$

When the angle of in-plane rotation ψ of the grating (with respect to the plane of incidence) is zero, or when it is very small, the grating equation may be written as

$$\sin \beta_{mx} = \sin \alpha + m\lambda\mathcal{F} \tag{7-15c}$$

$$\beta_{my} = 0 \tag{7-16c}$$

which is the familiar two-dimensional grating equation.

Angle of diffraction β_{mx} cannot exceed 90°. It is clear from Eqs. (7-15a) through (7-15c) that the number of diffraction orders that emerge within the range $\pm 90°$ is a very large number when $\mathcal{F}$ is small (i.e., for a coarse grating), and conversely the maximum value of m is small when $\mathcal{F}$ is large. The angle between neighboring diffraction orders is small for a coarse grating and large for a fine grating.

In a special case of interest (Fig. 7-10d), the zeroth diffraction order and its neighbor, the -1 order, are symmetrical with respect to the grating normal. If the angle of incidence is α, we have $\beta_0 = \alpha$ and $\beta_{-1} = -\alpha$. Equation (7-15c) reduces to

$$\sin \alpha = \frac{\lambda}{2}\mathcal{F} \tag{7-17}$$

This defines the angle of incidence α to achieve the special condition of symmetry.

Equivalency and Virtual Gratings

Equation (7-11) relates the fringe gradient, or frequency of fringes F, formed when two coherent beams intersect with half-angle θ. Equation (7-17) relates the angle of emergence α of the zero and -1 diffraction orders from a grating of frequency $\mathcal{F}$. The equations are the same. If we let $\theta = \alpha$, we find $F = \mathcal{F}$.

There exists a kinship—actually an equivalency—between fringe patterns and gratings. Figure 7-11 illustrates the relationship. In the volume of space where two coherent beams coexist, a steady-state formation of walls of constructive and destructive interference is generated. On the other hand, a real grating of the same pitch creates two coherent beams that coexist in space and create the same walls of interference in their region of intersection.

Remember the photographic plate $B-B$ exposed to the interference pattern in Fig. 7-5. The exposed and developed plate is a pattern of parallel, equally spaced bars governed by Eq. (7-11); it is a diffraction grating. If this diffraction grating is illuminated at a suitable angle $\alpha = \theta$, it will generate zero and -1 order beams equal to those that originally formed the grating. This is an equivalency relationship.

MOIRÉ INTERFEROMETRY 313

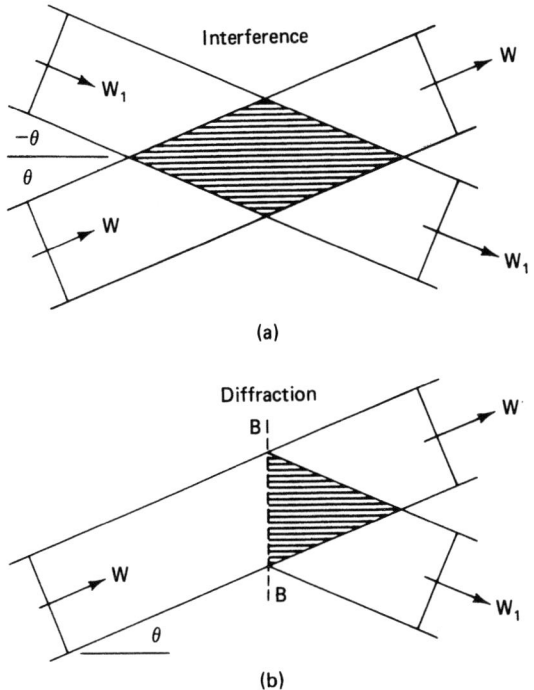

Figure 7-11 Virtual gratings formed (a) by two coherent beams and (b) by a real grating.

The three-dimensional zone of regularly spaced walls of interference represented in Fig. 7-11a or b will be called a *virtual grating*. Regardless of how it is created, this light possesses the same properties as light emerging from a real grating. The significant concept is that a virtual grating can be formed either by a real grating or by intersection of two coherent beams.

The equivalency is a powerful concept. Diffraction grating lines and regularly spaced interference fringes can be interchangeable entities.

7-5 Moiré Interferometry

Figure 7-12 is a schematic description of the moiré interferometry method we will study. A crossed-line diffraction grating is produced on the specimen. This is a high-reflection, symmetrical, phase-type grating of the sort shown in Fig. 7-9f; its frequency, $\mathscr{F} = f/2$, is high—for example, 1200 lines/mm (30,480 lines/in.) for all but two of the patterns exhibited in this chapter. The specimen grating is firmly attached to the specimen. When loads are applied to the specimen, the grating moves and deforms together with the specimen surface.

Two beams of coherent light illuminate the specimen grating obliquely from angles $+\alpha$ and $-\alpha$. This recreates the conditions of Section 7-2: the two beams generate walls of constructive and destructive interference (i.e., a virtual grating) in the zone of their intersection; the virtual grating is cut by the plane of the specimen surface, where an array of parallel and very closely

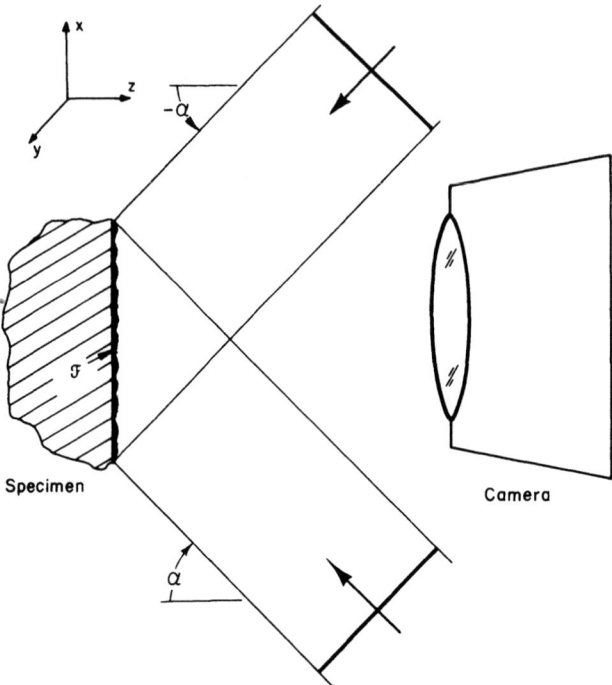

Figure 7-12 Schematic diagram of the essential elements of a moiré interferometry system.

spaced fringes are formed. These fringes are essentially bright and dark bars, and they act like the reference grating of geometrical moiré. This array of fringes, and also the walls of interference from which they are derived, are called a *virtual reference grating*. Their frequency is f, where

$$f = \frac{2}{\lambda} \sin \alpha \qquad (7\text{-}18)$$

This is a restatement of Eq. (7-11) in terms of f and α, symbols that represent the frequency of the virtual reference grating and the angle of the beams that create it.

The specimen grating and virtual reference grating interact to form a moiré pattern, which is viewed and photographed in the camera. This casual description repeats the essential description of geometrical moiré and serves to emphasize their common qualities. A more rigorous description of moiré interferometry will follow later. The casual description is conceptually useful and acceptable, however, since the governing equations of geometrical moiré and moiré interferometry are identical [55].

Note that the initial frequency of the specimen grating is half that of the reference grating. Conditions of Fig. 7-12 are directly analogous to geometric moiré with a fringe multiplication factor of 2, as described in Chapter 6. In Fig. 7-12, lines of the virtual reference grating are perpendicular to the x axis. Its interaction with the specimen grating produces the u_x displacement field. In practice, another pair of coherent incident beams can be used to form a reference grating perpendicular to the y axis, to interact with the corresponding array of lines of the crossed-line specimen grating and produce the u_y displacement field.

Representative patterns of moiré interferometry fringes (Figs. 7-13 and 7-14) depict the in-plane displacements of every point on the specimen surface as contour maps of equal-displacement fringes. Quantitatively, for each point in the fringe pattern,

$$u_x = \frac{1}{f}N_x \quad \text{and} \quad u_y = \frac{1}{f}N_y \tag{7-19}$$

where u_x, u_y are components of displacement in the x and y directions, respectively; N_x, N_y are fringe orders when lines of the reference grating are perpendicular to the x and y directions, respectively; and f is the frequency of the reference grating. We might note that these governing equations are identical for geometrical moiré and moiré interferometry.

Figures 7-13 and 7-14 are introduced here to illustrate qualities of moiré interferometry patterns. Reference grating frequency f was 2400 lines/mm (60,960 lines/in.) in both cases, which corresponds to a sensitivity ($1/f$) of 0.417 μm (16.4 μin.) per fringe order. If needed for special problems, N_x and N_y can be determined by interpolation to $\frac{1}{5}$ or $\frac{1}{10}$ of a fringe order, providing enhanced displacement resolution. In addition to high sensitivity and resolution, the patterns are characterized by high fringe contrast, very high spatial resolution of fringes in regions of large fringe gradients, and extensive range, or the ability to register very large absolute fringe orders. Why?

These desirable properties are characteristic of two-beam interference with coherent light. Indeed, moiré interferometry is a method of two-beam interference, as discussed in the next section.

A More Rigorous Explanation

The basic elements of a moiré interferometry system are illustrated again in Fig. 7-15. The phase-type specimen grating of frequency $\mathscr{F}_s$ intercepts two beams of coherent light which are incident at symmetrical angles $\pm \alpha$. One scheme for creating these two beams, A and B, is illustrated in Fig. 7-15a; where the beams are represented by corresponding rays. Beams A and B are separated by the beam splitter and redirected by mirrors to meet at the specimen surface. To negate effects of phase changes caused by reflections, the light should be polarized, with its plane of polarization either perpendicular or parallel to the plane of the diagram. The specimen grating diffracts the incident light such that rays represented by A'' and B'' emerge from each specimen point and propagate to the camera. Beams A'' and B'' interfere and generate an interference pattern on the film plane of the camera.

Figure 7-15a illustrates an interferometer. It functions like that represented by the box in Fig. 7-4, where the beam splitter is the entrance to the box and the specimen grating is the exit from the box. The qualities discussed in Section 7-2 for two-beam interference apply fully to moiré interferometry. Various optical arrangements can be used to produce incident beams A and B [29], but in each case the two beams are divided from a common beam, travel different paths, and meet again at the specimen. Two coherent beams emerge from the specimen grating with warped wavefronts; they coexist in space and generate optical interference.

Figure 7-15b illustrates additional details. At the top and bottom of the figure, diffracted beams stemming from incident beams A and B are illustrated by ray diagrams. These diffraction directions are prescribed by the grating equations (7-15) and (7-16).

When $\mathscr{F}_s = f/2$ and lines of the specimen and reference gratings are exactly parallel, light from beam A, diffracted in the first order of the specimen grating, emerges with angle $\beta = 0$ and with a wavefront represented by A'. Light from B that is diffracted in the -1 order also emerges

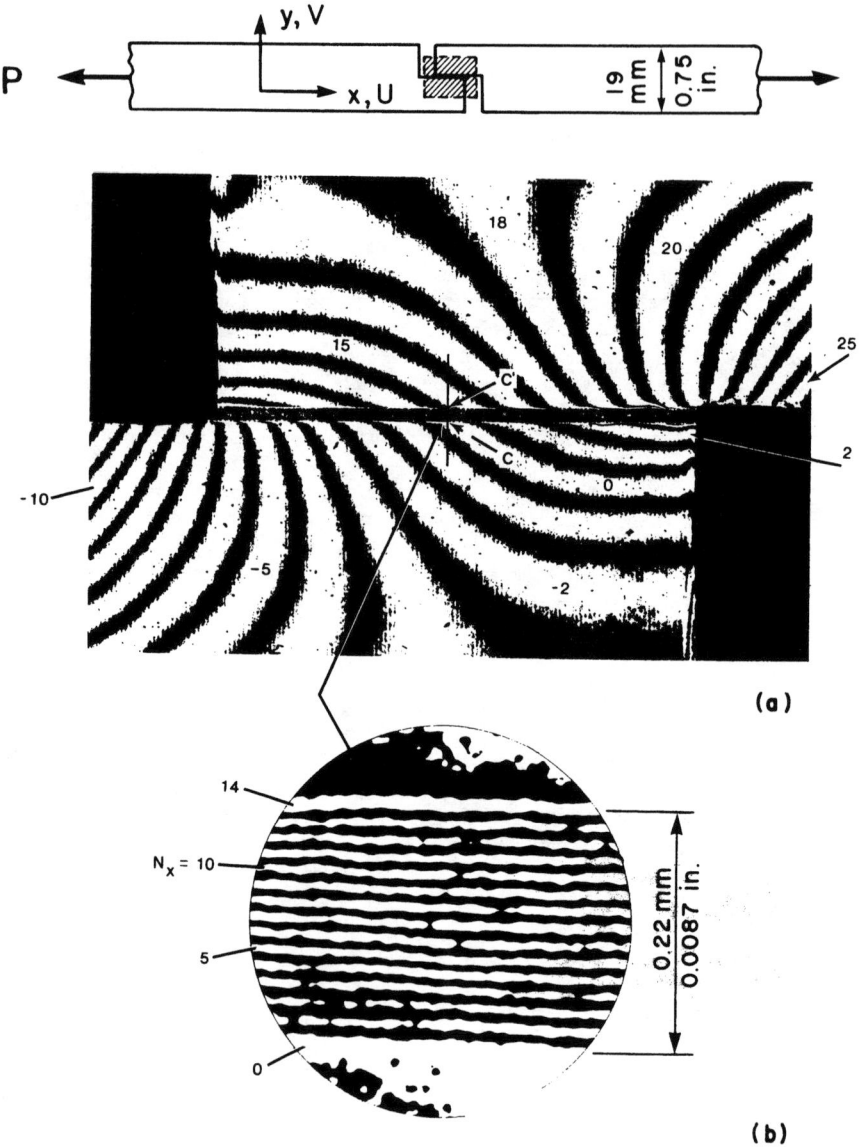

Figure 7-13 Moiré interferometry pattern of N_x, or the u_x displacement field, for a lap joint. The adherends are aluminum and the adhesive is a rubber-modified epoxy. (b) An enlargement of the zone surrounding line cc' in (a). $f = 2400$ lines/mm (60,960 lines/in.).

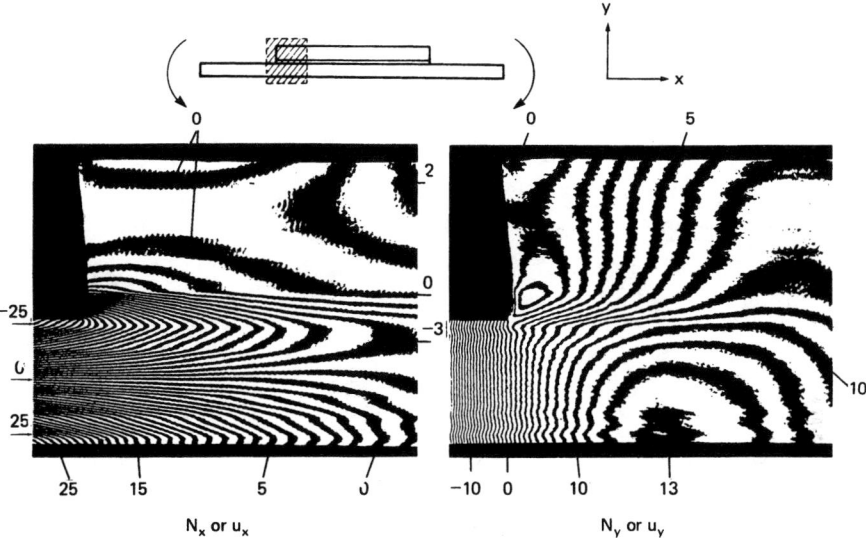

Figure 7-14 Patterns of N_x and N_y (or the u_x and u_y displacement fields) for a bonded joint between two quasi-isotropic graphite–epoxy members, subjected to bending. $f = 2400$ lines/mm.

along $\beta = 0$ and exhibits plane wavefront B'. Again, two coherent beams coexist in space, but their angle of intersection is zero. Their mutual interference produces a uniform intensity throughout the field, or as provided by the basic relationship for interference [Eq. (7-11)], a pattern of zero frequency—zero fringes/mm. This is called a *null* field.

Now let the specimen be subjected to forces that stretch it uniformly in the x direction such that the uniform normal strain ε_{xx} is a constant. The frequency of the specimen grating is thereby decreased to

$$\mathcal{F}_s = \frac{f/2}{1 + \varepsilon_{xx}}$$

Light from the first diffraction order of beam A does not emerge perpendicular to the grating. Instead, it emerges at an angle β_1 given by the grating equation (7-15c); the diffraction is two-dimensional and $\beta_{x1} = \beta_1$. Substituting the prevailing conditions into Eq. (7-15c),

$$m = 1 \quad \text{and} \quad \sin \alpha = -\frac{\lambda}{2} f \quad \text{[by Eq. (7-18)]}$$

we find

$$\sin \beta_1 = -\frac{\lambda f \varepsilon_{xx}}{2} \tag{7-20a}$$

This applies when ε_{xx} is small. Similarly, light from the -1 diffraction order of beam B emerges in the direction

$$\sin \beta_{-1} = \frac{\lambda f \varepsilon_{xx}}{2} \tag{7-20b}$$

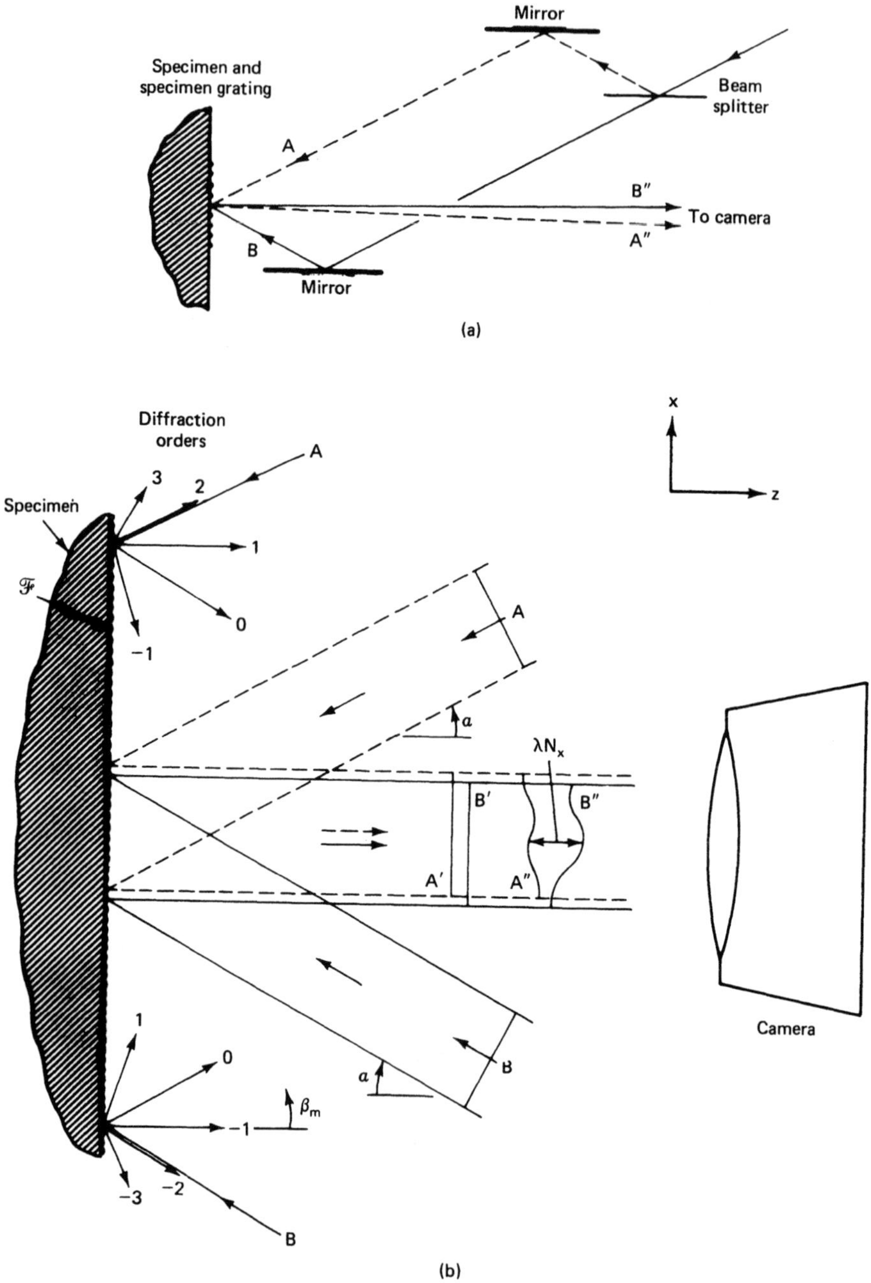

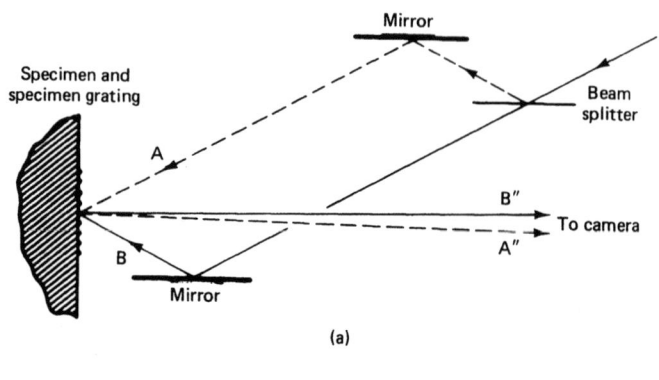

Figure 7-15 (a) Schematic diagram of an interferometer for moiré interferometry. (b) Diffraction by the specimen grating produces output beams with plane wavefronts A' and B' for the initial no-load condition of the specimen. Warped wavefronts A'' and B'' result from nonhomogeneous deformation of the specimen.

These two coherent beams propagate toward the camera with an angular separation of $|2\beta_1|$. The conditions of Fig. 7-4 are reproduced, and again the result is an interference pattern with uniformly spaced parallel fringes (parallel to the y axis) just as in Figs. 7-5 and 7-6. The fringe gradient F_{xx} (i.e., $\partial N_x/\partial x$) is proportional to the angle of intersection $|2\beta_1|$. This angle is very small when ε_{xx} is small; by substituting $\sin\theta = |\sin\beta_1|$ in Eq. (7-11), the magnitude of the fringe gradient is

$$F_{xx} = f\varepsilon_{xx} \tag{7-21a}$$

The fringe gradient (fringes/mm) is proportional to the reference grating frequency and the strain.

If the specimen grating is caused to rotate by an angle ψ (instead of stretching by ε_{xx}), a similar argument yields

$$F_{xy} = f\psi \tag{7-21b}$$

The fringes are perpendicular (essentially) to the lines of the reference grating and their gradient (F_{xy}, i.e., $\partial N_x/\partial y$) is proportional to the reference grating frequency and the rotation.

The resultant fringe gradient can be derived in another way. Equation (7-19) can be expressed in its incremental form,

$$\Delta u_x = \frac{1}{f}\Delta N_x \tag{7-19a}$$

where Δ signifies the changes of u_x and N_x that occur between two nearby points in the field. For uniform small strain, $\varepsilon_{xx} = \Delta u_x/\Delta x$, and therefore we may substitute $\Delta u_x = \Delta u\,\varepsilon_{xx}$ into Eq. (7-19a) to find

$$f\varepsilon_{xx} = \frac{\Delta N_x}{\Delta x} = F_{xx} \tag{7-22}$$

The fringe gradient $\Delta N_x/\Delta x$ is another name for F_{xx}, and we find that this derivation produces the same result as Eq. (7-21a). What does it mean? It corroborates the truth of Eq. (7-19).

Numerical evaluations of β_1 and F_{xx} are instructive. For light of the helium–neon laser ($\lambda = 632.8$ nm), a virtual reference grating of 2400 lines/mm and a normal strain of 0.1% (0.001 m/m), we find that

$$\beta_1 = -0.00076 \text{ rad} = -0.043°$$

and

$$F_{xx} = 2.4 \text{ fringes/mm} = 61.0 \text{ fringes/in.}$$

Angle β is tiny and, of course, the fringe frequency F_{xx} is 0.1% of the reference grating frequency f!

Next, let the specimen be subjected to nonhomogeneous deformation. This distorts the specimen grating such that its local frequency and orientation vary from point to point along the specimen. They vary as continuous functions, however, so that the directions of light diffracted from each point in the $+1$ and -1 diffraction orders also vary as continuous functions and create the slightly warped but continuous wavefronts depicted by A'' and B'' in Fig. 7-15. The camera collects these two beams of light and images A'' and B'' in its film plane with the same relative phase (or relative wavefront separation) as they had when they emerged from the specimen grating.

The two beams produce in the camera an interference pattern that is a contour map of the separation λN_x between these warped wavefronts. The warpages and the resultant wavefront separation are functions of in-plane displacements experienced by the specimen grating, and the interference fringe orders represent the displacements according to Eq. (7-19).

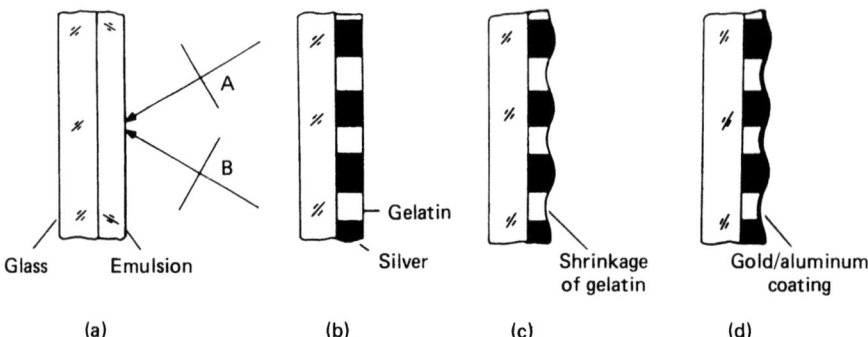

Figure 7-16 A mold for producing phase-type specimen gratings is generated optically on a high-resolution photographic plate: (a) expose photographic plate; (b) develop; (c) dry; (d) mirrorize.

7-6 Specimen Gratings

The method for producing the specimen gratings for the fringe patterns shown in this chapter is described first [45]. Other methods are referenced later.

Referring again to Fig. 7-15, it becomes clear that the most important quality is symmetrical diffraction efficiency. Then, for beams A and B of equal intensity, diffracted beams A'' and B'' will also be equal. The resultant fringe contrast will be optimum. A secondary requirement is high diffraction efficiency to enhance the intensities of A'' and B''. Of course any modest diffraction efficiency will suffice if the light source is powerful or the photographic exposure time is long. In addition, the grating applied to the specimen should be very thin to ensure high fidelity in transmitting the displacements at the surface of the specimen to the surface of the grating.

The specimen gratings are produced by an uncomplicated two-step procedure. First, a mold of the diffraction grating of required frequency is made by a photographic process. The surface of this mold has a regularly corrugated or furrowed surface. In the second step, this furrowed surface is reproduced on the surface of the specimen by a replication technique. Details of the process are illustrated in Figs. 7-16 through 7-18.

The Mold

The mold is made by the method outlined in Fig. 7-16. A high-resolution photographic plate is exposed to two intersecting beams of collimated, coherent light. The two intersecting beams form a steady-state or stationary interference pattern in space, governed by Eq. (7-11); it is a virtual grating. The photographic plate intercepts the virtual grating, and the emulsion is exposed in zones of constructive interference while remaining unexposed in zones of destructive interference. After the plate has been developed, the photographic emulsion consists of clear gelatin in the unexposed zones and gelatin containing silver crystals in the exposed zones. Upon drying, the gelatin shrinks, but its shrinkage is partially restrained by the silver crystals, resulting in the undulating surface illustrated in Fig. 7-16c. It is this undulation that transforms the surface of the photographic plate into a phase-type diffraction grating of frequency $\mathscr{F}_s$. The final step in producing the mold is to apply an ultrathin reflective coating of aluminum, or gold overcoated with aluminum, by evaporation (high-vacuum deposition). These molds can be made in advance by batch processing and stored for future use.

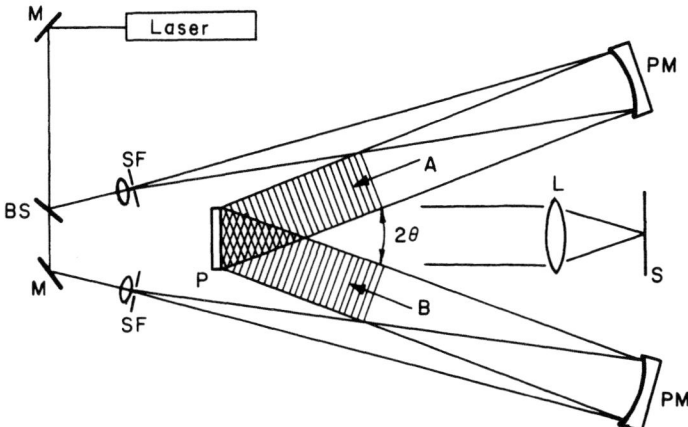

Figure 7-17 Optical arrangement to expose the photographic plate of Fig. 7-16 to a virtual grating of frequency $\mathcal{F}_s$: *M*, Mirror; *BS*, beam splitter; *SF*, spatial filter; *PM*, parabolic mirror; *L*, lens; *S*, screen.

The method is called a holographic process, and the result a *holographic grating*, because the dual-beam illumination and the high-resolution photosensitive plate are commonly used in holography. The interference pattern captured by the plate is a harmonic intensity distribution [Eq. (7-7)], assuring that the corrugations have a symmetrical profile.

A means of exposing the photographic plate is illustrated in Fig. 7-17. An easy method to establish the required angle 2θ is to insert at *P* a commercial reflective diffraction grating of half the desired specimen and grating frequency. Then two beams of light reflect back from the grating in its $+1$ and -1 diffraction orders. They are converged by lens *L* to two bright spots on the screen. Coarse adjustment is achieved by varying angle 2θ until the spots are superimposed. If lens *L* is then removed, a moiré pattern is seen on the screen, depicting the difference between the frequency of the virtual grating and twice that of the commercial grating. For fine adjustment, the angle is varied until the moiré pattern is reduced to a null field, whereupon the angle and virtual grating frequency are established to about 1 part in 10^5.

For many applications of moiré interferometry, a crossed-line specimen grating is required, one that has furrows perpendicular to orthogonal *x* and *y* directions. This is accommodated in a simple way. In the first step of the process, the photographic plate is exposed as illustrated in Fig. 7-16a, and then it is rotated 90° and exposed again to record the virtual grating in its orthogonal direction. Subsequent steps remain the same as for a unidirectional grating.

The use of high-resolution photographic plates for making specimen grating molds is described here because the process is easily accomplished in photomechanics laboratories. The surface relief or depth of furrows is limited, and the resultant diffraction efficiency is appreciably less than that obtainable by other methods. The diffraction efficiency is not a critical matter, however; for all the moiré interferometry patterns shown in this chapter, the specimen gratings were produced from molds made by the photographic plate technique. The use of photoresists instead of photographic emulsions is recommended for controllable furrow depth and maximum diffraction efficiency. Hutley [58] provides an excellent reference. Further opportunities lie in reactive ion-beam etching to create an extremely robust mold. Photoresist and etching techniques were used by Matsui et al. [59] to form special gratings.

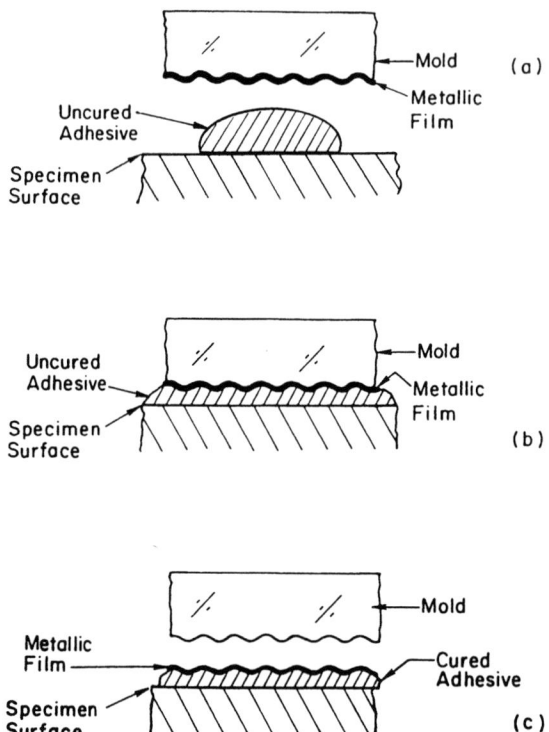

Figure 7-18 Steps in producing the specimen grating by a casting or replication process; the reflective metallic film is transferred to the specimen grating.

Replication

The grating is replicated on the surface of the specimen according to the steps shown in Fig. 7-18. A pool of liquid adhesive is poured on the specimen and squeezed into a thin film by pressing against the mold. Epoxy adhesives are suitable. After polymerization, the photographic plate is pried off—only a small prying force is required—leaving a reflective diffraction grating bonded to the surface of the specimen. The weakest interface in the system occurs between the gelatin of the photographic plate and the evaporated aluminum or gold, which accounts for the transfer of the reflective film to the specimen. The result is a reflective high-frequency, phase-type diffraction grating formed on the specimen. Its thickness is about 0.025 mm (0.001 in.).

Other methods have been used to produce phase-type specimen gratings for moiré. McKelvie and Walker [17] and Rowlands and Vallem [19] used a three-step process. In it, a relatively coarse amplitude grating was contact printed on a photoresist-coated plate to convert it to a phase grating called a master. A submaster was produced by replicating the phase grating surface in silicone rubber, much as in Fig. 7-18. The submaster was used to replicate the phase grating in epoxy on the specimen surface. Later, McKelvie, Walker, et al. [18,37,38] used a high-frequency holographic phase grating to produce the silicon rubber submaster. Silicone rubber of the two-component liquid type—not paste—has a distinct virtue for the intermediate step in that it is a nonadhesive replicating material. It permits easy separation from the master and also from the

MOIRÉ INTERFEROMETRY

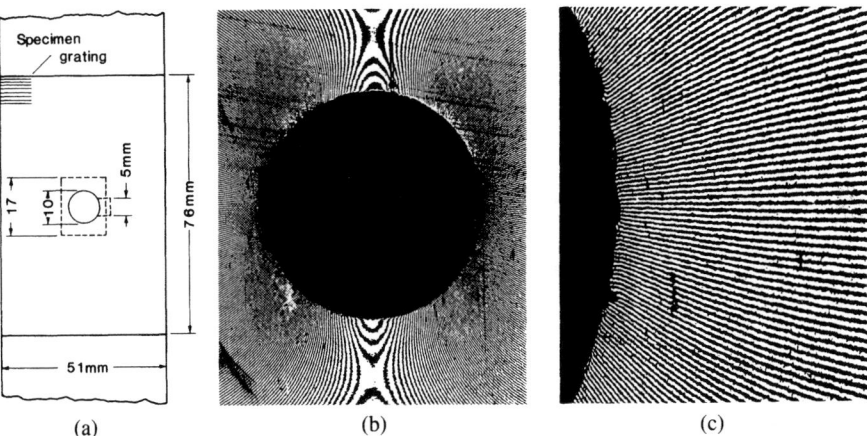

Figure 7-19 Demonstration of moiré interferometry at 97.6% of the theoretical limit of sensitivity. $f = 4000$ lines/mm (101,600 lines/in.).

final specimen grating; it bonds well, however, to specially printed surfaces. Post and Baracat [24] used silicone rubber by the method of Fig. 7-18 to produce high-frequency, low-reflectance specimen gratings.

Marchant et al. [12,13] produced high-frequency specimen gratings by coating the specimen itself with photoresist and exposing it to a virtual grating formed by two coherent beams.

7-7 Theoretical Limit

It is clear from physical reasoning and also from Eq. (7-18) that the theoretical upper limit for reference-grating frequency is approached as α (Fig. 7-15) approaches 90°. The theoretical limit is $f = 2/\lambda$, and the corresponding theoretical upper limit of sensitivity is a displacement of $\lambda/2$ per fringe order.

An experiment was conducted in which the specimen illustrated in Fig. 7-19a was prepared with a grating of 2000 lines/mm (50,800 lines/in.) [28]. The optical system depicted in Fig. 7-20 was used, where α was 77.4° and λ was 488.0 nm. This produced a virtual reference grating of 4000 lines/mm (101,600 lines/in.). For this wavelength, the theoretical limit is 4098 lines/mm, which means that the experiment was conducted at 97.6% of the theoretical limit of sensitivity.

The fringes were well resolved in the entire 51×76 mm (2×3 in.) area of the specimen grating. Enlargements of the very dense array of interference fringes around the hole are shown in Fig. 7-19b and c. The patterns show remarkable clarity of the interference fringes and verify that moiré interferometry is effective even very near its theoretical limit.

7-8 Optical Systems

Numerous different optical schemes can be contrived to form the virtual reference grating. Any means that brings coherent beams equivalent to A and B (Fig. 7-15) onto the specimen grating will suffice. A particularly simple means is illustrated in Fig. 7-20 [24]. Here, half the incident

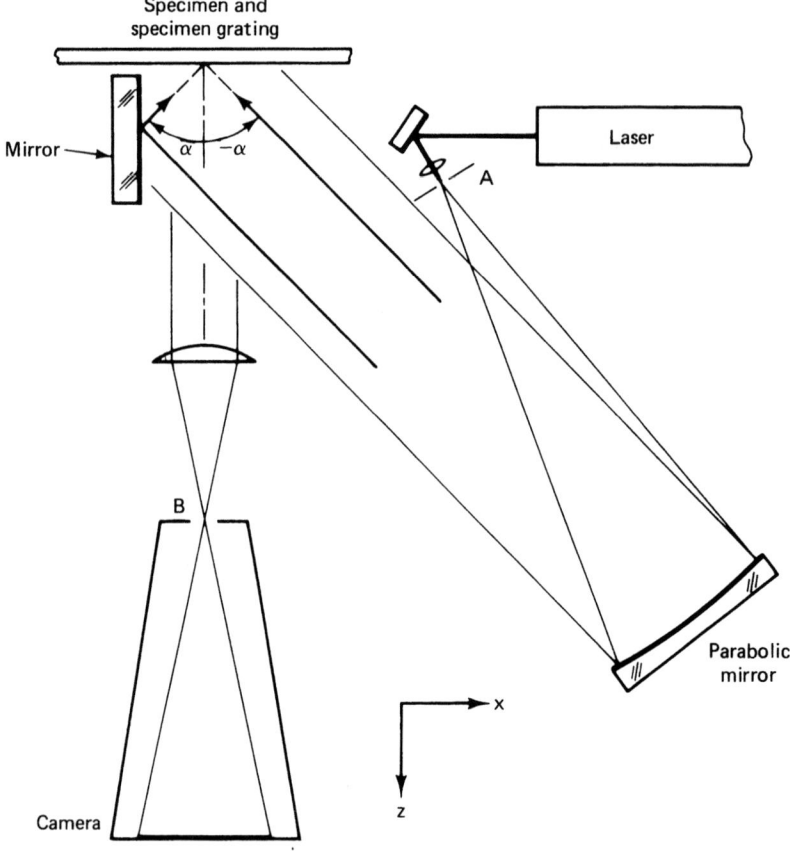

Figure 7-20 Optical arrangement for moiré interfermetry.

beam impinges directly on the specimen surface while the other half impinges indirectly in a symmetrical direction after reflection from a plane mirror. The entire optical system is shown schematically, including a lens that performs dual functions as a decollimating lens and an objective lens: it collects all the light that emerges essentially normal to the specimen surface and focuses the specimen surface onto the film plane of the camera. A parabolic mirror is shown as a means to form the collimated beam, but a collimating lens is a feasible alternative. Most of the patterns shown here were made with this optical arrangement or its four-beam version, shown in Fig. 7-21.

Ideally, the collimator and plane mirror should have sufficient optical quality to produce wavefronts that are plane to within a fraction of a wavelength. Much lower quality is acceptable, however, if the initial fringe pattern is to be subtracted from the load-induced fringe pattern, as discussed in Section 7-16 and Ref. 24.

For static analyses, the required laser power depends primarily on the diffraction efficiency of

MOIRÉ INTERFEROMETRY

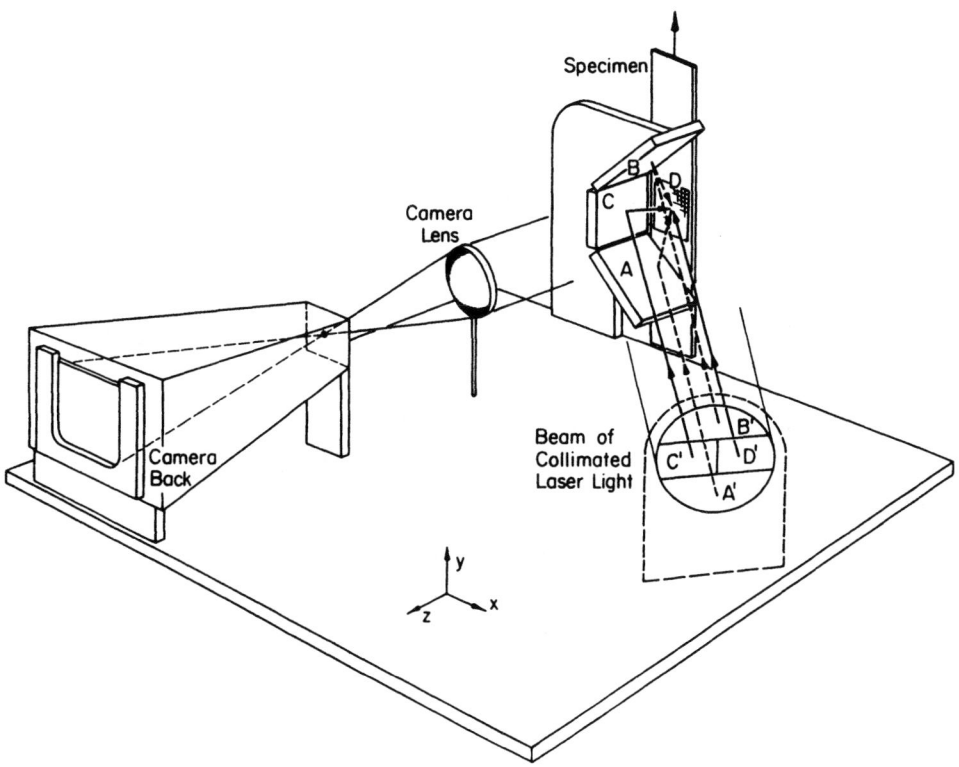

Figure 7-21 Four-beam optical arrangement to produce the N_x pattern with beams C' and D' and the N_y pattern with beams A' and B'.

the specimen grating, the magnification of the image, and the sensitivity of the film used to photograph the fringe pattern. Laser powers from 0.5 to 200 mW have been used successfully.

Adjustment

Angle α can be precisely adjusted by the following rather simple procedure. With the specimen in its unloaded (undeformed) condition, adjust the plane mirror while observing an aperture plate in plane A. Two bright dots will appear in plane A and they should be merged into one by adjusting the plane mirror; this adjusts the mirror perpendicular to the specimen. Attach a white card to aperture plate B and observe two bright dots on the card. They are from the two beams that form the interference pattern. These dots should be merged by adjusting the angle of incidence α and the parallelism of lines in the specimen and virtual reference gratings. Angle α can be adjusted by rotation or translation of the parabolic mirror. Parallelism can be adjusted by in-plane rotation of the specimen or by rotation of the plane mirror about an axis perpendicular to the specimen. Remove the white card and observe moiré interferometry fringes in the camera screen (using a magnifier if necessary). Make fine adjustments of α and parallelism to obtain a null field, or the closest approach to a null field that can be obtained. Additional details appear in Ref. 56.

Four-Beam Optical Arrangement

Generally, both the u_x and u_y displacement fields are required in an analysis. Various optical schemes can be arranged to provide both fields, and the four-beam system of Fig. 7-21 is an example [54]. Two adjustable mirrors, A and B, are added to the basic system of Fig. 7-20. Let light from sections C' and D' of the collimated laser beam be blocked by an opaque screen. Light from section A' strikes mirror A; each ray is reflected upward to lie in a plane parallel to the yz plane. The rays strike the specimen at an angle $+\alpha$ relative to the z axis. At the same time, light from section B' of the incident beam strikes mirror B and is reflected downward at an angle $-\alpha$. These two beams, from mirrors A and B, combine to form a virtual reference grating with its walls of interference and its lines perpendicular to the y axis. This reference grating interacts with the corresponding system of lines on the specimen grating to form the u_y displacement field viewed in the camera.

Next, let sections A' and B' of the incoming beam be blocked and let light from sections C' and D' reach the apparatus. After one part has been reflected from mirror C, two beams reach the specimen with angles $+\alpha$ and $-\alpha$ in the xz plane, just as in the case of Fig. 7-20. They combine to form a virtual reference grating with its lines perpendicular to the x axis and produce the u_x displacement field in the camera. Accordingly, the u_x and u_y fields can be viewed independently by blocking different portions of the incoming beam. As always, the camera lens and camera back are adjusted to focus an image of the specimen surface on the film plane.

In the example of corresponding u_x and u_y displacement fields shown in Fig. 7-14, the specimen was a beam with a bonded stiffener, subjected to pure bending. The beam and the stiffener were graphite–epoxy, and the adhesive was a relatively thick layer of epoxy. The distribution of in-plane displacements is clearly established near the critical portion of the joint. These tests were conducted for validation of companion analytical investigations, by comparisons of results from experimental and computational programs. Validation is exceedingly important for problems like this one, where theory predicts a singularity at the geometric discontinuity.

Figure 7-22 shows another example of u_x and u_y displacement fields. The specimen was an edge-notched aluminum tension coupon. After the specimen grating had been applied, it was fatigue loaded to initiate and advance a crack, which extends from the end of the notch. The patterns show residual displacement fields, that is, the displacements associated with plastic deformation and crack closure [49].

Rigid-Body Rotations

A special advantage accrues from use of the four-beam system. In practice, it is sometimes difficult to control the in-plane, rigid-body rotation of the specimen while applying external loads. When strains are determined from the displacement fields, such accidental rotations have virtually no influence on normal strains, but they strongly degrade the accuracy of shear-strain determinations. It is shown, however, that the shear strains remain independent of accidental rotations if the rotations of the u_x and u_y fields are identical [7]. With the four-beam system, the virtual reference gratings remain fixed with respect to each other. Rigid-body rotation of the specimen introduces fringe gradients $\partial u_x/\partial y$ and $\partial u_y/\partial x$ of identical magnitudes and opposite signs, and these extraneous gradients cancel when shear strains are calculated by Eq. (7-2).

Auxiliary Specimen Grating

Rigid-body rotations can be controlled in another way, one that is suitable for a two-beam optical system. With a two-beam system, the specimen (with its crossed-line specimen grating) and the

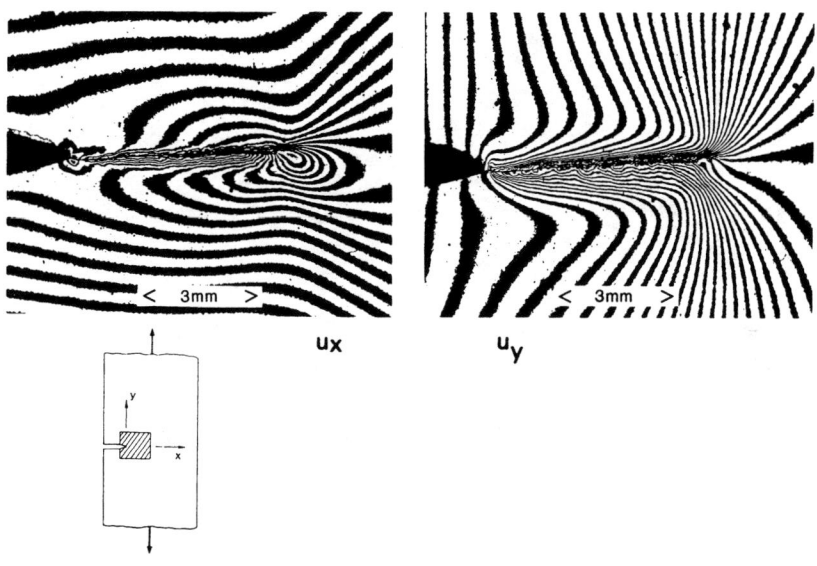

Figure 7-22 N_x and N_y, fringe patterns, depicting the u_x and u_y displacement fields around a fatigue crack in 7075-T6 aluminum. $f = 2400$ lines/mm.

loading fixture must be rotated through 90° to obtain u_x and u_y displacement fields. The problem is to provide means for establishing the precise rotation, to avoid extraneous fringes associated with excess or insufficient rotation of the specimen.

A good answer is to provide an extension of the specimen that projects from it in such a way that it remains unstressed and undeformed when the specimen is loaded. The specimen grating is replicated simultaneously on the specimen and the projection. The portion of the grating on the extension will be called an *auxiliary specimen grating*. It experiences rigid-body displacements identical to those of the specimen itself, and it can be used for alignment purposes.

In practice, the specimen would be loaded and its orientation adjusted to null the fringes in the auxiliary specimen grating. This can be done either by fine adjustment of the angular orientation of the specimen grating or by adjustment of the virtual reference grating. After recording the fringe pattern, the specimen is rotated 90° and fine adjustments are made again to null the fringes in the auxiliary specimen grating.

A good geometry for the auxiliary specimen extension is a circular disk connected along a small arc of the disk to a free boundary of the specimen. This assures negligible reinforcement of the specimen and negligible strain transfer to the extension. An alternative is a square extension connected along a length shorter than the side of the square. The extension can be made by machining it as a part of the specimen itself, or it can be cemented to the specimen. When the specimen grating is produced simultaneously on the specimen and extension, their mutual alignment

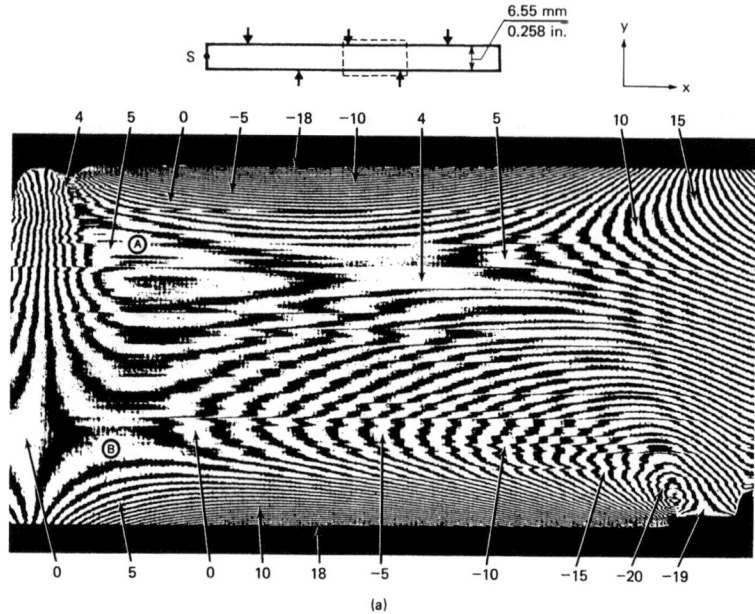

Figure 7-23 Fringe patterns of N_x in the portions of graphite–epoxy beams indicated by dashed boxes: (a) unidirectional composite; (b) quasi-isotropic composite; (c) shear strain along line B–B. $f = 2400$ lines/mm.

is achieved automatically, and the angle between x and y furrows of the gratings is identical in the specimen and extension. This identity of angles is the true requirement, rather than the unachievable objective of making the angles exactly 90°.

7-9 Quantitative Analysis

Fringe orders N_x are enumerated in Fig. 7-23. The two specimens are from a series of tests on composite members in five-point and three-point bending [51]. In Fig. 7-23a the specimen is a unidirectional composite, with fibers in the longitudinal (or horizontal) direction. Especially interesting features of the pattern are striations (i.e., very narrow horizontal fringes) and zigzag fringes. These are anomalies—deviations from otherwise smooth fringe contours. They occur where horizontal shear stresses ε_{xy} are large. Analysis showed that they indicate high localized shear strains in resin-rich zones between adjacent plies.

The specimen of Fig. 7-23b is a quasi-isotropic composite, with a repeating sequence of fiber directions of 45°, 0°, −45°, and 90°. The remarkable waviness of fringes and cyclic variations of fringe gradients in the pattern corresponds to the 4-ply sequence in the 48-ply specimen. The corresponding undulations of shear strains are shown. This results from two anomalous effects: (1) free-edge shear strains in regions of normal stresses, notably near the top and bottom of the

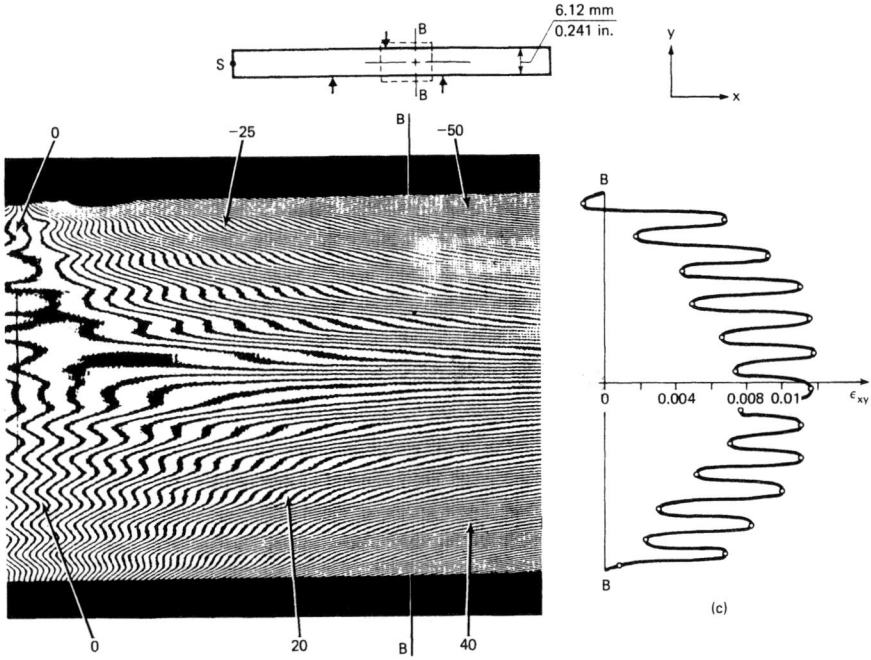

Figure 7-23 (*continued*)

specimen, and (2) shear-strain variations associated with the cyclic variations of shear compliance of individual plies in regions of shear stress, notably in the central portion of the bay [51].

Fringe Counting

When moiré is used for deformation studies, the location of the zeroth-order fringe is arbitrary, as discussed in Chapter 6. This is because rigid-body translations are unimportant for deformation analyses. Relative displacements are pertinent, and absolute fringe orders are not. In Fig. 7-23, a location under the central load was chosen as the zero-displacement datum and the fringe through that point was assigned zero order. The center of the bright fringe was chosen as zero, but this is arbitrary. The centers of dark fringes may be assigned intergral orders, instead. In fact, any intensity level between the maximum and minimum can be called zero order, and corresponding points on adjacent fringes become integral orders. Although this last option is seldom used, it can be useful when one feels compelled to locate the zero-displacement datum at a specific point on the specimen.

The rules of topography of continuous surfaces govern the order of fringes [20]. Adjacent fringes differ by plus or minus one fringe order, except in zones of local maxima or minima, where adjacent fringes may have equal fringe orders. Local-maxima and minima are usually distinguished by closed loops or saddle-shaped contours, as at A and B, respectively, in Fig. 7-23a. Fringes of unequal orders cannot intersect. To be correct, the fringe order at any point must be unique, independent of the path of the fringe count used to reach the point.

Usually, the sign of the fringe-order gradient is known a priori from knowledge of the physical behavior of the specimen. For example, in the lowermost portion of Fig. 7-23b, the stresses in the beam are tensile stresses and the strains are tensile (or positive). The gradient $\Delta u_x/\Delta x$ must be positive. Displacements—and fringe orders—must increase in this vicinity with coordinate x.

When the sign of the gradient is not known a priori, it can be determined easily by an experimental observation. First, establish the positive direction for coordinate x and displacement u_x. Then apply a tiny displacement to the (loaded or deformed) specimen while observing the fringe pattern. If the displacement is in the positive x direction, the fringe orders will all increase and as a result, the fringes will move in the direction of lower fringe orders. For example, if the beam of Fig. 7-23b is moved to the right by pressing a pencil lightly against it at point s, fringes along the bottom edge of the pattern will move toward the left. In a region of positive fringe gradient, fringes move in the direction opposite the rigid-body displacement of the specimen. For a negative fringe gradient, fringes move in the same direction as the displacement.

7-10 Bridges

Problems arise in which relative displacements of two separate elements of an assembly must be measured. Figure 7-24 is an example. The specimen is an adhesively bonded joint with the same dimensions and materials as that of Fig. 7-13. Soft bridges were used to span the gap between the aluminum adherends. Figure 7-24 shows u_x displacement fields for the same specimen, but for part a the applied load was about 25% of the failure load, while for part b it was about 70%. The fringes in the narrow adhesive zone are well defined in part a and there is no difficulty in establishing fringe orders on both sides of the joint. In part b, the gradient of u_x displacements in the adhesive was very much larger as a result of the increased load and also because of nonlinearity of adhesive properties, and the fringes were not resolved in the adhesive zone.

It remains possible, however, to determine the displacements across the adhesive by establishing the fringe orders in the aluminum adherends at the adherend/adhesive interfaces. This was done with the help of the bridges bonded to the aluminum adherends along the dashed lines drawn in Fig. 7-24a. The bridges allow a unique count of fringe orders on one side of the joint with respect to the other. While fringes in the bridges are not clearly reproduced in Fig. 7-24b, they were clearly visible on the photographic film and they were counted with the aid of magnification. In Fig. 7-24b, the change of fringe order across the adhesive is 59 fringes at the center, and the change increases to 63 fringes at the ends. Comprehensive results appear in Ref. 50.

The bridges were made from a soft epoxy, which had a modulus of elasticity of about 210 MPa (30,000 psi). With their low modulus and low-stiffness geometry, the bridges did not contribute in any significant amount to the load-bearing characteristics of the specimen. One bridge would have been sufficient for this specimen, but the second provided a reassuring check on the fringe count.

Soft, artificial bridges provide an effective scheme to quantify displacements in two-body problems. They allow accurate measurements of displacements for contact problems, where slippage occurs between the bodies and creates discontinuous displacements and discontinuous fringe counts across the contact line. Reference 53 is an example in which the relative displacements in a dovetail joint between a simulated turbine blade and disk were studied.

7-11 Anomalies and Subtraction of Uniform Gradients

When a specimen is loaded in tension, it exhibits a uniform array of moiré fringes—closely spaced, straight fringes—if the material behavior is uniform. Anomalous, or nonuniform behavior shows up as slight irregularities of the otherwise straight fringes.

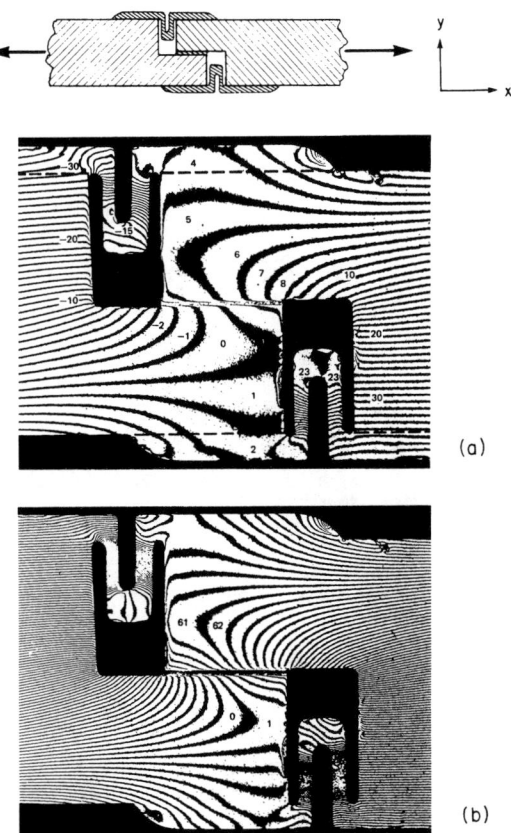

Figure 7-24 N_x fringe patterns for an adhesive joint; omega-shaped features are soft bridges to ensure a continuous path between the adherends for fringe counting. $f = 2400$ lines/mm.

Moiré interferometry offers a unique capability to analyze the nonuniformities. After the specimen is loaded in tension, the frequency of the virtual reference grating can be adjusted to cancel the uniform part of the fringe pattern. The remaining pattern shows the irregularities as bold variations of displacements across the sample [40].

To formalize the technique, let

$$F_{xx} = F_{xx0} + F_{xx1} \qquad F_{yy} = F_{yy0} + F_{yy1} \tag{7-23}$$

where subscripts 0 and 1 signify the uniform part of the fringe gradient and the remaining or nonuniform part, respectively. After the uniform part has been canceled by adjusting the virtual reference grating, the anomalous part of the strain field can be interpreted by

$$\varepsilon_{xx1} = \frac{1}{f}F_{xx1} \qquad \varepsilon_{yy1} = \frac{1}{f}F_{yy1} \tag{7-24}$$

An example is shown in Fig. 7-25a for a graphite–epoxy tensile specimen. It exhibits the

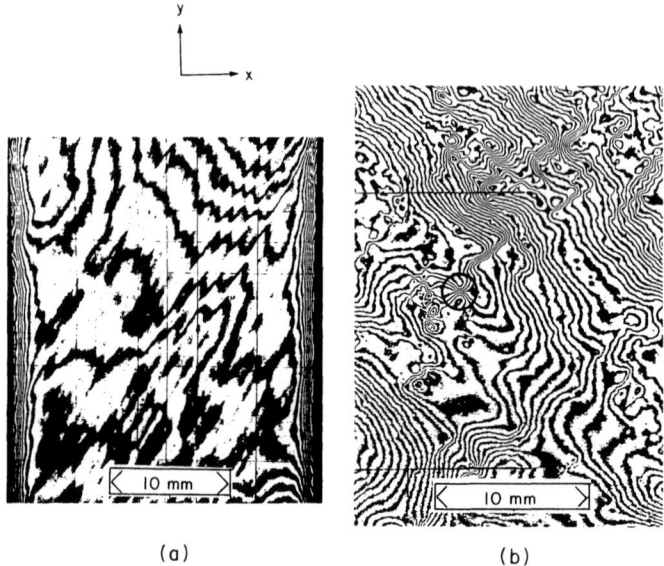

Figure 7-25 Nonuniform part of fringe patterns after the uniform part was subtracted: (a) N_x for graphite–epoxy tension member, showing free-edge effects; (b) N_y for plastically deformed copper tension member after unloading, showing strong heterogeneous deformations; the uniform plastic strain was 0.85%, while the heterogeneous strain in the circled zone was 0.53%, to produce a local maximum of 1.38% permanent strain. $f \approx 2400$ lines/mm.

transverse displacement field in the outermost ply after the fringe gradient $\mathbf{F}_{xx0}$ was subtracted. The gradient $\mathbf{F}_{xx0}$ was 5.8 fringes/mm (148 fringes/in.) and the fringes were so closely spaced that details were not obvious. However, the anomalous part, F_{xx1} is dramatically clear in Fig. 7-25a. The specimen exhibited so-called free-edge effects, manifest in the figure as fringe gradients near the specimen edges. By Eq. (7-24), the anomalous transverse strain ε_{xx1} was about 2400 μm/m near the edges.

Figure 7-25b shows the anomalous part of the residual displacement field for a plastically deformed copper tensile member. In this test, the specimen grating was replicated on the specimen in its virgin condition; the specimen was loaded into its small plasticity range and then unloaded; the unloaded specimen was installed in the optical system illustrated in Fig. 7-21, which was adjusted to cancel the uniform fringe gradient F_{yy0}, and the remaining pattern was photographed. It shows the nonuniform part of the displacement fringes and their gradient F_{yy1}. Substantial nonuniformities become evident in the permanent strain field; the peak anomalous strain occurs in the circled zone, where $\varepsilon_{yy1} = 0.0053$ m/m.

The technique can also be extended to shear strains. Anomalies in otherwise uniform shear strain fields can be detected by canceling the uniform part F_{xy0} (or F_{yx0}) by rotation of the specimen or the virtual reference grating.

This method is an exception to the general rule of metrology—that sensitivity diminishes as range is increased. Here the high sensitivity of moiré interferometry is retained, even though the actual displacements and fringe gradients may be very large.

MOIRÉ INTERFEROMETRY

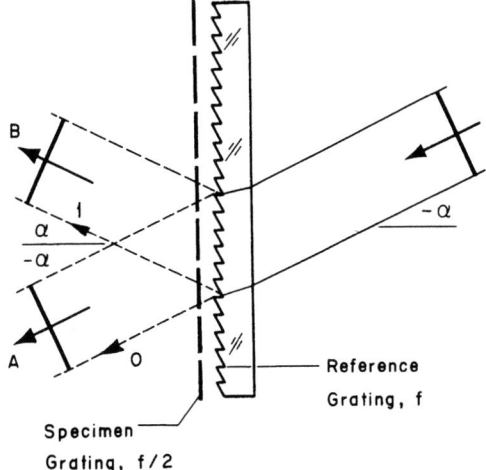

Figure 7-26 A virtual reference grating, produced by diffraction from a real reference grating, is intercepted by the specimen grating.

7-12 Alternative Optical Systems

The diagram of Fig. 7-15 is basic to moiré interferometry. Although the system of Fig. 7-20 is an attractive implementation, virtual reference gratings can be created by other means and incorporated into the basic system of Fig. 7-15. Figures 7-26 through 7-28 are examples.

A virtual reference grating can be formed by a real grating of the same frequency, as illustrated in Fig. 7-26. The zeroth and first diffraction orders are beams A and B, respectively, which

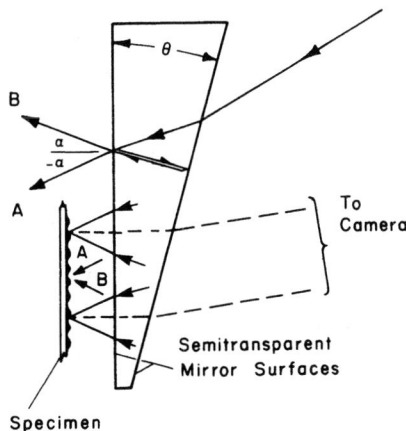

Figure 7-27 Virtual reference grating produced by a mirrorized wedge prism.

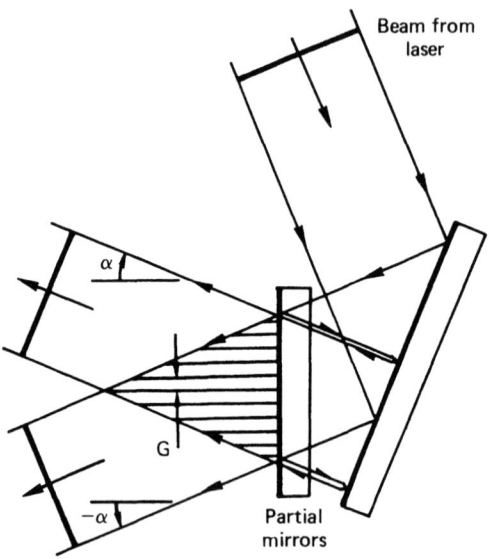

Figure 7-28 Virtual reference grating formed by adjustable partial mirrors.

correspond to the input beams of Fig. 7-15. They form a virtual reference grating in their zone of intersection. The specimen and specimen grating is located in this zone, as indicated in Fig. 7-26. Light diffracted from the specimen grating passes back through the real reference grating in its zero order to reach the camera. The arrangement is most closely analogous to that of geometrical moiré, with the specimen grating and real reference grating close together.

The angle of incidence α is prescribed by Eq. (7-18), and, as a consequence, the zeroth and first diffraction orders emerge in symmetrical directions. If the grating frequency f is in the range

$$\frac{2}{\lambda} > f \geq \frac{2}{3\lambda} \tag{7-25}$$

the diffraction angle will be so large that two and only two orders will emerge, the zeroth and first orders [23]. For light of the red helium–neon laser, this range is $3160 > f \geq 1054$ lines/mm or $80,280 > f \geq 26,760$ lines/in. This arrangement has been used for two applications [26,30], both with $f = 1200$ lines/mm (30,480 lines/in.). In the first application, a special advantage was the close proximity of specimen and reference gratings, which minimized thermal air currents in the gap between them. In the second, the special advantage was the precisely fixed reference grating frequency. Details relating to the use of real reference gratings are treated in these references.

Figure 7-27 illustrates another means for generating a virtual reference grating in a moiré interferometry system. A wedge-shaped prism of angle θ is coated with semitransparent mirror surfaces of approximately 60% reflectance. (Multilayer dielectric coatings are used to minimize absorption.) A portion of the incident ray penetrates both surfaces and emerges as A. Another portion is internally reflected once at each mirror surface and emerges as B. The same occurs for every ray of the incident beam, creating two intersecting beams. As in Fig. 7-15, two preferred beams emerge from the specimen grating, pass through the prism, and generate the moiré interferometry pattern seen in the camera.

MOIRÉ INTERFEROMETRY

Frequency f of the virtual grating depends on α, which in turn depends on wedge angle θ by

$$\sin \theta = \frac{1}{n} \sin \alpha = \frac{\lambda f}{2n} \qquad (7\text{-}26)$$

where n is the index of refraction of the prism material. For a frequency of 1200 lines/mm (30,480 lines/in.), red light of the He-Ne laser, and n near 1.5, we find that $\alpha = 22.3°$ and $\theta \approx 15°$.

Beam A is brighter than B, since it is not attenuated by reflection coefficients. If reflectance is 60%, the intensity of A is approximately three times that of B. Complete destructive interference is not possible, but very good fringe contrast is still obtained. From Eqs. (7-8) and (7-13), the contrast is 94%.

The wedge prism is even better adapted to thermal environments and was used for thermal expansion measurements [34]. Its surfaces are much more robust than the fragile surface of a real grating; it can be cleaned readily; and it can withstand much greater temperature extremes than commercial gratings. While angle θ of the wedge does not change with temperature, the refractive index does change and calibration is required.

A variation of the wedge prism is an air wedge, in which the semitransparent mirrors are coated on individual flat plates. They are mounted together in a fixture that permits fine adjustment of the air gap wedge angle. Such an arrangement provides an adjustable frequency f. It is most attractive for comparatively small frequencies [e.g., f = 200 to 1200 lines/mm (5000 to 30,000 lines/in.)].

Figure 7-28 is a variation of the air wedge. It has been used effectively for small fields of view in fracture mechanics studies [27,36,42,48]. Its special advantage is that the camera lens can be placed close to the specimen without blocking the incoming light, and therefore it permits large optical magnifications without very large lens-to-film distances.

Another optical system can be described in terms of Fig. 7-17. Let the specimen and specimen grating be located at P, where they intercept the virtual reference grating formed by beams A and B. Let the camera lens and camera be located between the incoming beams, in the region of L and S. This arrangement is attractive for relatively large fields of view.

7-13 Achromatic Systems

Again consider Fig. 7-15. A virtual reference grating of frequency f is formed when the angle 2α between beams A and B is given by the relationship of Eq. (7-18), rewritten here as

$$\frac{f}{2} = \frac{\sin \alpha}{\lambda} \qquad (7\text{-}18a)$$

The virtual reference grating is easily generated in monochromatic light, where λ is a fixed wavelength. It can be arranged for polychromatic light, too, by providing a system in which α varies in precise harmony with λ, wherein $(\sin \alpha)/\lambda$ is a constant. This harmony is provided in the +1 and −1 diffraction orders of an auxiliary grating of frequency $f/2$, as illustrated in the upper portion of Fig. 7-29. For these auxiliary grating conditions, namely angle of incidence equal to zero, grating frequency $\mathscr{F} = f/2$, diffraction order $m = 1$, and diffraction angle $\beta = \alpha$, the grating equation (7-15c) gives precisely the relationship of Eq. (7-18a) for every wavelength. This auxiliary grating, which spreads the light exactly as required, is called a *compensator* [21].

This subsystem can illuminate a moiré interferometry arrangement as also shown in Fig. 7-29. In the specimen space, a virtual reference grating of unique frequency f is produced by every wavelength of the polychromatic light. Geometric analysis shows that these virtual reference

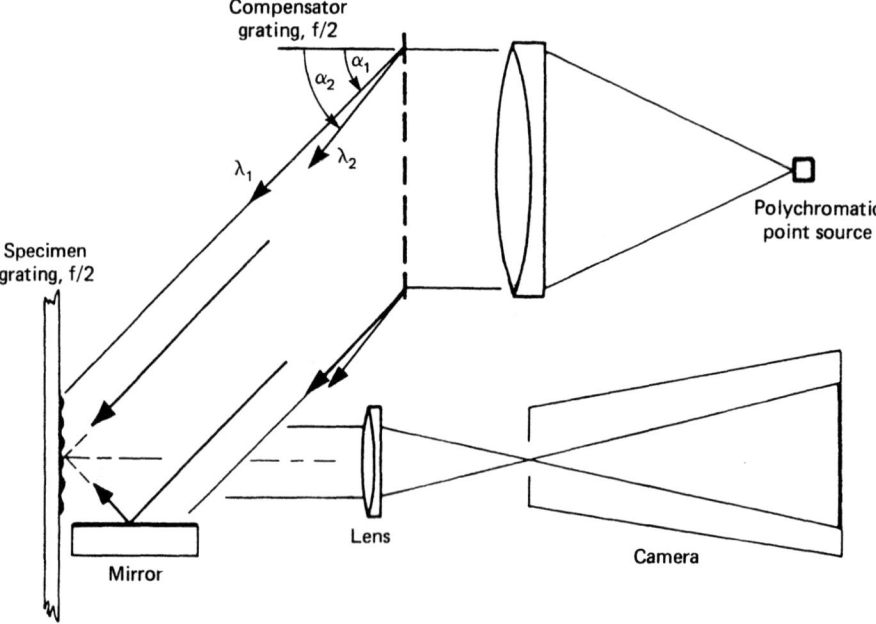

Figure 7-29 A compensator grating permits use of temporally noncoherent light.

gratings are in exact registration, with destructive interference occurring at identical locations for all wavelengths. The camera receives identical moiré patterns for every wavelength and records the scalar sum of the individual intensity distributions. The resulting pattern is the same as that produced with monochromatic light.

The width of the light source remains a critical parameter. In principle, the source must be a point or line when a virtual reference grating is employed. A laser diode appears to be a suitable source. Its size and convenient mechanical features make it especially attractive for portable instrumentation. Although it does not produce a very broad spectrum, its spectral bandwidth is sufficient to require compensation. Requirements on the width of the light source are relaxed when a real reference grating (Fig. 7-26) is used in close proximity to the specimen grating.

Examples of compact achromatic systems are illustrated in Fig. 7-30. In both parts a and b, the incoming beam is directed by a mirror (inclined at 45°) to strike the compensator grating. This grating, whose frequency is $f/2$, divides the beam into three paths, namely the 0, 1, and -1 diffraction orders. The zeroth-order light is wasted, but beams of the $+1$ and -1 orders are reflected in Fig. 7-30a by the vertical mirrors and recombine near the specimen to form a virtual reference grating of frequency f [60,61]. This interacts with the specimen grating (of frequency $f/2$) to form the moiré pattern, which is carried out of the system by the two beams that propagate essentially normal to the specimen. They are redirected by the 45° mirror into the camera lens and camera.

The compensator grating provides angles of diffraction in the $+1$ and -1 orders in accord with Eq. (7-18a). The frequency of the virtual reference grating is automatically the same for every wavelength.

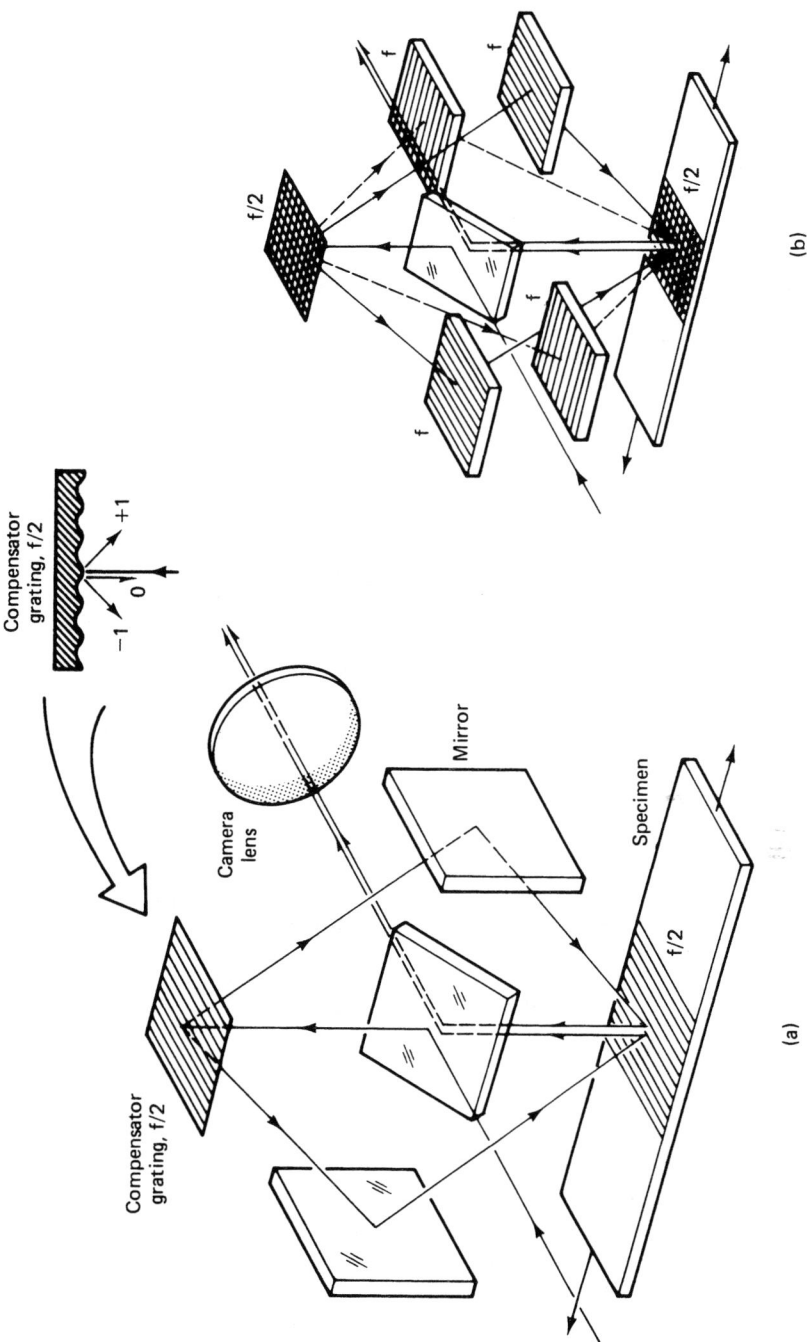

Figure 7-30 Optical arrangements for compact, achromatic moiré interferometers; the compensator allows use of noncoherent light: (a) two-beam system; (b) four-beam system. Both (a) and (b) allow temporally noncoherent light and (b) also allows spatially noncoherent light.

A four-beam system based on the design of Fig. 7-30a can be implemented by using a crossed-line compensator grating and two additional vertical mirrors. In this configuration, the additional vertical mirrors must be semitransparent to pass the input and output beams. An alternative would be to bring the beams in and out of the gaps between vertical mirrors, which would be done by rotating the inclined mirror by 45° about a vertical axis.

Another variation is illustrated in Fig. 7-30b. Although it is more costly to use gratings in place of the vertical mirrors, special advantages are accrued. In this case, a polychromatic light source of finite width (i.e., a spatially noncoherent source) can be used [61,62]. There are no strict coherence requirements for this moiré interferometer. In both schemes, a and b, the virtual reference grating is localized in a specific plane when a source of finite size is used, and the specimen grating must be located in that plane.

7-14 Vibration Control

Interferometry measures tiny displacements and is characteristically sensitive to vibrations. However, only certain elements of the system are highly sensitive. In Fig. 7-20, vibration of the plane mirror in the x direction would be devastating if the amplitude were of the order of the reference grating pitch, $1/f$. On the other hand, the amount of movement of the camera lens and camera back required to cause a noticeable blur depends on the smallest fringe separation in the image. This is usually two to three orders of magnitude larger than $1/f$, so sensitivity to vibrations of the camera is comparatively low.

A large degree of vibration control can be achieved by attaching a real reference grating to the specimen. This arrangement could minimize relative motions between the specimen grating and the reference grating, thus stabilizing the fringe pattern. This scheme is discussed in the next section.

7-15 Sensitivity to Out-of-Plane Motion

Moiré interferometry, as described here, is nearly insensitive to out-of-plane motion of the specimen and specimen grating. Referring to Fig. 7-15 or 7-20, it is completely insensitive to specimen translations in the z direction and to specimen rotations about the x axis. The moiré pattern does respond to rotations of the specimen about a line parallel to the y axis [17]. This is because the effective frequency of the specimen grating is the frequency projected onto the x-y plane. For a rotation about the y axis of magnitude Ψ, the effective frequency changes by $(1 - 1/\cos \Psi)$. A uniform extraneous fringe gradient, F_{xxe}, appears in the pattern, where the subscript e signifies extraneous or error. For small rotations Ψ,

$$F_{xxe} = -\frac{f\Psi^2}{2} \tag{7-27}$$

The error is always negative, corresponding to compressive strain. For $f = 2400$ lines/mm (60,960 lines/in.) and $\Psi = 0.01$ m/m, the extraneous gradient is 0.12 fringes/mm (3 fringes/in.). Whether this is negligible depends on the problem, particularly the magnitude of the fringe gradient caused by in-plane displacements. In the case of $\Psi = 0.001$ m/m, the extraneous gradient is only 0.0012 fringes/mm (0.03 fringe/in.).

In practice, rotations as small as $\Psi = 0.001$ m/m can be detected and corrected if necessary. Refer to Fig. 7-20 and remember the procedure for alignment. When $\Psi \neq 0$ (i.e., when the specimen grating and plane mirror are not mutually perpendicular), two separate dots of light appear in plane A at the focal plane of the collimating mirror (or collimating lens). Their separation is $2\Psi(FL)$, where FL is the focal length of the collimator. For example, if $FL = 1$ m and $\Psi = 0.001$ m/m, their separation becomes an easily observed 2 mm. The position of the specimen could be adjusted to reduce even this tiny rotation.

Another alternative is practical, too. Instead of adjusting the rotation of the specimen to eliminate the extraneous fringe gradient, the virtual reference grating can be adjusted to eliminate it. This requires an *auxiliary specimen grating*, as described in Section 7-8. Then, as extraneous fringes from out-of-plane rotations are developed, they can be nulled in the auxiliary specimen grating by adjusting the plane mirror of Fig. 7-20, or by translating (laterally) or rotating the parabolic mirror to change angle α.

Systems with Real Reference Grating

When a real reference grating is used, as in Fig. 7-26, it can be attached to the specimen for the purpose of vibration control. It will experience the same rigid-body, out-of-plane rotations as the specimen grating. The effective frequency is also the frequency projected onto the x-y plane and it experiences the same $(1 - 1/\cos \Psi)$ change as the specimen grating. No extraneous fringes are introduced.

An inconvenience of the arrangement of Fig. 7-26 is that the frequency of the virtual reference grating is not adjustable. The systems of Fig. 7-30 offer the same insensitivity to out-of-plane rotations, plus the additional advantage that the virtual reference grating can be varied by adjustment of the vertical mirrors (Fig. 7-30a) or by in-plane rotations of the gratings of frequency f (Fig. 7-30b). Such systems are amenable to mounting on the specimen to achieve a high degree of insensitivity to vibrations.

The issue can be generalized: when the optical system utilizes a real reference grating attached to the specimen, sensitivity to specimen motions is reduced.

7-16 Coping With The Initial Field

In general, the specimen grating will not have lines that are perfectly straight and uniformly spaced. Also, the optical elements used to form the reference grating will not be so accurate that a perfect reference grating is formed. The result is that a few fringes will usually appear in the field of view before the specimen is loaded. These must be subtracted from the pattern obtained after loading in order to determine the load-induced displacements. This can be done manually by subtracting the fringe orders at corresponding points in the load and no-load patterns. Alternatively, the subtraction can be performed simultaneously for the entire field of view by geometrical moiré.

There are at least two conditions, however, in which no subtraction is required. (1) When the initial pattern is sparse, but the full-load pattern exhibits a great number of fringes, the subtraction may be unnecessary to achieve the desired accuracy. (2) When the field of view is small, especially when it is so small that magnification is used in photographing the fringe pattern, the initial pattern seen in the field of view would usually be a small fraction of a fringe order and it could be neglected in most analyses.

When subtraction is required, the photographic transparencies (negatives) of the moiré interferometry patterns for the load and no-load conditions can be superimposed in registration. Alternatively, they can be photographed as a double exposure on the same film. In either event, the

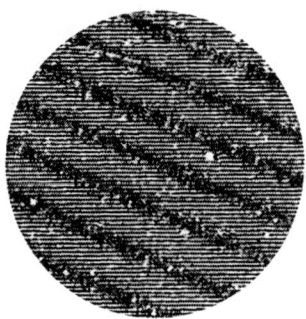

Figure 7-31 Enlarged view of a double-exposure negative of two moiré interferometry patterns: a no-load and a full-load pattern. Identical carrier patterns of 12 fringes/mm (300 fringes/in.) were added to the patterns.

two patterns interweave to produce a (geometrical) moiré pattern[1] that depicts the load-induced displacement field alone. The supermoiré fringes would not have good visibility, however, since the no-load pattern would usually be rather sparse. The superposition of closely spaced line arrays is required for clear supermoiré patterns. This is remedied by a carrier pattern.

Carrier Patterns and Optical Filtering

The number of fringes in the no-load or initial pattern can be made as large as desired by adjusting the apparatus. For example, if the specimen is given an in-plane rotation relative to the optical apparatus, or if the plane mirror of Fig. 7-20 is rotated about an axis parallel to z, a carrier pattern of rotation is introduced. An apparent displacement field of uniformly spaced fringes perpendicular to the specimen grating lines, called a *carrier pattern of rotation*, is added to the initial pattern. If the parabolic mirror is translated or rotated to change α, a *carrier pattern of extension* is added. This is characterized by uniformly spaced fringes parallel to the grating lines. It is convenient to use a carrier pattern of rather high frequency [e.g., 10 fringes/mm (250 fringes/in.) or more] if optical filtering is to be used later. Otherwise, a lower frequency is usually satisfactory [e.g., 2 fringes/mm (50 fringes/in.)].

Figure 7-31 is an enlarged view of a double exposure of two moiré interferometry patterns, a no-load (or initial) pattern and a full-load pattern. A carrier pattern of rotation of about 12 fringes/mm (300 fringes/in.) was identical in both patterns. As a result of deformation of the specimen, the full-load and initial patterns are different from each other, and therefore, fringes cross each other in a systematic way. They form a geometrical moiré pattern that depicts the difference of fringe orders present in the two constituent moiré interferometry patterns. This supermoiré pattern is the desired contour map of the load-induced displacements, independent of any distortions present in the no-load pattern.

The moiré fringes of Fig. 7-31 have relatively low visibility, but they can be transformed into fringes of high contrast by a simple optical filtering process. Each moiré fringe of Fig. 7-31 is comprised of an area of continuous tone (where the constructive interference fringes of the full-load pattern fall midway between the constructive interference fringes of the initial pattern) and an area of black and clear stripes (where constructive interference fringes of both patterns fall

---

[1] In Chapter 6, Parks calls this a *supermoiré* pattern. It has also been called *moiré of moiré* and *second-order moiré*.

upon each other in registration). The filtering process is possible because the striped areas diffract light while the continuous-tone areas do not.

The filtering apparatus is illustrated in Fig. 7-32, with the double-exposure negative (or two superimposed negatives) placed in the system. Light that passes through the striped areas is diffracted into several diffraction orders, while light that passes through the gray continuous-tone areas does not experience diffraction. The field lens converges the transmitted beams to separate points in its focal plane, numbered in Fig. 7-32 to correspond to diffraction orders from the striped areas. Of course, light from the continuous-tone areas converges to point 0. Since light from only one (nonzero) diffraction order is admitted into the camera, the image on the camera screen is bright at all points corresponding to striped areas, and it is black at all other points. With optical filtering, an enhanced contrast picture of the negative is made, using light of any nonzero diffraction order.

The result is seen in Fig. 7-33, in which the circled zone corresponds to Fig. 7-31. The enhancement of contrast is dramatic. The patterns depict the u_x and u_y displacement fields for the upper half of a composite tensile specimen with 15° off-axis fibers [24,41]. The loading condition was axial translation of the grips with (almost) no rotation of the grips. The initial field is canceled and load-induced fringes are displayed uniquely. The 15° orientation of the fibers makes the specimen asymmetrical. It deforms into a shallow S-shaped member, and the central region becomes a zone of uniform shear strain.

For effective optical filtering, the frequency of the carrier pattern should be large compared to the largest fringe gradient in the field of view. Because moiré interferometry produces pure two-beam interference, very-high-frequency carrier patterns are resolved in the film plane with high contrast. The ultimate limitation is the resolving power of the photographic film, which is sufficiently high for these purposes with most films.

What frequency should be used when calculating displacements? Equation (7-19) specifies $u_x = N_x/f$, where f is the frequency of the reference grating; more specifically, f is whatever reference grating frequency produces a null field when interacting with the undeformed specimen grating. However, when a carrier pattern of extension is used, the virtual grating frequency becomes $f + c$ (or $f - c$), where c is the carrier frequency. When a carrier pattern of rotation is used, the virtual grating is rotated, but its frequency remains f.

The answer is that Eq. (7-19) is always correct. The frequency that produces the null field should always be used, not the frequency that produces the carrier pattern. Perhaps the answer is intuitively predictable for this reason: the same pattern of load-induced fringes would be expected regardless of whether a carrier of extension or rotation is used.

7-17 Mechanical Differentiation For Normal and Shear Strains

Derivatives of displacements are required to calculate strains ε_{xx}, ε_{yy}, and ε_{xy}. Techniques of mechanical differentiation are available to produce whole-field contour maps of derivatives of displacements. The topic was introduced in Chapter 6 and is explored further here.

If two identical transparencies of a fringe pattern are superimposed in registration with each other, they appear as the original. However, if the top transparency is shifted, the two patterns interweave to form a pattern of geometrical moiré fringes. The effect is demonstrated in Fig. 7-34, where a moiré interferometry pattern of N_y is shifted by increments Δy and Δx. The geometrical moiré fringes are contour maps of $\Delta N_y/\Delta y$ and $\Delta N_y/\Delta x$ (i.e., contour maps of the finite-increment approximations to the fringe gradients F_{yy} and F_{yx}) [Eq. (7-22)]. This supermoiré technique is called *mechanical differentiation*.

In practice, the number of geometrical moiré fringes thus formed increases as the shifts Δx and

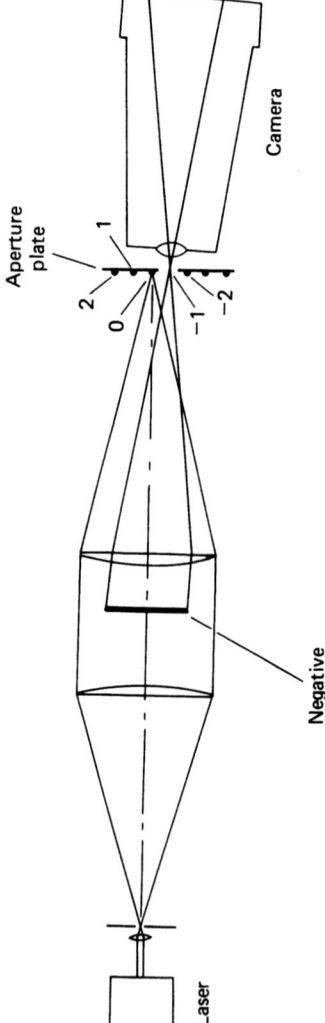

Figure 7-32 Optical filtering arrangement to enhance the contrast of fringes in the double-exposure negative of Fig. 7-31.

MOIRÉ INTERFEROMETRY 343

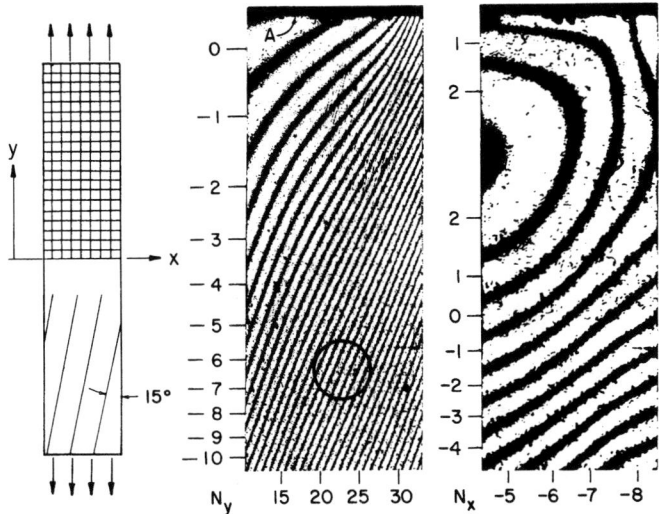

Figure 7-33 Result of optical filtering: circled region corresponds to Fig. 7-31. The initial or no-load fringes are canceled, and the patterns reveal load-induced displacements uniquely. The specimen is a graphite-polyimide composite with 15° off-axis fibers. $f = 1200$ lines/mm (30,480 lines/in.).

Δy increase. The shifts must be large enough to produce several fringes, but they must be small enough to ensure that the finite increments reasonably represent the true derivatives. Furthermore, the visibility of the geometrical moiré would be enhanced by adding identical carrier patterns to each of the superimposed and shifted fringe patterns.

For practical implementation, the double-exposure method can be used in conjunction with a movable filmholder. Let the camera back illustrated in Fig. 7-21 be replaced with that of Fig. 7-35. The latter is a filmholder mounted on an x, y micrometer stage. Referring to Fig. 7-21 (with the filmholder of Fig. 7-35), the experimental procedure is as follows. Install the specimen in the loading fixture mounted on the optical table. Adjust the system for null fields in the x and y directions. Load the specimen and photograph the u_x and u_y fields. Rotate the specimen to add a carrier pattern of rotation. Make an exposure of the u_x field (with carrier), shift the filmholder by an increment Δx, and make a second exposure on the same film. Repeat the double exposure on another film for the u_y field shifted by Δy. Repeat for double exposures of the u_y field shifted by Δx and by Δy. Optically filter these four films to obtain contour maps of $\Delta N_x/\Delta x$, $\Delta N_x/\Delta y$, $\Delta N_y/\Delta x$, and $\Delta N_y/\Delta y$.

This procedure was followed for a beam in three-point bending, illustrated in Fig. 7-36. The u_x and u_y displacement fields (i.e., the N_x and N_y fields) are shown for a small applied load; for a larger load, the fringes would be too closely spaced for reproduction on this page. The load was made about seven times greater for mechanical differentiation. Figure 7-37 shows the patterns of $\Delta N_x/\Delta x$, and $\Delta N_y/\Delta y$, after optical filtering. The mechanical shifting distances Δx and Δy were only 0.25 mm (0.010 in.) or 2% of the height of the beam. Patterns representing the cross-derivatives are shown in Fig. 7-38, where Δx and Δy was again 0.25 mm.

To obtain strains, the basic displacement and small-strain relationships can be written in incremental form:

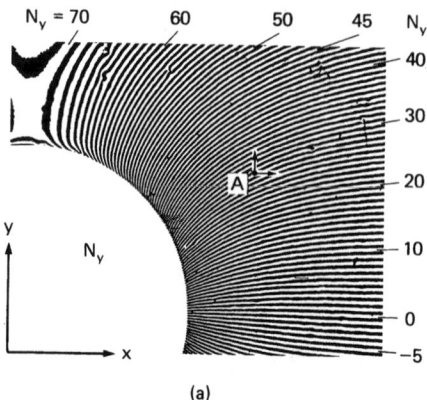

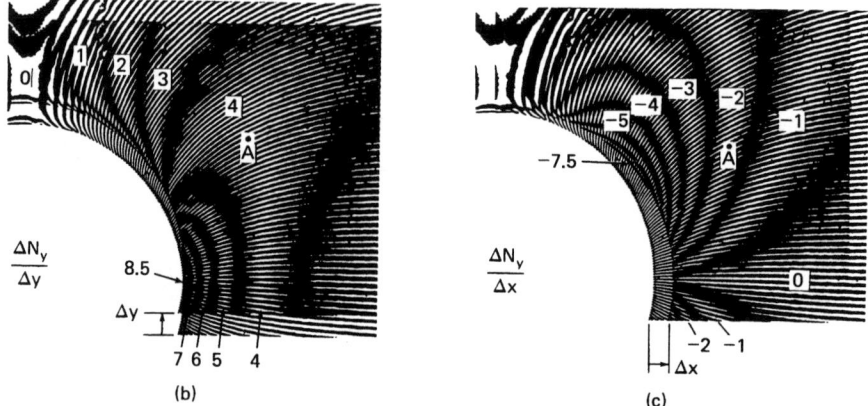

Figure 7-34 Demonstration of mechanical differentiation technique: (a) fringe pattern of N_y; (b) supermoiré pattern obtained by superimposing two transparencies of (a) with a shift Δy; (c) supermoiré pattern with a shift Δx. The supermoiré fringes of (b) and (c) approximate the derivatives $\partial N_y/\partial y$ and $\partial N_y/\partial x$, respectively.

$$\Delta u_x = \frac{1}{f} \Delta N_x \qquad \Delta u_y = \frac{1}{f} \Delta N_y \qquad (7\text{-}19b)$$

From this,

$$\varepsilon_{xx} = \frac{\Delta u_x}{\Delta x} = \frac{1}{f} \frac{\Delta N_x}{\Delta x}, \qquad \varepsilon_{yy} = \frac{1}{f} \frac{\Delta N_y}{\Delta y} \qquad (7\text{-}1a)$$

and

$$\varepsilon_{xy} = \frac{1}{f} \left(\frac{\Delta N_x}{\Delta y} + \frac{\Delta N_y}{\Delta x} \right) \qquad (7\text{-}2a)$$

When $\Delta x = \Delta y = \Delta$, a constant can be factored out, namely,

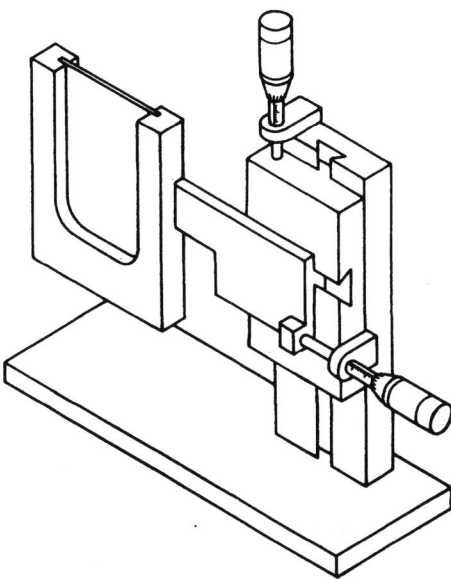

Figure 7-35 Camera back for the apparatus of Fig. 7-21, in which the film holder is mounted on a micrometer stage. It is used for shifting the film in the double-exposure technique of mechanical differentiation.

$$\frac{1}{f\Delta} = \text{sensitivity factor} \qquad (7\text{-}28)$$

which defines the sensitivity in strain per fringe order. In the examples here, $f = 2400$ lines/mm (60,960 lines/in.) and $\Delta = 0.25$ mm (0.010 in.), so the sensitivity factor is 0.00167 m/m per fringe order. The patterns of Fig. 7-37 can be interpreted as contour maps of normal strains ε_{xx} and ε_{yy} by multiplying the fringe orders by the sensitivity factor.

To obtain shear strain ε_{xy}, it is necessary to add the fringe orders of patterns in Fig. 7-38. This can be done on a whole-field basis by applying the concept of additive moiré described in Chapter 6. Figure 7-39a shows tracings of the centerlines of fringes in the two cross-derivative patterns of Fig. 7-38. Both parameters are labeled with their respective fringe orders. Each contour of the additive moiré is the locus of points of constant sum of these fringe orders. It is relatively easy to trace these contours as a continuous series of diagonals across the quadrilaterals in Fig. 7-39a. The result is Fig. 7-39b, which represents contours of the sum of the cross-derivatives. It is a contour map of shear strain ε_{xy}, when multiplied by the same sensitivity factor [46]. While manual tracing is involved, only a modest number of fringes are required. The method seems practical in this form.

These strain patterns are approximations in the same sense that electric-resistance strain gages or the finite-element computational method gives approximations of strain at a unique point. The derivative obtained by mechanical differentiation is a representative value over a line segment equal to the shifting distance. The shifting distance can usually be small, however, comparing favorably with strain gages and many finite-element analyses.

An invaluable asset is available in the u_x and u_y patterns. At any local point, the rate of change of fringe orders in x and y directions can be measured to determine the respective derivative. This

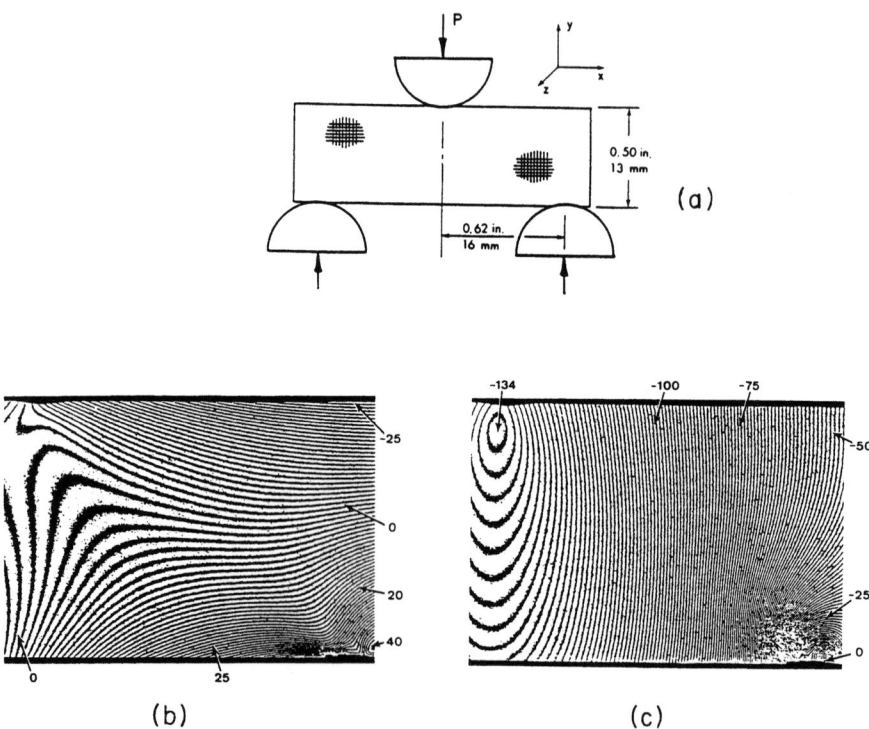

Figure 7-36 (a) Specimen in simple bending; (b) and (c) fringe patterns of N_x and N_y, respectively, for the right side of the specimen, and for small load P. $f = 2400$ lines/mm.

provides both a starting point for fringe-order assignments and a check on the fringe count. An easy check is available for positions of derivative fringes of zero order. These pass through the points where the displacement fringes (Fig. 7-36) lie parallel to the x or y axis: for example, $\partial u_x / \partial x = 0$, where the fringes in Fig. 7-36b are parallel to the x axis.

As a result of the carrier pattern of rotation, fringe orders in the cross-derivative fields are not necessarily whole integers at the centers of white fringes. This is because the terms being differentiated have two components: the load-induced displacements and the apparent displacements of the carrier pattern. The latter is a uniform fringe gradient, and its derivative is a constant. Accordingly, the pattern represents the desired derivative of the load-induced displacements plus a constant fringe order. This constant might be an integer, but it would usually be some integer plus a fraction. The constant integer goes unnoticed in the pattern, but the fraction changes the gray level of the integral-order fringes. In analyses where this fractional fringe accuracy is important, the position of an integral-order fringe (e.g., a zeroth-order fringe) could be determined locally from the displacement field and the gray-level correction obtained there could be applied as a constant thoughout the field. This correction might not be necessary for shear-strain calculations, however, if a four-beam optical arrangement is used. As pointed out in Section 7-8, fringes of rigid-body rotation (i.e., the carrier pattern) have no effect on the calculated shear strains if the rotations are identical for the u_x and u_y fields.

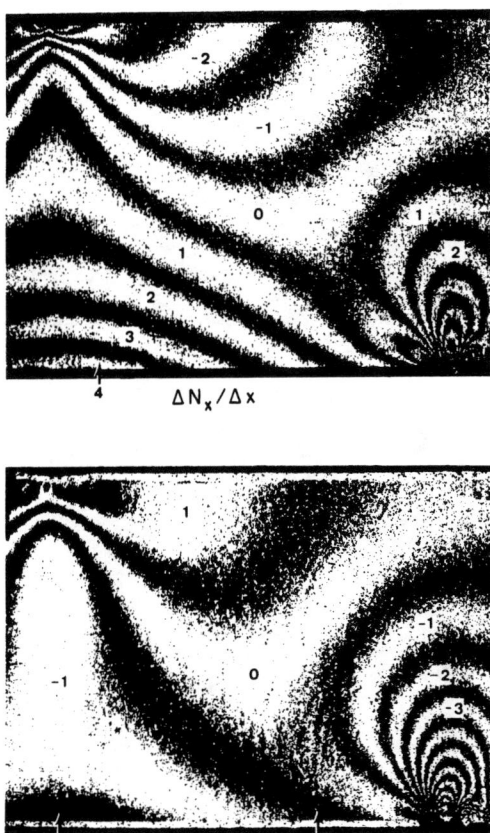

Figure 7-37 Patterns approximating contours of $\partial N_x/\partial x$ and $\partial N_y/\partial y$ by mechanical differentiation, for the beam of Fig. 7-36, $\Delta x = \Delta y = 0.25$ mm.

7-18 Vector Treatment

A vector is an artifice—a fiction—but a useful one. It does not exist, but it can be visualized in the mind and treated in drawings as a substitute for something that does exist. In this section, vectors will substitute for (1) moiré fringes, (2) grating lines, and (3) angles of emergence from the specimen grating. These quantities are intimately related to each other and to the deformation of the specimen. Relationships between the state of specimen deformation, the light emerging from the specimen, and the moiré interferometry fringes will be described quantitatively in terms of vectors.

Why is it useful to do this? Vectors do *not* provide a superior means to calculate strains from the fringe pattern. Instead, they help the experimentalist perceive details of the optical phenomenon, a knowledge she or he might use to apply the techniques more effectively and to extend the boundaries of the techniques. Vector representation was used in the development and explana-

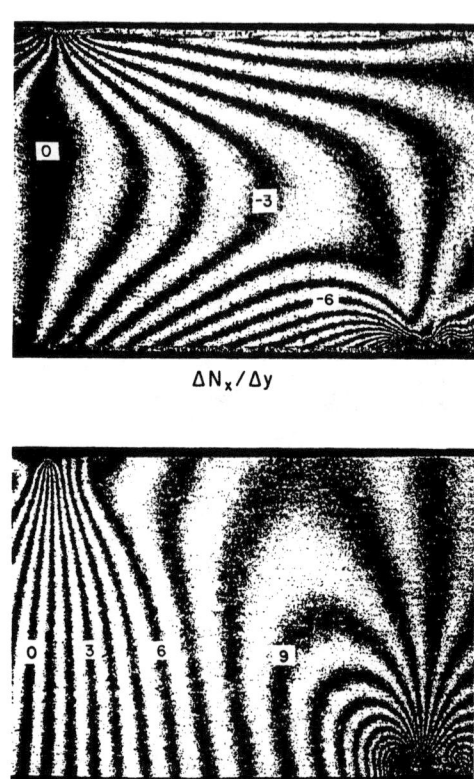

Figure 7-38 Patterns approximating cross-derivatives $\partial N_x/\partial y$ and $\partial N_y/\partial x$. $\Delta x = \Delta y = 0.25$ mm.

tion of the extension introduced in Section 7-19 [44]. Section 7-20 introduces another extension, one that relies on a knowledge of the directions of emergent light for its development and explanation [31].

Fringe Vectors

When one views an interference pattern on the surface of an object, the spatial frequency of the fringes (i.e., the fringe gradient) surrounding any point 0 is a vector quantity **F** [4,6,14,16]. The local frequency of the fringes, fringes per millimeter, is the magnitude of the vector; the perpendicular to the fringes is its line of action; the direction of increasing fringe order is its positive direction; and 0 is its point of application. Figure 7-40 illustrates vector properties of fringes near point 0. If N_x increases as shown, fringe vector $\mathbf{F}_x$ has the direction shown. Its magnitude is proportional to the fringe gradient, F_x, or the inverse of the local pitch of fringes in the N_x field.

Fringes parallel to the lines of the reference grating are always fringes of extension. They result from the difference between the frequency of the reference grating and (twice) the frequency of

MOIRÉ INTERFEROMETRY

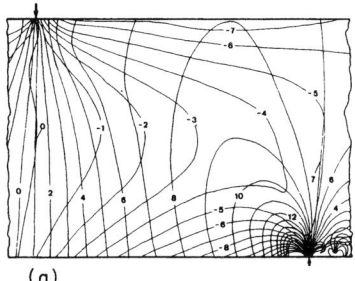

(a)

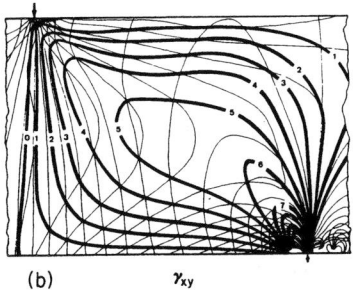

(b) γ_{xy}

Figure 7-39 (a) Centerlines of cross-derivative fringes of Fig. 7-38; (b) contours of the sum of cross-derivatives, $\partial N_x/\partial y + \partial N_y/\partial x$ (approximate).

the specimen grating. Their gradient is represented by the x component of $\mathbf{F}_x$, designed F_{xx}. Physically, the local pitch of the fringes in the x direction is $1/F_{xx}$.

Fringes perpendicular to the lines of the reference grating are fringes of rotation. They result from the angular difference between the lines of the reference and specimen gratings. Rigorously, fringes of rotation are perpendicular to the bisector of the angle between reference and specimen grating lines, but this is practically the y direction for the small deformations usually encountered in high-sensitivity moiré interferometry. Accordingly, fringes of rotation are represented by the y component of $\mathbf{F}_x$ and are designated by F_{xy}. The local pitch of fringes in the y direction is $1/F_{xy}$.

It is often convenient to interpret fringe patterns in terms of these fringe vector components. Strains are

$$\varepsilon_{xx} = \frac{1}{f} F_{xx} \qquad \varepsilon_{yy} = \frac{1}{f} F_{yy} \tag{7-29}$$

$$\varepsilon_{xy} = \frac{1}{f} (F_{xy} + F_{yx}) \tag{7-30}$$

if the initial fields were null or negligible compared to the load-induced fringes. Otherwise, the

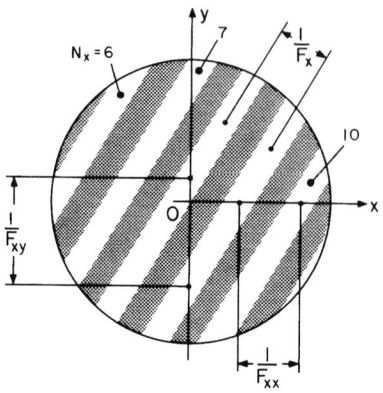

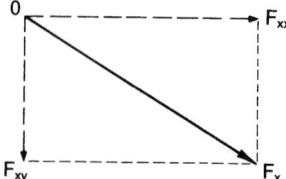

Figure 7-40 Vector $\mathbf{F}_x$ is a fringe vector; it depicts the direction and spatial gradient of fringe orders N_x near point O. Its components $\mathbf{F}_{xx}$ and $\mathbf{F}_{xy}$ represent the gradients of the extensional and rotational parts of u_x displacements at point O.

fringe vectors must be the differences taken from the corresponding load and no-load fields. To determine strains, fringe gradient data in x and y directions should be extracted from the fringe patterns.

Grating Vectors and Fringe Vectors

The kinship between gratings and interference fringes extends to their vector properties. The local spatial frequency (lines/mm) of any grating is a vector quantity, just like the local spatial frequency of a fringe pattern. *Grating vector* $\mathcal{F}$ represents the grating in a local region surrounding point 0. Vector $\mathcal{F}$ originates at point 0; it lies in the grating plane (i.e., the xy plane) perpendicular to the grating lines; its positive direction is the direction of increasing grating line index numbers (see Chapter 6 and Ref. 6); and its magnitude is the grating frequency $\mathcal{F}$. Obviously, the grating vector $\mathcal{F}_s$ for the specimen grating varies from point to point on the specimen, in harmony with the specimen deformation.

The spatial frequency of the reference grating surrounding point 0 is also a vector quantity, designated by $\mathbf{f}$. It originates at point 0 and lies perpendicular to the lines of the reference grating (or to the fringes comprising the grating if it is a virtual reference grating); its magnitude is the spatial frequency f. In the case of the reference grating, the vector is a constant, fixed in direction and magnitude.

For geometrical moiré, Rogers [16] explains a simple vectorial relationship between the vectors.

MOIRÉ INTERFEROMETRY

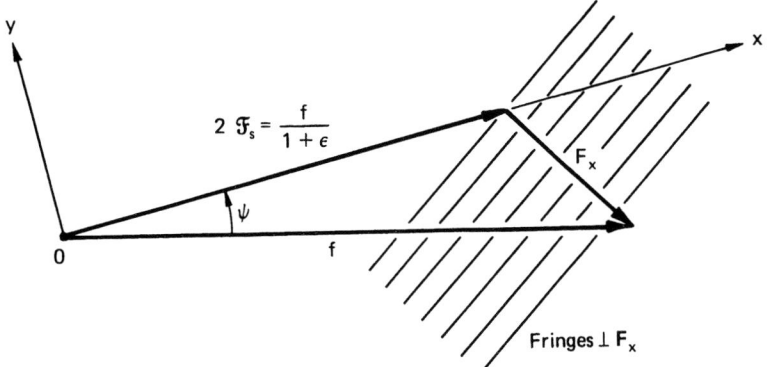

Figure 7-41 Vector diagram, in which the fringe vector $\mathbf{F}_x$ is the difference of the reference grating vector and twice the specimen grating vector. Here the x, y coordinate axes are assumed to rotate together with the specimen grating at point O.

It applies to moiré interferometry, too, since the governing equations are identical [55]. For moiré interferometry, the relationship is

$$\mathbf{F}_x = \mathbf{f} - 2\mathbf{F}_s \tag{7-31}$$

where the gratings produce the N_x fringe pattern. A similar relationship applies for $\mathbf{F}_y$. Accordingly, the fringe vector that defines the fringe gradient at any point 0 in the pattern is the difference of the reference grating vector and twice the specimen grating vector. The factor of 2 must be introduced because the initial frequency of the specimen grating—for a null field—is $f/2$. The order of subtraction is chosen such that extension of the specimen grating corresponds to a tensile strain [Eq. (7-22)].

The relationship is illustrated in Fig. 7-41 by a vector diagram in which the specimen grating rotation ψ and its extension ε are both greatly exaggerated. The fringe gradient surrounding point 0 is represented by $\mathbf{F}_x$, where the fringes lie perpendicular to the vector. It is clear that f, ψ, and ε fully define the local fringe gradients in the moiré pattern.

Local strains are intimately tied to the x and y components of the fringe vectors by Eq. (7-29) and (7-30). In Fig. 7-41, the x and y components of F_x provide the magnitudes of F_{xx} and F_{xy}, while a similar diagram for the N_y fringe pattern provides the remaining components.

Emergence Vectors

Another vector quantity can be defined, one that is very useful to help visualize the formation of fringe patterns. Figure 7-42 shows the directions (exaggerated) of emergent rays A and B when the grating is stretched in the x direction. They are diffracted by the specimen grating from incoming rays A' and B', respectively. Before the specimen was stretched, A and B were coincident with the z axis; the beams they represent had no angular separation and their mutual interference generated a null field. After stretching, the frequency of the specimen grating is decreased and angles of diffractions (with respect to the zero-order diffractions) decrease. Angles β_A, β_B, and β become finite. The lens reconverges beams represented by rays A and B, which intersect at a finite angle in the film plane.

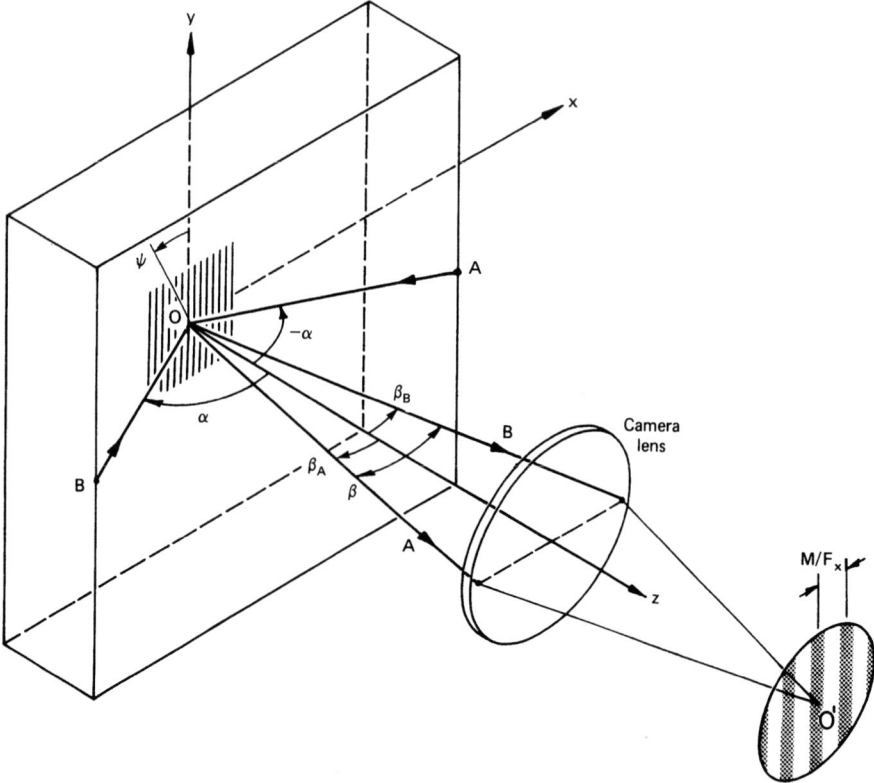

Figure 7-42 Emergent rays A and B when the specimen grating is stretched in the x direction. Angles β_A and β_B are greatly exaggerated.

The direction of emergent ray A in Fig. 7-42 depends on the disposition of the grating near point O. For specimen grating extension, emergence angle β_A lies in the horizontal (x-z) plane as shown; for small in-plane rotation ψ, β_A lies in the vertical plane; for small out-of-plane rotation of the specimen grating about the y axis, β_A lies in the horizontal plane; and for rotation about the x axis, β_A lies in the vertical plane. The same argument applies for emergent ray B.

We are keenly interested in these angles, since they determine the nature of the fringe pattern surrounding point O (or O'). The fringe gradient at O' in Fig. 7-42 (for $M = 1$) depends on the relative angular spread β between rays A and B by Eq. (7-11), which can be written in the notation of Fig. 7-42 as

$$F_x = \frac{2}{\lambda} \sin \frac{\beta}{2} \qquad (7\text{-}11a)$$

where $\beta = \beta_B - \beta_A$. The direction of fringes at O' depends on the directions of rays A and B; specifically, the fringes lie perpendicular to plane AOB.

We can represent the directions of rays A and B by vectors, such that their difference fully defines angle β and fringe gradient F_x. Let vector $\boldsymbol{\beta}_A^*$ be the projection on the xy plane of a unit vector lying in the direction of ray A; let $\boldsymbol{\beta}_B^*$ be the corresponding projection for ray B. Vectors $\boldsymbol{\beta}_A^*$ and $\boldsymbol{\beta}_B^*$ are called emergence vectors (Section 7-4) for rays A and B, respectively [55].

How does $\boldsymbol{\beta}_A^*$ and $\boldsymbol{\beta}_B^*$ respond to deformation of the specimen grating? The answer is remarkably simple. Vector $\boldsymbol{\beta}_A^*$ is the vector sum of its components:

$$\boldsymbol{\beta}_A^* = [\boldsymbol{\beta}_{\varepsilon xx}^* + \boldsymbol{\beta}_{\varepsilon xy}^* + \boldsymbol{\beta}_{\psi_R}^* + \boldsymbol{\beta}_{\psi_x}^* + \boldsymbol{\beta}_{\psi_y}^*]_A \tag{7-32}$$

where the terms on the right represent the fringe vector components caused by extension ε_{xx} (or normal strain), shear strain ε_{xy}, in-plane, rigid-body rotation ψ_R, and out-of-plane rotations Ψ_x and Ψ_y of the specimen grating, respectively—all for ray A. An equivalent expression applies for ray B. Table 7-2 gives the direction and magnitude of each of these emergence vector components. Thus $\boldsymbol{\beta}_A^*$ and $\boldsymbol{\beta}_B^*$ can be determined easily for any state of specimen grating deformation. Since these emergence vectors are projections of unit vectors in the directions of rays A and B, they establish the directions of rays A and B. Accordingly, the influence of specimen deformation on the rays that form the moiré pattern is determined.

Vector $\boldsymbol{\beta}^*$ is defined as the difference of emergence vectors,

$$\boldsymbol{\beta}^* = \boldsymbol{\beta}_B^* - \boldsymbol{\beta}_A^* \tag{7-33}$$

The contribution to $\boldsymbol{\beta}^*$ of every part of the specimen deformation is also given in Table 7-2. The vector sum $\boldsymbol{\beta}^*$ of all the nonzero contributions is represented in Fig. 7-43. Note that $\boldsymbol{\beta}_{\psi_y}^*$ is usually zero or negligible.

From Fig. 7-42 and the definition of the emergence vector, it can be shown [55] that

$$\frac{|\boldsymbol{\beta}^*|}{2} = \sin\frac{\beta}{2} \tag{7-34}$$

and therefore Eq. (7-11a) can be written in its vector form

$$\mathbf{F}_x = \frac{\boldsymbol{\beta}^*}{\lambda} \tag{7-35}$$

Figure 7-43 also illustrates $\mathbf{F}_x$ and the corresponding local fringe pattern: the x and y components of the fringe vector are obtained by dividing the corresponding emergence vector magnitudes listed in Table 7-2 by λ. They are

$$F_{xx} = f\varepsilon_{xx} \tag{7-21a, 7-29a}$$

and

$$F_{xy} = f\left(\psi_R - \frac{\varepsilon_{xy}}{2}\right) \tag{7-36}$$

Since the total grating rotation $\psi = \psi_R - (\varepsilon_{xy}/2)$,

$$F_{xy} = f\psi \tag{7-21b}$$

A similar analysis can be conducted for the emergence vectors for the N_y fringe pattern. In this case, components $\beta_{\varepsilon yy}^*$, $\beta_{\psi yx}^*$, and $\beta_{\psi_R}^*$ caused by positive strains and rotation, lie in the $+y$, and $+x$, and $+x$ directions, respectively.

The analysis is restricted to small strains and rotations. This is reasonable, as we might note from numerical considerations. If angle β (Fig. 7-42) becomes as large as 1°, the resultant fringe

Table 7-2 Emergence Vectors for the N_x Fringe Pattern

State of Deformation of Specimen Grating	Emergence Vector				Difference of Emergence Vectors $(\beta^* = \beta_B^* - \beta_A^*)$		Sum of Emergence Vectors $(\beta^{**} = \beta_A^* + \beta_B^*)$	
	Ray A		Ray B					
	Symbol, Direction	Magnitude	Symbol, Direction	Magnitude	Symbol, Direction	Magnitude	Symbol, Direction	Magnitude
Normal strain, ε_{xx}	$\boldsymbol{\beta}^*_{\varepsilon xxA}$ ←	$\dfrac{\varepsilon_{xx} f\lambda}{2}$	$\boldsymbol{\beta}^*_{\varepsilon xxB}$ →	$\dfrac{\varepsilon_{xx} f\lambda}{2}$	$\boldsymbol{\beta}^*_{\varepsilon xx}$ →	$\varepsilon_{xx} f\lambda$	$\boldsymbol{\beta}^{**}_{\varepsilon xx}$	0
Shear strain, ε_{xy}	$\boldsymbol{\beta}^*_{\varepsilon xyA}$ ↓	$\dfrac{\varepsilon_{xy} f\lambda}{2}$	$\boldsymbol{\beta}^*_{\varepsilon xyB}$ ↑	$\dfrac{\varepsilon_{xy} f\lambda}{2}$	$\boldsymbol{\beta}^*_{\varepsilon xy}$ ↑	$\varepsilon_{xy} f\lambda$	$\boldsymbol{\beta}^{**}_{\varepsilon xy}$	0
Normal strain, ε_{yy}		0		0	$\boldsymbol{\beta}^*_{\varepsilon yy}$	0	$\boldsymbol{\beta}^{**}_{\varepsilon yy}$	0
Rigid-body rotation, in-plane ψ_R	$\boldsymbol{\beta}^*_{\psi RA}$ ↑	$\dfrac{\psi_R f\lambda}{2}$	$\boldsymbol{\beta}^*_{\psi RB}$ ↓	$\dfrac{\psi_R f\lambda}{2}$	$\boldsymbol{\beta}^*_{\psi R}$ ↓	$\psi_R f\lambda$	$\boldsymbol{\beta}^{**}_{\psi R}$	0
Rigid-body rotation, out-of-plane, Ψ_x	$\boldsymbol{\beta}^*_{\Psi xA}$ ↑	$2\Psi_x$	$\boldsymbol{\beta}^*_{\Psi xB}$ ↑	$2\Psi_x$	$\boldsymbol{\beta}^*_{\Psi x}$	0	$\boldsymbol{\beta}^{**}_{\Psi x}$ ↑	$4\Psi_x$
Rigid-body rotation, out-of-plane, Ψ_y	$\boldsymbol{\beta}^*_{\Psi yA}$ →	$2\Psi_y + \dfrac{\Psi_y^2 f\lambda}{4}$	$\boldsymbol{\beta}^*_{\Psi yB}$ →	$2\Psi_y - \dfrac{\Psi_y^2 f\lambda}{4}$	$\boldsymbol{\beta}^*_{\Psi y}$	$-\dfrac{\Psi_y^2 f\lambda}{2}$	$\boldsymbol{\beta}^{**}_{\Psi y}$ →	$4\Psi_y$
Rigid-body translation, u_x, u_y, u_z		0		0	$\boldsymbol{\beta}^*_u$	0	$\boldsymbol{\beta}^{**}_u$	0

MOIRÉ INTERFEROMETRY

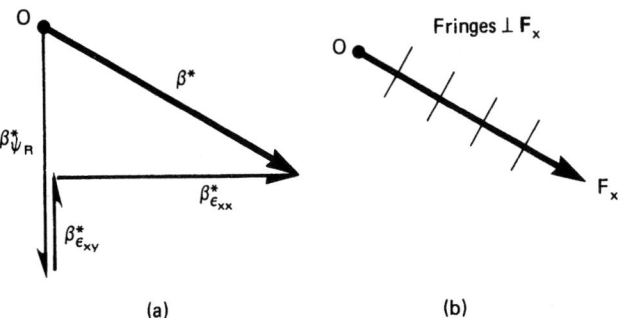

Figure 7-43 (a) Vector diagram showing the components and resultant of β^*, the difference of emergence vectors for rays A and B; (b) the corresponding fringe vector and fringe direction.

pattern (for $\lambda = 633$ nm and $M = 1$) has a gradient of 28 fringes/mm (700 fringes/in.), which is larger than that usually encountered in practice; the corresponding normal strain (for $\psi = 0$ and $f = 2400$ lines/mm) is only $\varepsilon_{xx} = 0.012$, or 1.2% strain. The general case is treated in Ref. 55.

7-19 ±45° Gratings

The optical configuration illustrated schematically in Fig. 7-44 was analyzed by means of emergence vectors [44]. The crossed-line specimen grating is produced on the specimen with its lines oriented 45° to the x and y axes. Illumination is by incoming coherent, collimated beams in directions KO, LO, and MO. Light from L is diffracted in the $+1$ diffraction order of the Y' specimen grating (i.e., the grating with lines perpendicular to y') and emerges in the direction OQ. Light from M is diffracted in the -1 order of grating X' to emerge in (essentially) the same direction. These two emergent beams produce a fringe pattern viewed by a camera.

The result is remarkable! When the specimen is deformed, the pattern is the u_x displacement field—the same pattern as normally observed, with specimen and reference gratings oriented perpendicular to the x axis. Similarly, the pattern of the u_y displacement field is generated by illuminating the $\pm 45°$ specimen gratings with coherent light from K and L.

The sensitivity—$1/f$ mm per fringe order—depends on the frequency of the virtual reference grating (formed by beams from K and L or L and M) by Eqs. (7-18) and (7-19), just as with X, Y gratings. Frequency of the X' and Y' specimen gratings must be $f/\sqrt{2}$ instead of $f/2$. Implementation of this scheme has merit for three-beam optical arrangement vis-à-vis four-beam arrangements like that of Fig. 7-21.

7-20 Determination of u_z

Moiré interferometry has been discussed in terms of extracting the in-plane displacements u_x and u_y. It is clear from the discussion of emergence vectors that the two emergent beams also carry information on out-of-plane displacements u_z. How can we extract u_z independently?

A contour map of out-of-plane displacements u_z would define the local gradients of displace-

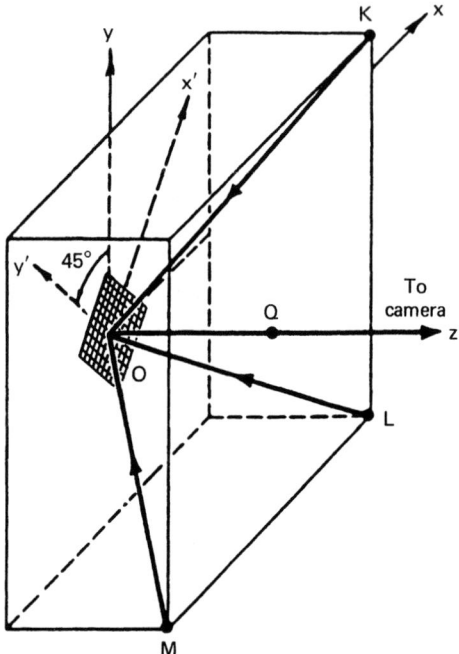

Figure 7-44 Specimen gratings oriented 45° to the x and y axes produces fringe patterns of N_x and N_y.

ments $\partial u_z/\partial x$ and $\partial u_z/\partial y$ everywhere in the field. These gradients are slopes Ψ_x and Ψ_y, respectively. From Table 7-2, Ψ_x and Ψ_y are proportional to (the x and y components of) the vector sum $\mathbf{B}^{**}$ of emergence vectors, where

$$\boldsymbol{\beta}^{**} = \boldsymbol{\beta}_A^* + \boldsymbol{\beta}_B^* \tag{7-37}$$

It follows that a fringe pattern in which the fringe gradients are everywhere proportional to $\mathbf{B}^{**}$ is a contour map of out-of-plane displacements u_z.

What is required is the pattern obtained when beams A and B are combined in such a way that their emergence vectors are added. Stated in another way, what is required is the pattern (or contour map) of the sum of wavefront warpages of beams A and B. A direct comparison of the wavefronts gives their difference, which is related uniquely to in-plane displacements. An indirect scheme that yields their sum is needed to produce the u_z pattern.

The optical arrangement of Fig. 7-45 can be used. An auxiliary partial mirror is added to the arrangement of Fig. 7-20 [31]. This produces a third beam, beam O, along the z axis. A carrier pattern of extension or rotation is introduced to cause an angular separation between emergent beams A and B. With the specimen in the no-load condition, beams A, O, and B pass through points in the camera aperture plane marked A, O, and B. Let the aperture be moved to block B, but admit A and O. These two beams, with different emergent directions, produce a carrier pattern which is photographed in the camera. Let the aperture be moved to admit beams O and B only, and photograph this carrier pattern on the same film for a double exposure.

What is the result? Assuming perfect optics and perfect alignment, the two superimposed carrier

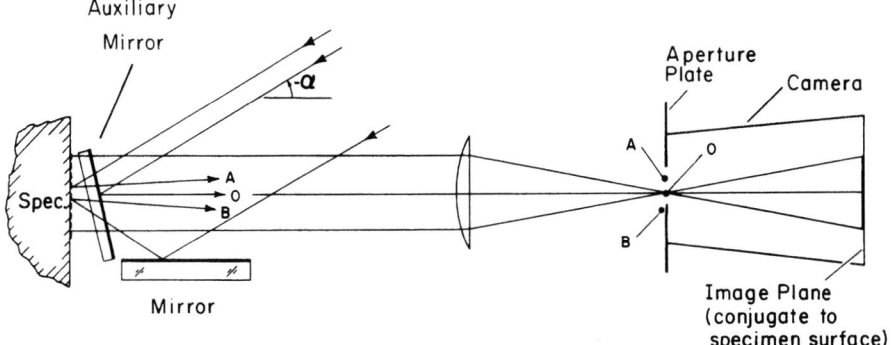

Figure 7-45 Optical system to produce fringe patterns of out-of-plane displacement u_z.

patterns recorded in the double exposure have the same frequency and same direction. They combine by geometrical moiré to give a null field. However, the carrier patterns have opposite signs. The angle between O and A is opposite to the angle between O and B, and the respective emergence vectors of A and B have opposite directions. Their sum is zero, and the null field of this supermoiré pattern represents the sum, or the additive moiré parameter (see Chapter 6).

When this double-exposure procedure is followed with the specimen in the loaded condition, beams A and B have warped wavefronts that carry both the in-plane and out-of-plane information. The supermoiré pattern formed in this double-exposure procedure, however, is the additive parameter; it depicts the out-of-plane displacements u_z. Optical filtering can be used to enhance its contrast. Out-of-plane displacements are extracted independently!

The optics and alignment will, more likely, not be perfect. Then the procedure can be conducted for no-load and load conditions and the respective films superimposed to yield the load-induced u_z displacement field. If this procedure is used, beam O should be displaced a bit from midway between A and B to create extra fringes in the no-load pattern and to improve visibility of the final load-induced moiré pattern. We will denote the fringes of this load-induced pattern by the symbol N^+.

Out-of-plane displacements are calculated by

$$u_z = \frac{\lambda}{2(1 + \cos \alpha)} N^+ \tag{7-38}$$

where α is the angle of incidence in Fig. 7-45. For the case of a virtual reference grating frequency of $f = 2400$ lines/mm (60,960 lines/in.) and light of wavelength $\lambda = 632.8$ nm (24.9 μin.), the angle of incidence is [by Eq. (7-18)] $\alpha = 49.4°$. The sensitivity for out-of-plane displacement measurements is 0.192 μm/fringe order (7.55 μin./fringe order), or 0.303λ/fringe order. This is about 67% greater than the $\lambda/2$ sensitivity of most interferometric methods for out-of-plane measurements, including holographic interferometry. Sensitivity approaches its limit, $\lambda/4$ per fringe order, as α is reduced.

In another version of this method, light from the three beams—A, O, and B in Fig. 7-45—is recorded simultaneously [31]. Three carrier patterns containing wavefront information can be separated by optical filtering to extract contour maps of in-plane displacements u_x (or u_y) and out-of-plane displacements u_z. It has the potential for simultaneous recording of transient in-plane and out-of-plane deformations.

7-21 Variations of Moiré Interferometry

Transmission Systems

The optical arrangements described here were reflection systems, suitable for opaque specimens. Each has a transmission counterpart, however, that can be used with transparent specimens. Referring to Fig. 7-15b, it is easy to visualize these changes: (1) the specimen is transparent; (2) the specimen grating is the same type of transparent phase grating as before, but without the reflective coating of aluminum or gold; and (3) the emergent beams and the camera are mirror images of those shown in the figure: beams with wavefronts A' and B' (or A'' and B'') pass through the specimen and enter the camera on the left side of the specimen.

With such an arrangement, the action is the same as for a reflection system, including diffraction by the specimen grating in $+1$ and -1 orders, formation of warped wavefronts from specimen deformation, and recording of their interference pattern in the camera. Transmission systems can be easier to construct and use because the optical elements are not as close together. Although attractive for special problems, they are limited to studies where the specimens are transparent and (usually) two-dimensional bodies.

Moiré-Fringe Multiplication

The sensitivity of moiré measurements depends on the frequency of the reference grating. At one stage of its evolution, high-frequency reference gratings were practical, but specimen gratings were restricted to low frequencies. They could be used together, and the ratio of reference grating frequency to specimen grating frequency was called the fringe multiplication factor [9–11].

The optical arrangements described here use a fringe multiplication factor of 2. Consider Fig. 7-15b again, but now let the initial frequency of the specimen grating be half that illustrated in the figure, or $\mathcal{F} = f/4$. Incoming beam A and the zeroth diffraction order would have the same directions shown. Twice as many diffraction orders would appear in the upper-left part of the figure, since the grating frequency was halved. From beam A, light of the $+2$ diffraction order would enter the camera; from B, light of the -2 order would enter the camera, combining to produce the same result as that obtained with the configuration shown in Fig. 7-15b. This is because the wavefront warpage introduced by a deformed specimen grating is proportional to the product of grating frequency and diffraction order.

The same scheme applies for any even factor. Moiré fringe multiplication by a factor of 60 has been reported in the literature [9], and higher factors have been demonstrated in the classroom. This requires high-quality specimen gratings, since the signal-to-noise ratio decreases as diffraction order increases. The lowest optical noise is achieved in the $+1$ or -1 diffraction orders, and these orders were chosen as the techniques evolved toward the present method of moiré interferometry. Only the transmission mode of moiré fringe multiplication is reported in the early literature, but the reflection mode is now an obvious counterpart.

Replication of Deformed Specimen Gratings

Techniques were introduced whereby a deformed specimen grating on a loaded structure was replicated and the replica was brought to the laboratory for analysis [17]. There it was installed in an optical system to record the moiré pattern. While the other methods described here are live, or real-time methods, the replication approach has attractive merit for special circumstances. This method has been used with low-frequency specimen gratings, fringe multiplication by high factors, and transmission systems. It can be used, of course, with high-frequency specimen gratings and either transmission or reflection optical systems.

Multiplication by One

An interesting method akin to those already described uses the same virtual reference grating for two separate purposes. First, it is used to print a specimen grating in a photosensitive coating on the specimen surface. Then it uses the same virtual grating as a reference grating [12, 13]. If the specimen surface is partially matte, light carrying the interference pattern can be observed from a convenient direction. However, speckle noise from the matte surface reduces fringe visibility. The method has potential for analysis of curved surfaces as well as flat surfaces.

7-22 Data Processing

A wealth of literature has emerged on computer-aided data processing of fringe patterns. Most of the systems contemplate relatively few fringes in the field of view and use sophisticated fringe interpolation, curve fitting, and differentiation procedures to wring out the best accuracy from available data.

Typically, moiré interferometry produces many fringes and an abundance of data. Carrier patterns (of low or modest frequencies) can easily be added to simplify fringe counting by eliminating fringe loops and thus make the fringe count monotonic. Consequently, computer-aided analysis would be much less sophisticated. Displacements can be graphed directly from the positions of integral fringe orders. Fringe derivatives and strains can be graphed directly from the distance between adjacent integral-order fringes (i.e., from $1/F_{xx}$ in Fig. 7-40) when fringes are closely spaced. Smoothing can be performed as the last step of the process, thus introducing minimal bias in treating random errors. With an abundance of fringes in the field of view, data processing is relatively easy and reliable.

Although whole-field visualization of deformations provides an extremely valuable guide, the engineer is usually interested in detailed data reduction for local zones only. Semiautomated, interactive schemes to determine the coordinates of fringes in selected zones would be sufficient in most instances.

7-23 Prospectus

Moiré interferometry is a whole-field quantitative optical method to determine in-plane displacement fields of opaque bodies. Strain fields are obtained indirectly. This real-time method of experimental mechanics is characterized by a remarkable set of qualities:

High sensitivity
Excellent fringe contrast
High spatial resolution
Extensive range
Pattern location coincident with specimen

Sensitivity corresponding to moiré with 2400 lines/mm (60,960 lines/in.) is readily achieved. This is a subwavelength sensitivity of 0.417 μm (16.4μin.) per fringe order. Resolution of 1/5 fringe order or 0.08 μm (3 μin.) is obtained by manual interpolation.

Specimen gratings are crossed-line, phase-type diffraction gratings applied by a replication technique. They are about 0.025 mm (0.001 in.) thick. The replication yields reflective gratings with specular (mirrorlike) qualities. They produce smooth wavefronts, which is the reason for low speckle noise and excellent fringe contrast.

The abundance of load-induced fringes makes strain determinations relatively easy and reliable. Graphical differentiation of displacement data or computer-aided analysis can be used. Whole-field mechanical differentiation is a practical alternative; it was demonstrated for normal and shear-strain determinations using a mechanical shifting distance of only 0.25 mm (0.010 in.).

What about the future? Very likely, we will see increasingly broad applications. This includes measurements for engineering design and verification; measurements for research analyses in solid mechanics, including two-body problems; observations and measurements for discovery—discovery of anomalous or unexpected behavior of materials, assemblies, and components. Hybrid experimental-computational analyses will proliferate.

The techniques will continue to evolve. We probably will see optical systems with tiny but powerful solid-state light sources. Portable systems well suited to mechanical laboratory and field environments will arrive. The real costs of experimental analyses will decline.

The author is an optimist on issues of technical progress: I predict that moiré interferometry will contribute superb and exciting advances beyond our current insight.

Acknowledgments

Much of the developmental work reported here on moiré interferometry techniques was sponsored by the National Science Foundation under grants ENG-7824609 and MEA-81092230. Applications illustrated in Figs. 7-13, 7-14, 7-17, and 7-23 were sponsored by NASA Langley Research Center under grants NAG-1-227, NAG-1-359, and NAG-1-291, respectively. This sponsorship and generous help of coworkers in the Department of Engineering Science and Mechanics of this university is gratefully acknowledged. I extend special acknowledgments and thanks to students and former students in photomechanics, who coauthored publications on this work and are largely responsible for advances described in this chapter, notably D. E. Bowles, W. A. Baracat, M. L. Basehore, E. M. Weissman, R. B. Watson, E. W. Brooks, G. Nicoletto, J. S. Epstein, S. J. Lubowinski, A. L. Highsmith, R. Czarnek, J. D. Wood, and D. Joh. I extend special thanks to Professor V. J. Parks for extensive and very helpful critiques of the chapter.

List of Symbols

	A	amplitude of light wave
	$\mathbf{A}$	unit vector defining the direction of an incident beam
	c	frequency of carrier pattern
	C	velocity of light in vacuum
	$\mathbf{D}_m$	unit vector defining the direction of the diffracted beam of order m
	$f, \mathbf{f}$	frequency of reference grating and its vector
	$F, \mathbf{F}$	fringe gradient and its vector
	F_x, F_y	fringe gradient in the N_x, N_y pattern, respectively
$F_{xx}, F_{xy}, F_{yy}, F_{yx}$		fringe gradient components $\partial N_x/\partial x$, $\partial N_x/\partial y$, $\partial N_y/\partial y$, $\partial N_y/\partial x$, respectively
	$\mathscr{F}, \mathscr{F}_s, \vec{\mathscr{F}}$	grating frequency, specimen grating frequency, and grating vector
	G	distance between adjacent fringes
	$\mathscr{G}$	grating pitch, or distance between adjacent lines or furrows
	I	intensity of light
	m	diffraction order
	n	index of refraction
	N	fringe order

N_x, N_y	moiré fringe order when lines of the reference grating are perpendicular to the x, y axis, respectively
N^+	fringe order in the out-of-plane displacement pattern
S	distance between wavefronts of equal phase
t	time
u_x, u_y, u_z	displacement components in Cartesian coordinates
W	wavefront
x, y, z	Cartesian coordinates
x', y'	coordinates in $\pm 45°$ directions
X, Y, X', Y'	specimen grating with lines perpendicular to x, y, x', y' axis, respectively
α	angle of incidence with respect to the grating normal
$\alpha^*, \alpha_x^*, \alpha_y^*$	incidence vector and its x, y components
β	angle of diffraction with respect to the grating normal
$\beta_m^*, \beta_{mx}^*, \beta_{my}^*$	emergence vector for the beam of diffraction order m, and its x, y components
β^*	the difference of emergence vectors for rays A and B
β^{**}	the sum of emergence vectors for rays A and B
Δ	change of; increment of
$\varepsilon_{xx}, \varepsilon_{yy}$	normal-strain component in Cartesian coordinates
ε_{xy}	shear-strain component in Cartesian coordinates
θ	angle of beam; angle between two beams
θ	angle of wedge prism
λ	wavelength of light
ϕ	phase angle of light wave
ψ	angle of in-plane rotation of specimen grating
ψ_R	rigid-body part of rotation ψ
Ψ_x, Ψ_y	angle of out-of-plane rotation about the x, y axis, respectively
ω	cyclic frequency of light waves
0, 1, 2	subscripts signifying beams or wavefronts; subscripts signifying diffraction orders
0, 1	subscripts signifying uniform and nonuniform part, respectively, of the strain field

References

1. J. Guild, *The Interference System of Crossed Diffraction Gratings*, Clarendon Press, Oxford, 1956.
2. J. W. Dally and W. F. Riley, *Experimental Stress Analysis*, 1st ed., McGraw-Hill, New York, 1965, 360.
3. C. A. Sciammarella, The Moiré Method, a Review, *Exp. Mech.* 22, no. 11 (1982), 418–433.
4. A. J. Durelli and V. J. Parks, *Moiré Analysis of Strain*, Prentice-Hall, Englewood Cliffs, NJ, 1970.
5. P. S. Theocaris, *Moiré Fringes in Strain Analysis*, Pergamon Press, Elmsford, NY, 1969.
6. C. A. Sciammarella and A. J. Durelli, Moiré Fringes as a Means of Analyzing Strains, *Proc. ASCE, J. Eng. Mech. Div.*, 87 (EM1), (1961), 51–74.
7. D. Post, The Moiré Grid-Analyzer Method for Strain Analysis, *Exp. Mech.*, 5, no. 11 (1965), 368–377.
8. A. J. Durelli and V. J. Parks, Moiré Fringes as Parametric Curves, *Exp. Mech.*, 7, no. 3 (1967), 97.
9. D. Post, Moiré Fringe Multiplication with a Non-symmetrical Double-Blazed Reference Grating, *Appl. Opt.*, 10, no. 4 (1971), 901–907.

10. D. Post and T. F. MacLaughlin, Strain Analysis by Moiré Fringe Multiplication, *Exp. Mech., 11*, no. 9 (1971), 408–413.

11. I. M. Daniel, R. E. Rowlands, and D. Post, Strain Analysis of Composites by Moiré Methods, *Exp. Mech., 13*, no. 6 (1973), 246–252.

12. N. Wadsworth, M. Marchant, and B. Billing, Real-Time Observation of In-Plane Displacements of Opaque Surfaces, *Opt. Laser Technol., 5*, no. 3 (June 1973), 119–123.

13. M. Marchant and S. M. Bishop, An Interference Technique for the Measurement of In-Plane Displacements of Opaque Surfaces, *J. Strain Anal., 9*, no. 1 (1974), 36–43.

14. K. A. Stetson, Homogeneous Deformations: Determination by Fringe Vectors in Hologram Interferometry, *Appl. Opt., 14*, no. 9 (1975), 2256–2259.

15. J. B. Abbis, M. J. Marchant, and A. C. Marchant, Recent Applications of Coherent Optics in Aerospace Research, *Opt. Eng., 15*, no. 3 (1976), 202–210.

16. G. L. Rogers, A Geometrical Approach to Moiré Pattern Calculations, *Opt. Acta, 24*, no. 1 (1977), 1–13.

17. J. McKelvie and C. A. Walker, A Practical Multiplied Moiré-Fringe technique, *Exp. Mech., 18*, no. 8 (1978), 316–320.

18. J. McKelvie, D. Pritty, and C. A. Walker, An Automatic Fringe Analysis Interferometer for Rapid Moiré Stress Analysis, *SPIE Proc., 164* (1978), 175–188.

19. R. E. Rowlands and J. H. Vallem, On Replication for Moiré-Fringe Multiplication, *Exp. Mech., 20*, no. 5 (1980), 167.

20. A. J. Durelli and A. Shukla, Identification of Isochromatic Fringes, *Exp. Mech., 23*, (March 1983), 111–119.

21. D. Post, Moiré Interferometry in White Light, *Appl. Opt., 18*, no. 24 (Dec. 15, 1979), 4163–4167.

22. A. McDonach, J. McKelvie, and C. A. Walker, Stress Analysis of Fibrous Composites Using Moiré Interferometry, *Opt. Lasers Eng., 1* (1980), 85–105.

23. D. Post, Optical Interference for Deformation Measurements—Classical, Holographic and Moiré Interferometry, in *Mechanics of Nondestructive Testing* (W. W. Stinchcomb, Ed.), Plenum Press, New York, 1980, 1–53.

24. D. Post and W. A. Baracat, High-Sensitivity Moiré Interferometry—A Simplified Approach, *Exp. Mech., 21*, no. 3 (1981), 100–104.

25. M. Basehore and D. Post, Moiré Method for In-Plane and Out-of-Plane Displacement Measurements, *Exp. Mech., 21*, no. 9 (1981), 312–328.

26. D. F. Bowles, D. Post, C. T. Herakovich, and D. R. Tenney, Moiré Interferometry for Thermal Expansion of Composites, *Exp. Mech., 21*, no. 12 (1981), 441–447.

27. C. W. Smith, D. Post, and G. Nicoletto, Prediction of Subcritical Crack Growth Data from Model Experiments, in *Developments in Theoretical and Applied Mechanics*, Vol. XI, University of Alabama Huntsville, April 1982, 167–179.

28. E. M. Weissman and D. Post, Moiré Interferometry Near the Theoretical Limit, *Appl. Opt., 21*, no. 9 (May 1, 1982), 1621–1623.

29. D. Post, Developments in Moiré Interferometry, *Opt. Eng., 21*, no. 3 (May 1982), 458–467.

30. R. B. Watson and D. Post, Precision Standard by Moiré Interferometry for Strain Gage Calibration, *Exp. Mech., 22*, no. 7 (1982), 256–261.

31. M. L. Basehore and D. Post, Displacement Fields (u, w) Obtained Simultaneously by Moiré Interferometry, *Appl. Opt., 21*, no. 14 (July 15, 1982), 2558–2562.

32. D. Post, High Sensitivity Displacement Measurements by Moiré Interferometry, *Proc. 7th Int. Conf. Exp. Stress Anal., Haifa, Israel*, August 1982, 397–408.
33. E. M. Weissman and D. Post, Full-Field Displacement Rosette by Moiré Interferometry, *Exp. Mech.*, 22, no. 9 (1982), 324–328.
34. M. W. Hyer, C. T. Herakovich, and D. Post, Thermal Deformations of Graphite Epoxy, in *1982 Advances in Aerospace Structures and Materials*, ASME, New York, 1982, 107–114.
35. R. Czarnek, D. Post, and C. T. Herakovich, Edge Effects in Composites by Moiré Interferometry, *Exp. Tech.*, January 1983, 18–21.
36. C. W. Smith, D. Post, G. Hiatt, and G. Nicoletto, Displacement Measurements Around Cracks in Three-Dimensional Problems by a Hybrid Experimental Technique, *Exp. Mech.*, 23, no. 1 (March 1983), 15–20.
37. C. A. Walker, J. McKelvie, and A. McDonach, Experimental Study of Inelastic Strain Patterns in a Model of a Tube–Plate Ligament Using an Interferometric Moiré Technique, *Exp. Mech.*, 23, no. 1 (March 1983), 21–29.
38. A. McDonach, J. McKelvie, P. MacKenzie, and C. A. Walker, Improved Moiré Interferometry and Applications in Fracture Mechanics, Residual Stress and Damaged Composites, *Exp. Tech.*, 23, no. 2 (June 1983), 20–24.
39. D. Post, Moiré Interferometry at VPI & SU, *Exp. Mech.*, 23, no. 2 (June 1983), 203–210.
40. D. Post, Moiré Interferometry for Damage Analysis of Composites, *Exp. Tech.*, 7, no. 7 (1983), 17–20.
41. M. P. Nemeth, C. T. Herakovich, and D. Post, On the Off-Axis Tensile Test for Unidirectional Composites, *Compos. Technol. Rev.*, 5, no. 2 (Summer 1983), 61–68.
42. C. W. Smith, D. Post, and G. Nicoletto, Experimental Stress-Intensity Factors in Three-Dimensional Cracked-Body Problems, *Exp. Mech.*, 23, no. 4 (December 1983), 378–381.
43. E. M. Weissman, D. Post, and A. Asundi, Whole-Field Strain Determination by Moiré Shearing Interferometry, *J. Strain Anal.*, 19, no. 2 (1984), 77–80.
44. R. Czarnek and D. Post, Moiré Interferometry with ±45-Deg Gratings, *Exp. Mech.*, 24, no. 1 (March 1984), 68–74.
45. M. L. Basehore and D. Post, High-Frequency, High-Reflectance Transferable Moiré Gratings, *Exp. Tech.*, 8, no. 5, 29–31 (1984).
46. D. Post, R. Czarnek, and D. Joh, Shear Strain Contours from Moiré Interferometry *Exp. Mech.*, 25, no. 3 (September 1985), 282–287.
47. D. Post, Moiré Interferometry for Deformation and Strain Studies, *Opt. Eng.*, 24, no. 4, (July–August 1985), 663–667.
48. C. W. Smith, D. Post, and J. S. Epstein, Algorithms and Restrictions in the Application of Optical Methods to Stress Intensity Factor Determination, *J. Theor. Appl. Fract. Mech.*, 2, no. 1 (September 1984), 81–89.
49. D. Post, R. Czarnek, and C. W. Smith, Patterns of u and v Displacement Fields Around Cracks in Aluminum by Moiré Interferometry, in *Applications of Fracture Mechanics to Materials and Structure* (G. C. Sih, E. Sommer, and W. Dahl, Eds.), Nijhoff, Hingham, MA, 1984, 699–708.
50. D. Post, R. Czarnek, J. D. Wood, D. Joh, and S. Lubowinski, Deformations and Strains in Adhesive Joints by Moiré Interferometry, NASA Contractor Report 172474, 1984; also D. Post, R. Czarnek, J. D. Wood, and D. Joh, Deformations and Strains in a Thick Adherend Lap Joint, *Adhesively Bonded Joints: Testing, Analysis and Design, ASTM STP 981* (W. S. Johnson, Ed.), American Society for Testing and Materials, Philadelphia, 1988, 107–118.
51. D. Post, R. Czarnek, D. Joh, and J. D. Wood, Deformation Measurements of Composite Multi-Span Beam Shear Specimens by Moiré Interferometry, NASA Contractor Report 3844, 1984.
52. J. D. Wood, Detection of Delamination Onset in a Composite Laminate Using Moiré Interferometry, *Compos. Technol. Rev.*, 7, no. 4 (Winter 1985), 121–128.
53. C. Ruiz, D. Post, and R. Czarnek, Moiré Interferometric Study of Dovetail Joints, *J. Appl. Mech.*, 52, no. 1 (March 1985), 109–114.
54. R. Czarnek and D. Post, Moiré Interferometry with ±45° Gratings Applied to the Dovetail Joint, *Proc. 1985 SEM Spring Conf. Experimental Mechanics*, Las Vegas June 9–14, 1985, 553–559.

55. A. Livnat and D. Post, The Governing Equations for Moiré Interferometry and Their Identity to Equations of Geometrical Moiré, *Exp. Mech. 25,* no. 4 (December 1985), 360–366.

56. D. Post, Moiré Interferometry for Composites, Section IV A-2, in *Manual on Experimental Methods for Mechanical Testing of Composites* (R. L. Pendleton and M. E. Tuttle, Eds.), Society for Experimental Mechanics, Bethel, CT, 1989, 67–80.

57. G. H. Spencer and M. V. R. K. Murty, General Ray-Tracing Procedure, *J. Opt. Soc. Am., 52,* no. 6 (1962), 672–678.

58. M. C. Hutley, *Diffraction Gratings,* Academic press, New York, 1982.

59. S. Matsui, K. Moriwaki, H. Aritome, S. Namba, S. Shin, and S. Suga, X-Ray Diffraction Grating for Synchrotron Radiation Spectroscopy: A New Fabrication Method, *Appl. Opt., 21,* no. 15 (Aug. 1, 1982), 2787–2793.

60. R. Czarnek, High Sensitivity Moiré Interferometry with Compact Achromatic Interferometer, *Optics Lasers Eng., 13* (1990), 100–115.

61. B. J. Chang, R. Alferness, and E. N. Leith, Space-Invarient Achromatic Grating Interferometers: Theory, *Appl. Opt., 14,* no. 7 (July 1975), 1592–1600.

62. E. N. Leith, G. Swanson, and S. Leon, Construction of Diffractive Optical Elements in Noncoherent Light, *SPIE, 503* (1984), 2–8.

CHAPTER

8

Holographic and Laser Speckle Interferometry

W. F. Ranson, M. A. Sutton, and W. H. Peters

University of South Carolina
Columbia, South Carolina

8-0 Introduction

This chapter presents a complete description of the principles of holography and speckle metrology that form the basic concepts in experimental mechanics. Of all the properties of holography and laser speckle, interferometry is quite possibly the most important in experimental mechanics. Both interferometry techniques permit the extension of interferometry type measurements of a diffuse reflecting object. The primary purpose of this chapter is to provide the reader with the basic concepts important to experimental mechanics. Since the initial discovery of holographic interferometry and later speckle metrology, numerous applications to deformable bodies have been presented in the literature, and space does not permit discussion of the numerous applications of these two techniques to experimental mechanics. Therefore, only the fundamentals are discussed, together with some experimental examples to illustrate the application of the theory of fringe formation. Insofar as possible, the formation of fringes related to surface deformation measurements is discussed independent of an orthogonal basis. Thus an attempt is made to present the equations in a general form. The particular experimental configuration for both holography and speckle metrology will dictate the measurement of a set of particular displacement components. These simple illustrative examples therefore demonstrate theoretical cases, and the interested reader should refer to the literature on experimental mechanics for contemporary applications.

Significant applications in holographic interferometry have been made since the development of the laser, Gabor's [1] original work, and the introduction of an off-axis reference beam by Leith and Upatnieks [2]. The concept of holographic interferometry applied to static problems has been more successful than dynamic problems because of the practical use of the helium–neon (He-Ne) laser. Most of the practices have been to measure surface displacement components of deformed bodies. Haines and Hildebrand [3] first developed a method to determine the surface deformation of solids. Ennos [4], Aleksandrov and Bonch-Bruevich [5], Solid [6], and Stroke et al. [7] extended the work of Haines and Hildebrand. The earlier important contributions included shearing interferometry, originally developed by Bryngdahl [8] and advanced by Saito et al. [9].

The first pulsed ruby holograms were obtained by Brooks et al. [10,11] and Jacobson and

McClung [12]. Earlier investigations of stress wave propagation included the works of Gottenburg [13] and Evensen and Aprahamian [14]. Applications of longitudinal impact of slender rods and transverse impact of flat plates were more successful in dynamic holographic interferometry. Powell and Stetson [15] originally developed time-average holography to measure resonant mode shapes and frequencies of transverse vibrations. This holographic technique has proven to be extremely useful, and probably the most significant engineering application is in the turbine engine industry, where mode shapes of turbine blades must be known to ensure that dangerous resonant conditions do not occur at operating engine speeds.

The history of specklelike phenomena predated the laser, although the laser was primarily responsible for the development of engineering applications of speckle interferometry. This technique has been utilized as an effective measure of surface deformation measurements in experimental mechanics. Leendertz in 1970 [16] established the foundation principles of speckle interferometry, proclaiming a sensitivity approximating holographic interferometry. Other investigators [17,18] contributed significantly to the early development of this technique. In this method, an object is illuminated by two beams that optically interfere. Surface displacement information is obtained by comparing the speckle patterns photographed before and after the deformation process. Archbold and Ennos [19] and later Duffy [20] proposed a different technique best described by the term "speckle photography." This method involves an illumination with a single divergent laser beam and recording double-exposure photographs of the speckle patterns.

Stetson [21] and later Hung [22] have presented thorough reviews of speckle methods to measure in-plane motions. Stetson divided the methods into speckle photography and speckle interferometry. Stetson defined speckle photography as the illumination of an object with a single divergent laser beam and photographing the response before and after the object is displaced. Speckle interferometry methods are defined as those which have an optical setup that corresponds to some type of interferometer. Therefore, there are two beams illuminating the object, and these beams optically interfere. In speckle interferometry the response must still be viewed or recorded both before and after the object is displaced.

Computer applications to laser speckle photography first appeared as methods to locate the minumum value of the fringes in order to calculate fringe spacing for pointwise filtering. Chambless and Broadway [23] demonstrated that digital filters were effective for the enhancement of the fringe patterns in speckle photography. The application of filters of this type is inexpensive and highly versatile when applied to noisy data such as speckle patterns. The term filter (or linear shift-invariant transform) is obtained by use of the Fourier transformation concept as the key to the design of the filters. When the speckle photograph is digitized by a camera interfaced to a computer, the data are discretized. For such discrete data, finite summation replaces the usual integrals and an appropriate fast Fourier transform is utilized as the linear filter. Then the discrete Fourier transform becomes a suitable tool when optical stress analysis is coupled to the computer.

In recent years direct processing of specklelike data has emerged as a technique in optical stress analysis. This technique is known as digital image processing and displacement analysis and is discussed by Chu et al. [24]. The principles of this method parallel the developments of speckle photography, in which small areas of the image are processed in order to determine object motion. Small subimages of a discrete recording of a speckle type of image are correlated for undeformed and deformed configurations of a body. This form of image correlation is defined as subpixel image restoration. The applications in optical stress analysis include the effect of image distortion due to the deformation of a body. Thus the digital image processing must include the effects of local rigid-body displacement, rigid-body rotation, and deformation. Parallels to speckle photography include the measurement of local rigid-body displacements only. Therefore, image correlation techniques will possibly extend the measurements of speckle photography to include the strain components. Neither method, however, will replace the other, because experiments to date

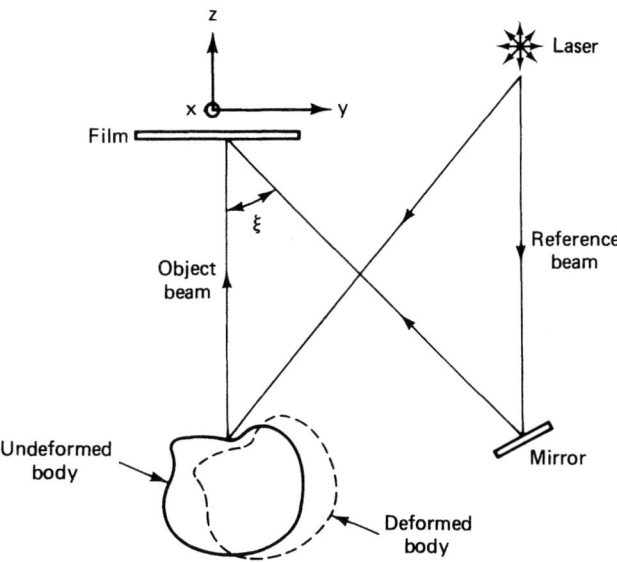

Figure 8-1 Off-axis hologram configuration for double-exposure holography.

indicate that speckle photography has greater sensitivity than image correlation. Therefore, problems that require a sensitive measurement technique will utilize the principles of holography and speckle photography.

Contemporary applications of the most important contributions of holography and laser speckle metrology in experimental mechanics are presented in Ref. 25. Current applications suggest future trends in experimental mechanics. The rapid development of computer technology will most certainly influence the applications of optical stress analysis. Improvements in data analysis of interference fringes and direct recording of speckle patterns represent a most promising area of development.

8-1 Basic Theory of Fringe Formation

Formation of a Double-Exposed Hologram

Consider an image-forming process as shown schematically in Fig. 8-1. This method of forming a double-exposed hologram utilizes an off-axis reference beam first developed by Leith and Upatnieks [2]. Other methods of double-exposed holograms have been developed [3,4]; however, only the off-axis reference beam will be considered here. Let the light reflected from the body on the hologram plane be given by

$$E_0 = \text{Re}\{B(x, y) \exp [i\theta(x, y)]\} \tag{8-1}$$

$B(x, y)$ is called the amplitude modulus and $\theta(x, y)$ is called the phase and both are, in general, functions of the coordinates of the hologram plane. The reference beam may be described as a plane wave with amplitude modulus A, which is constant with respect to x and y. This beam produces, at the hologram plane, a phase term $(2\pi/\lambda)\, x \sin \xi$, which results from the plane wave impinging on the hologram plane at some angle ξ. Leith and Upatnieks [2] showed that the reference beam can be represented by

$$E_R = \text{Re}\left[A \exp\left(\frac{i2\pi}{\lambda} x \sin \xi\right)\right] \tag{8-2}$$

Note that the light amplitude E will always be the real part of the complex function and that the subscripts O and R will denote the object and reference beams, respectively. Unless otherwise stated, it is understood that only the real part of the complex light amplitude has physical significance.

In holographic interferometry, two exposures must be made of the body. The first exposure is made of the body in some reference configuration, and a second exposure is made with the body deformed from the reference configuration.

For the first exposure, the total light amplitude at the hologram plane is the superposition of the two wavefronts from the reflected object beam and reference wave as given by

$$E_\text{I} = E_{O\text{I}} + E_R \tag{8-3}$$

For the second exposure the total light amplitude at the hologram plane is therefore

$$E_\text{II} = E_{O\text{II}} + E_R \tag{8-4}$$

The subscripts I and II denote the first and second exposures, respectively. In both exposures the reference will be of the form in Eq. (8-2). The off-axis reference beam is the same for both exposures in the formation process. However, if the light amplitude $E_{O\text{I}}$ is of the form in Eq. (8-1), then $E_{O\text{II}}$ may be conveniently expressed as

$$E_{O\text{II}} = \text{Re}[B(x, y) \exp[i(\theta + \Delta\theta)]] \tag{8-5}$$

where $\Delta\theta$ is the change in phase due to the deformation of the body. The phase change is determined from geometrical arguments of the deformation and configuration of the optical elements. An analytical expression for $\Delta\theta$ is underdeveloped "Phase Changes Due to Deformation of a Body."

The total light at the film plane for both exposures is

$$I_\text{film} = E_\text{I} \cdot E_\text{I}^* + E_\text{II} \cdot E_\text{II}^* \tag{8-6}$$

where E_I^* and E_II^* denote the complex conjugates of the light amplitudes. Substitution of Eqs. (8-4) and (8-5) into (8-6) yields

$$I_\text{film} = 2(B^2 + A^2) + AB \exp[i(\gamma - \theta)] \tag{8-7}$$
$$+ AB \exp[-i(\gamma - \theta)] + AB \exp[i(\gamma - \theta - \Delta\theta)] + AB \exp[-i(\gamma - \theta - \Delta\theta)]$$

where $\gamma = 2\pi/\lambda\, x \sin \xi$. Equation (8-7) describes the light intensity in the formation process of a double-exposed hologram. The information that contains the measure of object motion is the phase, $\Delta\theta$, which may be separated in the reconstruction process.

Reconstruction of a Double Exposure Hologram

The reconstruction process is illustrated in Fig. 8-2, and the light amplitude transmittance for the hologram is given in Ref. 2.

$$E_\text{trans} = I_\text{film} \cdot E_R \tag{8-8}$$

HOLOGRAPHIC AND LASER SPECKLE INTERFEROMETRY

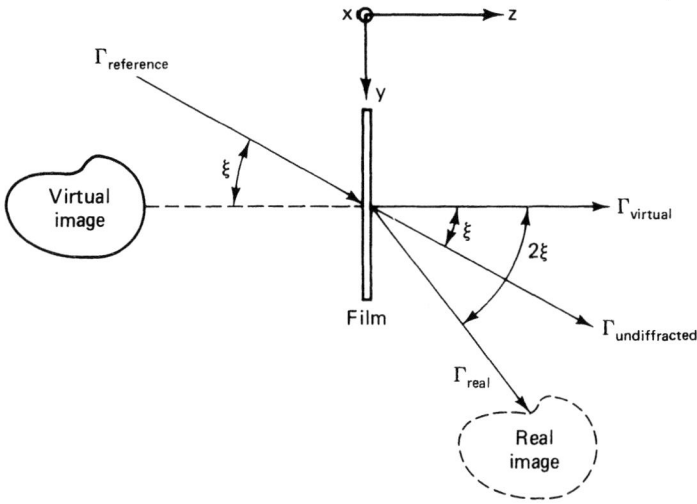

Figure 8-2 Reconstruction of an off-axis reference-beam hologram.

In the reconstruction process the light does not have to be the same as the reference beam in the formation process. Thus the hologram may be reconstructed with any sufficiently coherent light. This is particularly important in dynamic events using the pulsed ruby laser, since the light output from the ruby laser is only about 50 ns, and thus a continuous-operation laser must be used in the reconstruction process. Completing the multiplication indicated in Eq. (8-9), three distinct terms then may be separated out:

$$E_{\text{virtual}} = A^2B \exp(i\theta) + A^2B \exp[i(\theta + \Delta\theta)]$$
$$E_{\text{real}} = A^2B \exp[i(2\gamma - \theta)] + A^2B \exp[i(2\gamma - \theta - \Delta\theta)] \quad (8\text{-}9)$$
$$E_{\text{undiffracted}} = 2A(B^2 + A^2) \exp(i\gamma)$$

Since the phase change due to object motion is the quantity of interest, the undiffracted term does not contain any useful information. The term E_{virtual} is known as the virtual image and the term E_{real} is the real image when both contain the phase change information $\Delta\theta$. Therefore by suitable choices of the reconstruction angle the real and virtual images may be separated. The terms yield the same information; however, it is more popular to consider the virtual image. If the virtual image is photographed or observed, the intensity is given by

$$I_{\text{virtual}} = E_{\text{virtual}} \cdot E^*_{\text{virtual}} \quad (8\text{-}10)$$

Therefore, Eq. (8-10) becomes

$$I_{\text{virtual}} = 2A^4B^2(1 + \cos \Delta\theta) \quad (8\text{-}11)$$

An interference fringe will be defined as a locus of points when the intensity of the virtual image is zero. Therefore, a fringe will be determined where the term $1 + \cos \Delta\theta = 0$. As may be observed from Eq. (8-11), the phase change may be interpreted as a measure of object motion

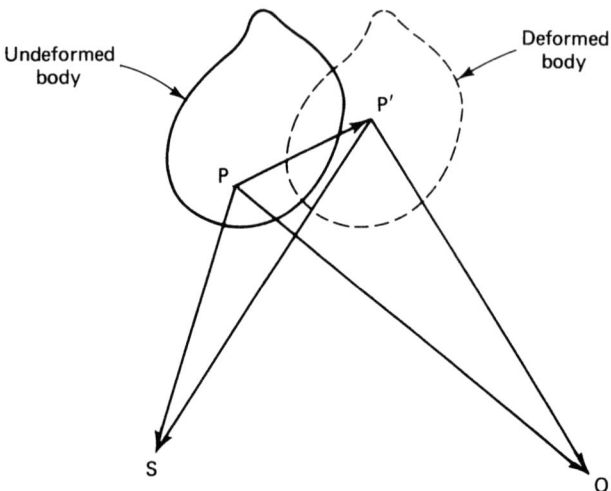

Figure 8-3 Geometry of fringe formation for a deformed body.

along lines of minimum light amplitude. The geometry of the phase change is described in the following paragraph.

Phase Change Due to Deformation of a Body

If the body is deformed between exposures of the hologram, the phase terms of the complex light amplitudes are different for the two exposures. This change in phase results from the motion of the body, which may be interpreted in terms of the surface deformation. The phase change, denoted as $\Delta\theta$ in Eq. (8-11), is determined from the change in the optical path length of the light. Refer to Fig. 8-3, the optical path length of the light for the body in the reference configuration is denoted by L_1 and is the distance

$$L_1 = |\mathbf{PS}| + |\mathbf{PO}| \qquad (8\text{-}12)$$

where S = light source
P = point on the surface of the body
O = point in the hologram plane

The distance $|\mathbf{PS}|$ between two points (S, P) is the length of the vector that connects them. Therefore, the distance $|\mathbf{PS}|$ is written in terms of the inner product of the point difference. The vector $\mathbf{PS}$ connecting the points P and S is defined as the point difference. Using the inner product rule, Eq. (8-12) may be written as

$$L_1 = (\mathbf{PS} \cdot \mathbf{PS})^{1/2} + (\mathbf{PO} \cdot \mathbf{PO})^{1/2} \qquad (8\text{-}13)$$

The point P on the surface of the body is deformed into the point P'. With the body in the deformed configuration, the optical path length of the light, denoted L_2, is the distance

$$L_2 = |\mathbf{P'S}| + |\mathbf{P'O}| \qquad (8\text{-}14)$$

The inner product of the point difference for the difference L_2,

$$L_2 = (\mathbf{P'S} \cdot \mathbf{P'S})^{1/2} + (\mathbf{P'O} \cdot \mathbf{P'O})^{1/2} \tag{8-15}$$

To find the phase change of the light, Eqs. (8-14) and (8-15) are used to determine the change in optical path length Δ, where

$$\Delta = L_1 - L_2 \tag{8-16}$$

The phase $\Delta\theta$ is related to Δ by $2\pi/\lambda$; therefore, the phase change is written as

$$\Delta\theta = \frac{2\pi}{\lambda}(L_1 - L_2) \tag{8-17}$$

This equation, although completely general, is not in a form convenient for numerical use. Let the point difference of the body before and after deformation be described in the following equation and illustrated schematically in Fig. 8-3.

$$\mathbf{PS} = \mathbf{P'S} + \mathbf{PP'} \tag{8-18}$$
$$\mathbf{PO} = \mathbf{P'O} + \mathbf{PP'}$$

In holographic interferometry, the displacement of points between exposures is small. Therefore, it will be assumed here that $PP'^2 \ll PP'$. The small deformation may be used to obtain an approximate expression for the phase change $\Delta\theta$.

$$\Delta\theta = \frac{2\pi}{\lambda}(\mathbf{ps} + \mathbf{po}) \cdot \mathbf{PP'} \tag{8-19}$$

The terms in parentheses in Eq. (8-19) are unit vectors in the directions $\mathbf{PS}$ and $\mathbf{PO}$. Therefore, in the following equations and illustrations, examples $\mathbf{ps}$ and $\mathbf{po}$ will denote unit vectors and components relative to an orthogonal basis will be given in terms of direction cosines.

Since the quantity that is observed is the intensity of the virtual image, or real image, Eq. (8-19) can be written as

$$I_{\text{virtual}} = C\left\{1 + \cos\left[\frac{2\pi}{\lambda}(\hat{ps} + \hat{po}) \cdot \overline{PP'}\right]\right\} \tag{8-20}$$

where $C = 2A^4B^2$.

For problems involving small deformations, the information contained on the hologram will yield phase changes that are related to the point difference $\mathbf{PP'}$. These results are still in a general form and some special cases will be considered. Also, since a reference configuration does not have to be an undeformed position, problems involving large deformations can be studied. The only requirement is that the change in configuration be small between exposures of the hologram.

To consider some applications of the holographic interferometry equations, it will be convenient to introduce a reference frame. Suppose that the origin of the reference frame is located at the point P as shown in Fig. 8-4.

Let the point difference have the following component representation:

$$\mathbf{PP'} = u_i \mathbf{e}_i$$

where $\mathbf{e}_i$ are unit vectors along the coordinate axis, Let the point difference $\mathbf{ps}$ have a set of direction cosines denoted by l_i, and $\mathbf{po}$ have a set of direction cosines denoted by m_i. The expression for the intensity of the virtual image is now

$$I_{\text{virtual}} = C\left\{1 + \cos\left[\frac{2\pi}{\lambda}(l_i u_i + m_i u_i)\right]\right\} \tag{8-21}$$

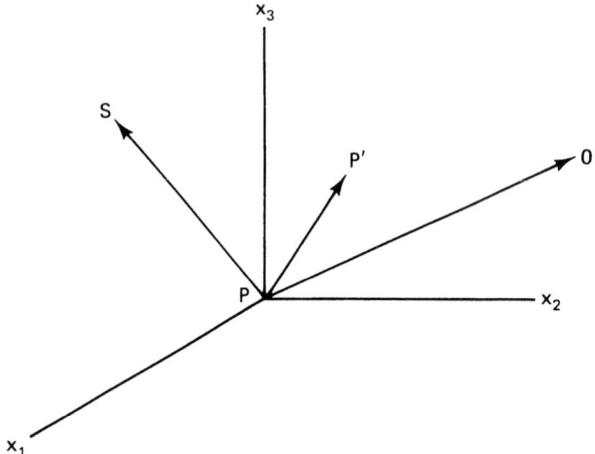

Figure 8-4 Orthogonal coordinate system.

Therefore, the information from the real or virtual images is dependent on the three components of the point difference **PP'**. Solid [6] and Ennos [4] state that if zero-order fringes can be found on the surface, three holograms are sufficient to determine the components u_i uniquely. Problems of this type usually involve a fixed point or a fixed boundary. If zero-order fringes cannot be observed, a single hologram technique as described by Aleksandrov and Bonch-Bruevich [5] must be used.

The phase change due to the deformation of a body is derived in terms of the inner product of the point difference and unit vectors describing the illumination and observation directions. The inner product is coordinate system dependent; therefore, experimental configurations will allow for component measurements of **PP'** depending on geometry.

Fringe Formation in Single-Beam Speckle Photography

Laser speckle patterns are a result of interference when an optically rough surface is illuminated with a coherent laser. Speckles were originally classified as objective or subjective. The latter are formed when the illuminated surface is through a lens or eye. Speckle size is therefore controlled by the numerical aperture of the lens. Single-beam subjective laser speckle photography has been successfully employed to measure surface deformation of solid bodies. The range of measurements is much larger than holographic interferometry, and in many experimental configurations is much easier to obtain.

The basic optical configuration from a single beam laser speckle method is illustrated schematically in Fig. 8-5. An object surface is illuminated by a laser beam. The location of the light source, which is denoted as S and P, is a point on the surface of the illuminated body. The procedure is similar to holographic interferometry, where two exposures for both undeformed and deformed configuration of the body are recorded on the same film.

The recorded light intensity of the first exposure is calculated in a similar manner to the equations in the preceding section:

$$I_1 = B^2(x, y) \qquad (8\text{-}22)$$

HOLOGRAPHIC AND LASER SPECKLE INTERFEROMETRY 373

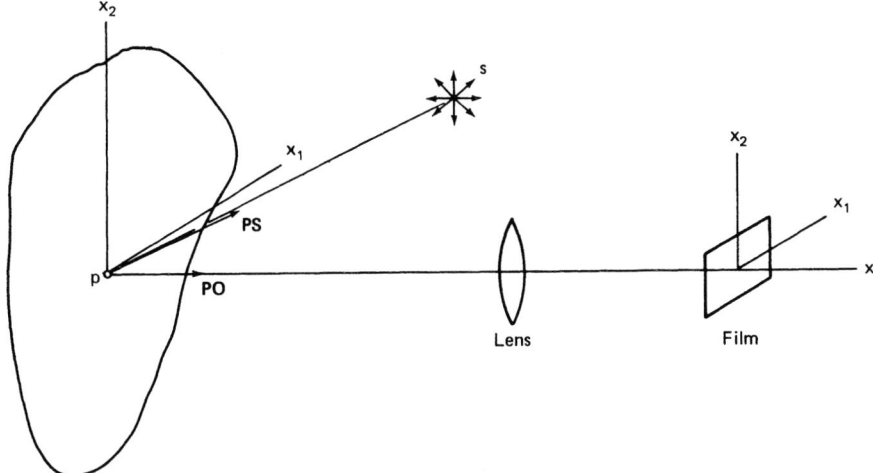

Figure 8-5 Schematic diagram for experimental arrangement in single-beam speckle photography.

If the body is deformed between exposures, the image on the film plane will be shifted to new coordinates x' and y'. The light amplitude in the deformed configuration may be expressed as

$$I_{\text{II}} = B^2(x', y') \tag{8-23}$$

The surface displacement vector **PP**′ may be expressed in the component representation

$$\mathbf{PP}' = u_p\mathbf{e}_x + v_p\mathbf{e}_y + w_p\mathbf{e}_z \tag{8-24}$$

Then the film-plane displacement vector can be written as

$$[\mathbf{PP}']_{\text{film}} = M[u_p\mathbf{e}_x + v_p\mathbf{e}_y] \tag{8-25}$$

where M is the magnification of lens, typically $M < 1$ and $u_{pf} = MU_p$, $v_{pf} = MV_p$.

Coordinates for the film plane for the second exposure are $x' = x + u_f$ and $y' = y + v_f$. Now the expression for the total intensity is

$$I_{\text{total}} = B^2(x, y) + B^2(x + u_f, y + v_f) \tag{8-26}$$

Interference fringes are obtained by taking optically the Fourier transform of the amplitude transmission function of the processed photographic transparency. This procedure may be obtained in several methods, depending on the data analysis. The amplitude transmission function $g(x, y)$ is assumed linear in the ranges of interest and may be expressed as

$$g(x, y) = a + bI_{\text{total}} \tag{8-27}$$

where a and b are constants.

The light amplitude in the Fourier transform plane, which is proportional to the transform of $g(x, y)$ times a quadratic-phase factor, is expressed in

$$G(\omega_1, \omega_2) = \exp\left[\frac{id(\omega_1^2 + \omega_2^2)}{2K}\right] F[g(x, y)] \tag{8-28}$$

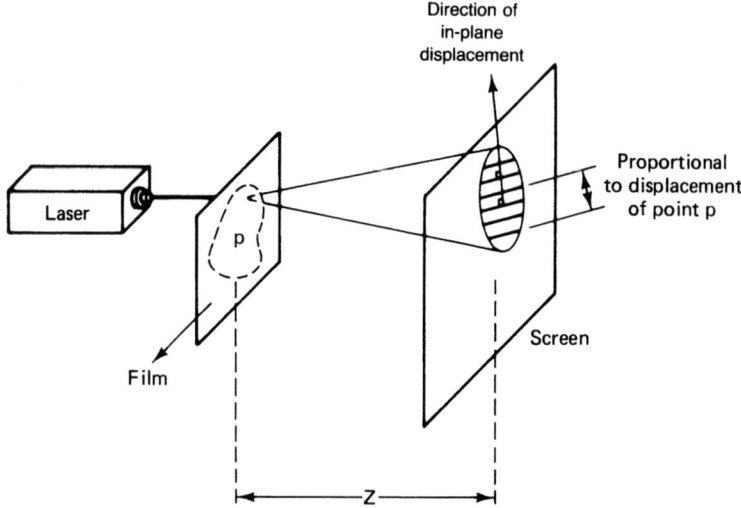

Figure 8-6 Data analysis for point filtering in single-beam speckle photography.

$F[g(x, y)]$ denotes the Fourier transform of $g(x, y)$. In the transform plane, the vectorial spatial-frequency variable is defined as

$$\boldsymbol{\omega} = \omega_1 \mathbf{e}_{xs} + \omega_2 \mathbf{e}_{ys}, \qquad \omega_1 = \frac{KX_s}{d} \qquad \omega_2 = \frac{KY_s}{d} \tag{8-29}$$

The coordinates in the transform plane are denoted as (X_s, Y_s) and constant $K = 2\pi/\lambda$.

Equation (8-28) has a final form as

$$G(\omega_1, \omega_2) = \exp\left[\frac{id(\omega_1^2 + \omega_2^2)}{2K}\right][a\delta(\omega_1, \omega_2) + bF(I_{\text{total}})] \tag{8-30}$$

The delta function represents the (idealized) point focus of the illuminating beam. It contributes to the amplitude only at the center ($\omega_1 = \omega_2 = 0$); in practice, it is a small area around this point. Outside this small area the intensity in the transform plane takes the form

$$I(\omega_1, \omega_2) = 2b^2|Ap(\boldsymbol{\omega}) * Ap(\boldsymbol{\omega})|^2[1 + \cos(\boldsymbol{\omega} \cdot \mathbf{u}_f)] \tag{8-31}$$

Terms that relate to the interference effects in Eq. (8-31) will be considered in the data analysis. Fringes will be defined when $I(\omega_1, \omega_2) = 0$ or when $[1 + \cos(\boldsymbol{\omega} \cdot \mathbf{u}_f)] = 0$. Fringes are determined by the technique of pointwise filtering, as illustrated in Fig. 8-6. For a small region of illumination of the transparency in the neighborhood of P, the displacement is assumed constant and fringes are formed when

$$\boldsymbol{\omega} \cdot \mathbf{u}_{pf} = (2n - 1)\pi \qquad n = 1, 2, 3, \ldots \tag{8-32}$$

For most problems of interest only the first fringe is observed; therefore,

$$X_s u_{pf} + Y_s v_{pf} = \frac{\lambda d}{2} \tag{8-33}$$

HOLOGRAPHIC AND LASER SPECKLE INTERFEROMETRY

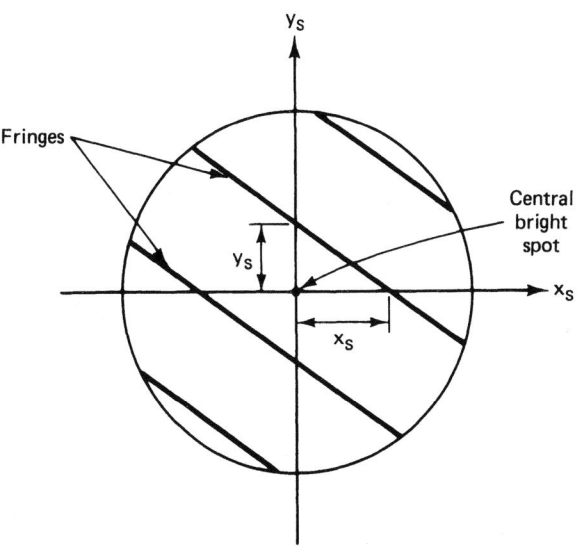

Figure 8-7 Schematic for determining the in-plane displacement component single-beam speckle photography.

The schematic for determining components in Eq. (8-33) is illustrated in Fig. 8-7 by measuring the distances X_s and Y_s between the center and points of intersection of the first fringe and X_s, Y_s, respectively.

Fringe Formation in Shearing Speckle Interferometry

This section presents the general theory of shearing speckle interferometry. A schematic diagram of the experimental apparatus for shearing speckle interferometry is shown in Fig. 8-8. Thus a point p is imaged at (x, y) and $(x + \Delta x, y + \Delta y)$ in the film plane, and a neighboring point p_1 is also imaged at (x, y) in the film plane. To obtain fringe data, a double-exposure technique is used, where a photograph is taken of the body in some reference configuration. After deformation, another exposure is recorded on the same film.

With the body in the reference configuration, the light amplitudes at the film plane for both points p and p_1 can be described in the following manner:

$$E_p = A_p(x, y) \exp[i\theta_p(x, y)] \tag{8-34}$$
$$E_{p1} = A_{p1}(x, y) \exp[i\theta_{p1}(x, y)]$$

where $A(x, y)$ is the amplitude modulus and $\theta(x, y)$ is the phase and both are, in general, functions of the coordinates of the film plane. The total intensity for the first exposure can be written as

$$I_1 = A_p^2 + A_{p1}^2 + 2A_pA_{p1} \cos(\theta_p - \theta_{p1}) \tag{8-35}$$

If the body is deformed and the point p and p_1 are displaced to points p' and p_1', the image of

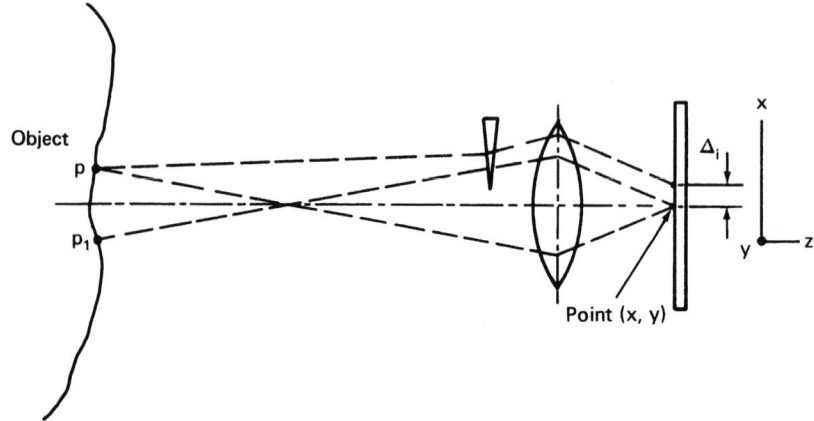

Figure 8-8 Schematic diagram for experimental arrangement in shearing speckle interferometry.

point p on the film plane will be shifted to new coordinates x' and y'. The light amplitudes in the deformed configuration can be written in the following form (Fig. 8-9):

$$E'_p = A_p(x', y') \exp[i\theta_p(x', y') + \Delta\theta_p]$$
$$E'_{p1} = A_{p1}(x', y') \exp[i\theta_{p1}(x', y') + \Delta\theta_{p1}]$$
(8-36)

$\Delta\theta_p$ and $\Delta\theta_{p1}$ account for the phase change in each field due to motion of the object relative to the illumination field. So the total intensity expression for both exposures is

$$I_{total} = A_p^2 + A_{p1} + 2A_pA_{p1}\cos[\theta_p(x, y) - \theta_p(x, y)] + A_p^2 + A_{p1}$$
$$+ 2_{A_pA_{p1}} \cos[\psi(x', y') + \Delta\theta]$$
(8-37)

$$\psi(x', y') = \theta_p(x', Y') - \theta_{p1}(x', y') \qquad \Delta\theta = \Delta\theta - \Delta\theta_{p1}$$

Interference fringes are obtained by taking optically the Fourier transform of the amplitude transmission function of the transparency. This can be done by several methods, depending on

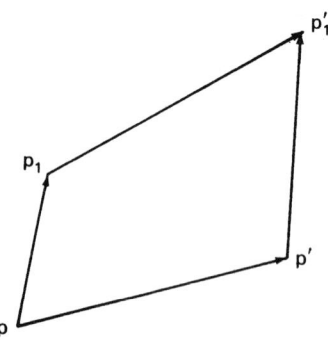

Figure 8-9 Displacement vector neighbor points P and P'.

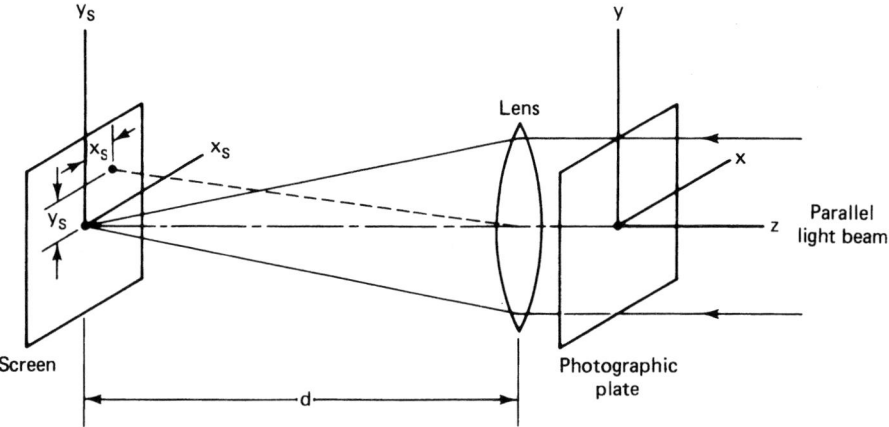

Figure 8-10 Optical Fourier transform for fringe observation in shearing speckle interferometry.

the data analysis. The amplitude transmission function $g(x, y)$ is linear for the ranges of interest and can be expressed as

$$g(x, y) = A + bI_{\text{total}} \tag{8-38}$$

where a and b are constants. Substituting the expression for I_{total} from Eq. (8-37) into Eq. (8-38), an expression for the amplitude transmission is obtained:

$$\begin{aligned} g(x, y) = A + b\{&A_p^2 + A_{p1}^2 + 2A_p A_{p1} \cos[|\theta|(x, y)] \\ &+ A_p(x + u_j, y + v_f) + A_p(x + u_f, y + v_f) \\ &+ [2A_p(x + u_f, y + v_f) A_p(x + u_f, y + v_f) \\ &\times \cos[\psi(x + u_f, y + v_f) + \Delta\theta]\} \end{aligned} \tag{8-39}$$

The procedure developed in the section "Fringe Formation in Speckle Metrology" is used to express the light amplitude $G(\omega_1, \omega_2)$ in the Fourier transform plane (Fig. 8-10), which is proportional to the Fourier transform of $g(x, y)$ times a quadratic phase factor, that is,

$$G(\omega_1, \omega_2) = \exp\left[\frac{id(\omega_1^2 + \omega_2^2)}{2K}\right] F[g(x, y)] \tag{8-40}$$

$F[g(x, y)]$ denotes the Fourier transform of the function $g(x, y)$, and d is the distance between the transparency and the screen. In the transform plane, the vectorial spatial-frequency variable is defined as

$$\boldsymbol{\omega} = \omega_1 \mathbf{e}_{xs} + \omega \mathbf{e}_{ys}$$

$$\omega_1 = \frac{KX_s}{d}$$

$$\omega_2 = \frac{KY_s}{d}$$

The coordinates in the transform plane are denoted as (x_s, y_s) and the constant $k = 2\pi/\lambda$, where λ is the wavelength of the light.

Substituting $g(x, y)$ from Eq. (8-39) into Eq. (8-40) yields

$$G(\omega_1, \omega_2) = \exp\left[\frac{id(\omega_1^2 + \omega_2^2)}{2K}\right] a\gamma(\omega_1, \omega_2) + bF(I_{total}) \tag{8-41}$$

where $\delta(\omega_1, \omega_2)$ is the delta function, defined as

$$\delta(\omega_1, \omega_2) = \exp[-i(\omega_1, x + \omega_2, y) dx\, dy]$$

The delta function represents the (idealized) point focus of the illuminating beam. It contributes to the amplitude $G(\omega_1, \omega_2)$ only at the center ($\omega_1 = \omega_2 = 0$); in practice it is a small area around this point. Outside this small area the intensity $|G(\omega_1, \omega_2)|^2$ in the transform plane is

$$I(\omega_1, \omega_2) = b^2 |F(I_{total})|^2 \tag{8-42}$$

Using the identity

$$\cos(\psi + \Delta\theta) = \cos\psi \cos\Delta\theta - \sin\psi \sin\Delta\theta$$

in the expression of I_{total}, Eq. (8-42), the Fourier transform of I_{total} can be written as

$$\begin{aligned}F(I_{total}) &= A_p(\omega) * A_p(\omega) * [1 + \exp[i(\omega \cdot u_f)]] \\&+ A_{p1}(\omega) * A_{p1}(\omega) + A_{p2}(\omega) * A_{p2}(\omega) \exp[i(\omega \cdot u_f)] \\&+ 2A_p(\omega) * A_{p1}(\omega) * C(\omega) + 2A_p(\omega) * A_p(\omega) * C(\omega) \exp[i(\omega \cdot u_f)] \\&\cos\Delta\theta - 2A_p(\omega) * A_{p1}(\omega) * S(\omega) \exp[i(\omega \cdot u_f)] \sin\Delta\theta\end{aligned} \tag{8-43}$$

$F(\cdot)$ indicates the two-dimensional Fourier transform with respect to the variable u_f; A_p, A_{p1}, and A_{p2} are the Fourier transforms of the functions a_p, a_{p1}, and a_{p2}, respectively; $*$ denotes the convolution integral, such as

$$A_p(\omega) * A_{p1}(\omega) * C(\omega) = F[A_p(x, y) A_{p1}(x, y) \cos\psi(x, y)]$$

8-2 Holographic Interferometry Applications

Surface Deformation Measurements

A basic experimental configuration for double-exposure holographic interferometry is illustrated in Fig. 8-11. This off-axis reference-beam system must be a stable interferometry system in order to isolate the optical elements from the surroundings. A laser may be either continuous or pulsed, depending on the problem. Surface preparation of models is not difficult for holography, and generally any diffuse reflecting surface is sufficient, which is a particular advantage over classical interferometry because the same material may be used for model and prototype in holography. The reconstruction process is usually accomplished by placing the processed transparency in the original reference beam for observation of the virtual or real image. The virtual image is viewed by observation through the transparency. On the other hand, the real image is viewed on a white background by projecting the reference beam through the photographic plate. If a piece of film is used instead of a white background, the real image may be recorded without further optical elements. Some problems in reconstruction can exist due to the fact that the interference fringes do not focus at the same plane as the model.

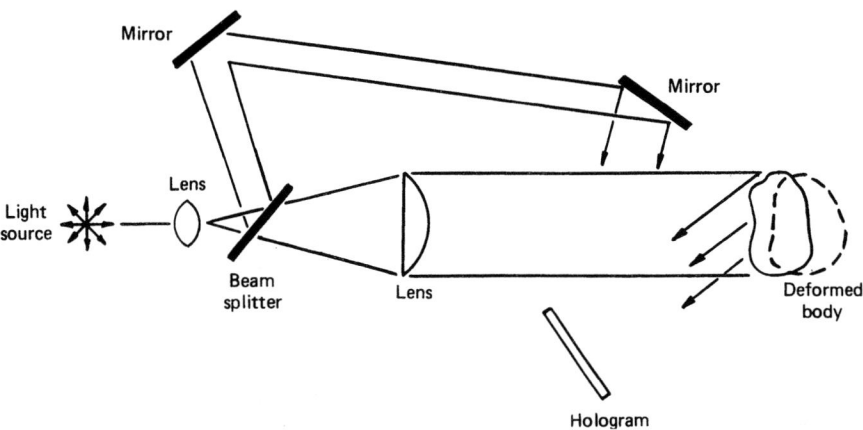

Figure 8-11 Schematic diagram for double-exposure holographic interferometry.

Equation (8-20) is the basic equation used to interpret the formation of fringes in holographic interferometry as a measure of surface deformation. Although the off-axis reference beam is used in the experiments, different displacement components may be obtained by changing the optical configuration of illumination and observation unit vectors. The surface displacements of a cantilever beam with a transverse end load may be used as an illustration. For an experimental geometry shown in Fig. 8-12 the set of direction cosines are $l_1 = 0$, $l_2 = -1.0$, $l_3 = 0$, $m_1 = 0$, $m_2 = -\cos\theta$, $m_3 = \sin\theta$. Therefore, Eq. (8-21) is a measure of displacement components in the y and z directions. The formation of fringes is a measure of the lateral deformation of the beam. Figures 8-13 and 8-14 offer a comparison of the experimental and theoretical fringes for this example. If the experimental geometry is changed as illustrated in Fig. 8-15, the fringes as predicted by Eq. (8-21) are a measure of the transverse deflection of the beam. For this geometry the direction cosines are $l_1 = 1$, $l_2 = 0$, $l_3 = 0$, $m_1 = \cos\xi$, $m_2 = 0$, $m_3 = \sin\xi$. Figure 8-16 is a photograph of the virtual image in which these fringes are a measure of the displacement.

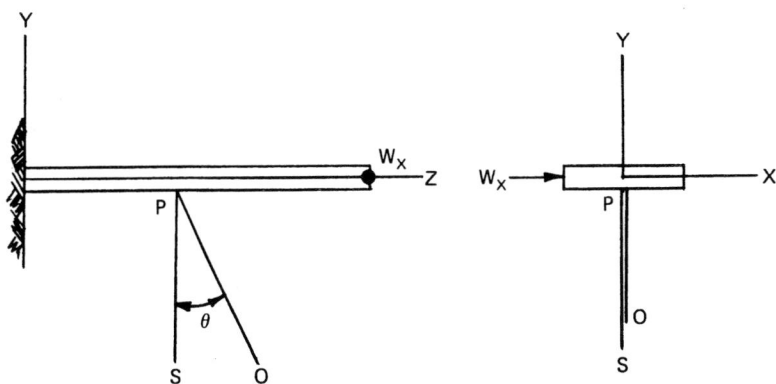

Figure 8-12 Geometric configuration used to measure the lateral deflection of a cantilever beam.

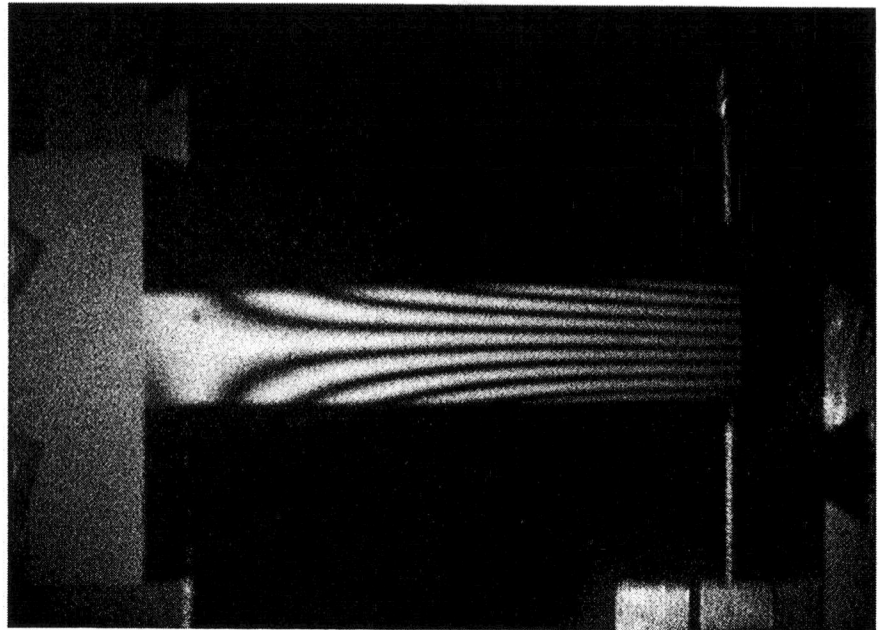

Figure 8-13 Holographic reconstruction of the virtual image of the cantilever beam.

Double-exposure dynamic holography utilizes the same basic component of surface experimental geometry; however, the problems are generally more complex due to the requirements of a coherent laser with short pulse duration. Difficulties of a ruby laser for use in holography and requirements are discussed by Holloway [26]. However, even with the difficulty of pulse holography, the data are amenable to solution of many important problems in stress-wave propagation. The essential requirement for wave propagation studies is a single-frequency laser. A short-pulse-duration pulse ruby laser has been demonstrated to be a suitable light source capable of controlled light output of 50-ns duration. Thus the pulse duration is satisfactory and sufficiently fast to stop photographically the motion in most materials.

Figure 8-17 is an example of the virtual image of the central impact of a flat aluminum plate. This example illustrates the high-quality interference fringes that may be obtained with pulsed holography. Figure 8-18 shows the central impact of a flat composite plate. Edge impact loading of an aluminum plate is illustrated in Fig. 8-19.

Nondestructive Evaluation *(NDE)*

Sensitivity of the formation of fringes with small surface deformations has been very useful in the nondestructive testing of solid bodies. The presence of cracks or subsurface voids in composites, and broken fibers in tires, are only a few of the many examples of the application of holography to NDE. Although the fringes are interpolated quantitatively for the components of surface deformation, NDE applications usually require only a qualitative evaluation. If the fringes are formed for a geometry without geometrical discontinuities, distortions of fringe patterns may be used as

HOLOGRAPHIC AND LASER SPECKLE INTERFEROMETRY 381

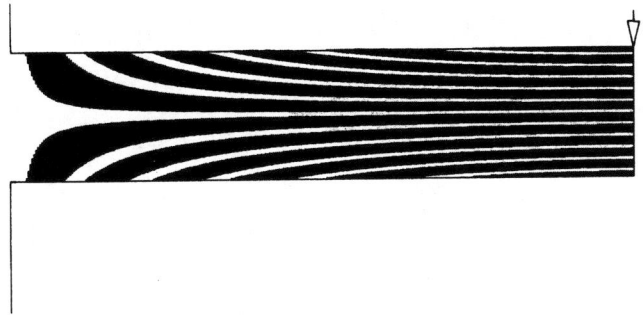

Figure 8-14 Theoretical fringes for comparison to Fig. 8-13.

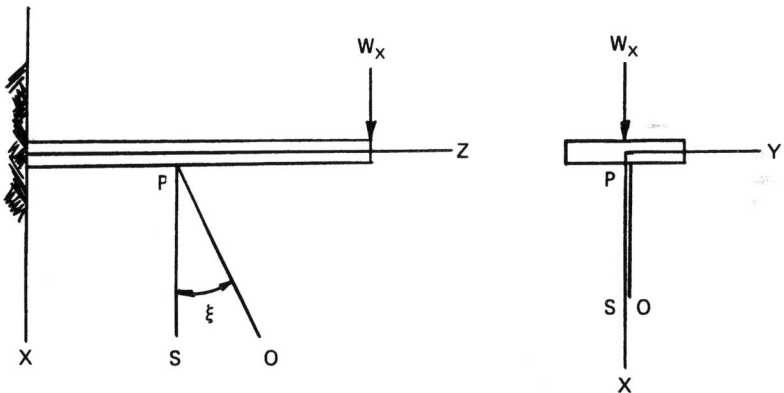

Figure 8-15 Geometrical configuration used to measure the transverse deflection of a cantilever beam.

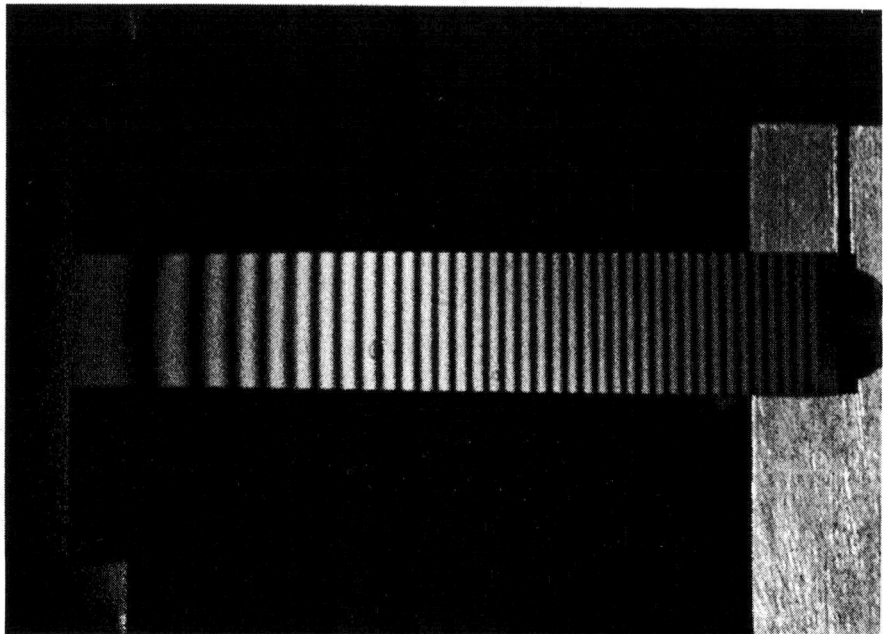

Figure 8-16 Holographic reconstruction of the virtual image of the cantilever beam.

a detection of any discontinuities. The qualitative evaluation of holography is therefore useful for NDE.

The detection of flaws in composites has been one of the major applications of holographic NDE. Several illustrations will be used to demonstrate the potential of holography to evaluate deformation qualitatively. An off-axis holographic experiment used to measure surface deformations of a circular cylinder is illustrated schematically in Fig. 8-20. This schematic illustrates only one possible geometry that defines the unit vectors **ps** and **po**. In general, the hologram will record fringes in relation to the general formulation of surface deformation. However, some simplifications can be made in many problems; and with careful selection of the geometry, the amount of data can be reduced. Direction cosines for the cylinder are given by $l_1 = 0$, $l_2 = 0$, $l_3 = 1$, $m_1 = \sin \theta$, $m_2 = 0$, $m_3 = \cos \theta$. This particular geometry assumes that the surface of the cylinder is illuminated with collimated light parallel to the three directions. The mechanical loading for a uniform cylinder, usually internal pressure, produces a symmetrical deformation with respect to the longitudinal axis of the cylinder. This geometry and loading will therefore produce a fringe pattern that will predict a set of parallel lines along the longitudinal axis of the cylinder neglecting the end effects. Therefore, the expected fringe patterns for cylinders without discontinuities will produce a pattern of parallel straight fringes. Also, qualitative evaluation requires only a knowledge of geometry and wavelength of the laser, and measurements are made independent of material properties.

The mechanical loading of a capped-end, thin-wall composite cylinder was applied by internal pressure of 206 kPa (30 psi) and fringes observed qualitatively. Figure 8-21 is a hologram of a

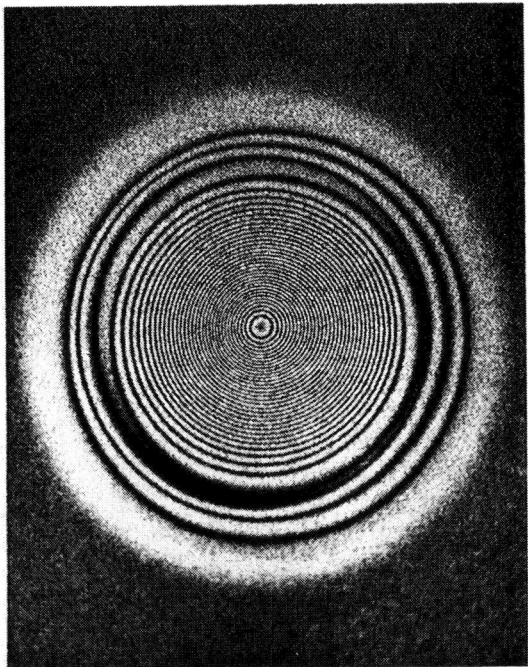

Figure 8-17 Virtual image of the transverse impact of an aluminum plate 12 μs after impact.

thin composite cylinder, and an internal spot delamination is easily observed in the central region of the cylinder. Since detection of a flaw is desired, the anomaly of the fringe patterns illustrates the effects of discontinuities. This type of fringe pattern is characteristic of symmetrical deformation in regions removed from the flaw. The small circular pattern in the center is approximately the same size as the flaw. Figure 8-22 is a reconstruction of a cylinder with a more complex fringe pattern. However, a delamination along the entire length of the cylinder is observed. This pattern may be very difficult to analyze for a quantitative description of the fringes, although qualitative evaluation of a flaw is rather easy.

Time-Average Holography

Object motion during the exposure time may produce an interference fringe pattern similar to the double-exposure method; however, the interpretation of the data yields a technique known as time-average holography. This problem formulation is based on previous results; the basic experimental configuration was illustrated in Fig. 8-1. During the exposure of the film plate, the object beam may be represented as in the equation

$$E_0 = A_0 \exp[i(x, y) + \phi_0 \sin \omega t] \tag{8-44}$$

where the phase is modified by $\phi_0 \sin \omega t$, which represents the motion of the body during the

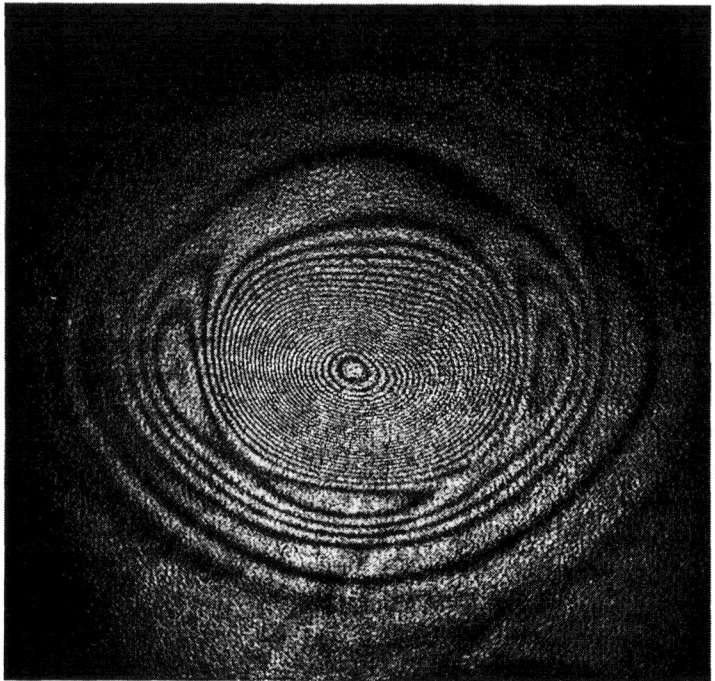

Figure 8-18 Virtual image of the transverse impact of a unidirectional composite plate 30 μs after impact.

exposure time. Since the object beam is combined with the reference beam, the intensity during the film exposure is

$$I_{\text{total}} = \frac{1}{t_1} \int_0^{t_1} E_0 \cdot E_0^* \, dt \tag{8-45}$$

where t_1 is the exposure time.

The intensity of the reconstructed virtual image may be obtained by substitution of Eq. (8-45) into Eq. (8-10). The resulting expression for this image is therefore

$$I_{\text{virtual}} = A_0^2 A_R^4 t_1^2 J_0^2 \left[(\mathbf{ps} + \mathbf{po}) \cdot \mathbf{PP'} \right] \tag{8-46}$$

where J_0 is the Bessel function and $\phi = (\mathbf{ps} + \mathbf{po}) \cdot \mathbf{PP'}$ is the phase change due to object motion. A fringe order is defined when the intensity of $I_{\text{virtual}} = 0$, which occurs for zeros of J_0, which corresponds to the maximum amplitude of vibration of a point on the surface of the object.

Equation (8-46) is the basic equation for the formation of fringes in time-average holography. However, this equation generally does not represent the most useful form. The argument of ϕ when expressed in terms of components of an orthogonal coordinate system provides a component representation most useful in time-average holography. When **PS**, **PO**, and **PP'** are coplanar, as shown in Fig. 8-4, and $\theta_1 = \theta_2$, a fringe is observed when

$$d = \frac{\theta_0 \phi_0 \lambda}{4\pi \cos \theta} \tag{8-47}$$

HOLOGRAPHIC AND LASER SPECKLE INTERFEROMETRY

Figure 8-19 Virtual image of an aluminum plate with edge loading 16 μs after impact.

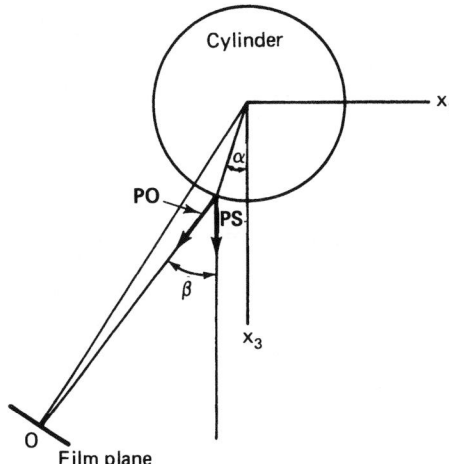

Figure 8-20 Geometrical configuration used to measure symmetrical deformation of a circular cylinder.

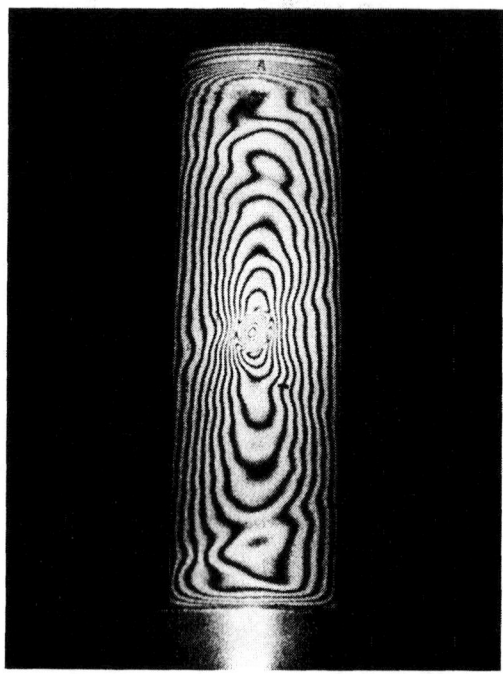

Figure 8-21 Virtual image of a thin-walled composite cylinder with a small circular delamination in the central portion of the cylinder.

An off-axis reference beam experiment for time-average holography was used to determine the formation of fringes for vibrating turbine blades as an example of the time-average method. A turbine blade was mounted in a fixture and a small speaker was used as the source of excitation for the blade. At each natural frequency the standing-wave pattern produced the fringe pattern for the time-average holography. Figures 8-23 and 8-24 illustrate the high-quality fringe pattern obtained utilizing this method. These two examples only illustrate the techniques; the reader may realize the vast number of engineering applications of time-average holography.

8-3 Speckle Metrology Applications

Speckle Photography

The original contributions of speckle interferometry in the early 1980s provided the fundamental principles for applications to numerous problems. The excellent review article by Parks [25] identified some of the most important contributions of speckle photography. Among these are the research of Stetson [27–29], Adams and Maddux [30], Chiang [31,32], Cloud [33], Barker and Fourney [34], and Duffy [35], and workers at the National Physical Laboratory [17] and Loughborough University of Technology [16]. The research activity coupled with computer applications of data analysis will provide significant contributions to experimental mechanics in future years.

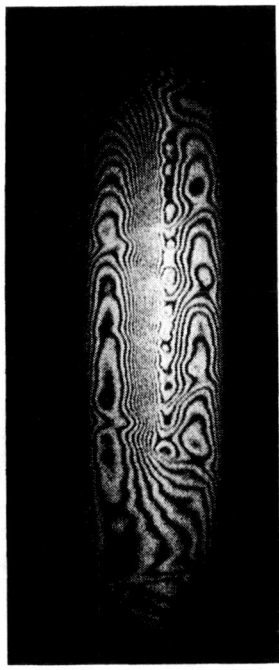

Figure 8-22 Virtual image of a thin-walled composite cylinder with a delamination along the longitudinal section of the cylinder.

An illustration of the measurement capability is given in a simple experimental example. If the geometry of Fig. 8-12 is used as a basis for a double-exposure speckle image, the deflection of the beam is easily measured. The difference in this method from that of the holography experiment occurs in the location of observation and illumination. Observation and illumination directions are interchanged and the recording is done by a lens and camera. Displacement components are measured as components along the x direction (denoted as the U component). Note that although the experiment is similar to holography, the component of measurement is the transverse deflection. A typical fringe pattern with pointwise filtering is illustrated in Fig. 8-25. This process is completed for as many points along the beam as desired. Figure 8-26 compares the measured data with the theoretical transverse deflection of the beam.

The sensitivity and range of speckle photography are discussed in Refs. 24 and 25. The sensitivity of speckle photography measurements as defined by Chiang [36] is the smallest measurable displacement resolvable by a lens according to the Rayleigh criterion. Parks has presented an example which illustrates that the sensitivity of the measurement is 36 μm. In this example Parks assumes a 5000-Å wavelength and 4° half-angle subtended by the camera lens. Also, measurement accuracy to one-tenth of a fringe is assured in this example. The range of measurement is defined as the upper bound on the measurement capability. If one assumes that displacement in the presence of strain is the quantity of interest, the question is difficult to answer. Equation (8-32) assumes that the displacement field in the area of illumination is constant. For problems in experimental mechanics, this is not generally the case and strains are superimposed on the

Figure 8-23 Virtual image of a time-average hologram of a turbine blade.

displacements. Chiang illustrates the effect of strain on the fringe contrast and shows that decorrelation results when strains are superimposed on the speckle patterns. At the present time, the upper bound for measurement is subject to discussion, and a practical approach is to record a succession of photographs such that only small increments occur for each double-exposure speckle photograph.

Shearing Speckle Interferometry

The experimental arrangement for the example problem used to illustrate the shearing speckle measurements is shown in Fig. 8-27. An optical wedge is placed in front of the camera at O. Also, for this example $u_p = v_p = 0$ while $w_p \neq 0$ and $A_p = A_{p1}$.

Derivative information is related to minimum of $I(\omega_1, \omega_2)$ [Eq. (8-42)], which will be minimum—upon substitution of Eq. (8-43)—when $(1 + \cos \Delta\theta)$ and $\sin \Delta\theta$ are both zero. Therefore, $\Delta\theta = (2n - 1)\pi$ will define a fringe. The phase change can be written in the form

$$\Delta\theta = \frac{2\pi}{\lambda} (\mathbf{ps} + \mathbf{po}) \cdot (\mathbf{P_2P_2'} - \mathbf{PP'}) \tag{8-48}$$

where

$$\mathbf{ps} = \begin{bmatrix} l_s \\ m_s \\ n_s \end{bmatrix} \quad \text{and} \quad \mathbf{po} = \begin{bmatrix} l_o \\ m_o \\ n_o \end{bmatrix}$$

Figure 8-24 Virtual image of a time-average hologram of a turbine blade with different natural frequency than Fig. 8-23.

representing the directions from point P on the object to the light source and camera, respectively. Equation (8-48) can be written in the following form:

$$\Delta\theta = \frac{2\pi}{\lambda} \left[(l_s + l_o) \left(\frac{\partial u}{\partial x} \Delta x + \frac{\partial u}{\partial y} \Delta y + \frac{\partial u}{\partial z} \Delta z \right) \right.$$
$$+ (m_s + n_o) \left(\frac{\partial v}{\partial x} \Delta x + \frac{\partial v}{\partial y} \Delta y + \frac{\partial v}{\partial z} \Delta z \right)$$
$$\left. + (n_s + n_o) \left(\frac{\partial w}{\partial x} \Delta x + \frac{\partial w}{\partial y} \Delta y + \frac{\partial w}{\partial z} \Delta z \right) \right] \quad (8\text{-}49)$$

Fringe order can be determined from the expression

$$\lambda \left(N - \frac{1}{2} \right) = (l_s + l_o) \left(\frac{\partial u}{\partial x} \Delta x + \frac{\partial u}{\partial y} \Delta y + \frac{\partial u}{\partial z} \Delta z \right)$$
$$+ (m_s + n_o) \left(\frac{\partial v}{\partial x} \Delta x + \frac{\partial v}{\partial y} \Delta y + \frac{\partial v}{\partial z} \Delta z \right)$$
$$+ (n_s + n_o) \left(\frac{\partial w}{\partial x} \Delta x + \frac{\partial w}{\partial y} \Delta y + \frac{\partial w}{\partial z} \Delta z \right) \quad (8\text{-}50)$$

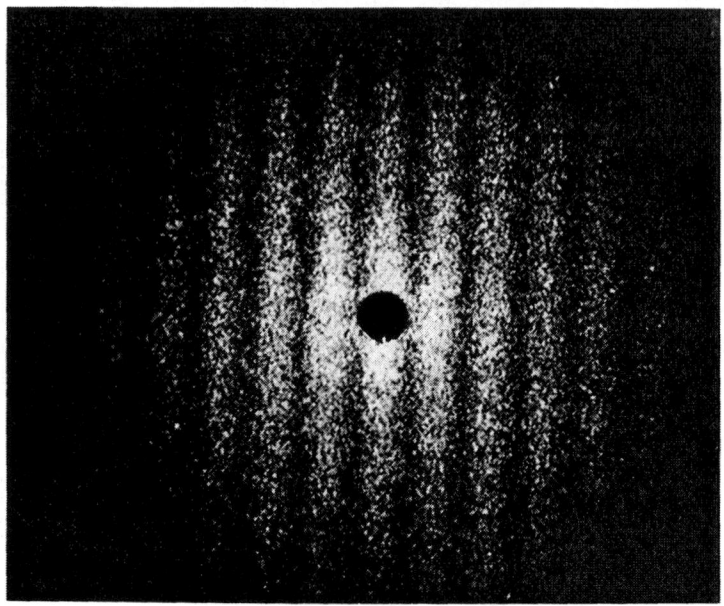

Figure 8-25 Laser speckle photograph.

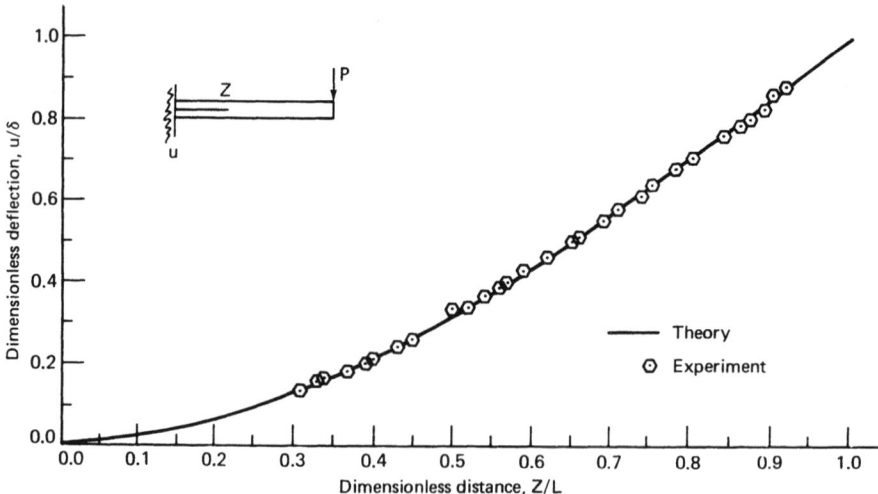

Figure 8-26 Comparison of theory and experiment for laser speckle photography measurements of the deflection of a cantilever beam.

HOLOGRAPHIC AND LASER SPECKLE INTERFEROMETRY

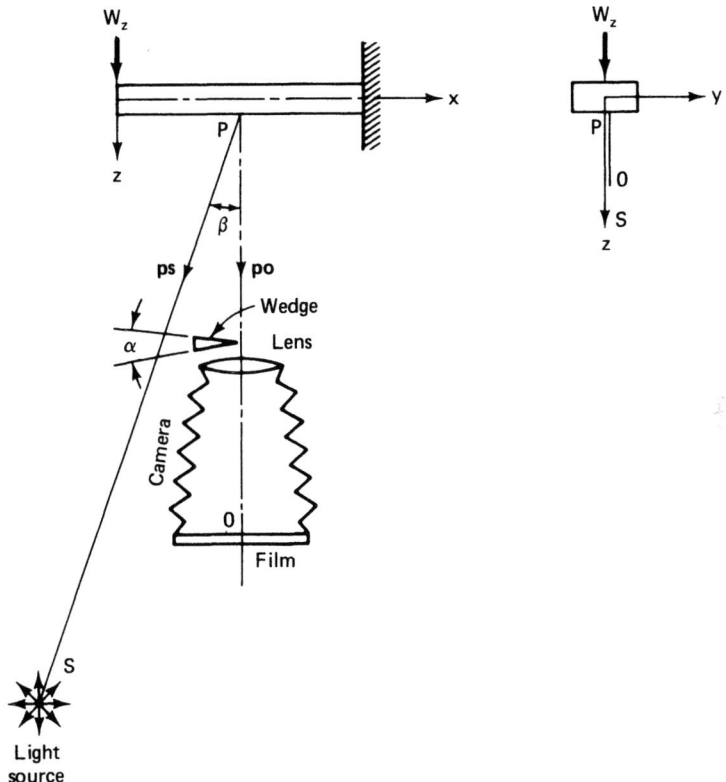

Figure 8-27 Geometrical configuration of the shearing speckle experimental for a cantilever beam with a transverse end load.

Fringe-order data for the shearing speckle are obtained by taking optically the Fourier transform of the double-exposed transparency as shown in Fig. 8-10. Data for this example are the following:

$$l_o = m_o = n_o = 1$$

$$l_s = \sin \beta \quad m_s = 0 \quad n_s = \cos \beta$$

For a shear of value $\Delta x = 2.3 \times 10^{-3}$ m (0.0899 in.), SI units in the negative x direction only, and assuming that the only nonzero displacement component is w_p, the fringe order n is expressed as

$$N = (1 + \cos \beta) \frac{\partial w}{\partial x} \frac{\Delta x}{\lambda} + \frac{1}{2} \tag{8-51}$$

where the wavelength of the laser light used $\lambda = 6328$ Å. A comparison of the theoretical and experimental results for the case of free-end deflection of 128 (0.005-in.) units is given in Fig. 8-28.

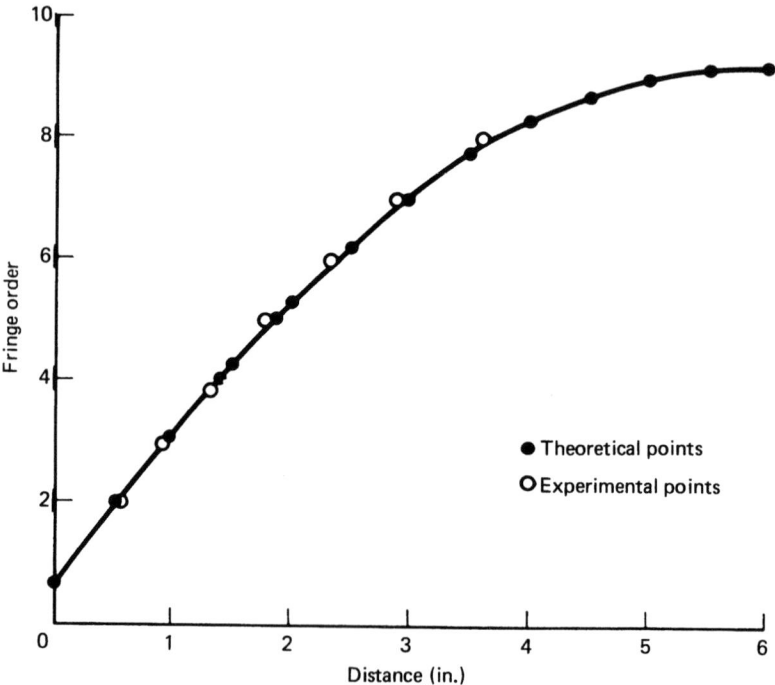

Figure 8-28 Comparison of theory and experiment.

8-4 Image-Processing Applications in Speckle Metrology

Computer vision technology is one of the fastest-expanding technologies today, and any attempt here to describe the system will certainly be obsolete in the near future. However, if the current trends continue, certainly, the impact of this technology will provide for significant changes in experimental mechanics. Although computing capability will improve our ability to solve larger problems, several basic concepts should remain constant. The equations of holographic interferometry and laser speckle photography, which predict the formation of fringes for deformable bodies, are the fruits of a mature science. Applications may change as computers are integrated into this area. However, direct processing of discrete optical data will still require a camera interfaced to a host computer. If we assume that these two areas will continue to form the basic elements of computer technology, image processing may be discussed as a basic concept, and improvements in computer technology may not necessarily change these basic concepts.

Fringe formation in holography and speckle metrology appear as a noisy pattern that is a result of the coherent illumination. Data-processing applications of the fringe patterns usually require the location of the minimum value in the presence of optical noise [37–40]. Therefore, the applications discussed in the next section are concerned with the processing of fringes in the presence of a noisy background. Specific engineering applications are not included.

Direct recording of speckle images in discrete forms with a camera interfaced to a host computer represent the other basic problem [41,42]. The data processing is analogous to the principles in

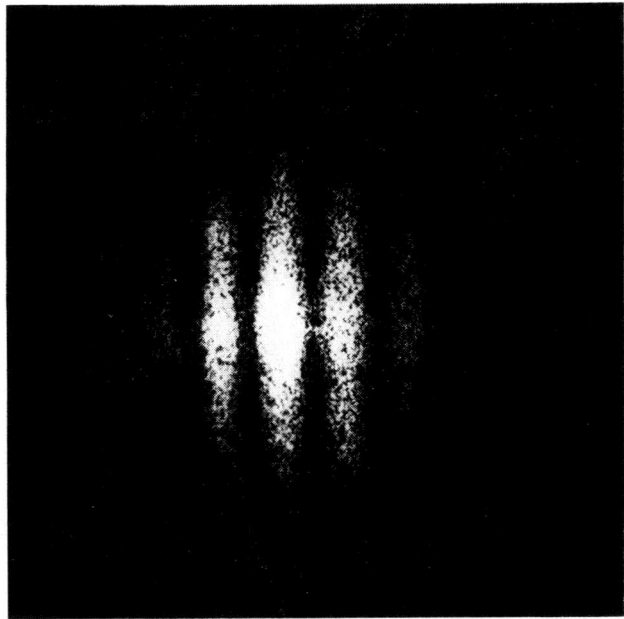

Figure 8-29 (a) Photograph of speckle photography pattern; (b) Digitized scan of a horizontal line of data.

white-light speckle photography and will be presented independent of the computer applications. Each method may be significantly different, with the magnitude of problem applications.

Image Processing of Speckle Photography Fringe Patterns

A single line of image data recorded in host computer from a speckle photograph is illustrated in Fig. 8-29a. As may be observed from the graphical presentation of a single line (Fig. 8-29b), the fringe pattern is noisy due to the coherent illumination of the transparency. Improvements in the data analysis may be obtained through highly effective digital filters. The term "linear filter," or "linear shift-invariant transform," was first applied to the speckle fringes by Chambless and Broadway [23]. These authors presented the background information of the design of a digital filter that utilizes the calculations of a fast Fourier transform. Basically, this method calculates the discrete transform, and the digital filter is used as a smoothing operation that cuts off the higher-order harmonics. The inverse transform of the filtered data is a smooth representation of the original data with the elimination of the coherent optical noise.

Figure 8-29b is a graphical representation of a single line of discrete data and the inverse transform with different levels of digital filters. As may be observed, a 256-harmonic discrete transform is an accurate representation of the data (Fig. 8-30). If the first 160 harmonics are used to obtain the inverse transform, a smoother representation is obtained (Fig. 8-31). This process is then continued until the optimal filter is obtained. Calculations suggest that a filter that retains only the first 15 harmonics is nearly optimum for a digital filter for this type of data, as illustrated in Fig. 8-32. Further experiments have verified that this filter provides accurate results of speckle photography data.

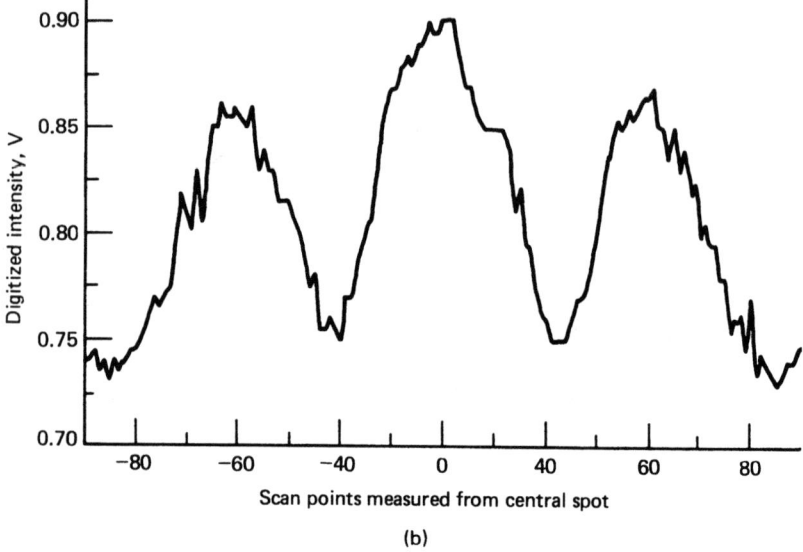

Figure 8-29 (*continued*)

Data recorded in the example discussed in Section 8-3 were used as an example of digital filtering of speckle patterns. A digital filter retaining only the first 15 harmonics was constructed to calculate fringe spacing from the smooth inverse transform. Five lines of video were recorded, and the average of these lines represented the data. The spacing from the smooth data was then compared with a manual scan as recorded by an observer. Figure 8-33 is a comparison of the manual scan, computer scan, and the theory for the cantilever beam example. As may be observed, a digital filter produces results as accurate as those that can be obtained by a skilled observer.

The fast Fourier transform used as a filter for speckle data is one-dimensional and is amenable to this type of data. However, coherent noise exists in all full-field applications of holography, shearing speckle interferometry, and full-field speckle interferometry. A two-dimensional fast Fourier transform applied in the same manner may be optimum for filtering full-field data. Until now, computing time was prohibitive except on large-scale mainframe computers. However, with improvements today, this form of automated data analysis may now be a feasible concept of image processing of optical data.

Digital Image of Speckle Patterns [24]

White-light speckle photography, developed primarily by Chiang et al. [36], is directly analogous to the digital imaging of speckle patterns. The frequency content of the discrete image is small compared to the frequency content of a laser speckle image. The direct recording of a white-light speckle image in a computer using direct imaging of a camera interfaced to the computer is defined as a digital speckle pattern. Data analysis, although analogous to speckle interferometry, is sufficiently different to require a discussion in order to complete the computer applications.

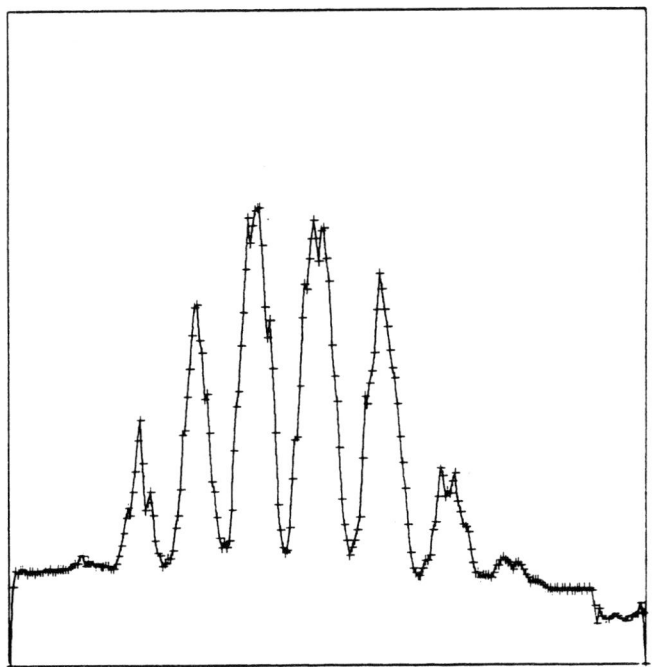

Figure 8-30 Plot of fringe pattern intensity using 256 harmonics.

Given the digital speckle images of the global response of the body before and after deformation, the local object motion from the global reference and deformed configuration of a body must be calculated. Thus the problem is geometrical in nature and initially is not concerned with the external forces necessary to cause the deformation.

Suppose that the brightness levels of a digital speckle image of a body surface can be described as a function $f(x, y)$ described in terms of the unstrained coordinates x, y, as illustrated schematically in Fig. 8-34. Now suppose that the body is deformed with respect to the reference configuration, which can be described as a function $f^*(x^*, y^*)$ described in terms of the deformed coordinates x^*, y^*, which is also illustrated schematically in Fig. 8-34. If the displacement functions are substituted into the representation of the deformed image function, then $f^*(x^*, y^*)$ is given by

$$f^*(y^*, y^*) = f^*(x + u, y + v)$$

where

$$x^* = x + u \qquad (8\text{-}52)$$
$$y^* = y + u$$

The digital speckle image surface is assumed to be parallel to the plane of the scattering surface.

Since the deformed image is not just a uniform translation in the x, y directions, local object motion in the presence of geometric distortion (strain) has to be considered. Figure 8-35 illustrates the local motion and the deformation of a subset image.

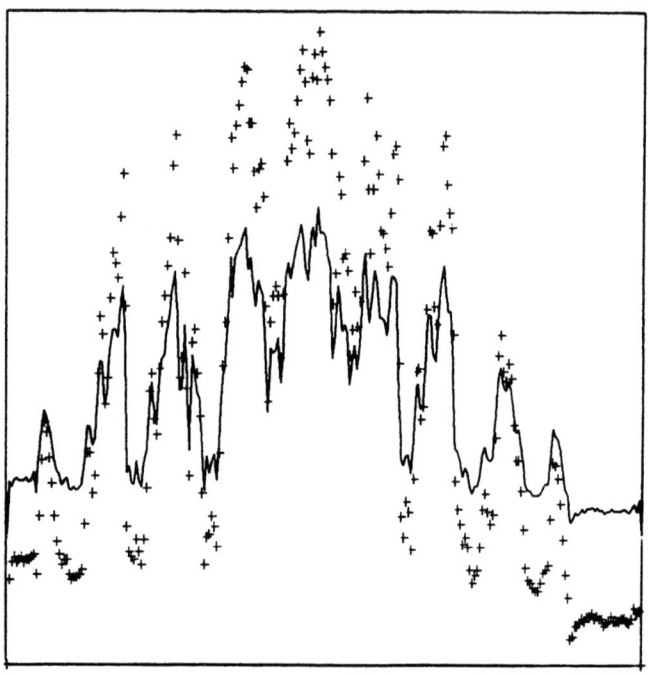

Figure 8-31 Plot of fringe pattern intensity using the first 160 harmonics. The smooth curve is the inverse transform of the digital filter data.

A digital image is fundamentally discrete; thus the gray levels are defined only at integral values of coordinates x, y. Pixels in the undeformed image, however, may move to position between pixels and the deformed image. Therefore, interpolation or surface fitting is required for the gray levels of the deformed image.

Suppose that a subset image is in the neighborhood surrounding a point of interest P_0. Assume that this subset image is so small that a line element in the undeformed configuration is deformed to a line element in the deformed one without warping, as shown in Fig. 8-35. A small rectangular subset image with center P_0 is moved and deformed to a parallelogram with center P_0.

Let the local coordinates x, y be assigned to the midpoint P_0 of the undeformed subset image. Assume that the displacement components of P_0 to P_0^* due to the deformation are u_0 and v_0. Two sides of the rectangle in the undeformed image intersect positive x and y axes at points P_a and P_b, respectively. Transformation of the three control points P_0, P_a, and P_b, due to deformation is illustrated in Fig. 8-36. The midpoint $P_0(0, 0)$ displaced to $P_0^*(u_0, v_0)$, while the midpoints of the sides $p_a(x_a, y_a)$ and $P_b(x_b, y_b)$ are displaced to $P_a^*(x_a^*, y_a^*)$ and $P_b^*(x_b^*, y_b^*)$, respectively.

To calculate the coordinates of an arbitrary point in the subset image after deformation, a spatial transformation function based on the control grid interpolation has to be determined. Three control points P_0, P_a, and P_b, which form three vertices of a rectangle in the undeformed image, are specified to make up an undeformed control grid. Pixels are located exactly at the nodes of the grid. The undeformed control grid maps into a control grid of contiguous parallelogram in the deformed image. The rectangle vertices map directly to the vertices of a corresponding parallel-

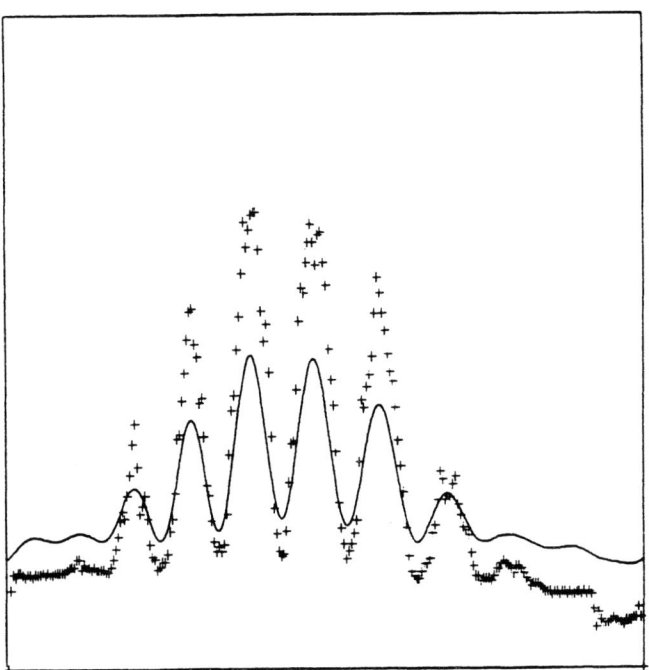

Figure 8-32 Plot of fringe pattern intensity using the first 15 harmonics. The smooth curve is the inverse transform of the digital filter data.

ogram. Points inside an input rectangle map to points within the corresponding output parallelogram. Similarly for line element P_0P_b,

$$x_b^* = \frac{\partial u}{\partial y} Y_b + u_0$$

$$y_b^* = \left(1 + \frac{\partial w}{\partial y}\right) y_b + v_0 \tag{8-53}$$

Compare equations for control points $P_a(x_a, 0)$ and $P_b(0, y_b)$ and note that $c_0 = u_0$ and $d_0 = v_0$; we obtain

$$C_1 = 1 + \frac{\partial u}{\partial x} \quad d_1 = \frac{\partial v}{\partial x}$$

$$C_2 = \frac{\partial u}{\partial y} \quad d_2 = 1 + \frac{\partial v}{\partial y} \tag{8-54}$$

Substitution of Eq. (8-52) in Eq. (8-53) yields

$$x^* = \left(1 + \frac{\partial u}{\partial x}\right)x + \frac{\partial u}{\partial y} y + u_0$$

$$y^* = \frac{\partial v}{\partial x} x + \left(1 + \frac{\partial v}{\partial y}\right) y + v_0 \tag{8-55}$$

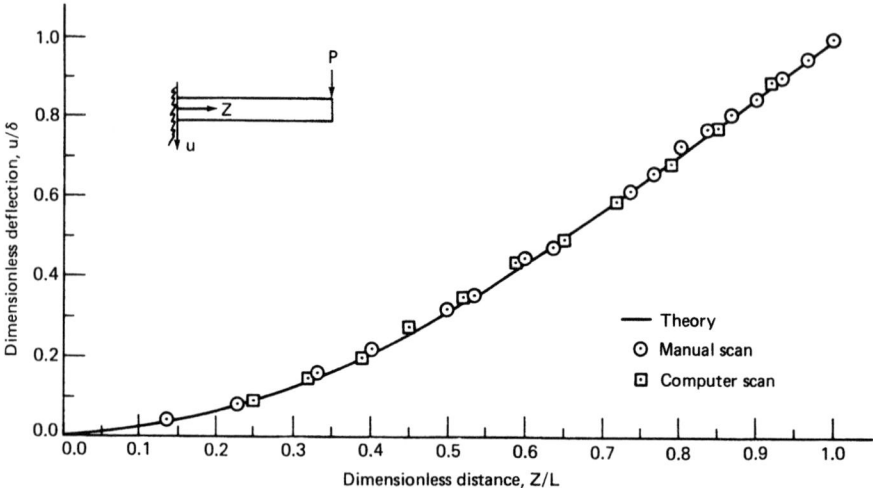

Figure 8-33 Comparison of digital filtered fringe spacing and spacing obtained by an observer.

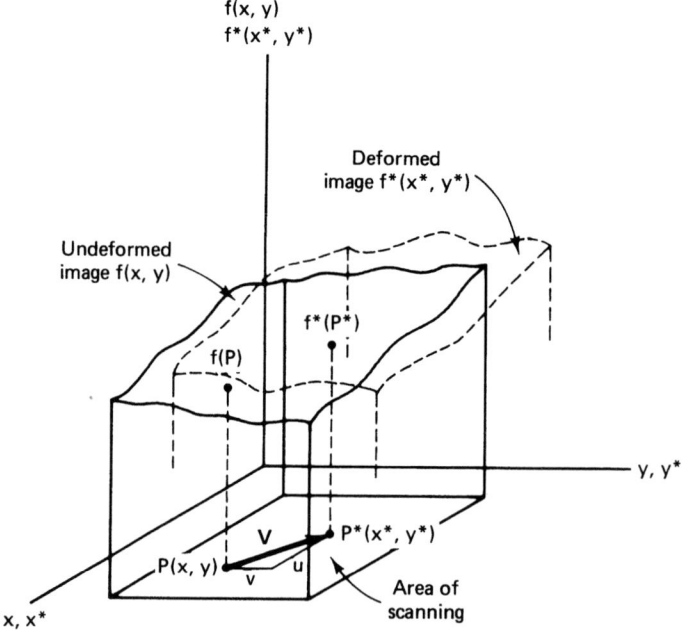

Figure 8-34 Gray-level representation of deformed and undeformed digital speckle images.

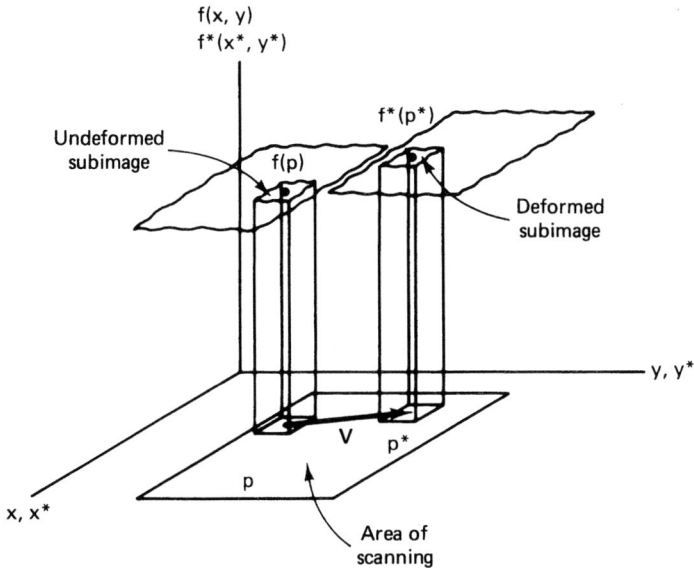

Figure 8-35 Local motion and distortion of a subset image.

Equation (8-54) describes the spatial transformation of a rectangle control grid in the undeformed image to a parallelogram in the deformed image.

As discussed previously, the pixels in the undeformed image map to fractional positions in the deformed image, generally falling in the space between four deformed pixels. Interpolation is necessary to determine the gray level of the deformed image.

As for the case of first-order interpolation, the bilinear function is required because fitting a plane through four points is an overconstrained problem. Bilinear interpolation produces more desirable results, with a slight increase in programming complexity and execution time. Let $g(x, y)$ be a function that is known at the vertices of the unit square. The value $g(x, y)$ at an arbitrary point inside the square can be obtained by fitting a hyperbolic paraboloid through the four known points. This paraboloid is defined by the linear equation [24]

$$g(x, y) = ax + by + cxy + d \qquad (8\text{-}56)$$

where $\quad a = g(1, 0) - g(0, 0)$
$\qquad\quad b = g(0, 1) - g(0, 0)$
$\qquad\quad c = g(1, 1) + g(0, 0) - g(0, 1) - g(1, 0)$
$\qquad\quad d = g(0, 0)$

The schematic illustration of the bilinear interpolation is shown in Fig. 8-37.

The piecewise bilinear function given by Eq. (8-56) interpolates four adjacent pixel neighborhoods and forms surfaces that match in amplitude at the neighborhood boundaries but do not match in slope. Therefore, the piecewise bilinear interpolation will form a continuous surface whose derivatives have discontinuities at the neighborhood boundaries. The undesired effects produced by the slope discontinuities of bilinear function involving magnification may be improved by the bicubic splines.

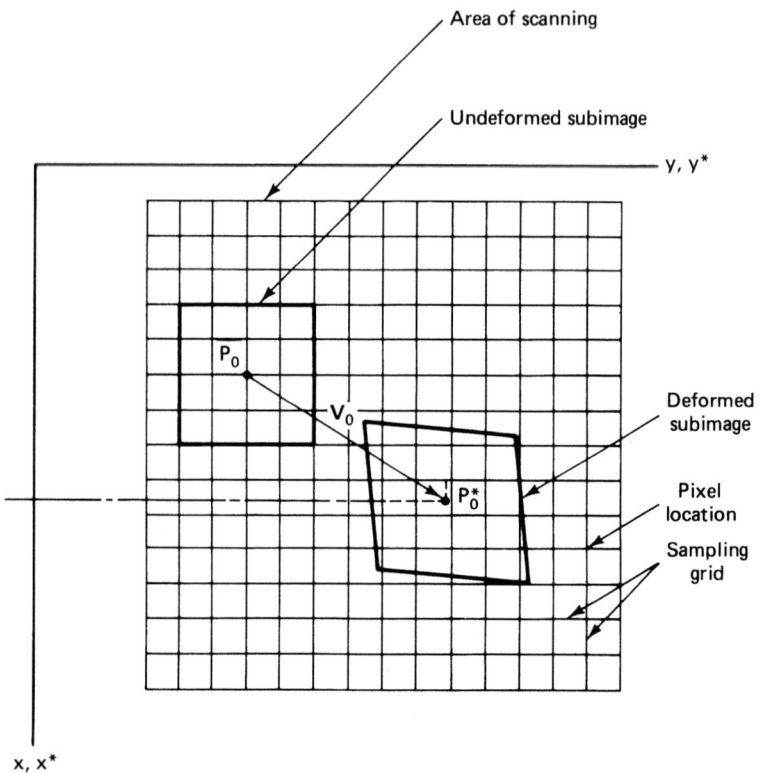

Figure 8-36 Deformation of a subimage in a sampling grid.

The cross-correlation function for two sets of random data describes the general dependence of the value of one set of data on the other. Consider the pair of digital speckle images before and after deformation of an object body. The gray-level representations for the undeformed and deformed images are $f(x, y)$ and $f^*(x^*, y^*)$, respectively. Substitution of displacements into the deformed expression of digital images gives

$$f^*(x^*, y^*) = f^*(x + u, y + v) \tag{8-57}$$

where $x^* = x + u(x, y)$
$y^* = y + v(x, y)$

To measure the local in-plane displacement components u and v, the images $f(x, y)$ and $f^*(x^*, y^*)$ must be correlated. The least-squares cross-correlation function is defined in the equation

$$C(u, v) = \int_M [f(x, y) - f^*(x + u, y + v)]^2 \, dx \, dy \tag{8-58}$$

where M is the scanning area of the image considered. If $C(u, v)$ is evaluated for all possible values of the functions u and v, $C(u, v)$ will be a minimum when $f^*(x + u, y + v)$ is transformed

HOLOGRAPHIC AND LASER SPECKLE INTERFEROMETRY

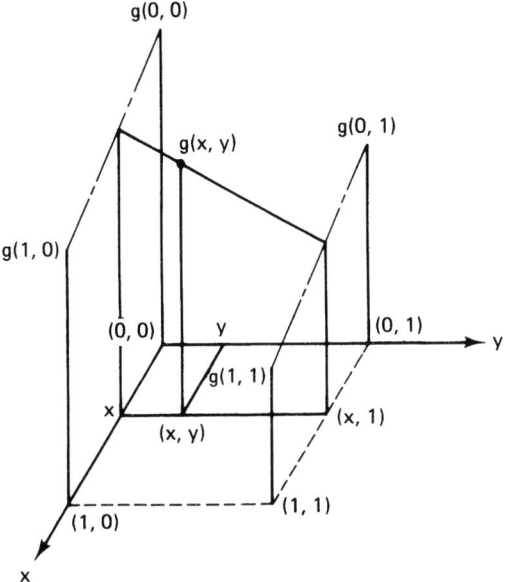

Figure 8-37 Bilinear interpolation.

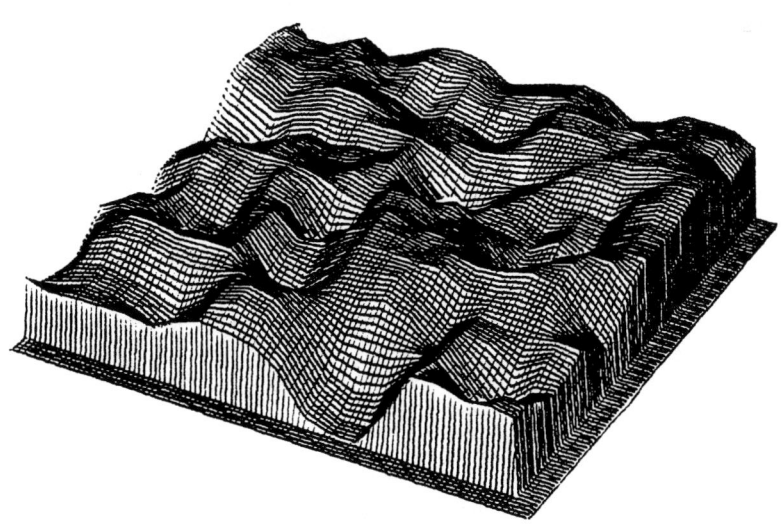

Figure 8-38 Bilinear subimage.

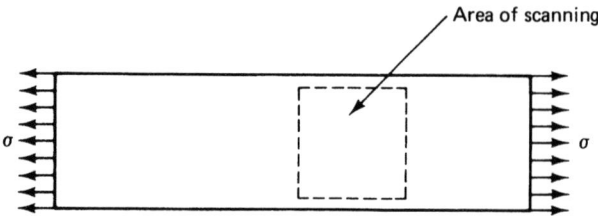

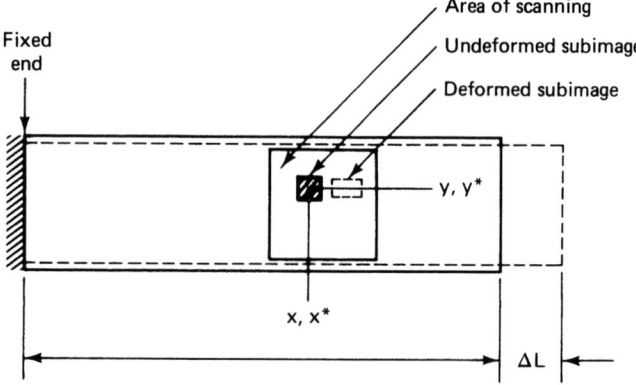

Figure 8-39 Deformed and undeformed configuration of a tensile specimen.

to match $f(x, y)$. Equation (8-58) is not applied in a straightforward manner because u and v are functions of the unstrained coordinates x and y, which are unknown quantities to be determined.

Consider some neighborhood of a point P_0 in the undeformed image as illustrated in Fig. 8-36. After deformation, P_0 is displaced to a point P_0^* and its neighborhood is also distorted, due to effect of deformation gradients. Suppose that these gradients are uniform within the subset image; Eq. (8-58), which governs the deformation of a subset image, can be applied to derive the displacement function u and v as follows:

$$u = u_x x + u_y y + u_0$$
$$v = v_x x + v_y y + v_0 \tag{8-59}$$

where u_0 and v_0 are the displacement components of P_0, which is set to be the origin of the local coordinates x, y, as shown in Fig. 8-37. Substitution of Eq. (8-59) into Eq. (8-58) yields

$$C(u_0, u_0, u_x, u_y, u_y)$$
$$= \int_{\delta M} [f(x, y) - f(X + u_0 + u_x x + v_y y, + v_0 + v_x x + v_y y)]^2 \, dx \, dy \tag{8-60}$$

where M is the scanning area of the subset image surrounding the point of interest. Note that the correlation function C is a function of displacement components u_0, v_0 and the deformation

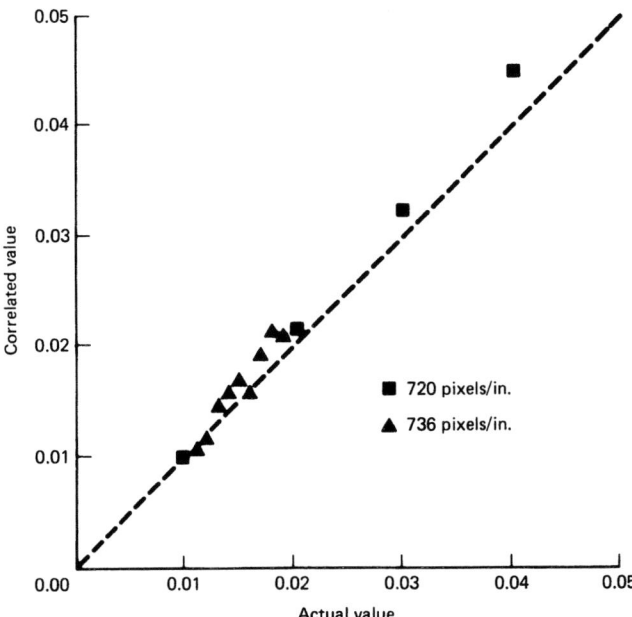

Figure 8-40 Correlated strain values versus theoretical strain values for a uniform tensile test.

gradients, which are assumed to be uniform within the subset image (Fig. 8-38). The correct values of u_0, v_0 and the deformation gradients corresponding to the object deformation will result in a minimum value of the correlation function C. This correlation procedure computes the displacements and their gradients at a point, and by repeating this procedure at other points, enough information may be obtained about the deformation for the particular problem of interest.

Illustration of the image correlation technique is illustrated by a uniform tension test as shown schematically in Fig. 8-39. This figure also defines the coordinate system for the undeformed and deformed configurations of a subimage in the neighborhood of a point of interest within the area of scanning. The images before and after deformation of the tension specimen correspond to stretched amount ΔL. Local rigid-body translations as well as deformation gradients at any point of interest were determined by correlation.

Strain components may be considered uniform in the area of interest and the only nonzero components are $\partial u/\partial x$ and $\partial v/\partial y$. Results for the uniform strain test are shown in Fig. 8-40. Correlated values of $\partial v/\partial y$ agree with the theoretical values within a reasonable accuracy (less than 10% error) for most of the data, ranging from 0.01 to 0.03.

The correlation of deformation gradient terms $\partial u/\partial x$, $\partial v/\partial y$ is very sensitive to small changes in the minimum of the correlation value. As a possible improvement to the calculation of strains, higher-order interpolations such as bicubic splines may be used instead of the first-order interpolation. Higher-order interpolations may reduce the sensitivity of the gradient terms and therefore increase the accuracy of strain measurements.

8-5 Summary and Future Trends

Since the introduction of the laser in the early 1960s, coherent-wave interferometry has made significant contributions to experimental mechanics. If the last 20 years are an indication, research in experimental mechanics certainly has a bright future. The future, quite surely, will be coupled to the computer. For example, the limitations of holography and speckle metrology occur in the time required to analyze the data. The formation of fringes is an easy task compared to the reduction of fringe information stress and strain.

In the past, computer vision required large mainframe computers to process large volumes of data. However, desktop workstations perform this task today and at a fraction of the cost. Color graphics coupled to vision make it possible to present fringe-type data in color. Color fringe patterns should be the tool of the experimentalist and not regarded as a thing of the past.

References

1. D. Gabor, *Nature, 161* (1948), 777; *Proc. R. Soc. London, A197* (1949), 454.

2. E. Leith and J. Upatnieks, *J. Opt. Soc. Am., 52* (1962), 1123.

3. K. A. Haines and B. P. Hildebrand, Surface-Deformation Measurement Using the Wavefront Reconstruction Technique, *Appl. Opt. 5*, no. 4 (1966), 595–602.

4. A. E. Ennos, Measurement of In-Plane Surface Strain by Hologram Interferometry, *J. Sci. Instrum.* (*J. Phys. E*), Ser. 2, 1., 1968.

5. E. B. Aleksandrov and A. M. Bonch-Bruevich, Investigation of Surface Strains by the Hologram Technique, *Sov. Phys. Tech. Phys., 12,* no. 2 1967.

6. J. E. Solid, Holographic Interferometry Applied to Measurement of Small Static Displacements of Diffusely Reflecting Surfaces, *Appl. Opt., 8* (1969), 1587–1595.

7. G. W. Stroke, G. M. Brown, and R. M. Grant, Theory of Holographic Interferometry, *J. Acoust. Soc. Am., 4*, no. 5 (1969), 1166–1179.

8. Olof Bryngdahl, Shearing Interferometry by Wavefront Reconstruction, *J. Opt. Soc. Am., 58* (July 1968).

9. H. Saito, I. Yamaguchi, and K. Hachimine, An Application of Holographic Interferometry to Stress Analysis of Elastic Bending Plate, Institute of Physical Chemistry, Research Paper 1802, October 1970.

10. R. E. Brooks, L. O. Heflinger, and R. F. Wuerker, Pulsed Laser Holograms, *J. Quantum Electron., 2* (1966), 275–279.

11. R. E. Brooks, L. O. Heflinger, R. F. Wuerker, and A. R. Briones, Holography Photography of High-Speed Phenomena with Conventional and Q-Switched Ruby Lasers, *Appl. Phys. Lett., 7* (1965), 92–94.

12. A. D. Jacobson and F. J. McClung, Holograms with Pulsed Laser Illumination, *Appl. Opt., 4* (1965), 1509–1510.

13. W. G. Gottenburg, Some Applications of Holographic Interferometry, *Exp. Mech., 8* (1969), 281–285.

14. D. A. Evensen and R. Aprahamian, Applications of Holography to Vibrations, Transient Response, and Wave Propagation, TRW Systems Group Report AM 70-11, May 1970.

15. R. L. Powell and K. A. Stetson, Interferometric Vibration Analysis by Wavefront Reconstruction, *J. Opt. Soc. Am., 5,* no. 12 (1965), 1593–1598.

16. J. A. Leendertz, Interferometric Displacement Measurement on Scattering Surfaces Utilizing Speckle Effect, *J. Phys. E: Sci. Instrum., 3* (1970), 214–218.

17. E. Archbold, J. M. Burch, and A. E. Ennos, Recording of In-Plane Surface Displacement by Double-Exposure Speckle Photography, *Opt. Acta, 17* (1970), 883–898.

18. J. N. Butters and J. A. Leendertz, A Double-Exposure Technique for Speckle Pattern Interferometry, *J. Phys. E: Sci. Instrum., 4* (1971), 277–279.
19. E. Archbold and A. E. Ennos, Displacement Measurement from Double-Exposure Laser Photography, *Opt. Acta, 19* (1972), 253–271.
20. D. E. Duffy, Measurement of Surface Displacement Normal to the Line of Sight, *Exp. Mech., 14* (September 1974), 378–384.
21. K. A. Stetson, A Review of Speckle Photography and Interferometry, *Opt. Eng., 14,* no. 5 (1975), 482–489.
22. Y. Y. Hung, Displacement and Strain Measurement, Chapter 4 in *Speckle Metrology* (R. K. Erf, Ed.), Academic Press, New York, 1978.
23. D. A. Chambless and J. A. Broadway, Digital Filtering of Speckle-Photography Data, *Exp. Mech., 19,* no. 8 (1979), 286–289.
24. T. C. Chu, W. F. Ranson, M. A. Sutton, and W. H. Peters, Applications of Digital Image Correlation Techniques to Experimental Mechanics, *Exp. Mech., 25,* no. 3 (1985), 232–244.
25. V. J. Parks, The Range of Speckle Metrology, *Exp. Mech., 20* (June 1980), 181–191.
26. D. H. Holoway, Application of Holographic Interferometry to Stress Wave and Crack Propagation Problems, *Opt. Eng., 21,* no. 3 (1982).
27. K. A. Stetson, A Rigorous Treatment of the Fringes of Hologram Interferometry, *Optic, 4* (1969), 385–400.
28. K. A. Stetson, New Design for Laser-Speckle Interferometry, *Opt. Laser Technol., 2* (1970), 179–181.
29. K. A. Stetson, Analysis of Double-Exposure Speckle Photography with Two-Beam Illumination, *J. Opt. Soc. Am., 64,* no. 6 (1974), 857–861.
30. F. D. Adams and G. E. Maddux, Synthesis of Holography and Speckle Photography to Measure 3-D Displacements, *Appl. Opt., 13,* no. 2 (1974), 219.
31. F. P. Chiang and R. P. Khetan, Strain Analysis by One-Beam Laser. Speckle Interferometry—2: Multiaperture Method, *Appl. Opt., 18,* no. 13 (July 1979).
32. F. P. Chiang and G. Jaisingh, On the Influence of Strain in One-Beam Laser Speckle Interferometry, presented to the Society for Experimental Stress Analysis, San Francisco, May 22, 1979.
33. G. Cloud, Practical speckle Interferometry for Measuring In-Plane Deformation, *Appl. Opt., 14,* no. 4 (1975), 878–884.
34. D. B. Barker and M. E. Fourney, Displacement Measurements in the Interior of 3-D Bodies Using Scattered-Light Speckle Patterns, *Exp. Mech., 16,* no. 6 (1976), 209–214.
35. D. E. Duffy, Measurement of Surface Displacement Normal to the Line of Sight, *Exp. Mech., 14,* no. 9 (1974), 378–384.
36. F. P. Chiang (Ed.), Coherent Optical Technique and Experimental Mechanics, *Opt. Eng., 21,* no. 3 (1982).
37. Y. Y. Hung, C. P. Hu, and C. E. Taylor, Speckle-Shearing Interferometric Camera—A Tool for Measurement of Derivatives of Surface-Displacement, *Proc. SPIE, 41* (1973), 169–175.
38. Y. Y. Hung, C. P. Hu, and C. E. Taylor, Speckle Moiré Interferometry: A Tool for Complete Measurement of In-Plane Surface Displacement, in *Developments in Theoretical and Applied Mechanics, 7* (SECTAM VII), Catholic University of America, Washington, DC, 1974.
39. Y. Y. Hung and C. E. Taylor, Measurement of Slopes of Structural Deflections by Speckle-Shearing Interferometry, *Exp. Mech., 14,* no. 7 (1974), 281–285.
40. Y. Y. Hung, A Speckle-Shearing Interferometer: A Tool for Measuring Derivatives of Surface Displacements, *Opt. Commun., 11,* no. 2 (1974), 132–135.
41. W. H. Peters and W. F. Ranson, Digital Imaging Techniques in Experimental Stress Analysis, *Opt. Eng., 21,* no. 3 (1982), 427–432.
42. M. A. Sutton, W. J. Wolters, W. H. Peters, W. F. Ranson, and S. R. McNeil, Determination of Displacements Using an Improved Digital Correlation Method, *Image Vision Comput., 1,* no. 3 (1983), 133–139.

CHAPTER

9

Shadow Optical Method of Caustics

Jörg F. Kalthoff

Experimentelle Mechanik
Ruhr-Universität Bochum
Bochum, Germany

9-0 Introduction

The shadow optical method of caustics is a relatively new experimental technique in stress–strain analysis introduced by Manogg [1,2] in 1964. The method is sensitive to stress gradients and therefore is an appropriate tool for quantifying stress concentration problems. Manogg originally used the method for investigating crack-tip stress intensifications. The technique was extended later by Theocaris [3–5], Rosakis [6,7], and the author and his colleagues [8–12] to different conditions of loading and material behavior in static as well as dynamic situations. Shadow optical images of test specimens under loading in general are characterized by very simple geometric patterns, which can easily be evaluated. Because of the simplicity of shadow patterns, the method can also be applied successfully for investigating rather complex phenomena, such as transient problems. Despite the complexity that may be inherent in the problems to be investigated, the clearness of the recordings generated allows the derivation of reliable, informative data. The author and his colleagues have applied the caustic technique to the investigation of various problems of practical interest in the field of fracture dynamics, in particular to the behavior of propagating and subsequently arresting cracks and the behavior of cracks under different impact loading conditions.

This chapter illustrates the basic physical principle of the shadow optical method of caustics and presents the mathematical description of the imaging process. Caustic evaluation formulas are presented for various stress–strain concentration problems. Experimental techniques for the generation and recording of shadow patterns in static and dynamic situations are discussed with special emphasis on laboratory test conditions. The applicability of the technique for investigating problems of practical relevance is demonstrated by several examples.

9-1 Physical Principle

Stresses in a solid alter the optical properties of the solid. Tensile stresses reduce the thickness of the body due to Poisson's effects and they also reduce the refractive index of the material,

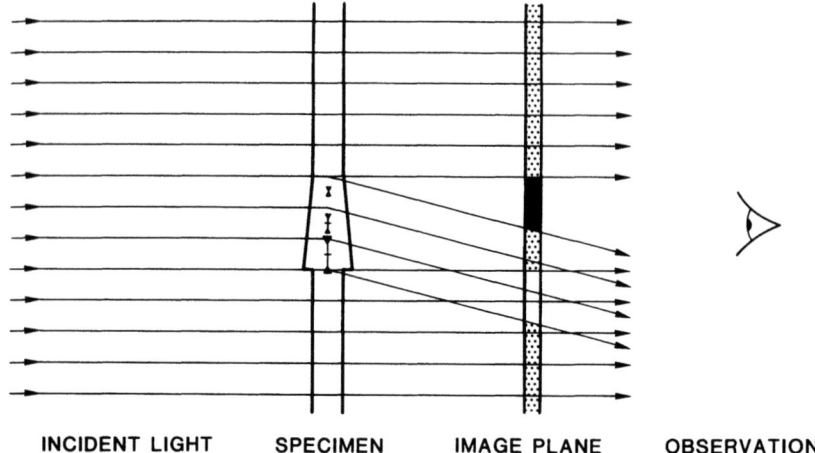

Figure 9-1 Principle of shadow optical imaging process (schematically).

since the material becomes optically less dense. The reverse situation applies for compressive stresses. These changes in the optical properties are utilized in the shadow optical method of caustics to make stress distributions in solids visible.

Model Considerations for Shadow Optical Images

The principle of the shadow optical imaging process is illustrated in Fig. 9-1. Considered is a plate, the middle section of which is subjected to linearly increasing compressive stresses. According to this stress distribution, the thickness of the plate and the refractive index of the material increase linearly in this section of the plate. Both effects have the same influence on the deflection of light rays. Here the two effects are combined and substituted in a simple model having the same net influence on the light deflection (i.e., a prism). The plate is illuminated from the left side by a parallel incident light beam. The stress-free parts of the plate are traversed by the light rays without deflection; in the middle section of the plate, however, the light rays are deflected (in this simplified model by the same angle). Consequently, the light distribution in an arbitrary plane (image plane or reference plane) behind the specimen (i.e., at the right side of the specimen) is not uniform any more. Certain areas are not hit by light rays and thus appear dark, whereas other areas are hit by light rays twice and thus appear as areas of increased brightness. The light distribution can be made visible on a screen. The resulting pattern represents a quantitative description of the stress distribution in the plate.

Shadow optical light distributions can be observed in different ways, in transmission or in reflection arrangements, as real or as virtual images. In transmission arrangements the observation direction is opposite to the direction of the illuminating light beam; in reflection arrangements it is in the same direction. Real images are obtained in reference planes located ahead of the specimen—when looking in the observation direction; virtual images are obtained in reference planes located behind it. The different possibilities of observation and the resulting light distributions are explained in Fig. 9-2. Two characteristic examples are considered which illustrate typical but simplified compressive and tensile stress concentration problems. A plate in its middle

SHADOW OPTICAL METHOD OF CAUSTICS

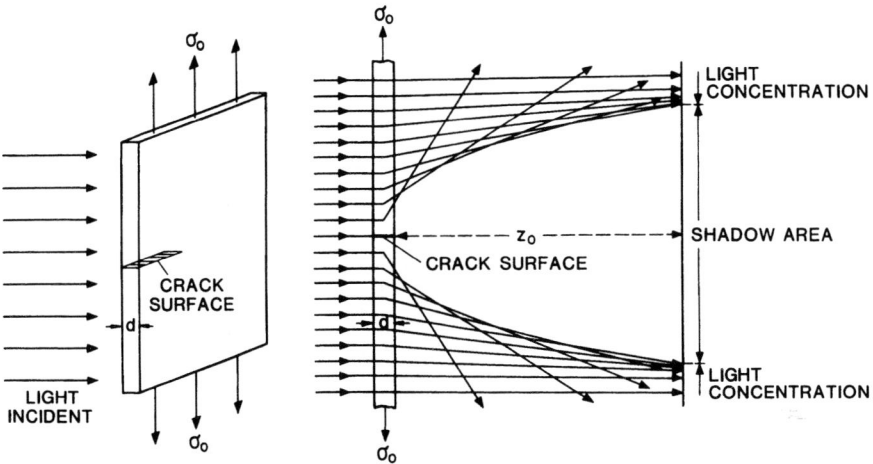

Figure 9-3 Light-ray distribution for a realistic stress concentration problem (schematically).

the image points move away from the central point again. The shadow optical image therefore exhibits a sharp boundary line between the area of darkness and the surrounding area of light concentration. This boundary line between the two areas is called the caustic curve ("caustic" is Greek for "focal line").

These particular light rays, which directly hit the caustic curve in the image plane, are of special importance; they are called the "initial" light rays. The locus of all points where the initial light rays traverse the object plane (specimen) is called the initial curve. Light rays traversing the object plane at points that compared with the initial curve are closer to (or farther away from) the crack tip are deflected more steeply (or more shallowly) than the initial light rays. Consequently, in both cases the image points lie outside the caustic curve. The caustic curve in the image plane is the direct image of the initial curve in the object plane. In the mathematical sense of the theory of imaging, this mapping process of the initial curve onto the caustic curve is in general not reversible and does not represent a one-to-one correspondence.

Figure 9-4 shows examples of experimentally observed shadow optical images and the resulting caustic curves for various stress concentration problems. These are: a 60° notch subjected to compressive and tensile loading (Fig. 9-4a), a compressive edge load acting on a half-plane (Fig. 9-4b), and a crack subjected to tensile (mode I) loading (Fig. 9-4c). The shadow optical pictures were photographed with various materials under the different observation modes specified in the figure.

The derived interpretation rules of shadow optical light distributions are verified by these examples. For example, the virtual shadow pattern of the notch under compression is identical to the real pattern of the notch under tension (Fig. 9-4a). The same is true for the real pattern of the notch under compression and the virtual pattern of the notch under tension. In transmission, the real shadow pattern of the notch under tension shows a distinct region of central darkness, whereas the real patterns of the notch under compression and of the compressive edge load exhibit central regions of increased brightness (Fig. 9-4a and b, right photographs). The reverse applies for the virtual patterns (Fig. 9-4a and b, left photographs). The shape of the real crack-tip caustic observed in transmission is the same as that of the virtual crack-tip caustic observed in reflection (Fig. 9-4c).

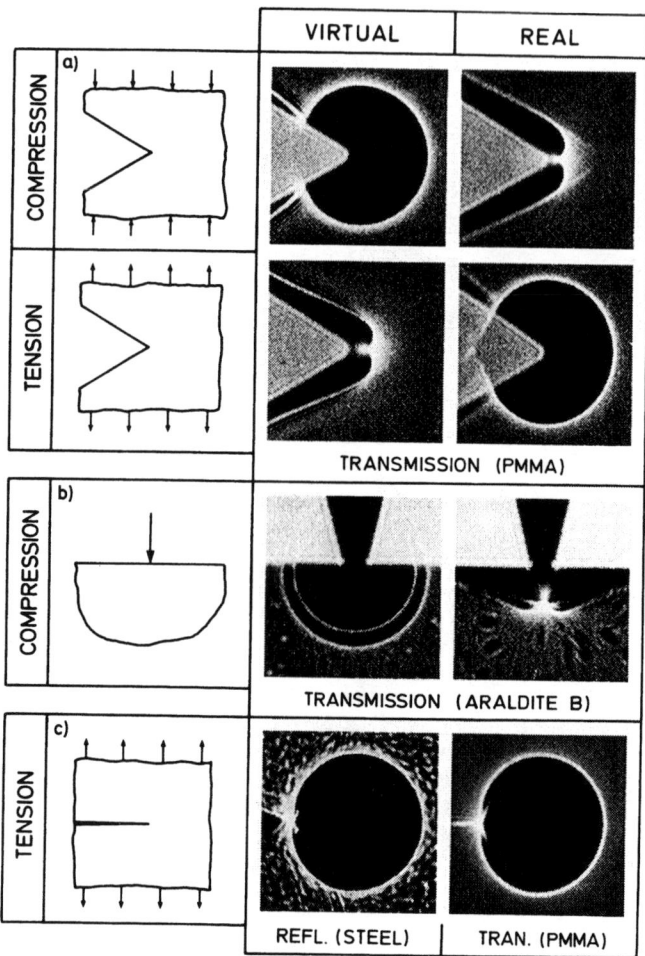

Figure 9-4 Experimentally observed shadow patterns.

9-2 Quantitative Description of Shadow Optical Images

General Mapping Equations

The mapping of the object plane E (specimen) onto a shadow optical image plane (or reference plane) E' by a parallel incident light beam is considered in Fig. 9-5. The schematic representation applies for a transmission arrangement and observation of a real shadow optical image (i.e., $z_0 > 0$). A notched specimen subjected to tensile loading is considered. The following quantitative description, however, applies quite generally for any kind of stress concentration problem or observation mode. The Cartesian (x, y) and polar (r, ϕ) coordinate systems are used in the object

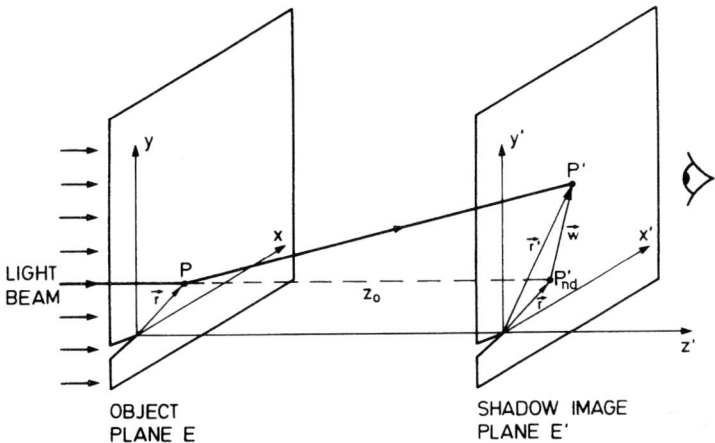

Figure 9-5 Light-ray deflection.

plane, while (x', y') and (r', ϕ') are used in the image plane. Tensile stresses are defined as positive.

A light ray is considered which traverses the object plane E at the point P with the distance $\mathbf{r}$ from the origin of the coordinate system. In an unloaded specimen this light ray would not experience any light deflection in the object plane and thus would hit the image plane E' at the point P'_{nd}, with the distance $\mathbf{r}'_{nd} = \mathbf{r}$ from the origin. However, the stress distribution in a specimen subjected to loading would deflect the light ray. Consequently, the light ray will hit the image plane E' displaced by the vector $\mathbf{w}$ at the point $P'(\mathbf{r}')$ with

$$\mathbf{r}' = \mathbf{r} + \mathbf{w} \tag{9-1}$$

The direction and magnitude of the displacement vector $\mathbf{w}$ are controlled by the change in optical path length which the light ray experiences in the object plane. Figure 9-6 illustrates the simplified one-dimensional case of stresses and light deflections in the y direction. The planar wavefront of an impinging light beam is distorted when traversing the specimen as a result of changes in the thickness of the plate and in the refractive index of the material. The local retardation of the distorted wavefront with regard to an equivalent wavefront that did not pass through the specimen, denoted s, is given by

$$s = [n_l(y) - 1]\, d_l(y) \tag{9-2}$$

where $n_l(y)$ = local refractive index of the material and $d_l(y)$ = local thickness of the specimen. As is readily derived from Fig. 9-6, $w = z_0[\partial \Delta s(y)/\partial y]$ and, consequently, for the general two-dimensional case (see also Ref. 13)

$$\mathbf{w} = z_0\, \mathbf{grad}\, \Delta s(r, \phi) \tag{9-3}$$

When an arbitrary observation mode is considered, Δs is written in the general form

$$\Delta s = (n - 1)\Delta d_{\text{eff}} + d_{\text{eff}}\Delta n \tag{9-4}$$

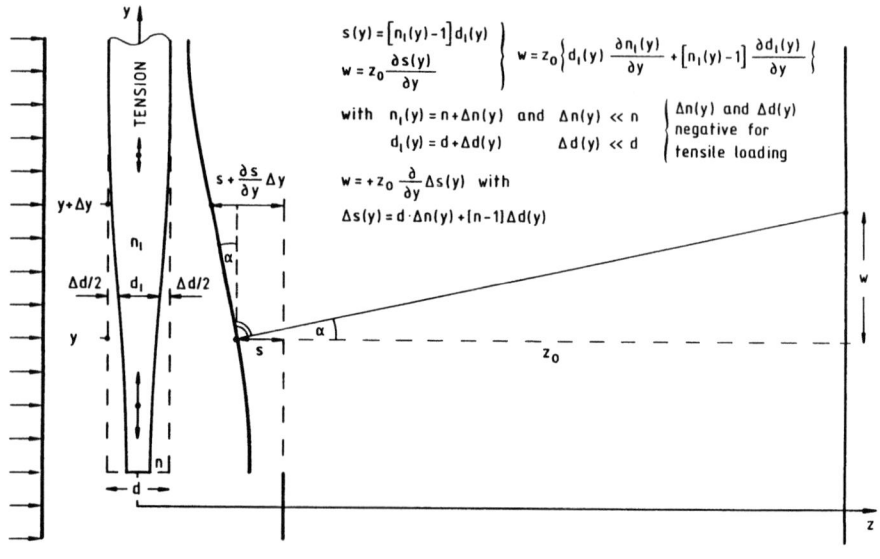

Figure 9-6 Derivation of the light deflection w (one-dimensional consideration).

where s = optical path length, retardation of wavefront
d_{eff} = effective thickness of the plate
n = refractive index $\Big\}$ for transparent specimens
$d_{\text{eff}} = d$
$n = -1$ $\Big\}$ for nontransparent specimens in reflection
$d_{\text{eff}} = d/2$
d = actual thickness of the specimen

Furthermore, for real images in transmission or reflection $z_0 > 0$, and for virtual images in transmission or reflection $z_0 < 0$.

The formal treatment of the reflection case equivalent to the transmission case results from the following consideration. In transmission, the surface deformations at both sides of the specimen and the change in the refractive index along the total thickness of the specimen contribute to the change in the optical path length. In reflection, only the deformation of the illuminated front surface determines the change in the optical path length (see Fig. 9-7). Thus the reflection case is obtained from the more general transmission case by formally setting the refractive index $n = -1$ (change in the direction of the light rays and $\Delta n = 0$) and by using an effective thickness $d_{\text{eff}} = d/2$ (consideration of the deformation at one surface only).

Furthermore, changes Δn in the refractive index due to the principal stresses σ_1, σ_2, σ_3, are described by the Maxwell–Neumann law. For the general case of a birefringent, transparent material

$$\Delta n_1 = A\sigma_1 + B(\sigma_2 + \sigma_3) \qquad \Delta n_2 = A\sigma_2 + B(\sigma_1 + \sigma_3) \tag{9-5}$$

SHADOW OPTICAL METHOD OF CAUSTICS

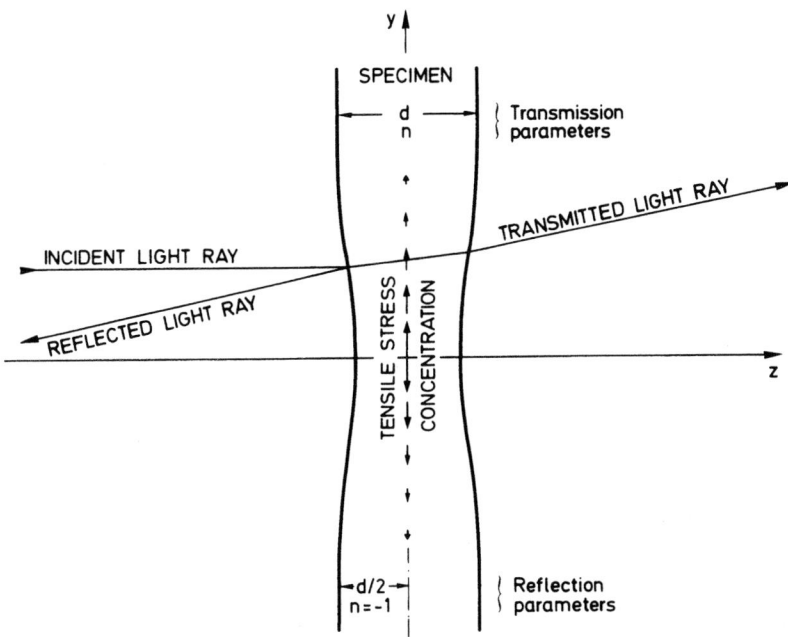

Figure 9-7 Relevant parameters for transmission and reflection of light rays.

where A and B are material constants. For optically isotropic, nonbirefringent materials $A = B$ and $\Delta n_1 = \Delta n_2 = \Delta n$. For reflection $A = B = 0$. Changes Δd_{eff} due to the prevailing stresses are described by Hooke's law.

$$\Delta d_{\text{eff}} = \left[\frac{1}{E}\sigma_3 - \frac{v}{E}(\sigma_1 + \sigma_2)\right] d_{\text{eff}} \tag{9-6}$$

where $\sigma_3 = 0$ for plane stress
$\Delta d_{\text{eff}} = 0$ for plane strain
E = Young's modulus
v = Poisson's ratio

With Eqs. (9-5) and (9-6), Eq. (9-4) can be written as

$$\Delta s_1 = d_{\text{eff}}(a\sigma_1 + b\sigma_2) \qquad \Delta s_2 = d_{\text{eff}}(a\sigma_2 + b\sigma_1) \tag{9-7}$$

where $a = A - (n - 1)v/E$, $b = B - (n - 1)v/E$ for plane stress
$a = A + vB$, $b = B + vB$ for plane strain

Equation (9-7) can be rearranged in a more convenient form in terms of the sum and the difference of the principal stresses as

$$\Delta s_{1/2} = cd_{\text{eff}}[(\sigma_1 + \sigma_2) \pm \lambda(\sigma_1 - \sigma_2)] \tag{9-8}$$

where

$$c = \frac{A+B}{2} - \frac{(n-1)v}{E} \qquad \lambda = \frac{A-B}{A+B-2(n-1)v/E} \qquad \text{for plane stress}$$

$$c = \frac{A+B}{2} + vB \qquad \lambda = \frac{A-B}{A+B+2vB} \qquad \text{for plane strain}$$

The constant c describes the change in the optical path length obtained with a specific material for a certain stress situation. Thus the constant c is a quantitative measure of the resulting shadow optical effect and therefore is called the "shadow optical constant." Influences on the change of the optical path length due to anisotropy effects of the material ($A \neq B$) are described by the coefficient λ. Numerical values for the material constants A, B, n, and the deduced shadow optical constant c and the anisotropy coefficient λ are given for various materials in Table 9-1. Note that the shadow optical constant c is negative for transmission arrangements but positive for reflection arrangements.

Equations (9-1), (9-3), and (9-8) describe the mapping of the object plane onto the shadow optical image plane for arbitrary stress distributions $\sigma_{1,2}(r, \phi)$. The specific mapping equation for a special stress concentration problem is obtained by inserting into the general equation (9-8) the individual formula for the stress distribution of the particular stress concentration problem considered.

Mapping Equations and Caustics for Specific Examples

Three stress concentration problems will be discussed simultaneously in a comparative manner: (a) a compressive edge load P acting on a half-plane; (b) a circular hole in a plate subjected to a biaxial stress field p, q; and (c) a crack in a plate under tensile load with mode I stress intensity factor K_I.[2] Graphical presentations of the three problems considered are given in Fig. 9-8.

The linear elastic stress concentration fields for the three examples are given by the following equations:

(a) Figure 9-8a:

$$\sigma_r = \frac{2P}{\pi} \frac{\sin \phi}{r}$$

$$\sigma_\phi = 0$$

$$\tau_{r\phi} = 0$$

(b) Figure 9-8b (formulas are also given in Chapter 1 of this book):

$$\sigma_r = \frac{p+q}{2}\left(1 - \frac{R^2}{r^2}\right) - \frac{p-q}{2}\left(1 - 4\frac{R^2}{r^2} + 3\frac{R^4}{r^4}\right) \cos 2\phi$$

$$\sigma_\phi = \frac{p+q}{2}\left(1 + \frac{R^2}{r^2}\right) + \frac{p-q}{2}\left(1 \qquad\quad + 3\frac{R^4}{r^4}\right) \cos 2\phi \qquad (9\text{-}9)$$

$$\tau_{r\phi} = \qquad\qquad\qquad\qquad \frac{p-q}{2}\left(1 + 2\frac{R^2}{r^2} - 3\frac{R^4}{r^4}\right) \sin 2\phi$$

---

[2] For the definition of the stress intensity factor K, see textbooks on fracture mechanics (e.g., Ref. 14 and Chapters 1 and 19 of this book).

Table 9-1 Constants for Caustic Evaluation

Material	Elastic Constants		Refractive Index	General Optical Constants		Shadow Optical Constants				Effective Thickness, d_{eff}
	Young's Modulus (MN/m²)	Poisson's Ratio, ν				For Plane Stress		For Plane Strain		
				A (m²/N)	B (m²/N)	c (m²/N)	λ	c (m²/N)	λ	
					Transmission					
Optically anisotropic										
Araldite B	3,660[a]	0.392[a]	1.592	-0.056×10^{-10}	-0.620×10^{-10}	-0.970×10^{-10}	-0.288	-0.580×10^{-10}	-0.482	d
CR-39	2,580	0.443	1.504	-0.160×10^{-10}	-0.520×10^{-10}	-1.200×10^{-10}	-0.148	-0.560×10^{-10}	-0.317	d
Plate glass	73,900	0.231	1.517	$+0.0032 \times 10^{-10}$	-0.025×10^{-10}	-0.027×10^{-10}	-0.519	-0.017×10^{-10}	-0.849	d
Homalite 100	4,820[a]	0.310[a]	1.561	-0.444×10^{-10}	-0.672×10^{-10}	-0.920×10^{-10}	-0.121	-0.767×10^{-10}	-0.149	d
Optically isotropic										
PMMA	3,240	0.350	1.491	-0.530×10^{-10}	-0.570×10^{-10}	-1.080×10^{-10}	~ 0	-0.750×10^{-10}	~ 0	d
					Reflection					
All materials	E	ν	-1	0	0	$2\nu/E$	0	—	—	$d/2$

[a] Dynamic values.

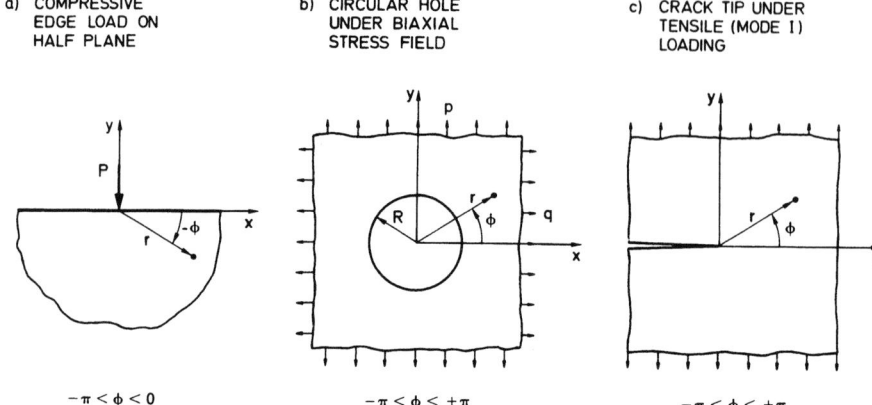

Figure 9-8 Typical stress concentration problems.

(c) Figure 9-8c (formulas are also given in Chapter 1 of this book):

$$\sigma_r = \frac{K_I}{\sqrt{2\pi r}} \frac{1}{4}\left(5\cos\frac{\phi}{2} - \cos\frac{3}{2}\phi\right)$$

$$\sigma_\phi = \frac{K_I}{\sqrt{2\pi r}} \frac{1}{4}\left(3\cos\frac{\phi}{2} + \cos\frac{3}{2}\phi\right)$$

$$\tau_{r\phi} = \frac{K_I}{\sqrt{2\pi r}} \frac{1}{4}\left(\sin\frac{\phi}{2} - \sin\frac{3}{2}\phi\right)$$

With these stress distributions and Eq. (9-8) (for $\lambda = 0$)[3], the mapping equations (9-1) and (9-3) become

(a) $\quad x' = r\cos\phi - \dfrac{2P}{\pi} z_0 c d_{\text{eff}} r^{-2} \sin 2\phi$

$\quad\quad y' = r\sin\phi + \dfrac{2P}{\pi} z_0 c d_{\text{eff}} r^{-2} \cos 2\phi$ $\quad$ for $-\pi < \phi < 0$

(b) $\quad x' = r\cos\phi - 4 z_0 c d_{\text{eff}} R^2 (p-q) r^{-3} \cos 3\phi$

$\quad\quad y' = r\sin\phi - 4 z_0 c d_{\text{eff}} R^2 (p-q) r^{-3} \sin 3\phi$ $\quad$ for $-\pi < \phi < \pi$ $\quad$ (9-10)

(c) $\quad x' = r\cos\phi - \dfrac{K_I}{\sqrt{2\pi}} z_0 c d_{\text{eff}} r^{-3/2} \cos\dfrac{3}{2}\phi$

$\quad\quad y' = r\sin\phi - \dfrac{K_I}{\sqrt{2\pi}} z_0 c d_{\text{eff}} r^{-3/2} \sin\dfrac{3}{2}\phi$ $\quad$ for $-\pi < \phi < \pi$

The complete family of the deflected light rays forms a shadow space behind the object plane. Its surface is an envelope to the light rays and is called the caustic surface. The intersection of

---

[3] For simplicity, only the isotropic case is considered in this context.

this surface with the image plane forms the caustic curve. This caustic curve is a multivalued, singular solution of Eqs. (9-1) and (9-3) (i.e., the mapping of points along the caustic curve is not reversible). Thus a necessary and sufficient condition for the existence of the caustic curve is obtained if the Jacobian of Eqs. (9-1) and (9-3) vanishes, that is,

$$\frac{\partial x'}{\partial r}\frac{\partial y'}{\partial \phi} - \frac{\partial x'}{\partial \phi}\frac{\partial y'}{\partial r} = 0 \tag{9-11}$$

The coordinates r, ϕ of points P that fulfill Eq. (9-11) form the initial curve in the object plane, and the mapping of this initial curve onto the image plane is the caustic curve.

Consequently, application of Eq. (9-11) to the mapping Eqs. (9-10) gives the equation of the initial curves

(a) $r = \left[\dfrac{4}{\pi}|z_0||c|d_{\text{eff}}P\right]^{1/3} \equiv r_0$ for $-\pi < \phi < 0$

(b) $r = [12|z_0||c|d_{\text{eff}}R^2(p-q)]^{1/4} \equiv r_0$ for $-\pi < \phi < \pi$ (9-12)

(c) $r = \left[\dfrac{3}{2}\dfrac{K_1}{\sqrt{2\pi}}|z_0||c|d_{\text{eff}}\right]^{2/5} \equiv r_0$ for $-\pi < \phi < \pi$

For the three examples considered, the initial curves are circles around the center point of stress concentration with fixed radii r_0. With the mappings Eqs. (9-10) the caustic curves are finally obtained as images of the initial curves [Eqs. (9-12)] and are given as

(a) $x' = r_0(\cos\phi - \text{sgn}(z_0c)\tfrac{1}{2}\sin 2\phi)$
$y' = r_0(\sin\phi + \text{sgn}(z_0c)\tfrac{1}{2}\cos 2\phi)$ for $-\pi < \phi < 0$

(b) $x' = r_0(\cos\phi - \text{sgn}(z_0c)\tfrac{1}{3}\cos 3\phi)$
$y' = r_0(\sin\phi - \text{sgn}(z_0c)\tfrac{1}{3}\sin 3\phi)$ for $-\pi < \phi < +\pi$ (9-13)

(c) $x' = r_0(\cos\phi - \text{sgn}(z_0c)\tfrac{2}{3}\cos\tfrac{3}{2}\phi)$
$y' = r_0(\sin\phi - \text{sgn}(z_0c)\tfrac{2}{3}\sin\tfrac{3}{2}\phi)$ for $-\pi < \phi < +\pi$

Mathematically, the caustic curves are generalized epicycloids. The caustics are shown graphically in Fig. 9-9 for negative and positive signs of sgn(z_0c); that is, for transmission ($c < 0$) or reflection ($c > 0$) arrangements and for real ($z_0 > 0$) or virtual ($z_0 < 0$) images.

An illustrative picture of the complete distribution of deflected light rays in the image plane is obtained in a simple manner by considering the mapping of rays that traverse the object plane along lines $\phi = $ const. The images of those lines obtained with Eqs. (9-10) for the three examples considered are shown in Fig. 9-10. The caustic curves appear as envelopes to the obtained families of image lines (see Fig. 9-9 for comparison).

For the quantitative evaluation of caustics, a length parameter between characteristic points on the caustic curve is defined (e.g., the maximum diameter of the caustic curve given by the distances D in Fig. 9.9). These distances are related to the radii of the initial curve by the equations

(a) $D = 2.6r_0$

(b) $D = 2.67r_0$ (9-14)

(c) $D = 3.17r_0$

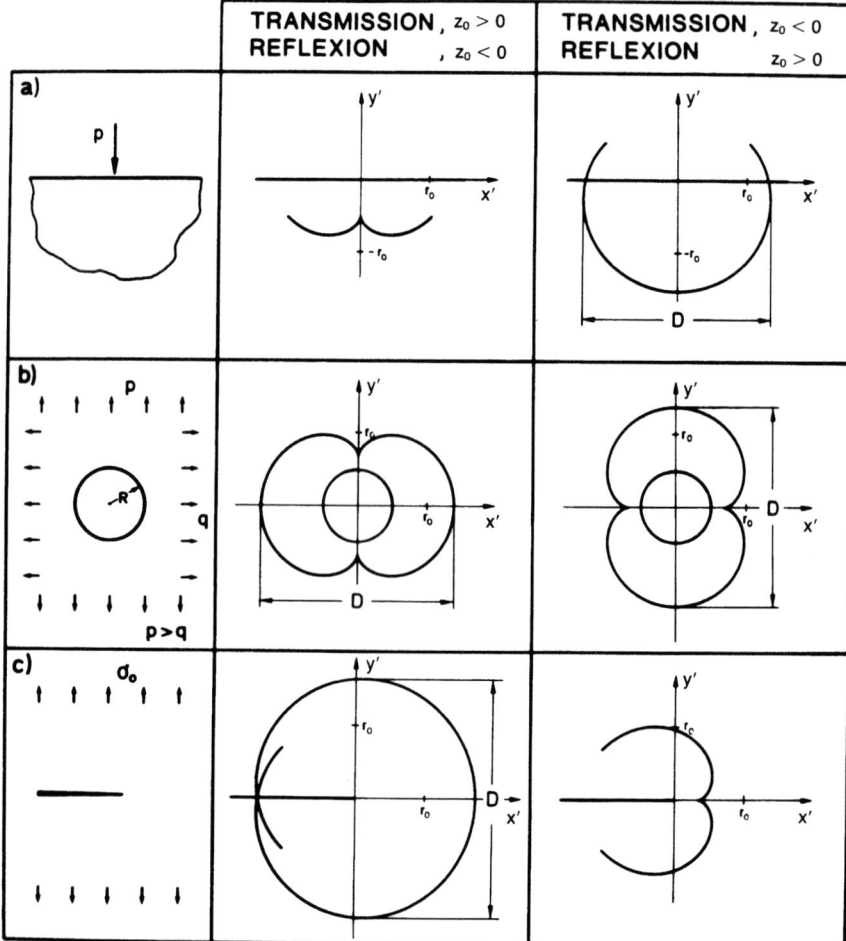

Figure 9-9 Caustic curves for the stress concentration problems considered.

With Eqs. (9-12) and (9-14) a quantitative formula is obtained in each case relating the size of the shadow optical pattern with the generating load parameter:

(a) $\quad P = \dfrac{\pi}{4(2.6)^3 |z_0| \, |c| d_{\text{eff}}} D^3$

(b) $\quad p - q = \dfrac{1}{12(2.67)^4 |z_0| \, |c| d_{\text{eff}} R^2} D^4 \qquad (9\text{-}15)$

(c) $\quad K_I = \dfrac{2\sqrt{2\pi}}{3(3.17)^{5/2} |z_0| \, |c| d_{\text{eff}}} D^{5/2}$

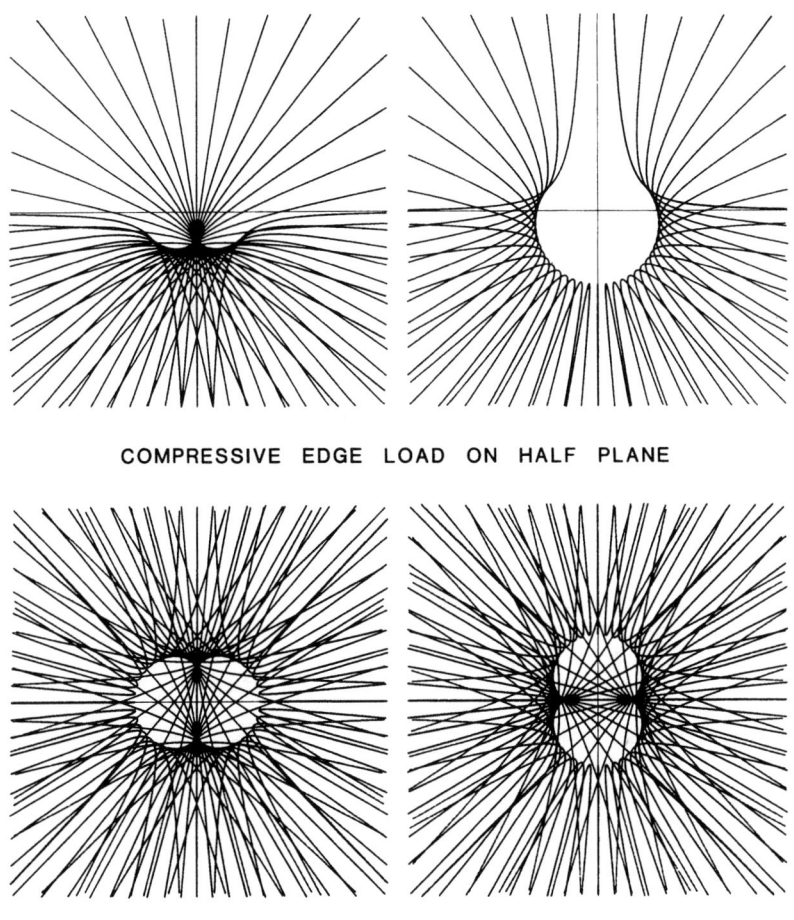

Figure 9-10 Shadow optical light distributions for the stress concentration problems considered. Left side: sgn($z_0 c$) negative. Right side: sgn($z_0 c$) positive.

Thus, with Eq. (9-15) and the distance D measured in an experimentally obtained caustic, the generating load parameter can be determined: that is, the magnitude of the compressive edge load P, or the difference of the biaxial stresses $p - q$, or the crack-tip stress intensity factor K_I.

It is also possible to use characteristic length parameters of the caustic curves other than those defined in Fig. 9-9, or to use caustic curves obtained in reference planes of opposite sign. However, to obtain evaluation formulas that are reliable and sufficiently accurate, it is advantageous to evaluate that caustic and to select that length parameter between two characteristic points on the caustic curve which represents the largest value. Furthermore, both characteristic points should be points on the caustic curve itself. The use of a characteristic point that does not lie on the caustic (e.g., a point that is related to the boundary of the specimen) should be

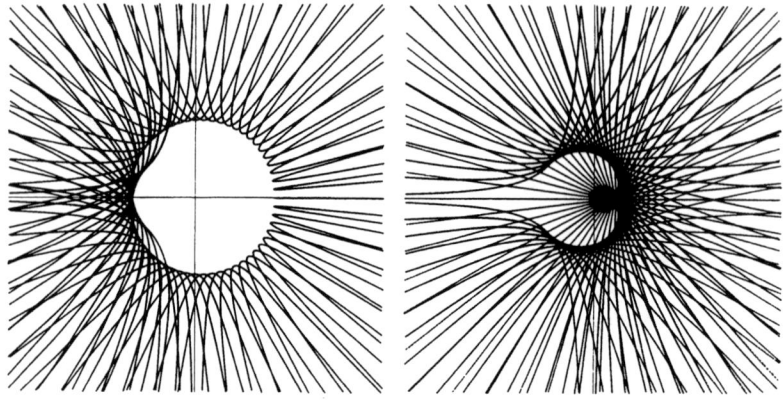

CRACK UNDER MODE I LOADING

Figure 9-10 *continued*

avoided. Because of the shadow optical imaging process, such a point is not sharply reproduced and thus cannot be located with sufficient accuracy. Also, some points that lie on the caustic curve may first appear as suitable characteristic points, but a more detailed consideration shows that they are not appropriate (e.g., the cusps of the caustic curves in Fig. 9-9). The plots of the light-ray distributions (Fig. 9-10) indicate that these points can be located experimentally only with reduced accuracy, as a result of the focal-point character of the cusps. The characteristic length parameters D defined in Fig. 9-9 were selected on the basis of the criteria discussed above.

When transparent materials are considered that are optically anisotropic (i.e., birefringent: $\lambda \neq 0$), we must take into account the anisotropy term in Eq. (9-8) that became zero in the treatment of the isotropic case (which was the only case considered so far). As a result, the caustic curves for the three examples considered split up into double caustic curves. The consideration of the anisotropy term renders the mathematical analysis more complex. For the crack problem, the relevant equations shall be given in this context:

A straightforward but lengthy calculation [15] analogous to the isotropic case yields for the mapping equations

$$x' = r \cos \phi - \frac{K_I}{\sqrt{2\pi}} z_0 c d_{\text{eff}} r^{-3/2} \left[\cos \frac{3}{2}\phi \pm \frac{\lambda}{4} 3 \sin 2\phi \, \text{sgn}(\phi) \right]$$
$$y' = r \sin \phi - \frac{K_I}{\sqrt{2\pi}} z_0 c d_{\text{eff}} r^{-3/2} \left[\sin \frac{3}{2}\phi \pm \frac{\lambda}{4} (-1 - 3 \cos 2\phi) \, \text{sgn}(\phi) \right]$$
(9-16)

The radius of the initial curve is given by the expression

$$r_0 = \left(\frac{3}{2} \frac{K_I}{\sqrt{2\pi}} |z_0| |c| d_{\text{eff}} B_I \right)^{2/5}$$
(9-17)

SHADOW OPTICAL METHOD OF CAUSTICS

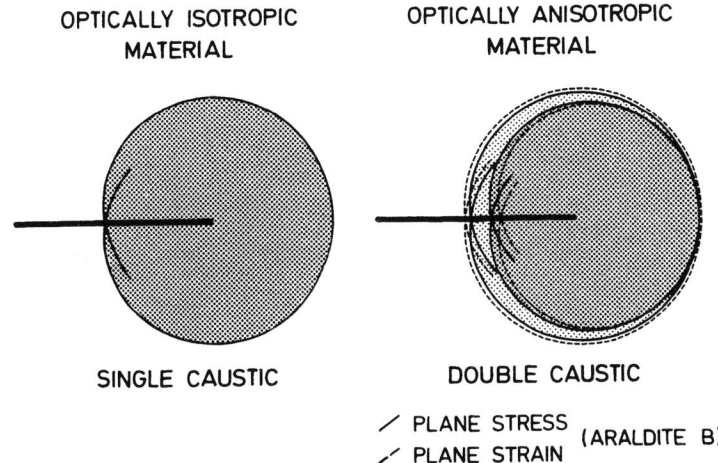

Figure 9-11 Mode I single caustic for optically isotropic material and double caustic for optically anisotropic material.

with

$$B_I = \left| \pm \frac{\lambda}{4} \sin \phi \operatorname{sgn}(\phi) - \left[1 \pm \operatorname{sgn}(\phi) \frac{\lambda}{4} \left(7 \sin \frac{\phi}{2} - \sin \frac{3}{2}\phi \right) \right. \right.$$
$$\left. \left. + \left(\frac{\lambda}{4}\right)^2 \left(\frac{25}{2} + \frac{7}{2} \cos 2\phi \right) \right]^{1/2} \right| \quad (9\text{-}18)$$

Because of the plus and minus signs of the λ-affected term in the expression of B_I, two initial curves result. Furthermore, these two initial curves are not represented by an exact circle, as in the isotropic case. But, the shape of these initial curves is almost identical to a circle; that is, deviations from an exact circle are within the thickness of the line the circle would be drawn with for all practically relevant values of the anisotropy coefficient λ.

With Eqs. (9-16) and (9-17) the caustic curves are obtained as

$$x' = r_0 \left(\cos \phi - \operatorname{sgn}(z_0 c) \frac{2}{3} B_I^{-1} \left[\cos \frac{3}{2}\phi \pm \frac{\lambda}{4} 3 \sin 2\phi \operatorname{sgn}(\phi) \right] \right)$$
$$y' = r_0 \left(\sin \phi - \operatorname{sgn}(z_0 c) \frac{2}{3} B_I^{-1} \left[\sin \frac{3}{2}\phi \pm \frac{\lambda}{4} (-1 - 3 \cos 2\phi) \operatorname{sgn}(\phi) \right] \right) \quad (9\text{-}19)$$

with B_I given by Eq. (9-18).

For $\lambda = 0$ these equations reduce to the expressions derived for the isotropic case: see Eq. (9-13c). As discussed before in the context of the initial curves, the two signs in Eq. (9-18) now define two caustic curves, not one, as obtained in the isotropic case. Thus, the additional consideration of the anisotropy term in the basic mapping equations results in a split-up of the single caustic curve obtained for the isotropic case into a double caustic curve. Figure 9-11 compares crack-tip caustics for the optically anisotropic material Araldite B to the corresponding single

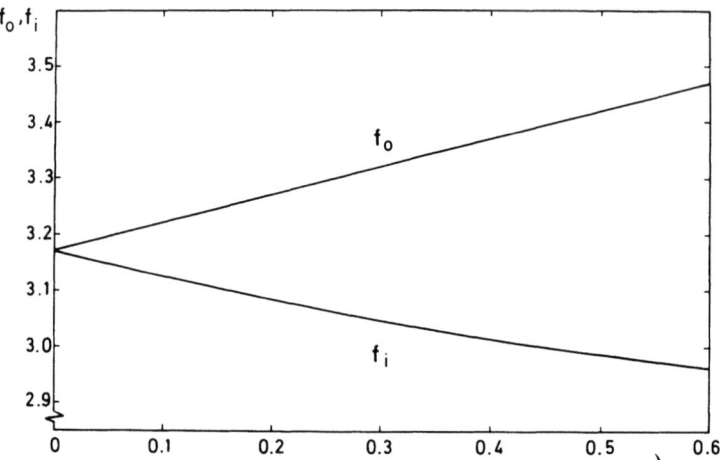

Figure 9-12 Numerical factors $f_{o,i}$ for evaluating anisotropic mode I crack-tip caustics.

caustic. Since the coefficient of anisotropy λ depends on the state of stress ($|\lambda|$ is larger for plane strain than for plane stress; see Table 9-1) the split-up of the double caustic is larger for plane strain than for plane stress.

The outer as well as the inner caustic can be used for determining the stress-intensity factor. The evaluation formula for the crack-tip stress-intensity factor is then

$$K_1 = \frac{2\sqrt{2\pi}}{3f_{o,i}^{5/2} |z_0| |c| d_{\text{eff}}} D_{o,i}^{5/2} \qquad (9\text{-}20)$$

where $D_{o,i}$ = outer, inner diameter of the crack-tip caustic
$f_{o,i}$ = numerical factor for evaluation of the outer, inner caustic

Numerical values of the factors $f_{o,i}$ that lie around the value 3.17 for the single caustic are given in Fig. 9-12 as a function of the anisotropy coefficient λ. Anisotropic evaluation formulas for the other two examples considered (see Fig. 9-8) can be derived in a analogous manner.

9-3 Crack-Tip Caustics

Most applications of the shadow optical method of caustics have so far been in the field of fracture mechanics. Consequently, the theory of caustics around crack tips has been developed to the greatest degree. Some of the results that are of more general interest and can be transferred to other caustic problems as well are presented in this section.

Mode I, Mode II, and Mode III Caustics

Three possible loading conditions exist for cracks (see Fig. 9-13): tensile loading (mode I), in-plane shear loading (mode II), and antiplane shear loading (mode III). Any arbitrary loading condition can be represented by a superposition of these three fundamental types of loading. For the sake of completeness, both positive and negative stress intensity factors are considered (for negative values of K_I, slits with a finite opening are considered instead of cracks to allow compressive stress concentrations to build up).

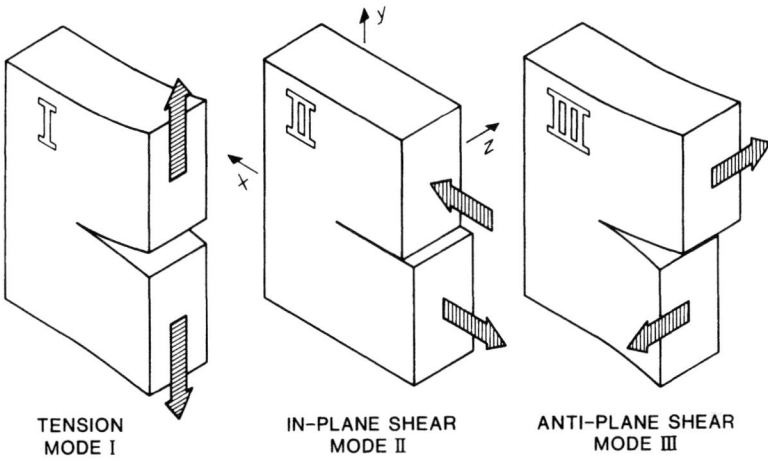

Figure 9-13 Fundamental crack-tip loading conditions: tension, mode I; in-plane shear, mode II; antiplane shear, mode III.

Since both in-plane and out-of-plane loading states are discussed in this context, it is necessary to consider the complete stress and displacement fields in all three directions [14,16].[4] These are given by

Mode I:

$$\sigma_r = \frac{K_I}{\sqrt{2\pi r}} \tfrac{1}{4}(5 \cos \tfrac{1}{2}\phi - \cos \tfrac{3}{2}\phi)$$

$$\sigma_\phi = \frac{K_I}{\sqrt{2\pi r}} \tfrac{1}{4}(3 \cos \tfrac{1}{2}\phi + \cos \tfrac{3}{2}\phi)$$

$$\tau_{r\phi} = \frac{K_I}{\sqrt{2\pi r}} \tfrac{1}{4}(\sin \tfrac{1}{2}\phi + \sin \tfrac{3}{2}\phi)$$

$$\sigma_z = v(\sigma_r + \sigma_\phi) \qquad \tau_{rz} = \tau_{\phi z} = 0$$

$$u = \frac{K_I}{G} \sqrt{\frac{r}{2\pi}} [\cos \tfrac{1}{2}\phi (1 - 2v + \sin^2 \tfrac{1}{2}\phi)]$$

$$v = \frac{K_I}{G} \sqrt{\frac{r}{2\pi}} [\sin \tfrac{1}{2}\phi (2 - 2v - \cos^2 \tfrac{1}{2}\phi)]$$

$$w = 0$$

---

[4] Formulas are also given in Chapters 1 and 19 of this book.

Mode II:

$$\sigma_r = \frac{K_{II}}{\sqrt{2\pi r}} \tfrac{1}{4}(-5 \sin \tfrac{1}{2}\phi + 3 \sin \tfrac{3}{2}\phi)$$

$$\sigma_\phi = \frac{K_{II}}{\sqrt{2\pi r}} \tfrac{1}{4}(-3 \sin \tfrac{1}{2}\phi - 3 \sin \tfrac{3}{2}\phi)$$

$$\tau_{r\phi} = \frac{K_{II}}{\sqrt{2\pi r}} \tfrac{1}{4}(\cos \tfrac{1}{2}\phi + 3 \cos \tfrac{3}{2}\phi)$$

$$\sigma_z = v(\sigma_r + \sigma_\phi) \qquad \tau_{rz} = \tau_{\phi z} = 0 \tag{9-21}$$

$$u = \frac{K_{II}}{G}\sqrt{\frac{r}{2\pi}}\,[\sin \tfrac{1}{2}\phi (2 - 2v + \cos^2 \tfrac{1}{2}\phi)]$$

$$v = \frac{K_{II}}{G}\sqrt{\frac{r}{2\pi}}\,[\cos \tfrac{1}{2}\phi (-1 + 2v + \sin^2 \tfrac{1}{2}\phi)]$$

$$w = 0$$

Mode III:

$$\tau_{xz} = -\frac{K_{III}}{\sqrt{2\pi r}} \sin \tfrac{1}{2}\phi \qquad \sigma_x = \sigma_y = \sigma_z = \tau_{xy} = 0$$

$$\tau_{yz} = +\frac{K_{III}}{\sqrt{2\pi r}} \cos \tfrac{1}{2}\phi$$

$$w = \frac{K_{III}}{G}\sqrt{\frac{2r}{\pi}} \sin \tfrac{1}{2}\phi \qquad u = v = 0$$

where G is the shear modulus of the material:

$$G = \frac{E}{2(1 + v)}$$

In addition, u, v, and w are the displacements in x, y, and z directions and K_I, K_{II}, and K_{III} are the stress intensity factors for mode I, mode II, and mode III, respectively. The stress σ_2 and the displacements u and v in the equations for mode I and mode II are given for the case of plane strain but can be easily changed to plane stress by taking $\sigma_2 = 0$ and by replacing v by $v/(1 - v)$ in the equations for u and v.

Since the equations for the stress fields for the mode I and the mode II cases are similar in form, the procedure for obtaining mode II caustics is analogous to the one described in the preceding section for the mode I case. For mode III, however, the in-plane stresses and displacements are zero and changes in the optical path length according to Eq. (9-8) would be zero. Physically, the out-of-plane loading does not produce changes in the thickness of the plate or (for the transmission case) in the refractive index of the material. Thus a shadow optical effect on the basis of the previously discussed considerations does not result. The out-of-plane displacement $w(x, y)$, however, can be utilized for generating a shadow optical image, but only in reflection. The change in optical path length in this case is simply given by

$$\Delta s = -2w(x, y) \tag{9-22}$$

SHADOW OPTICAL METHOD OF CAUSTICS

A shadow optical effect due to the out-of-plane displacement w does not result for transmission arrangements, since the surfaces of the plate, although nonplanar, remain parallel. Consequently, light rays traversing the plate are not deflected but only slightly displaced.

Assuming isotropic material behavior ($\lambda = 0$) first, the mapping equations for the three fracture modes are:

Mode I:

$$x' = r \cos \phi - \frac{K_I}{\sqrt{2\pi}} z_0 c d_{\text{eff}} \, r^{-3/2} \cos \tfrac{3}{2}\phi$$

$$y' = r \sin \phi - \frac{K_I}{\sqrt{2\pi}} z_0 c d_{\text{eff}} \, r^{-3/2} \sin \tfrac{3}{2}\phi$$

Mode II:

$$x' = r \cos \phi + \frac{K_{II}}{\sqrt{2\pi}} z_0 c d_{\text{eff}} \, r^{-3/2} \sin \tfrac{3}{2}\phi \qquad (9\text{-}23)$$

$$y' = r \sin \phi - \frac{K_{II}}{\sqrt{2\pi}} z_0 c d_{\text{eff}} \, r^{-3/2} \cos \tfrac{3}{2}\phi$$

Mode III:

$$x' = r \cos \phi - 2\frac{K_{III}}{\sqrt{2\pi}} \frac{z_0}{G} r^{-1/2} \sin \tfrac{1}{2}\phi$$

$$y' = r \sin \phi + 2\frac{K_{III}}{\sqrt{2\pi}} \frac{z_0}{G} r^{-1/2} \sin \tfrac{1}{2}\phi$$

Application of Eq. (9-11) to these mapping equation gives the equations of the initial curves

Mode I:

$$r = \left[\frac{3}{2} \frac{|K_I|}{\sqrt{2\pi}} |z_0| |c| d_{\text{eff}}\right]^{2/5} \equiv r_0$$

Mode II:

$$r = \left[\frac{3}{2} \frac{|K_{II}|}{\sqrt{2\pi}} |z_0| |c| d_{\text{eff}}\right]^{2/5} \equiv r_0 \qquad (9\text{-}24)$$

Mode III:

$$r = \left[\frac{|K_{III}|}{\sqrt{2\pi}} \frac{|z_0|}{G}\right]^{2/3} \equiv r_0$$

As in the examples discussed previously, the initial curves are circles around the origin with fixed radius r_0. With the mapping Eqs. (9-23) the caustic curves are obtained as images of the initial curves [Eqs. (9-24)]. The caustic equations are:

Mode I:

$$x' = r_0(\cos \phi - \text{sgn}\,(K_I z_0 c) \tfrac{2}{3} \cos \tfrac{3}{2}\phi)$$

$$y' = r_0(\sin \phi - \text{sgn}\,(K_I z_0 c) \tfrac{2}{3} \sin \tfrac{3}{2}\phi)$$

Mode II:

$$x' = r_0(\cos \phi + \text{sgn}\,(K_{II} z_0 c) \tfrac{2}{3} \cos \tfrac{3}{2}\phi) \qquad (9\text{-}25)$$

$$y' = r_0(\sin \phi - \text{sgn}\,(K_{II} z_0 c) \tfrac{2}{3} \sin \tfrac{3}{2}\phi)$$

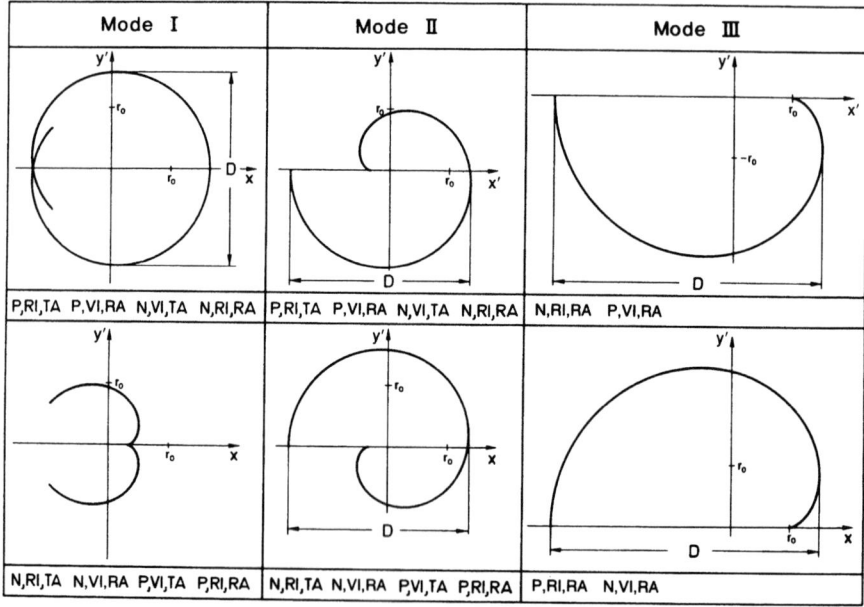

Figure 9-14 Crack-tip caustic curves for mode I, mode II, and mode III loading. P, N = Positive, negative loading; RI, VI = real ($z_0 > 0$), virtual ($z_0 < 0$) image; TA, RA = transmission ($c < 0$), reflection ($c > 0$) arrangement.

Mode III:

$$x' = r_0(\cos \phi - \text{sgn}\,(K_{III}z_0 c)2 \sin \tfrac{1}{2}\phi)$$

$$y' = r_0(\sin \phi + \text{sgn}\,(K_{III}z_0 c)2 \cos \tfrac{1}{2}\phi)$$

Graphical presentations of caustic curves for the three fracture modes and the different possible observation modes are given in Fig. 9-14. In contrast to the mode I caustics, the mode II and the mode III caustics are asymmetric. Graphical presentations of light distributions for the mode II and the mode III shadow images are given in Fig. 9-15. For the mode I light distributions, see Fig. 9-10.

It can easily be seen that the characteristic length parameters D defined in Fig. 9-14 are related to the radii of the initial curves by:

Mode I: $\quad D = 3.17 r_0$

Mode II: $\quad D = 3.02 r_0$ \hfill (9-26)

Mode III: $\quad D = 4.5 r_0$

Consequently, with Eqs. (9-24) and (9-26) the evaluation formulas for the three fracture modes become:

Mode I:

$$|K_I| = \frac{2\sqrt{2\pi}}{3(3.17)^{5/2}|z_0|\,|c|d_{\text{eff}}} D^{5/2}$$

SHADOW OPTICAL METHOD OF CAUSTICS

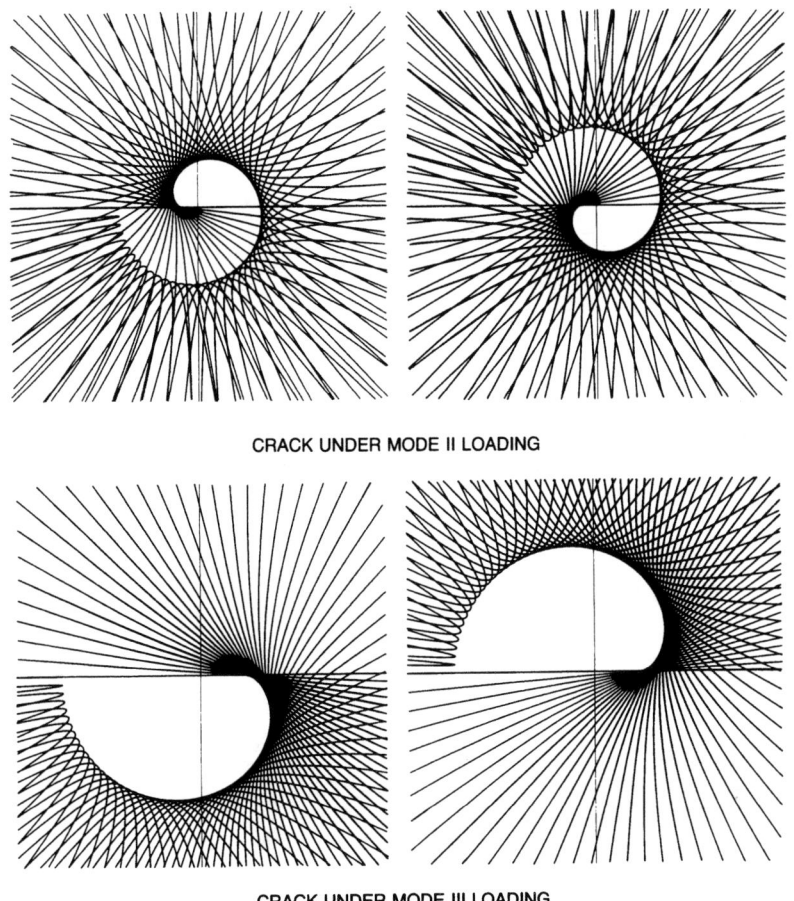

CRACK UNDER MODE II LOADING

CRACK UNDER MODE III LOADING

for reflection only

Figure 9-15 Shadow optical light distributions: crack under mode II loading; crack under mode III loading (for reflection only). Left side, sgn(Kz_0c) negative. Right side, sgn(Kz_0c) positive.

Mode II:

$$|K_{II}| = \frac{2\sqrt{2\pi}}{3(3.02)^{5/2}|z_0| \, |c|d_{\text{eff}}} D^{5/2} \tag{9-27}$$

Mode III:

$$|K_{III}| = \frac{\sqrt{2\pi}G}{(4.5)^{3/2}|z_0|} D^{3/2}$$

The equivalent equations for the anisotropic case ($\lambda \neq 0$) are given for completeness as well. Since the mode III crack problem by definition applies to the isotropic case only (see Table 9-1), only the mode II crack problem is considered:

As shown in detail in Ref. 17, the mapping equations are

$$x' = r \cos \phi + \frac{K_{II}}{\sqrt{2\pi}} z_0 c d_{eff} r^{-3/2} \left[\sin \frac{3}{2}\phi \pm \frac{\lambda}{8} \frac{-7 \cos \phi - 9 \cos 3\phi}{\sqrt{3 \cos^2 \phi + 1}} \right]$$

$$y' = r \sin \phi - \frac{K_{II}}{\sqrt{2\pi}} z_0 c d_{eff} r^{-3/2} \left[\cos \frac{3}{2}\phi \pm \frac{\lambda}{8} \frac{13 \sin \phi + 9 \sin 3\phi}{\sqrt{3 \cos^2 \phi + 1}} \right]$$

(9-28)

and the initial curves are obtained as

$$r_0 = \left(\frac{3}{2} \frac{|K_{II}|}{\sqrt{2\pi}} |z_0||c|d_{eff} B_{II} \right)^{2/5}$$

(9-29)

with

$$B_{II} = \left| \pm \frac{\lambda}{8} A^{3/2} \left(\frac{49}{12} \pm 15 \cos 2\phi + \frac{9}{4} \cos 4\phi \right) + \text{sgn}(z_0 c) R^{1/2} \right|$$

(9-30)

where

$$A = (3 \cos^2 \phi + 1)^{-1}$$

(9-31)

and

$$R = 1 \pm \frac{\lambda}{8} A^{3/2} \left(-\frac{403}{6} \sin \frac{\phi}{2} + 75 \sin \frac{3}{2}\phi - 15 \sin \frac{5}{2}\phi \right.$$
$$\left. + \frac{63}{4} \sin \frac{7}{2}\phi + \frac{9}{4} \sin \frac{9}{2}\phi \right)$$
$$+ \left(\frac{\lambda}{8} \right)^2 \frac{A^3}{288} (796643 + 1035720 \cos 2\phi + 292572 \cos 4\phi$$
$$+ 9720 \cos 6\phi - 5103 \cos 8\phi)$$

(9-32)

As a result of the plus and minus signs of the λ-affected terms in Eqs. (9-30) and (9-32), two initial curves are obtained, similar to the anisotropic case of the mode I problem. In contrast to the mode I problem, though, the initial curves of the mode II problem are no longer represented by circles. The initial curves show marked deviations from the shape of an exact circle; in particular, a discontinuity in the curve is obtained at the fracture surfaces.

With Eqs. (9-28) and (9-29), the caustic curves are obtained as

$$x' = r_0 \left\{ \cos \phi + \frac{2}{3} \text{sgn}(z_0 c) B_{II}^{-1} \left[\sin \frac{3}{2}\phi \pm \frac{\lambda}{8}(-7 \cos \phi - 9 \cos 3\phi) A^{1/2} \right] \right\}$$

$$y' = r_0 \left\{ \sin \phi - \frac{2}{3} \text{sgn}(z_0 c) B_{II}^{-1} \left[\cos \frac{3}{2}\phi \pm \frac{\lambda}{8}(13 \sin \phi + 9 \sin 3\phi) A^{1/2} \right] \right\}$$

(9-33)

with B_{II} given by Eqs. (9-30) through (9-32).

Figure 9-16 shows anisotropic mode II crack-tip double caustics according to Eq. (9-33) for the material Araldite B ($\lambda = -0.288$). Similar to the equivalent mode I problem (see Fig. 9-11), the split-up of the double caustic is larger for plane strain than for plane stress.

Because of the complex geometry of the anisotropic mode II crack-tip caustic and also because of possible experimental disturbances of the caustic curves near the fracture surfaces (e.g., due

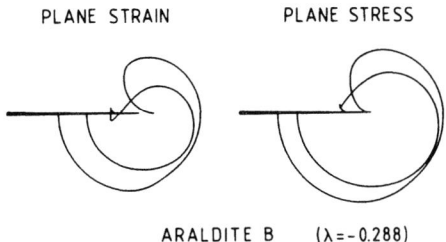

Figure 9-16 Mode II double caustics for optically anisotropic ($\lambda \neq 0$) material: plane strain and plane stress.

to residual stresses introduced into the material by the machining process if saw cuts are used instead of real cracks), it is advantageous to take more than one size parameter from the caustics and to use an overdeterministic approach for determining the mode II stress-intensity factor. The seven size parameters defined in Fig. 9-17 have proven to yield results that are the least affected by eventual experimental disturbances. By correlating these size parameters, namely the diameters D with the average value $\bar{r}_0$ of the two corresponding $\bar{r}_0$ values of the initial curves (because the initial curves are not circular, these two values are principally different, although the differences are not very large, but they are not identical as in the isotropic cases or practically identical as in the anisotropic mode I case), relations of the kind $D = f\,\bar{r}_0$ hold, similar to Eqs. (9-14), or (9-26). Numerical values of the f-factors for the seven size parameters defined in Fig. 9-17 are given in Fig. 9-18 as a function of the anisotropy coefficient λ. In particular, the numerical values of the factors f_{oh} and f_{ih} associated with the size parameter in the horizontal direction lie around 3.02 for the corresponding size parameter of the single caustic in the isotropic case. With the f-factors, the mode II stress-intensity factor is obtained as

$$|K_{II}| = \frac{2\sqrt{2\pi}}{3 f_{o,i}|z_0||c|d_{\text{eff}}} D_{o,i}^{5/2} \qquad (9\text{-}34)$$

Results for cracks under mixed-mode (mode I/mode II) loading and isotropic material behavior ($\lambda = 0$) are obtained by a superposition of the stress field equations and consequently of the mapping equations for the pure mode I and mode II cases. Caustic curves for different ratios of the mode II to the mode I stress-intensity factor $\mu = K_{II}/K_I$ are shown in Fig. 9-19a. Depending on the K_{II}/K_I ratio, all intermediate stages between pure mode I and pure mode II caustics are possible.

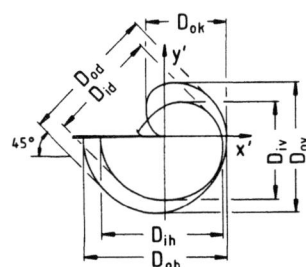

Figure 9-17 Definition of size parameters D of anisotropic ($\lambda \neq 0$) mode II crack-tip caustics.

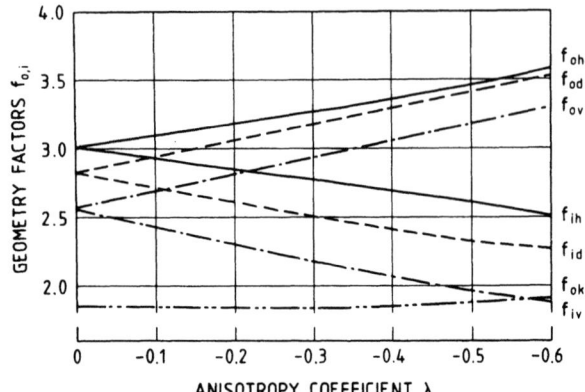

Figure 9-18 Numerical factors $f_{o,i}$ for evaluating anisotropic ($\lambda \neq 0$) mode II crack-tip caustics.

From the two diameters D_{max} and D_{min}, defined in Fig. 9-19b, the two stress-intensity factors K_I and K_{II} are determined [18,19]. With Fig. 9-20, resulting from Eqs. (9-25) for the complete mode I/mode II caustic curve, the stress intensity factor ratio $\mu = K_{II}/K_I$ is first determined from the measured value $(D_{max} - D_{min})/D_{max}$. According to Fig. 9-21, this μ value then defines the value of the numerical factor f, describing the relationship between the characteristic length parameter D_{max} and the radius r_0 of the initial curve, $D_{max} = fr_0$. Thus the absolute value of the mode I stress-intensity factor K_I can be determined from the measured distance D_{max} as

a) MIXED MODE CAUSTICS

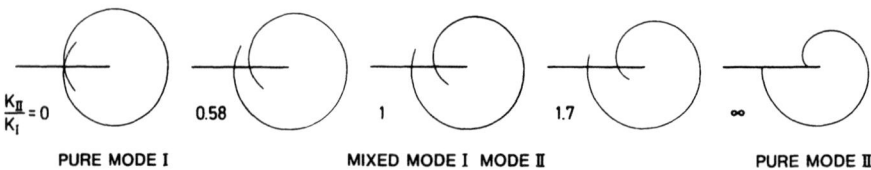

b) DEFINITION OF D_{max} and D_{min}

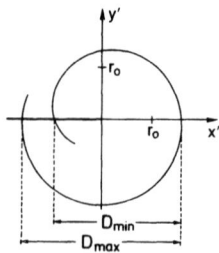

Figure 9-19 Mixed-mode crack-tip caustics. (a) caustics; (b) definition of D_{max} and D_{min}.

SHADOW OPTICAL METHOD OF CAUSTICS

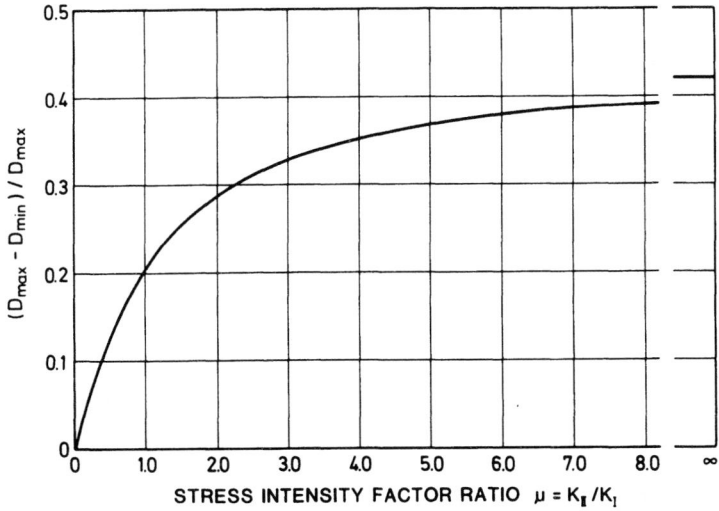

Figure 9-20 Determination of the stress intensity factor ratio μ from isotropic ($\lambda = 0$) mixed-mode crack-tip caustics.

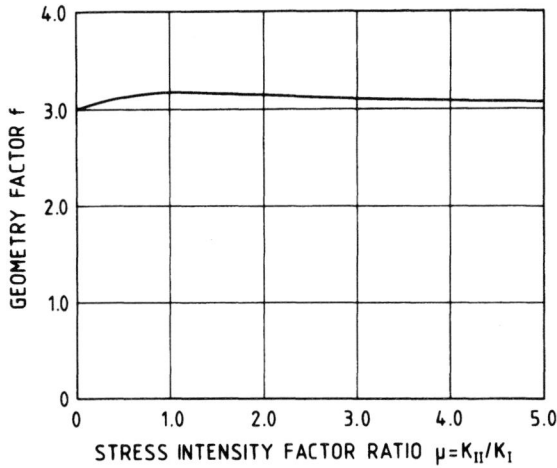

Figure 9-21 Numerical factor f for determination of stress-intensity factor from isotropic ($\lambda = 0$) mixed-mode crack-tip caustics.

$$K_I = \frac{2\sqrt{2\pi}}{3f^{5/2}|z_0| |c|d_{\text{eff}}} D_{\text{max}}^{5/2} \frac{1}{\sqrt{1+\mu^2}} \qquad (9\text{-}35)$$

and

$$K_{II} = \mu K_I$$

For optically anisotropic materials ($\lambda \neq 0$) the mapping equations are [20]

$$x' = r\cos\phi - z_0 c d_{\text{eff}} \frac{1}{\sqrt{2\pi}} r^{-3/2} \left\{ K_I \cos\frac{3}{2}\phi - K_{II} \sin\frac{3}{2}\phi \pm \left(\frac{\lambda}{8}\right) T^{-1/2} \left[K_I^2 (3\cos\phi \right. \right.$$
$$\left. \left. - 3\cos 3\phi) + K_I K_{II}(-4\sin\phi + 12\sin 3\phi) + K_{II}^2(7\cos\phi + 9\cos 3\phi) \right] \right\}$$

$$y' = r\sin\phi - z_0 c d_{\text{eff}} \frac{1}{\sqrt{2\pi}} r^{-3/2} \left\{ K_I \sin\frac{3}{2}\phi - K_{II} \cos\frac{3}{2}\phi \pm \left(\frac{\lambda}{8}\right) T^{-1/2} \left[K_I^2 (3\sin\phi \right. \right.$$
$$\left. \left. - 3\sin 3\phi) - K_I K_{II}(4\cos\phi + 12\cos 3\phi) + K_{II}^2(13\sin\phi + 9\sin 3\phi) \right] \right\} \qquad (9\text{-}36)$$

with

$$T = K_I^2 \sin^2\phi + 2K_I K_{II} \sin 2\phi + K_{II}^2 (3\cos^2\phi + 1)$$

It is recognized from Eq. (9-36) that the anisotropy of the material yields quadratic and coupled terms in the stress-intensity factors that do not exist in the isotropic case. The caustic curve for different ratios of the mode II to mode I stress-intensity factor $\mu = K_{II}/K_I$ is shown in Fig. 9-22 for the material Araldite B ($\lambda = -0.288$) and plane-stress conditions. With the seven size parameters originally defined for anisotropic pure mode II caustics (see Fig. 9-17) used now in an analogous manner for anisotropic mixed-mode (mode I/mode II) caustics as well, the two stress-intensity factors K_I and K_{II} are obtained from the caustics, following in principle the procedure described for evaluating isotropic mixed-mode caustics but using the overdeterministic approach. That is, from the measured value $(D_{oh} - D_{ok})/D_{oh}$, first, the stress intensity factor ratio $\mu = K_{II}/K_I$ is determined with the help of Fig. 9-23; then numerical values of the f-factors for the seven size parameters D are obtained for the prevailing μ values with the help of Fig. 9-24. In particular, note that for the special case $\mu = 0$ (i.e., a pure mode I caustic), the values of f_{iv} and f_{ov} become identical to the values for f_i and f_o of Fig. 9-12 for $\lambda = -0.288$, and, furthermore, that for the special case $\mu = \infty$ (i.e., a pure mode II caustic), the values of f_{ih} and f_{oh} become identical to the equivalent values in Fig. 9-18 for $\lambda = -0.288$. With the determined values of the μ ratios and the f-factors, the two stress-intensity factors K_I and K_{II} are then obtained via the equations

$$K_I = \frac{2\sqrt{2\pi}}{3f_{o,i}^{5/2}|z_0| |c|d_{\text{eff}}} D_{o,i}^{5/2} \frac{1}{\sqrt{1+\mu^2}} \qquad (9\text{-}37)$$

and

$$K_{II} = \mu K_I$$

Dynamic Caustics

Two cases of dynamic fracture problems are considered: stationary cracks under dynamic loading, and propagating cracks under stationary external loading.

For a stationary crack subjected to a dynamically applied load, the near-field stress distribution is the same as for a statically loaded crack [21]. The stress intensity factor, however, becomes a

SHADOW OPTICAL METHOD OF CAUSTICS

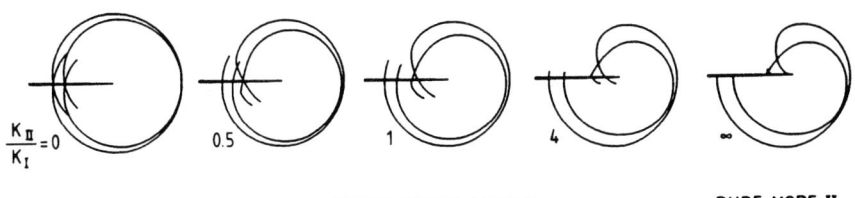

Figure 9-22 Anisotropic ($\lambda \neq 0$) mixed-mode crack-tip caustics.

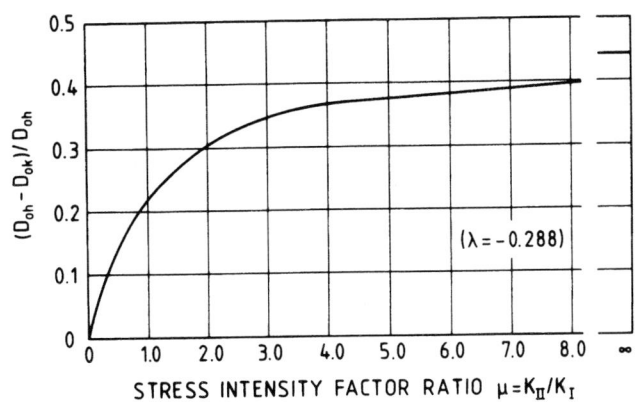

Figure 9-23 Determination of the stress-intensity factor ratio μ from anisotropic ($\lambda \neq 0$) mixed-mode crack-tip caustics.

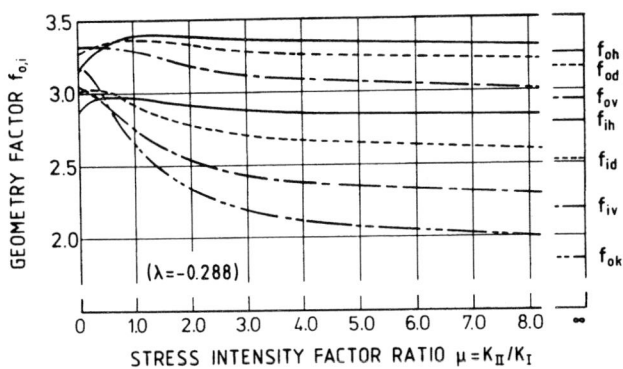

Figure 9-24 Numerical factors for the determination of stress-intensity factors from anisotropic ($\lambda \neq 0$) mixed-mode crack-tip caustics.

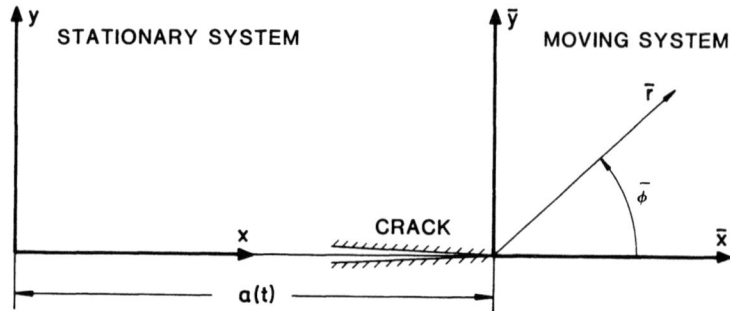

Figure 9-25 Coordinate system at the tip of a moving crack.

function of time. Consequently, since the dependencies from the radial distance r and the angle ϕ are identical for the dynamic and static cases, the results presented in the preceding subsection also apply to dynamically loaded cracks, with the stress-intensity factors K_I, K_{II}, or K_{III} replaced by $K_I(t)$, $K_{II}(t)$, or $K_{III}(t)$.

For a propagating crack, the near-field stress distribution differs from that for a stationary crack due to inertia effects (see, e.g., Refs. 22 and 23).

In a coordinate system $(\bar{x}, \bar{y})$ or $(\bar{r}, \bar{\phi})$ with its origin in the moving crack tip (see Fig. 9-25).

$$x = a(t) + \bar{x} = a(t) + \bar{r} \cos \bar{\phi}$$
$$y = \bar{y} = \bar{r} \sin \bar{\phi}$$
(9-38)

where a is the crack length, the near-field stress distribution for a propagating crack with an instantaneous crack velocity $v = da/dt$ is given as

$$\sigma_{\bar{x}} = \frac{K_I}{\sqrt{2\pi\bar{r}}} \frac{1 + \alpha_2^2}{4\alpha_1\alpha_2 - (1 + \alpha_2^2)^2} \left[(1 + 2\alpha_1^2 - \alpha_2^2) p(\bar{\phi}, \alpha_1) - \frac{4\alpha_1\alpha_2}{1 + \alpha_2^2} p(\bar{\phi}, \alpha_2) \right]$$

$$\sigma_{\bar{y}} = \frac{K_I}{\sqrt{2\pi\bar{r}}} \frac{1 + \alpha_2^2}{4\alpha_1\alpha_2 - (1 + \alpha_2^2)^2} \left[-(1 + \alpha_2^2) p(\bar{\phi}, \alpha_1) + \frac{4\alpha_1\alpha_2}{1 + \alpha_2^2} p(\bar{\phi}, \alpha_2) \right] \quad (9\text{-}39)$$

$$\tau_{\bar{x}\bar{y}} = \frac{K_I}{\sqrt{2\pi\bar{r}}} \frac{1 + \alpha_2^2}{4\alpha_1\alpha_2 - (1 + \alpha_2^2)^2} \alpha_1 [q(\bar{\phi}, \alpha_1) - q(\bar{\phi}, \alpha_2)]$$

where

$$p(\bar{\phi}, \alpha_j) = \frac{[\cos \bar{\phi} + (\cos^2 \bar{\phi} + \alpha_j^2 \sin^2 \bar{\phi})^{1/2}]^{1/2}}{(\cos^2 \bar{\phi} + \alpha_j^2 \sin^2 \bar{\phi})^{1/2}}$$

$$q(\bar{\phi}, \alpha_j) = \frac{[-\cos \bar{\phi} + (\cos^2 \bar{\phi} + \alpha_j^2 \sin^2 \bar{\phi})^{1/2}]^{1/2}}{(\cos^2 \bar{\phi} + \alpha_j^2 \sin^2 \bar{\phi})^{1/2}}$$
(9-40)

and

$$\alpha_j = \left(1 - \frac{v^2}{c^2}\right)^{1/2} \quad j = 1, 2$$
(9-41)

with

Longitudinal wave speed:

$$c_1 = \sqrt{\frac{E}{\rho}} \sqrt{\frac{1-v}{(1+v)(1-2v)}}$$

(9-42)

Transverse wave speed:

$$c_2 = \sqrt{\frac{E}{\rho}} \sqrt{\frac{1}{2(1+v)}}$$

With this stress distribution, the mapping equations with regard to the moving coordinate system are

$$\bar{x}' = \bar{r} \cos \bar{\phi} - \frac{K_1}{\sqrt{2\pi}} z_0 c d_{\text{eff}} \bar{r}^{-3/2} F^{-1} G_1(\alpha_1, \bar{\phi})$$

$$\bar{y}' = \bar{r} \sin \bar{\phi} - \frac{K_1}{\sqrt{2\pi}} z_0 c d_{\text{eff}} \bar{r}^{-3/2} F^{-1} G_2(\alpha_1, \bar{\phi})$$

(9-43)

where

$$F = \frac{4\alpha_1 \alpha_2 - (1 + \alpha_2^2)^2}{(\alpha_1^2 - \alpha_2^2)(1 + \alpha_2^2)}$$

(9-44)

and

$$G_1(\alpha_1, \bar{\phi}) = \frac{-1}{\sqrt{2}} (g^{1/2} + \cos \bar{\phi})^{-1/2} (g^{-1/2} - g^{-1} \cos \bar{\phi} - 2g^{-3/2} \cos^2 \bar{\phi})$$

$$G_2(\alpha_1, \bar{\phi}) = \frac{1}{\sqrt{2}} (g^{1/2} + \cos \bar{\phi})^{-1/2} (\alpha_1^2 g^{-1} \sin \bar{\phi} - \alpha_1^2 g^{-3/2} \sin 2\bar{\phi})$$

(9-45)

with

$$g = 1 + (\alpha_1^2 - 1) \sin^{-2} \bar{\phi}$$

(9-46)

It can be easily shown by numerical calculations (see the more detailed discussion in Ref. 9) that the influence of α_1 on the functions $G_1(\alpha_1, \bar{\phi})$ and $G_2(\alpha_1, \bar{\phi})$ is negligibly small for all crack velocities of practical relevance (i.e., $v < 0.3c_1$). In particular, the error made by neglecting the dependency of α_1 on G_1 and G_2 is small in comparison to that on the factor F. Thus with an accuracy sufficient for engineering purposes these functions can be approximated by

$$G_1(\alpha_1, \bar{\phi}) \approx \cos \tfrac{3}{2}\bar{\phi}$$

$$G_2(\alpha_1, \bar{\phi}) \approx \cos \tfrac{3}{2}\bar{\phi}$$

(9-47)

With these approximations the equation for the initial curve is obtained in a manner analogous to the static considerations [see Section 9-2, Eq. (9-12c)] as

$$\bar{r} = \left[\frac{3}{2} \frac{K_I}{\sqrt{2\pi}} |z_0| |c| d_{\text{eff}} F^{-1} \right]^{2/5} \equiv \bar{r}_0$$

(9-48)

and consequently the evaluation formula becomes

$$K_I = \frac{2\sqrt{2\pi} \, F}{3(3.17)^{5/2} |z_0| |c| d_{\text{eff}}} D^{5/2}$$

(9-49)

Thus the stress-intensity evaluation formula for a propagating crack is the same as that for a

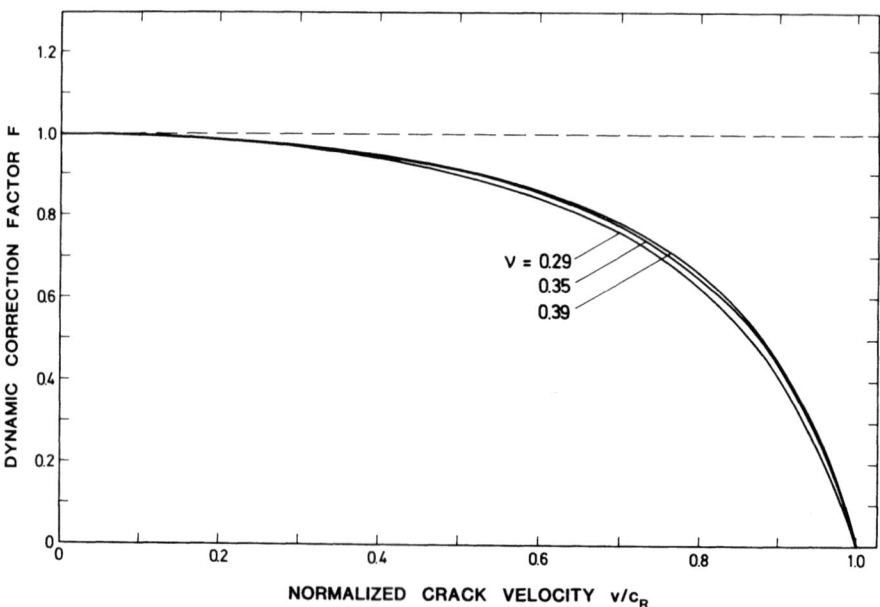

Figure 9-26 Correction factor for evaluating dynamic crack-tip caustics.

stationary crack [Eq. (9-15c)] except for a correction factor $F(v)$. This factor accounts for velocity effects on the r and ϕ distribution of the dynamic stress field for a propagating crack. The factor F as a function of crack velocity v is given by Fig. 9-26. F is less than 1, but for practically relevant crack velocities it is nearly equal to 1.

Figure 9-27 shows crack-tip caustics for difference crack velocities but with a fixed stress intensity factor. The crack velocities are normalized by the Rayleigh wave speed $c_R \approx (0.862 + 1.14v)/[\sqrt{2}(1 + v)^{3/2}]\sqrt{E/\rho}$. The dynamic caustics are shown in comparison to the corresponding caustic for a stationary crack. For all crack velocities, the shape of the dynamic caustic is practically the same as for the stationary caustic, but the size of the caustic increases slightly with increasing velocity according to the correction factor F.

For optically anisotropic caustics the situation becomes more complicated because of the additional anisotropy term in the equations and the resulting formation of double caustics. It can be shown, however, that the outer caustic increases with crack velocity in a similar manner to the single caustic for optically isotropic materials [9]. The increase in size of the inner caustic with crack velocity is somewhat larger than that for the single caustic. Thus the isotropic correction factor $F(v)$ given by Eq. (9-44) also represents a good approximation for double caustics if applied to the outer caustic.

Influence of Higher-Order Effects on Caustics

So far only the near-field stress distributions around crack tips have been considered. These stress field solutions are strictly valid as $r \to 0$; that is, the range of applicability of these solutions is restricted to the direct vicinity of the crack tip. In the intermediate- and far-field regions, additional higher-order terms of the stress field solution will have to be included. Since the caustic is the

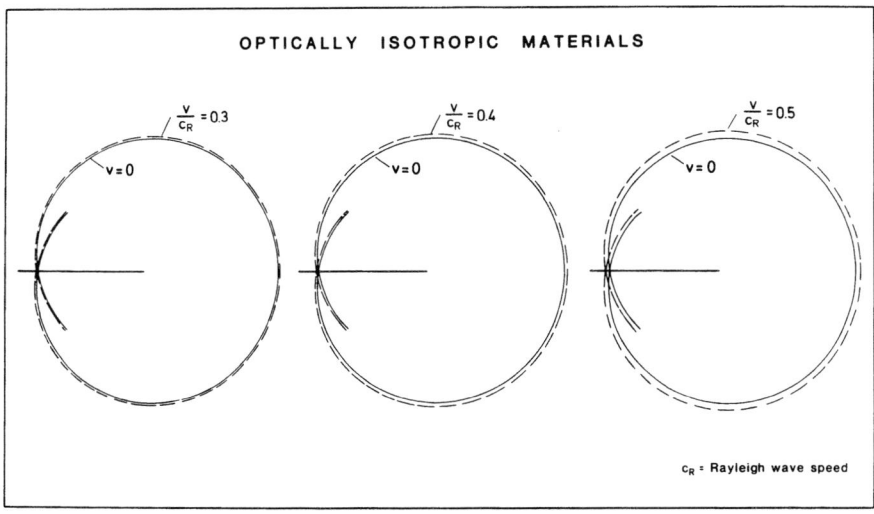

Figure 9-27 Caustics for propagating cracks: optically isotropic materials. c_R, Rayleigh wave speed.

image of the initial curve with finite radius r_0 around the crack tip, these higher-order terms may have an influence on the formation of the caustic. In general, no practical problems are encountered for long cracks in large specimens. However, for short cracks and for cracks approaching the boundary of the specimen, the possible influence of higher-order terms must be considered. The complete solution of the crack-tip stress field is given by a series of terms, with the first-order term representing the near-field stress distribution. For tensile (mode I) loading the first six terms are [24]

$$\sigma_r = \frac{K_I}{\sqrt{2\pi r}} \frac{1}{4} \{5 \cos \tfrac{1}{2}\phi - \cos \tfrac{3}{2}\phi\} + a_2 \cos^2 \phi$$
$$+ r^{1/2} a_3 \{3 \cos \tfrac{1}{2}\phi + \cos \tfrac{5}{2}\phi\} + r a_4 \{\cos \phi + 3 \cos 3\phi\}$$
$$+ r^{3/2} a_5 \{\cos \tfrac{3}{2}\phi + 3 \cos \tfrac{7}{2}\phi\} + r^2 a_6 \{+ 2 \cos 4\phi\}$$

$$\sigma_\phi = \frac{K_I}{\sqrt{2\pi r}} \frac{1}{4} \{3 \cos \tfrac{1}{2}\phi + \cos \tfrac{3}{2}\phi\} + a_2 \sin^2 \phi$$
$$+ r^{1/2} a_3 \{5 \cos \tfrac{1}{2}\phi - \cos \tfrac{5}{2}\phi\} + r a_4 \{3 \cos \phi - 3 \cos 3\phi\} \qquad (9\text{-}50)$$
$$+ r^{3/2} a_5 \{7 \cos \tfrac{3}{2}\phi - 3 \cos \tfrac{7}{2}\phi\} + r^2 a_6 \{2 \cos 2\phi - 2 \cos 4\phi\}$$

$$\tau_{r\phi} = \frac{K_I}{\sqrt{2\pi r}} \frac{1}{4} \{\sin \tfrac{1}{2}\phi + \sin \tfrac{3}{2}\phi\} - a_2 \tfrac{1}{2} \sin 2\phi$$
$$+ r^{1/2} a_3 \{\sin \tfrac{1}{2}\phi - \sin \tfrac{5}{2}\phi\} + r a_4 \{\sin \phi - 3 \sin 3\phi\}$$
$$+ r^{3/2} a_5 \{5 \sin \tfrac{3}{2}\phi - 5 \sin \tfrac{7}{2}\phi\} + r^2 a_6 \{\sin \phi - 2 \sin 4\phi\}$$

In general, the most important term in expanding the range of applicability of a first-order

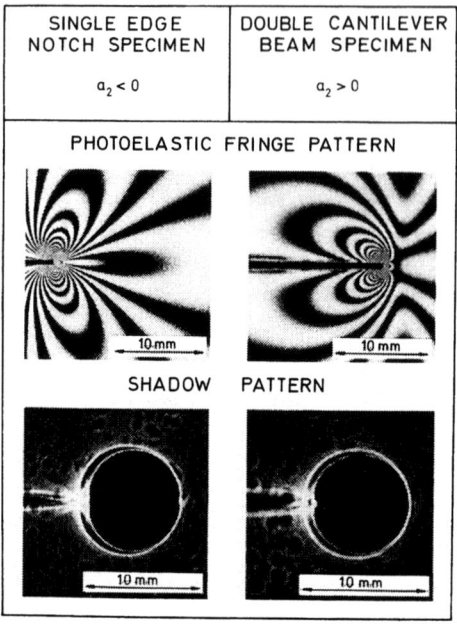

Figure 9-28 Influence of the second-order term of the crack-tip stress field equations on photoelastic fringes and shadow optical caustics.

solution is the term following (i.e., the second-order term). This term represents a constant stress in the x direction, $\sigma_x = a_2$, which is superimposed on the singular near-field stress distribution. However, the shadow optical effect is sensitive to stress gradients only. Constant stresses do not give rise to changes in the optical path length [see Eqs. (9-1), (9-3), and (9-8)]. Thus the second-order term does not have any influence on the formation of the caustic.[5] In practical applications this can be a significant advantage of the shadow optical method of caustics over other stress optical techniques (e.g., photoelastic, moiré, or holographic techniques) that directly measure stresses or strains. Figure 9-28 illustrates the influence of the second-order term on the shadow optical and the photoelastic pictures of a crack-tip stress distribution. For single-edge-notch (SEN) specimens, the sign of the coefficient a_2 is negative, causing the isochromatic fringes in the photoelastic picture to lean forward, whereas in double-cantilever-beam (DCB) specimens the sign of a_2 is positive, so that the isochromatic fringes lean backward. These changes in the isochromatic patterns can cause severe difficulties in the evaluation procedure of photoelastic crack-tip fringe patterns. The shadow optical pictures, however, are identical, independent of the sign of the second-order coefficient.

Terms of orders higher than 2 have an influence on the formation of the caustic. Figure 9-29 shows the resulting changes by including the next four higher-order terms. For arbitrary values of the individual coefficients and both signs (i.e., positive and negative coefficients), changes in the shape of the caustics are shown in comparison to the near-field caustic. Since in practice the

---

[5] This statement applies to single caustics obtained with isotropic materials. Slight influences of the second order can, however, result from the anisotropy term ($\lambda \neq 0$) in Eq. (9-8) when double caustics are considered.

SHADOW OPTICAL METHOD OF CAUSTICS

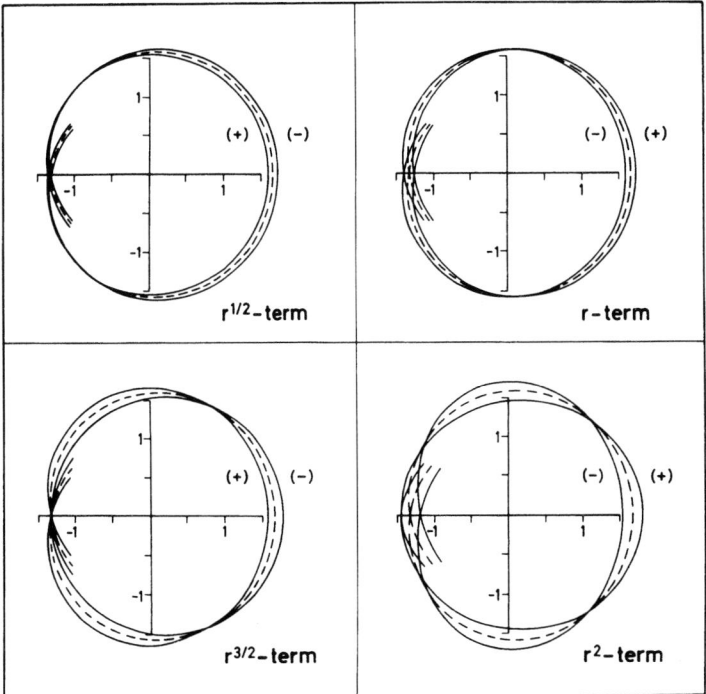

Figure 9-29 Influence of higher-order terms of the crack-tip stress field equations on the caustic curve: (a) $r^{1/2}$ term; (b) r term; (c) $r^{3/2}$ term; (d) r^2 term.

coefficients of the higher-order terms are generally not known, a quantitative estimate of the influence of higher-order effects on the caustics is difficult to make. For practical applications it will be appropriate to compare the shape of the experimentally observed caustic with the theoretically calculated near-field caustic. Deviations of a significant amount would be an indication of the nonnegligible influence of the higher-order terms. In these cases caution must be used in evaluating stress-intensity factors from such caustics.

Elastic–Plastic Caustics

For cracks in ductile materials (e.g., structural steels) the linear theory of elasticity is no longer applicable. The crack-tip stress field is described by elastic–plastic relations. The stresses are bounded by the yield stress, and consequently the strains near the crack tip become unbounded when the crack tip is approached. Thus, instead of an elastic stress concentration problem, a plastic strain concentration problem is encountered. But analogous to elastic stress concentrations, caustics also result for plastic strain concentrations [6,7].

With regard to an application to steels, the following consideration applies to the reflection case only and virtual images ($z_0 < 0$). Because of the deformation of the specimen surface, the change in the optical path length of light rays according to Eq. (9-4) is given by

$$\Delta s = -\Delta d \tag{9-51}$$

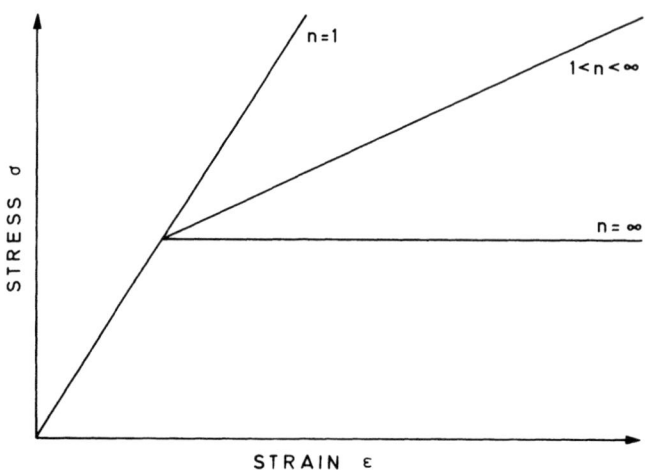

Figure 9-30 Elastic–plastic stress–strain behavior.

where d is the total thickness of the specimen. The reduction in specimen thickness is obtained from the plastic strain ε as

$$\Delta d = \varepsilon_{zz} d \qquad (9\text{-}52)$$

Elastic strains ε^e are neglected relative to the plastic strains ε^p, since $\varepsilon^e \ll \varepsilon^p$. As a result of the incompressibility of the material under plastic deformation (constancy of volume),

$$\varepsilon_{zz} = -(\varepsilon_{rr} + \varepsilon_{\phi\phi}) \qquad (9\text{-}53)$$

Consequently, the mapping equations (9-1) and (9-3) take the form

$$\mathbf{r}' = \mathbf{r} + z_0 d\ \mathbf{grad}(\varepsilon_{rr} + \varepsilon_{\phi\phi}) \qquad (9\text{-}54)$$

The strains in the near-field region around a crack tip have been calculated by Hutchinson [25,26], and Rice and Rosengren [27]. For an elastic perfectly-plastic material [i.e., a non-hardening material with $n \to \infty$ (see Fig. 9-30)], the strains are unbounded in a fan region ($-79.7° < \phi < +79.7°$) ahead of the crack tip (see Fig. 9-31). Outside this fan region the strains are bounded by finite values. Quantitatively, the strains are given by the following expressions:

$$\left.\begin{array}{l} \varepsilon_{rr} = 0 \\ \varepsilon_{\phi\phi} = \dfrac{J}{2.57\sigma_0 r}\dfrac{\sqrt{3}}{2}\cos\phi \end{array}\right\} -79.7° < \phi < +79.7°$$

$$\left.\begin{array}{l} \varepsilon_{rr} = \text{bounded} \\ \varepsilon_{\phi\phi} = \text{bounded} \end{array}\right\} \begin{array}{l} -180° < \phi < -79.7° \\ +79.7° > \phi < +180° \end{array} \qquad (9\text{-}55)$$

where J is the so-called J integral and σ_0 is the tensile yield stress. With Eqs. (9-55), the mapping equations (9-54) become

$$\left.\begin{array}{l} x' = r\cos\phi - \dfrac{Jz_0 d\sqrt{3}}{2.57\sigma_0(2)} r^{-2}\cos\phi \\ y' = r\sin\phi - \dfrac{Jz_0 d\sqrt{3}}{2.57\sigma_0(2)} r^{-2}\sin\phi \end{array}\right\} -79.7° < \phi < +79.7° \qquad (9\text{-}56)$$

SHADOW OPTICAL METHOD OF CAUSTICS

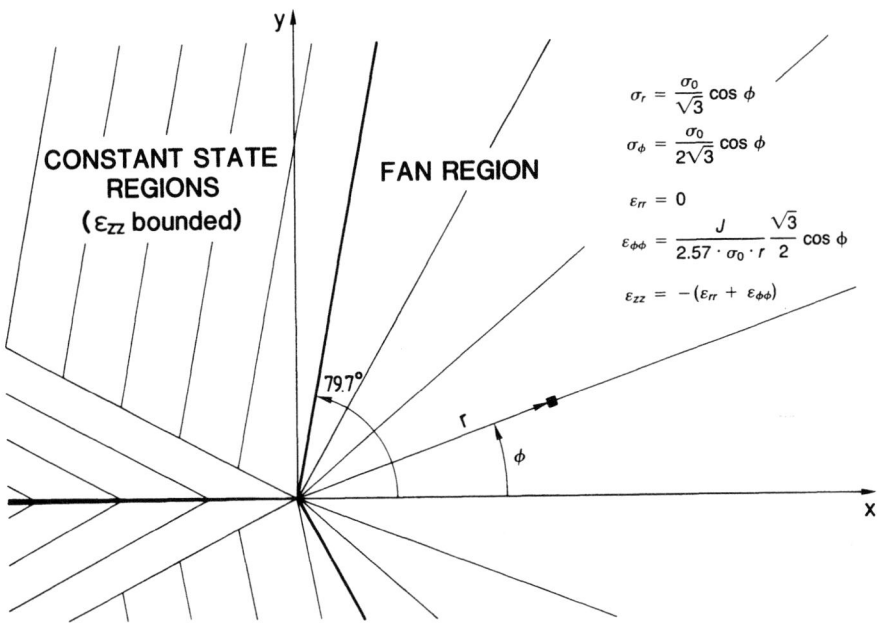

Figure 9-31 Elastic–plastic stress–strain field at the tip of a crack: Hutchinson–Rice–Rosengren.

The bounded strains outside the fan region ($-79.7° < \phi < +79.7°$) cannot contribute to the caustic. Application of Eq. (9-11) yields the initial curve

$$r = \left[\frac{\sqrt{3}J|z_0|d}{2.57\sigma_0}\right]^{1/3} \equiv r_0 \tag{9-57}$$

which again is a circle of fixed radius r_0 around the crack tip (but defined only for segment $-79.7° < \phi < 79.7°$). The mapping of the initial curve (9-57) onto the reference plane by Eq. (9-56) gives the equations of the caustic curve:

$$\left. \begin{array}{l} x' = r_0(\cos\phi + \operatorname{sgn}(z_0)\tfrac{1}{2}\cos 2\phi) \\ y' = r_0(\sin\phi + \operatorname{sgn}(z_0)\tfrac{1}{2}\sin 2\phi) \end{array} \right\} \quad -79.7° < \phi < +79.7° \tag{9-58}$$

The caustic curve in a reference plane $z_0 < 0$ is graphically shown in Fig. 9-32. Analogous to the elastic case, the characteristic length parameter is defined as the maximum diameter D of the caustic in the y direction, which is related to the radius of the initial curve by

$$D = 2.38 r_0 \tag{9-59}$$

With Eqs. (9-57) and (9-59), one obtains

$$J = \frac{\sigma_0}{13.5|z_0|d} D^3 \tag{9-60}$$

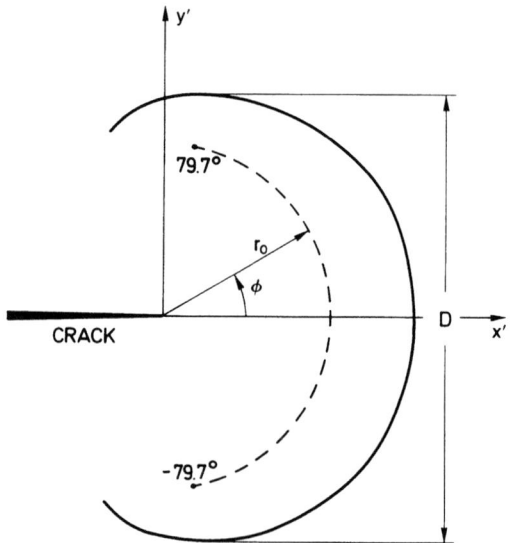

Figure 9-32 Elastic–plastic crack-tip caustic ($n = \infty$). (From Ref. 6.)

This equation allows the determination of the J-integral value from elastic–plastic crack-tip caustics generated by an initial curve located within the elastic–plastic stress–strain field around the crack tip.

For strain-hardening materials, $1 < n < \infty$ (see Fig. 9-30), the strain fields become more complicated and the caustics can no longer be calculated analytically. A numerical calculation of the deflections of an array of light rays around the crack tip, however, gives a picture of the light distribution in the reference plane. Results obtained for $n = 25, 9, 3$, and 1 are shown in Fig. 9-33 [7]. The caustic curve becomes visible as the boundary curves between areas of zero density (shadow) areas and high density (light concentration)].

It can be seen that the caustic curve for $n = 25$ resembles the analytically determined caustic curve for nonhardening materials, $n = \infty$. Furthermore, for the limiting case $n = 1$ (i.e., an elastic material), the caustic curve obtained is identical to the elastic crack-tip caustic curve. The caustic curve for $n = 3$ and $n = 9$ represent intermediate stages between these two limiting cases.

Figure 9-34 shows an experimentally observed elastic–plastic shadow pattern photographed on a tool steel, showing a low rate of strain hardening. Despite the formation of slide lines, which necessarily disturb the reflection conditions of the polished specimen surface, the shadow pattern is of high quality. The resulting caustic curve is in good agreement with the theoretically derived caustic curve and demonstrates the applicability of the shadow optical method of caustics for investigating plastic strain concentration problems.

9-4 Experimental Techniques

Experimental arrangements used for shadow optical investigations are generally quite simple. Special sophisticated equipment is not needed. The only essentials are a suitable light beam for illumination of the specimen and a device for recording the shadow patterns.

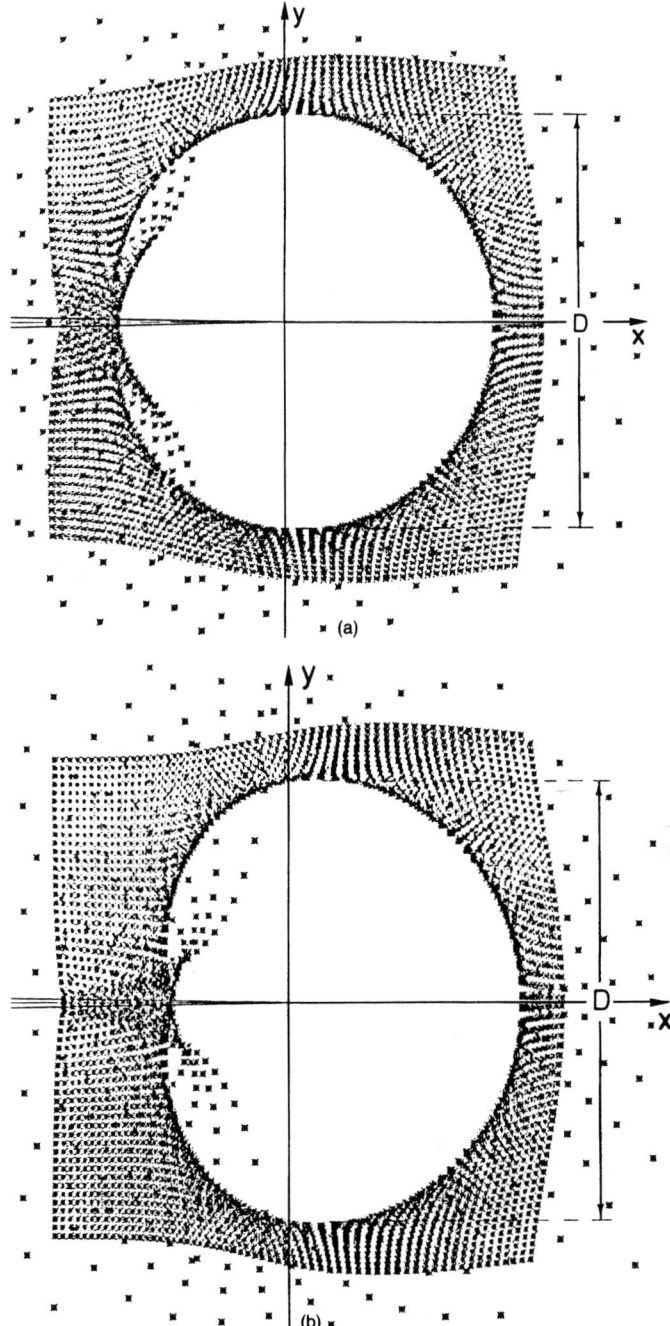

Figure 9-33 Shadow optical light distributions for a crack in an elastic–plastic material: (a) $n = 1$; (b) $n = 3$; (c) $n = 9$; (d) $n = 25$. (From Ref. 7.)

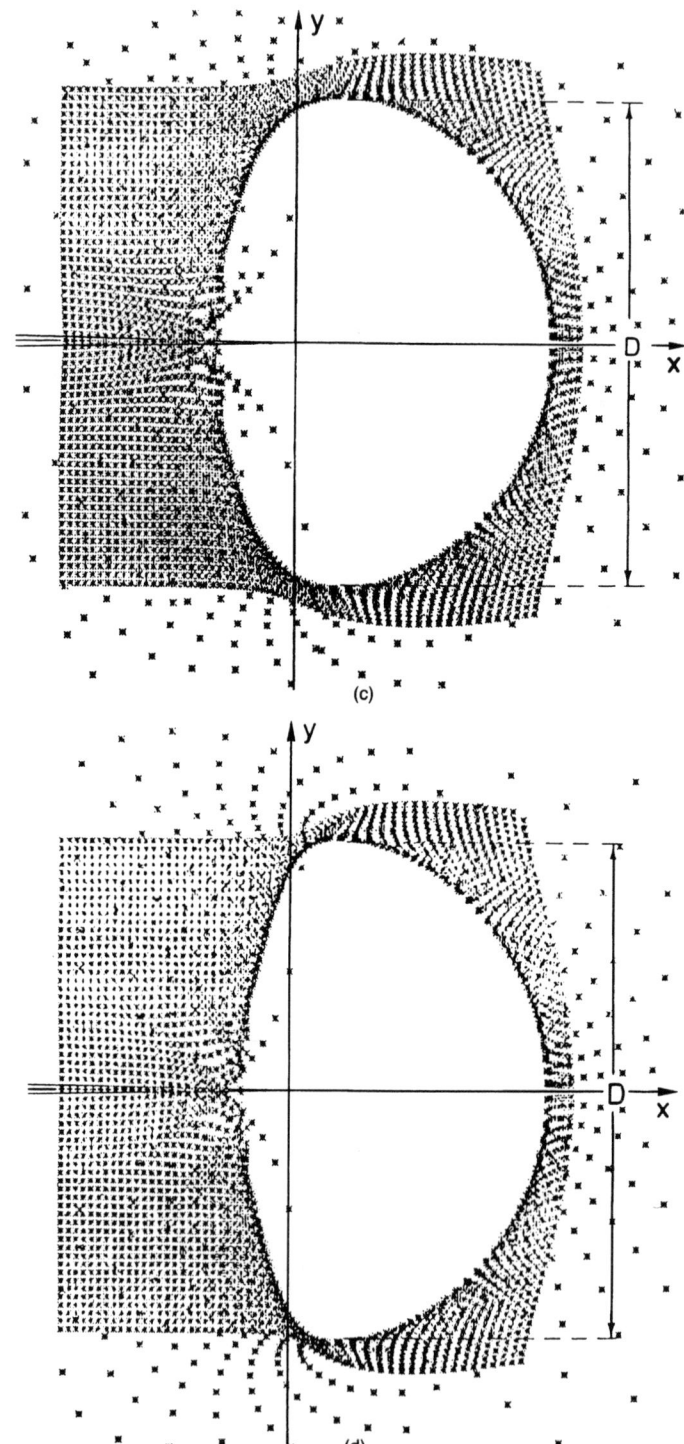

Figure 9-33 *continued*

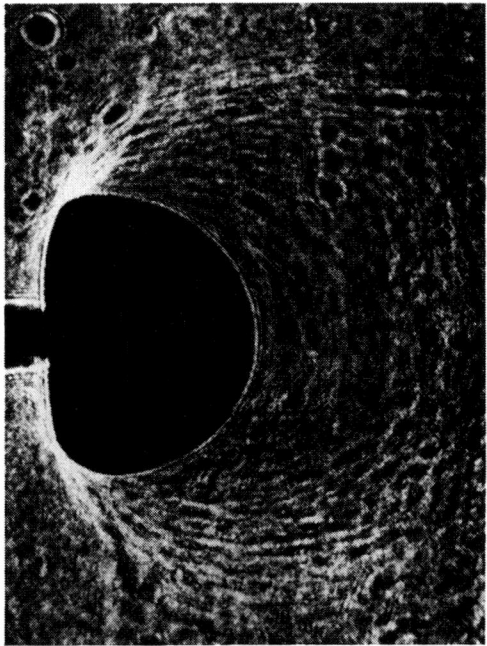

Figure 9-34 Experimentally observed elastic–plastic shadow pattern. (From Ref. 6.)

Illumination of Specimens and Recording of Shadow Patterns

In the description of the physical principle of the shadow optical method, a parallel light beam has been considered. Shadow optical pictures, however, can also be obtained with divergent or convergent light beams. The parallel light beam has been considered for simplicity reasons in the theoretical considerations. In practice parallel light beams are seldom used. Shadow optical arrangements with divergent, parallel, and convergent light beams are schematically shown in Fig. 9-35. The illustrations show the transmission case, but they apply in an analogous manner to the reflection case also. Regardless of what arrangement is utilized, the light beam has to fulfill only one, but a very stringent, requirement: the light beam has to be exceptionally parallel, divergent, or convergent. To achieve this property, the light source must have the essential features of a point source, that is, a small aperture and a large distance from the object. If these conditions are not sufficiently fulfilled, the shadow optical pattern gets a blurry "nonsharp" appearance. The boundary line between the regions of darkness and light concentration (i.e., the caustic curve) is no longer represented by a distinctly marked line. Thus the quantitative evaluation of shadow patterns would be difficult or even erroneous.

Real caustics can be recorded directly on a photographic film positioned in the reference plane. This direct recording of shadow patterns is possible in transmission arrangements as well as in reflection arrangements. In the latter case the light beam has to be slightly tilted with regard to the normal of the specimen to get a separation of the reflected beam from the impinging beam. The direct recording of shadow patterns, however, is seldom used in practice. Shadow patterns are generally recorded with a photographic camera (e.g., a conventional 35-mm camera). When

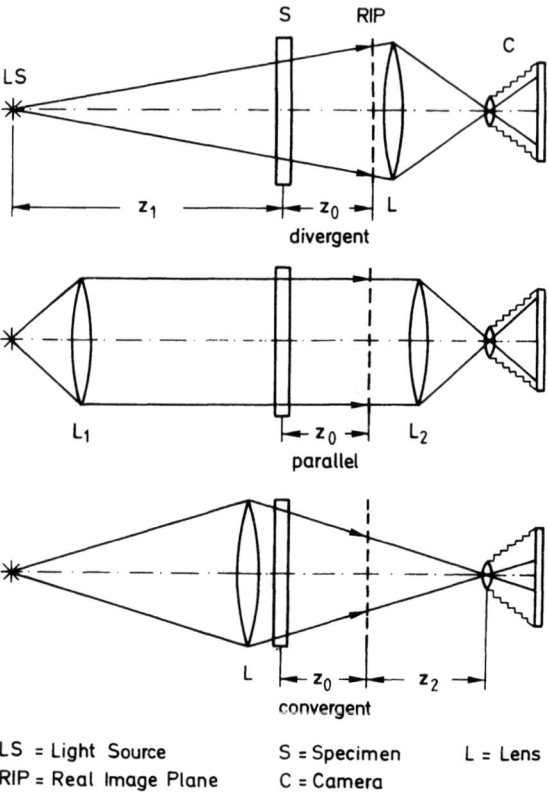

Figure 9-35 Shadow optical arrangements with divergent, parallel, and convergent light beams.

shadow optical arrangements with divergent or parallel light beams are utilized, an additional field lens is needed to focus the light into the lens of the camera. In the case of convergent light beams, the camera is positioned directly in the focal point of the light beam. Recording arrangements for divergent, parallel, and convergent light beams are shown together with the respective illumination arrangements in Fig. 9-35. For the registration of shadow optical pictures in a certain reference plane, the photographic camera is focused on this special plane. The use of a screen positioned in the reference plane is not needed. The recording of shadow patterns with a photographic camera has an advantage in that not only real but also virtual shadow patterns can be registered. In the latter case it is only necessary to focus the camera on the respective virtual reference plane. (For reasons of simplicity, the virtual image planes are not marked in Fig. 9-35.)

Scaling Factor for Nonparallel Light Beams

When shadow optical arrangements with divergent or convergent light beams are utilized, the mapping equations and the caustic curves for parallel light beams no longer apply in their simple

SHADOW OPTICAL METHOD OF CAUSTICS

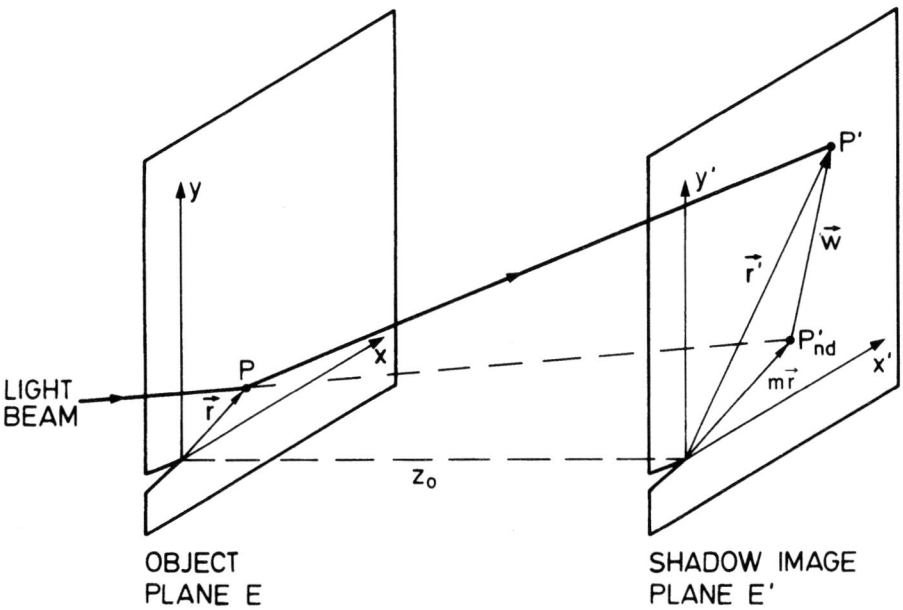

Figure 9-36 Light-ray deflection for a nonparallel light beam.

form but have to be modified. Figure 9-36 illustrates the mapping process of a nonparallel beam. As in Fig. 9-5, Section 9-2, a transmission arrangement with the observation of a real shadow image is considered. A nondeflected ray traversing the object plane at the point $P(\mathbf{r})$ would not hit the image plane at the point P'_{nd} with $\mathbf{r}'_{nd} = \mathbf{r}$, as is the case for a parallel beam (see Section 9-2 and Fig. 9-5), but it would hit the image plane at P'_{nd} with $\mathbf{r}'_{nd} = m\mathbf{r}$ (see Fig. 9-36), where:

For divergent light beams:

$$m = \frac{z_1 + z_0}{z_1}$$

For convergent light beams:

$$m = \frac{z_2}{z_2 + z_0} \qquad (9\text{-}61)$$

z_1 and z_2 are defined as the distances from the light source to the specimen and from the camera to the image plane, respectively (see Fig. 9-35). z_1 and z_2 are always positive. Since for real images $z_0 > 0$, $m > 1$ for divergent beams and $m < 1$ for convergent beams. Consequently, the basic shadow optical mapping Eq. (9-1) for nonparallel light beams has to be modified to

$$\mathbf{r}' = m\mathbf{r} + \mathbf{w} \qquad (9\text{-}62)$$

With this modified mapping equation, the relations between the size of the shadow pattern and the respective generating load parameter for the three examples considered in Section 9-2 (see Fig. 9-8) are

(a) $$P = \frac{1}{m^2} \frac{\pi}{4(2.6)^3 |z_0| \, |c|d_{\text{eff}}} D^3$$

(b) $$p - q = \frac{1}{m^3} \frac{1}{12(2.66)^4 |z_0| \, |c|d_{\text{eff}} R^2} D^4 \qquad (9\text{-}63)$$

(c) $$K_I = \frac{1}{m^{3/2}} \frac{2\sqrt{2\pi}}{3(3.17)^{5/2} |z_0| \, |c|d_{\text{eff}}} D^{5/2}$$

As can easily be seen, for divergent and parallel arrangements ($m \geq 1$) the characteristic length parameter D steadily increases in size with increasing distance z_0 between specimen and reference plane. This is not true for convergent arrangements if large distances z_0 are utilized, since the scaling factor m decreases with increasing z_0. For a fixed distance $z_0 + z_2$ between the specimen and the photographic camera, the largest caustic is obtained in a reference plane with a distance z_0 given by

(a) $z_0 = \tfrac{1}{3}(z_0 + z_2)$

(b) $z_0 = \tfrac{1}{4}(z_0 + z_2)$ $\qquad (9\text{-}64)$

(c) $z_0 = \tfrac{2}{5}(z_0 + z_2)$

For virtual instead of real shadow images, Eqs. (9-61) through (9-63) apply in the same form. But since for virtual images $z_0 < 0$, the scaling factors m take different values, $m < 1$ for divergent beams and $m > 1$ for convergent beams. Caution must now be used with divergent arrangements, since the size of the caustic becomes smaller with increasing distance $|z_0|$, when large values of $|z_0|$ are utilized.

Instrumentation

Shadow optical arrangements with convergent light beams are generally utilized in practice since they require only one lens and since the view field obtained with this lens is largest. Instead of lenses, some shadow optical arrangements use concave mirrors. Figure 9-37 shows a shadow optical arrangement with a concave mirror compared to an arrangement with a lens. Concave mirrors have the advantage that they are available with larger apertures and larger focal lengths than lenses. Thus larger view fields can be obtained. Furthermore, large distances between the light source and the specimen can be realized. As a consequence, the quality of the illuminating light beam is improved and the distance z_0 between the specimen and the reference plane can be varied over a larger range (see also the preceding subsection).

All arrangements using mirrors, as well as reflection arrangements with nontransparent specimens, require light beams that are slightly tilted with respect to the optical axis. Consequently, the influence of astigmatism due to the tilting of the light beams has to be considered. Errors in the evaluation procedure are minimized if the following precautions are taken. Optical arrangements should be used for which the direction of the characteristic length parameter of the caustic is perpendicular to one of the focal lines. This focal line and the corresponding focal length should be used in the evaluation procedure.

To determine experimentally the absolute size of the caustic curve that is recorded with a photographic camera, it is appropriate to photograph together with the caustic a scale that is positioned in the real or virtual image plane. The length parameter D of the caustic is then easily

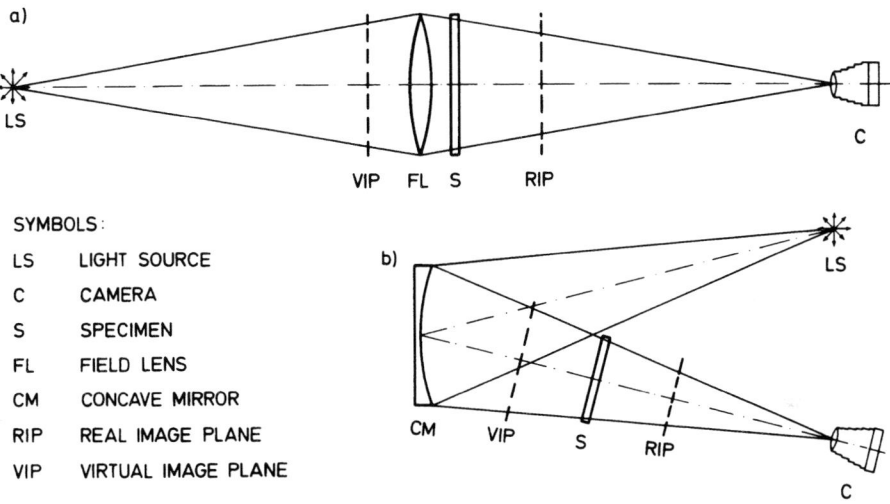

Figure 9-37 Shadow optical arrangements utilizing (a) a field lens and (b) a concave mirror.

determined by comparison with the scale. The scale can also be mounted on the specimen. Since a shadow optical image is photographed, the scale is not sharply reproduced, but the symmetry of the obtained line patterns of the scale nevertheless allows a sufficiently accurate evaluation. For arrangements with scaling factors $m \neq 1$, the size of the caustic determined by comparison with such a scale must, however, be corrected, since the caustic exists in the reference plane, whereas the scale is positioned in the object plane. Consequently, the real size of the caustic is obtained by multiplication with the corresponding scaling factor m.

Application of the shadow optical method for investigating dynamic processes requires the recording of time-dependent shadow patterns with high-speed cameras. Figure 9-38 shows the experimental arrangements of a Cranz–Schardin 24-spark high-speed camera for recording shadow patterns in transmission or in reflection. Basically, the Cranz–Schardin system consists of a photographic recording unit that contains 24 individual cameras with optical axes that meet in the object. The light for illuminating the specimen is generated by 24 sparks, which are assembled in a special light unit. The light emitted from one individual spark is focused into the lens of the corresponding individual camera via a lens or a mirror; the latter is used in Fig. 9-38. The sparks are triggered at different predetermined times (minimum picture interval time = 1 μs) via an electronic control unit. For simplicity, only two of the total 24 light beams are shown in Fig. 9-38. A picture of the experimental setup is given in Fig. 9-39.

Model Materials and Preparation of Specimens

Specimens for shadow optical investigations have to meet relatively high standards concerning their optical quality. In particular, care must be taken that the specimens used in transmission arrangements are made from materials that are free of local variations in density and in thickness. Specimens used in reflection arrangements, on the other hand, must have an optically planar surface at the illuminated side. If these conditions are not sufficiently fulfilled, disturbing influences on the shadow patterns may appear, resulting in erroneous interpretations.

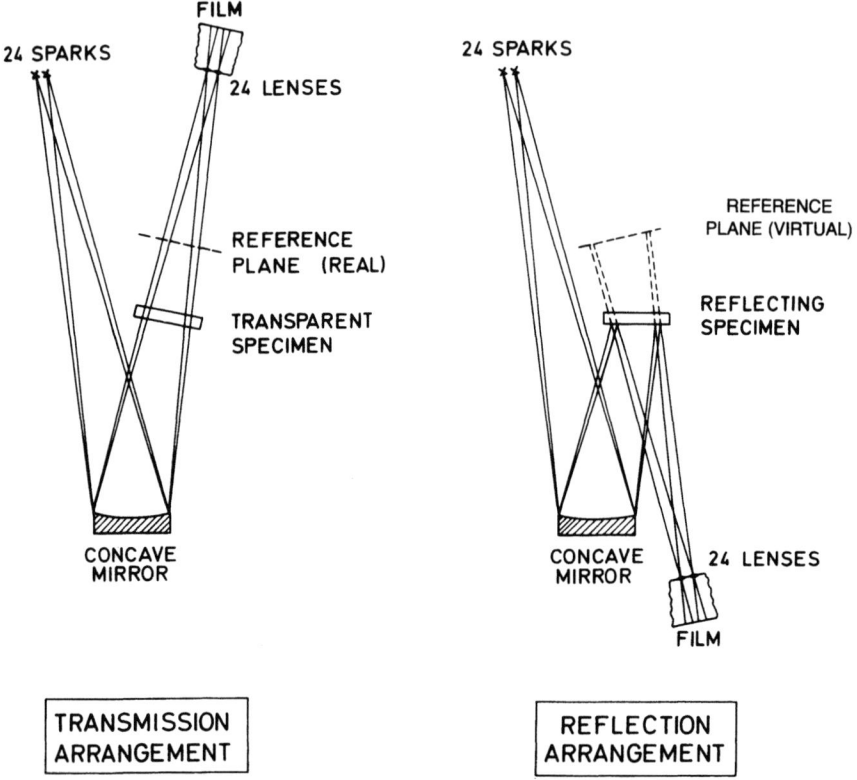

Figure 9-38 Cranz–Schardin 24-spark high-speed camera in transmission and reflection arrangements.

Various model materials are available for machining transparent specimens (see Table 9-1): polymethylmethacrylate (PMMA) has a high shadow optical constant c. It is almost optically isotropic ($\lambda \simeq 0$) and thus yields single caustic curves. The material, however, exhibits strong viscoelastic effects, so that for time-dependent loads its relaxation behavior must be considered. For example, shadow patterns obtained with PMMA do not immediately disappear when the applied load is removed. Depending on the time during which the load has been active, the shadow patterns can decay over time ranges as long as several minutes. The materials Homalite 100 and CR-39 also show a high shadow optical sensitivity and relatively small effects of anisotropy (i.e., double caustics become visible only in shadow optical arrangements of high resolution). Similar to PMMA, the material CR-39 has the disadvantage of being very viscoelastic. Homalite 100, however, shows considerably smaller viscoelastic effects. A material well suited for shadow optical investigations is the epoxy resin Araldite B: its shadow optical sensitivity is high, but viscoelastic effects are negligible. The material is highly anisotropic and thus exhibits clearly pronounced double caustics (see also the next subsection).

In reflection arrangements the shadow optical sensitivity is controlled by the size of the surface deformation produced by the generating load and thus is determined by the elastic properties of the material (i.e., Young's modulus E and Poisson's ratio ν). The deformation at the rear surface

SHADOW OPTICAL METHOD OF CAUSTICS

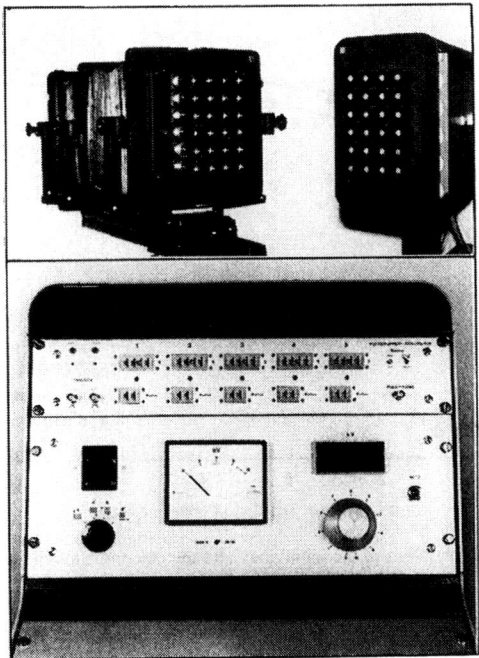

Figure 9-39 Cranz–Schardin 24-spark high-speed camera.

of the specimen and changes in the density of the material (resulting in changes of the refractive index for transparent materials) do not contribute to the shadow optical effect in reflection. Therefore, reflection arrangements are generally less sensitive than transmission arrangements that integrate over all the effects discussed above. The front surface of the specimen must be optically planar and mirrored, at least in the localized area that generates the caustic. For steel specimens, best results are obtained by subsequent grinding, lapping, and polishing of the front surface of the specimen.

Influences of Local Plasticity and State of Stress

Stresses or strains obtained in stress concentration problems can be very high, theoretically often unbounded. Thus in real materials plastically deformed regions result. For the investigation of linear problems of elasticity it is therefore necessary to assure that these plasticity effects do not have a disturbing influence on the formation of the caustic. In practice, this is usually achieved by utilizing shadow optical arrangements with an initial curve being larger and lying outside the plastically deformed region. The caustic is then generated from areas where the linear elastic stress–strain relations remain valid. This requirement is realized experimentally by choosing a sufficiently large distance $|z_0|$ between the specimen and the reference plane so that

$$r_0 > r_{pl} \tag{9-65}$$

where r_{pl} is the size of the plastically deformed region.

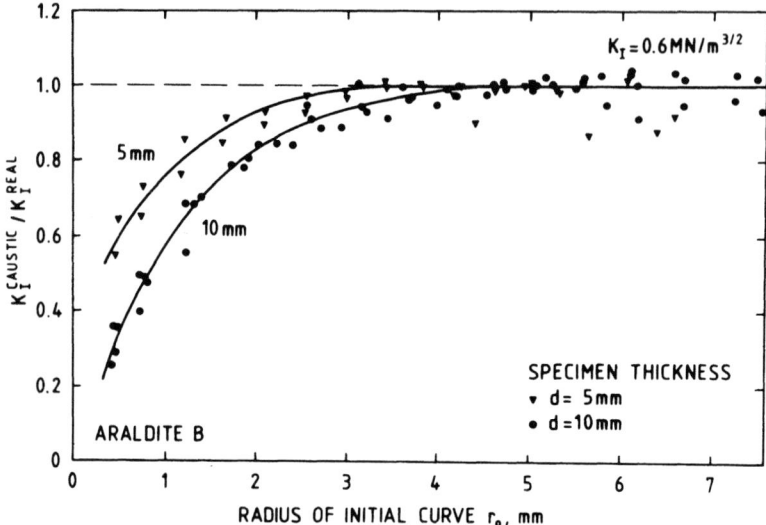

Figure 9-40 Influence of the radius of the initial curve on the determination of stress-intensity factors.

The values of the shadow optical constant c and of the anisotropy coefficient λ are different for plane stress and plane strain (see Table 9-1). Therefore, in Eqs. (9-15), (9-20), (9-24), (9-27), (9-35), (9-37), and (9-49) for the evaluation of caustics, those values must be utilized that apply for the state of stress that prevails along the initial curve. Usually, this is a state of plane stress. With very high stress gradients, however, even in plates a state of plane stress cannot develop in the localized area around stress concentration points due to the very large differences in the constraints. Consequently, instead of plane stress a three-dimensional state of stress applies. If the initial curve falls within this region of three-dimensional state of stress, naturally the determination of stress-intensity factors by evaluation formulas that are based on the assumption of plane-stress conditions becomes erroneous. Figure 9-40 [28] shows stress-intensity factors determined by the caustic technique $K_I^{CAUSTIC}$ in relation to the mechanically applied stress-intensity factors K_I^{REAL}, which were derived from experiments in which the size of the initial curve was varied by varying the distance z_0 between the specimen and the reference plane [see, e.g., Eq. (9-12)]. Only if the radius of the initial curve becomes larger than a certain value, which obviously depends on the thickness d of the specimen, do the experimentally determined stress-intensity factors $K_I^{CAUSTIC}$ agree with the real stress intensity-factors applied K_I^{REAL}; that is, only then do plane-stress conditions prevail. Thus, special care must be taken in the evaluation of caustics that the distance $|z_0|$ between the specimen and the reference plane is chosen sufficiently large that the initial curve lies outside the region of three-dimensional state of stress,

$$r_0 > r_{ps} \qquad (9\text{-}66)$$

where r_{ps} is the smallest radius around the center of stress concentration outside which a state of plane stress applies.

With optically anisotropic materials the correct choice of the distance $|z_0|$ between the specimen and the reference plane can be easily verified experimentally. If the experimentally observed split-up of the double caustic, $(D_o - D_i)/D_i$, agrees with the theoretically predicted split-up for plane-

SHADOW OPTICAL METHOD OF CAUSTICS

stress conditions, the initial curve lies outside the region of mixed state of stress (i.e., in the region of plane stress). However, if the experimentally observed split-up is larger, the initial curve lies within the region of mixed state of stress and an evaluation of the caustic would be erroneous.

The preceding validity requirements [Eqs. (9-65) and (9-66)] of course apply in an analogous manner to any technique for determining stress-intensity factors from crack-tip stress fields.

9-5 Applications

Some typical applications of the shadow optical method of caustics for investigating dynamic problems of practical interest are presented in this section. All the pictures reported were photographed with a Cranz–Schardin 24-spark high-speed camera. The shadow patterns either were recorded in transmission arrangements with specimens made from the model material Araldite B or photographed in reflection arrangements with specimens made from the high-strength maraging steel X2 NiCoMo 18-9-5.[6]

Dynamic Loading of Bend Specimens

The history of the load input into a specimen that is impacted by a falling knife edge is investigated. An experiment is performed with a bend specimen measuring 412 mm × 72 mm × 10 mm made from the material Araldite B. The support span is 260 mm. The wedge of 5.3 kg mass impacts the specimen at a velocity of 1 m/s. The knife-edge tip radius ($\rho_0 = 1$ mm) is small compared to the specimen dimensions and thus produces a pointlike loading condition. The virtual compression caustic formed around the point of impact is photographed with a transmission arrangement in a reference plane at $z_0 < 0$. A schematic illustration of the experimental setup is given in Fig. 9-41.

A series of shadow optical photographs is shown in Fig. 9-42. The times given with the photographs were measured from the moment of first contact between the wedge and the specimen. Quantitative data on the impact load $P_c(t)$, which were obtained according to Eq. (9-15a) from the recorded caustics, are given in Fig. 9-43. In addition to the shadow optically determined load values P_c, the load curve $P_H(t)$ is given, which is measured by a strain gage near the tip of the striking knife edge. The two measurements are in good agreement. It is recognized that the load input into the specimen is not given by a steadily increasing function of time. The load shows an oscillating behavior with an overall increasing tendency. This is caused by eigenvibrations of the specimen that are excited by the impact process.

Cracks Under Impact Loading

Impact tests with precracked bend specimens loaded in a pendulum or a drop-weight test machine are widely used for measuring dynamic material strength values. The response of the specimen at the crack tip during the impact event is investigated (see also Refs. 29–31). An experiment is performed with a precracked bend specimen subjected to dynamic loading by a drop weight. The experimental arrangement is shown in Fig. 9-44. The specimen, made from the high-strength steel

---

[6] Produced by Krupp Stahl AG and designated Krupp HFX 6358. Nominal composition: 18% Ni, 9% Co, 4.8% Mo, and <0.03% C. Heat treatment: 480°C for 4 hours in air. This steel is similar to the American designation 18 Ni maraging grade 300.

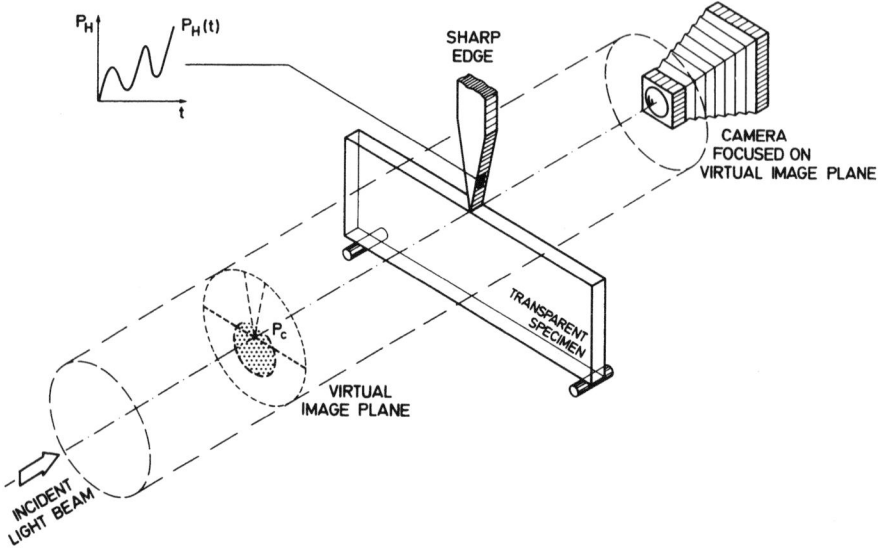

Figure 9-41 Experimental arrangement for investigating the load input into a specimen by a falling knife edge.

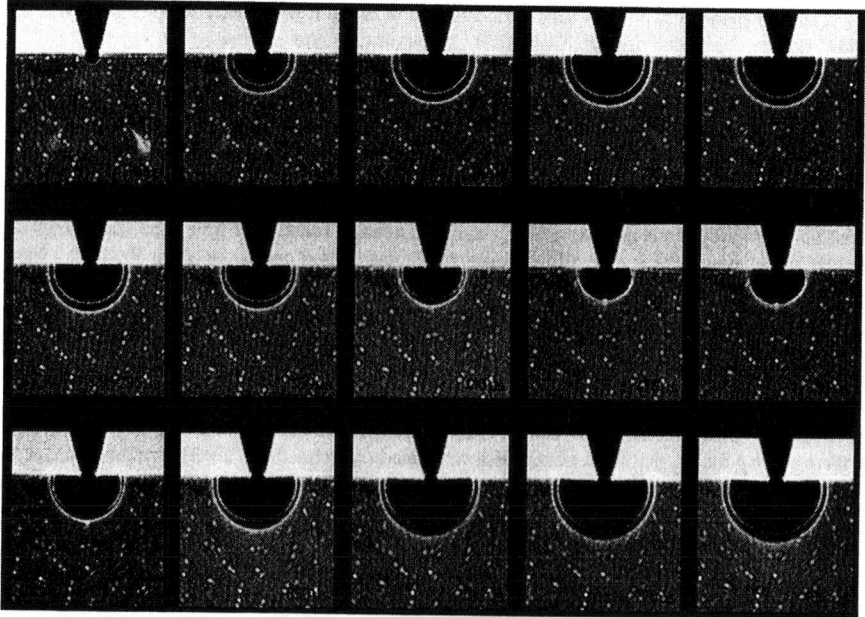

Figure 9-42 Series of shadow optical photographs for a knife edge impacting the edge of a plate (virtual images photographed in transmission with an Araldite B specimen).

SHADOW OPTICAL METHOD OF CAUSTICS

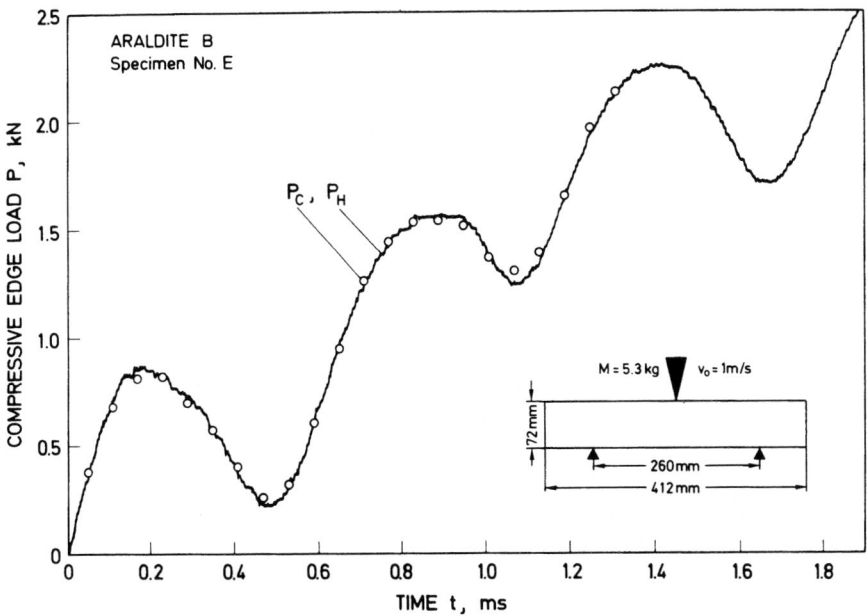

Figure 9-43 Load input into a specimen by a falling knife edge.

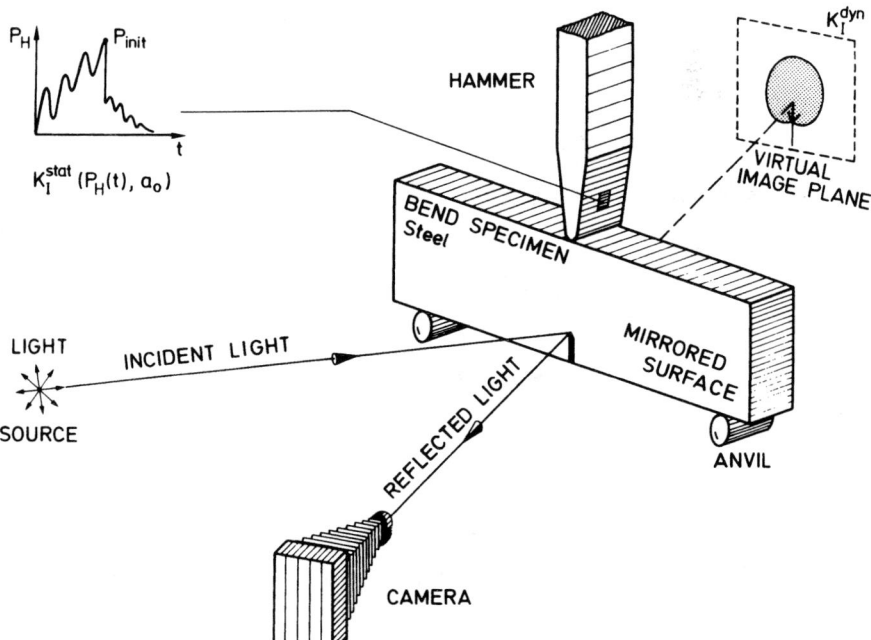

Figure 9-44 Experimental arrangement for investigating the behavior of cracks under impact loading.

X2 NiCoMo 18-9-5, measures 330 mm × 60 mm × 10 mm. The specimen is supported by anvils with an opening of 240 mm. The initial crack in the specimen has a length of 20 mm and is slightly blunted to increase the load-carrying capacity of the specimen and thus the observation time before failure. Dynamic loading is achieved by a drop weight of 90-kg mass impacting the specimen at a velocity of 0.5 m/s.

A series of 12 virtual shadow optical pictures, photographed in reflection with $z_0 < 0$, is presented in Fig. 9-45. Only the central part of the impacted specimen is shown. Quantitative data obtained according to Eq. (9-15c) from such shadow patterns are given in Fig. 9-46. The shadow optically determined dynamic stress-intensity factors K_I^{dyn} are plotted as a function of time. In addition, the stress-intensity factors $K_I^{stat}(P_H)$ are shown. These values were determined from the load signal P_H registered at the instrumented tup of the striking hammer, utilizing a conventional stress-intensity factor formula [32] for a precracked specimen under an equivalent quasi-static loading

$$K_I^{stat} = \frac{P_H S}{dW^{3/2}} \frac{3(a/W)^{1/2}\{1.99 - (a/W)(1 - a/W)[2.15 - 3.93(a/W) + 2.7(a^2/W^2)]\}}{2(1 + 2a/W)(1 - a/W)^{3/2}}$$

(9-67)

where P_H = load registered at the tup of the striking hammer
d = specimen thickness
S = support plan
W = specimen width
a = crack length

The times in Fig. 9-46 are given in absolute units and also in relative units by normalization with the time τ, where τ is the period of the oscillation of the impacted specimen. Values of τ are obtained from the approximation formula [33]

$$\tau = \frac{1.68(SWdCE)^{1/2}}{c_0}$$

(9-68)

where S = support span
W = specimen width
d = specimen thickness
C = specimen compliance
E = Young's modulus
c_0 = sound wave speed (5000 m/s for steel)

The K_I^{stat} values show a strongly oscillating behavior, whereas the actual dynamic stress-intensity factors K_I^{dyn} show a more steadily increasing tendency. In the time range $t < 3\tau$, these differences are very pronounced. The differences become smaller with increasing time, but even for times larger than 3τ the influences of dynamic effects obviously have not vanished and there are still marked differences between K_I^{stat} and K_I^{dyn}.

The data indicate that influences of dynamic effects on the crack-tip stress-intensity factor must, in principle, be taken into account for measuring dynamic material strength data in impact tests. In particular, these dynamic influences can be very large for specimens failing in a brittle manner in the early time range. Only for larger crack initiation times resulting for specimens that fail in a more ductile manner will an evaluation procedure based on a quasi-static analysis of the dynamic test yield sufficiently reliable material strength data.

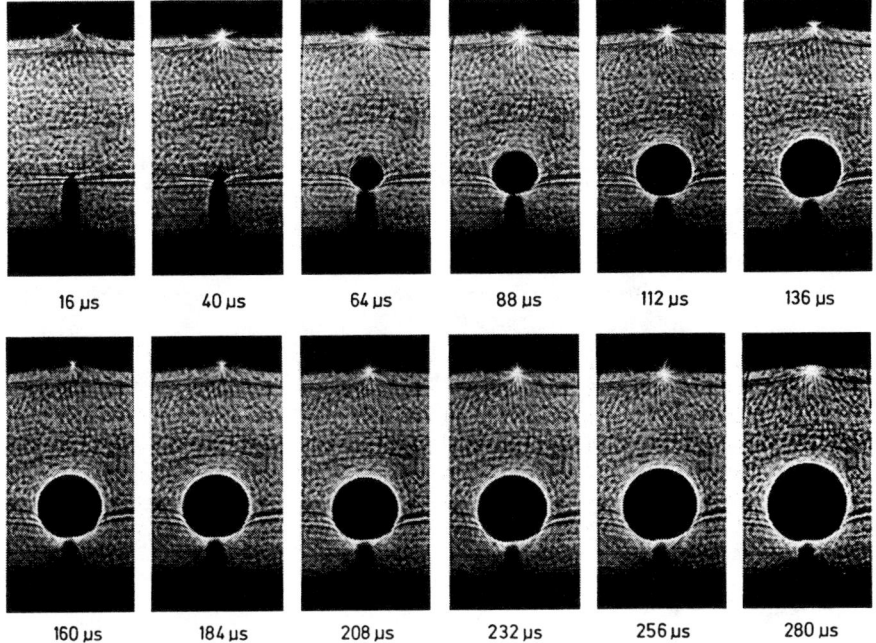

Figure 9-45 Series of shadow optical photographs of a crack under impact loading (virtual images photographed in reflection with a high-strength steel specimen).

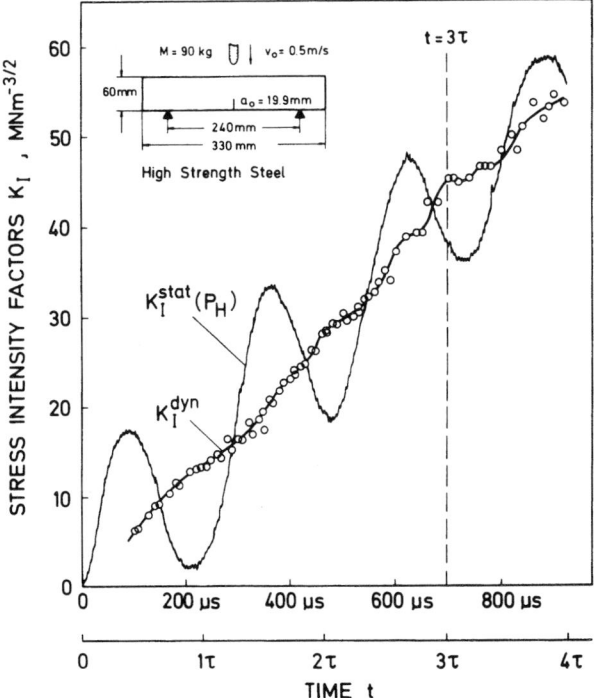

Figure 9-46 Stress-intensity factors for a crack under impact loading by a drop weight.

Propagating Cracks

The dependence of the crack-tip stress-intensity factor K_I from the crack velocity is investigated. The experimental arrangement is shown in Fig. 9-47. In a single-edge-notch (SEN) specimen under uniform tensile loading, a propagating crack is initiated from a preexisting notch. With increasing crack length and reduction of the remaining ligament, the crack-tip stress intensification increases and the crack therefore accelerates. The experiment was performed with a specimen made from Araldite B.

Figure 9-48 shows real shadow optical patterns photographed in transmission with $z_0 > 0$. Only three of the 24 pictures are reproduced here, showing the early crack propagation phase. The momentary crack velocity is given with each picture. Quantitative data obtained according to Eq. (9-20) from such photographs are given in Fig. 9-49. The dynamic stress-intensity factor K_I^{dyn} is shown as a function of the crack velocity v. For small and intermediate crack velocities ($v < 250$ m/s) the stress-intensity factor increases only slightly with crack velocity, but in the higher crack velocity range ($v > 250$ m/s) the increase in K_I^{dyn} with v becomes very steep and K_I^{dyn} obviously approaches unbounded values. The maximum possible crack velocity is therefore limited. For the material Araldite B, the resulting terminal crack velocity is about 400 m/s. (The highest terminal crack velocity theoretically possible [22] is the Rayleigh wave speed c_R. For Araldite B, $c_R = 970$ m/s.)

Arresting Cracks

Similar to the crack initiation toughness K_{Ic} (i.e., the critical stress-intensity factor for failure of the material), the crack arrest toughness K_{Ia} is the critical stress-intensity factor at the moment of arrest. K_{Ia} describes the ability of a material to arrest a propagating crack. Crack arrest toughness values are measured in special laboratory experiments. To get information on the stress condition at the tip of an arresting crack, the general behavior of a propagating and subsequently arresting crack is investigated (see also Refs. 34–37). One speculation is that the stress condition at arrest must be static, since the crack velocity is zero at the moment of arrest [38].

Arrest of a propagating crack is achieved in a specimen with a decreasing stress-intensity field [e.g., a wedge-loaded double-cantilever-beam (DCB) specimen (Fig. 9-50)]. The propagating crack is initiated from a preexisting notch of length a_0 at a critical displacement 2δ. The notch tip is blunted in order to store sufficient elastic energy in the specimen before crack initiation so that the crack can accelerate to high velocities. During crack propagation the displacement 2δ stays constant because of the stiffness of the wedge-loading system. The specimen, however, becomes more compliant because of the increase of crack length. Consequently, the crack-tip stress intensification decreases with increasing crack length and the crack finally comes to arrest at an arrest crack length a_a. The stress-intensity factory history for propagating and subsequently arresting cracks is investigated with DCB specimens measuring 321 mm × 127 mm × 10 mm made from Araldite B.

A series of six real shadow optical patterns photographed in transmission with $z_0 > 0$ is reproduced in Fig. 9-51. The momentary crack-tip position is given with each photograph. It can be seen that with increasing crack length the shadow patterns decrease in size, indicating decreasing stress-intensity factors for the advancing crack. Quantitative data obtained from such photographs with the evaluation formula (9-20) are shown in Fig. 9-52. Results for cracks initiated at different values of the stress-intensity factor at initiation (K_{Iq}) are summarized in the figure. Values of the dynamic stress-intensity factor K_I^{dyn} (experimental points) are shown as a function of crack length a, together with the corresponding static stress-intensity factor (K_I^{stat}) curve. The static values were

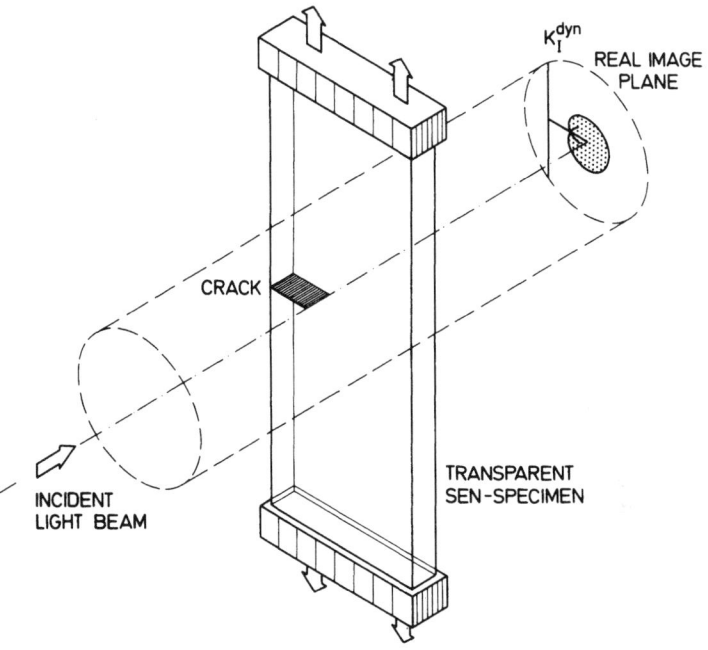

Figure 9-47 Experimental arrangement for investigating the behavior of propagating cracks.

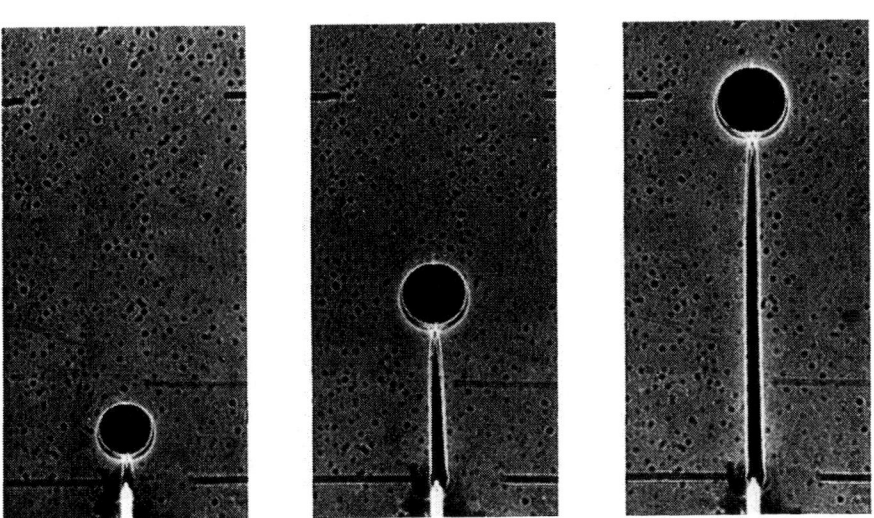

Figure 9-48 Series of shadow optical photographs of a propagating crack (real images photographed in transmission with an Araldite B specimen).

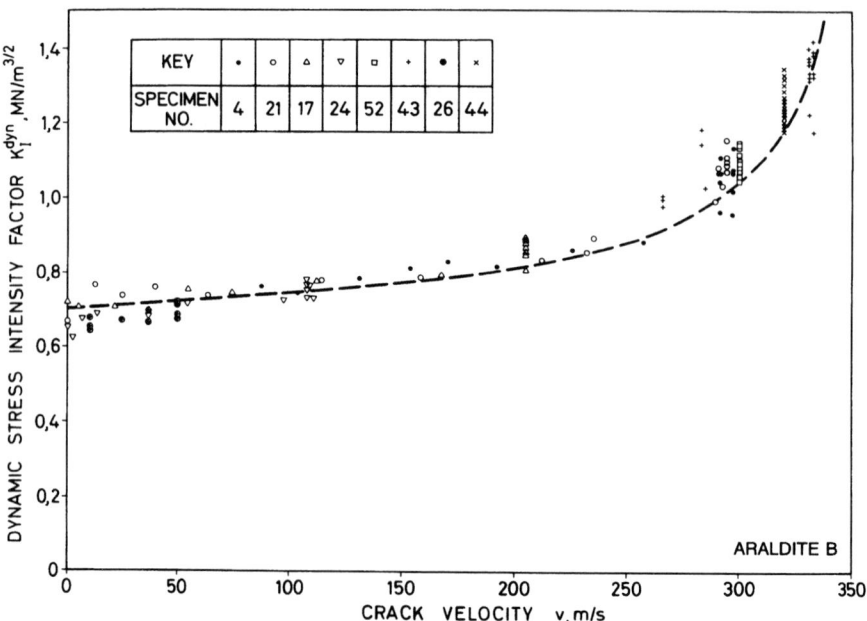

Figure 9-49 Dynamic stress-intensity factors for propagating cracks.

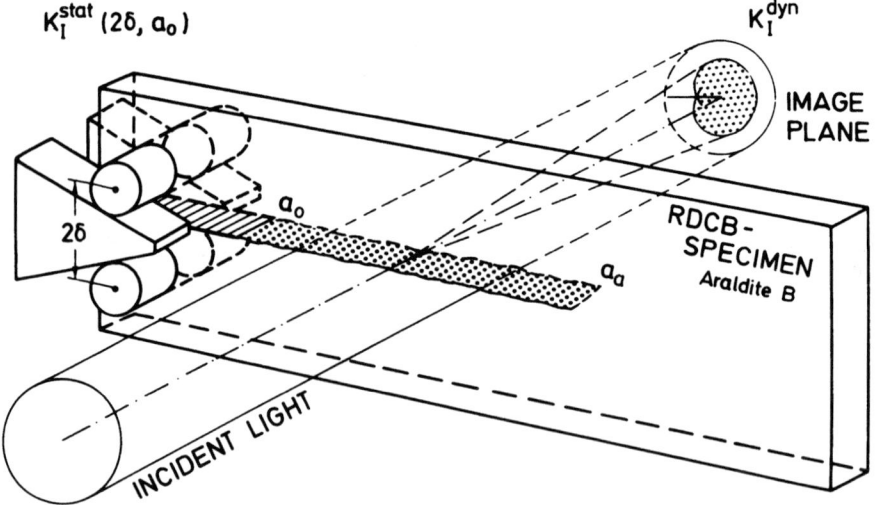

Figure 9-50 Experimental arrangement for investigating the behavior of an arresting crack. RDCB, double-cantilever beam.

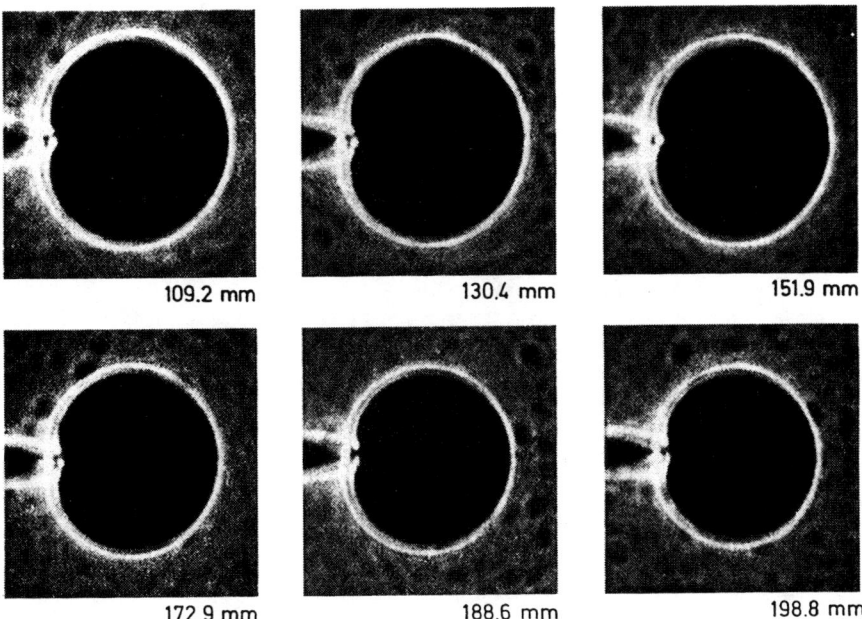

Figure 9-51 Series of shadow optical photographs of a propagating and subsequently arresting crack (real images photographed in transmission with an Araldite B specimen).

determined from the measured critical displacement 2δ utilizing the conventional stress intensity factor formula [32] for a stationary crack:

$$K_I^{stat} = \frac{\sqrt{3}}{4}\frac{EH^{3/2}}{a^2} 2\delta \frac{1 + 0.64(H/a)}{1 + 1.92(H/a) + 1.22(H/a)^2 + 0.39(H/a)^3} \quad (9\text{-}69)$$

where E = Young's modulus
$2H$ = height of the specimen
a = crack length

In addition to the stress-intensity factors, the measured crack velocities are given in the lower part of the diagram. The following characteristics of the crack arrest process can be deduced from the results.

At the beginning of the crack propagation phase the dynamic stress intensity factor K_I^{dyn} is smaller than the corresponding static value K_I^{stat}. At the end of the crack propagation phase the dynamic stress-intensity factor K_I^{dyn} is larger than the corresponding static value K_I^{stat}. Only after arrest does the dynamic stress-intensity factor K_I^{dyn} approach the static stress-intensity factor at arrest (K_{Ia}^{stat}). Differences between the dynamic and the static stress-intensity factor curves become smaller for cracks that are initiated at lower K_{Iq} values and thus propagate at lower velocities. The dynamic effects obviously decrease with decreasing velocity, as one might expect.

For two experiments the behavior of the dynamic stress-intensity factor in the postarrest phase is shown in Fig. 9-53. The dynamic stress-intensity factors K_I^{dyn} are plotted as functions of time

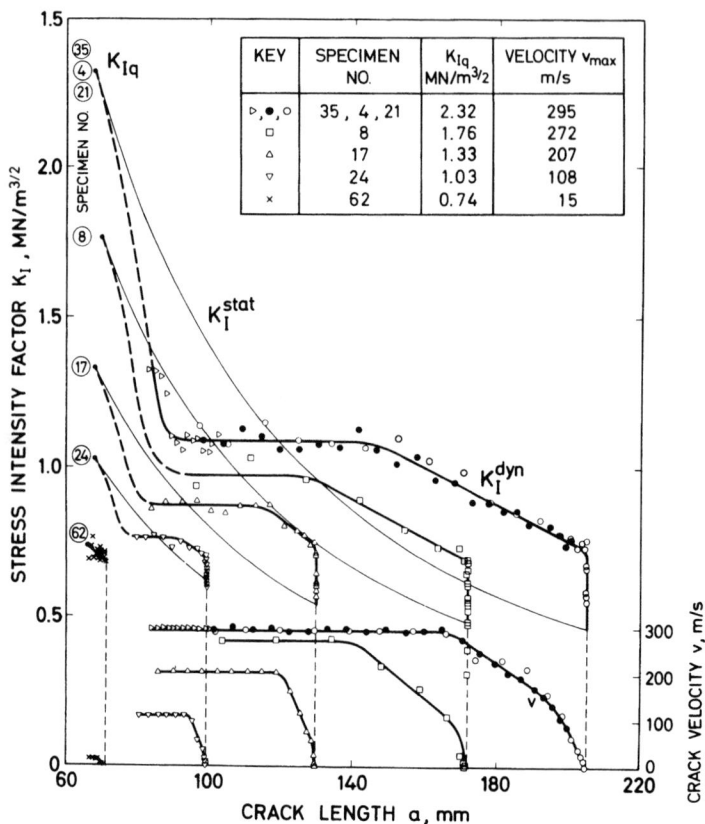

Figure 9-52 Stress-intensity factors for propagating and subsequently arresting cracks.

t. It is recognized that K_I^{dyn} oscillates around the value of the static stress-intensity factor at arrest K_{Ia}^{stat}. Only for sufficiently large times after arrest does the dynamic stress intensity factor approach the static value.

The data indicate that a dynamic state of stress prevails at the moment of arrest, although the crack velocity is zero at this moment. A static state of stress is obtained only a long time after arrest. This behavior is in accordance with the concept of recovered kinetic energy [39]. Elastic waves are generated by the propagating crack; thus kinetic energy is radiated into the specimen. At the finite boundaries of the specimen, these elastic waves are reflected and thus can later interact with the crack tip again and contribute to the stress-intensity factor. This process is illustrated in Fig. 9-54, showing the shadow optical picture of a fast-propagating crack (1000 m/s) in a high-strength steel specimen. The photograph was taken in a reflection arrangement with $z_0 < 0$. In addition to the crack-tip shadow pattern, it shows the generation of elastic waves by the propagating crack, the reflection of these waves at the boundaries of the specimen, and the subsequent interaction of the reflected waves with the crack tip.

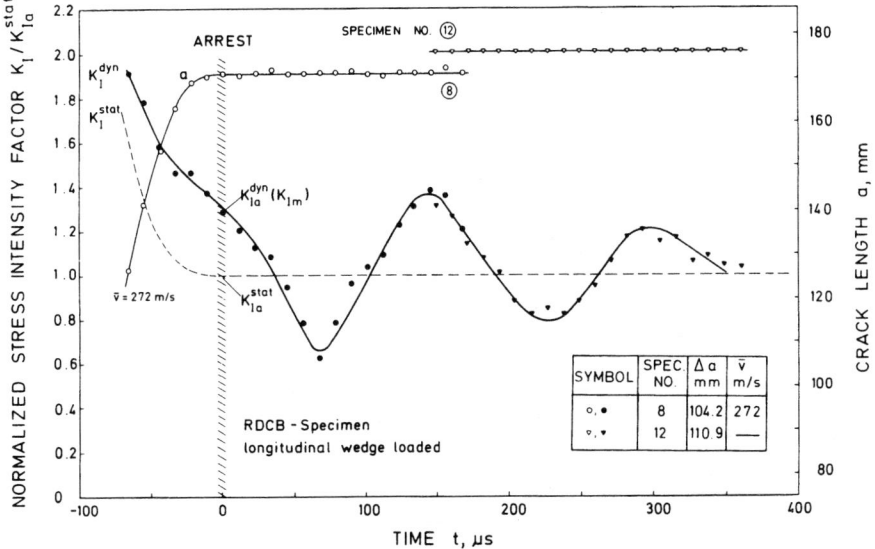

Figure 9-53 Stress-intensity factor behavior in the postarrest phase.

The observed findings are summarized in Fig. 9-55. The data indicate that dynamic effects do have an influence on the crack arrest process. Therefore, in principle, these effects have to be taken into account for measuring the crack arrest toughness K_{Ia}. A static analysis can yield correct arrest toughness data only if the crack propagation velocity prior to arrest is sufficiently small.

Stress Field in a Specimen Under Projectile Loading

The stress condition in a specimen that is dynamically loaded by an impinging projectile is investigated (see also Ref. 40). The loading arrangement is shown in Fig. 9-56. A projectile impinges on a specimen twice the length of the projectile. Both projectile and specimen are made from the same material, Araldite B. The impact process generates a compressive wave, which propagates into the specimen, thus creating a compressive state of stress. At the free end of the specimen, the compressive wave is then reflected as a tensile wave. Similar processes take place in the projectile. A compressive wave is generated in the projectile which afterward is reflected at its free end. These two tensile waves, after having traveled the same distance, meet in the middle of the specimen and create a tensile state of stress, which then spreads over the entire specimen. Subsequent wave reflection processes cause compressive stresses again to result for later times, and so on.

The stress history in the specimen has been made visible by the shadow optical hole field technique. A row or an array of small holes is drilled into the specimen. The stresses at the position of the hole generate a stress concentration around the hole which is registered by the shadow optical technique, here applied in transmission with $z_0 > 0$. According to the results presented in Section 9-2 (see, in particular, Fig. 9-9), a tensile stress field will generate a shadow pattern with the intersection line pointing in the direction of the applied stress field, whereas for

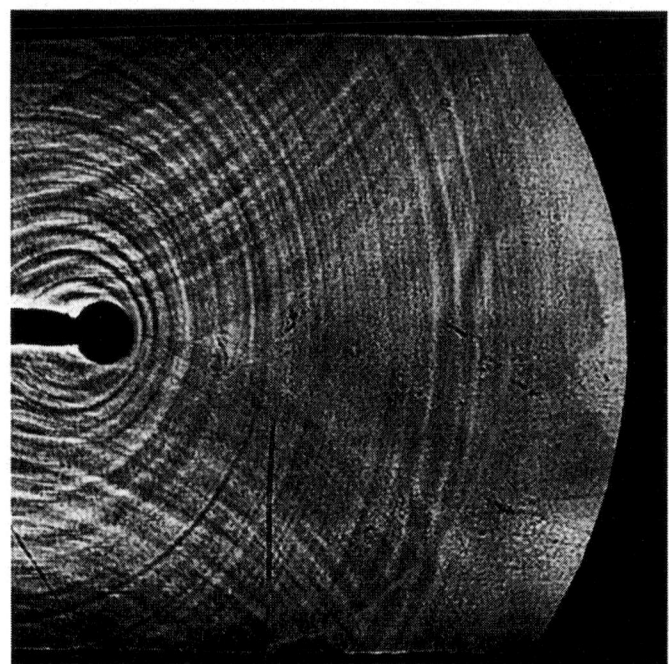

Figure 9-54 Shadow optical photograph of a fast-propagating crack in a high-strength steel specimen (virtual image photographed in reflection).

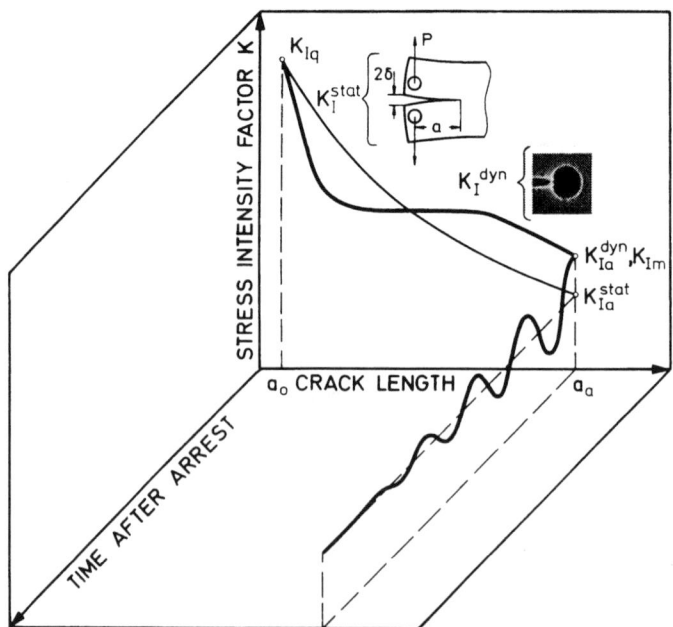

Figure 9-55 Schematic representation of crack arrest behavior.

SHADOW OPTICAL METHOD OF CAUSTICS

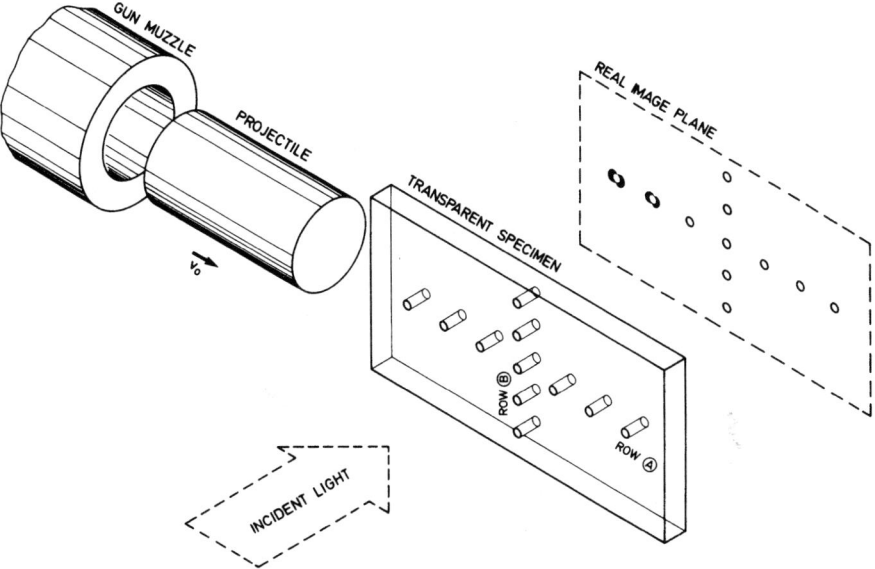

Figure 9-56 Experimental arrangement for investigating the stress history in a target plate impacted by a projectile.

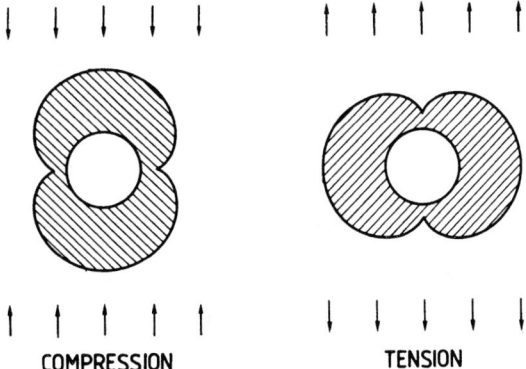

Figure 9-57 Shadow patterns around circular holes under (a) compression and (b) tension. Real images ($z_0 > 0$) in transmission.

a compressive stress field the intersection line is oriented perpendicular to the stress field, as shown in Fig. 9-57. Furthermore, the size of the caustic is a measure of the magnitude of the stress.

A series of 24 shadow optical pictures of a row of holes (1 mm diameter) positioned in length direction of the specimen (see Fig. 9-56, row A) is reproduced in Fig. 9-58. The recording times of the photographs are given with each picture; the time at which compression changes into tension has been set equal to zero. The first pictures show the propagation of the compressive wave into the specimen (frames 1–5), creating a compressive loading field (frames 6–12). The compressive stresses are then reduced in magnitude and change into tensile stresses, as indicated by the change in the direction of the shadow patterns (frames 13–14). A tensile stress field is observed for the following time interval (frames 15–22); later, the stresses change into compression again (frames 23–24).

Figure 9-59 shows the results for a row of three holes positioned at half-length and at $\frac{1}{4}$, $\frac{1}{2}$, and $\frac{3}{4}$ of the width of the specimen (see Fig. 9-56, row B). In addition to the shadow optical recordings, the stresses were directly measured by a strain gage located between two of the holes. The shadow optical pictures are shown in Fig. 9-59 in comparison to the strain-gage result. It is recognized that the shadow optical patterns change their directions and their sizes analogous to the strain-gage signal. In particular, between frames 13 and 15 the compression shadow patterns change into tension shadow patterns, in agreement with the strain-gage measurements. The last shadow optical picture again indicates compressive stresses. The data demonstrate the applicability of the shadow optical hole field technique for providing a simple overview of stress distributions in specimens.

Double-Crack Configuration Loaded by a Tensile Stress Pulse

The stress-intensity factor history of two mutually interacting parallel cracks under dynamic loading by an asymmetrically impinging tensile stress pulse in now investigated (see also Refs. 40 and 41). A specimen made from Araldite B measuring 400 mm × 100 mm × 10 mm contains two single-edge notches of length $a_0 = 25$ mm and distance $d = 20$ mm. The holding fixture consists of two blocks; one is at a fixed position, the other is suddenly accelerated by an impinging projectile in a special loading device (i.e., $s = 0$ for $t < t_0$ but $s \neq 0$ for $t > t_0$). Thus a tensile stress pulse is initiated and propagates into the specimen. The crack that is hit first is denoted crack A; the other crack, which is hit somewhat later, is denoted crack B. The experimental arrangement is shown in Fig. 9-60.

Figure 9-61 shows a series of six real shadow optical pictures, photographed in transmission with $z_0 > 0$. The recording times are given with each picture. The time when the tensile stress pulse reaches the centerline between the two cracks A and B is set equal to zero in the figure. For early times crack A exhibits the larger shadow pattern (i.e., the larger stress-intensity factor) since crack A is hit first by the stress pulse. The caustic indicates a pure mode I loading, whereas crack B shows a mixed-mode (mode I/mode II) loading due to disturbances of the stress pulse resulting from the previous interaction with crack A. For later times the situation changes. Both caustics are of the same size, and even later the crack B caustic becomes larger than the crack A caustic, which indeed becomes smaller. Due to the interaction processes of the two crack-tip stress fields, crack A now also shows a mixed-mode loading type. Quantitative data on the mode I stress-intensity factor histories for a crack configuration of similar geometry (crack length $a_0 = 15$ mm, crack distance $d = 20$ mm) are shown in Fig. 9-62. The stress-intensity factor curves $K_I^A(t)$ and $K_I^B(t)$ of the two cracks oscillate around each other. Only for early times is the stress-intensity factor of crack A larger than for crack B, $K_I^A > K_I^B$. At later times the crack B stress-intensity

SHADOW OPTICAL METHOD OF CAUSTICS

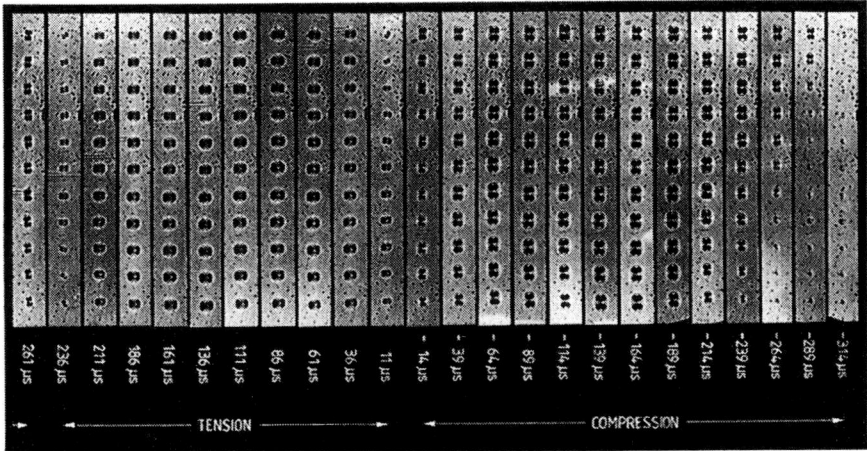

Figure 9-58 Series of shadow optical photographs of a specimen with a row of holes impacted by a projectile (real images photographed in transmission with an Araldite B specimen).

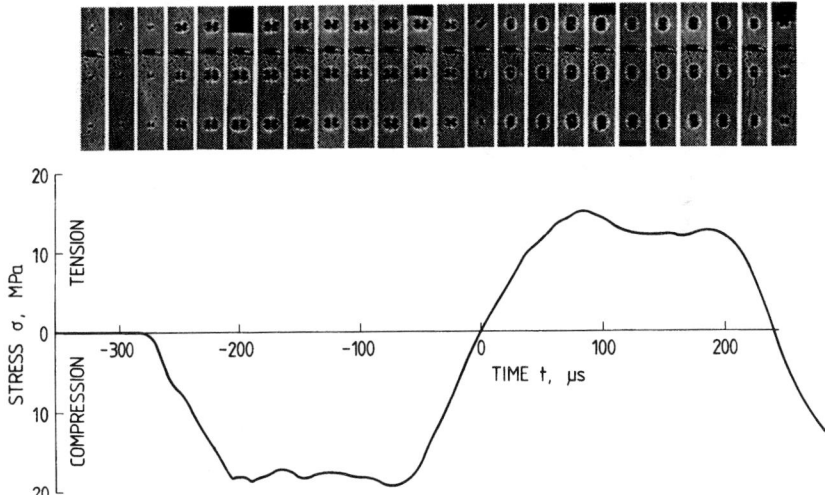

Figure 9-59 Series of shadow optical photographs for a specimen with a row of holes impacted by a projectile (real images photographed in transmission with an Araldite B specimen). Comparison with strain-gage signal.

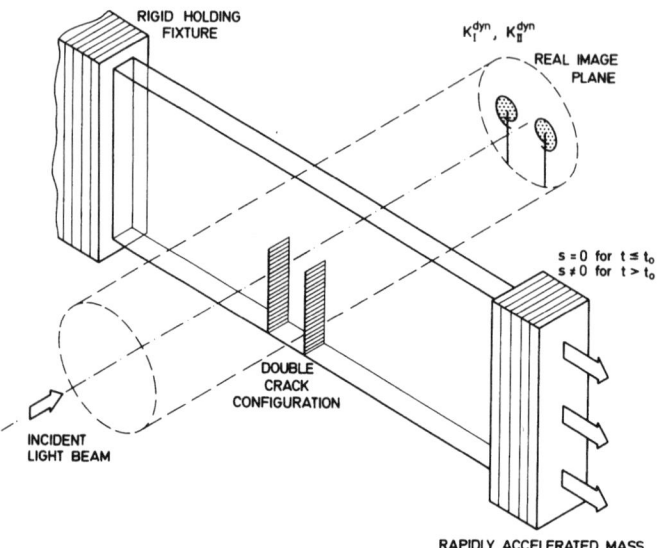

Figure 9-60 Experimental arrangement for investigating the interaction of double-crack configuration loaded by an asymmetrically impinging stress pulse.

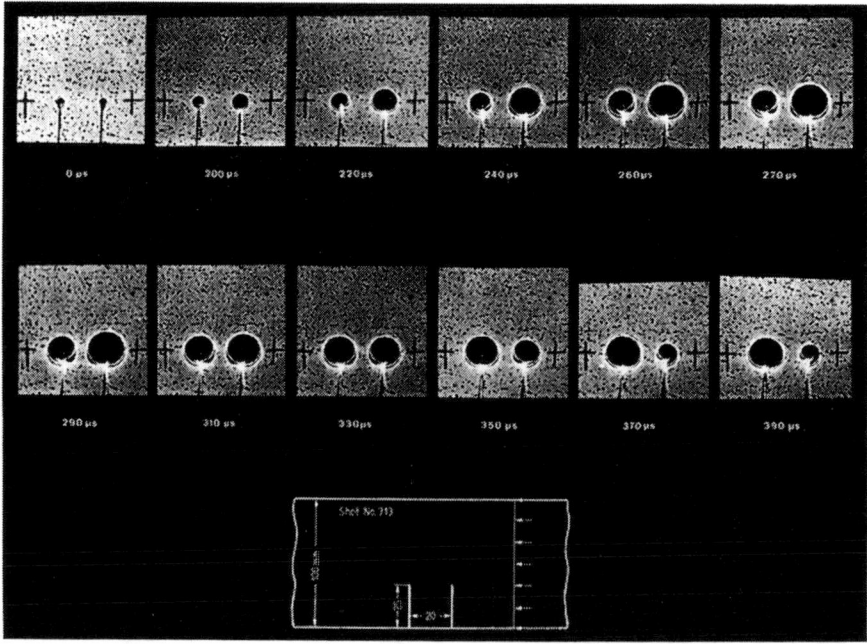

Figure 9-61 Series of shadow optical photographs of a dynamically loaded double-crack configuration (real images photographed in transmission with an Araldite B specimen).

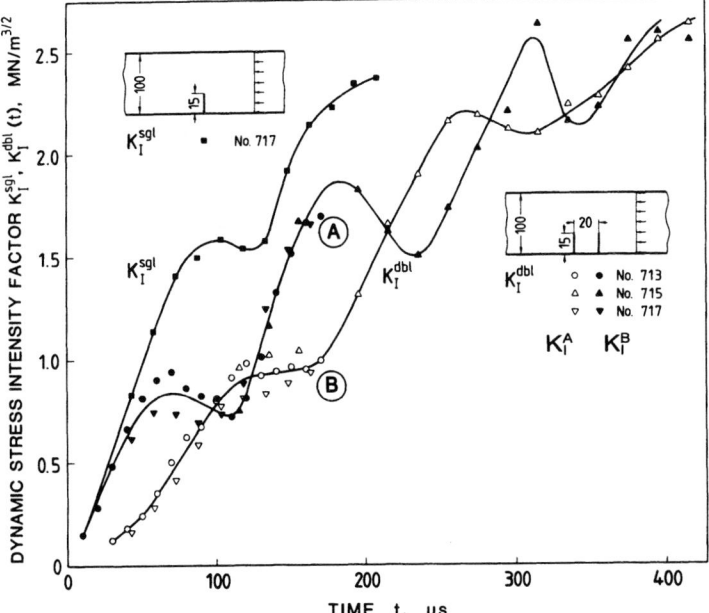

Figure 9-62 Stress-intensity factor history for dynamically loaded double-crack configuration.

factor can be larger than the crack A stress-intensity factor (e.g., at the time intervals 95 $\mu s <$ $t <$ 125 μs, 215 $\mu s < t <$ 285 μs, etc.). For large times the stress-intensity factors of both cracks approach the same values. Thus, despite the more exposed position of crack A toward the impinging stress pulse, the stress-intensity factor of the hidden crack B can for certain time intervals be larger than for crack A. For any times the stress-intensity factors K_I^A and K_I^B of the double-crack configuration are smaller than the stress-intensity factor K_I^{sgl} of an equivalent single crack; only for very early times, that is, before an interaction between the two cracks A and B can take place, K_I^A is identical with K_I^{sgl}, as one might expect.

9-6 Summary

An overview has been given on the shadow optical method of caustics. The basic physical principles of the method, the mathematical description of shadow optical imaging processes, and the derivation of caustic evaluation formulas have been presented. The realization of the technique in the laboratory has been discussed, and the applicability of the method for investigating practical problems has been demonstrated.

It has been shown that the shadow optical method of caustics is based on the deflection of light rays due to stress gradients. The method is therefore very well suited for the investigation of any kind of stress–strain problem that shows large gradients (e.g., stress concentration problems around holes, notches, cracks, contact areas, etc.). For typical examples the shadow optical formulation is presented. It is shown that the shadow optical patterns generated are simple and easy to evaluate. In general, only one characteristic length parameter is taken from the measured caustic in order to determine the generating load parameter.

Figure 9-63 Comparison of (a) shadow optical and (b) photoelastic crack-tip patterns.

Most other experimental techniques of stress analysis are based on effects that are directly proportional to stresses and strains. The specific advantages and disadvantages of the shadow optical method of caustics due to these differences in the dependence from the stress become evident by a comparison of a shadow optical pattern and a photoelastic pattern obtained for the same crack-tip stress field shown in Fig. 9-63. The shadow spot represents a direct quantitative measure of the stress intensification at the crack tip. The photoelastic pattern is more complicated than the shadow optical pattern due to the larger number of isochromatic fringes. Thus the evaluation of photoelastic stress concentration patterns naturally becomes a laborious task. In particular, in the near-field region around the crack tip the reduced resolution makes an evaluation difficult, and the derived results are necessarily of reduced accuracy. Furthermore, data on the loading condition in the direct vicinity of the crack tip can be obtained only by extrapolation of data measured at distances farther away from the crack tip. The shadow optical picture does not suffer from these disadvantages. The simplicity and clearness of the shadow optical picture result from the fact that stress distributions with small variations of stress do not become visible. This, however, also implies certain disadvantages. Far-field distributions around stress concentrations (e.g., around the crack tip in Fig. 9-63) do not result in shadow optical effects because of the small stress gradients and therefore cannot be investigated. The photoelastic pattern, however, yields accurate information in particular in this far-field range. The shadow optical method of caustics and the photoelastic method of isochromatic fringes are therefore not to be considered as competitive experimental tools that can be used with the same success and efficiency in obtaining the same kind of data. Depending on the specific property of interest and problem to be investigated, one or the other measuring techniques is more appropriate. Both techniques have their specific ranges of applicability and complement each other.

Because of its inherent simplicity and the ease in evaluation of the recorded light patterns, the shadow optical method of caustics yields reliable data even if the problem to be investigated is very complicated. The technique is therefore very well suited for investigating complex phenomena (e.g., dynamic processes). Most applications of the shadow optical method of caustics so far have been in the field of fracture dynamics. Fast variations in crack-tip stress intensifications can be measured easily and accurately. Shadow optical analyses are available not only for the most important case of cracks under tensile loading, but also for shear types of loading and mixed-mode loading conditions. The applicability of the evaluation formulas derived for the investigation of dynamic fracture problems has been discussed. In addition to the analyses of linear elastic stress intensifications, the shadow optical formulation has been developed for elastic–plastic strain concentration fields around cracks in materials that show ductile behavior. Caustic evaluation formulas are presented for determining the relevant fracture mechanics parameters, that is, the linear elastic stress-intensity factors K_I, K_{II}, and K_{III} or the elastic–plastic J-integral value.

The applicability of the shadow optical method of caustics for investigating problems of practical relevance in the field of fracture dynamics has been demonstrated by several examples. For many other stress concentration problems, the shadow optical method of caustics is an appropriate tool for investigation.

List of Symbols

a	crack length
a, b	elasto-optical constants
A, B	material constants in Maxwell–Neumann's law
$a_{2,3,4}\ldots$	coefficients of higher-order terms in crack-tip stress distribution
c	shadow optical constant
c_0	sound wave speed
c_1	longitudinal wave speed
c_2	transverse wave speed
C	compliance of specimen
d	specimen thickness
d	distance between two cracks in double-crack configuration
d_{eff}	effective thickness of the specimen
D	characteristic length parameter for caustic evaluation
$D_{o,i}$	outer, inner characteristic length parameter
$D_{\max,\min}$	maximum, minimum length parameter of mixed-mode caustic
e	index characterizing elastic behavior
E	Young's modulus
f	numerical factor for caustic evaluation
$f_{o,i}$	numerical factor for evaluating outer, inner caustic
G	function
G	Lamé's constant, $G = E/2(1 + v)$
H	height of specimen
I_n	numerical factor in the Hutchinson–Rice–Rosengren stress field equations
J	J integral
K	stress-intensity factor
n	refractive index
η	strain-hardening coefficient
O	origin of coordinate system
P	edge load, unit [N/m]
p	index characterizing plastic behavior
p, q	biaxial stresses in y, x direction
r, ϕ	polar coordinate system in object plane (specimen)
r', ϕ'	polar coordinate system in image (reference) plane
$\overline{r}, \overline{\phi}$	polar coordinate system at the tip of a moving crack
r_0	radius of initial curve

r_{pl}	radius of plastically deformed region around the crack tip
r_{ps}	smallest radius around the center of stress concentration outside which a state of plane stress exists
R	radius of circular hole
S	support span
s	optical path length
t	time
u, v, w	displacements in x, y, z direction
ν	crack velocity
W	width of the specimen
x, y	Cartesian coordinate system in object plane (specimen)
x', y'	Cartesian coordinate system in image (reference) plane
$\bar{x}, \bar{y}$	Cartesian coordinate system at the tip of a moving crack
z	direction of optical axis
z_0	distance between object plane (specimen) and image (reference) plane
α	velocity dependent factor
ε	strain
λ	coefficient of anisotropy
μ	ratio of the mode II to mode I stress intensity factor, K_{II}/K_I
υ	Poisson's ratio
ρ	density of the material
ρ	notch-tip radius
ρ	wedge-tip radius
σ	normal stress
σ_0	tensile yield stress
τ	shear stress
τ	period of the oscillation of impacted specimen

References

1. P. Manogg, Anwendung der Schattenoptik zur Untersuchung des Zerreissvorgangs von Platten, Dissertation, Freiburg, Germany, 1964.

2. P. Manogg, Schattenoptische Messung der spezifischen Bruchenergie während des Bruchvorgangs bei Plexiglas, *Proc. Int. Conf. Phys. Non-Crystalline Solids*, Delft, The Netherlands, 1964, 481–490.

3. P. S. Theocaris and N. Joakimides, Some Properties of Generalized Epicycloids Applied to Fracture Mechanics, *J. Appl. Mech.*, 22 (1971), 876–890.

4. P. S. Theocaris, The Reflected Caustic Method for the Evaluation of Mode III Stress Intensity Factor, *Int. J. Mech. Sci.*, 23 (1981), 105–117.

5. P. S. Theocaris, Stress Concentrations at Concentrated Loads, *Exp. Mech.*, 13 (1973), 511–528.

6. A. J. Rosakis and L. B. Freund, Optical Measurement of the Plastic Strain Concentration at a Tip in a Ductile Steel Plate, *J. Eng. Mater. Technol.*, 104 (1982), 115–125.

7. A. J. Rosakis, C. C. Ma, and L. B. Freund, Analysis of the Optical Shadow Spot Method for a Tensile Crack in a Power-Law Hardening Material, *J. Appl. Mech.*, 50 (1983), 777–782.

8. J. F. Kalthoff, J. Beinert, and S. Winkler, Analysis of Fast Running and Arresting Cracks by the Shadow-Optical Method of Caustics, *IUTAM Symp. Opt. Methods Mech. Solid* (A. Lagarde, Ed.), University of Poitiers, France, Sept. 10–14, 1979, Sijthoff-Noordhoff, Alphen aan den Rijn, The Netherlands, 1980, 497–508.

9. J. Beinert and J. F. Kalthoff, Experimental Determination of Dynamic Stress Intensity Factors by Shadow Patterns, in *Mechanics of Fracture*, Vol. 7, *Experimental Fracture Mechanics* (G. C. Sih, Ed.), Nijhoff, Hingham, MA, 1981, 280–330.

10. J. F. Kalthoff, Stress Intensity Factor Determination by Caustics, *Proc. Int. Conf. Experimental Stress Analysis*, organized by Japan Society of Mechanical Engineers (JSME) and American Society for Experimental Stress Analysis (SESA), Honolulu-Maui, May 23–29, 1982, 1119–1126.

11. J. F. Kalthoff, W. Böhme, and S. Winkler, Analysis of Impact Fracture Phenomena by Means of the Shadow Optical Method of Caustics, *Proc. 7th Int. Conf. Experimental Stress Analysis*, organized by SESA, Haifa, Israel, Aug. 23–27, 1982, 148–160.
12. J. F. Kalthoff, Experimental Fracture Dynamics, in *Crack Dynamics in Metallic Materials*, CISM (International Center for Mechanical Sciences), Udine, Italy, Course 310 (J. R. Klepazko, Ed.), Springer-Verlag, New York, 1990, 69–253.
13. M. Born and E. Wolf, *Principles of Optics*, Pergamon Press, Elmsford, NY, 1970.
14. D. Broek, *Elementary Engineering Fracture Mechanics*, Nijhoff, Hingham, MA, 1982.
15. R. Podleschny and J. F. Kalthoff, Geschlossene Lösung anisotroper Modus-I Kaustiken, Technische Mitteilung RUB EM T 91/1, Experimentelle Mechanik, Ruhr-Universität Bochum, February, 1991.
16. P. C. Paris and G. C. Sih, Stress Analysis of Cracks, *Fracture Toughness Testing and Its Applications*, ASTM STP 381, American Society for Testing and Materials, Philadelphia, 1965, 30–83.
17. R. Podleschny and J. F. Kalthoff, Verbesserte Bestimmung bruchmechanischer Schub-(Modus II)-Spannungsintensitätsfaktoren mit Hilfe des schattenoptischen Kaustikenverfahrens, *Proc. 13. GESA-Symposium, Experimentelle Mechanik in Forschung und Praxis*, Bremen, May 10–11, 1990, VDI-Berichte 815, VDI-Verlag, Düsseldorf, 1990, 323–335.
18. P. S. Theocaris, Complex Stress Intensity Factors of Bifurcation Cracks, *J. Mech. Phys. Solids*, 20 (1972), 265–279.
19. U. Seidelmann, Anwendung des schattenoptischen Kaustikenverfahrens zur Bestimmung bruchmechanischer Kennwerte bei überlagerter Normal- und Scherbeanspruchung, Bericht 2/76 des Fraunhofer-Instituts für Festkörpermechanik, Freiburg, 1976.
20. R. Podleschny and J. F. Kalthoff, A Novel Arrangement for Mixed-Mode Impact Testing—Characterization by Shadow Optics, *Proc. Int. Conf. Mixed-Mode Fracture and Fatigue*, Vienna, July 15–17, 1991; ESIS Publication 14. MEP (Inst. Mech. Engs.), London, (1993), 61–73.
21. G. C. Sih, *Handbook of Stress Intensity Factors*, Institute of Fracture and Solid Mechanics, Lehigh University, Bethlehem, PA, 1973.
22. L. B. Freund, Crack Propagation in an Elastic Solid Subjected to General Loading—I. Constant Rate of Extension, *J. Mech. Phys. Solids*, 20 (1972), 129–140.
23. J. F. Kalthoff, Zur Ausbreitung und Arretierung schnell laufender Risse, *Fortschritt-Berichte der VDI-Zeitschriften*, Reihe *18*, no. 4, VDI-Verlag, Düsseldorf, 1978, 1–95.
24. M. L. Williams, On the Stress Distribution at the Base of a Stationary Crack, *J. Appl. Mech.*, 24 (1957), 109–114.
25. J. W. Hutchinson, Singular Behavior at the End of a Tensile Crack in a Hardening Material, *J. Mech. Phys. Solids*, *16* (1968), 13–31.
26. J. W. Hutchinson, Plastic Stress and Strain Fields at a Crack Tip, *J. Mech. Phys. Solids*, *16* (1968), 337–347.
27. J. R. Rice and G. F. Rosengren, Plane Strain Deformation Near a Crack Tip in a Power-Law Hardening Material, *J. Mech. Phys. Solids*, *16* (1968), 1–12.
28. U. Soltész and J. Beinert, Bestimmung des Spannungszustandes in der Umgebung einer Rißspitze mit einem schattenoptischen Verfahren, Wissenschaftlicher Bericht W 6/81, Fraunhofer-Institut für Werkstoffmechanik, Freiburg, December, 1981.
29. J. F. Kalthoff, S. Winkler, and J. Beinert, The Influence of Dynamic Effects in Impact Testing, *Int. J. Fract.*, 13 (1977), 528–531.
30. J. F. Kalthoff, W. Böhme, S. Winkler, and W. Klemme, Measurements of Dynamic Stress Intensity Factors in Impacted Bend Specimens, CSNI-Specialist Meeting on Instrumented Precracked Charpy Testing, Electric Power Research Institute, Palo Alto, CA, Dec. 1–3, 1980, 1–17.
31. J. F. Kalthoff, S. Winkler, W. Böhme, and W. Klemm, Determination of the Dynamic Fracture Toughness K_{Id} in Impact Tests by Means of Response Curves, *Proc. 5th Int. Conf. Fract.*, Cannes, March 29–April 3, 1980, in *Advances in Fracture Research*, Pergamon Press, Elmsford, NY, 1981, 363–373.
32. ASTM E 399, Standard Test Method for Plane-Strain Fracture Toughness of Metallic Materials, *Annual Book of ASTM Standards*, Part 10, American Society of Testing and Materials, Philadelphia, 1983.
33. D. R. Ireland, Critical Review of Instrumented Precracked Charpy Testing, *Proc. Int. Conf. Dyn. Fract. Toughness*, London, July 5–7, 1976, 47–62.

34. J. F. Kalthoff, S. Winkler, and J. Beinert, Dynamic Stress Intensity Factors for Arresting Cracks in DCB Specimens, *Int. J. Fract., 12* (1976), 317–319.

35. J. F. Kalthoff, J. Beinert, and S. Winkler, Measurements of Dynamic Stress Intensity Factors for Fast Running and Arresting Cracks in Double-Cantilever-Beam Specimens, in *Fast Fracture and Crack Arrest,* ASTM STP 627 (G. T. Hahn and M. F. Kanninen, Eds.), American Society for Testing and Materials, Philadelphia, 1977, 161–176.

36. J. F. Kalthoff, J. Beinert, S. Winkler, and W. Klemm, Experimental Analysis of Dynamic Effects in Different Crack Arrest Test Specimens, In *Crack Arrest Methodology and Applications,* ASTM STP 711 (G. T. Hahn and M. F. Kanninen, Eds.), American Society for Testing and Materials, Philadelphia, 1980, 109–127.

37. J. F. Kalthoff, Bruchdynamik laufender und arretierender Risse, In *Internationales Seminar über Bruchmechanik, Schadensanalyse für die Praxis, Bruchsicherheit,* (H. P. Rossmanith, Ed.), Vienna, June 12–13, 1980, K 1-22; published in *Grundlagen der Bruchmechanik* (H. P. Rossmanith, Ed.), Springer-Verlag, New York, 1982, 191–219.

38. P. B. Crosley and E. J. Ripling, Characteristics of a Run-Arrest Segment of Crack Extension, in *Fast Fracture and Crack Arrest,* ASTM STP 627 (G. T. Hahn and M. F. Kanninen, Eds.), American Society for Testing and Materials, Philadelphia, 1977, 203–227.

39. G. T. Hahn et al, Critical Experiments, Measurements and Analyses to Establish a Crack Arrest Methodology for Nuclear Pressure Vessel Steels, Reports BMI—1937, 1959, 1985 prepared for U.S. Nuclear Regulatory Commission, Battelle Columbus Laboratories, Ohio, 1975–1978.

40. J. F. Kalthoff and S. Winkler, Fracture Behavior Under Impact, First and Second Annual Report prepared for U.S. ARO European Research Office, London, IWM Reports W 8/82 and W 10/82, Fraunhofer-Institut für Werkstoffmechanik, Freiburg, 1983.

41. J. F. Kalthoff, On Some Current Problems in Experimental Fracture Dynamics, in *Workshop on Dynamic Fracture* (W. G. Knauss, Ed.), California Institute of Technology, Pasadena, CA, Feb. 17–18, 1983, 11–35.

CHAPTER

10

Optical Heterodyning

Karl A. Stetson
United Technologies Research Center
East Hartford, Connecticut

10-0 Introduction

Most interferometric methods of optical metrology are pictorial and present data in the form of an image in which light and dark fringes may be photographed or observed by eye. Although fine for qualitative inspection, this often presents difficulties for quantitative measurement. Although fringes define exact units of displacement, the center of a fringe in an image may not be located more precisely than perhaps $\frac{1}{10}$ of a fringe spacing. Also, it is often practical to obtain data only at the locations of the fringe centers; displacement at other locations has to be inferred by interpolation.

The recent advances in electronic computing create additional problems for utilizing fringe data. Optical metrology is often employed to confirm the predictions of computer modeling of a test object. If quantitative comparisons are to be done efficiently, both theoretical and experimental values should be available as numerical data on a computer. Furthermore, optical data often require extensive numerical processing if they are to be converted into mechanical displacement or strain, and this is also most efficiently done on a computer. For all the foregoing reasons, it is desirable to convert the pictorial data of interferograms into digital electronic form.

Perhaps the most versatile way to accomplish this is by what is called heterodyne (or more precisely, homodyne) interferometry. The concept is easily understood in terms of a conventional interferometer in, for example, the Mach–Zehnder configuration. Suppose that a device is placed in one of the beams that continuously advances the phase of light passing through it. The fringes at the output will advance continuously across the field of view and the irradiance at any location will fluctuate sinusoidally. The time phase of these fluctuations can be measured by a photodetector and phase meter, and it will be directly proportional to fringe order. Fringe order, in turn, is proportional to the change in optical path length that the interferometer is being used to measure.

There are many advantages to the heterodyne interferometer. The motion of the fringes eliminates the need to find fringe centers, and therefore the photodetector can make a measurement anywhere in the output field. Modern electronic phase meters can measure the phase of good signals to better than 0.1°, which corresponds to locating a fringe center to 1 part in 3600.

Furthermore, instruments equipped with interface connectors for computer-controlled operation and transfer of data are available and greatly facilitate the electronic processing of the interferometer output.

This chapter presents a discussion of optical heterodyning from the point of view of coherent optical metrology. It will begin with a discussion of equipment and basic techniques. This will be followed by a summary of applications involving holography and speckle photogrammetry. It will conclude with a discussion of concomitant systems.

10-1 Basic Technology

Heterodyne Generating Systems

There are basically three practical systems for generating heterodyne phase modulation in a beam of light. For low-frequency modulation, a simple system may be constructed from a rotating half-wave plate followed by a quarter-wave plate. The theory of this device has been well described [1], and it will suffice here to state that it converts a beam of linearly polarized light into two collinear, orthogonally polarized beams that differ in optical frequency by four times the rotation frequency of the half-wave plate. These may be separated by a polarization beam splitter and used in an interferometer. The half-wave plate must be free of wedge so that it does not wobble the propagation direction of the beam. Thin mica wave plates, which are mounted with no substrates, are available that serve the purpose very well. The half-wave plate must rotate without vibration at a constant speed to ensure an unmodulated heterodyne frequency. A hollow-shafted synchronous motor offers the best solution for this. The quarter-wave plate must be adjusted relative to the polarization beam splitter to eliminate mixing of the frequency-shifted beams.

A few hundred hertz is probably the practical limit to the frequencies obtainable by a single rotating half-wave plate. Higher frequencies make it easier to separate the heterodyne signals of the interferometer from low-frequency noise created by mechanical vibrations. These can be obtained mechanically by a rotating radial grating. With gratings of modest dimensions rotating at quite reasonable speeds, even megahertz frequency differences can be obtained between the plus and minus first orders. Speed control, centering, and uniformity of the grating, however, become much more important with this system than with the rotating half-wave plate.

Perhaps the most widely used devices for heterodyne generation are acousto-optic modulators. These employ a traveling acoustic wave in a material such as quartz to diffract light. The width of the acoustic beam, together with the Bragg effect, can limit the diffraction to zero order and a single first order with excellent efficiency. The primary problem with these devices is that they generate extremely high offset frequencies, 40 to 100 MHz. To reduce the offset to lower values, it is common to use two such modulators set to a small frequency difference, one in each beam of the interferometer.

All these systems will perform better with light from a laser than with light from a thermal source. First, their apertures are generally small and their modulation is angularly dependent, which means that lasers provide the most efficient means for passing reasonable amounts of energy through them. Second, both wave plates and gratings have wavelength dependence, so that spectral purity is important. It is not surprising, therefore, that heterodyne interferometry is generally a coherent optical technique.

Phase Meters

A variety of instruments are available for electronic phase measurement; however, it is good to bear in mind that phase per se is often only proportional to displacement and strain, which comprise

OPTICAL HETERODYNING

the real output. Electronic instruments are available with internal microprocessors for mathematical operations such as offset, normalizing, scaling, and averaging, and these can be set up to convert the phase measurement directly into the desired output. These features, together with access for computer control, are important considerations for any system where large amounts of data are to be processed. The need for computer control extends even to the scanners that move the detectors across the output field, and to any other repetitive operation.

Some additional consideration must be given to dealing with the problem of phase changes greater than 360°. A simple phase meter will reset itself each time this transition occurs and thereby lead to ambiguities in the output. The way in which this problem is solved will depend on the operation of the particular phase meter used. In general, either the transitions are detected and counted, or a gate control is used to offset the sampling of the two signals.

Scintillation Noise

Heterodyne interferometry has some intrinsic advantages over image scanning, for example by a TV camera, in that it is not directly affected by changes in image brightness nor by fringe contrast. There is, however, an indirect effect resulting from the speckles formed by the diffuse coherent fields found in holography and speckle metrology. Two randomly diffuse fields will interfere to form a single field with a random speckle pattern similar to that of either field alone. When the phase of one field advances continuously, as in heterodyne interferometry, the speckles all scintillate at the heterodyne frequency. These scintillations will be randomly phased, and if the detector sampling the field accepts enough speckles, these uncorrelated signals will add to zero. A small detector may be required, however, to increase the spatial resolution of the mechanical measurement, and the scintillation signal may then be significant.

When two coherent diffuse fields are perfectly correlated, their individual speckle patterns match perfectly and they can interfere with unity contrast. This seldom happens, and usually the field can be divided into a correlated portion that gives macroscopic fringes and an uncorrelated portion that gives speckle scintillation. The scintillation forms a random background signal that perturbs the phase due to the macroscopic fringes and causes errors in the measurements. This phenomenon has been thoroughly analyzed for hologram interferometry [2] and the results applied to speckle photogrammetry [3]. In practice, the scintillation effect may establish the practical limits to metrology by heterodyne interferometry.

10-2 Holographic Applications

Stored Fringe Patterns

The applications of heterodyne techniques to hologram interferometry have been discussed extensively by Dändliker in Ref. 2, to which the reader is referred for a complete exposition of the topic. This section reviews material from Ref. 2 with supplementary commentary.

Heterodyne techniques can be applied only to an interferometer that operates concomitantly (i.e., in real time). A conventional double-exposure or time-average hologram reconstructs an image that contains a permanent interference pattern as an integral part of the reconstructed field. Access to the separate components interfering within this field must be obtained before heterodyne techniques can be applied. This can be done for double-exposure holograms by recording the two exposures with separate reference beams. When these are duplicated as reconstruction beams, the hologram creates from each an independent image field whose phase is directly related to its reconstruction beam. If the relative phase of the two reconstruction beams is now swept contin-

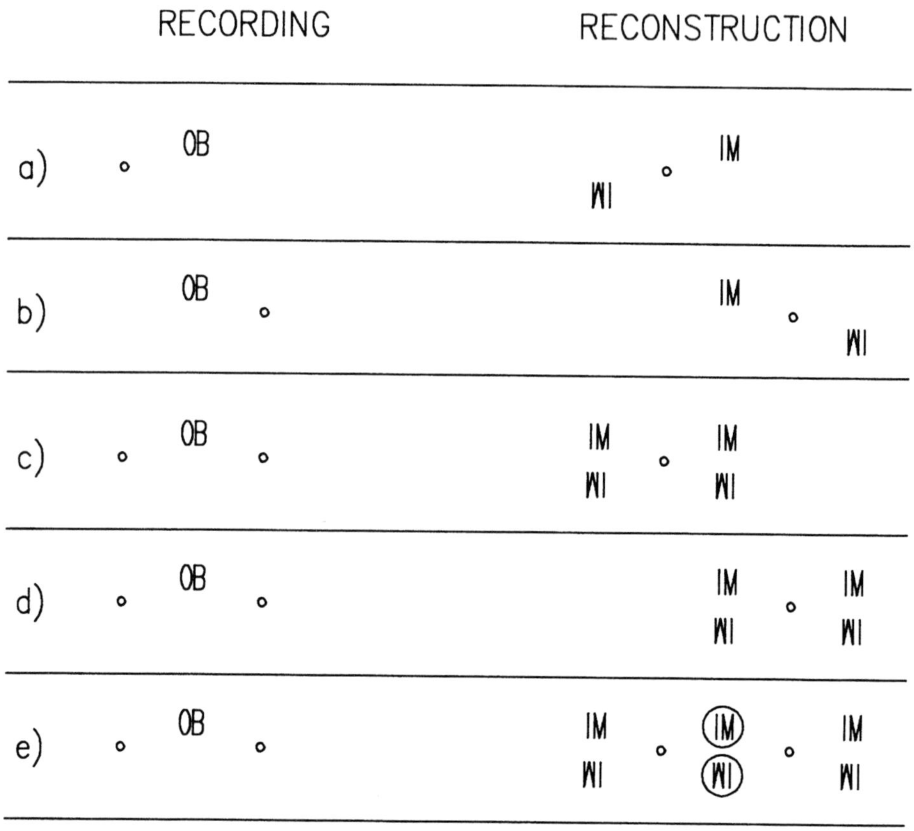

480   HANDBOOK ON EXPERIMENTAL MECHANICS

	RECORDING	RECONSTRUCTION
a)	○ OB	○ IM WI
b)	OB ○	IM ○ WI
c)	○ OB ○	IM ○ IM WI WI
d)	○ OB ○	IM ○ IM WI WI
e)	○ OB ○	IM ○ (IM) ○ IM WI (WI) WI

Figure 10-1 Tabulation of recording and reconstruction geometries for two-reference-beam holography.

uously by a heterodyne generator, the fringes in the reconstruction will scan across the image field. If the two reference beams are duplicated exactly, the two images will overlap with perfect fidelity, and the remaining fringes will depend only on the deformations the object experienced between exposures.

Some care must be taken in setting up the original reference beams because the hologram has the capacity to reconstruct a true and conjugate image for each of the two exposures for a total of four images. Hence eight images result from two reconstruction beams, and care must be taken as to how these overlap. Figure 10-1 shows in tabular form how this occurs. Figure 10-1a, on the left, shows a hologram recorded with the reference beam to the lower left of the object, and at the right, it reconstructs two images. The true image, IM, lies to the upper right of the reconstruction beam and the conjugate, WI, lies to the lower left (note that the conjugate image WI is upside down and backward from the true image IM). In Fig. 10-1b the reference beam lies to the lower right and the resulting hologram reconstructs IM to the upper left and WI to the lower right. Both beams are present during the recording for Fig. 10-1c, d, and e. On the right-hand side of Fig. 10-1c, the hologram reconstructs four images from the left reconstruction beam: IM

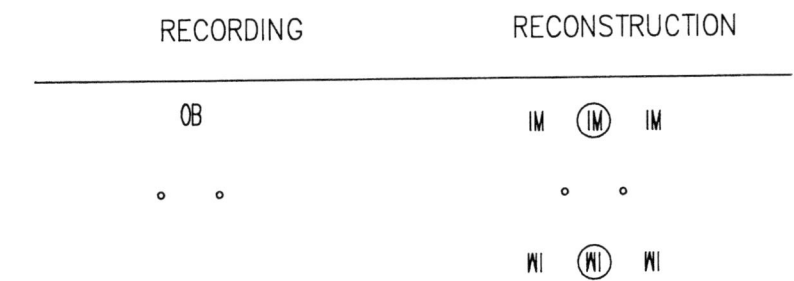

Figure 10-2 Image overlap for a hologram with closely spaced reference beams.

upper left and right, and WI lower left and right. The situation in Fig. 10-1d is the same except that the reconstruction beam to the right is used. In Fig. 10-1e, both reconstruction beams are present and eight images occur; however, the two upper and two lower images in the center overlap to produce interference fringes. These pairs of images have been circled in Fig. 10-1e for emphasis.

The original reference beams should be arranged to keep the central pairs of images from overlapping during reconstruction; otherwise, errors may occur. This can be done by keeping the two reference beams reasonably close together and well separated from the object, as shown in Fig. 10-2. This, however, brings the two pairs of noninterfering images close to pairs of interfering images. These overlapping images reduce the contrast of the interfering fringes and generate scintillation noise. The magnitude of the resulting errors have been evaluated in Ref. 2 as a function of image parameters.

Another problem can arise if a substantial portion of the object does not move between the exposures. The nonlinearities of the holographic film can cause cross talk between the two recordings, which results in spurious fringes that can also cause errors. A discussion of these effects is well beyond the scope of this chapter but is available in the literature [4].

When two reconstruction beams are used to reconstruct a double-exposure hologram, fringes will result from any changes in angles and curvatures of the reference beams. Even if these are too small to generate visible fringes, the heterodyne method of readout will easily pick them up and add false information to the data. If the hologram was removed for photographic processing, it must be repositioned precisely with respect to the reconstruction beams. Even distortions of the photographic medium may show up, particularly if it is a gelatin emulsion. In-situ hologram processing provides perhaps the most straightforward solution to this problem. Thermoplastic recording systems are particularly suited to this work because of their high diffraction efficiency and dry, rapid processing. If conventional photographic materials are used, they should be bleached to form efficient, low-noise, phase holograms to maximize the amount of light on the photodetectors. It has been shown in the literature [5] that repositioning errors are reduced if the reference beams are kept close together. There exists a tradeoff between repositioning errors and increased scintillation noise, and the choice must be made with regard to the actual application.

If the fringes formed by the object deformation do not localize on the surface of the object, their contrast will be reduced when the detectors scan the focused image. This increases scintillation noise and reduces accuracy. If the fringes are localized on a plane that is slightly displaced from the object surface, it may be practical to move the detectors to where the fringes are localized and scan the defocused image of the object. This will increase the area of the object sampled by the detector, but the improvement in the signal may be more important. In many cases, the fringes may be localized along a line that is not in the field of view. This results from displacements of the object that are transverse to the direction of observation in addition to various rotations, strains, or shears. Here it may be desirable to alter deliberately the angle of one reconstruction beam relative to the other to make the two images coincide as much as possible, and for such manipulations collimated reference beams are desirable.

Conventional time-average holograms of a vibrating object are not easily adapted to this readout scheme because of the continuous motion of the object. This can be overcome, however, by recording stroboscopic holograms that sample the object field at only one instant in the vibration cycle. Phasor vibrations (i.e., vibrations that circulate on an object surface) may go unnoticed in the reconstruction of a time-average hologram. With the added sensitivity of heterodyne readout, however, they become quite significant. This may be either a benefit or a hindrance, depending on the application.

Concomitant Fringe Patterns

The concomitant, or real-time, hologram interferometer is intrinsically suited to heterodyne readout. The heterodyne generator may be introduced between the object and reference beams in the recording setup. The hologram is recorded with no frequency offset and reconstructed with a frequency offset. In-situ photographic processing may be particularly important in this configuration, where only one field is perturbed by distortions of the emulsion. In the double-exposure technique both hologram reconstructions are affected similarly. Less light may be available at the photodetectors because of diffuse scattering by the object itself, and the combination of high-efficiency bleached holograms and a high-power laser is very desirable.

When a double-exposure hologram is reconstructed with two reference beams, it is usually possible to keep both passing through the same general volume of air to minimize the effects of air turbulence. In the concomitant hologram interferometer, this is more difficult to do, and the air along the beam paths may need to be enclosed to keep it from introducing jitter into the phase output of the system.

One advantage of the concomitant hologram interferometer is that it may be used to measure object deformation as a function of time. This may be particularly attractive if a number of detector channels are available either in parallel or by serial multiplexing. With this configuration, simultaneous spatial and temporal deformations could be evaluated.

The concomitant hologram interferometer is also adaptable to a simple form of vibration readout that uses a low-frequency heterodyne signal [6]. The detector output is monitored with the heterodyne generator running at a frequency substantially lower than that of the vibration, and a count is made of the average number of positive-going (or negative-going) zero crossings per vibration cycle over a substantial time period. As the time period increases, this average converges to the peak-to-peak vibration amplitude in cycles of an interference fringe.

The method can be implemented by connecting the detector output to a counter and the vibration signal to the input for an external clock. The minimum detectable amplitude is that required to generate local maxima and minima on top of the sine wave generated by the heterodyne signal. This is the heterodyne frequency in cycles divided by the vibration frequency in radians. It is not difficult to resolve one-thousandth of a wavelength or better with this method. Note that only the vibration magnitude is important to this detection scheme, not the phase of the vibration. Noise, however, will show up as an additive background if it is sufficient to cause local maxima and minima, and it cannot be removed by band-pass filtering.

10-3 Speckle Photogrammetry

When a diffusely reflecting object is illuminated by laser light and imaged by an optical system, random speckles will modulate the image field. If the object is well focused, the image speckles will appear to move as if attached to the object surface, to first-order approximation. For this reason, laser speckles can serve as markers for points on the object surface, and measurements of speckle displacements can be used to measure object displacements.

The use of specklegrams (i.e., photographs of specific patterns) in optical metrology is well documented [7]. In one of the most common techniques, double-exposure specklegrams are probed by a narrow beam. Fringes appear in the halo of light diffracted away from the readout beam (by the recorded speckle pattern), which can be used to measure object displacements. These fringes result from interference between the two fields scattered by the two components of the double-exposure specklegram.

To apply optical heterodyning to this technology [8] it is necessary to separate the two interfering fields, just as with double-exposure holograms. The most practical way to do this is to record the two specklegram exposures on separate photographic plates. The two plates are then introduced into the two beams of an interferometer, where the resulting halo patterns are combined to form fringes. If the two plates are moved jointly so that a new region is illuminated, any change in the fringe pattern is a measure of the differential image displacement. Such a system functions as a photocomparator, and it is logical to refer to this metrology as speckle photogrammetry. It is very simple, of course, to apply optical heterodyning to the interferometer by frequency shifting the two beams involved. Photodetectors in the output field will then convert the moving fringes to sinusoidal signals suitable for phase measurement.

With the two exposures recorded on separate plates, it is difficult to determine the absolute displacement of the object. One of the most practical applications of heterodyne speckle photogrammetry, however, is strain measurement, for which absolute displacement is not important. If an object surface is strained between the recording of two specklegrams, the second speckle pattern will be distorted with respect to the first. This will show up under the photocomparator as a directional magnification of one pattern relative to the other. One-dimensional strain will equal the relative image displacement divided by the distance between the points examined. Strain in two dimensions is a tensor that may be coupled with in-plane rotations. Relative image displacements in two dimensions for two directions of scan on the photographs are required to specify it.

A typical interferometer for this purpose is shown in Fig. 10-3, which has numerous features to help improve accuracy. The symmetrical arrangement of six mirrors makes it possible to have an equal path length for both beams while keeping both specklegrams in the same plane and equidistant from the output beam splitter. The three input mirrors and beam splitter may be mounted together as one unit, and so may the three output mirrors and beam splitter. It is desirable to mount the input beam splitter and $M2$ on a horizontal translation stage so that the pair may be moved toward $M1$ or $M3$ to equalize the path lengths. The output beam splitter and $M5$ may ride on a similar stage to allow movement toward $M4$ or $M6$. The output beam splitter should be constructed, as shown, from two identical etalons with the reflecting surface in the center. In this way it affects both the reflected and transmitted halos equally. It is desirable that all optics on the output side of the interferometer be of the highest affordable surface flatness to minimize any residual fringes they might introduce into the halo interference.

The performance of this system is limited, in general, by the scintillation noise described previously for heterodyne hologram interferometry. It arises from the facts that the halos themselves contain speckles inverse in size to the diameter of the beam passing through each specklegram, and a finite number of them are collected by a detector in the halo. Analysis of this problem [3] has shown that it creates a root-mean-square variation in strain values proportional to the square of the ratio of speckle size in the specklegram to gage length. The constant of proportionality is related to the fringe contrast and the percentage of the halo employed in the measurement.

Correct alignment of the interferometer will improve fringe contrast and increase accuracy. The most difficult problem is to keep the beams all in one plane, especially with six adjustable

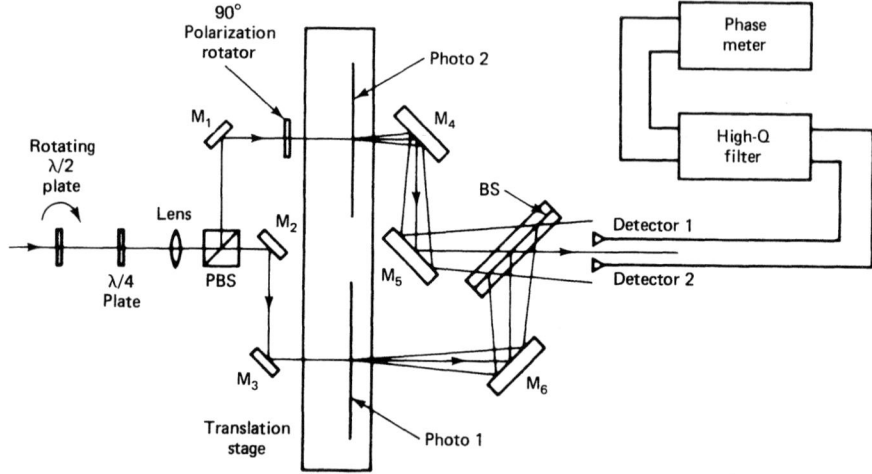

Figure 10-3 Interferometric comparator for heterodyne readout of specklegram halos. *M1* through *M6* are mirrors, *PBS* is a polarization beam splitter, and *BS* is a symmetrically compensated beam splitter.

mirrors and two adjustable beam splitters. In aligning such a system, it is very useful to have the interferometer on a flat supporting surface with a movable reference mirror to retroreflect the beam after each mirror. Care must also be taken to note whether the input beam splitter has any wedge that deviates the transmitted beam, so that this can be taken into account.

The sensitivity provided by heterodyne readout in speckle photogrammetry makes it important to deal with the problem created by the spherical perspective of most lenses. Spherical perspective makes the image of an object that moves toward the lens appear to grow in size, and such motion generates a false isotropic apparent strain. Telecentric lens systems can be designed to eliminate this problem, but their size must be greater than that of the object under study.

Application of speckle photogrammetry to measurement of strain on nickel-based alloys at high temperature has been reported [9]. The interferometric photocomparator may also be applied to white-light photographs of randomly textured objects [i.e., so-called white-light speckle photography (or perhaps "speck" photograph)]. In principle, even electron microscope photographs and aerial photographs can be analyzed with this system.

10-4 Concomitant Strain-Measuring Systems

Heterodyne Speckle Interferometry

A number of concomitant strain-measuring systems can be augmented by optical heterodyning, and two of these are discussed in this and the following sections. The first is one of the earliest forms of speckle interferometry introduced by Leendertz [10] (see Chapter 8). Two beams from a laser illuminate an object surface at equal and opposite angles to its surface normal. A lens, whose axis is centered on the illuminated region, forms an image of the object surface which is recorded on a photographic plate. The plate is replaced to the location it occupied during exposure and serves as a random moiré mask for the image field. As the object moves equally toward one

OPTICAL HETERODYNING

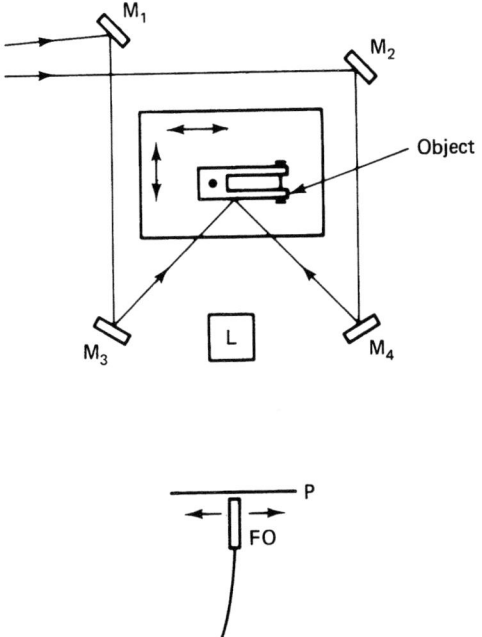

Figure 10-4 Setup for heterodyne speckle interferometry. *M1* through *M4* are mirrors, *L* is a lens, *FO* is a fiber-optic probe, and *D* is a detector pair.

illumination beam and away form the other (i.e., sideways), the relative phase of the two reflected fields varies cyclically. The resulting speckle pattern in the image field will cyclically correlate and decorrelate with the recorded pattern. This results in a sinusoidal fluctuation of light transmitted through the mask. Strain of the object surface can generate visible fringes in the light transmitted through the mask, which in turn can be used to measure the strain value.

The system described above is a concomitant interferometer and is suited to heterodyne readout. After the recording has been made and replaced, a relative frequency shift is introduced between the two illumination beams. This causes the cyclic correlation to occur at the heterodyne frequency which can be detected as a sinusoidal fluctuation of light anywhere in the image field. If strain is present, the phase of the fluctuation will vary across the image, and the differential phase between two points can be measured by a pair of photodetectors and a phase meter. This will measure the differential displacement between the two object points where the detectors are located and, normalized by the detector separation, this becomes a measure of strain.

A setup used to test this technique is shown in Fig. 10-4. The object was an aluminum fork with a screw adjustment to draw the ends together, as shown in Fig. 10-5. Two wire strain gages were fastened to the surface of the bar to provide reference strain measurements. A pair of 0.6 mm-diameter core optical fibers, shown behind the photographic plate in Fig. 10-4, provided a

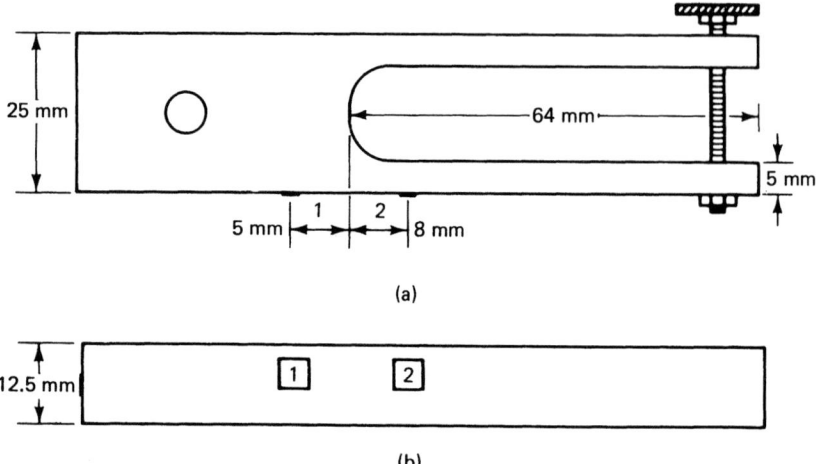

Figure 10-5 Illustration of the test object in Fig 10-4: (a) top view, (b) front view. Strain gages are located at points 1 and 2.

convenient means to sample two small, nearby areas in the image and channel the light to the detectors. Figure 10-6 shows results obtained in the actual operation of this system. Strains are measured along a 40-mm length of the surface by means of two sequential photographs, A and B, between which the object was translated 15 mm so that a new region could be examined. In each case, after the photograph was replaced, the screw was tightened so as to provide a 500 microstrain reading at gage 2. The strain readings as determined by speckle interferometry (continuous plot) show quite good agreement with the gages at the gage locations. Because of its ability to measure strain at many locations, however, the optical method shows the actual strain contour that results from the fillet inside the fork.

The practical limits of this technique are set in one aspect by scintillation effects. Small, closely spaced detector apertures are desirable to get fine resolution of strain details. Small apertures, however, accept only a limited number of the speckles formed by the lens, all of which scintillate in random phase. The mask removes a certain percentage of this randomness but not all of it. The remainder is suppressed by accepting a sufficient number of speckles. This number will increase with lens aperture, but this will correspondingly decrease the distance the object may translate without decorrelation. It is helpful to mount the object on an x,y translation stage to compensate for the translations due to the load. Consequently, the detector apertures cannot be reduced too far without introducing random errors into the measurements. These can be seen in Fig. 10-6 as the irregularity of the optical strain measurements.

Low light levels also limit the use of this technique. Considerable light is lost by the diffuse scattering of the object, and the photographic transparency further reduces the light available to the detectors. If small detector apertures are used, only a small fraction of the light in the image field reaches the detectors and detector noise is a serious problem. It was for that reason that the 40-mm region of the object in Fig. 10-6 was done as two sections rather than as one.

Heterodyne Optical Strain Sensor

In some situations it is practical to sacrifice the scanning ability of heterodyne speckle interferometry to gain more reliable strain measurements at a single point on the object. This is particularly true

OPTICAL HETERODYNING

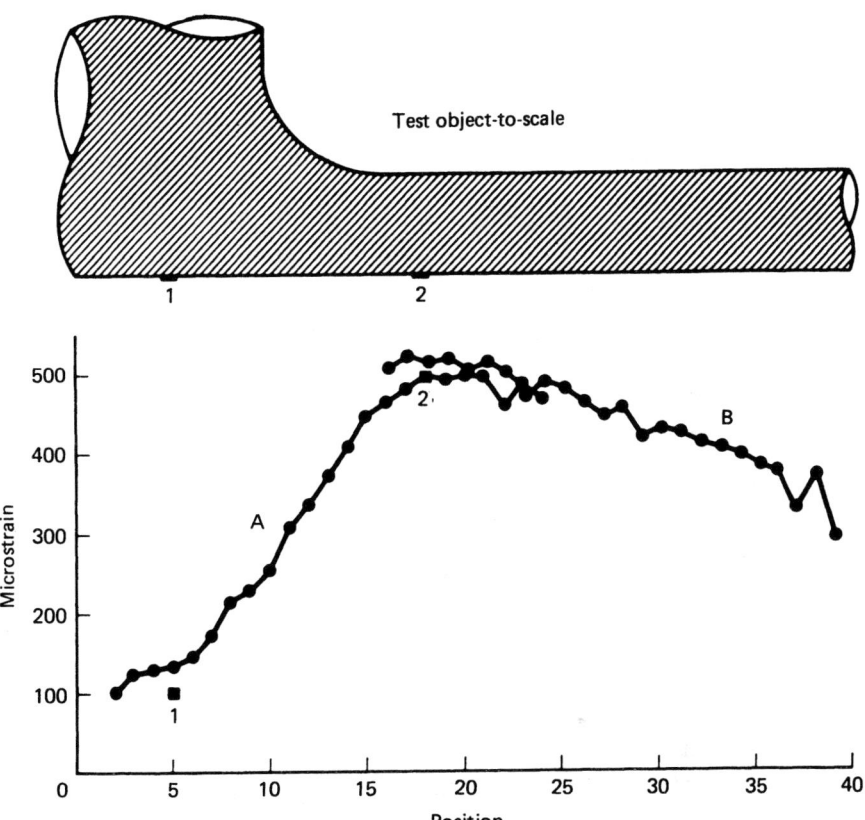

Figure 10-6 Plot of strain versus distance along the object surface for cantilever bending. The two squares are readings from the strain gages; the circles are measurements via heterodyne speckle interferometry.

when the object is at temperatures higher than strain gages can tolerate and is subject to the robust stresses of a tensile testing machine, for example. An ingenious optical technique has been developed by Sharpe (see Chapter 13) in which the angular deviations of light reflected by small indentations in the object surface are observed. An alternative system using similar principles developed by this author requires no surface indentations and makes use of optical heterodyning [11].

The device, which is called a heterodyne optical strain sensor, is shown in Fig. 10-7. Two input beams are focused to small spots, about 0.1 mm in diameter, on the object surface at a spacing of 1 to 2 mm. Light scattered at equal and opposite angles to the surface normal is collected by two lens systems after being deviated to the normal, for convenience, by a pair of prisms. The lens pairs are prealigned so that the object surface and the recording plate are at conjugate focal planes (Fourier transforms pairs). Polarizers are used to remove light orthogonally polarized by the object surface. The two fields are recorded on a single photographic plate, which is relocated to serve as a pseudomoiré mask. Each recorded field has a randomly modulated fringe pattern of the type shown in Fig. 10-8. When a frequency offset is introduced between the two illumination

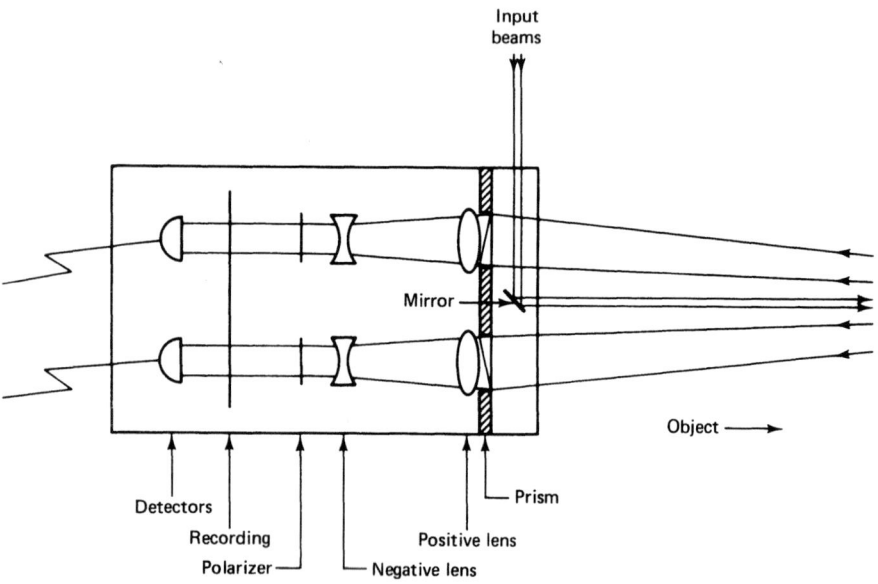

Figure 10-7 Diagram of the heterodyne optical strain sensor.

beams, the fringes move with respect to the recording and the light transmitted becomes sinusoidally modulated. The two transmitted fields are detected, and the phase difference of the two signals is measured. If the object surface is strained, the phase difference will change in a positive or negative direction depending on the direction of the fringe scan. When normalized by the appropriate factors, this phase change is an unambiguous measure of strain.

This system provides excellent utilization of light because all the laser output is focused onto two object spots. Scintillation effects are not set by the speckles associated with each illuminated spot but rather by the fringes created by the pair of spots. The illumination beams may be translated to follow the object under test. Its narrow acceptance angle (8°) and long working distance (500 mm) make this instrument well suited for high-temperature operation. This has been verified for this system at divisions of the author's corporation, and measurements of strain up to 10,000 microstrain have been recorded with accuracies on the order of 1%. Finally, the system of vibration detection described in connection with concomitant fringe patterns is easily applied to this strain sensor. A sum or difference counter is necessary to combine the two outputs, however, depending on whether the sinusoidal detector signals are out of phase or in phase, respectively.

10-5 Summary and Future Possibilities

Optical heterodyning is a practical technique for converting fringe data into electronic form, and it has been demonstrated to improve greatly the accuracy of quantitative fringe analysis. Resolution of fringe order to 1 part in 1000 is routinely achieved with properly constructed systems. With optical fields that are coherent but diffuse, the technique is usually limited by the random scintillation of uncorrelated speckles. The problem is alleviated by increasing the detector aperture,

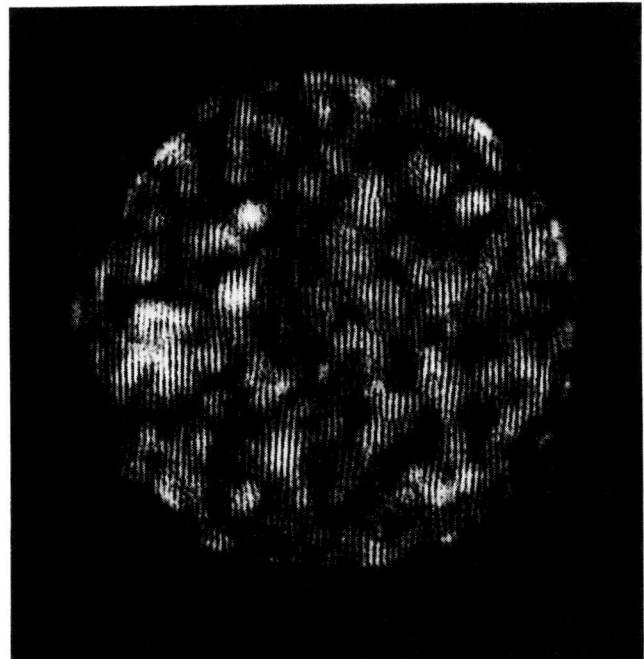

Figure 10-8 Typical pattern recorded at the back focal plane of the heterodyne optical strain sensor.

which lowers the geometrical resolution of the measurement. Numerous practical systems have been set up for measurement of displacement, vibration, and strain that can be applied to routine engineering measurements.

If demand for this technology grows in proportion to its potential capabilities, a number of developments can be foreseen. The first is likely to be the production of commercially heterodyne-generating systems specifically designed for holography and speckle applications. Such systems must now be assembled from components. Detector and signal-conditioning technology may also become commercially adapted to these applications. Again, usable systems must currently be put together from components, the compatibilities of which must be evaluated by the user, who may not be a specialist in electronics. The actual optical systems employed in heterodyne holography or speckle interferometry are essentially conventional and need little specialization.

Finally, heterodyne technology seems likely to make possible many new innovations in holography and speckle metrology. The researcher who investigates this technology should expect to find a considerable number of undiscovered phenomena available for exploitation.

References

1. G. E. Sommargren, Up/Down Frequency Shifter for Optical Heterodyning, *J. Opt. Soc. Am.*, 65 (1975) 960–961.

2. R. Dändliker, Heterodyne Holographic Interferometry, in *Progress in Optics* Vol. XVII (E. Wolf Ed.), North-Holland, Amsterdam, 1972, Chap. 1.

3. K. A. Stetson, Effect of Scintillation Noise in Heterodyne Speckle Photogrammetry, *Appl. Opt.*, *23* (1984), 920–923.

4. J. Katz and E. Marom, Effects of Nonlinear Recording in Holographic Interferometry, *J. Opt. Soc. Am.*, *69* (1979), 696–705.

5. R. Dändliker, R. Thalmann, and J. F. Willemin, Fringe Interpolation by Two-Beam Holographic Interferometry: Reducing Sensitivity to Hologram Misalignment, *Opt.Commun.*, *42* (1982), 301–306.

6. K. A. Stetson, Method of Vibration Measurements in Heterodyne Interferometry, *Opt. Lett.*, *7* (1982), 233–234.

7. R. K. Erf (Ed.), *Speckle Metrology*, Academic Press, New York, 1978.

8. G. B. Smith and K. A. Stetson, Heterodyne Readout of Specklegram Halo Interference Fringes, *Appl. Opt.*, *19* (1980), 3031–3033.

9. K. A. Stetson, The Use of Heterodyne Speckle Photogrammetry to Measure High-Temperature Strain Distributions, in *Holographic Data Non-destructive Testing* (D. Vukicevic, Ed.), *Proc. SPIE*, *370*, 1983, 46–55.

10. J. A. Leendertz, Interferometric Displacement Measurement of Scattering Surfaces Utilizing Speckle Effect, *J. Sci. Instrum.* (J. Phys. E), *3* (1970), 214–218.

11. K. A. Stetson, A. Heterodyne Optical Strain Sensor, *Proc. 29th Int. Instrum. Symp. Instrum. Soc. Am.*, May 1983, 65–69.

CHAPTER

11

Brittle Coating

Part I Stresses and Strains in Brittle Coatings

A. J. Durelli

The Catholic University of America
Washington, D.C.

11-0 Introduction

If the elastic constants of the coating and the specimen were the same, the same state of stress would be found on the common layer of the coating and the specimen.[1] Since these constants are different, even the sign of the stress in the coating may differ from that in the specimen. The developments that follow are taken to a large extent from a previous publication [1], where they were presented in greater detail.

At the time the coating is cured, it develops, in general, residual tensile stresses that will help those produced by the external loading to crack the coating. Sometimes these residual stresses are sufficiently high to crack the coating without any external loading. The theoretical analysis conducted in the following paragraphs takes this into account, assuming that the residual stresses in the coating are the same as the residual stresses in the calibration strip. This can be assured by keeping the specimen and the calibration strips together during the curing period.

11-1 State of Stress in Coating

On the top surface of the coating (Fig. 11-1) there are no external loads; therefore, the stresses perpendicular to this surface are zero. For the layer immediately below the top surface, these stresses will be zero, or close to zero. Without appreciable error, it can be assumed that a state of plane stress exists in the coating. Assume also perfect adhesion between the coating and the base specimen. The strains in the coating will be then identical to the strains in the specimen, and the principal directions of the stresses in the coating will also coincide with those of the

---

[1] The word "specimen" is used in this chapter to indicate the material under the coating.

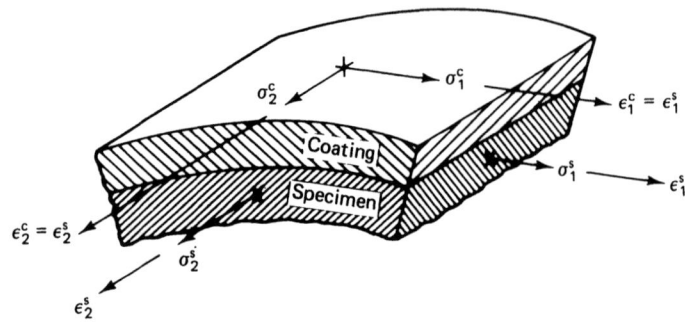

Figure 11-1 Stresses and strains in the coating and the specimen before any crack is opened.

specimen. The relations between the stresses in the coating and the stresses in the specimen can be derived from the following way. By Hooke's law,

$$E_s\varepsilon_1^s = \sigma_1^s - \nu_s\sigma_2^s \qquad E_s\varepsilon_2^s = \sigma_2^s - \nu_s\sigma_1^s$$
$$E_c\varepsilon_1^c = \sigma_1^c - \nu_c\sigma_2^c \qquad E_c\varepsilon_2^c = \sigma_2^c - \nu_c\sigma_1^c \qquad (11\text{-}1)$$

where E_c = modulus of elasticity of the coating
E_s = modulus of elasticity of specimen
ε_1^c = principal strain in coating in direction of ε_1^s produced by external load
ε_1^s = algebraically larger principal strain in specimen
ε_2^c = principal strain in coating in direction of ε_2^s produced by external load
ε_2^s = algebraically smaller principal strain in specimen
ν_c = Poisson's ratio of brittle coating
ν_s = Poisson's ratio of specimen
σ_1^c = principal stress in coating in direction of σ_1^s produced by external load; it is the difference between the total stress and the residual stress
σ_1^s = algebraically larger principal stress in specimen
σ_2^c = principal stress in coating in direction of σ_2^s produced by external load; it is the difference between the total stress and the residual stress
σ_2^s = algebraically smaller principal stress in specimen

The sign convention used is positive ε for elongation and positive σ for tension.

Since perfect adhesion between base metal and coating requires that

$$\varepsilon_1^c = \varepsilon_1^s \quad \text{and} \quad \varepsilon_2^c = \varepsilon_2^s \qquad (11\text{-}2)$$

Eqs. (11-1) and (11-2) can be solved to find $\sigma_1^c \sigma_2^c$ in terms of σ_1^s and σ_2^s.

$$\sigma_1^c = \frac{E_c}{E_s(1 - \nu_c^2)}[(1 - \nu_c\nu_s)\sigma_1^s + (\nu_c - \nu_s)\sigma_2^s] \qquad (11\text{-}3a)$$

$$\sigma_2^c = \frac{E_c}{E_s(1 - \nu_c^2)}[(1 - \nu_c\nu_s)\sigma_2^s + (\nu_c - \nu_s)\sigma_1^s] \qquad (11\text{-}3b)$$

$$\sigma_1^c - \sigma_2^c = \frac{E_c}{E_s}\frac{1 + \nu_s}{1 + \nu_c}(\sigma_1^s - \sigma_2^s) \qquad (11\text{-}3c)$$

BRITTLE COATING

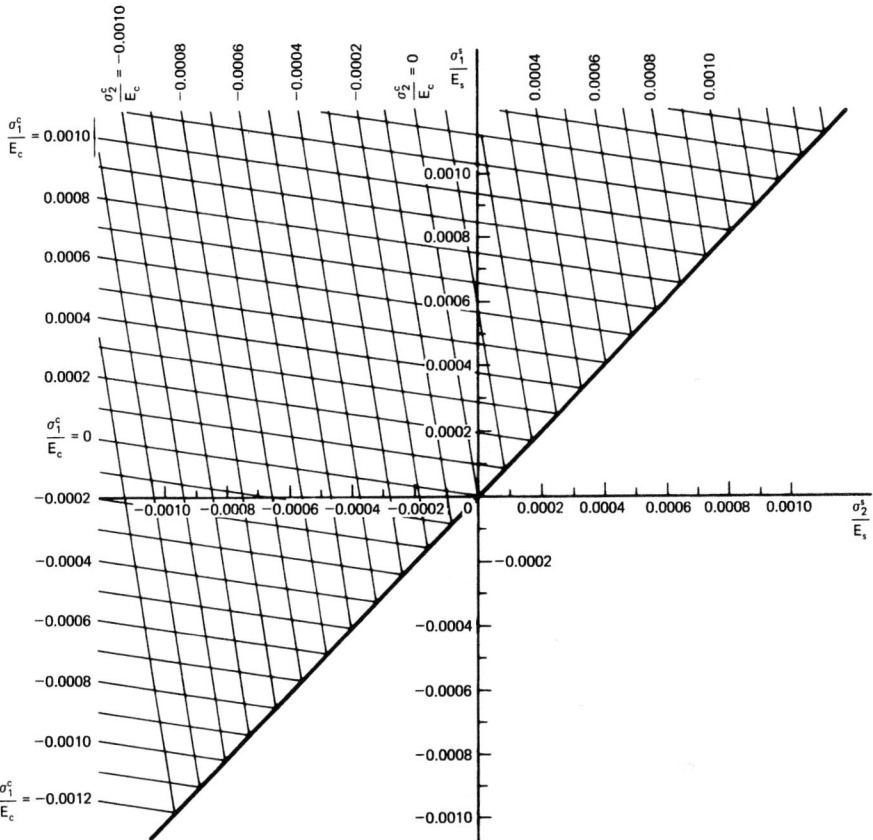

Figure 11-2 Principal stresses in the coating as function of the principal stresses in the specimen. $\nu_c = 0.42$ and $\nu_s = 0.29$.

Since σ_1^s has been defined as the algebraically larger principal stress, $\sigma_1^s - \sigma_2^s$ is always positive. Since the values of modulus of elasticity and Poisson's ratio are always positive, Eq. (11-3c) shows that $\sigma_1^c - \sigma_2^c$ is always positive. In other words, σ_1^c, which has been defined as the principal stress in the coating in the direction of σ_1^s, is always algebraically larger than σ_2^c.

Equations (11-3) are the basic equations governing the relation between the stresses in the coating and the stresses in the specimen. For the typical case where $\nu_2 = 0.42$, $\nu_1 = 0.29$, the graphs of these basic equations are represented in Fig. 11-2. For other values of ν_c and ν_s the graph of the basic equations will take the same general shape.

It is common practice to determine the sensitivity of the brittle coating by means of cantilever calibration strips as shown in Fig. 11-3, which are not necessarily made out of the same material as the specimen. The smaller principal stress in the metal strip σ_2^s, is zero. If we call A the point on the cantilever calibration strip where the cracks start to appear, then from Eqs. (11-3a) and

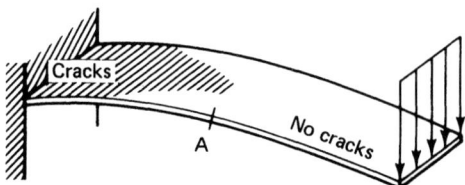

Figure 11-3 Cantilever calibration strip, used to determine the strain sensitivity of the coating. The deflection at the free end of the strip is known. The strain at A is called the strain sensitivity ε_s^d.

(11-3b), by setting σ_2^s equal to zero, by using the relation $\sigma_1^s = E_s \varepsilon_s^d$, and by changing ν_s to ν_a, we obtain

$$\sigma_1^c = \frac{1 - \nu_c \nu_a}{1 - \nu_c^2} E_c \varepsilon_s^d$$

$$\sigma_2^c = \frac{\nu_c - \nu_a}{1 - \nu_c^2} E_c \varepsilon_s^d$$

(11-4)

where ν_a is the Poisson's ratio of the material of the calibration strip and ε_s^d is the larger principal strain at A (Fig. 11-3), commonly called the strain sensitivity in direct loading of the brittle coating used.

11-2 Law of Failure of Brittle Coatings

Historically, in the first published paper on the subject, O. Dietrich and E. Lehr assumed that the brittle coatings they used failed according to the maximum tensile strain law. This law states that if, and only if, the larger principal strain reaches a critical value, cracks perpendicular to the direction of this principal strain will appear. This assumption, however, was not substantiated by their experiments. Studies [2, 3] on Stresscoat soon showed that in a simple compression test, the lateral tensile strain ε_1^c (Fig. 11-4) may be several times larger than the critical tensile strain ε_s^d required to crack the coating (Fig. 11-3), and yet no cracks will appear. In fact, no cracks have yet been reported on brittle coatings in carefully controlled simple compression tests. This phenomenon proves that the maximum tensile strain law is not suitable. To describe completely the law of failure of a coating, variables such as manufacturer's coating number, speed of loading, and gradients of principal stresses must be considered. Since the maximum tensile stress law of failure is generally adopted for brittle materials, it seems plausible that this law should be used as a first approximation of the failure phenomena of brittle coatings. In this chapter it will be assumed that the brittle coating fails when the maximum tensile stress in the coating reaches a critical value. The simplicity of this law helps in the understanding of the coating-failure phenomenon and in the analysis of the data. Experimental data indicate that the error introduced by assuming this law probably would not be large in many cases of ordinary application. Part II of the chapter shows how the same approach used here could be applied to other laws of failure when more experimental data prove them to represent the phenomenon more accurately.

11-3 Determination of Ultimate Strength of Coating

As the coating changes from the liquid state to the solid state, a hydrostatic tension due to shrinkage will be produced. In some cases, this tension may be larger than the ultimate strength of the

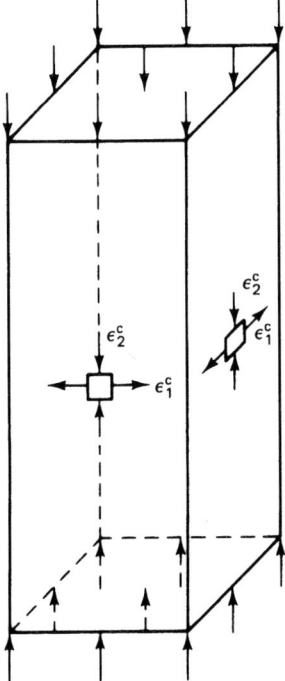

Figure 11-4 Simple compression on a prismatic column. The lateral tensile strain ε_1^c of the coating is several times larger than the strain sensitivity ε_s^d, yet no cracks appear.

coating and may produce random cracks indicating no preferential direction. In most cases, however, the hydrostatic tension is not sufficient to crack the coating. Some time after the coating solidifies, its creep property will relieve some of this hydrostatic tension. When the curing has been done at a temperature higher than room temperature, the coating must be further cooled to the test temperature. If this cooling takes place suddenly, the induced hydrostatic tension in the coating may again be larger than the ultimate strength and random cracks will appear. In properly conducted tests, however, this cooling is very slow, so that a large part of the hydrostatic tension set up by the temperature drop is relieved. Still, an appreciable amount of residual tension denoted by σ_R^c probably remains.

The maximum tensile stress law of failure assumes that the coating will crack when the maximum principal stress in the coating is larger than, or equal to, a critical value σ_u, defined as the ultimate strength of the coating. For the calibration strip (Fig. 11-3), if the coating starts to crack at point A, then σ_u must be equal to the sum of the residual stress σ_R^c and the applied stress σ_1^c. Using the first of Eqs. (11-4) to obtain the value of σ_1^c, we have

$$\sigma_u = \frac{1 - \nu_c \nu_a}{1 - \nu_c^2} E_c \varepsilon_s^d + \sigma_R^c \qquad (11\text{-}5)$$

The strain sensitivities ε_s^d for a particular brittle coating at 24°C and 50% relative humidity as commonly determined on aluminum cantilever calibration strips are approximately 106×10^{-5}

and 70 × 10^{-5}, respectively. The values of ν_c and E_c at 24°C as determined by Stokey [4] are approximately 0.45 and 1500 MPa, and 0.42 and 2000 MPa, respectively. Using these values and taking ν_a as 0.33, Eq. (11-5) gives the values of $\sigma_u - \sigma_R^c$ as 1.68 and 1.44 MPa.

The strain sensitivity of a brittle coating that is free of any residual stress is estimated by Ellis [5] to be about 350 × 10^{-5}. Using this value in Eq. (11-5), the ultimate strength comes out to be of the order of 6.2 MPa. This implies that the residual stress σ_R^c is of the order of 4.7 MPa in the example above.

The values of E_c given above are for completely dried coatings. At the time of test, the outer surface of the coating will generally be completely dried. Since it is the outer layer that fails, the value of E_c for the completely dried coating seems a legitimate one to use in our calculations. The ultimate strength determined above will obviously be the ultimate strength of the outer layer.

11-4 Analysis of Brittle-Coating Data Using Maximum Tensile Stress Law of Failure

Cracks Perpendicular to σ_1 by Direct Loading

Assuming the maximum tensile stress law of failure, the coating will crack when the value of σ_1^c as calculated from Eqs. (11-3) is larger than, or equal to, $\sigma_u - \sigma_R^c$. Hence cracks perpendicular to σ_1^c will be impending when

$$\frac{E_c}{E_s(1 - \nu_c^2)} [(1 - \nu_c \nu_s)\sigma_1^s + (\nu_c - \nu_s)\sigma_2^s] = \sigma_u - \sigma_R^c \tag{11-6}$$

In properly conducted tests, residual stress σ_R^c present in the coating on the calibration strip and in the coating on the specimen may be taken as the same.

We shall define σ_o^d as the minimum σ_1^s required to crack the coating in the direct-loading test when σ_2^s is zero. Then from Eq. (11-6), by substituting σ_o^d for σ_1^s and setting σ_2^s equal to zero, we obtain

$$\frac{E_c}{E_s(1 - \nu_c^2)}(1 - \nu_c \nu_s)\sigma_0^d = \sigma_u - \sigma_R^c$$

or

$$\sigma_0^d = \frac{E_s}{E_c} \frac{1 - \nu_c^2}{1 - \nu_c \nu_s}(\sigma_u - \sigma_R^c) \tag{11-7}$$

From Eqs. (11-5) and (11-7), we can eliminate $\sigma_u - \sigma_R^c$ and determine the value of σ_0,

$$\sigma_0^d = \frac{1 - \nu_c \nu_a}{1 - \nu_c \nu_s} E_s \varepsilon_s^d \tag{11-8}$$

Equations (11-5) and (11-8) are to be used for direct loading. For relaxation loading, the technique of which is explained later, the strain at A on the calibration strip shown in Fig. 11-4 will be denoted by ε_s^r. Equations (11-5) and (11-8) will then be

$$\sigma_u = \frac{1 - \nu_c \nu_a}{1 - \nu_c^2} E_c \varepsilon_s^r + \sigma_R^c \tag{11-9}$$

$$\sigma_0^r = \frac{1 - \nu_c \nu_a}{1 - \nu_c \nu_s} E_s \varepsilon_s^r \tag{11-10}$$

where σ_0^r is the minimum σ_1^s required to crack the coating in the relaxation loading test, where σ_2^s is zero.

BRITTLE COATING

For the special case where the calibration strip material has the same Poisson's ratio as the test specimen, so that $\nu_a = \nu_s$, Eqs. (11-8) and (11-10) reduce to

$$\sigma_0^d = E_s \varepsilon_s^d \tag{11-11}$$

$$\sigma_0^r = E_s \varepsilon_s^r \tag{11-12}$$

The value of ν_c is in the neighborhood of 0.44. If the calibration strips are of aluminum and the specimen is of steel, so that $\nu_a = 0.33$, $\nu_s = 0.29$, the ratio $(1 - \nu_c \nu_a)/(1 - \nu_c \nu_s)$ is 0.975. If we use Eqs. (11-11) and (11-12) instead of Eqs. (11-8) and (11-10), there will be an error of 2.5%.

Equation (11-6) gives the relation that must exist between the applied stresses σ_1^s and σ_2^s and the ultimate strength σ_u of the coating when cracks are impending. This ultimate strength σ_u is related in Eq. (11-7) to σ_0^d, the minimum uniaxial tensile stress in the specimen required for cracking the coating. By eliminating σ_u from Eqs. (11-6) and (11-7), we obtain the relation that must exist between σ_1^s, σ_2^s, and σ_0^d when a crack in the coating perpendicular to σ_1^s is impending:

$$\frac{\sigma_1^s}{\sigma_0^d} + \frac{\nu_c - \nu_s}{1 - \nu_c \nu_s} \frac{\sigma_2^s}{\sigma_0^d} = 1 \tag{11-13}$$

Equation (11-13) gives us a measure of the effect of the algebraically smaller principal stress σ_2^s on the cracking of the coating perpendicular to the algebraically larger principal stress σ_1^s. It will be called the biaxiality equation. Whenever the combinations of σ_1^s and σ_2^s are such that the sum of the quantities on the left side of Eq. (11-13) is equal to or larger than unity, the stress σ_1^c in the coating will be equal to or larger than the maximum stress that the coating can take, and cracks perpendicular to σ_1^s will appear. For the special case where σ_2^s vanishes, it follows from the biaxiality equation that σ_1^s must be equal to σ_0^d for the cracks to appear. This checks with the definition of σ_0^d.

Cracks Perpendicular to σ_2^s by Direct Loading

Immediately after the coating has cracked in the direction perpendicular to σ_1^s, ε_1^c becomes less than ε_1^s at all points in the coating, except the common layer with the specimen. For any point at the edge of the cracks (point P in Fig. 11-5 and points P_1 and P_2 in Fig. 11-6), both the residual stress σ_R^c and the applied stress σ_1^c in the direction of σ_1^s are reduced to zero. For any point not at the edge of the cracks (point Q in Fig. 11-5), these stresses are also reduced, but not necessarily to zero. Further increase of the load may cause the stress in the direction of σ_2^s at some points to reach the ultimate strength σ_u. Cracks perpendicular to σ_2^s will now originate there. At the instant when these secondary cracks are impending, the stresses in the coating at the points where the secondary cracks originate are given by

$$\sigma_1^c = k(\sigma_u - \sigma_R^c) \qquad \sigma_2^c = \sigma_u - \sigma_R^c \tag{11-14}$$
$$\varepsilon_1^c < \varepsilon_1^s \qquad \varepsilon_2^c = \varepsilon_2^s$$

so that k is defined as the ratio of σ_1^c to $\sigma_u - \sigma_R^c$ at the instant when the cracks perpendicular to σ_2^s are just going to appear and at the point where these cracks originate. In the above, we have neglected the change of residual stress in the direction of σ_2^s due to the opening up of the cracks perpendicular to σ_1^s. However, this would be taken into account by the experimental determination of k. As explained later, k may not be a constant and may depend on thickness of the coating and other factors.

To obtain the relation between σ_1^s, σ_2^s, and σ_0^d when the cracks perpendicular to σ_2^s are opening

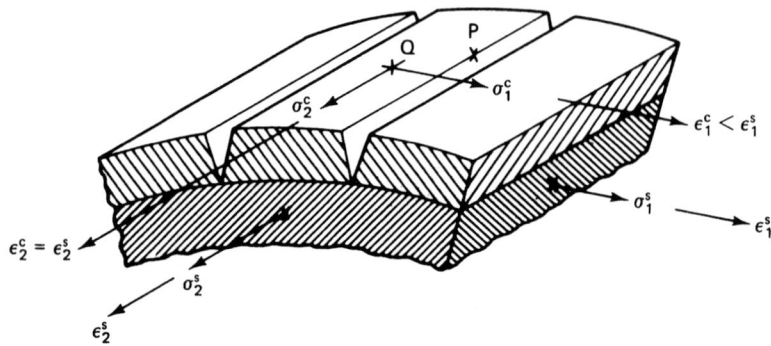

Figure 11-5 Stresses and strains in the coating and the specimen after the cracks perpendicular to σ_1^s are opened.

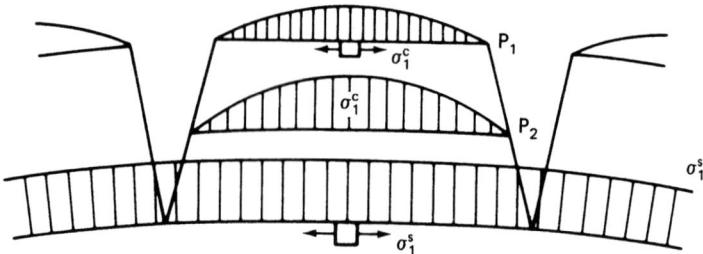

Figure 11-6 Distribution of σ_1^s after the coating is cracked perpendicular to σ_1^s.

up, we proceed as follows. From the second and fourth of Eqs. (11-1) and the first and last of Eqs. (11-14), we obtain

$$\sigma_2^c = k\nu_c(\sigma_u - \sigma_R^c) + \frac{E_c}{E_s}(\sigma_2^s - \nu_s\sigma_1^s) \tag{11-15}$$

By Eq. (11-15) and the second of Eqs. (11-14), we obtain

$$\frac{E_c}{E_s}(\sigma_2^s - \nu_s\sigma_1^s) = (1 - k\nu_c)(\sigma_y - \sigma_R^c) \tag{11-16}$$

By Eqs. (11-7) and (11-16), we conclude that cracks perpendicular to σ_2^s will be impending when

$$\frac{1 - \nu_c^2}{(1 - k\nu_c)(1 - \nu_c\nu_s)}\frac{\sigma_2^s}{\sigma_0^d} - \nu_s\frac{1 - \nu_c^2}{(1 - k\nu_c)(1 - \nu_c\nu_s)}\frac{\sigma_1^s}{\sigma_0^d} = 1 \tag{11-17}$$

Equation (11-17) gives the condition under which additional cracks perpendicular to σ_2^s will be opened after those cracks perpendicular to σ_1^s have been already open. It states that if the combinations of σ_1^s and σ_2^s are such that the sum of the left-side quantities of Eq. (11-17) is equal to or larger than unity, then the stresses in the coating in the direction of σ_2^s will be equal to or greater than the maximum stress that the coating can take, and cracks perpendicular to σ_2^s will appear in addition to those perpendicular to σ_1^s which already exist. This equation is applicable only after the coating has been cracked in the direction perpendicular to σ_1^s.

Let the load on a specimen be increased slowly. Call P the load on the specimen when cracks perpendicular to σ_1^s begin to appear. Then the stresses in the specimen σ_1^s and σ_2^s corresponding to the load P must satisfy the biaxiality equation (11-13). After the cracks perpendicular to σ_1^s opened up, the load on the specimen is increased slowly again. Call αP the load on the specimen when cracks perpendicular to σ_2^s begin to appear. Within the elastic limit of the material of the specimen, stresses in the specimen are always proportional to the load. The principal stresses in the specimen when the external load is αP are therefore $\alpha \sigma_1^s$ and $\alpha \sigma_2^s$. These are the stresses occurring when the second crack is impending and must satisfy Eq. (11-17), in which σ_1^s and σ_2^s are replaced by $\alpha \sigma_1^s$ and $\alpha \sigma_2^s$, respectively,

$$\frac{1 - \nu_c^2}{(1 - k\nu_c)(1 - \nu_c \nu_s)} - \frac{\sigma \sigma_2^s}{\delta_0^d} - \nu_s \frac{1 - \nu_c^2}{(1 - k\nu_c)(1 - \nu_c \nu_s)} - \frac{\alpha \sigma_1^s}{\sigma_0^d} = 1 \qquad (11\text{-}18)$$

Equations (11-13) and (11-18) can be solved simultaneously to give σ_1^s and σ_2^s,

$$\frac{\sigma_1^s}{\sigma_0^d} = \frac{1 - \nu_c \nu_s}{1 - \nu_s^2}\left(1 - \frac{1 - k\nu_c}{\alpha}\frac{\nu_c - \nu_s}{1 - \nu_c^2}\right)$$

$$\frac{\sigma_2^s}{\sigma_0^d} = \frac{1 - \nu_c \nu_s}{1 - \nu_s^2}\left(\nu_s + \frac{1 - k\nu_c}{\alpha}\frac{1 - \nu_c \nu_s}{1 - \nu_c^2}\right) \qquad (11\text{-}19)$$

$$\frac{\sigma_2^s}{\sigma_1^s} = \frac{\nu_s + \dfrac{1 - k\nu_c}{\alpha}\dfrac{1 - \nu_c \nu_s}{1 - \nu_c^2}}{1 - \dfrac{1 - k\nu_c}{\alpha}\dfrac{\nu_c - \nu_s}{1 - \nu_c^2}}$$

where σ_1^s and σ_2^s are the principal stresses corresponding to the load P producing the cracks perpendicular to σ_1^s; α is always larger than unity. σ_0^d is given by Eq. (11-8).

Cracks Produced by Relaxation

A technique called relaxation loading is often employed in brittle-coating analysis. Here the coating is cured while the maximum load P is maintained on the specimen, so that only a hydrostatic residual stress σ_R^c is present in the coating while the specimen is under load P. When this maximum load P is later completely released, the induced stresses in the brittle coating are identical to those induced by a load of $-P$ on the specimen. Similarly, the stresses induced in the brittle coating as a result of a partial release of the initial high load P to the final low load $P - \beta P$ will be identical to those caused by a loading of $-\beta P$ in the specimen. Here a relaxation load of $-\beta P$ will denote a release of the initial load P to the final load $P - \beta P$. Obviously, the initial load P will be completely released if β is equal to unity.

It should be pointed out that the terms "direct loading" and "relaxation loading" are entirely relative. We can call one loading the direct loading and its reversed loading the relaxation loading. We can also call the reversed loading the direct loading and the unreversed loading the relaxation loading. In other words, a direct loading can be any loading. A relaxation loading is a reversed direct loading.

Suppose that a coating of sensitivity ε_s^d starts to crack in a direction perpendicular to σ_1^s when the load is P. Then the stresses in the specimen, σ_1^s and σ_2^s, corresponding to the load P, must satisfy the biaxiality equation (11-13). This cracked coating is then removed and a new coating of a strain sensitivity ε_s^r is applied and cured. Suppose that this new coating starts to crack at the same point in a direction perpendicular to the first crack under the relaxation load $-\beta P$. Within the elastic limit of the material of the specimen, stresses in the specimen are proportional to the load so that the stresses in the specimen when the load is $-\beta P$ are $-\beta\sigma_1^s$ and $-\beta\sigma_2^s$. These are the stresses occurring when the second crack is impending and must again satisfy the biaxiality equation (11-13). However, the algebraically larger principal stress is now $-\beta\sigma_2^s$, the algebraically smaller stress is now $-\beta\sigma_1^s$, and the uniaxial load σ_0^r required to crack the coating is now $(1 - \nu_c\nu_a)/(1 - \nu_c\nu_s) E_s\varepsilon_s^r$, as given by Eq. (11-10), instead of $(1 - \nu_c\nu_a)/(1 - \nu_c\nu_s)E_s\varepsilon_s^d$, as given by Eq. (11-8).

Now $-\beta\sigma_2^s$, $-\beta\sigma_1^s$, and $(\varepsilon_s^r/\varepsilon_s^d)\sigma_0^d$ will be the algebraically larger principal stress, algebraically smaller principal stress, and the minimum uniaxial tensile stress in the specimen required for cracking the coating used in the relaxation-loading test. Substituting these new values of σ_1^s, σ_2^s, and σ_0^d into Eq. (11-13) yields

$$\frac{-\beta\varepsilon_s^d}{\varepsilon_s^r}\frac{\sigma_2^s}{\sigma_0^d} + \frac{\nu_c - \nu_s}{1 - \nu_c\nu_s}\frac{-\beta\varepsilon_s^d}{\varepsilon_s^r}\frac{\sigma_1^s}{\sigma_0^d} = 1 \tag{11-20}$$

Equations (11-13) and (11-20) can be solved simultaneously to give

$$\frac{\sigma_1^s}{\sigma_0^d} = \frac{(1 - \nu_c\nu_s)^2}{(1 - \nu_c^2)(1 - \nu_s^2)}\left(1 + \frac{\varepsilon_s^r}{\beta\varepsilon_s^d}\frac{\nu_c - \nu_s}{1 - \nu_c\nu_s}\right)$$

$$\frac{\sigma_2^s}{\sigma_0^d} = \frac{-(1 - \nu_c\nu_s)^2}{(1 - \nu_c^2)(1 - \nu_s^2)}\left(\frac{\nu_c - \nu_s}{1 - \nu_c\nu_s} + \frac{\varepsilon_s^r}{\beta\varepsilon_s^d}\right) \tag{11-21}$$

where β = the initial high load in relaxation test minus low load in relaxation test producing crack perpendicular to σ_2^s divided by the direct load P producing crack perpendicular to σ_1^s
σ_1^s, σ_2^s = principal stresses corresponding to direct load P
σ_0^d = minimum uniaxial tensile stress in specimen required to crack coating used in direct loading, given by Eq. (11-8).

β is always larger than zero. The coating used in relaxation loading is assumed to have the same Poisson's ratio as the coating used in direct loading.

11-5 Failure Chart of the Coating When σ_1^s Reaches Ultimate Tensile Strength

The failure chart shown in Fig. 11-7 represents in a convenient way the failure phenomenon of the coating for the case of $\nu_c = 0.42$, $\nu_s = 0.33$, and $k = 0.6$. The abscissa and the ordinate are the dimensionless quantities σ_2^s/σ_0^d and σ_1^s/σ_0^d In the special case where the material of the cantilever calibration strip has the same Poisson's ratio as the material of the test specimen, these quantities would be simply $\sigma_2^s/E_s\varepsilon_s^d$, and $\sigma_1^s/E_s\varepsilon_s^d$ respectively. Each point on the chart corresponds to a specific combination of the principal stresses σ_1^s and σ_2^s. By definition, σ_1^s is always larger than or equal to σ_2^s; hence all possible combinations of the principal stresses are represented by points confined in the half-plane to the upper left of the line $N'N$, where $\sigma_1^s = \sigma_2^s$. The biaxiality equation (11-13) is represented by the straight line AB. This line always intercepts the σ_1^s/σ_0^d axis at unity. Its slope depends on the Poisson's ratios of the coating ν_c and of the specimen ν_s. If ν_c and ν_s are equal, the slope is equal to zero and the line AB will be a horizontal line. Equation (11-17) is represented by the straight line CDE. The position of this line depends on the values of ν_c, ν_s, and k.

For any combinations of the principal stresses σ_1^s and σ_2^s corresponding to points inside zone

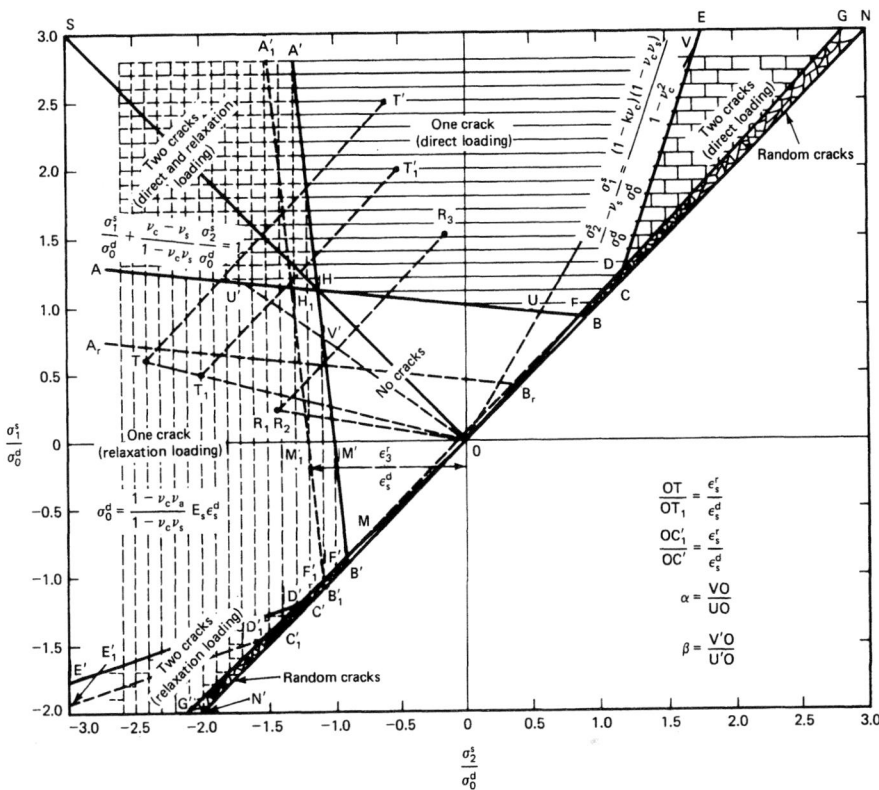

Figure 11-7 Failure chart for the coating. Maximum tensile stress law of failure assumed. Poisson's ratio in specimen $\nu_s = 0.33$, Poisson's ratio in coating $\nu_c = 0.42$, $k = 0.6$, and angle $GON = 1.5°$.

$N'BA$, the sum of the quantities on the left side of the biaxiality equation (11-13) is less than unity, so that no cracks perpendicular to σ_1^s would appear. For any combinations of the principal stresses σ_1^s and σ_2^s corresponding to points inside zone ABN, the sum of the quantities on the left side of the biaxiality equation (11-13) is larger than unity, so that cracks would open up perpendicular to σ_1^s. For any combinations of the principal stresses σ_1^s and σ_2^s represented by points inside zone ECN, the sum of the quantities on the left side of Eq. (11-17) is equal to or larger than unity, so that additional cracks perpendicular to σ_2^s will appear.

For any values of the principal stresses σ_1^s and σ_2^s represented by points on the line $N'N$, these two principal stresses are equal. By Eqs. (11-3), the two principal stresses in the coating, σ_1^c and σ_2^c, are also equal to each other. The state of stress in both the specimen and the coating is therefore hydrostatic. For those points on the line $N'B$, the sum of the quantities on the left side of the biaxiality equation is less than unity; the stress in the coating σ_1^c, in the direction of σ_1^s, is less than the maximum stress the coating can take; and no cracks will appear. For those points on the line BN, the sum above is larger than unity, so that σ_1^c is larger than the maximum stress the coating can take. Being equal to σ_1^c, or σ_2^c at the same time also exceeds that limiting stress. In fact, since the state of stress is hydrostatic, the stress in any direction is equal to σ_1^c

and larger than the limiting stress that the coating can take. Random cracks will therefore appear. Theoretically, random cracks will occur only for the state of stress represented by the points exactly on the line BN. Practically, any state of stress inside a zone GFBN will produce random cracks. It seems reasonable to assume that random cracks will appear when σ_2^s reaches a certain percentage of σ_1^s. Hence the line FG defining the zone of random cracks can be assumed to be a straight line passing through the origin. The angle that the line FG makes with the line ON should be determined experimentally.

In a relaxation test, the sign of the load is reversed so that all stresses change sign. For instance, if the stresses at a point in a specimen under a direct load P are given as $\sigma_x = 25$ MPa, $\sigma_y = -100$ MPa, and $\tau_{xy} = 0$, so that $\sigma_1^s = \sigma_x = 25$ MPa and $\sigma_2^s = \sigma_y = -100$ MPa, then the stresses at the same point under a relaxation load $-P$ must be $\sigma_1^s = -\sigma_y = 100$ MPa and $\sigma_2^s = -\sigma_x = -25$ MPa.

If the strain sensitivities of the coatings used in the direct and relaxation loadings are equal so that σ_0^d for the two loadings are also equal, the stresses due to the loadings P and $-P$ are always represented by two points on the failure chart symmetrically located with respect to the 45° line OS. For instance, if T represents the stresses due to a direct load P, then T' will represent the stresses due to a relaxation load $-P$.

Suppose that the stresses at a point under the direct load P correspond to points inside zone $A'B'N'$, where $A'B'$ and AB are symmetrically located with respect to OS; then the stresses at the same point under the relaxation load $-P$ will correspond to points in the zone ABN, and one crack will appear. For this reason, zone $A'B'N'$ is marked "one-crack zone" and covered with dashed lines to indicate that one crack will appear when the load is reversed. Since the zone AHA' lies inside ABN and $A'B'N'$, it is marked "two-crack zone" to indicate that cracks will appear under both the direct load and the relaxation load. In the zone BHB' the points before and after the reversal of load all lie under the line AB; hence it is marked "no-crack zone" to indicate that no cracks will appear under either the direct load or the relaxation load. Similarly, lines $F'G'$ and $C'E'$ are drawn so that these two lines and lines FG and CE are symmetrically located with respect to the line OS. Zones $E'D'G'$ and $G'F'B'N'$ will be the zones where two cracks and random cracks, respectively, will appear upon reversal of load.

Where the strain sensitivities of the coatings used in the direct loading and the relaxation loading are different, the stresses due to direct and relaxation loadings are no longer represented at points symmetrically located with respect to line OS. For instance, if in the example above we have $\varepsilon_s^d = 60 \times 10^{-5}$, $\varepsilon_s^r = 72 \times 10^{-5}$, and $E_s = 69,000$ MPa, then $\sigma_0^d = 41$ MPa and $\sigma_0^d = 50$ MPa. Under the direct load P the stresses are given by $\sigma_1^s/\sigma_0^d = 0.60$ and $\sigma_2^s/\sigma_0^d = -2.40$. Under the relaxation load $-P$ the stresses are given by $\sigma_1^s/\sigma_0^d = 2.00$ and $\sigma_2^s/\sigma_0^d = -0.50$.

On the failure chart the point T_1', representing the stresses due to the relaxation loading, is obtained from the point T, representing the stresses due to the direct loading, by two displacements. The first is a radial displacement TT_1 toward or away from the origin so that the ratio of OT to OT_1 is the same as the ratio of ε_s^r to ε_s^d. The second is a reflection of T_1T_1' about line OS. If we draw a line $A_1'M_1'B_1'$ parallel to $A'M'B$ and with the ratio OM_1'/OM' equal to the ratio $\varepsilon_s^r/\varepsilon_s^d$, then for any point inside zone $A_1'B_1'N'$ representing the stresses due to a direct loading P, the corresponding point representing the stresses due to the relaxation loading $-P$ will lie inside the zone ABN, and one crack will appear. The line of demarcation of the several zones is therefore $A_1'B_1'$ instead of $A'B'$. Similarly, the line of demarcation $C'E'$ also shifts to its new position $C_1'E_1'$, where $C_1'E_1'$ is parallel to $C'E'$ and the ratio OC_1'/OC' is equal to the ratio $\varepsilon_s^r/\varepsilon_s^d$.

The location of the lines AB and CE depends only on a ν_c, ν_s, and k, and not on the sensitivity of the coating. The uniaxial stress is shown in Eq. (11-8) to be directly proportional to both the strain sensitivity of the coating and the modulus of elasticity of the specimen. For different combinations of ν_c, ν_s, and k, the lines of demarcation AB and CE, will occupy slightly different

BRITTLE COATING

positions from those shown in Fig. 11-8. The general shape of those zones of no crack, one crack, two orthogonal cracks, and random cracks, however, will still be the same. The different positions of the line AB corresponding to four combinations of ν_c, and ν_s are shown on a larger scale in Fig. 11-8.

11-6 Formula for the Calculation of "Apparent" Stress

The failure chart and all the formulas in this chapter have been derived from the assumption that the brittle coating fails when the maximum tensile stress in the coating reaches the ultimate strength. If, in addition to this, a second assumption is made (i.e., that the value of σ_2^s is zero), the biaxiality equation (11-13) will give $\sigma_1^s = \sigma_0^d$, so that Eq. (11-8) becomes

$$\sigma_1^s = \sigma_0^d = \frac{1 - \nu_c \nu_a}{1 - \nu_c \nu_s} E_s \varepsilon_s^d \qquad (11\text{-}22)$$

The stress σ_1^s obtained from Eq. (11-22) will be termed "apparent" stress. On the failure chart, this apparent stress is represented as a point on the σ_1^s/σ_0^d axis at which the value of the ordinate is unity. Where σ_2^s/σ_0^d is known to be zero or small compared with unity, the apparent stress represents a good estimate of the true stress. The error of this estimate increases as the absolute magnitude of the biaxiality ratio σ_2^s/σ_1^s becomes larger and larger, reaching a value of about 15% when the ratio σ_2^s/σ_1^s is ± 1.

11-7 Cracking of Coating by Direct Load Followed by Perpendicular Cracking by Higher Direct Load

We shall consider in more detail the analysis of the data of a coating cracking in two orthogonal directions. In Fig. 11-9, let ABC be the isoentatic[2] for the external load P. Let DEF be the isoentatic for the external load $2P$. At this amount of load, cracks perpendicular to σ_2^s also appear. Let the dashed lines GHI and LMN be the isoentatics for the second crack for the loads $2P$, $3P$, respectively. Then the stresses at A, B, and C, due to the load P, and the stresses at D, E, and F, under the load $2P$, must correspond to points located on line AF of Fig. 11-7. The stresses at G, H and I, due to the load $2P$, and the stresses at L, M, and N, due to the load $3P$, must correspond to points on line DE of Fig. 11-7. Further, the stresses at the points of intersection O_1, O_2, O_3, and O_4 of solid and dashed isoentatics due to the load P are uniquely determined from Eqs. (11-19) by putting α equal to 3 and by using the known values of ε_s^d, E_s, ν_a, ν_c, ν_s, and k. This unique set of principal stresses is represented at point U on line AF shown in Fig. 11-7. Point U may also be determined graphically by drawing a straight line from the origin O to intersect the two lines AF and DE at U and V so that the ratio of the length OV to the length OU is 3. Similarly, the value of α can be estimated for other points, such as B in Fig. 11-9, and both principal stresses uniquely determined graphically or by Eqs. (11-19).

The positions of the points on the line AF corresponding to four combinations of ν_c, ν_s, and $\alpha/(1 - k\nu_c)$ are shown on a large scale in Fig. 11-8. For other combinations of ν_c, ν_s, and $\alpha/(1 - k\nu_c)$, the positions of these points can be obtained by interpolation.

---

[2] Isoentatic P is the locus of the ends of the cracks when the applied external load is P. As a first approximation, isoentatics are frequently considered as isobars, or the loci of points at which the value of σ is the same.

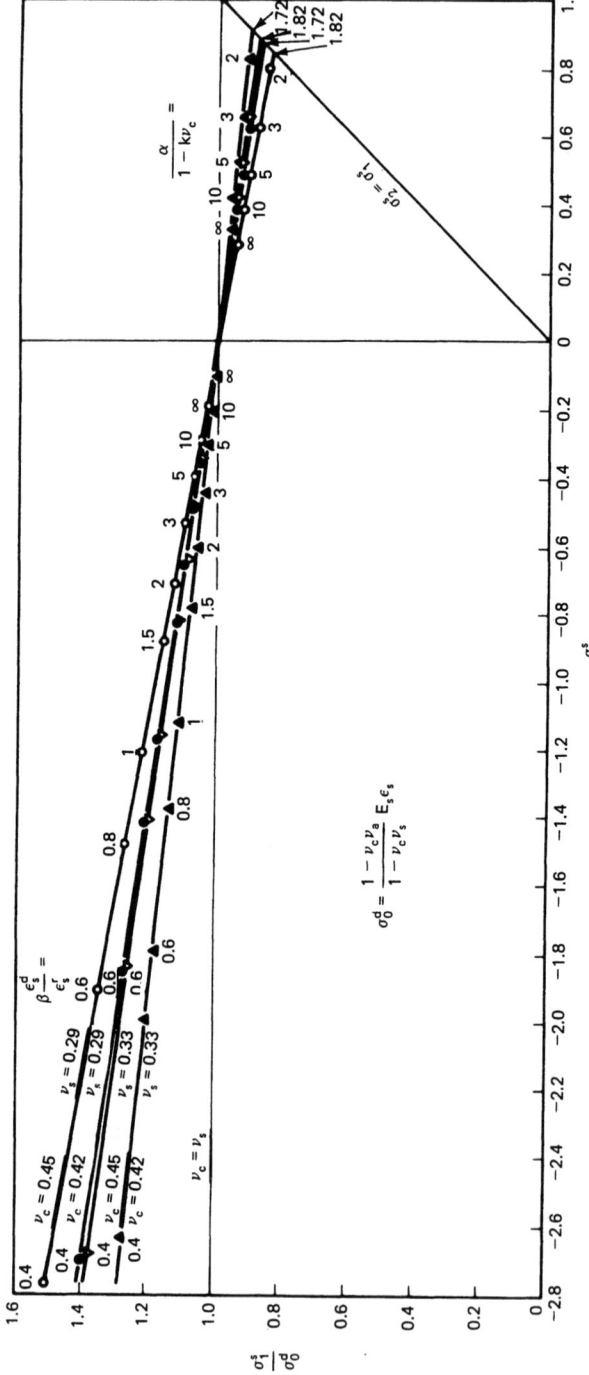

Figure 11-8 Chart for the determination of both principal stresses, assuming the maximum tensile stress law of failing in the coating. Values shown on the portions of curves to the right and to the left of the σ_1^s/σ_0^d axis are those of $\alpha/(1 - k\nu_c)$ and $\beta(\varepsilon_s^d/\varepsilon_s^f)$, respectively.

BRITTLE COATING

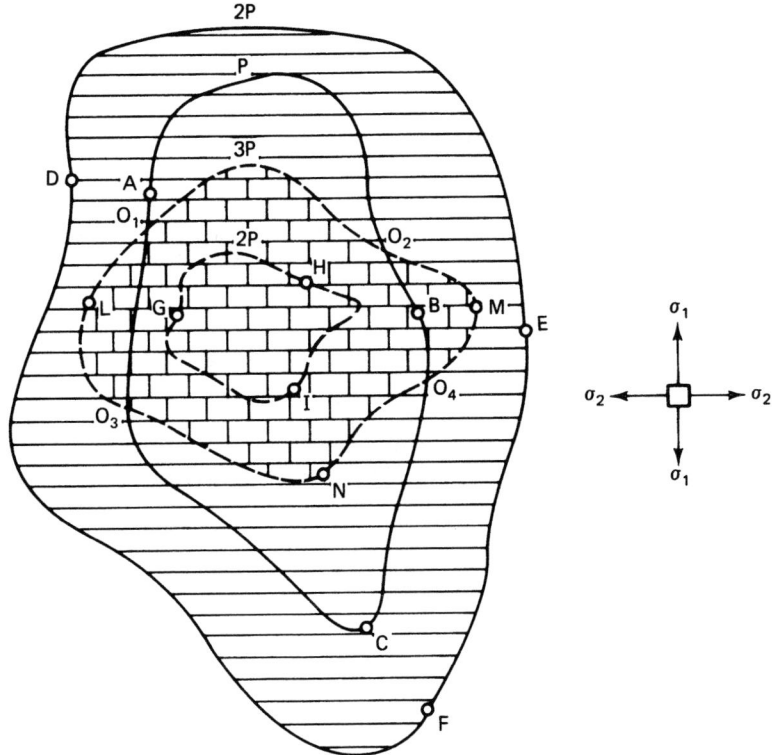

Figure 11-9 Example of isoentatics for cracks perpendicular to σ_1 (solid lines) and for the second cracks perpendicular to σ_2 (dashed lines).

To demonstrate the foregoing analysis, we shall solve the following example. At a certain point A of a steel specimen, cracks in one direction begin to appear when the external load is 6700 N. At the same point A, cracks perpendicular to the above-mentioned cracks begin to appear when the load reaches 13,000 N. The strain sensitivity σ_s^d of the coating is obtained from the aluminum cantilever calibration strips 60×10^{-5}. The values of the constants are given as $\nu_a = 0.33$, $\nu_c = 0.44$, $\nu_s = 0.29$, $E_s = 21{,}700$ MPa, and $k = 0.5$. The two principal stresses are required in the specimen at point A when the external load is 4400 N. We shall assume that the coating cracks according to the maximum tensile stress law of failure.

To solve our example, we first calculate the value of σ_0^d by substituting the given values of ν_a, ν_c, ν_s, E_s, and ε_s^e in Eq. (11-8),

$$\sigma_0^d = \frac{1 - \nu_c\nu_a}{1 - \nu_c\nu_s}E_s\varepsilon_s^d = 122 \text{ MPa}$$

By substituting the given values of ν_c, ν_s, and k into Eqs. (11-19) and noting that $\alpha = 13{,}000/6700 \approx 2.0$, we have

$$\frac{\sigma_1^s}{\sigma_0^d} = \frac{1 - \nu_c \nu_s}{1 - \nu_s^2}\left(1 - \frac{1 - k\nu_c}{\alpha}\frac{\nu_c - \nu_s}{1 - \nu_c^2}\right) = 0.883$$

$$\frac{\sigma_2^s}{\sigma_0^d} = \frac{1 - \nu_c \nu_s}{1 - \nu_s^2}\left(\nu_s + \frac{1 - k\nu_c}{\alpha}\frac{1 - \nu_c \nu_s}{1 - \nu_c^2}\right) = 0.679$$

or

$$\sigma_1^s = 110 \text{ MPa} \quad \text{and} \quad \sigma_2^s = 83 \text{ MPa}$$

The same analysis can be conducted graphically. In Fig. 11-8 we selected the second line corresponding to $\nu_c = 0.42$ and $\nu_s = 0.29$. On this line we locate the point where $\alpha/(1 - k\nu_c) = 2.56$. The coordinates of this point are 0.90 and 0.68, giving the values of σ_1^s/σ_0^d and σ_2^s/σ_0^d, respectively. By interpolation, for $\nu_c = 0.44$, $\nu_s = 0.29$, $\alpha/(1 - k\nu_c) = 2.56$, the values of σ_1^s/σ_0^d and σ_2^s/σ_0^d are therefore 0.89 and 0.68, respectively. This checks with 0.88 and 0.68 calculated above. These principal stresses are those in the specimen at point A corresponding to an external load of 6700 N. The principal stresses in the specimen at point A corresponding to an external load of 4500 N are therefore $\sigma_1^s = 72$ MPa tension and $\sigma_2^s = 55$ MPa tension. Since the directions of σ_1^s and σ_2^s are always perpendicular and parallel, respectively, to the direction of the first crack, the stresses occurring at point A on the surface of the specimen are solved completely.

11-8 Direct Load Cracking Followed by Perpendicular Cracking by Relaxation Load

The following is an analysis of the data obtained from a relaxation test. In Fig. 11-10 let the thick lines ABC and DEF be the isoentatics for the direct loads P and $2P$, respectively. Let the lines GHI and LMN be the isoentatics due to the relaxation loading $-0.3P$ and $-0.6P$, respectively. The relaxation-loading test is run after the coating of the direct-loading test is removed and a new coating is put on the specimen. Again, the stresses at A, B, and C, due to the load P, and the stresses at D, E, and F, due to the load $2P$, must correspond to points located on line AF of Fig. 11-7. The stresses at G, H, and I, due to the relaxation load $-0.3P$, and the stresses at L, M, and N, due to the relaxation load $-0.6P$, must correspond to points on line $F'A'$ of Fig. 11-7. Further, the stresses at the points of intersection O_1, O_2, O_3, and O_4, of thick and thin isoentatics, due to the direct load P, are uniquely determined from Eqs. (11-21) by putting $\beta = 0.6$ and by using the known values of ε_s^d, ε_s^r, E_s, ν_a, ν_c, and ν_s. This unique set of principal stresses is represented as point U' on line AF shown in Fig. 11-7. Point U' may also be determined graphically by drawing a straight line through the origin O to intersect the two lines AF and $A'F'$ at U' and V' so that the ratio of the length OV' to the length OU' is $\beta(\varepsilon_s^d/\varepsilon_s^r)$. Similarly, the value of β can be determined for other points, such as B in Fig. 11-10, and both principal stresses uniquely determined by Eqs. (11-21). The positions of the points on the line AF corresponding to four combinations of ν_c, ν_s, and $\beta(\varepsilon_s^d/\varepsilon_s^r)$ are shown on a large scale in Fig. 11-8. For other combinations of ν_c, ν_s, and $\beta(\varepsilon_s^d/\varepsilon_s^r)$, the positions of these points can be obtained by interpolation.

It should be pointed out that the relaxation-loading test is conducted on an uncracked coating, not on the coating that was already cracked in the direct-loading test. Otherwise, the cracks would make ε_1^c different from ε_1^s so that the analysis above would no longer be valid.

BRITTLE COATING

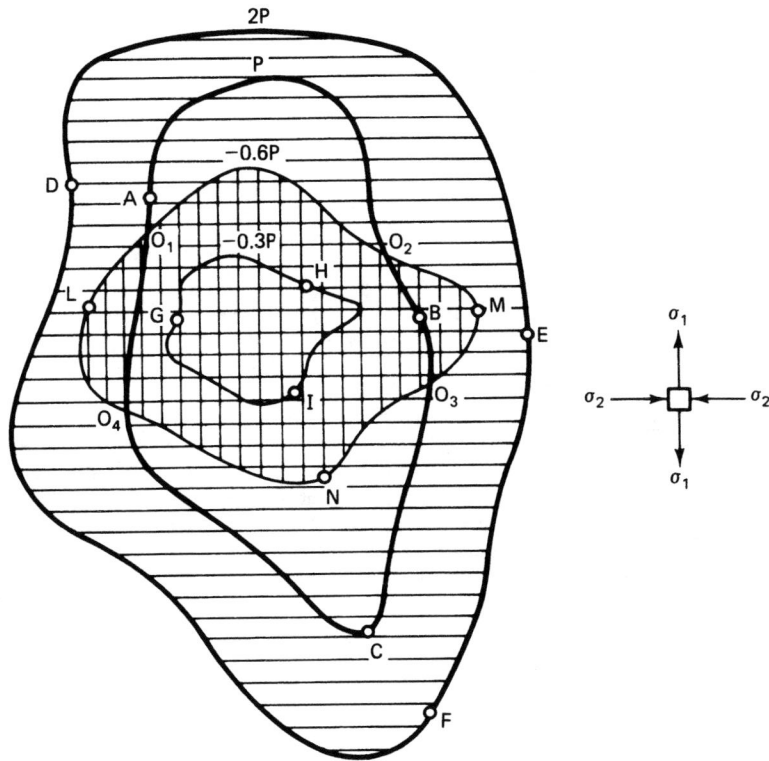

Figure 11-10 Example of isoentatics due to direct loading (thick lines) and to relaxation loading (thin lines).

11-9 Cracking by Direct Load Followed by Perpendicular Cracking by Highest Direct Load and Highest Relaxation Load at Different Location

When cracks perpendicular to σ_1^s are obtained, but no cracks perpendicular to σ_2^s appear under the highest possible direct load and relaxation load corresponding to certain values of $\alpha/(1 - k\nu_c)$ and $\beta(\varepsilon_s^d/\varepsilon_s^r)$, negative information can still be obtained. The state of stress must correspond to points on the failure line AF between these values of $\alpha/(1 - k\nu_c)$ and $\beta(\varepsilon_s^d/\varepsilon_s^r)$ corresponding to the highest loads that fail to crack the coating in the direction perpendicular to σ_2^s.

To demonstrate the statement above, we shall solve the following example. At a certain point A of a steel specimen, cracks in one direction began to appear when the external load was 6700 N. At the same point A, no cracks perpendicular to the above-mentioned cracks appeared at the highest load of 15,500 N. A separate relaxation test in which the initial load of 15,500 N was completely released also showed no cracks at A perpendicular to the above-mentioned crack. The strain sensitivities of the coatings used in the direct loading and the relaxation loading as determined on aluminum calibration strips are $\varepsilon_s^d = 60 \times 100^{-5}$ and $\varepsilon_s^r = 70 \times 10^{-5}$. The values of the constants are given as $\nu_s = 0.29$, $\nu_a = 0.33$, $\nu_c = 0.44$, $E_s = 207{,}000$ MPa, and $k = 0.5$. The upper limit and the lower limit of both principal stresses at point A are required when the external load is 4500 N.

To solve this problem, we first calculate the value of σ_0^d from Eq. (11-8),

$$\sigma_0^d = \frac{1 - \nu_c \nu_a}{1 - \nu_c \nu_s} E_s \varepsilon_s^d = 121.89 \text{ MPa}$$

Next, we note that a crack did appear at point A at a load of 6700 N. Hence, at the load, the principal stresses at A must correspond to a point located on the line $\nu_c = 0.44$, $\nu_s = 0.29$ in Fig. 11-8. (This line is not shown in Fig. 11-8, but its position can be obtained there by interpolation.) On this line, a point that corresponds to $\alpha/(1 - k\nu_c) = 2.99$ can be located by interpolation. Now for any point that lies to the right of this point, $\alpha/(1 - k\nu_c)$ is less than 2.99 or α is less than 2.33, using our given values, $k = 0.5$ and $\nu_c = 0.44$. A direct load of 2.33 times the load that cracks the coating perpendicular to σ_1^s should therefore produce a crack perpendicular to σ_2^s. Since the 15,500-N load, which is 2.33 times the 6700-N load, did not succeed in cracking the coating perpendicular to σ_2^s, our stresses must be represented by points on the line $\nu_c = 0.44$, $\nu_s = 0.29$ to the left of the point $\alpha/(1 - k\nu_c) = 2.99$. Similarly, any point to the left of the point $\beta(\varepsilon_s^d/\varepsilon_s^r) = 2.00$ or $\beta = 2.33$ would mean that a crack perpendicular to σ_2^s should be produced by the release of the 15,600 N-load. Since this did not happen in our given data, our stresses must lie to the right of the point $\beta(\varepsilon_s^d/\varepsilon_s^r) = 2.00$. The upper and lower limits of σ_1^s/σ_0^d and σ_2^s/σ_0^d are therefore the coordinates of the points $\alpha/(1 - k\nu_c) = 2.99$ and $\beta(\varepsilon_s^d/\varepsilon_s^r) = 2.00$ on the line $\nu_c = 0.44$, $\nu_s = 0.29$, or

$$0.88 < \frac{\sigma_1^s}{\sigma_0^d} < 1.13 \quad \text{and} \quad -0.70 < \frac{\sigma_2^s}{\sigma_0^d} < 0.62$$

or

$$107 \text{ MPa tension} < \sigma_1^s < 138 \text{ MPa tension}$$
$$-86 \text{ MPa compression} < \sigma_2^s < 75 \text{ MPa tension}$$

The best estimate for the values of σ_1^s and σ_2^s will therefore be the mean of the two extremes:

$$\left. \begin{array}{l} \sigma_1^s = 81.5 \text{ MPa} \\ \sigma_2^s = -3.5 \text{ MPa} \end{array} \right\} \text{ under the external load of 7 MPa}$$

The apparent stress calculated from Eq. (11-22) is

$$\sigma_1^s = 81 \text{ MPa}$$

which is practically equal to the mean, 81.5 MPa, obtained above. Hence, for this example, the apparent stress calculated by the simple equation (11-22) represents the best estimate for σ_1^s. This is generally true of the stress σ_1^s at any point where the coating is cracked perpendicular to σ_1^s by a direct load but not cracked perpendicular to σ_2^s by the highest direct load and the highest relaxation load. Similarly, it is also generally true that the apparent-stress formula represents the best estimate for σ_2^s at any point where the coating is cracked perpendicular to σ_2^s by a relaxation load, but not cracked perpendicular to σ_1^s by the highest relaxation load and the highest direct load.

BRITTLE COATING

11-10 Determination of Second Principal Stress by Graphical Integration

When the values of σ_1^s and ρ_1 are known at every point on an isostatic S_2 (Fig. 11-11) and the value of σ_2^s is known at one point on this isostatic, it is possible to determine the value of σ_2^s at any point on this isostatic by integrating the Lamé–Maxwell equations,

$$\sigma_2^s = \int \frac{\sigma_2^s - \sigma_1^s}{\rho_1} dS_2$$

In experimental stress analysis, integration is necessary to calculate the value of σ_2^s. From brittle-coating data the value of σ_1^s is not determined exactly, but rather an approximation to σ_1^s, which was called apparent stress. Using this apparent stress, the integration may be performed in the following manner. The isostatic S_2 is first divided into a number of short segments ΔS_2 of equal length (Fig. 11-11). The following quantities are then calculated in the order stated.

1. $(\sigma_2^s - \sigma_1^s)/\rho_1$ at P, the point of known σ_2^s
2. Change of σ_2^s between P and Q, by the formula

$$\Delta\sigma_2^s = \frac{\sigma_2^s - \sigma_1^s}{\rho_1} \Delta S_2$$

where $(\sigma_2^s - \sigma_1^s)/\rho_1$ are those at P

3. σ_2^s at Q by the relation σ_2^s at $Q = (\sigma_2^s$ at $P) + \Delta\sigma_2^s$ calculated in step 2
4. Change of σ_2^s between Q and R by the formula

$$\Delta\sigma_2^s = \frac{\sigma_2^s - \sigma_1^s}{\rho_1} \Delta S_2$$

where $(\sigma_2^s - \sigma_1^s)/\rho_1$ are those at Q
5. σ_2^s at R, by the relation σ_2^s at $R = (\sigma_2^s$ at $Q) + \Delta\sigma_2^s$ calculated in step 4

The calculations can be continued until the values of σ_2^s at all points to the right of P are obtained. The values of σ_2^s at all points $Q', R', \ldots$ to the left of P can be calculated in a similar way. If a point is reached at which the value of σ_2^s is known, the value obtained through integration can be checked against this known value and the error distributed along the path of integration.

The radius of curvature ρ_1 can be measured on the curves S_1 by two methods. Where ρ_1 is short, circles of different radii can be drawn on the curve by a compass. The one that fits the curve closest at the point of tangency will give the best estimate of ρ_1. Where ρ_1 is long, the base l and height h of a short segment of the curve can be measured (Fig. 11-12) and ρ_1 calculated from the formula

$$\rho_1 = \frac{l^2}{8h} + \frac{h}{2}$$

11-11 Effect of Refrigeration

Generally, no cracks can be obtained in a large portion of the surface of the test specimen because even under the highest possible load, the stresses in these areas are not large enough to crack the coating. These stresses correspond to points in the no-crack zone ABM of Fig. 11-7. The techniques usually employed to obtain crack patterns in these areas are refrigeration or dye etching, or both.

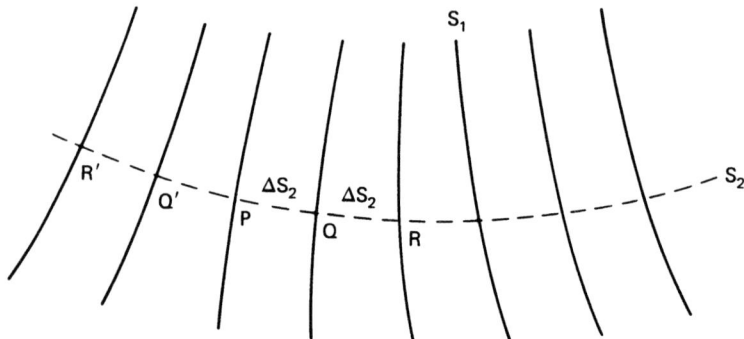

Figure 11-11 Graphical integration of Lamé–Maxwell equations. Division into a number of equal segments ΔS_2 of an isostatic S_2.

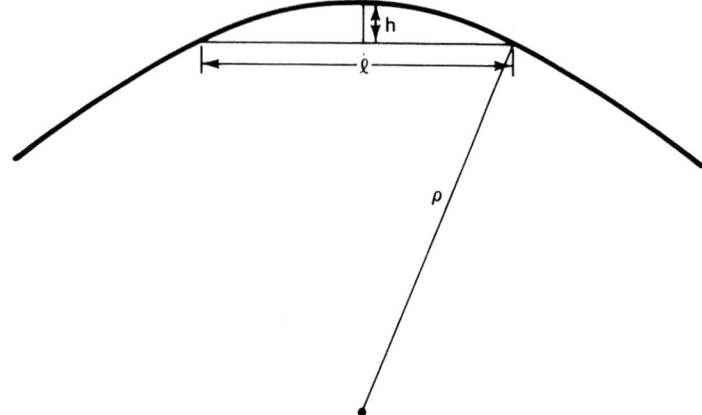

Figure 11-12 Graphical determination of the radius curvature of any curve. The radius of curvature ρ is given approximately by $l^2/8h + h/2$.

Refrigeration produces two separate effects. First, it changes the ultimate tensile strength of the coating from σ_u^0 at test temperature t_0 to σ_u^r at the lowered temperature t_r. Second, it induces a hydrostatic tension in the coating. This tension, denoted by σ_H^c, can be computed as follows:

The contraction per unit length of the coating due to the temperature drop from t_0 to t_r is

$$\gamma_c(t_0 - t_r)$$

and the contraction per unit length of the specimen due to the temperature drop from t_0 to t_s is

$$\gamma_s(t_0 - t_s)$$

where γ_c and γ_s are the coefficients of lineal thermal expansion of the coating and the specimen, respectively.

BRITTLE COATING

The expansion per unit length of the coating due to the hydrostatic tension σ_H^c at temperature t_r is given by the corresponding hydrostatic strain,

$$(1 - \nu_c)\frac{\sigma_H^c}{E_c^r}$$

where E_c^r is the modulus of elasticity of the brittle coating at temperature t_r. The contraction of the test specimen due to the hydrostatic compression is negligible.

Assuming perfect adhesion between coating and specimen, we have

$$\gamma_c(t_0 - t_r) - \gamma_s(t_0 - t_s) = (1 - \nu_c)\frac{\sigma_H^c}{E_c^r}$$

or

$$\sigma_H^c = \frac{E_c^r}{1 - \nu_c}[\gamma_c(t_0 - t_r) - \gamma_s(t_0 - t_s)] \tag{11-23}$$

This fictitious hydrostatic tension in the specimen σ_H^s, which would induce the same amount of stress in the coating as refrigeration, can be computed by substituting E_c^0 for E_c, σ_H^c for σ_1^c, and σ_H^s for both σ_1^s and σ_2^s in Eq. (11-3a),

$$\sigma_H^c = \frac{E_c^0}{E_s(1 - \nu_c^2)}(1 - \nu_c\nu_s + \nu_c - \nu_s)\sigma_H^s$$

or

$$\sigma_H^s = \frac{E_s}{E_c^0}\frac{1 - \nu_c}{1 - \nu_s}\sigma_H^c$$

where E_c^0 is the modulus of elasticity of the brittle coating at temperature t_0.

Once the constants σ_u^0, σ_u^r, γ_c, γ_s, ν_c, and ν_s, have been determined for the particular coating and test specimen used, the effect of refrigeration can be evaluated.

Refrigeration is represented on the failure chart by two separate displacements. The point is first displaced toward or away from the origin until it is at σ_0^r/σ_0^0 times its initial distance from the origin. It is the displaced upward to the right along a 45° line until the increase of the ordinate or abscissa is equal to σ_H/σ_0^r. The final location will then dictate whether there will be no crack, one crack, two orthogonal cracks, or random cracks after refrigeration.

The radial displacement represented by the change of σ_0^d computed by the change of $(\sigma_u - \sigma_0^d)/E_c$ is probably always small compared with the displacement along the 45° line due to σ_H^s, as the example above illustrates. In the following discussion, σ_0^0/σ_0^r therefore is assumed to be unity and refrigeration is treated as a single displacement upward to the right along a 45° line.

Returning to the failure chart, let MF be a line parallel to NN'. For any combination of principal stresses represented by points inside the zone $MFBN'$, refrigeration will move them across the line FB into the zone of random cracks. For those represented by the points inside the zone AFM, refrigeration will move them across the line AF representing a crack perpendicular to σ_1. Further refrigeration will move them across the line CE representing the second crack orthogonal to the first. The effect of refrigeration on relaxation loading is similar to that on direct loading (i.e., a displacement upward to the right along a 45° line).

From Eqs. (11-23) and (11-24) we have

$$\sigma_H^s = \frac{E_c^r}{E_c^0}\frac{E_s}{1-\nu_s}[\gamma_c(t_0 - t_r) - \gamma_s(t_0 - t_s)]$$

or

$$\frac{\sigma_H^s}{\sigma_0^d} = \frac{E_c^r}{E_c^0}\frac{E_s}{\sigma_0^d(1-\nu_s)}[\gamma_c(t_0 - t_r) - \gamma_s(t_0 - t_s)]$$

Refrigeration can therefore be represented on the failure chart as a shift of the line AB to its new position A_r, B_r so that A_rB_r is parallel to AB and the distance BB_r is equal to $\sqrt{2}$ times the right-hand quantity of Eq. (11-25). A_rB_rN will now be the zone of no crack, and A_rB_rN' will now be the zone of one or more cracks. The length of the shift, BB_r, depends, of course, on the degree of refrigeration.

Examples of experimental determinations of the location of the line OG in the failure chart and of the value of k can be found in Ref 1, mentioned at the beginning of the chapters.

11-12 Applications to The Analysis of Stress Fields

Figure 11-13 gives the crack pattern on the outside surface of a pressure vessel. The no-crack zone, the one-crack family zone, and the two-crack family zone, for σ, are clearly shown. Knowing that $\varepsilon_S = 0.0006$, it will be a straightforward operation to determine the largest tensile principal stress in a large part of the field. The region where two of these are largest and equal to each other is also shown.

Figure 11-14 shows a detail of the pattern on a tank shoe and Fig. 11-15 the detail of the pattern on a landing gear. Figures 11-16 and 11-17 show the surface of a large-diameter steel pressure vessel with a longitudinal weld.

It should be obvious to the reader that the coating method has other possibilities than the location of a strain gage. No other method can give in larger zones of the field, the tensor, in one photograph.

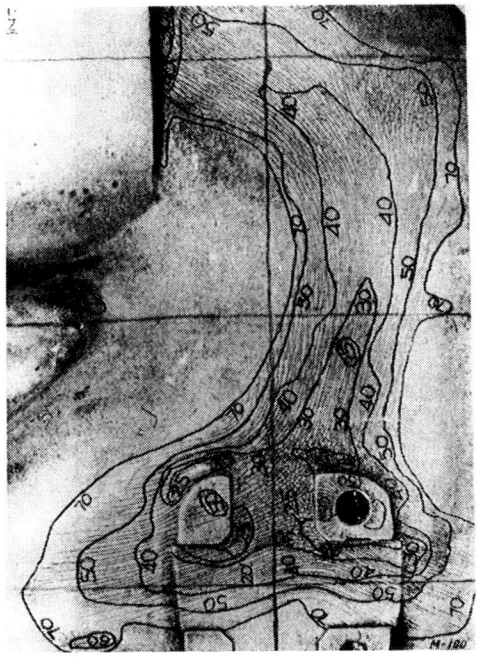

Figure 11-13 Brittle-coating pattern on a pressure vessel. Figures on isoentatics correspond to internal pressure (psi $\times$ 10^2).

Figure 11-14 Brittle-coating pattern on a portion of a tank shoe.

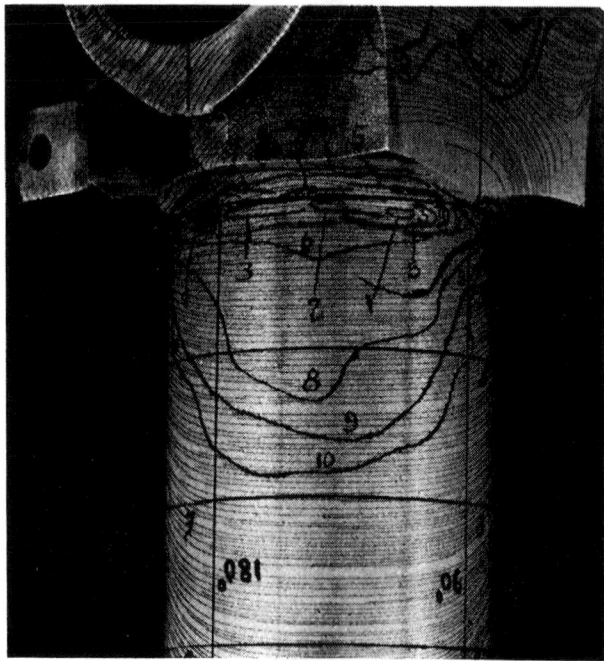

Figure 11-15 Brittle-Coating Pattern on a Landing Gear Component. For a load of 1 lb

Code No.	Stress (psi)
0	13.3
1	12.2
2	11.0
3	9.72
4	8.58
5	7.41
6	6.27
7	5.29
8	4.38
9	3.65
10	3.31

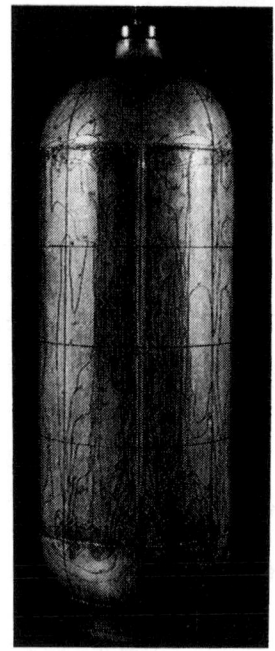

Figure 11-16 Overall view of the brittle-coating crack pattern on a steel pressure vessel.

BRITTLE COATING

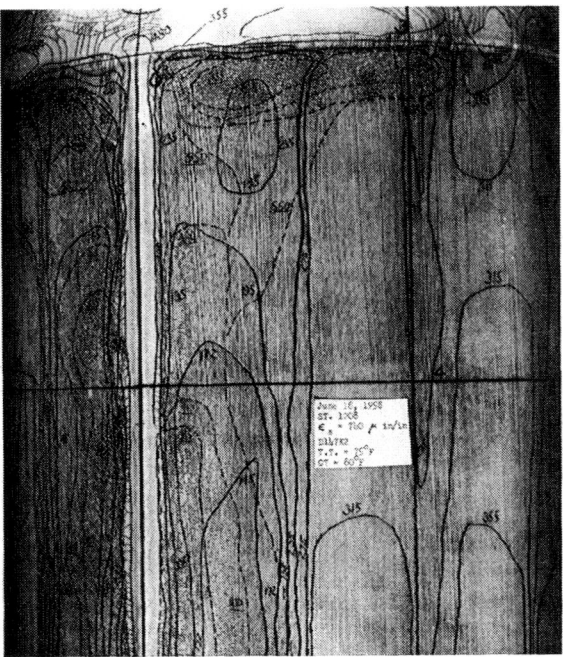

Figure 11-17 Brittle-coating crack pattern over the forward section of the cylindrical shell near the longitudinal weld: solid isoentatics give maximum principal stresses. Dashed isoentatics give maximum principal stresses. ST, 1208; ε_s = 740 μin./in.; D147K2; TT = 75°F, CT = 80°F.

Part II Application of Brittle Coating

John Hall

Micro Engineering II
Upland, California

Ferdi Stern*

F. B. Stern Company
Wayland, Massachusetts

11-13 Test Procedures

Applying Resin-Based Brittle Coating

Since most industrial applications of the brittle-coating technique are confined to the use of commercially available brittle coatings, some discussion of the practical details of these coatings is in order. The use of brittle coating for quantitative determination on surface strains requires a specialized technique; and while technique is important with all experimental stress analysis methods, it is especially so with brittle coating. Inattention to the many details, such as surface preparation, has been found by experience to reduce accuracy and may easily result in the loss of the desired data or, more seriously, in the loss of confidence in the method.

Selection of Coating for Test Temperature

There are many variables that affect the sensitivity of brittle coating. Some of the most important ones are the brittle-coating number, the curing conditions of the coating, the coating thickness, the temperature and humidity at the time of test, and the loading history of the coating.

The sensitivity of the coating varies with the grade (number) of the coating, where lower sensitivity is expected for lower numbers of the coating. The sensitivity of the coating also decreases with increased humidity. Thus the coating number must be selected for the temperature and humidity expected at the time of test. Charts that give the coating sensitivity as a function of temperature and humidity are provided by coating manufacturers. Figure 11-18 shows some typical relationships. Coating numbers that are selected from these charts can be expected to start cracking at 700 to 800 μm/m (μin./in.) of strain. A coating number two numbers above the chart value should be used for a more sensitive coating, where a threshold strain sensitivity of 500 to 600 μm/m (μin./in.) is required. A coating four numbers higher than the chart value, which will yield strain sensitivity of about 300 to 400 μm/m (μin./in.), is about as far as it is possible to go without

---

*Deceased

BRITTLE COATING

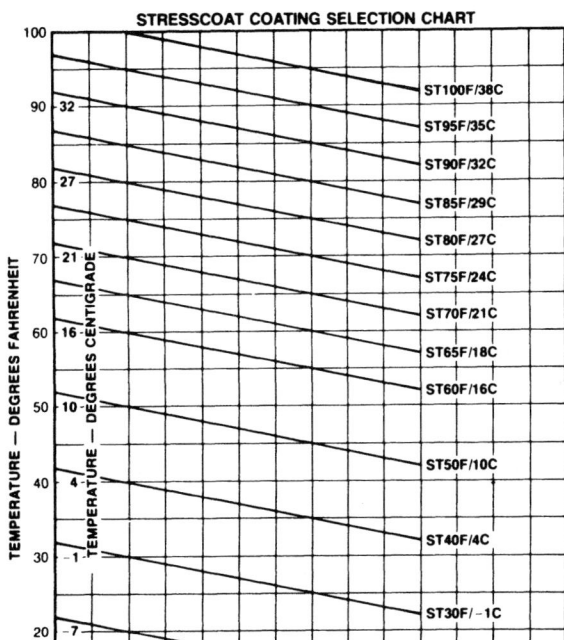

Figure 11-18 Coating sensitivity as a function of temperature and humidity.

experiencing random cracking in the coating during its cure. This random cracking is known as "crazing." Where lower strain sensitivity is desired, a softer coating, two numbers lower than the chart value, is used to start cracking at strains of approximately 1000 μm/m (μin./in.)

Coating Thickness

The strain sensitivity of the coating increases with an increase in coating thickness. To avoid variation in the coating strain sensitivity, it is imperative to spray a uniform thickness of coating onto the structural component. Buildup of coating thickness at reentrant corners and edges of the structural components should be avoided. Experimental results show that a coating thickness of 0.1 to 0.13 mm (0.004 to 0.005 in.) is most desirable in producing a uniform coating sensitivity. Calibration strips that are used to determine the threshold strain sensitivity of the coating in use must also be sprayed with the same thickness. In actual practice, six or more calibration strips are sprayed at the same time as the test part. Color is a good guide to coating thickness.

Curing Coating

The strain sensitivity of the coating is also influenced by the curing temperature. As a rule of thumb, the coating should be dried at least 6°C (10°F) above testing conditions, but below 52°C (125°F) and by all means at a uniform temperature. Particular care should be taken to reduce the temperature slowly to the testing temperature after curing is completed.

The coated structural component should then be moved to the testing spot an hour or two before starting the test, to allow it to come to temperature equilibrium. The calibration strips should be left immediately adjacent to the parts at all times during curing and testing, preferably being in contact with them. Calibration strips must undergo the same conditions as the test part. Rapid cooling should be avoided, as thermal crazing may occur; similarly parts and calibration bars should be protected from cool air currents, which may cause localized crazing.

Creep of Coating

Since the major constituent of brittle coating is wood resin, the coating is very susceptible to creep under sustained load. Typically, the strain sensitivity of the coating will be reduced by more than 50% for a load duration of 1000 seconds. It is therefore important that the load duration be as short as possible. Moreover, all efforts should be made to duplicate the loading history in the calibration strips with that of the structural component (see Fig. 11-19).

Ceramic Coatings

Calibrated ceramic-based coatings for experimental stress analysis yield accurate stress measurements in the temperature range 10 to 316°C (50 to 600°F). The sensitivity of commercially available coatings varies from 200 to 1800 μm/m (μin./in.), depending on the coefficient of expansion of the metal on which the coating is applied. Changes in temperature and humidity and the presence of oil do not affect test results appreciably. Tests may be conducted outdoors for indefinite lengths of time, in areas where abrasion by air or dust prevents use of resin-based materials, or under liquid environments.

Principle of Operation

Successful use of the ceramic coatings depends on a controlled difference between the coefficient of expansion of the coating and that of the base metal. If the coefficient of expansion of the coating is greater than that of the base metal, residual tension will be "locked" in the coating when it cools from the firing temperature. This process develops a coating that cracks at low strain levels. If, on the other hand, the coefficient of expansion of the coating is less than that of the base metal, residual compression will be locked in the coating, which then requires a higher strain level to initiate crack patterns.

Coating Selection and Calibration

Figure 11-20 is a selection chart for choosing a ceramic coating for use with specific materials. Carbon steels, for example, have a coefficient of expansion of 12 to 14 μm/m/°C (7 to 8 μin./in./°F), straight chromium steels from 10 to 12 μm/m/°C (5.5 to 6.5 μin./in./°F), and the austenitic chromium–nickel steels from 16 to 18 μm/m/°C (9 to 10 μin./in./°F). This coefficient of thermal expansion is a mean value for the range 27 to 318°C (80 to 600°F). Since data on coefficient of expansion for a particular metal may be difficult to obtain, a calibration bar is usually made from

BRITTLE COATING

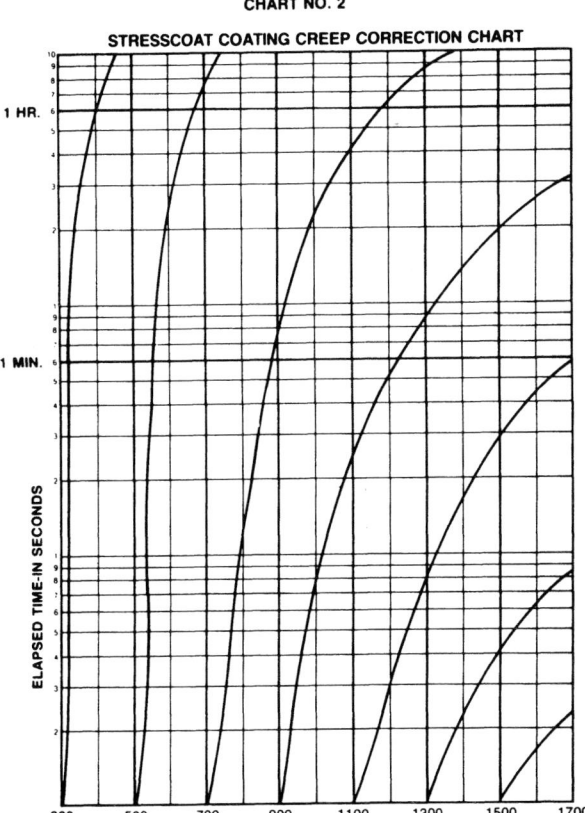

Figure 11-19 Stresscoat coating creep correction chart.

the same material as the test structure, and the strain sensitivity of the coating is determined from it. Strain sensitivity is not appreciably affected by temperature variations to 318°C (600°F), as shown in Fig. 11-21.

Surface Preparation, Firing of Coating, and Detection of Crack Pattern

The test specimen should be prepared by a slight roughening of the surface such as sandblasting or vapor-honing, followed by a thorough cleaning by solvent used in the coating.

Ceramic coatings are a fine-grit material mixed with a volatile solvent carrier, which is sprayed onto the test part. It is then air dried to a soft powder. The coating material should be well stirred and shaken in the container to provide a homogeneous mixture that can be sprayed. The material is sprayed from a distance of 7.6 to 17.7 cm (3 to 7 in.), as thick and wet as possible without causing running. The coating is air dried for 30 minutes, then placed in an oxidizing furnace for

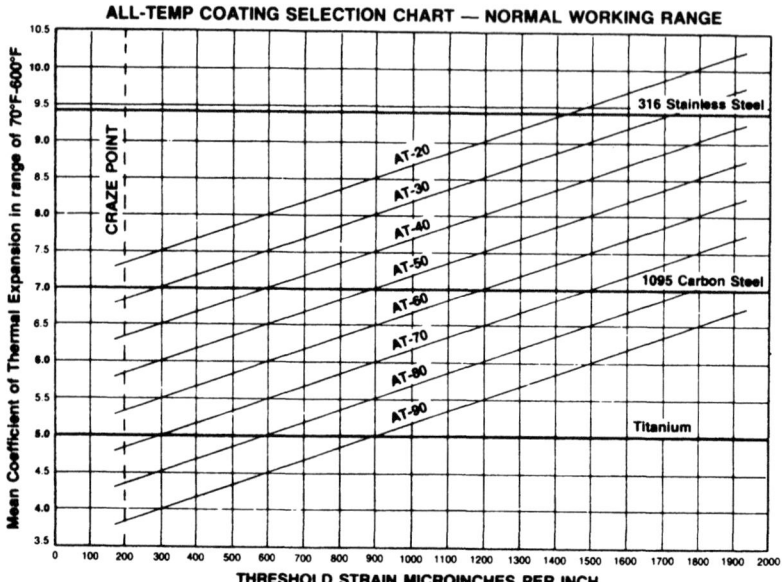

Figure 11-20 Selection chart for ceramic coatings.

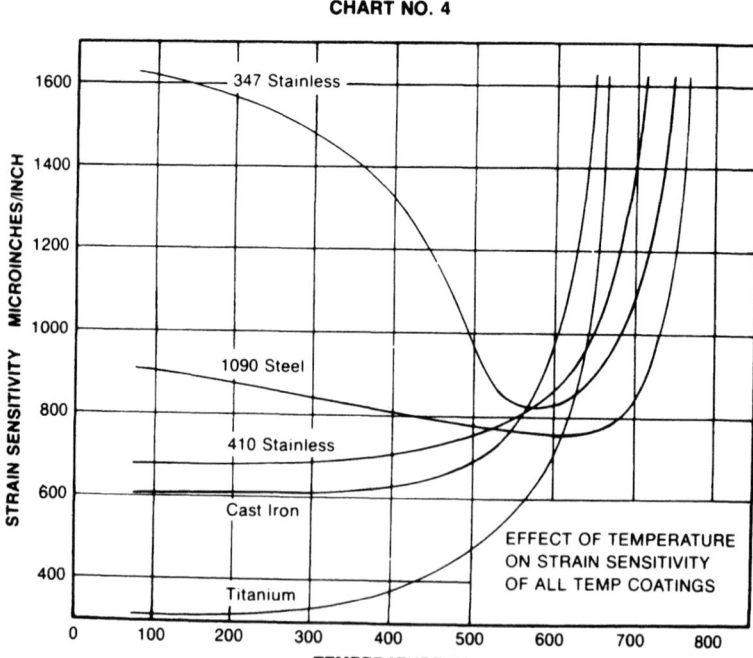

Figure 11-21 Effect of temperature on strain sensitivity.

about 15 minutes at 510 to 516°C (950 to 1050°F) until the coating appears glossy, and then air cooled to the test temperature. Coating thickness of 0.08 to 0.3 mm (0.003 to 0.012 in.) has little effect on the sensitivity of the coating.

The test piece may now have the load applied and cracks will form. To detect these cracks, the Statiflux[3] system must be used. This system consists of white charged particles that are sprayed on the coating. Wherever there is a crack, the powder will build up at the crack and provide a visual indication. A typical crack pattern developed by this technique is shown in Fig. 11-22. Calibration of the coating sensitivity is achieved by coating a standard calibration bar. The calibration bar must be of the same material as the test item.

11-14 Applications

Tensile Strains

Brittle coatings crack at right angles to the principal tension strain. Figure 11-23 shows the example of a typical coated crane hook under tensile load.

Compressive Strains by Relaxation Loading

Tests have been conducted using static loads held for a period of time allowing the coating to creep. The load is released and the former compression areas will put the coating into tension, thereby producing cracks (see Fig. 11-24).

Impact Loading

Tests have been conducted using impact loads. Often when the strain levels are above 900 μm/m (μin./in.) the test requires the technique of electrolyte and charged particles to show cracks. Figure 11-25 offers a particularly vivid illustration of impact loading.

Centrifugal Force Loading

Figure 11-26 shows patterns produced in brittle coating on a disk spun at high speeds.

Locating Strain Gages

Figure 11-27 shows a boom-crane tip that was subjected to brittle-coating test to locate areas of high stress. Strain gages are placed at right angles to the cracks, as determined from coating indications.

11-15 Summary

Resin-based, strain-indicating brittle coatings are used to locate high stress concentrations and as an aid to placement of resistance-type strain gages. They are also used to demonstrate location and direction of principal tension strains.

---

[3] A process whereby special electrostatically charged powder is applied to a surface with powder built up at the cracks.

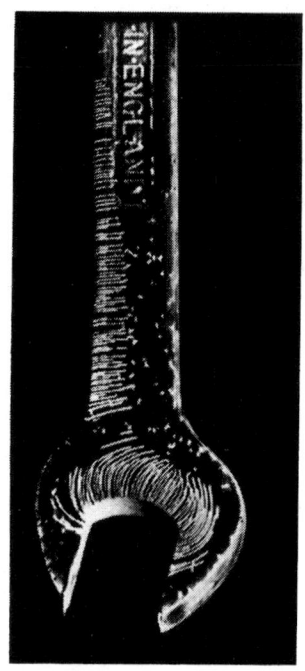

Figure 11-22 Crack pattern developed by the Statiflux system.

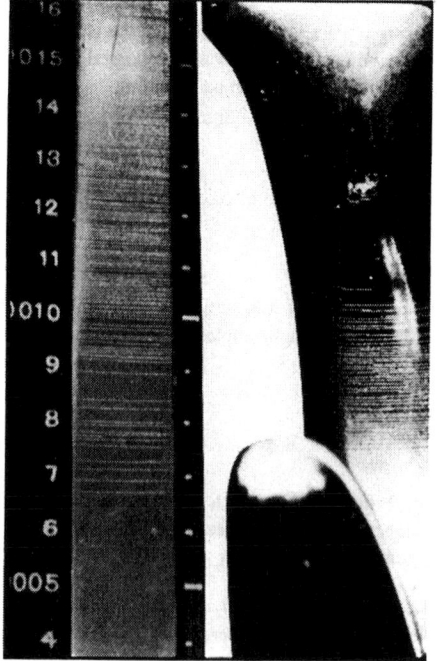

Figure 11-23 Typical coated crane hook under tensile load.

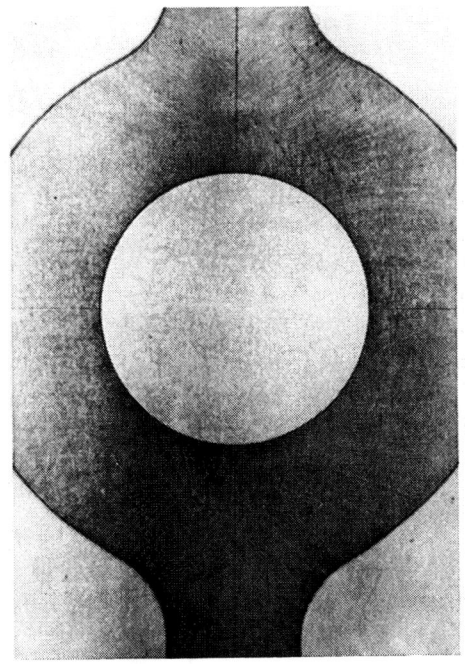

Figure 11-24 Tension-produced cracks in coating.

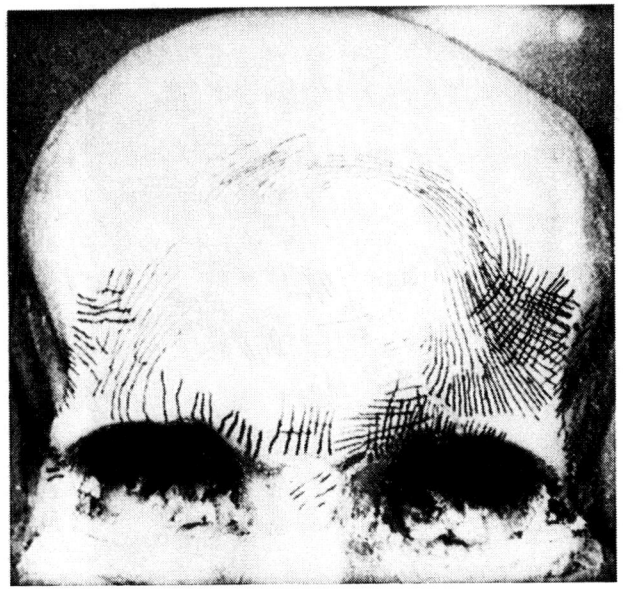

Figure 11-25 Skull showing the effect of a baseball's impact.

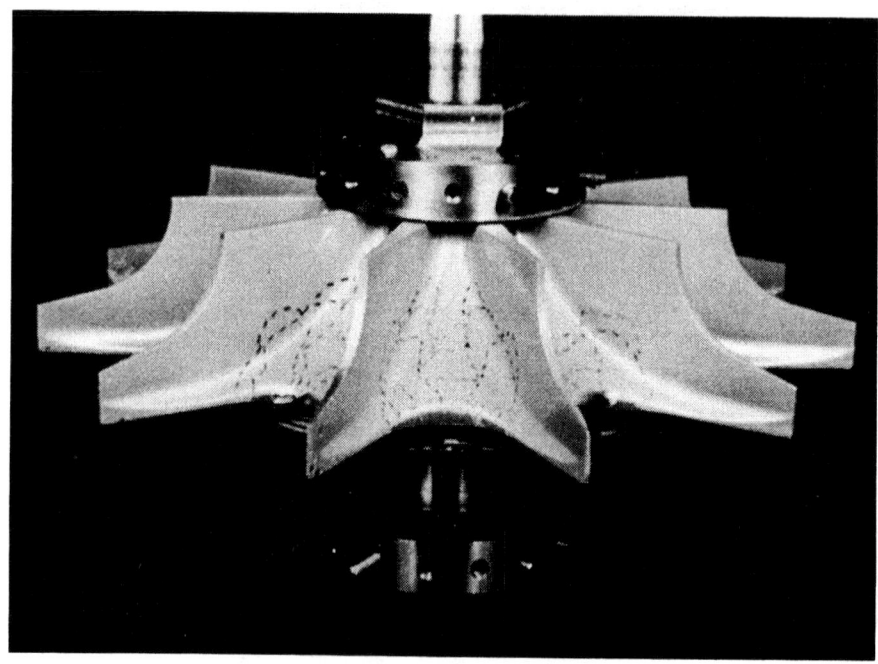

Figure 11-26 Patterns produced in brittle coating on a disk spun at high speeds.

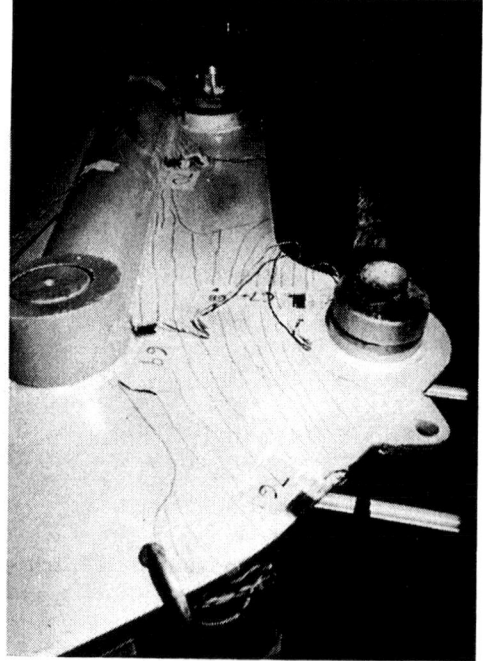

Figure 11-27 Boom-crane tip subjected to brittle-coating test.

Ceramic-based strain-indication brittle coatings are used in environments that are hostile to resin-based coatings. They are used in long-term tests and also in product quality control.

References

1. A. J. Durelli, E. A. Phillips, and C. H. Tsao, *Introduction to the Theoretical and Experimental Analysis of Stress and Strain*, McGraw-Hill, New York, 1958.
2. A. J. Durelli and T. N. deWolf, Law of Failure of Stresscoat, *Proc. Soc. Exp. Stress Anal.*, 6, no. 2 (1948), 68–83.
3. A. J. Durelli, R. H. Jacobson, and S. Okubo, Further Studies of Some Properties of Stresscoat, *Proc. Soc. Exp. Stress Anal.*, 13, no. 1 (1955), 35.
4. W. F. Stokey, Elastic and Creep Properties of Stresscoat, *Proc. Soc. Exp. Stress Anal.*, 10, no. 1, (1952), 179–186.
5. G. Ellis, Resinous Composition for Determining the Strain Concentration in Rigid Articles, U.S. patent 2,428,559, Oct. 7, 1947.

For further information on stresses and strains in brittle coatings see also the following papers.

6. A. J. Durelli, Determination of Stresses on Free Boundaries by Means of Isostatics *Semi-Annual East. Photoelasticity Conf.*, 15 (1942), 32–42.
7. A. J. Durelli, Experimental Determination of the Isostatic Lines, *J. Appl. Mech.*, 9 (1942), A155–A160.
8. A. J. Durelli, Stress Analysis Research by Means of Brittle Coatings *Armour Res. Found. Rep.*, 10 (1947), 7–9.
9. A. J. Durelli, What Kind of Information Does Brittle Coating Give? *Prod. Eng.*, 19, (June 1948), 86–91; (July 1948), 133–150.
10. A. J. Durelli and T. N. DeWolf, Law of Failure of Stresscoat, *Armour Res. Found. Rep.*, 6 (1949), 68–83.
11. A. J. Durelli, Heat-Treated Brittle Coating Increases Sensitivity, *Prod. Eng.*, 22 (1951), 144–147.
12. A. J. Durelli and S. Okubo, Influence of Strain Gradient on Brittle Coating Sensitivity, *Prod. Eng.*, 24 (1953), 136–137.
13. A. J. Durelli and C. H. Tsao, Use of Brittle Coating Data in Stress Analysis, *Soc. Exp. Stress Anal.*, 11 (1953), 181–196.
14. A. J. Durelli and S. Okubo, Crack Density Studies in Stresscoat, *Soc. Exp. Stress Anal.*, 11 (1953), 153–160.
15. A. J. Durelli, S. Okubo, and R. H. Jacobson, Study of Some Properties of Stresscoat, *Soc. Exp. Stress Anal.*, 12 (1955), 55–76.
16. A. J. Durelli, S. Okubo and R. H. Jacobson, Further Studies of Stresscoat Properties, *Soc. Exp. Stress Anal.*, 13 (1956), 35–58.
17. A. J. Durelli and J. W. Dally, Some Properties of Stresscoat Under Dynamic Loading, *Soc. Exp. Stress Anal.*, 15 (1957), 43–56.
18. A. J. Durelli, S. Okubo, and J. W. Dally, A Dynamic Strain Calibration Device, *Soc. Exp. Stress Anal.*, 15 (1958), 67–72.
19. A. J. Durelli, J. W. Dally, and V. J. Parks, Further Studies of Properties of Stresscoat Under Dynamic Loading, *Soc. Exp. Stress Anal.*, 15 (1958), 57–66.
20. A. J. Durelli and J. W. Dally, Variables Affecting Brittle Coating in Stress Analysis, *Prod. Eng.*, 30 (1959), 302–305.
21. A. J. Durelli, A. S. Kobayashi, and K. E. Hofer, Lueder's Lines Detection by Means of Brittle Coatings, *Exp. Mech. Soc. Exp. Stress Anal.*, 1 and 18 (1961), 91–101.
22. A. J. Durelli, V. J. Parks, and K. E. Hofer, Extending the Stress-Analysis Range of Brittle Coatings, *Exp. Mech. Soc. Exp. Stress Anal.*, 2 and 19 (1962), 137–141.

CHAPTER

12

Nondestructive Evaluation

James E. Doherty

Illinois Tool Works
Glenview, Illinois

12-0 Introduction

NDE (nondestructive evaluation), sometimes referred to as NDI (nondestructive inspection) or NDT (nondestruction testing), has in recent years become a widely used tool not only for detecting flaws, but also for characterizing and evaluating them as to type and size. This change in emphasis from detection to characterization came about because of an interest in estimating the remaining life of components. The change paralleled the development of fracture mechanics and other life management systems, which require quantitative information about the severity of flaws.

While many new and enhanced NDE methods were developed in response to the new quantitative focus, many were based on the traditional nondestructive methods of penetrant, magnetic inspection, eddy-current, ultrasonic, and x-ray NDE. The objectives of this chapter are to review the basic elements and principles of these most commonly used NDE methods, to indicate new developmental features that have occurred to extend the capability of these methods, and to indicate the direction of future developments. Several more exhaustive reviews of NDE methods are available [1-6].

12-1 Penetrant Inspection

Penetrant inspection is the most widely used NDE method [6,7]. It is fast, simple, and inexpensive. The primary advantage is that it can be used on many shapes and usually does not require a large amount of training to implement. The method is applicable only to the detection of surface-connected defects and, in particular, is most effective for detecting closed flaws such as cracks. Open blemishes, gashes, dents, and other similar surface anomalies are relegated to the conditional visual inspection. The method is applicable to components made from many materials, ranging from simple steels to the most exotic superalloys used in jet engine components.

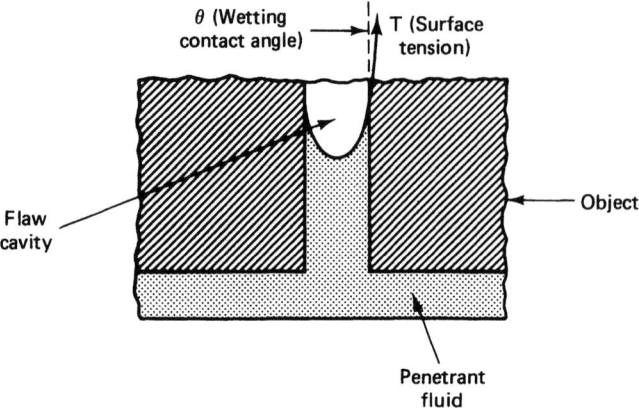

Figure 12-1 Schematic of the penetrant process.

Physical Principles

Liquid penetrant inspection depends mainly on the ability of a liquid to be absorbed preferentially into small cavities that are open to the surface. Typically, the cavities of interest are exceedingly small, at least in one dimension, and are not easily visible. The primary factors that determine the propensity for a liquid to enter a cavity depend on the relative ratios of surface tension and the ability of the liquid to wet the surface. Figure 12-1 illustrates simply that the ability of the penetrant to wet the wall of a cavity draws the fluid into the cavity. The surface tension T and the wetting contact angle are the physical parameters most important to the penetrant process. Good penetrant fluids have small surface contact angles and high surface tension. Other forces not shown include cavity pressures due to the cavity being closed off from the external environment and gravity, which is not very important. The ability of the penetrant to wet the surface is by far the most important property that determines the success of a particular penetrant system. Similarly, other factors, such as cleanliness (the presence of a material with poor wetting characteristics), can significantly reduce the effectiveness of a penetrant to indicate surface flaws. For this reason, the success of penetrant inspection often clearly depends on cleaning and removing foreign matter from the surface to be inspected.

The viscosity of the penetrant liquid is not a factor in the basic principles of capillary absorption. Viscosity is related only to the rate at which liquid can flow. This, however, does not mean that viscosity is not important to the penetrant process; the rate of flow of a liquid into very fine and small cavities can be very slow if the viscosity is high. Simply put, the Reynolds number of the flow in small cavities even at very low flow rates can be high.

The process of penetrant inspection begins with the application of the penetrant fluid. The next step is the removal of excess penetrant from the surface being inspected. Once this has been done, the same forces that drew the penetrant into the flaw cavities draw the penetrant back onto the freshly cleaned surface. The material that has bled onto the surface is seen as a penetrant indication. In many instances, this bleed-out process is enhanced by the use of a fine-particle powder placed on the surface. This powder, called the developer, aids in the removal of penetrant from the surface flaw or discontinuity. The developer literally draws penetrant from the flaw onto the surface, thereby increasing the ability for the penetrant to be observed. If contrasting colors are used, the developer can increase the contrast of the indication. The increased sensitivity caused

NONDESTRUCTIVE EVALUATION

by the application of the developer does reduce the resolution of two closely spaced indications. Also, if too much developer is applied (too thick), the penetrant in the flaw will be exhausted before it has significantly saturated the developer film, and contrast along with sensitivity will be reduced.

In general, the sensitivity of a particular penetrant system is determined by selecting the viscosity and the wetting characteristics of the penetrant fluid. High-viscosity fluids are more convenient for detecting larger size and grosser flaws. They are less susceptible to accidental removal during the process of cleaning excess penetrant after penetrant application. Also, such low-sensitivity penetrants are frequently used in situations where the surface is rough. Rough surfaces will act as penetrant sinks and thus offer the potential of obscuring flaws of interest if a high-sensitivity system is used.

Practical Penetrant Processes

Several practical penetrant processes have been devised for different applications. The selection of a specific approach depends on the size and the location of the object to be inspected, the types of defect to be found, and the surface condition of the component to be inspected. Other practical considerations are the number and rate at which components must be inspected. Many penetrant processes are designed to handle large numbers of components, and the penetrant inspection line is similar to many other manufacturing processes in the use of continuous movement and automatic parts-handling equipment.

There are essentially three basic types of liquid penetrant system: water washable, postemulsifiable, and solvent removable. Each system has its own particular characteristics that make it an optimum choice in certain applications. The water-washable system is quick, cheap, and easily automated. The postemulsifiable system is expensive and automatable, and is more controllable. The solvent-removable system is very easy for hand operation, and usually is highly sensitive. Figure 12-2 summarizes the process elements included in each of the basic penetrant processes.

Specifications and Standards

The classification of penetrant materials use by the federal government and in the manufacture of U.S. government material is done by the Air Force. Materials are classified by manufactured product name into four levels, 1 through 4, with level 4 the most sensitive. The Qualified Products List (QPL) is available from the Materials at Laboratory Wright-Patterson Air Force Base. The QPL classification of sensitivity has been generally adopted worldwide by all manufacturers and users of aerospace equipment. Some original equipment manufacturers have their own requirements above and beyond the Air Force requirements; owners and users of equipment requiring penetrant inspection should follow manufacturers' recommendations. Specific standards and specifications are listed in Table 12-1.

Environmental protection standards apply to the waste products of penetrant inspection, in particular to the wash water, which contains oil. These standards vary by state and local government; in all cases the user is fully responsible for proper disposal. Generally, spent wash water cannot be disposed of down the drain without pretreatment. Penetrant manufacturers now offer equipment for concentrating the objectional material so that the bulk of the waste water can be reused or put down the drain. The residue is disposed of as hazardous waste.

Novel Penetrant Systems

Several novel penetrant systems have been developed for particular applications. Although some are in continued use to meet special inspection needs, most are not generally used either because

Penetrant system	Step									
	Clean and dry	Apply penetrant and wait 0	Emulsify	Wash 0	Clean	Dry, then dry develop	Dry, then nonaqueous develop	Aqueous develop and dry	Develop 0	Inspect
Water-washable	X	X		X		X or	X or	X	X	X
Post-emulsifiable	X	X	X	X		X or	X or	X	X	X
Solvent-removable	X	X			X		X		X	X

Figure 12-2 Schematic of basic liquid penetrant processes. A choice of the development method is indicated for the water-washable and postemulsifiable systems. A status check using fluorescent lights is often used to assure that the step has been completed satisfactorily at locations indicated by 0.

of their expense or because they detect only certain types of flaws. Three of these are the krypton exposure technique (KET) [8], the differential absorption technique [9], and the use of radiography to enhance the visibility of penetrant fluids [10].

The KET uses radioactive krypton gas for a penetrant rather than a fluid. In this process a component is placed in a vessel, which is evacuated and back-filled with radiographic krypton gas. After a suitable dwell time, the gas is removed by evacuation and the components are brought back to ambient pressure. Indication detection proceeds by the application of special sprayed-on photographic emulsions. After a suitable exposure time, these emulsions are developed, usually in situ, in normal photographic processes. The presence of indications is then identified by the location of darkened areas in the emulsion. The advantage of the method is that it is sensitive to locating large internal cavities that are only barely connected to the surface. In this way, it complements liquid penetrant, which would produce only a very small indication. The advantage of KET in finding large subsurface defects, such as might be produced in the casing process, is obvious. Although the KET system had difficulty during early development stages because it produced spurious indications, it has been found that the inspection method works best where some residual oxide is present. It appears that krypton gas is preferentially absorbed inside cavities whose surfaces have slightly oxidized. For this reason, it is more effective on components such as nickel alloys, where high-temperature exposure can produce internal oxidation, in comparison to titanium alloys, where oxidation begins by absorption rather than by oxide film formation.

Table 12-1 Liquid Penetrant Inspection Standards and Specifications

ASTM Standard
    E165 Liquid penetrant inspection
Government specifications
    MIL-I6866 (FSC-6850)   Inspection penetrant method
    MIL-I9684 (FSC 6850)   Inspection penetrants, nondestructive testing
    MIL-I9445 (FSC 6635)   Inspection unit, fluorescent penetrant, general specifications for
AMS-SAE specifications
    AMS 2645   Fluorescent penetrant inspection
    AMS 2646   Contrast-dye penetrant inspection
    AMS 3155   Oil-fluorescent penetrant, solvent soluble
    AMS 3156   Oil-fluorescent penetrant, water soluble
    AMS 3158   Solution-fluorescent penetrant, water base

The differential absorption technique is used for identifying indications where some fluid is absorbed into the surface in a nonuniform manner. This method detects primarily distributed conditions rather than isolated flaws. The method can identify nonuniform surface situations of various types, and in many cases, the readout is simply dependent on variation in coloring of the surface. A novel application of this method has been developed to inspect for surface damage in some composite materials. In this application, a low-viscosity fluid carrying a heavy element is allowed to penetrate into the surface. The presence of large amounts of subsurface damage and porosity can then be detected using x-ray spectroscopy to identify those areas where large amounts of penetrant fluid is absorbed.

Heavy-element-containing liquids also have been used for the inspection of composite materials by means of radiographic methods for locating indications. Typical applications are the detection of subsurface delaminations in composite materials. In these cases, heavy-halogen-containing fluids are allowed to penetrate into the surface. Their presence is detected through radiography because they are a heavier absorber of radiographic energy than are the composite materials being inspected. This method can also detect internal surface-connected flaws in low-atomic-number metals such as aluminum. The penetrant is applied within the cavity, removed, and then detected in flaws using radiography. The particular disadvantage of this method is that the high-density penetrant fluids are usually dangerous to human beings.

12-2 Magnetic Inspection

In magnetic inspection, flaws are identified by detecting disturbed magnetic fields [11]. The method has evolved into several approaches; however, traditionally, magnetic particle has been the most widely used method for magnetic inspection. In magnetic particle inspection, the disturbed magnetic fields are detected using small magnetic particles, which outline a flaw. An extension of magnetic particle inspection is called flux leakage [12] or magnetic perturbation. The disturbed fields produced by flaws are detected using small magnetometers such as a coil or Hall device. This adaptation is more quantitative than the classic approach because it is implemented using instrumentation and has the capability of being automated. An extension of the method, applicable to nonferromagnetic materials, detects the magnetic fields produced when electric currents pass near flaws. This method, called electric-current perturbation [13], uses the same type of magnetometer as used by the flux leakage method.

Magnetic Particle Inspection Methods

A magnetic field flux is induced either by causing the direct passage of current or by using induced current (see Table 12-2). Figure 12-3 shows that flaw detection capability is dependent on the relative orientation of the flaw and the direction of current flow. This limitation occurs because the field induced by the current must be perpendicular to the flaw plane if flux leakage is to occur. Magnetic currents are applied using several approaches. These are the use of coils (single or multiple loop), yokes, central conductors, or direct contact. Each of these methods is used to particular advantage in several different applications; see Table 12-2.

Excitation Current

Both alternating and direct currents can be used in magnetic inspection. Direct-current excitation has an advantage over alternating-current excitation in that current is distributed across the cross section of the part, causing sensitivity to defects slightly below the surface. The skin effect limits

Table 12-2 Methods for Magnetizing Materials

	Applications	Features	Limitations
Yokes	Large surface area parts; localized inspection	No electrical contact; portable	Time consuming
Coils	Small to medium-size parts; usually long parts	No electric contact; magnetize all surfaces	Depends on L/D ratio
Central conductors	Short hollow parts; long tubular parts	No electric contact, all surfaces magnetized; ideal for establishing residual field	Large conductors might be required; off-center conductor used for large-diameter parts
Direct contact			
Head and clamp shots	Solid casting and forgings, long parts	Fast, whole surface magnetized; useful for complex shapes	Contact; arc-burn potential; high currents
Prods	Surface flaw detection; welds; large castings and forgings	Easy field alignment; portable	Small inspection area; arc-burn potential

alternating currents to a zone very close to the surface. An advantage of alternating current over direct current is that the equipment is cheaper. Typically, indications deeper than 50 mils are not detectable using alternating-current excitation.

Some applications require that parts be demagnetized after magnetic inspection. Methods for demagnetization usually employ alternating currents at successively lower intensity, as would be produced by moving an object through a coil. Demagnetization is a consideration (1) when subsequent operation may allow abrasive magnetic particles to adhere to the surface, or (2) when arc welding is to be performed, because strong residual magnetic fields can deflect arcs away from the point of application.

Magnetic Flux Leakage/Magnetic Perturbation

For the magnetic flux leakage method, an extension of the magnetic particle method, leakage fields are detected using magnetometers. An advantage of this method is that it can be more quantitative; as the output of the inspection represents a measurement of voltages, it does not depend on the qualitative assessments of visual readings common in magnetic particle inspection. In the magnetic flux leakage method, the magnetometer is usually scanned in a regular fashion over the surface on which flaws are to be detected. Magnetometers based on the Hall principle

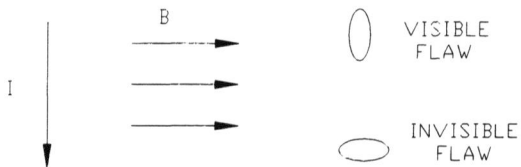

Figure 12-3 Relationship of detectable flaws to magnetic field orientation.

NONDESTRUCTIVE EVALUATION

have an advantage over magnetometers based on the coil principle, in that the former measure the absolute value of the magnetic field. Another advantage is that they do not have to move to detect the presence of field as does a coil. A disadvantage of Hall devices is that they are sometimes difficult to make in very small sizes.

An extension of the magnetic method to detect flaws in nonferromagnetic materials has been developed. This method detects flaws by the presence of magnetic fields produced by the perturbed electric currents near the flaw. The basic principles of inducing currents for this method are similar to those used for magnetic methods.

12-3 Eddy-Current Inspection

Eddy-current inspection is an old method being used with increasing frequency. This revitalized interest is due to the possibility of using this approach to inspect nonferromagnetic materials and to detect small flaws, often in complicated geometries. Traditionally, the method has been used to inspect tubular goods and rod stock for fabrication defects such as seams, laps, and cracks. The method is regularly used to sort dissimilar metals and detect differences in their composition and microstructure. Key uses along these lines are the measurement of hardness of various kinds of steel, coating thicknesses, and flaw detection [14–16]. In situations where several of the pertinent independent variables can be controlled, the method can suitably measure electrical conductivity, magnetic permeability, grain size, and even the wall thickness of thin stock. The method can be applied with different equipment based on slightly different principles. It is used in both manual and automatic inspections.

Physical Principles of Operation

The basic concept behind eddy-current inspection is easy to understand; however, because of different arrangements of coils, different types of equipment, and the use of complex number notations in formal method descriptions, eddy-current inspection is often confusing. The eddy-current method is based on the principle of electromagnetic induction; flaws (and other conditions) are detected because they change the currents induced by an induction coil [16]. Figure 12-4 shows that in some cases these induction coils can be encircling coils, as would be used to inspect tubes and rods, or nearby coils, as used for the inspection of planar surfaces.

The electrical description of the induction coil and the currents in the material is one of two coils with a mutual inductance coupling. Any change in this system can be detected as a change in the electrical parameters of the induction coil. The parameter that most completely describes these changes is the electrical impedance. Eddy-current instruments detect a change in material condition or flaws by indicating a change in impedance or a change in an element of the impedance [17–19].

During an inspection, the locus of the impedance vector of an ideal eddy-current system as it moves on the impedance plane is important. Figure 12-5 shows the response of an eddy-current system in an impedance plane when the conductivity of the test material is varied from 100% of that of standard copper to 0% (air). For an almost pure conductor, the impedance is almost an entirely inductive resistive load, whereas in air, where the material has 0 conductivity, the impedance represents an inductive reactance load [15]. Other materials with varying conductivities exist as intermediate cases.

Similar diagrams can be drawn for other changes in material conditions. For example; at a given conductivity, a locus of curves can be shown for the motion of the impedance on the impedance plane as the inductive coupling is reduced. An example (Fig. 12-6) occurs when a

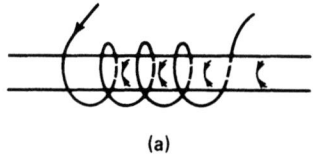

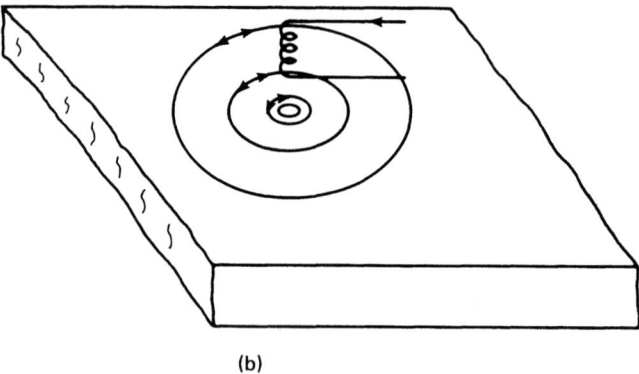

Figure 12-4 Schematic of induced eddy currents using: (a) encircling coils and (b) nearby coils.

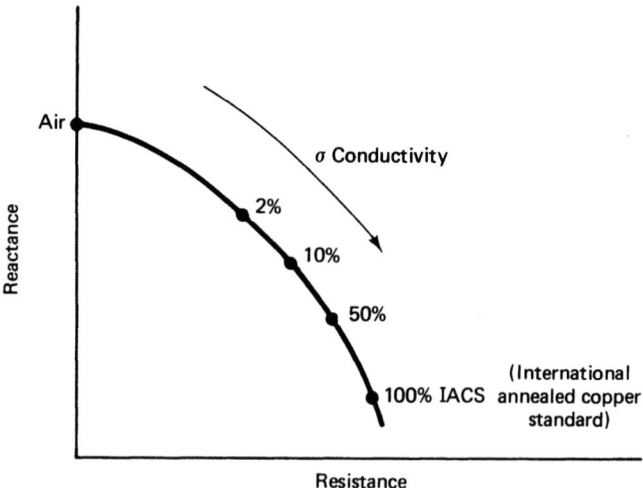

Figure 12-5 Schematic showing locus of impedance when the object conductivity changes.

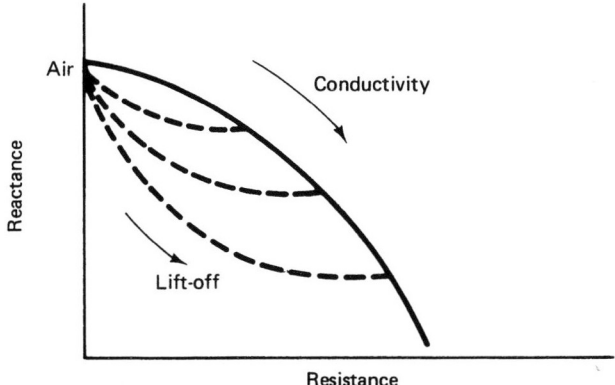

Figure 12-6 Schematic showing effect of lift-off on coil object impedance.

nearby, or pancake, coil is lifted off from the surface until it has the response of pure air. These curves, called lift-off curves (in some other applications called fill-factor curves), represent changes in the performance of the eddy-current system as the degree of coupling or mutual inductance is varied.

The preceding examples of how changes in an inductance coil system cause changes in the impedance provide only a general understanding of the basic principles of eddy current because real flaws such as cracks will cause impedance changes with both reactive and resistive components. Therefore, the locus of motion on the impedance plane can be complex. Because lift-off can never be completely separated from effects due to changes in other conditions, the spacing between the eddy-current coil and the inspection part must be controlled during any inspection.

Geometrical effects are not the only important ones present in an eddy-current response. For instance, induced eddy currents are not uniformly distributed throughout the thickness of a component, but are distributed according to the physical laws that describe the well-known skin effect. This means that below the surface the amplitude of induced currents exponentially decays with depth. The decay is scaled by the skin depth, which is directly proportional to the square root of the ratio of resistivity of the material and the product of the magnetic permeability and the applied frequency. The dependence of the skin depth on the frequency is both a limitation and a benefit because it allows for selecting the optimum skin depth for detection of flaws of interest. Figure 12-7 graphically shows that 50% of the current density in eddy currents is within one skin depth of the surface. The phase also varies similarly in that one skin depth represents a π phase change between the induced current and the drive current. This characteristic allows thickness of sheet stock to be accurately measured as a phase change if a detector coil can be placed on an opposite side of the sheet from the eddy generator.

Rules of thumb based on empirical studies for detecting the best frequency for flaw detection have been developed; the maximum sensitivity to surface-connected, cracklike defects occurs when the crack depth is nearly equal to one skin depth. The current distribution due to the skin-depth phenomenon is also important in determinating the effective sensitivity of the method to subsurface flaws. Again, empirical studies have shown that (globular) subsurface equivalency-type defects can be detected if their depth-to-diameter ratio is less than 3 [14].

Skin depth is dependent on frequency, the parameter that describes the inherent sensitivity of a particular eddy-current inspection. Figure 12-8 shows how skin depth varies with frequency for

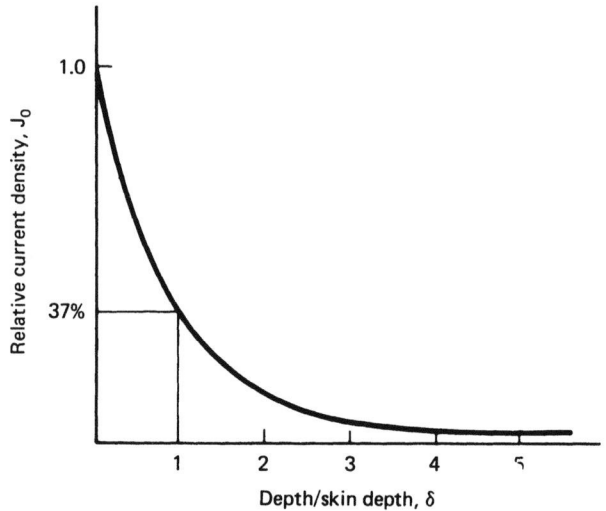

Figure 12-7 Normalized variation in eddy-current density with depth.

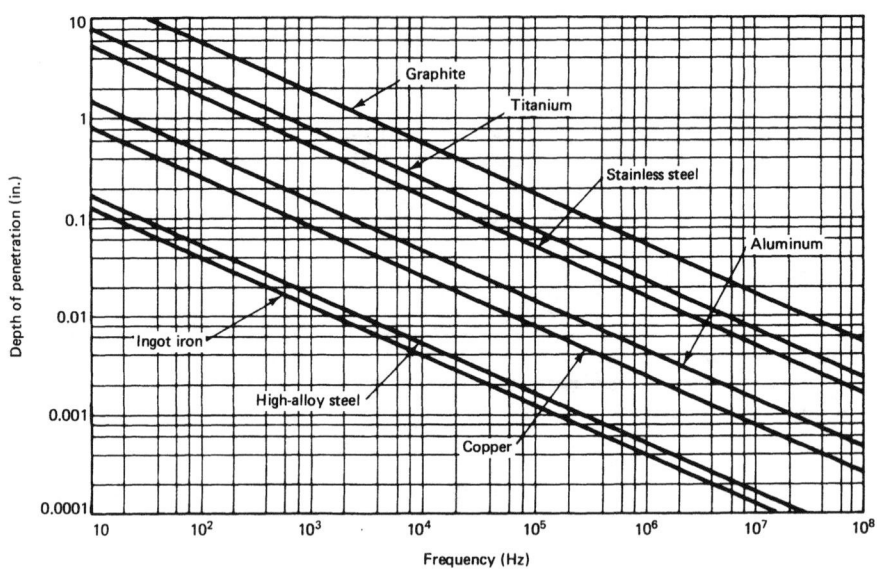

Figure 12-8 Variation of skin depth with frequency for several common materials.

NONDESTRUCTIVE EVALUATION

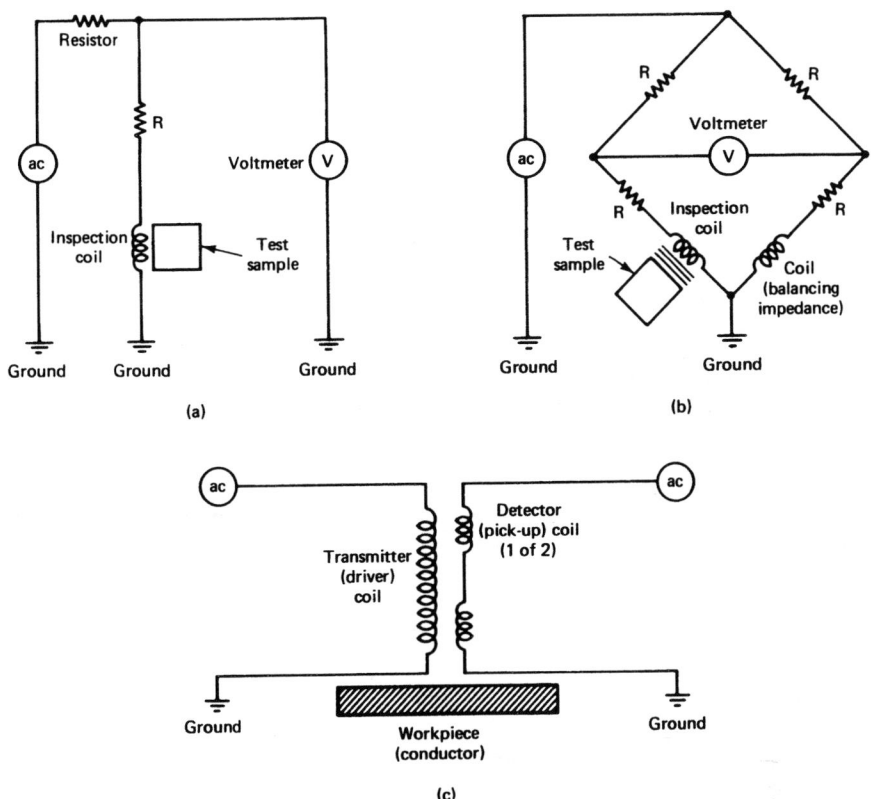

Figure 12-9 Schematic eddy-current instruments: (a) simple amplitude instrument; (b) impedance bridge system; (c) induction bridge system.

seven common materials. As shown, for a given frequency of operation, the sensitivity of the method can vary over several orders of magnitude because material properties affect skin depth.

Eddy-Current Instruments

Eddy-current instruments can be classified into three general types [15]. One type measures amplitude of the impedance and simply looks for changes in the magnitude of the impedance. The second, a bridge-type instrument, can be made to display the impedance on the impedance plane, allowing independent measurements of phase and amplitude components of the impedance. The third type is the induction bridge instrument, which uses the transformer coupling between the part and two receiver coils to detect a change in the phase and amplitude of the voltage. Since the induction bridge is typically built into the probe itself, instrumentation is very simple (see schematic Fig. 12-9c).

Figure 12-9 shows simple amplitude instruments. Many very successful instruments have been built on these principles. The instruments tend to be very small, lightweight, and battery operated;

and if properly designed, they can be very sensitive. In addition to the simple circuitry shown, these instruments often include special lift-off compensating circuitry. The operating frequency is not constant because the resonating frequency of the tank circuit is dependent on the combined inductance of the inspection on the test sample. Bucking voltages and other types of device are often used to enable the voltage meter to operate over six ranges at high sensitivities. This type of instrument is often very popular for manual, portable operations, and it has been used for determining parts, material sorting, and material composition as well as for detecting flaws in aerospace components. The range of operating frequency can be chosen by selecting proper electrical parameter values for the eddy-current probe according to the case in question.

Figure 12-9b shows schematically an impedance bridge type of eddy-current instrument. Among the several varieties of the instruments, the differences are dependent on how balance is achieved. The options available are the use of a balancing coil similar to the drive coil with and without a reference sample. Some of these instruments do not operate at perfect balance and use bucking voltages in the voltmeter circuit for the voltmeter to operate at high sensitivity. In general, impedance plane instruments operate over a wider frequency range than do the amplitude-type devices. Many high-frequency instruments, however, do have a tendency to roll off in sensitivity at the high frequencies, thus showing uniform sensitivity to small surface flaws with increasing frequency. Some of the newer high-performance instruments pay more attention to providing uniform performance over the entire band of operation.

Impedance plane instruments include a phase-shifting circuitry to allow for arranging the display such that lift-off or other geometrical factors produce horizontal deflections. This arrangement allows a decoupling of lift-off information from flaw information to some degree. Figure 12-6 showed that this lift-off deflection direction is not necessarily perpendicular to a conductivity deflection direction or, for that matter, any deflection produced by a flaw. Therefore, when this method is used to reduce sensitivity to lift-off, the sensitivity to the flaws of interest can be reduced. Figure 12-6 also showed that at a given operating frequency, sensitivity to surface flaws can vary by as much as two orders of magnitude, depending on the conductivity of the material.

Some misconception exists as to the bandwidth of the output of eddy-current instruments. Although some instruments can often operate at frequencies as high as 10 MHz, the information contained in the output of the instruments rarely exceeds a few kilohertz. The maximum output response frequency is often considerably lower. In the case of manual operation, this is sometimes an advantage, because the instrument has an inherent high-frequency cutoff, filtering out the effect of rapid lift-off variations produced by hand scanning over surfaces. On the other hand, limited bandwidth can be bothersome if high sensitivity is to be achieved in high-scanning-speed automated systems.

Inspection Coils

Inspection coils take a variety of shapes. As already indicated, there are two basic kinds: those designed primarily to operate on one side of the test object and those designed to encircle the test object. Differential coil systems are often used to desensitize the inspection system to gross variations in products. Although these systems offer increased sensitivity to small, isolated flaws, they can produce confusing results for long or distributed flaws. In this case, they will essentially detect only the edges or the fringes of a large flaw, being differential in nature.

Normally, in the inspection of flat surfaces for cracks or other discontinuities, simple axial or pin-type coils are used, as shown in Fig. 12-10. Methods to increase the sensitivity of coils by using ferrite and other magnetic field flux-conducting materials have been developed. Horseshoe-shaped ferrites, used to concentrate the flux, are commonly used in many high-sensitivity applications [17]. Where high sensitivity is desired, the aim is to concentrate the flux into as small a

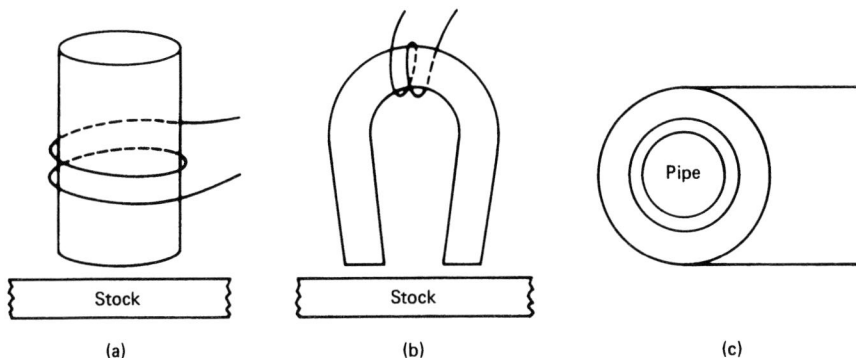

Figure 12-10 Eddy-current coil configurations showing shielded and horseshoe-type probe coils; (a) pin coil with ferrite core; (b) horseshoe ferrite core; (c) encircling coil.

volume as possible. As intuition indicates, the relative sensitivity of the system will be dependent on the relative volume or cross section of the flaw to that of the kernel of the flux density. Generally, schemes that are designed to concentrate the flux into small volumes limit the penetration of the flux density even when the skin depth is relatively large. Empirical rules of thumb usually scale both the eddy-current penetration and the flaw sensitivity to the cross-sectional diameter of the kernel flux density. In some applications, shields and other high- and low-reluctance materials are used to concentrate the flux density. Figure 12-10 also shows the use of ferrite cores to force the flux density in a special probe into a small area.

Advanced Methods

Several advanced eddy-current methods are currently under development: YIG sphere or microwave method [20], pulsed eddy current [21], and remote-field eddy current [22].

The YIG sphere method generates very high (10^{10} Hz) eddy currents by using the resonant properties of a small YIG sphere. The method has high sensitivity to small surface flaws because of the negligible skin depth at the operation frequency. Crack sizing is possible in slightly open cracks because surface currents propagate over the crack faces.

The pulsed eddy-current method was developed long ago to detect deep flaws. Recent work has extended the method to the detection and sizing of small (<1 mm deep) surface flaws. In concept, the method appears similar to pulse-echo ultrasonics; flaws are detected by the presence of signals reflected by flaws. Flaw depth is measured as a shift in time in the response from deeper flaws.

The remote-field eddy-current method is dependent on electromagnetic propagation phenomena not anticipated by classic models. Particularly suited to the inspection of tubular pipes (both ferromagnetic and nonferromagnetic materials), the method can be used to detect changes in wall thickness, axial flaws, and circumferential flaws. The method is particularly attractive because it is insensitive to lift-off.

12-4 Radiographic Inspection

Radiographic inspection [3,23] is one of the oldest NDE inspection methods. Its primary advantage is that it can detect the existence of flaws in the volume as well as some surface flaws. However,

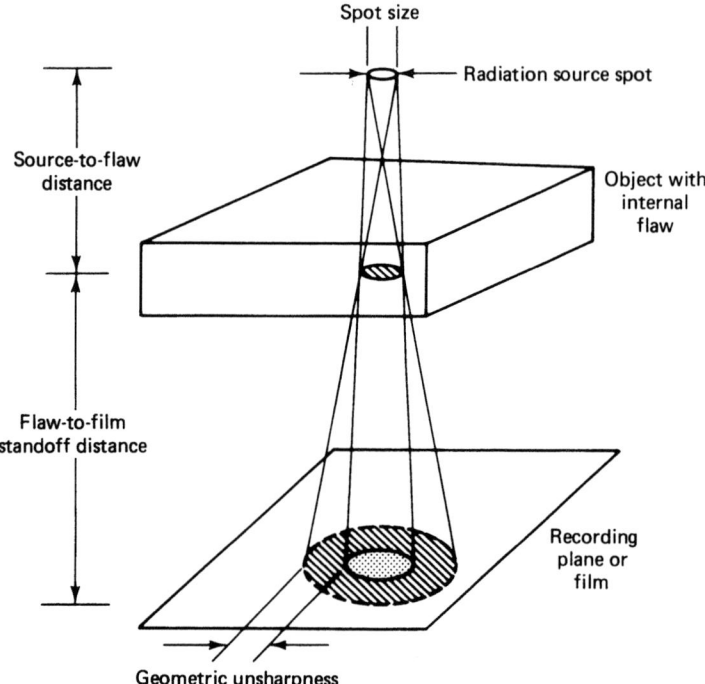

Figure 12-11 Schematic of typical x-ray setup. Geometric unsharpness exists because of finite source spot size.

it is much less effective in detecting planar-type flaws unless careful attention is paid to the orientation of the x-ray beam relative to the flaw orientation. The method also can be expensive to implement. The costs are normally associated with the necessary capital equipment and purchasing, processing, and storing film. Larger users of the x-ray process reduce costs by reclaiming used silver. A traditional advantage of the radiographic method is that it produces a permanent record for reference at a later time. This advantage has become less unique as other NDE methods become implemented in computer-aided formats capable of storing inspection data.

The recurrent costs associated with film in recent times have been addressed by the availability of real-time x-ray systems. These systems have high initial costs, but operational costs are lower when a large number of similar test objects are to be examined. These systems meet the requirement for permanent records because images can be stored on videotape, although radiographic sensitivity is less than or equal to film.

Figure 12-11, a schematic of the arrangement in a typical x-ray environment, shows that the source has a finite size and that a geometric unsharpness is produced on the film because of the size and the spacing of the film and object. X-ray sources have finite size because of the necessity of cooling the target, since most of the energy generated by the electron beam is thermal. The amount of geometric unsharpness can be calculated simply from geometric relationships. Unsharpness is equal to the source size times the ratio of the standoff object film distance to the source distance. It is clear that the geometric unsharpness can be reduced to a small value by

Table 12-3 Radiographic Exposure Equivalent Factors[a]

Material	50 kV	100 kV	150 kV	220 kV
Magnesium	0.6	0.6	0.5	0.08
Aluminum	1.0	1.0	0.12	0.18
Titanium	—	—	0.45	0.35
Steel	—	12	1.0	1.0
Copper	—	18	1.6	1.4
Lead	—	—	14	12

[a] The reference material is aluminum for 50 and 100 kV and is steel for 150 and 220 kV. The exposure time for a thickness of a given material is the exposure of the equivalent thickness of the reference material. The equivalent thickness is the product of the thickness of the material of interest and the equivalence factor.

increasing the source–object distance. However, an additional concern factor is that x-ray sources allow an inverse-square radiation law; increasing the source–object distance by 2 will cause an increase in the exposure by a factor of 4.

Although several types of x-ray source are used, the most common one is the x-ray tube. This tube generates x-rays by bombarding a target, usually tungsten, with energized electrons. The excitation energy of the electrons is a measure of the highest x-ray energy that can be produced. X-ray tubes produce a broad spectrum of energy, with the maximum intensity usually somewhat below the maximum energy. Higher x-ray energies are required for the radiography of denser materials and thicker objects. Table 12-3 shows the equivalent exposure needed to radiograph common materials. Such rules of equivalence can also be developed for any specific material of interest.

The filament current in an x-ray generator is a measure of x-ray output. Higher x-ray output can be used to reduce exposure time for a given acceleration voltage. Special x-ray systems using very high filament current densities for very short times have been developed to radiograph objects in rapid motion. These special units can successfully expose many times without melting their x-ray targets. Extremely high energy sources have been developed for x-raying thick sections of material quickly. These systems are commonly based on the use of an electron linear accelerator to produce the high-energy electrons to bombard the target and produce high-energy x-rays. Such high-energy sources are, for the most part, not transportable, although a portable system called the miniature linear accelerator was developed under the sponsorship of the Electric Power Research Institute.

The exposure needed to establish proper densities on x-ray film can be calculated for different x-ray setups using the following principles. The exposure time is linearly related to the milliampere excitation of the filament or the time. Exposure increases as the square of the source film distance. Tables of equivalency of these factors can be found in the open literature for different materials. Exposure charts can be determined easily from standard material thickness for given film speeds.

Gamma-ray sources are also used in radiography. The advantages of gamma-ray sources are their simplicity, compactness, and easy transportability. They are regularly used to radiograph thick sections and components not easily transported to a primary source. In contrast to x-ray tubes, the energy of gamma-ray sources is quantized at fixed energy levels associated with the disintegration of radioactive atoms. The intensity of the source depends on the amount of radioactive material in the source. Because of the decay, the amount of radiation is continually reduced, with half-lives shown in Table 12-4.

Thermal neutrons are also used in radiography. Neutrons and x-rays obey different absorption rules; and, in general, neutron absorption does not depend on density; unlike x-ray absorption, moreover, it does not generally increase with atomic number. Neutron radiography is commonly

Table 12-4 Half-Life of Radioactive Sources

Source	Half-Life	Gamma-Ray Energy
Thulium-170	127 days	0.084 and 0.54
Indium-192	70 days	0.137–0.651
Cesium-137	33 years	0.66
Cobalt-60	5.3 years	1.17 and 1.33

used to detect changes or the presence of low-atomic-number materials. This type of evaluation can be done in the presence of higher-atomic-number materials such as steel because of the low absorption capacity of steel. The detection of corrosion products, primarily the detection of hydrogen and oxygen, is one important use for neutron radiography.

There are three kinds of neutron radiographic source: primary reactor sources, which, by their nature, are large, expensive, and immobile; somewhat less expensive californium-intensified sources, which are smaller and have low output and less well collimated beams; and finally, cogeneration systems based on the excitation of neutrons by high-energy electrons in certain materials. The latter process is consistent with more modern approaches where portable sources have been developed.

Radiographs are typically recorded on radiographic film, which, unlike photographic film, generally has emulsion on two sides to increase the propensity for image formation. The double emulsion increases the number of image-forming photons that will be absorbed in the film. The amount of darkening, called the density, depends on exposure. The density is defined as the log of the transmittance, which is equal to the log of the ratio of the incident light intensity to the transmitted light intensity, when the film is placed on a viewing box.

Different film has different exposure characteristics. Figure 12-12 shows the characteristic curves of three different films. Some films have higher speed; that is, less exposure is required to darken the film. As with photographic film, high-speed radiographic film sacrifices some image definition to obtain the speed. Some films have higher contrast than others. Typically, the contrast is determined by the rate at which the density increases with exposure. Film Z has a lower film contrast than film X, particularly in the region where the film characteristic is essentially linear. Figure 12-12 shows that the contrast can be a nonlinear function of thickness. Most good radiographs are taken in the linear range of thickness under evaluation because in this range the highest film contrast can be obtained. This is illustrated schematically in Fig. 12-12, which shows that a small flaw can be most readily seen in the place where the contrast is changing most rapidly. As is clear from Fig. 12-12, high-contrast film, can evaluate only small ranges in thickness at a given set of exposure.

Radiographic Screens

When x-ray beams strike a film, usually less than 1% of the energy is absorbed. Radiographic screens are used as a means to increase the efficiency of exposing the film and to decrease exposure time. Two types of screen are used—lead and fluorescent.

For radiography in the range of 150 to 400 kV, lead-foil screens are used in contact with the film. The lead foil increases the photographic activity of the film because it generates electrons when x-rays are absorbed by the screen, and these electrons act to enhance or activate the silver compound in the film. Contrast is also improved because the lead screen tends to absorb lower-energy scattered radiation more than it does the primary beam. Intensification produced by radio-

NONDESTRUCTIVE EVALUATION 543

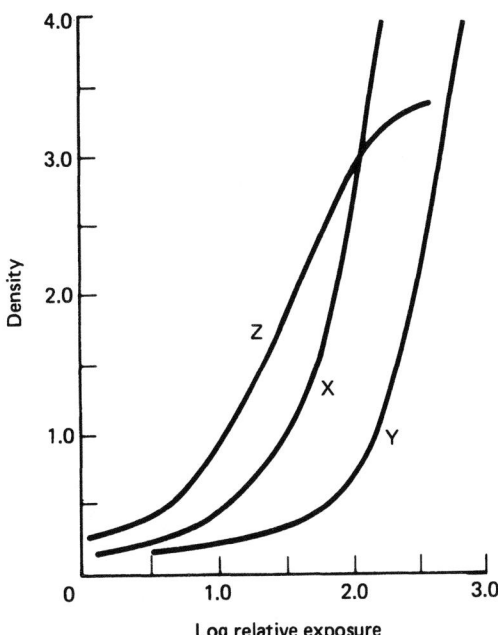

Figure 12-12 Relative exposure characteristics for three types of film.

graphic lead screens can occur only if sufficient energy is in the x-ray beam. Intensification screens can also be made from lead oxide or lead oxide–coated paper. In these cases, the screens are usually factory sealed with the film in light-tight envelopes. An advantage of this type of package is that it is particularly easy to keep a film clean. A disadvantage is that the lower lead density makes for less effectiveness in removing scattered radiation, as discussed below.

Fluorescent screens, which work by emitting light radiation when struck by x-rays, are also used to intensify exposure. Significant reduction in exposure times can be achieved when these screens are used. This advantage is offset, however, by the disadvantage that these screens typically give poor definition in the radiograph compared to exposures made with lead screens. This poor definition is associated with the spreading of light emitted from the screen. Fluorescent screens also produce what is called screen mottling or a texture in the film. Screen mottling is associated with statistical variations in the number of photons absorbed in one area of the screen versus another.

Scattered radiation can be bothersome and can reduce subject contrast. Because of this, some attention must be given to obtain high-quality radiographs. Scattered radiation can originate from the environment and is produced when the primary beam interacts with walls, floors, and other structures. Significant reduction in scattered radiation can be accomplished using lead shields. Lead shields are mounted in the back of the film cassette to control scattered radiation coming from the floor or other supporting structures supporting the film. Masks and diaphragms are also used as a means to prevent contact by radiation coming from other directions.

The standard means to control quality of radiographic exposure is to use penetrameters. A penetrameter, a standard test piece included in every radiograph, is made from the same material as the specimen being radiographed. It has a simple geometric form with a hole or cylinder (wire) whose dimensions are in some simple relationship to the thickness of the part being tested. Penetrameters with holes are commonly used in the United States and are described by ASTM standards. The ASTM plaque-type penetrameter contains three holes of diameters t, $2t$, and $4t$, where t is the thickness of the penetrameter. Radiographic sensitivities can be specified by the penetrameter thickness in terms of a percentage of the specimen thickness and the diameter of the hole. For example, 2-$2t$ is the resolution of a hole whose diameter is twice the thickness of the penetrameter and the penetrameter is 2% of the specimen thickness. The 2-$2t$ quality level is the most commonly specified for routine radiography; however, in some critical components, a higher level of 1-$2t$ or even 1-$1t$ may be required. The ASTM wire penetrameters are similar to the European-type penetrameters.

12-5 Ultrasonic Inspection

Ultrasonic inspection is one of the more widely used nondestructive testing methods. Originally, it was established to provide volumetric inspection to detect flaws on the inside of forgings and castings. It has since become an important method for inspecting welds and has been extended to detect surface flaws such as cracks [24,25,26]. The equipment for ultrasonic testing evolved from that used in the ultrasonic delay lines of early radar systems.

The primary advantages of ultrasonic testing are the ability to penetrate solid objects and to detect flaws deep within the part. In many cases, ultrasonic inspection has greater sensitivity than radiographic inspection to volumetric flaws. Additional advantages over radiography are that ultrasonics can be used to determine the depth, or distance, to the flaw and to detect cracklike flaws. Another important advantage is the ability to conduct a nonhazardous inspection. Ultrasonic inspection can be applied almost immediately after repair or fabrication welding. Although ultrasonic inspection is not widely used during fabrication in the United States, it is used in Europe both during fabrication and for evaluation of in-service weldments.

The disadvantages of the ultrasonic method are related mostly to limited sensitivity produced by material and physical properties such as grain size and surface roughness. Because ultrasonic inspection is performed using beams, some component geometries can limit accessibility to flaws in some locations.

This section reviews the physical principles of ultrasonic inspection and the practical processes used to implement it. Included are a discussion of some important codes and standards used in controlling the ultrasonic method, and finally, some advanced application methods.

Basic Principles

The ultrasonic method depends on the interaction of ultrasonic waves with flaws. These flaws can have different configurations, but in all cases they must constitute impedance discontinuities to be able to interact with the waves. An impedance discontinuity is a change in either local density or elastic properties. The processes of interaction include those most often encountered in wave-propagation situations: refraction, scattering, defraction, reflection, and absorption, as discussed later in this subsection.

Ultrasound propagates as bulk, surface, and plate waves. Bulk waves propagate in a three-dimensional medium, which is essentially infinite in extent. Surface waves propagate along the surface between high- and low-acoustic-impedance materials and at a velocity close to the longitudinal velocity in the bulk material. Plate or Lamb waves are often encountered in ultrasonic

Table 12-5 Acoustic Impedance of Several Materials

Material	Impedance
Air	0.00004
Aluminum	1.75
Brass	3.3
Copper-110	4.2
Ethylene glycol	0.18
Glass	1.2–1.5
Glycerine	0.24
Iron, cast	2.5–4.0
Mercury	1.95
Nylon	0.18–0.28
Oil	0.15
Plexiglas	0.32
Steel	
Carbon	4.7
Stainless	4.1–4.7
Teflon	0.3
Titanium	2.75
Water	0.149

inspections. The many complex plate-wave mode shapes are similar to the modes of vibration of a drum or a large sheet.

Although recognizing the subtleties of the different wave phenomena is not always important, in many practical inspection cases involving complex geometry, the process of mode conversion (the change from one propagating mode to another) must be understood if results are to be interpreted properly.

Reflection and transmission of waves at interfaces is the most commonly encountered ultrasonic process. An interface can be the boundary of a flaw or of a component being inspected. The governing physical property in reflection and transmission is the acoustic impedance, which is the product of the density, velocity, and area of the impedance discontinuity. The area is not considered because impedance calculations usually involve the transmission or characterization of waves through interfaces. In these cases, the area of the impedance discontinuity is larger than the ultrasonic beam and therefore is typically taken as 1. The commonly tabulated values of acoustic impedance are simply the product of the velocity times the density. Table 12-5 lists the acoustic impedance of several typical materials of interest.

The relative values of the acoustic impedance for the materials on either side of an interface determine the amount of sound transmitted through the interface. The amount of transmitted energy in a normally incident beam is equal to the square of the ratio of the sum and difference of the impedances of the two mediums. Energy is conserved at the interface, so energy that is not transmitted through the interface is reflected from it. This principle has practical significance because many ultrasonic inspections use a liquid medium to transmit the ultrasound from the transducer to the object to be inspected. Because liquids have a relatively low acoustic impedance compared to metals, for many materials inspected under water, only 20% of the incident energy transmits through the interface. Thus the insertion loss associated with ultrasonic inspection is typically high.

Refraction is another acoustic-wave process associated with interfaces. It occurs because of differences between the velocity of sound in two materials. The propagation direction of ultrasonic beams changes when a beam is incident on an interface at an angle. These interactions are described by Snell's law, which applies even if mode conversion takes place. It states that the ratio of the sine of the angles of incidence and deflection is equal to the ratio between the velocities of two

Table 12-6 Critical Angles to Establish 45° Shear Wave (Measured from Normal)

	Immersion Testing (in Water)	Contact Testing (Acrylic Wedge)
Aluminum	13.5	25
Brass	23	44
Iron, cast	15–25	28–50
Steel		
Low carbon	14.5	26.5
Stainless (302)	15	28
Titanium	14	26

media. The applicability of Snell's law is that it can be used to predict the beam incident angle necessary to obtain any desired angle of propagation in the object of interest.

Although theory exists to describe the relative amplitudes for mode conversion as the wave travels through the interface, these theories are not usually evaluated. Interesting cases occur at certain critical incident angles. At these angles, the propagation direction of the refracted longitudinal wave will approach 90°, leaving only a mode-converted shear wave to propagate into the second medium. Many inspections are done at this critical angle or higher in order to reduce the number of propagating beams in the second medium. This eases interpretation of reflected return information in complex structures. Table 12-6 gives the angle needed to establish the 45° shear wave in common materials.

A third important physical process involving ultrasonic waves is attenuation, which, as in all waves, involves absorption and scattering. Absorption occurs because of interactions at the atomic level between the propagating wave and irregularities of the crystal structure. Ultrasonic inspection of materials with high absorption can be troublesome. The measurement of absorption is of interest only when it is being used as a means to assess special atomic-level processes. Because many of these absorption processes are thermally activated, sound-wave absorption is temperature dependent and usually increases with frequency.

The most important component of attenuation involves scattering. Scattering occurs because the materials are not homogeneous. Discontinuities, such as grain boundaries, twin boundaries, inclusions, porosity, and multiphase structures, constitute interfaces that can disperse ultrasonic energy. Slight differences in acoustic impedance among discontinuities smaller than one-third to one-fifth the wavelength usually appear to be homogeneous to the propagating wave and do not cause scattering. Return signals from scattering, called coherent noise, increase in intensity as the discontinuity size increases and exceeds the wavelength of interest. The scattering affects the application of the ultrasonic method in many practical cases, particularly in the testing of components with large grain sizes. One such case is cast stainless steel, where the large-scale variable microstructure produces all manner of random reflections that raise havoc with the propagating beam. On the other hand, grain size may not necessarily be the most important factor. Many cast materials that have a large grain size undergo phase transformation at lower temperatures, thereby reducing the scale of the microstructure. Thus these materials can be inspected because of the fine scale of the resultant transformed structure.

Defraction is a wave phenomenon involving the interaction of edges with beams. In ultrasonic inspection, the most important or common occurrence of defraction is associated with the ultrasonic transducer. The transducer, acting as a port or piston, emanates a pressure pulse. Because this piston is finite in size, the defraction effects from its edges can interfere with the main beam and produce nonlinear changes in the amplitude immediately in front of the transducer piston. Figure

NONDESTRUCTIVE EVALUATION

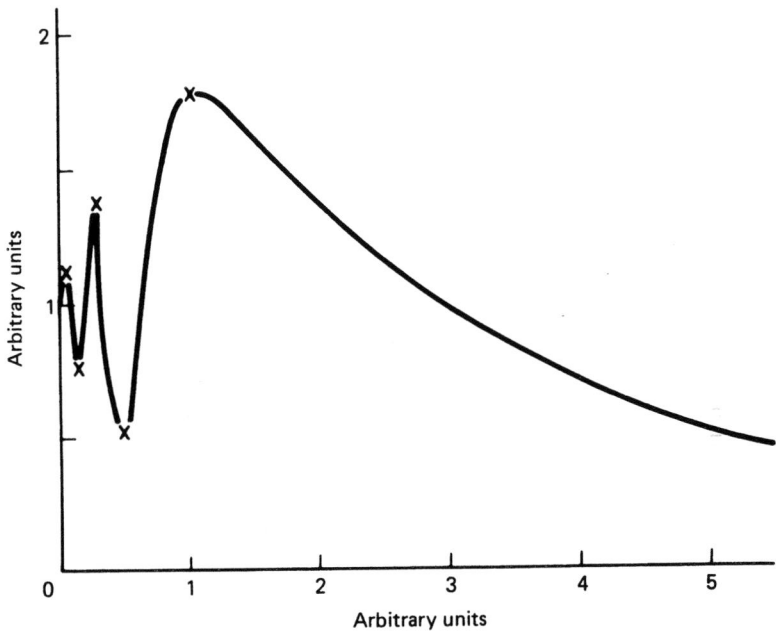

Figure 12-13 Variation of sound amplitude in the near field. The horizontal scale depends on wavelength and crystal diameter.

12-13 shows the variations of sound amplitude in the near field close to the transducer front. Because of the widely varying amplitude in this near field, a standoff distance is usually used to assure that the area to be inspected is in the far field.

Because defraction effects are produced by interference, they can be described by simple geometric relationships between the wavelength and the diameter of the transducer. The beginning of the far-field zone can be calculated because the wavelength is small in comparison to the transducer diameter. The beginning of the far field for circular pistons is the area of the aperture divided by two times the wavelength. In the far-field zone, the transducer appears to be a point radiator and the amplitude begins to decrease as the reciprocal of the square of the distance. Defraction also happens when flaws interact with the beam. Equivalent near-field and far-field effects can be observed in the beam reflected by the flaw.

This discussion has addressed only defraction from main beams that are perpendicular to the aperture. In addition, other beams, called side lobes, are produced by the finite size of the aperture. Although usually ignored, in some circumstances side lobes can emanate energy and produce various ultrasonic reflections that might be interpreted as flaw reflections.

Practical Inspection Methods

Practical ultrasonic testing methods are concerned with the way ultrasonic beams propagate through test materials and the methods by which the ultrasonic information is displayed. The two primary methods for controlling ultrasonic beams propagating through media are pulse echo and through transmission. The pulse-echo method is a single-sided approach whereby the transmitted and

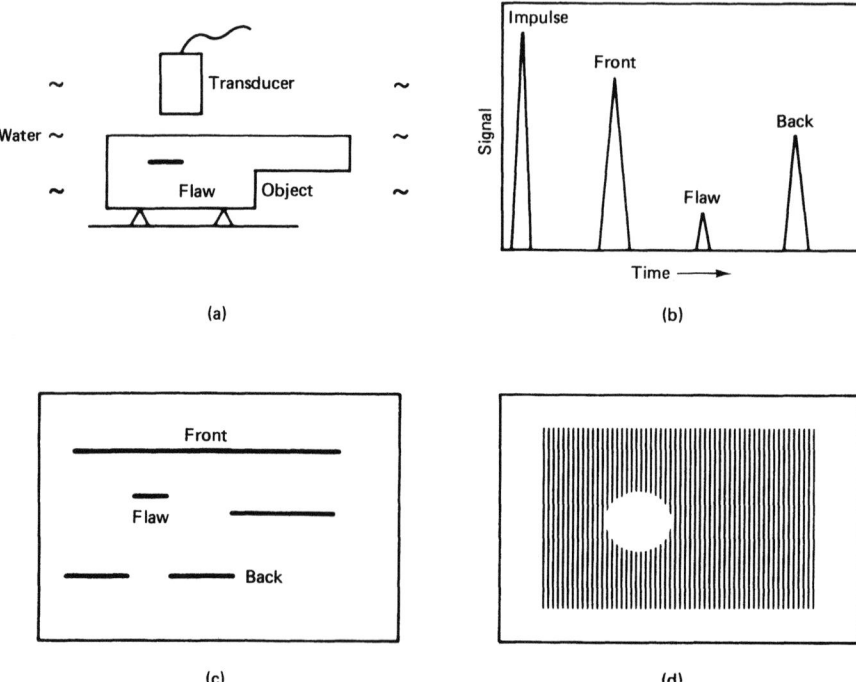

Figure 12-14 Common ultrasonic data displays: (a) object to be inspected; (b) A-scan (one-dimensional) fixed transducer; (c) B-scan (one-dimensional) cross section; (d) C-scan (two-dimensional) plan view.

received signals occur on the same side and are produced and received by the same transducer. This method is reminiscent of radar, or sonar, where items of interest or interfaces are detected by the presence of echoes at a given range. In a through-transmission mode, the object to be inspected is located between the transmitter and the receiver. Objects of interest usually appear as shadows or an increase in loss of beam transmission. This method is good for inspecting laminates and other multilayer structures for interface flaws. A variance of the through-transmission is performed by using a single transducer on one side. In this method, a signal reflected from an interface beyond the object of interest is monitored, and flaws are indicated as shadows or variations in the amplitude.

The common displays for ultrasonic information are the A-, B-, and C-scans (see Fig. 12-14). The A-scan is simply an amplitude-versus-time representation of the received information. The display can be scaled in distance so that flaw depth can be read directly. The B-scan is a collection of A-scans that shows all flaws in a cross section. The C-scan is a plan view of returned signals from a given range thickness. Conventionally, the C-scan is a binary display, where the information above a certain threshold is shown in a contrasting black or white shade. A C-scan can have graduated shade scales, and it is now a common feature with computer-aided systems for both B- and C-scans to be displayed in color graphic modes.

The basic equipment typically used for ultrasonic inspection consists of a scanning device and a transducer for sending the sound, which includes pulsers, receivers, and gates for identifying received signals in certain areas. Ultrasonic instruments appear in the traditional manual format as well as with advanced computer-controlled models. All instruments contain a pulser circuit producing impulses to excite the ultrasonic transducer. Classically, these pulsers were simple high-voltage spikes ranging from 100 to 1000 V. They operated at repeat frequencies, typically in the range 400 to 1200 pulses per second. Some modern pulsers use square-wave pulsers consisting of two interface impulses connected together at a predetermined delayed time. Square-wave pulsers increase performance because they enhance the efficiency of the conversion from electrical to acoustical energy in the transducer.

The receiver is the second important element of ultrasonic instruments. Instrument receivers usually provide a capability to select the band of operation that is believed to enhance signal-to-noise ratios. In many modern applications, filtering (band selection) is not used because of the good noise factor of modern instruments. Receivers also contain an attenuation/gain control, so the amplitude of return signals can be measured accurately. The receiver characteristic of interest is its linearity. This is because classical inspection requirements have been based on the principle that the magnitude of the received signal is a measure of size of the flaw. An important control on the receiver, typically termed the reject, is an electronic device used to effectively eliminate small, low-level signals from the instrument's display screen. Care must be taken in exercising the use of reject because it can affect the linearity of the instrument.

Most instruments contain gates for monitoring the presence or size of signals of interest. The simplest gates report the presence of a signal exceeding a given amplitude in a certain time frame representing a specific thickness of material. Multiple simple gates can be used to zone the material, giving a more refined indication of flaw depth. Many modern gates have the capability to report both the depth and amplitude, or size of indication signals, that can occur in a depth range.

Many large-scale ultrasonic systems contain scanning systems of various types. These scanning systems are used to transport the ultrasonic transducer over a fixed path to allow the evaluation of a large object quickly and thoroughly. In many cases scanning is very time-consuming because of the high resolution required. In modern industrial systems, computer-controlled scanning devices are integrated with the instrument and produce complex displays of large components.

Probably the most important element of an ultrasonic system is the ultrasonic transducer, because it produces the electrical-mechanical conversion. Like many other electrical mechanisms, it is a nonlinear device whose performance quite often depends on the details of a specific transducer. Performance standards and characteristics of transducers have not yet developed into industry-wide formats, although one has been proposed. Characteristics are usually measured and determined by individual users, so that procurement and acceptance of transducers depend on individual performance characteristics determined by a specific user. The performance characteristics recognized to be important are pulse shape, power, frequency, and bandwidth.

The shape of the pulse produced by transducer is also important in determining the other characteristics. Classically, two types of transducer exist: those with very short pulses and those with longer pulses. The short-pulse output of a transducer, shown in Fig. 12-15, is typical of wideband, low-power transducers. The output of a long-pulse transducer, also shown in Fig. 12-15, is representative of high-power, narrow-bandwidth transducers. The selection of a particular characteristic depends on the specific needs of the inspection. In through-transmission inspection, high-powered, narrowband transducers can be used and, in fact, are often indicated where thick, absorptive materials must be evaluated, such as some composites. On the other hand, where small, isolated flaws must be detected in high-strength materials, wideband transducers are indicated because resolving smaller indications require short pulses.

As mentioned earlier in the discussion on ultrasonic principles, the frequency at which an inspection is conducted is important in the performance of the inspection. Often, the stipulated frequency on the label of the transducer is not the frequency at which it operates. When the

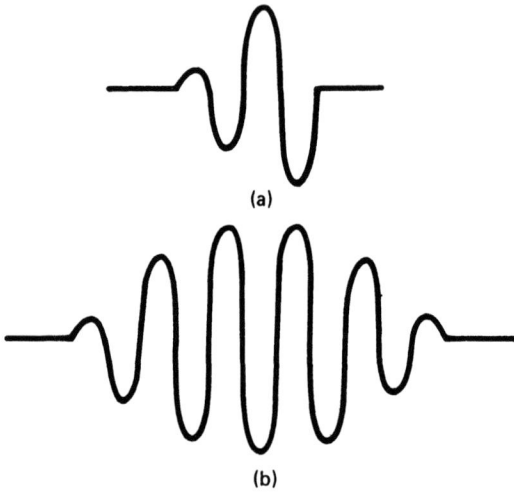

Figure 12-15 Output traces of two typical transducers: (a) high damping; (b) low damping.

inspection may be frequency dependent, frequency must be measured and must be a part of the inspection of the transducer specification.

Transducers come in two basic types: contact or immersion. Contact transducers are used in manual inspections. They usually contain a plastic wedge to produce angle beams or to stand the transducer off the surface. The standoff distance assures that the element is operating in the far-field zone or, in some cases, removes the effect of some interferences that can occur in the transmit signal. Contact transducers are particularly effective when applied on rough surfaces because the layer of couplant has a tendency to smooth out and reduce the surface irregularities.

Immersion transducers are designed to be used with a body immersed in a water bath and are most often employed on automatic systems. Another immersion inspection performed with the same type of transducer is one employing water columns or bubblers. In these cases, a jet of water provides the ultrasonic couplant between the transducer and the inspection object.

Transducers also come in focused and unfocused modules. Focused modules are used to (1) correct for the effect of curve surfaces, which can distort ultrasonic beams, and (2) increase the acoustic intensity in the volume, in particular in the near-surface zone, where it is necessary to improve detection reliability. Also, focused transducers provide a means to describe flaw shape, which is particularly useful when the flaw shape is larger than the beam diameter. The flaw size is determined by the area producing reflected energy. Another benefit of focusing is that the near-field zone is contracted, thus allowing a shorter standoff distance in some inspection procedures.

For most practical inspections, transducers are circular in projection. However, sometimes it is convenient to use long, rectangular transducers called paintbrush transducers if a large volume of materials must be inspected quickly. In such a device, sacrificed spatial resolution may be acceptable when high-speed inspection is required and only an indication of the presence of a flaw is necessary.

A new type of transducer, called the phased array, has become available. In this transducer, shown schematically in Fig. 12-16, the ultrasonic beam can be steered and focused depending on the time relationship between when the various elements are excited. The design of modern arrays usually includes varying the size of the element and the spacing between the elements to minimize

NONDESTRUCTIVE EVALUATION

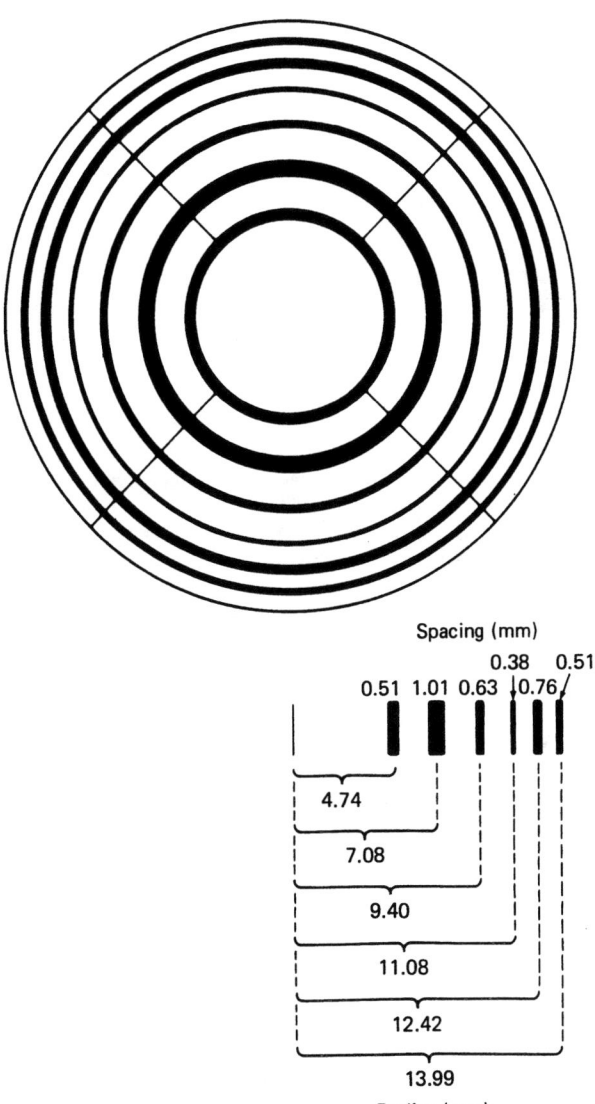

Figure 12-16 Phased-array transducer.

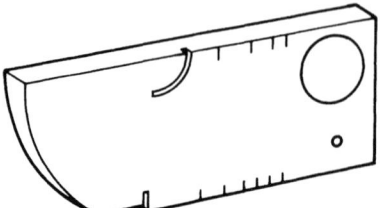

Figure 12-17 The International Institute of Welding reference block.

and reduce the impact of grating effects. Phase array systems typically are limited to frequencies below 5 MHz because of the cost of the very-high-speed electronic devices needed for higher frequencies.

Inspection Standards

Inspection standards typically are developed from reflectors produced in a material equivalent to the material to be inspected. The approach evolved because of a convention that the size of the flaw is proportional to the return signal. This is generally true for flaws that are smaller than the beam size and similar in geometry and type. Accordingly, inspection standards typically constitute a group of reference blocks or pieces of material with standard reflectors such as flat-bottomed holes or side-drilled holes located at various depths within the material. The approach developed because of historical differences between the instrumentation and performance of transducers. If reflectors are properly standardized in amplitude using these reference blocks, their relative reflectance can be determined at any given depth using any piece of equipment.

The success of this approach depends on the assumptions that the material in the reference blocks has acoustic characteristics identical to those of the material under test and that the geometry of the standardized reflectors is indeed uniform from test block to test block or set of test blocks. To address this problem, the U.S. National Institute of Science and Technology has available a standard set of reference blocks for aluminum.

Another reference block commonly used to calibrate contact inspection is the International Institute of Welding (IIW) block. The IIW block, which is constructed of low-carbon steel in the normalized condition, is grain size number 8. Figure 12-17 shows this block.

New Technology

Now technical interest is in transforming nondestructive testing from a simple, binary (good/bad) flaw-detection technology into a support system for lifetime management. Ultrasonic testing in response to this challenge has become more quantitative. The most significant change in ultrasonic methods has been the evolution away from simple amplitude methods for determining flaw size [27]. New methods have been developed that use the time between received signals as the means to measure flaw size. A specific example is shown in Fig. 12-18, which indicates the principles of the tip-diffracted method for the length of cracks [28]. Crack size is related to the time difference between the signal specularly reflected from the crack and the signal diffracted from the crack tip. By evaluating the crack from two different directions, crack tilt or orientation can be determined. Variants of this principle can be used to size cracks oriented near the surface or in the interior.

Over the last few years, considerable effort has been directed at evolving ultrasonic schemes

NONDESTRUCTIVE EVALUATION

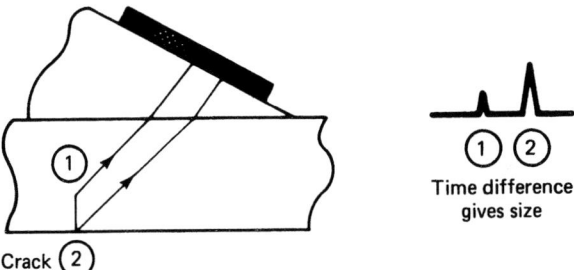

Figure 12-18 Principles of the satellite-pulse method for sizing cracks.

to image internal volumes of material. Two of these receiving considerable attention are acoustic holography [29] and synthetic aperture focusing [30]. The latter, outlined in Fig. 12-19, allows the reconstruction of image data. Imaging methods have the advantage over simple presentations such as A-, B-, and C-scans because they use spatially averaged information in image construction. They also provide an increase in signal-to-material noise ratio. The resolution is limited by the wavelength of the ultrasound. Because in most cases flaws of interest have features smaller than the wavelength, they produce incomplete images. For example, as might be anticipated from the discussion on ultrasonic principles, cracks appear as two separate scattering centers, one from each tip. No information is provided from the (intervening) surfaces unless a specular reflection of that surface occurs because, for most cracks, the surface roughness is smaller than the wavelength. Also, the production of images is significantly time-consuming. In addition, in situations other than very simple geometry, it is believed that the reading of ultrasonic images still requires significant interpretation skills.

12-6 Summary

This chapter is a condensation into a few pages of the very large, rapidly developing field of NDE technology; as a consequence, some may find it rather brief. For those readers, we recommend

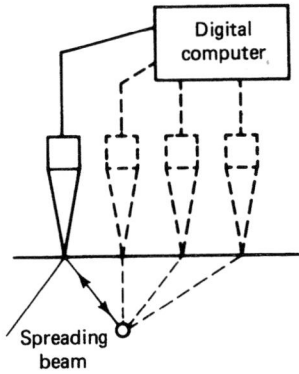

Figure 12-19 Schematic of synthetic aperture imaging method. The lens is simulated by focusing artificially; one sensor serves many locations; and collected signals are added numerically.

the vast amount of literature available for specific NDE methods. Major changes in the field are anticipated in the future, particularly in the use of more sophisticated instrumentation to implement the methods. Beyond this, continued evolution of NDE as an engineering tool is expected. NDE is no longer just a yes-or-no, good-or-bad evaluation tool, but is fully integrated into manufacturing and application engineering approaches; and NDE methods will increasingly be joined with component remaining life prediction and process control systems. Coupled with these transformations is the awareness of the statistical nature of inspection. It is now firmly accepted that NDE assessment is not a quantized black-or-white process; there are no unique inspection levels at which the detection capability changes from zero to 100%.

Inspection performance is a continuous, variable process; and flaw detectability is an estimable parameter that can be described as a distribution function for a range of conditions. The more sophisticated inspection systems now being developed incorporate the statistical nature of NDE performance. In these cases, lifetime prediction, or component performance, is expressed in statistical terms such as risk of failure.

It should be pointed out that because of space limitations, this chapter has not discussed several well-known NDE inspection methods. These include visual inspection, acoustic emission, and the leak test, which are regularly used and have a significant place in the inspection repertoire. Visual inspection is very important to many industries, which have developed and do follow detailed visual inspection procedures. Acoustic emission, a once-popular method because it offered to provide detailed flaw information from a single sensor, is now used less extensively because this expectation was not fulfilled. However, the method is employed commercially to detect cracks in some on-line pressure vessels and to monitor mechanical tests in an attempt to identify the first incidence of cracking. Leak testing as an inspection method is important to the semiconductor industry, as it can be used to detect leaks in hermetically sealed packages.

In conclusion, it should be mentioned that one significant source for NDE information is generally accessible to all workers for a minimal cost. This source is the Nondestructive Testing Information Center (NTIAC). NTIAC, a Department of Defense information analysis center located at and operated by the Texas Research Institute in Austin, maintains a database of current NDE literature, government reports, and unpublished documents that can be acquired. The NTIAC database can also be used to obtain bibliographic studies at a nominal cost. Furthermore, it publishes state-of-the-art reports, handbooks, symposium proceedings, and a free newsletter.

References

1. R. C. McMaster (Ed.), *Nondestructive Testing Handbook,* Vol. 1 Ronald Press, New York, 1959.

2. Nondestructive Evaluation and Quality Control, *Metals Handbook,* 9th ed., Vol. 17, American Society for Metals, Metals Park, OH, 1989.

3. *Nondestructive Testing Handbook,* 2nd ed., American Society for Nondestructive Testing, Columbus, OH. *Leak Testing,* Vol. 1, July 1982; *Liquid Penetrant,* Vol. 2, December 1982; *Radiography,* Vol. 3, January 1985; *Electromagnetic Testing,* Vol. 4, September 1986; *Acoustic Emission Testing,* Vol. 5, February 1988; *Magnetic Particle Testing,* Vol. 6, June 1989; *Ultrasonic Testing,* Vol. 7, May 1991. Additional volumes on *Visual Testing, Inferred Testing,* and *Special Testing Methods* are in preparation.

4. *Nondestructive Evaluation,* Report NMAB-252, National Academy of Sciences, AD-692491, June 1969.

5. *Annual Book of ASTM Standards,* Park 31, American Society for Testing and Materials, Philadelphia, 1972.

6. *Nondestructive Testing: Methods, Techniques and Their Applications,* Report DDC-TAS-71-58-1, Defense Documentation Center, AD-733850, December 1971.

7. W. D. Rummel, S. J. Mullen, B. K. Christner, F. B. Ross, and R.E. Muthart, Reliability of Nondestructive Inspection

(NDI) of Aircraft Engine Components, SA-ALC/MM8151, National Information Services, Washington, DC, AD-A155320.

8. J. Glatz, Detecting Micro Defects with Gas Penetrants, *Met. Prog.*, February 1985, p. 18.

9. R. C. McMasters, Nondestructive Testing, 26th Edgar Marburg Lecture, *Proc. ASTM, 52* (1952), 617.

10. F. H. Chang et al., Real-Time Characterization of Damage Growth in Graphite/Epoxy Laminates, *J. Compos. Mater., 10* (July 1976).

11. E. H. Rodgerd and C. P. Merhib, *A Report Guide to Magnetic Particle Testing Literature*, Report AMRA-MS-65-04, Army Materials Research Agency, AD-617758, June 1965.

12. C. G. Gardner and J. R. Barton, Recent Advances in Magnetic Field Methods of Nondestructive Evaluation for Aerospace Applications, *Advanced Technology for Production of Aerospace Engines*, AGARD-CP-64-70, September 1970, 18.1–18.9.

13. R. E. Beissner and M. J. Sablik, Theory of Electric Current Perturbation Probe Optimization, in *Review of Progress in Quantitative NDE*, Vol. 3 (D. O. Thompson and D. E. Chimenti, Eds.), Plenum Press, New York, 1984, 633–641.

14. *Eddy Current Testing, Classroom Training Handbook*, CT-6-5, 2nd ed., General Dynamics/Convair Division, San Diego, CA, 1979.

15. Nondestructive Inspection and Quality Control, *Metals Handbook*, 8th ed., Vol. 11, American Society for Metals, Metals Park, OH, 1976, 75–92.

16. Eddy Current Examination Method for Installed Nonferromagnetic Steam Generator Heat Exchanger Tubing, ASME Boiler and Pressure Vessel Code, Section V, Article 8, Appendix 1, American Society of Mechanical Engineers, New York, 1978.

17. H. L. Libby, *Introduction to Electromagnetic Nondestructive Test Methods*, Wiley-Interscience, New York, 1971.

18. R. C. McMaster (Ed.), *Nondestructive Testing Handbook*, Vol. II, Ronald Press, New York, 1963, 36.1–42.74.

19. H. S. Jackson, *Introduction to Electric Circuits*, 2nd ed., Prentice-Hall, Englewood Cliffs, NJ, 1965.

20. J. M. Prince and B. A. Auld, Development of Active Microwave Ferromagnetic Resonance Eddy Current Probes and Associated Signal Processing Methods, in *Review of Progress in Quantitative Nondestructive Evaluation*, Vol. 3 (D. O. Thompson and D. E. Chimenti, Eds.), Plenum Press, New York, 1984, 1369–1376.

21. D. L. Waidelich, Pulsed Eddy Currents, in *Research Techniques in Nondestructive Testing* (R. S. Sharpe, Ed.), Academic Press, London, 1970.

22. T. Kikuta, J. L. Fisher, S. N. Rowland, and W. D. Jolly, Far-Field Eddy Current Model for Carbon Steel Gas Pipes, *Proc. 4th Int. Symp. Offshore Mech. Arctic Eng.*, Dallas, Feb. 17–22, 1985.

23. R. A. Quinn and C. C. Sigl (Eds.), *Radiography in Modern Industry*, 4th ed., Eastman Kodak Company, Rochester, NY, 1980.

24. J. Krautkrammer, *Ultrasonic Testing of Materials*, Springer-Verlag, Berlin, 1969.

25. *Nondestructive Testing: Ultrasonics*, Rep. DDC-TAS-71-55-1, Defense Documentation Center, AD-733700, November 1971.

26. D. Jolly and D. Burton, Advanced Ultrasonic Transducer Performance and Produceability, U.S. Air Force Contract F33615-85-C-5036 Final Report, WRDC-TR-90-8030, May 1991.

27. *Guide to Ultrasonic Testing Literature*, Vol. 6, Report AMMRC-MS-69-03, Army Materials and Mechanics Research Center, AD-689455, April 1969.

28. G. J. Gruber, Characterization of Flaws in Piping Welds Using Satellite Pulses, *Mater. Eval.* April 1984.

29. M. Ahmed, K. Y. Want, and A. F. Metherell, Holography and Its Application to Acoustic Imaging, *Proc. IEEE, 67*, no. 4 (1979).

30. C. F. Schueler, H. Lee, and G. Wade, Fundamentals of Digital Ultrasonic Imaging, *IEEE Trans. Sonics Ultrason., SU-31*, no. 4 (July 1984).

CHAPTER

13

High-Temperature Strain/Displacement Measurement

William N. Sharpe, Jr.

Department of Mechanical Engineering
The Johns Hopkins University
Baltimore, Maryland

13-0 Introduction

When the test temperature exceeds 260°C (500°F), strain measurement becomes infinitely more challenging. There are numerous requirements for measurements in the range of 260 to 540°C (500 to 1000°F), but this chapter focuses on temperatures between 540 and 815°C (1000 and 1500°F). Power-plant components such as piping and inlet nozzles operate at the lower end of this range. Steam turbines and air-breathing jet engines have components that experience temperatures at the upper end and higher. Increased concern for efficiency is pushing the demands on the design and the material to new limits; this in turn requires strain measurements in the development phase to validate a prototype and in the operational phase to monitor performance. Suffice it to say that accurate and reliable means of measuring strains between 540 and 815°C would enable significant advances in the development of power plants of various kinds.

The foil or wire strain gage has become widely and successfully used over the last 40 years and has perhaps generated a false sense of security as to the ease of making accurate and reliable strain measurements with very high resolution. Gages, adhesives, and coatings are readily available that permit even the novice to obtain reasonable results on a first attempt. That is far from the case at high temperature. The novice would be lucky to get any reliable data at all on the first attempt.

One pays little attention to the difference between static and dynamic strain measurement at room temperature under ordinary conditions, but entirely different approaches to the two conditions are often taken at high temperature. Figure 13-1 illustrates the two kinds of strain—with arbitrary strain and time scales. If the test item is a high-pressure steam line, the time scale may be on the order of days; if it is a jet engine turbine blade, it may be seconds. One would probably use one kind of gage in the first case and another kind in the second. It is perhaps useful to specify some of the characteristics of the "ideal" high-temperature strain gage:

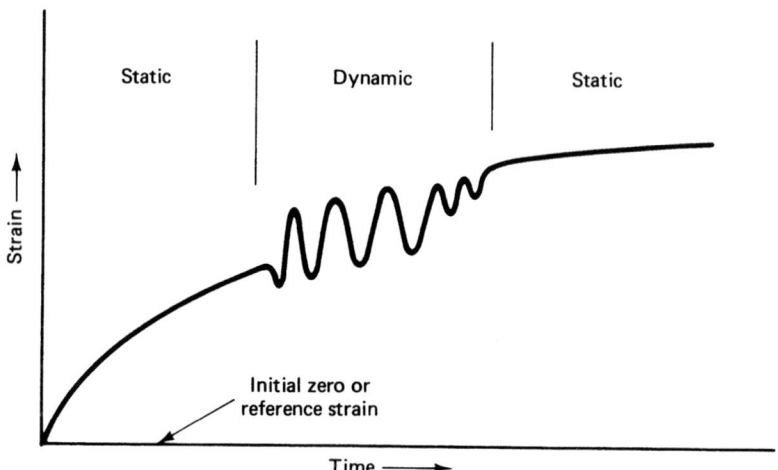

Figure 13-1 Schematic of static and dynamic strain (arbitrary scales).

    Temperature range: 20 to 815°C

                Size: 3 mm × 3 mm

              Range: ±10,000 microstrain

       Resolution: 100 microstrain

               Error: ±5%

                Life: 10,000 hours

  Acceleration loads: 100,000 g

Note that with the exception of the temperature range, these requirements are easily met at room temperature by foil gages.

In spite of this rather grim prognosis, strains are measured successfully under the foregoing conditions. But this is accomplished in laboratories that have built up a considerable expertise over the years. Furthermore, they have often developed and manufactured their own gages for their particular application. The companies supporting those laboratories are in the business of developing new engines, power plants, or components. They find the expense of testing prototypes to be quite cost effective compared with the costs of building several units which are run to failure in order to identify problem areas. Clearly, high-temperature strain measurement is a specialized activity requiring great care and skill.

In 1973 the author conducted a review of techniques for measuring strains at high temperatures, reported in Ref. 1. Resistance strain gages were surveyed; they were used almost exclusively for dynamic strain measurement. Capacitance strain gages were just appearing on the scene. In the intervening years, there have been no startling advances in either kind of gage: just a steady improvement in techniques and a gaining of considerable experience. A more recent review [2] describes some of these advances. This chapter covers wire resistance strain gages and capacitance strain gages in some detail because they are the two major techniques for high-temperature measurement at this time. Extensometers for material property testing at high temperature are then discussed; this is followed by brief discussions of various specialized optical techniques.

13-1 Resistance Strain Gages

The earliest form of resistance strain gage was a grid of fine wire; users manufactured their own gages and developed their own techniques for application. Many laboratories today take the same approach for high-temperature strain measurement. The three common types of resistance strain gage are illustrated in Fig. 13-2. Figure 13-2a shows the familiar wire gage. Wire on the order of 0.02 mm in diameter is wound into a grid pattern and fastened to the component surface with special adhesives. This version is used primarily for dynamic strains. The encapsulated version (Fig. 13-2b) places the active wire element inside a sealed tube, which is then welded to the component. With the improved protection against oxidation, this version is used for longer-term static strain measurement. A relatively new type of resistance gage is shown in Fig. 13-2c. This thin-film gage is deposited directly onto the component over a layer of insulating material. This very small gage is used exclusively for dynamic work.

The resistance R of a strain-gage element is a function of time and temperature as well as strain:

$$R = f(\varepsilon, T, t) \tag{13-1}$$

The gage factor is

$$F = \frac{1}{R} \frac{\partial f}{\partial \varepsilon}\bigg|_{\varepsilon_0, T_0, t_0} \tag{13-2}$$

and is determined experimentally by mounting the gage on a calibrated deflection apparatus and subjecting the gage to a known strain at various temperatures. Procedures for this are described in the American Society for Testing and Materials Standard E251.

The apparent strain is measured by mounting the gage on an unloaded test coupon and observing the "strain" output as the coupon is heated. The slope of this curve at a particular temperature is given by

$$\frac{1}{RF} \frac{\partial f}{\partial T}\bigg|_{\varepsilon_0, T_0, t_0} \tag{13-3}$$

This apparent strain can be several thousand microstrain over a large temperature range simply because of the mismatch in thermal expansion between the gage wire and the component material.

Drift of a gage is the change in output over time with no change in strain in the test item; its rate at a particular strain level is given by

$$\frac{1}{RF} \frac{\partial f}{\partial t}\bigg|_{\varepsilon_0, T_0, t_0} \tag{13-4}$$

It may be difficult to separate the apparent strain and the drift at very high temperatures. If the test article is a thin-gage material, additional concerns arise; these are discussed by Hudson [3].

It is quite natural to try to extend the familiar resistance strain-gage technology to high temperatures, but the record over the years has not been good. Some of the problems with resistance gages are [4] excessive drift, poor corrosion resistance, high or unstable coefficient of thermal expansion, large gage factor change with temperature, large and unstable apparent strain, low electrical resistivity (causing low output), and poor fatigue strength. With this list of problems,

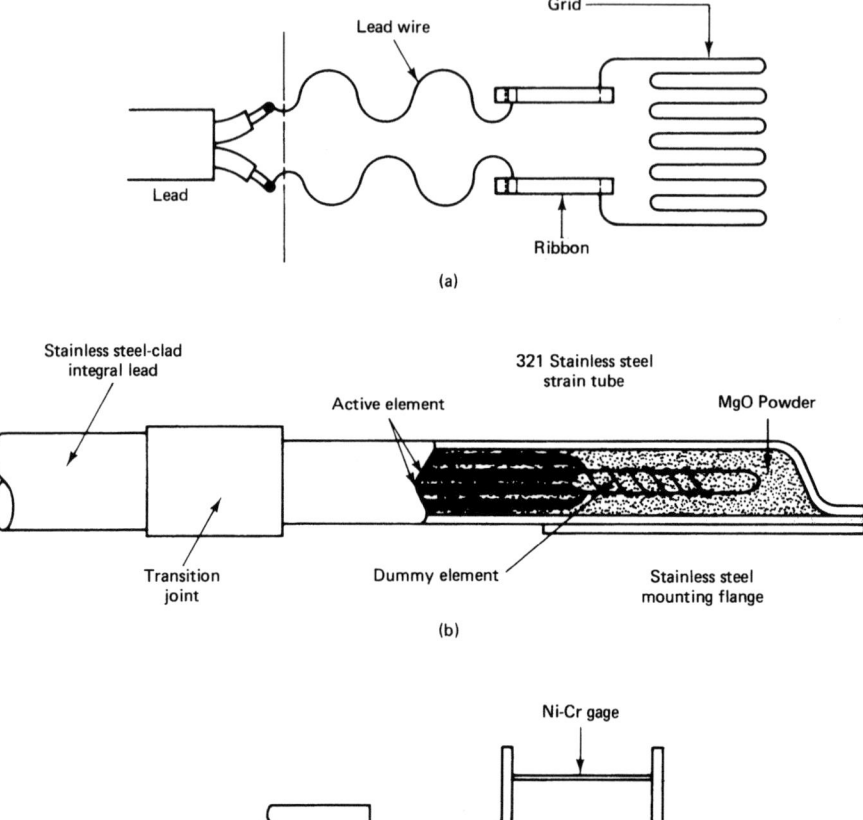

Figure 13-2 (a) Schematic of wire gage; (b) schematic of encapsulated wire gage; (c) schematic of thin-film gage.

it is a wonder that these gages are used at all; yet with proper care and experience, acceptable measurements can be made.

The distinction between static and dynamic strain measurement becomes very important at high temperature, and the specifications of the measurement system are also quite different. A static strain gage usually can be larger because it is used on larger components; a gage length of 25 mm is often acceptable. Creep strains are often important under static loading, so the range needs to be larger for static sensors. Clearly, the life can be smaller for dynamic measurements; a few hours may be sufficient. And dynamic strain measurement is often desired at higher temperatures. A jet engine is a good example of the measurement problem. The central disk runs at temperatures around 650°C and experiences slowly varying centrifugal loads. The turbine blades operate at 815 to 1100°C and vibrate with changing engine conditions. These differences between static and dynamic behavior make it convenient to separate the following discussion along those lines.

Dynamic Strain Measurement

Weise and Foster [5] and Prosser [6] describe techniques and procedures for accurate, reliable strain measurement on turbine blades. Their reports exemplify the magnitude of the effort required to develop a capability for accurate high-temperature strain measurement.

The wire from which the gage is made is obviously important, and researchers have been searching for years for improved alloys [1]. Combinations of nickel–chromium, iron–chromium–aluminum, platinum–tungsten, and platinum–nickel have been used with varying alloy ratios and small doses of other elements. Weise and Foster [5] evaluated various candidates and selected an alloy of 75% Ni, 20% Cr, and 4.5% Al for a temperature up to 815°C. This wire, which was originally 20 μm in diameter, was formed into the grid shape and then flattened to a thickness of 15 μm in a specially made press. The flattening process made the gage easier to handle without destroying the grid pattern. The gage was then heated at 566°C for 2 hours to stabilize the material and reduce scatter in the gage factor. This particular alloy and its treatment were chosen after an extensive test program.

Other wire alloys are used; Raymondo [7] used an Fe-Cr-Al alloy and found it usable to 870°C. Wu et al [8] made single-element gages of a specially developed alloy of Fe, Cr, Al, V, Ti, and Y. Their dual-element gages contained an active element fabricated from wire made of P, W, Re, and Ni, and a compensating element made from Pt and Ir. These are actually two gages in one installation; the two elements have differing coefficients of thermal expansion and are connected in adjacent arms of a Wheatstone bridge. A ballast resistor is added in the compensating arm and can be adjusted to minimize the apparent strain. Both gages are reported to be usable to 700°C. Recent work [9] has extended the upper limit to 800°C. Lemcoe [10] developed a gage using an Fe-25/Cr-7.5/Al alloy that was usable to 1035°C; however, it had a very high apparent strain. Drenner et al. [11] conducted an extensive survey of wire gages in 1970; their report contains 343 references. Wire gages are commercially available that are made of alloys suitable for use at high temperature. Transition metal compounds have recently been examined by Lei et al. [12], who concluded that the Group IV metal carbides were the best candidates.

The adhesive system used to attach the wire gate to the component must insulate the wire from the material, provide protection from the atmosphere (which may include eroding particles and corrosive gases moving at high speeds), and transmit the strain in the component to the wire. Considering the different thermal coefficients of expansion of the component, wire, and adhesive, this is no trivial requirement, especially when one realizes that rapid thermal shocks may be present. The common attachment systems for wire gages are ceramic cements and the flame-spray technique. The latter technique, also known as the Rokide process, is one in which a high-purity aluminum oxide rod is fed into a flame-spray gun, where it is melted in an oxyacetylene flame. The molten material is sprayed onto the component surface to form an insulating layer 50 to 100 μm thick. The wire gage is then laid on the precoat and attached finally with another layer of flame-sprayed material. The wire gage is usually held to its shape by two or three thin strips of tape. The attaching adhesive layer (ceramic cement or flame spray) bonds the portions of the gage not covered by the tape strips. The strips are then removed and the final coating applied. Needless to say, handling small gages made from wire only 25 μm thick is a delicate process. Commercially available gages are received with this handling tape (or its equivalent) in place.

Weise and Foster [5] investigated the various gage attachment systems and concluded that the flame-spray system is superior for use at temperatures up to 815°C. Their final attachment procedure is: grit-blast the component surface with size 120 aluminum oxide, apply an adhesion layer of flame-sprayed nickel-aluminide from 25 to 75 μm thick, apply a base coat of flame-sprayed alumina to about the same thickness, tape down the gage and flame-spray an alumina coating 50 to 100 μm thick, remove the tape strips, and spray on an alumina coating up to 125 μm thick. The final installation has a total thickness of not more than 375 μm (0.015 in.). Ceramic adhesive systems are sometimes preferable at lower temperatures.

Not surprisingly, many of the failures of wire gages occur where the wire of the active element is joined to the larger lead wire that goes to the instrumentation. Weise and Foster [5] bond the active gage wire to a Karma (76 Ni, 20 Cr, Fe, Al) ribbon 25 μm thick by 375 μm wide using ultrasonic welding. These short pieces of ribbon are attached to the intermediate leads of 50 μm Pt-10 Ni wire by ultrasonic welding; these intermediate leads are formed in a convoluted pattern (see Fig. 13-2a) for improved fatigue life. The primary lead wires are 30-gage chromel–alumel in a duplex insulation of carded asbestos with a tightly woven glass cover.

One has to be careful of the attachment of the primary lead wires in view of the thermal transients and dynamic loads. The dynamic loads are especially troublesome in turbines. Lead wires are usually attached by tack welding thin metal shim stock in place over them. However, one must be careful that these welds do not reduce the fatigue life of the component.

The instrumentation used with wire gages is basically the same as for foil gages at lower temperature except that one has to be able to balance out the apparent strain, which can be larger than the range of many units.

What is different from room-temperature gages is that it is necessary to test and evaluate the strain-gage installation before usage on the actual component. According to Prosser [6], the largest contributor to measurement error is the uncertainty in gage factor. Gages need to be installed on test specimens that can be subjected to known strains at various temperatures. The test specimens need to be of the same material as the component to be tested, and the gages need to be installed in exactly the same fashion as they will be on the test article. Gage installations need to be tested not only for the gage factor, but for erosion, acceleration forces (this requires a spin test facility), and fatigue. All this testing beforehand is necessary to permit evaluation of the measurement uncertainty when the final test is conducted.

Figure 13-3 [6] shows the variation in gage factor of several gages up to 815°C. The narrow scatter band is a reflection of the care in use of standardized procedures. One would expect to know the gage factor of a similar installation to within 5%. Gages and lead wires installed according to his procedures can withstand an acceleration of 40,000 g and the erosion of flowing air in a jet engine. Furthermore, they can survive a cyclic strain loading of 3000 microstrain peak-to-peak for 10 million cycles. Most of Prosser's tests were of jet engine turbine blades, and he recommends checking out the installation on an individual blade by subjecting it to vibration in a mechanical shaker. This will eliminate installations that are not good enough before the blade is inserted for the final test.

A newer resistance strain gage, the thin-film gage, is similar in concept to the familiar foil gage except that the resistive metal is deposited directly by a photo fabrication process on an insulating substrate applied to the test article. The primary advantage of the thin-film gage is its thinness—25 μm as opposed to the 400 μm of a complete wire gage installation. This thinness is important when the gage is to be attached to a turbine blade because a thicker gage can disturb the airflow pattern and also perhaps change the mechanical response of the blade.

Stowell and Weise [13] have developed thin-film gages for use on turbine blades at temperatures up to 540°C (with a goal of 650°C). The most difficult problem to resolve has been the insulating substrate. Since it is so thin, it is susceptible to failure during testing. For example, carbide

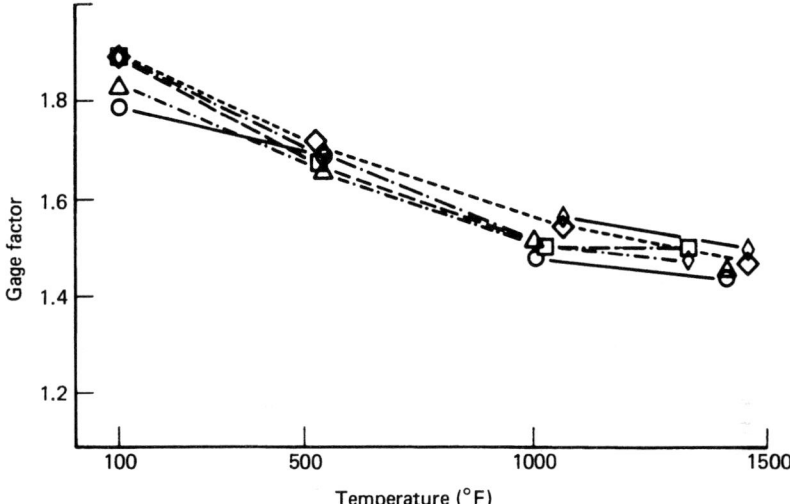

Figure 13-3 Typical gage factor uncertainty and variation with temperature for wire gages. Each symbol represents one of several gages installed on a test specimen. (From Ref. 6.)

particles extruding from Inconel 718 can push through the deposited insulating layer. Sputtered alumina is the choice of these investigators for the substrate, but it has to be applied in two to four layers to avoid the development of porosity during the sputtering process.

After the substrate has been applied, a 2000-Å-thick layer of nichrome is sputtered over the area. This is covered with a layer of gold 1 to 2 μm thick. A photoresist mask of the gage and tabs is then baked onto the gold layer and developed. Selective etchants are used to remove the excess gold and nichrome and then to remove the gold layer from the active part of the gage. The protective top coat of alumina is then sputtered on. Leads are attached by thermocompression bonding of small gold ribbons to the gold-plated tabs; these ribbons are wrapped around the final lead wire. The installation receives a final protective coating of ceramic cement.

Gage factors are determined in much the same fashion as for foil gages (i.e., applied to a constant-moment beam and evaluated at temperature). The scatter at room temperature is small, 5% or less. It is interesting to note that a thin-film gage can be evaluated by gluing a foil gage directly on top of the installation; the foil gage can easily be removed before the test at high temperature. Stowell and Weise [13] report that thin-film gages have proven more reliable than wire gages in turbine tests, and they expect that thin-film gages will eventually be less expensive.

Static Strain Measurement

For long-term strain measurement, the encapsulated strain gage shown schematically in Fig. 13-2b is used. This type of gage, carrying the name Ailtech, is manufactured by the Electronic Instrumentation Division of the Eaton Corporation. The active strain element (the single loop of wire) and the coiled dummy element are made from either a Pt-W or a Ni-Cr alloy. This dual-element gage forms half of a Wheatstone bridge; a temperature-compensation resistor is provided by the manufacturer for reduced apparent strain output. The two elements are hermetically sealed in a metal tube and insulated from it by highly compacted ceramic powder, which also serves to transfer the strain from the component to the active wire element. This strain tube is integrally attached to the stainless steel–covered lead–one purchases the gage plus a lead approximately 3 m long. The strain tube is normally made from stainless steel, but other materials can be used. Gage lengths range from 5 to 25 mm.

These gages are attached by spot welding the flanges, which are attached to the strain tube by the manufacturer, to the test article. This is a very easy process and is also used to fasten down the lead wire tube. Since these gages are most often used on rather large structures, there is little problem of fatigue life reduction because of the welding.

Nightingale and Twa [14] investigated the capabilities of Ailtech model SG425 gages for measuring low-level strains at temperatures up to 482°C (900°F). These gages are recommended for static usage to 482°C with intermittent temperatures up to 650°C. Their concern was the variability between individual gages, and they devised methods to precalibrate the as-received gages. A clamping fixture was built that can be used to clamp the gage for gage factor determination or for apparent strain measurements. From tests of eight gages from the same lot, they found an average difference of 3.5% from the manufacturer's gage factor, with one gage being 9% different. On the other hand, they found that the change in gage factor with temperature followed the manufacturer's prediction. Their recommended procedure for calibrating an individual gage is: (1) bond the gage to a calibration bar with epoxy adhesive and perform load cycles at room temperature to determine the gage factor; (2) use that gage factor in conjunction with the manufacturer's data to predict gage factors at other temperatures; and (3) clamp the gage to an unloaded piece of the same material as the test article and temperature cycle the assembly to obtain the apparent strain curve. These procedures will yield gage factors for an individual gage with a relative uncertainty of 4% up to 482°C. Similar procedures should be useful for higher temperatures.

Hayes [15] investigated the possibility of extending the encapsulated gages to higher temperatures for measurement of low-frequency vibration of components in a fluidized-bed power plant. The gage manufacturer performed a special stabilization treatment at 815°C to improve the high-temperature performance. Three types of gages—SG125 (Ni-Cr, $\frac{1}{4}$ bridge), SG325 (Ni-Cr, $\frac{1}{2}$ bridge), and SG425 (Pt-W, $\frac{1}{2}$ bridge)—were evaluated by attaching them to uniaxial test specimens that were loaded inside a resistance furnace mounted in an electrohydraulic test machine. The specimen strains were also monitored by two high-temperature extensometers. Numerous load and temperature cycles were applied; specimens were held at temperature for approximately 70 hours. The apparent strains at 650°C were on the order of 30,000 microstrain for the SG125, 2000 for the SG325, and 5000 for the SG425. It should be noted that the strains imposed by loading were small—on the order of 300 microstrain. A major concern in these experiments was the uniformity of loading in the specimen (i.e., the lack of bending strains). The measured strain values were compared with the strain calculated from the measured load and handbook values of the elastic modulus at temperature for the type 304 stainless steel specimen. The SG325 gages showed an average error of −9% at 650°C and, surprisingly, −5% at 843°C. The SG425 gages showed an average error of +9% at 650°C, but were not evaluated at the higher temperatures. The SG125 gages had an error of +6% at 650°C and 22% at 843°C. The general conclusion is that reliable low-frequency strain data can be obtained at these temperatures for the moderately long times required.

Smith [16] used a specially designed constant-moment loading apparatus that could be mounted inside a laboratory furnace to evaluate encapsulated gages at 593°C (1100°F). Gages were calibrated in place, using foil gages, which were removed prior to the long-term test. He tested SG425-type gages and thermally preconditioned SG125 gages. Figure 13-4 is a plot of the drift response of four of the gages over 1000 hours. These four show the best results; data for eight more gages are included in the report. The SG125 gages are found to have limited applicability for these longer times. A potential problem with this type of gage is thermal effects on the lead wires

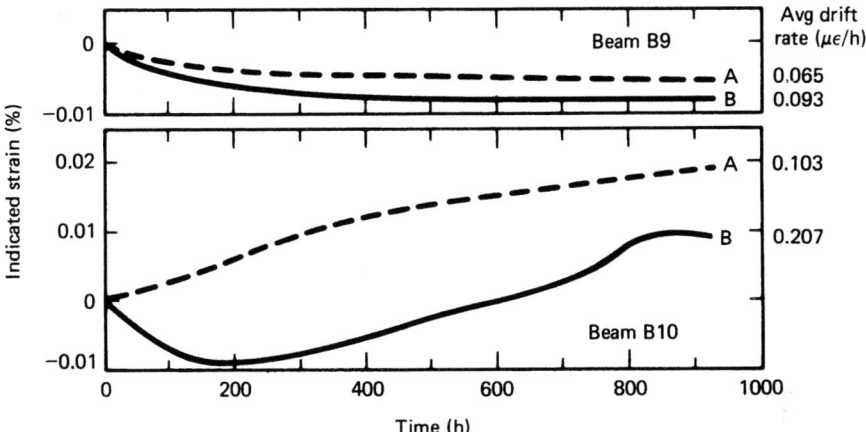

Figure 13-4 Drift response for four Ailtech encapsulated wire gages. Two gages were mounted on beam 9 and two on beam 10. (From Ref. 16.)

because of slight variations in characteristics of the lead wires. Smith [16] recommends in-place calibrations to evaluate this effect as well as to establish the gage factor.

In the presence of thermal transients, the temperature of the encapsulated gage may be quite different from that of the structure. This could lead to serious errors in the apparent strain computation. Adams [17] has investigated this effect by using a specially manufactured temperature sensor in the same type of enclosure as an encapsulated gage. He concludes that if one measures the temperature in this fashion, appropriate corrections to the apparent strain can be made for transients as high as 56°C per second. A similar approach is described by Englund [18].

Carbon, carbon-coated, and carbon composite materials are used in high-temperature components, and strain measurement requires special considerations. Szafranski [19] gives a detailed report on early tests of two kinds of wire gages on silicon carbide–coated carbon–carbon composites. Transient, apparent strain, and loading tests were conducted to temperatures as high as 800°C, with failure of the adhesive bonds being the main problem. Free-filament Pt-W gages with 6-mm gage length were evaluated on graphite specimens [20]; the results showed their behavior to be "reasonably accurate" up to 500°C. Lanius et al. [21] describe a free-filament strain gage and temperature gage developed for use on carbon composites at lower maximum temperatures of 370°C. The 6.35-mm-long gages were attached with either ceramic cement or flame spray and were designed to produce low apparent strain and hysteresis on the carbon composite.

13-2 Capacitance Strain Gages

The concept of building a strain transducer that converts the change in distance between two gage points into a change in capacitance is not new. In fact, the first handbook [22] contains an entire chapter on electric-capacitance gages. The capacitance C between two parallel plates of area A separated by a distance d is

$$C = \frac{Ak}{3.6\pi d} \tag{13-5}$$

where k is the dielectric constant (equal to 1.0 for air). Various clever ideas based on varying either A or d have led to the development in the last two decades of capacitance gages specifically for high-temperature strain measurement. Geometrical stability at temperature is more easily achieved than metallurgical stability, and this fact underlies the advantages that capacitance gages have (in principle) for long-term, high-temperature measurement. The changes in capacitance of these gages is very small, as is the resistance change in wire gages, but the electronic circuitry is more complicated and expensive. Recent advances in electronics have been important to the development of capacitance gages.

There are two gages commercially available at this time. One is a variable capacitance gage, referred to as the CERL-Planer gage after its originators, G. V. Planer Ltd. in conjunction with the Central Electricity Research Laboratory of Great Britain (see Ref. 23). The other is referred to as the Boeing gage [24] and operates on a differential capacitance basis. These two gages and the results of evaluative tests are discussed below. There are other types of capacitance gages [25,26], and Bloss [27] has developed a capacitance-based extensometer for use in calibrating high-temperature strain gages up to 815°C at rates of 55°C per second.

CERL-Planer Gage

Figure 13-5a is a schematic of the Planer gage; it consists of two springlike arches of different radii with capacitance plates mounted between the peaks of the arches. The two arches are welded together during manufacture of the gage, and the gage itself is spot-welded to the test component. Relative displacement between the two ends of the gage produces a change in the air gap between the electrodes and thus a change in capacitance. The gage is made from a highly stable alloy whose coefficient of thermal expansion is close to that of the stainless or ferritic steels widely used at high temperature. The shortest gage length is 19 mm. An installed gage is shown in Fig. 13-5b; such a gage normally would be covered with a little stainless steel box to protect it from physical damage and to keep dust from accumulating between the electrodes.

The capacitance varies nonlinearly from approximately 1.3 to 0.4 pF over a tensile strain of 10,000 microstrain. An individual calibration curve at room temperature is furnished with each gage. However, it is easy to clamp the gage to a calibration device for an on-site calibration. In fact, the gages can often be removed, recalibrated, and used again. If there are significant differences between the way the gages are installed for calibration and the way they are installed for use, large errors may arise [28]. The small capacitance changes are measured with a transformer ratio arm bridge; the two electrodes are insulated from the grounded gage and structure. A typical instrument will operate at a carrier frequency of 1.6 kHz at approximately 50 V and permit a resolution of 1 microstrain. The high-temperature coaxial cable leads are attached to the gage electrodes via the small coiled ribbons shown in Fig. 13-5b. Cable lengths as long as 100 m have been used.

The CERL-Planner gage has been used extensively, primarily in Great Britain, at temperature up to 650°C. Average drift rates for tests lasting 1000 hours or more are on the order of 0.02 microstrain per hour, although the rate can be an order of magnitude higher during the first 100 hours. Response to temperature transients as fast as 0.5°C per second is adequate, and the mechanical resonant frequency of the gage is above 1 kHz.

Boeing Gage

The Boeing gage consists of two cylindrical inner capacitor plates mounted on (but insulated from) a small rod and passing through an outer annular capacitor plate. The rod is attached to the test article at one end, and the outer capacitor plate is also attached to it. When the component is

HIGH-TEMPERATURE STRAIN/DISPLACEMENT MEASUREMENT

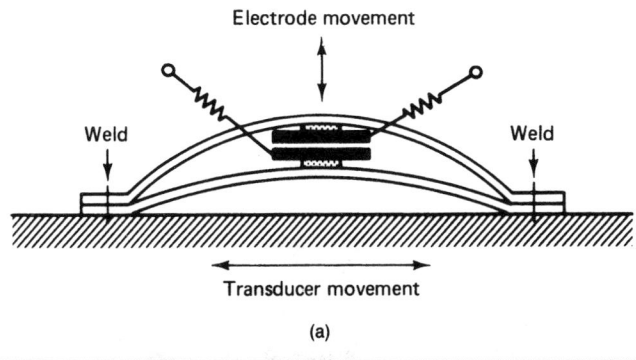

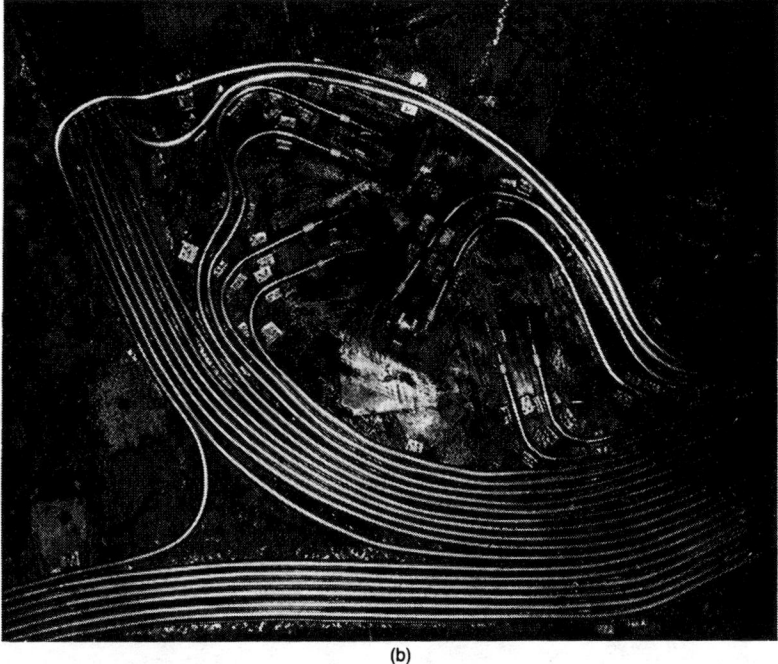

Figure 13-5 (a) Schematic of the Planer gage. (b) Photograph of an installation of four gages on a steam turbine case. Note the stainless steel–covered lead wires. (Courtesy of S. Wnuk.)

strained, the inner cylindrical plates move with respect to the outer annular plate; the area of the capacitor increases for one inner plate and decreases for the other one. This is then a linear and differential gage. Figure 13-6 presents a schematic of the gage and a photograph of a typical installation.

The rod carrying the inner plates can be made of the same material as the test piece; this reduces the apparent strain. However, if the test article is a composite, it may be difficult to choose a rod material with the correct coefficient of expansion [19]. One can see from Fig. 13-6 that this rod

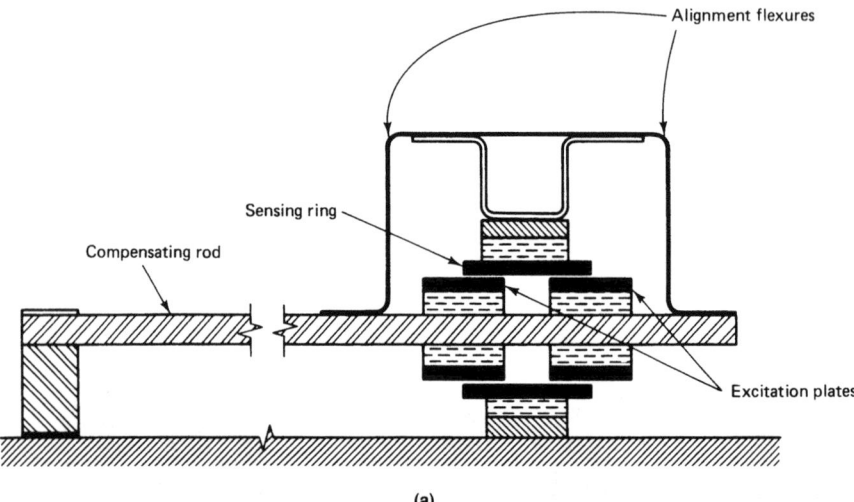

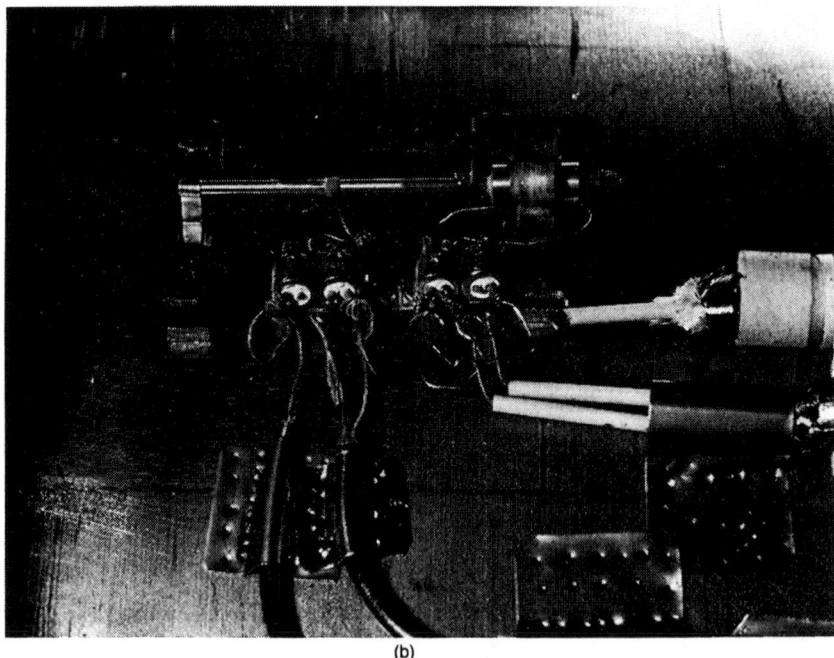

Figure 13-6 (a) Schematic and (b) photograph of a Boeing gage. The gage length in (b) is 25 mm, and the lead wires on the left are to thermocouples mounted on the compensating rod and on the component. (Courtesy of J. E. Smith.)

HIGH-TEMPERATURE STRAIN/DISPLACEMENT MEASUREMENT

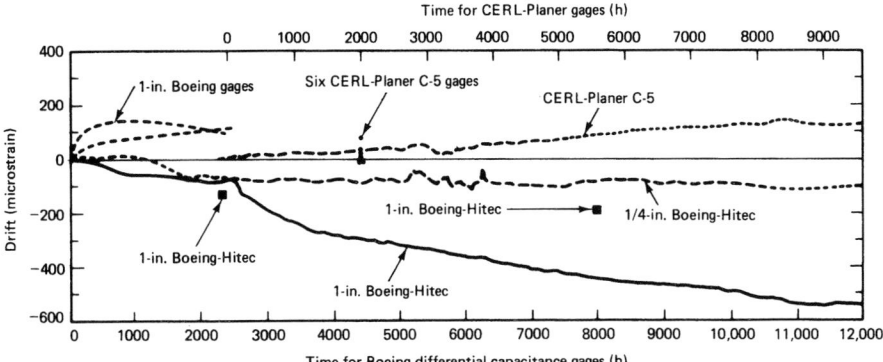

Figure 13-7 Drift of capacitance gages. The continuous results are for 12,000 hours at 593°C. The 1-in. Boeing gage continuous data for 2000 hours and the data at discrete times are from other tests. (From Ref. 16.)

sits some distance (approximately 3 mm) above the component surface. It is therefore usually necessary to pay attention to the temperature difference between the rod and the component while the test article is heating up. For this reason, the gage is equipped with thermocouples attached to the rod and to the component. The end of the rod is attached to the test article by spot-welding the attachment ribbons, and the outer annular capacitor plate is attached in the same fashion. A flexure bracket attached to the rod rides over the annular capacitor and serves to maintain concentricity of the capacitor plates. The gage is supplied in an alignment frame which is removed after the gage is applied.

Electrically, the gage is half of a capacitance bridge. An instrument module is available from the gage manufacturer (Hitec Corp., Westford, MA); it provides the excitation (7 V rms at 3.4 kHz) to the two capacitors and converts the difference into a voltage proportional to strain. A full-scale output of 5 V can correspond to 2500 microstrain. High-temperature coaxial cables are used for this capacitance gage also.

The Boeing gage is available in gage lengths of either 6.4 or 25 mm, and is 9 mm high (to the top of the flexure bracket). As would be expected, the strain range is very large—160,000 and 40,000 microstrain for the two gage lengths, respectively. This device has been successfully used to 815°C, although most experience has been gained at 650°C. It must be mounted on a flat surface. Shims can be used to mount a Boeing gage on a curved surface, but care must be taken to avoid or account for bending effects. These gages have been used for thermal transients as fast as 17°C per second with appropriate corrections for the temperature differences.

Smith [16] conducted a long-term study of the drift and gage factor stability of capacitance gages following the same procedures as used for the Ailtech gage (discussed earlier). One Boeing gage of each gage length and one CERL-Planer gage were tested for over 10,000 hours on unstrained beams at 593°C; the beams were periodically loaded to check the gage factor. Smith recommends that periodic measurements of the total capacitance of the individual capacitors and of the insulation resistance of the leads be made for diagnostic purposes if a gage should fail.

The drift results are presented in Fig. 13-7, taken from Smith's report [16]. Excellent performance is shown there, with only slightly more than 100 microstrain drift in 12,000 hours for the 6-mm Boeing gage. The Planer gage and the larger Boeing gage show a somewhat higher (but quite acceptable) drift. Data from other studies are included in Fig. 13-7 and generally support the results.

Gage factor stability with time was evaluated occasionally and variations of less than 2% were found over an 11,000-hour period. Furthermore, the difference between gage factors at 22 and 593°C was less than 5%. One then concludes that these capacitance gages are accurate, stable strain transducers for use on large components at elevated temperatures. The Boeing gage has the advantage of a larger strain range and better measurement of transient thermal response, whereas the CERL-Planer gage is cheaper, smaller, and more rugged.

13-3 Electromechanical Extensometers

The properties of materials for use at high temperature must be determined, and this requires a whole new set of specifications for the strain measurement system. The test specimens can be designed to accommodate the measurement system, and the more controllable laboratory environment makes the task more manageable. Uniaxial tensile tests are quite a bit easier than cyclic tension–compression or biaxial–loading experiments. Cyclic tests typically use an "hourglass" shape for the specimen to avoid buckling problems. This requires a very short gage length technique for axial strain measurement, so the diametral strain is often measured and converted to axial strain—a procedure that presumes the material to be completely isotropic, with a well-specified value of Poisson's ratio. Tensile creep experiments can use a cylindrical specimen, but the long-term stability of the measurement system is a major concern.

Electromechanical extensometers are devices that are clamped or otherwise fastened to two gage locations on the sample so that the relative motion between them can be transmitted to some sensing device. In high-temperature work, the sensing device (resistive-, inductive-, or capacitive-based) is located outside the hot region and operates at a very moderate temperature. The arms transmitting the specimen displacement to the sensor are often made of quartz, which has a thermal coefficient of expansion near zero. The MTS Systems Corporation of Minneapolis manufactures extensometers of this type [29]. Excellent reviews [30–34] cover the various facets of high-temperature material property measurements, including discussions of extensometers.

Each laboratory tends to have its own particular loading and heating system and therefore develops its own extensometer to fit its unique needs. Extensometers are usually custom-made and incorporate features peculiar to the material at hand and information desired. However, Schmale and Cross [35] describe a commercially available system complete with furnace, water-cooled grips, and diametral extensometer.

One can expect that a notch cut into a specimen for the purpose of attaching the extensometer is likely to have a deleterious effect under low-cycle fatigue testing at high temperature. The solution to this problem is to machine a specimen with two ridges, which delineate the gage length over which strain is measured. But such a ridge is still a stress concentration, and a great deal of study has gone into what the proper shape of the ridge should be (see Chapters 14 and 15 of Ref. 31 and also Ref. 30). This problem is one reason for the popularity of diametral extensometers.

In addition to the details of attachment to the specimen and transmission of the displacement outside the hot region, electromechanical extensometers differ in their choice of sensing device. Characteristics of extensometers presented in the technical literature prior to 1973 are given in Ref. 1; the sensing device was either a linear-variable-differential transformer or a strain-gage-based displacement transducer. A typical diametral extensometer has a gage length of 6 mm and a resolution of less than 10 microstrain. Operating temperatures could be as high as 980°C and accuracies of better than 1% are reported. Reference 30 contains detailed descriptions of typical

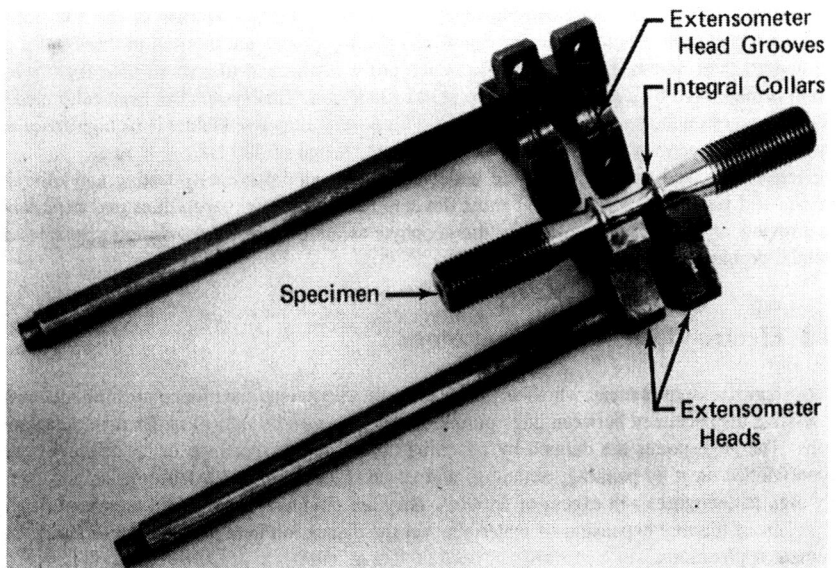

Figure 13-8 Disassembled mechanical extensometer linkage. (From Ref. 36.)

instruments. There has been a recent trend toward capacitive-based sensors, and some more recent electromechanical extensometers are described below to give an indication of modern capabilities.

A typical axial extensometer for use up to 605°C was used by Cowles et al. [36]; it has a gage length of 12 mm and uses rods running parallel to the specimen axis to transmit the displacement outside the furnace. Two (to correct for bending) capacitance-based proximity detectors are used to measure displacement outside the furnace. The relative uncertainty is estimated to be 2% for strains larger than 100 microstrain. A photograph of this device is presented in Fig. 13-8. That extensometer required integral collars machined with the specimen; more recent devices have dispensed with the collars.

Keusseyan and Li [37] use high-temperature capacitance probes to measure the displacement between platens into which the specimen is fastened, but the gage length is ill-defined in this arrangement. Walters and Hales [38] measured axial strain with the sensor being a capacitance gage mounted inside the furnace between two knife edges forced against the specimen, Yeakley and Lindholm [39] developed a novel ring-shaped capacitive transducer that fits onto the specimen inside the furnace; they extended this concept to a biaxial transducer usable to 927°C.

The low-density-silica tiles for insulating space vehicles are brittle and require a very sensitive strain measurement system capable of operating at quite high temperatures. Duggan [40] developed a capacitance-based biaxial transducer capable of resolving on the order of 1 microstrain at 1100°C. Jordan and Chan [41] describe a two-axis extensometer for tension–torsion testing of tubular specimens at 649°C.

Lemcoe [42] used the special resistance gages he developed to construct a biaxial transducer consisting of three superalloy flexure pins to which the gages are attached. This extensometer, which is approximately 60 mm tall, is pressed against the specimen and can measure strains as large as 1% at 760°C. An induction-based extensometer is described by Yazdi et al. [43]; it has a displacement range of 150 mm and has been evaluated at 500°C. A novel approach by Sumner

[44] requires drilling holes in each end of the specimen—with the bottoms of the holes at the transition to the uniform cross section. Small nickel-alloy probes are inserted in these holes and held against their bottom by a locking nut. Each probe is attached to a small optical slit whose motion is monitored by a position-sensitive photovoltaic cell. This system has been calibrated via foil strain gages on the specimen and used at 700°C. A very attractive feature is its high-frequency response, which permits strain measurement at a loading rate of 500 Hz.

Electromechanical extensometers are widely used in material property testing and can yield accurate and reliable data. Each user must develop his or her own capabilities and experience, and a review of the literature will make the neophyte aware of some of the design principles and potential problems.

13-4 Electro-Optical Extensometers

Electro-optical extensometers, which are also used primarily in material property testing, determine the relative displacement between gage points on the specimen by optical rather than mechanical means. The gage points are defined by attaching "flags" to the specimen or by creating optical discontinuities on it by painting, abrading, and so on. Electro-optical extensometers are used at very high temperatures—in excess of 2500°C. They are obviously very useful for measuring the coefficient of thermal expansion of materials, but the discussion here is restricted to strain measurement applications.

Targets are important considerations in the measurement system incorporating an electro-optical extensometer. Early extensometers were simply traveling microscopes whose cross hairs were manually aligned with the targets. A very sharply defined target is required, and one can imagine the additional difficulties involved in looking into a glowing red- to white-hot furnace. These same problems are present with an electro-optical transducer because it must also precisely locate the target. Gaal [45] presents an extensive evaluation of target configurations and fiducial marks up to 2500°C. Interestingly enough, he presents his results via photographs at temperature. The conclusion is that a wedge with the straight front face toward the observer is the most sharply defined at 2500°C. Fiducial marks on the specimen lose their contrast when the specimen becomes luminous. A novel way to avoid this problem has been put forth by Jordan et al. [46]. Small x-ray fluorescent targets are used and scanned with an x-ray beam; the technique has been demonstrated at 600°C.

One type of optical extensometer uses a laser beam, which is swept past the targets at a constant rate. The time between intensity changes at the targets is proportional to the length of the gage section. One example is described by Babcock and Hochstein [47], who used an argon laser because it has a shorter wavelength than the predominant portion of the blackbody radiation. Furthermore, narrowband optical filters are effective in blocking the background radiation. Their graphite specimen was resistively heated, and the gage section was delineated by a thin sputtered film of refractory metal. The laser beam was moved by an acousto-optic diffraction cell at the rate of 40 kHz. Strains on the order of 100 microstrain with rise times of longer than 1 ms were measured.

Commercial optical trackers are used for noncontacting displacement measurement (typical instruments are manufactured by Optron of Woodbridge, CT, and Physitech of Huntington Valley, PA). An optical discontinuity is focused through a telescope onto the cathode of a photomultiplier tube and held in position there by an electromagnetic feedback system (basically an oscilloscope in reverse). The feedback signal is proportional to the displacement. An extensometer version follows two targets through the same telescope via an image splitter. Thompson et al. [48] used such a differential optical tracker and argon laser illumination to measure strains in graphite specimens at 2200°C; the targets were also graphite and fastened to the specimen. They report an accuracy of ± 100 microstrain over a 25-mm gage length, but caution that uniformity of illumination and alignment of the target on the photomultiplier tube are critical. Calfo and Bizon [49] used a similar arrangement for strain measurement on the leading edge of a turbine blade subjected to a thermal cycle in a test rig. The targets were platinum–rhodium wires mounted on the leading edge and backlighted to quartz–iodine lamps. The thermal cycle lasted 4 minutes and heated the blade to 1100°C—generating strains as large as 1.27%. The resolution was 400 microstrain, and results were reproducible within ±4%.

Marion [50] describes in detail an electro-optical extensometer usable to 2700°C on resistively heated graphite specimens. The targets are small nodules of graphite cement approximately 2 mm high which are filed to obtain a sharp edge. They are illuminated with an argon laser and tracked by digital-line-scan cameras. These cameras are a linear array of 1024 photodiodes 0.43 mm wide and spaced 25 μm apart—coupled to a lens system. They are scanned by a controller circuit, which can have a scan rate as fast as 1 kHz. The displacement range is 4.24 mm, and the uncertainty is 4 μm; the gage length is of course variable depending on the specimen configuration.

A newer type of measurement device is the laser-based gaging system. A sheet of laser light (usually He-Ne) is produced by a small laser and suitable optics in the transmitter. It passes over the object to be measured and into a photodiode array in the receiver. While designed primarily for automatic dimension measurement in production and quality control, these devices can also be used to measure the distance between targets at elevated temperatures. Their resolution (on the order of 0.1 μm), stability, and relatively low cost make them attractive electro-optical extensometers.

An optical extensometer that does not require targets simply tracks the motion of two points on the specimen that are illuminated with a split laser beam. The surface must be reflective, and rigid-body motion is a concern, as is turbulence of the air near the heated specimen. A commercial version has been used to measure the stress–strain curve of a niobium alloy at 1100°C; it has also measured ±2000 microstrain cyclic strains at 538°C [51].

When very high temperatures and fast frequency response are required, electro-optical extensometers may well be the answer. They are different from electromechanical extensometers in that one purchases the optical system but still must develop the targets and the lighting. However, they are similar in that they are used primarily for the determination of material properties.

13-5 Optical Techniques

Whole-field optical techniques such as moiré and speckle interferometry can be extended to higher temperatures, although the number of applications to date is limited. Since the principles and techniques are discussed in earlier chapters, only the special considerations associated with high-temperature measurement will be covered here. In addition, a laser-based interferometric technique for measuring strains at a point on a real-time basis will be briefly described.

If one photographs a moiré grating on the specimen at temperature and subsequently processes it optically, the main difference in high-temperature work is the grating itself. Cloud et al. [52] discuss some of the details of applying 40-line/mm gratings for use up to 750°C. After a photoresist grating has been applied to the specimen, a vacuum-deposited coating of platinum, gold, or even aluminum can be applied. This produces a relief grating, and the photoresist may or may not burn away or vaporize at temperature. If this grating is lightly polished, the metal coating over the photoresist is removed, and the photoresist can be washed away with solvent. The result is a metal relief grating on the specimen surface. The metal coating can be applied by electroplating—a trickier process, but one that requires only simple equipment.

Gratings can be etched onto the specimen by electroetching or chemical milling after the photoresist is applied. Considerable practice and expertise are required to prevent undercutting the photoresist. Sciammarella and Rao [53] used this technique to study the deformation near the tip of a crack in a stainless steel ring at 593°C. They used a 12-line/mm grating because of the large deformations generated.

Cloud et al. [52] also describe a very simple technique in which the high-temperature grating is made from refractory paint. The paint is sprayed over the photoresist grating and rubbed down to the surface of the photoresist. The photoresist can then be washed away with solvent or left in place to vaporize at temperature. A high-contrast grating is produced; in fact, the paint can be colored to improve contrast. A grating applied in this fashion to a superalloy specimen has been used for strain measurements up to 1370°C [54].

Speckle interferometry suffers from the decorrelation caused by air movement near a heated specimen, but this problem can be eliminated by using a pulsed laser. Chiang et al. [55] used a laser with a pulse length of a approximately 60 ns in conjunction with the one-beam laser speckle method to measure transient strains in an aluminum plate heated with a propane torch. They took specklegrams approximately every 17 seconds, which was the pumping time of the ruby laser. Optical processing yielded the complete strain field over a limited region. Although the maximum temperature at which strain was measured was only 468°C, in principle the technique could be extended to higher temperatures since it is entirely noncontacting. Blackbody radiation could become a problem, but a different type of laser and interference filters could be used. Another approach uses a TV camera to record the fringe patterns [56,57]. The air turbulence near the specimen does not cause severe problems; since the fringe patterns are recorded continuously, one can identify the disturbances caused by air movement. This technique, known as electronic speckle pattern interferometry (ESPI), has been used to measure strain–displacement at temperatures as high as nearly 3000°C.

The author has used a laser-based interferometric technique to measure strains over a gage length of 100 μm on notched superalloy specimens at 650°C [58]. The principle underlying this method of measuring in-plane relative displacements is quite simple. Two very small indentations are placed approximately 100 μm apart on the specimen surface with a Vickers hardness tester. Upon illumination with a laser interference fringe patterns are produced because of the path differences between light rays reflecting from the sides of the indents. Motion of these fringes is proportional to the relative displacement between the two points. Fringe motion is monitored on a real-time basis by scanning mirrors, which are controlled by a minicomputer, and photomultiplier tubes. The minicomputer does the appropriate calculations and outputs a voltage signal proportional to strain to a monitoring oscilloscope. It also controls the electrohydraulic test machine and stores the load-strain data for later analysis.

The specifications of this interferometric strain/displacement gage (ISDG) as it was used for the experiments at 650°C are: gage length, 100 μm; range, 3% strain; resolution, 200 microstrain; sampling rate, 10 per second; and relative uncertainty, ±3%. Figure 13-9 is an electron micrograph of a pair of indentations in a notched Inconel 718 specimen cycled in a tension–compression at 650°C with a 2-minute hold in tension and a 2-minute hold in a compression; 253 cycles were run, and the entire test took 17 hours. Note the microcracks that developed: the indentations in this oxidation-resistant material are still reflective. Various loading spectra were used on other specimens, and the local cyclic strains were recorded for use in evaluation of cyclic plasticity theories [58].

The ISDG technique has also been used to measure crack-opening displacements near fatigue crack tips at 650°C [59]. Because of its high resolution (0.02 μm), threshold values for crack growth under creep and fatigue conditions can be obtained quickly. Dynamic displacements have been measured in precracked ceramic specimens at 1000°C by placing indentations in platinum tabs on either side of the crack tip [60]. The results were used in a hybrid experimental-numerical

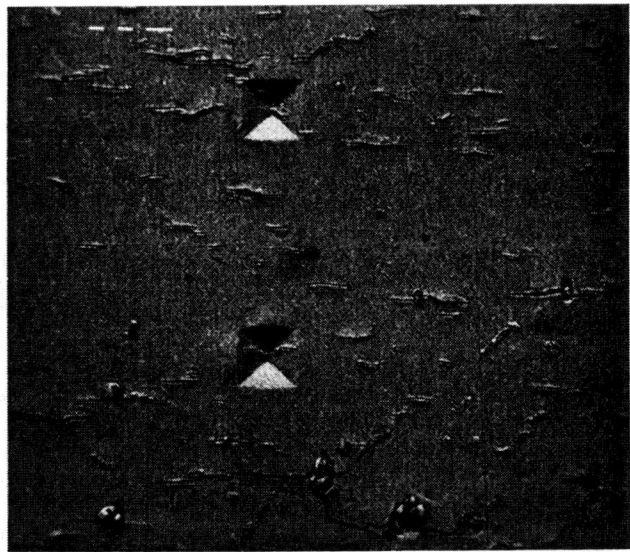

Figure 13-9 Indentations after cyclic testing at 650°C for 17 hours. The two indentations are 100 μm apart. (From Ref. 58.)

procedure to determine the dynamic fracture toughness. The technique has also been used to measure stress–strain curves of 304L stainless steel at 800°C in a high-rate servohydraulic machine [61]. The author has used the IDG with linear diode arrays as fringe sensors to record creep strain in notched specimens of a zirconium alloy at 250°C for 570 hours [62]. Stability of the measurement system is dependent only on the mechanical stability of the diode mounts, since fringe position, not intensity, is recorded. This approach has potential for creep testing at higher temperatures.

A technique operating on optical diffraction instead of interference has been used by Pih and Liu [63] to measure strain in ceramics at 1370°C. Targets are attached to the specimen to form a thin slit through which a laser beam shines. The fringes in the diffraction pattern are monitored to measure the change in slit width and therefore the change in length of the specimen.

The availability of digital video cameras has enabled the development of a full-field displacement strain measurement based on a correlation integral image recognition algorithm. Images before and after deformation are compared to determine the relative displacement field. This technique, known as digital image correlation, has potential for high temperature measurements, and preliminary measurements at 600°C have been made [64].

Another noncontacting, nontarget optical technique has been developed by Yamaguchi [65]. A laser beam is split into two beams, which are focused onto a small area on the specimen; each beam has the same angle with respect to the normal to the specimen. The speckle patterns generated by each beam interfere and the resulting fringes are monitored to determine the strain. Lant [66] describes an application of this technique to measure strains at 450°C.

13-6 Summary

The measurement of strains in test specimens or components at temperatures above 540°C is quite a bit more difficult than at room temperature. However, advances over the past 20 years enable one to measure static and dynamic strains at temperatures up to 815°C with commercially available or well-developed sensors. Strain measurements at higher temperatures can be made but require very specialized techniques.

Wire resistance strain gages applied with either ceramic cement or plasma sprays are best used for dynamic strain measurement. A new approach is the sputtered thin-film gage, which may well become widely used as the application techniques are developed. The encapsulated wire gage is more suitable for static long-term measurements but is limited to temperatures below 650°C. In all cases, the attachment of leads, the application of protective coating of the gage, and so on, require very specialized and carefully developed procedures.

Capacitance strain gages are becoming more widely used because they show excellent stability at temperatures up to 815°C. Their large strain range makes them especially suitable for long-term creep measurements; however, their size makes the measurement of rapid thermally or mechanically induced strains more difficult.

Strain measurement on test specimens is usually accomplished with electromechanical or electro-optical extensometers. Commercial units are available, but many researchers develop their own specialized systems. Mechanical extensometers are successfully used to 1095°C, while optical extensometers are employed at higher temperatures.

There are various optical techniques for making point or whole-field strain measurements. However, they are still restricted to specialized research activities.

In conclusion, one can make strain measurements with reasonable validity in the range of 540 to 815°C. However, such measurements are not trivial, and one must be prepared to invest a considerable amount of time in learning the intricacies of the various techniques. Room-temperature strain measurements are easy, but the same is not true at high temperature. In the literature, the successful measurements are the results of many years of effort and experience.

References

1. William N. Sharpe, Jr., Strain Gages for Long-Term High-Temperature Strain Measurement, *Exp. Mech.*, 15, no. 12 (1975), 482–488.
2. C. O. Hulse, High Temperature Static Strain Gage Development Contract, NASA Contractor Report CR-180811, 1987.
3. L. D. Hudson, Recent Experiences with Elevated-Temperature Foil Strain Gages with Application to Thin-Gage Materials, *Proc. 6th Annu. Conf. Hostile Environments and High Temperature Measurements,* Society for Experimental Mechanics, 1989, 68–81.
4. W. A. Stange, Advanced Techniques for Measurement of Strain and Temperature in a Turbine Engine, *AIAA/SAE/ASME 19th Jet Propulsion Conf.*, Seattle, WA, AIAA-83-1296, 1983.
5. R. A. Weise and J. H. Foster, High Temperature Strain Gage System for Application to Turbine Engine Components, AFWAL-TR-80-2126, 1981.
6. J. C. Prosser, Rotating Strain Gauge Instrumentation for Gas Turbine Engines, *BSSM-SESA Int. Conf.*, Edinburgh, 1981.
7. Philip Raymondo, Static-Strain Measurements on Gas-Turbine Combustor Liners, *Exp. Tech.*, 8, no. 2 (1984), 46–49.
8. Tsen-tai Wu, Liang-cheng Ma, and Lin-bao Zhao, Development of Temperature-Compensated Resistance Strain Gages for Use to 700°C, *Exp. Mech.*, 21, no. 3 (1981), 117–123.
9. L.-C. Ma, T.-T. Wu, and L.-B. Zhao, Development of Temperature-Compensated Resistance Strain Gages for Use to 800°C, *Exp. Mech.*, 30, (1990), 17–19.
10. M. M. Lemcoe, Development of Strain Gages for Use to 1311 K (1900°F), NASA CR-132485, 1974.

11. D. C. Drenner, C. M. Jackson, N. Crites, and J. E. Sorenson. Topical Report on LMFBR Instrumentation, National Technical Information Service, TID-25816, November 1970.

12. J. F. Lei, H. Okimura, and J. O. Brittain, Evaluation of Some Thin Film Transition Metal Compounds for High Temperature Resistance Strain Gauge Application, *Mat. Sci. Eng, A111* (1989), 145–154.

13. W. R. Stowell and R. Weise, Application of Thin Film Strain Gages and Thermocouples for Measurement on Aircraft Engine Parts, *AAIA/SAE/ASME 19th Jet Propulsion Conf.*, Seattle, WA., AIAA-83-1982, 1983.

14. C. M. Nightingale and G. J. Twa, Evaluation of Precalibration Techniques for Ailtech SG425 Strain Gages, presented at Western Regional Strain Gage Committee, February 1982.

15. J. K. Hayes, Evaluation of High Temperature Weldable Strain Gages, Combustion Engineering, Inc., CENC-1591, April 1983.

16. J. E. Smith, Assessment of Current High-Temperature Strain Gages, Oak Ridge National Laboratories, ORNL/TM-7025, December 1979.

17. P. H. Adams, Transient Temperature Response of Strain Gauges, *BSSM-SESA Int. Conf.*, Edinburgh, 1981.

18. D. R. Englund, The Dual Element Method of Strain Gauge Temperature Compensation, *Proc. 4th Annu. Conf. Hostile Environments and High Temperature Measurements*, Society for Experimental Mechanics, 1987, 40–43.

19. F. J. Szafranski, Thermo-Structural Measurements in a SiC Coated Carbon–Carbon Hypersonic Glide Vehicle, *Proc. 6th Annu. Conf. Hostile Environments and High Temperature Measurements*, Society for Experimental Mechanics, 1989, 45–54.

20. R. Steindler, High-Temperature Strain-Gage Behavior on Carbon Materials, *Exp. Mech., 28*, (1988), 244–248.

21. S. J. Lanius, R. G. Brasfield, and S. P. Wnuk, Development of a Strain/Temperature Gauge and Attachment System for Use on Carbon Composites at Elevated Temperatures, *Proc. 4th Annu. Conf. Hostile Environments and High Temperature Measurements*, Society for Experimental Mechanics, 1987, 53–59.

22. M. Hetényi (Ed.), *Handbook of Experimental Stress Analysis*, Wiley, New York, 1950.

23. B. E. Noltingk, Measuring Static Strains at High Temperatures, *Exp. Mech., 15*, no. 10 (1974), 420–423.

24. Darrell R. Harting, Evaluation of a Capacitive Strain Measuring System for Use to 1500°F, ISA ASI 75251, 1975, 289–297.

25. O. Larry Gillette and James L. Mullineaux, Development of High-Temperature Capacitance Strain Gauges, *ISA Trans., 8*, no. 1 (1969), 52–61.

26. E. B. Norris and L. M. Yeakley, Development of System for Monitoring In-Service Strains in Power Plant Piping, National Technical Information Service, PB 257 752, January 1976.

27. R. L. Bloss, An Extensometer for Use as a Laboratory Standard at Temperatures to 1500°C, *ISA Trans., 10*, no. 3 (1971), 242–249.

28. R. Fidler, The Calibration of CERL-Planer Capacitance Strain Gauges, *Strain*, 1986, 171–177.

29. R. Meyer, Axial-Torsional Extensometer for Use to 1200°C, *Proc. 4th Annu. Conf. Hostile Environments and High Temperature Measurements*, Society for Experimental Mechanics, 1987, 82–93.

30. *Manual on Low Cycle Fatigue Testing*, ASTM Spec. Tech. Pub. 465, American Society for Testing and Materials, Philadelphia, 1969.

31. M. S. Loveday, M. F. Day, and B. F. Dyson, *Measurement of High Temperature Mechanical Properties of Materials*, Her Majesty's Stationery Office, London, 1982.

32. M. H. Hirschberg, Elevated Temperature Fatigue Testing of Metals, NASA Tech. Memo. 82745, December 1981.

33. M. A. McGaw and P. A. Bartolotta, The NASA Lewis Research Center High Temperature Fatigue and Structures Laboratory, *Proc. 4th Annu. Conf. Hostile Environments and High Temperature Measurements*, Society for Experimental Mechanics, 1987, 12–29.

34. G. A. Hartman, L. P. Zawada, and S. M. Russ, Techniques for Elevated Temperature Testing of Advanced Ceramic Composite Materials, *Proc. 4th Annu. Conf. Hostile Environments and High Temperature Measurements*, Society for Experimental Mechanics, 1988, 31–38.

35. David T. Schmale and Robert W. Cross, Techniques for Elevated Temperature Mechanical Testing, NUREG/CR-2793, Sandia National Laboratories, Albuquerque, NM, June 1982.

36. B. A. Cowles, D. L. Sims, J. R. Warren, and R. V. Miner, Jr., Cyclic Behavior of Turbine Disk Alloys at 650°C, Trans. ASME, 102, October 1980.

37. R. L. Keusseyan and Che-yu Li, Precision Strain Measurement at Elevated Temperatures Using a Capacitance Probe, J. Test. Eval., 9, no. 3 (1981), 214–217.

38. D. J. Walters and R. Hales, An Extensometer for Creep-Fatigue Testing at Elevated Temperatures and Low Strain Ranges, J. Strain Anal., 16, no. 2 (1981), 145–147.

39. L. M. Yeakley and U. S. Lindholm, Development of Capacitance Strain Transducers for High-Temperature and Biaxial Applications, Exp. Mech., 14, no. 8 (1974), 331–336.

40. Michael F. Duggan, Combined-Loaded Testing of the Space-Shuttle Insulation at 1100°C, Exp. Mech., 20, no. 10 (1980), 350–356.

41. E. H. Jordan and C. T. Chan, A Unique Elevated-Temperature Tension–Torsion Fatigue Test Rig, Exp. Mech., 27, (1987), 172–183.

42. M. M. Lemcoe, Development of High-Temperature Biaxial-Strain Transducer for Use to 1033 K (1400°F), Exp. Mech., 19, no. 2 (1979), 56–62.

43. A. R. Yazdi, W. E. Deeds, and C. V. Dodd, Temperature-Compensated Induction Extensometer, Rev. Sci. Instrum., 49, no. 12 (1978), 1684–1687.

44. G. Sumner, Measurement of Dynamic and Static Strain in Small Specimens at Elevated Temperatures, J. Strain Anal., 9, no. 2 (1974), 130–134.

45. P. S. Gaal, Some Experimental Aspects of High-Temperature Thermal Expansion Measurements by Optical Telescopes, High Temp. High Pressures, 2 (1972), 49–57.

46. E. H. Jordan, D. M. Pease, H. Canistraro, and D. Brew, X-Ray Beam Method for Displacement Measurement in Hostile Environments, Proc. 6th Annu. Conf. Hostile Environments and High Temperature Measurements, Society for Experimental Mechanics, 1989, 27–30.

47. S. G. Babcock and P. A. Hochstein, High-Strain-Rate Testing of Rapidly Heated Conductive Materials to 7000°F, Exp. Mech., 10, no. 2 (1970), 78–83.

48. R. A. Thompson, W. E. Jorgenson, and M. L. Callabresi, An Optical Technique Measurement of Material Specimens at Elevated Temperatures, SAND75-8261, Sandia Laboratories, Albuquerque, NM, September 1975.

49. Frederick D, Calfo and Peter T. Bizon, Experimental Determination of Transient Strain in a Thermally-Cycled Simulated Turbine Blade Utilizing a Non-Contact Technique, NASA TM-73886, Lewis Research Center, January 1978.

50. Robert H. Marion, A New Method of High-Temperature Strain Measurement, Exp. Mech., 18, no. 4 (1978), 134–140.

51. D. Voorhes, G. Wyntjes, and M. Hercher, A Non-Contact Real-Time Laser Extensometer for Hostile Environments, Proc. 5th Annu. Conf. Hostile Environments and High Temperature Measurements, Society for Experimental Mechanics, 1988, 53–58.

52. G. Cloud, R. Radke, and J. Peiffer, Moiré Gratings for High Temperatures and Long Times, Exp. Mech., 19, no. 10 (1979), 19N-21N.

53. C. A. Sciammarella and M. P. K. Rao, Failure Analysis of Stainless Steel at Elevated Temperatures, Exp. Mech., 19, no. 11 (1979), 389–398.

54. G. Cloud and M. Bayer, High Temperature Moiré, Proc. 5th Annu. Conf. Hostile Environments and High Temperature Measurements, Society for Experimental Mechanics, 1987, 35–39.

55. F. P. Chiang, R. Anastasi, J. Beatty, and J. Adachi, Thermal Strain Measurement by One-Beam Laser Speckle Interferometry, Appl. Opt., 19, no. 16 (1980), 2701–2704.

56. C. A. Sciammarella, G. Bhat, and Y. Shao, Measurement of Strains at High Temperature by Means of a Portable Holographic Moiré Camera, Proc. Conf. Hostile Environments and High Temperature Measurements, Society for Experimental Mechanics, 1989, 21–26.

57. J. T. Malmo, O. J. Løkberg, and G. A. Slettemoen, Interferometric Testing at Very High Temperatures by TV Holography (ESPI), *Exp. Mech., 28*, 1988, 315–321.
58. W. N. Sharpe, Jr., and M. Ward, Benchmark Cyclic Plastic Notch Strain Measurements, *J. Eng. Mater. Technol., 105* (October 1983), 235–241.
59. W. N. Sharpe, Jr., A New Optical Technique for Rapid Determination of Creep and Fatigue Thresholds at High Temperature, AFWAL-TR-84-4028, Air Force Wright Aeronautical Laboratories, Wright-Patterson Air Force Base, Ohio, April 1984.
60. K. H. Yang and A. S. Kobayashi, A Hybrid Procedure for Dynamic Characterization of Ceramics at Elevated Temperatures, *Proc. Conf. Hostile Environments and High Temperature Measurements*, Society for Experimental Mechanics, 1989, 41–44.
61. E. A. Fuchs and D. R. Williams, Strain Measurement at Elevated Temperatures and High Strain Rates, *Proc. 4th Annu. Conf. Hostile Environments and High Temperature Measurements*, Society for Experimental Mechanics, 1988, 42–48.
62. W. N. Sharpe, Jr. and H. Zeng, Creep Strain Measurements at Notch Roots in Zirconium Pressure-Tube Material, *Proc. Third Conf. Fossil Plant Inspections*, Electric Power Research Institute, Baltimore, Maryland, August, 1991.
63. H. Pih and K. C. Liu, Laser Diffraction Methods for High-Temperature Strain Measurements, *Exp. Mech., 31*, 1991, 60–64.
64. J. L. Turner, S. S. Russell, and M. A. Sutton, Application of Digital Image Analysis to Strain Measurement of Elevated Temperature, *Proc. Conf. Hostile Environments and High Temperature Measurements*, Society for Experimental Mechanics, 1989, 31–34.
65. I. Yamaguchi, A Laser-Speckle Strain Gauge, *J. Physics E, 14*, 1981, 1270–1273.
66. C. T. Lant, High Temperature Optical Strain Measurement System, *Proc. 4th Annu. Conf. Hostile Environments and High Temperature Measurements*, Society for Experimental Mechanics, 1988, 8–11.

CHAPTER

14

Thermoelastic Stress Analysis

Brian J. Rauch and Robert E. Rowlands

Department of Engineering Mechanics
University of Wisconsin—Madison
Madison, Wisconsin

14-0 Introduction

Under adiabatic and reversible conditions, a cyclically loaded, isotropic structure experiences in-phase temperature variations that are proportional to the change in the sum of the principal stresses. Thermoelastic stress analysis (TSA, thermoelasticity) uses an infrared radiometer to measure the local temperature fluctuations and relates these changes to the associated dynamic stresses by thermodynamic principles. This approach differs from dissipative methods, such as vibrothermography, which associate local temperature variations with dissipated, as opposed to stored, energy [1]. TSA is a full-field, noncontacting technique that determines stress information for actual structures in their operating environments with a sensitivity similar to that of strain gages.

It is well known that the temperature of a gas decreases when the gas is expanded and increases when compressed. A similar effect occurs in solids. The temperature changes associated with elastic deformations in a solid are very small and until recently were considered to be of negligible importance. However, recent advances in infrared photodetector technology have made it possible to measure these temperature variations efficiently and reliably. Moreover, commercial TSA equipment is now available. In 1982 a British company, Ometron Ltd., began manufacturing and marketing the first commercial system, the SPATE 8000 [SPATE is an acronym for stress pattern analysis (by measurement of) thermal emission]. The SPATE 8000 and a subsequent improved version, the SPATE 9000 (Fig. 14-1), have been used extensively in the United States, the United Kingdom, and Australia during the past decade.

As with any new technology, the advances have been rapid and the challenges substantial. TSA is an emerging technique that to date has been used successfully for the evaluation and validation of design concepts, fracture mechanics, damage detection, fatigue monitoring, and residual stress analysis.

14-1 History and Theoretical Foundations

The first observation of the thermoelastic effect is attributed to Weber in 1830 [2]. He observed that when a vibrating wire receives a sudden increase in tension, it experiences a delay in the change of its fundamental frequency. He hypothesized that this delay was due to the increased

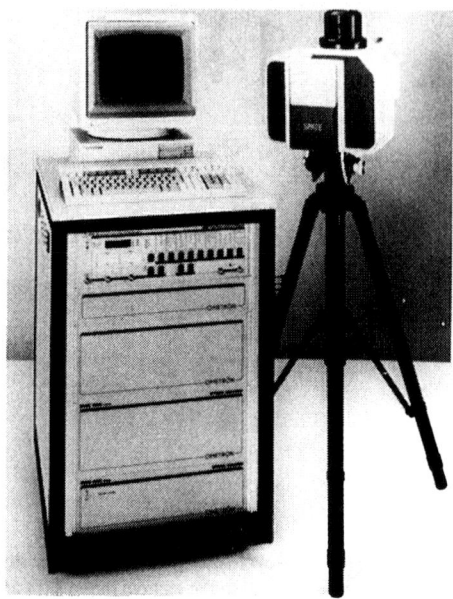

Figure 14-1 The SPATE 9000 thermoelastic stress analysis system. (Courtesy of Ometron Ltd.)

magnitude of stress causing a transient temperature change in the wire. The thermoelastic effect was given a theoretical foundation in 1853 by William Thomson [3], later known as Lord Kelvin. Thomson derived a linear relationship between the temperature change of a solid and the change in the sum of the principal stresses for isotropic materials. Compton and Webster confirmed this theory experimentally in 1915 [4].

The theoretical considerations of thermoelasticity remained relatively dormant until the 1930s when Zener published a series of articles on the internal friction of solids [5–8]. Rocca and Bever analytically investigated nonlinear effects and measured the thermoelastic response in iron and nickel near the Curie point [9]. During the 1950s Biot used irreversible thermodynamics to extend the theory to include anisotropic, viscoelastic, and plastic material responses [10–13].

In 1967 Belgen [14,15] conducted the first noncontacting thermoelastic stress analysis measurements. He used an infrared radiometer to relate the stresses to the small temperature changes measured in a vibrating cantilevered beam. Mountain and Webber developed the first prototype of the commercial SPATE system for a British Ministry of Defense project. They published early results of their work in 1978 [16].

The theoretical basis for thermoelastic stress analysis rests on the first and second laws of thermodynamics. Potter and Greaves recently derived a thermoelastic relationship for reversible, adiabatic thermodynamic events, which includes anisotropic material behavior, mean stresses, temperature-variant elasticity, and expansion coefficients [17]. Such behavior can be expressed as

$$\rho C_\varepsilon \frac{dT}{T} = \left[\frac{\partial C_{ijkl}}{\partial T} (\varepsilon_{kl} - \alpha_{kl} \Delta T) - C_{ijkl} \left(\alpha_{kl} + \Delta T \frac{\partial \alpha_{kl}}{\partial T} \right) \right] d\varepsilon_{ij} \qquad (14\text{-}1)$$

where ρ is the material density, C_ε the specific heat for constant deformation, T the temperature, C_{ijkl} the elasticity tensor, α_{kl} the thermal expansion coefficient tensor, and ε_{kl} the strain tensor.[1] Equation (14-1) assumes the standard summation convention; that is, repeated indices in a term indicate summation, with the indices running from 1 to 3.

Equation (14-1) is restricted to elastic deformations because its derivation includes the Duhamel–Neumann constitutive equation for linearly elastic thermoelasticity. Special forms of this equation have been derived to include irreversible events. Most notable is work by Enke and Sandor to include the effects of cyclic plasticity [18].

While Eq. (14-1) is cumbersome to work with, in most instances many of the terms are of negligible significance. For an isotropic material having temperature-independent elastic and thermal expansion coefficients, this relationship simplifies to

$$\Delta T = \frac{-\alpha}{\rho C_\varepsilon} T \Delta(\sigma_p + \sigma_q + \sigma_r)$$
$$= -K_m T \Delta(\sigma_p + \sigma_q + \sigma_r) = -K_m T \Delta(\sigma_x + \sigma_y + \sigma_z) \quad (14\text{-}2)$$

where σ_p, σ_q, σ_r, and σ_x, σ_y, σ_z are the principal and normal stresses, respectively. K_m is known as the thermoelastic material constant. Equation (14-2) is recognized as the classical TSA equation derived by Lord Kelvin. For plane-stress problems, σ_z is equal to zero and is omitted from Eq. (14-2).

Neither a state of pure shear stress nor a static load produces any thermoelastic response for an isotropic material. Therefore, interpretation of thermoelastic stress analysis results may require some insight from the loading and geometry of the component.

For zero mean stresses, Eq. (14-2) provides isopachic information, hence locates the center of Mohr's circle. If there is a mean stress about some nonzero cyclic loading, TSA provides only the change in isopachics. There are higher-order effects to Eq. (14-1) for a nonzero mean stress. This topic of nonzero mean stress is the subject of current research in the evaluation of residual stresses and is discussed in Section 14-6 (Applications).

For a plane-stressed orthotropic material whose stiffness and thermal expansion coefficients are independent of temperature, Eq. (14-1) may be expressed as

$$\Delta T = \frac{-1}{\rho C_\varepsilon} T(\alpha_{11}\Delta\sigma_{11} + \alpha_{22}\Delta\sigma_{22}) \quad (14\text{-}3)$$

where σ_{11} and σ_{22} are the normal stresses in the directions of material symmetry. This equation can be used advantageously if interpreted cautiously. The temperature information of Eq. (14-3) is no longer proportional to the isopachics but is a linear combination of the changes in the normal stresses in the directions of material symmetry.

The application of Eq. (14-3) for the quantitative stress analysis of composite materials can be quite difficult. To date many of the applications of thermoelastic stress analysis to composite materials have been more qualitative than quantitative, and the concept has been used primarily for damage detection and monitoring.

14-2 Equipment

A main component of a thermoelastic stress analysis system (Fig. 14-2) is a raster-scanning infrared detector capable of measuring the small temperature changes associated with the thermoelastic

---

[1] A list of the symbols used in this chapter precedes the References.

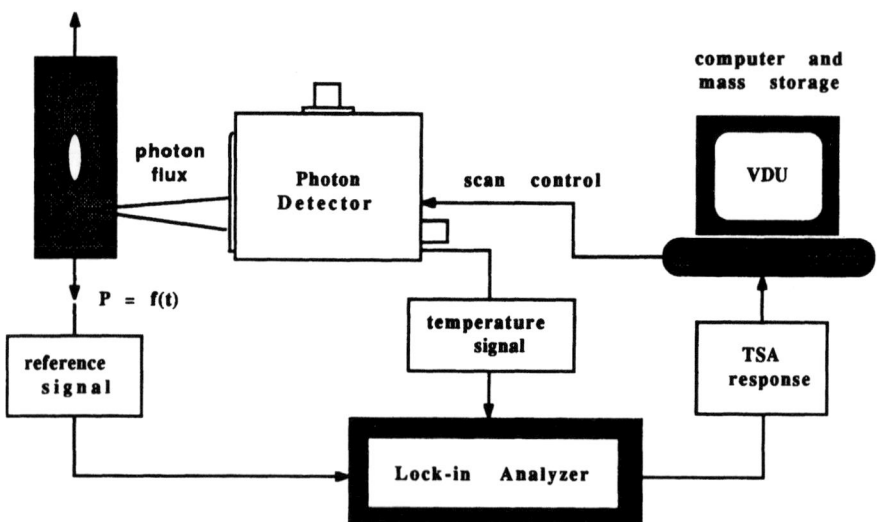

Figure 14-2 Schematic of typical thermoelastic stress analysis system.

effect (change in stresses). The infrared detector acts as a transducer, which converts the incident radiant energy into electrical signals. The commercial SPATE system uses a photon emission detector (photodetector) instead of a thermal detector largely because of its excellent sensitivity [19].

Infrared photon detectors have a semiconductor component that can be characterized by the band-gap energy required to excite the semiconductor to produce an output signal. Band-gap energy is the minimum energy required to cause an excited carrier to leave the valence energy band and enter the mobile conduction band. This occurs if the incident photons from the target specimen have sufficient energy [19]. Photons of a maximum wavelength and less will have the required energy to excite the semiconductor (i.e., the shorter wavelengths have higher energies).

The voltage output from a photodetector is proportional to the rate of incident photons. The photon emittance from a component being analyzed, known as the spectral radiant photon emittance, is an important parameter to consider when selecting a detector [20]. The detector will exhibit optimum sensitivity if the maximum number of photons striking the detector have sufficient energy to excite the semiconductor.

Planck's law relates the emittance of a blackbody to the wavelength and the temperature of the radiating body by

$$\phi_\lambda = \frac{2c}{\lambda^4}\left(\frac{1}{\exp(hc/\lambda kT) - 1}\right) \tag{14-4}$$

where ϕ_λ is the spectral radiant photon emittance, c the speed of light, λ the wavelength of light, h Planck's constant, and k Boltzmann's constant. At room temperature the peak photon emittance occurs at a wavelength of 12 μm. At higher temperatures the peak emittance shifts to shorter wavelengths [20].

The SPATE system uses a mercury-doped cadmium–telluride ($Hg_{1-x}Cd_xTe$) photovoltaic photon detector. By varying x of the semiconductor chemical composition, the required band-gap

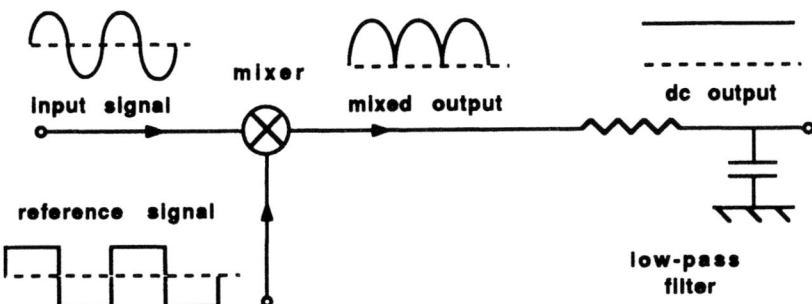

Figure 14-3 Schematic of simplified lock-in analyzer.

energy of the detector can be controlled. The maximum wavelength to which the detector will respond depends on the band-gap energy and therefore also on semiconductor composition. The SPATE system approximately has $x = 0.2$, for a band-gap energy of 0.1 eV, and it will respond to a peak photon wavelength of 12.4 μm. This detector therefore works well at room temperatures.

Photon detectors require liquid nitrogen cooling (77 K) to minimize background radiation effects and to improve the specific detectivity (signal-to-noise ratio). Unwanted radiation can be further reduced by decreasing the detector's angle of view. A cryogenic aperture placed directly in front of the detector will not reduce the target signal and can improve the specific detectivity as much as an order of magnitude compared to systems without such an aperture [21].

An ideal photon detector would be equally responsive to all photon wavelengths having sufficient energy to excite the detector. The Hg:CdTe detector is sensitive to wavelengths shorter than 12.4 μm. However, as a result of absorption and other dissipative effects from the atmosphere, a photodetector is restricted to two discrete windows of operation, 3 to 5 and 8 to 14 μm. A detector designed for the lower range has advantages for elevated-temperature testing. However, a detector chosen for this range would not be responsive to the peak spectral photon emittance wavelength at room temperature.

Assuming that surface emissivity does not vary with wavelength, the Stefan–Boltzmann law for photon detection expresses the total radiant photon emittance ϕ as [20]

$$\phi = eBT^3 \tag{14-5}$$

where e is the surface emissivity and B is the Stefan–Boltzmann constant. The detector output voltage S is linearly related to the total radiant photon emittance as $S = R\phi + R\phi_b$, where R is the photodetector responsivity and ϕ_b the incident photon rate for background emittance. Differentiating this expression with respect to time and substituting the result into Eq. (14-2) (assuming ϕ_b does not vary with time), the relationship between the detector output voltage and the first stress invariant I_1 for isotropic materials becomes

$$\frac{dS}{dt} = \frac{-3eRB\alpha T^3}{\rho C_\varepsilon} \frac{dI_1}{dt} \tag{14-6}$$

The lock-in analyzer of Fig. 14-3 is the signal-processing unit that extracts the thermoelastic information from the inherently noisy detector output signal. Typical output from a photon detector for high cyclic stress in a steel specimen may be a 0.020-V peak-to-peak signal buried in a large-bandwidth, 0.800-V peak-to-peak noise [19].

A lock-in analyzer may be described in a simplified manner (Fig. 14-3) as a series-connected signal mixer and a low-pass filter. The analog instrument mixes a reference signal at the load frequency with the detector output to extract the actual thermoelastic response [22]. The reference signal can originate from a function generator, load transducer, strain gage, or any other source that is in phase with the loading frequency.

Mixing two signals of distinct frequencies, say ω_1 and ω_2, produces an output signal with frequency content $\omega_1 + \omega_2$ and $\omega_1 - \omega_2$. The output signal from the photon detector will include the actual thermoelastic response at the reference frequency ω_r and other broadband frequencies due to the inherent noise. Mixing the thermoelastic response with the reference frequency will cause a portion of the resultant output to be a dc signal ($\omega_r - \omega_r = 0$) and a signal at twice the original frequency ($\omega_r + \omega_r = 2\omega_r$). Noise components are also mixed with the reference frequency, but their contribution to the dc signal is negligible.

The low-pass filter of the lock-in analyzer is used to recover the dc signal from the mixer output. The filter removes the portion of the signal above a cutoff frequency. Some detector noise that is close to the reference frequency will pass through the filter. The lock-in analyzer then normalizes the dc signal from the reference signal amplitude to a value proportional to the actual measured thermoelastic output.

The lock-in analyzer used with the SPATE system separates the detector output into two components: one in phase and the other 90° out of phase with respect to the reference signal. An adiabatic structural response is in phase with the reference loading frequency. If heat conduction occurs in regions of high stress gradients, or if the loading frequency is insufficient for adiabatic response, a phase shift from the reference frequency will occur and will be apparent as an out-of-phase detector signal. A quick confirmation of adiabatic response is to check for the absence of any out-of-phase detector signal.

The output-noise bandwidth of a low-pass filter is inversely proportional to the filter time constant. It is advantageous to select the filter time constant as large as possible to minimize the noise bandwidth. However, TSA rapidly scans a component with a small dwell time per measurement. A two-pole, low-pass filter takes 6.6 time constants to settle within 99% of a new value. Too long a time constant for a small dwell time will cause information to be "dragged" from previous locations. It is often advantageous to do most of the noise reduction with the analog-to-digital (A/D) conversion of the lock-in analyzer output instead of with the lock-in analyzer's low-pass filter.

Thermoelastic stress analysis can also be conducted without a lock-in analyzer. Harwood and Cummings [23,25] at the National Engineering Laboratory (NEL) in East Kilbride, U.K., have pioneered variable-amplitude (stationary random signal) TSA by replacing the lock-in analyzer with a fast Fourier transform (FFT) analyzer and have used entirely digital signal processing. They have devised techniques for measuring power spectral density, frequency-response functions, and performing modal-type analyses.

The TSA computer system of Fig. 14-2 performs three main functions. First, the scan area is selected and the mirrors are controlled to focus the component's photon emission onto the detector. Second, the computer samples the lock-in analyzer output signal (A/D conversion) and stores thermoelastic data from the component. Finally, the computer is used as a video display unit (VDU) to provide immediate information about the testing progress as well as to perform post-processing and data-reduction operations.

14-3 Test Materials and Methods

Thermoelastic stress analysis studies can be conducted on virtually any structure or material that is subjected to a sufficient dynamic load. The stress sensitivity for an isotropic material depends

Table 14-1 Common Engineering Materials Stress Sensitivity

Material	K_m (Pa^{-1})	$\Delta\sigma$, $T = 293$ K (MPa)	$\Delta\varepsilon$ Corresponding Strain Sum
Steel	3.5×10^{-12}	1	3.5×10^{-6}
Aluminum	8.8×10^{-12}	0.4	4.0×10^{-6}
Titanium	3.5×10^{-12}	1	6.5×10^{-6}
Epoxy	6.2×10^{-11}	0.055	11.0×10^{-6}
Magnesium alloy	1.405×10^{-11}	0.22	—
Glass	3.85×10^{-12}	0.91	—

on the thermoelastic constant K_m and the ambient temperature [Eq. (14-2)]. The minimum temperature sensitivity reported for the photodetector with the commercial SPATE system is 0.001°C. With this resolution, the available stress sensitivity for some common engineering materials has been determined, as listed in Table 14-1 [26]. These values compare favorably with the sensitivity available from strain gages. Moreover, the technique has the advantage of being full-field and noncontacting.

Thermoelastic stress analysis investigations have been performed on many different materials, components, and structures. Among the materials tested are structural steels, aluminum and magnesium alloys, boron aluminum, zirconium, titanium, ceramics, copper, brass, Plexiglas, epoxy, polyester, rubber, wood, brick, concrete, reinforced plastics, and human and animal bone.

A basic thermoelastic experiment requires little preparation compared to most stress measurement techniques. Vibration isolation is unnecessary, and the method is well suited for use in industrial settings. The SPATE equipment has shown some difficulty in producing consistent data when operated in extremely noisy environments (≥ 120 dBa). An acoustic hood has been developed at the National Engineering Laboratory in East Kilbride to alleviate this problem [19].

The only component preparation required for TSA is possibly a thin flat black surface coating. A high-emissivity coating is desirable for two reasons [19]. First, the black coating more closely approximates a blackbody, which enhances the photodetector sensitivity by increasing the photon flux from the structure. Second, the coating provides the uniform surface emissivity necessary for quantitative stress analysis. Most nonmetallic specimens do not require coatings, with the notable exceptions of brick and concrete. A surface coating is required for metallic materials because they tend to have both low and variable emissivity. Commercial flat black paints available in aerosol cans are well suited as surface coatings and often have emissivities of 0.9 and greater. Krylon Ultra Flat Black has been evaluated as having a surface emissivity of approximately 0.94 [26]. Commercial high-temperature paints are also suitable for elevated-temperature testing.

Current signal-processing capabilities allow loading frequencies between 0.5 Hz and 20 kHz. The minimum suitable frequency for a structure depends on the material properties, which affect heat transfer and the adiabatic loading requirements. The question of allowable loading frequencies for adiabatic response is addressed in the next section, Calibration and Experimental Considerations.

A computer is typically used to control the scanning and rate of data acquisition. It is often necessary to scan only a small portion of a structure. The recording time can be significantly reduced if it is not necessary to scan the entire component.

With the commercial SPATE system, the region of a structure that is to be scanned is indicated by a visual channel that accompanies the photodetector. The computer determines the scanning rate by controlling the detector mirrors and the sample time per point. Scans may require a few minutes for simple regions to 8 hours or more for precise measurements of structures with complex

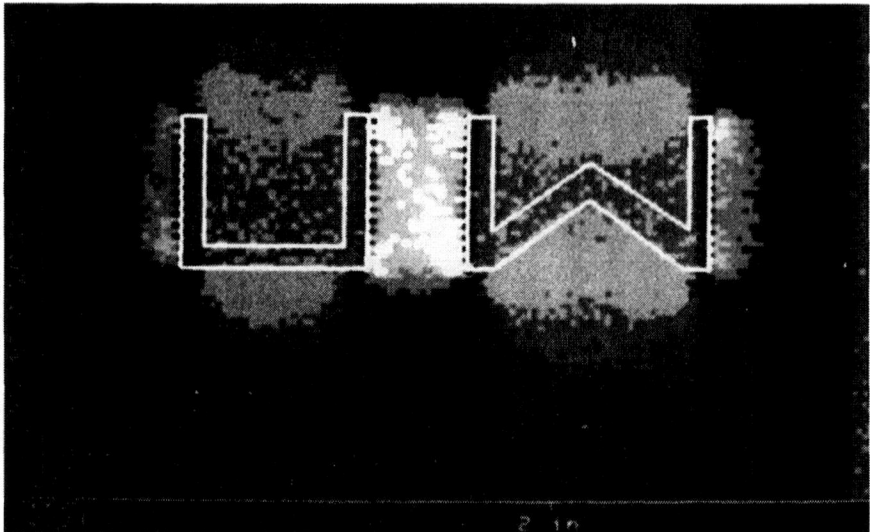

Figure 14-4 Isopachic stress contours of an aluminum plate loaded vertically in tension with the letters "UW" cut out. The letters are highlighted for clarity.

geometry or large stress gradients. The current SPATE system has a spatial resolution of 0.5 mm from a detector distance of 0.5 m. However, the manufacturer markets a lens attachment to improve this resolution to 0.15 mm for an 8-mm-diameter scan area.

A sample scan is shown in Fig. 14-4. The aluminum plate is 3 in. wide by $\frac{1}{8}$ in. thick with the letters UW machined through its thickness. The plate was loaded uniaxially at 20 Hz with an 800-lb mean load (2100 psi) and a 120-lb sinusoidal load (320 psi). A thin coat of a standard commercial flat black paint was applied to enhance the surface emissivity. The region scanned (3 in. × 2 in.) had a data acquisition time of approximately 1 hour. While the letters are highlighted for clarity, the sensitivity of the thermoelastic data is evident around the regions of stress concentration. Also, the inherent noise associated with thermoelastic measurements can be seen.

Postprocessing of recorded thermoelastic data may take many forms. Typically, the raw thermoelastic data are displayed in full-field, color-coded stress maps showing isopachic contours. Various filter and statistical operations can be performed on the data to produce more coherent results [27]. The ability to separate the thermoelastically recorded stress data into individual stress components is discussed in the next section.

Of course this section does not cover all the experimental methods associated with thermoelastic stress analysis, but only the most basic technique. More advanced techniques include the random-excitation method of Harwood and Cummings, mentioned earlier, and residual stress testing, fracture mechanics, and high-temperature applications. The latter will be discussed in the Applications section.

14-4 Calibration and Experimental Considerations

Quantitative analysis necessitates calibrating the thermoelastic signal. The calibration procedure may take either of two different forms. One concept is based on the theoretical relationship between the specimen material properties and measurement system characteristics. The other approach is based on comparing the measured thermoelastic signal with either a theoretical solution or a strain-gage response. Calibration methods described presently are for the classical thermoelastic equation [Eq. (14-2)]; slight modifications are necessary to include the orthotropic effects from Eq. (14-3).

At least for isotropic materials, the theoretical calibration approach relates the predicted digitally sampled photodetector voltage to the change in the sum of the principal stresses. The calibration constant is derived by combining the thermoelastic constant K_m (which includes material properties), detector and lock-in analyzer sensitivities, and various signal attenuation correction factors. These individual factors form a calibration constant based on thermoelastic theory and the physics of infrared detection. This approach has limited usefulness because it requires an accurate knowledge of many individual factors that are difficult to determine. However, this is a useful way to determine whether the classical thermoelastic equation (14-2) is the appropriate model or whether other factors (material nonlinearities, mean stress effects, etc.) need to be included in the theory [Eq. (14-1)]. Calibration measurements of this type are very well detailed in reports by Gutta [28] and Harwood and Cummings [29].

Experimental calibration is most reliably accomplished by preparing a specific calibration specimen using the same material, loading frequency, and ambient conditions as the test structure. The calibration specimen typically employs a geometry and loading for which the state of stress is known theoretically. Examples include a beam subjected to four-point bending, a disk diametrically compressed, or an axially loaded plate. Strain gages are located in several regions of low stress gradients, and their outputs are compared to the thermoelastic response signals (proportional to $\Delta T/T$) to determine a calibration factor. Several values should be averaged and compared to theoretical solutions [i.e., Eq. (14-2)] to minimize the effects of scatter and to provide a reliable value of K_m.

It is important to address what load frequency is acceptable for adiabatic material response. Adiabatic response requires that a balance be maintained between the thermal and mechanical energy; that is, the process must be reversible. The minimum loading frequency to accomplish this depends on the thermal conductivity and stress gradients in the structure. Most metals are safely adiabatic for frequencies above 2 Hz, but other factors such as the coating and system electronics may introduce concern below 10 Hz. Higher loading frequencies usually can improve the detail in a scan. For instance, experience indicates that about 20 Hz is a good loading frequency for minimum signal noise when analyzing aluminum components. For materials of decreasing thermal conductivity, however, the necessary frequency for adiabaticity increases. This is particularly true with nonmetallic components. A case in point is polyvinyl chloride (PVC), which may require a loading frequency of 100 Hz for an adiabatic response [30].

The presence or absence of nonadiabatic response fortunately can be observed by checking the phase of the response signal with respect to the load signal. If conduction is occurring in the component or coating, a detectable phase shift will occur between the reference and loading signals.

Coatings on the component are used to provide an enhanced and uniform surface emissivity. However, they introduce potential difficulties under certain conditions. A coating can act as an insulating layer and thereby drag down the true thermoelastic response or cause the detector signal to lag the loading frequency [30]. This effect becomes more pronounced with increased coating thickness and loading frequency. However, these difficulties can be ignored for a coating thickness between 20 and 30 μm and a loading frequency range of 5 to 200 Hz. Most thermoelastic stress analyses are conducted under these conditions.

The photodetector of a thermoelastic system does not focus on a mathematical point of zero dimensions, but instead focuses on an finite-sized spot in space. This situation can cause difficulty,

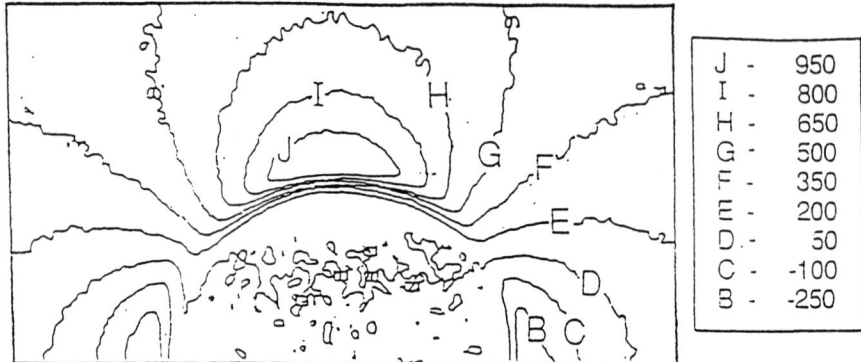

Figure 14-5 Isopachic stress contours around a central hole in a plate horizontally loaded. The contour lines near the edge of the hole illustrate the edge effect.

particularly on the edges of a specimen. An edge effect occurs because the detector "sees" a spot that is partly on the highly stressed specimen and partly on the stress-free background. The quality of an edge signal is further reduced by the cyclic motion of the structure, which provides different data to the detector from the different spatial positions. Maximum stresses often occur at an edge, and the lack of reliable thermoelastic edge data inhibits the ability to obtain such stresses accurately. This is one of the aspects that has caused many TSA analyses to be more qualitative than quantitative.

Figure 14-5 shows a clear example of the edge effect: measured isopachic contours appear around a central hole in a plate that is horizontally loaded. At the top of the hole, lower-valued contour lines are closer to the edge. This incorrectly suggests that the stress decreases as one approaches the hole. Away from the edge of the hole, the contour lines are reliable.

Fortunately, all is not lost with respect to the edge effect. Accurate stress concentration factors can be calculated by using the more reliable data slightly away from the edge [31].

Determining individual stress components from measured thermoelastic data is an important topic of contemporary research. To date, methods have been developed only for relatively simple, specific geometries. These include the work by Ryall et al. (described in the Application section of this chapter), Huang, Rowlands, et al. [31–36], Stanley [37] and Boyle and Hamilton [38]. Quantitative thermoelastic stress analysis has been hampered by noisy data and unreliable edge information. Most two-dimensional analyses employ equilibrium-based techniques that lead to boundary value problems. The edge difficulties obviously complicate this type of analysis and therefore have promoted smoothing and extrapolation techniques to enhance the measured edge information.

14-5 Composite Materials

Composite materials provide an interesting challenge and opportunity for thermoelastic stress analysis. As Eq. (14-3) demonstrates, the thermoelastic response for orthotropic media is not proportional to the sum of the change of the principal stresses, but rather a linear combination of the change in the stress components in the directions of material symmetry. Also, Wong [39] has shown that the diffusive heat transfer between plies of a laminate gives rise to a loading-frequency effect. These features complicate thermoelastic stress analysis of composite materials.

The effects of heat diffusion must be considered when working with laminated composite materials. While each ply of a laminate may be considered to be homogeneous, the material system as a whole may not [40]. Convective heat transfer from the surface of the component may be ignored, but diffusive heat transfer due to the varying stresses in the different plies can be significant. This effect is strongly dependent on the loading frequency.

The heat-transfer phenomenon described above changes the fundamental principle that the thermoelastic concept measures only the surface stress on a component. At lower loading frequencies, subsurface stress levels participate more in the thermoelastic response and several plies of a laminate may contribute. At higher frequencies, the response more nearly reflects the surface ply response. This interply heat transfer becomes negligible at sufficiently high frequencies. However, such frequencies depend on the constituent materials and may be too large for most conventional loading apparatuses. Surface coatings further complicate the thermoelastic response of composites, and their effects must be considered.

The diffusive heat transfer between composite plies produces both difficulties and opportunities. Since the surface temperature is no longer solely dependent on the response of the surface ply, subsurface effects may be investigated by applying various loading frequencies. Wong has also used this heat transfer to derive a potential stress separation scheme which is not applicable to isotropic materials [39]. Feng et al. [36] have evaluated the magnitude of the individual stresses thermoelastically across the center section of a tensile composite plate containing a central, circular hole.

Because of the difficulty of determining quantitative stress information for composite materials thermoelastically, most such analyses have consisted of qualitative, rather than quantitative nondestructive evaluation (NDE) [41,42]. The thermoelastic detection and characterization of damage by comparing TSA scans can successfully predict failure and regions of delamination, as well as identify locations of crack initiation. Quantitative NDE has been the goal of many researchers, and considerable effort has been expended in determining a damage parameter for predicting remaining life in structural components [43].

14-6 Applications

Partially because of its nonintrusive, full-field characteristics, thermoelastic stress analysis has proven to be a very diverse and effective engineering tool. Some of the applications are unique and deserve further elaboration, particularly those to residual stress analysis, fracture mechanics, and high-temperature testing. Other current and relevant engineering applications of thermoelastic stress analysis are described in the literature [44–53].

Residual Stress Analysis

The nondestructive determination of residual stresses is an active and emerging topic of thermoelastic research. Presently two main approaches of quantitative residual stress analysis are being pursued. The first technique, known as TERSA (thermal evaluation for residual stress analysis) [54], uses a laser to heat a statically loaded component. A mechanical chopper in the photon signal path provides a pseudodynamic signal. The other approach, elaborated upon below, is based on the second-order mean stress effects found in Eq. (14-1) [55].

Although in its early stages of development, the evaluation of residual stresses by the method of Ryall et al. has been successfully applied to the thermoelastic determination of mean stresses. An interesting by-product of this method is the determination of the individual stress components from the isopachic data. So far this approach has been limited to simple specimen geometries.

For isotropic materials, Eq. (14-1) can be rewritten as [56]

$$\rho C_\varepsilon \frac{\dot{T}}{T} = -\left[\alpha + \left(\frac{\nu}{E^2}\frac{\partial E}{\partial T} - \frac{1}{E}\frac{\partial \nu}{\partial T}\right)\sigma_{kk}\right]\dot{\sigma}_{kk} + \left(\frac{(1+\nu)}{E^2}\frac{\partial E}{\partial T} - \frac{1}{E}\frac{\partial \nu}{\partial T}\right)\sigma_{ij}\dot{\sigma}_{ij} \qquad (14\text{-}7)$$

where standard indicial notation is used. Integrating this expression and omitting higher-order terms shows that the thermoelastic signal observed at the loading frequency is proportional to [55]

$$I_1 - C_1 I_1^2 + C_2 \sigma_p \sigma_q \qquad (14\text{-}8)$$

Constants C_1 and C_2 are dependent on material properties. For a body subjected to dynamic plus static loadings, the thermoelastic signal then may be expressed in radial coordinates as proportional to the quantity

$$\sigma_{rr}^{(d)} + \sigma_{\theta\theta}^{(d)} - 2C_1(\sigma_{rr}^{(d)} + \sigma_{\theta\theta}^{(d)})(\sigma_{rr}^{(s)} + \sigma_{\theta\theta}^{(s)}) + C_2(\sigma_{rr}^{(d)}\sigma_{\theta\theta}^{(s)} + \sigma_{\theta\theta}^{(d)}\sigma_{rr}^{(s)} - 2\sigma_{r\theta}^{(d)}\sigma_{r\theta}^{(s)}) \qquad (14\text{-}9)$$

where the superscripts s and d refer to the static and dynamic stress amplitudes, respectively.

For two-dimensional elasticity, the equations of equilibrium $\sigma_{ij,j} = 0$, and compatibility, $\nabla^2 \sigma_{ij} = 0$, imply the existence of a scalar potential function Θ such that $\nabla^4 \Theta = 0$. In polar coordinates, the stresses are then given by

$$\sigma_{rr} = \frac{1}{r}\frac{\partial \Theta}{\partial r} + \frac{1}{r^2}\frac{\partial^2 \Theta}{\partial \theta^2} \qquad \sigma_{\theta\theta} = \frac{\partial^2 \Theta}{\partial r^2}$$

$$\sigma_{r\theta} = -\frac{\partial}{\partial r}\left(\frac{1}{r}\frac{\partial \Theta}{\partial \theta}\right) \qquad (14\text{-}10)$$

The form of the stress function Θ depends on the geometry and type of problem considered. These functions are infinite series with unknown coefficients that may be truncated after a sufficient number of terms to describe a solution with acceptable accuracy. The method chooses the same potential functions for the static and dynamic fields, differing only by the undetermined coefficients of the series terms. The problem is now reduced to determining the unknown series coefficients and evaluating Eq. (14-10) to determine the individual stress components. This is accomplished by substituting the potential functions and Eqs. (14-10) into Eq. (14-9) and performing a nonlinear least-squares fit to the measured thermoelastic data.

Using the determined coefficients, the individual components of the static and dynamic stresses are evaluated by Eqs. (14-10). Analyses of this type also may be used to smooth the measured thermoelastic data with the continuous function from the least-squares estimation to reconstruct isopachic results.

Ryall applied this concept to an aluminum plate with a central hole and subjected to an applied static load of 45 kN plus a sinusoidal dynamic load amplitude of 15 kN at 19 Hz. Figure 14-6 show the resolved patterns of dynamic tangential stress and the static shear stress. While the results presented are for static and dynamic stress patterns, current research involves modifying this technique to evaluate residual stresses.

Fracture Mechanics

Thermoelasticity has been used to determine stress-intensity factors and crack-tip velocities. Fracture mechanics requires high stress sensitivity and fine spatial resolution because of the localized

THERMOELASTIC STRESS ANALYSIS 593

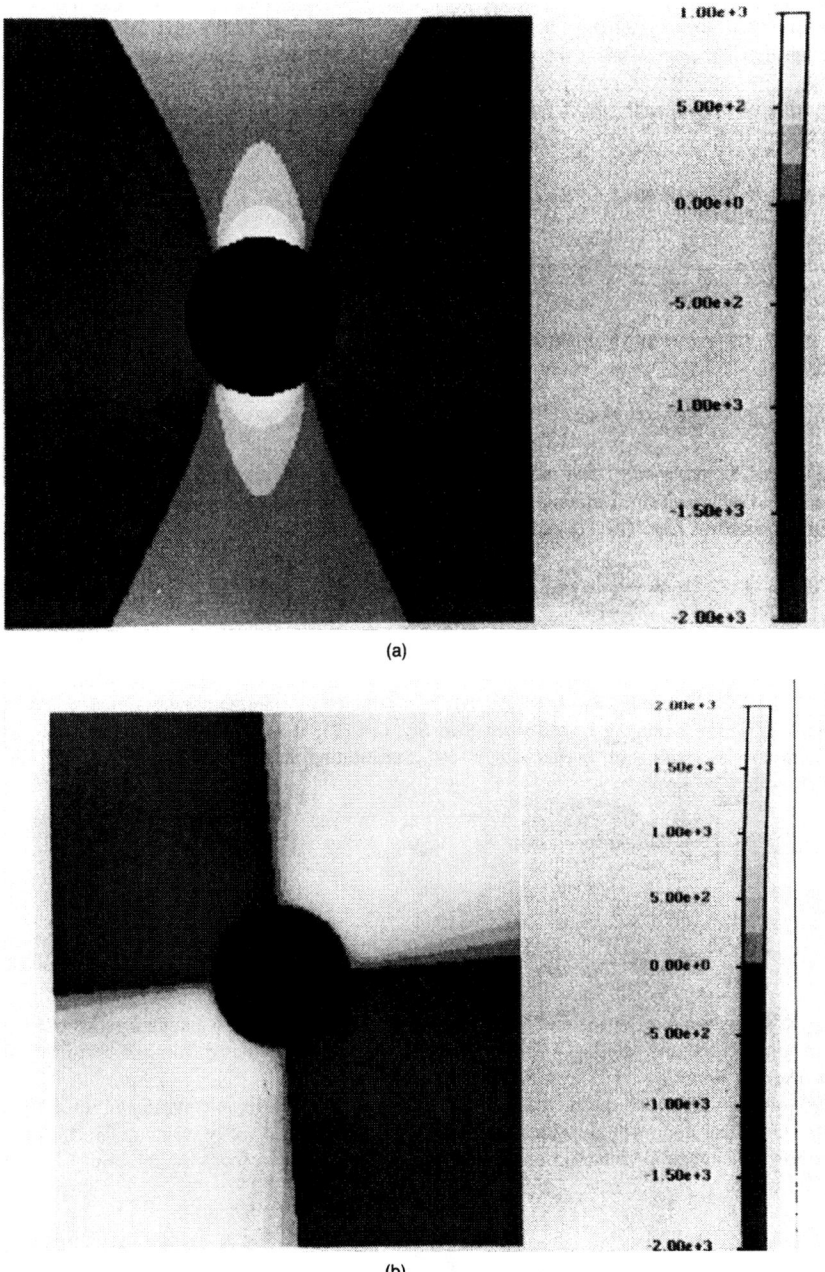

Figure 14-6 (a) Dynamic tangential stress and (b) static shear stress from an aluminum plate with a central hole loaded vertically. (Courtesy of T. G. Ryall, Aerospace Research Laboratory, Melbourne, Australia.)

regions of high stress gradients. Stanley and Chan [57,58], and Leaity and Smith [59] have published detailed papers about the methodology and limitations of thermoelasticity to fracture analysis.

Ignoring higher-order terms, Westergaard's equations for the elastic normal stresses near the crack tip are [60]

$$\sigma_x = \frac{K_I}{\sqrt{2\pi r}} \cos \tfrac{1}{2}\theta \, (1 - \sin \tfrac{1}{2}\theta \sin \tfrac{3}{2}\theta)$$

$$\quad - \frac{K_{II}}{\sqrt{2\pi r}} \sin \tfrac{1}{2}\theta \, (2 + \cos \tfrac{1}{2}\theta \cos \tfrac{3}{2}\theta)$$

$$\sigma_y = \frac{K_I}{\sqrt{2\pi r}} \cos \tfrac{1}{2}\theta \, (1 + \sin \tfrac{1}{2}\theta \sin \tfrac{3}{2}\theta) \qquad (14\text{-}11)$$

$$\quad + \frac{K_{II}}{\sqrt{2\pi r}} \sin \tfrac{1}{2}\theta \cos \tfrac{1}{2}\theta \cos \tfrac{3}{2}\theta$$

where K_I and K_{II} are mode I and mode II stress intensity factors and r, θ are polar coordinates with origin at the crack tip. Defining σ_m as the sum of the principal stresses (the thermoelastically measured quantity), Eqs. (14-11) can be added to obtain

$$\sigma_m = \sigma_x + \sigma_y = \frac{2K_I}{\sqrt{2\pi r}} \cos \frac{\theta}{2} - \frac{2K_{II}}{\sqrt{2\pi r}} \sin \frac{\theta}{2} \qquad (14\text{-}12)$$

This expression can be used to determine the stress intensity factors in either single- or mixed-mode loadings.

For mode I loading only ($K_{II} = 0$), K_I may be evaluated by scanning along a line parallel to the crack [57]. By replacing r with $y/\sin \theta$ in Eq. (14-12), it may be shown that $\sigma_{m,\max}$, the maximum normal stress sum, occurs at $\theta = 60°$. Substituting these results into Eq. (14-12) and squaring both sides provides

$$\sigma^2_{m,\max} = \left[\frac{2K_I^2}{\pi} \sin 60° \cos^2(30°) \right] \frac{1}{y}$$

or

$$y = \left[\frac{3\sqrt{3}}{4\pi} K_I^2 \right] \frac{1}{\sigma^2_{m,\max}} \qquad (14\text{-}13)$$

where y is the vertical distance from the crack to the line scanned. By scanning a series of lines parallel to the crack and plotting y versus $1/\sigma^2_{m,\max}$, K_I can be evaluated from the best-fit slope of the resulting line.

There are two thermoelastic methods of determining the crack-tip velocity da/dN for a propagating crack. The first method for finding crack-tip velocity is by directly scanning the crack and monitoring the change in distance between the peak thermoelastic responses. Values of C and m of the Paris law

$$\frac{da}{dN} = C(\Delta K)^m \qquad (14\text{-}14)$$

can be determined from measured values of ΔK, over the loading cycle. However, such an approach has not been reliable.

The second approach is to scan parallel to the crack in a region containing the fracture. The crack propagation will cause the resulting thermoelastic data to be "skewed" instead of the symmetric pattern that would result from a stationary crack. Stanley and Chan have detailed [57] how to relate this skewing to the crack-tip velocity.

Thermoelastic stress analysis is an effective technique for experimental fracture mechanics. However, the method is not without some limitations [59]. The mean stresses are associated with second-order thermoelastic effects. Consequently, neither they nor residual stresses are accounted for in stress-intensity factors. It is impractical to measure crack-tip velocities less than 10^{-5} mm/cycle. Also, it does not seem possible to account for crack closure effects during fatigue crack growth.

High-Temperature Stress Analysis

High-temperature thermoelasticity is another promising area of research. The noncontacting feature of the technique, and the ability to perform through a furnace window if necessary, are important here. Enke [61] has shown that thermoelasticity is able to maintain its stress sensitivity at elevated temperatures. Theoretically, there is no upper temperature limit to the technique, and measurements have been successfully made at temperatures in excess of 1000°C.

The only currently available commercial thermoelastic system is the SPATE 9000. The discussion of high-temperature analysis will therefore focus on this unit. The primary difficulties with high-temperature analyses occur because the photodetector is designed for use at room-temperature ambient conditions.

Equation (14-5) shows that photon emittance is highly temperature dependent. At elevated temperatures the emittance can saturate the photodetector. Photodetector saturation implies that the detector no longer returns to a neutral state upon removal of the incident radiation [20]. This problem can be alleviated by adding an attenuating filter to the system to reduce the incident radiation to an acceptable level.

As stated in the Equipment section of this chapter, the SPATE photodetector is designed to operate in the radiation wavelength range of 8 to 14 μm, to take advantage of peak spectral radiant emittance of structures at room temperature. However, at elevated temperature testing conditions, the window at 3 to 5 μm is advantageous. Fortunately, the SPATE detector will respond to this window if a notch filter is inserted into the optical path. Enke suggests some inexpensive laboratory materials that may perform the task sufficiently well.

Not all temperatures are equally desirable for thermoelastic testing. Certain temperature ranges are advantageous because the maximum photon flux falls within an acceptable atmospheric transmission window. If possible, testing should have an ambient condition around 170°C, between 460 and 540°C or between 620 and 950°C [19].

14-7 Summary

Thermoelastic stress analysis has achieved considerable popularity during the past decade, and its use continues to grow rapidly. Potential new uses for thermoelastic stress analysis emerge as the technique develops. Fundamental research in this area continues to produce significant advancements as software, hardware, and signal-processing techniques are refined.

Thermoelasticity offers several advantages compared with other methods. The technique provides full-field, noncontacting stress information with resolution similar to a strain gage and at small spatial increments. Virtually any type of material component or structure can be analyzed over a broad frequency range with little setup and specimen preparation. However, along with its advantages, thermoelasticity has limitations. Primary difficulties associated with this method include the general requirement of cyclic loading for adiabatic heat transfer, the unreliable edge data, and the fact that the output signal involves a linear combination of the stresses, not individual stress components. These features have tended to make TSA results, at least until recently, more qualitative than quantitative. This is particularly true with respect to composite materials, but research continues in this area.

Thermoelasticity is still in its infancy. Current research is addressing the method's shortcomings and producing advancements quickly. Continuing efforts are enabling this concept to become a practical tool for quantitative engineering applications.

Acknowledgment

The authors thank Dr. Neal Enke of Ford Motor Company for his valuable suggestions, comments and careful proofreading of this chapter.

List of Symbols

	B	Stefan–Boltzmann constant
	c	speed of light
	C_{ijkl}	fourth-order material elasticity tensor
	C_ε	specific heat for constant deformation
	e	emissivity
	h	Planck's constant
	I_1	first stress invariant $= \sigma_p + \sigma_q + \sigma_r$
	k	Boltzmann's constant
	K_m	thermoelastic constant
	R	photodetector responsivity
	S	photodetector output voltage
	t	time
	T	absolute temperature
	α_{ij}	thermal expansion coefficient tensor
	ε_{ij}	strain tensor
	Θ	stress functions
	λ	wavelength of light
	ν	Poisson's ratio
	ρ	material density
	σ_{ij}	stress tensor
	σ_m	stress sum $= \sigma_x + \sigma_y$
$\sigma_p, \sigma_q, \sigma_r$		principal stresses, $\sigma_p \geq \sigma_q \geq \sigma_r$
$\sigma_x, \sigma_y, \sigma_z$		normal stresses
	ϕ	total radiant photon emittance
	ϕ_b	incident photon rate for background emittance
	ϕ_λ	spectral radiant photon emittance

References

1. C. E. Bakis, and K.L. Reifsnider, Nondestructive Evaluation of Fiber Composite Laminates by Thermoelastic Emission, in *Review of Progress in Quantitative NDE Plenum* (D. O. Thompson and D. E. Chimenti Ed.), Williamsburg, VA, June 21–22, 1987, 1109–1116.
2. W. Weber, Über die specifische Warme fester Korper insbesondere der Metalle, *Ann. Phys. Chem.*, 96 (1830), 177–213.
3. W. Thomson, (Lord Kelvin) On the Dynamical Theory of Heat, *Trans Soc. Edinburgh*, 20 (1853), 261–283.
4. K. T. Compton, and D. B. Webster, Temperature Changes Accompanying the Adiabatic Compression of Steel. Verification of W. Thomson's Theory to a Very High Accuracy, *Phys. Rev.*, 5 (1915), 159–166.
5. C. Zener, Internal Friction in Solids. I. Theory of Internal Friction in Reeds, *Phys. Rev.*, 52 (1937) 230–235.
6. C. Zener, Internal Friction in Solids. II. General Theory of Thermoelastic Internal Friction, *Phs. Rev.*, 53 (1938), 90–99.
7. C. Zener, Internal Friction in Solids. IV. Relation Between Cold Work and Internal Friction, *Phys. Rev.*, 53 (1938), 582–586.
8. C. Zener, Internal Friction in Solids. V. General Theory of Macroscopic Eddy Currents, *Phys. Rev.*, 53 (1938), 1010–1013.
9. R. Rocca, and M. B. Bever, The Thermoelastic Effect in Iron and Nickel (as a Function of Temperature), *Trans. AIME*, 188 (1950), 327–333.
10. M. A. Biot, On Anisotropic Viscoelasticity, *J. Appl. Phys.*, 25 (1954), 1385–1391.
11. M. A. Biot, Plasticity and Consolidation in a Porous Anisotropic Solid, *J. Appl. Phys.*, 26 (1955), 182–185.
12. M. A. Biot, Irreversible Thermodynamics with Application to Viscoelasticity, *Phys. Rev.*, 97 (1955), 1463–1469.
13. M. A. Biot, Thermoelasticity and Irreversible Thermodynamics, *J. Appl. Phys.*, 27 (1956), 240–253.
14. M. H. Belgen, Structural Stress Measurements with an Infrared Radiometer, *ISA Trans.*, 6 (1967), 49–53.
15. M. H. Belgen, Infrared Radiometric Stress Instrumentation Application Range Study, NASA Report CR-1067, 1968.
16. D. S. Mountain, and J. M. B. Webber, Stress Pattern Analysis by Thermal Emission (SPATE), *Proc. Soc. Photoopt. Inst Eng.*, 164 (1978), 189–196.
17. R. T. Potter, and L. J. Greaves, The Application of Thermoelastic Stress Analysis Techniques to Fibre Composites, *Proc. SPIE*, 817, (1987), 134–146.
18. N. F. Enke, and B. I. Sandor, Cyclic Plasticity Analysis by Differential Infrared Thermography, *Proc. VI Int. Cong. Experimental Mechanics*, Portland, OR, June 6–10, 1988, 836–842.
19. N. Harwood, and W. M. Cummings, *Thermoelastic Stress Analysis*, Bristol, Adam Hiler, (1991).
20. N. F. Enke, Thermographic Stress Analysis of Isotropic Materials, Ph.D. thesis, University of Wisconsin—Madison, 1989.
21. R. J. Keyes, *Optical and Infrared Detectors*, 2nd ed., Springer-Verlag, New York, 1980.
22. EG&G Princeton Applied Research, *A Lock-In Primer*, EG&G, Princeton, NJ 1986.
23. N. Harwood, and W. M. Cummings The Theoretical Basis of the Use of Random Excitation Signals for Thermoelastic Stress Analysis, *2nd Int. Conf. Stress Analysis by Thermoelastic Techniques*, SPIE, Vol. 731 (B. Gasper, Ed.), London, 1986.
24. N. Harwood, and W. M. Cummings, Experimental Stress Analysis Under Modal Conditions Using an Infrared Technique, *Proc. 5th Int. Modal Analysis Conference*, Vol. 1, London, April 1987, 399–405.
25. N. Harwood, and W. M. Cummings, Frequency-Domain Analysis Techniques Developed for SPATE Response Signals, *Proc. SPIE*, 1084 (1989), 143–158.
26. D. E. Oliver, Stress Pattern Analysis by Thermal Emission, in *Handbook of Experimental Mechanics*, 1st ed., (A. S. Kobayashi, Ed.), Prentice-Hall, Englewood Cliffs, NJ, 1986.

27. B. R. Boyce, and D. E. Oliver, A Computerized Toolbox for the New Breed of Imaging Transducer, *Proc. VI Int. Cong. Experimental Mechanics*, Portland, OR, June 6–10, 1988, 819–824.

28. R. F. Gutta, *Stress Analysis Comparison Between Calculations, Conventional Strain Gages and Thermoelastic Technique*, General Electric Company, Schenectady, NY, 1990.

29. N. Harwood, and W. M. Cummings, Calibration and Qualitative Assessment of the SPATE Measurement System, National Engineering Laboratory, UK, NEL Report 705, 1986.

30. J. McKelvie, Consideration of the Surface Temperature Response to Cyclic Thermoelastic Heat Generation, *Proc., SPIE, 731* (1987), 44–53.

31. Y. M. Huang, R. E. Rowlands, and J. R. Lesniak, Simultaneous Stress Separation, Smoothing of Measured Thermoelastic Information, and Enhanced Boundary Data, *Exp. Mech., 30* (1990), 398–403.

32. Y. M. Huang, M. H. Abdelmohsen, D. Lohr, Z. Feng, R. E. Rowlands and P. Stanley, Determination of Individual Stress Components from SPATE Isopachics Only, *Proc. VI Int. Cong. Experimental Mechanics*, Portland, OR, June 6–10, 1988, 578–584.

33. Y. M. Huang, H. Abdelmohsen, and R. E. Rowlands, Determination of Individual Stresses Thermoelastically, *Exp. Mech., 30* (1990), 88–94.

34. Y. M. Huang, and R. E. Rowlands, Quantitative Stress Analysis Based on the Measured Trace of the Stress Tensor, *J. Strain Anal., 26* (1991), 55–63.

35. Y. M. Huang, and R. E. Rowlands, Experimental Thermoelastic Solution of Boundary-Value Stress Analysis, submitted for publication, 1993.

36. Z. Feng, D. Zhang, R. E. Rowlands, and B. I. Sandor, Thermoelastic Determination of Individual Stress Components in Loaded Composites, *Experimental Mechanics, 32(2)*, June (1992), 89–95.

37. P. Stanley, Stress Separation from SPATE Data for a Rotationally Symmetrical Pressure Vessel, *Proc. SPIE 1084* (1989), 72–83.

38. J. T. Boyle, and R. Hamilton, A Method of Thermographic Stress Separation, *4th Conference on Applied Solid Mechanics* (A. R. S. Porter, Ed.) Elsevier, Amsterdam, 1991.

39. A. K. Wong, A Non-Adiabatic Thermoelastic Theory and the Use of SPATE on Composite Laminates, *Int. Conf. Experimental Mechanics*, Copenhagen, Aug. 20–24, 1990.

40. A. K. Wong, A Non-Adiabatic Thermoelastic Theory for Composite Laminates, RAE Tech. Report TR90007, 1990.

41. C. E. Bakis, and K. L. Reifsnider, Nondestructive Evaluation of Fiber Composite Laminates by Thermoelastic Emission, *Review of Progress in Quantitative NDE, Plenum*, (D. O. Thompson and D. E. Chimenti, Eds.), Williamsburg, VA, June 21–22, 1987, 1109–1116.

42. C. E. Bakis, H. R. Yih, W. W. Stinchcomb, and K. L. Reifsnider, Damage Initiation and Growth in Notched Laminates Under Reversed Cyclic Loading, in *Composite Materials: Fatigue and Fracture*, (P. A. Lagace, Ed.), American Society for Testing and Materials, Philadelphia, 1989, 66–83.

43. M. Heller, S. Dunn, J. F. Williams, and R. Jones, Thermomechanical Analysis of Composite Specimens, *Compos. Struct., 11* (1988), 309–324.

44. C. E. Bakis, R. A. Simonds, L. W. Vick, and W. W. Stinchcomb, Matrix Toughness, Long-Term Behavior, and Damage Tolerance of Notched Graphite Fiber-Reinforced Composite Materials, in *Composite Materials: Testing and Design*, Vol. 9 (P. Garbo, Ed.), American Society for Testing and Materials, Philadelphia 1990, 349–370.

45. R. G. Bream, B. C. Gasper, B. E. Lloyd, S. W. J. Page, Application of Thermoelastic Stress Measurement to Both Modal Analysis and the Dynamic Behavior of Electrical Power Plant Structures, *Proc. SPIE, 817* (1987), 109–133.

46. J. L. Duncan, and A. K. Mackenzie, A Comparative Study of Stress Distributions in Human Bone Under Simple and Complex Loading Conditions and as Modified by Insertion of a Metallic Prosthesis, *Proc. SPIE, 1084*, (1989), 111–113.

47. K. Kageyama, K. Ueki, and M. Kikuchi, Fatigue Damage Analysis of Notched Carbon/Epoxy Laminates by Thermoelastic Emission and Three-Dimensional Finite Element Methods, *Proc. 7th Int. Conf. Composite Materials*, Beijing, August 1989.

48. J. R. Lesniak, Internal Stress Measurements *Proc. VI Int. Cong. Experimental Mechanics*, Portland, OR, June 6–10, 1988, 825–829.

49. A. J. Loader, W. B. Turner, and N. Harwood, Stresses in Vehicle Chassis Joints: A Comparison of SPATE with other Analysis Techniques *Proc. SPIE, 731* (1987), 149–153.
50. J. M. Madakacherry, M. Murphy, and R. Aiken, Thermoelastic Stress Analysis of an Automobile Engine, *Society for Experimental Mechanics Conf..*, New Orleans, June 1986, 931–938.
51. S. A. Rogers, Fatigue Cracking of Cooling Water Pipes, *Proc. SPIE, 1084* (1989); 119–128.
52. B. I. Sandor, and D. Zhang, Thermographic Stress Analysis of Bonded Joints, *ASNT Fall Conf.*, Valley Forge, PA, Oct. 9–13, 1989, 90–97.
53. D. Zhang, and B. I. Sandor, Thermographic Stress Analysis of Damage Evolution in Composite Joints (extended abstract only), *1989 SEM Spring Conf. Experimental Mechanics*, Cambridge, MA, May 1989, 13–17.
54. D. S. Mountain, and G. P. Cooper, Thermal Evaluation for Residual Stress Analysis (TERSA)— A New Technique for Assessing Residual Stress, *Proc. SPIE, 1084*, (1989) 103–110.
55. T. G. Ryall, P. M. Cox, and N. F. Enke, On the Determination of Dynamic and Static Stress Components from Experimental Thermoelastic Data, *Mechanics of Materials, Vol. 14, No. 1*, 1992, 47–59.
56. A. K. Wong, R. Jones, and J. G. Sparrow, Thermoelastic Constant or Thermoelastic Parameter?, *J. Phys. Chem. Solids, 48* (1987), 749–753.
57. P. Stanley, and W. K. Chan, The Determination of Stress Intensity Factors and Crack-Tip Velocities from Thermoelastic Infrared Emissions, *Proc. Int. Conf. Fatigue of Engineering Materials and Structures*, Vol. 1, Sheffield, September 1986, 105–114.
58. P. Stanley, and W. K. Chan, Mode II Crack Studies Using the SPATE Technique, *Proc. Int. Conf. Experimental Mechanics*, New Orleans, 1986, 916–923.
59. G. P. Leaity, and R. A. Smith, The Use of SPATE to Measure Residual Stresses and Fatigue Crack Growth, *Fatigue Fract. Eng. Mater. Struct., 12* (1989), 271–282.
60. G. C. Sih, On the Westergaard Method of Crack Analysis, *Int. J. Fract, 2* (1966), 628–631.
61. N. F. Enke, High Temperature Stress Analysis, Theory and Applications of Thermographic Stress Analysis, workshop notes of ASTM Committee Week, Atlanta, November 1988.

CHAPTER
15

Similitude, Modeling, and Dimensional Analysis

Donald F. Young

Iowa State University
Ames, Iowa

15-0 Introduction

Due to the general complexity of many engineering problems, it is often necessary to obtain solutions through experimentation, and it is, of course, highly desirable that the experimental results be as widely applicable as possible. It is in this regard that the concept of *similitude* is frequently used; that is, with properly designed and executed experiments, results obtained on one system can be used to describe the behavior of other similar systems. Experiments are commonly carried out in the laboratory utilizing a *model* to develop empirical formulations or to make specific predictions of certain characteristics of a given prototype. To use a model in this manner it is first necessary to establish certain well-defined relationships between the model and the "other" system. These relationships are called *similarity requirements, modeling laws,* or *model design conditions*, and they can be obtained using the principles of similitude. In this chapter these principles are developed, primarily through the use of *dimensional analysis*, and a variety of specific applications are described. In the following section certain basic concepts that provide the basis for dimensional analysis are considered. More detailed discussions of the various topics included in this chapter can be found in a number of texts [1–15].

15-1 Dimensions, Units, and Equations

Observations made during the course of an experiment have two general characteristics: *qualitative* and *quantitative*. The qualitative aspect of the observation serves to identify the nature, or type, of observation (e.g., length, time, stress, velocity, etc.), whereas the quantitative aspect provides a numerical measure of the observation. The quantitative characteristic requires both a number and a standard by which various observations can be compared. A standard for a length might be a meter or an inch, for time an hour or second, and for mass a slug or kilogram. Such standards are called *units*, and several systems of units are in common use. The qualitative characteristic is conveniently described in terms of certain *primary quantities*, such as length L, time T, mass M, and temperature Θ.[1] These primary quantities can then be used to provide a qualitative description of any other *secondary quantity*, for example, area $\doteq L^2$, velocity $\doteq LT^{-1}$, density $\doteq ML^{-3}$, and so on, where the symbol $\doteq$ is used to indicate the *dimensions* of the secondary quantity in terms of the primary quantities. The primary quantities are also referred to as *basic dimensions*.

For a wide variety of problems involving mechanical systems, only the three basic dimensions L, T, and M are required. Alternatively, L, T, and F could be used, where F is the basic dimension of force. Since Newton's law states that force is equal to mass times acceleration, it follows that $F \doteq MLT^{-2}$ or $M \doteq FT^2L^{-1}$. Thus, secondary quantities expressed in terms of M can be expressed in terms of F through the relationship above. For example, stress σ is a force per unit area, so that $\sigma \doteq FL^{-2}$, but an equivalent dimensional equation is $\sigma \doteq ML^{-1}T^{-2}$. Table 15-1 provides a list of dimensions for a number of quantities commonly used in experimental mechanics.

The general notion of "dimensions" dates back many centuries, with references to dimensions found in the works of ancient Greek scholars, but the foundations of dimensional analysis seem to have been first proposed by Fourier in the early nineteenth century in connection with his studies on the theory of heat. As outlined in a historical review of dimensional analysis by Macagno [16], further developments followed sporadically through the studies by Lord Rayleigh (1877–1878), Carvallo (1891), Vaschy (1892), Riabouchinsky (1911), Buckingham (1914), and Bridgman (1922).

Although there have been many long discourses on the philosophical aspects of dimensions, it is not necessary to pursue these discussions in this brief introduction. Nevertheless, as stated by Langhaar [3], "Dimensions serve a mathematical purpose. They are a code for telling us how the numerical value of a quantity changes when the basic units of measurement are subjected to prescribed changes. This is the only characteristic of dimensions to which we need to ascribe significance in the development of dimensional analysis." Thus, for some secondary quantity, such as velocity, which has the dimensions of LT^{-1}, it follows that a change in the basic units used in the measurement of length and time will yield a different numerical value for the velocity. For example, if $V = 50$ mph, a change to the Système International (SI) can be made by the following procedure:

$$V = (50 \text{ mi/h}) \frac{(5280 \text{ ft/mi})(0.3048 \text{ m/ft})}{3600 \text{ s/h}} = 22.35 \text{ m/s}$$

It is clear from this specific example that for any quantity that has dimensions, a change in the units used in the measurement of the basic dimensions will give a corresponding change in the numerical value of the quantity, and this change is related to the dimensions of the quantity. But *the ratio of the magnitudes of two like quantities is independent of the size of the units used in the measurement of each*, and this is a basic postulate on which dimensional analysis is based. This postulate applies to all commonly used measuring systems. For example, if the length and width of a table are 3 m and 1 m, respectively, the length/width ratio is 3, and if alternatively the length and width are measured in terms of feet, the ratio is still equal to 3. Based on this postulate it can be shown [1] that the dimensions of any secondary quantity S can be expressed in the form $S \doteq F^\alpha L^\beta T^\gamma \cdots$, where the exponents have values such as 0, 1, 2, an so on.

The second postulate on which dimensional analysis is based states that *the form of an equation that describes a certain phenomenon is independent of the size of units used in the measurement of the quantities involved*. For example, the normal stress σ, in a bar having a cross-sectional area A, loaded with an axial load P, is given by the equation $\sigma = P/A$, and this equation is valid regardless of the units selected for P and A. Such equations are called *complete equations*.

---

[1] A list of the symbols used in this chapter precedes the References.

SIMILITUDE, MODELING, AND DIMENSIONAL ANALYSIS

Table 15-1 Dimensions Associated with Common Physical Quantities

	Force System	Mass System
Acceleration	LT^{-2}	LT^{-2}
Angle	$F^0L^0T^0$	$M^0L^0T^0$
Angular acceleration	T^{-2}	T^{-2}
Angular velocity	T^{-1}	T^{-1}
Coefficient of thermal expansion	Θ^{-1}	Θ^{-1}
Density	$FL^{-4}T^2$	ML^{-3}
Energy	FL	ML^2T^{-2}
Force	F	MLT^{-2}
Frequency	T^{-1}	T^{-1}
Heat	FL	ML^2T^{-2}
Length	L	L
Mass	$FL^{-1}T^2$	M
Modulus of elasticity	FL^{-2}	$ML^{-1}T^{-2}$
Moment of a force	FL	ML^2T^{-2}
Moment of inertia (area)	L^4	L^4
Moment of inertia (mass)	FLT^2	ML^2
Momentum	FT	MLT^{-1}
Poisson's ratio	$F^0L^0T^0$	$M^0L^0T^0$
Power	FLT^{-1}	ML^2T^{-3}
Pressure	FL^{-2}	$ML^{-1}T^{-2}$
Specific heat	$L^2T^{-2}\Theta^{-1}$	$L^2T^{-2}\Theta^{-1}$
Specific weight	FL^{-3}	$ML^{-2}T^{-2}$
Spring constant (linear)	FL^{-1}	MT^{-2}
Strain	$F^0T^0T^0$	$M^0L^0T^0$
Stress	FL^{-2}	$ML^{-1}T^{-2}$
Surface coefficient of heat transfer	$FL^{-1}T^{-1}\Theta^{-1}$	$MT^{-3}\Theta^{-1}$
Surface tension	FL^{-1}	MT^{-2}
Temperature	Θ	Θ
Time	T	T
Thermal conductivity	$FT^{-1}\Theta^{-1}$	$MLT^{-3}\Theta^{-1}$
Torque	FL	ML^2T^{-2}
Velocity	LT^{-1}	LT^{-1}
Viscosity (dynamic)	$FL^{-2}T$	$ML^{-1}T^{-1}$
Viscosity (kinematic)	L^2T^{-1}	L^2T^{-1}
Work	FL	ML^2T^{-2}

In the following section it will be shown that a complete equation is *dimensionally homogeneous* (i.e., the dimensions of all terms in the equation are the same). For example, the equation for the velocity V of a uniformly accelerated body is

$$V = V_0 + at \tag{15-1}$$

where V_0 is the initial velocity, a the acceleration, and t the time interval. In terms of dimensions the equation is

$$LT^{-1} \doteq LT^{-1} + LT^{-1} \tag{15-2}$$

and thus Eq. (15-1) is dimensionally homogeneous.

Some equations that are known to be valid contain constants having dimensions. The equation for the distance d traveled by a freely falling body can be written as

$$d = 16.1t^2 \tag{15-3}$$

and a check of the dimensions reveals that the constant must have the dimensions of LT^{-2} if the equation is to be dimensionally homogeneous. Actually, Eq. (15-3) is a special form of the well-known equation from physics for freely falling bodies,

$$d = \frac{gt^2}{2} \tag{15-4}$$

in which g is the acceleration of gravity. Equation (15-4) is dimensionally homogeneous and valid in any system of units. For $g = 32.2$ ft/s² the equation reduces to Eq. (15-3) and thus Eq. (15-3) is valid only for the system of units using feet and seconds. Equations that are restricted to a particular system of units can be denoted as *restricted homogeneous equations,* as opposed to complete equations, which are *general homogeneous equations.* The preceding discussion indicates one rather elementary, but important application of dimensional analysis: the determination of the generality of a given equation simply based on a consideration of the dimensions of the various terms in the equation.

The two postulates described in this section serve as the basis for developing the fundamental theorem of dimensional analysis. Since these postulates are so broad and nonrestrictive in scope, it is apparent that dimensional analysis will be a powerful tool that can be universally applied to the great variety of problems encountered in experimental mechanics.

15-2 Theory of Dimensional Analysis[2]

Typically, in any given problem involving several variables, u, v, w, and so on, it can be presumed that there is some functional relationship between the variables such that

$$H(u, v, w, \ldots) = 0 \tag{15-5}$$

The second postulate, given in Section 15-1, states that the form of the specific equation that expresses this relationship should not depend on the system of units used in the measurement of the various quantities involved [i.e., Eq. (15-5) is a complete equation]. It follows from this postulate that quantities involved in a certain phenomenon must combine in a particular manner, as demonstrated by the following analysis. The quantities that appear in a given problem will be generally referred to as variables, although some, such as the acceleration of gravity, may be constant in any given problem. Thus, when the term "variable" is used, it should be interpreted to mean any quantity, including dimensional and nondimensional constants.

In terms of basic dimensions the variables can be expressed as

$$u \doteq F^{u_1}L^{u_2}T^{u_3} \qquad v \doteq F^{v_1}L^{v_2}T^{v_3} \qquad w \doteq F^{w_1}L^{w_2}T^{w_3} \qquad \text{etc.}$$

where it has been assumed for simplicity that only three basic dimensions are required. One of the variables is now selected, such as u, which involves the basic dimension F and a new set of variables

$$u \qquad v' = \frac{v}{(u)^{v_1/u_1}} \qquad w' = \frac{w}{(u)^{w_1/u_1}}, \qquad \text{etc.}$$

---

[2] The derivation presented in this section follows that given by E. B. Wilson, Jr., in his book *An Introduction to Scientific Research,* McGraw-Hill, New York, 1952, pp. 322-324.

SIMILITUDE, MODELING, AND DIMENSIONAL ANALYSIS

is defined. Note that for the new primed variables the basic dimension F has been eliminated, that is,

$$v' \doteq L^{v'_2}T^{v'_3} \qquad w' \doteq L^{w'_2}T^{w'_3}, \qquad \text{etc.}$$

Equation (15-5) can now be written as

$$H[u, v'(u)^{v_1/u_1}, w'(u)^{w_1/u_1}, \ldots] = 0 \qquad (15\text{-}6)$$

The system of units used to measure the basic dimension F is now changed. However, the function H can be influenced only through the variable u, since the primed variables are independent of F. Since Eq. (15-6) is presumed to be a complete equation, it should not depend on the system of units used. Thus the situation exists in which u is changing but the function H is not. Therefore, the function H, written in terms of the primed variables, must not depend on u, so that

$$H[u, v'(u)^{v_1/u_1}, w'(u)^{w_1/u_1}, \ldots] = P(v', w', \ldots) = 0 \qquad (15\text{-}7)$$

The process is now repeated to eliminate the basic dimension L by defining a new set of variables

$$v' \qquad w'' = \frac{w'}{(v')^{w'_2/v'_2}}, \qquad \text{etc.}$$

The double-primed variables will now be independent of the basic dimension L and Eq. (15-7) can be written as

$$P[v', w''(v')^{w'_2/v'_2}, \ldots] = 0 \qquad (15\text{-}8)$$

The same argument as before is used; that is, the system of units for L is changed, but the function P must not change, and it follows that

$$P[v', w''(v')^{w'_2/v'_2}, \ldots] = Q(w'', \ldots) = 0 \qquad (15\text{-}9)$$

This procedure is used for each basic dimension to arrive at the result

$$H(u, v, w, \ldots) = \phi(\pi_1, \pi_2, \ldots) = 0 \qquad (15\text{-}10)$$

in which the final variables π_1, π_2, and so on, are dimensionless and fewer in number than the original number of variables by the number of basic dimensions. This result may be summarized as follows:

> If an equation involving k variables is to be a complete equation, it can be reduced to a relationship among $k - r$ independent dimensionless products, where r is the number of basic dimensions required to describe the variables.

The dimensionless products are commonly referred to as *pi terms*, and the statement above, the *Buckingham pi theorem*[3] (or simply the "*pi theorem*"). Buckingham used the symbol π to represent a dimensionless group, and this notation is still widely used.

---

[3] As noted in Section 15-1, a number of investigators contributed to the development of dimensional analysis, and it is not clear why the name Buckingham is often linked with the theorem since his papers followed several earlier publications on the subject. Buckingham's publications, however, clearly indicated how dimensional analysis could be used in many practical problems and stimulated interest in the United States. See, for example, E. Buckingham, On Physically Similar Systems: Illustrations of the Use of Dimensional Equations, *Phys. Rev.*, **4** (1914), 345–376.

It is to be noted that Eq. (15-10) can be solved explicitly for any one of the pi terms

$$\pi_1 = \phi(\pi_2, \pi_3, \ldots) \tag{15-11}$$

and each of the pi terms will be of the general form $q_i/u^{a_i}v^{b_i}w^{c_i} \cdots$, where q_i is any variable in the original list of variables. Thus Eq. (15-11) can also be expressed as

$$q_i = u^{a_i}v^{b_i}w^{c_i} \cdots \phi(\pi_2, \pi_3, \ldots) \tag{15-12}$$

Since the function ϕ involves only dimensionless terms, it follows that all terms on the right-hand side of Eq. (15-12) will have the same dimensions as q_i on the left-hand side of the equation. It is therefore clear that all such equations will be dimensionally homogeneous.

15-3 Some Applications of Dimensional Analysis

Although the procedure for obtaining pi terms may appear to be rather long and tedious, it actually is not and can be applied in a simple, systematic manner. There are several procedures that can be used to develop pi terms, and three commonly used techniques are described below.

Repeating Variables

An examination of the steps followed in the derivation of the pi theorem reveals that what is actually done is equivalent to selecting a number of the original variables (equal to the number of basic dimensions required for the problem), and subsequently using these variables to eliminate the dependence of the remaining variables on the basic dimensions. As noted, the resulting dimensionless groups can be expressed in the form $q_i/u^{a_i}v^{b_i}w^{c_i} \cdots$, where the variables $u, v, w, \ldots$ are the *repeating variables* and the q_i's represent the remaining variables. The exponents $a_i, b_i, c_i, \ldots$ are determined so that the combination is dimensionless. The necessary steps for determining pi terms using the concept of repeating variables are as follows:

1. List all variables that are involved in the problem, and express each of the variables in terms of basic dimensions.
2. Select a number of repeating variables, where the number required is equal to the number of basic dimensions needed to describe all the variables. All required basic dimensions must be included within the group of repeating variables, and each repeating variable must be dimensionally independent of the others (i.e., the dimensions of one repeating variable cannot be reproduced by some combination of products of powers of the remaining repeating variables). Since the repeating variables will generally appear in more than one pi term, do not use the dependent variable as a repeating variable.
3. Form a dimensionless group by dividing one of the nonrepeating variables by the product of the repeating variables, each raised to an exponent that will make the combination dimensionless.
4. Repeat step 3 for each of the remaining nonrepeating variables. The required number of dimensionless products will be equal to the number of original variables minus the number of basic dimensions as prescribed by the pi theorem.

As an example of this procedure, the problem of the pressure drop per unit length along a smooth tube in which a viscous, incompressible fluid is flowing will be considered. The variables

SIMILITUDE, MODELING, AND DIMENSIONAL ANALYSIS

are specified, based on the experimenter's knowledge of the problem. For this example the pressure drop is assumed to be a function of four variables, that is

$$\Delta p = f(d, \mu, \rho, V) \qquad (15\text{-}13)$$

and the corresponding dimensions are:

$\Delta p \doteq FL^{-3}$ pressure drop per unit length

$d \doteq L$ pipe diameter

$\mu \doteq FL^{-2}T$ viscosity

$\rho \doteq FL^{-4}T^2$ density

$V \doteq LT^{-1}$ velocity

Since the number of basic dimensions is three in this example, three repeating variables are required and ρ, d, and V are selected for this purpose.

The repeating variables are now combined with one of the remaining variables to form a pi term; for example, let

$$\pi_1 = \frac{\Delta p}{\rho^a d^b V^c}$$

Since the pi terms must be dimensionless, it follows that

$$\frac{FL^{-3}}{(FL^{-4}T^2)^a(L)^b(LT^{-1})^c} \doteq F^0 L^0 T^0$$

and therefore

$$FL^{-3} \doteq (FL^{-4}T^2)^a(L)^b(LT^{-1})^c$$

The exponents on the corresponding basic dimensions on each side of the equation must be equal, and therefore the following three equations must be satisfied:

$1 = a$ for F

$-3 = -4a + b + c$ for L

$0 = 2a - c$ for T

It follows that $a = 1$, $b = -1$, and $c = 2$, and π_1 is therefore

$$\pi_1 = \frac{d\Delta p}{\rho V^2}$$

The process is repeated with the remaining variable so that

$$\pi_2 = \frac{\mu}{\rho^a d^b V^c}$$

and the required values for the exponents are $a = 1$, $b = 1$, and $c = 1$. The resulting pi term is

$$\pi_2 = \frac{\mu}{\rho d V}$$

According to the pi theorem there should be two pi terms for this problem, since there are five variables and three basic dimensions. The dimensionless products π_1 and π_2 which have been determined represent a suitable set and indicate that this problem can be studied using the new variables

$$\frac{d\Delta p}{\rho V^2} = \phi\left(\frac{\mu}{\rho V d}\right) \tag{15-14}$$

or alternatively,

$$\frac{d\Delta p}{\rho V^2} = \phi_1\left(\frac{\rho V d}{\mu}\right) \tag{15-15}$$

in which the dimensionless variable on the right-hand side of the equation is recognized as the Reynolds number.

Alternative Procedure

A different approach to the determination of pi terms can be developed by recognizing that the pi terms always consist of products of the variables raised to appropriate powers. If in a given problem there are k variables, $u_1, u_2, \ldots, u_k$, then any pi term is of the form $u_1^{x_1} u_2^{x_2} \cdots u_k^{x_k}$, where the x's are chosen so that the combination is dimensionless. Also, each of the variables can be expressed in terms of their basic dimensions in the form

$$u_i \doteq F^{a_i} L^{b_i} T^{c_i}$$

where a_i, b_i, and c_i are known for any variable. Thus the dimensions for any pi term can be expressed as

$$(F^{a_1} L^{b_1} T^{c_1})^{x_1} (F^{a_2} L^{b_2} T^{c_2})^{x_2} \cdots (F^{a_k} L^{b_k} T^{c_k})^{x_k}$$

To make the product dimensionless, the x's must be selected so that the dependence of the product on each basic dimension is eliminated. For example, to eliminate F,

$$a_1 x_1 + a_2 x_2 + \cdots + a_k x_k = 0 \tag{15-16}$$

Similarly, to eliminate L,

$$b_1 x_1 + b_2 b_2 + \cdots + b_k x_k = 0 \tag{15-17}$$

and to eliminate T,

$$c_1 x_1 + c_2 x_2 + \cdots + c_k x_k = 0 \tag{15-18}$$

Thus, to make the product dimensionless, these three algebraic equations must be satisfied simultaneously.

There will be as many equations as basic dimensions, and the number of unknown x's will equal the number of variables. Any set of x's that satisfies this system of equations will yield a dimensionless product. Since there will always be more unknowns than equations, a unique solution will not exist. However, from the theory of linear algebraic equations it is known that the number of linearly independent solutions is equal to the number of unknown x's minus the rank of the matrix of coefficients, where this matrix is of the form

$$\begin{bmatrix} a_1 & a_2 & \cdots & a_k \\ b_1 & b_2 & \cdots & b_k \\ c_1 & c_2 & \cdots & c_k \end{bmatrix}$$

SIMILITUDE, MODELING, AND DIMENSIONAL ANALYSIS

and the *rank* is the highest-order, nonzero determinant that is contained within the matrix. This matrix is called the *dimensional matrix*. The rank is usually equal to the number of basic dimensions. Since each independent solution will give a pi term, it can be concluded that the number of independent dimensionless products obtained from this method is equal to the number of variables minus the rank of the dimensional matrix (number of basic dimensions). This result is, of course, consistent with the pi theorem.

To illustrate this procedure with a specific example, the problem of pressure drop along a tube will again be considered. Since the variables are Δp, ρ, d, V, and μ, the pi terms are of the form

$$(\Delta p)^{x_1}(\rho)^{x_2}(d)^{x_3}(V)^{x_4}(\mu)^{x_5}$$

and in terms of basic dimensions

$$(FL^{-3})^{x_1}(FL^{-4}T^2)^{x_2}(L)^{x_3}(LT^{-1})^{x_4}(FL^{-2}T)^{x_5}$$

For this combination to be dimensionless, three auxiliary equations must be satisfied:

$$x_1 + x_2 + 0 + 0 + x_5 = 0 \quad \text{for } F \tag{15-19}$$

$$-3x_1 - 4x_2 + x_3 + x_4 - 2x_5 = 0 \quad \text{for } L \tag{15-20}$$

$$0 + 2x_2 + 0 - x_4 + x_5 = 0 \quad \text{for } T \tag{15-21}$$

The dimensional matrix can, therefore, be written as

	Δp	ρ	d	V	μ
F	1	1	0	0	1
L	-3	-4	1	1	-2
T	0	2	0	-1	1

The rank of the matrix is three and there will be two linearly independent solutions for the x's.

There are three equations [Eqs. (15-19), (15-20), and (15-21)] available for solving for five unknowns, and arbitrary values can be assigned to two of the unknowns. Many combinations are possible. For example, let

$$x_1 = 1$$

$$x_5 = 0$$

Equations (15-19) through (15-21) then become

$$1 + x_2 = 0$$

$$-3 - 4x_2 + x_3 + x_4 = 0$$

$$2x_2 - x_4 = 0$$

and it then follows that $x_2 = -1$, $x_3 = 1$, $x_4 = -2$. From this result

$$\pi_1 = (\Delta p)^1 (\rho)^{-1}(d)^1(V)^{-2}(\mu)^0$$

and therefore

$$\pi_1 = \frac{d\Delta p}{\rho V^2}$$

For the second pi term, let $x_1 = 0$ and $x_5 = 1$, and it follows that $x_2 = -1$, $x_3 = -1$, and $x_4 = -1$, so that

$$\pi_2 = \frac{\mu}{\rho d V}$$

This result is the same as that obtained previously. However, the same pi terms were obtained by judicious selection of the assumed values for two of the x's. An infinite number of correct pi terms exist, since there is an infinite number of combinations of x's that could be assumed. The only guide at this point is to keep the pi terms as simple as possible. This can be achieved by setting assumed values equal to unity or zero. It can be shown that although there are an infinite number of pi terms possible, once a suitable set has been found, all the rest can be formed by products of this set raised to various powers.

Forming Pi Terms by Inspection

The two methods described above provide a step-by-step procedure which, if executed properly, will provide a correct and complete set of pi terms. Although these methods are simple and straightforward, they are rather tedious, particularly for problems in which large numbers of variables are involved. Since the only restrictions placed on the pi terms are that they be (1) dimensionless, (2) independent, and (3) correct in number, it is possible to simply form them by inspection, without resorting to one of the more formal procedures. The following example will be used to illustrate this technique.

The end deflection of a cantilever beam having a rectangular cross section and subjected to a concentrated load at its free end is to be considered. The deflection is assumed to be a function of the following variables:

$$\delta = f(b, d, l, E, P) \tag{15-22}$$

where $\delta \doteq L$ end deflection
 $b \doteq L$ beam width
 $d \doteq L$ beam depth
 $l \doteq L$ beam length
 $E \doteq FL^{-2}$ modulus of elasticity
 $P \doteq F$ load

Application of the pi theorem indicates that four pi terms are required. Let π_1 contain the variable of interest, which is δ in this example. Now divide δ by some other variable having the dimension of length, such as l, to form the pi term

$$\pi_1 = \frac{\delta}{l}$$

Now select a second variable b, and again divide by l to obtain

$$\pi_2 = \frac{b}{l}$$

Similarly,

$$\pi_3 = \frac{d}{l}$$

SIMILITUDE, MODELING, AND DIMENSIONAL ANALYSIS

Note that independence of each pi term from those preceding it can be assured by incorporating a new variable in each pi term (i.e., π_1 contains δ, π_2 contains b, and π_3 contains d,) so that it is clear that these pi terms are independent. The final pi term can be formed by dividing P by El^2 so that

$$\pi_4 = \frac{P}{El^2}$$

Actually, El^2/P or P/Eb^2 or P/Ed^2 would be suitable, and the final choice is arbitrary. This problem can now be studied using the pi terms

$$\frac{\delta}{l} = \phi\left(\frac{b}{l}, \frac{d}{l}, \frac{P}{El^2}\right) \tag{15-23}$$

Although this procedure is equivalent to the first method described, it is less structured. But with a little practice, the pi terms can be readily formed by inspection, and this method offers an attractive alternative to the more formal procedures.

It may be concluded that on the basis of dimensional considerations, problems can be simplified and described in terms of dimensionless variables. The application of dimensional analysis provides three significant advantages.

1. The number of variables to be considered is always reduced.
2. The new dimensionless variables enable experiments to be performed more economically and efficiently.
3. Relationships developed on the basis of dimensionless variables are general and are not restricted to any particular system of units.

The usefulness of dimensional analysis in a variety of applications will be illustrated in more detail in subsequent sections, along with considerations of difficulties and pitfalls associated with this powerful tool.

15-4 Theory of Models

The term "model" is widely used in many different contexts. For example, models may be conceptual, in the sense that they are used to visualize or understand phenomena, and they may not in fact resemble the actual system. Frequently, this type of model is intended to be an instructional tool. Although conceptual models may be quite useful, the term "model" as commonly used in engineering implies an entity that provides more detailed information. The "engineering model" generally conforms to the following definition. *A model is a representation of a physical system that may be used to predict the behavior of the system in some desired respect.* The physical system for which the predictions are to be made is called the *prototype*. Note that the term "predict" appears in this definition, which implies that the model is more than a conceptual model (i.e., it can be used to make accurate, quantitative predictions of the behavior of the prototype).

There are two broad classes of engineering models: *physical models* and *mathematical models*. The physical model resembles the prototype in appearance but is usually of a different size, may involve different materials, and frequently operates under loads, temperatures, and so on, that differ from those of the prototype. The mathematical model consists of one or more equations that describe the behavior of the system of interest. The equations are based on certain basic laws and principles, and usually involve simplifying assumptions. The number of required assumptions can often be reduced by the use of a physical model, since it represents the phenomenon of interest on a different scale. The most complex problems can be studied with the aid of physical models, and major engineering projects involving structures, aircraft, ships, air and water pollution, rivers, harbors, dams, and so on, frequently involve the use of physical models. With the successful development of a valid model, it is possible to predict the immediate and future behavior of the prototype under a set of specified inputs, and to examine a priori the effect of various possible design modifications. Usually, a model is smaller than the prototype, and therefore less expensive and more easily handled in the laboratory; and, if necessary, tests on models can be carried to destruction, whereas this type of testing on a prototype is generally not practical. Also, experimental observations made on models can provide basic data that are useful in formulating general laws or empirical relationships. In the remainder of this chapter the principles underlying the design of physical models will be presented, along with examples taken from several different engineering disciplines.

Model Design Conditions

The theory of models can be readily developed using the principles of dimensional analysis. It has been shown that any given problem can be described in terms of a set of pi terms as

$$\pi_1 = \phi(\pi_2, \pi_3, \ldots, \pi_n) \tag{15-24}$$

In formulating this relationship, only a knowledge of the general nature of the physical phenomenon, and of the variables involved, is required. Specific values for variables, size of components, types of materials, and so on are not needed to perform the dimensional analysis. Thus Eq. (15-24) applies to any system that is governed by the same variables. If Eq. (15-24) describes the behavior of a particular prototype, a similar relationship can be written for a model of this prototype, that is,

$$\pi_{1m} = \phi_m(\pi_{2m}, \pi_{3m}, \ldots, \pi_{nm}) \tag{15-25}$$

where the form of the function will be the same as long as the same phenomenon is involved in both the prototype and the model.

The pi terms can be developed so that π_1 contains the variable that is to be predicted from observations made on the model. Therefore, if the model is designed and operated under the following conditions

$$\pi_{2m} = \pi_2$$
$$\pi_{3m} = \pi_3 \tag{15-26}$$
$$\vdots$$
$$\pi_{nm} = \pi_n$$

then with the presumption that the form of ϕ is the same for model and prototype, it follows that

$$\pi_1 = \pi_{1m} \tag{15-27}$$

Equation (15-27) is the desired *prediction equation* and indicates that the measured value of π_{1m} obtained with the model will be equal to the corresponding π_1 for the prototype as long as the other pi terms are equal. The conditions specified by Eqs. (15-26) provide the model design conditions, also called similarity requirements or modeling laws.

SIMILITUDE, MODELING, AND DIMENSIONAL ANALYSIS

As a simple example of the procedure, consider the beam deflection problem of the preceding section. A dimensional analysis of this problem indicated that four pi terms are involved, that is,

$$\frac{\delta}{l} = \phi\left(\frac{b}{l}, \frac{d}{l}, \frac{P}{El^2}\right) \tag{15-23}$$

Since this relationship applies to both prototype and model, Eq. (15-23) is assumed to govern the prototype, with a similar relationship

$$\frac{\delta_m}{l_m} = \phi_m\left(\frac{b_m}{l_m}, \frac{d_m}{l_m}, \frac{P_m}{E_m l_m^2}\right) \tag{15-28}$$

for the model. The design conditions are therefore

$$\frac{b_m}{l_m} = \frac{b}{l} \qquad \frac{d_m}{l_m} = \frac{d}{l} \qquad \frac{P_m}{E_m l_m^2} = \frac{P}{El^2}$$

The size of the model cross section is obtained from the first two design conditions, and they indicate that the pertinent model lengths must be scaled as

$$b_m = \frac{l_m}{l}b \quad \text{and} \quad d_m = \frac{l_m}{l}d \tag{15-29}$$

The remaining design condition yields the required model loading,

$$P_m = \frac{E_m}{E}\left(\frac{l_m}{l}\right)^2 P \tag{15-30}$$

It is to be noted that this model design requires not only geometric scaling, but also the correct scaling of the load. This result is typical of most model designs (i.e., there is more to the design than simply scaling the geometry). With the design conditions satisfied, the prediction equation becomes

$$\delta = \frac{l}{l_m}\delta_m \tag{15-31}$$

Thus measured model deflections must be multiplied by the ratio of the prototype length to model length to obtain the corresponding prototype deflection.

Scales

It is clear from the foregoing example that the ratio of the like quantities for prototype and model naturally arise from the design conditions (e.g., the design condition involving the beam width and length requires that $b/b_m = l/l_m$). The ratio l/l_m or b/b_m is the *length scale* or the *length scale factor*. For most models there will be only one length scale, and all lengths are scaled in accordance with this scale. There are, however, other scales such as the force scale P/P_m, and the modulus of elasticity scale E/E_m. In fact, a scale can be defined for each of the variables in the problem. Thus, it is actually meaningless to talk about the "scale" of a model without specifying which scale. In this chapter the length scale will be designated n_l, and other scales as n_P, n_E, and so on, where the subscript indicates the particular scale. Also, the ratio of the prototype value of the model value will be defined as the scale (rather than the inverse). With this definition, the length scale will generally be greater than unity since the model is usually smaller than the prototype.

The number of scales that can be arbitrarily designated in a given problem is equal to the number of basic dimensions required to describe the variables. All other scales are then expressible in terms of these specified scales. For example, derived quantities such as area A, moment of inertia I, velocity V, and pressure p, can be determined in terms of the length, time, and force scales so that

$n_A = n_l^2$ area

$n_I = n_l^4$ moment of inertia

$n_V = n_l n_t^{-1}$ velocity

$n_p = n_F n_l^{-2}$ pressure

It is apparent that the basic procedure for the design of a model is straightforward, but the validity of the design depends on the correctness of the dimensional analysis. If important variables are omitted in the formulation of the problem, the list of required model design conditions will be incomplete and, therefore, $\pi_1 \neq \pi_{1m}$. In some problems it may not be possible to satisfy one or more of the design conditions, as a result of practical difficulties related to material selection, size limitations, and so on, and again $\pi_1 \neq \pi_{1m}$. These models are classified as *distorted models* and may still be useful if the distortion can somehow be taken into account. A further discussion of distorted models is given in a following section.

15-5 Structural Models

The use of models to determine motion, deformation, fracture, buckling, and so on in various types of structures is commonplace. Although for many of these problems the required information can be obtained from analytical or numerical solutions, there remains a large class of structural problems that can best be studied through physical models. In some circumstances structural models represent one of the simplest classes of models from which accurate predictions of prototype behavior can be obtained. This is particularly true if the materials exhibit linear elastic behavior (i.e., if they obey Hooke's law). Conversely, some structural modeling problems may be exceedingly difficult (e.g., problems involving inelastic material behavior, fracture, dynamic loading, etc.). Although it is not possible to cover the entire spectrum of pertinent structural problems that have been, or could be, studied by means of models, a brief account of the more salient aspects of structural modeling, along with commonly encountered difficulties, is given in this section.

Elastic Structures Under Static Loading

Since material properties generally play an important role in problems involving structures, it is necessary to prescribe a constitutive relationship that relates applied forces to deformations. This is usually accomplished through the use of stress–strain equations which, for *linear elastic* material behavior, are characterized by two constants, the modulus of elasticity E and Poisson's ratio ν.[4] Many common structural materials are completely defined by these two material properties if the stresses do not exceed the proportional limit.

Consider a structure that is subjected to one or more external loads designated as P and P_i, where P_i is equivalent to a set of loads $P_1, P_2, \ldots$. The geometry of the structure can be described by a set of lengths l and λ_i, where l is some characteristic length and λ_i represents all other lengths

---

[4] Alternatively, E and G, the shear modulus, or G and ν, could be used since $G = E/[2(1 + \nu)]$. In general, however, two elastic constants are required to characterize the material properties for linear elastic behavior.

required to define the geometry. For the present it is assumed that the material behavior is linear elastic, and the effect of the weight of the structure is not considered.

With these conditions any stress component σ at a point x_i, can be expressed in the form

$$\sigma = f(l, x_i, \lambda_i, P, P_i, E, \nu) \tag{15-32}$$

Dimensional analysis now yields

$$\frac{\sigma l^2}{P} = \phi\left(\frac{x_i}{l}, \frac{\lambda_i}{l}, \frac{P_i}{P}, \frac{P}{El^2}, \nu\right) \tag{15-33}$$

Similarity requirements are obtained by making the pi terms on the right side of Eq. (15-33) equal between model and prototype. Equality of the first two pi terms, x_i/l and λ_i/l, requires that *geometric similarity* be maintained between model and prototype, not only with regard to shape but also with respect to prescribed coordinates.

The loading scale is established from the relationship

$$\frac{P}{P_m} = \frac{E}{E_m} n_l^2 \tag{15-34}$$

Note that the model and prototype materials need not be the same, but the elastic moduli scale and the length scale fix the loading scale. All additional loads P_i must be in the same ratio, that is

$$\frac{P_i}{P_{im}} = \frac{P}{P_m} \tag{15-35}$$

The last pi term in Eq. (15-33) imposes the rather stringent similarity requirement that Poisson's ratio be equal for model and prototype materials. The possible relaxation of this restriction is discussed in a subsequent section.

If all the aforementioned design conditions are satisfied, the prediction equation is

$$\frac{\sigma l^2}{P} = \frac{\sigma_m l_m^2}{P_m} \tag{15-36a}$$

and when combined with Eq. (15-34), it yields

$$\frac{\sigma}{\sigma_m} = \frac{E}{E_m} \tag{15-36b}$$

Note that stresses at corresponding points will be the same in model and prototype only for the case in which the model and prototype are constructed of the same materials.

Since any displacement component u or strain component ε will be a function of the same variables given in Eq. (15-33), it follows that the same model design conditions are required for displacements and strains as for stresses. The corresponding prediction equations become

$$\frac{u}{u_m} = n_l \tag{15-37}$$

and

$$\varepsilon = \varepsilon_m \tag{15-38}$$

that is, the displacements scale varies as the length scale, whereas the strains are equal in model and prototype. Similarly, for the prediction of statically indeterminate forces R,

$$\frac{R}{R_m} = \frac{P}{P_m} \tag{15-39}$$

It should be emphasized that *these scaling laws for elastic structures are valid for both small and large deformations, as long as the material in both model and prototype obeys Hooke's law.* Loads of other types (e.g., line, surface, and volume loads, or boundary displacements) can be readily incorporated into the analysis.

Although only structures constructed of a single material have been considered, the analysis can be readily extended to *composite structures* if all materials involved in the structure obey Hooke's law. In this case the original list of variables given in Eq. (15-32) is expanded to include the various moduli E_1, E_2, E_3, and so on, and Poisson's ratios ν_1, ν_2, ν_3, and so on. Additional design conditions (of the form $E_1/E_2 = E_{1m}/E_{2m}$, etc. and $\nu_1 = \nu_{1m}$, etc.) imply that the ratios of the moduli of elasticity for the various components in the structures must be equal, and Poisson's ratio for corresponding components must be equal. A potential difficulty that may be encountered in models of composite structures is the lack of similar behavior at material interfaces, where the bonding mechanisms may not be properly scaled [15].

Elastic Structures with Small Deformations

Many structures are "stiff" and the deformations are very small. In this case it is known that stresses, strains, displacements, and indeterminate forces are *linear functions of the loads*, and *the principle of superposition can be applied* (i.e., each load can be treated separately, with the effect of combined loading being the sum of the individual effects). These results are a consequence of the fact that the basic equations of classical elasticity, which govern this type of problem, are linear.

The previously developed general expression for stress [Eq. (15-33)] indicated that

$$\frac{\sigma l^2}{P} = \phi\left(\frac{x_i}{l}, \frac{\lambda_i}{l}, \frac{P_i}{P}, \frac{P}{El^2}, \nu\right) \tag{15-33}$$

However, for a stiff structure the stress must be a linear function of the load, and therefore $\sigma l^2/P$ cannot be a function of P/El^2, so that

$$\frac{\sigma l^2}{P} = \phi_1\left(\frac{x_i}{l}, \frac{\lambda_i}{l}, \frac{P_i}{P}, \nu\right) \tag{15-40}$$

and thus for small deformations the stresses do not depend on the modulus of elasticity. A similar argument shows that statically indeterminate reactions can be expressed as

$$\frac{R}{P} = \phi_2\left(\frac{x_i}{l}, \frac{\lambda_i}{l}, \frac{P_i}{P}, \nu\right) \tag{15-41}$$

For strains and displacements

$$\varepsilon = \frac{P}{El^2} \phi_3\left(\frac{x_i}{l}, \frac{\lambda_i}{l}, \frac{P_i}{P}, \nu\right) \tag{15-42}$$

and

$$\frac{u}{l} = \frac{P}{El^2} \phi_4\left(\frac{x_i}{l}, \frac{\lambda_i}{l}, \frac{P_i}{P}, \nu\right) \tag{15-43}$$

These equations indicate that strains and displacements vary inversely with the modulus of elasticity for stiff structures.

The principle of superposition allows the effect of each load to be considered separately. For example, if a structure is loaded with a series of loads, $P_1, P_2 \ldots, P_n$, the displacement due to P_1, is given by the expression

$$\frac{u_1}{l} = \frac{P_1}{El^2} \Phi\left(\frac{x_i}{l}, \frac{\lambda_i}{l}, \nu\right) \tag{15-44}$$

and for the other loads

$$\frac{u_2}{l} = \frac{P_2}{El^2} \Phi_2\left(\frac{x_i}{l}, \frac{\lambda_i}{l}, \nu\right)$$
$$\vdots \tag{15-45}$$
$$\frac{u_n}{l} = \frac{P_n}{El^2} \Phi_n\left(\frac{x_i}{l}, \frac{\lambda_i}{l}, \nu\right)$$

The displacement due to the combined loads is then given by the equation

$$u = u_1 + u_2 + \cdots + u_n \tag{15-46}$$

The various functions $\Phi_1, \Phi_2, \ldots, \Phi_n$ generally will not be the same. It is clear that models designed on the basis of the small deformation assumption offer considerable simplification in the manner in which they need to be tested.

Gravitational Loading

In the foregoing analyses of structural models, the effect of the weight of the structure has not been considered. For many structures, however, the weight is important, and, in fact, the "dead load" may be the dominant load. To account for the effect of weight, the additional variable, specific weight of the material w, needs to be included as a pertinent variable. For example, in Eq. (15-32) for stress, if w is included, an additional pi term, such as wl^3/P, is required. It now follows that the additional design condition is

$$\frac{wl^3}{P} = \frac{w_m l_m^3}{P_m}$$

and the required specific weight for the model is

$$w_m = \frac{P_m}{P} n_l^3 w \tag{15-47}$$

Since

$$\frac{P_m}{P} = \frac{1}{n_l^2} \frac{E_m}{E}$$

then

$$w_m = \frac{E_m}{E} n_l w \tag{15-48}$$

This design condition imposes some rather severe restrictions on the materials that can be used for the model. In particular, the same materials cannot be used for model and prototype, since this would require that the length scale be unity. For length scales greater than unity with

$E_m = E$, the model material must be heavier than that of the prototype. If it is desired that $w_m = w$ for $n_l > 1$, then $E_m < E$ (i.e., the model material would have to be less stiff).

For practical reasons it is often necessary to have the model and prototype constructed of the same material. Thus $E_m = E$ and the design condition given by Eq. (15-48) will not be satisfied for $n_l > 1$ unless the weight of the model can be increased. It may be possible to add a distributed external load to simulate the increased weight. This procedure may provide an appropriate effect for other members but generally not for the one on which the load is applied. Also, although not easily accomplished, it is possible to increase the weight by testing the model in an acceleration field that can be increased over and above the normal acceleration of gravity. For example, the model could be tested in a centrifuge in which the speed of rotation can be varied to obtain the acceleration that will give the desired weight [17].

For small displacements the principle of superposition can be utilized so that the effect of the dead load can be obtained independently of applied external loads and the results subsequently added to the effects due to external loading. Thus, if the weight is the only load, then

$$\sigma_w = f(l, x_i, \lambda_i, w, E, \nu) \tag{15-49}$$

so that

$$\frac{\sigma_w}{E} = \phi\left(\frac{x_i}{l}, \frac{\lambda_i}{l}, \frac{wl}{E}, \nu\right) \tag{15-50}$$

For small displacements

$$\frac{\sigma_w}{E} = \frac{wl}{E} \phi_1\left(\frac{x_i}{l}, \frac{\lambda_i}{l}, \nu\right) \tag{15-51}$$

and therefore

$$\frac{\sigma_w}{wl} = \phi_1\left(\frac{x_i}{l}, \frac{\lambda_i}{l}, \nu\right) \tag{15-52}$$

For geometrically similar structures having the same Poisson's ratio, the right-hand side of Eq. (15-52) is a constant for stresses evaluated at corresponding points, so that

$$\sigma_w = c_1 wl \tag{15-53}$$

where the constant c_1 can be obtained from a model study. Stresses due to external loads can be determined as described previously from Eq. (15-40) expressed as

$$\sigma_P = c_2 \frac{P}{l^2} \tag{15-54}$$

where c_2 is a constant for geometrically similar structures having the same Poisson's ratio. The total stress is obtained as a sum of the stresses due to weight and external loading, so that

$$\sigma_t = \sigma_w + \sigma_P \tag{15-55}$$

A similar analysis can be made for other characteristics of interest.

Effect of Poisson's Ratio

In the most general case, stresses, strains, displacements, and indeterminate forces for structures fabricated of materials obeying Hooke's law depend on Poisson's ratio ν, as well as the modulus of elasticity E. Since ν is a dimensionless constant, it follows that to have an undistorted model, in the general case, Poisson's ratio must be the same for model and prototype structures. This is a stringent requirement, and unless the same materials are used for model and prototype, it is difficult to satisfy. In certain types of model study—for example, stress analysis using photoelastic models—Poisson's ratio is usually not the same for model and prototype, and the effect of Poisson's ratio on the stress distribution must be assessed. For *plane elasticity* problems, some general guidelines can be given. For plane-stress or plane-strain problems involving simply connected bodies (no holes in body) for which the body forces are zero or constant or vary linearly with position, the stress distribution is known to be independent of Poisson's ratio. Similar problems involving multiply connected bodies containing a hole or holes can also be modeled without regard to Poisson's ratio if the resultant force on the boundary of each hole is zero. It has also been shown that if there are unbalanced forces on internal boundaries, the stress is *linearly* related to Poisson's ratio [18]. Thus, if two models having different Poisson's ratios are tested, the results of these tests can be used to predict the stresses in a geometrically similar prototype having a Poisson's ratio different from that of the models. Nevertheless, for the more general three-dimensional problems, Poisson's ratio may be a significant factor and must be considered. There are no general rules in this case, and it must be left to the experience and knowledge of the experimenter to decide on the relative importance of this material property.

Dynamic Loading

In the previous sections the loads were assumed to be static (i.e., they are applied slowly so that inertial effects are not significant). In many important applications, however, the loading is developed rapidly so that this assumption is not valid. To illustrate this type of problem, consider the prediction of strains in an elastic structure under the influence of a dynamic pressure loading over some part of the surface of the structure. As before, let ε represent one component of strain at the position x_i, and let the geometry of the system be described by a set of characteristic lengths, l and λ_i, where the λ_i's also include the required spatial coordinates of the loading. If the material obeys Hooke's law, two elastic constants E and ν are required, as previously discussed. However, since the structure is under dynamic loading, an additional material property, its mass density ρ, must be included in the analysis.

Let the pressure at any point be described in dimensionless form as

$$\frac{p}{p_0} = \psi\left(\frac{t}{l\sqrt{\rho/E}}\right) \tag{15-56}$$

where the quantity $l\sqrt{\rho/E}$ has the dimension of time and can be combined with t to form a dimensionless time variable, and p_0 is a reference pressure. The functional relationship for the strain can therefore be written as

$$\varepsilon = f(l, x_i, \lambda_i, E, \nu, \rho, p_0, t) \tag{15-57}$$

where it is tacitly implied that the form of the pressure function Ψ is the same in both the model and prototype. In dimensionless terms Eq. (15-57) can be written as

$$\varepsilon = \phi\left(\frac{x_i}{l}, \frac{\lambda_i}{l}, \nu, \frac{p_0}{E}, \frac{t}{l\sqrt{\rho/E}}\right) \tag{15-58}$$

The three pi terms x_i/l, λ_i/l, and ν yield the similarity requirements previously considered (i.e., geometric similarity must be maintained, and Poisson's ratio for model and prototype materials must be the same). The pressure scale is established from the condition

$$\frac{p_0}{E} = \frac{p_{0m}}{E_m} \tag{15-59}$$

which indicates that if the same material is used in model and prototype, the pressures at scaled locations and times must be the same. The time scale for the problem is established from the condition

$$\frac{t}{l\sqrt{\rho/E}} = \frac{t_m}{l_m\sqrt{\rho_m/E_m}} \qquad (15\text{-}60)$$

and, in general, corresponding times in the model and prototype will be different. If the same materials are used in model and prototype, the time scale will equal the length scale. Since the length scale is generally greater than unity, it follows that when modeling with the same materials, corresponding times in the model will be shorter than for the prototype. Thus, in a problem of this type, the model and prototype loading pressure–time relationship, when expressed in terms of p/p_0 and the dimensionless time variable must be identical, whereas in terms of real time, the model and prototype pressure–time relationship must be different. From a practical point of view, this is a common problem encountered in dynamic testing, since it is difficult to generate a properly scaled model loading. If all similarity requirements are met, it follows that $\varepsilon = \varepsilon_m$, where the strains in the model and prototype are measured at corresponding times based on the time scale. A similar analysis can be made for the prediction of stresses, displacements, forces, and so on.

When dealing with structural dynamics problems, the velocity V and acceleration a at certain locations within the structure may be of interest, as well as a vibratory frequency ω. With the establishment of the time scale, as in Eq. (15-60), the appropriate scales for these quantities can be readily expressed in terms of the time and length scales as follows:

$$\frac{V}{V_m} = n_l n_t^{-1} \quad \text{velocity} \qquad (15\text{-}61)$$

$$\frac{a}{a_m} = n_l n_t^{-2} \quad \text{acceleration} \qquad (15\text{-}62)$$

$$\frac{\omega}{\omega_m} = n_t^{-1} \quad \text{frequency} \qquad (15\text{-}63)$$

Nonlinear Material Behavior

In the example of structural models considered thus far, the material was assumed to obey Hooke's law and was characterized by two material properties, such as E and ν. If the stress–strain relationship for the material is nonlinear, the pertinent properties cannot be so readily identified, and the development of similarity requirements is further complicated. For the static, nonlinear case it can be assumed that any stress component can be expressed as

$$\frac{\sigma}{E} = \phi(\varepsilon_i, \gamma_j) \qquad (15\text{-}64)$$

during any monotonically increasing loading phase, where E in some characteristic modulus having the dimensions of stress, ε_i the strain components, and γ_j a set of j dimensionless coefficients. For example, in a simple uniaxial tension or compression test, let

$$\sigma = E\varepsilon + E_1\varepsilon^2 + E_2\varepsilon^3 + \cdots$$

and thus

$$\frac{\sigma}{E} = \varepsilon + \frac{E_1}{E}\varepsilon^2 + \frac{E_2}{E}\varepsilon^3 + \cdots \qquad (15\text{-}65)$$

SIMILITUDE, MODELING, AND DIMENSIONAL ANALYSIS

which is the form of Eq. (15-64) with the ratios E_1/E, E_2/E, and so on, corresponding to the γ_j's. As far as dimensional analysis is concerned, for a material characterized in this manner, the pertinent material properties are E and the dimensionless γ_j's. The use of this set of material properties will not alter the design conditions from those obtained for linear elastic materials. To arrive at this conclusion, it is tacitly implied that the form of the constitutive relationship as given by Eq. (15-64) is the same for model and prototype materials, which requires that the dimensionless stress–strain curves (σ/E versus ε) be the same for model and prototype materials. The obvious way to satisfy this condition is to use the same materials in model and prototype. Although, in principle, the same materials are not required, it is very difficult (and usually impractical) to satisfy the required conditions related to the nonlinear constitutive equations if different materials are used.

An additional consideration arises if the material is loaded beyond the elastic range. Since in the plastic range the strain is not a single-valued function of stress, the loading history becomes important. Thus, for model studies that extend into the plastic range, the model loads must follow the same pattern as the corresponding prototype loads. For the usual case in which $E = E_m$, it follows that model loads must be scaled as $P_m/l_m^2 = P/l^2$ with identical loading patterns. It can be concluded that *models designed and operated on the basis of the similarity requirements for linear elastic structures can be used for the study of structures involving nonlinear material behavior, including plastic action, if the model and prototype materials have the same dimensionless stress–strain relationships and the same pattern of loading.*

For models used to study cracking and fracture, special attention must be given to so-called scale effects. Since the initiation of fracture may depend on the material microstructure, which would not be scaled even if the same material were used for model and prototype, cracking and fracture may not be accurately predicted. A discussion of this complex aspect of modeling is given in Ref. 15.

For dynamically loaded structures with very rapid loading rates, the stresses developed in the structure may be a function not only of the strains but also of the rate at which the strain develops. In this case let any stress component be expressed in the form

$$\sigma = Ef(\varepsilon_i, \gamma_j, \dot{\varepsilon}_k, \eta_l) \tag{15-66}$$

where E is a characteristic modulus; ε_i and γ_j are strains and dimensionless coefficients, respectively, as defined previously; $\dot{\varepsilon}_k$ denotes a set of strain rates; and η_l represents one or more material constants, which involve the basic dimension of time. The subscripts i, j, k, l simply indicate that, in general, the stress may depend on several strains, strain rates, and material constants. Neither the specific number of quantities nor the form of the function is important for the present purpose. If, for example, one or more of the η's has the basic dimension of T, a pi term of the form η/t will be required. This pi term will lead to the design condition

$$\frac{\eta}{t} = \frac{\eta_m}{t_m} \tag{15-67}$$

with the corresponding time scale

$$\frac{t}{t_m} = \frac{\eta}{\eta_m} \tag{15-68}$$

If the same materials are used for model and prototype ($\eta = \eta_m$), the time scale is unity. However, this result is in conflict with the time scale $t/t_m = n_l$, which arises from another design condition [see Eq. (15-60)]. Thus, if strain-rate effects are important, it is not possible to use the same materials in model and prototype without distortion. It is to be noted that if the same materials

Table 15-2 Some Common Fluid-Flow Variables and Dimensionless Groups

Variable	Dimensionless Group	Name	Significance (force ratio)
Acceleration of gravity, g	$\dfrac{\rho V^2}{K}$	Cauchy number[a]	$\dfrac{\text{inertia}}{\text{compressibility}}$
Bulk modulus, K Characteristic length, l	$\dfrac{p}{\rho V^2}$	Euler number	$\dfrac{\text{pressure}}{\text{inertia}}$
Density, ρ	$\dfrac{V^2}{gl}$	Froude number	$\dfrac{\text{inertia}}{\text{gravitational}}$
Pressure, p (or pressure drop) Speed of sound, c	$\dfrac{V}{c}$	Mach number[a]	$\dfrac{\text{inertia}}{\text{compressibility}}$
Surface tension, ξ	$\dfrac{\rho V l}{\mu}$	Reynolds number	$\dfrac{\text{inertia}}{\text{viscous}}$
Velocity, V Viscosity, μ	$\dfrac{\rho V^2 l}{\xi}$	Weber number	$\dfrac{\text{inertia}}{\text{surface tension}}$

[a] Since the speed of sound in a fluid is equal to $c = \sqrt{K/\rho}$, it follows that the Cauchy number is equal to the square of the Mach number. Hence either number may be used as an index of the relative effects of inertia and compressibility.

are not used, it is necessary to actually determine the specific form of the constitutive relationships for the model and prototype materials, and the relationships must be of the same form. This is an exceedingly difficult task, and generally constitutive relationships for inelastic behavior (with or without strain-rate effects) are not available.

15-6 Fluid-Flow Models

Fluid motion in most circumstances is exceedingly complicated, and practical problems involving moving fluids often require the use of empirical data for their solution. It is therefore not surprising that dimensional analysis and fluid-flow models have played an important role in the field of hydraulics and fluid mechanics [19,20]. Applications of interest vary widely and can include diverse problems related to flow in pipes, sediment transport in rivers, biological flows, wind-induced forces on structures, and aerodynamic flutter. Although the basic concepts of similitude previously discussed also serve as the basis for fluid-flow modeling, the variables involved and the resulting similarity requirements are significantly different from those found in structural models.

Table 15-2 lists a number of variables that commonly arise in fluid-flow systems. This list is obviously not exhaustive but indicates important variables, particularly fluid properties. Application of dimensional analysis leads to the development of corresponding dimensionless groups, or pi terms, as indicated in Table 15-2. These groups occur so frequently that special names are associated with them, as indicated. Also, as will be demonstrated in Section 15-9, these dimensionless groups can be given a physical interpretation, and each actually represents an index of the ratio of two forces that occur in the flow field. For example, the Reynolds number is an index of the ratio of the inertia forces to the viscous forces in the fluid, and it is well known that for very small Reynolds

SIMILITUDE, MODELING, AND DIMENSIONAL ANALYSIS

numbers the flow is dominated by viscous effects, whereas for very large Reynolds numbers viscous forces are small compared with inertial forces. This physical interpretation can be quite useful in the design of experiments and in the interpretation of test results.

Similarity requirements are obtained by equating corresponding dimensionless groups; for example, for problems in which the Reynolds number is important it follows that a model design condition is

$$\frac{\rho V l}{\mu} = \frac{\rho_m V_m l_m}{\mu_m} \tag{15-69}$$

and therefore the velocity scale is

$$\frac{V}{V_m} = \frac{\rho_m}{\rho} \frac{\mu}{\mu_m} \frac{l_m}{l} \tag{15-70}$$

If the fluid properties are the same for model and prototype then

$$V_m = \frac{l}{l_m} V \tag{15-71}$$

which indicates the model velocity must be higher than that of the prototype if the model is smaller in size than the prototype. Reynolds number similarity is often an important design condition for fluid-flow models and one that is not always easily satisfied. Additional design conditions can be obtained from the other dimensionless groups in Table 15-2. Fortunately, not all the groups are generally required in a given problem, and each specific application must be considered individually. The following examples illustrate how some of these dimensionless groups arise naturally in the analysis of certain types of problem. For all the examples the fluid is assumed to be incompressible and Newtonian.

Flow Through Pipes

One of the simplest, yet very important flow problems is concerned with flow through straight circular pipes. Of particular interest is the pressure drop Δp, which occurs over a length l in a pipe of diameter d in which a fluid flows with velocity V. In an earlier example, it was assumed that the pipe wall was smooth, but in general the wall will be rough with an average roughness height h. The pertinent fluid properties are fluid density ρ and fluid viscosity μ, so that

$$\Delta p = f(l, d, h, \rho, \mu, V) \tag{15-72}$$

and from the pi theorem

$$\frac{\Delta p}{\rho V^2} = \phi\left(\frac{l}{d}, \frac{h}{d}, \frac{\rho V d}{\mu}\right) \tag{15-73}$$

For a fully developed flow, the pressure drop should be directly proportional to the length, so that

$$\frac{\Delta p}{\rho V^2} = \frac{l}{d} \phi_1\left(\frac{h}{d}, \frac{\rho V d}{\mu}\right) \tag{15-74}$$

Equation (15-74) can be written as

$$\Delta p = \phi_2\left(\frac{h}{d}, \frac{\rho V d}{\mu}\right) \frac{l}{d} \frac{\rho V^2}{2} \tag{15-75}$$

where the function ϕ_2 is the "friction factor," which must be determined experimentally even for

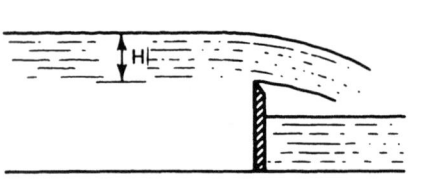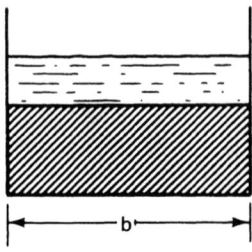

Figure 15-1 Flow over sharp-edged weir.

this relatively simple flow (unless the flow is laminar). It is clear from Eq. (15-74) that similarity requirements include not only the equality of Reynolds numbers, but also an "equivalent" scaled roughness h/d. Since roughness is difficult to quantify, this last condition is not easily satisfied.

Open-Channel Flow

Consider the flow of a liquid over a wide, sharp-edged weir as illustrated in Fig. 15-1. It is desired to relate the average velocity V of the fluid flowing over the weir crest to the head H. For open-channel, or free-surface, flow, the velocity depends on the acceleration of gravity g. Also, since there is a free surface, surface tension ξ may be significant in addition to the usual fluid properties, ρ and μ. Thus

$$V = f(H, g, \xi, \rho, \mu) \tag{15-76}$$

and therefore

$$\frac{V^2}{gH} = \phi\left(\frac{\rho V H}{\mu}, \frac{\rho V^2 H}{\xi}\right) \tag{15-77}$$

where the three dimensionless groups are the Froude number, the Reynolds number, and the Weber number, respectively. It can easily be shown that if the same fluid is used in a model and a prototype for this problem, the design conditions resulting from Eq. (15-77) cannot be simultaneously satisfied. However, for H relatively large, neither viscous nor surface tension effects are significant and therefore

$$\frac{V^2}{gH} = C \tag{15-78}$$

where C is a constant.[5] Since the discharge Q is equal to VbH, it follows that

$$Q = \sqrt{Cg}\, bH^{3/2} \tag{15-79}$$

which is a well-known formula for this type of weir. Models involving open-channel flow are usually designed on the basis of the Froude criteria, but they are frequently distorted, since the Reynolds numbers are not matched if the same fluid is used in model and prototype (see Section 15-8).

---

[5] The same result would be obtained if initially only the three variables, V, H, and g were considered. Note that in a problem in which there is only one pi item, that pi term must be equal to a constant.

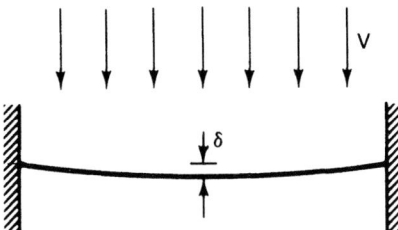

Figure 15-2 Flow past an elastic wire.

Fluid–Structure Interaction

As a final example, the flow past an elastic wire stretched between two rigid supports as shown in Fig. 15-2 will be considered. The variable of interest is the deflection of the wire δ at the center of the span. The external force acting on the wire is due to the drag of the fluid, which will be a function of the fluid velocity V, density ρ, and viscosity μ, as well as the wire diameter D and length l. The important property of the wire for static deflections is the modulus of elasticity E, so that

$$\delta = f(V, \rho, \mu, D, l, E) \qquad (15\text{-}80)$$

and therefore

$$\frac{\delta}{l} = \phi\left(\frac{D}{l}, \frac{\rho V D}{\mu}, \frac{\rho V^2}{E}\right) \qquad (15\text{-}81)$$

For this system the model and prototype must be geometrically similar ($D/l = D_m/l_m$), operated at the same Reynolds number, and in addition

$$\frac{\rho V^2}{E} = \frac{\rho_m V_m^2}{E_m} \qquad (15\text{-}82)$$

If the same fluid is used in both systems, the velocity scale is determined from Reynolds number similarity (i.e., $n_V = n_l^{-1}$) and therefore from Eq. (15-82) $n_E = n_l^{-2}$. Since typically the length scale is greater than unity, $E_m > E$ and the model wire must be stiffer than the prototype to maintain similarity.

Although this example has been treated as a statics problem, it is known that vortices may be shed from the wire at a regular frequency, thereby causing the wire to vibrate, with large amplitudes of vibration possible if the vortex-shedding frequency matches the natural frequency of the wire. If this phenomenon is to be correctly modeled, some measure of the mass of the wire must be included (mass per unit length or density) which will yield an additional pi term and design condition. It can be shown that this additional design condition would require the densities of the material used for the wires to be the same in both model and prototype (if the same fluid is used) even though their elastic moduli are different.

15-7 Thermal Models

Another important class of problems arises in which heat transfer is involved. Applications in this field also vary widely and include investigations related to thermal stresses, temperature distributions in flowing fluids, effects of fires, and so on. Although the basic concepts of dimensional analysis and modeling are still applicable to thermal models, an additional basic dimension, that of temperature Θ, is required to describe some of the variables that enter these problems. Heat is generally not considered a basic dimension since it has the dimensions of work (FL), although this concept has long evoked numerous discussions, starting with the work of Lord Rayleigh [21] in 1915.

Additional common material properties and variables to be considered in thermal modeling include the coefficient of thermal expansion β, the surface coefficient of heat transfer h, specific heat c, and thermal conductivity k. Corresponding dimensions for these quantities are given in Table 15-1. As in the case of fluid-flow models, special names are frequently associated with the pi terms that often arise in dealing with thermal models, and some common dimensionless groups include:

$$\text{Grashof number} \quad GR = \frac{\rho^2 g \beta \Delta \theta l^3}{\mu^2}$$

$$\text{Nusselt number} \quad Nu = \frac{hl}{k}$$

$$\text{Péclet number}[6] \quad Pe = \frac{\rho c V l}{k}$$

$$\text{Prandtl number} \quad Pr = \frac{c\mu}{k}$$

To illustrate the nature of similarity requirements for thermal models, two simple examples will be considered. Additional examples and more detailed discussions of this subject can be found in Refs. 3, 10, and 22.

Heat Transfer in Pipe Flow

A classical problem in heat transfer is concerned with the rate at which heat is transmitted through the wall (per unit area) of a smooth-walled pipe through which a liquid is flowing. For this *forced-convection* problem, the variable of interest is the heat-transfer coefficient h, since by definition the rate of heat transfer per unit area is $h\Delta\theta$, where $\Delta\theta$ is the difference between the temperature of the pipe surface and some average fluid temperature. For fully developed turbulent flow

$$h = f(d, V, \rho, \mu, c, k) \tag{15-83}$$

where d and V refer to the pipe diameter and average velocity, respectively, and ρ, μ, c, and k refer to the fluid properties of density, viscosity, specific heat, and thermal conductivity, respectively. Since there are seven variables and four basic dimensions, the pi theorem indicates that three pi terms are required and a suitable set is

$$\frac{hd}{k} = \phi\left(\frac{\rho V d}{\mu}, \frac{c\mu}{k}\right) \tag{15-84}$$

which indicates that the Nusselt number is a function of the Reynolds number and the Prandtl number. Thus similarity requirements for this type of problem include equality of the Reynolds

---

[6]The Péclet number is the product of the Prandtl number and the Reynolds number.

number and the Prandtl number. Numerous experiments have shown that heat-transfer data can be successfully correlated on the basis of Eq. (15-84). If, in addition, *free convection* is considered, the Nusselt number will also become dependent on the Grashof number, which is an important dimensionless group for heat-transfer problems in which buoyant forces are significant.

Transient Heat Transfer to Immersed Bodies

Consider a solid body, having a characteristic length l and initial temperature θ_0, suddenly immersed in a liquid bath of temperature θ_b. Assume that the bath is large and stirred so that its temperature remains essentially constant. Of interest is the temperature at some specified point in the body, at any time after immersion. As discussed by Langhaar [3], the pertinent variables in this problem are related by the equation

$$\theta - \theta_0 = f(l, \lambda_i, \rho c, k, h, \theta_b - \theta_0, t) \tag{15-85}$$

where θ is the body temperature at some specified point, at time t, and λ_i represents a set of length terms that define the shape of the body as well as the coordinates of the point of interest. It is to be noted that heat transfer processes are governed by temperature differences, and $\theta - \theta_0$ and $\theta_b - \theta_0$ are the pertinent temperature variables. Dimensional analysis yields

$$\frac{\theta - \theta_0}{\theta_b - \theta_0} = \phi\left(\frac{\lambda_i}{l}, \frac{hl}{k}, \frac{kt}{\rho c l^2}\right) \tag{15-86}$$

The dimensionless group hl/k has the same form as the Nusselt number but is usually called the *Biot number* when the thermal conductivity refers to a solid rather than a fluid. The last dimensionless group, which is the *Fourier number*, establishes the time scale (i.e., if the same material is used for model and prototype, then $t_m = t/n_l^2$). Since n_l is typically much greater than unity, it follows that for this type of thermal model, corresponding events occur more rapidly in the model than in the prototype. The other model design conditions include geometric similarity and equality of Biot numbers.

15-8 True, Adequate, and Distorted Models

There is obviously an advantage in keeping the number of variables in any given problem to a minimum. For each variable that is eliminated, a corresponding pi term is eliminated, and thus the number of design conditions reduced. In some problems it is possible to combine variables to cause this reduction. For example, in the beam deflection problem the cross-sectional variables considered were beam width and depth, but if it is assumed that only bending deflections are significant, it is known that the pertinent cross-sectional variable is actually the moment of inertia I, which can be used to replace the width and depth. The corresponding design condition thus becomes

$$I_m = \frac{I}{n_l^4} \tag{15-87}$$

This condition provides for much greater flexibility in the selection of the model geometry and allows structural elements having rather complicated cross sections to be modeled with simpler model geometries (such as rectangles). Although a model designed on this basis may be suitable for the prediction of deflection, it would not generally give correct results for predicting other variables, such as bending stresses, which depend on the depth of the cross section as well as the moment of inertia. Models that may be suitable for predicting a particular variable, or class of variables, but cannot be used for other predictions, are called *adequate models* to distinguish them from the more general class of *true models*. Invariably, true models are geometrically similar to the prototype, whereas for adequate models this is not necessarily the case. Nevertheless, adequate models are useful and commonplace, and the possibility of simplifying a model design by a judicious combination of variables, or by neglecting variables that do not have a strong influence on the phenomenon of interest, should be given serious consideration in the early stages of a model design. Unfortunately, a frequently encountered difficulty occurs in modeling when one or more design conditions, known to be important, cannot be satisfied. A model for which this is the case is called a *distorted model*. Distorted models often arise because suitable materials cannot be found for the model. The classic example of such a model occurs in the study of open-channel flows in which both the Reynolds number and the Froude number are involved.

The velocity scale based on the Froude criteria is

$$\frac{V}{V_m} = \sqrt{\frac{l}{l_m}} = \sqrt{n_l} \tag{15-88}$$

if the model and prototype are operated in the same gravitational field. However, the velocity scale based on Reynolds number similarity is

$$\frac{V}{V_m} = \frac{\mu}{\mu_m} \frac{\rho_m}{\rho} \frac{l_m}{l} \tag{15-89}$$

and therefore to satisfy both conditions

$$\frac{\mu/\rho}{\mu_m/\rho_m} = (n_l)^{3/2} \tag{15-90}$$

where the ratio μ/ρ is the kinematic viscosity. Although there is no inherent impediment in satisfying this design condition, it may be quite difficult, if not impossible, from a practical viewpoint to find a suitable model fluid, particularly for large length scales. For problems involving rivers, harbors, and so on, for which the prototype fluid is water, the models are also relatively large, so that the only practical model fluid is water. However, in this case Eq. (15-90) will not be satisfied and a distorted model will result. Generally, models of this type are distorted and are designed on the basis of the Froude number, with the Reynolds numbers different in model and prototype. Distorted models can be successfully used, but results obtained from this type of model are obviously more difficult to interpret than those obtained with true or adequate models.

Established, standard techniques for handling distortion are not available, and essentially each distorted model design must be considered individually. Techniques that have been used can be broadly grouped into three categories:

1. Neglect certain variables (and therefore pi terms) that lead to the distortion, with the understanding that predictions made by the model will be approximate. This is a common technique, and its suitability depends on the use to be made of the model, as well as past experience, to ascertain the degree of error to be expected.
2. Determine the effect of the distortion analytically. For some applications it may be possible to deduce the likely effect of the distortion through a consideration of the equations that govern the phenomenon, or through physical reasoning. Of course, if a complete analysis were available, there would be no need for a model study. It may, however, be possible to obtain a partial solution, which will indicate the effect of the distortion, even though the complete solution is not available.

3. Determine the effect of the distortion empirically. For a distorted model the implication is that $\pi_1 \neq \pi_{1m}$, where π_1 contains the variable to be predicted, since one or more of the independent pi terms are not equal between model and prototype (e.g., $\pi_2 \neq \pi_{2m}$). To indicate the degree of distortion, let $\pi_{2m} = \alpha \pi_2$, where α is the *distortion factor*, and the prediction equation becomes $\pi_1 = \delta \pi_{1m}$, where δ is the *prediction factor*. If prototype measurements of π_1 can be obtained corresponding to the model predictions of π_1, δ can be determined as a function of α. The assumption here is that some prototype data are available to determine empirically the effect of distortion, with the model subsequently used to predict the behavior of the prototype under conditions for which data are not available. Alternatively, it may be possible to run a series of tests involving several models, with one of the models considered to be a prototype, to determine the nature of the relationship between the prediction and distortion factors. From basic model theory

$$\frac{\pi_1}{\pi_{1m}} = \frac{\phi(\pi_2, \pi_3, \ldots, \pi_n)}{\phi(\pi_{2m}, \pi_{3m}, \ldots, \pi_{nm})} \tag{15-91}$$

and therefore for a distorted model

$$\delta = \frac{\phi(\pi_2, \pi_3, \ldots, \pi_n)}{\phi(\alpha \pi_2, \pi_3, \ldots, \pi_n)} = \phi_1(\alpha, \pi_2, \pi_3, \ldots, \pi_n) \tag{15-92}$$

where it is assumed that only one pi term is distorted. Unfortunately, as demonstrated by Eq. (15-92), the prediction factor may be a function not only of the degree of distortion as specified by α but also of the specific values of the other prototype pi terms. Thus, empirically determined relationships accounting for distortion may be quite limited in applicability and must be carefully interpreted when any generalizations are to be made.

15-9 Similarity Laws From Differential Equations

The use of dimensional analysis to obtain similarity laws requires only a knowledge of the variables that influence the phenomenon of interest. Although the simplicity of this approach is attractive, it must be recognized that the omission of one or more important variables will lead to serious errors in a model design. Thus, the determination of the variables to be considered is an essential first step in any dimensional analysis, and this step requires a good understanding of the basic laws governing the phenomenon. An alternative approach is available if the equations (usually differential equations) governing the phenomenon are known. As will be demonstrated by the following examples, similarity laws can be readily developed from the governing equations, even though it may not be possible to obtain analytic solutions to the equations.

Damped Vibratory Motion

To illustrate the method, the motion of the damped, simple spring–mass system of Fig. 15-3 will be considered. For viscous damping with a linear spring the well-known differential equation describing the free, vibratory motion of this system is

$$m\frac{d^2x}{dt^2} + c\frac{dx}{dt} + kx = 0 \tag{15-93}$$

where m is the mass, c the damping coefficient, k the spring constant, x the displacement, and t the time. Typical initial conditions would be $x = x_0$ and $dx/dt = v_0$ at $t = 0$. Although the

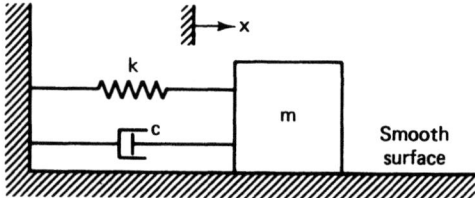

Figure 15-3 Damped, simple spring–mass system.

solution to this particular problem can easily be obtained, similarity requirements can be deduced without obtaining the solution in the following manner. Rewrite the equation in dimensionless form by defining new dependent and independent variables; that is, let

$$\bar{x} = \frac{x}{x_0} \quad \text{and} \quad \bar{t} = \frac{t}{\sqrt{m/k}}$$

It follows that Eq. (15-93) can then be expressed in the form

$$\frac{d^2\bar{x}}{d\bar{t}^2} + \frac{c}{\sqrt{mk}} \frac{d\bar{x}}{d\bar{t}} + \bar{x} = 0 \tag{15-94}$$

with the initial conditions

$$\bar{x} = 1 \quad \frac{d\bar{x}}{d\bar{t}} = \frac{v_0}{x_0}\sqrt{\frac{m}{k}}$$

Thus, the dimensionless displacements in two systems will be equal,

$$\frac{x}{x_0} = \frac{x_m}{x_{0m}} \tag{15-95}$$

at the corresponding times

$$\frac{t}{\sqrt{m/k}} = \frac{t_m}{\sqrt{m_m/k_m}} \tag{15-96}$$

if

$$\frac{c}{\sqrt{mk}} = \frac{c_m}{\sqrt{m_m k_m}} \quad \frac{v_0}{x_0}\sqrt{\frac{m}{k}} = \frac{v_{0m}}{x_{0m}}\sqrt{\frac{m_m}{k_m}} \tag{15-97}$$

These four relationships constitute the similarity requirements for this problem and are, of course, the same as those obtained by dimensional analysis starting with the list of variables

$$x = f(m, c, k, x_0, v_0, t) \tag{15-98}$$

This method, based on a knowledge of the governing equations, involves essentially two steps:

1. Express the dependent and independent variables in dimensionless form using appropriate characteristic lengths, times, velocities, and so on, for the system.
2. Rewrite the equations in terms of these dimensionless variables so that only dimensionless groups appear in the equations. Boundary and initial conditions must also be expressed in

SIMILITUDE, MODELING, AND DIMENSIONAL ANALYSIS

dimensionless form. The dimensionless groups that arise naturally from this procedure will correspond to the pi terms obtained from a dimensional analysis.

Fluid Motion

As a second, more complicated example, consider the steady motion of an incompressible, Newtonian fluid. The governing differential equations are the continuity equation

$$\frac{\partial u}{\partial x} + \frac{\partial v}{\partial y} + \frac{\partial w}{\partial z} = 0 \tag{15-99}$$

where u, v, and w are velocity components in the x, y, and z directions, respectively, and three equations of motion, where, for example, the equation written in the vertical y direction is

$$\rho\left(u\frac{\partial v}{\partial x} + v\frac{\partial v}{\partial y} + w\frac{\partial v}{\partial z}\right) = -\frac{\partial p}{\partial y} - \rho g + \mu\left(\frac{\partial^2 v}{\partial x^2} + \frac{\partial^2 v}{\partial y^2} + \frac{\partial^2 v}{\partial z^2}\right) \tag{15-100}$$

In this equation ρ and μ are fluid density and viscosity, respectively, and g is the acceleration of gravity. Two similar equations can be written in the x and z directions. This system of nonlinear, partial differential equations cannot, in general, be solved analytically.

Let l be some characteristic length, p_0 a characteristic pressure, and V a characteristic velocity, and define a new set of dimensionless variables

$$\bar{x} = \frac{x}{l} \quad \bar{y} = \frac{y}{l} \quad \bar{z} = \frac{z}{l} \quad \bar{p} = \frac{p}{p_0}$$

$$\bar{u} = \frac{u}{V} \quad \bar{v} = \frac{v}{V} \quad \bar{w} = \frac{w}{V}$$

Equation (15-99) and (15-100) can now be written as

$$\frac{\partial \bar{u}}{\partial \bar{x}} + \frac{\partial \bar{v}}{\partial \bar{y}} + \frac{\partial \bar{w}}{\partial \bar{z}} = 0 \tag{15-101}$$

and

$$\underbrace{\frac{\rho V^2}{l}}_{\text{Inertia}}\left(\bar{u}\frac{\partial \bar{v}}{\partial \bar{x}} + \bar{v}\frac{\partial \bar{v}}{\partial \bar{y}} + \bar{w}\frac{\partial \bar{v}}{\partial \bar{z}}\right) = -\underbrace{\frac{p_0}{l}\frac{\partial \bar{p}}{\partial \bar{y}}}_{\text{Pressure}} - \underbrace{\rho g}_{\text{Gravity}} + \underbrace{\frac{\mu V}{l^2}}_{\text{Viscous}}\left(\frac{\partial^2 \bar{v}}{\partial \bar{x}^2} + \frac{\partial^2 \bar{v}}{\partial \bar{y}^2} + \frac{\partial^2 \bar{v}}{\partial \bar{z}^2}\right) \tag{15-102}$$

The quantities multiplying the dimensionless variables can be interpreted as indices for the various forces as indicated. Now divide by one of these indices, such as the one representing the inertia force, to obtain

$$\bar{u}\frac{\partial \bar{v}}{\partial \bar{x}} + \bar{v}\frac{\partial \bar{v}}{\partial \bar{y}} + \bar{w}\frac{\partial \bar{v}}{\partial \bar{z}} = -\left[\frac{p_0}{\rho V^2}\right]\frac{\partial \bar{p}}{\partial \bar{y}} - \left[\frac{V^2}{gl}\right]^{-1} + \left[\frac{\rho Vl}{\mu}\right]^{-1}\left(\frac{\partial^2 \bar{v}}{\partial \bar{x}^2} + \frac{\partial^2 \bar{v}}{\partial \bar{y}^2} + \frac{\partial^2 \bar{v}}{\partial \bar{z}^2}\right) \tag{15-103}$$

The dimensionless terms in brackets are the Euler number $p_0/\rho V^2$, the Froude number V^2/gl, and the Reynolds number $\rho Vl/\mu$. It is apparent how these dimensionless groups arise naturally from the equations, and how they can be interpreted as indices representing the ratio of forces. For example, Reynolds number represents the ratio of the inertia force to the viscous force and the

Froude number the ratio of the inertia force to the gravity force. Two similar dimensionless equations of motion are obtained for the other two coordinate directions (without the Froude number, since gravity acts only in the vertical direction).

These equations can now be used to establish similarity requirements, and from Eq. (15-103) it follows that if the Euler number, the Froude number, and the Reynolds number are equal in two geometrically similar systems, then the flow fields will be *kinematically similar* ($\bar{u} = \bar{u}_m$, $\bar{v} = \bar{v}_m$, $\bar{w} = \bar{w}_m$) and *dynamically similar* (similar pressure distributions as well as the various force ratios being equal). It can further be shown that for flows without a free surface, the Froude number can be omitted, and if there is no possibility of cavitation, the Euler number can also be omitted as a similarity requirement.

The use of governing equations to obtain similarity laws is a powerful alternative to dimensional analysis. This approach has the advantage that the variables and assumptions are clearly defined. In addition, a physical interpretation of the dimensionless groups can often be obtained, and it may be possible to simplify a problem through an order-of-magnitude analysis based on the relative values of the dimensionless groups [8].

15-10 Summary

The use of dimensional analysis, both as a tool for correlating data and as a basis for the design of models, is well established and thoroughly documented in the literature. In this chapter an attempt has been made to provide a concise, broad overview of the subject, with particular emphasis on the use of dimensional analysis for the establishment of similarity requirements to be used in the design of models. A major strength of dimensional analysis is that it can be applied to highly complex phenomena with a knowledge of only the variables involved. This relative simplicity of the method, however, represents a weakness, since omission of one or more important variables generally will not be detected by dimensional analysis, and errors in a model design are not likely to be detected until prototype behavior is found to be not as predicted. It is therefore necessary to have a good understanding of the fundamental physical laws underlying the phenomenon involved. The problem of selecting pertinent variables can be relieved through the use of governing equations to obtain similarity requirements in those cases for which the equations are known and the assumptions that usually accompany such equations are acceptable. Regardless of the method utilized, the concept of similitude is a powerful one that can be used in the planning and interpretation of experiments throughout a broad spectrum of engineering disciplines.

List of Symbols

a	acceleration
A	area
b	beam width
c	sound velocity
c	specific heat
c	viscous damping coefficient
d	beam depth
d	distance traveled
d	pipe diameter
D	wire diameter

E	modulus of elasticity
F	primary quantity, force
g	acceleration of gravity
G	shear modulus
h	roughness height
h	surface coefficient of heat transfer
H	fluid depth
I	area moment of inertia
k	spring constant
k	thermal conductivity
K	bulk modulus
l	beam length
l	characteristic length
L	primary quantity, length
m	mass
M	primary quantity, mass
n	model scale
p	pressure
P	load
Q	fluid discharge
r	number of basic dimension
R	indeterminate force
t	time
T	primary quantity, time
u	component of displacement
u, v, w	Cartesian components of fluid velocity
u, v, w	general variables
V	velocity
w	specific weight
x	displacement
x, y, z	Cartesian coordinates
α	distortion factor
β	coefficient of thermal expansion
γ	dimensionless material property
δ	deflection
δ	prediction factor
ε	component of strain
$\dot{\varepsilon}$	component of strain rate
η	material property
θ	temperature
Θ	primary quantity, temperature
λ	length
μ	fluid viscosity
ν	Poisson's ratio
ξ	surface tension
π	dimensionless group
ρ	density
σ	component of stress
ω	frequency

References

1. P. W. Bridgman, *Dimensional Analysis*, Yale University Press, New Haven, CT, 1922.
2. G. Murphy, *Similitude in Engineering*, Ronald Press, New York, 1950.
3. H. L. Langhaar, *Dimensional Analysis and Theory of Models*, Wiley, New York, 1951.
4. H. E. Huntley, *Dimensional Analysis*, Macdonald, London, 1952.
5. W. J. Duncan, *Physical Similarity and Dimensional Analysis: An Elementary Treatise*, Edward Arnold, London, 1953.
6. L. I. Sedov, *Similarity and Dimensional Methods in Mechanics*, Academic Press, New York, 1959.
7. D. C. Ipsen, *Units, Dimensions, and Dimensionless Numbers*, McGraw-Hill, New York, 1960.
8. S. J. Kline, *Similitude and Approximation Theory*, McGraw-Hill, New York, 1965.
9. V. J. Skoglund, *Similitude—Theory and Applications*, International Textbook, Scranton, PA, 1967.
10. W. E. Baker, P. S. Westline, and F. T. Dodge, *Similarity Methods in Engineering Dynamics—Theory and Practice of Scale Modeling*, Hayden (Spartan Books), Rochelle Park, NJ, 1973.
11. E. S. Taylor, *Dimensional Analysis for Engineers*, Clarendon Press, Oxford, 1974.
12. E. de St. Q. Issacson and M. de St. Q. Issacson, *Dimensional Methods in Engineering and Physics*, Wiley, New York, 1975.
13. D. J. Schuring, *Scale Models in Engineering*, Pergamon Press, Elmsford, NY, 1977.
14. E. Szucs, *Similitude and Modeling*, Elsevier, New York, 1980.
15. G. M. Sabnis, H. G. Harris, R. N. White, and M. S. Mirza, *Structural Modeling and Experimental Techniques*, Prentice-Hall, Englewood Cliffs, NJ, 1983.
16. E. O. Macagno, Historico-Critical Review of Dimensional Analysis, *J. Franklin Inst.*, 292 (1971), 391–402.
17. E. Fumagalli, *Statical and Geomechanical Models*, Springer-Verlag, New York, 1973, 33–57.
18. J. Dundurs, Dependence of Stress on Poisson's Ratio in Plane Elasticity, *Int. J. Solids Struct.*, 3 (1967), 1013–1021.
19. M. S. Yalin, *Theory of Hydraulic Models*, Macmillan, London, 1971.
20. J. J. Sharp, *Hydraulic Modelling*. Butterworths, London, 1981.
21. Lord Rayleigh, The Principles of Similitude, *Nature*, 95 (1915), 66–68; see also comment by D. Riabouchinsky (p. 591) and reply (p. 644).
22. W. M. Rohsenow and H. Y. Choi, *Heat, Mass, and Momentum Transfer*, Prentice-Hall, Englewood Cliffs, NJ, 1961, 427–443.

CHAPTER
16

Experimental Modal Analysis

Randall J. Allemang and David L. Brown

Structural Dynamics Research Laboratory
Department of Mechanical, Industrial, and Nuclear Engineering
University of Cincinnati
Cincinnati, Ohio

16-0 Introduction

Experimental modal analysis refers to the process of determining the modal parameters (frequencies, damping factors, modal vectors, and modal scaling) of a linear, time-invariant system by way of an experimental approach. Naturally, the modal parameters may be determined by analytical means, such as finite-element analysis, and one of the common reasons for experimental modal analysis is the verification of the results of the analytical approach. Often, though, an analytical model does not exist and the modal parameters determined experimentally serve as the model for future evaluations such as structural modifications. Predominantly, experimental modal analysis is used to explain a dynamics problem, vibration or acoustic, that is not obvious from intuition, analytical models, or similar experience.

The history of experimental modal analysis began in the 1940s with work oriented toward measuring the modal parameters of aircraft so that the problem of flutter could be accurately predicted. At that time, transducers to measure dynamic force were primitive and the analog nature of the approach yielded a time-consuming process that was not practical for most situations. With the advent of digital minicomputers and the fast Fourier transform (FFT) in the 1960s, the modern era of experimental modal analysis was born. Today, experimental modal analysis represents an interdisciplinary field that brings together the signal conditioning and computer interaction of electrical engineering, the theory of mechanics, vibrations, acoustics, and control theory from mechanical engineering, and the parameter estimation approaches of applied mathematics [1-5].

Experimental Modal Analysis Overview

The process of determining modal parameters from experimental data involves several phases. While these phases can be, in the simplest cases, very abbreviated, experimental modal analysis depends on the understanding of the basis for each phase. As in most experimental situations, the success of the experimental modal analysis process depends on having very specific goals for the test situation. Such specific goals affect every phase of the process in terms of reducing the errors associated with that phase. While there are several ways of breaking down the process, one possible delineation of these phases is as follows:

Modal analysis theory
Experimental modal analysis method
Modal data acquisition
Modal parameter estimation
Modal data presentation

Modal analysis theory refers to that portion of classical vibrations that explains, theoretically, the existence of natural frequencies, damping factors, and mode shapes for linear systems. This theory includes lumped-parameter, or discrete, models as well as continuous models. This theory also includes real normal modes as well as complex modes of vibration as possible solutions for the modal parameters. This phase of the experimental modal analysis process will not be repeated here but is summarized in textbooks on analytical and experimental modal analysis and on vibration. The relationship of transforms to vibration theory is very important to the comprehension of modern experimental modal analysis methods. Particularly, since discrete Fourier transforms are often involved in the modal data acquisition phase, this aspect of modal theory is critical.

Assuming a basic understanding of classic vibration theory and, in particular, the role of modal parameters in structural dynamics, the theoretical basis for the experimental modal analysis methods is the next concern. Experimental modal analysis methods involve the theoretical relationship between measured quantities and the classic vibration theory often represented as matrix differential equations. All modern methods trace from the matrix differential equations but yield a final mathematical form in terms of measured data. Such measured data can be raw input and output data in the time or frequency domains or some form of processed data (e.g., impulse-response or frequency-response functions). Most current methods involve processed data such as frequency-response functions in the estimation of modal parameters. In summary, though, the experimental modal analysis method establishes the form of the data that must be acquired.

The modal data acquisition phase involves the practical aspects of acquiring the data that will be required to serve as input to the modal parameter estimation phase. Therefore, a great deal of care must be taken to assure that the data match the requirements of the theory as well as the requirements of the numerical algorithm involved in the modal parameter estimation. The theoretical requirements involve concerns such as system linearity as well as time invariance of system parameters. The numerical algorithms are more concerned with the bias errors in the data and with any overall dynamic range considerations.

The modal parameter estimation phase is concerned with the practical problem of estimating the modal parameters, based on a choice of mathematical model as justified by the experimental modal analysis method, from the measured data. Problems that occur at this phase most often arise from violations of assumptions used to justify earlier phases. Serious theoretical problems such as nonlinear considerations will cause serious problems in the estimation of modal parameters and may completely invalidate the experimental modal analysis approach. Serious practical problems such as bias errors resulting from the digital signal processing will cause similar problems but are a function of data acquisition techniques, which can be altered to minimize such errors. The modal parameter estimation phase is that point in the experimental modal analysis process where the errors of all earlier work are accumulated. Unfortunately, the cause of the error is often attributed to a faulty modal parameter estimation algorithm rather than to faulty theoretical assumptions or a faulty modal data acquisition procedure.

The final phase of the process involves modal data presentation. This may simply be the numerical tabulation of the frequency, damping, and modal vectors along with the associated geometry of the measured degrees of freedom. More often, modal data presentation involves the

EXPERIMENTAL MODAL ANALYSIS

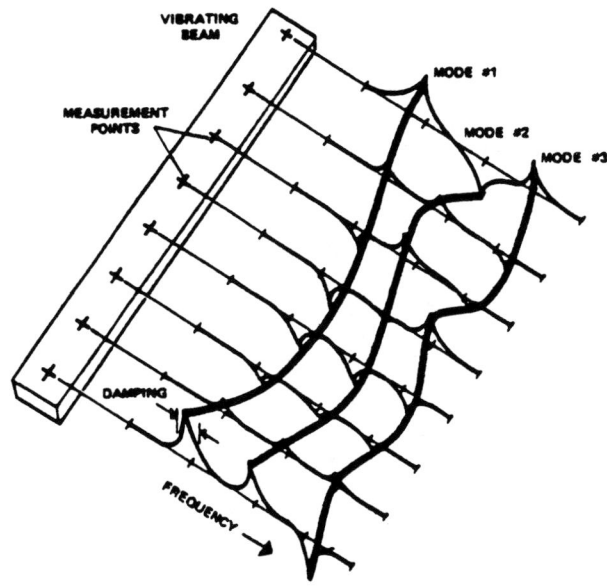

Figure 16-1 Experimental modal analysis example.

plotting and animation of such information. This involves the additional information required to construct a three-dimensional representation of the test object. Primarily, this requires a display sequence involving the order in which the degrees of freedom will be connected. Newer approaches to the plotting and animation of the modal vectors involve hidden line calculations, which require surface definition as well. Modal data presentation, therefore, is that process of providing a physical interpretation of the modal parameters.

Figure 16-1 is an example of all phases of the process. First of all, a continuous beam is being evaluated for the first few modes of vibration. Modal analysis theory explains that this is a linear system and that the modal vectors of this system should be real normal modes. The experimental modal analysis method that has been used is based on the frequency-response function relationships to the matrix differential equations of motion. At each measured degree of freedom, the imaginary part of the frequency-response function for that measured response degree of freedom and a common input degree of freedom is superimposed perpendicular to the beam. Naturally, the modal data acquisition involved in this example involves the estimation of frequency-response functions for each degree of freedom shown. Note that the frequency-response functions are complex-valued functions and that only the imaginary portion of each function is shown. One method of modal parameter estimation suggests that for systems with light damping and widely spaced modes, the imaginary part of the frequency-response function, at the damped natural frequency, may be used as an estimate of the modal coefficient for that response degree of freedom. The damped natural frequency can be identified as the frequency of the positive and negative peaks in the imaginary part of the frequency-response functions. The damping can be estimated from the sharpness of the peaks. In this very simple way, the modal parameters have been estimated. Modal data presentation for this case is shown as the lines connecting the peaks. While animation is possible, a reasonable interpretation of the modal vector can be gained in this simple case from plotting alone.

16-1 Modal Analysis Theory

Modal analysis theory has not really changed over the last century. The significant difference has come in how this thoery relates to the experimental world. The advances of the last 40 years with respect to measurement and analysis capabilities have caused a reevaluation of what aspects of the theory relate to the practical world of testing. With this in mind, the aspect of transform relationships has taken on renewed importance, since digital forms of the integral transforms are in constant use. The theory from the vibrations point of view involves a more thorough understanding of how the structural parameters of mass, damping, and stiffness relate to the impulse-response function (time domain), the frequency-response function (Fourier domain), and the transfer function (Laplace domain) for having single and multiple degrees of freedom.

Transform Relationships

One of the keys to understanding experimental modal analysis involves the comprehension of the relationships between the different domains used to describe the dynamics of a structural system. Historically, this has involved the time, frequency (Fourier), and Laplace domains. These relationships, with respect to a structural system, are the integral transforms (Fourier and Laplace) that reflect the information contained by the governing differential equations transformed to each domain. It is important to note that these are integral relationships and that the governing differential equations represent continuous relationships in each domain. As the digital approach to the measurement of data is considered, similar relationships between time, frequency, and Z domains can be formed that reflect the information contained by the governing finite-difference equations in each domain. The relationships in this discrete case are represented by discrete transforms (discrete Fourier tranform, Z transform). Whether the concept is approached from the continuous or the discrete case, the idea of the transform relationship is of primary importance.

The concept of a transform is obviously very special. In general, there are four criteria for a mathematical process to be designated as a transform. First of all, the transform process must involve a change of independent variable that represents the change from one domain to another. Second, the transformed variable must be the variable of interest. Third, the transform process must be computationally very simple but unique. Finally, there must be no loss of information in the transform process. The last criterion is often the true distinction between a transform and some form of parameter estimation. Since, with respect to modal analysis, the concepts of transforms and parameter estimation are both important, this distinction is a very important one.

Transform Properties

To process data into measurements efficiently and without error, the properties of transforms must be well understood. Digital data always are truncated to a limited time period and often are multiplied by a time-varying function to enhance certain characteristics in the data. This sort of alteration of the data in the time domain has distinctly unique but often bewildering effects in another domain if the properties of the transform are not considered.

While the complete list of transform properties for each transform can be found in numerous applied mathematics handbooks, several properties should be noted because of their frequent application in digital signal processing. First of all, scalar multiplication is commutative within and between domains. Therefore, calibration of signals using constant calibration factors can be performed without detrimental effect in any domain. Second, the multiplication of two functions in one domain is equivalent to the convolution of the transformed functions in the resulting domain. This property is essential for explaining the effects of weighting functions applied in the time domain as a possible way of minimizing the effects of the bias error that will occur as a result of the truncation of the data in the time domain (leakage error). Third, differentiation in the time domain with respect to the independent time variable is equivalent to multiplication in the transformed domain by the transformed variable. This property gives a very simple algebraic way to convert data from acceleration to velocity or displacement in the frequency of Laplace domains via multiplication. Finally, the shift theorem, which involves multiplying the time-domain data by a complex function of a shift frequency $[cos(\omega t) + j\, sin(\omega t)]$ is the basis for most frequency-shifted Fourier transform algorithms (zoom) used in digital signal analyzers today.[1]

Degrees of Freedom

The development of any theoretical concept in the area of vibrations, including modal analysis, depends on an understanding of the concept of the number of degrees of freedom N of a system. This concept is extremely important to the area of modal analysis, since the number of modes of vibration of a mechanical system is theoretically equal to the number of degrees of freedom. From a practical point of view, the relationship between this theoretical definition of the number of degrees of freedom and the number of *measurement degrees of freedom* is often confusing. For this reason, we review the concept of degree of freedom as a preliminary to the modal analysis material that follows.

To begin with, the basic definition that is normally associated with the concept of the number of degrees of freedom involves the following statement: the number of degrees of freedom for a mechanical system is equal to the number of independent coordinates (or minimum number of coordinates) required to locate and orient each mass in the mechanical system at any instant in time. As this definition is applied to a point mass, three degrees of freedom are required, since the location of the point mass involves knowing the x, y, and z translations of the center of gravity of the point mass. As this definition is applied to a rigid-body mass, six degrees of freedom are required, since θ_x, θ_y, and θ_z rotations are required in addition to the x, y, and z translations, if both the orientation and location of the rigid-body mass are to be defined at any instant in time. This concept is represented in Fig. 16-2. As this definition is extended to any general deformable body, it should be obvious that the number of degrees of freedom can now be considered to be infinite. While this is theoretically true, it is quite common, particularly with respect to finite-element methods, to view the general deformable body in terms of a large number of physical

---

[1] A list of the symbols and notational conventions used in this chapter precedes the References

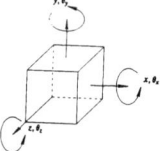

Figure 16-2 Degrees of freedom of a rigid body.

points of interest with six degrees of freedom for each of the physical points. In this way, the infinite number of degrees of freedom can be reduced to a large but finite number.

When measurement limitations are imposed on this theoretical concept of the number of degrees of freedom of a mechanical system, the difference between the theoretical number of degrees of freedom N and the number of measurement degrees of freedom begins to evolve. Initially, for a general deformable body, the number of degrees of freedom N can be considered to be infinite or equal to some large finite number if a limited set of physical points of interest is considered, as discussed in the preceding paragraph. The first measurement limitation that needs to be considered is that there is normally a limited frequency range that is of interest to the analysis. For example, most dominant structural modes of vibration for an automobile would be located between 0 and 200 Hz. As this limitation is considered, the number of degrees of freedom of this system that are of interest is now reduced from infinity to a reasonable finite number. The next measurement limitation that needs to be considered involves the physical limitation of the measurement system in terms of amplitude. A common limitation of transducers, signal-conditioning, and data acquisition systems results in a dynamic range of 80 to 100 dB (10^4 to 10^5) in the measurement. This means that the number of degrees of freedom is reduced further as a result of the dynamic range limitations of the measurement instrumentation. Finally, since few rotational transducers exist at this time, the normal measurements that are made involve only translational quantities (displacement, velocity, accelerations, force) and thus do not include rotational effects, or rotational degrees of freedom. In summary, even for the general deformable body, the theoretical number of degrees of freedom that are of interest is limited to a very reasonable finite value ($N = 1-50$). Therefore, this number of degrees of freedom is the number of modes of vibration that are of interest.

Finally, then, the number of *measurement degrees of freedom* can be defined as the number of physical locations at which measurements are made times the number of measurements made at each physical location. For example, if x, y, and z accelerations are measured at each of 100 physical locations on a general deformable body, the number of measurement degrees of freedom would be 300. It should be obvious that since the physical locations are chosen somewhat arbitrarily and certainly without exact knowledge of the modes of vibration that are of interest, there is no specific relationship between the number of degrees of freedom N and the number of measurement degrees of freedom. In general, to define N modes of vibration of a mechanical system, the number of measurement degrees of freedom must be equal to or larger than N. Note also that even though the number of measurement degrees of freedom is larger than N, this is not a guarantee that N modes of vibration can be found from the measurement degrees of freedom. The measurement degrees of freedom must include physical locations that allow a unique determination of the N modes of vibration. For example, if none of the measurement degrees of freedom are located on a portion of the mechanical system that is active in one of the N modes of vibration, portions of the modal parameters for this mode of vibration cannot be found.

In the development of the single and multiple degree of freedom information in the following, the assumption is made that a minimal set of measurement degrees of freedom ($=N$) exists that

EXPERIMENTAL MODAL ANALYSIS

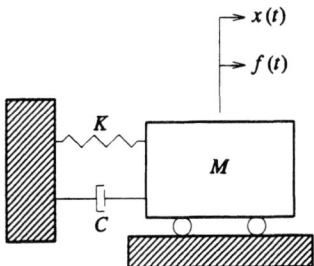

Figure 16-3 Single degree of freedom system.

will allow for N modes of vibration to be determined. In reality, the number of measurement degrees of freedom is normally chosen much larger than the number of modes of vibration that are to be determined; this is because prior knowledge of the modes of vibration is not available. If the set of measurement degrees of freedom is large enough and if the measurement degrees of freedom are distributed uniformly over the general deformable body, the N modes of vibration will normally be found. The measurement degrees of freedom are defined experimentally as those locations (and directions) at which inputs are applied to the system N_i together with those locations (and directions) at which outputs are measured N_o. Since the number of inputs N_i normally is small and normally is a subset of the number of outputs N_o, the number of measurement degrees of freedom normally is equal to the number of outputs or response transducers used in the experimental test.

Single Degree of Freedom Systems

To understand modal analysis, complete familiarity with single degree of freedom systems is necessary. In particular, single degree of freedom systems, as presented and evaluated in the time, frequency (Fourier), and Laplace domains, serve as the basis for many of the models that are used in modal parameter estimation. This single degree of freedom approach is obviously trivial for the modal analysis case, since a modal vector cannot be defined. The true importance of this approach results because the multiple degree of freedom case can be viewed as simply a linear superposition of single degree of freedom systems.

Time Domain: Impulse Response Function

The general mathematical representation of a single degree of freedom (SDOF) system, as shown in Figure 16-3, is expressed by Eq. (16-1):

$$M \ddot{x}(t) + C \dot{x}(t) + K x(t) = f(t) \tag{16-1}$$

From differential equation theory, the transient response of this SDOF system to a transient force in the form of a theoretical impulse, can be assumed to be in the following form:

$$x(t) = Ae^{\lambda_1 t} + Be^{\lambda_2 t} \tag{16-2}$$

The characteristic frequencies in this solution, λ_1 and λ_2, are determined from the characteristic equation of Eq. (16-1). This yields characteristic frequencies of the following form:

$$\lambda_{1,2} = -\frac{C}{2M} \pm \sqrt{\left[\frac{C}{2M}\right]^2 - \left[\frac{K}{M}\right]} \qquad (16\text{-}3)$$

With respect to the preceding result, critical damping, damping ratio, and damping factor can now be defined. *Critical damping*, C_c, is the damping that reduces the radical in the solution to the characteristic equation to zero. This form of representing the damping is a physical approach and, therefore, involves the appropriate units for equivalent viscous damping.

$$C_c = 2M\sqrt{\frac{K}{M}} \qquad (16\text{-}4)$$

The fraction of critical damping, or *damping ratio*, ζ, is the ratio of the actual system damping to the critical system damping. This approach to the description of damping is dimensionless since the units have been normalized.

$$\zeta_1 = \frac{C}{C_c} \qquad (16\text{-}5)$$

Systems are often classified depending on their damping ratios. That is:

Overdamped system $\zeta_1 > 1$
Critically damped system $\zeta_1 = 1$
Underdamped system $\zeta_1 < 1$

The roots of the overdamped system yield two real roots that lie on the real axis; if damping were to increase, the roots would move apart. For the case of the critically damped system, there is only one real root, again lying on the real axis. The underdamped system yields two complex roots, one being the complex conjugate of the other. As the system is altered (damping and frequency changes), these roots stay in the second and third quadrant of the *s* plane. It should be pointed out that if any roots of the characteristic equation were to lie to the right of the imaginary (frequency) axis, the system would be unstable.

For most real structures, unless active damping systems are present, the damping ratio is rarely greater than 10%. For this reason, all further discussion is restricted to underdamped systems $\zeta < 1$. With reference to Eq. (16-3), this means that the two roots $\lambda_{1,2}$ wil always be complex conjugates. Also, it can be proven that the two coefficients A and B will be complex conjugates of each other. For an underdamped system, the roots of the characteristic equation can be written as:

$$\lambda_1 = \sigma_1 \pm j\omega_1 \qquad (16\text{-}6)$$

where σ_1 = damping factor
ω_1 = damped natural frequency
Ω_1 = undamped natural frequency = $\sqrt{\sigma_1^2 + \omega_1^2}$

The roots of the characteristic equation in Eq. (16-3) can also be written as

$$\lambda_1, \lambda_1^* = -\zeta_1 \Omega_1 \pm \Omega_1 \sqrt{1 - \zeta_1^2} \qquad (16\text{-}7)$$

EXPERIMENTAL MODAL ANALYSIS

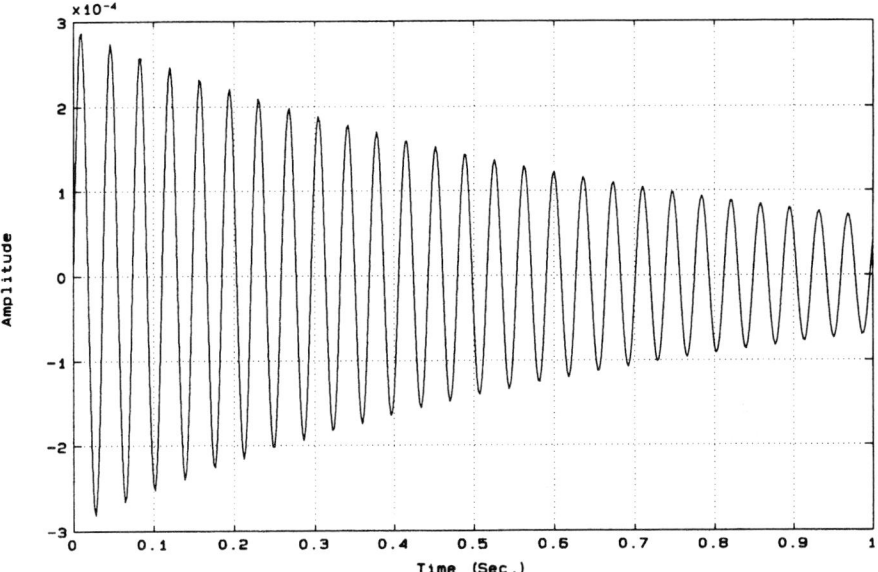

Figure 16-4 Time domain: SDOF impulse-response function.

From the equation above the definition of *damping factor, damped natural frequency,* and *undamped natural frequency* can be noted. The real part of a root of the characteristic equation is the *damping factor*, since it is this parameter that describes the exponential decay or growth of the harmonic. This parameter has the same units as the imaginary part of the root of the characteristic equation, which is defined as the *damped natural frequency*. This yields a situation in which damping is described in terms of hertz or radians per second. The *undamped natural frequency* is the magnitude of the complex root; when the damping is zero, the undamped and damped natural frequencies are equal.

The transient response of the single degree of freedom system can be determined from Eq. (16-1), assuming that the initial conditions are zero and that the system excitation $f(t)$ is a unit impulse. The response of the system $x(t)$ to such a unit impulse is known as the *impulse-response function* $h(t)$ of the system. Therefore:

$$h(t) = A\, e^{\lambda_1 t} + A^*\, e^{\lambda_1^* t} \tag{16-8}$$

$$h(t) = e^{\sigma_1 t}\, [A\, e^{(+j\omega_1 t)} + A^*\, e^{(-j\omega_1 t)}] \tag{16-9}$$

Thus, the magnitude A controls the amplitude of the impulse response, the real part of the pole is the decay rate, and the imaginary part of the pole is the frequency of oscillation. Figure 16-4 illustrates the impulse-response function for a single degree of freedom system.

Frequency Domain: Frequency-Response Function

Equation (16-1) is the time-domain representation of the system in Fig. 16-3. An equivalent equation of motion may be determined for the Fourier or frequency (ω) domain. This representation has the advantage of converting a differential equation to an algebraic equation. This is accomplished by taking the Fourier transform of Eq. (16-1). Thus, Eq. (16-1) becomes:

$$[-M\omega^2 + jC\omega + K]X(\omega) = F(\omega) \tag{16-10}$$

Equation (16-10) is an equivalent representation of Eq. (16-1) in the Fourier domain. If the system forcing function $F(\omega)$ and its response $X(\omega)$ are known, the system characteristic $H(\omega)$ can be calculated. That is:

$$H(\omega) = \frac{X(\omega)}{F(\omega)} = \frac{1}{-M\omega^2 + jC\omega + K} \tag{16-11}$$

The quantity $H(\omega)$ is known as the *frequency-response function* of the system. In other words, a frequency-response function relates the Fourier transform of the system input to the Fourier transform of the system response. The denominator of Eq. (16-10) is known as the characteristic equation of the system and is the same characteristic equation as that found in the solution of Eq. (16-1) in the time domain. Note that the characteristic values of this complex equation are in general complex even though the equation is a function of a real-valued independent variable (ω). With this in mind, it can be proven that the characteristic values of this equation, known as the complex roots of the characteristic equation or the complex poles of the system, are the same as those found from Eq. (16-1).

The frequency-response function $H(\omega)$ can now be rewritten as a function of the complex poles by using the factored form of the polynomial equation as follows:

$$H(\omega) = \frac{1/M}{(j\omega - \lambda_1)(j\omega - \lambda_1^*)} \tag{16-12}$$

Since the frequency-response function is a complex-valued function of a real-valued independent variable (ω), the frequency-response function, as shown in Fig. 16-5a or b, is represented by a pair of curves.

Laplace Domain: Transfer Function

Just as in the preceding case for the frequency domain, the equivalent information can be presented in the Laplace domain by way of the Laplace transform. The only significant difference in the development concerns the fact that the Fourier transform is defined from negative infinity to positive infinity while the Laplace transform is defined from zero to positive infinity with initial conditions. The Laplace representation, also, has the advantage of converting a differential equation to an algebraic equation. The development using Laplace tranforms begins by taking the Laplace transform of Eq. (16-1). Thus, Eq. (16-1) becomes, assuming zero initial conditions:

$$[Ms^2 + Cs + K]X(s) = F(s) \tag{16-13}$$

Therefore, using the same logic as in the frequency-domain case, the transfer function can be defined just as the frequency-response function was defined earlier.

$$H(s) = \frac{X(s)}{F(s)} = \frac{1}{Ms^2 + Cs + K} \tag{16-14}$$

The quantity $H(s)$ is defined as the *transfer function* of the system. In other words, a transfer function relates the Laplace transform of the system input to the Laplace transform of the system response. The transfer function can also be written

$$H(s) = \frac{1/M}{s^2 + (C/M)s + (K/M)} \tag{16-15}$$

Note that Eq. (16-14) is valid under the assumption that the initial conditions are zero.

EXPERIMENTAL MODAL ANALYSIS

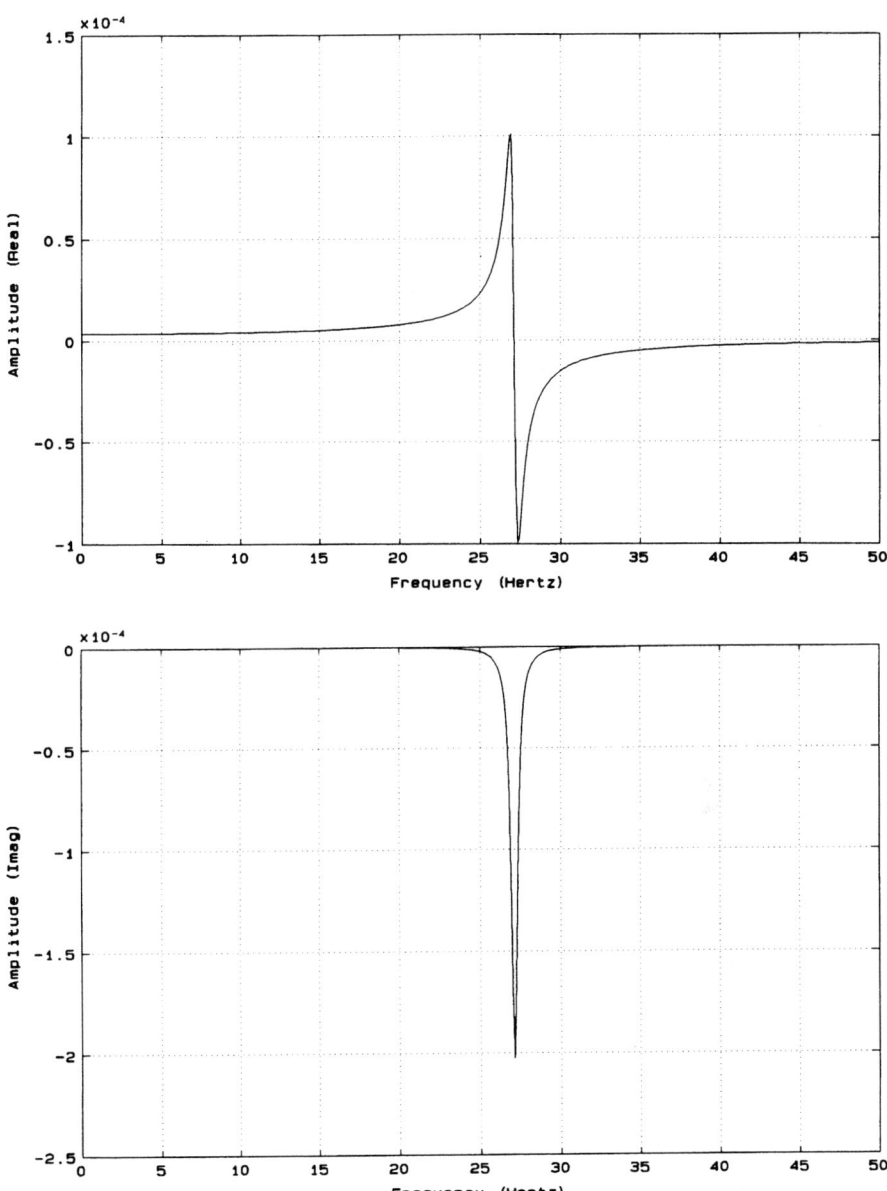

Figure 16-5a Frequency domain: SDOF frequency-response functions.

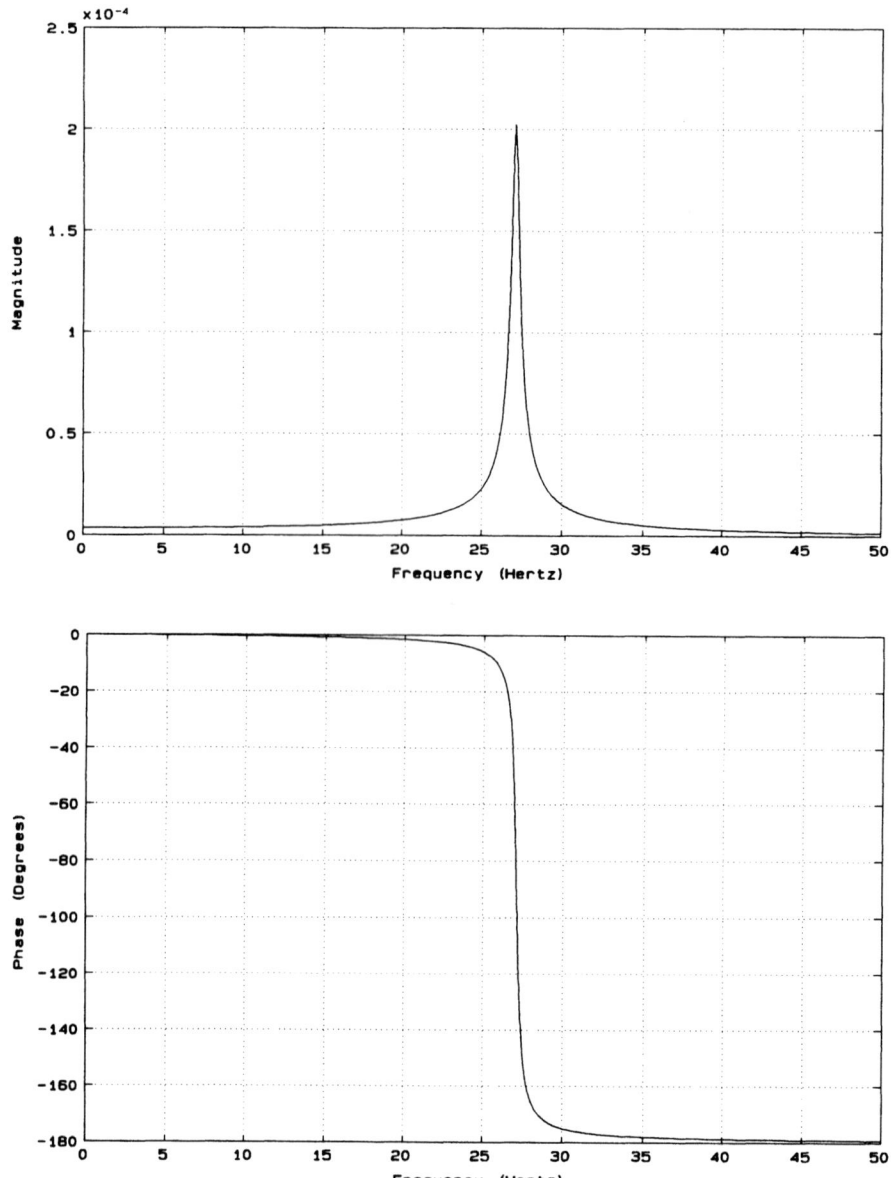

Figure 16-5b Frequency domain: SDOF frequency-response functions.

EXPERIMENTAL MODAL ANALYSIS

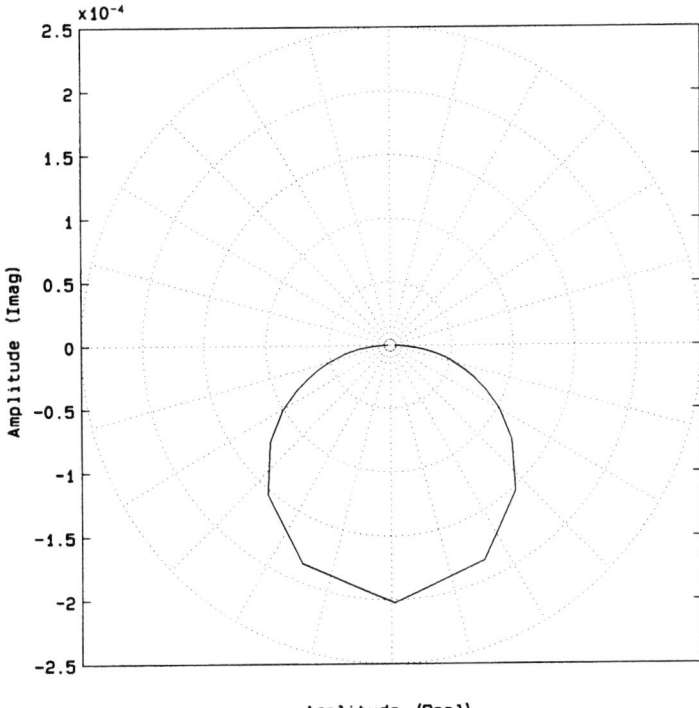

Figure 16-5c Frequency domain: SDOF frequency-response functions.

The denominator term is once again referred to as the characteristic equation of the system. As noted in the preceding two cases, the roots of the characteristic equation are given in Eq. (16-3).

The transfer function $H(s)$ can now be rewritten, just as in the frequency-response function case, as

$$H(s) = \frac{1/M}{(s - \lambda_1)(s - \lambda_1^*)} \tag{16-16}$$

Since the transfer function is a complex-valued function of a complex independent variable s, the transfer function is represented, as shown in Fig. 16-6a or b, as a pair of surfaces. Remember that the variable s in Eqs. (16-13) through (16-16) is a complex variable; that is, it has a real part and an imaginary part. Therefore, it can be viewed as a function of two variables that represent a surface.

The definitions of undamped natural frequency, damped natural frequency, damping factor, percent of critical damping, and residue are all relative to the information represented by Fig. 16-6. The projection of this information onto the plane of zero amplitude yields the information as shown in Fig. 16-7.

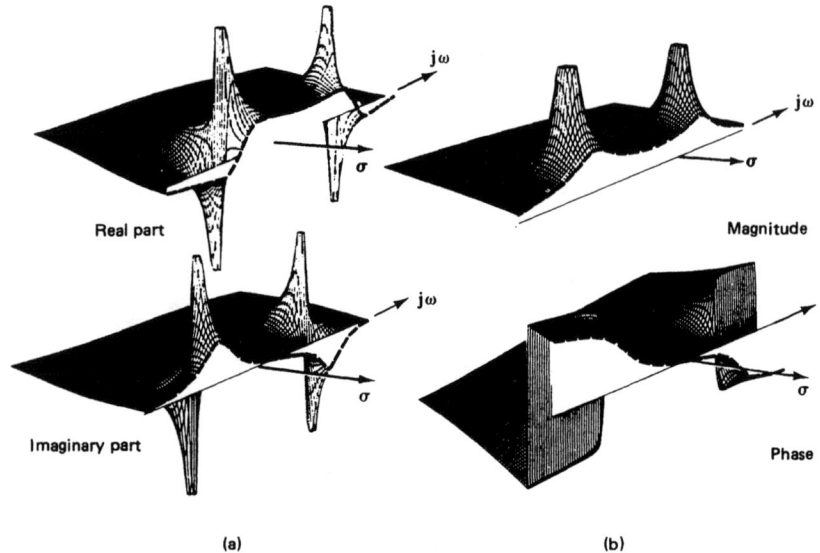

Figure 16-6 Laplace domain: SDOF transfer functions.

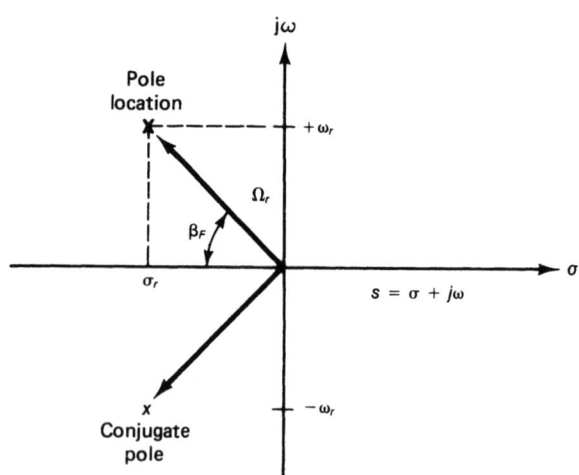

Figure 16-7 Laplace domain projection.

EXPERIMENTAL MODAL ANALYSIS

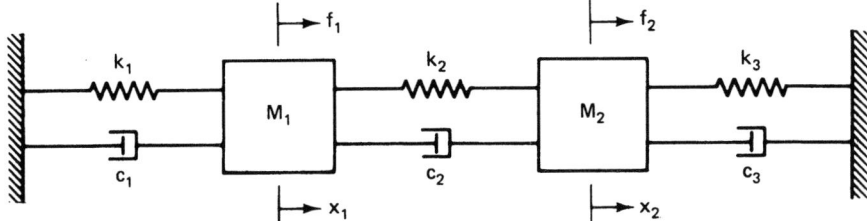

Figure 16-8 Multiple degree of freedom system.

The concept of *residues* can now be discussed in terms of the partial fraction expansion of the transfer function equation. Equation (16-16) can be expressed in terms of partial fractions:

$$H(s) = \frac{1/M}{(s - \lambda_1)(s - \lambda_1^*)} = \frac{A}{(s - \lambda_1)} + \frac{A^*}{(s - \lambda_1^*)} \tag{16-17}$$

The residues of the transfer function are defined as being the constants A and A^*. The terminology and development of residues comes from the evaluation of analytic functions in complex analysis. The residues of the transfer function are directly related to the amplitude of the impulse-response function. In general, the residue A can be a complex quantity. As shown for a single degree of freedom system, the residue A is purely imaginary.

At this point, it can be noted that the Laplace transform formulation is simply the general case of the Fourier transform development if the initial conditions are zero. The frequency-response function is the part of the transfer function evaluated along the $s = j\omega$ axis. From an experimental point of view, when one talks about measuring a transfer function, the frequency-response function is actually being estimated via the discrete Fourier transform.

Multiple Degree of Freedom Systems

The real application of modal analysis concepts begin when a continuous, nonhomogeneous structure is described as a lumped-mass, multiple degree of freedom system. At this point, the modal frequencies, the modal damping, and the modal vectors, or relative patterns of motion, can be found via an estimate of the mass, damping, and stiffness matrices or via the measurement of the associated frequency-response functions. The two degree of freedom system, shown in Fig. 16-8, is the most basic example of a multiple degree of freedom system. This example is a useful means for discussing modal analysis concepts, since a theoretical solution can be formulated in terms of the mass, stiffness, and damping matrices or in terms of the frequency-response functions.

The equation of motion for the system in Fig. 16-8, using matrix notation, is as follows:

$$\begin{bmatrix} M_1 & 0 \\ 0 & M_2 \end{bmatrix} \begin{Bmatrix} \ddot{x}_1 \\ \ddot{x}_2 \end{Bmatrix} + \begin{bmatrix} (C_1 + C_2) & -C_2 \\ -C_2 & (C_2 + C_3) \end{bmatrix} \begin{Bmatrix} \dot{x}_1 \\ \dot{x}_2 \end{Bmatrix}$$

$$+ \begin{bmatrix} (K_1 + K_2) & -K_2 \\ -K_2 & (K_2 + K_3) \end{bmatrix} \begin{Bmatrix} x_1 \\ x_2 \end{Bmatrix} = \begin{Bmatrix} f_1 \\ f_2 \end{Bmatrix}$$

or more concisely,

$$[M]\{\ddot{x}\} + [C]\{\dot{x}\} + [K]\{x\} = \{f\} \tag{16-18}$$

The process of solving Eq. (16-18) when the mass, damping, and stiffness matrices are known is shown in most classical texts concerning vibrations.

The development of the frequency-response function solution for the multiple degree of freedom case is shown in some detail in the section entitled Frequency-Response Function Method. This development relates the mass, damping, and stiffness matrices to a transfer function model involving multiple degrees of freedom. Since the transfer function model can be considered to be the general case of the frequency-response function, this demonstrates that the same modal information can be determined whether the mass, damping, and stiffness matrices are known or whether the associated frequency-response functions are measured. Just as in the analytical case, where the ultimate solution can be described in terms of one degree of freedom systems, the frequency-response functions between any input and response degree of freedom can be represented as a linear superposition of the single degree of freedom models derived earlier.

As a result of the linear superposition concept, the equations for the impulse-response function, the frequency-response function, and the transfer function for the multiple degree of freedom (MDOF) system are defined as follows:

Impulse-response function:

$$h_{pq}(t) = \sum_{r=1}^{N} A_{pqr} e^{\lambda_r t} + A_{pqr}^* e^{\lambda_r^* t} \tag{16-19}$$

Frequency-response function:

$$H_{pq}(\omega) = \sum_{r=1}^{N} \frac{A_{pqr}}{j\omega - \lambda_r} + \frac{A_{pqr}^*}{j\omega - \lambda_r^*} \tag{16-20}$$

Transfer function:

$$H_{pq}(s) = \sum_{r=1}^{N} \frac{A_{pqr}}{s - \lambda_r} + \frac{A_{pqr}^*}{s - \lambda_r^*} \tag{16-21}$$

where t = time variable
 ω = frequency variable
 s = laplace variable
 p = measured degree of freedom (response)
 q = measured degree of freedom (input)
 r = modal vector number
 A_{pqr} = residue
 $A_{pqr} = Q_r \psi_{pr} \psi_{qr}$
 Q_r = modal scaling factor
 ψ_{pr} = modal coefficient
 λ_r = system pole
 N = number of modal frequencies

It is important to note that the residue A_{pqr} in Eqs. (16-19) through (16-21) is the product of the modal deformations at the input q and response p degrees of freedom and a modal scaling factor for mode r. Therefore, the product of these three terms is unique, but each of the three terms by themselves is not unique.

Figures 16-9 and 16-10 represent impulse-response and frequency-response functions for the relatively simple two degree of freedom system shown in Fig. 16-8. Note that the impulse-response function shown in Fig. 16-9 is difficult to interpret visually for the individual modal parameters, in contrast to Fig. 16-4. The frequency-response function for the situation of input and response at the same measurement degree of freedom (same location and direction) is represented in Fig. 16-10a through c. This frequency-response function is known as a *driving point* measurement.

EXPERIMENTAL MODAL ANALYSIS

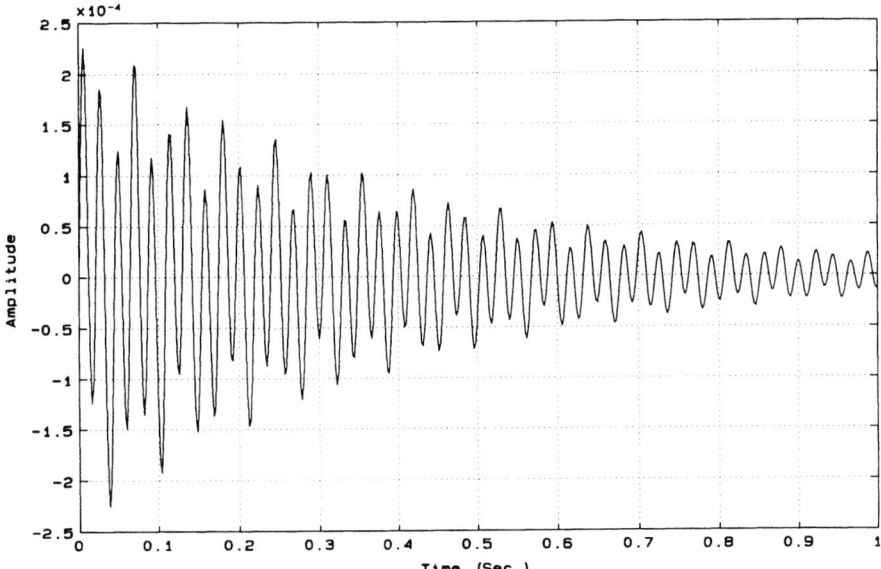

Figure 16-9 Time domain: MDOF impulse-response function.

The frequency-response function for the situation of input and response at different measurement degrees of freedom (different locations and/or directions) appears in Fig. 16-10d though f. This frequency-response function is known as a *cross-point* measurement.

Damping Mechanisms

To evaluate multiple degree of freedom systems that are present in the real world, the effect of damping on the complex frequencies and modal vectors must be considered. Many physical mechanisms are needed to describe all the possible forms of damping that may be present in a particular structure or system. Some of the classical types are structural damping, viscous damping, and Coulomb damping. It is generally difficult to ascertain which type of damping is present in any particular structure. Indeed most structures exhibit damping characteristics that result from a combination of all three types above, plus others.

It will suffice to say that whenever a structure is modeled with a particular form of damping—for example, viscous—the damping model is an equivalent model to whatever type of damping may actually be present.

Rather than consider the many, different physical mechanisms, the probable location of each mechanism, and the particular mathematical representation of the mechanism of damping that is needed to describe the dissipative energy of the system, we use a model that is concerned only with the resultant mathematical form. This model represents a hypothetical form of damping that is proportional to the system mass or stiffness matrix. Therefore:

$$[C] = \alpha[M] + \beta[K] \tag{16-22}$$

For this special form of damping, the coordinate transformation that diagonalizes the system mass

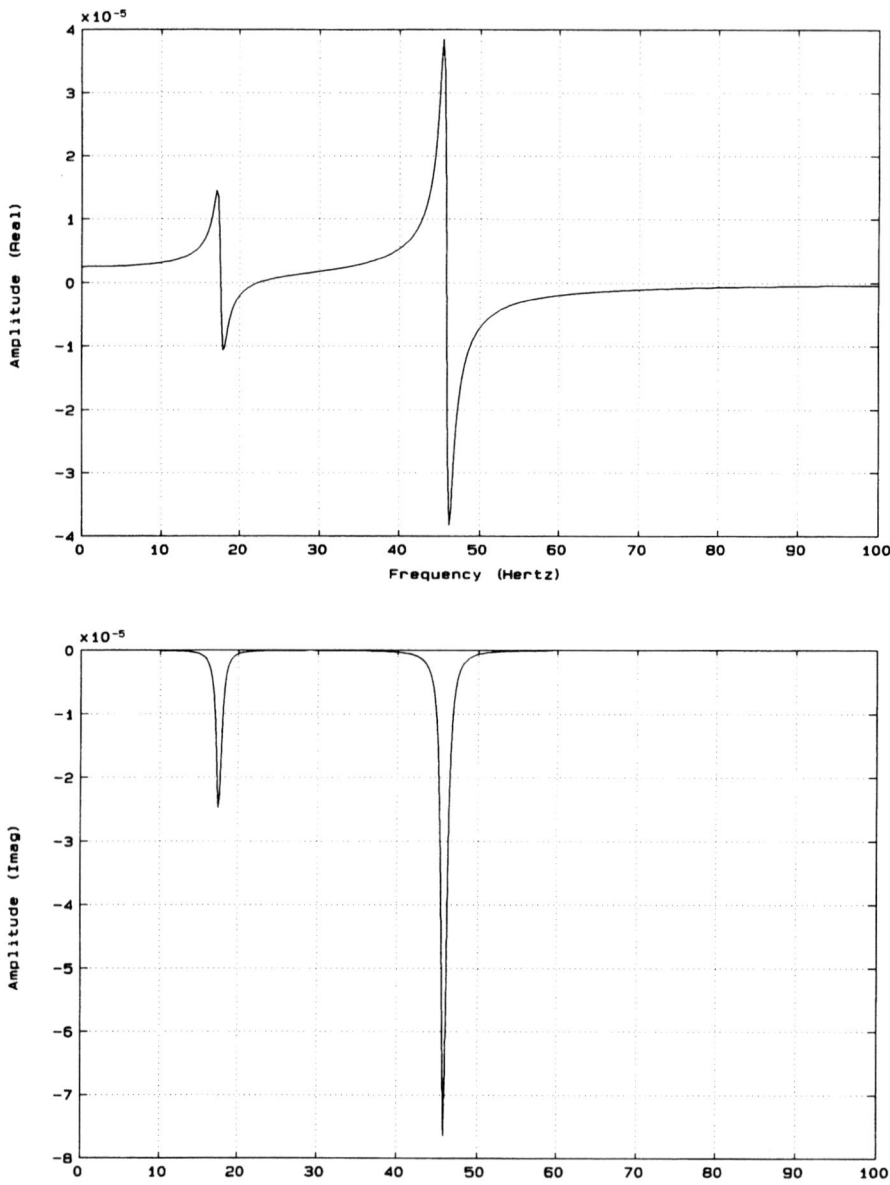

Figure 16-10a Frequency domain: MDOF frequency-response function. Input and response at same degree of freedom (real/imaginary format).

EXPERIMENTAL MODAL ANALYSIS

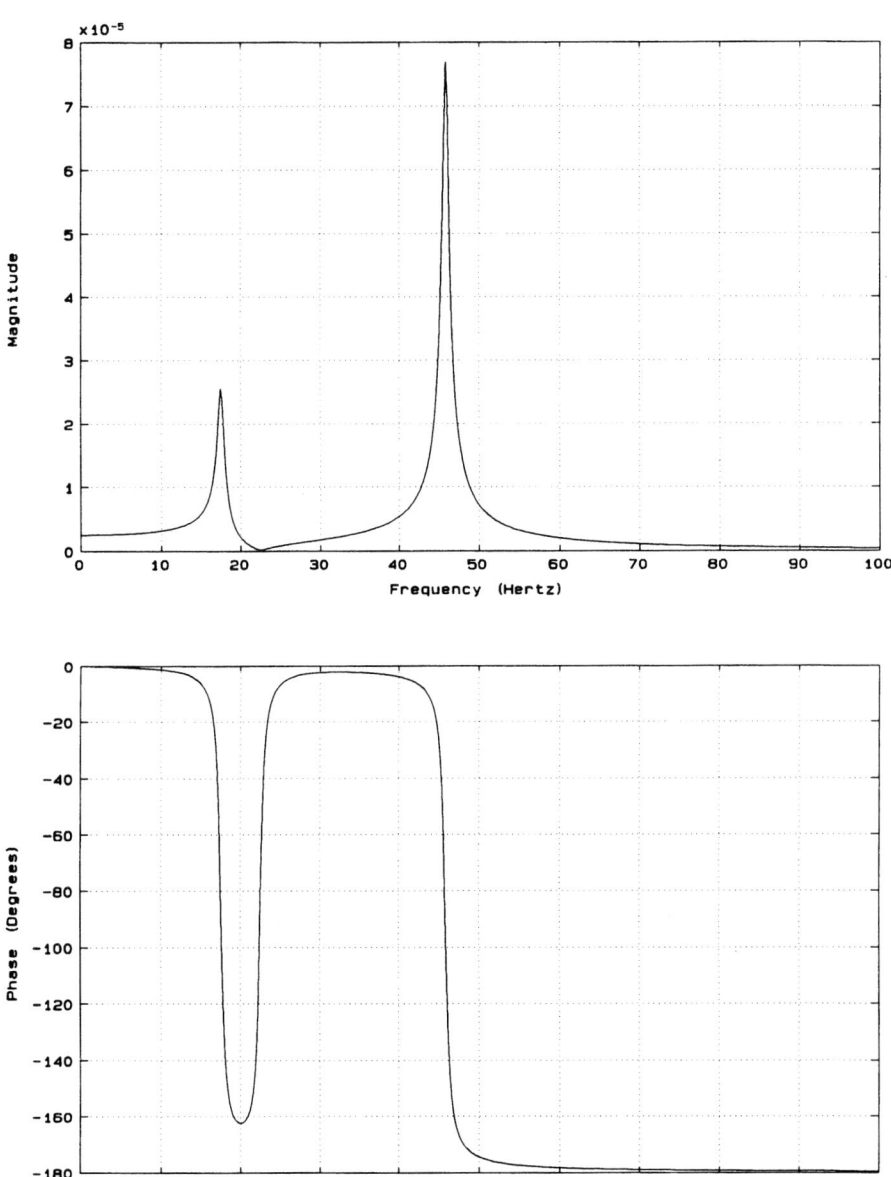

Figure 16-10b Frequency domain: MDOF frequency-response function. Input and response at same degree of freedom (magnitude/phase format).

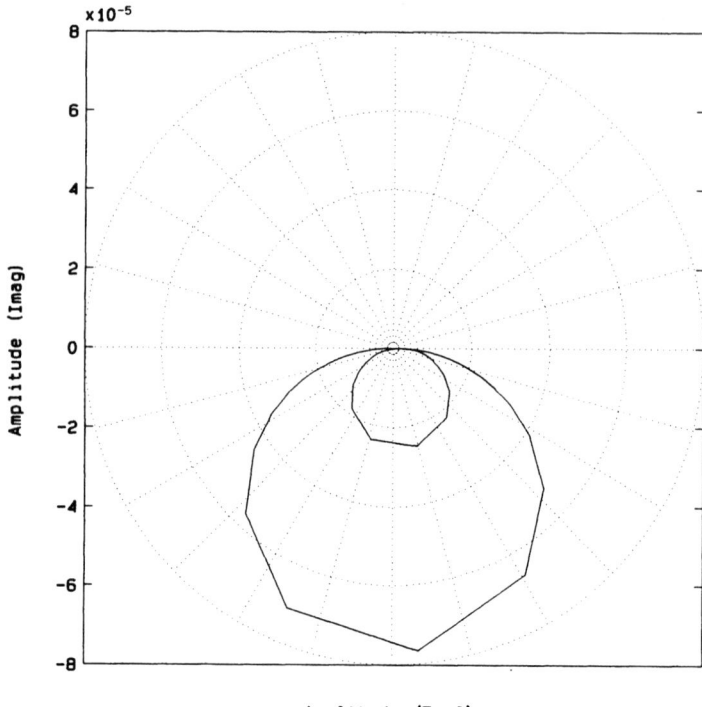

Amplitude (Real)

Figure 16-10c Frequency domain: MDOF frequency-response function. Input and response at same degree of freedom.

and stiffness matrices will also diagonalize the system-damping matrix. Therefore, when a system with proportional damping exists, that system of coupled equations of motion can be transformed to a system of equations representing an uncoupled system of single degree of freedom systems that are easily solved. The transformation matrix that transforms the physical coordinates $\{x\}$ to the coordinates that will uncouple the original equations of motion is the *modal matrix* $[\Psi]$. The modal matrix is the matrix of modal vectors, where each column of the modal matrix is a different modal vector. The transformation relationship can be stated as follows:

$$\{x\} = [\Psi] \{q\} \qquad (16\text{-}23)$$

Substituting Eq. (16-23) into Eq. (16-1) yields:

$$\lceil M \rfloor \{\ddot{q}\} + \lceil C \rfloor \{\dot{q}\} + \lceil K \rfloor \{q\} = \{f'\} \qquad (16\text{-}24)$$

The resulting motion coordinate in Eq. (16-24) $\{q\}$ is known as the *principal coordinate* or the *modal coordinate* for this system.

Real and Complex Modal Vectors

Modal vectors are a representation of the relative pattern of motion that is associated with each modal frequency and do not have a unique absolute value. In this sense, the modal vectors have

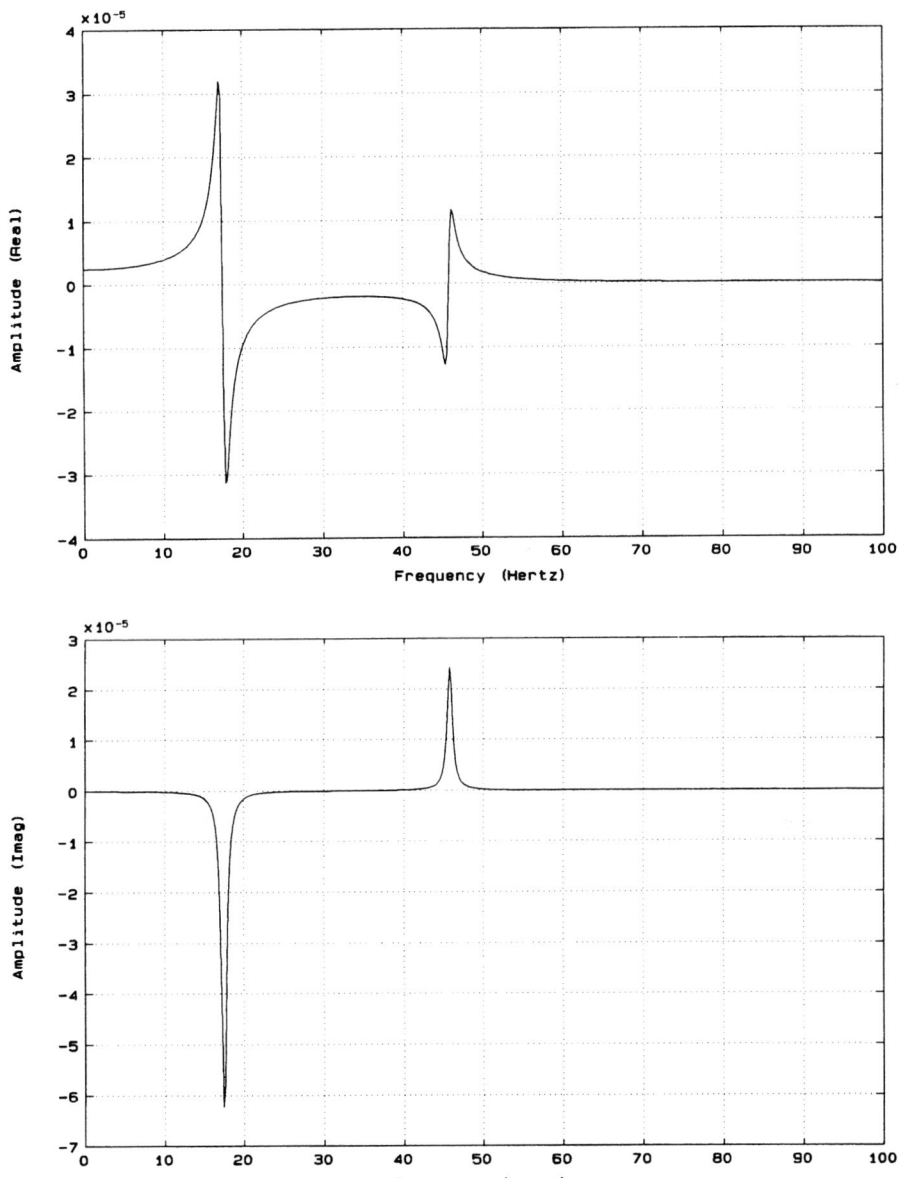

Figure 16-10d Frequency domain: MDOF frequency-response function. Input and response at different degrees of freedom (real/imaginary format).

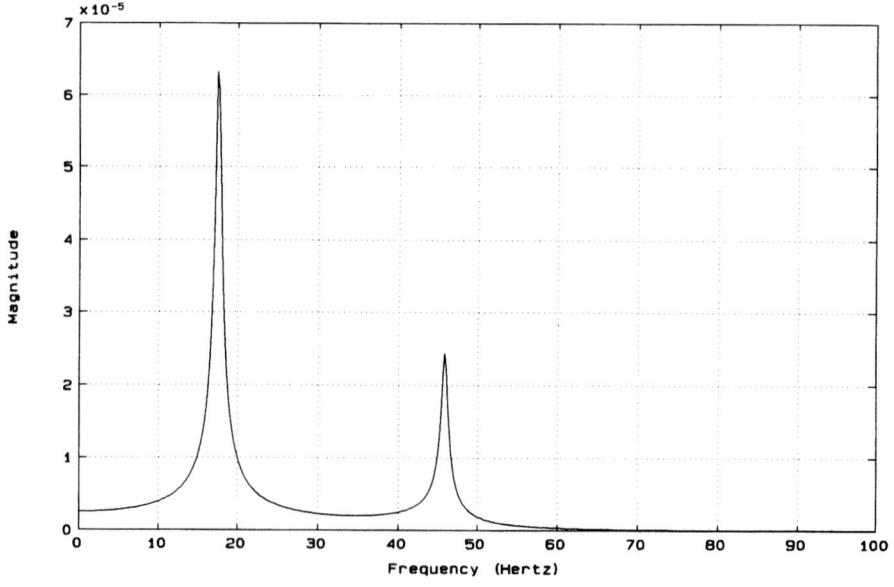

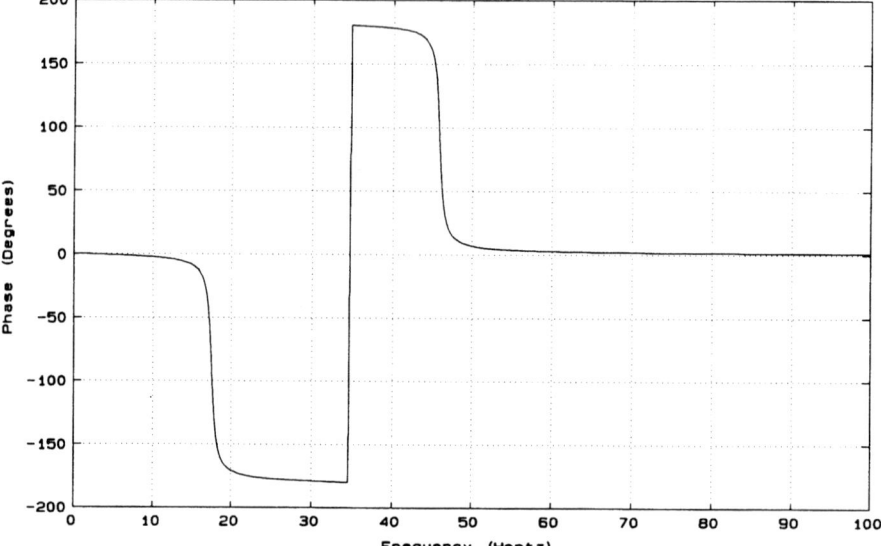

Figure 16-10e Frequency domain: MDOF frequency-response function. Input and response at different degrees of freedom (magnitude/phase format).

EXPERIMENTAL MODAL ANALYSIS

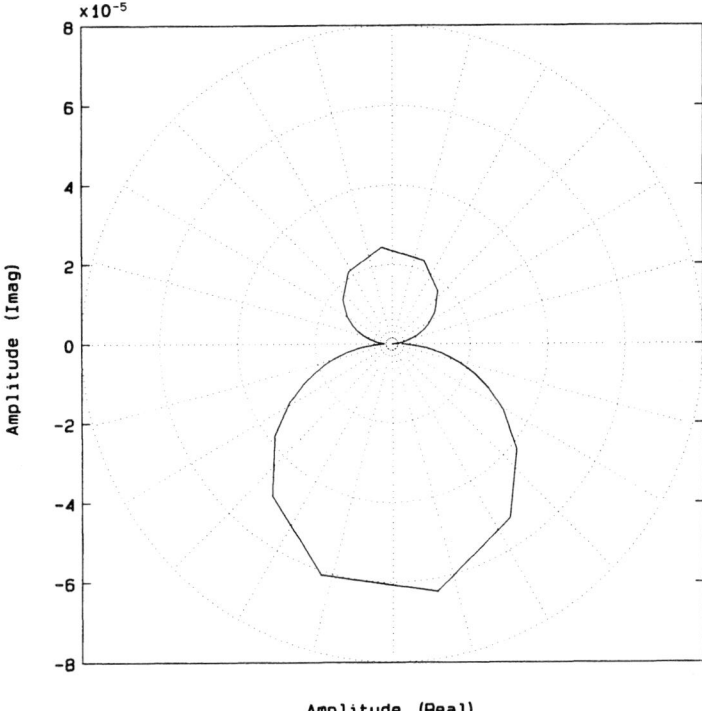

Figure 16-10f Frequency Domain: MDOF frequency response function. Input and response at different degrees of freedom (Complex format).

the characteristics of the eigenvector in an eigenvalue–eigenvector problem. Modal vectors that are found from the solution of Eq. (16-1) or from experimental testing usually are *real modal vectors*. Real modal vectors, or *normal modes,* are modal vectors that can be normalized so that the numerical pattern representing the motion can be described by real numbers (magnitudes with phase angles of ±180°) alone. Real modal vectors will always occur when dealing with undamped or proportionally damped systems. When dealing with systems with damping that cannot be described mathematically by Eq. (16-22), in other words, nonproportionally damped systems, *complex modal vectors* will result. Complex modal vectors are modal vectors that cannot be normalized to a numerical pattern described by only real numbers. If a complex modal vector is normalized so that one of the positions, or modal coefficients, is a real number, some or all of the remaining modal coefficients will be complex valued.

Modal Scaling

Traditionally, modal scaling has been defined in terms of the mass matrix and the modal vectors arising from the undamped, or proportionally damped, equations of motion that can be reduced to the form of Eq. 16-24.

$$\{\psi\}_r^T [M] \{\psi\}_r = M_r \tag{16-25}$$

Note that this definition involves modal vectors that are real or normal modal vectors and that the definition of *modal mass* from this analytical point of view is not unique. Equation (16-25) clearly indicates that even though the mass matrix is an absolute quantity, since the modal vectors can be scaled arbitrarily, the modal mass depends on the scaling, or normalization, of the modal vectors. From an experimental point of view, the modal mass can be defined in a consistent manner. Rather than defining modal mass in terms of an absolute quantity A_{qqr}, modal mass is defined based on Q_r, which is a relative quantity. If scaling of $\{\psi\}_r$ is changed, Q_r will be altered accordingly. While this concept of modal mass being a relative quantity (relative to scaling) is inconsistent with the engineering view of mass being an absolute quantity, this approach is consistent with the analytical definition of modal mass.

Therefore, the rth modal mass of a multi-degree-of-freedom system is defined as

$$M_r = \frac{1}{j2\, Q_r \omega_r} = \frac{\psi_{pr}\psi_{qr}}{j2\, A_{pqr}\omega_r} \qquad (16\text{-}26)$$

where M_r = modal mass
Q_r = scaling constant
ω_r = damped natural frequency

Note also that in the case of the single degree of freedom with the scaled modal coefficient equal to unity, the modal mass that is computed with Eq. (16-26) is equal to the physical mass of the single degree of freedom system. This leads to the same condition for a multiple degree of freedom case. If the largest scaled modal coefficient is equal to unity, Eq. (16-26) will also compute a quantity of modal mass that has physical significance. The physical significance is that the quantity of modal mass computed under these conditions will be a number between zero and the total mass of the system. Therefore, under this scaling condition, the modal mass can be viewed as the amount of mass that is participating in each mode of vibration. Obviously, for a translational rigid-body mode of vibration, the modal mass should be equal to the total mass of the system.

Note that the modal mass defined in Eq. (16-26) is developed in terms of displacement over force units. If measurements, and therefore residues, are developed in terms of any other units (velocity over force or acceleration over force), Eq. (16-26) will have to be altered accordingly.

Once the modal mass is known, the modal damping and stiffness can be obtained through the following equations:

$$C_r = 2\sigma_r M_r = \text{modal damping for mode } r \qquad (16\text{-}27)$$

$$K_r = (\sigma_r^2 + \omega_r^2)M_r = \text{modal stiffness for mode } r \qquad (16\text{-}28)$$

For the case of complex modes of the system (nonproportional damping), the preceding definition is inconsistent with the theoretical approach to solving for the eigenvalues and eigenvectors of the system. For the case of complex modes, an alternate form of modal scaling has become widely used based upon the definition of *modal A*. The modal A scaling factor is an equivalent concept to modal mass that relates the scaled modal vectors and the residues determined from the measured frequency-response functions. This relationship is as follows:

$$M_{A_r} = \frac{\psi_{pr}\psi_{qr}}{A_{pqr}} \qquad (16\text{-}29)$$

Note that this definition of modal A is the reciprocal of the Q_r scaling factor defined earlier and is developed in terms of displacement over force units. If measurements, and therefore residues, are developed in terms of any other units (velocity over force or acceleration over force), Eq. (16-29) will have to be altered accordingly just as in the case of the definition of modal mass.

EXPERIMENTAL MODAL ANALYSIS

For proportionally damped systems, a relationship between the modal mass and the modal A scaling factors can be determined.

$$M_{A_r} = \pm j2M_r\omega_r \qquad (16\text{-}30)$$

16-2 Experimental Modal Analysis Methods

Over the past 40 years, at least four general categories of experimental modal analysis methods have been identified:

Sinusoidal input–output model
Frequency-response function
Damped complex exponential response
General input–output model

Historically, the modal characteristics of mechanical systems have been estimated by techniques that fall into either the first or second category. The experimental modal analysis methods that fall into the last two categories are composite approaches that utilize elaborate parameter estimation algorithms based on structural models. This section reviews and provides references for the current work in each of these areas. A brief review of the theory is also provided for the frequency-response function method, since it is the basis for the majority of experimental modal analysis activity at this time.

To evaluate and improve any approach to experimental modal analysis, the relative merits of all viable techniques must be well understood. Thus many articles have been written to try to compare and contrast one method with another. Unfortunately, most of these comparisons have been heavily concerned with differences that are a function of specific implementations of the various techniques. These comparisons were also potentially biased by the test engineers themselves, whose expertise tended to be restricted to one of the areas of testing. Since each method involves very special testing awareness, this sort of analysis has limited value.

In the evaluation of experimental modal analysis methods, the differences in the theoretical approach are obviously of prime concern. Since most experimental modal analysis methods involve similar theoretical basis, the only significant areas of difference concern the concept of real versus complex modal vectors and the explicit measurement of the input. The debate over the need to describe complex-valued modes of vibration may never end. Certainly, the concept of a complex mode, since it contains a real mode as a special case, appears to be the most general case. Likewise, some experimental modal analysis methods do not require the measurement of the input. While this can be advantageous when the implicit nature of the input is known or assumed, it seems prudent, when the input can be measured, to do so.

Beyond the direct theoretical differences, though, several key evaluation considerations, may or may not be a direct function of the theory. The availability of confidence factors, the potential for implementation, stability, and precision of the solution algorithm, the sensitivity to random and/or bias errors in the measured data, and the need for operator expertise may control the ability to estimate valid modal parameters. Specifically, through knowledge of these aspects with respect to other experimental modal analysis methods, the use of each of the approaches may be enhanced as a result of this transfer of technology between the methods.

System Assumptions

Three basic assumptions concerning any structure are made in order to perform an experimental modal analysis. First, the structure is assumed to be *linear*. This means that the response of the structure to any combination of forces, simultaneously applied, is the sum of the individual responses to each of the forces acting alone. For a wide variety of structures this is a very good assumption. When a structure is linear, its behavior can be characterized by a controlled-excitation experiment in which the forces applied to the structure have a form convenient for measurement and parameter estimation rather than being similar to the forces that are actually applied to the structure in its normal environment. For many important kinds of structure, however, the assumption of linearity is not valid. In these cases it is hoped that the linear model that is identified provides a reasonable approximation of the structure's behavior.

The second basic assumption is that the structure is *time invariant*. This essentially means that the parameters to be determined are constants. In general, a system that is not time invariant will have components whose mass, stiffness, or damping depends on factors that are not measured or are not included in the model. For example, some components may be temperature dependent. In this case the temperature of the component is viewed as a time-varying signal; hence, the component has time-varying characteristics. Therefore, the modal parameters that would be determined by any measurement and estimation process would depend on the time (by this temperature dependence) that any measurements were made. If the structure that is tested is changing with time, measurements made at the end of the test period would determine a set of modal parameters different from the measurements made at the beginning of the test period. In other words, the measurements made at the two different times will be inconsistent, violating the assumption of time invariance.

The third basic assumption is that the structure is *observable*. This means that the input–output measurements that are made contain enough information to generate an adequate behavioral model of the structure. Structures and machines that have loose components, or more generally, degrees of freedom of motion that are not measured, are not completely observable. Consider describing the motion of a partially filled tank of liquid when complicated sloshing of the fluid occurs. Sometimes enough measurements can be made to render the system observable under the form chosen for the model, and sometimes no realistic amount of measurements will suffice until the model is changed. This assumption is particularly relevant because the data normally describe an incomplete model of the structure. This occurs in at least two different ways. First, the data normally are limited to a minimum and maximum frequency as well as a limited frequency resolution. Second, no information is available relative to local rotations as a result of a lack of transducers available in this area.

Many other assumptions can be made with respect to the development of the theory concerning a modal analysis method. Maxwell's reciprocity is most often assumed but generally is not enforced in the development of the algorithm. Current research on parameter estimation algorithms is concerned with enforcing this condition. The repeated-root problem is also of great interest now. In a repeated-root problem, one complex root or eigenvalue occurs more than once in the characteristic equation. Each root with the same value has an independent modal vector or eigenvector. This situation can be detected only by the use of multiple inputs or references. Detection is critical in developing a complete modal model so that arbitrary input–output information can be synthesized. While theoretical repeated roots are of debatable significance, the acquisition of data is normally restricted to a fixed-frequency resolution, which gives the possibility of a number of eigenvalues between the observed frequencies. This situation represents a very real problem and is referred to as a pseudo-repeated-root problem.

Excitation Assumptions

The primary assumption concerning the excitation of a linear structure is that the excitation is observable. Whenever the excitation is measured, this assumption simply implies that the measured

EXPERIMENTAL MODAL ANALYSIS

characteristic properly describes the actual input characteristics. For the case of multiple inputs, the different inputs often must be uncorrelated for the computational procedures to yield a solution. In most cases this means only that the multiple inputs must not be perfectly correlated at any frequency. As long as the excitation is measured, the validity of these limited assumptions can be evaluated.

Currently, there are a number of techniques that can be used to estimate modal characteristics from response measurements with no measurement of the excitation. If this approach is used, the excitation assumptions are much more imposing. Obviously, if the excitation is not measured, accurate estimates of generalized parameters such as modal mass, stiffness, or damping cannot be generated. Even under the assumption that the estimation of these parameters is not required, all these techniques have one further restriction: an assumption must be made concerning the characteristics of the excitation of the system. Usually, one assumes that the autospectrum of the excitation signal is sufficiently smooth over the frequency interval of interest.

In particular, the following assumptions about the excitation signal can be used:

1. The excitation is impulsive. The autospectrum of a short pulse (time duration much smaller than the period of the greatest frequency of interest) is nearly uniform, or constant in amplitude, and largely independent of the shape of the pulse.
2. The excitation is white noise. White noise has an autospectrum that is uniform over the bandwidth of the signal.
3. The excitation signal is a step. A step signal has an autospectrum that decreases in amplitude in proportion to the reciprocal of frequency. The step signal can be viewed as the integral of an impulsive signal.
4. The excitation is a Wiener–Levy signal. This signal can be treated as the integral of a white-noise signal and has an autospectrum similar to that of the step signal.
5. There is no excitation. This is called the free-response or free-decay situation. The structure is excited to a condition of nonzero displacement, or nonzero velocity, or both. Then the excitation is removed, and the response is measured during free decay. This kind of response can be modeled as the response of the structure to an excitation signal that is a linear combination of impulsive and step signals.

When the excitation autospectrum is uniform, the autospectrum of the response signal is proportional to the square of the modulus of the frequency-response function. In this case, the poles of the response spectrum are the poles of the frequency response, which are the parameters of the system resonances. If the autospectrum is not uniform, the excitation spectrum can be modeled as an analytic function, to a precision comparable to typical experimental error in the measurement of spectra. In this model, the excitation spectrum has poles that account for the nonuniformity of the spectrum amplitude. The response signal, therefore, can be modeled by a spectrum that contains zeros at the zeros of the excitation and the zeros of the frequency response, and contains poles at the poles of the excitation and at the poles of the frequency response. It is obviously important that the force spectrum have no poles or zeros that coincide with poles of the frequency response. Figure 16-11 presents a typical frequency-response function measurement situation that occurs when a electrodynamic shaker (voltage feedback) is used to excite a structral system. It is clear that distinctly different results will occur if the response autospectrum is used to estimate the frequency and damping as compared to the frequency-response function.

For transient inputs, such as an impact or step relaxation, the assumption of smooth excitation spectra is generally true; but for operating inputs or inputs generated by an exciter system, care must be taken to ensure that the input force spectrum is smooth. This is especially true for tests performed using a hydraulic or an electromechanical exciter, because the system being analyzed may "load" the exciter system (the structure's impedance is so low that the desired force level

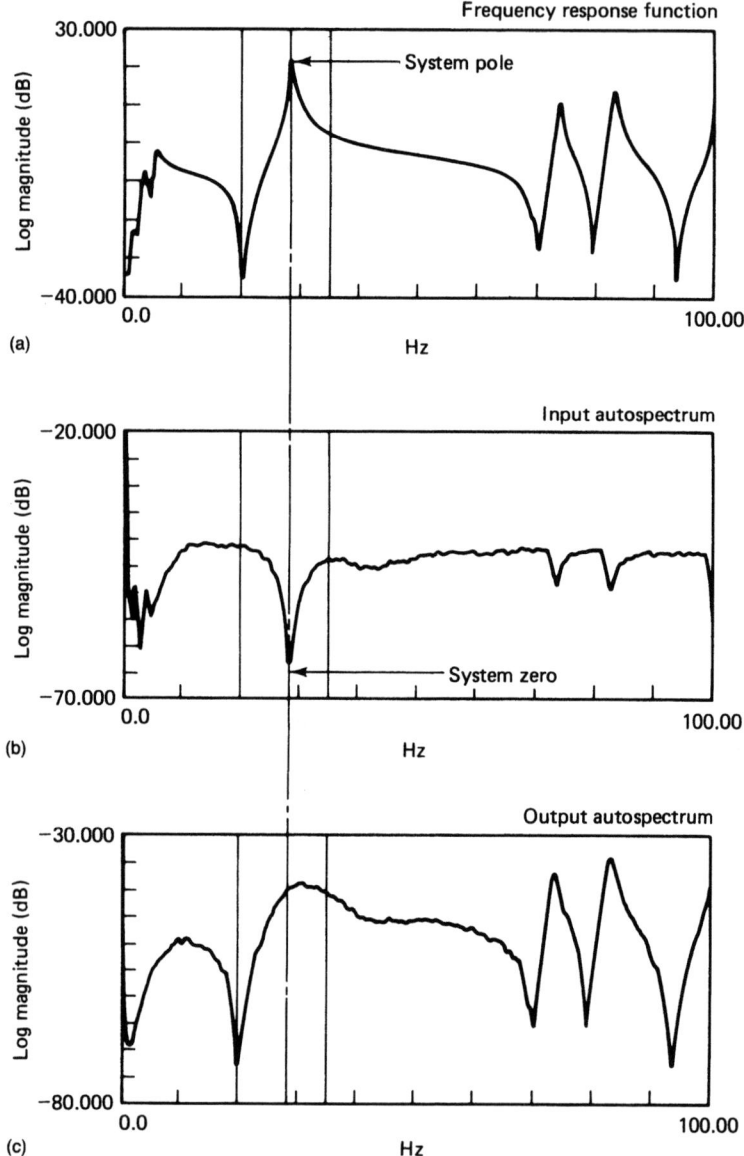

Figure 16-11 Measurement example consisting of (a) frequency-response function; (b) input autospectrum; (c) output autospectrum.

EXPERIMENTAL MODAL ANALYSIS

cannot be achieved within the constraint of small motion), and this causes a nonuniformity in the input force spectrum.

To determine the characteristics of the system from the response, the response must have the same poles as the frequency response, or the analysis process must correct for the zeros and poles of the excitation. If the force input spectrum has a zero in the frequency range of interest, the pole location measured from the response spectrum will not match that of the frequency response. Presently, there is a great deal of interest in determining modal parameters from measured response data taken on operating systems (e.g., turbulent flow over an airfoil, road inputs to automobiles, and environmental inputs to proposed large space structures). For these cases, care must be taken not to confuse poles that are system resonances with those that exist in the output spectrum as a result of inputs.

In general, the poles of the response include those of the frequency response and of the input spectrum. Therefore, if the force is not measured, it is not possible without some prior knowledge about the input to determine whether the poles of the response are truly system charcteristics. If no poles or zeros exist in the force spectrum in the frequency range of interest, then any poles in the response in this range must be a result of the system characteristics. Obviously, when the excitation can be measured, it is prudent to do so.

Sinusoidal Input–Output Method

Methods covered in the sinusoidal input–output category involve excitation that consists of only one frequency during the observation period. While excitation involves a single frequency, the response will initially involve many frequencies as a result of the initiation of the excitation (transient). Even after this initial transient, the response may contain energy at more than one frequency because of the harmonic distortion of the excitation caused by system nonlinearities. This harmonic distortion is normally removed by filtering the response before the data are processed, leaving a single frequency of information in both the input and output signals.

The sinusoidal input–output methods require minimal data acquisition capabilities but generally involve more sophistication in the test setup or in the postprocessing of the acquired data. Since only a single frequency is present in the input and output signals, time domain methods or at most, small blocksize fast Fourier transform (FFT) methods can be used to determine signal amplitudes. Depending on the approach used, the data will yield the modal parameters somewhat directly or will require considerable postprocessing. In the forced normal mode approach, the configuration of the test setup (location and phasing of multiple exciters) yields the modal parameters somewhat directly. In the forced response decomposition approach, the test setup is very general, but the modal parameters are found using an elaborate postprocessing procedure.

Forced Normal Mode Excitation Method.

The forced normal mode excitation method of experimental modal analysis is the oldest approach to the estimation of dynamic structural parameters. The first method to use the application of multiple inputs in the estimation of modal parameters, this method is still used in the aerospace industry for ground vibration testing of aircraft structures. This method was originally outlined in an article by Lewis and Wrisley in 1950 [6] and begins with the matrix form of the differential equation for the system being tested, Eq. (16-1).

$$[M]\{\ddot{x}\} + [C]\{\dot{x}\} + [K]\{x\} = \{f\} \qquad (16\text{-}31)$$

Very simply stated, Lewis and Wrisley found that a number of exciters, utilizing a common frequency and monophase amplitudes, could be tuned to exactly balance the dissipative forces in a structure. This is represented by Eqs. (16-32) and (16-33) and occurs when the phase lag angle ϕ lags the input force by 90° at every response location.

$$\{f\} = \{F\} \sin(\omega t) \qquad \{x\} = \{X\} \sin(\omega t - \phi) \qquad (16\text{-}32)$$

$$\{f\} = [C] \{\dot{x}\} \qquad (16\text{-}33)$$

When the force balance is achieved, the differential equations of motion describing the structure can be reduced to the undamped homogeneous differential equations of motion at that particular frequency. This is represented by Eq. (16-34).

$$[M] \{\ddot{x}\} + [K] \{x\} = \{0\} \qquad (16\text{-}34)$$

To more thoroughly explain this phenomena, De Veubeke in 1956 published an article [7] that explains the theoretical basis for this testing in terms of characteristic phase lag theory. Further development of the practical application of this theory was enhanced by the concept of effective number of degrees of freedom by Trail-Nash [8]. This concept explains that the required number of exciters is a function of the effective number of degrees of freedom, not the total number of degrees of freedom. The effective number of degrees of freedom is a function of modal density and damping. Finally, Asher utilized the determinant of the real part of the frequency-response matrix to locate damped natural frequencies and determine the effective number of degrees of freedom [9].

Most of the work done since 1958 has been concerned with improvements in the implementation of the method, primarily the force appropriation [10–19]. One of the most advanced implementations of this method involves approximately 500 channels of data acquisition and co/quad analysis equipment controlled from a minicomputer. Extensive tuning criteria are utilized as well as real-time animated displays of the modal vector and the out-of-phase response. This can be particularly useful for optimum exciter location as well as for force appropriation. Once a modal vector has been tuned using a 90° phase lag angle criterion, the excitation frequency can be varied with no change in modal vector. Theoretically, this can be used as a check to determine whether the modal vector has been adequately tuned.

In addition to this potential confidence check, the excitation can be removed from the system once a modal vector has been tuned. If the modal vector contains only responses due to a single mode of vibration, the exponential decay at all response positions should contain only the excitation frequency, and the envelope of exponential decay should give an accurate estimate of the system damping.

The forced normal mode excitation method works well in the presence of proportional damping but theoretically does not include the concept of complex modes of vibration nor the concept of repeated and multiple roots. (Proportional damping is a mathematical approach to the description of damping that states that the damping matrix resulting from whatever damping mechanism is present is proportional to the mass matrix, the stiffness matrix, or to some linear combination of the two.) As a result of this theoretical limitation, the practical application of the 90° phase lag criterion is normally applied only to within ±10°. Likewise, added difficulty is encountered in evaluating the exponential decay purity as well as force appropriation. Much work has been done on automated tuning algorithms to alleviate this. These algorithms alter excitation magnitude and phase to try to achieve 90° phase lag criteria under severe impedance-matching situations. Unfortunately, the location of the excitation cannot be evaluated automatically in this process [16–19].

EXPERIMENTAL MODAL ANALYSIS

Forced Response Decomposition Method.

The forced response decomposition method uses an array of exciters (multiple input) to excite the system into a forced response at a single frequency. While the magnitude of each input and the phasing between inputs may be chosen randomly or according to some particular regime, the inputs are held constant during the observation period. Therefore, after the initial transient decays, the response is a steady-state forced response of this system. A forced response vector is created by using all output points of interest simultaneously. This is represented by Eqs. (16-35) thrugh (16-37).

$$[M]\{\ddot{x}\} + [C]\{\dot{x}\} + [K]\{x\} = \{f\} \tag{16-35}$$

$$\{f\} = \{F\} \sin(\omega_k t) \tag{16-36}$$

$$\{x\} = \{X\} \sin(\omega_k t - \phi_k) \tag{16-37}$$

The forced response vector that is generated must be a linear superposition of the modal vectors of the system (expansion theorem). Since in a given frequency range of interest, N modal vectors will contribute to the response, the individual modal vectors cannot be determined from one forced response vector. If N or more independent forced response vectors can be generated, the N modal vectors can be determined. A large number of independent forced response vectors can be generated by two approaches. First of all, at a single frequency, many different input vectors can be generated with randomly chosen magnitudes and relative phasing. Each of these choices will yield a potentially independent forced response vector. Second, this process can be repeated for different frequencies which will, once again, yield potentially independent forced response vectors. Since the degree of independence among the forced response vectors is unknown, and since the best estimate of the modal vectors is desired, many more forced response vectors, compared to the number of expected modal vectors (N), are acquired.

The postprocessing of the forced response vectors involves using a singular value decomposition of the data spanned by the forced response vectors. If N_V forced response vectors are acquired (where $N_V >> N$), the number of significant singular values in the $N_V \times N_V$ data space is an indication of the number of contributing modal vectors in the data. The singular vectors, associated with the significant singular values, provide a transformation matrix to transform the forced response vectors to modal vectors. Frequency and damping values are found in a second-stage solution process. If frequency-response functions or the input vectors required to force a specific normal mode are desired, these characteristics can be found at this time as well.

The forced response decomposition method, in one form or another, is currently receiving much research attention [20–27]. While the methods have not been used much to this point, they are very attractive because data acquisition requirements are minimal (when the data acquisition has been specifically designed for this method) and because sophisticated postprocessing techniques, which require minimal computational power (micro- or minicomputers) are available.

Frequency-Response Function Method

The frequency-response function method of experimental modal analysis is the most commonly used approach to the estimation of modal parameters. This method originated as a testing technique as a result of the use of frequency-response functions in the forced normal mode excitation method to determine natural frequencies and effective number of degrees of freedom [28–33]. With the advent of the computer and minicomputer, the frequency-response function method became a separate, viable technique [34–60].

In this method, frequency response functions are measured using excitation at single, or multiple, points. The relationships between the input F and the response X for both single and multiple inputs are shown in Eqs. (16-38) through (16-40).

Single-input relationship:

$$X_p = H_{pq} F_q \tag{16-38}$$

$$\begin{Bmatrix} X_1 \\ X_2 \\ \cdot \\ \cdot \\ X_p \end{Bmatrix} = \begin{Bmatrix} H_{1q} \\ \cdot \\ \cdot \\ \cdot \\ H_{pq} \end{Bmatrix} F_q \tag{16-39}$$

Multiple input relationship:

$$\begin{Bmatrix} X_1 \\ X_2 \\ \cdot \\ \cdot \\ X_p \end{Bmatrix}_{N_o \times 1} = \begin{bmatrix} H_{11} & \ldots & H_{1q} \\ H_{21} & & \\ \cdot & & \\ \cdot & & \\ H_{p1} & \ldots & H_{pq} \end{bmatrix}_{N_o \times N_i} \begin{Bmatrix} F_1 \\ F_2 \\ \cdot \\ \cdot \\ F_q \end{Bmatrix}_{N_i \times 1} \tag{16-40}$$

The excitation F may be narrowband or broadband, random or deterministic [46, 47, 50, 56, 58]. The frequency-response functions are used as input data to algorithms that estimate modal parameters using a frequency-domain model. Through the use of the fast Fourier transform, the Fourier transform of the frequency-response function, the impulse-response function, can be calculated for use in modal parameter estimation algorithms involving time-domain models.

Most of the work over the last 10 years has focused on the simultaneous measurement of multiple columns of the frequency-response function matrix. This work has involved establishing the numerical and excitation requirements for solving the relationship identified in Eq. (16-40), developing alternate estimation algorithms, and developing modal parameter estimation algorithms that are matched to this new data acquisition/estimation procedure. This work has revolutionized experimental modal analysis testing for several reasons. First of all, to be sure that all modal vectors have been found experimentally, a number of excitation (reference) points must be utilized, either one at a time or simultaneously. This minimizes the possibility of exciting the system at or near a node of one of the modal vectors, would provide inaccurate estimates of that modal vector. Second, multiple columns (or rows) of the frequency response function matrix are necessary for the detection of repeated or pseudorepeated (close) modal frequencies. The ability to measure, detect, and identify the presence of repeated or close modes was not generally possible prior to this research. Finally, the increased number of measurements per measurement cycle obtained using multiple inputs does not have an adverse effect on the time required to acquire frequency-response function data. Therefore, a more complete set of data, allowing for a more complete dynamic model of the system to be validated, is possible in the same measurement time.

Theory

If all or part of the elements of the frequency-response matrix can be measured, each column will contain information that can be used to estimate modal vectors. Since the frequency-response matrix is considered to be symmetric in accordance with the Maxwell–Betti relations, each row will contain the information needed to estimate modal vectors.

To obtain estimates of the modal vectors, the frequency-response functions are used as input in a parameter estimation scheme based on Eq. (16-41) and (16-42) or Eq. (16-43) and (16-44), the time-domain equivalent formulation involving the impulse response-function [61–63].

EXPERIMENTAL MODAL ANALYSIS

Single Reference:

$$H_{pq}(\omega) = \sum_{r=1}^{N} \frac{A_{pqr}}{j\omega - \lambda_r} + \frac{A^*_{pqr}}{j\omega - \lambda_r^*} \tag{16-41}$$

Multiple Reference:

$$[H(\omega)] = \sum_{r=1}^{N} \frac{[A_r]}{j\omega - \lambda_r} + \frac{[A_r^*]}{j\omega - \lambda_r^*} \tag{16-42}$$

Often Eq. (16-41) is altered by an assumption of real modes, a specific damping mechanism, or known system poles. Under such assumptions, the evaluation of Eq. (16-41) for the remaining modal parameters becomes much simpler. Most current research and development in the area of the frequency-response function method involves the modal parameter estimation algorithms that are related to the time or frequency-domain models equivalent to Eq. (16-41) and (16-42) and/or Eq. (16-43) and (16-44). Much of this work involves algorithms that utilize as much of the redundant information within multiple rows and columns of the frequency-response function matrix as is possible. The application of the polyreference approach [61–63] to impulse-response functions is a primary example.

Damped Complex Exponential Method

The damped complex exponential response methods of experimental modal analysis are now receiving considerable attention. These methods are normally formulated to utilize data corresponding to the free decay of a system generated by the release of an initial condition, but they apply quite generally to impulse-response function data as well. Since impulse-response function data are scaled to include the forcing condition, use of this method on impulse-response function data yields properly scaled modal parameters that can be used to calculate generalized mass and stiffness, which would not be possible if free-decay responses were used. Even so, the formulation of the impulse-response function generally involves the computation of the frequency-response function via a fast Fourier transform, potentially imposing bias errors, which may degrade the estimation of the modal parameters.

Single Reference:

$$h_{pq}(t) = \sum_{r=1}^{N} A_{pqr} e^{\lambda_r t} + A^*_{pqr} e^{\lambda_r^* t} \tag{16-43}$$

Multiple Reference:

$$[h(t)] = \sum_{r=1}^{N} [A_r] e^{\lambda_r t} + [A_r^*] e^{\lambda_r^* t} \tag{16-44}$$

While the current implementation of damped complex exponentials is certainly new, the basis of much of the work was formulated in the eighteenth century by Prony [64]. Currently, the three approaches that have demonstrated capability are the Ibrahim time-domain approach, the polyreference approach, and the eigensystem realization algorithm approach, which are discussed in the following sections. The first two approaches, as well as other similar formulations [65–71], have been explained in a unified damped complex exponential theory based strongly on the polyreference theoretical approach [72–83]. While the current explanation of the Ibrahim time-domain approach does not include the possibility of using multiple initial conditions in a single solution formulation, the unified dampled complex exponential theory includes this as part of the general case.

Ibrahim Time-Domain Approach

One practical implemention of the damped complex exponential method is the Ibrahim time-domain (ITD) method [65–71], developed to extract the modal parameters from damped complex exponential response information. Digital free-decay response data are measured at various points on the structure. If response data from all the selected measurement positions cannot be obtained simultaneously because of equipment restrictions, a common position is retained between measurement groups. A recurrence matrix is created from the free-decay data, and the eigenvalues of this matrix are exponential functions of the poles of the system, from which the poles are easily computed. The eigenvectors of the recurrence matrix are response residues, from which the mode shapes are determined.

The damped complex exponential response method is rather straightforward in application if the necessary data acquisition hardware, computer facilities, and software are available. This approach computes the poles and residues based on a specific initial vibration condition of the structure. A number of different initial conditions can be established, analogous to the practice of using several exciter positions in ordinary single-input modal surveys, until all the important modes have been excited. All the modes cannot be established from one exciter position, nor can all the modes be determined from one initial condition.

Although this technique is based on free-decay data, the procedure also can be used with operating inputs if the free decay is computed from the operating inputs by using "random-decrement" averaging [67, 68] or from measured auto- and cross-correlation functions. Again, it should be emphasized that this can be done only if there are no poles or zeros in the input spectrum in the frequency range of interest.

An additional development with respect to this technique is the concept of modal confidence factor [56]. The modal confidence factor is a complex number calculated for each identified mode of the structure, while undergoing an exponential decay form of vibration test. The modal confidence factor is based on the relation of the modal deflection at a particular measurement point to the modal deflection at that same measurement point at any time earlier or later in the free-decay response. Therefore, if modal vectors are estimated from exponential decay data and two separate estimates are calculated from sets of data taken some fixed time Δt apart, the relationship between the measured second estimate of the modal vector and the calculated estimate of the modal vector based on the measured first estimate of the modal vector is the modal confidence factor. The purpose of the modal confidence factor is to provide an indicator for determining whether an estimated modal vector is real or computational.

Polyreference Approach

A more recent implementation of the damped complex exponential approach to experimental modal analysis is the polyreferecne approach developed by Vold et al. [61]. Most of the comments relative to the ITD approach can be repeated with respect to the polyreference approach. Of particular importance, once again, is the increased requirement for computational power and the desirability of acquiring all response data simultaneously to reduce time-invariance problems.

The polyreference approach utilizes all measured damped complex exponential information, from all references or initial conditions, simultaneously in the estimation of modal frequencies. Additionally, the polyreference approach breaks new ground in the estimation of a single modal coefficient for each measurement degree of freedom in the presence of multiple initial conditions or multiple inputs. This characteristic is shared by the frequency-response function method when the polyreference approach is used as the parameter estimation algorithm and with some of the approaches within the mathematical input–output model methods. The formulation of the algorithm

EXPERIMENTAL MODAL ANALYSIS

such that constraints are included to account for redundant information is an advantage but requires that the total data set be acquired so as to match this assumption. This means that the data acquisition best matches the analysis procedure when all data can be acquired simultaneously.

While much of the work utilizing the polyreference approach is quite recent, the evaluation of the method based on comparisons between experimentally measured and synthesized frequency-response functions is quite impressive compared to other modal parameter estimation approaches.

Eigensystem Realization Algorithm Approach

The Eigensystem realization algorithm (ERA) approach is a very recent technique that is basically an extended version of the Ho–Kalman system realization algorithm [75]. The ERA algorithm was developed at NASA-Langley Research Center under an interdisciplinary effort involving structural dynamics and controls. This method is similar to the other damped complex exponential methods in that it involves solutions of a matrix eigenvalue problem. The ERA approach is most like the polyreference time-domain approach in that multiple initial conditions can be involved in the matrix formulation. Therefore, repeated roots or eigenvalues can be identified with this approach as in the polyreference time domain approach. Other significant attributes of this approach include the extensive use of accuracy indicators to assess effects of noise and nonlinearities as well as rank information provided by singular value decomposition techniques [72–83].

The ERA approach is based on well-established realization (state-space) theory using the concepts of controllability and observability. The approach determines a complete state-space model based on the important principles of minimal realization theory attributed to Ho and Kalman. The Ho–Kalman procedure uses a sequence of real matrices known as Markov parameters (impulse-response functions) to construct a state-space representation of a linear system. The ERA approach begins with a block data matrix formulated from damped complex exponential functions, such as free-decay responses. This block data matrix is similar to a general Hankel matrix and includes information from several initial conditions and a weighted set of damped complex exponential functions. The weighted set of functions means that points of interest or points with large response can be emphasized without loss of capabilitiy of the method. The state-space matrices are found from the block data matrix by factorization of the block data matrix using singular value decomposition. Based on the rank evaluation of the block data matrix in this factorization procedure, a state-space set of matrices can be formulated based on the reduced order. Eigenvalues and eigenvectors of this reduced-order, state-space model are then found. Accuracy indicators such as the rank of the block data matrix, modal amplitude coherence, modal phase collinearity, and data reconstruction are used to identify the final set of modal parameters.

The ERA approach is a very recent but promising method that demonstrates extensive use of accuracy indicators. Several preliminary studies comparing multiple reference alogrithms indicate good agreement among methods, but the identification of nonrealistic modal parameters is still a significant problem. The ERA approach, through the extensive development and use of accuracy indicators, attempts to deal with this part of the identification problem more commmpletely than most other approaches. The accuracy indicators utilized in the ERA approach are already being applied to several other approaches with simliar success.

Mathematical Input–Output Model Method

The experimental modal analysis methods included in the category of mathematical input–output model methods are the approaches that generally involve input and response data independently without the need for creating auto- and cross-moment functions. There is no other restriction with regard to time- or frequency-domain models, effective number of degrees of freedom, and so on.

On this basis, two approaches are currently in use that can be described in this fashion. First of all, the autoregressive moving-average approach is a time-domain formulation that utilizes a pole-zero model as the basis for the description of the system characteristics. While this model is appropriate, it cannot be easily constrained to account for known system information. Additionally, although the current application of this technique does not involve multiple inputs, the theoretical background for the multiple-input case is well developed.

The other approach currently in use involves a reduced structural matrix model for the basis of the description of the system characteristics. This model involves the reduced mass, stiffness, and damping matrices with regard to the measured degrees of freedom. The model easily accounts for constraints such as known elements in the mass, stiffness, and damping matrices or known characteristics of the distribution within the matrices such as symmetric or banded characteristics. The method also incorporates the multiple-input case routinely as known terms in the forcing vector of the matrix differential equation that serves as the mathematical model. These two approaches to experimental modal analysis will be briefly described in the following subsections.

Autoregressive Moving-Average Approach

One approach to estimating the modal characteristics from input–output data is the autoregressive moving average procedure. This method has been applied to the determination of structural parameters by Gersch [84–87] and Pandit [88]. With this technique the response data are assumed to be cause by a white random noise input to the structure. The technique computes the best statisitical model of the system in terms of its poles (from the autoregressive part) and zeros (from the moving-average part), as well as statistical confidence factors on the parameters. It has been primarily used to estimate the characteristics of buildings being excited by wind forces. The data used in the computational process are the autocorrelation functions of the responses measured at various points on the structure. Since in the general case the inputs are not measured, the modal vectors are determined by referencing each response function to a single response to provide relative magnitude and phase information.

Single Reference:

$$\sum_{i=0}^{m} a_i\, x(t - i\Delta t) = \sum_{i=0}^{n} b_i\, f(t - i\Delta t) \tag{16-45}$$

Multiple Reference:

$$\sum_{i=0}^{m} [a_i]\, [x(t - i\Delta t)] = \sum_{i=0}^{n} [f(t - i\Delta t)]\, [b_i] \tag{16-46}$$

The solution for the autoregressive moving-average coefficients proceeds in a two-stage, least squares fashion in the Gersch solution. In the first stage a "long" autoregressive model is solved linearly by using the Yule–Walker equations. This process uses output covariance functions to determine the autoregressive coefficients based on a determination of the order of the autoregressive model. The second stage involves setting up an equivalent moving-average model for the output involving convolution of the impulse-response function and the input function. This procedure also involves computations using covariance functions and results in the least-squares computation of the moving-average coefficients.

Since the solution for the autoregressive moving-average coefficients and thus the structural parameter estimates are statistically based, statistical confidence factors, called coefficients of variation, for the natural frequencies and damping can be easily calculated. These coefficients represent the ratio of the standard deviation of each parameter with respect to the actual parameter.

EXPERIMENTAL MODAL ANALYSIS

Obviously, the random response function method is complicated, and users vary in their choice of a solution procedure. For further details or information concerning varied solution approaches, technical articles can provide further reference [5,84,88].

Reduced Structural Matrix Approach

Over the last 10 years, there has been increasing interest in being able to estimate reduced structural (mass, stiffness, and damping) matrices from experimental data. Most of these methods are based on an indirect approach utilizing the estimated modal parameters to synthesize the reduced matrices [89–92].

This approach has not generated the anticipated results for a number of reasons. First of all, regardless of the approach used, the solution for the reduced matrices is not unique. There are many combinations of matrix relationships that can be generated from the given set of estimated modal parameters. Second, the reduced frequency range of the modal parameter estimates means that the matrices will be weighted to represent an incomplete model [90]. Third, the limitation of the precision of the modal parameter estimates as a result of commonly accepted experimental error tends to desensitize the process of estimating the reduced matrices. Finally, the problem of invalid modal parameter estimates obviously will result in invalid estimates of reduced matrices.

Time domain:

$$[M]\{\ddot{x}\} + [C]\{\dot{x}\} + [K]\{x\} = \{f\} \tag{16-47}$$

Frequency domain:

$$-\omega^2 [M]\{X\} + j\omega [C]\{X\} + [K]\{X\} = \{F\} \tag{16-48}$$

An algorithm developed in Germany by Link and Vollan [89] attempts to use frequency-domain input and response data to directly estimate the reduced matrices. This method has been designated identification of structural system parameters (ISSPA). Leuridan used this formulation as a starting point and has published encouraging results [91–93] using the same general approach for the estimation of modal parameters referred to as the direct system parameter identification (DSPI) method. Since modal parameters are found as a result of the solution of the eigenvalue problem using the reduced matrix estimations, the process of estimating the reduced matrices may represent the ultimate goal in experimental modal analysis. If this process could be correlated with a purely theoretical finite-element approach, the engineering design cycle would be complete.

Since there are many more known pieces of information (input and output information at different times) than unknowns that must be estimated, the solution is a function of the pseudoinverse procedure chosen. Leuridan and Vold have evaluated pseudoinverse numerical procedures utilizing the normal equations, the least-squares approach, and the Householder reflections approach. Since the system matrix that results is often ill-conditioned, the Householder reflections approach yields more numerical precision for a given computational word size but at a sacrifice in speed.

Link and Vollan formulate the pseudoinverse approach on the basis of a singular value decomposition procedure under the restriction that the rank of the data-dependent matrices be equal to the effective number of degrees of freedom. This effective number of degrees of freedom is dependent on the number of theoretical system poles in the frequency range of interest, the accuracy of the measured data, and the computational precision of the computer with respect to the solution algorithm used.

Once the unknown elements of the mass, stiffness, and damping matrices have been found, the modal parameters are estimated from the $[M]$, $[C]$, and $[K]$ matrices by way of a complex eigenvalue–eigenvector solution algorithm such as the QR algorithm.

A confidence of validity check of the frequencies, damping factors, and modal vectors can be performed using a back-substitution procedure. The dynamic response is calculated and compared to the original measured response. The agreement between these responses is regarded as a measure of the accuracy of the estimated modal parameters.

16-3 Modal Data Acquisition

Acquisition of data that will be used in the formation of a modal model involves many important technical concerns. The primary concern is the digital signal processing or the converting of analog signals into a corresponding sequence of digital values that accurately describe the time-varying characteristics of the inputs to and responses from a system. Once the data have become available in digital form, the most common approach is to transform them from the time domain to the frequency domain by use of a discrete Fourier transform algorithm. Since such an algorithm involves discrete data over a limited time period, there are large potential problems with this approach that must be well understood.

Digital Signal Processing

To determine modal parameters, the measured input (excitation) and response data must be processed and put into a form that is compatible with the test and modal parameter estimation methods. As a result, digital signal processing of the data is a very important step in structural testing. This is one of the technology areas where a clear understanding of the time–frequency–Laplace domain relationships is essential. The conversion of the data from the time domain into the frequency and Laplace domains is important in both the measurement process and subsequently in the parameter estimation process.

Digital signal processing of the measured input and response data is used for the following reasons:

1. *Condensation*. In general, the amount of measured data tremendously exceeds the information present in the desired measurements (frequency response, unit impulse response, coherence function, etc.) Therefore, digital signal processing is used to condense the data.
2. *Measurements*. The measurements that will be used in the modal parameter estimation process must be estimated. Since there are many excitation, measurement, and modal parameter estimation procedures, there is likewise a large number of digital signal-processing options.
3. *Noise reduction*. Signal processing is used to reduce the influences of noise in the measurement process. The types of noise have been classified for convenience as follows:
    a. *Noncoherent noise*. This noise is due to electrical noise on the transducer signals, unmeasured excitation sources, etc., which are noncoherent with respect to the measured input signals or to some other signal used in the averaging process. Zero mean noncoherent noise can be removed by averaging with respect to a reference signal. This reference signal could be the input signal in terms of a spectrum averaging process or a synchronization or trigger signal in terms of cyclic averaging or random decrement process.
    b. *Signal-processing noise*. The signal processing itself may generate noise. For example, "leakage" is a classic source of noise when using fast Fourier transforms for computing frequency-domain measurements. This type of noise is reduced or eliminated by using completely observed time signals (periodic or transient), by using windows of various types, or by increasing the frequency resolution.
    c. *Nonlinear noise*. If the system is nonlinear measurements of free decay, frequency response,

EXPERIMENTAL MODAL ANALYSIS

or unit impulse function may be distorted, causing problems later during the estimation of modal parameters. Nonlinear distortion noise can be eliminated by linearizing the test structure before testing or by randomizing the input signals to the structure. This will cause the nonlinear distortion noise to become noncoherent with respect to the input signal. The nonlinear noise can then be averaged from the data in the same manner as ordinary noncoherent noise.

Historically, frequency-response function or unit impulse-response function has been the most important measurement estimated. There has been a great deal of research in the area of signal processing for determining these measurements [20–27,45–60,94–96]. Much of the research in this area conducted over the last 15 years has been involved in understanding the digital signal-processing implications on the ultimate estimate of measurement characteristics. Future work will establish the statistical error bounds on the ultimate estimates of the modal parameters that are derived from the measurements. In this way, importance can be properly given to the relative errors in the different phases of the measurement process (calibration, instrumentation, data acquisition, estimation algorithm, etc.). The process of representing an analog signal as a series of digital values is a basic requirement of modern digital signal-processing analyzers. In practice, the goal of the analog-to-digital conversion (ADC) process is to obtain the conversion while maintaining sufficient accuracy in terms of frequency, magnitude, and phase. When dealing strictly with analog devices, this concern was satisfied by the performance characteristics of each individual analog device. With the advent of digital signal processing, the performance characteristics of the analog device are only the first criteria of consideration. The characteristics of the analog-to-digital conversion now become of prime importance.

This process of analog-to-digital conversion involves two separate concepts, each of which is related to the dynamic performance of a digital signal-processing analyzer. *Sampling* is the part of the process related to the timing between individual digital pieces of the time history. *Quantization* is the part of the process related to describing an analog amplitude as a digital value. Primarily, sampling considerations alone affect the frequency accuracy, while both sampling and quantization considerations affect magnitude and phase accuracy.

Sampling

Sampling is the process of recording the independent variable of an analog process. This can be done in an absolute sense, where the independent variable is in terms of time. Quite often, particular advantage can be gained if the sampling process proceeds in a relative sense when this independent variable is in terms of some event. This relative approach is the basis of the processing of data related to rotating equipment, where the event is a revolution of a shaft. In either case, the same theories apply to the sampling process. In the relative approach, there is simply a change of variable associated with the independent axis (time versus event).

The process of sampling arises from the need to describe analog time histories in a digital fashion. This can be done, in general, by recording a digitized amplitude and a reference time of measurement or in the more common method of recording amplitudes at uniform increments of time (Δt). Since all analog-to-digital converters sample at constant sampling increments during each sample period, all further discussion is restricted to this case.

Two theories or principles apply to the process of digitizing analog signals and recovering valid frequency information. Shannon's sampling theorem states, very simply, the following:

$$F_{\text{samp}} \geq F_{\text{max}} \times 2.0 \tag{16-49}$$

Obviously, this theorem has to do with the maximum frequency that can be described accurately.

Table 16-1 Digitization Equations ($F_{samp} = 2 F_{max}$)

Sampling Parameter	Automatically Determines	Blocksize Determines
Δt	$F_{max} = \dfrac{1}{2\Delta t}$	$T = N\Delta t$ $\Delta f = \dfrac{1}{N\Delta t}$
F_{max}	$\Delta t = \dfrac{1}{2F_{max}}$	$T = N\Delta t$ $\Delta f = \dfrac{1}{N\Delta t}$
Δf	$T = \dfrac{1}{\Delta f}$	$\Delta t = \dfrac{T}{N}$ $F_{max} = \dfrac{N}{2}\Delta f$
T	$\Delta f = \dfrac{1}{T}$	$\Delta t = \dfrac{T}{N}$ $F_{max} = \dfrac{N}{2}\Delta f$

It should be noted that this equation establishes the limit that can be used to digitize a signal and still identify a certain maximum frequency component. For reasons involving the practical limitation of the analog filters used prior to any digitization, the sampling frequency is often chosen to be greater than twice the maximum frequency of interest. In this case, Eq. (16-49) still applies as stated by the inequality. Therefore, to understand the application of Shannon's sampling theorem in any digital signal-processing situation, the sampling frequency, the maximum frequency, and the *observed or displayed frequency* must be considered. The observed frequency is always less than the maximum frequency, as defined by the equality in Eq. (16-49), to account for the roll-off characteristic of any filters used in the analog (or digital) signal processing. Note that in this situation the resulting frequency information must be displayed appropriately to avoid viewing the data that are considered to be invalid.

Rayleigh's criterion or principle, first formulated in the field of optics, has to do with the resolution of two closely related spaced frequency components. For a time record of T seconds, the lowest frequency component measurable is

$$\Delta f = \frac{1.0}{T} \tag{16-50}$$

It should be emphasized that Eq. (16-50) is always true regardless of the absolute frequency involved.

With these two principles in mind, the selection of sampling parameters can be summarized as shown in Table 16-1. Note that for this case, the equality in Eq. (16-49) has been used. Also

EXPERIMENTAL MODAL ANALYSIS

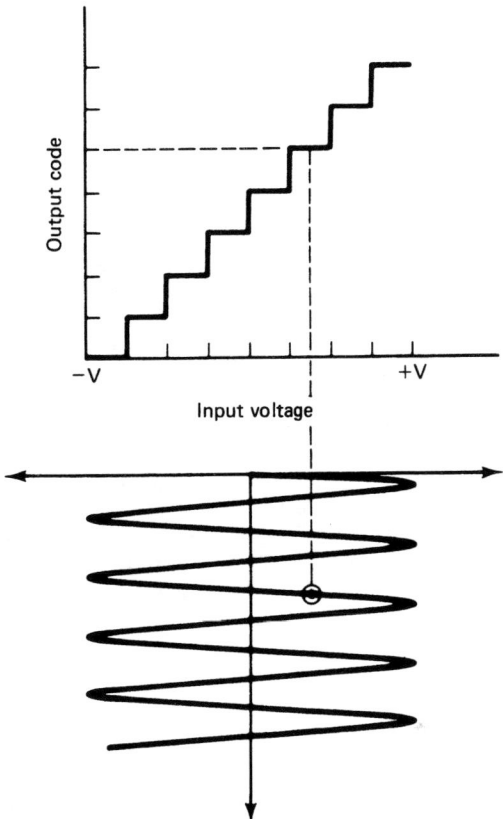

Figure 16-12 Quantization: a three-bit analog-to-digital converter.

note that the term "blocksize" refers to the number of time domain samples acquired contiguously for each average. The blocksize is normally an integer power of 2, for example, $2^{10} = 1024$.

Quantization

Quantization refers to the conversion of a specific analog value of amplitude to the nearest discrete value available in the analog-to-digital converter. This process involves representing a range of voltage by a fixed number of integer steps. Normally, the range of voltage is chosen to be between positive and negative limits for a given voltage limit. The number of discrete levels is a function of the number of bits in the analog-to-digital conversion. An ADC with an "M-bit" converter is able to determine signal amplitude within one part in 2 raised to the Mth power; 10- and 12-bit converters are very common. An example of a three-bit converter is shown in Fig. 16-12.

The most important consideration with respect to optimum quantization depends on the concept of *actual* word size versus *effective* word size. The actual word size is the number of bits available in each word in the ADC or computer. The effective word size is the number of bits used in each word in any operation in the ADC or computer. The first consideration is that of dynamic range. This has to do with the input to the ADC in the digitization process. Since the actual word size of the ADC is fixed, the dynamic range of the ADC (60 dB for 10 bits, 72 dB for 12 bits) is meaningful only if the quantization of an input signal of interest involves all the bits of the actual word. There are two situations in which this is not true. First, if the input ranges for the ADC are not automatically or manually set to the optimum position, some loss of dynamic range will occur. This means that the maximum level of the data should not be less than half the input voltage range. Second, if the signal has more dominant information content outside the band of interest, a significant portion of the dynamic range will be used to describe the unwanted characteristic. This will reduce the potential dynamic range available to the portion of the signal in which there is interest. Some common examples of this are a large mean value offset or large harmonic component (e.g., 60 Hz) as shown in Fig. 16-13. Both these situations cause the effective word size of the ADC to be smaller than the actual word size of the ADC with respect to the information of interest.

This dynamic range consideration with respect to quantization is particularly important when a multiplexer is used to obtain a large number of channels of data in parallel. Since a multiplexer configuration often involves only one ADC channel for a number of multiplexer channels, the dynamic range of all the channels must be similar; otherwise the effective word size for many of the channels will be much less than the actual word size, whereupon the signals must be amplified prior to digitization to ensure that each channel has approximately the same dynamic range. Naturally, this amplification factor must be taken into account in the final calibration of the data.

Discrete Fourier Transform

The discrete Fourier transform algorithm is the basis for the formulation of any frequency-domain function in modern data acquisition systems. In terms of an integral Fourier transform, for a function to be evaluated it must exist for all time in a continuous sense. For the realistic measurement situation, data are available in a discrete sense over a limited time period. Figures 16-14 and 16-15 represent schematically what the discrete Fourier transform is doing with respect to the time-domain data. The discrete Fourier transform, therefore, is based on a set of assumptions concerning this discrete sequence of values. The assumptions can be reduced to two situations that must be met by every signal processed by the discrete Fourier transform algorithm. The first assumption is that the signal must be a totally observed transient with respect to the time period of observation. If this cannot be done, then the signal must be composed only of harmonics of the time period of observation. If one of these two assumptions is not met by any discrete history processed by the discrete Fourier transform algorithm, the resulting spectrum will contain bias errors accordingly. Much of the data processing that is considered with respect to acquisition of data for the formulation of a modal model revolves around an attempt to assure that the input and response histories match one of these two assumptions. For a more complete understanding of the discrete Fourier transform algorithm and the associated problems, a number of good references explain all the pertinent details [74,75,81].

Integral Fourier transform (forward):

$$X(f) = \int_{-\infty}^{\infty} x(t) e^{-j2\pi ft} \, dt \qquad (16\text{-}51)$$

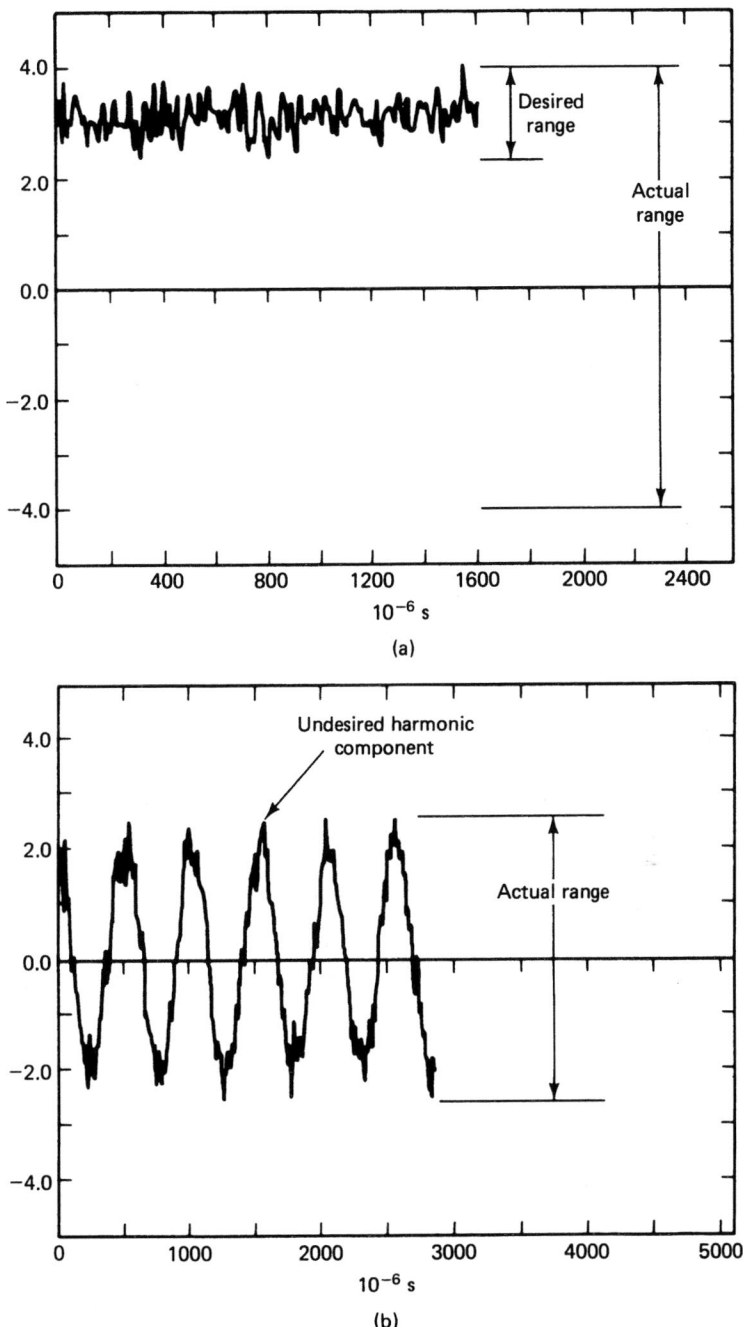

Figure 16-13 Quantization: ADC input optimization with (a) large mean value offset and (b) large harmonic component.

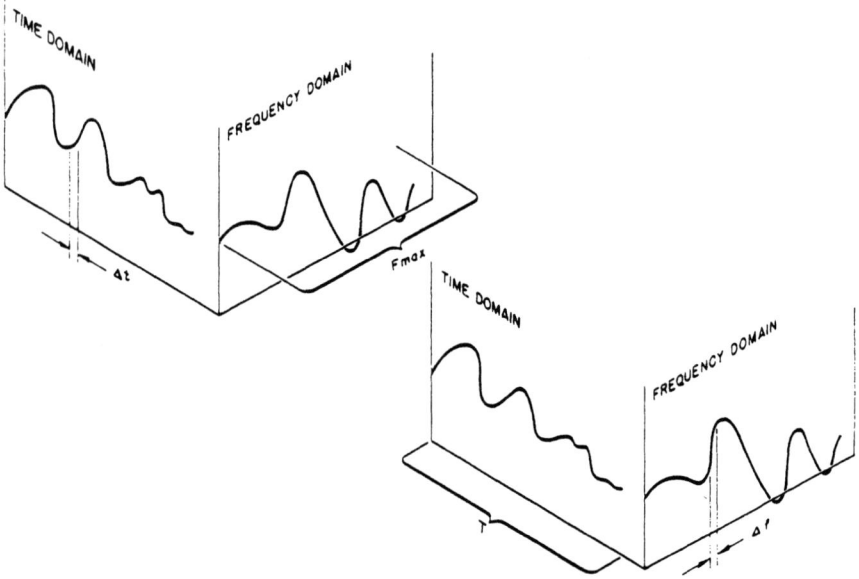

Figure 16-14 Time- and frequency-domain relationships.

Integral Fourier transform (reverse):

$$x(t) = \int_{-\infty}^{\infty} X(f) e^{j2\pi ft} \, df \tag{16-52}$$

Discrete Fourier transform (forward):

$$X(f_k) = \sum_{n=0}^{N-1} x(t_n) e^{-j2\pi f_k t_n} \tag{16-53}$$

$$X(f_k) = \sum_{n=0}^{N-1} x(t_n) e^{-j2\pi(kn/N)} \tag{16-54}$$

Discrete Fourier transform (reverse):

$$x(t_n) = \sum_{k=0}^{N-1} X(f_k) e^{-j2\pi(kn/N)} \tag{16-55}$$

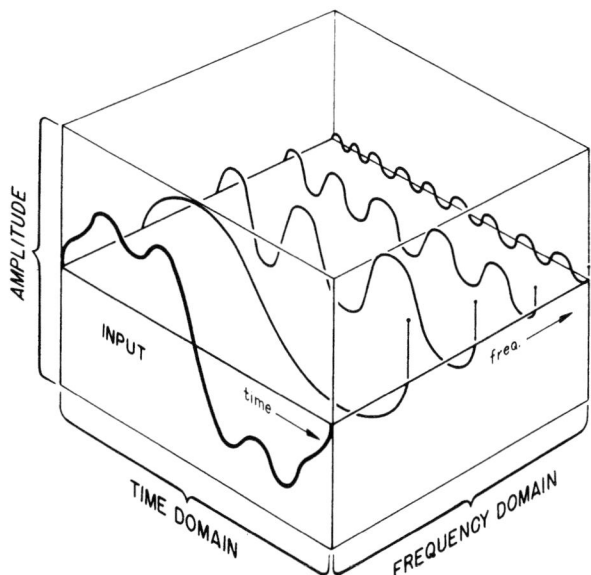

Figure 16-15 Discrete Fourier transform concept.

where N = blocksize (power of 2)
Δt = time spacing
Δf = frequency spacing
$n = 0,1,2, \ldots, N - 1$
$k = 0,1,2, \ldots, N - 1$
$\Delta f = 1/T$
$f_k = k\Delta f$
$T = N\Delta t$
$t_n = n\Delta t$

Errors

The accurate measurement of frequency-response functions depends heavily on the reduction of errors stemming from the digital signal processing. To take full advantage of experimental data in the evaluation of experimental procedures and verification of theoretical approaches, errors in measurement, generally designated noise, must be reduced to acceptable levels.

With respect to the frequency-response function measurement, the errors in the estimate are generally grouped into two categories: variance and bias. The variance portion of the error is due to random deviations of each sample function from the mean. Statistically, then, if sufficient sample functions are evaluated, the estimate will closely approximate the true function with a high degree of confidence. The bias portion of the error, on the other hand, is not necessarily reduced as a result of having many samples. The bias error is due to a system characteristic or measurement procedure consistently resulting in an incorrect estimate. Therefore, the expected value is not equal to the true value. Examples of this are system nonlinearities or digitization errors such as aliasing or leakage. With this type of error, knowledge of the form of the error is vital in reducing the resultant effect in the frequency-response function measurement.

Specifically, two general categories of problems exist which may cause significant error even when great care has been taken to deal with inaccurate system assumptions and obvious measurement mistakes. The first category is concerned with the limitations of using finite information. Any measurement instrument is limited in time resolution, or frequency bandwidth. However, sampling a signal at discrete times also introduces a form of amplitude error (called aliasing) that converts high-frequency energy to lower frequencies. This source of error would be classified as a bias. Thus, the time resolution and frequency bandwidth parameters are generally dictated by an antialiasing filter in front of the sampler. The shape of this filter influences the in-band accuracy and the stop-band rejection characteristics of the instrument. Obviously, filters are not perfect, and there is no such thing as absolute rejection. Strong signals with potential aliasing often are present to some extent. Another form of amplitude error is associated with the quantization of the analog signal to a digital signal. Since only discrete amplitude levels are possible, the amplitude will often be in error. This source of error normally has a Gaussian distribution and therefore is part of the variance portion of the total error.

In analog to time resolution limits, there is always a limit on frequency resolution. This is ultimately determined by the total effective time over which coherent data are collected. The effect of this finite collection time is the introduction of another type of nonlinear error (called leakage), which converts energy at each frequency into energy within a relatively narrow band nearby. This type of error is controlled to some extent by weighting (or windowing) the original time-domain data. However, this type of error will always cause a considerable bias in any portion of a measurement that is sufficiently close to a strong signal. In the situation of excitation of lightly damped structures, this leakage error, compared to all other sources of error, is usually the largest bias error, often much greater than the variance error.

Leakage Error

Leakage error is basically due to a violation of an assumption of the fast Fourier transform algorithm, namely that the true signal is periodic within the sample period used to observe the sample function. When both input and output are totally observable (transient input with completely observed decay output within the sample period) or are harmonic functions of the time period of observation T, there will be no contribution to the bias error due to leakage.

Leakage is probably the most common and, therefore, the most serious digital signal-processing error. Unlike aliasing and many other errors, the effects of leakage can only be reduced, not completely eliminated. The leakage error can be reduced by any of the following methods:

Cyclic averaging
Periodic excitation
Increase in frequency resolution
Frequency-response function estimation algorithm
Windowing or weighting functions

To understand how each of these methods reduces the leakage error, the origin of leakage must be well understood.

EXPERIMENTAL MODAL ANALYSIS

The discrete Fourier transform algorithm assumes that the data to be transformed are periodic with respect to the frequency resolution of the sampling period. Since, in general, the real world does not operate on the basis of multiples of some arbitrary frequency resolution, this assumption introduces an error known as leakage.

The following is one approach (there are many other equivalent presentations) used to explain leakage in terms of convolution.

The concept of multiplication and convolution represents a transform pair with respect to Fourier and Laplace transforms. More specifically, if two functions are multiplied in one domain, the result is the convolution of the two transformed functions in the other domain. Conversely, if two functions are convolved in one domain, the result is the multiplication of the two transformed functions in the other domain. When a signal is observed in the time domain with respect to a limited observation period T, the signal that is observed can be viewed as the multiplication of two infinite time functions as shown in Fig. 16-16a and b. The resulting time-domain function is, in the limit, the signal that is processed by the Fourier transform, which is shown in Fig. 16-16c. Therefore, by this act of multiplication, the corresponding frequency domain functions of Fig. 16-17a and b will be convolved to give the result equivalent to the Fourier transform of Fig. 16-16c. In this way, the difference between the infinite and the truncated signal can be evaluated theoretically.

To evaluate the frequency-domain result, the concept of convolution must be considered. First of all, the integral equation representing the convolution of two time-domain signals $x(t)$ and $y(t)$ can be given by Eq. (16.56).

$$Z(\phi) = \int_{-\infty}^{\infty} X(\omega) Y(\phi - \omega) d\omega \tag{16-56}$$

Therefore, the evaluation of the convolution of two functions is a function as well. The value of the new function can be viewed as the integration (or summation) over all frequencies of the product of the two frequency-domain functions, where one function has been shifted in frequency. This result is shown in Fig. 16-17. For simplicity, only the amplitude results are given. For a complete understanding of the bias error, a discussion of the phase effects is also needed. Note that Fig. 16-17a is the Fourier transform of Fig. 16-16a, Fig. 16-17b is the Fourier transform of Fig. 16-16b, and Fig. 16-17c is the convolution of Fig. 16-17a and b.

For the practical case, the resulting function shown in Fig. 16-17c is not continuous but occurs in a digital sense. For this case Eq. (16-56) can be adjusted accordingly as follows:

$$Z(\phi) = \sum_{i=-N}^{+N} X(i\Delta\omega) Y(\phi - i\Delta\omega) \tag{16-57}$$

where $\phi = k\Delta\omega$
$k = -N \leftrightarrow +N$

With this in mind, two cases must be evaluated with respect to whether the theoretical harmonic chosen for the example is periodic with respect to the window period T. These cases are shown in Fig. 16-18 and 16-19. Note that harmonic signal was chosen for this example, since any other signal can be thought to be simply a linear sum of such harmonics. Since the Fourier transform is also linear in this sense, the result shown is valid for any theoretical signal satisfying the Dirichlet conditions.

Therefore, when an analog signal is digitized in a Fourier analyzer, the analog signal has been multiplied by a function of unity (for a period of time T) in the time domain. This results in a

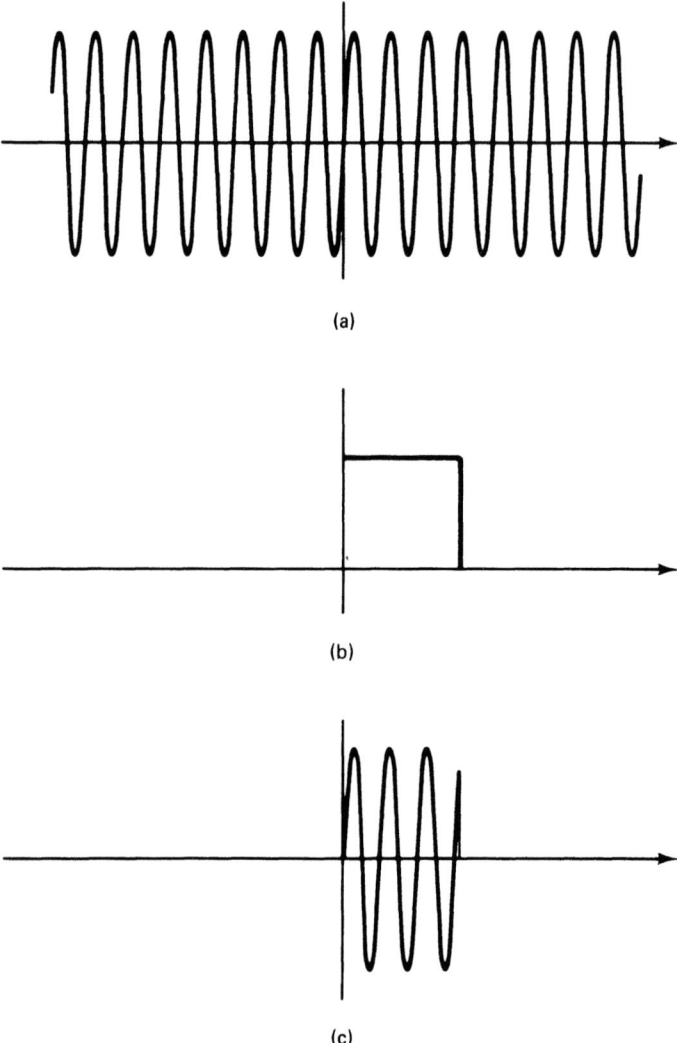

Figure 16-16 Time-domain function: (a) theoretical harmonic; (b) theoretical window; (c) multiplication of signals.

convolution of the two signals in the frequency domain. This process of multiplying an analog signal by some sort of weighting function is loosely referred to as *windowing*. Whenever a time function is sampled, the transform relationship between multiplication and convolution must be considered. Likewise, whenever an additional weighting function such as a Hanning window is used, the effects of such a window can be evaluated in the same fashion.

EXPERIMENTAL MODAL ANALYSIS

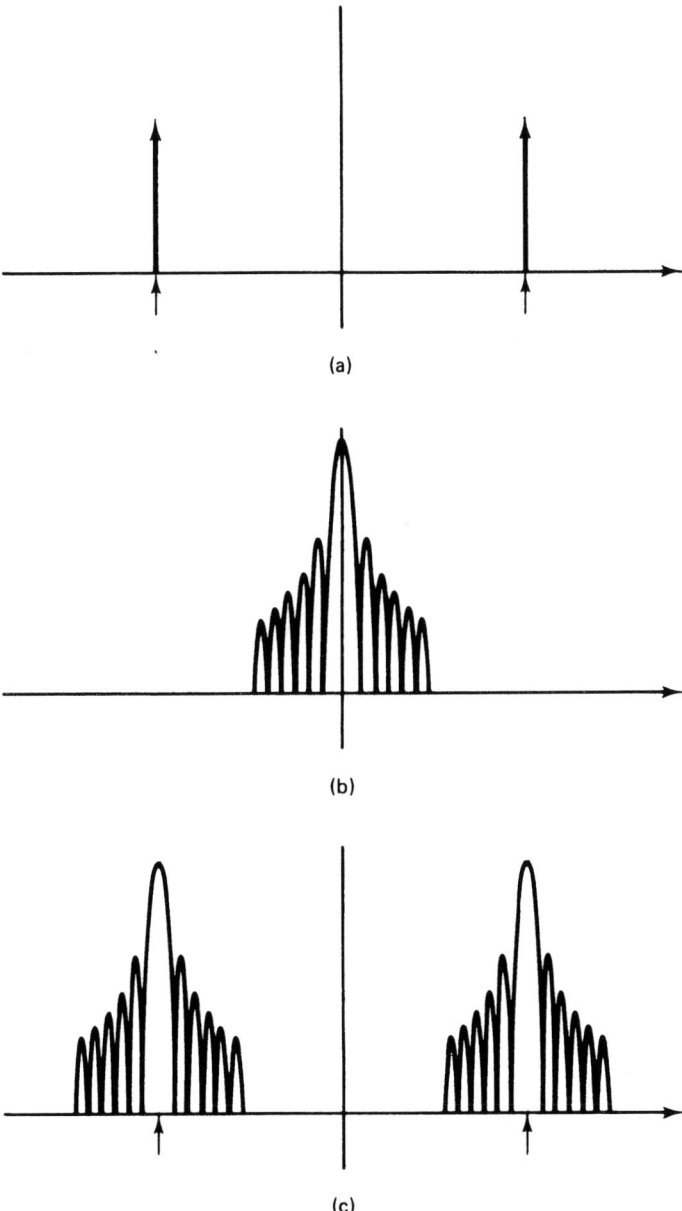

Figure 16-17 Frequency domain: (a) theoretical harmonic; (b) theoretical window; (c) convolved signals.

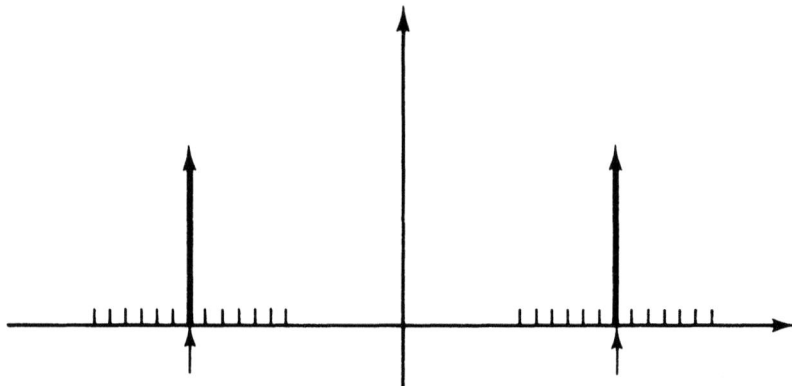

Figure 16-18 Frequency domain: periodic signal.

Aliasing Error

If frequency components larger than half the sampling frequency occur in the analog time history, amplitude and frequency errors will result. These errors are a result of the inability of the Fourier transform to decide which frequencies are within the analysis band and which frequencies are outside. The problem is explained graphically in Fig. 16-20.

A summary of what happens to signals above the F_{max} defined by the equality in Eq. (16-49) is shown graphically in Fig. 16-21. This demonstrates that a signal above F_{max} will appear after being digitized as a frequency below F_{max}. This serious error is controlled by using analog filters prior to digitization, to low-pass only the information below F_{max}. Naturally, since filters have a limited out-of-band rejection, the positioning of the cutoff frequency of the filters must be made with respect to the F_{max} and the roll-off characteristic of the filter.

Most modern digital data acquisition systems use a single low-pass analog filter with an ADC set at a single, high-speed sampling frequency rather than multiple filters and multiple sampling rates in the ADC. For example, the maximum frequency of interest for must structural and acoustic

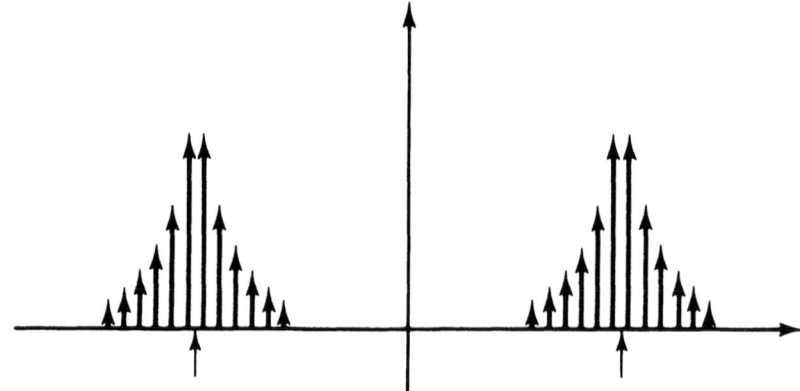

Figure 16-19 Frequency domain: nonperiodic signal.

Figure 16-20 Aliasing example.

measurement situations is below 25 kHz. Therefore, 25 kHz can be used as the maximum frequency (f_{max}). For this case, the sampling frequency (f_{samp}) is normally chosen to be 100 kHz.

While this choice of sampling frequency is higher than the minimum identified by Shannon's sampling theorem, it obeys the inequality of Shannon's sampling theorem and provides for a conservative sampling situation. With this sampling criterion, the low-pass analog filter (with 80 to 100 dB/octave roll-off) can be set at 25 kHz to eliminate aliasing effects. Note that since the sampling frequency is chosen higher than necessary (4× compared to 2×), this provides a guard band of frequency to prevent any aliasing effects from contaminating the digitized data below the maximum frequency. Since most structural and acoustic measurement situations do not require information from 0 Hz to 25 kHz, further reduction of the digitized information into an appropriate frequency band of interest can be achieved through the use of digital filtering.

Other Errors

To understand that the possibility for error originates at every step in the measurement process, a number of other errors must be discussed. While all the following errors are possible, many are a function of the way in which the ADC hardware operates or often malfunctions. Many of these errors can be evaluated by performing a histogram on a signal with known characteristics.

Quantization Error. This is the difference between the actual analog signal and the measured digitized value. Normally it is plus or minus half of one part in 2 raised to the Mth power. Since this error is a random event, averaging will minimize the effect on the resulting measurements.

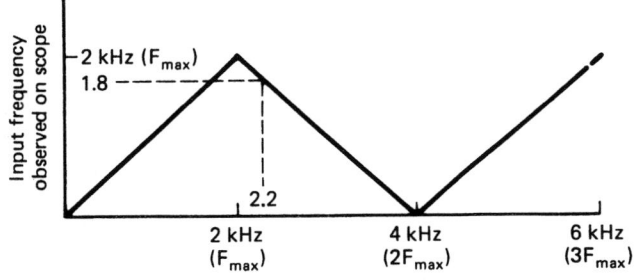

Figure 16-21 Aliasing contribution.

Differential Nonlinearity. This type of error results if the roundoff that occurs to cause quantization error is not regular (some of the spacing between counts varies). Differential nonlinearity causes the "noise" from quantization error to be biased.

Bit Dropout. One bit in the ADC may never be set. Obviously, any sample requiring this bit in the ADC word to be set will be in error and the error will be biased. A similar problem exists if one bit of the ADC word is always set.

Reference Voltage. The reference voltage used by the ADC may drift within or between sample periods. Since this drift is not known or measured, this error will cause a bias in any resulting estimate of the dependent variable.

Overload and Overload Recovery. The ADC may take several sampling increments to recover when it is overloaded. This is normally a problem under severe overloads, but if it occurs, the result will be a bias in the estimate of the amplitude.

Aperture–Clock Jitter. The value of the amplitude recorded does not correspond to the assumed instant in time t. This type of error will resulting a bias in the estimate of time and frequency parameters.

Digitizer Noise. The random setting of plus or minus one bit when the input is zero is referred to as digitizer noise. This error may become dominant in transient excitation since a large part of the observed histories may be very small or actually zero, as in the case of impact testing. This error can be controlled to some extent by averaging and the use of special window functions.

Transducer Considerations

The transducer considerations are often the most overlooked aspect of the experimental modal analysis process. Miscalculations involving the actual type and specifications of the transducers, mounting of the transducers, and calibration of the transducers will often be some of the largest sources of error.

Transducer specifications are concerned with the magnitude and frequency limitations that the transducer is designed to meet. This involves the measured calibration at the time the transducer was manufactured, the frequency range over which this calibration is valid, and the magnitude and phase distortion of the transducer, compared to the calibration constant over the range of interest. The specifications of any transducer signal conditioning must be included in this evaluation.

Transducer mounting involves evaluation of the mounting system to ascertain whether the mounting system has compromised any of the transducers specifications. This normally involves the possibility of relative motion between the structure under test and the transducer. Very often, the mounting systems that are convenient to use and allow ease of alignment with orthogonal reference axes are subject to mounting resonances that result in substantial relative motion between the transducer and the structure under test in the frequency range of interest. Therefore, selection of the mounting system depends heavily on the frequency range of interest and on the test conditions. Test conditions are factors such as temperature, roving or fixed transducers, and surface irregularity. Table 16-2 gives a brief review of many common transducer mounting methods.

Transducer calibration refers to the actual engineering unit per volt output of the transducer and signal-conditioning system. Calibration of the complete measurement system is needed to verify that the performance of the transducer and signal-conditioning system is proper. Obviously, if the measured calibration differs widely from the manufacturer's specifications, the use of that particular transducer and signal-conditioning path should be questioned. Also, certain applications,

EXPERIMENTAL MODAL ANALYSIS

Table 16-2 Transducer Mounting Methods

Method	Frequency Range (Hz)	Main Advantage	Main Disadvantage
Hand-held	20–1,000	"Quick look"	Poor measuring quality for long sample periods
Putty	0–200	Alignment, ease of mounting	Low frequency range; creep problem during measurement
Wax	0–2,000	Ease of application	Temperature limitations; frequency range limited by wax thickness; alignment
Hot glue	0–2,000	Quick setting time, good alignment	Temperature-sensitive transducers
Magnet	0–2,000	Quick setup	Requires magnetic material; alignment; bounce a problem during impact; surface preparation important
Adhesive film	0–2,000	Quick setup	Alignment; flat surface
Epoxy–cement	0–5,000	Mounts on irregular surface; alignment	Long curing time
Stud mount	0–10,000	Accurate alignment if carefully machined	Difficult setup
		Approximate frequency ranges; depends on transducer mass and contact conditions	

such as impact testing, involve slight changes in the transducer system (such as adding mass to the tip of an instrumented hammer) that affect the associated calibration of the transducer.

Ideally, on-site calibration should be performed both before and after every test to verify that the transducer and signal-conditioning system are operating as expected. The calibration can be performed using the same signal-processing and data analysis equipment that will be used in the data acquisition. A number of calibration methods can be used to calibrate the transducer and signal conditioning. Some of these methods yield a calibration curve, with magnitude and phase, as a function of frequency, while others simply estimate a calibration constant. Most of the current calibration methods are reviewed in Table 16-3. Note that some of the methods are appropriate for field calibration, while others are more suited for permanent installations in calibration laboratories [97–103].

Frequency-Response Function Estimation

The definition of the frequency response function is directly related to the concept of the transfer function. The frequency-response function is the complex ratio of output response to input excitation as a function of frequency for a single-input, single-output situation. The definition of transfer function is essentially the same, except that the transfer function is a function of the complex Laplace variable s. For the specific case of the real part of the Laplace variable restricted to zero, the transfer function becomes equal to the frequency-response function.

Structurally, frequency-response functions are normally used to describe the input–output force–displacement relationships of any system. Most often, the system is assumed to be linear and time invariant, although this is not necessary. When assumptions of linearity and time invariance are not valid, the measurement of frequency-response functions also depend on the independent

Table 16-3 Calibration Methods

Method		Remarks
	Calibration methods	
Inversion test		Constant — Can only be used with transducer that has stable dc output; calibration against local earth's gravity
Comparison method	Transducer / Reference transducer	Frequency response — Calibration against reference transducer
Reciprocity method	Transducer / Mass / Exciter	Constant and/or frequency response — Calibration against mass-loaded shaker
Drop method	1 g	Constant — Calibration against local earth and gravity; used for ac-coupled transducers
Ratio method	F → M; $\dfrac{a}{F} = \dfrac{1}{m}$ (Rigid mass)	Frequency-response ratio — Calibration against known a/F for a rigid mass

variables of time and input. In this way, a conditional frequency-response function is measured as a function of independent variables in addition to frequency.

The computation of the theoretical frequency-response function depends on the transformation of data from the time of the frequency domain. The Fourier transform is used for this computation, just as the Laplace transform is used to transform from the time domain to the Laplace domain. Unfortunately, though, the integral Fourier transform definition requires time histories from negative infinity to positive infinity. Since this is not possible experimentally, the computation is performed digitally using a fast Fourier transform algorithm based on only a limited time history. In this way the theoretical advantages of the Fourier transform can be implemented in a digital computation scheme.

#EXPERIMENTAL MODAL ANALYSIS

Table 16-4 Summary of Frequency-Response Function Estimation Models

		Assumed Location of Noise	
Technique	Solution Method[a]	Force Inputs	Response
H_1	LS	No noise	Noise
H_2	LS	Noise	No noise
H_v	TLS	Noise	Noise

[a] LS, least squares; TLS, total least squares.

Frequency-Response Function Estimation

The primary difference in the methods used to estimate the frequency-response function information is in the assumption of where the noise enters the measurement problem. Table 16-4 summarizes this characteristic for the three methods that have been widely used.

Based on the assumed location of the noise entering the estimation process, Equations (16-58) through (16-60) represent the corresponding model for the H_1, H_2, and H_v estimation procedures.

H_1 technique:

$$[H]_{N_o \times N_i} \{F\}_{N_i \times 1} = \{X\}_{N_o \times 1} - \{\eta\}_{N_o \times 1} \tag{16-58}$$

H_2 technique:

$$[H]_{N_o \times N_i} \{\{F\}_{N_i \times 1} - \{v\}_{N_i \times 1}\} = \{X\}_{N_o \times 1} \tag{16-59}$$

H_v technique:

$$[H]_{N_o \times N_i} \{\{F\}_{N_i \times 1} - \{v\}_{N_i \times 1}\} = \{X\}_{N_o \times 1} - \{v\}_{N_o \times 1} \tag{16-60}$$

Equation (16-61) presents the solution of Eq. (16-58) for a case involving six inputs or excitations. Note that in all methods, the inversion of a matrix is involved. Therefore, the inputs (references) that are used must not be fully correlated so that the inverse will exist. Extensive evaluation tools (principal force analysis) have been developed to detect and avoid this condition [25–27].

H_1 technique:

$$\begin{bmatrix} H_{p1} \\ H_{p2} \\ \cdot \\ \cdot \\ H_{p6} \end{bmatrix} = \begin{bmatrix} GFF_{11} & \cdots & GFF_{61} \\ GFF_{12} & & \cdot \\ \cdot & & \cdot \\ \cdot & & \cdot \\ GFF_{16} & \cdots & GFF_{66} \end{bmatrix}^{-1} \begin{bmatrix} GXF_{p1} \\ GXF_{p2} \\ \cdot \\ \cdot \\ GXF_{p6} \end{bmatrix} \tag{16-61}$$

Single-Input Example (H_1, H_2)

The frequency-response function can be computed in several ways through the use of the fast Fourier transform algorithm. In each case, the desired frequency-response function H of the system in Fig. 16-22 is a function of frequency and input and output positions, and all noise inputs (K,L,M,N) are assumed to be zero. Therefore, the system of Fig. 16-22 is reduced to that of Fig. 16-23. In all cases, only an estimate of the frequency-response function is possible. No additional notation is used to denote that the frequency-response functions are estimates.

The most reasonable, and most common approach to the estimation of frequency response is

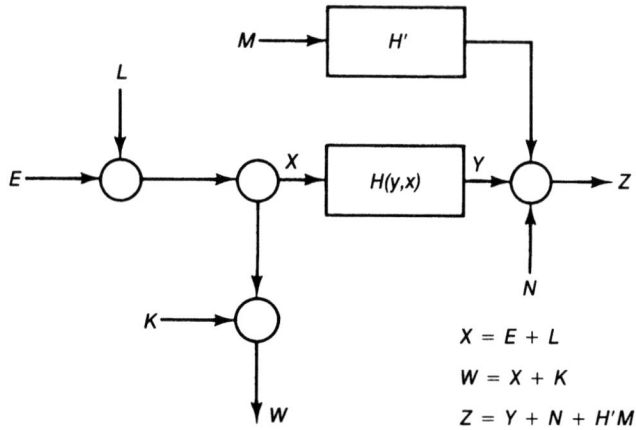

Figure 16-22 System model: with possible noise inputs.

by use of a least-squares technique. This standard technique for estimating parameters of noisy signals is based in this case on the auto and cross-power spectra.

Currently there is much interest in the formulation of the frequency-response function, in a least-squares sense, based on minimizing the noise on the input, the noise on the output, or the noise in a vector sense [91]. While the use of each approach can be justified on the basis of the location and form of the noise, it should be noted that as the noise is eliminated, each approach yields the same answer. With this in mind, there are two common formulations of frequency response based on power spectra.

The estimate of cross-power and auto power spectra for our model in Figs. 16-22 and 16-23 can be defined as follows:

$$G_{XF} = \sum_{i=1}^{N_{avg}} X_i F_i^* \tag{16-62}$$

$$G_{FX} = \sum_{i=1}^{N_{avg}} F_i X_i^* \tag{16-63}$$

$$G_{FF} = \sum_{i=1}^{N_{avg}} F_i F_i^* \tag{16-64}$$

$$G_{XX} = \sum_{i=1}^{N_{avg}} X_i X_i^* \tag{16-65}$$

where F^* = complex conjugate of F
X^* = complex conjugate of X

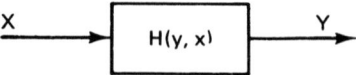

Figure 16-23 System model: without noise inputs.

EXPERIMENTAL MODAL ANALYSIS

The historical formulation of the frequency-response function, often referred to as H_{1xF}, tends to minimize the noise on the output. This formulation is

$$H_{1xF} = \frac{G_{XF}}{G_{FF}} \tag{16-66}$$

The more recent formulation in terms of implementation of the frequency-response function, often referred to as H_{2xF}, tends to minimize the noise on the input. This formulation is

$$H_{2xF} = \frac{G_{XX}}{G_{FX}} \tag{16-67}$$

Note that in both formulations, the phase information is preserved in the cross-power spectrum term.

In actual practice, the estimated values of the cross-power and auto power spectra depend on the form of the input signal (deterministic, random, etc.) Statistically, therefore, the power spectrum estimates are not unique unless the input is stationary. Even so, with nonstationary or deterministic inputs, the ratio of power spectra used in Eq. (16-66) and (16-67) will be unique and, therefore, valid within the limitations of the frequency content of the input and linearity of the model.

One important consideration of the two formulations for frequency-response function estimation is the behavior of each formulation in the presence of a bias error such as leakage. In both cases, the estimate will differ from the expected value, particularly in the region of a resonance or antiresonance. For example Eq. (16-66) tends to underestimate the value at resonance, while Eq. (16-67) tends to overestimate it.

In addition to the attractiveness of computing estimates of H using the H_1 or H_2 algorithm in terms of the minimization of the error, the availability of cross-power and auto power spectra allows the determination of another function, called ordinary coherence, which can express a level of confidence in the frequency-response function estimate.

$$\gamma_{XF}^2 = \frac{|G_{XF}|^2}{G_{FF} \, G_{XX}} = \frac{G_{FX} \, G_{XF}}{G_{FF} \, G_{XX}} \tag{16-68}$$

The quantity γ_{XF}^2 is called the scalar or *ordinary coherence function* and is a frequency-dependent, real value between zero and one. The coherence function can be thought to describe the division of output power into coherent and incoherent parts with respect to the input.

When the coherence is zero, the output is caused totally by sources other than the measured input. In general, then, the coherence can be a measure of the degree of noise contamination in a measurement. Thus, with more averaging, the estimate of coherence may contain less variance, therefore giving a better estimate of the noise energy in a measured signal. This is not the case, though, if the low coherence is due to bias errors such as nonlinearities, multiple inputs, or leakage. In all these cases, the estimated coherence function will approach, in the limit, the expected value of coherence at each frequency, depending on the type of noise present in the structure and measurement system. Note that with more averaging, the estimated value of coherence will not increase; the estimated value of coherence always approaches the expected value from the upper side.

The coherence function indicates the degree of causality in a frequency-response function. If the coherence is equal to one at any specific frequency, the system is said to have perfect causality at that frequency. In other words, the measured response power is caused totally by the measured input power (or by sources that are coherent with the measured input power). A coherence value less than unity at any frequency indicates that the measured response power is greater than that due to the measured input. This is because some extraneous noise also contributes to the output power. It should be emphasized, however, that low coherence values do not necessarily imply poor estimates of the frequency-response function, simply that more averaging is needed for a reliable result.

Two special cases of low coherence are worth particular mention. The first situation occurs when there is a *leakage bias error* in one or both of the input and output measurements. This causes the coherence in the area of the peaks of the frequency response to be less than unity. This error can be reduced by the use of weighting functions or by cyclic averaging. The second situation occurs when there is a significant *propagation time delay* between the input and output, as may be the case with acoustic measurements. If a propagation delay of length t is compared to a sample function length of T, a low estimate of coherence will be estimated as a function of the ratio t/T [83]. This propagation delay causes a bias error in the frequency response and should be removed prior to computation if possible.

Multiple-Input Example (H_1)

Multiple-input estimation of frequency-response functions is desirable for several reasons. The principal advantage is the increase in the accuracy of estimates of the frequency-response functions. During single-input excitation of a system, there may exist large differences in the amplitudes of vibratory motion at various locations because of the dissipation of the excitation power within the structure. This is especially true when the structure has heavy damping. Small nonlinearities in the structure consequently cause errors in the measurement of the response. With multiple-input excitation, the vibratory amplitudes across the structure typically are more uniform, with a consequent decrease in the effect of nonlinearities.

A second reason for improved accuracy is the increase in consistency of the frequency-response functions compared to the single-input method. When a number of exciter systems are used, the elements from columns of the frequency-response function matrix corresponding to those exciter locations are being determined simultaneously. With the single-input method, each column is determined independently, and it is possible for small errors of measurement due to nonlinearities and time-dependent system characteristics to cause a change in resonance frequencies, damping, or mode shapes among the measurements in the several columns. This is particularly important for the advanced analysis methods, such as the polyreference method [23,24] and the reduced matrix estimation method [44,71–73, 82], which use frequency-response functions for multiple columns simultaneously.

Another advantage of the simultaneous measurement of a number of columns of the frequency-response function matrix is the ability to use a linear combination of frequency-response functions in the same row of the matrix, to enhance specific modes of the system. This technique is analogous to the forced normal mode excitation experimental modal analysis in which a structure is excited by a forcing vector that is proportional to the modal vector of interest. For this analysis, the coefficients of a preliminary experimental modal analysis are used to weight the frequency-response functions, so that the sum emphasizes the modal vector that is sought. The revised set of conditioned frequency-response functions is analyzed to improve the accuracy of the modal vector. A simple example of this approach for a structure with approximate geometrical symmetry would be to excite at two symmetric locations. The sum of the two frequency-response functions at a specific response location should enhance the symmetric modes. Likewise, the difference of the two functions should enhance the antisymmetric modes.

An additional advantage of the multiple-input excitation is a reduction of the test time. In general, using multiple-input estimation of frequency response functions permits frequency-response functions to be obtained for all input locations in approximately the same time needed to acquire a set of frequency-response functions for one of the input locations, using a single-input estimation method.

EXPERIMENTAL MODAL ANALYSIS

The theoretical basis of multiple-input frequency-response function analysis is well documented in a number of sources [74–81]. Some of the more recent work in the development of the theory has been concerned with a concise matrix formulation of ordinary, partial, and multiple coherence [76–78,81]. While much has been written about multiple-input theory, very little experimental work has been documented. The experimental work that has been documented is primarily concerned with acoustic measurements. The application of the multiple-input theory to experimental modal analysis apparently had not been seriously investigated. Prior to the availability of fast Fourier analysis equipment, the multiple-input estimation approach was not very practical.

It also needs to be noted that this application of multiple-input/output theory represents a very special case compared to the general case. For this case, everything about the inputs is known or can be controlled. The number of inputs, the location of the inputs, and the characteristics of the inputs are controlled by the test procedure. For the general case, none of these characteristics may be known.

Consider the case of N_i inputs and N_o outputs measured during a modal test on a dynamic system as shown in Fig. 16-24. The model assumed for the dynamics is:

$$X_p = \sum_{q=1}^{N_i} H_{pq} F_q + N_p \tag{16-69}$$

where X_p = spectrum of the pth output
F_q = spectrum of the qth input
H_{pq} = frequency-response function of output p with respect to input q
N_p = spectrum of the "noise"; that part of X_p not linearly related to F_q

Least-Squares Estimation (H_1)

The least-squares method for determining the frequency-response functions is derived by minimizing the magnitudes of the output noise spectra. The solution of this minimization process provides the matrix equation

$$[GXF] = [H][GFF] + [Z] \tag{16-70}$$

where $[GXF]$ is an $N_o \times N_i$ frequency-dependent matrix, called the input/output-cross-spectrum matrix;
$[GFF]$ is the $N_i \times N_i$ frequency-dependent matrix, called the input-cross-spectrum matrix;
$H]$ is the $N_o \times N_i$ frequency-response function matrix; and
$[Z]$ is the $N_o \times N_i$ frequency-dependent matrix, called the noise-cross-spectrum matrix.

In the experimental procedure, the input and response signals are measured, and the averaged cross-spectra and autospectra necessary to create the $[GXF]$ and $[GFF]$ matrices are computed. If the computation of ordinary, multiple, or partial coherence functions will be required, the diagonal elements of the output-cross-spectrum matrix $[GXX]$ must be computed also.

The least-squares estimate of the frequency-response matrix is computed assuming that the noise-cross-spectrum matrix averages to zero, and so the formal solution yields:

$$[H] = [GXF][GFF]^{-1} \tag{16-71}$$

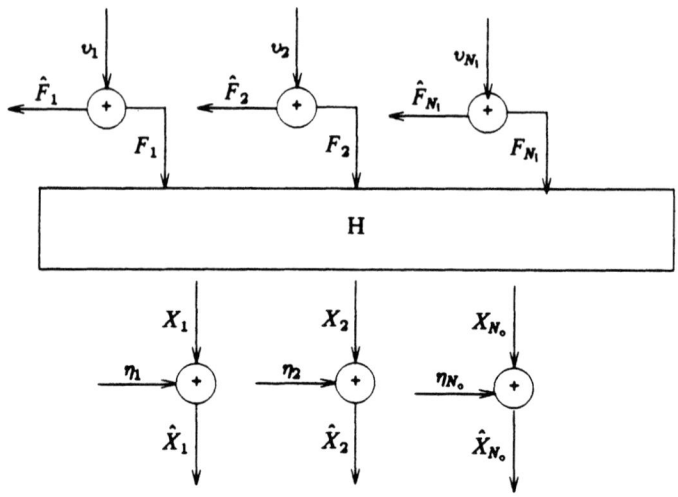

$F = \hat{F} - v$ Actual input
$X = \hat{X} - \eta$ Actual output
$\hat{X}_p$ = Spectrum of the p-th output, measured
$\hat{F}_q$ = Spectrum of the q-th input, measured
H_{pq} = Frequency response function of output p with respect to input q
v_q = Spectrum of the noise part of the input
η_p = Spectrum of the noise part of the output

Figure 16-24 System model: multiple inputs.

Note that the input-cross-spectrum matrix must be inverted, at least implicitly, at every frequency in the analysis range. This means that the "computational load" on the analysis system is greater than for the single-input case, in which only the reciprocal of a single-input autospectrum is computed.

Equation (16-71) is valid regardless of whether the various inputs are correlated. Unfortunately, in a number of situations the input-cross-spectrum matrix $[GFF]$ may be singular for specific frequencies or frequency intervals. When this happens, the inverse of $[GFF]$ will not exist and Eq. (16-71) cannot be used to solve for the frequency-response function at those frequencies or in those frequency intervals. A computational procedure that solves Eq. (16-71) for $[H]$ should therefore monitor the rank the matrix $[GFF]$ that is to be inverted and, ideally, provide direction on how to alter the input signals or use the available data when a problem exists.

The concept of the coherence function, as defined for single-input measurement, needs to be

EXPERIMENTAL MODAL ANALYSIS

expanded to include the variety of relationships that are possible for multiple inputs. Ordinary coherence is defined as the correlation coefficient describing the linear relationship between any two single spectra. This is exactly the coherence function that is defined for single-input measurement. Great care must be taken in the interpretation of ordinary coherence when more than one input is present. The ordinary coherence of an output with respect to an input can be much less than unity even though the linear relationship between inputs and outputs is valid, because of the influence of the inputs.

Partial coherence is defined as the ordinary coherence between a conditioned output and another conditioned output, between a conditioned input and another conditioned input, or between a conditioned input and a conditioned output. The output and input are conditioned by removing contributions from other input(s). The removal of the effects of the other input(s) is formulated on a linear least-squares basis. The order of removal of the inputs during "conditioning" has a definite effect on the partial coherence if some of the inputs are mutually correlated. There will be a partial coherence function for every input/output, input/input, and input/output combination for all permutations of conditioning. The usefulness of partial coherence, especially between inputs, for experimental modal analysis would appear to be of limited value, since all inputs are, ideally, uncorrelated. The inputs may however be mutually related without any two of them being fully coherent. Except for the dual-input case, the use of ordinary coherence between inputs is therefore insufficient. The computation procedure, to be presented, suggests the additional verification of a limited number of partial coherences among all inputs. These partial coherences are generated as intermediate results, and indicate, independent of the particular permutation of conditioning, whether a least-squares solution is possible.

Multiple coherence is defined as the correlation coefficient describing the linear relationship between an output and all known inputs. There will be a multiple coherence function for every output. Multiple coherence can be used to evaluate the importance of unknown contributions to each output. These unknown contributions can be measurement noise, nonlinearities, or unknown inputs. Particularly, as in the evaluation of ordinary coherence, a low value of multiple coherence near a resonance often means that the "leakage" error is present in the frequency-response function. Unlike the ordinary coherence function, a low value of multiple coherence is not expected at antiresonances. The antiresonances for different response locations occur at the same frequency. Though one response signal may have a poor signal-to-noise ratio at its antiresonance, other inputs will not at the same frequency.

The ordinary coherence function can be formulated in terms of the elements of the matrices defined earlier. The ordinary coherence function between the pth output and the qth input can be computed from the following formula:

$$\gamma_{pq}^2 = \frac{|[GXF]|^2}{[GFF]_{qq} [GXX]_{pp}} \tag{16-72}$$

The formulation of the equations for the multiple coherence functions can be simplified from the normal computational approach to the following equation.

$$MCOH_p = \sum_{k=1}^{N} \sum_{q=1}^{N} \frac{H_{pk} \, GFF_{kq} \, H_{pq}^*}{GXX_p} \tag{16-73}$$

If the multiple coherence of the pth output is near unity, then the this output is well predicted from the set of inputs using the least-squares frequency-response functions.

Of the variety of situations that can cause difficulties in the computation of the frequency-response functions, the most troublesome is the case of coherent or partially coherent inputs. Unfortunately, there are a number of situations in which the input-cross-spectrum matrix $[GFF]$ may be singular at specific frequencies or frequency intervals. When this happens, the inverse of

[GFF] will not exist and Eq. (16-71) cannot be used to solve for the frequency-response function at those frequencies or in those frequency intervals. First, one of the input autospectra may be zero in amplitude over some frequency interval. When this occurs, all the cross-spectra in the same row and column in the input-cross-spectrum matrix [GFF] will also be zero over that frequency interval. Consequently, the input-cross-spectrum matrix [GFF] will be singular over that frequency interval. Second, two or more of the input signals may be fully coherent over some frequency interval. This can be evaluated and documented by computing the ordinary and conditioned partial coherences among the inputs. Last, numerical problems, which cause the computation of the inverse to be inexact, may be present. This can happen when an autospectrum is near zero in amplitude, when the cross-spectra have large dynamic range with respect to the precision of the computer, or when the matrix is ill-conditioned because of nearly redundant input signals.

Of the variety of situations that can cause difficulties in the computation of the frequency-response functions, the highest potential for trouble is the case of coherent inputs. If two of the inputs are fully coherent, there are no unique frequency-response functions associated with those inputs. This can be measured by computing the ordinary and conditioned partial coherence functions among all inputs. Although the signals used as inputs to the exciter systems must be uncorrelated random inputs, the response of the structure at resonance, combined with the inability to completely isolate the exciter systems from this response, will result in the ordinary or conditioned partial coherence functions with values other than zero, particularly at the system poles. As long as the coherence functions are not unity at these frequencies, Eq. (16-71) can be solved uniquely for the frequency-response functions. A unique solution is obtained when [GFF] is invertible. Note that the auto- and cross-spectra involved in the calculation of the multiple-input case for the estimation of frequency-response functions should be computed from analog time data that have been digitized *simultaneously*. If data are not processed in this manner, many more averages will be required to reduce the variance on each individual auto- and cross-spectrum, and the multiple-input approach to the estimation of frequency-response functions will be less efficient.

Error Reduction Methods

Several factors contribute to the quality of actual measured frequency-response function estimates. Some of the most common sources of error are measurement mistakes. With a proper measurement approach, most errors of this type, such as overloading the input and extraneous signal pickup via ground loops or strong electric or magnetic fields nearby, can be avoided. Violations of test assumptions are often the source of another inaccuracy and can be viewed as a measurement mistake. For example, frequency response and coherence functions have been defined as parameters of a linear system. Nonlinearities will generally shift energy from one frequency to many new frequencies, in a way that may be difficult to recognize. The result will be a distortion in the estimates of the system parameters, which may not be apparent unless the excitation is changed. One way to reduce the effect of nonlinearities is to randomize these contributions by choosing a randomly different input signal for each of the N_a averages. Subsequent averaging will reduce these contributions in the same manner that random noise is reduced. Another example involves control of the system input. One of the most obvious requirements is to excite the system with energy at all frequencies for which measurements are expected. It is important to be sure that the input signal spectrum does not have "holes" where little energy exists. Otherwise, coherence will be very low, and the variance on the frequency-response function will be large.

Assuming that the system is linear, the excitation is proper, and obvious measurement mistakes are avoided, some amount of noise will present in the measurement process. Noise is a general designation describing the difference between the true value and the estimated value. A more exact designation is to view noise as the total error comprised of two terms, variance and bias. Each of these classifications is merely a convenient grouping of many individual errors that cause a specific kind of inaccuracy in the function estimate. The variance portion of the error essentially has a Gaussian distribution and can be reduced by any form of synchronization in the measurement or analysis process. The bias or distortion portion of the error causes the expected value of the estimated function to be different from the true value. Normally, bias errors are removed if possible, but if the form and the source of a specific bias error are known, many techniques may be used to reduce the magnitude of the specific bias error.

Five different approaches can be used to reduce the error involved in frequency-response function measurements in current fast Fourier transform analyzers. First of all, the *use of different frequency-response function estimation algorithms* (H_v compared to H_1) will reduce the effect of the leakage error on the estimation of the frequency-response function computation. The use of *averaging* can significantly reduce errors of both variance and bias and is probably the most general technique in error reduction in frequency-response function measurement. *Selective excitation* is often used to verify nonlinearities or to randomize characteristics. In this way, bias errors due to system sources can be reduced or controlled. The *increase of frequency resolution* through the zoom fast Fourier transform can improve the frequency-response function estimate primarily by reduction of the leakage bias error due to the use of a longer time sample. The zoom fast Fourier transform by itself is a linear process and does not involve any specific error reduction characteristics compared to a baseband fast Fourier transform. Finally, the *use of weighing functions (windows)* is widespread, and much has been written about their value [74,75,81,93,104]. Primarily, weighting functions compensate for the bias error (leakage) caused by the analysis procedure.

Signal Averaging

The averaging of signals is normally viewed as a summation or weighted summation process in which each sample function has a common abscissa. Normally, the designation of "history" is given to sample functions with the abscissa of absolute time and the designation of "spectrum" is given to sample functions with the abscissa of absolute frequency. The spectra are normally generated by Fourier transforming the corresponding history. To generalize and consolidate the concept of signal averaging as much as possible, the case of relative time could also be considered. In this way "relative history" may be discussed with units of the appropriate event rather than seconds, and a "relative spectrum" becomes the corresponding Fourier transform, with units of cycles per event. This concept of signal averaging is used widely in structural signature analysis where the event is a revolution. This kind of approach simplifies the application of many other concepts of signal relationships such as Shannon's sampling theorem and Rayleigh's criterion of frequency resolution.

The process of signal averaging as it applies to frequency-response functions is simplified greatly by the intrinsic uniqueness of the frequency-response function. Since the frequency-response function can be expressed in terms of system properties of mass, stiffness, and damping, it is reasonable to conclude that in most realistic structures, the frequency-response functions are considered to be constants just like mass, stiffness, and damping. This concept means that when formulating the frequency-response function using Eq. (16-58), (16-59), or (16-60), the estimate of frequency response is intrinsically unique, as long as the system is linear. In general, the auto- and cross-power spectra are statistically unique only if the input is stationary and sufficient averages have been taken. Nevertheless, the estimate of frequency response is valid whether the input is stationary, nonstationary, or deterministic.

The concept of the intrinsic uniqueness of the frequency-response function also permits greater freedom in the testing procedure. Each function can be derived as a result of a separate test or as the result of different portions of the same continuous test situation. In either case, the estimate of frequency-response function will be the same as long as the time history data have been acquired simultaneously for the auto- and cross-power spectra that are used in any computation for frequency-response or coherence function.

The approaches to signal averaging vary only in the relationship between the sample functions used. Since the Fourier transform is a linear function, there is no theoretical difference in the use of histories or spectra. (Practically, though, there are precision considerations, which will be discussed later). With this in mind, the signal averaging useful to frequency-response function measurements can be divided into three classifications:

Asynchronous
Synchronous
Cyclic

These three classifications refer to the trigger and sampling relationships between sample functions.

Asynchronous Signal Averaging

The classification of asynchronous signal averaging refers to the case in which no known relationship exists between individual sample functions except for the intrinsic uniqueness of the frequency-response function. In this case the least-squares approach to the estimate of frequency response must be used, since no other way of preserving phase and improving the estimate is available. In this situation, the trigger for digitization (sampling and quantization) takes place in a random fashion dependent only on the equipment availability. The digitization is said to be in a free-run mode.

Synchronous Signal Averaging

The synchronous classification of signal averaging adds the constraint that each sample function be initiated with respect to a system input. This fact, together with the intrinsic uniqueness, would allow the frequency-response function to be formed as a summation of ratios of Y divided by X, since phase is preserved. Even so, the reduction of variance and the value of coherence available with the power spectra approach would preclude the use of any other technique in most cases. The ability to synchronize the initiation of digitization allows for use of nonstationary or deterministic inputs, with a resulting increased signal-to-noise ratio and reduced leakage. Both these improvements in the frequency-response function estimate are due to more of the input and output being observable in the limited time window.

The synchronization takes place as a function of a trigger signal occurring in the input (internally) or in some event related to the input (externally). An example of an internal trigger would be the case of an impulsive input used to estimate the frequency response. All sample functions would be initiated when the input reached a certain amplitude and slope. A similar example of an external trigger would be the case of the impulsive excitation to a speaker used to trigger the estimate of frequency response between two microphones in the sound field. Again, all sample functions would be initiated when the trigger signal reached a certain amplitude and scope.

Cyclic Signal Averaging

The cyclic classification of signal averaging involves the added constraint that the digitization be coherent between sample functions. This means that the exact time (absolute or relative) between each pair of sample functions is used to enhance the signal-averaging process. Rather than trying to keep track of elapsed time between sample functions, the normal procedure is to allow no time to elapse between successive sample functions. This process can be described as a comb digital filter in the frequency domain, with the teeth of the comb at frequency increments dependent on the periodic nature of the sampling with respect to the event measured. The result is an attenuation of the spectrum between the teeth not possible with other forms of averaging.

This form of signal averaging is very useful for filtering periodic components from a noisy signal, since the teeth of the filter are positioned at harmonics of the frequency of the sampling reference signal. This is of particular importance in signature applications, where it is desirable to extract signals connected with various rotating members. This same form of signal averaging is particularly useful for reducing leakage during frequency-response measurements and also has been used extensively for evoked response measurements in biomedical studies.

A very common application of cyclic signal averaging is in the area of analysis of rotating structures. In such a signature analysis application, the peaks of the comb filter are positioned to match the fundamental and harmonic frequencies of a particular rotating shaft or component. This is particularly powerful, since in one measurement it is possible to filter all the possible frequencies generated by the rotating member from a given data signal. With a zoom Fourier transform type of approach, potentially only one shaft frequency at a time can be examined depending on the zoom power necessary to extract the shaft frequencies from the surrounding noise.

In the application of cyclic signal averaging to frequency-response function estimates, the corresponding fundamental and harmonic frequencies are now simply the frequency resolution Δf and integer multiples of Δf. In this case, the spectra between each Δf is reduced, with an associated reduction of the bias error called leakage.

The implementation of cyclic signal averaging proceeds in a manner easily applicable to most fast Fourier transform analyzers. The cyclic averaged inputs and outputs are normally computed by simply summing successive time records. The important requirement of the successive time records is that no data be lost. Therefore, these successive time records could be laid end to end to create a digitized time record of length N_cT. The cyclic averaged records are then created by simply adding each time record of length T together in a block mode.

While the basic approach to cyclic averaging involves using the data weighted uniformly over the total sample time N_cT, the benefits that can be gained by using weighting functions can also be applied. The application of a Hanning window to the successive time records before the summation occurs yields an even greater reduction of the bias error. Therefore, for frequency-response measurements, Hanning-weighted signal averaging should drastically reduce the leakage errors that can exist when using broadband random excitation techniques to measure frequency response. Figure 16-25 shows the results of a series of measurements performed by measuring the frequency response of a very lightly damped automotive frame. The value of N_c indicates the number of cyclic time records averaged together, and N_a is the number of asynchronous auto- and cross-spectrum averages: a total of N_cN_a time records were sampled, thus making the same amount of independent information statistically available in each averaging case. Note that the total time required to acquire the measurement is the same for every case and no change has been made in the placement of the transducers or the form of the excitation. Clearly, the measurement using cyclic averaging with Hanning weighting show a significant reduction of the bias error. An interesting point is that the data near the antiresonance also are drastically improved as a result of the sharp roll-off of the line shape of the Hanning-weighted averaging.

Overlapping Time Records

At least two common averaging techniques use histories that may or may not overlap. In both cases, the averaging techniques involve enhancement of the data by processing random data histories. The first case is that of overlap processing, which uses individual sample histories that

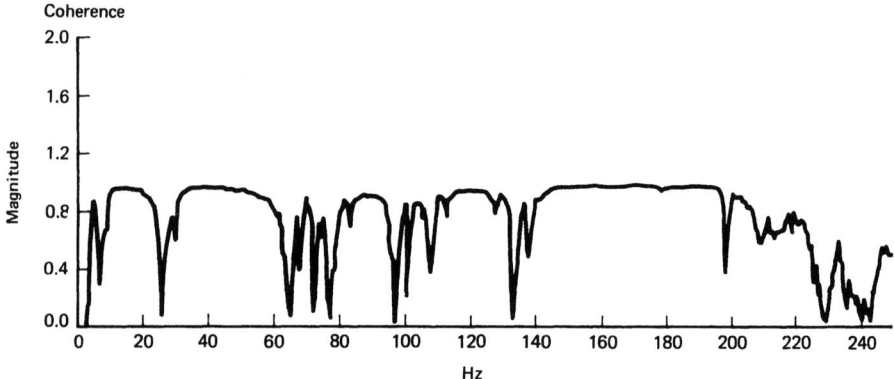

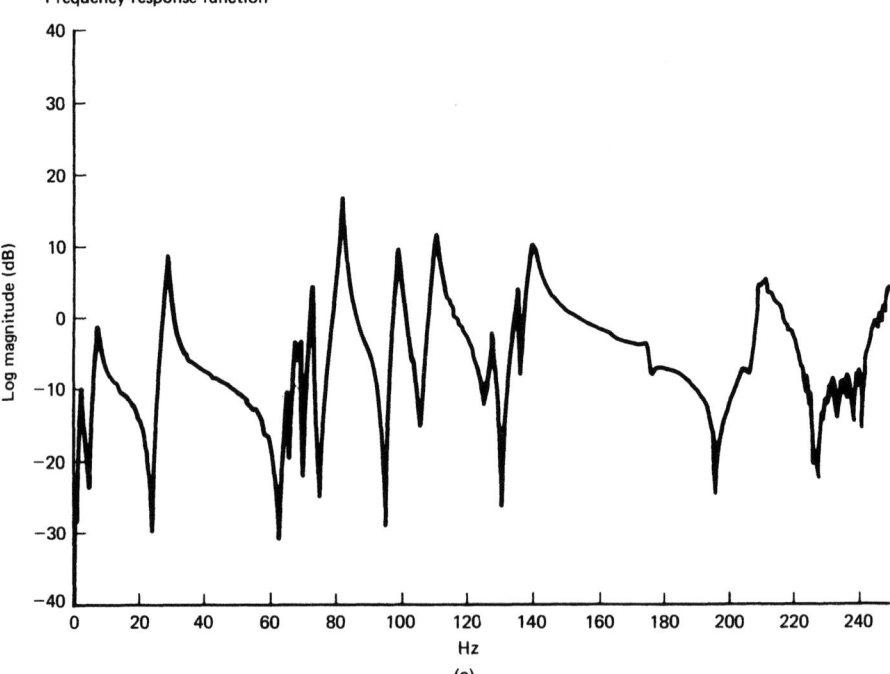

Figure 16-25 Asynchronous averaging (a) $N_c = 1$, $N_a = 80$; (b) asynchronous averaging with Hann: $N_c = 1$, $N_a = 80$; (c) cyclic averaging $N_c = 4$, $N_a = 20$; (d) cyclic averaging with Hann: $N_c = 4$, $N_a = 20$.

EXPERIMENTAL MODAL ANALYSIS

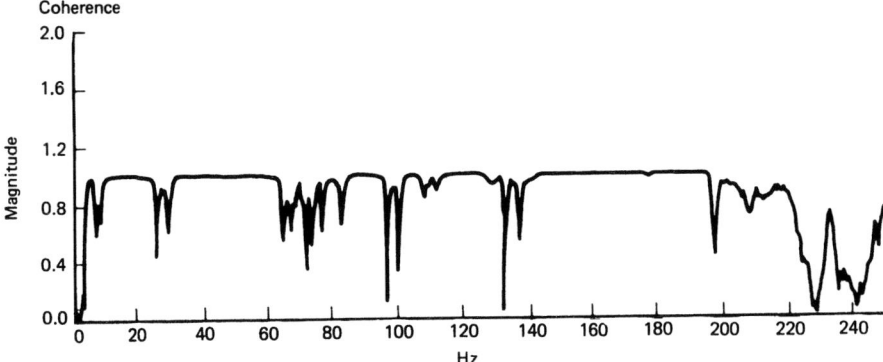

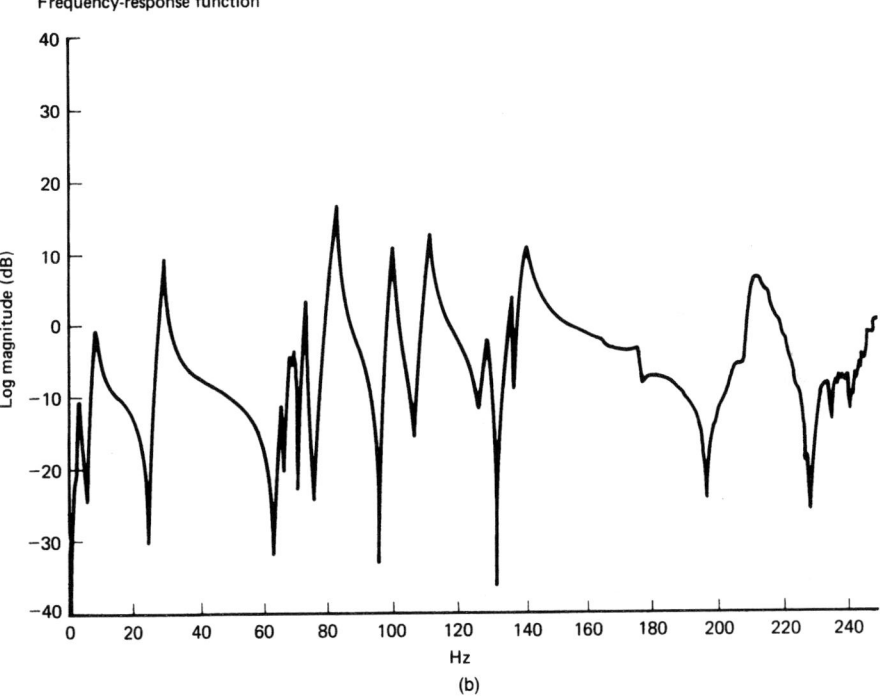

Figure 16-25 (*continued*)

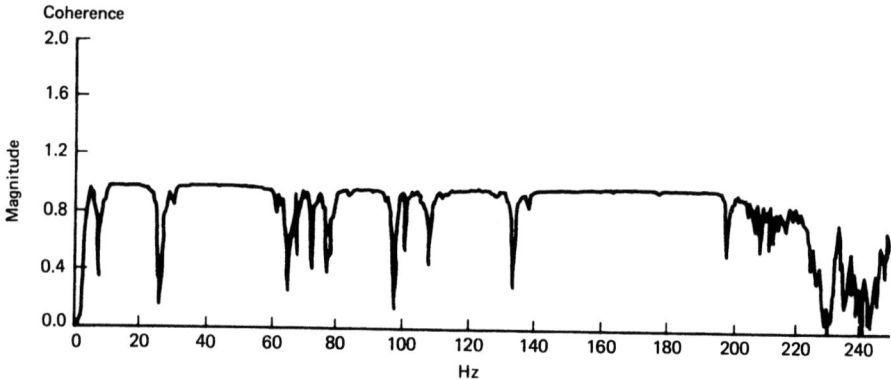

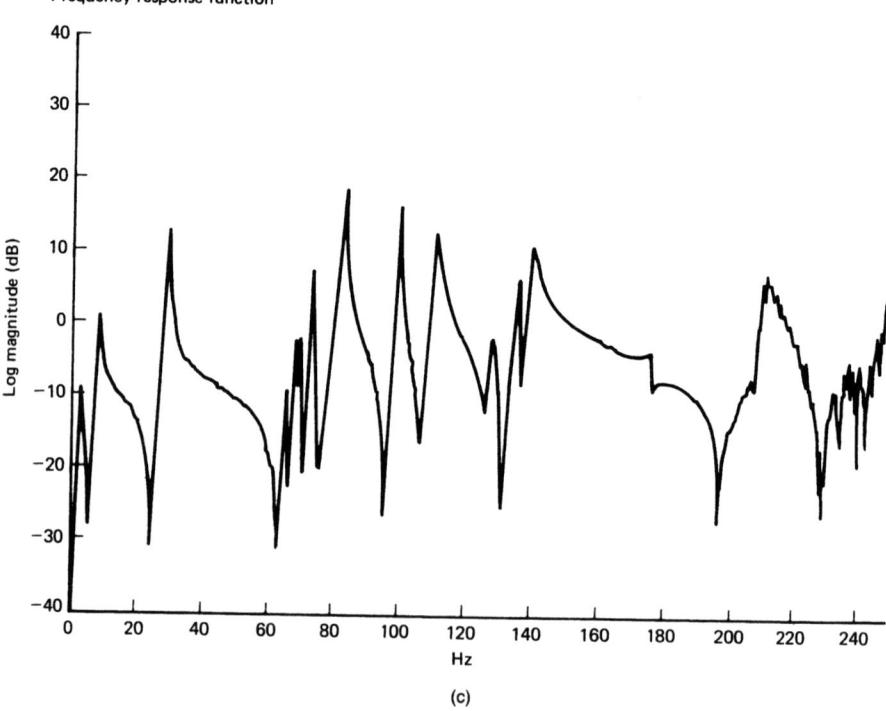

(c)

Figure 16-25 (*continued*)

EXPERIMENTAL MODAL ANALYSIS

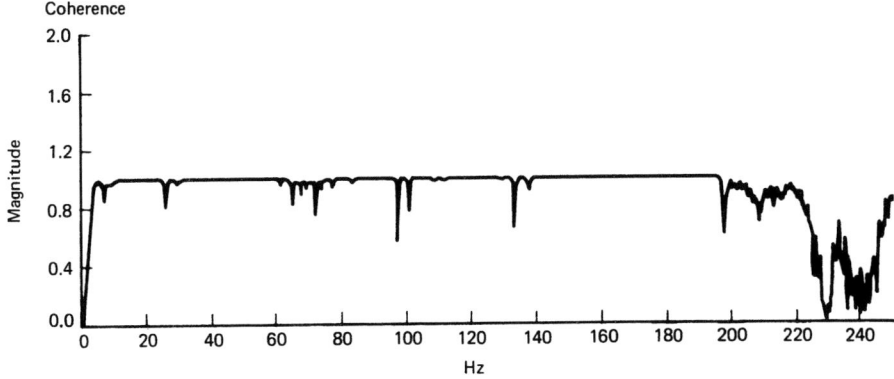

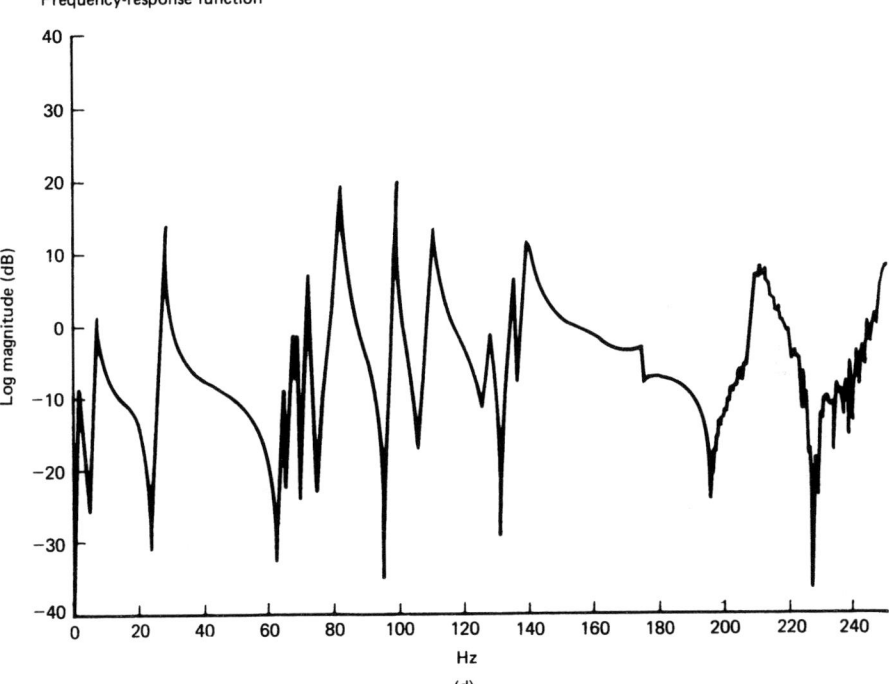

(d)

Figure 16-25 (*continued*)

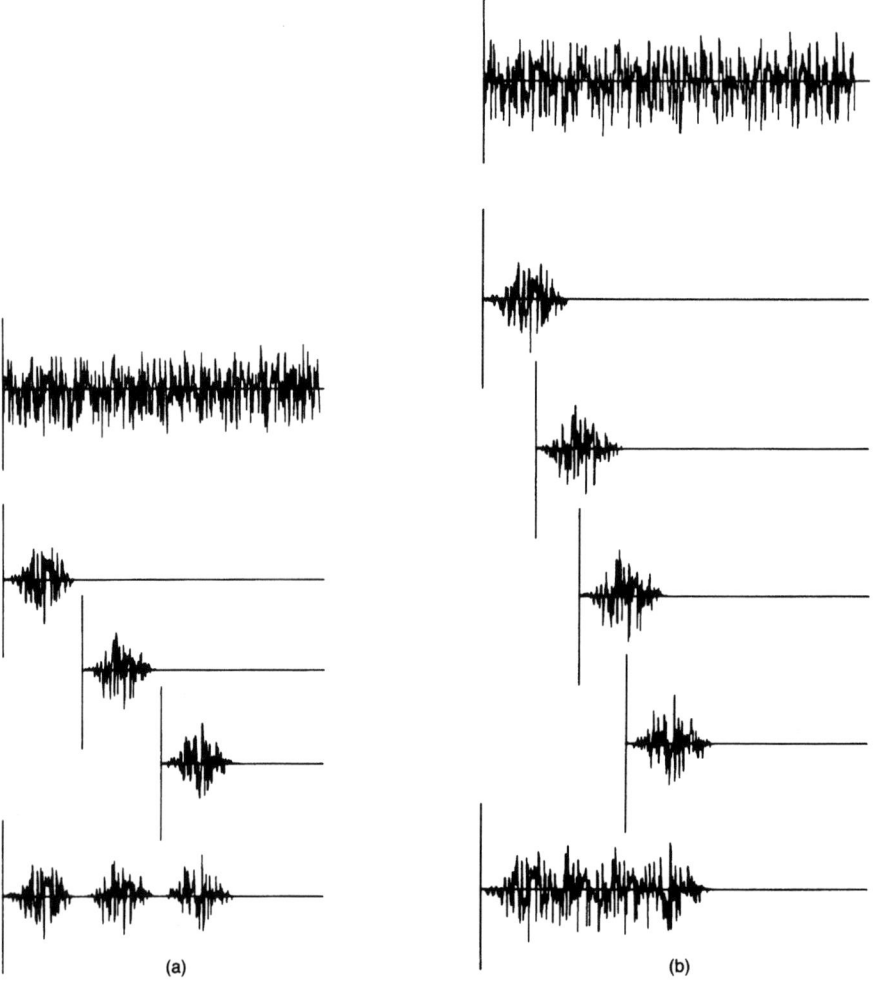

Figure 16-26 Overlap processing: (a) zero overlap; (b) 50% overlap.

are not totally independent of one another. The dependence that occurs results from each successive history starting before the preceding history ends. For the general case, where the time data are not weighted in any fashion, it should be obvious that this averaging procedure does not involve any new data and, therefore, statistically does not improve the estimation process. In the special case of weighting functions, this technique can use data that otherwise are ignored. Figure 16-26a is an example of a data record that has been weighted to reduce the leakage error using a Hanning weighting function. The data prior to 20% of each sample period and after 80% of each sample period is nearly eliminated by the use of Hanning window. Using an overlap factor of at least 20 to 30% in Fig. 16-26b involves these data once again in the averaging process.

The second case involving overlapping histories is that of random decrement analysis [63,64,84,85]. This process involves the overlapping of histories to enhance the deterministic portion of the random record. In general, the random response data can be considered to be made up of a deterministic part and a random part. Averaging in the time domain, the random part can be reduced if a trigger signal with respect to the information of interest exists. In the preceding discussions, this trigger signal has been a function of the input (asynchronous or synchronous averaging) or of the sampling frequency (cyclic averaging). More generally, though, the trigger function can be any function with characteristics related to the response history. Specifically, then, the random decrement technique relies on the assumptions that the deterministic part of the random response signal itself contains a free-decay step and impulse-response functions and can be used as the trigger function. Therefore, by starting each history at a specific value and slope of the random response function, characteristics related to the deterministic portion of the history will be enhanced.

There are three specific cases of random decrement averaging that represent the limiting results of its use [86]. In the first case, each starting value is chosen when the random response history reaches a specific constant level, with alternating slopes for each successive starting value. The random decrement history for this case becomes the free-decay step-response function. An example of this case for the first few averages is shown in Fig. 16-27.

In the second case, each starting value is chosen when the random response history crosses the zero axis with positive slope. The random decrement history for this case becomes the free-decay, positive impulse-response function.

In the third case, each starting value is chosen when the random response history crosses zero with negative slope. The random decrement history for this case becomes the free decay negative impulse response function.

Therefore, in each of these cases, the random decrement technique acts like a notched digital filter with pass bands at the poles of the trigger function. This tends to eliminate spectral components not coherent with the trigger function.

If a secondary function is utilized as the trigger function, only the history related to the poles of the secondary function will be enhanced by this technique. If the trigger function is sinusoidal, the random decrement history will contain information related only to that sinusoid. Likewise, if the trigger function is white noise, the random decrement history will be a unit impulse function at time zero. One useful example of this concept was investigated for conditioning random response histories so that information unrelated to the theoretical input history is removed. In this situation, the theoretical input history serves as the trigger function. The random decrement history formed on the basis of this trigger function represents the random response function that would be formed if the theoretical input history were truly the system input. In reality, the measured input history may vary due to noise, impedance mismatch, and so on.

Excitation

Inputs that can be used to excite a system to determine frequency-response functions belong to one of two classifications. The first classification is that of a random signal. Signals of this form can be defined only by their statistical properties over some time period. Any subset of the total time period is unique, and no explicit mathematical relationship can be formulated to describe the signal. Random signals can be further classified as stationary or nonstationary. Stationary random signals are a special case: the statistical properties of the random signals do not vary with respect

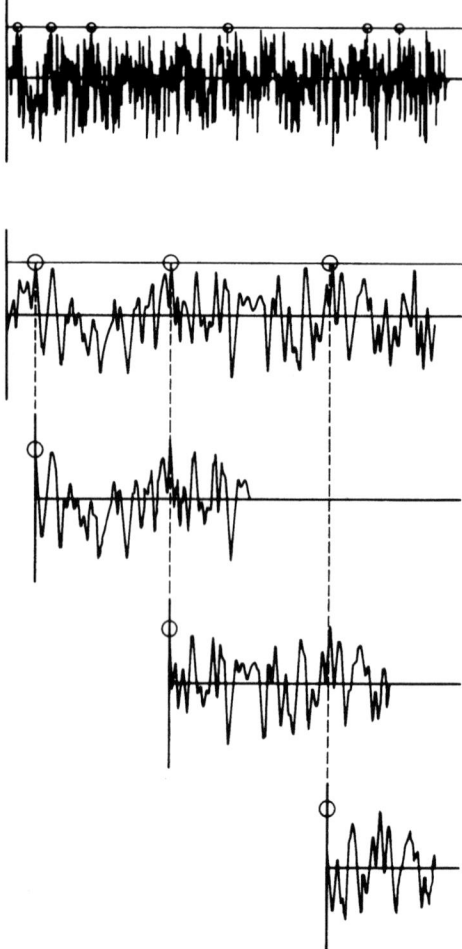

Figure 16-27 Random decrement averaging.

to translations with time. Finally, stationary random signals can be classified as ergodic or nonergodic. A stationary random signal is ergodic when a time average on any particular subset of the signal is the same for any arbitrary subset of the random signal. All the random signals commonly used as input signals fall into the category of ergodic, stationary random signals.

The second classification of inputs that can be used to excite a system in order to determine frequency-response functions is that of a deterministic signal. Signals of this form can be represented in an explicit mathematical relationship. Deterministic signals are further divided into periodic and nonperiodic classifications. The most common inputs in the periodic deterministic signal designation are sinusoidal, while the most common inputs in the nonperiodic deterministic designation have a transient form.

The choice of input to be used to excite a system in order to determine frequency-response functions depends on the characteristics of the system, the characteristics of the parameter estimation, and the expected use of the data. The characterization of the system is primarily concerned with the linearity of the system. As long as the system is linear, all input forms should give the same expected value. Naturally, though, all real systems have some degree of nonlinearity. Deterministic input signals result in frequency-response functions that depend on the signal level and type. A set of frequency-response functions for different signal levels can be used to document the nonlinear characteristics of the system. Random input signals, in the presence of nonlinearities, result in a frequency-response function that represents the best linear representation of the nonlinear characteristics for a given level of random signal input. For small nonlinearities, use of a random input will not differ greatly from the use of a deterministic input.

The characterization of the parameter estimation is concerned primarily with the type of mathematical model being used to represent the frequency-response function. Generally, the model is a linear summation based on the modal parameters of the system. Unless the mathematical representation of all nonlinearities is known, the parameter estimation process cannot properly weight the frequency-response function data to include nonlinear effects. For this reason, random input signals are prevalently used to obtain the best linear estimate of the frequency-response function when a parameter estimation process using a linear model is to be utilized.

The expected use of the data entails consideration of the degree of detailed information required by any postprocessing task. For experimental modal analysis, this can range from implicit modal vectors needed for troubleshooting to explicit modal vectors used in an orthogonality check. As more detail is required, input signals, both random and deterministic, will need to match the system characteristics and parameter estimation characteristics more closely. In all possible uses of frequency-response function data, the conflicting requirements of accuracy, equipment availability, testing time, and testing cost will normally reduce the possible choices of input signal.

With respect to the reduction of the variance and bias errors of the frequency-response function, random or deterministic signals can be used most effectively if the signals are periodic with respect to the sample period or totally observable with respect to the sample period. If either of these criteria is satisfied, regardless of signal type, the predominant bias error, leakage, will be eliminated. If these criteria are not satisfied, the leakage error may become significant. In either case, the variance error will be a function of the signal-to-noise ratio and the amount of averaging.

Many signals are appropriate for use in experimental modal analysis. Some of the most commonly used signals are described in the following sections [42,43,87–91]. For excitation signals that require the use of a shaker. Figure 16-28a shows a typical test configuration; Fig. 16-28b shows a typical test configuration suitable for use with an impact form of excitation. The advantages and disadvantages of each excitation signal are summarized in Table 16-5.

Slow-Swept Sine

The slow-swept sine signal is a periodic deterministic signal with a frequency that is an integer multiple of the FFT frequency increment. Sufficient time is allowed in the measurement procedure for any transient response to the changes in frequency to decay so that the resultant input and response histories will be periodic with respect to the sample period. Therefore, the total time needed to compute an entire frequency-response function will be a function of the number of frequency increments required and the system damping.

Periodic Chirp

The periodic chirp is a fast-swept sine signal that is a periodic deterministic signal and is formulated by sweeping a sine signal up or down within a frequency band of interest during a single sample

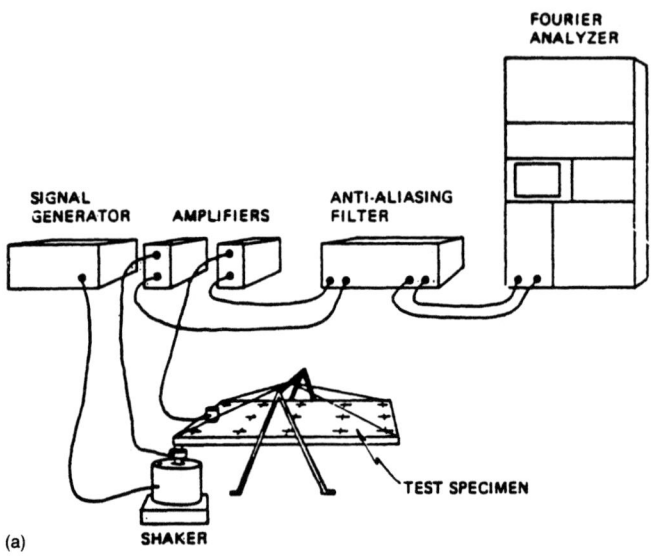
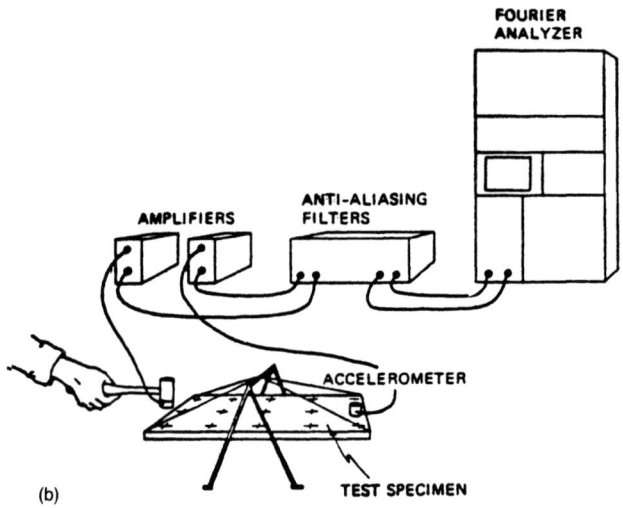

Figure 16-28 Typical test configurations: (a) shaker; (b) impact hammer.

EXPERIMENTAL MODAL ANALYSIS

Table 16-5 Summary of Excitation Signals

	Sine Steady State	True Random	Pseudo-random	Periodic Random	Fast Sine	Impact[b]	Burst[b] Sine	Burst[b] Random
Minimum leakage	No	No	Yes	Yes	Yes	Yes	Yes	Yes
Signal-to-noise ratio	Very high	Fair	Fair	Fair	High	Low	High	Fair
Rms-to-peak ratio	High	Fair	Fair	Fair	High	Low	High	Fair
Test measurement time	Very long	Good	Very good	Fair	Fair	Very good	Very good	Very good
Controlled frequency content	Yes	Yes[a]	Yes[a]	Yes[a]	Yes[a]	No	Yes[a]	Yes[a]
Controlled amplitude content	Yes	No	Yes[a]	No	Yes[a]	No	Yes[a]	No
Removes distortion	No	Yes	No	Yes	No	No	No	Yes
Characterize nonlinearity	Yes	No	No	No	Yes	No	Yes	Yes

[a] Requires additional equipment or special hardware.
[b] Transient in analyzer window.

period. Normally, the fast-swept sine signal is made up of only integer multiples of the FFT frequency increment. This signal is repeated without change, so that the input and output histories will be periodic with respect to the sample period.

Impact (Impulse)

The impact signal is a transient deterministic signal formed by applying an input pulse to a system lasting only a very small part of the sample period. The width, height, and shape of this pulse determine the usable spectrum of the impact. Briefly, the width of the pulse determines the frequency spectrum, while the height and shape of the pulse control the level of the spectrum. Impact signals have proven to be quite popular because they allow the freedom of applying the input with some form of instrumented hammer. While the concept is straightforward, the effective utilization of an impact signal is very involved [87,88].

Step Relaxation

The step relaxation signal is a transient deterministic signal formed by releasing a previously applied static input. The sample period begins at the instant the release occurs. This signal is normally generated by the application of a static force through a cable, which is then cut or allowed to release through a shear-pin arrangement [90].

Pure Random

The pure random signal is an ergodic, stationary random signal that has a Gaussian probability distribution. In general, the frequency content of the signal contains all frequencies (not just integer multiples of the FFT frequency increment), but the signal may be filtered to include only information in a frequency band of interest. The measured input spectrum of the pure random signal will be altered by any impedance mismatch between the system and the exciter.

Pseudorandom

The pseudorandom signal is an ergodic, stationary random signal consisting only of integer multiples of the FFT frequency increment. The frequency spectrum of this signal has a constant amplitude with random phase. If sufficient time is allowed in the measurement procedure for any transient response to the initiation of the signal to decay, the resultant input and response histories will be periodic with respect to the sample period. The number of averages used in the measurement procedure is a function of the reduction of the variance error only. In a noise-free environment, only one average may be necessary.

Periodic Random

The periodic random signal is an ergodic, stationary random signal consisting only of integer multiples of the FFT frequency increment. The frequency spectrum of this signal has random amplitude and random phase distribution. Since a single history will not contain information at all frequencies, a number of histories are necessary. For each average, an input history is created with random amplitude and random phase. The system is excited with this input in a repetitive cycle until the transient response to the change in excitation signal decays. The input and response histories should then be periodic with respect to the sample period and are recorded as one average in the total process. With each new average, a new history, uncorrelated with earlier input signals, is generated so that the resulting measurement will be completely randomized.

Random Transient (Burst Random)

The random transient signal is neither a completely transient deterministic signal nor a completely ergodic, stationary random signal but contains properties of both signal types. The frequency spectrum of this signal has random amplitude and random phase distribution and contains energy throughout the frequency spectrum. The difference between this signal and the periodic random signal is that the random transient history is truncated to zero after some percentage of the sample period (normally 50 to 80%). The measurement procedure duplicates the periodic random procedure but without the need to wait for the transient response to decay. The point of truncation of the input history is chosen so that the response history decays to zero within the sample period. Even for lightly damped systems, the response history will decay to zero very quickly as a result of the damping provided by the exciter system trying to maintain the input at zero. This damping, provided by the exciter system, is often overlooked in the analysis of the characteristics of this signal type. Since this measured input, although not part of the generated signal, includes the variation of the input during the decay of the response history, the input and response histories are totally observable within the sample period and the system damping is unaffected.

Increased Frequency Resolution

An increase in the frequency resolution of a frequency-response function affects measurement errors in several ways. Obviously, finer frequency resolution allows more exact determination of the damped natural frequency of each modal vector. The increased frequency resolution means that the level of a broadband signal is reduced. The most important benefit of increased frequency resolution, though, is a reduction of the leakage error. Since the distortion of the frequency-response function due to leakage is a function of frequency spacing, not frequency, the increase in frequency resolution will reduce the true bandwidth of the leakage error centered at each damped natural frequency. To increase the frequency resolution, the total time per history must be increased in direct proportion. The longer data acquisition time will increase the variance error problem when transient signals are used for input, as well as emphasizing any nonstationary problem with the data. The increase of frequency resolution often calls for multiple acquisition and/or processing of the histories to obtain an equivalent frequency range. This additional work will increase the data storage and documentation overhead as well as extending the total test time.

There are two approaches for increasing the frequency resolution of a frequency-response function. The first approach involves increasing the number of spectral lines in a baseband measurement. The advantage of this approach is that no additional hardware or software is required. Often, FFT analyzers do not have the capability to alter the number of spectral lines used in the measurement. The second approach involves the reduction of the bandwidth of the measurement while holding the number of spectral line constant. If the lower frequency limit of the bandwidth is always zero, no additional hardware or software is required. Ideally, though, for an arbitrary bandwidth, hardware and/or software to perform a frequency-shifted, or digitally filtered, FFT will be required.

The frequency-shifted FFT process for computing the frequency-response function has additional characteristics pertinent to the reduction of errors. Primarily, more accurate information can be obtained on weak spectral components if the bandwidth is chosen to avoid strong spectral components. The out-of-band rejection of the frequency-shifted FFT is better than most analog filters that could be used in a measurement procedure to attempt to achieve the same results. Additionally, the precision of the resulting frequency-response function will be improved as a result of processor gain inherent in the frequency-shifted FFT calculation procedure. An example of the improvement of the frequency-response function using a frequency-shifted FFT can be seen in Fig. 16-29.

Weighting Functions

Weighting functions, or data windows, probably constitute the most common approach to the reduction of the leakage error in the frequency-response function. While weighting functions are sometimes desirable and necessary, weighting functions are often used when one of the other approaches to error reduction would give superior results. Averaging, selective excitation, and increasing the frequency resolution all act to reduce the leakage error by the elimination of the cause of the error. Weighting functions, on the other hand, attempt to compensate for the leakage error after the fact. This compensation for the leakage error causes an attendant distortion of the frequency and phase information of the frequency-response function, particularly in the case of closely spaced, lightly damped system poles. This distortion is a direct function of the width of the main lobe and the size of the side lobes of the spectrum of the weighting function. Examples of some common weighting functions are given in Fig. 16-30. Complete details concerning these and many other weighting functions are available from many sources [74,75,93,104].

Weighting functions may be applied to all classifications of signal averaging. The most common case is a weighting function equal to the inverse of the number of averages used in the estimate. When this weight is used, the individual power spectra can be weighted at the end of the signal averaging or as an ongoing procedure referred to as stable averaging. This type of weighting introduces no further distortion in the frequency-response function estimate; however, it does not act to compensate for the leakage error.

16-4 Modal Parameter Estimation

Modal parameter estimation is a special case of system identification in which a priori model of the system is known to be in the form of modal parameters. Therefore, regardless of the form of input–output data acquired, the form of the model used to represent the experimental data can be

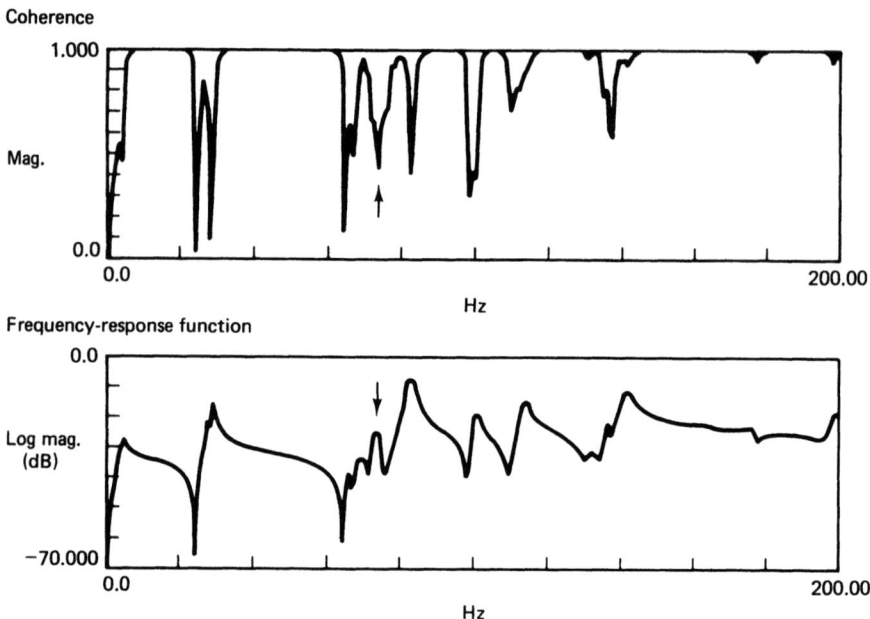
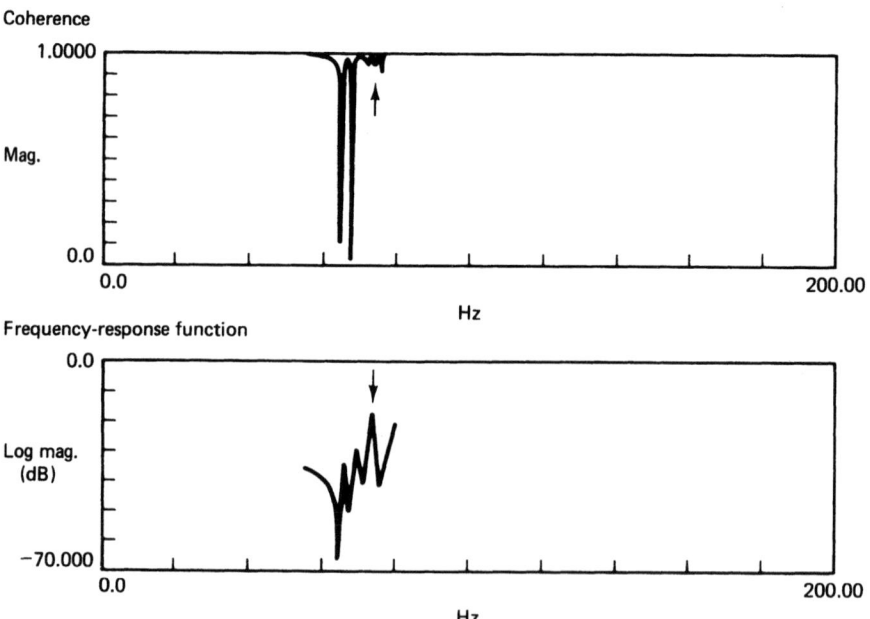

Figure 16-29 Increased frequency resolution.

EXPERIMENTAL MODAL ANALYSIS

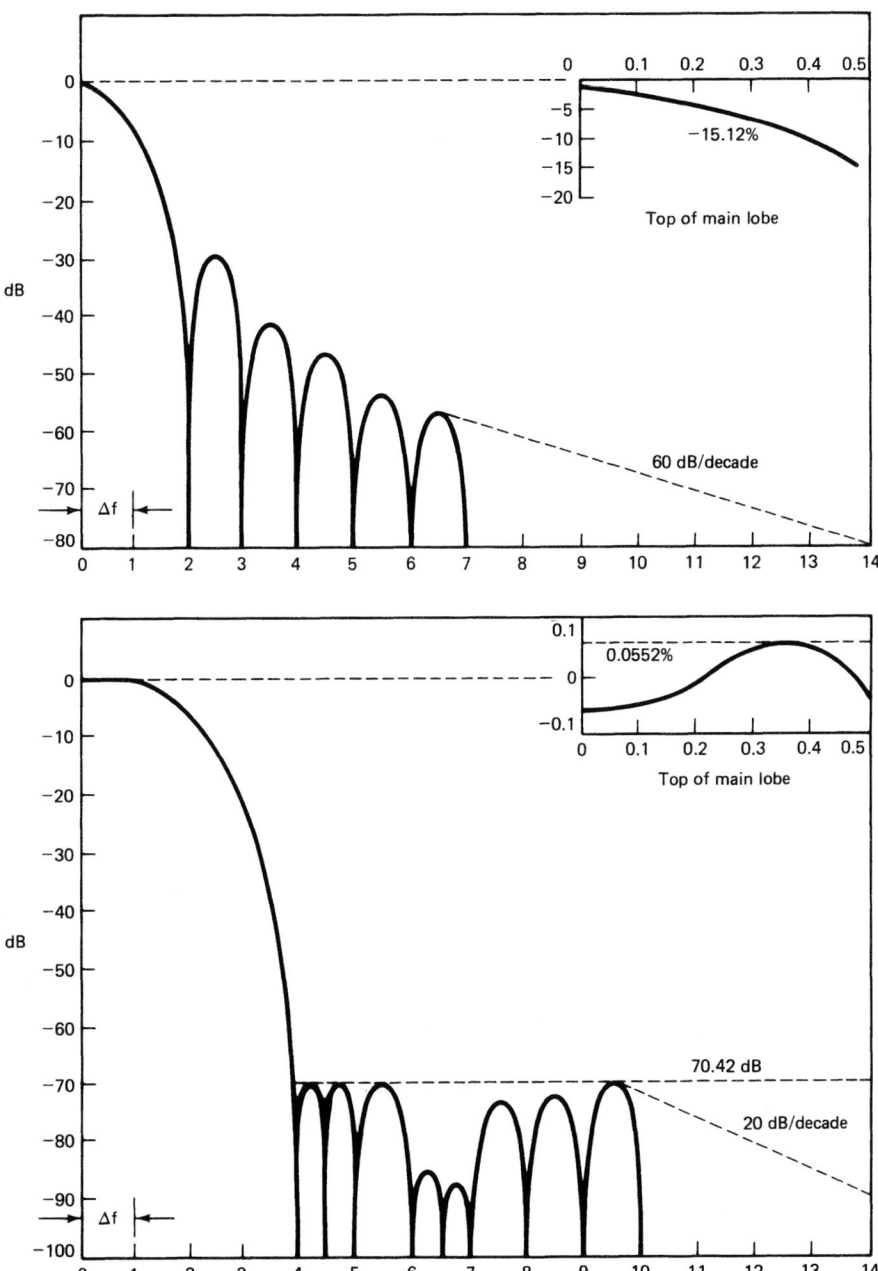

Figure 16-30 Typical weighting function: (a) Hann window; (b) flat-top window.

stated in a modal model in either time, frequency, or Laplace domain using temporal and spatial information. Research into modal parameter estimation represents the largest body of work impacting the area of structural testing over the last 10 to 15 years.

Research in modal parameter estimation over the last 10 years has yielded many algorithms that are being used privately or sold as a part of commercial software. While most of these individual algorithms are well understood, the comparison of one algorithm to another has become one of the current thrusts of this research effort. This work attempts to characterize different classes of modal parameter estimation techniques in terms of similarities and differences rather than performance. The modal parameter estimation process involves a greatly overdetermined problem; thus the estimates of modal parameters resulting from different algorithms (performance) will not be the same as a result of differences in the modal model and model domains differences in how the algorithms use the data, differences in the way the data are weighted or condensed, and differences in user expertise.

Nevertheless, the solution to many difficult problems in modal parameter estimation have been advanced over the last 10 years and much is now known about comparing and contrasting different modal parameter estimation algorithms. The problem of repeated and multiple (weakly coupled) roots is now well understood and solvable. The solution also explains much of the difficulty that users encountered during the 1970s and early 1980s when applying single-reference, multiple degree of freedom modal parameter estimation algorithms. Insufficient data were available to resolve repeated or pseudorepeated modal frequencies with those early algorithms. Detection and identification of repeated and multiple roots are critical in development of a complete modal model so that arbitrary input–output information can be synthesized. While great strides have been made in this area, further work is needed to adequately describe the multiple-root problem.

Much of the recent work has been focused on the development of a conceptual understanding of modal parameter estimation technology. This understanding involves the ability to conceptualize the data and the data domain, the evaluation of the order of the problem, the condensation of the data, and a common parameter estimation theory that can serve as the basis for developing any of the algorithms in use today. The following subsections review these concepts as applied to the current modal parameter estimation methodology.

Therefore, modal parameter estimation involves the use of simplistic or sophisticated approaches to the estimation of frequency, damping, and modal coefficients from the measured data. The measured data can be in relatively raw form in terms of force and response data in the time or frequency domain or in a processed form such as frequency-response or impulse-response functions. Most modal parameter estimation is based on the measured data being the frequency-response function or the equivalent impulse-response function, typically found by inverse Fourier transforming the frequency-response function. Regardless of the form of the measured data, the modal parameter estimation techniques can be further divided into several categories according to the manner in which the data are used in the various algorithms:

Single measurement, single degree of freedom (SDOF) approximations
Single measurement, multiple degree of freedom (MDOF) approximations
Multiple measurement, SDOF approximations
Multiple measurement, MDOF approximations
Multiple reference, SDOF approximations
Multiple reference, MDOF approximations

Since the single degree of freedom equations are simply special cases of the multiple degree of freedom equations, all theoretical discussions will be made in terms of the multiple degree of freedom case.

EXPERIMENTAL MODAL ANALYSIS

Concept: Mathematical Models

The most general model that can be used is one in which the elements of the mass, damping, and stiffness matrices are estimated based on measured forces and responses. Effectively, this means that the model that is used is the matrix differential equation like that shown in Eq. (16-74) for time-domain data or in Eq. (16-75) for frequency-domain data.

$$[M]\{\ddot{x}\} + [C]\{\dot{x}\} + [K]\{x\} = \{f\} \quad (16\text{-}74)$$

$$-[M]\{X\}\omega^2 + j[C]\{X\}\omega + [K]\{X\} = \{F\} \quad (16\text{-}75)$$

If Eq. (16-74) or (16-75) is used as the model for parameter estimation, the elements of the unknown matrices must first be estimated from the known force and response data measured in the domain of time or frequency. Once the matrices have been estimated, the modal parameters can be found by the solution of the classic eigenvalue–eigenvector problem. Note that as a result of truncation of the data in terms of frequency content, limited numbers of degrees of freedom, and measurement errors, the matrices found by Eq. (16-74) or (16-75) will not, in general, be directly comparable to matrices determined from a finite-element approach. Instead, the matrices that are estimated will simply yield valid input–output relationships and valid modal parameters. This is because there is an infinite number of sets of mass, damping, and stiffness matrices that will yield the same modal parameters over a reduced frequency range limited to the dynamic range of the measurements. For this reason, Eq. (16-74) or (16-75) is often premultipled by the inverse of the mass matrix so that the elements of the two matrices [A] and [B] are estimated.

$$[I]\{\ddot{x}\} + [A]\{\dot{x}\} + [B]\{x\} = \{f\} \quad (16\text{-}76)$$

$$[I]\{X\}\omega^2 + j[A]\{X\}\omega + [B]\{X\} = \{F\} \quad (16\text{-}77)$$

Many new methods are being developed around the basic models described in Eqs. (16-74) through (16-77). More commonly, though, the existing modal parameter estimation methods used in commercial modal analysis systems use a model based on measured frequency-response or impulse-response functions. While the exact model used as the basis for modal parameter estimation varies, almost all models used in conjunction with frequency-response function data can be described by a general model in the Laplace domain or an equivalent general model in the time domain. Note that the general model in the Laplace domain (transfer function) includes the frequency-domain model (frequency-response function) as a subset. In studies carried out by Klosterman [34], Van Loon [35], and Richardson [39], a derivation is given for the general formula of the transfer function of a multiple degree of freedom system with viscous or hysteretic damping. For general viscous damping, the transfer function for a multiple degree of freedom mechanical system can be written as shown earlier.

Single reference:

$$H_{pq}(\omega) = \sum_{r=1}^{N} \frac{A_{pqr}}{j\omega - \lambda_r} + \frac{A^*_{pqr}}{j\omega - \lambda^*_r} \quad (16\text{-}78)$$

Multiple reference:

$$[H(\omega)] = \sum_{r=1}^{N} \frac{[A_r]}{j\omega - \lambda_r} + \frac{[A^*_r]}{j\omega - \lambda^*_r} \quad (16\text{-}79)$$

$$[H(\omega)] = \sum_{r=1}^{2N} \frac{[A_r]}{j\omega - \lambda_r} \quad (16\text{-}80)$$

$$[H(\omega)] = [\psi] \lceil \Lambda \rfloor [L] \quad (16\text{-}81)$$

$$\lceil \Lambda \rfloor = \text{diagonal matrix} \left(\frac{1}{j\omega - \lambda_r}\right)$$

Other forms of this model are commonly found where certain assumptions or mathematical relationships are used. For example, an equivalent model can be found when the common denominator of Eq. (16-79) is formed yielding a polynomial numerator and polynomial denominator of maximum order 2N. The denominator polynomial is then a function of the system poles. Often an assumption is made concerning the modal vectors being normal (real) rather than complex. This reduces the number of unknowns that must be estimated by N.

Several common modal parameter estimation algorithms are based on a time-domain model. In these cases, the general model represents the Fourier transform equivalent of the frequency-response function, the impulse-response function. Therefore, a mathematical expression for the impulse-response function can be obtained by a Fourier transform of Eq. (16-78):

Single reference:

$$h_{pq}(t) = \sum_{r=1}^{N} A_{pqr} e^{\lambda_r t} + A^*_{pqr} e^{\lambda^*_r t} \tag{16-82}$$

Multiple reference:

$$[h(t)] = \sum_{r=1}^{N} [A_r] e^{\lambda_r t} + [A^*_r] e^{\lambda^*_r t} \tag{16-83}$$

$$[h(t)] = \sum_{r=1}^{2N} [A_r] e^{\lambda_r t} \tag{16-84}$$

$$[h(t)] = [\psi] [e^{\lambda_r t}] [L] \tag{16-85}$$

$$[e^{\lambda_r t}] = \text{diagonal matrix } (e^{\lambda_r t})$$

Concept: Residuals

Continuous systems have an infinite number of degrees of freedom but, in general, only a finite number of modes can be used to describe the dynamic behavior of a system. The theoretical number of degrees of freedom can be reduced by using a finite frequency range (f_a, f_b). Therefore, for example, the frequency response can be broken up into three partial sums, each covering the modal contribution corresponding to modes located in the frequency ranges $(0, f_a)$, (f_a, f_b) and (f_b, ∞) as shown in Fig. 16-31.

In the frequency range of interest, the modal parameters can be estimated to be consistent with Eq. (16-78). In the lower and higher frequency ranges, residual terms can be included to handle modes in these ranges. In this case, Eq. (16-86) can be written as

$$H_{pq}(\omega) = R_{F_{pq}} + \sum_{r=1}^{N} \frac{A_{pqr}}{\omega - \lambda_r} + \frac{A^*_{pqr}}{\omega - \lambda^*_r} + R_{I_{pq}}(\omega) \tag{16-86}$$

where ω = frequency variable
p = measured degree of freedom (response)
q = measured degree of freedom (input)
r = modal vector number
A_{pqr} = residue
λ_r = system pole
N = number of modal frequencies
$R_{F_{pq}}$ = residual flexibility
$R_{I_{pq}}(\omega)$ = residual inertia

EXPERIMENTAL MODAL ANALYSIS

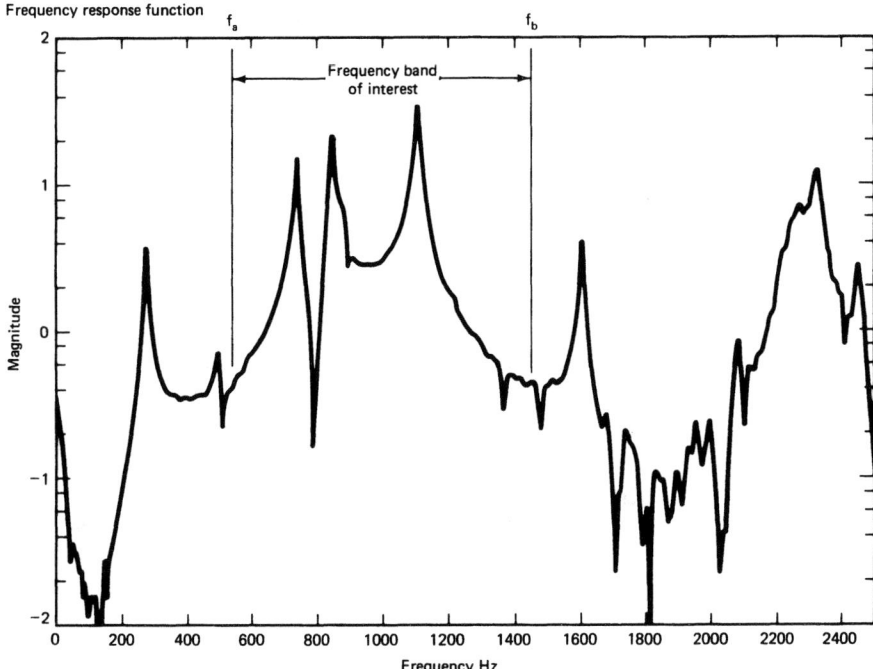

Figure 16-31 Frequency response function: partial ranges.

In many cases the lower residual term is called the inertia restraint, or residual inertia, and the upper residual term is called the residual flexibility. Note that in this common formulation of residuals, both terms are real-valued quantities. The lower residual is a term reflecting the inertia or mass of the lower modes and is an inverse function of the frequency squared. The upper residual is a term reflecting the flexibility of the upper modes and is constant with frequency.

In this case, the form of the residual is based on a physical concept of how the system poles below and above the frequency range of interest will affect the data in the range of interest. As the system poles below and above the range of interest are located in the proximity of the boundaries, these effects are not the simple real-valued quantities noted in Eq. (16-86). In these cases, residual modes may be included in the model to partially account for these effects. When this is done, the modal parameters that are associated with these residual poles have no physical significance but may be required to compensate for strong dynamic influences from outside the frequency range of interest. Using the same argument, the lower and upper residuals can take on any mathematical form that is convenient as long as the lack of physical significance is understood. Mathematically, power functions of frequency (zero, first, and second order) are commonly used within such a limitation. In general, the use of residuals is confined to frequency-response function models. This is primarily because of the difficulty of formulating a reasonable mathematical model and solution procedure in the time domain for the general case that includes residuals.

Concept: Model Order (Single Reference, Single Measurement)

The order of the model refers to the number of unknowns that must be estimated in the model. In the modal parameter estimation case, this refers to the frequency, damping, and complex modal coefficient for each mode of vibration at every measurement degree of freedom plus any residual terms that must be estimated. Therefore, the order of the model is directly dependent on the number of modal frequencies, N, that are to be estimated. For example, for a single frequency-response function with N modes of vibration, assuming that no residuals were required, $4N$ unknowns must be estimated. For cases involving real measured data, the order of the model is an extremely important decision, since the estimates of modal parameters are affected by the choice of the order of this model. The difficulty in such cases arises from the inability to be certain that the correct order of the model has been chosen. Obviously if the number of modes of vibration is more or less than N, modes of vibration will be found that do not exist physically or modes of vibration will be missed that actually do occur physically. In addition, though, the values of frequency, damping, and complex modal coefficient for the physically realizable modes of vibration will be affected by an inappropriate choice of the number of modes of vibration.

The number of modes of vibration is normally chosen between one and an upper limit, depending on the memory limitations of the computational hardware. The true number of system poles is a function of the frequency range of the measurements used to estimate the model parameters. By observing the number of peaks in the frequency-response function, the minimum number of system poles can be estimated. This estimate, though, is normally low based upon poles occurring at nearly the same frequency (pseudorepeated roots), limits on dynamic range, and poorly excited modes. For these reasons, the estimate of the correct order of the model is often in error. While it is not necessarily obvious, when the order of the model is other than optimum, the estimate of the modal parameters will be in error.

Many of the parameter estimation techniques that are used will assume that only one mode exists in the range of interest and all the other modes appear as residual terms. For this case, Eq. (16-86) can be rewritten as

$$H_{pq} = R_{F_{pq}} + \frac{A_{pqr}}{s - \lambda_r} + \frac{A^*_{pqr}}{s - \lambda^*_r} + R_{I_{pq}}(\omega) \tag{16-87}$$

Concept: Model Order (Multiple Reference, Multiple Measurements)

The estimation of an appropriate model order is the most important problem encountered in modal parameter estimation. This problem is complicated by the formulation of the parameter estimation model in the time or frequency domain, due to a single or multiple reference formulation of the modal parameter estimation model, and by the effects of random and bias errors on the modal parameter estimation model. The basis of the formulation of the correct model order can be seen by expanding the second-order equations of motion to a higher-order model ($2N$). This is a necessary process required to handle the case of spatial information truncated to a size smaller than the number of eigenvalues in the measured data. This concept can be developed in several ways. One method is to start with the equation of motions and, for example: Laplace transform these equation to determine the characteristic equation in matrix form.

$$s^2 [M] \{X(s)\} + s [C] \{X(s)\} + [K] \{X(s)\} = \{0\} \tag{16-88}$$

This characteristic equation is a matrix polynomial of second order, which can be partitioned as follows:

EXPERIMENTAL MODAL ANALYSIS

$$\begin{bmatrix} [M_{11}] & \cdots & [M_{1n}] \\ [M_{21}] & & [M_{2n}] \\ \cdot & & \cdot \\ \cdot & & \cdot \\ [M_{n1}] & \cdots & [M_{nn}] \end{bmatrix} s^2 +$$

$$\begin{bmatrix} [C_{11}] & \cdots & [C_{1n}] \\ [C_{21}] & & [C_{2n}] \\ \cdot & & \cdot \\ \cdot & & \cdot \\ [C_{n1}] & \cdots & [C_{nn}] \end{bmatrix} s +$$

$$\begin{bmatrix} [K_{11}] & \cdots & [K_{1n}] \\ [K_{21}] & & [K_{2n}] \\ \cdot & & \cdot \\ \cdot & & \cdot \\ [K_{n1}] & \cdots & [K_{nn}] \end{bmatrix} = 0 \qquad (16\text{-}89)$$

This partitioned equation can be expanded to a higher-order matrix polynomial and put in a generic form as follows:

$$[A]_{2n} s^{2n} + [A]_{2n-1} s^{2n-1} + [A]_{2n-2} s^{2n-2} + \cdots + [A]_0 = 0 \qquad (16\text{-}90)$$

Note that the size of the matrices $[A]$ is the same as the size of the partitioned submatrices in the preceding equation. Also note that each $[A]$ matrix involves a matrix product and summation of several $[M_{kl}]$, $[C_{kl}]$, and $[K_{kl}]$ submatrices.

This form of the characteristic equation can also be obtained directly by Laplace transforming a higher-order differential equation but with a smaller matrix size equal to the partition size. This higher-order differential equation will have the same eigenvalues as the original second-order equation.

The limit of this process would be to reduce the size of the matrices to a single order or to a scalar equation.

$$\alpha_N s^N + \alpha_{N-1} s^{N-1} + \alpha_{N-2} s^{N-2} + \cdots + \alpha_0 = 0 \qquad (16\text{-}91)$$

The eigenvalues of this scalar characteristic equation would be the same as the original second-order matrix polynomial equation.

Therefore, the concept of model order involves the order of the matrix coefficients involved in the modal parameter estimation model and the order of the polynomial, or finite-difference, terms in the modal parameter estimation model. There are a significant number of procedures for aiding these decisions and selecting the appropriate estimation model and much research effort has been expended over the last 10 years on this effort [80,82,91,105–107]. Procedures for estimating the appropriate matrix size and model order represent another difference between various estimation procedures.

Concept: Solution Procedure

With respect to Eq. (16-78) through (16-85), which represent most of the general models for almost all modal parameter estimation, a number of important aspects relating to the solution for modal parameters can be noted. Since most modal parameter estimation deals with frequency-response or impulse-response function models, the discussion will be specific to these cases. First of all, Eqs. (16-78) and (16-82) represent a nonlinear model in terms of the unknown modal parameters. This can be noted from the unknowns in the numerator and denominator of Eq. (16-78) and the unknowns as the argument of the transcental functions of Eq. (16-82). The nonlinear aspect of the model must be addressed in one of two ways. The first approach is to use an iterative solution procedure to solve the nonlinear estimation problem. This approach allows all modal parameters to vary according to a constraint relationship until an error criterion reaches an acceptably low value. The second approach involves separating the nonlinear estimation problem into two linear estimation problems. For the case of structural dynamics, the common technique is to estimate $2N$ frequencies and damping values in a first stage and then to estimate the $2N$ modal coefficients plus any residuals in a second stage.

The iterative approach to the solution of the nonlinear estimation approach was one of the first techniques used. The problems associated with iterative approaches are well known. First of all, a set of starting values must be chosen to initiate the sequence. Naturally, the number and the value of these starting values affect the final result. Poor initial estimates can lead to problems of convergence. For these reasons, close operator supervision normally is required for successful use of this technique.

The alternative to the iterative approach is to reformulate the nonlinear problem into a number of linear stages so that each stage is stable. Almost all algorithms in use at this time involve this concept to some degree. For example, with respect to Eq. (16-78), if all system poles can be found, the estimation of the complex residues can proceed in a linear fashion. This procedure of separating the nonlinear problem into a two-stage linear problem is a common technique for most estimation methods today.

The actual data used in the estimate of the modal parameters also have importance to the results. Based on the choice of the order of the model N, there are $4N$ modal parameters to be estimated. If residuals, in one form or another, are also included, the number of modal parameters to be estimated will be slightly higher. The common approach to solving for these unknown modal parameters is to find an equation involving known information for every unknown that will be found. In this case, the frequency-response function or impulse-response function that has been measured provides the known information, and Eq. (16-78) or (16-82) can be repeated for different frequencies or time values to obtain the sufficient number of equations. These equations, for the linear case, can then be solve simultaneously for the unknown modal parameters. As an illustration of this relationship, consider a common modal parameter situation in which the number of modes in the frequency range of interest is between 1 and 30. Assuming the highest order model means that slightly more than 120 modal parameters must be estimated. From a single frequency-response measurement, 1024 known values of the function will be available (512 complex values at successive values of frequency). Obviously, many more equations, based on the known values of frequency response, can be formed than are needed to find the unknown modal parameters. The most obvious solution is simply to choose enough equations to solve for the modal parameters. The problem arises in determining what part of the known information is to be involved in the solution. Obviously, as different portions of the known data (e.g., data near a resonance compared to an antiresonance) are used in the solution, the estimates of modal parameters will vary. Since the quality of the data is marginal, the variance can be quite large. When the modal parameters that are estimated appear to be nonphysical, this variance is often the reason.

To address this problem, all or a large portion of the data can be used if an averaging type of solution procedure is used. One well-known procedure is the formulation of the problem in a way that serves to minimize the squared error between the data and the estimated model. This least-squares error approach is the most commonly used technique in the area of modal parameter estimation at this time. Note that as long as there are many more known pieces of information

EXPERIMENTAL MODAL ANALYSIS

than unknowns that must be estimated, many more equations can be formed than are needed to solve for the unknowns. The least-squares error approach to the solution allows for all the redundant data to be used to estimate the modal parameters in a computationally efficient manner. The least-squares error approach normally can be derived directly from the linear equations using a normal equations or pseudoinverse approach. In general, this procedure does not increase the memory or computational requirements of the computational hardware significantly. It is important to stress that any solution procedure that can be used is only estimating a "best" solution based on the choice of the model, the order of the model, and the known, measured data used in the model.

Concept: Global Modal Parameters

The concept of global modal parameters simply means that there is only one answer for each modal parameter, and the modal parameter estimation solution procedure enforces this constraint. Every frequency-response or impulse-reponse function measurement theoretically contains the information that is represented by the characteristic equation, the modal frequencies and damping. If individual measurements are treated in the solution procedure independent of one another, there is nothing to guarantee that a single set of modal frequencies and damping will be generated. Likewise, if more than one reference is measured in the data set, redundant estimates of the modal vectors can be estimated unless the solution procedure uses all references in the estimation process simultaneously. Most of the current modal parameter estimation algorithms estimate the modal frequencies and damping in a global sense, but very few estimate the modal vectors in a global sense.

Concept: Modal Participation Factors

Modal participation factors are a result of multiple-reference, modal parameter estimation algorithms, and they relate how well each modal vector is excited from each of the reference locations. The combination of the modal participation factors (vector) and the modal coefficients (vector) for a given mode give the residue information (matrix) for that mode. In general, these two vectors represent portions of the right and left eigenvectors associated with the structural system for that specific mode of vibration. Normally, the system can be assumed to be reciprocal, and the right and left eigenvectors, and therefore the modal participation vector and the modal vector, will be proportional to one another. Under this assumption, the modal participation vector can be used in a weighted least-squares error solution procedure to estimate the modal vectors in the presence of multiple references. Theoretically, for reciprocal systems, these modal participation factors should be in proportion to the modal coefficients of the reference degrees of freedom for each modal vector. The use of modal participation factors in a solution procedure allows for the possible constraint concerned with Maxwell's reciprocity between the reference degrees of freedom to be evaluated. Most multiple-reference, modal parameter estimation methods estimate modal participation factors as part of the first-stage estimation of global modal frequencies and damping.

Concept: Data Domain

Modal parameters can be estimated from a variety of different measurements that exist as discrete data in different data domains (time, frequency, $Z,$). These measurements, which can include free decays, forced responses, frequency responses, and unit impulse responses, can be processed one at a time or in partial or complete sets simultaneously. The measurements can be generated with no measured inputs, a single measured input, or multiple measured inputs. The data can be measured individually or simultaneously. In other words, there is a tremendous variation in the types of measurement and in the types of constraint that can be placed on the testing procedures used to acquire the data.

In the area of experimental modal analysis is it very important to develop a conceptual understanding of the mathematical association among time, frequency, and Laplace domain relationships. The measurements and data processing are performed or represented in the time, frequency, and/or Laplace domain.

The Fourier and Laplace transforms are the mechanisms for moving from one domain to another. In general, the Fourier transform is more important in the measurement process, while the Laplace transform is more important in the modal parameter estimation, or data reduction process, from a theoretical point of view. In the actual measurement and data reduction processes, these two transforms are often implemented in a numerical sense. In these cases the fast Fourier transform and Z transform are used in place of the Fourier transform and the Laplace transform, respectively. In reality the FFT and the Z transform are simply discrete versions of the continuous transforms.

Another important concept in experimental modal analysis, and particularly modal parameter estimation, involves the relationships between the temporal (time and/or frequency) information and the spatial information. Data can then be reduced as a function of the superposition of the underlying temporal characteristics (modal frequencies) or as a function of the superposition of the underlying spatial characteristic (modal vectors). The time domain is the starting point for any analysis or experiment in the modal analysis area. Analytically, the equation of motions are formulated in the time domain, and experimentally, the measurements are made as function of time. This information can be transformed into the frequency or the Laplace domain using the Fourier or the Laplace transforms or their discrete counterparts.

The matrix equation of motion for a general multi-degree-of-freedom system can be written as

$$[M]\{\ddot{x}(t)\} + [C]\{\dot{x}(t)\} + [K]\{x(t)\} = \{f(t)\} \tag{16-92}$$

In this equation, the following assumptions, already discussed, have been made concerning the system:

Linear
Time invariant ($[M]$ $[C]$, and $[K]$ are constant)
Observable
Viscous damping

Fourier transforming the equation of motion:

$$-\omega^2 [M]\{X(\omega)\} + j\omega [C]\{X(\omega)\} + [K]\{X(\omega)\} = \{F(\omega)\} \tag{16-93}$$

Laplace transforming the equation of motion:

$$s^2 [M]\{X(s)\} + s [C]\{X(s)\} + [K]\{X(s)\} = \{F(s)\} \tag{16-94}$$

While Eqs. (16-92) through (16-94) can serve as the basis for a system model for a structural testing situation, normally data are acquired in the form of some intermediate measurement to take advantage of averaging (condensation). Historically, this practice was necessary because data storage space was limited, and it is still attractive due to data evaluation methodologies. The solutions for the foregoing representations of the system are the responses of the system due to an arbitrary forcing function. In terms of most modal parameter estimation algorithms, solutions equivalent to the solution of the homogeneous matrix equation of motion are desirable. This can be conceptualized as the response of the system to a unit impulse forcing function. Using this concept for the forcing function, Eqs. (16-95) through (16-97) can be formulated:

Time domain:

$$[h(t)] = \sum_{r=1}^{N} [A_r] e^{\lambda_r t} + [A_r^*] e^{\lambda_r^* t} \tag{16-95}$$

Frequency domain:

$$[H(\omega)] = \sum_{r=1}^{N} \frac{[A_r]}{j\omega - \lambda_r} + \frac{[A_r^*]}{j\omega - \lambda_r^*} \qquad (16\text{-}96)$$

Laplace domain:

$$[H(s)] = \sum_{r=1}^{N} \frac{[A_r]}{s - \lambda_r} + \frac{[A_r^*]}{s - \lambda_r^*} \qquad (16\text{-}97)$$

Equations (16-95) through (16-97) also can be expressed in terms of their eigenvalue–eigenvector solution as follows:

Time domain:

$$[h(t)] = [\psi] \lceil e^{\lambda_r t} \rfloor [L] \qquad (16\text{-}98)$$

Frequency domain:

$$[H(\omega)] = [\psi] \lceil \Lambda \rfloor [L] \qquad (16\text{-}99)$$

Laplace domain:

$$[H(s)] = [\psi] \lceil \Lambda \rfloor [L] \qquad (16\text{-}100)$$

where $[\psi]$ is a matrix of left eigenvectors that correspond to the modal matrix, $[L]$ is a matrix of right eigenvectors and is often referred to as the modal participation matrix, and $[\Lambda]$ is the diagonal eigenvalue matrix and has the form $e^{\lambda_r t}$ along the diagonal for the time domain, $1/(j\omega - \lambda_r)$ for the frequency domain, and $1/(s - \lambda_r)$ for the Laplace domain.

The equations above are in matrix form. Individual elements (the impulse-response functions, the frequency-response functions, or the transfer functions) can be represented as follows:

Impulse-response function:

$$h_{pq}(t) = \sum_{r=1}^{N} A_{pqr} e^{\lambda_r t} + A_{pqr}^* e^{\lambda_r^* t} \qquad (16\text{-}101)$$

Frequency-response function:

$$H_{pq}(\omega) = \sum_{r=1}^{N} \frac{A_{pqr}}{j\omega - \lambda_r} + \frac{A_{pqr}^*}{j\omega - \lambda_r^*} \qquad (16\text{-}102)$$

Transfer function:

$$H_{pq}(s) = \sum_{r=1}^{N} \frac{A_{pqr}}{s - \lambda_r} + \frac{A_{pqr}^*}{s - \lambda_r^*} \qquad (16\text{-}103)$$

where A_{pqr} is the residue between the response ponit p and the input point q, r is the index on the mode number, and λ_r is the rth eigenvalue.

It should be noted that the unit impulse-response function, the frequency-response function, and the transfer function are merely transform pairs and contain exactly the same information.

Several important observations can be made from the mathematical relations just developed.

1. The equations of motion are linear, second-order differential equation, and the solution can be constructed from the superposition of eigensolutions.
2. The eigensolutions given in Eqs. (16-98) through (16-100) consist of the product of three characteristic functions. The first set of characteristic functions is the eigenvector of the system that corresponds to columns of the modal matrix. This is spatial information, which depends

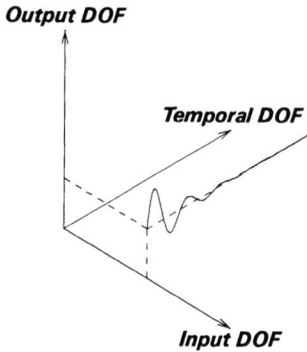

Figure 16-32 Conceptualization of modal characteristic space.

on the geometry of the test object or, in other words, the location and direction of the measured points. It includes both translational and rotational coordinates. The second set of characteristics consist of the temporal functions defined by the eigenvalues of the system. "Temporal" refers to a variable that is a function of time, but in this case the definition is extended to include frequency and the Laplace variables, since they are transforms of time. These characteristics are damped exponential functions in the time domain and single degree of freedom responses in the frequency and Laplace domains. The third set of characteristics corresponds to a matrix of eigenvectors. If the mass, stiffness, and damping matrices are symmetric, this set of characteristics is proportional to the transpose of those defined in the modal matrix and consists of spatial information. If the matrices are not symmetric, then the two spatial characteristics are simply the left and right eigenvectors.

3. Any arbitrary solution to the equation of motion consists of a superposition of these sets of characteristics.

From a conceptual viewpoint, the solutions of the equations of motion can be visualized as occupying a volume with the coordinate axis defined in terms of the three sets of characteristics. Two axes of the conceptual volume correspond to spatial information and the third axis to temporal information. The spatial coordinates are in terms of the input and response points of the system. The temporal axis is either time or frequency, depending on the domain of the measurements. These three axes define a volume that is referred to as the "characteristic space" as represented in Fig. 16-32. This space or volume represents all possible solutions to the equations of motion as expressed by Eqs. (16-98) through (16-100). Information parallel to one of the axes consists of a solution composed of the superposition of the characteristics defined by that axis. The other two characteristics determine the scaling of each term in the superposition.

Concept: Data Subspace

Any structural testing procedure measures a subspace of the total possible data subspace available. Modal parameter estimation algorithms then may use all this subspace or may choose to further limit operations to a subsequent subspace. This is not a problem, since it is possible to estimate

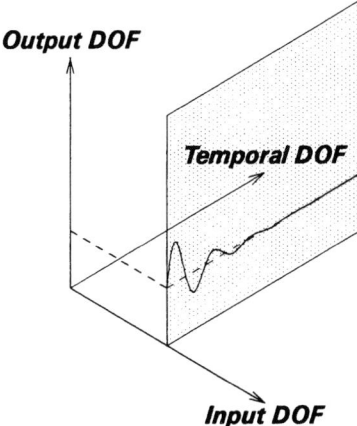

Figure 16-33 Characteristic space representation: single measurement.

the total space by experimentally measuring a subspace that samples all three characteristics. Experimentally, any point in this space can be measured. The particular subspace measured and the weighting of the data within the subspace are the main differences between the various modal parameter estimation procedures that have been developed historically.

In general, the amount of measured subspace information greatly exceeds that which is necessary to solve for the unknown modal characteristics. Another major difference between the various modal parameter estimation procedures is the type of condensation algorithms used to reduce to the data to match the number of unknowns: for example, least squares (LS), singular value decomposition (SVD), and so on. This will be discussed in a later section. As is the case with any overspecific solution procedure, there is no unique answer. The answer that is obtained depends on the data selected, the weighting of the data, and the unique set of algorithms used in the solution process. As a result, the answer is the "best" answer depending on the objective functions associated with the algorithms being used. Historically, this point has created some confusion, since many users expect different methods to give the same answer.

As mentioned earlier, various methods differ in their use of subspace. As might be expected, the selection of the subspace has a significant influence on the results. To estimate all the modal parameters, the subspace must encompass a region that includes contributions of all three characteristics. A classic example is the necessity to use multiple reference data in estimating "repeated roots."

In the past, many modal techniques used information (subspace) where only one or two characteristics were varied. For example, in the early 1970s, the modal parameter estimation methods fit only one unit impulse function or one frequency response at a time. With reference to the characteristic space of the problem, this is represented by Fig. 16-33. In this case, only the temporal characteristic is used and as might be expected, only temporal characteristics (eigenvalues) can be estimated rom the single measurement. In practice, multiple measurements were taken, and the spatial information was extracted from the multiple measurements by successive estimation of the residues for each model from each measurement.

Later, data acquisition equipment became more sophisticated and modal parameter estimation algorithms evolved, techniques that fit the data in a plane of the characteristic space began to appear. For example, this corresponds to the data taken at a number of response points but from

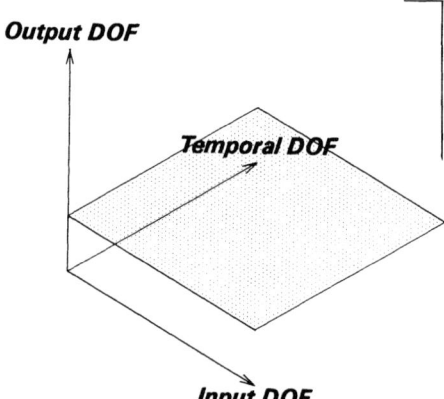

Figure 16-34 Characteristic space representation: column(s) of measurements.

a single excitation point or reference. This representation of several column(s) of measurements is shown in Fig. 16-34. For this case, representing a single input (reference), it is not possible to compute repeated roots and it is difficult to separate closely coupled modes.

Recent modal methods utilize data taken at a large number of response points due to excitation at a small number of input points. This representation of several row(s) of the potential measurement matrix is shown in Fig. 16-35.

This is the best possible measurement condition. It defines a subspace that includes the contributions of all the characteristics. This allows the best possibility of estimating all the important modal parameters. In general, this data subspace should allow a low-order model to be used. Whether a low-order model can be used depends on whether the spatial information uniquely defines the eigenvectors.

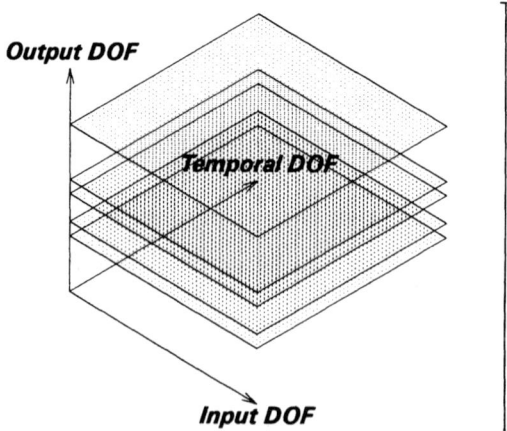

Figure 16-35 Characteristic space representation: row(s) of measurements.

EXPERIMENTAL MODAL ANALYSIS

It should be obvious that the data that define the subspace must be consistent, for the algorithms to estimate accurate modal parameters. This requirement triggered the need to measure all the data simultaneously and has led to recent advancements in data acquisition, digital signal processing, and instrumentation designed to facilitate this measurement problem.

Concept: Data Condensation

Several important concepts should be delineated in the area of data condensation algorithms. Condensational algorithms are used to reduce the measured data to match the number of unknowns in the modal parameter estimation algorithms. There are a large number of condensation algorithms available. Based on the modal parameter estimation algorithms in use today, the three of algorithm most often used are:

Least Squares: Least squares (LS), or weighted least squares (WLS), is used to minimize the squared error between the measured data and the estimation model. Historically, this is one of the most popular procedures for performing a pseudoinverse solution to an overspecified system. The main advantage of this method is computational speed and ease of implementation, while the major disadvantage is lack of numerical precision.

Transformations: A large number of transformations can be used to reduce the data. In the transformation methods, the measured data are reduced by approximation by way of the superposition of a set of significant vectors. The number of significant vectors is equal to the amount of independent measured data. This set of vectors is used to approximate the measured data and serve as input to the parameter estimation procedures.

Singular Value Decomposition: Singular value decomposition (SVD) is an example of one of the more popular transformation methods. The major advantage of these methods is numerical precision; the disadvantage lies in the computational speed and memory requirements associated with SVD.

Coherent Averaging: In this popular method for reducing the data, the data are weighted by performing a dot product between the data and a weighting vector (spatial filter). Information in the data that is not coherent with the weighting vectors is averaged out of the data. The method, often referred to as a spatial filtering procedure, has both speed and precision. To achieve precision, however, a good set of weighting vectors is necessary. In general, the optimum weighting vectors are connected with the solution, which is unknown. It should be noted that least squares is an example of a noncoherent averaging process.

The least squares and the transformation procedures tend to weight the modes of vibration that are well excited. This can be a problem when trying to extract modes that are not well excited. The solution is to use a weighting function for condensation which tends to enhance the mode of interest. This can be accomplished in a number of ways:

1. In the time domain, a spatial filter or a coherent averaging process can be used to filter the response to enhance a particular mode or set of modes. For example, by averaging the data from two symmetric exciter locations, the symmetric modes of vibration can be enhanced. A second example is using only the data in a local area of the system to enhance local modes. The third method consists of using estimates of the modes shapes as weighting functions to enhance particular modes.
2. In the frequency domain, the data can be enhanced in the same manner as the time domain. Additionally enhancement can be achieved by weighting the data in a frequency band near the natural frequency of the mode of interest.

Obviously, the type of condensation has a significant influence on the results of the parameter estimation process.

Concept: Model Order Determination

Much of the work concerned with modal parameter estimation since 1975 has involved methodology for determining the correct model order for the modal parameter model. Since the number of unknowns in the modal parameter model is a function of N, determining the model order essentially reduces to estimating N. As has always been the case, the minimum model order can be estimated easily by counting the number of peaks in the frequency-response function in the frequency band of analysis. This is a minimum estimate of N, since the frequency-response function measurement may be at a node of one or more modes of the system, repeated roots may exist, and/or the frequency resolution of the measurement may be too coarse to observe modes that are closely spaced in frequency. Several measurements can be observed, and a tabulation of peaks existing in any or all measurements can be used as a more accurate minimum estimate of N. A more automated procedure for including the peaks that are present in several frequency-response functions is to observe the summation of frequency response function power. This function represents the autopower or automoment of the frequency-response functions summed over a number of response measurements and is normally formulated as follows:

$$H_{\text{power}}(\omega) = \sum_{p=1}^{N_o} H_{pq}(\omega) H_{pq}^*(\omega) \tag{16-104}$$

These simple techniques are extremely useful but do not provide an accurate estimate of model order when repeated roots exist or when modes are closely spaced in frequency. For these reasons, an appropriate estimate of the order of the model is of prime concern and is the single most important problem in modal parameter estimation.

A number of techniques have been developed as guides or aids to the user attempting to determine a reasonable estimate of the model order for set of representative data. Much of the skill involved in modal parameter estimation entails the use of these tools. Most of the techniques that have been developed allow the user to establish a maximum model order to be evaluated (in many cases, this is set by the memory limits of the computer algorithm). Data are acquired based on an assumption that the model order is equal to this maximum. In a sequential fashion, the data are evaluated to determine whether a model order less than the maximum will describe the data sufficiently. This is the point at which the user's judgment and the use of various evaluation aids become important. One of the simplest techniques is to synthesize an impulse-response function or a frequency-response function and compare it to the measured function to see whether modes have obviously been missed. This curve-fitting procedure is also used as a measure of the overall success of the modal parameter estimation procedure. The difference between the two functions can be quantified and normalized to give an indicator of the degree off it. Obviously, a comparison can be poor for many reasons, an incorrect model order simply being one of the possibilities.

Error Chart

Another method that has been used to indicate more directly the correct model order is the error chart. Essentially, the error chart is a plot of the error in the model as a function of increasing model order. The error in the model is a normalized quantity that represents the ability of the model to predict data that were not involved in the estimate of the model parameters. For example, when using measured data in the form of an impulse-response function, only a small percentage of the total number of data values are involved in the estimate of modal parameters. If the model is estimated based on 10 modes, only 4 × 10 data points are required, at a minimum, to estimate the modal parameters. The error in the model can then be estimated by the ability of the model to predict the next several data points in the impulse-response function compared to the measured

EXPERIMENTAL MODAL ANALYSIS

data points. For this case of 10 modes and 40 data points, the error in the model would be calculated from the predicted and measured data points 41 through 50. When the model order is insufficient, this model error will be large; when the model error reaches the "correct" value, however, further increase in the model order will not result in a further decrease in the error. Figure 16-36 is an example of an error chart.

Stability Diagram

A further enhancement of the error chart is the stability diagram. The stability diagram is developed in the same fashion as the error chart and involves tracking the estimates of frequency, damping, and possibly modal participation factors as a function of model order. As the model order is increased, more and more modal frequencies are estimated but, hopefully, the estimates of the physical modal parameters will stabilize as the correct model order is found. For modes that are very active in the measured data, the modal parameters will stabilize at a very low model order. For modes that were poorly excited in the measured data, the modal parameters may not stabilize until a very high model order is chosen. Nevertheless, the nonphysical (computational) modes will not stabilize at all during this process and can be sorted out of the modal parameter data set more easily. Note that inconsistencies (frequency shifts, leakage errors, etc.) in the measured data set will obscure the stability and render the stability diagram difficult to use. Normally, a tolerance, in percentage, is given for the stability of each of the modal parameters that are being evaluated. Figure 16-37 is an example of a stability diagram. In Fig. 16-37, a summation of the frequency-response function power is plotted on the stability diagram for reference.

Mode Indication Function

Mode indication functions (MIF) are normally real-valued, frequency-domain functions that exhibit local minima at the natural frequencies of real normal modes. One mode indication function can be plotted for each reference available in the measured data. The primary mode indication function will exhibit a local minimum at each of the natural frequencies of the system under test. The secondary mode indication function will exhibit a local minimum at repeated or pseudorepeated roots of order 2 or more. Additional mode indication functions yield local minima for successively higher orders of repeated or pseudorepeated roots of the system under test.

The development the mode indication function is based on finding a force vector $\{F\}$ that will excite a normal mode at each frequency in the frequency range of interest. [108] If a normal mode can be excited at a particular frequency, the response to such a force vector will exhibit the 90° phase lag characteristic. Therefore, the real part of the response will be as small as possible, particularly when compared to the imaginary part or the total response. To evaluate this possibility, a minimization problem can be formulated as follows:

$$\min_{\|F\|=1} \frac{\{F\}^T [H_{\text{Real}}]^T [H_{\text{Real}}]\{F\}}{\{F\}^T ([H_{\text{Real}}]^T [H_{\text{Real}}] + [H_{\text{Imag}}]^T [H_{\text{Imag}}]) \{F\}} = \lambda \qquad (16\text{-}105)$$

This minimization problem is similar to a Rayleigh quotient, and it can be shown that the solution to the problem is found by finding the smallest eigenvalue $\lambda_{\min}$ and the corresponding eigenvector $\{F\}_{\min}$ of the following problem:

$$[H_{\text{Real}}]^T [H_{\text{Real}}] \{F\} = \lambda ([H_{\text{Real}}]^T [H_{\text{Real}}] + [H_{\text{Imag}}]^T [H_{\text{Imag}}]) \{F\} \qquad (16\text{-}106)$$

The eigenvalue problem above is formulated at each frequency in the frequency range of interest. Note that the result of the matrix product $[H_{\text{Real}}]^T[H_{\text{real}}]$ and $[H_{\text{Imag}}]^T[H_{\text{Imag}}]$ in each case is

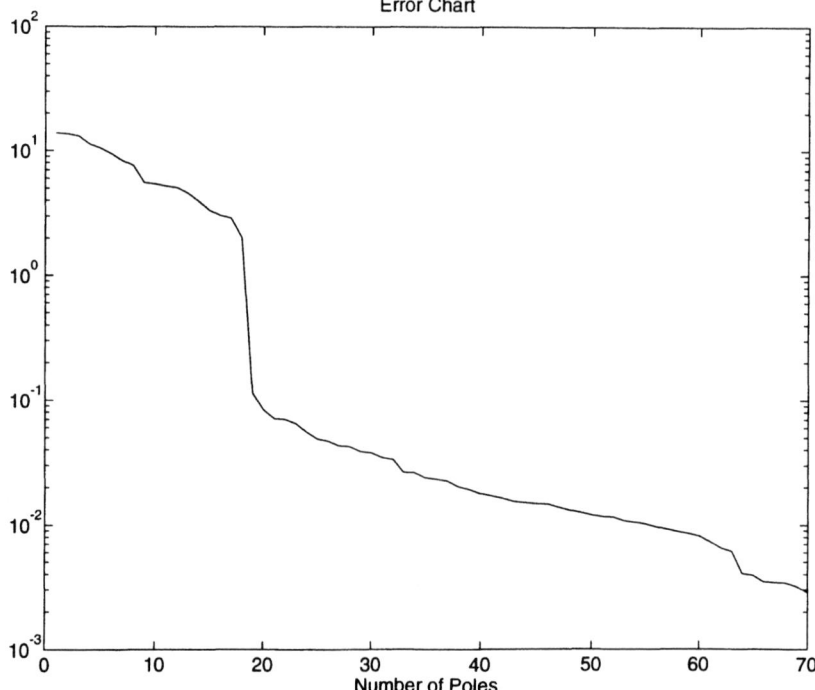

Figure 16-36 Model order determination: error chart.

a square, real-valued matrix of size equal to the number of references in the measured data $N_i \times N_i$. The resulting plot of a mode indication function for a multiple reference case can be seen in Fig. 16-38.

Rank Estimation

A more recent model order evaluation technique involves the estimate of the rank of the matrix of measured data. An estimate of the rank of the matrix of measured data gives a good estimate of the model order of the system. Essentially, the rank is an indicator of the number of independent characteristics contributing to the data. While the rank cannot be calculated in an absolute sense, the rank can be estimated from the singular value decomposition of the matrix of measured data. For each mode of the system, one singular value should be found by the SVD procedure. The SVD procedure finds the largest singular value first and then successively finds the next largest. The magnitudes of the singular values are used in one of two different procedures to estimate the rank. The concept that is used is that the singular values should go to zero when the rank of the matrix is exceeded. For theoretical data, this will happen exactly. For measured data, because of random errors and small inconsistencies in the data, the singular values will not become zero but will be very small. Therefore, the rate of change of the singular values is used as an indicator

EXPERIMENTAL MODAL ANALYSIS

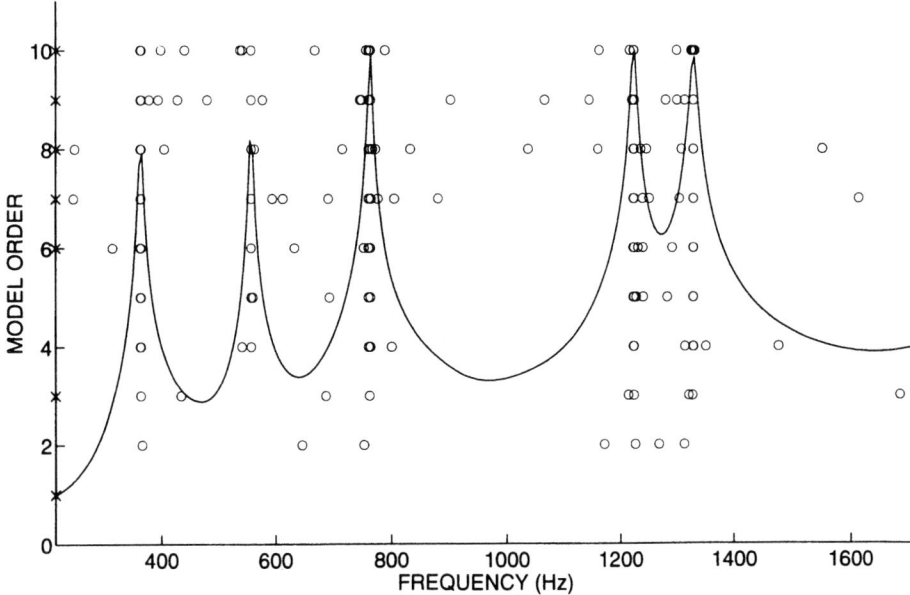

Figure 16-37 Model order determination: stability diagram.

rather than the absolute values. In one approach, each singular value is divided by the first (largest) to form a normalized ratio. This normalized ratio is treated much like the error chart, and the appropriate rank (model order) is chosen when the normalized ratio approaches an asymtote. In another similar approach, each singular value is divided by the preceding singular value, forming a normalized ratio that will be approximately 1 if the successive singular values are not changing in magnitude. When there is a rapid decrease in the magnitude of the singular value, the ratio of successive singular values drops (or peaks if the inverse of the ratio is plotted) as an indicator of rank (model order) of the system. Figure 16-39 and 16-40 are examples of these procedures for rank estimation.

Concept: Fundamental Model

In light of the discussion above, it ha become apparent over the past several years that most of the modal parameter estimation processes available could have been developed by starting from a general matrix polynomial formulation of the differential equation. The general matrix polynomial formulation of the differential equations is given below in the time, frequency, and Laplace (assuming zero initial conditions) domains.

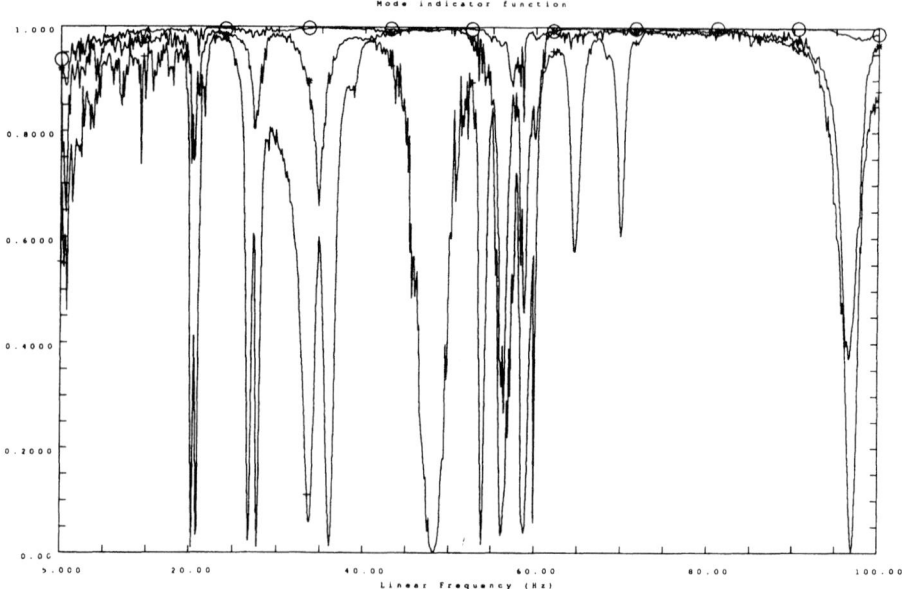

Figure 16-38 Model order determination: mode indication function.

Time domain:
$$[A]_n \frac{d^n x(t)}{dt^n} + [A]_{n-1} \frac{d^{n-1} x(t)}{dt^{n-1}} + \cdots + [A]_0 x(t) =$$
$$[B]_m \frac{d^m f(t)}{dt^m} + [B]_{m-1} \frac{d^{m-1} f(t)}{dt^{m-1}} + \cdots + [B]_0 f(t) \quad (16\text{-}107)$$

Frequency domain:
$$[[A]_n (j\omega)^n + [A]_{n-1}(j\omega)^{n-1} + \cdots + [A]_0]\, X(\omega) =$$
$$[[B]_m (j\omega)^m + [B]_{m-1}(j\omega)^{m-1} + \cdots + [B]_0]\, F(\omega) \quad (16\text{-}108)$$

Laplace domain:
$$[[A]_n (s)^n + [A]_{n-1}(s)^{n-1} + \cdots + [A]_0]\, X(s) =$$
$$[[B]_m (s)^m + [B]_{m-1}(s)^{m-1} + \cdots + [B]_0]\, \{F(s)\} \quad (16\text{-}109)$$

In terms of sampled data, the time-domain matrix differential equation reduces to a set of finite-difference equations.

Sampled time domain:
$$[[A]_n x_n + [A]_{n-1} x_{n-1} + \cdots + [A]_0 x_0] =$$
$$[[B]_m f_m + [B]_{m-1} f_{m-1} + \cdots + [B]_0 f_0] \quad (16\text{-}110)$$

EXPERIMENTAL MODAL ANALYSIS

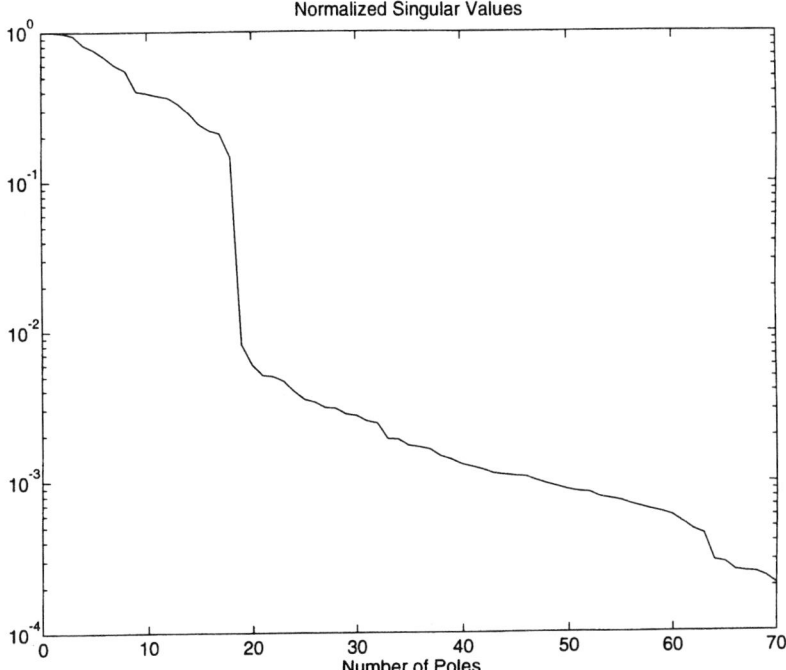

Figure 16-39 Model order determination: rank estimation. Normalized singular values versus number of poles.

The solution to the equation above corresponds to a sampled form of the solution for the continuous equations. In other words, it is a sampled unit impulse-response function.

The Z transform of the finite-difference equation corresponds to a discrete Laplace transform and gives an equation equivalent in form to the Laplace transform of the continuous equations of motion.

Sampled frequency domain (Z domain):

$$[[A]_n z^n + [A]_{n-1} z^{n-1} + \cdots + [A]_0] X(z) = \\ [[B]_m z^m + [B]_{m-1} z^{m-1} + \cdots + [B]_0] F(z) \tag{16-111}$$

The Z-transform formulation is important in rederiving many of the existing time-domain parameter algorithms. Historically, many of the modal parameter estimation methods were derived using clever mathematical manipulations. As a result, it is difficult to see the common threads that run through most of the algorithms. Using a common starting point to recreate the various methods makes many of the advantages and disadvantages of the methods clear.

The models above represent a fundamental model that can be used to explain or rederive any modal parameter estimation algorithm in use at the present time. These models correspond to autoregressive moving-average (ARMA) models that were developed as a set of finite-difference

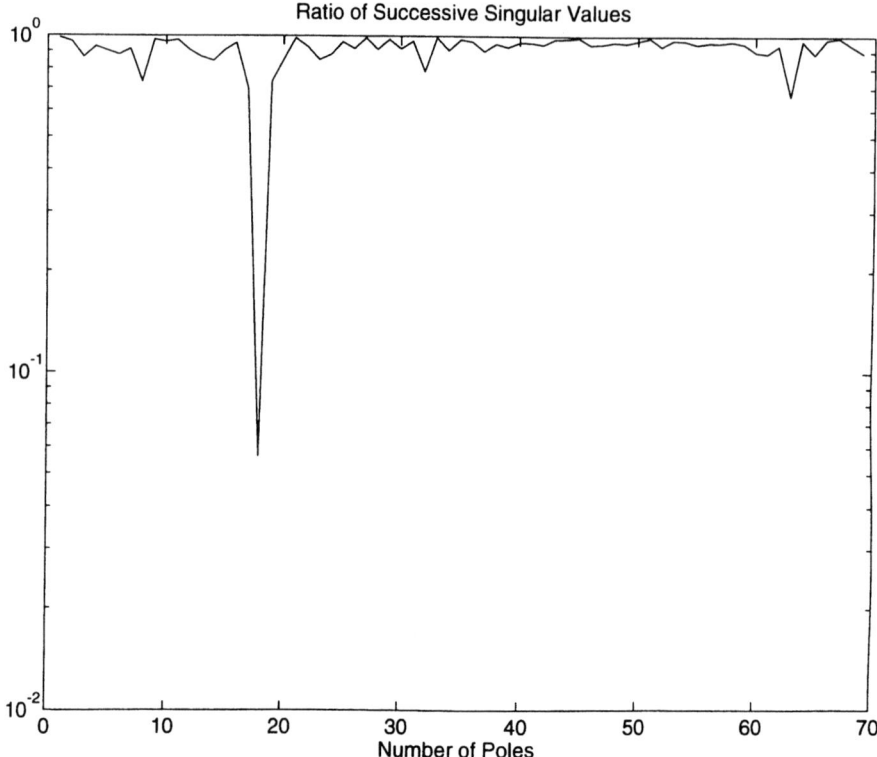

Figure 16-40 Model order determination: rank estimation. Ratio of successive singular values versus number of poles.

equations and were specific to the time domain. However, with the general use of both the Fourier transform and the Laplace transform, the information from the frequency and Laplace domains is used in an interchangeable manner. While the historical definition of an ARMA model involves a time-domain, finite-difference form, for the purposes of the following discussion, the general matrix polynomial model has been expanded from the finite-difference formulation to include all the polynomial forms that will result when the ARMA model is transformed into the frequency or Z domains.

The general matrix polynomial model equations are broken into three general classifications (shown in the Laplace domain):

High-order model:

$$\sum_{r=0}^{n} [[A]_r s^r] X(s) = \sum_{r=0}^{m} [[B]_r s^r] F(s) \qquad (16\text{-}112)$$

Low order model:

$$[[A]_2 s^2 + [A]_1 s + [A]_0] X(s) = [[B]_1 s + [B]_0] F(s) \tag{16-113}$$

Zeroth-order model:

$$[A]_0 X(s) = F(s) \tag{16-114}$$

The high-order model is typically used when the system is undersampled in the spatial domain. For example, the limiting case occurs when only one point is measured on the structure. For this case, the left-hand side of the general matrix polynomial model corresponds to a scalar polynomial equation with the order equal to or greater than the number of measurable eigenvalues.

The number of terms on the right-hand side can be zero for the homogeneous case and greater than the order of the right-hand side when the forcing function is complicated. One of the obvious difficulties in using this type of formulation is determining the order of the right-hand and left-hand polynomials.

The low-order model is used when the spatial information is complete. In other words, the number of physical coordinates is greater than the number of measurable eigenvalues. For this case, the order of the left-hand side is 2 and the right-hand side can be less than or greater than 2, depending on the number of residual modes in the data. Residual modes are modes whose eigenvalue or natural frequency is out of the frequency band of analysis, and they make a significant contribution in the frequency range of interest. If the number of physical coordinates is greater than the number of measurable eigenvalues, the number of physical coordinates is reduced by using a transformation to be equivalent to the approximate number of measurable eigenvalues. SVD or some other orthogonal transformation is often used to reduce the spatial information. In most cases even when the spatial information must be condensed, it is necessary to use a model greater than 2 to compensate for distortion errors or noise in the data and to compensate for the case of the location of the transducers being insufficient to define the structure totally.

The zeroth order model corresponds to a case of neglect of the temporal information, with only the spatial information used. These methods directly estimate the eigenvectors as a first step. In general, these methods are programmed to process data at a single temporal condition or variable. In this case, the method is essentially equivalent to the single degree of freedom methods, which have been used with frequency-response functions. In other words, the zeroth-order matrix polynomial model compared to the higher-order matrix polynomial models is similar to the comparison between the SDOF and MDOF methods used historically in modal parameter estimation.

Concept: Modal Model Validation

Once the modal parameters have been determined, several procedures exist that allow for the modal model to be validated. Some of the procedures that are used are:

Measurement synthesis
Visual verification (animation)
Finite-element analysis
Modal vector orthogonality
Modal vector consistency (modal assurance criterion)
Modal modification prediction

Modal complexity
Modal phase collinearity and mean phase deviation

All these methods depend on the evaluation of an assumption concerning the modal model. Unfortunately, the success of the validation method only defines the validity of the assumption; the failure of the modal validation generally does not identify the cause of the problem.

Measurement Synthesis

The simplest validation procedure is to compare the data synthesized from the modal model with the measured data. This is particularly effective if the measured data were not part of the data used to estimate the modal parameters. This approach serves as an independent check of the modal parameter estimation process.

Visual Verification

Another relatively simple method of modal model validation is to evaluate the modal vectors visually. While this can be accomplished from plotted modal vectors superimposed on the undeformed geometry, the modal vectors are normally animated (superimposed on the undeformed geometry), to permit quick assessment of the modal vector. Particularly, modal vectors are evaluated for physically realizable characteristics such as discontinuous motion or out-of-phase problems. Often, rigid-body modes of vibration are evaluated to determine scaling (calibration) errors or invalid measurement degree-of-freedom assignment or orientation. Naturally, if the system under test is believed to be proportionally damped, the modal vectors should be normal modes, and this characteristic can be quickly observed by viewing an animation of the modal vector.

Finite-Element Analysis

The results of a finite-element analysis of the system under test can provide another method of validating the modal model. While the problem of matching the number of analytical degrees of freedom N_a to the number of experimental degrees of freedom N_e causes some difficulty, the modal frequencies and modal vectors can be compared visually or through orthogonality or consistency checks. Unfortunately, when the comparison is not acceptable, the question of error in the experimental model versus error in the analytical model cannot be easily resolved. Generally, assuming minimal errors and sufficient analysis and test experience, reasonable agreement can be found in the first 10 deformable modal vectors; agreement for higher modal vectors will be more difficult.

Modal Vector Orthogonality

Another method that has been used historically to validate an experimental modal model is the weighted orthogonality check. In this case, the experimental modal vectors are used together with a mass matrix normally derived from a finite-element model to evaluate orthogonality. The experimental modal vectors are scaled so that the diagonal terms of the modal mass matrix are unity. With this form of scaling, the off-diagonal values in the modal mass matrix are expected to be less than 1.0 (10% of the diagonal terms).

EXPERIMENTAL MODAL ANALYSIS

Theoretically, for the case of proportional damping, each modal vector of a system will be orthogonal to all other modal vectors of that system when weighted by the mass, stiffness, or damping matrix. In practice, these matrices are made available by way of a finite-element analysis, and normally the mass matrix is considered to be the most accurate. For this reason, any further discussion of orthogonality will be made with respect to mass matrix weighting. As a result, the orthogonality relations can be stated as follows:

For $r \neq s$:

$$\{\psi_r\} [M] \{\psi_s\} = 0 \tag{16-115}$$

For $r = s$:

$$\{\psi_r\} [M] \{\psi_s\} = M_r \tag{16-116}$$

Experimentally, the result of zero for the cross-orthogonality (Eq. 16-115) is rarely achieved, but values up to one-tenth of the magnitude of the generalized mass of each mode are accepted. It is a common procedure to form the modal vectors into a normalized set of mode shape vectors with respect to the mass matrix weighting. The accepted criterion in the aerospace industry, where this confidence check is made most often, is for all the generalized mass terms to be unity and all cross-orthogonality terms to be less than 0.1. Often, even under this criterion, an attempt is made to adjust the modal vectors so that the cross-orthogonality conditions are satisfied [109–111].

In Eqs. (16-115) and (16-116), the mass matrix must be an $N_o \times N_o$ matrix corresponding to the measurement locations on the structure. This means that the finite-element mass matrix must be modified from whatever size and distribution of grid locations are required in the finite-element analysis to the $N_o \times N_o$ square mix corresponding to the measurement locations. This normally involves some sort of reduction algorithm as well as interpolation of grid locations to match the measurement situation [112–115].

When Eq. (16-115) is not sufficiently satisfied, one (or more) of three situations may exist. First, the modal vectors may be invalid, perhaps because of measurement error or problems with the modal parameter estimation algorithms. This is a very common problem and many times contributes to the problem. Second, the mass matrix may be invalid. Since the mass matrix is not easily related to the physical properties of the system, invalidity here probably contributes significantly to the problem. Third, the reduction of the mass matrix may be invalid [114,115]. This can certainly be a realistic problem and can cause severe errors. The most obvious example of this situation would be a relatively large amount of mass reduced to a highly flexible measurement location, such as the center of an unsupported panel. In such a situation the measurement location is weighted very heavily in the orthogonality calculation of Eq. (16-115) but may represent only incidental motion of the overall modal vector.

In all probability, all three situations contribute to the failure of cross-orthogonality criteria on occasion. Failure to satisfy the orthogonality conditions does not indicate where the problem originates. From an experimental point of view, it is important to try to develop methods that indicate confidence that the modal vector is or is not part of the problem.

Modal Vector Consistency

Since the residue matrix contains redundant information with respect to a modal vector, the consistency of the estimate of the modal vector under varying conditions such as excitation location or modal parameter estimation algorithms can be a valuable confidence factor to be used in the evaluation of the experimental modal vectors.

The common approach to estimation of modal vectors from the frequency-response function method is to measure a complete row or column of the frequency-response function matrix. This will give reasonable definition to the modal vectors that have a nonzero modal coefficient at the excitation location and can be completely uncoupled with the forced normal mode excitation method. When the modal coefficient at the excitation location of a modal vector is zero (very small with respect to the dynamic range of the modal vector) or when the modal vectors cannot be uncoupled, the estimation of the modal vector will contain potential bias and variance errors. In such cases additional rows and/or columns of the frequency-response function matrix are measured to detect potential problems.

The function of the *modal scale factor* (MSF) is to provide a means of normalizing all estimates of the same modal vector. When two modal vectors are scaled similarly, elements of each vector can be averaged (with or without weighting), differenced, or sorted to provide a best estimate of the modal vector or an indication of the type of error vector superimposed on the modal vector. In terms of modern, multiple-reference modal parameter estimation algorithms, the modal scale factor is a normalized estimate of the modal participation factor between two references for a specific mode of vibration.

The function of the *modal assurance criterion* (MAC) is to provide a measure of consistency between estimates of a modal vector. This provides an additional confidence factor in the evaluation of a modal vector from different excitation locations. The modal assurance criterion also constitutes a method of determining the degree of causality between estimates of different modal vectors from the same system.

The modal scale factor and the modal assurance criterion also provide a method of easily comparing estimates of modal vectors originating from different sources. The modal vectors from a finite-element analysis can be compared and contrasted with those determined experimentally as well as modal vectors determined by way of different experimental or modal parameter estimation methods. In this approach, methods can be compared and contrasted in order to evaluate the mutual consistency of different procedures rather than estimating the modal vectors specifically.

The residue matrix for mode r of a system can be represented as follows for a multiple-input (reference) case:

$$[A_r] = Q_r \begin{bmatrix} \psi_1\psi_1 & \psi_1\psi_2 & \psi_1\psi_3 & \cdots & \cdots & \cdots & \psi_1\psi_q \\ \psi_2\psi_1 & \psi_2\psi_2 & \psi_2\psi_3 & \cdots & \cdots & \cdots & \psi_2\psi_q \\ \psi_3\psi_1 & \psi_3\psi_2 & \psi_3\psi_3 & \cdots & \cdots & \cdots & \psi_3\psi_q \\ \cdots & \cdots & \cdots & \cdots & \cdots & \cdots & \cdots \\ \psi_p\psi_1 & \psi_p\psi_2 & \psi_p\psi_3 & \cdots & \cdots & \cdots & \psi_p\psi_q \end{bmatrix}_{N_o \times N_i} \quad (16\text{-}117)$$

This representation for the residue matrix demonstrates that each element of the residue matrix for modal vector r consists of the product of a scaling constant, the modal coefficient of the excitation location, and the modal coefficient of the response location. Therefore, for each column, the residue vector $\{A_{qr}\}$ can be written in terms of the common modal vector $\{\psi_r\}$ as follows:

$$\{A_{pqr}\} = Q_r\, \psi_{pr}\, \{\psi_r\} \quad (16\text{-}118)$$

From Eqs. (16-117) and (16-118) it is easy to see that each row or column is simply the same modal vector times the modal coefficient of the common response location or excitation location respectively. Therefore, the proportionality constant between rows c and d or columns c and d can be referred to as a modal scale factor equal to:

$$MSF_{cdr} = \frac{\psi_{cr}}{\psi_{dr}} \quad (16\text{-}119)$$

If Eq. (16-119) is used as the basis for a model to calculate a least-squares error estimate of the

EXPERIMENTAL MODAL ANALYSIS

proportionately constant between rows or columns of the residue matrix, the model is linear as follows:

$$A_{cqr} = MSF_{cdr} A_{dqr} \tag{16-120}$$

In vector notation, this would be:

$$\{A_{cr}\} = MSF_{cdr} \{A_{dr}\} \tag{16-121}$$

All the elements of the modal vectors in rows/columns c and d exhibit the relationship stated in Eq. (16-121). The value of the modal scale factor is to be calculated to minimize the sum of the squared errors between corresponding elements of each modal vector [47,49]. All or part of each modal vector can be used in such a calculation. If some elements are consciously excluded, obviously a form of weighted least-squares error estimation is involved. The modal scale factor is defined, according to this approach, as follows:

$$MSF_{cdr} = \frac{\{A_{cr}\}^H \{A_{dr}\}}{\{A_{dr}\}^H \{A_{dr}\}} \tag{16-122}$$

Equation (16-122) implies that the modal vector of row/column d is the reference to which the modal vector of row/column c is compared. In the general case, modal vector c can be considered to be made of two parts. The first part will be the part correlated with modal vector d. The second part will be the part that is not correlated with modal vector d and will be made up of contamination from other modal vectors and of any random contribution. This error vector will be considered to be noise. If the modal assurance criterion is defined as a scalar constant relating the portion of the automoment of the modal vector that is linearly related to the reference modal vector, then the following equation can be developed:

$$MAC_{cdr} = \frac{|\{A_{cr}\}^H \{A_{dr}\}|^2}{\{A_{cr}\}^H \{A_{cr}\}\{A_{dr}\}^H \{A_{dr}\}} \tag{16-123}$$

Alternatively, the modal assurance criterion can be formulated as follows:

$$MAC_{cdr} \frac{\{A_{cr}\}^H \{A_{dr}\}\{A_{dr}\}^H \{A_{cr}\}}{\{A_{cr}\}^H \{A_{cr}\}\{A_{dr}\}^H \{A_{dr}\}} \tag{16-124}$$

The modal assurance criterion is a scalar constant relating the causal relationship between two modal vectors. The constant will take on values from zero, representing no consistent correspondence, to one, representing a consistent correspondence. In this manner, if the modal vectors under consideration truly exhibit a consistent relationship, the modal assurance criterion should approach unity and the value of the modal scale factor can be considered to be reasonable.

The value of the modal assurance criterion can indicate the validity of the modal scale factor. While certain implications of the modal assurance criterion depend on the calculations involving rows or columns, some general discussion is applicable to all cases.

The modal assurance criterion can take on values between zero and one. If the modal assurance criterion has a value near zero, this is an indication that the modal vectors are not consistent. This can be due to any of the following reasons:

1. The system is nonstationary. This condition can occur whenever the system is undergoing a change in mass or stiffness during the testing period.
2. The system is nonlinear. System nonlinearities will appear differently in frequency-response functions generated from different exciter positions or excitation signals. The modal parameter estimation algorithms also will not handle the different nonlinear characteristics in a consistent manner.

3. There is noise on the reference modal vector. This case is the same as noise on the input of a frequency-response function measurement. No amount of signal processing can remove this type of error.
4. The modal parameter estimation is invalid. Even if the frequency-response functions measurements contain no errors, the modal parameter estimation may not be consistent with the data. For example, the modal parameter estimation algorithm may use a complex system pole model when only real-valued system poles exist.
5. The modal vectors are from linearly unrelated mode shape vectors. Hopefully, since the different modal vector estimates are from different excitation positions, this measure of inconsistency will imply that the modal vectors are orthogonal.

Obviously, if the first four reasons can be eliminated, the interpretation of the modal assurance criterion can be similar to that for an orthogonality calculation.

If the modal assurance criterion has a value near unity, this is an indication that the modal vectors are consistent. It does not necessarily mean that they are correct. The modal vectors can be consistent for any of the following reasons:

1. The modal vectors have been incompletely measured. This situation can occur whenever too few response stations have been included in the experimental determination of the modal vector. For example, two symmetric modes of the wings of an airplane will appear to be identical if only response stations at the wing tips are used to define the modal vectors.
2. The modal vectors are the result of a forced excitation other than the desired input. This situation would obtain if, during the measurement of the frequency-response function, a rotating piece of equipment with an unbalance were present in the system being tested.
3. The modal vectors are primarily coherent noise. Since the reference modal vector may be arbitrarily chosen, this modal vector may not be one of the true modal vectors of the system. It could be simply a random noise vector or a vector reflecting the bias in the modal parameter estimation algorithm. In any case, the modal assurance criterion will reflect only a causal relationship to the reference modal vector.
4. The modal vectors represent the same modal vector with different arbitrary scaling. If the two modal vectors being compared have the same expected value when normalized, the two modal vectors should differ only by the complex-value scale factor, which is a function of the common modal coefficients between the rows or columns.

Therefore, if the first three reasons can be eliminated, the modal assurance criterion indicates that the modal scale factor is the complex constant relating the modal vectors and that the modal scale factor can be used to average, difference, or sort the modal vectors.

It is very important to notice that the modal assurance criterion can indicate only consistency, not validity. If the same errors; random or bias, exist in all modal vector estimates, this will not be delineated by the modal assurance criterion. Invalid assumptions are normally the cause of this sort of potential error. Even though the modal assurance criterion is unity, the assumptions involving the system or the modal parameter estimation techniques are not necessarily correct. The assumptions may cause consistent errors in all modal vectors under all test conditions verified by the modal assurance criterion.

Modal Modification Prediction

The use of a modal model to predict changes in modal parameters caused by a perturbation (modification) of the system is becoming more of a reality as more measured data are acquired simultaneously. In this validation procedure, a modal model is estimated based on a complete modal test. This modal model is used as the basis for predicting a perturbation to the system that was tested, such as the addition of a mass at a particular point on the structure. Then the mass is added to the structure and the perturbed system is retested. The predicted and measured data or modal model can be compared and contrasted as a measure of the validity of the underlying modal model.

Modal Complexity

Modal complexity is a variation of the use of sensitivity analysis in the validation of a modal model. When a mass is added to a structure, the modal frequencies either should be unaffected or should shift to a slightly lower frequency. Modal overcomplexity is a summation of this effect over all measured degrees of freedom for each mode. Modal complexity is particularly useful for the case of complex modes in an attempt to quantify whether the mode is genuinely a complex mode, a linear combination of several modes, or a computational artifact. The mode complexity is normally indicated by the *mode overcomplexity value* (*MOV*), which is the percentage of response points that actually would cause the damped natural frequency to decrease when a mass is added compared to the total number of response points. A separate MOV is estimated for each mode of vibration, and the ideal result should be 1.0 (100%) for each mode.

Modal Phase Collinearity and Mean Phase Deviation

For proportionally damped systems, each model coefficient for a specific mode of vibration should differ by 0° or 180°. The *modal phase collinearity* (MPC) is an index expressing the consistency of the linear relationship between the real and imaginary parts of each modal coefficient. This concept is essentially the same as the ordinary coherence function with respect to the linear relationship of the frequency-response function for different averages or the modal assurance criterion with respect to the modal scale factor between modal vectors. The MPC should be 1.0 (100%) for a mode that is essentially a normal mode. A low value of MPC indicates a mode that is complex (after normalization) and is an indication of a nonproportionally damped system or errors in the measured data and/or modal parameter estimation.

Another indicator that defines whether a modal vector is essentially a normal mode is the *mean phase deviation* (MPD). This index is the statistical variance of the phase angles for each mode shape coefficient for a specific modal vector from the mean value of the phase angle. The MPD is an indication of the phase scatter of a modal vector and should be near 0° for a real, normal mode.

Summary

Modal parameter estimation is probably one of the most misunderstood aspects of the total experimental modal analysis process. In no way can the modal parameter estimation method compensate for basic errors made in the measurement of the data used in the estimation process. Since most modal parameter estimation methods are mathematically intimidating, many users do not fully understand the ramifications of the decisions made during the measurement stages, or later in the modal parameter estimation process. Table 16-6 is an attempt to categorize current modal parameter estimation methods based on the general concepts introduced in this section. This sort of overview of modal parameter estimation should be used simply as a guide toward further study and understanding of the details of the individual modal parameter estimation methods.

Table 16-6 Summary of Modal Parameter Estimation Methods

Method	Time, Frequency, or Spatial Domain	Single or Multiple Degrees of Freedom	Modal Parameter Estimation Characteristics				
			Global Modal Frequencies and Damping Factors	Repeated Modal Frequencies and Damping Factors	Global Modal Vectors	Global Modal Participation Factors	Residuals
Quadrature amplitude	Frequency	SDOF	No	No	No	No	No
Kennedy–Pancu circle fit	Frequency	SDOF	No	No	No	No	Yes
SDOF polynomial	Frequency	SDOF	Yes/no	No	No	No	No
Nonlinear frequency domain	Frequency	MDOF	No	No	No	No	Yes
Complex exponential	Time	MDOF	No	No	No	No	No
Least-squares complex exponential (LSCE)	Time	MDOF	Yes	No	No	No	No
Ibrahim time domain (ITD)	Time	MDOF	Yes	No	Yes	No	No
Multireference Ibrahim time domain (MITD)	Time	MDOF	Yes	Yes	Yes/no	No/yes	No
Eigensystem realization algorithm (ERA)	Time	MDOF	Yes	Yes	Yes	Yes	No
Orthogonal polynomial	Frequency	MDOF	Yes	No	No	No	Yes
Multireference orthogonal polynomial	Frequency	MDOF	Yes	Yes	Yes	Yes	Yes
Polyreference time domain	Time	MDOF	Yes	Yes	No	Yes	No
Polyreference frequency domain	Frequency	MDOF	Yes	Yes	Yes	Yes	Yes
Time-domain direct parameter identification	Time	MDOF	Yes	Yes	Yes	Yes	No
Frequency-domain direct parameter identification	Frequency	MDOF	Yes	Yes	Yes	Yes	Yes
Multi-MAC	Spatial	SDOF	No	Yes	Yes	No	No
Mulit-Mac/Complex Mode Indicator Function (CMIF)/ enhanced FRF	Spatial	MDOF	Yes	Yes	Yes	Yes	No

16-5 Summary

Experimental modal analysis involves the combination of several evolving technologies. In the measurement stage, the technology of digital signal processing and discrete transforms is of primary importance in terms of making accurate measurements and understanding the relative significance of the errors that remain. The understanding of this technology often is the limiting factor the successful estimation of modal parameters. In the estimation stage, methods are changing rapidly as a result of the transfer of existing technology from other fields in which estimation methods have been more common over the past several decades. Finally, the impact of the technology concerned with solid-state electronics on both the data acquisition hardware and the computational hardware has made routinely accessible advanced measurement and estimation techniques that were impractical 5 years ago. These, and many other factors, are the reasons behind the great interest in experimental modal analysis as simply one more part of an integrated computer-aided engineering approach to the solution of problems in structural dynamics.

List of Symbols

Matrix Notation

$\{..\}$	braces enclose column vector expressions
$\{..\}^T$	row vector expressions
$[..]$	brackets enclose matrix expressions
$[..]^H$	complex conjugate transpose, or Hermitian transpose, of a matrix
$[..]^T$	transpose of a matrix
$[..]^{-1}$	inverse of a matrix
$[..]_{q \times p}$	size of a matrix: q rows, p columns
$\lceil .. \rfloor$	diagonal matrix

Operator Notation

A^*	complex conjugate
$\dot{x}$	first derivative with respect to time of dependent variable x
$\ddot{x}$	second derivative with respect to time of dependent variable x
$\sum_{i=1}^{n} A_i B_i$	summation of $A_i B_i$ from $i = 1$ to n

Roman Alphabet

A_{pqr}	residue for response location p, reference location q, of mode r
C	damping
e	base e (2.71828 . . .)
F	input force
F_q	spectrum of qth reference
GFF	autopower spectrum of reference
GFF_{qq}	autopower spectrum of reference q
GFF_{ik}	cross-power spectrum of reference i and reference k
GFX	cross-power spectrum of reference/response
GXF	cross-power spectrum of response/reference
GXX	autopower spectrum of response
GXX_{pp}	autopower spectrum of response p

$h(t)$	impulse-response function
$h_{pq}(t)$	impulse-response function for response location p, reference location q
$H(s)$	transfer function
$H(\omega)$	frequency response function; when no ambiguity exists, H is used instead of $H(\omega)$
$H_{pq}(\omega)$	frequency response function for response location p, reference location q; when no ambiguity exists, H_{pq} is used instead of $H_{pq}(\omega)$
$H_1(\omega)$	frequency-response function estimate with noise assumed on the response; when no ambiguity exists, H_1 is used instead of $H_1(\omega)$
$H_2(\omega)$	frequency-response function estimate with noise assumed on the reference; when no ambiguity exists, H_2 is used instead of $H_2(\omega)$
$H_S(\omega)$	scaled frequency-response function estimate; when no ambiguity exists, H_S is used instead of $H_S(\omega)$
$H_v(\omega)$	frequency-response function estimate with noise assumed on both reference and response; when no ambiguity exists, H_v is used instead of $H_v(\omega)$
$[I]$	identity matrix
j	$\sqrt{-1}$
K	stiffness
L	modal participation factor
M	mass
M_r	modal mass for mode r
M_{A_r}	modal A scaling factor for mode r
$MCOH$	multiple coherence function
N	number of modes
N_i	number of references (inputs)
N_o	number of responses (outputs)
p	output, or response point (subscript)
q	input, or reference point (subscript)
r	mode number (subscript)
R_I	residual inertia
R_F	residual flexibility
s	Laplace domain variable
t	independent variable of time (seconds)
t_k	discrete value of time (seconds) $t_k = k \Delta t$
T	sample period
x	displacement in physical coordinates
X	response
X_p	spectrum of pth response
z	Z-domain variable

Greek Alphabet

Δf	discrete interval of frequency (Hz or cycles/s)
Δt	discrete interval of sample time (seconds)
ζ	damping ratio
ζ_r	damping ratio for mode r
η	noise on the output
λ_r	rth complex eigenvalue, or system pole $\lambda_r = \sigma_r + j\omega_r$
$\lceil \Lambda \rfloor$	diagonal matrix of poles in Laplace domain
ν	noise on the input

σ	variable of damping (rad/s)
σ_r	real part of the system pole, or damping factor, for mode r
$\{\psi\}$	scaled eigenvector or modal vector
ψ_{pr}	scaled pth response of a complex modal vector for mode r
$\{\psi\}_r$	scaled complex modal vector for mode r
$[\Psi]$	scaled complex modal vector matrix
ω	variable of frequency (rad/s)
ω_r	imaginary part of the system pole, or damped natural frequency, for mode r (rad/s) $\omega_r = \Omega_r\sqrt{1 - \zeta_r^2}$
Ω_r	undamped natural frequency (rad/s) $\Omega_r = \sqrt{\sigma_r^2 + \omega_r^2}$

References

1. L. D. Mitchell, A Perspective View of Modal Analysis Keynote Address, 6th International Modal Analysis Conference, *Proc. Int. Modal Analysis Conf.*, 1988, 5 pp.

2. H. Vold, Insight, Not Numbers, Keynote Address, 7th International Modal Analysis Conference, *Proc. Int. Modal Analysis Conf.*, 1989, 4 pp.

3. R. S. Pappa, Identification Challenges for Large Space Structures, Keynote Address, 8th International Modal Analysis Conference, *Proc. Int. Modal Analysis Conf.*, 1990, 9 pp.

4. H. G. Natke, Survey on the Identification of Mechanical Systems, *Proc. Road-Vehicle Systems and Related Mathematics*, 1987, 69–116.

5. R. J. Allemang, and D. L. Brown, *Experimental Modal Analysis and Dynamic Component Synthesis*, AFWAL-TR-87-3069, Vols. I–IV, 1987.

6. R. C. Lewis, and D. L. Wrisley, A System for the Excitation of Pure Natural Modes of Complex Structures, *J. Aeronaut. Sci.*, 17, 11 (1950), 705–722.

7. B. F. De Veubeke, A Variational Approach to Pure Mode Excitation Based on Characteristic Phase Lag Theory, AGARD, Report 39, 1956, 35 pp.

8. R. W. Trail-Nash, On the Excitation of Pure Natural Modes in Aircraft Resonance Testing, *J. Aeronaut. Sci.*, 25, no. 12 (1958), 775–778.

9. G. W. Asher, A Method of Normal Mode Excitation Utilizing Admittance Measurements, Dynamics of Aeroelasticity, *Proc. Inst. Aeronautical Sciences*, 1958, 69–76.

10. F. J. Hawkins, An Automatic Resonance Testing Technique for Exciting Normal Modes of Vibration of Complex Structures, IUTAM Symposium, *Progrés Recents de la Mécanique des Vibrations Linéaires*, 1965, 37–41.

11. F. J. Hawkins, GRAMPA—An Automatic Technique for Exciting the Principal Modes of Vibration of Complex Structures, Royal Aircraft Establishment, RAE-TR-67-211, 1967.

12. G. A. Taylor, D. R. Gaukroger, and C. W. Skingle, MAMA—A Semiautomatic Technique for Exciting the Principal Modes of Vibration of Complex Structures, Aeronautical Research Council, ARC-R/M-3590, 1967, 20 pp.

13. R. R. Craig, Jr. and Y. W. T. Su, On Multiple-Shaker Resonance Testing, *AIAA J.*, 12, no. 7 (1974), 924–931.

14. M. Feix, An Iterative Self-Organizing Method for the Determination of Structural Dynamic Characteristics, European Space Agency, ESA-TT-232 (N76-331183), 1975, 36–60.

15. P. Ibanez, Force Appropriation by Extended Asher's Method, SAE Paper 760873, 1976, 16 pp.

16. G. Morosow, Exciter Force Apportioning for Modal Vibration Testing Using Incomplete Excitation, Ph.D, dissertation, Department of Civil, Environmental, and Architectural Engineering, University of Colorado at Boulder, 1977, 132 pp.

17. G. Morosow, and R. S. Ayre, Force Appropriation for Modal Vibration Testing Using Excitation, *Shock Vib. Bull.*, 48, part 1 (1978), 39–48.

18. A. M. Kabe, Mode Shape Identification and Orthogonalization, USAF Space Division Report SD-TR-88-93, 1988, 25 pp.

19. A. M. Kabe, An Efficient Direct Measurement Mode Survey Test Procedure, USAF Space Division Report SD-TR-89-01, 1988, 34 pp.

20. R. Williams, and H. Vold, The Multi-Phase Step-Sine Method for Experimental Modal Analysis, *J. Anal. Exp. Modal Anal.*, 1, no. 2 (1986), 25–34.

21. R. Fillod, G. Lallement, and J. L. Piranda, Identification Using a Variable Phase Multipoint Excitation, *Proc. 11th Int. Seminar Modal Analysis*, Katholieke Universiteit te Leuven, 1986, 20 pp.

22. F. Lembregts, P. Sas, and H. Van der Auweraer, Integrated Stepped Sine Modal Analysis, *Proc. 11th International Seminar Modal Analysis*, Katholieke Universiteit te Leuven, 1986, 12 pp.

23. H. Van der Auweraer, P. Vanherck, P. Sas, and R. Snoeys, Accurate Modal Analysis Measurements with Programmed Sine Wave Excitation, *Mech. Syst. Signal Process.*, 1, no. 3 (1987), 295–300.

24. F. Lembregts, H. Van der Auweraer, and J. Leuridan, Integrated Stepped-Sine System for Modal Analysis, *Mech. Syst. Signal Process*, 1, no. 4 (1987), 415–424.

25. A. J. Severyn, M. Lally, R. Rost, and D. L. Brown, Hardware Considerations for Spatial Domain Sine Testing, *Proc. 12th Int. Seminar Modal Analysis*, Katholieke Universiteit te Leuven, 1987, 8 pp.

26. F. Deblauwe, C. Y. Shih, R. Rost, and D. L. Brown, Survey of Parameter Estimation Algorithms Applicable to Spatial Domain Sine Testing, *Proc. 12th Int. Seminar Modal Analysis*, Katholieke Universiteit te Leuven, 1987, 15 pp.

27. M. Lally, D. L. Brown, A. J. Severyn, and H. Vold, System Support for Spatial Sine Testing, *Proc. Int. Modal Analysis Conf.*, 1988, 620–628.

28. C. V. Stahle, and W. R. Forlifer, Ground Vibration Testing of Complex Structures, *Flight Flutter Testing Symposium*, NASA-SP-385, 1958, 83–90.

29. C. V. Stahle, Jr., Phase Separation Technique for Ground Vibration Testing, *Aerospace Eng.*, July 1962, 8 pp.

30. J. W. Pendered, and R. E. D. Bishop, A Critical Introduction to Some Industrial Resonance Testing Techniques, *J. Mech. Eng. Sci.*, 5, no. 4 (1963), 368–378.

31. J. W. Pendered, and R. E. D. Bishop, Extraction of Data for a Subsystem from Resonance Test Results, *J. Mech. Eng. Sci.*, 5, no. 4 (1963), 368–378.

32. J. W. Pendered, and R. E. D. Bishop, The Determination of Modal Shapes in Resonance Testing, *J. Mech. Eng. Sci.*, 5, no. 4 (1963), 379–385.

33. R. E. D. Bishop and G. M. L. Gladwell, An Investigation into the Theory of Resonance Testing, *Phil. Trans. R. Soc. London, Ser. A*, 225, A-1055 (1963), 241–280.

34. A. Klosterman, On the Experimental Determination and Use of Modal Representations of Dynamic Characteristics, Ph.D. Dissertation, Department of Mechanical Engineering, University of Cincinnati, 1971, 184 pp.

35. Patrick Van Loon, MOdal Parameters of Mechanical Structures, Ph.D. dissertation, University of Leuven, Belgium, 1974, 183 pp.

36. A. Klosterman, and R. Zimmerman, Modal Survey Activity via Frequency Response Functions, SAE Paper 751068, 1975.

37. R. W. Potter, A General Theory of Modal Analysis for Linear Systems, *Shock Vib. Digest*, 7, no. 11 (1975).

38. E. Sloane, and B. McKeever, Modal Survey Techniques and Theory, SAE Paper 751067, 1975, 27 pp.

39. M. Richardson, Modal Analysis Using Digital Test Systems, *Seminar on Understanding Digital Control and Analysis in Vibration Test Systems* (Part 2), 1975, 43–64.

40. B. K. Wada, Modal Test—Measurement and Analysis Requirements, SAE Paper 751066, 1975, 17 pp.

41. K. Ramsey, Effective Measurements for Structural Dynamics Testing Part I, *Sound Vib.*, November 1975.

42. K. Ramsey, Effective Measurements for Structural Dynamics Testing Part II, *Sound Vib.*, April 1976.

43. N. Mirimand, J. F. Billaud, F. Leleux, and J. P. Krenevez, Identification of Structural Modal Parameters by Dynamic Tests at a Single Point, *Shock Vib. Bull, 46,* part 5 (1976), 197–212.

44. E. L. Peterson, and A. L. Klosterman, Obtaining Good Results from an Experimental Modal survey, *Society of Environmental Engineers, Symposium,* London, 1977, 22 pp.

45. C. V. Stahle, Modal Test Methods and Applications, *J. Environ. Sci.,* January/February 1978, 4 pp.

46. G. D. Carbon, D. L. Brown and R. D. Zimmerman, Survey of Excitation Techniques Applicable to the Testing of Automotive Structures, SAE Paper 770029, 1977.

47. R. J. Allemang, Investigation of Some Multiple Input/Output Frequency Response Function Experimental Modal Analysis Techniques, Ph.D. dissertation, Department of Mechanical Engineering, University of Cincinnati, 1980, 358 pp.

48. R. D. Zimmerman, D. L. Brown, and R. J. Allemang, Improved Parameter Estimation Techniques for Experimental Modal Analysis, Final Report: NASA Grant NSG-1486, NASA-Langley Research Center, 23 pp., 1982.

49. R. J. Allemang, and D. L. Brown, A Correlation Coefficient for Modal Vector Analysis, *Proc. International Modal Analysis Conf.,* 1982, 110–116.

50. R. J. Allemang, R. W. Rost, and D. L. Brown, Dual Input Estimation of Frequency Response Functions for Experimental Modal Analysis of Aircraft Structures, *Proc. Int. Modal Analysis Conf.,* 1982, 333–340.

51. G. D. Carbon, D. L. Brown, and R. J. Allemang, Application of Dual Input Excitation Techniques to the Modal Testing of Commercial Aircraft, *Proc. Int. Modal Analysis Conf.,* 1982, 559–565.

52. R. J. Allemang, D. L. Brown, and R. W. Rost, Dual Input Estimation of Frequency Response Functions for Experimental Modal Analysis of Automotive Structures, SAE Paper 821093, 1982.

53. R. J. Allemang, R. W. Rost, and D. L. Brown, Multiple Input Estimation of Frequency Response Functions: Excitation Considerations, ASME Paper 83-DET-73, 1983, 11 pp.

54. K. B. Elliott, and L. D. Mitchell, Improved Frequency Response Function Circle Fits, *Modal Testing and Model Refinement,* ASME AMD, 59, 1983, 63–76.

55. R. J. Allemang, D. L. Brown, and R. W. Rost, Multiple Input Estimation of Frequency Response Functions for Experimental Modal Analysis, U.S. Air Force Report Number AFATL-TR-84-15, 1984, 185 pp.

56. R. W. Rost, Investigation of Multiple Input Frequency Response Function Estimation Techniques for Modal Analysis, Ph.D. dissertation, Department of Mechanical Engineering, University of Cincinnati, 1985, 219 pp.

57. H. Vold, J. Crowley, and G. Rocklin, A Comparison of H_1, H_2, H_v, Frequency Response Functions, *Proc. Int. Modal Analysis Conf.,* 1985, 272–278.

58. C. Van Karsen, A Survey of Excitation Techniques for Frequency Response Function Measurement, M. Sci. thesis, University of Cincinnati, Department of Mechanical Engineering, 1987, 81 pp.

59. L. D. Mitchell, R. E. Cobb, J. C. Deel, and Y. W. Luk, An Unbiased Frequency Response Function Estimator, *J. Anal. Exp. Modal Anal., 3,* no. 1 (1988), 12–19.

60. R. E. Cobb, and L. D. Mitchell, A Method for the Unbiased Estimate of System FRFs in the Presence of Multiple-Correlated Inputs, *J. Anl. Exp. Modal Anal. 3,* no. 4 (1988), 123–128.

61. H. Vold, J. Kundrat, T. Rocklin, and R. Russell, A Multi-Input Modal Estimation Algorithm for Mini-Computers, SAE Paper 820194, 1982, 10 pp.

62. L. Zhang, H. Kanda, D. L. Brown, and R. J. Allemang, A Polyreference Frequency Domain Method for Modal Parameter Identification, ASME Paper 85-DET-106 1985, 8 pp.

63. F. Lembregts, R. Snoeys, and J. Leuridan, Multiple Input Modal Analysis of Frequency Response Functions Based on Direct Parameter Identification, *Proc. 10th Int. Seminar Modal Analysis,* Part IV, Katholieke Universiteit te Leuven, 1985.

64. R. Prony, Essai Experimental et Analytique sur les Lois de la Dilatibilé des Fluides Elastiques et sur celles de la Force Expansive de la Vapeur de l'Eau et de la Vapeur de l'Alkool, à Differentes Températures, *J. Ec. Polytech. (Paris), 1,* no. 2 [Floreal et Prairial, An. III] (1795), 24–76.

65. S. R. Ibrahim, and E. C. Mikulcik, A Method for the Direct Identification of Vibration Parameters from the Free Response, *Shock Vib. Bull, 47,* part 4 (1977), 183-198.

66. S. R. Ibrahim, Modal Confidence Factor in Vibration Testing, *Shock Vib. Bull, 48,* part 1 (1978), 65-75.

67. S. R. Ibrahim, and E. C. Mikulcik, A Method for the Direct Identification of Vibration Parameters from the Free Response, *Shock Vib. Bull., 47,* part 4 (1977), 183-198.

68. S. R. Ibrahim, The Use of Random Decrement Technique for Identification of Structural Modes of Vibration, AIAA Paper 77368, 1977, 10 pp.

69. R. S. Pappa, and S. R. Ibrahim, A parametric Study of the "ITD" Modal Identification Algorithm, *Shock Vib. Bull., 51,* part 3 (1981), 43-72.

70. R. S. Pappa, Some Statistical Performance Characteristics of the "ITD" Modal Identification Algorithm, AIAA Paper 82-0768, 1982, 19 pp.

71. K. Fukuzono, Investigation of Multiple-Reference Ibrahim Time Domain Modal Parameter Estimation Technique, M.Sc. thesis, University of Cincinnati, Department of Mechanical Engineering, 220 pp., 1986.

72. R. S. Pappa, and J. N. Juang, Galileo Spacecraft Modal Identification Using an Eigensystem Realization Algorithm, AIAA Paper 84-1070-CP, 1984, 18 pp.

73. R. S. Pappa, and J. N. Juang, An Eigensystem Realization Algorithm (ERA) for Modal Parameter Identification, NASA-JPL Workshop, Identification and Control of Flexible Space Structures, Pasadena, CA, June 1984, 20 pp.

74. J. N. Juang and R. S. Pappa, Effects of Noise on ERA-Identified Modal Parameters, AAS Paper AAS-85-422, August 1985 (Vail, CO), 23 pp.

75. B. L. Ho, and R. E. Kalman, Effective Construction of Linear State-Variable models from Input/Output Data, *Proc. 3rd Ann. Allerton Conf. Circuit and System Theory,* 1965, 449-459.

76. J. N. Juang, and R. S. Pappa, An Eigensystem Realization Algorithm (ERA) for Modal Parameter Identification and Model Reduction, NASA/JPL Workshop on Identification and Control of Flexible Space Structures, Pasadena, CA, June 1984, 20 pp.

77. R. S. Pappa, and J. N. Juang, Galileo Spacecraft Modal Identification Using an Eigensystem Realization Algorithm AIAA Paper 84-1070, 1984; *J. Astronaut. Sci.,* January/March 1985.

78. J. N. Juang, and R. S. Pappa, An Eigensystem Realization Algorithm for Modal Parameter Identification and Model Reduction, *AIAA J. Guid. Control Dyn., 8,* no. 4 (1985), 620-627.

79. R. S. Pappa, The Eigensystem Realization Algorithm, *Advanced Modal Analysis Seminar,* University of Cincinnati, 1985, 106 pp.

80. J. N. Juang, Mathematical Correlation of Modal Parameter Identification Methods via System Realization Theory, *J. Anal. Exp. Modal Anal., 2,* no. 1 (1987), 1-18.

81. J. N. Juang and R.S. Pappa, A Comparative Overview of Modal Testing and System Identification for Control of Structures, *Shock Vib. Digest, 20,* no. 6 (1988), 4-15.

82. R. W. Longman and J. N. Juang, Recursive Form of the Eigensystem Realization Algorithm for System Identification, *AIAA J. Guid. Control Dyn., 12,* no. 5 (1989), 647-652.

83. M. Bergman, R. W. Longman, and J. N. Juang, Variance and Bias Computation for Enhanced System Identification, *IEEE, 28th Conf. Decision and Control,* 1989, 8 pp.

84. W. Gersch, On the Achievable Accuracy of structural System Parameter Estimates, *J. Sound Vib., 34,* no. 1 (1974), 63-79.

85. W. Gersch, and S. Luo, Discrete Time Series Synthesis of Randomly Excited Structural System Responses, *J. Acous. Soc. Am. A, 51,* no. 1 (1972), 402-408.

86. W. Gersch, and D. R. Sharpe, Estimation of Power Spectra with Finite-Order Autoregressive Models, *IEEE Trans. Autom. Control, AC-18* (August 1973), 367-369.

87. W. Gersch, and D. A. Fouth, Least Squares Estimates of Structural System Parameters Using Covariance Function Data, *IEEE Trans. Autom. Control, AC-19,* no. 6 (December 1974), 898-903.

88. S. M. Pandit, and H. Suzuki, Application of Data Dependent Systems to Diagnostic Vibration Analysis, ASME Paper 79-DET-7, September 1979, 9 pp.

89. M. Link, and A. Vollan, Identification of Structural System Parameters from Dynamic Response Data, *Z. Flugwiss.* 2, no. 3 (1978), 165–174.

90. A. Berman, and W. G. Flannelly, Theory of Incomplete Models of Dynamic Structures, *AIAA J.*, 9, no. 8 (1971), 1481–1487.

91. J. Leuridan, D. Brown, and R. Allemang, Direct System Parameter Identification of Mechanical Structures with Application to Modal Analysis, AIAA Paper 82-0767 *Proc. 23rd Structures, Structural Dynamics Materials Conf.*, Part 2, 1982, 548–556.

92. J. Leuridan, and J. Kundrat, Advanced Matrix Methods for Experimental Modal Analysis—A Multi-Matrix Method for Direct Parameter Extraction, *Proc. Int. Modal Analysis Conf.*, 1982, 192–200.

93. J. Leuridan, Some Direct Parameter Model Identification Methods Applicable for Multiple Input Modal Analysis, Ph.D. dissertation, Department of Mechanical Engineering, University of Cincinnati, 1984, 384 pp.

94. D. Formenti, R. J. Allemang, and R. W. Rost, Analytical and Experimental Modal Analysis, University of Cincinnati, Modal Analysis Course Notes, 1978 (rev. 1990).

95. M. Rades, Analysis of Measured Structural Frequency Response Data, *Shock Vib. Digest*, 14, no. 4 (1982), 21–32.

96. R. J. Allemang, and D. L. Brown, Experimental Modal Analysis and Dynamic Component Synthesis, AFWAL-TR-87-3069, Vol. III, *Modal Parameter Estimation*, 1987, 200 pp.

97. F. Lembregts, J. Leuridan, L. Zhang, and H. Kanda, Multiple Input Modal Analysis of Frequency Response Functions Based on Direct Parameter Identification, *Proc. 4th Int. Modal Analysis, Conf.*, 1986, 598.

98. L. G. Kelly, *Handbook of Numerical Methods and Applications*, Addison-Wesley, Reading, MA,.

99. H. Vold, Orthogonal Polynomials in the Polyreference Method, *Proc. 11th Int. Seminar Modal Analysis*, Katholieke Universiteit de Leuven, 1986.

100. H. Van Der Auweraer, and J. M. Leuridan, Multiple Input Orthogonal Polynomial Parameter Estimation, *Proc. 11th Int. Seminar Modal Analysis*, Katholieke Universiteit de Leuven, 1986.

101. H. Van der Auweraer, and J. Leuridan, Multiple Input Orthogonal Polynomial Parameter Estimation, *Mech. Syst. Signal Process*, 1, no. 3 (1987), 259–272.

102. G. M. Lee, M. W. Trethewey, Application of the Least Squares Time Domain Algorithm for Efficient Modal Parameter Estimation, *Mech. Syst. Signal Process*, 2, no. 1 (1988), 21–38.

103. D. Capecchi, Difference Models for Identification of Mechanical Linear Systems in Dynamics, *Mech. Syst. Signal Process*, 3, no. 2 (1989), 157–172.

104. H. Vold, and T. Rocklin, The Numerical Implementation of a Multi-Input Modal Estimation Algorithm for Mini-Computers, *Proc. Int. Modal Analysis Conf.*, 1982, 542–548.

105. M. Rades, System Identification Using Real Frequency Dependent Modal Characteristics, *Shock Vib. Digest*, 18, no. 8 (1986).

106. G. R. Tomlinson, Detection, Identification, and Quantification of Nonlinearity in Modal Analysis—A Review, *Proc. Int. Modal Analysis Conf.*, 1986, 837–843.

107. T. Vinh, A. Haoui, and Y. Chevalier, Extension of Modal Analysis to Nonlinear Structures by Using the Hilbert Transform, *Proc. 2nd Int. Modal Analysis Conf.*, 1984, 852–857.

108. R. Williams, J. Crowley, and H. Vold, The Multivariate Mode Indicator Function in Modal Analysis, *Proc. 3rd Modal Analysis Conf.*, 1985 66–70.

109. S. I. Gravitz, An Analytical Procedure for Orthogonalization of Experimentally Measured Modes, *J. Aero/Space Sci.*, 25, (1958), 721–722.

110. J. McGrew, Orthogonalization of Measured Modes and Calculation of Influence Coefficients, *AIAA J.*, 7, no. 4 (1969), 774–776.

111. W. P. Targoff, Orthogonality Check and Correction of Measured Modes, *AIAA J.*, 14, no. 2 (1976), 164–167.

112. R. J. Guyan, Reduction of Stiffness and Mass Matrices, *AIAA, Jr.*, 3, no. 2 (February 1965), 380.

113. B. Irons, Structural Eigenvalue Problems: Elimination of Unwanted Variables, *AIAA J.*, *3*, no. 5 (May 1965), 961–962.

114. B. Downs, Accurate Reduction of Stiffness and Mass Matrices for Vibration Analysis and a Rationale for Selecting Master Degrees of Freedom, ASME Paper 79-DET-18, 1979, 5 pp.

115. J. D. Sowers, Condensation of Free Body Mass Matrices Using Flexibility Coefficients, *AIAA J.*, *16*, no. 3 (March 1978), 272–273.

CHAPTER
17

Hybrid Experimental-Numerical Stress Analysis

Albert S. Kobayashi

University of Washington
Seattle, Washington

17-0 Introduction

One of the frustrations of an experimental stress analyst is the lack of a universal experimental procedure that solves all problems. Referring to his second principle, Durelli states that "seldom does one method give a complete solution, with the most efficiency" [1]. Examples of this second principle are seen in photoelastic-coating and brittle-coating techniques that require additional strain-gage testing in locations of high stress concentrations identified by these two techniques. The hybrid experimental-numerical technique is an aberration of the above where numerical stress analysis replaces the second experimental method.

Early applications of the hybrid experimental-numerical stress analysis technique were limited to separation of two- and three-dimensional states of stress in photoelastic specimens. The first stress invariant obtained through a finite-difference solution to the compatibility equation and the maximum shear stress $(\sigma_1 - \sigma_2)/2$ distribution provided by the isochromatics yielded the two planar components of the principal stresses of σ_1 and σ_2. [2–4]. Figure 17-1 illustrates the application of this solution procedure of separating the two stress components in a ring subjected to diametral compression [4]. The shear-difference method [2,5] and Filon's method [6,7], on the other hand, used isochromatic and isoclinic data to integrate the equilibrium equations along a straight line and a stress trajectory, respectively. The latter two single-purpose numerical techniques provided only the stresses along a specified integration path. Figure 17-2 illustrates the application of Filon's method in separating the two stress components in a disk subjected to diametral compression.

In contrast to the above, the modern supercodes based on the finite-element, boundary element, and finite-difference methods yield the complete states of stress, strain, and displacement for the given constitutive relations and boundary and initial conditions. The uncertainty or lack of knowledge in these given conditions, however, limited the accuracy and the otherwise voluminous outputs of these supercodes. Inaccurate numerical modeling procedures generated results with obvious errors and are credited for the resurgence of three-dimensional photoelasticity in the 1970s. The hybrid experimental-numerical stress analysis technique of today reduces, if not eliminates,

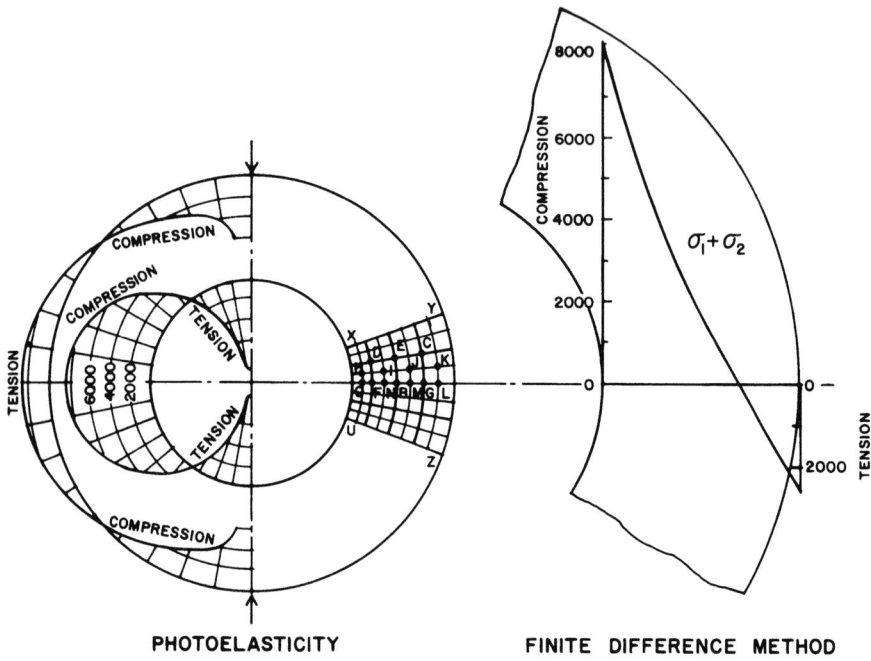

Figure 17-1 Circular ring under diametrical compression. (From Ref. 4.)

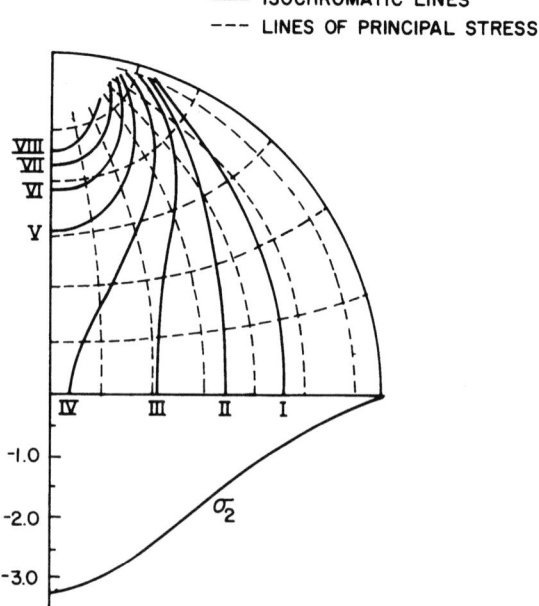

Figure 17-2 Disk under diametrical compression. Isochromatic fringe order shown by roman numerals. (From Ref. 6.)

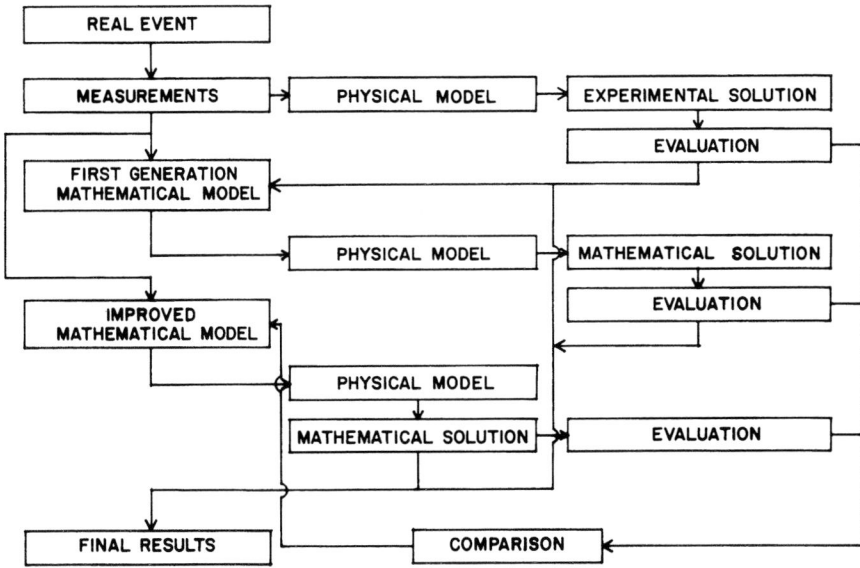

Figure 17-3 Flowchart of hybrid solution. (Adapted from Fig. 1 of Ref. 9.)

the foregoing uncertainties in prescribed input conditions by using experimentally determined boundary and initial conditions. The output from the otherwise proven numerical techniques is either the constitutive relation or the complete states of displacement, strain, and stress that cannot be readily extracted through the use of a single experimental technique in stress analysis. Thus the hybrid experimental-numerical technique is an extremely efficient stress analysis technique that often provides more information than needed.

17-1 Hybrid Experimental-Numerical Stress Analysis

In hybrid experimental-numerical stress analysis, as described in the preceding section, data are generated through both experimental and numerical procedures. The use of highly sophisticated numerical procedures for processing experimental data does not fall within this category due to the passive role of the former. Although this book does not contain a separate chapter on numerical processing of experimental data, the importance of numerical processing is emphasized throughout with the unavoidable and extremely effective uses of minicomputers in almost every experimental procedure. The full potential of hybrid experimental-numerical stress analysis, on the other hand, is yet to be developed and its procedure has been formalized only recently [8,9].

Laerman [8,9] noted that experimental data are interpreted through a physical model of the real event which can equally be interpreted through mathematical and physical models. A combination of mathematical and physical models, which augment each other, thus "yields a better knowledge of the structure reactions and the most reliable results in stress analysis." Figure 17-3 shows schematically the sequence of data transmission and feedback in the hybrid technique. The physical model in this figure represents the modeling necessary to convert the experimental measurements to meaningful results, such as the constitutive relation for converting strain-gage data to stresses. This flowchart accounts for the inevitable updating of the first-generation mathematical model, which is bound to be simple, through comparative study of related mathematical and experimental results. The outcome of this comparative study is an improved mathematical model, which yields complementary results to those obtained experimentally. The mathematical modeling in Fig. 17-3 encompasses classical modeling, which is often limited in scope, as well as numerical modeling of modern times.

In the following, the utility of the hybrid experimental-numerical stress analysis technique will be demonstrated by some stress analysis problems involving two- and three-dimensional structural components, biomechanics, and fracture mechanics.

17-2 Elastic Analysis of Structural Components

Using Stress Equations of Compatibility

The numerical techniques used in modern hybrid techniques for structural analysis are vastly superior to their predecessors since they provide the entire states of stress, strain, and displacements. As a straightforward extension of the classical hybrid technique for stress analysis, Rao [10] used measured temperature and surface traction data to solve, by the finite-difference method, the Beltrami–Michell stress equations of compatibility interior to an axisymmetric solid. Figure 17-4a shows the end retaining ring, which is shrink-fitted to the two ends of the rotor and is used to contain the end loops of rotor windings, in a turbogenerator. The distributions of hoop stress, which is generated by shrink fitting and the centrifugal force, obtained by the hybrid technique, three-dimensional frozen-stress photoelasticity, and a two-dimensional analog, are shown in Fig. 17-4b. The utility of this hybrid technique is demonstrated by the author's remark that "the time needed for the analysis is smaller than that required by the time-consuming and tedious shear-difference methods." [10].

Jacobs, on the other hand, extended the approach above [11] to orthotropic plane problems [12] and applied the procedure to separate the interfacial stresses of a composite prism under axial compression [13], as shown in Fig. 17-5. In this study, the isochromatic patterns of the center slice of the three-dimensional photoelastic model were used to determine the boundary stresses and maximum shear stress along the interface. Stress equilibrium and strain compatibility equations along the interface were used to obtain the σ_{zz} and σ_{yy} stress distributions shown in Fig. 17-6. Excellent agreement between the computed maximum shear stress $\sigma_{zz} - \sigma_{yy}$ and that obtained directly by three-dimensional photoelasticity is noted.

Using the Three-Dimensional Stress Function

Figure 17-7 shows a water turbine wheel and its curvilinear finite-difference grid representation, which was analyzed by Barishpolsky [14]. Frozen-stress photoelasticity was used to determine the stress tensor on the complex boundaries. These boundary values were input to two curvilinear finite-difference equations of the Laplace and Poisson equations involving the six stress components of three-dimensional linear elasticity, where the number of equations equaled the number of nodes and thus reduced the computational time three- to sixfold over standard finite-difference codes. The procedure is extended to steady-state, three-dimensional problems, where measured surface temperature must be input in addition to the measured surface tractions [15].

The third example in this elastic analysis involves a large deflection of a circular plate that is supported by a yielding foundation [16]. The resultant nonlinear plate equation is difficult to solve

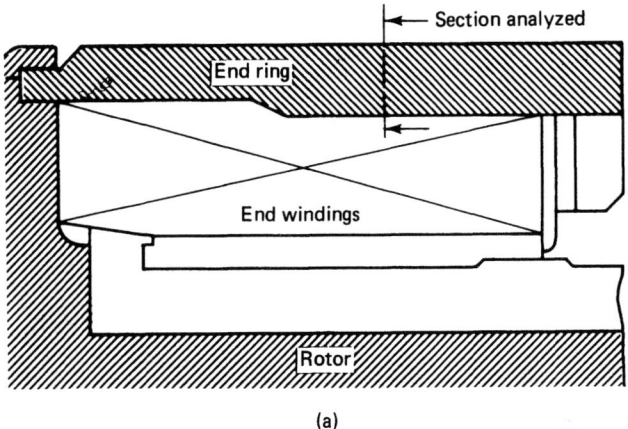

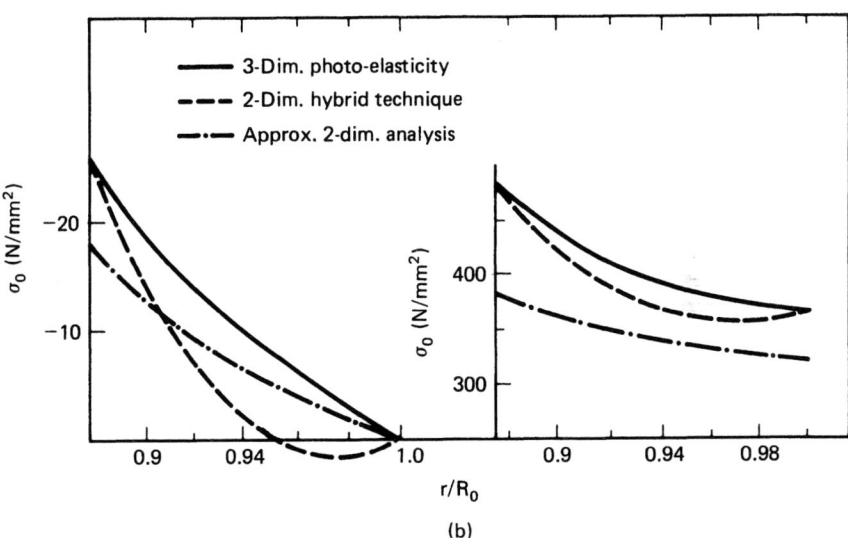

Figure 17-4 (a) End ring assembly on rotor; (b) hoop stress in end ring due to centrifugal forces. (From Ref. 10.)

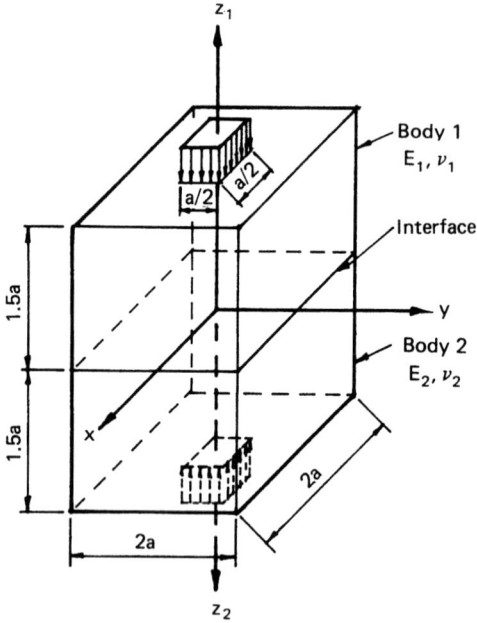

Figure 17-5 Composite prism subjected to partial compression. (From Ref 13.)

even by numerical means. Also, the supporting boundary conditions cannot be determined, since friction forces between the plate and the support are unknown. Laerman used Ligtenberg's reflection moiré method to determine the slope of defected surface of a PSMI plate subjected to a concentrated load as shown in Fig. 17-8 [16]. Numerical differentiation of the slope data then yielded the plate bending moments. The membrane stresses were determined by reflection photoelasticity. The remaining unknowns in the two plate equilibrium equations with geometric nonlinearity were the contact and friction forces. The friction forces were obtained by direct numerical analysis, while the contact forces were obtained numerically after the governing second-order partial differential equation had been converted into a finite-difference equation. Figure 17-9 shows the distributions of in-plane, contact and friction stresses p and t, respectively. The utility of the hybrid technique is demonstrated by the complete solution for the otherwise complex plate problem in Fig. 17-8.

Using the Boundary Element Method

While the specialized codes used in the hybrid techniques above are computationally more efficient by design, off-the-shelf codes in finite-element and boundary element methods are often used for sheer expediency. For an elastostatic problem, the boundary element method is more computationally efficient and natural, since the input data consist of experimentally determined boundary displacements and tractions. When used together with double-exposure, laser speckle interferometry, "the measured surface displacements become the input data needed in the boundary element method to calculate the traction vectors at specified points on the boundary" [17], as well as in the interior of the body. Moslehy and Ranson [17] demonstrated the utility of this hybrid technique

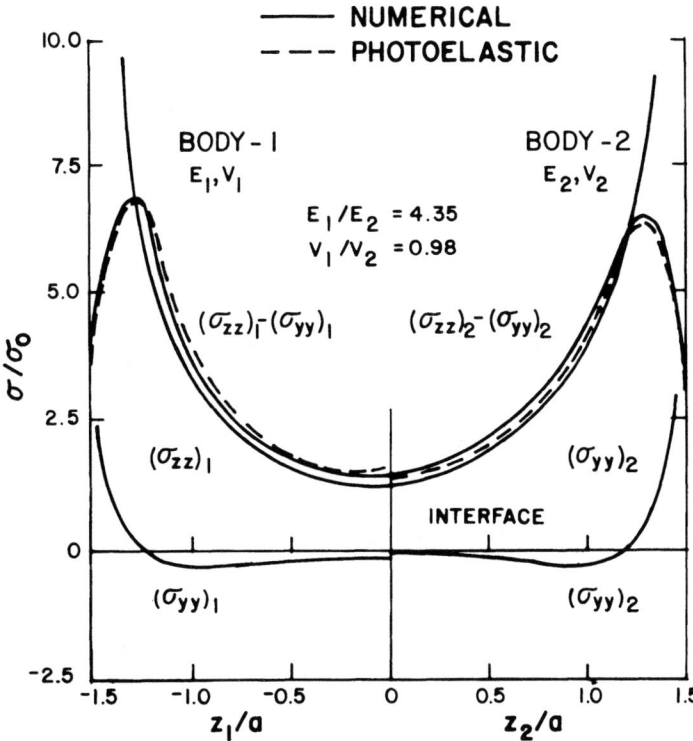

Figure 17-6 Distribution of σ_{zz}, σ_{yy} and $(\sigma_z - \sigma_y)$ along the axis $x = 0$, $y = 0$ and comparison with the corresponding photoelastic results. (From Ref. 13.)

by the excellent agreements in theoretically and experimentally obtained stresses interior to a cantilever beam, as shown in Fig. 17-10, with a transverse end load. In a similar application of the hybrid technique, Balas et al. [18] used the double-aperture, laser speckle interferometry and demonstrated the advantage of the hybrid technique by analyzing only the region of interest of a plate-stiffened frame. In this case, the recorded displacements were input to a simplified boundary, which is represented by dashed lines, of the frame structure shown in Fig. 17-11. The boundary element method was used to determine the stress distributions along the three cross sections shown in Fig. 17-12.

As a variation in the above-mentioned hybrid technique, Umeagukwu et al. [19] used two-dimensional photoelasticity together with a boundary element code to optimize the filets in a double-notch tension plate as shown in Fig. 17-13. The interior principal stresses obtained by the hybrid technique were used to distribute the load more evenly along the net section and resulted in a better understanding of the filet optimization problem.

Corneoscleral Envelope

The interocular pressure of the human eye is maintained at a nearly constant level of 15 to 20 mm Hg through a complex physiological system involving the mechanical, biochemical, and

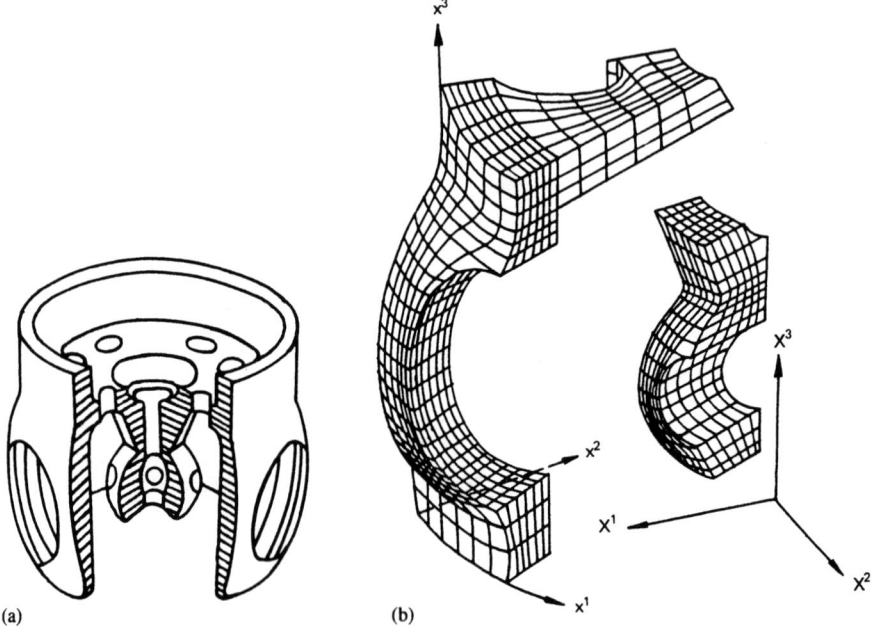

Figure 17-7 (a) Model of the working wheel of a water turbine; (b) finite-difference network. (From Ref 14.)

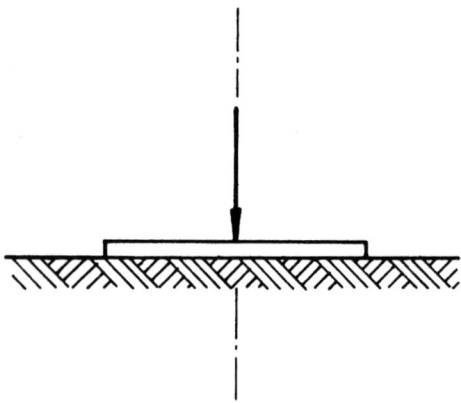

Figure 17-8 Centrally loaded plate supported on yielding subgrade. (From Ref. 16.)

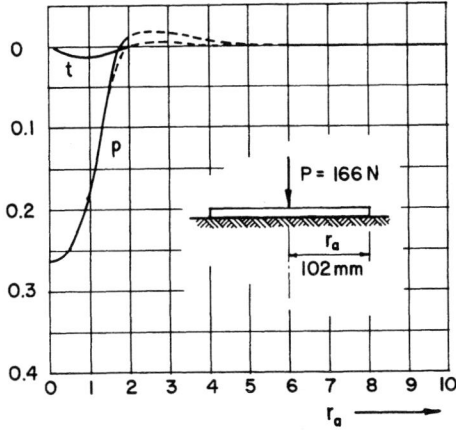

Figure 17-9 Contact and friction stresses p and t. (From Ref. 16.)

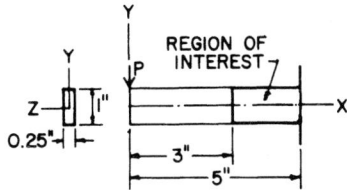

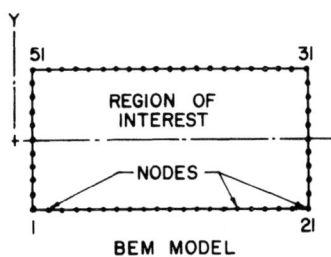

Figure 17-10 Bending of a cantilever beam with a transverse end load. (From Ref 17.)

neurological responses of the eye [20]. When the outflow of ocular fluid is restricted by pathological conditions, the ensuing increase in interocular pressure eventually results in glaucoma, which is the direct cause of 13.5% of the blindness in the United States [21]. Tonometry monitors this interocular pressure by measuring the exterior mechanical response of the cornea, which is indented or flattened by a tonometer plunger. The tonometer reading is thus affected by the mechanical response of the pressurized corneoscleral envelope, which is essentially a pressure vessel containing the optical and neurological components.

The mechanical properties of the cornea and sclera are difficult to obtain because of the small size, delicacy, and natural curvature of these tissues. The commonly used ocular rigidity [22],

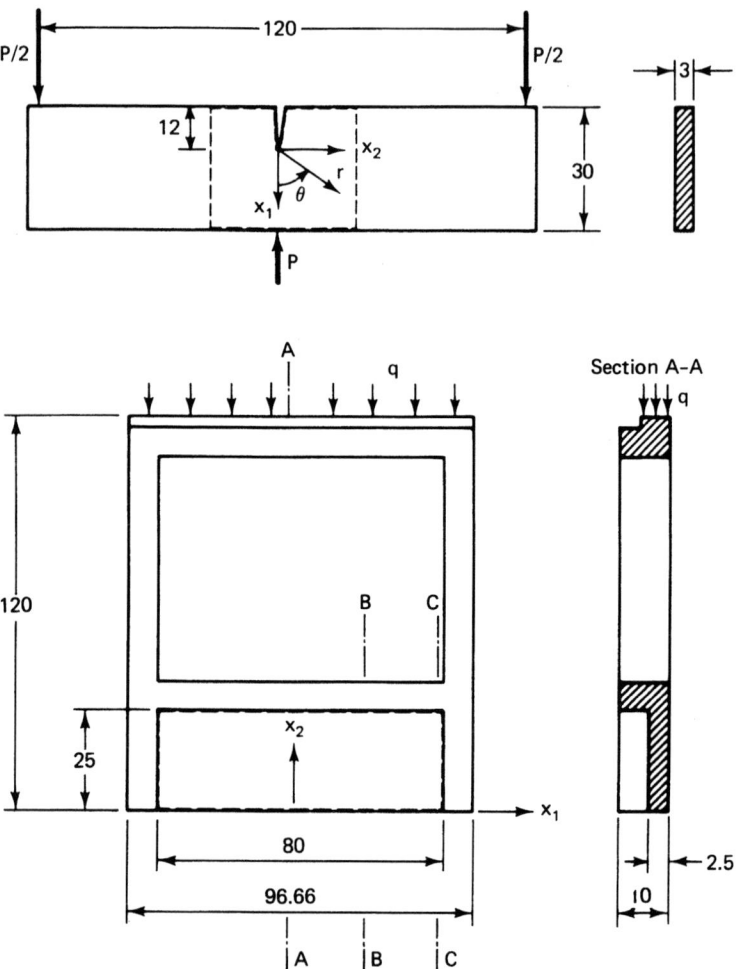

Figure 17-11 Plate-stiffened frame. (From Ref. 18.)

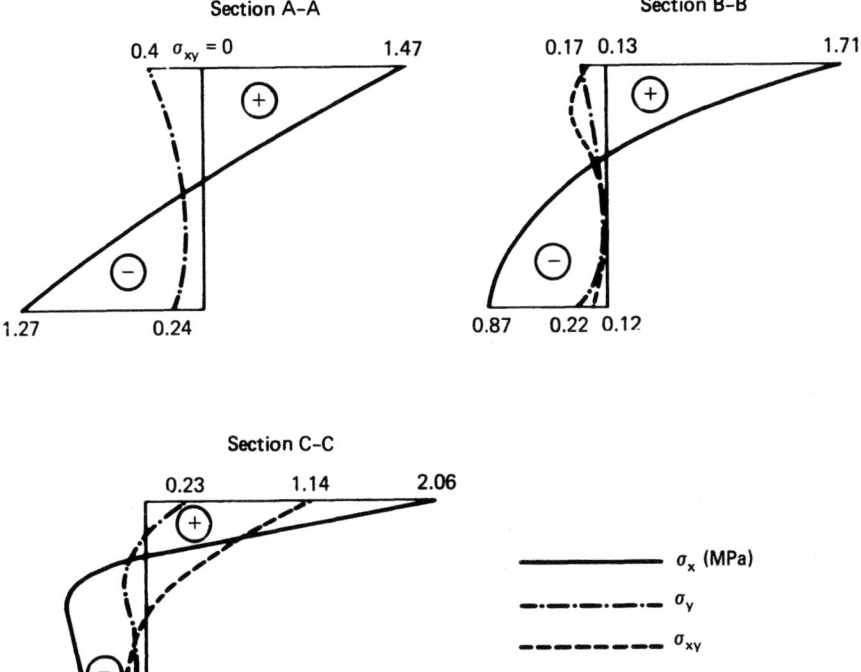

Figure 17-12 Stresses in plate-stiffened frame. (From Ref. 18.)

which relates the pressure and volume of the corneoscleral envelope, is a global coefficient and is not suitable for analyzing the local deformation process under tonometer loading. Simple tension testing of excised strips of the cornea [23] and the sclera [24] yielded erroneous modulus of elasticity and Poisson's ratio by the loosening of the collagen fibrils from the soft mucopolysaccharide at the excised edges. To overcome the deficiencies of these global and local approaches, Woo et al. [25,26] developed a hybrid experimental-numerical procedure for determining the local mechanical properties of an intact corneoscleral envelope.

Woo's experimental procedure consisted of measuring the cornea and sclera deformations as well as the volume changes of pressurized anterior segments of enucleated human eyes. A flying spot scanner, as shown in Fig. 17-14, was used to measure the relative motions of two white targets on the cornea or sclera, which were mounted on a McEwen-type chamber [27]. Woo's numerical procedure consisted of matching, through trial and error, the measured and computed deformations and volume changes. A pressurized axisymmetric finite-element model of the anterior segment of the corneoscleral envelope was used to execute the finite-element code in its application mode for this purpose. The resultant isotropic, trilinear, elastic stress–strain relations obtained for this analog model of the corneoscleral envelope are shown in Fig. 17-15. The trilinear stress–strain relationships were incorporated into a finite-element model of the total eye, which was used to calculate the nonlinear intraocular pressure–volume relation. The lack of bending rigidity in

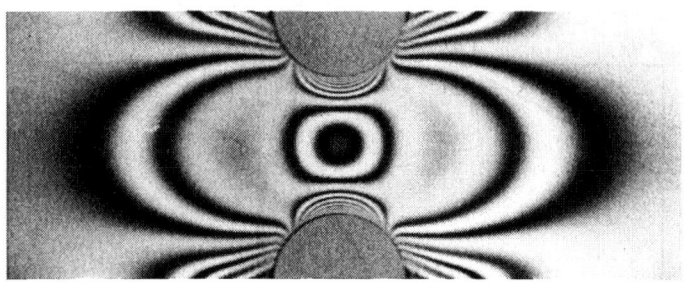

OPTIMIZED GEOMETRY

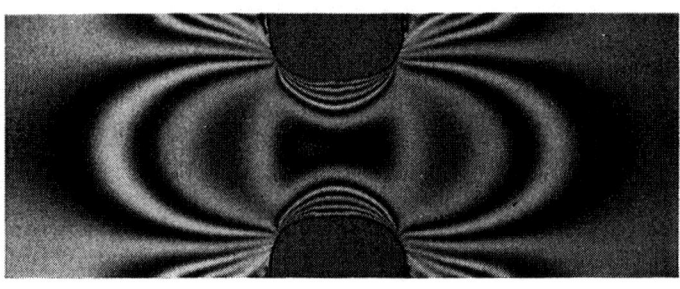

UNOPTIMIZED GEOMETRY

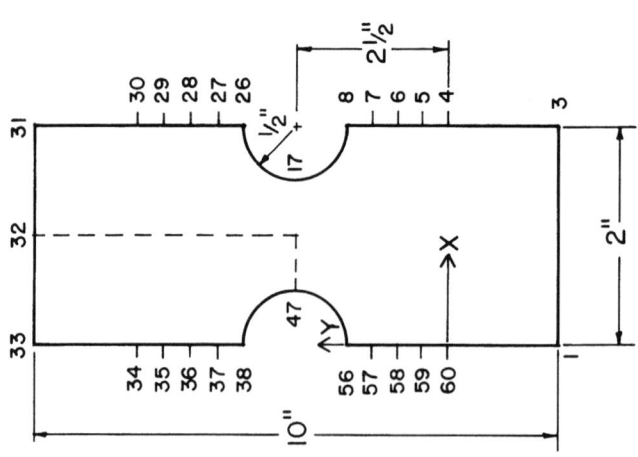

Figure 17-13 Optimization of external notched plate. (From Ref. 19.)

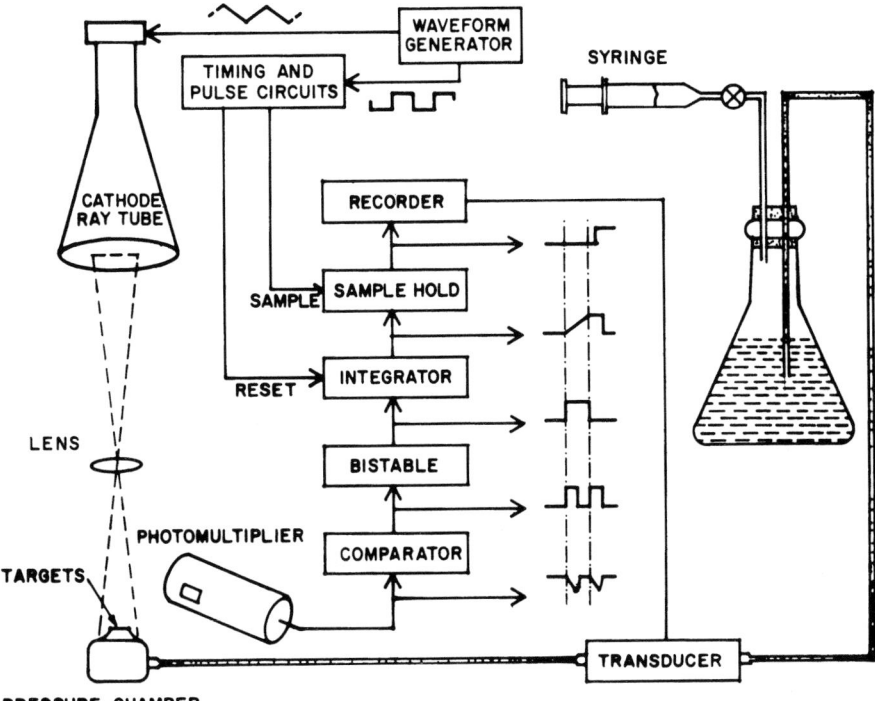

Figure 17-14 Block diagram of flying spot scanner. (From Ref. 25.)

the cornea under the tonometer probe was compensated for by artificially reducing the bending stiffness of the finite elements in the compression region. With this modification, excellent correlations between the calculated [28] applanation tonometer responses and published experimental results [29], as shown in Fig. 17-16, were obtained.

The membrane shell elements, which were later used to construct the corneoscleral envelope, shown in Fig. 17-17, removed the above-mentioned artificial reduction in bending rigidity in the solid elements used by Woo. Woo's experimental data [25] were reevaluated by this membrane shell model, which yielded a slightly different distribution of elastic moduli along the corneoscleral shell. Such differences demonstrate the inevitable interdependence of experimental data and numerical modeling of the hybrid experimental-numerical technique, where the finite-element model is used as an analog model of the experiment.

17-3 Elastic–Plastic Fracture Mechanics

Fracture parameters governing elastic, elastic–plastic, and dynamic fracture, with the exception of geometric quantities, such as crack-opening displacements and crack-tip opening angles, cannot be measured directly. In practice, even the geometric parameters above are difficult to quantify and are often computed by using analog models of the crack. Strain-energy release rate and stress-

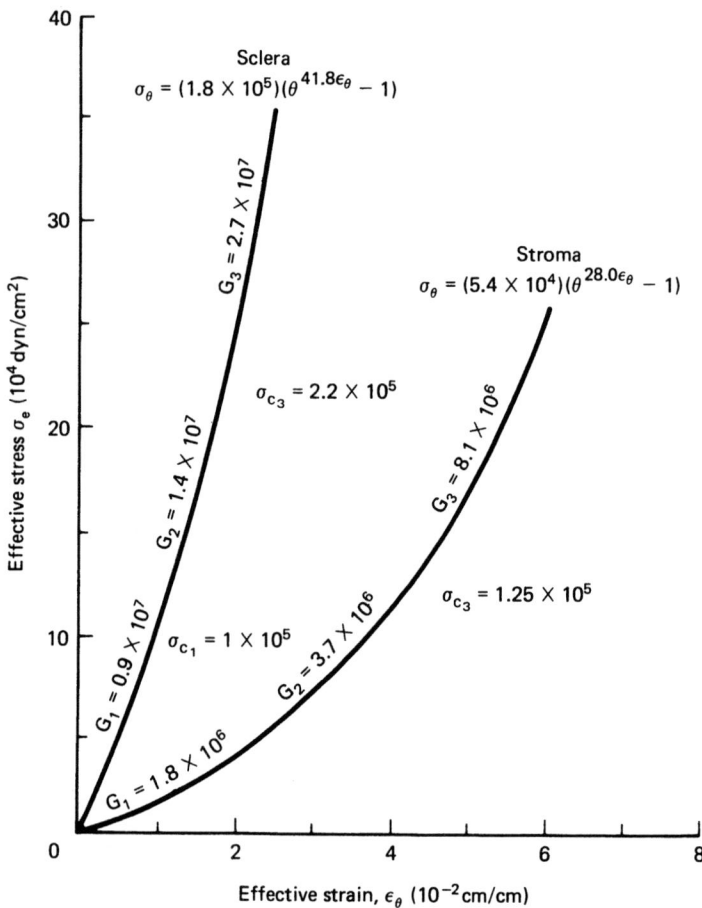

Figure 17-15 Trilinear and experimental stress–strain relation. (From Refs. 25 and 26.)

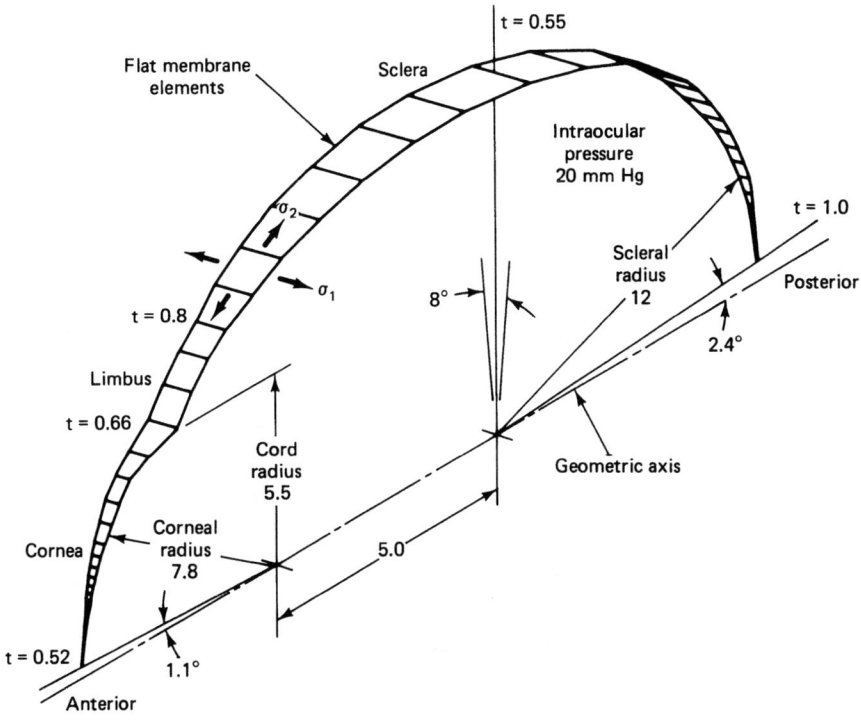

Figure 17-16 Applanation force/intraocular pressure versus applanation area. (From Ref. 28; data points from Ref. 29.)

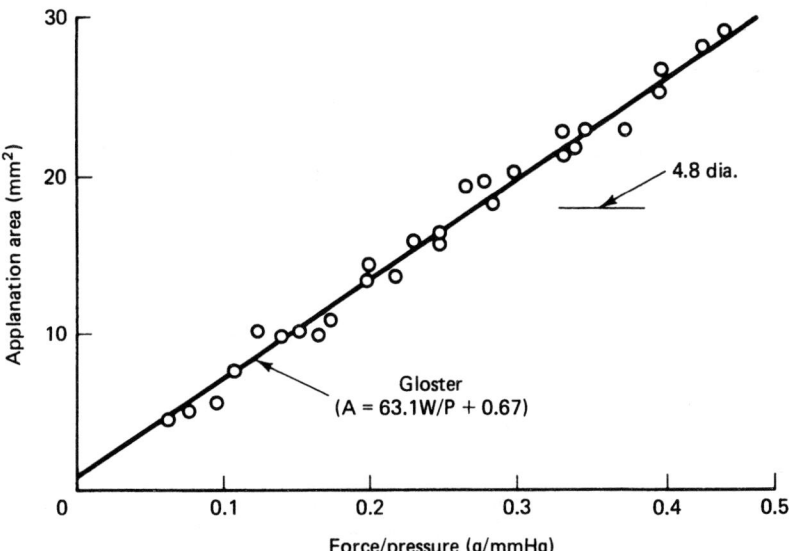

Figure 17-17 Membrane model finite-element arrangement. t, thickness; all dimensions in millimeters.

intensity factor in linear elastic fracture mechanics, which constitute a well-established analog model of the crack, can be computed accurately by using modern numerical codes. The various fracture packages for these codes have been verified by a benchmark problem [30] and thus should provide correct numerical solutions to well-defined boundary value problems. Once the strain-energy release rate or stress-intensity factor has been determined, the onset of brittle fracture can be predicted if the critical values of these quantities are known. Their elastic–plastic extension, the J integral, has also been used with some success in predicting the onset of ductile fracture. Laws governing other fracture phenomena, such as stable crack growth under large-scale yielding, are being investigated through empirical correlations of fracture data with computed fracture parameters.

An approach that has been used to establish a stable crack growth criterion is to input actual crack-growth data as additional boundary values to an elastic–plastic finite-element code. Kanninen et al. [31] used the finite-element code in its "generation mode" to study stable crack growth and instability of A533 steel and 2219-T87 aluminum center-crack and compact specimens. A similar approach was used by Shih et al. [32], who studied stable crack growth and instability of A533-B compact specimens. The experimentally determined load-line displacement versus crack-length relation, as shown in Fig. 17-18, was used to simulate crack extension in the two-dimensional finite-element model shown in Fig. 17-19.

Two sets of elastic–plastic analyses based on J_2 deformation and J_2 flow theories of plasticity were conducted. Figure 17-20 shows excellent agreements between the measured and computed applied load versus load-line displacement relations obtained by these two numerical analyses. The computed fracture parameters included the crack-opening displacement (COD), the crack-opening angle (COA), the J integral, and the rate of change of the J integral, dJ/da. Since the fracture criterion for stable crack growth must be independent of specimen geometry and crack extension, these fracture parameters were then scrutinized for constancy during crack extension. Typical dJ/da and COA variations with crack extensions obtained by Shih et al. are shown in Figs. 17-21 and 17-22, respectively. Both Kaninnen and Shih concluded from their hybrid experimental-numerical investigations that the COA was an attractive fracture criterion for stable crack growth in the presence of large-scale yielding.

The studies above demonstrate the utility of the hybrid experimental-numerical technique in extracting candidate parameters that cannot be obtained directly from experimental or numerical analysis alone. The hybrid experimental-numerical technique provided computed fracture parameters, such as J and dJ/da, under actual test conditions and not under assumed test conditions, which normally would have been prescribed in pure numerical analysis. The technique also yielded numerically consistent COD and COA, which in theory are more measurable but in practice are difficult to determine. The elastic–plastic finite-element codes with fracture packages, including those mentioned above, have yet to be subjected to the rigorous scrutiny leveled on the elastic codes. The wide variations in the J integrals of the mid-1970s [33] hopefully have been reduced, if not eliminated, in the elastoplastic finite-element codes of today.

17-4 Dynamic Fracture

The state of science on dynamic fracture mechanics studied with dynamic photoelasticity was presented by J. W. Dally in his 1979 William M. Murray Lecture [34]. He noted that the crack-tip state of stress provided by dynamic photoelasticity and dynamic caustic techniques will continue to enhance our understanding of the complex phenomena of dynamic crack propagation. Dynamic fracture studies by these techniques, however, are limited to photoelastic polymers and to plane-stress problems when photoelastic coatings or caustics are used. The hybrid experimental-numerical

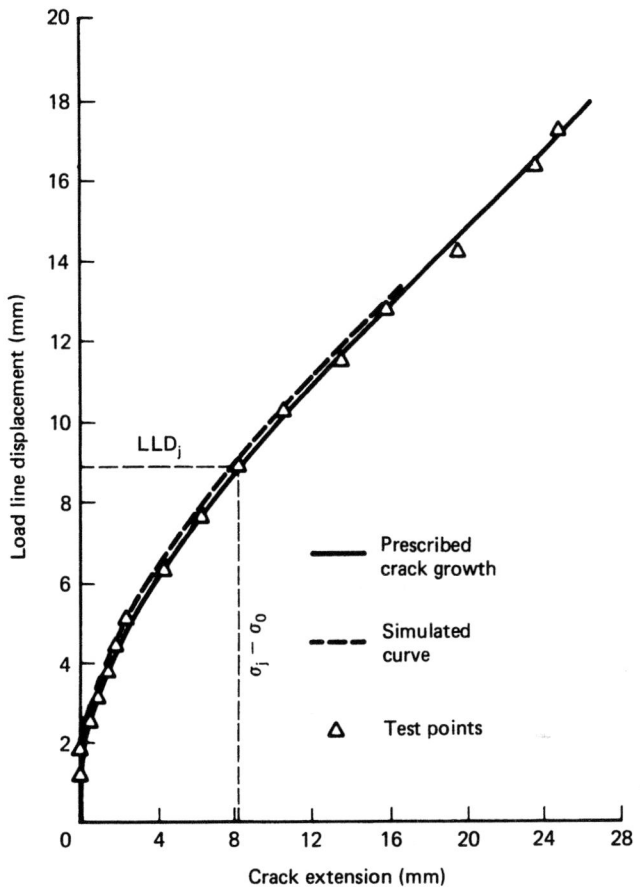

Figure 17-18 Prescribed and simulated load-line displacement versus crack extension for 4T compact specimen. (From Ref 32.)

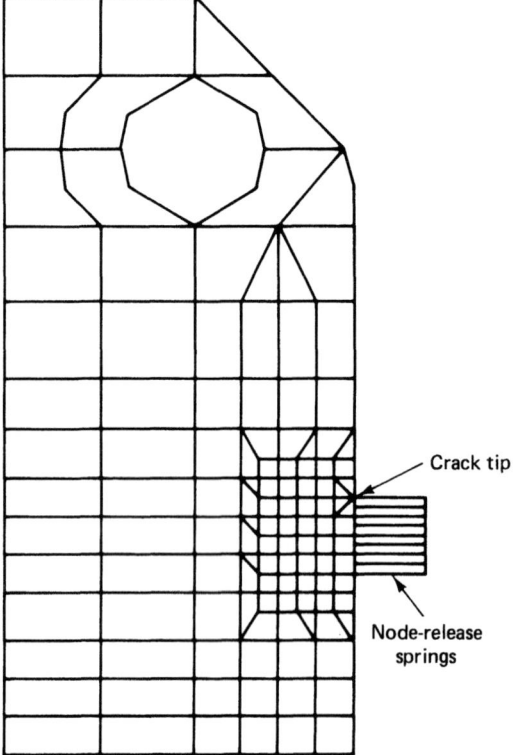

Figure 17-19 Finite-element model for 4T compact specimen. (From Ref 32.)

technique, when used with the generation mode of the finite-element or finite-difference method, will extract dynamic fracture parameters in opaque materials as well as in non-plane-stress problems. These dynamic codes, unlike the well-studied static codes, required verification prior to use in dynamic fracture mechanics. Fracture dynamic results generated by various two-dimensional elastodynamic finite-difference codes [35,36] and finite-element codes [37,38] have been compared [39] with dynamic caustic results of fracturing polymeric specimens [40,41]. Similar verification studies have been conducted with dynamic photoelasticity [42].

The verified numerical code can also be used to check ancillary results deduced from the original experimental results, such as the variation in input work, which cannot be easily measured, during the fracture process. Numerical analysis also provides the transient energy partition for the input boundary and initial conditions. Such energy partition can then be used to check the hypothesis used in deducing the experimentally determined energy partition. The legend of Fig. 17-23 shows an internally notched, semicircular photoelastic specimen that was loaded with end rotation and shear deformation [43]. The reported dynamic fracture toughness versus crack-velocity relation [34] was used as a dynamic fracture criterion to execute a dynamic finite-element code in its application mode, which yielded the crack propagation and dynamic stress intensity factor histories [44], which are in good agreement with the numerical results. Having verified the numerical modeling of the photoelastic experiment, the energies during crack propagation [45] were computed

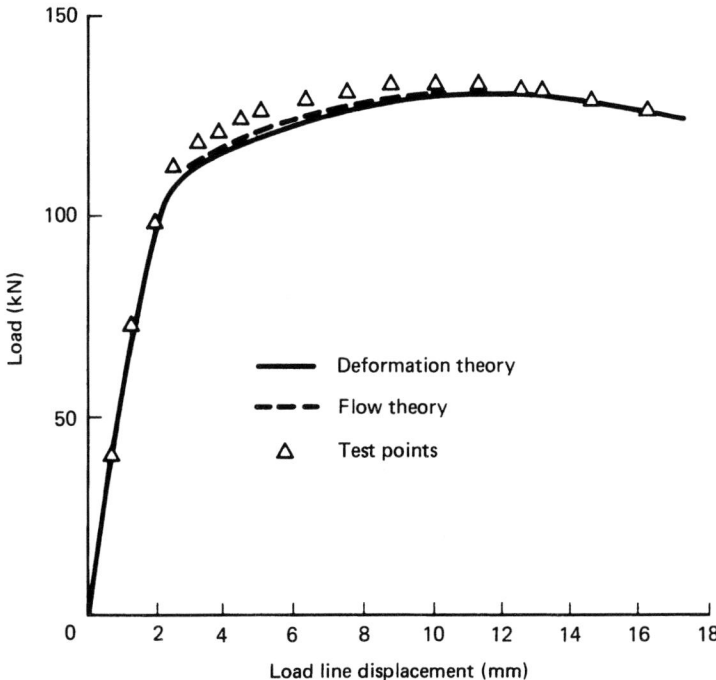

Figure 17-20 Applied load versus load-line displacement for 4T compact specimen, 25% side-grooved, $W - a_0 = 40$ mm (1.593 in.). (From Ref. 32.)

and plotted as shown in Fig. 17-24. The internal consistency in the computed energy partition verifies the basic postulates of negligible viscoelastic damping and negligible energy dissipation at the finite specimen boundaries during the dynamic crack propagation period.

A relatively simple application of the hybrid technique is the determination of the dynamic stress-intensity factor in an impacted notch bend specimen [46]. Measured time variations in the striker load were input to the finite-element model of a dynamic finite-element code, which was then used to compute the time variations in the dynamic finite-element code. This code was then used to compute the time variations in the dynamic stress-intensity factor [47]. The numerical code was also verified by comparing the computed and measured dynamic strains near the crack tip, as shown in Fig. 17-25. Figure 17-26 shows the variations in dynamic and corresponding static stress-intensity factors with time prior to the crack propagation. These results show the inadequacy of the static stress-intensity factor, which was computed by using a static formula and the instantaneous striker load. It also indicates the futility of interpreting such impact fracture responses without the use of proper dynamic analysis [47]. As a verification of codes, Fig. 17-27 shows the agreement between three independent dynamic fracture analyses of another impacted three-point bend specimen [48].

The effectiveness of the hybrid experimental-numerical technique in dynamic fracture can be demonstrated by the extraction of the dynamic fracture toughness K_{ID} in structural materials. Kanazawa et al. [49] used the measured crack velocities to drive a dynamic finite-difference code in its generation mode to analyze the crack propagation and arrest behaviors in 15-mm-thick

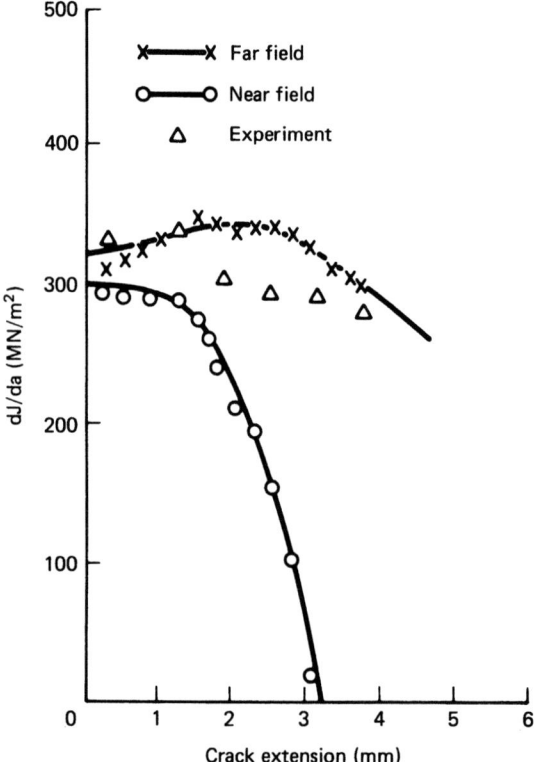

Figure 17-21 dJ/da for compact specimen T-61, $a/W = 0.801$. (From Ref. 32.)

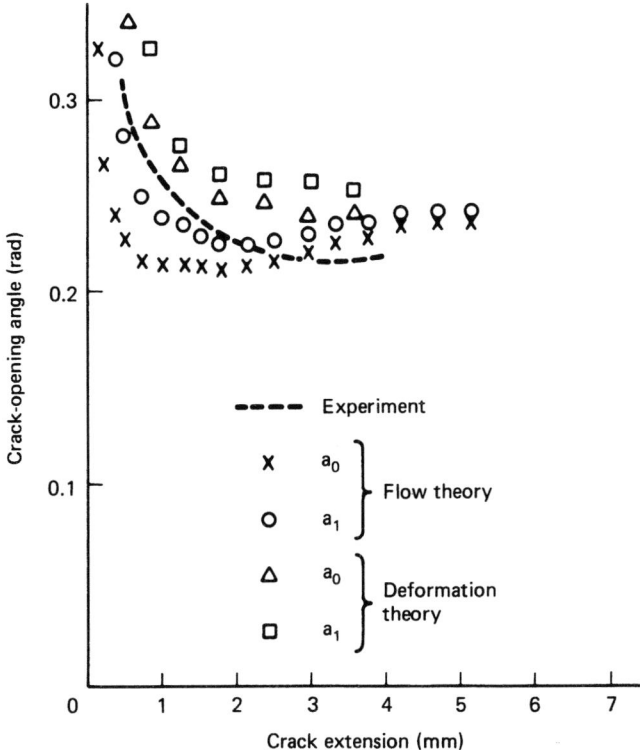

Figure 17-22 Crack-opening angle versus crack extension for 4T compact specimen. (From Ref. 32.)

structural steel. Figure 17-28 shows the double-tension specimens, which were fractured at a constant temperature of -40 to $-61°C$ and at an increasing temperature gradient to room temperature. Under the latter temperature gradient, the crack was arrested because of increasing toughness. Crack velocities were measured by crack gages with a rung spacing of 30 mm. Transient strains, which were also measured at the strain-gage location shown in Fig. 17-28, were used to check the numerical analysis. Variations in the static and dynamic stress-intensity factors K_S and K_D, with crack propagations, as shown in Fig. 17-29, were determined by the finite-difference codes. Figure 17-30 shows the resultant dynamic fracture toughness versus crack-velocity relation for this series of dynamic fracture testing.

Figure 17-31 shows a wedge-loaded, modified-tapered double-cantilever-beam (WL-MTDCB) specimen that was fabricated from plate glass [50]. The specimen was 25% side-grooved to guide the propagating crack. The flexible, long, tapered beam sections were designed to lessen the friction with the silicon carbide loading pin. The specimen was wedge-loaded to fracture in a 500-kg Instron testing machine, and the crack extension history was recorded by a Krak-Gage and associated instrumentation [50]. Figure 17-32 shows typical crack length versus time data, which are characterized by the unambiguous initial period of crack acceleration and have not been observed in dynamic fracture of metals and photoelastic polymers. The average of the two data sets, which is represented by the solid curve in Fig. 17-32, was used to drive a dynamic finite-element code in its generation mode. The resultant K_I^{dyn} as well as the static stress-intensity factor K_I^{stat}, which

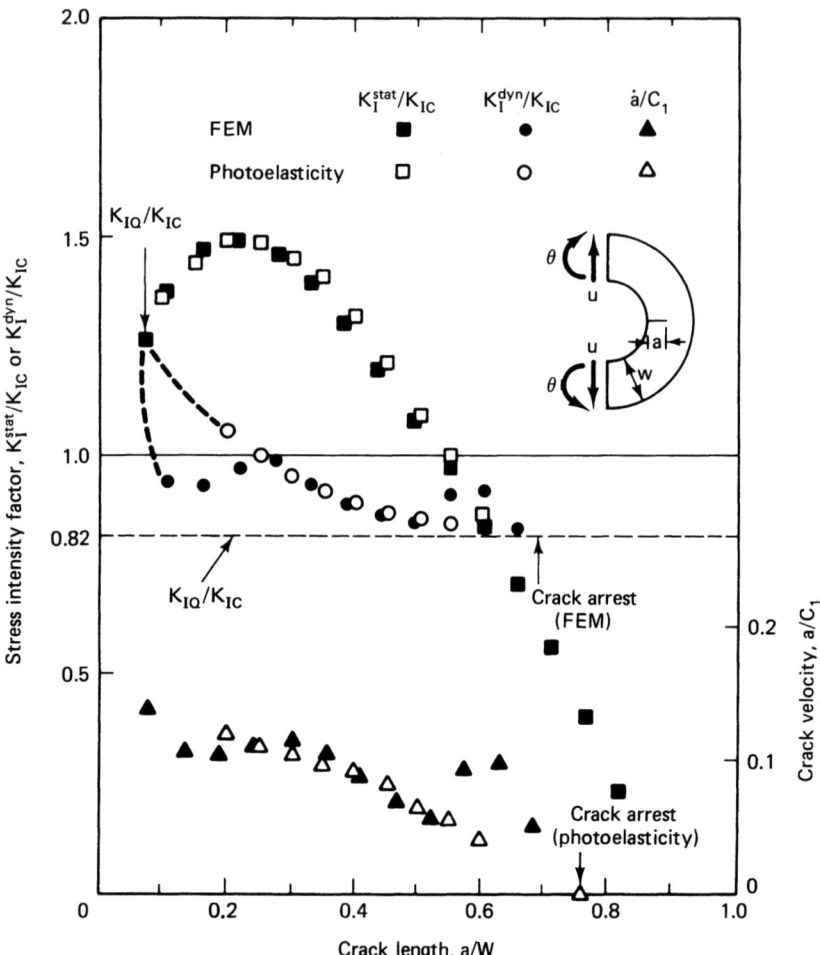

Figure 17-23 Stress-intensity factors and crack velocities of internally notched semicircular Homalite-100. Specimen subjected to end rotation and shear formation. FEM, finite-element method. (From Refs. 43 and 44.)

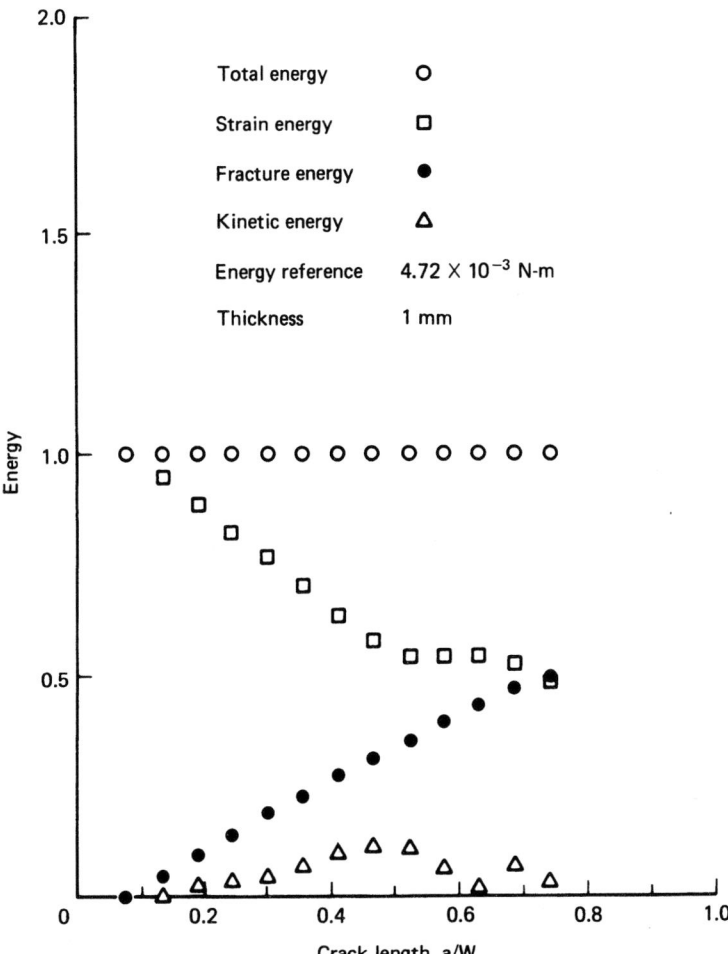

Figure 17-24 Energies in internally notched, semicircular Homalite-100. Specimen subjected to end rotation and shear deformation. (From Ref. 44.)

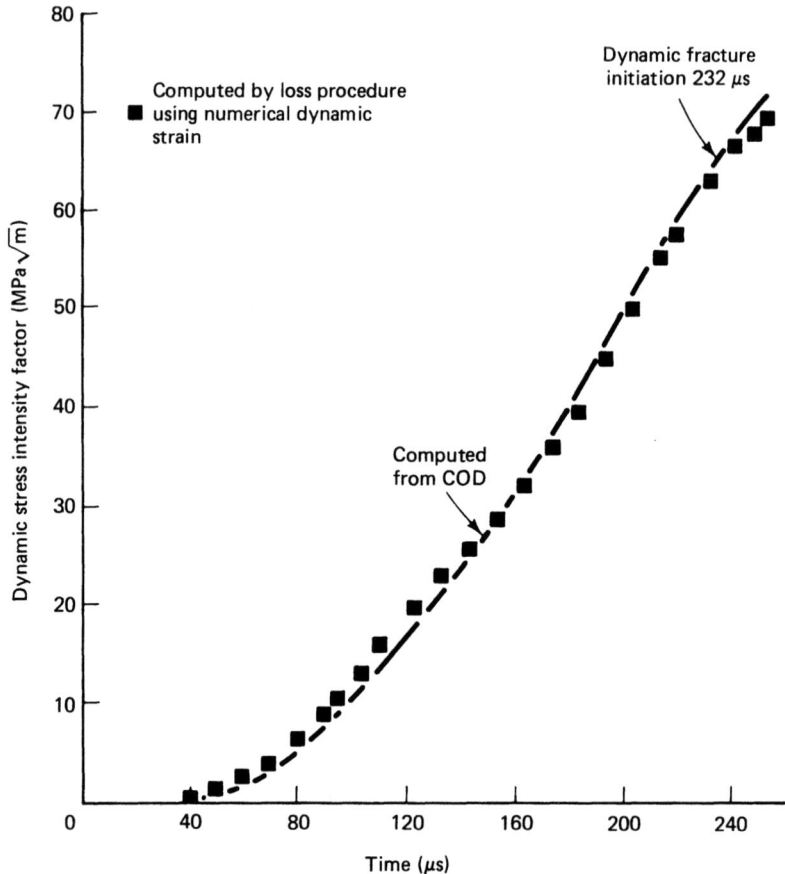

Figure 17-25 Dynamic strain at location 1, A533B bend specimen. (From Ref. 47.)

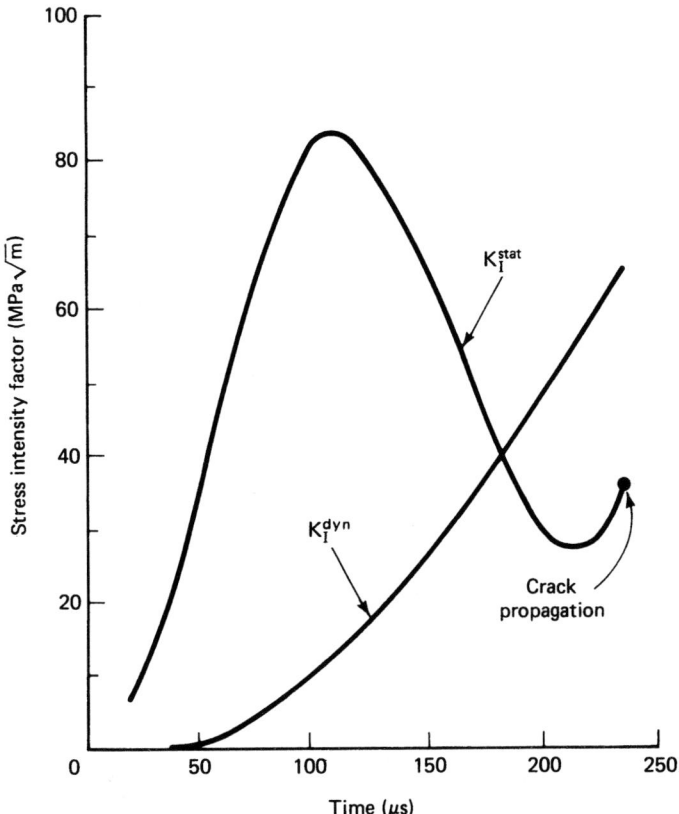

Figure 17-26 Stress intensity factors of an impacted A533B steel notched bend specimen. $L = 229$; $W = 51$; $B = 25$; $a = 25$ mm. (From Ref. 47.)

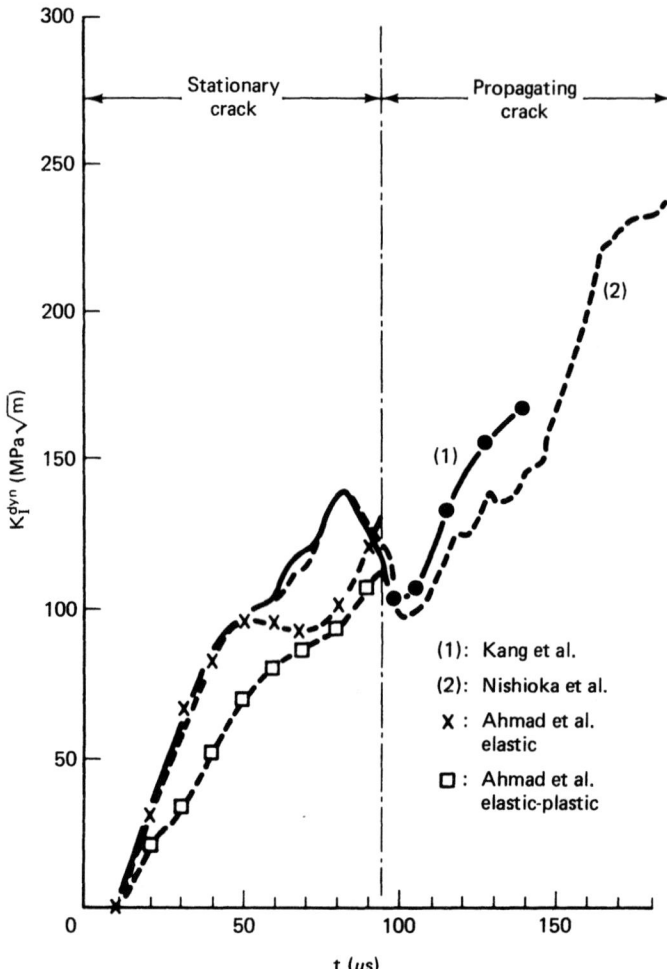

Figure 17-27 Dynamic stress intensity factor of a dynamic tear test specimen. (From Ref. 48.)

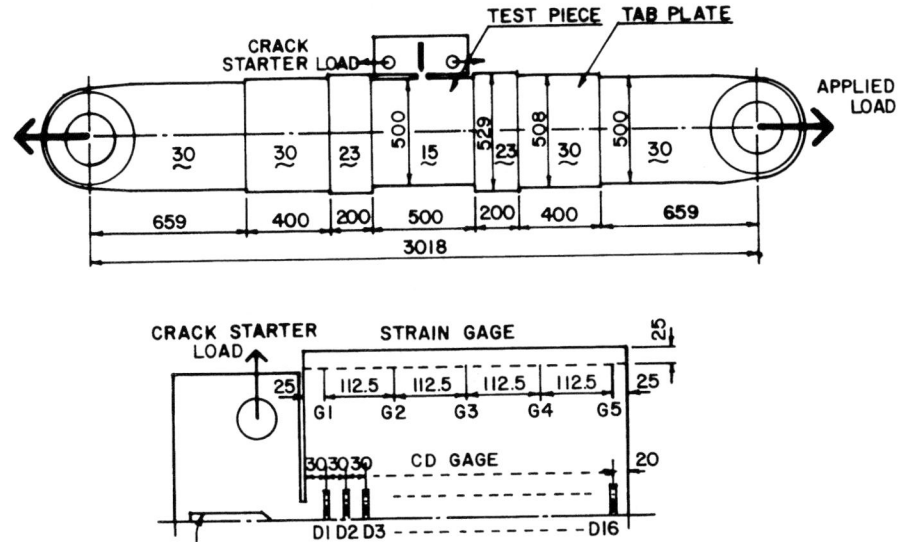

Figure 17-28 Specimens and position of crack-detector gages and strain gages. (From Ref. 49.)

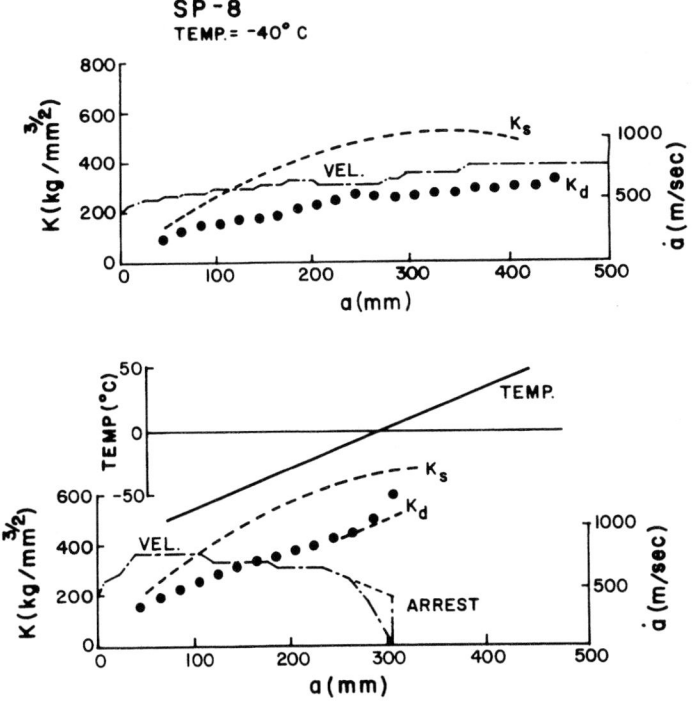

Figure 17-29 Variation of dynamic fracture toughness with crack extension. (From Ref. 49.)

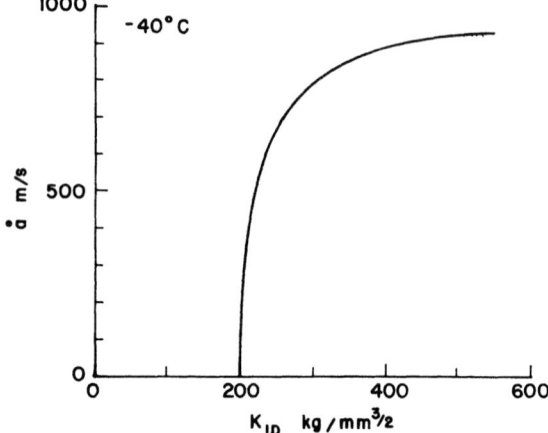

Figure 17-30 K_{ID} versus a KAS steel. (From Ref. 49.)

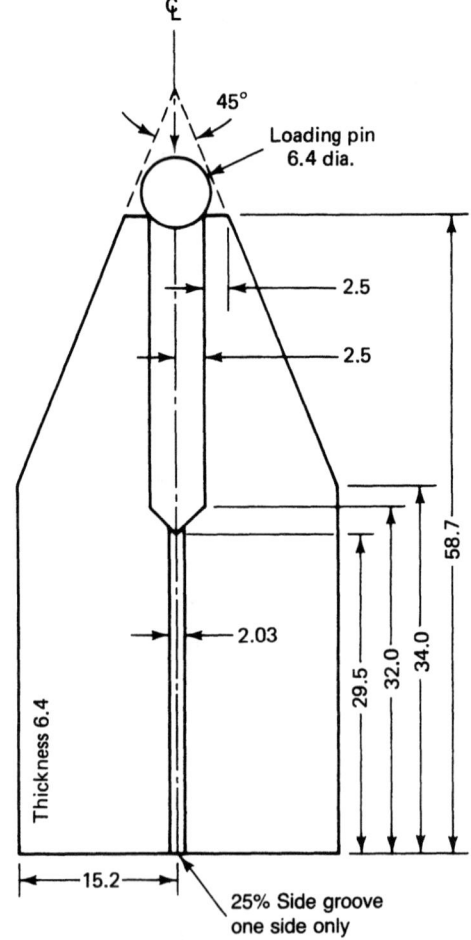

Figure 17-31 Glass WL-MTDCB specimen (all dimensions in millimeters). (From Ref. 50.)

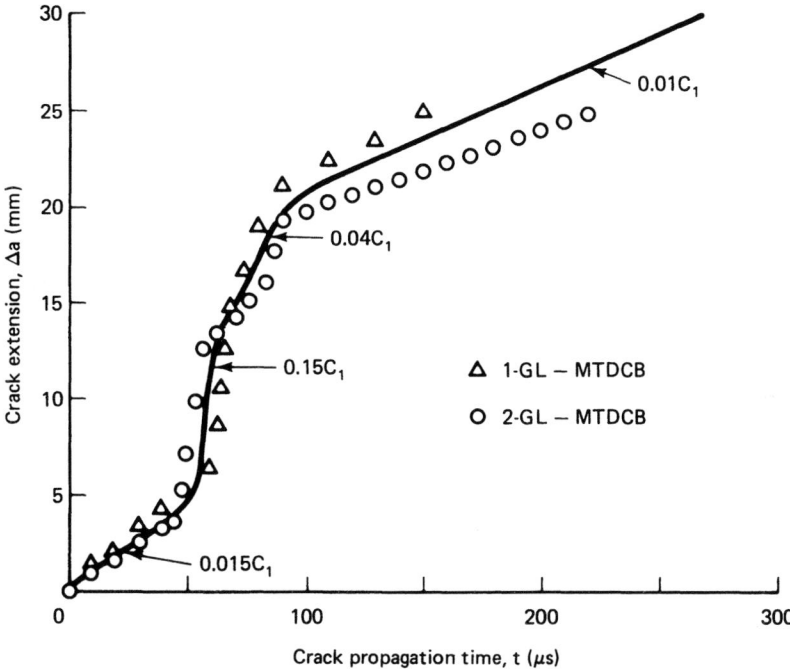

Figure 17-32 Crack extension versus time of fracturing WL-MTDCB glass specimen. (From Ref. 50.)

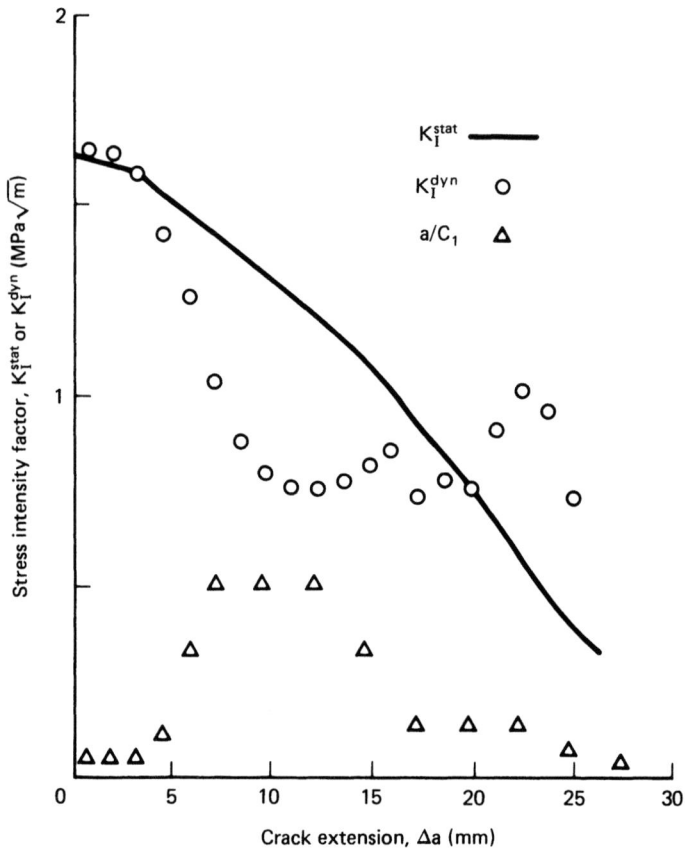

Figure 17-33 Stress-intensity factors and crack velocities in a fracturing WL-MTDCB glass specimen. (From Ref. 50.)

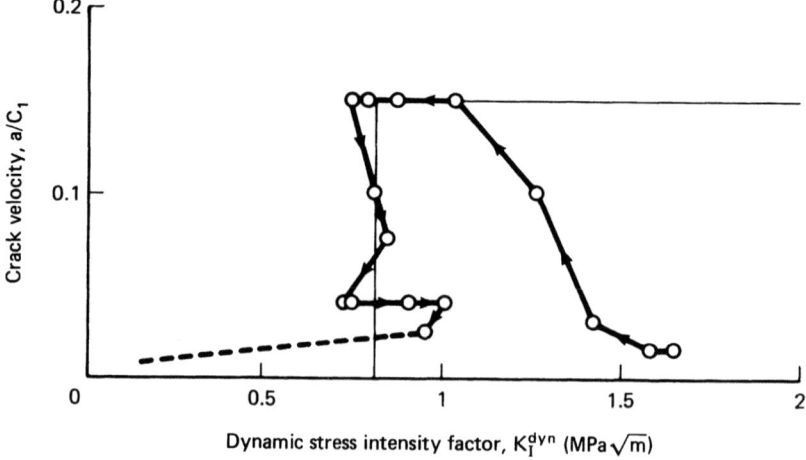

Figure 17-34 Dynamic fracture toughness versus crack-velocity relation for glass. (From Ref. 50.)

was also computed by finite-element analysis, are shown in Fig. 17-33. Although it is not obvious from Fig. 17-33, unlike the dynamic fracture of metals and polymers, the crack never arrested in these and other ceramic WL-MTDCB specimens [51]. Thus the K_I^{dyn} versus a curve in Fig. 17-34 should continue past the nominal static fracture toughness $K_{IC} = 0.73\text{MPa }\sqrt{m}$ as indicated by the dashed line. Notable is the lack of the gamma-shaped K_I^{dyn} versus a curve commonly observed in metals and polymers.

17-5 Summary

The hybrid experimental-numerical technique yields reliable information that cannot be obtained by the use of the experimental or numerical technique alone. The utility of the hybrid experimental-numerical technique in experimental mechanics is demonstrated by case studies in two- and three-dimensional stress analysis, biomechanics, and fracture mechanics.

References

1. A. J. Durelli, *Applied Stress Analysis*, Prentice-Hall, Englewood Cliffs, NJ 1967, 6.
2. J. W. Dally and W. F. Riley, *Experimental Stress Analysis*, McGraw-Hill, New York, 1978 459–467.
3. G. H. Lee, *An Introduction to Experimental Stress Analysis*, Wiley, New York, 1950, 202–208.
4. M. M. Frocht, *Photoelasticity*, Vol. 2, Wiley, New York 1948, 238–332.
5. M. M. Frocht, *Photoelasticity*, Vol. 1, Wiley, New York, 1941, 252–286.
6. E. G. Coker and L. N. G. A. Filon, *Treatise on Photoelasticity*, Cambridge University Press, London, 1931, 143–145.
7. A. J. Durelli and W. F. Riley, *Introduction to Photomechanics*, Prentice-Hall, Englewood Cliffs N J, 1965, 185–186.
8. K. H. Laerman, Recent Developments and Further Aspects of Experimental Stress Analysis in the Federal Republic of Germany and Western Europe, *Exp. Mech.*, 21 (February 1981), 49–57.
9. K. H. Laerman, Über das Prinzip der hybriden Technik in der experimentallen Spannungsanalyse, *Messen + Pruen/Automatik*, April 1983, 184–190.
10. G. V. Rao, Experimental-Numerical Hybrid Techniques for Body-Force and Thermal Stress Problems—Applications in Power Industry, *Proc. 1982 J. Conf. Exp. Mech.*, SESA, 1982, 398–404.
11. K. Chandrashekhara and K. A. Jacobs, Experimental-Numerical Hybrid Technique for Two-Dimensional Stress, *Strain*, 13, no. 4 (1977), 25–31.
12. K. Chandrashekhara and K. A. Jacobs, Experimental-Numerical Hybrid Technique for Stress Analysis of Orthotropic Composites, in *Development in Composite Material*, Applied Science Publishers, Barking, Essex, England, 1977, 67–83.
13. K. A. Jacobs and K. Chandrashekhara, Photoelastic Analysis a Composite Prism Subjected to Partial Compression, *ASME J. Appl. Mech.*, 45 (June 1978), 436–438.
14. B. M. Barishpolsky, A Combined Experimental and Numerical Method for the Solution of Generalized Elasticity Problem, *Exp. Mech.*, 20 (October 1980), 345–349.
15. B. M. Barishpolsky, Development of a Combined Experimental and Numerical Method for the Thermoelastic Analysis, *Proc. 1981 SESA Spring Meeting*, 1981.
16. K. H. Laerman, Hybrid Analysis of Plate Problems, *Exp. Mech.*,21 (October 1981), 386–388.
17. F. A. Moslehy and W. F. Ranson, Laser Speckle Interferometry and Boundary Integral Techniques in Experimental

Stress Analysis, in *Developments in Theoretical and Applied Mechanics*, Vol. X (J. E. Stoneking, Ed.) University of Tennessee, Chattanooga, 1980, 473–492.

18. J. Balas, J. Sladek, and M. Drzik, Stress-Analysis by Combination of Holographic Interferometry and Boundary Integral Method, *Exp. Mech. 23* (June 1983), 196–202.

19. I. Umeagukwu, W. Peters, and W. F. Ranson, The Experimental Boundary Integral Method in Photoelasticity, in *Developments in Theoretical and Applied Mechanics*, Vol. XI (T. J. Chung and G. R. Karr, Eds.), University of Alabama, Huntsville, 1982, 181–191.

20. R. Collins and R. van der Werft, *Mathematical Models of the Dynamics of the Human Eye*, Springer-Verlag, Berlin, 1980, 1.

21. J. M. Enoch, Causes and Costs of Visual Impairment, *Vision and Its Disorders*, NINDB Monograph No. 4, HEW, 1967, 23–28.

22. J. S. Friedenwald, Contribution to the Theory and Practice of Tonometry, *Am. J. Ophthalmol., 20* (1937), 985–1024.

23. G. N. Nyquist, Rheology of the Cornea: Experimental Techniques and Results, *Exp. Eye Res., 7* (1968), 183–188.

24. J. Gloster, E. S. Perkins, and M. Pommiesr, Extensibility of Strips of Sclera and Cornea, *Br. J. Ophthalmol. 41* (1957), 103–110.

25. S. Woo, A. S. Kobayashi, W. A. Schlegal, and C. Lawrence, Nonlinear Material Properties of Intact Cornea and Sclera, *Exp. Eye Res., 14* (1972) 29–39.

26. S. Woo, A. S. Kobayashi, W. A. Schlegal, and C. Lawrence, Mathematical Model of the Corneo-Scleral Shell as Applied to Pressure–Volume Relations and Applanation Tonometry, *Ann. Biomed. Eng. 1* (September 1972), 87–98.

27. R. St. Helen and W. K. McEwen, Rheology for the Human Eye, *Am. J. Ophthalmol., 51* (1961), 539–548.

28. A. S. Kobayashi, C. A. Brown, and A. F. Emery, Viscoelastic Response of the Corneo-Sclera Shell Under Tonometer Loading, ASME Preprint 75-WA/Bio-2, 1975.

29. J. Gloster, Tonometry and Tonography, *Int. Ophthalmol. Clin., 5* (1965).

30. J. J. McGowan (Ed.), *A Critical Evaluation of Numerical Solutions to the "Benchmark" Surface Flaw Problem*, Society for Experimental Stress Analysis, 1980.

31. M. F. Kanninen, E. F. Rybicki, R. B. Stonesifer, D. Broek, A. R. Rosenfield, C. W. Marschall, and G. T. Hahn, Elastic-Plastic Fracture Mechanics for Two-Dimensional Stable Crack Growth and Instability Problems, in *Elastic-Plastic Fractures* (J. D. Landes, J. A. Begley, and G. A. Clarke, Eds.), ASTM STP 668, American Society for Testing and Materials, Philadelphia, 1979, 121–150.

32. C. F. Shih, H. G. deLorenzi, and W. R. Andrew, Studies on Crack Initiation and Stable Crack Growth, in *Elastic-Plastic Fractures* (J. D. Landes, J. A. Begley, and G. A. Clarke, Eds.), ASTM STP 668, American Society for Testing and Materials; Philadelphia, 1979, 65–120.

33. W. K. Wilson and J. R. Osias, A Comparison of Finite Element Solutions for an Elastic–Plastic Problem, *Int, J. Fract., 14* (1978), R95–R108.

34. J. W. Dally, Dynamic Photoelastic Studies of Fracture, *Exp. Mech.*, 19 (1979), 349–361.

35. M. Shumely and M. Perl, the SMF2D Code for Proper Simulation of Crack Propagation, in *Crack Arrest Methodology and Applications* (G. T. Hahn and M. F. Kanninen, Eds.), ASTM STP 711, American Society for Testing and Materials, Philadelphia, 1980, 54–69.

36. C. H. Popelar and P. C. Gehlen, "Modeling of Dynamic Crack Propagation: II. Validation of Dynamic Analysis, *Int. J. Fract., 15* (1979), 159–177.

37. L. Hodulak, A. S. Kobayashi, and A. F. Emery, A Critical Examination of a Numerical Fracture Dynamic Code, in *Fracture Mechanics* (P. C. Paris, Ed.), ASTM STP 700, American Society for Testing and Materials, Philadelphia, 1980, 174–188.

38. T. Nishioka and S. M. Atluri, Numerical Analysis of Dynamic Crack Propagation: Generation and Prediction Studies, *Eng. Fract. Dyn., 16*, no. 3 (1982), 303–332.

39. L. Hodulak, A. S. Kobayashi, and A. F. Emery, Influence of Dynamic Fracture Toughness on Dynamic Crack Propagation, *Eng. Fract. Mech., 13* (1980), 85–93.
40. J. F. Kalthoff, J. Beinert, and S. Winkler, Measurements of Dynamic Stress Intensity Factors for Fast Running and Arresting Cracks in Double Cantilever-Beam Specimens, in *Fast Fracture and Crack Arrest* (G. T. Hahn and M. F. Kanninen, Eds.), ASTM STP 627, American Society for Testing and Materials, Philadelphia, 1977, 161–176.
41. J. F. Kalthoff, J. Beinert, and S. Winkler, Influence of Dynamic Effects on Crack Arrest, Institute für Festkorpemechanik, EPRI RP 1022-1, IKFM 404012, August 1978.
42. A. S. Kobayashi, K. Seo, J.-Y. Jou, and Y. Urabe, Dynamic Analysis of Homalite-100 and Polycarbonate Modified Compact Tension Specimens, *Exp. Mech, 20* (September 1980), 73–79.
43. J. W. Dally, A. Shukla, and T. Kobayashi, A Dynamic Photoelastic Study of Crack Propagation in a Ring Specimen, in *Crack Arrest Methodology and Applications* (G. T. Hahn, and M. F. Kanninen, Eds.), ASTM STP 711, American Society for Testing and Materials, Philadelphia, June 1980, 161–177.
44. A. S. Kobayashi, A. F. Emery, and B.-M Liaw, Dynamic Fracture of Three Specimens, *Fracture Mechanics,* (14), Vol. II, *Testing and Application* (J. C. Lewis and G. Sines, Eds.), ASTM STP 791, American Society for Testing and Materials, Philadelphia, 1983, II-251–II-265.
45. A. Shukla and J. W. Dally, A Photoelastic Study of Energy Loss During a Fracture Event, *Exp. Mech.,* 21 (April 1981), 163–168.
46. S. Mall, A. S. Kobayashi, and F. J. Loss, Dynamic Fracture Analysis of Notch Bend Specimen, in *Crack Arrest Methodology and Applications* (G. T. Hahn and M. F. Kanninen, Eds.), ASTM STP 711, American Society for Testing and Materials, Philadelphia, 1980, 70–85.
47. A. S. Kobayashi, M. Ramulu, and S. Mall, Impacted Notch Bend Specimens, *ASME J. Pressure Vessel Technol., 104,* (February 1982) 25–30.
48. A. S. Kobayashi, Numerical Analysis in Fracture Mechanics, in *Application of Fracture Mechanics to Materials and Structures* (G. C. Sih, E. Sommer, and W. Dahl, Eds.), Nijhoff, Hingham, MA, 1984, 27–56.
49. T. Kanazawa, S. Machida, T. Teramoto, and H. Yoshinari, Study on Fast Fracture and Crack Arrest, *Exp. Mech., 21* (February 1981), 78–88.
50. A. S. Kobayashi, A. F. Emery, and B.-M. Liaw, Dynamic Fracture Toughness of Glass, in *Fracture Mechanics of Ceramics,* Vol. 6 (R. C. Bradt, G. G. Evans, D. P. H. Hasselman, and F. Lange, Eds.), Plenum Press, New York, 1983.
51. A. S. Kobayashi, A. G. Emery, and B.-M. Liaw, Dynamic Fracture Toughness of Reaction Bonded Silicon Nitride, *J. Am. Ceram. Soc., 66,* no. 2 (1983), 151–155.

CHAPTER

18

Residual Stresses

Robert E. Rowlands
University of Wisconsin—Madison
Madison, Wisconsin

18-0 Introduction

"Darn it—it failed" is a phrase uttered too often by engineers and consumers upon viewing the catastrophic results of our inadequate technology. Many service failures of structural or machine components result not from stresses due to applied loads but from residual stresses. The associated cost of replacement and loss of production is prohibitive and often results in adverse customer relations, to say nothing of the potential for loss of life or limb. Residual stresses exist in a body free from external force or constraint. They frequently form during fabrication operations, such as casting, rolling, welding, stamping, or forging. Their presence in metal components such as the Liberty Bell[1] (Fig. 18-1) is well documented. They also occur in fiber-reinforced composites due to the matrix solidifying around the reinforcement, and from different plies of a stacked laminate having different coefficients of expansion (contraction) in different directions.

Residual stresses are as detrimental in causing flow or fracture as stresses due to live loads. They are usually very undesirable for the following reasons: (1) they are difficult to measure, and (2) they add to stresses due to applied loads. Advantageous compressive residual surface stresses are sometimes introduced intentionally, as with shot peening, quenching, tempered glass, and so on.

Residual stresses can be extremely troublesome because of difficulty in measuring them nondestructively; the unpredictability of their magnitude, sense, and direction; their adverse ability to combine with stress corrosion, environmental, and fatigue situations; and difficulty in removing them. Residual stresses should be evaluated in the actual member under operating conditions to assess their effect on the intended service capability of the component. Although some modeling of residual stresses has been done, such approaches are normally less than ideal. For quality

---

[1] The Liberty Bell was recast several times and was first rung in Philadelphia in 1753. Cracking was noticed in 1828. Attempts were made to arrest the crack growth by drilling holes and removing material at the crack tip. This action was unsuccessful and the crack continued to grow, due to residual stresses. It cracked beyond repair during ringing in 1846, at which time it was removed from service and now resides in Independence Hall [1].

Figure 18-1 The Liberty Bell. (Courtesy of Measurements Group, Inc.)

control during manufacture, techniques for determining residual stresses should be usable on a production-line basis. Although progress has been made in measuring residual stresses, extensive efforts remain necessary to develop adequate nondestructive, cost-effective, and expedient techniques.

Notwithstanding the widespread occurrence of residual stresses and their importance in determining component reliability, the subject receives vastly inadequate attention. Quantitative consideration of residual stresses remains one of the least addressed topics of mechanical or structural design. This is extremely unfortunate. Our need to address the subject is proportional to the product of the technical importance of residual stresses times the extent of our lack of knowledge of and inability to measure them. It is hoped that the availability of this chapter, together with other review articles on residual stresses [1–13], will stimulate at least some coverage of the topic in engineering curricula.

Partly because the formation of residual stresses is covered extensively elsewhere [1–3,5,9], and partly through space limitations, the present treatment emphasizes methods for measuring residual stresses. Only sufficient comments are devoted to the formation and removal of residual stresses to attempt to place the various aspects into perspective. Although every effort has been made to provide a generous bibliography, the literature on the topic is extensive and diffuse, and no attempt has been made here to be "all-inclusive," Techniques covered include hole boring as applied to metals, composites, and rock; ultrasonic methods of Rayleigh surface waves and acoustoelasticity; x-rays; electrical-resistance strain gages; nuclear acoustic resonance; photoelasticity; and finite elements.

18-1 Hole-Drilling Methods

The hole-drilling technique involves monitoring the change in stresses (or strains) produced when a hole is drilled into a member containing a residual stress. It is one of several material-removal

RESIDUAL STRESSES

methods. The change in strain may be measured using photoelasticity, brittle-coating, or electrical strain gages. Knowing the magnitude and direction of relaxed strains, the size of the hole, and material properties, the residual stresses can be predicted by classical and/or empirical analysis. The method is often only semidestructive, in that the hole produced may be insignificant or repaired subsequently by, for example, inserting a bolt or welding. However, the repair may affect the stress field. Blind rather than through-holes are sometimes used, although interpretation of the residual stresses may be more complicated. Drilling of holes of various depths can be employed to assess the variation of the stresses with depth.

The hole-drilling method can be used to survey residual stresses throughout the surface of large components. With photoelastic or brittle coatings, the associated pattern can be used effectively to estimate the magnitude and direction of biaxial stresses. Although used most easily with flat structures, the technique is useful with fillets, welds, castings, and rolled shapes. Commercial equipment is now available. As with many of the methods for evaluating residual stresses, the hole-drilling technique can be more qualitative than quantitative. In addition to being the least destructive of the mechanical methods, results can be obtained rapidly and economically. Unlike some concepts, the technique permits evaluation of the residual stresses essentially at a point.

Numerous variations of the hole-drilling technique have been proposed and used, along with a variety of analyses. Several of the methods are summarized here. The reader is also directed to review articles on this topic [3,6,7,10].

Mathar's Method (1934)

The hole-drilling method appears to have been proposed first by J. Mathar of the University of Aachen [14]. He based his technique on the fact that the shape of a circular hole drilled into a stressed (say, residual) member will change due to the stresses. A record of the shape change is a measure of the residual stresses present. If the initial stress field were unidirectional, its magnitude could be deduced from measuring the diametral change in one direction. A biaxial stress field would typically require measuring at least three diametral changes—thereby providing information on the magnitude and direction of the residual stresses.

Mathar recorded diametral changes with a mirror extensometer of the Martens type and used a calibration based on the theoretical stress analysis in a loaded plate [15–18]. He used Kirsch's solution [15] for infinite plates and the approximation for finite-width plates by Willheim and Leon [16]. In such studies one may find Mesmer's analysis [17] convenient if the principal directions are known, or Campus's [18] generalization when the principal directions are not known. Mathar's calibration accounted for changes in hole depth. Drill pressure was maintained constant by a spring. He also suggested that if only one extensometer were available and the stress field varied slowly, three small neighboring holes be drilled and the diametral change monitored in a different orientation each time.

Mathar applied his method to determine residual stresses in wide-flange beams, welds, and bridge sections. He proposed using the technique on concrete dams, on tunnel walls, and in deep shafts with coring to determine the stresses in rock. The latter approaches hole-boring concepts employed currently and as described in Section 18-2.

Soete and Vancrombrugge's Method (1950)

Soete and Vancrombrugge employed a concept similar to that of Mathar except that they used electrical-resistance strain gages around the hole rather than measuring hole-diameter changes with an extensometer. Although they did not have particularly small gages, the tiny gages now available

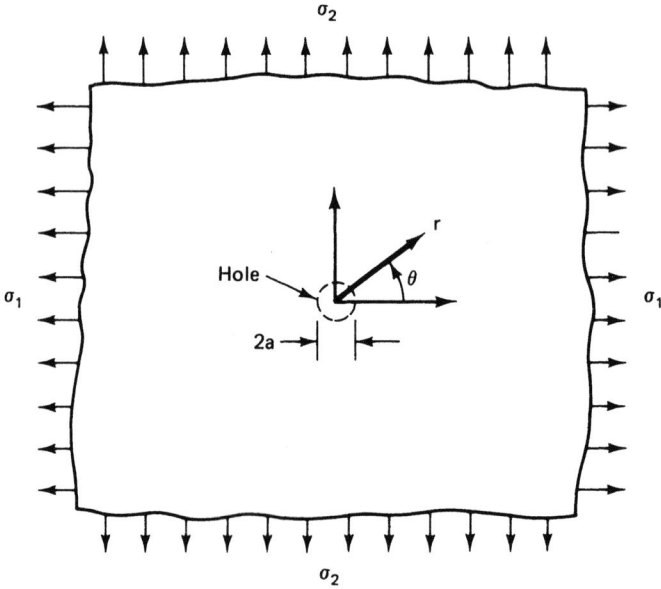

Figure 18-2 Geometry for method by Soete and Vancrombrugge. (From Ref. 20.)

permit use of shallow holes of very small diameter, thereby reducing the destructiveness. Moreover, use of small gages and small holes allows applications in regions of sharp stress gradient.

Soete and Vancrombrugge [19,20] assumed the structure to be an infinite, homogeneous, isotropic elastic plate under biaxial plane stress (Fig. 18-2). Call the original state of residual stress σ. The effect of opening up (drilling) in this plate a stress-free hole of finite radius a is equivalent to applying on the circumference of radius a equal and opposite normal and shearing stresses, which existed at the free field. The resulting stress field σ'' is just the Kirsch solution [15] for a biaxially loaded infinite plate containing a hole of radius a. For the free-field case (no hole), the residual stresses on any circumference are

$$\sigma_r = \frac{\sigma_1 + \sigma_2}{2} + \frac{\sigma_1 - \sigma_2}{2} \cos 2\theta$$
$$\sigma_\theta = \frac{\sigma_1 + \sigma_2}{2} - \frac{\sigma_1 - \sigma_2}{2} \cos 2\theta \tag{18-1}$$

Moreover, the stresses σ'' anywhere in the biaxially loaded plate with the hole of radius a are

$$\sigma_r'' = \frac{\sigma_1 + \sigma_2}{2}\left(1 - \frac{a^2}{r^2}\right) + \frac{\sigma_1 - \sigma_2}{2}\left(1 + \frac{3a^4}{r^4} - \frac{4a^2}{r^2}\right)\cos 2\theta$$
$$\sigma_\theta'' = \frac{\sigma_1 + \sigma_2}{2}\left(1 + \frac{a^2}{r^2}\right) - \frac{\sigma_1 - \sigma_2}{2}\left(1 + \frac{3a^4}{r^4}\right)\cos 2\theta \tag{18-2}$$

Since the stress field beyond the radius a goes from σ to σ'' due to the drilling,

$$\sigma' = \sigma'' - \sigma \tag{18-3}$$

RESIDUAL STRESSES

is the change in stress (at $r \geq a$) associated with the hole drilling. From Eqs. (18-1) through (18-3), the change in stresses are

$$\sigma'_r = -\frac{\sigma_1 + \sigma_2}{2}\frac{a^2}{r^2} + \frac{\sigma_1 - \sigma_2}{2}\left(\frac{3a^4}{r^4} - \frac{4a^2}{r^2}\right)\cos 2\theta \quad (18\text{-}4)$$

$$\sigma'_\theta = \frac{\sigma_1 + \sigma_2}{2}\frac{a^2}{r^2} - \frac{\sigma_1 - \sigma_2}{2}\frac{3a^4}{r^4}\cos 2\theta$$

and the change in radial strains recorded during drilling would be[2]:

$$\varepsilon'_r = \frac{1}{E}(\sigma'_r - \nu\sigma'_\theta) \quad (18\text{-}5)$$

If the radial strain gradient is not insignificant, the actual strain monitored by a radial gage would be the integrated average, that is

$$\varepsilon'_m = \frac{1}{r_2 - r_1}\int_{r_1}^{r_2}\varepsilon'_r\, dr \quad (18\text{-}6)$$

where the radial difference $(r_2 - r_1)$ is the grid length of the gage. Substituting Eqs. (18-4) and (18-5) into (18-6) and carrying out the integration produces

$$\varepsilon'_m = \frac{1}{E}\frac{2a^2}{r_1 r_2}\cos 2\theta\left[-1 + \frac{1+\nu}{4}\frac{a^2(r_1^2 + r_1 r_2 + r_2^2)}{r_1^2 r_2^2}\right](\sigma_1 - \sigma_2)$$
$$- \frac{1+\nu}{2E}\frac{a^2}{r_1 r_2}(\sigma_1 + \sigma_2) \quad (18\text{-}7)$$

or

$$\varepsilon'_m = \frac{A'}{E}(\sigma_1 + \sigma_2) + \frac{B'}{E}(\sigma_1 - \sigma_2)\cos 2\theta \quad (18\text{-}8)$$

where

$$A' = -\frac{1+\nu}{2}\frac{a^2}{r_1 r_2} \quad (18\text{-}9)$$

and

$$B' = \frac{2a^2}{r_1 r_2}\left[-1 + \frac{1+\nu}{4}\frac{a^2(r_1^2 + r_1 r_2 + r_2^2)}{r_1^2 r_2^2}\right] \quad (18\text{-}10)$$

Angle θ is measured from the σ_1 direction (Fig. 18-2). If the principal directions of the residual stresses σ are known, Eqs. (18-7) and (18-8) demonstrate that radial gages at two different orientations are adequate to evaluate the individual principal residual stresses, σ_1 and σ_2. If the principal directions of the original residual stresses are not known, it will be necessary to monitor three radial gages during the drilling procedure. The previous equations must then be developed in terms of σ_x, σ_y, and τ_{xy} or else σ_1, σ_2, and the principal direction. As is frequently the case when the principal directions are not known, a three-element strain-gage rosette is used whose elements are at angles θ, $\theta + 45°$, and $\theta - 45°$. Equation (18-7) may then be solved for the residual strains at the hole location but in the undrilled plate to be

---

[2] A list of the symbols used in this chapter precedes the References.

$$\varepsilon_0 = K_1\varepsilon_0' + K_2(\varepsilon_+' + \varepsilon_-')$$
$$\varepsilon_+ = K_1\varepsilon_+' + K_2(\varepsilon_+' + \varepsilon_-') \qquad (18\text{-}11)$$
$$\varepsilon_- = K_1\varepsilon_-' + K_2(\varepsilon_+' + \varepsilon_-')$$

where

$$K_1 = \frac{B}{B'} \qquad K_2 = \frac{AB' - A'B}{2A'B'} \qquad A = \frac{1-\nu}{2} \qquad B = \frac{1+\nu}{2} \qquad (18\text{-}12)$$

ε_0', ε_+', and ε_-' are the measured strains at angle θ, $\theta + 45°$, and $\theta - 45°$, respectively, and ε_0, ε_+, and ε_- are the residual strains at θ, $\theta + 45°$, and $\theta - 45°$, respectively. The residual stresses are then obtained from these residual strains of Eqs. (18-11) and Hooke's law.

Soete and Vancrombrugge applied their technique to the determination of residual stresses in both butt and fillet welds. Although sometimes violated in practice, this method assumes the following:

1. Plane stress—stresses do not vary with depth; as such, the method is not applicable for general three-dimensional geometries.
2. Plate is infinite with respect to the size of hole drilled.
3. Hole is drilled completely through the plate.
4. Residual stresses relax linearly elastically.

The method does not necessitate any supplementary calibration and it will handle any ratio of σ_1/σ_2. In that very small holes can be utilized, the technique need not necessarily be destructive. The concept can provide residual stresses essentially at a point.

Riparbelli's Method (1950)

A main contribution of Riparbelli's work was effective use of electrical-resistance strain gages and the manner of connecting them into the Wheatstone bridge advantageously [21]. Like previous investigators, he assumed that the hole was drilled into an infinite plate and thereby used the Kirsch stress analysis of Eqs. (18-2). A calibration technique is used in actual problems.

For a biaxial state of stress such that the principal directions are known, the residual stresses were related to the measured strains ε_1' and ε_2' (associated with drilling) as follows:

$$\sigma_1 = E\frac{K_1\varepsilon_1' - K_2\varepsilon_2'}{K_1^2 - K_2^2}$$
$$\sigma_2 = E\frac{K_1\varepsilon_2' - K_2\varepsilon_1'}{K_1^2 - K_2^2} \qquad (18\text{-}13)$$

Sensitivity factors K_1 and K_2 are obtained experimentally under applied loading.[3] K_1 and K_2 depend on factors such as the material, geometry, and the hole and gage dimensions. For SR-4 gages mounted about the drilled hole of a plane-stressed aluminum plate and connected into a Wheatstone bridge in the balanced concept of Fig. 18-3, calibrated values of $K_1 = 0.460$ and $K_2 = -0.029$ were obtained. Riparbelli applied his method to welded plates.

---

[3] Unless stated otherwise, the A's, B's, C's and K's are general coefficients, and their use with one analysis reviewed here need not have any relationship with that of any other analysis.

RESIDUAL STRESSES

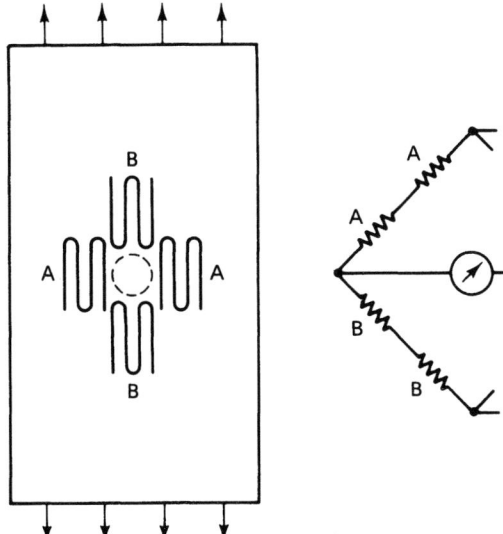

Figure 18-3 Strain-gage orientation and bridge configuration for Riparbelli's method. (From Ref. 21.)

Boiten and Ten Gate's Method (1952)

Soete and Vancrombrugge [20] and Riparbelli [21] integrated the radially varying strain effect over the length of the strain gages. Boiten and Ten Gate [22] went one step further and averaged the effect over the total area of the grid (double integration of varying strain over the length and width of the gage). They also included the transverse sensitivity influence of the gage. At least for plane-stress problems they again made use of the Kirsch analysis [15].

Boiten and Ten Gate developed their analysis for flexed plates by bonding three-element rosettes to both the top and bottom faces of the plate. They took care to ensure coincident gage location on both faces of the plate.

Kelsey's Method (1956)

Whereas most previous investigations assumed that the stress field was uniform with depth, Kelsey [23] proposed a method to measure residual stresses that vary with depth below the surface. The method provides both the magnitude and direction of the principal residual stresses. Results indicate that stresses can be evaluated to a depth below the surface of about half the hole diameter. The method again involves drilling between a cluster of three or four radially located electrical-resistance strain gages. A calibration bar identical in material, gage size and location, and hole to the structure of interest size is used.

For depth-dependent varying stresses, it is assumed that as the hole depth is increased by a small distance Δz, the average stress σ existing over this incremental depth will produce identical surface strain change $\Delta \varepsilon$ in the structure of concern as in the calibration specimen. This surface strain change $\Delta \varepsilon$ is assumed to be proportional to the average stress σ over this increment. Assuming this to be true under uniaxial or biaxial loading, one may then write the increments of surface strains $\Delta\varepsilon_l$ and $\Delta\varepsilon_t$ in terms of the average subsurface stress over the depth increment as

$$\Delta\varepsilon_l = \frac{1}{E}(K_1\sigma_l - \nu K_2\sigma_t)$$
$$\Delta\varepsilon_t = \frac{1}{E}(K_1\sigma_t - \nu K_2\sigma_l)$$
(18-14)

By inverting Eqs. (18-14), the average stresses over the particular Δz become

$$\sigma_l;\ \sigma_t = \frac{E}{K_1^2 - \nu K_2^2}\{[K_1(\Delta\varepsilon_l) + \nu K_2(\Delta\varepsilon_t)];\ [K_1(\Delta\varepsilon_t) + \nu K_2(\Delta\varepsilon_l)]\}$$
(18-15)

l and t are orthogonal, but not necessarily principal directions. If these are not the principal directions, the surface strains must be recorded in at least three directions. The rosette equations, together with Hooke's law, can then be used to evaluate the principal subsurface stresses and their directions.

The validity of Eqs. (18-14) was demonstrated experimentally. K_1 and K_2 are empirical constants of proportionality which, from Eqs. (18-14), are related to the subsurface stresses averaged over the increment and the monitored surface strain increments by

$$K_1 = \frac{E}{\sigma_l^2 - \sigma_t^2}[\sigma_l(\Delta\varepsilon_l) - \sigma_t(\Delta\varepsilon_t)]$$
$$K_2 = \frac{-E}{\nu(\sigma_l^2 - \sigma_t^2)}[\sigma_l(\Delta\varepsilon_t) - \sigma_t(\Delta\varepsilon_l)]$$
(18-16)

K_1 and K_2 are obtained from a supplementary calibration test. They can be evaluated from Eqs. (18-16) by employing calibration specimens of known stress (typically) uniaxial tensile stress specimens) distribution E and ν, drilling the hole, and monitoring the surface strain changes. Having evaluated the K's, one can evaluate the residual stresses for an unknown situation by Eqs. (18-15). In actuality, Kelsey employed "live stresses" and obtained the $\Delta\varepsilon$'s associated with the hole introduction by loading undrilled specimens and measuring the full-load strains, unloading, drilling, then reloading and again recording the strains. The difference in strain fields represents the same effect as drilling the loaded plate.

Kelsey evaluated the influence of employing holes of varying diameters and depths relative to the thickness of structure. He also substantiated experimentally the validity of evaluating the K's under uniaxial loading and then using them in a biaxial stress field. Kelsey checked the reliability of his hole-drilling concept by correlating residual stress results in a thick plate with those obtained by removing layers of material.

Typical results are demonstrated in Figs. 18-4 and 18-5. Figure 18-4 illustrates surface strains recorded on aluminum plates subjected to uniaxial stress and bending (nonuniform stress) as a function of depth of drilling. Agreement between subsurface strains obtained by hole drilling and those from removing layers is demonstrated in Fig. 18-5. Kelsey notes that while the layer technique required 75 worker-hours, the hole-drilling method only took 10 worker-hours.

The following were among the significant conclusions of Kelsey's research:

1. Although additional effort is needed to assess the effect of machining techniques, the method is independent of machining stresses as long as the same material and machining procedure are used for the actual structure and for the calibration specimens.
2. The method appears to be reliable for determining residual stress profiles as a function of depth.

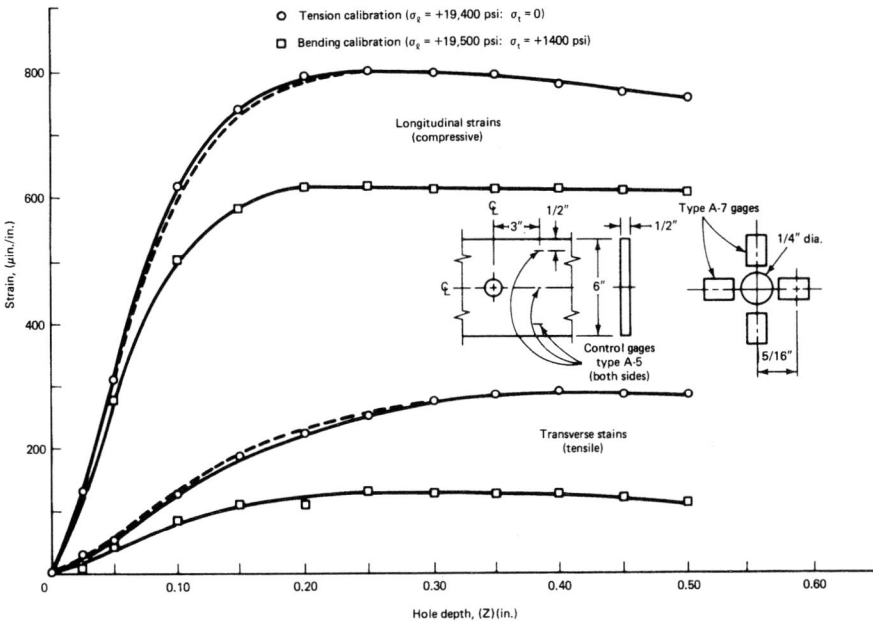

Figure 18-4 Strains relieved by drilling hole in thick test bar subjected to uniform and nonuniform stress fields. (From Ref. 23.)

3. The method is limited to determining residual stresses in planes normal to the hole axis (i.e., essentially parallel to the surface).
4. Response is effective up to a depth equal to about half the hole diameter.
5. A reliable method of determining residual stresses without the need for "through-the-thickness" holes fosters the nondestructiveness of the concept.

Nisida and Takabayashi's Method (1965)

Nisida and Takabayashi [24] used birefringent coating to measure residual stresses by the hole-drilling method. The structure of interest is coated with a sheet of photoelastic material through which a hole is drilled into the metal. Residual stresses in the latter are evaluated from the fringe pattern associated with the released stresses. The photoelastic effect is correlated experimentally with the residual principal stresses. The method was demonstrated by application to welds and steel plates.

Under biaxial stress, the residual stresses σ_1 and σ_2 are determined from the expressions

$$\sigma_1 = \frac{E_m f_\sigma}{4h_c E_c} N_A \quad \text{and} \quad \sigma_2 = \frac{E_m f_\sigma}{4h_c E_c} N_B \tag{18-17}$$

where f_σ is the photoelastic coefficient of the coating; E_m and E_c are the elastic moduli of the structural member and photoelastic coating materials, respectively; h_c is the coating thickness; and N_A and N_B are the photoelastic fringe orders at points A and B of Fig. 18-6. The location of points A and B on the hole boundary give the directions of principal stresses σ_1 and σ_2, and can be

794 HANDBOOK ON EXPERIMENTAL MECHANICS

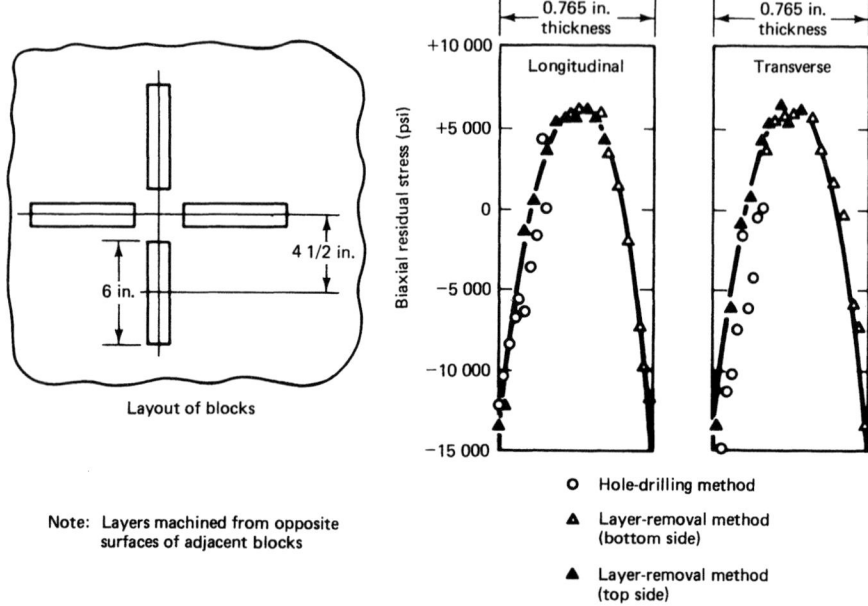

Figure 18-5 Comparison of subsurface residual stresses obtained by hole-drilling and layer-removal techniques. (From Ref. 23.)

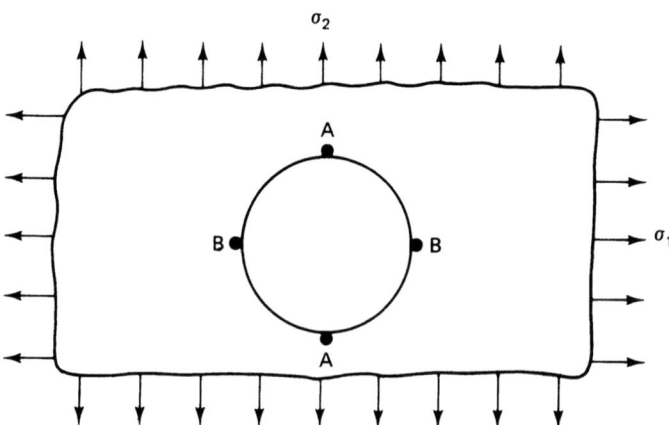

Figure 18-6 Location of photoelastic birefringence reading at drilled hole for residual stress determination.

RESIDUAL STRESSES

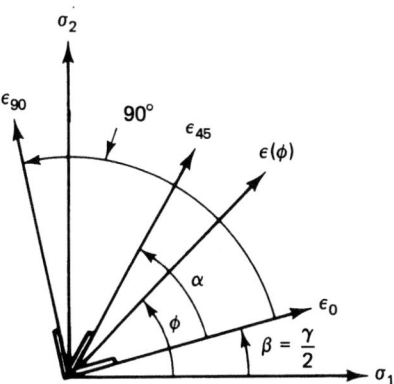

Figure 18-7 Coordinate system for hole-drilling technique of Rendler and Vigness (From Ref. 25.)

determined by symmetry of the fringe pattern. Experience demonstrates that the birefringence associated with the released stresses diminishes rapidly to zero away from the hole (~2 or 3 diameters), so the area of the coating need not be large.

Rendler and Vigness's Method (1966) [25,26]

Kelsey's method [23] involves two calibration constants which in reality are material dependent. Consequently, his concept necessitates a companion calibration specimen and evaluation. Rendler and Vigness [25] enhanced the hole-drilling concept of evaluating residual stresses by eliminating the need for individual, supplementary calibration tests. This improvement was achieved by assuming (and subsequently justifying) a specific functional form for the distribution of surface-strain change accompanying the release of subsurface residual stresses. It is shown that the two constants employed are valid for all isotropic, elastically responding situations. Only a knowledge of the elastic modulus E and Poisson's ratio ν is needed. Like Kelsey's method, the present approach is valid for partial holes.

Somewhat similarly to Kelsey, Rendler and Vigness assumed a surface strain change ε given by

$$\varepsilon(\phi) = [K(\phi)]\sigma \tag{18-18}$$

in terms of released residual stress σ at some distance below the surface. Angle ϕ is measured from the direction of principal stress σ (Fig. 18-7). $K(\phi)$ was selected of the form

$$K(\phi) = A + B \cos 2\phi \tag{18-19}$$

where coefficients A and B will be discussed subsequently. If linear strains are monitored along the three radial directions $\phi = \beta$, $\phi = \beta + 45°$, and $\phi = \beta + 90°$, the residual stress field at the bottom of the hole, in terms of measured surface-strain releases, is

$$\sigma_1 = \frac{\varepsilon_0(A + B \sin \gamma) - \varepsilon_{45}(A - B \cos \gamma)}{2AB(\sin \gamma + \cos \gamma)} \tag{18-20a}$$

$$\sigma_2 = \frac{\varepsilon_{45}(A + B \cos \gamma) - \varepsilon_0(A - B \sin \gamma)}{2AB(\sin \gamma + \cos \gamma)} \tag{18-20b}$$

$$2\beta = \gamma = \tan^{-1} \frac{\varepsilon_0 - 2\varepsilon_{45} + \varepsilon_{90}}{\varepsilon_0 - \varepsilon_{90}} \tag{18-20c}$$

where β is the angle between ε_0 and σ_1. Strains ε_0, ε_{45}, and ε_{90} of Eqs. (18-20) correspond to surface strains measured along $\alpha = 0°$, $\alpha = 45°$, and $\alpha = 90°$ of Fig. 18-7. Gupta noted that the correct quadrant for angle γ of Fig. 18-7 must be determined from consideration of the numerator and denominator of Eq. (18-20c) to ensure that $\sigma_1 \geq \sigma_2$ in Eqs. (18-20) [26].

A and B are related to isotropic elastic properties by

$$A = \frac{1}{2E}(k_1 - \nu k_2)$$
$$B = \frac{1}{2E}(k_1 + \nu k_2) \tag{18-21}$$

where the same k_1 and k_2 are valid for all isotropic, elastic materials. If E and ν are known, one calibration to determine k_1 and k_2 will thus suffice for all isotropic elastic materials. Moreover, by putting $\sigma_2 = 0$ and applying a $\sigma_1 = \sigma \neq 0$ to a specimen,

$$A = \frac{\varepsilon_0 + \varepsilon_{90}}{2\sigma} \qquad B = \frac{\varepsilon_0 - \varepsilon_{90}}{2\sigma} \tag{18-22}$$

From Eq. (18-22) and known E and ν, the k's were obtained from a steel specimen:

$$k_1 = -0.2015 \quad \text{and} \quad k_2 = -0.238 \tag{18-23}$$

Substituting these k's of Eqs. (18-23) and (18-21) into Eqs. (18-20) yields

$$\sigma_1 = \frac{\varepsilon_{45}(1 - 2\cos\gamma) - \varepsilon_0(1 + 2\sin\gamma)}{0.952(\sin\gamma + \cos\gamma)} \frac{E}{\nu}$$
$$\sigma_2 = \frac{\varepsilon_0(1 - 2\sin\gamma) - \varepsilon_{45}(1 + 2\cos\gamma)}{0.952(\sin\gamma + \cos\gamma)} \frac{E}{\nu} \tag{18-24}$$

where γ is given by Eq. (18-20c). Provided similitude is maintained, A and B of Eqs. (18-22) are independent of hole size. Coefficient 0.952 of Eqs. (18-24) is therefore independent of the isotropic, elastic material, provided dimensional similitude with the original calibration geometry is maintained. Equations (18-24) assume using a 45° rosette. Although no dependence of strain sensitivity on either plate thickness or plate width was observed, the authors suggested maintaining holes such that structural boundaries are at least eight hole diameters from the hole centerline and the plate thickness at least four hole diameters.

Bathgate's Method (1968)

Drawing heavily on the developments by Soete and Vancrombrugge [19,20], Kelsey [23], and Rendler and Vigness [25], Bathgate [27] developed expressions relating the principal residual stresses and principal orientation at the base of a blind hole to the measured surface-strain relaxations around the drilled hole. The method is supposedly suitable for evaluating residual stresses that vary in magnitude and/or direction with depth.

Cordiano and Salerno's Method (1969)

Cordiano and Salerno developed a hole-drilling method particularly applicable to welds [28]. They assumed the following for long welds: (1) the neighboring residual stresses are in a state of plane stress; (2) the principal stresses are parallel and perpendicular to the weld directions (σ_1 parallel to the weld); (3) the residual stresses approach zero some distance from the weld; and (4) the

RESIDUAL STRESSES

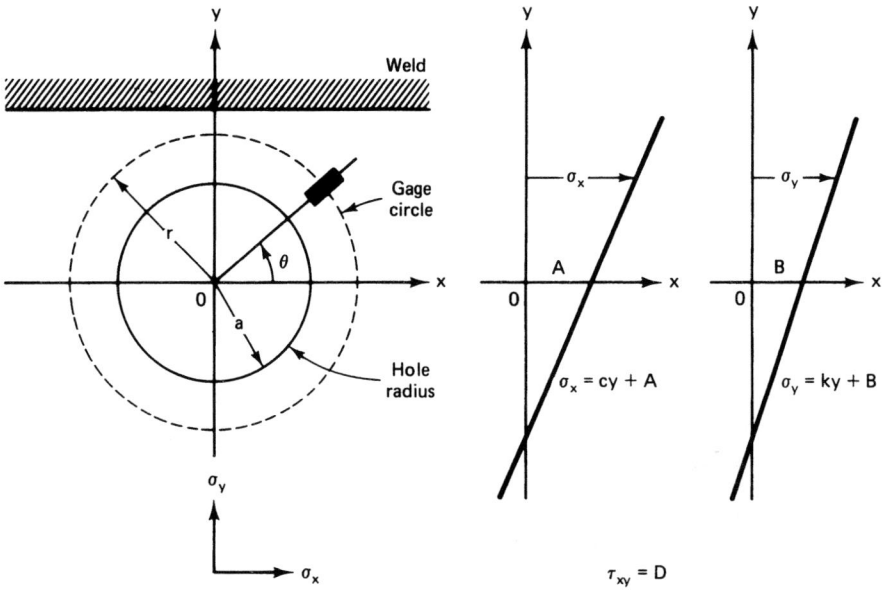

Figure 18-8 Coordinate system and stress distribution assumed by Cordiano and Salerno. (From Ref. 28.)

stresses increase linearly as the weld is approached. For the geometry of Fig. 18-8, the stress field was assumed to be of the form

$$\sigma_x = cy + A$$
$$\sigma_y = ky + B \qquad (18\text{-}25)$$
$$\tau_{xy} = D$$

where, A, B, D, c, and k are constants to be evaluated. Although the stresses of Eqs. (18-25) do not satisfy equilibrium identically, they were found to agree well with observation. For this case, the relaxed radial strain ε_r, measured at radius r and at orientation θ (Fig. 18-8) due to drilling a hole of radius a at the origin, is given by

$$\begin{aligned}
E(\varepsilon_r) = &-\frac{A+B}{2}(1+\nu)\frac{a^2}{r^2} - \frac{A-B}{2}\left[4\frac{a^2}{r^2} - 3(1+\nu)\frac{a^4}{r^4}\right]\cos 2\theta \\
&-\frac{cr}{4}(1+\nu)\frac{a^4}{r^4}\sin\theta + \frac{kr}{4}\left[(\nu-3)(\nu+1)\frac{a^2}{r^2} + \nu(1+\nu)\frac{a^4}{r^4}\right]\sin\theta \\
&-\frac{(c-k)r}{4}\left[(5+\nu)\frac{a^4}{r^4} - 4(1+\nu)\frac{a^6}{r^6}\right]\sin 3\theta \\
&- D\left[4\frac{a^2}{r^2} - 3(1+\nu)\frac{a^4}{r^4}\right]\sin 2\theta
\end{aligned} \qquad (18\text{-}26)$$

Since there are five coefficients, A, B, c, k, and D, to be evaluated, at least five independent measurements of ε_r are to be made in five radial directions θ. The first two terms of Eq. (18-26) represent the relaxation strain in a biaxial-stress field where A and B are the principal stresses. The next three terms of Eq. (18-26) represent the relaxation strain corresponding to the linear variation of the principal stresses with y, where c and k are proportionality constants that give a measure of the strain gradient, and the last term represents the relaxation strain due to the shear stress D. Once A, B, c, k, and D have been determined, the residual stresses are evaluated from Eqs. (18-25). Cordiano and Salerno used excesses of strain gages and data, plus least squares, to obtain a "best-fit" evaluation of the constants of Eqs. (18-25) and (18-26).

This theory was used to determine residual stresses in the vicinity of welds and as influenced by shot peening and surface grinding. Residual stresses in the welds were found to approach the yield strength of the material. It was also found that residual stresses were reduced 25% by grinding and 50% by shot peening.

This method of evaluating residual stresses involves through-the-plate holes, assumes plane stress, and does not necessitate a supplemental calibration specimen.

Bert's Method (1968)

Bert et al. [29,30] extended the concepts of Rendler and Vigness to fiber-reinforced composites. If 1 and 2 represent the in-plane orthotropic axes of material symmetry (not necessarily principal directions of stress or strain), and the 3-direction is normal to the plate (Fig. 18-9), then

$$\varepsilon_{11} = \frac{\sigma_{11}}{E_{11}} - \frac{\nu_{21}\sigma_{22}}{E_{22}}$$

$$\varepsilon_{22} = \frac{\sigma_{22}}{E_{22}} - \frac{\nu_{12}\sigma_{11}}{E_{11}} \quad (18\text{-}27)$$

$$\gamma_{12} = \frac{\tau_{12}}{G_{12}} \quad \text{and} \quad \tau_{12} = \sigma_{12}$$

where $\nu_{12}E_{22} = \nu_{21}E_{11}$. The authors assumed a relaxed radial strain distribution around a drilled hole, given by

$$\varepsilon(\alpha) = [A + B\cos(2\alpha)]\sigma_{11} + [A + B\cos(2\alpha + 180°)]\sigma_{22} + [C\sin(2\alpha)]\tau_{12} \quad (18\text{-}28)$$

where A, B, and C are three independent coefficients that must be determined experimentally. By evaluating Eq. (18-28) simultaneously at each of $\alpha = 0°$, $45°$, and $90°$, and then solving for the stresses, one obtains

$$\sigma_{11} = \frac{(A+B)\varepsilon_0 - (A-B)\varepsilon_{90}}{4AB}$$

$$\sigma_{22} = \frac{(A+B)\varepsilon_{90} - (A-B)\varepsilon_0}{4AB} \quad (18\text{-}29)$$

$$\tau_{12} = \frac{2\varepsilon_{45} - \varepsilon_0 - \varepsilon_{90}}{2C}$$

For uniaxial loading of rectangularly orthotropic materials ($\sigma_{22} = \tau_{12} = 0$), these equations provide

$$A = \frac{\varepsilon_0 + \varepsilon_{90}}{2\sigma_{11}} \quad \text{and} \quad B = \frac{\varepsilon_0 - \varepsilon_{90}}{2\sigma_{11}} \quad (18\text{-}30)$$

RESIDUAL STRESSES

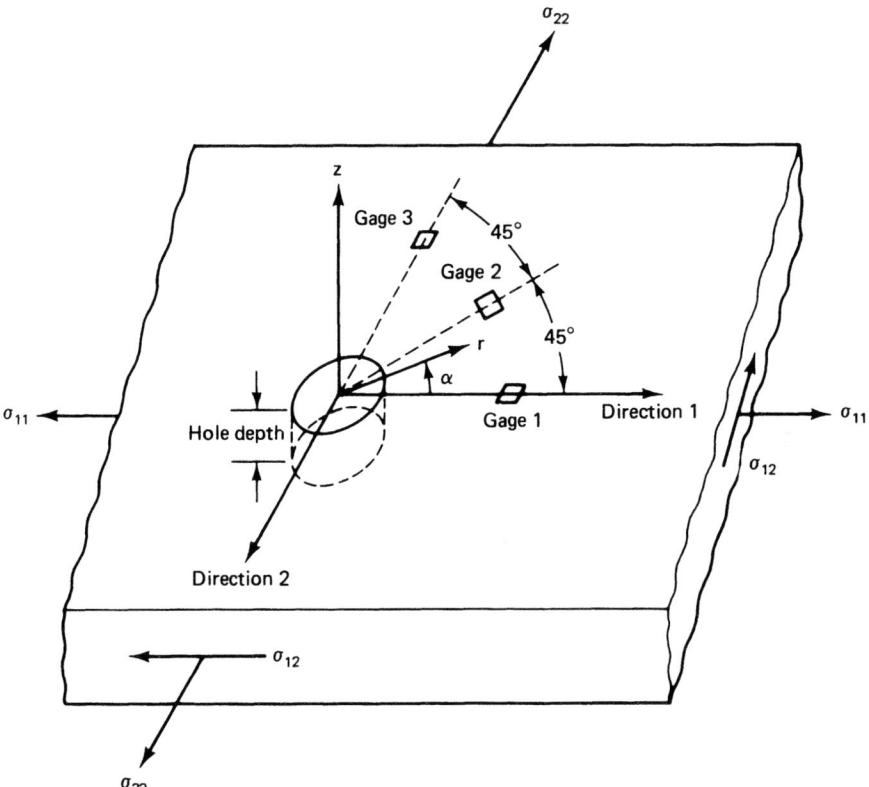

Figure 18-9 Schematic diagram showing Bert's coordinate system, stress components, hole geometry, and strain-gage locations. (From Ref. 29.)

Moreover, C can be obtained from a pure shear test as

$$C = \frac{2\varepsilon_{45} - \varepsilon_0 - \varepsilon_{90}}{2\tau_{12}} \qquad (\sigma_1 = \sigma_2 = 0) \tag{18-31}$$

Once A, B, and C have been determined by calibration, Eqs. (18-29) can be employed to evaluate the residual stresses in orthotropic plates by the hole-drilling technique and from measured strains ε_0, ε_{45}, and ε_{90}.

The method was applied to fiberglass composite plates. Through-the-thickness holes were employed. Constants A and B were determined from uniaxial tests, while C was evaluated from a shear test. The similitude conditions for measurements at a particular ratio of hole depth to diameter, which exist for thick materials, did not prevail in thin plates. Calibration tests must therefore be made for each combination of strain-gage size and plate thickness. Through-the-thickness holes are also desirable. Again, plane-stress and infinite plates (with respect to hole diameters) were assumed.

Shewchuk's Method (1976)

Another set of authors presented a modified hole-drilling method in which no supplemental calibration specimen is necessary [31,32]. For an infinite, isotropic plate containing a small hole and subjected to uniaxial tension σ_x, the Kirsch solution in terms of nondimensional hole diameter $d/r = \lambda$ (d = hole diameter, r = radial distance to strain gage) is

$$\sigma_r = \frac{\sigma_x}{2}\left[\left(1 - \frac{\lambda^2}{4}\right) + \left(1 - \lambda^2 + \frac{3}{16}\lambda^4\right)\cos 2\theta\right]$$

$$\sigma_\theta = \frac{\sigma_x}{2}\left[\left(1 + \frac{\lambda^2}{4}\right) - \left(1 + \frac{3}{16}\lambda^4\right)\cos 2\theta\right] \tag{18-32}$$

$$\tau_{r\theta} = \frac{-\sigma_x}{2}\left[\left(1 + \frac{\lambda^2}{2} - \frac{3}{16}\lambda^4\right)\sin 2\theta\right]$$

Using these equations and Hooke's law, the authors related the relaxed strain ε' to the residual stress by the following expressions:

$$\varepsilon'_r = K_r\frac{\sigma_x}{E} \quad \text{and} \quad \varepsilon'_\theta = K_\theta\frac{\sigma_x}{E} \tag{18-33}$$

K_r and K_θ are the relaxation coefficients (at any angle θ) in the radial and tangential directions, respectively. They are given by

$$K_r = -\frac{\nu + 1}{8}\lambda^2 - \left[\frac{\lambda^2}{2} - \frac{3(\nu + 1)}{32}\lambda^4\right]\cos 2\theta$$

$$K_\theta = -\frac{\nu + 1}{8}\lambda^2 + \left[\frac{\lambda^2}{2} - \frac{3(\nu + 1)}{32}\lambda^4\right]\cos 2\theta \tag{18-34}$$

These nondimensional relaxation coefficients depend on the point of measurement (λ, θ) and Poisson's ratio ν. Figures 18-10 and 18-11 illustrate the variation of K_r and K_θ with θ for aluminum. For residual stresses in aluminum, the technique appears to be accurate within $\pm 5\%$. This is comparable with x-ray results.

Hole Preparation

In 1973 Bush and Kromer [33] found that conventional hole drilling using rotating cutters could introduce unacceptable machining stresses. This, among other considerations, prompted investigators to seek superior techniques for machining stress-free holes [33–39]. Electric-discharge machining (EDM) was rejected because the equipment is not portable, and stress-free holes were not achieved [33]. Electrochemical machining is equally nonportable. Lasers are expensive and impractical. However, Bush and Kromer [33] reported, and others substantiated [34–38], the usefulness of abrasive jet machining. This process directs a stream of air containing fine abrasive particles (typically, 50-μm aluminum oxide) against the workpiece, thereby eroding a hole of desired quality. The method is very acceptable and convenient. Beaney and Procter [34,35,38] have used the technique extensively. They subsequently developed a device having an orbiting nozzle to improve concentricity of the hole.

A stress-free, flat-bottomed hole having a sharp edge is desired for blind holes. Moreover, holes with incrementally varying depths are sought, to permit the monitoring of the residual stresses as a function of depth. These features motivate the machining of holes by end mills or drills. Flaman [39] conducted an extensive comparison of hole-drilling methods using conventional

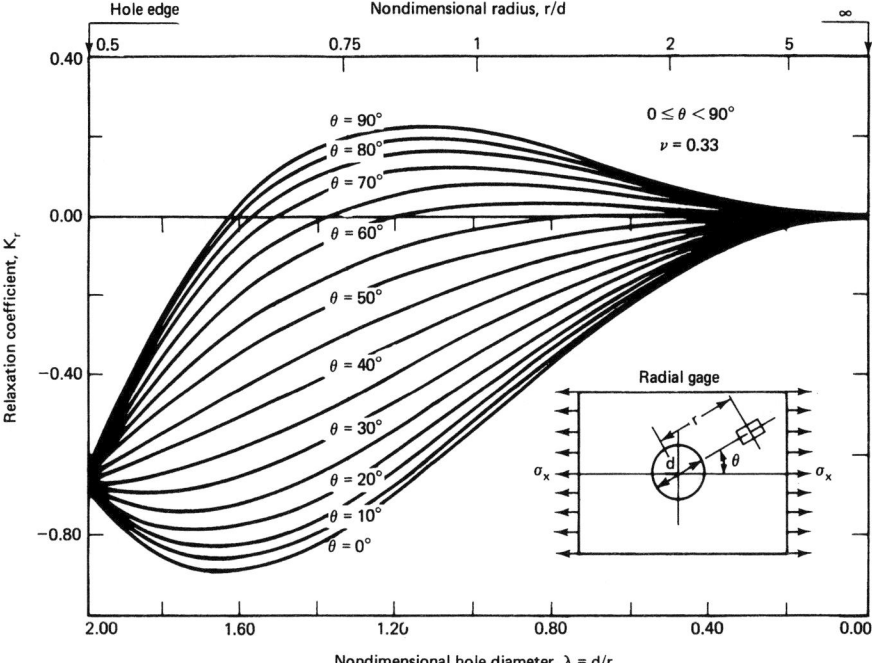

Figure 18-10 Values of radial relaxation coefficient K_r. (From Ref. 31.)

low-speed end mills and an ultra-high-speed drill (350,000 to 400,000 rpm) on mild steel, stainless steel, nickel, aluminum, and copper. The ultra-high-speed drill generally produced superior holes.

Hole Misalignment

Standard hole-drilling techniques for determining residual stresses assume the hole to be drilled concentrically with the center of the strain gages. This suggests the use of some precision guide for locating and drilling the hole. In practice, the hole is often not concentric with the gages. Sandifer and Bowie [40] and Ajovalasit [41] analyzed the effects of the stress (strain) predictions of such eccentricities. The reader is referred to these articles for details. Ajovalasit demonstrates that errors in predicted stresses can be as high as 80% if hole misalignment exists and is not accounted for.

Numerical Analyses

Residual stress determination by the hole-drilling technique typically uses a Kirsch-type solution by which to interpret relaxed strains in terms of stress. The following assumptions usually are made: (1) geometric boundaries are (infinitely) far from the hole, and (2) surface strains around holes of finite depth (blind hole) are similar to those around a through-hole in a similarly thin plate. The validity of such assumptions has been investigated using finite elements [42–44].

Schajer [42] and Manning and Flaman [43] investigated numerically the effect on the calibration coefficients of blind holes of varying depth. While A and B of Eqs. (18-20) are material dependent,

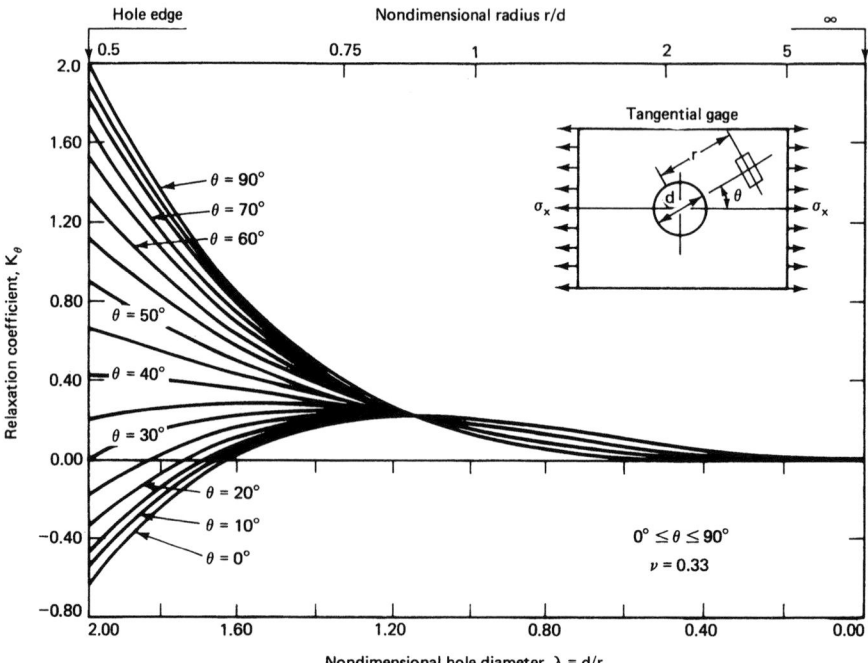

Figure 18-11 Values of tangential relaxation coefficient K_θ. (From Ref. 31.)

Rendler and Vigness [25] introduced two "material-independent" coefficients, k_1 and k_2. Beaney and Procter [34] subsequently considered the group term

$$\frac{\nu k_2}{k_1} = \frac{B - A}{B + A} \tag{18-35}$$

as the second constant instead of k_2. Also in a quest to find material-independent coefficients, Schajer suggested using the following:

$$C_1 = \frac{2EA}{1 + \nu} \tag{18-36a}$$

$$C_2 = 2EB \tag{18-36b}$$

Both Schajer [42], and Manning and Flaman [43] developed finite-element solutions of blind holes of varying depth in loaded plates to assess the validity of coefficients A and B of Eqs. (18-20) [or C_1 and C_2 of Eqs. (18-36)] (Fig. 18-12). The horizontal coordinate z/r_m of Fig. 18-12 is the hole depth divided by the mean gage radius of the rosette used (r_m = 2.56 mm). In addition to the individual finite-element predictions by Schajer and Manning, Fig. 18-12 contains values of A and B as measured by Rendler and Vigness [25]. Both numerical predictions agree well with experiment. Schajer used quadrilateral elements having linear nodal interpolation. He employed models of at least two widely different mesh sizes, which provided essentially identical results. In their analysis, Manning and Flaman used the cyclic symmetry option of the MSC/NASTRAN

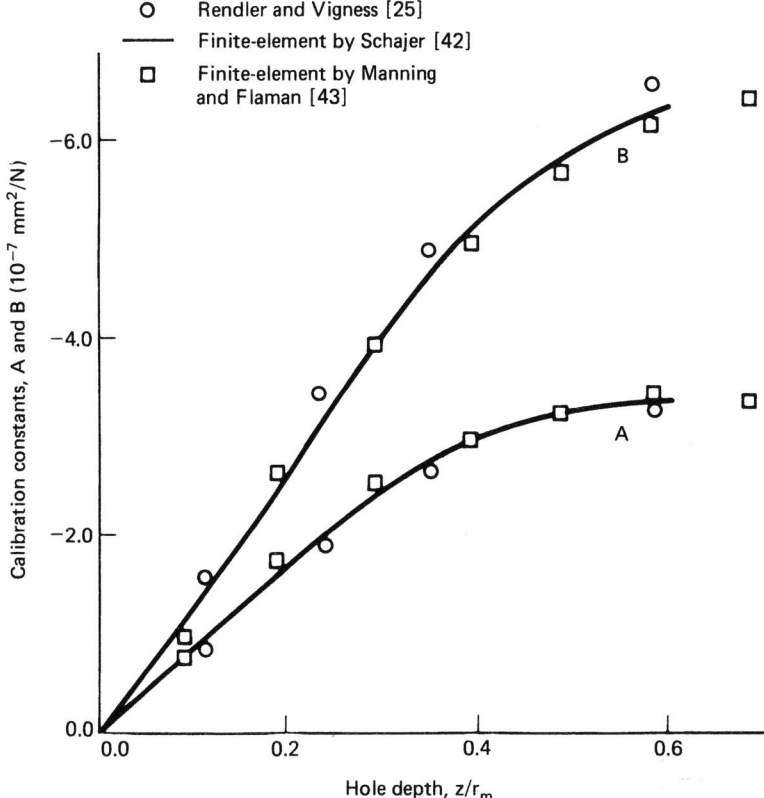

Figure 18-12 Comparison of experimentally and numerically evaluated calibration coefficients A and B of analysis by Rendler and Vigness [Eqs. 18-20]. (From Ref. 43.)

program. While the two numerical solutions of Fig. 18-12 agree well with each other, as well as with the experimental values, differences are possibly associated with how the strains were evaluated numerically. Schajer assumed constant strain across the gage in the radial direction, whereas Manning provided for a totally varying strain under the gage.

Figures 18-13 and 18-14 demonstrate the variations of C_1 and C_2 of Eqs. (18-36) with hole depth and as evaluated by Schajer (r_m = mean gage radius of rosette). The "theoretical" values shown in Figs. 18-13 and 18-14 are approached asymptotically for large hole depths. These data show that increasing the relative hole radius a/r_m, increases the magnitude of the curves but has little effect on their shapes. This implies that the character of measured relaxed strains depends on the gage layout and only the sensitivity depends on the hole diameter. Schajer observed that both C_1 and C_2 vary by less than 2% over $.025 \leq \nu \leq 0.35$. Moreover, for a hole whose depth equals its diameter, k_1 is constant within 2% and $\nu k_2/k_1$ varies only between 0.32 to 0.27 over the same change in Poisson's ratio. These small variations in constants qualify them as essentially "material independent." In practice, $\nu k_2/k_1$ is often taken equal to 0.3 [34,44].

Newwar et al. [31,32] employed relaxation coefficients [Eqs. (18-34)] in the hole-drilling analysis. Newwar and Miller subsequentially conducted a finite-element analysis to study the

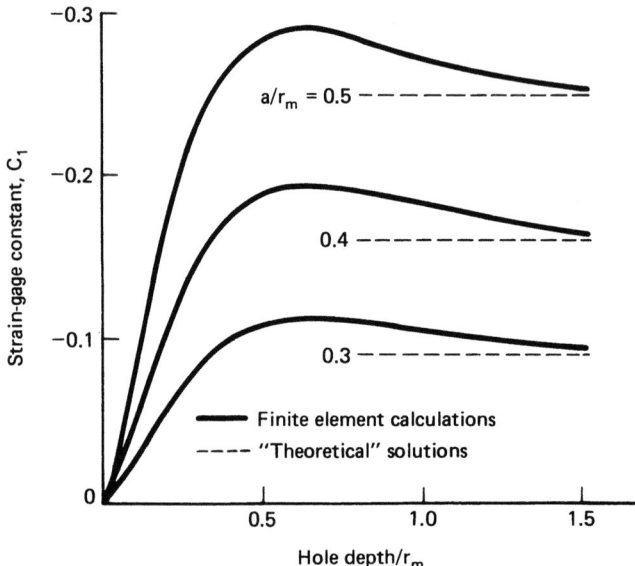

Figure 18-13 Numerically predicted coefficient C_1 of Eq. (18-36a) as a function of relative hole depth. (From Ref. 42.)

variation of such coefficients as a function of relative hole depth and distance of the gages from the hole [44]. Numerically, using STRUDL version V3MO, they modeled the blind hole with a very fine mesh in the immediate vicinity of the hole and at the top surface to allow for small depth changes associated with incremental drilling. They found that the relative hole diameter (hole diameter-to-gage radius) plays a significant role as the plate thickness increases.

Implementation

After our survey of many of the analytical developments of the hole-drilling concept, practical aspects of the method are in order. References 45–48 are informative in this respect. A standard on this technique has been issued by ASTM [47].

Once a commercial three-element rosette of the hole-drilling type has been applied, one carefully drills a hole [~1/16 in. (1.6 mm) diameter] in the center of the rosette and to a depth equal to or slightly greater than that of the diameter. It is advisable to monitor the strains incrementally as the hole is drilled. A dummy-gage rosette should be used. With reference to Fig. 18-15, surface radial strains relaxed by drilling are related to the relieved residual principal stresses σ_1 and σ_2 by [47]

$$\varepsilon_r(\alpha) = A + B \cos 2\alpha)\sigma_1 + (A - B \cos 2\alpha)\sigma_2 \tag{18-37}$$

RESIDUAL STRESSES

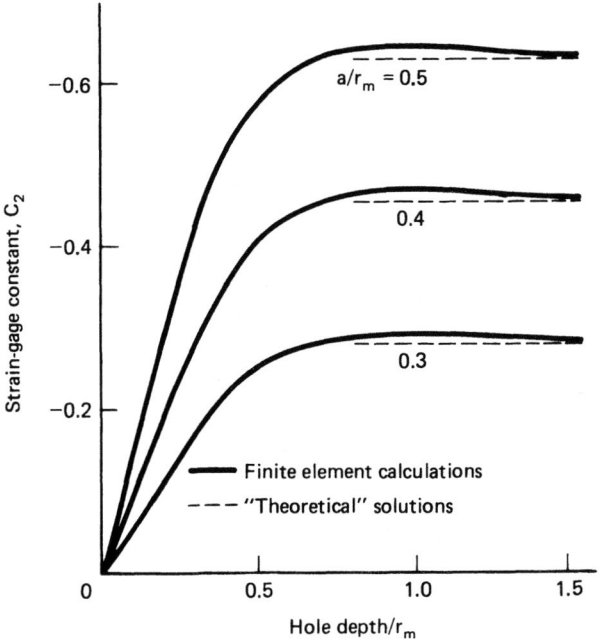

Figure 18-14 Numerically predicted coefficient C_2 of Eq. (18-36b) as a function of relative hole depth. (From Ref. 42.)

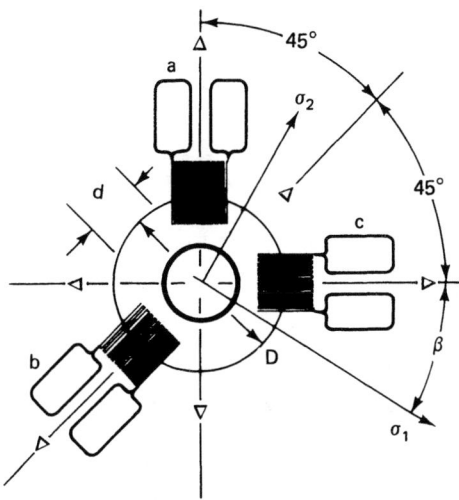

Figure 18-15 Typical three-element strain-gage rosette layout for hole-drilling technique. (From Ref. 47.)

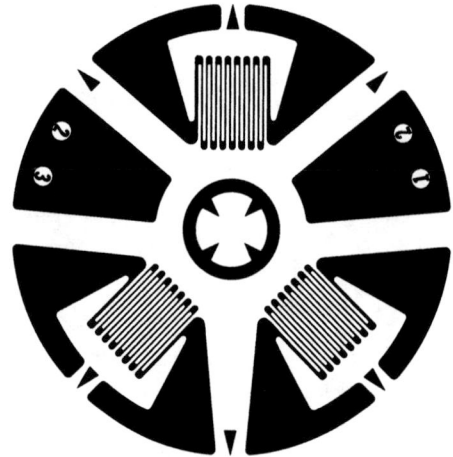

Figure 18-16 Commercial three-element strain-gage rosette for hole drilling. (Courtesy of Measurements Group, Inc.)

where $A = -(1 + \nu)/2Er_0^2$
$B = -[(1 + \nu)/2Er_0^2][4/(1 + \nu) - 3/r_0^2]$
α = angle between σ_1 and $\varepsilon_r(\alpha)$
E, ν = elastic properties
$r_0 = D/d$, normalized hole diameter (Fig. 18-15) $(2.5 \leq r_0 \leq 3.4)$

The principal residual stresses relieved and principal direction β are given by (Fig. 18-15)

$$\sigma_1, \sigma_2 = \frac{\varepsilon_a + \varepsilon_c}{4A} \pm \frac{\sqrt{2}}{4B}\sqrt{(\varepsilon_a - \varepsilon_b)^2 + (\varepsilon_b - \varepsilon_c)^2} \tag{18-38}$$

$$\tan 2\beta = \frac{\varepsilon_c - 2\varepsilon_b + \varepsilon_a}{\varepsilon_c - \varepsilon_a} \tag{18-39}$$

The hole should be machined incrementally in approximately 10 steps, reading the strains at each interval. It is convenient to record strain versus depth for each gage. Hole machining can be done with a drill, end mill, or abrasive jet. However, "good" holes having flat bottoms should be used. Care should be taken to minimize machining stresses and not to record any thermal strains.

Commercial equipment is available for applying the hole-drilling method. Figure 18-16 shows a three-element strain-gage rosette (Micro-Measurements Group, Vishay, Inc.). A machining guide assembly is also available from Measurements Group, Vishay, Inc., as are other components for the technique.

18-2 Hole Boring—Rock Mechanics [49–54]

The safe design and construction of items such as mines, subway systems, and nuclear waste repositories require determining residual stresses in rock. Several techniques have been developed

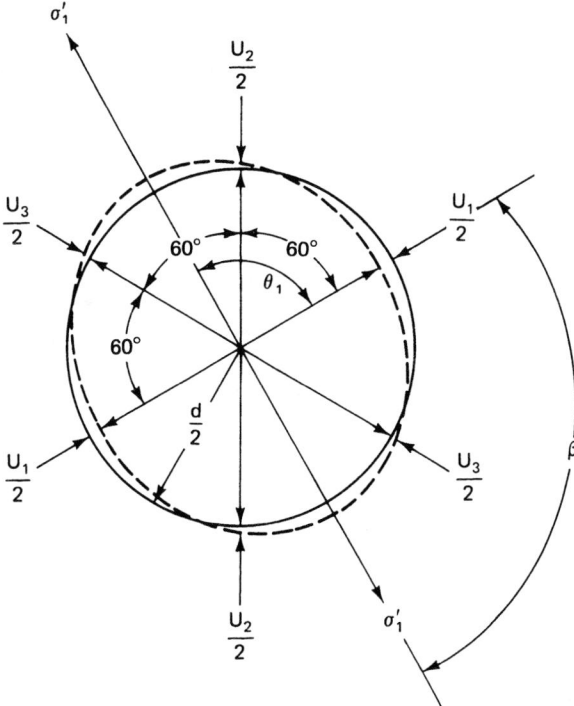

Figure 18-17 A 60° deformation rosette.

to measure such stresses in situ, the most prevalent of which involve monitoring the diametral changes of a borehole during overcoring. The methods typically assume that the rock behaves in a linear elastic manner and again use Kirsch's solution for a hole in a loaded infinite plate.

The concept involves boring a circular hole in the rock, into which is located a device capable of measuring the diametral changes of this hole upon overcoring with a concentric core bit. Overcoring of the original hole relieves the residual stresses (which had been present and are related to strains or changes in the inside diameter of the cylinder).

Assuming plane stress (in the plane normal to the axis of the hole), linear-elastic isotropic response, the in-plane principal stresses (original residual stresses) σ_1' and σ_1' and the principal direction β are given by [50,54]

$$\sigma_1', \sigma_2' = \frac{E}{6d}\left\{(U_1 + U_2 + U_3) \pm \frac{\sqrt{2}}{2}[(U_1 - U_2)^2 + (U_2 - U_3)^2 + (U_3 - U_1)^2]^{1/2}\right\}$$
(18-40)

$$\beta = \frac{1}{2}\tan^{-1}\frac{3(U_2 - U_3)}{2U_1 - U_2 - U_3}$$
(18-41)

where angle β is measured from diametral direction change U_1 to σ_1', U_1, U_2, and U_3 are three measured diametral changes at 60° intervals (Fig. 18-17). Although several diametral deformation gages have been proposed, probably the U.S. Bureau of Mines (USBM) gage is used most widely

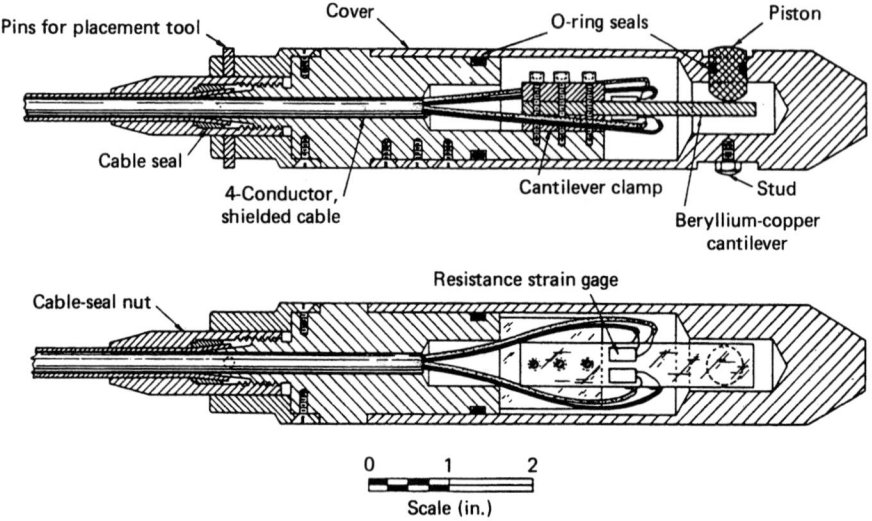

Figure 18-18 U.S. Bureau of Mines borehole deformation gage. (From Ref. 51.)

(Fig. 18-18) [51]. The deformation transducer employs three pistons (at 60° intervals), which contact the wall of the inner hole. Each piston activates a separate beryllium–copper cantilever beam on which four strain gages are mounted and connected as a four-arm Wheatstone bridge. Bridge output is typically conveyed to the surface through a shielded cable. The instrument is relatively stable over long time periods and is comparatively insensitive to temperature or other environmental factors. Based on a rock modulus of $E = 21$ GPa (3×10^6 psi), the gage would respond to the stresses as small as 0.1 MPa (13 psi). The instrument normally fits a 3.8-cm (1.5-in.)-diameter hole. Variations of the USBM borehole gage have been proposed, including those equipped to determine the longitudinal stress in non-plane-stress situations.

A somewhat different borehole gage for determining residual stresses in rock in situ includes monitoring during overcoring the response of three three-element strain-gage rosettes bonded to the wall of a hole [52]. All six components of residual stress can be determined from these strain readings.

Other methods of measuring residual stresses in rock include photoelastic meters [53,54], hydraulic fracturing, sonic techniques, vibrating wire gages, rigid-inclusion stress gages, and hydraulic cells or flat jacks [49,54]. Photoelastic meters include wedge-loaded photoelastic disks located inside a hole during overboring. Hydraulic fracturing involves pressurizing a hole or cavity in the rock and predicting the residual stresses from the known hydraulic pressure when rock fracture occurs. Sonic methods depend on establishing the stress dependence of the sonically determined material properties E and ν. Hydraulic flat jacks are thin-walled, fluid-filled metal bladders. Their use involves bonding two strain gages on a rock, after which a slot is cut in the rock between the gages and into which an empty flat jack is placed. By pressurizing the flat jack to return the gages to their initial readings, present before the slot was machined, one can interpret the original residual normal stress in the rock as equal to the hydraulic pressure in the flat jack.

18-3 Ultrasonic (Acoustical) Techniques

Ultrasonics is the discipline associated with the propagation of high-frequency mechanical waves in the range beyond the audible. The field is sometimes referred to as "ultrasound." Ultrasonic waves or vibrations can be produced in any medium—solid, liquid, or gas. Engineering applications of ultrasonics often employ frequencies in the high-kilohertz through megahertz range. Ultrasonic waves exhibit regular features of common mechanical waves, including dispersion, transmission, reflection, diffraction, interference, and attenuation. They can be longitudinal or transverse in form, or a combination thereof. Moreover, ultrasonic surface or Rayleigh waves can occur. There are mechanical equivalents of polarizers and wave plates, such that plane or circularly polarized ultrasonic waves of reasonably pure frequencies (monochromatic) can be generated. Using ultrasonic similarities of visible light waves, ultrasonic (or acoustical) holographic techniques have been proposed for internally analyzing three-dimensional objects [55].

Ultrasonic stress analysis is based on the relationship between wave propagation velocity and the higher-order elastic constants, which are stress dependent. The method involves introducing ultrasonic longitudinal, transverse-shear, or surface Rayleigh waves into the material of interest and measuring their velocity change with stress. Practical difficulties encountered in executing the experimental procedures include (1) the need to transmit effective waves of suitable characteristics into the component being analyzed, (2) the measurement of very small stress-induced velocity changes, and (3) the need to separate the effects of velocity changes of stress from those of material microstructure, anisotropy, inhomogeneity, and so on. While both the x-ray and ultrasonic techniques of residual stress analysis are nondestructive, ultrasonic methods appear to hold greater potential for three-dimensional and subsurface applications.

Surface-Wave Techniques

When a solid is subjected to a mechanical impact (or pulse) or train of pulses, stress waves of various types move out through the material. Two mechanical phenomena propagate in an infinite medium or far from boundaries: the fastest or dilational (irrotational) wave in which the particle motion is parallel to the direction of event propagation and the slower transverse shear (distortional, equivoluminal) waves whose particle motion is transverse to the direction of wave propagation. In bounded media, Rayleigh (surface) waves are also produced at the surface. Whereas acoustoelasticity (Section 18-4) typically utilizes transverse mechanical waves, Rayleigh waves can be employed to investigate residual surface stresses.

Rayleigh waves are similar to the gravitational waves produced on the surface of a liquid [56,57]. Particle motion of surface or Rayleigh waves vibrates along elliptical orbits with one component parallel to the direction of motion (the latter being parallel to the surface) and the other perpendicular thereto and normal to the surface. In solids, this elliptical motion is retrograde to the direction of wave propagation at the surface, becoming prograde at approximately 0.5-wavelength below the surface. The major diameter of the ellipse is normal to the surface and has a length approximately 1.5 times that of the minor diameter. Rayleigh-wave amplitudes decrease rapidly (exponentially) with depth, becoming negligible 1.5 wavelengths below the surface. Rayleigh waves are slightly slower than shear waves. These waves are nondispersive; that is, their propagation velocity is frequency independent and depends only on the elastic constants of the material, and indirectly on the state of stress through stress-dependent changes in the higher-order elastic constants. A change in Rayleigh-wave velocity can be used to measure the magnitude of surface stresses. The technique is potentially very useful for residual stresses, which often are a maximum near the surface. The depth-penetration dependence on wavelength affords the possibility

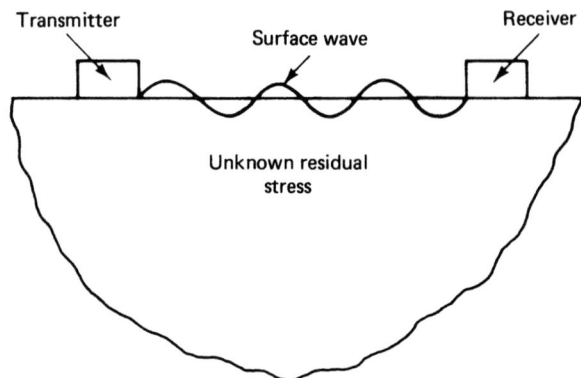

Figure 18-19 Schematic of Rayleigh-wave technique.

of obtaining the stress gradient with depth by varying the frequency (or wavelength λ) of the Rayleigh waves.

Rayleigh-wave determination of surface stresses typically employs separate transmitting and receiving transducers[4] applied to the surface of the component (Fig. 18-19). The transducers are normally cross-cut crystals, which produce longitudinal mechanical waves when electrically oscillated. Rayleigh surface waves are produced in the component through refraction by ensuring that the longitudinal waves from the transducer enter the component at the appropriate incidence angle as determined by Snell's law [58],

$$\frac{\sin \phi_1}{\sin \phi_2} = \frac{\nu_1}{\nu_2} \qquad (18\text{-}42)$$

where ϕ_1 is the angle of the incident (or receiving) longitudinal wave measured from the normal to the surface, ϕ_2 the orientation from the normal to the refracted wave, ν_1 the transducer incidence (received) longitudinal wave velocity, and ν_2 the velocity in the component.

Receiving and transmitting transducers are similar. Surface-wave velocity is determined by measuring the time for the ultrasonic wave to traverse the gage length defined by the distance between the two transducers. The average stress (over, say, a depth of $\lambda/2$ to λ) is then determined from the stress–velocity calibration for the material. In this respect the method acts like a strain gage. A coupling agent is often used between the transducer and the component being interrogated. For residual stress analysis, a given material must be calibrated by measuring the relative change in acoustic travel time as a function of applied stress.

Transmitting and receiving transducers are normally made of a piezoelectric material [quartz, ammonium dihydrogren phosphate (ADP), barium titanate, lead zirconate titanate (PZT), etc.] whose dimensions change when subjected to a voltage, and vice versa. Magnetostrictive transducers are also sometimes used. A magnetostrictive material is one whose dimensions change when

---

[4] In this case, the transducer converts the electrical or magnetic energy into mechanical vibrations, or waves (or vice versa). A horn or stereo speaker is an acoustical transducer that converts the electrical waves into audible air waves.

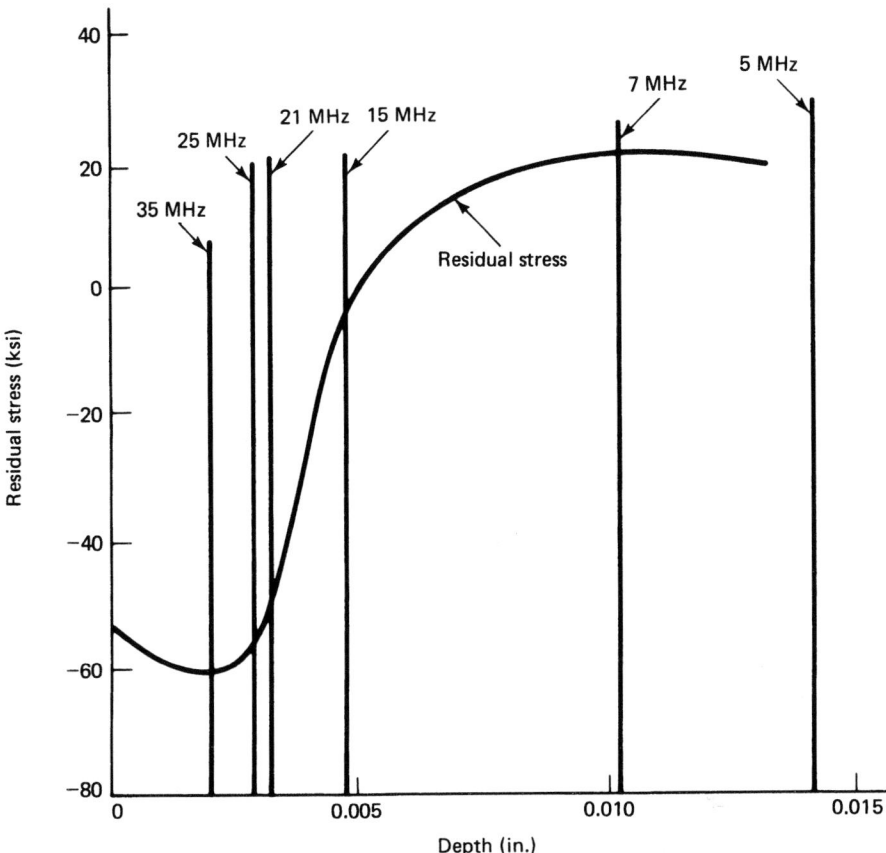

Figure 18-20 Location of stresses determined by Rayleigh waves at various frequencies relative to a typical shot-peened, induced residual stress field. (From Ref. 59.)

subjected to a magnetic, rather than electric, field. Iron and nickel are common core materials of magnetostrictive transducers. Transducer design is a fairly well-developed science, and a wide variety of models are available commercially.

Stress gradients normal to a surface can be evaluated by exciting the transmitting transducer at various frequencies. The Rayleigh-wave velocity responds to the stress in a surface layer of the material, this layer having a thickness of the order of the wavelength λ, where $c = f\lambda$. By changing the frequency f of the induced Rayleigh wave, one measures the sonic influence of the stress extending over varying depths from the surface. This enables the stress gradient normal to the surface to be determined. Figure 18-20 illustrates how residual stresses near the surface can be scanned by Rayleigh waves using transmitting frequencies of 5 and 7 MHz operated at their fundamental and higher frequencies [59]. In this case the depth of scan is based on an effective layer for stress analysis purposes of 0.6λ. These penetration depths are compared in the figure with a typical residual stress profile as might be produced by commercial shot peening.

Although Rayleigh-wave techniques are attractive for evaluating surface residual stress, improved electronic equipment having adequate resolutions is needed. Applications to polyaxial states of stress, composite materials, and curved surfaces are also warranted.

Velocity Measurements

Since a stress change of 100 psi (0.69 MPa) in metals produces an ultrasonic velocity change of only a few parts in 10^6, practical utilization of the concepts necessitates the ability to measure time changes in nanoseconds. The following electronic techniques are used to make the time measurements: (1) sing around, (2) pulse superposition, and (3) pulse echo. For residual stress measurements, one normally must establish separately a stress–velocity calibration. This is typically done using applied loads (stresses), and the velocity dependence due to live or residual stresses is assumed to be identical.

The sing-around method involves transmitting the sound energy such that it "sings around" the system (mechanical gage length plus electronic circuitry) at a repetition rate determined by the ultrasonic transit time between transducers. This is accomplished by retriggering the transmitter by an electric signal from the receiver. A refined version of the sing-around method effectively increases the ultrasonic path length by retriggering the transmitter from an echo among the multiple reflections. Changes in transient time, hence acoustic velocity, are determined by the change in sing-around frequency measured by an electronic counter.

Numerous pulse-superposition concepts have been employed to measure velocity changes. In this approach, the time required for the signal to traverse the gage length under stress is compared with that necessary to traverse some calibrated electronic delay line, or else the phase difference to these two signals is compared. Variations of this concept have been employed by McSkimin [60] and Benson [61].

At least as originally proposed, the pulse-echo method consists of utilizing a single transducer as both transmitter and receiver. An acoustical wave train is generated electrically within the specimen by the transducer. The magnitude of successive echoes received by the transducer attenuates exponentially with echo number for a particular stress level. A stress change alters the pulse-echo pattern, the change being a measure of the change in stress magnitude.

Because stress-induced changes in ultrasonic velocity are small, appreciable efforts have been devoted to developing electrical systems of adequate resolution. McSkimin [60] achieved a resolution of 1 part in 10^6, while Forgacs [62] developed a sing-around technique capable of 1 part in 10^7. A pulse-echo system capable of resolving time changes of 40 ps was reported by Chick et al. [63]. Additional velocity measuring techniques are described in Ref. 64–67. McKannan claimed stress sensitivity down to 0.9 MPa (100 psi) in aluminum [58].

Figure 18-21 shows the stress–velocity calibration of glass obtained using a phase-shift technique having a time resolution of 4 ns, which corresponds to a stress resolution in aluminum of 300 psi (2 MPa) [68].

18-4 Acoustoelasticity

When polarized light (transverse shear vibration) is incident on an initially isotropic but now stressed (now optically anisotropic) photoelastic material, the wave breaks into two plane-polarized rays, which propagate along the principal directions (stress or strain) with different velocities (Chapter 5). This velocity difference causes the two rays to exit the photoelastic medium out of phase with each other. At locations where the phase difference is 180°, destructive interference occurs and half-order fringes occur. Where the phase shift is 360° (corresponding to a full wave-

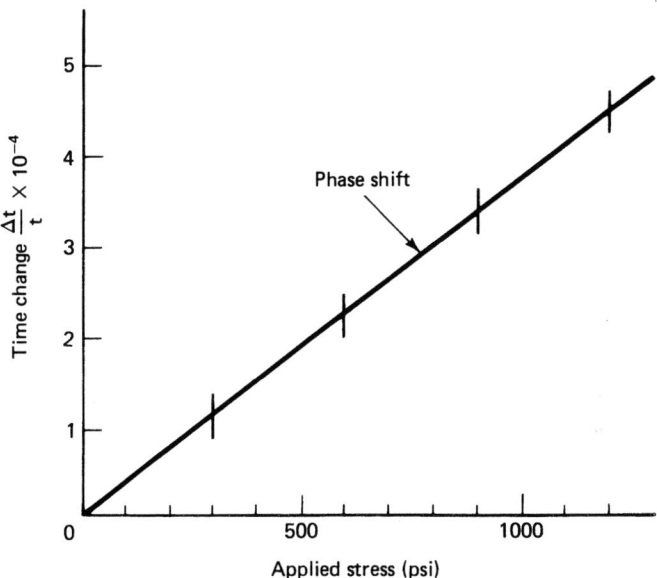

Figure 18-21 Stress–velocity calibration of glass (phase-shift technique). (From Ref. 68.)

length), constructive interference is present and an integral fringe occurs. One thereby obtains light and dark isochromatic fringes.

Acoustoelasticity is like photoelasticity except that the light waves are replaced normally by transverse ultrasonic or acoustic shear waves (~2 to 10 MHz, typically) [69–90]. The acoustical waves divide into components and propagate along the principal directions of a stressed medium with a velocity difference (phase shift) proportional to the stress difference

$$(\sigma_1 - \sigma_2) \propto (n_1 - n_2) \propto (\nu_1 - \nu_2) \tag{18-43}$$

where n is the index of refraction, ν the propagation velocity, and the subscripts 1 and 2 denote principal directions. Potential advantages of acoustoelasticity include

1. Actual structures rather than models can be analyzed. This could lead to in-situ field applications.
2. Both residual stresses and those due to applied loads can be measured. As such the concept has been applied to both elastic and inelastic stresses.
3. Like optical photoelasticity, the technique is conceptually applicable to two- and three-dimensional structures under static, dynamic, or inelastic response.

Disadvantages and/or shortcomings include:

1. Results are influenced by metallurgical effects, such as forming anisotropy and grain structures.
2. At least currently, the technique is essentially a point-by-point analysis.
3. Since acoustical birefringence results from both initial (residual) and load-imposed stresses, the situation must be compared with a stress-free situation.

Table 18-1 Comparison of Photoelasticity and Acoustoelasticity (from Ref. 76)

	Photoelasticity (plastic)	Acoustoelasticity (metal)
Speed of sensing waves in unstressed medium, v_0	2×10^8 m/s	3×10^3 m/s
Frequency, f	5×10^{14} Hz	10^7 Hz
Wavelength, λ	0.0004 mm	0.3 mm
Fringe order at proportional limit, N	50 cm^{-1}	0.2 cm^{-1}
Stress per fringe order (thickness = 2.54 cm)	0.4 MPa	825 MPa

4. The method has limited resolution, although attempts have been made to address this topic. Based on a 2.54-cm thickness, the relatively long wavelength of acoustical waves ($f \sim 10^7$ Hz and $\lambda \sim 0.33$ mm for acoustoelasticity compared with $f \sim 10^{14}$ Hz and $\lambda \sim 0.0004$ mm for photoelasticity) requires a stress of 825 MPa to produce an acoustical relative retardation of one wavelength in metal, while only 0.4 MPa produces a photoelastic relative retardation of one wavelength in plastic [76] (Table 18-1).
5. Sophisticated circuitry and electronics are needed (must measure time changes of the order of nanoseconds).
6. Variability of transducer coupling to a structure can introduce questions of accuracy and reproducibility.

Although the acoustoelastic method is still in its infancy, the typical approach is to introduce transverse-shear waves into the structure from an emitting transducer contacting the component. One may use a transmitting transducer on the front face and a separate receiving transducer on the back face. Reflective acoustoelasticity employs a common transducer as transmitter and receiver by receiving the waves reflected from the back side of the component. Reflective acoustoelasticity is similar to reflective photoelasticity, where the captured light is reflected off the back face of the birefringent coating. Transducers are available that emit polarized acoustical shear waves akin to plane-polarized light. Principal directions can be obtained acoustically by rotating a single transducer (or transmitter and receiver together) until a maximum signal is obtained—indicative of the isoclinic orientation. One typically measures absolute rather than relative velocities along the principal directions. Acoustoelasticity evaluated stresses in a loaded plate with a hole are illustrated in Fig. 18-22.

Much of the available research on acoustoelasticity has concerned simple stress fields, analytical relationships between wave velocities and the stresses, and electronic techniques for measuring accurately the wave velocities. However, recent efforts have involved two-dimensional states of stress [76,86], inelastic strains [70,71,88], initially anisotropic materials, use of longitudinal as well as transverse-shear waves [88,89], the shear-difference technique of stress separation [86], and fracture problems [90]. Numerous applications involve residual stresses [70,72,74,77,78,83–85,87,88]. Blinka and Sachse report an acoustoelastic method for determining principal-stress differences as small as 0.25 MPa (35 psi) in metal [79].

Besides x-ray, acoustical techniques are almost the only method able to measure residual stress totally nondestructively. Their potential for measuring subsurface stresses, and for stress-analyzing actual components in the field without use of models, is extremely attractive. Acoustoelasticity should be developed to its fullest capabilities. Objectives should include improving sensitivity and assessing the potential for three-dimensional, transient, and thermoelastic problems, and those involving composite materials. The abilities to provide full-field analysis, perhaps using arrays of transducers and some means of electrical-to-visual display, would be extremely advantageous.

RESIDUAL STRESSES

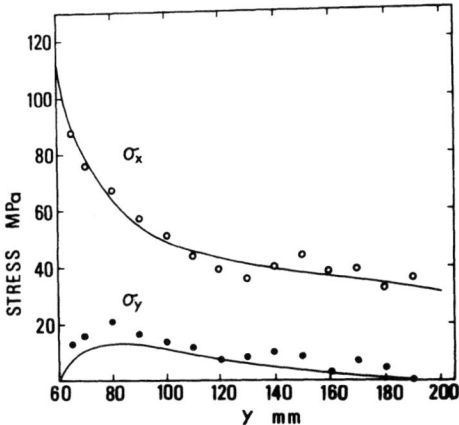

Figure 18-22 Acoustoelastically determined stresses along the horizontal axis of symmetry of a vertically loaded metal plate containing a central circular hole. (From Ref. 86.)

18-5 X-Ray [91–106]

X-ray techniques provide one of the very few nondestructive methods of measuring residual stresses. However, even x-ray determination of subsurface stresses by removing material may involve destructive techniques. Unlike most other methods, x-rays permit direct measurement of the loaded state without the need of an unloaded measurement. This renders the method very suitable for measuring residual, as well as load-induced, stresses.

X-rays, like gamma, ultraviolet, infrared, and microwave radiation, are portions of the electromagnetic spectrum. As with mechanical waves, x-rays can exhibit interference, reflections, and diffractions. Their relatively short wavelength ($\lambda \sim Å = 10^{-8}$ cm $= 10^{-4}$ μm; $f \sim 10^{18}$ Hz based on $f\lambda = c = 3 \times 10^{10}$ cm/s) enables them to penetrate objects such as metals, which are normally opaque to visible light.

As with other techniques, the inability to measure stress causes one to measure strains from which stresses are evaluated through the constitutive relationships. The x-ray technique utilizes the interatomic spacing (and changes thereof) of certain lattice planes as the gage length. X-ray stress analysis has a resolution of 20 to 35 MPa (3000 to 5000 psi). This compares to ~2 MPa (300 psi) for strain gages on metals. X-rays are used to measure surface stresses (0.0001 to 0.001 in. penetration). Subsurface stresses and stress gradients are determined by removing surface layers.

Although space does not permit reviewing the extensive literature on x-ray stress analysis, the reader is referred to the summary-type articles of Refs. 91–99.

Principles

X-ray stress (strain) analysis is based on the Bragg diffraction equation

$$n\lambda = 2d \sin \theta \tag{18-44}$$

where n is an integer (diffraction order), λ the wavelength of the incident x-ray, d the interplanar spacing of the polycrystalline material under consideration, and θ the Bragg angle (θ = angle of

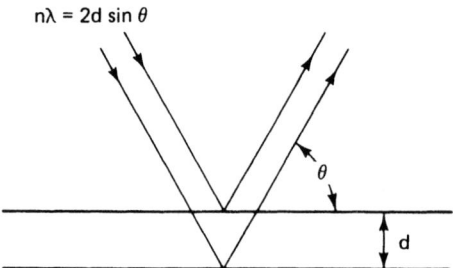

Figure 18-23 Bragg diffraction at point p.

incidence = angle of diffraction) (Fig. 18-23). A change in the interplanar spacing Δd will produce a corresponding change in the Bragg angle $\Delta\theta$, such that lattice strain can be expressed from

$$\frac{\Delta d}{d_0} = \varepsilon = -(\cot\theta)\Delta\theta = -(\cot\theta)\frac{\Delta(2\theta)}{2} \tag{18-45}$$

where d_0 is the unstrained lattice spacing, ε is the mechanical strain, and $\Delta\theta$ is the change in Bragg diffraction orientation due to the strain. X-ray diffraction angles θ are measured from the atomic plane, whereas in optics incident and reflected angles are measured typically from the normal to the surface (Fig. 18-23).

Diffraction occurs when Eq. (18-44) is satisfied [i.e., for a combination of λ, d and θ which satisfies Eq. (18-44)]. X-ray stress analysis equipment consisting of an x-ray source (λ) and the capability for measuring angles θ and their changes with orientations ψ is called a diffractometer.

Stress and strain at a point p under investigation can be obtained from x-ray data with the aid of the stress and strain ellipsoids (Fig. 18-24). Principal strains, ε_1 and ε_2, are assumed to lie in the plane of the material and ε_3 normal to the plane. From strain transformation and the angles of Fig. 18-24, the normal strain $\varepsilon_{\phi\psi}$ in the z-x plane is given by [92]

$$\varepsilon_{\phi\psi} = \varepsilon_1\alpha_1^2 + \varepsilon_2\alpha_2^2 + \varepsilon_3\alpha_3^2 \tag{18-46}$$

where the direction cosines are given by

$$\begin{aligned}\alpha_1 &= \cos\phi\sin\psi \\ \alpha_2 &= \sin\phi\sin\psi \\ \alpha_3 &= \cos\psi = \sqrt{1 - \sin^2\psi}\end{aligned} \tag{18-47}$$

Similarly, the stress component σ_ϕ in the plane of the material ($\psi = \pi/2$) can be written in terms of the principal stresses σ_1, σ_2, and in-plane angle ϕ, that is,

$$\sigma_\phi = \sigma_1\cos^2\phi + \sigma_2\sin^2\phi \tag{18-48}$$

Substituting Hooke's law

$$\begin{aligned}\varepsilon_1 &= \frac{1}{E}[\sigma_1 - \nu(\sigma_2 + \sigma_3)] \\ \varepsilon_2 &= \frac{1}{E}[\sigma_2 - \nu(\sigma_1 + \sigma_3)] \\ \varepsilon_3 &= \frac{1}{E}[\sigma_3 - \nu(\sigma_2 + \sigma_1)]\end{aligned} \tag{18-49}$$

RESIDUAL STRESSES 817

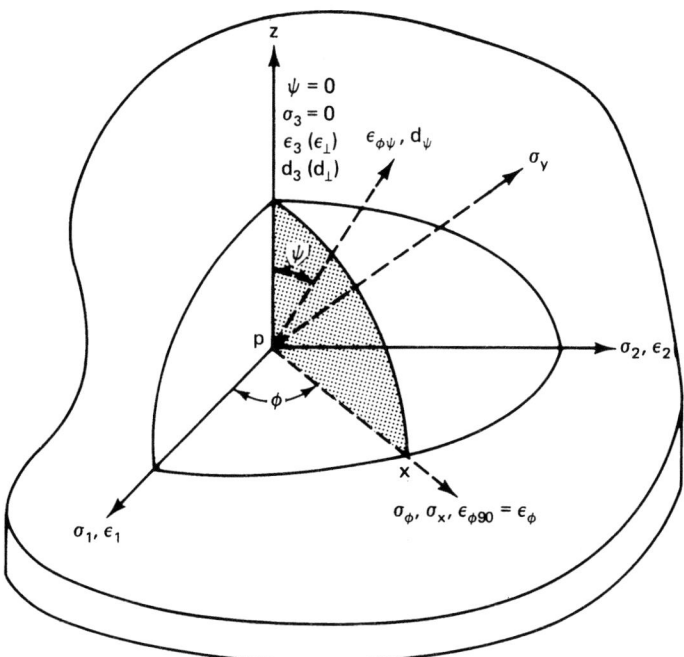

Figure 18-24 Stress and strain ellipsoids at point p.

into Eq. (18-46) yields

$$\varepsilon_{\phi\psi} = \frac{1+\nu}{E}(\sigma_1\alpha_1^2 + \sigma_2\alpha_2^2 + \sigma_3\alpha_3^2) - \frac{\nu}{E}(\sigma_1 + \sigma_2 + \sigma_3) \quad (18\text{-}50)$$

At a stress-free surface ($\sigma_3 = 0$), Eqs. (18-47) and (18-50) combine to give

$$\varepsilon_{\phi\psi} = \frac{1+\nu}{E}(\sigma_1 \cos^2\phi + \sigma_2 \sin^2\phi)\sin^2\psi - \frac{\nu}{E}(\sigma_1 + \sigma_2) \quad (18\text{-}51)$$

such that Eqs. (18-48) and (18-51) can be expressed as[5]

$$\varepsilon_{\phi\psi} = \frac{1+\nu}{E}\sigma_\phi \sin^2\psi - \frac{\nu}{E}(\sigma_1 + \sigma_2) \quad (18\text{-}52)$$

This is a basic mechanical equation of the x-ray diffraction method of stress analysis. It relates lattice strain in the arbitrary orientation defined by angles ϕ and ψ of Fig. 18-24 to in-plane stress. It remains to relate the principal-stress term of Eq. (18-52) to the normal strain and to express strains in terms of diffraction quantities.

The atomic planes defining the gage length are perpendicular to the incident and diffracted

---

[5] Note that knowledge of the principal direction ϕ is not necessary.

beam bisector, while it is the normal strain along this bisector that is being measured at point p of Figs. 18-23 and 18-24. If $d_\perp$ is the measured spacing of atomic planes parallel to the surface in the strained state, the strain $\varepsilon_\perp$ normal to the surface ($\psi = 0$) is

$$\varepsilon_\perp = \varepsilon_{\phi 0} = \frac{d_\perp - d_0}{d_0} \tag{18-53}$$

From Eqs. (18-46), (18-47), and (18-49) at $\psi = 0$, and assuming a stress-free surface,

$$\varepsilon_\perp = \varepsilon_{\phi 0} = -\frac{\nu}{E}(\sigma_1 + \sigma_2) \tag{18-54}$$

Equations (18-52) through (18-54) give

$$\varepsilon_{\phi\psi} = \frac{1 + \nu}{E} \sigma_\phi \sin^2\psi + \varepsilon_\perp \tag{18-55}$$

or, upon rearranging to solve for the in-plane stress,

$$\sigma_\phi = (\varepsilon_{\phi\psi} - \varepsilon_\perp)\frac{E}{(1 + \nu)\sin^2\psi} \tag{18-56}$$

However, from Eq. (18-53),

$$\varepsilon_{\phi\psi} - \varepsilon_\perp = \frac{d_\psi - d_0}{d_0} - \frac{d_\perp - d_0}{d_0} = \frac{d_\psi - d_\perp}{d_0} \approx \frac{d_\psi - d_\perp}{d_\perp} \tag{18-57}$$

Combining Eqs. (18-56) and (18-57) gives

$$\sigma_\phi = \sigma_x = \frac{d_{\psi x} - d_\perp}{d_\perp}\frac{E}{(1 + \nu)\sin^2\psi} = K^*\frac{d_{\psi x} - d_\perp}{d_\perp} \tag{18-58}$$

which is the in-plane stress at point p in any x direction at angle ϕ from σ_1 (Fig. 18-24). Similarly, the stress in the y direction is

$$\sigma_y = \sigma_{\phi+90} = K^*\frac{d_{\psi y} - d_\perp}{d_\perp} \tag{18-59}$$

The stress-diffraction coefficient

$$K^* = \frac{E}{(1 + \nu)\sin^2\psi} \tag{18-60}$$

is usually evaluated in terms of bulk elastic properties. However, for best accuracy it should be determined experimentally by x-ray.

Angle ϕ does not appear in Eq. (18-58). This is fortunate, since the in-plane principal directions are not known a priori. The unstrained atomic spacing d_0 is also unnecessary, eliminating the need for measurements in the unstrained (unstressed) condition. The latter aspect makes x-ray very suitable for determining residual stresses.

Equation (18-58) demonstrates that determination of an in-plane stress component σ_ϕ requires at least two diffraction readings in the same z-x plane of Fig. 18-24: one at $\psi = 0$ (normal to the surface) to provide $d_\perp$ and a second reading at $\psi \neq 0$ (typically $45° \leq \psi \leq 60°$) to obtain d_ψ. Diffraction determination of $d_\perp$, and $d_{\psi\text{I}}$, $d_{\psi\text{II}}$, and $d_{\psi\text{III}}$ in separate planes defined by $\phi = \phi_\text{I}$, $\phi = \phi_\text{II}$, $\phi = \phi_\text{III}$, determines three in-plane stresses $\sigma_{\phi\text{I}}$, $\sigma_{\phi\text{II}}$, $\sigma_{\phi\text{III}}$, from which the principal stresses σ_1, σ_2 and their directions can be obtained. This concept is similar to the use of three-element strain-gage rosettes. Lattice spacing and their changes are determined by x-ray using Eqs. (18-44) and (18-45). From Eqs. (18-45), (18-57), and (18-58), one obtains

$$\sigma_\phi = (2\theta_\perp - 2\theta_\psi)\frac{\cot\theta}{2}\frac{E}{(1+\nu)\sin^2\psi}\frac{\pi}{180} \qquad (18\text{-}61)$$

$$= K'\cot\theta(2\theta_\perp - 2\theta_\psi) \qquad (18\text{-}62)$$

with angle θ expressed in degrees. The $\cot\theta$ term is sometimes considered constant such that Eq. (18-62) becomes

$$\sigma_\phi = K(2\theta_\perp - 2\theta_\psi) = K\Delta(2\theta) \qquad (18\text{-}63)$$

where $\theta_\perp$ and θ_ψ are the measured Bragg diffraction angles at $\psi = 0$ and $\psi \neq 0$, respectively. Using Eq. (18-63) and the two-tilt method is conceptually simple. However, superior values of σ are obtained by plotting d_ψ of Eq. (18-58) versus $\sin^2\psi$ for several values of ψ and determining the best slope of this straight line $\partial(d_\psi)/\partial(\sin^2\psi) = \sigma d_\perp[(1+\nu)/E]$ by least squares. Figure 18-25 illustrates the schematic operation of an x-ray diffractometer and the shift in the diffracted beam between $\psi = 0$ and $\psi \neq 0$.

Influencing Variables, Equipment, Applications, and Advances

X-ray diffraction results can be affected by grain size, depth of ray penetration, accuracy in determining the diffraction angle θ, material anisotropy, and material texture [92]. Consequently, although the method possesses numerous advantages, its use requires careful application. This can be illustrated as follows: while mechanical lapping may be used to remove successive layers during in-depth stress measurements, the lapping itself can introduce substantial surface stresses. This is especially vexing because it is recognized that about 60% of the diffracted beam comes from the outer 5-μm (0.0002-in.) layer of material [96].

X-rays are being used for a variety of industrial applications, including determination of residual stresses resulting from manufacturing processes (heat treating, carburizing, shot peening, machining, etc.) and operating conditions such as fatigue [93]. Figure 18-26 illustrates use of x-rays to determine the residual-stress profile below the surface of a steel due to grinding. The agreement between the x-ray-determined stresses and those evaluated otherwise is good. A variety of x-ray stress analysis equipment is available commercially [93,94,98,103].

Contemporary advances in nonvisible radiation stress analysis include the practical determination of three-dimensional states of stress [104], conversion of diffracted x-ray patterns into light that is transmitted via fiber optics and then electro-optically intensified for conversion into an electronic signal for refinement and interpretation [105], utilization of x-ray concepts but with neutron radiation [106], x-ray measurement of grinding stresses in metal composites [96], as well as applications to ceramics [100].

18-6 Photomechanical Techniques

Photomechanical stress analyses have not found, nor are they expected to find, widespread utilization in determining residual stresses of metallic components. Photoelastic and brittle coatings and moiré do find limited use with the hole-drilling method [107–117]. Nevertheless, the ease, accuracy, and expediency of electrical strain gages are difficult to surpass in such applications. While photoelastic coatings have been used to evaluate plane-stress cases, Redner used three-dimensional stress freezing, slicing, and fringe multiplication to determine the residual-stress gradient normal to a surface [107]. Additional birefringent and oblique-incident methods for

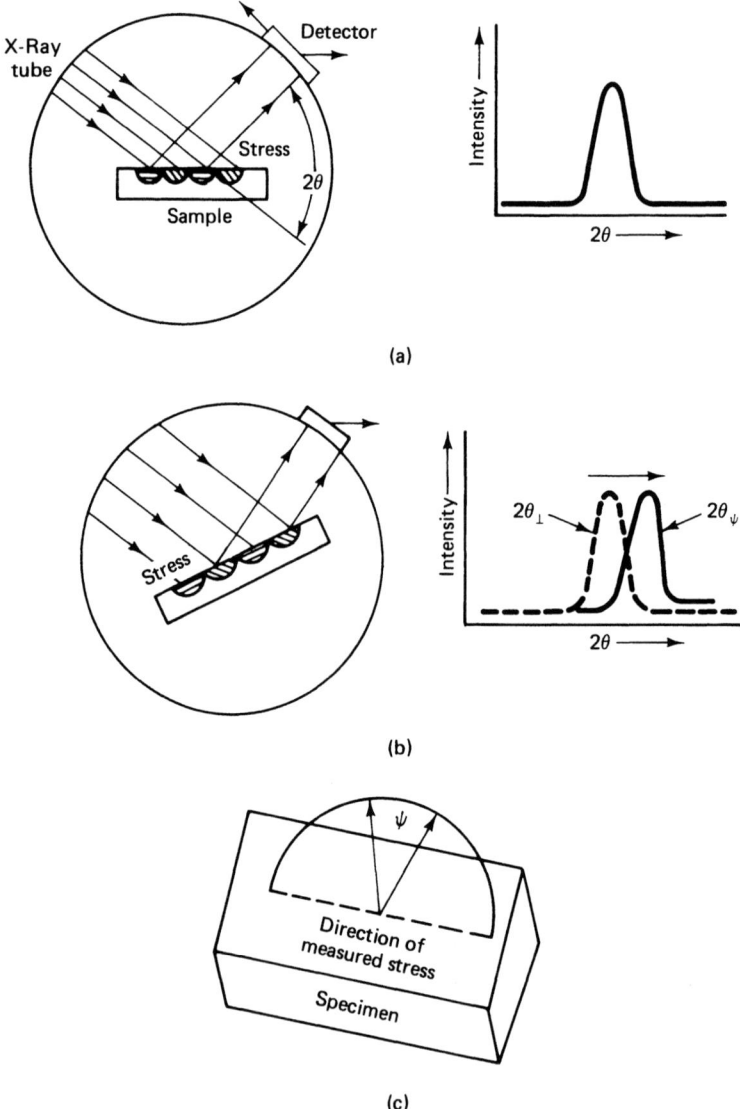

Figure 18-25 Schematic of x-ray diffractometer and shift in diffracted beam. (a) Schematic of a diffractometer. The incident beam diffracts x-rays of wavelength λ from planes that are parallel to the sample's surface and satisfy Bragg's law. The diffracted beam is recorded as intensity versus scattering angle by a detector moving with respect to the specimen. (b) After the specimen has been tilted through angle ψ, diffraction occurs from the same planes but from other grains. The peak occurs at different angles, $2\theta_\psi$. (c) After the specimen has been tilted, the stress is measured in a direction that is the intersection of the circle of tilt and the surface of the specimen. (From Ref. 93.)

RESIDUAL STRESSES

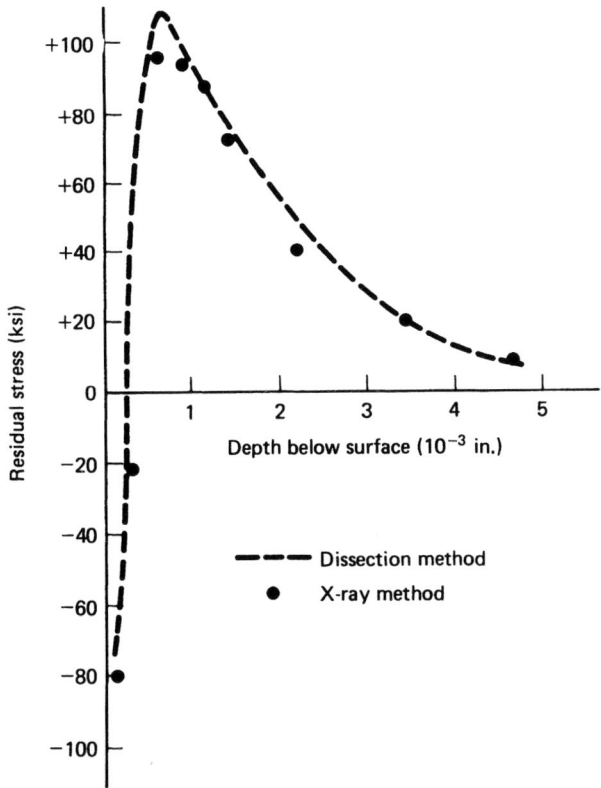

Figure 18-26 Residual stress as a function of depth below the ground surface in a heat-treated steel. (X-ray measurements by Koistinen and Marburger from Ref. 102, dissection measurement by Letner from Ref. 101. Note inability of dissection method to measure stress right at the surface.) (From Ref. 94, p. 465, Fig. 16.11; reprinted with permission.)

analyzing residual stresses are discussed in Refs. 108–113. Techniques such as moiré could be useful to diagnose residual stresses associated with welds [117].

18-7 Numerical Analyses

Inelastic and elevated-temperature polyaxial constitutive response, and uncertain temperature gradients and boundary conditions, complicate the finite-element analysis of residual stresses in homogeneous and composite materials. In addition to the blind-hole numerical analyses discussed previously [42–44], Refs. 110, 118, and 119 provide informative results. Frick et al. [110] presented a numerical technique to compute, from measured surface strains, the internal residual stresses in isotropic and orthotropic homogeneous or laminated structures. Their analysis is verified experimentally. Migliore presented a nonlinear finite-element procedure with incremental displacement loading/unloading to predict the residual stresses in metal sheets such as automotive body components [118]. Reference 119 describes the numerical determination of residual stresses in welded pipe.

18-8 Relieving Residual Stresses

Residual stresses are easily introduced but can be difficult to remove. Annealing is probably the most prevalent means of relieving residual stresses. However, the process can be expensive and incomplete. Large components are difficult to handle. Stress relief can cause dimensional changes.

18-9 Summary, Discussion, and Future

Residual stresses are of great concern. They are usually unwanted, difficult to evaluate, and expensive to remove. Their presence can be detrimental in large items such as transportation vehicles or nuclear reactors components [120], and in tiny electronic "thin films." The seriousness of their presence, combined with our insufficient ability to measure and/or remove them, mandates that the discipline receive increased attention. Nondestructive methods such as ultrasonics, magnetics, nuclear acoustic resonance [121], Barkhausen [122,123], and radiation (x-ray, neutrons) techniques should be emphasized. The ability to have both local and full-field visual display would be advantageous. Progress will necessitate the combined talents of capable individuals experienced in stress analysis, manufacturing, materials science, and instrumentation, and supported by adequate facilities. Contemporary technology provides reasonably reliable stress analyses due to live loads, expedient computer-assisted design methods, and robotic manufacturing abilities. Significant but unknown residual stresses could negate these capabilities.

Acknowledgments

Helpful conversations with J. B. Cohen of Northwestern University, S. Redner of Measurements Group, Inc., and C. O. Ruud of Pennsylvania State University are acknowledged.

List of Symbols

	a	hole radius
A, A', B, B', C, c		coefficients
	c	coating, as a subscript
	c	velocity
	d	hole diameter
	d	interatomic spacing
	d_0	interatomic spacing of unstrained material
	E	Young's modulus
	f	frequency
	f_σ	material stress fringe value
	h	thickness
k, K, K_i		coefficients
	l	longitudinal coordinate, as a subscript
	m	structural member, as a subscript

	N	isochromatic fringe order
	n	index of refraction
	n	diffraction order
	r	radial coordinate
	r_m	mean radius of strain-gage rosette
	t	tangential coordinate, as a subscript
	U_i	diametral change of bore hole
	v	velocity
	x, y, z	Cartesian coordinates
$\alpha, \beta, \gamma, \theta, \phi, \psi$		angles
	α_i	direction cosines
	$\Delta(\theta), \Delta(2\theta)$	change in Bragg diffraction angle
	Δ_ε	change in strain
	γ	shear strain
	ε'	relaxed strain
	ε	strain
	ε_m	average strain
	θ	polar coordinate
	λ	wavelength
	λ	ratio of hole diameter to radius of strain rosette, $\lambda = d/r$
	ν	Poisson's ratio
	σ	normal stress
	τ	shear stress

References

1. J. O. Almen and P. H. Black, *Residual Stresses and Fatigue in Metals*, McGraw-Hill, New York, 1963.

2. O. J. Horger, Residual Stresses, in *Handbook of Experimental Stress Analysis* (M. Hetényi, Ed.), Wiley, New York, 1950, 459–478.

3. J. A. Haigren, T. C. Huang, and E. I. Blount, Measurement of Residual Stress in Review, Report at the SAE Golden Anniversary Meeting, Detroit, January 1955.

4. Evaluation of Methods for Measurement of Residual Stresses, SAE Report TR-147, March 1960.

5. C. S. Barrett, A Critical Review of Various Methods of Residual Stress Measurement, *Proc. Soc. Exp. Stress Anal.*, II, no. 1, (1945), 147–156.

6. Methods of Residual Stress Measurement, Society of Automotive Engineering, Report J936, New York, 1965.

7. N. Tebedge, G. A. Alpsten, and L. Tall, Measurement of Residual Stresses—A Study of Methods, Fritz Engineering Laboratory Report 337.8, Lehigh University, February 1971.

8. C. O. Ruud, A Review of Nondestructive Methods for Residual Stress Measurement, *J. Met., 33* (July 1981), 35–40.

9. J. F. Throop and H. S. Reemsnyder, Residual Stress Effects in Fatigue, ASTM-STP 776, American Society for Testing and Materials, Philadelphia, 1981.

10. R. J. Glover and J. M. Boag, A Review of Mechanical Methods for Residual Stress Measurement, Report 82-404-K, Ontario Hydro (Research Division), Toronto, Canada, 1982.

11. Residual Stress—The Nemesis of the Liberty Bell, *Epsilonics*, Micromeasurements, Inc., Raleigh, NC, December 1983.

12. J. J. Lynch, The Measurement of Residual Stresses, in *Residual Stress Measurements* (R. G. Treuting et al., Eds.), American Society for Metals, Metals Park, OH, 1952, 42–96.

13. L. J. Vande Walle, (Ed.), *Residual Stress for Designers and Metallurgists*, American Society for Metals, Metals Park, OH, 1981.

14. J. Mathar, Determination of Initial Stresses by Measuring the Deformations Around Drilled Holes, *Trans. ASME*, 56 (1934), 249–254.

15. G. Kirsch, Theory of Elasticity and Application in Strength of Materials, *Z. Vev. Dtsch. Inge.*, 42, no. 29 (1898), 797–807.

16. A. Willheim and Z. Leon, *Z. Math. Phys.*, 64 (1916), 233.

17. G. Mesmer, Fliesserscheinungen beim Spannungsmessverfahren, *Arch. Eisenhuettenws.*, 10, no. 2 (August 1936), 59–63.

18. F. Campus, Research Studies and Considerations for Welded Structures, *Sci. Lettr.*, Liège, 1946, 103–109.

19. W. Soete, Measurement and Relaxation of Residual Stresses, *Sheet Met. Ind.*, 26, no. 266 (June 1949), 1269–1281.

20. W. Soete and R. Vancrombrugge, An Industrial Method for the Determination of Residual Stresses, *Proc. SESA*, VIII, no. 1 (1950), 17–28.

21. C. Riparbelli, A Method for the Determination of Initial Stresses, *Proc. SESA*, VIII, no. 1, (1950), 173–196.

22. R. G. Boiten and W. Ten Cate, A Routine Method for the Measurement of Residual Stresses in Plates, *Appl. Sci. Res.*, 3, Sec. A, no. 2 (1952), 317–343.

23. R. A. Kelsey, Measuring Nonuniform Residual Stresses by the Hole Drilling Method, *Proc. SESA*, XIV, no. 1 (1956), 181–194.

24. M. Nisida and H. Takabayashi, Thickness Effects in "Hole Method" and Application of Method of Residual Stress Measurement, *Sci. Pap. IPCR*, 59, no. 2 (April 1965), 78–86.

25. N. J. Rendler and I. Vigness, Hole-Drilling Strain-Gage Method of Measuring Residual Stresses, *Exp. Mech.*, 6, no. 12 (1966), 577–586.

26. B. Gupta, Hole-Drilling Technique: Modifications in the Analysis of Residual Stresses, *Exp. Mech.*, 13, no. 1 (1973), 45–48.

27. R. G. Bathgate, Measurement of Non-Uniform Biaxial Residual Stresses by the Hole-Drilling Method, *Strain*, 4, no. 2, (April 1968), 20–29.

28. H. V. Cordiano and V. L. Salerno, Study of Residual Stresses in Linearly Varying Biaxial-Stress Fields, *Exp. Mech.*, 9, no. 1 (1969), 17–24.

29. C. W. Bert and G. L. Thompson, A Method for Measuring Planar Residual Stresses in Rectangularly Orthotropic Materials, *J. Compos. Mater.*, 2, no. 4 (1968), 244–253.

30. B. R. Lake, F. J. Appl, and C. W. Bert, An Investigation of the Hole-Drilling Technique for Measuring Residual Stress in Rectangularly Orthotropic Materials, *Exp. Mech.* 10, no. 6 (1970), 233–239.

31. A. M. Newwar, K. McLachlan, and J. Shewchuk, A Modified Hole-Drilling Technique for Determining Residual Stresses in Thin Plates, *Exp. Mech.*, 16, no. 6 (1976), 226–232.

32. A. M. Newwar and J. Shewchuk, On the Measurement of Residual-Stress Gradients in Aluminum-Alloy Specimens, *Exp. Mech.*, 18, no. 7 (1978), 269–276.

33. A. J. Bush and F. J. Kromer, Simplification of the Hole-Drilling Method of Residual Stress Measurements, *ISA Trans.*, 12, no. 3 (1973), 249–259.

34. E. M. Beaney and E. Procter, A Critical Evaluation of the Centre-Hole Technique for the Measurement of Residual Stresses, *Strain*, 10, no. 1 (January 1974), 7–14.

35. E. M. Beaney, Accurate Measurement of Residual Stress on any Steel Using the Centre-Hole Method, *Strain*, 12, no. 3 (July 1976), 99–105.

36. J. E. Bynum, Modifications to the Hole-Drilling Technique of Measuring Residual Stresses for Improved Accuracy and Reproducibility, *Exp. Mech.*, 21, no. 1 (1981), 21–30.

37. F. Witt, F. Lee, and W. Rider, A Comparison of Residual Stress Measurements Using Blind Hole, Abrasive Jet and Trepan Ring Methods, *Exp. Tech.*, 7, no. 2 (February 1983), 41–45.
38. E. Procter and E. M. Beaney, Recent Developments in Center-Hole Techniques for Residual-Stress Measurements, *Exp. Tech.*, 6, no. 6 (December 1982), 10–15.
39. M. T. Flaman, Brief Investigation of Induced Drilling Stresses in the Center-Hole Method of Residual-Stress Measurement, *Exp. Mech.*, 22, no. 1 (1982), 26–30.
40. J. P. Sandifer and G. E. Bowie, Residual Stress by Blind-Hole Method with Off-Center Hole, *Exp. Mech.*, 18, no. 5 (1978) 173–179.
41. A. Ajovalasit, Measurement of Residual Stresses by the Hole-Drilling Method: Influence of Hole Eccentricity, *J. Strain Anal.*, 14, no. 4 (1979), 171–178.
42. G. S. Schajer, Application of Finite-Element Calculations to Residual Stress Measurements, *ASME Trans., J. Eng. Mater. Tech.*, 103, no. 2 (April 1981), 157–163.
43. B. W. Manning and M. T. Flaman, Finite-Element Calculations of Calibration Constants for Determination of Residual Stresses with Depth by the Hole-Drilling Method, Ontario Hydro (Research Division) Report 82-88-K, Toronto, Canada, 1982.
44. A. M. Newwar and J. A. Miller, A Finite-Element Model of the Blind-Hole Drilling Technique, *Proc. J. SESA–JSME Conf. Exp. Mech.*, Hawaii, 1982, 696–702.
45. W. R. Delameter and T. C. Mamaros, Measurement of Residual Stresses by the Hole-Drilling Method, Sandia–Livermore Report SAND77-8006, April 1977.
46. Measurements of Residual Stresses by Hole-Drilling Strain-Gage Method, Bulletin TN503-1, Measurements Group of Vishay, Inc., Raleigh, NC, 1985.
47. Determining Residual Stresses by the Hole-Drilling Strain-Gage Method, ASTM Standard E837-81, American Society for Testing and Materials, Philadelphia, 1981.
48. S. Redner and C. C. Perry, Factors Affecting the Accuracy of Residual Stress Measurements Using the Blind-Hole Drilling Technique, *Proc. 7th Int. Conf. Exp. Stress Anal.*, Haifa, Israel, August 1982, 604–616.
49. C. Fairhurst, Methods of Determining In Situ Rock Stresses at Great Depths, Technical Report 1-68 to U.S. Corps of Engineers, Missouri River Division, Omaha, NE, February 1968.
50. L. Obert, Determination of the Stress in Rock—A State of the Art Report, presented at the 69th Annu. American Society for Testing and Materials Meeting, Atlantic City, NJ, June 1966.
51. L. Obert, R. H. Merrill, and T. A. Morgan, Borehole Deformation Gage for Determining the Stress in Mine Rock, U.S. Bureau of Mines Report 5978, 1962.
52. E. R. Leeman and D. J. Hayes, A Technique for Determining the Complete State of Stress in Rock Using a Single Borehole, *1st Int. Conf. Rock Mech.*, Lisbon, Portugal, Vol. II, 1966, 17–24.
53. A. Robert, I. Hawkes, F. T. Williams, and R. K. Dhir, A Laboratory Study of the Photoelastic Stressmeter, *Int. J. Rock Mech.*, 1, no. 3 (May 1964), 451–458.
54. L. Obert and W., I. Duvall, *Rock Mechanics and the Design of Structures in Rock*, Wiley, New York, 1967.
55. B. P. Hildebrand and B. B. Brenden, *An Introduction to Acoustical Holography*, Plenum Press, New York, 1972.
56. A. P. Van den Heuvel, Surface-Wave Electronics, *Sci. Technol.*, (January 1969), 52–60.
57. H. Kolsky, *Stress Waves in Solids*, Dover, New York, 1963.
58. E. C. McKannan, Ultrasonic Measurement of Stress in Aluminum, in Non-destructive Testing: Trends and Techniques, NASA SP-5082, 1966.
59. R. E. Rowlands, Nondestructive Stress Analysis by Ultrasonic Techniques, IITRI Document 69-475D, 1969.
60. H. J. McSkimin, Variations of the Ultrasonic Pulse-Superposition Method for Increasing the Sensitivity of Delay-Time Measurements, *J. Acoust. Soc. Am.*, 37, no. 5 (1965), 864–871.
61. R. W. Benson, Development of Non-destructive Methods for Determining Residual Stresses and Fatigue Damage in Metals, Final Report NAS8-20208 by Robert W. Benson & Associates, Inc., to NASA, March 1968.

62. R. L. Forgacs, Improvements in the Sing-Around Techniques for Ultrasonic Velocity Measurements, *J. Acoust. Soc. Am.*, *32*, no. 12, (1960), 1697–1698.

63. B. B. Chick, G. P. Anderson, and R. Truell, Improvement in Ultrasonic Velocity Measuring Techniques, Final Report Contract 66-0150d, Metals Research Lab., Brown University, Providence, RI, 1967.

64. N. P. Cedrone and D. R. Curran, Electronic Pulse Method for Measuring the Velocity of Sound in Liquids and Solids, *J. Acoust. Soc. Am.*, *26*, no. 6 (November 1954), 963–966.

65. F. Buckens, The Velocity of Rayleigh Waves Along a Prestressed Semi-Infinite Medium Assuming a Two-Dimensional Anisotropy, *Ann. Geofis. (Rome)*, *11*, no. 2 (1958), 99–112.

66. A. Myers, L. MacKinnon, and F. E. Hoare, Modifications to Standard Pulse Techniques for Ultrasonic Pulse Measurements, *J. Acoust. Soc. Am.*, *31*, no. 2 (1959).

67. J. R. Chapman, Velocity-Stress Dependence of Surface Waves in Aluminum Utilizing Ultrasonic Techniques, M.S. thesis, Department of Electrical Engineering, Vanderbilt University, Nashville, TN, August 1962.

68. I. M. Daniel and R. E. Rowlands, Ultrasonic Techniques for Nondestructive Stress Analysis, IITRI Report D1014, January 1969.

69. R. W. Benson and V. J. Raelson, Acoustoelasticity, *Prod. Eng.*, *30*, no. 24 (July 1959), 56–59.

70. F. R. Rollins, Study of Methods for Nondestructive Measurement of Residual Stresses, WADC Technical Report, Wright-Patterson AFB, Ohio, December 1959, 59–561.

71. R. W. Benson, Ultrasonic Stress Analysis, *Ultrason. News*, *6*, no. 14 (1962).

72. F. R. Rollins, Ultrasonic Methods for Nondestructive Measurement of Residual Stresses, WADC Technical Report 61-42, Parts I and II, Wright-Patterson AFB, Ohio, May 1961 and January 1963.

73. R. T. Smith, Stress-Induced Anisotropy in Solids—The Acousto-elastic Effect, *Ultrasonics*, July–September 1963.

74. D. I. Crecraft, The Measurement of Applied and Residual Stress in Metals Using Ultrasonic Waves, *J. Sound Vib.*, *5*, no. 1 (1967), 173–192.

75. D. I. Crecraft, Ultrasonic Measurement of Stresses, *Ultrasonics*, *6* (1968), 117–121.

76. N. N. Hsu, Acoustical Birefringence and the Use of Ultrasonic Waves for Experimental Stress Analysis, *Exp. Mech.*, *14*, no. 5 (1974), 169–176.

77. P. J. Noronda and J. J. Wert, An Ultrasonic Technique for the Measurement of Residual Stress, *J. Test. Eval.*, *3* (1975), 147–152.

78. P. J. Noronda, J. R. Chapmann, and J. J. Wert, Residual Stress Measurement and Analysis Using Ultrasonic Techniques, *J. Test. Eval.*, *1*, no. 3 (May 1973), 209–214.

79. J. Blinka and W. Sachse, Application of Ultrasonic-Pulse-Spectroscopy Measurements to Experimental Stress Analysis, *Exp. Mech.*, *16*, no. 12 (1976), 448–453.

80. F. Bach and V. Askegaard, General Stress-Velocity Expressions in Acoustoelasticity, *Exp. Mech.*, *19*, no. 2 (February 1979), 69–75.

81. G. S. Kino, Acoustical Imaging of Stress Fields, *J. Appl. Phys.*, *50* (1979), 2607–2613.

82. Y. Iwashimizu and K. Kubomura, Stress-Induced Rotation of Polarization Directions in Elastic Waves in Slightly Anisotropic Materials, *Inst. J. Solids, Struct.*, *9* (1973), 99–114.

83. H. Fukuoka, H. Toda, and T. Yamane, Acoustoelastic Stress Analysis of Residual Stress in a Patch-Welded Disk, *Exp. Mech.*, *18*, no. 7 (1978), 277–280.

84. M. P. Scott, D. M. Barnett, and D. B. Ilic, The Nondestructive Determination of Residual Stresses in Extruded Billets from Acoustoelastic Measurements, *IEEE Ultrason. Symp. Proc.*, New Orleans, 1979.

85. G. C. Johnson, On the Applicability of Acoustoelasticity for Residual Stress Determination, *J. Appl. Mech.*, *48*, (December 1981), 791–795.

86. K. Okada, Acoustoelastic Determination of Stress in Slightly Orthotropic Materials, *Exp. Mech.*, *21*, no. 12 (1981), 461–466.

87. H. Fukuoka, H. Toda, and H. Naka, Nondestructive Residual-Stress Measurement in a Wide-Flanged Beam by Acoustoelasticity, *Exp. Mech., 23*, no. 3 (1983), 120–128.

88. W. Y. Lu, Residual Stress Evaluation by Ultrasonics in an Elastic–Plastic Material, *Proc. Spring Conf. SESA*, Cleveland, 1983, 77–83.

89. R. H. Bergman and R. A. Shahbender, Effect of Statically Applied Stresses on the Velocity of Propagation of Ultrasonic Waves, *J. Appl. Phys., 29* (1958), 1736.

90. G. S. Kino, Acoustic Measurements of Stress Fields and Microstructure, *J. Nondestr. Eval., 1*, no. 1 (1980), 66–67.

91. The Measurement of Stress by X-Ray, Report TR-182 prepared by the X-Ray Subcommittee of Division 4—Residual Stresses and Fatigue—of the SAE Iron and Steel Technical Committee.

92. Residual Stress Measurement by X-Ray Diffraction, SAE Report J784a, prepared by the SAE Iron and Steel Technical Subcommittee, Warrendale, PA, 1971.

93. M. R. James and J. B. Cohen, The Measurement of Residual Stresses by X-Ray Diffraction Techniques, *Treatise Mater. Sci. Technol., 19A* (1980), 1–62.

94. B. D. Cullity, *Elements of X-Ray Diffraction,* 2nd ed., Addison-Wesley, Reading, MA, 1978.

95. J. B. Cohen, P. Georgopoulos, and C. Noyan, Workshop Notes on X-Ray Measurement of Residual Stresses, Northwestern University, Evanston, IL, 1983.

96. D. N. French and B. A. MacDonald, Experimental Methods of X-Ray Stress Analysis, *Exp. Mech., 9*, no. 10 (1969), 456–462.

97. C. O. Ruud, A Review of Nondestructive Methods for Residual Stress Measurement, *J. Met., 33*, no. 7 (July 1981), 35–40.

98. C. O. Ruud, X-Ray Analysis and Advances in Portable Field Instrumentation, *J. Met., 31*, no. 6 (June 1979), 10–15.

99. P. S. Prevey, A Comparison of X-Ray Diffraction Residual Stress Measurement Methods on Machined Surfaces, *Adv. in X-ray Anal., 19* (1976), 709–724.

100. C. O. Ruud and C. P. Gazzara, Residual Stress Measurements in Alumina and Silicon Carbide, *J. Am. Ceram. Soc., 68* (1985).

101. H. R. Letner, Residual Grinding Stresses in Hardened Steel, *Trans. ASME, 77* (1955), 1089.

102. P. Koistinen and R. E. Marburger, *Trans. ASM, 51* (1959), 537–555.

103. M. James and J. B. Cohen, PARS—A Portable X-Ray Analyzer for Residual Stresses, *ASTM J. Test. Eval., 6*, no. 2 (1978), 91–94 (U.S. Patent 4,095,103).

104. H. Dölle and J. B. Cohen, Residual Stresses in Ground Steel, *ASM AIME Metall. Trans. A, 11A* (January 1980), 159–164.

105. C. O. Ruud, P. S. DiMascio, and D. M. Melcher, Nondestructive Residual-Stress Measurement on the Inside Surface of Stainless-Steel Pipe Weldments, *Exp. Mech., 24*, no. 2 (June 1984), 162–168.

106. M. J. Schmank and A. D. Krawitz, Measurement of a Stress Gradient Through the Bulk of an Aluminum Alloy Using Neutrons, *ASM AIME Metall. Trans. A, 13A* (June 1982), 1069–1076.

107. S. Redner, Photoelastic Investigation of Stresses Relieved During Shallow Hole Drilling in Residual Stress Measuring Technology, *Proc. Int. Conf. Methods Exp. Mech.*, Prague, 1974, 342–348.

108. S. Redner and W. E. Nichola, Measurement of Residual Strains and Stresses in Transparent Materials, presented at SESA meeting, Cleveland, May 1983.

109. A Shinohara and M. Nakazawa, Stress Analysis in Textile Fiber by Means of Micro-Photoelasticity, presented at SESA meeting, Hawaii, May 1982.

110. R. P. Frick, G. A. Gurtman, and H. D. Meriwether, Experimental Determination of Stresses in an Orthotropic Material, *J. Mater., 2* (December 1967), 719–748.

111. R. W. Ansevin, The Nondestructive Examination of Surface Stresses in Glass, *Mater. Eval.* (March 1967), 58–64.

112. Photoelastic Measurements of Birefringence and Residual Strains in Transparent or Translucent Plastic Materials, ASTM Standard D4093-82, American Society for Testing and Materials, Philadelphia, 1982.
113. M. Nisida, A Method for Measuring Stresses on Metal Surface with Photoelasticity, *Sci. Pap. IPCR, 59*, no. 2 (1965), 69–77.
114. C. W. Gadd, Residual Stress Indications in Brittle Lacquer, *Proc. SESA, IV*, no. 1 (1946), 74–77.
115. A. G. Tokarcik and M. H. Polzin, Quantitative Evaluation of Residual Stresses by the Stresscoat Drilling Technique, *Proc. SESA, IX*, no. 2 (1952), 195–207.
116. H. R. Letner, Application of Optical Interference to the Study of Residual Surface Stresses, *Proc. SESA, X*, no. 2 (1953), 23–36.
117. A. MacDonach, J. McKelvie, P. MacKenzie, and C. A. Walker, Improved Moiré Interferometry and Applications in Fracture Mechanics, Residual Stress and Damaged Composites, *Exp. Tech., 7*, no. 6 (1983), 20–24.
118. H. Migliore, Finite-Element Prediction of Residual Stress in Formed Sheet Metal, presented at SESA meeting, Chicago, May 1975.
119. E. F. Rybicki, D. W. Schmueser, R. W. Stonesifer, J. J. Groom, and H. W. Mishler, A Finite-Element Model for Residual Stresses and Deflections in Girth-Butt Welded Pipe, *ASME J. Pressure Vessel Technol., 100* (August 1978), 256–262.
120. J. Druez and A. Bazergui, Through-Thickness Measurement of Residual Stresses in Thin Tubes, *Exp. Mech., 23*, no. 2 (June 1983), 211–216.
121. D. K. Hsu, R. G. Leisure, and G. A. Matzkanin, Investigation of Residual Stress Using Nuclear Acoustic Resonance, *Proc. SESA Meeting*, Dearborn, MI, May 1981, 79–83.
122. C. O. Ruud, P. S. DiMarscio, and J. J. Yavelak, Comparison of Three Residual-Stress Measurement Methods on a Mild Steel Bar, *Exp. Mech., 25*, no. 4 (December 1985), 338–343.
123. G. A. Matzkanin, R. E. Beissner, and C. M. Teller, The Barkhausen Effect and Its Application to Nondestructive Evaluation, *NTIAC-79-2* (October 1979).

CHAPTER

19

Composite Materials

Isaac M. Daniel

Northwestern University
Evanston, Illinois

19-0 Introduction

A composite is a material system consisting of two or more phases on a macroscopic scale whose properties are superior to those of the constituent materials acting independently. One of the phases is usually discontinuous and stiffer and is called reinforcement, whereas the less stiff one is continuous and is called matrix. The properties of a composite depend on the properties of the constituents, geometry, and distribution of the phases. One of the most important parameters is the volume fraction or concentration of the reinforcement. The distribution of the reinforcement is a measure of homogeneity, whereas its orientation affects the anisotropy of the system.

Composites have unique advantages over monolithic materials, such as high strength, high stiffness, long fatigue life, low density, and adaptability to the intended function of the structure. Additional improvements can be realized in corrosion resistance, wear resistance, appearance, temperature-dependent behavior, thermal insulation, thermal conductivity, and acoustic insulation. The basis for the high performance of composite materials lies in the high specific strength (strength over density) and high specific stiffness (modulus over density) and in the anisotropic and heterogeneous character of the material. The latter provides the composite system with many degrees of freedom for optimum configuration of the material.

Composites can be broadly classified into particulate composites, discontinuous fiber composites, and continuous fiber composites. In fibrous composites the fibers are stiffer and stronger and carry most of the load, whereas the role of the matrix is to support, protect, and transfer the stresses to the fibers. The most advanced composites used in the most critical applications are continuous fiber composites. The basic building block of such composites is a single layer of unidirectional fibers in the matrix called a lamina or ply. A multidirectional laminate consists of several plies or laminae stacked on top of each other at various orientations.

The concept of fibrous reinforcement is very old, starting with straw-reinforced clay bricks, steel-reinforced concrete, and developing into fiber-reinforced polymeric resins and metals. Applications abound, including pipes, containers, boats, aircraft and aerospace structures, automotive components, and sports equipment. The rapid progress in composites technology in recent years is due to several factors, including significant advances in materials science in the areas of fibers and polymers, requirements for high-performance materials in the aerospace industry, and development of sophisticated numerical methods of analysis. Prospects for the future are bright as the costs of the basic materials and fabrication decrease, and as new high-volume applications, such as in the automotive industry, appear.

The study of composites is a philosophy of material design along with structural design. The process is not one of purely structural analysis, but one of structural synthesis and evaluation. It is a science and technology requiring close interaction among structural design and analysis, materials science, and process engineering. The scope of composite materials technology consists of the following tasks: (1) investigation of basic characteristics of the constituent and composite materials, (2) optimum material configuration for given service environment, (3) development of effective and efficient fabrication procedures, (4) development of analytical procedures for determining material properties and predicting structural behavior, and (5) development of effective experimental methods for material characterization, stress analysis, failure analysis, and nondestructive evaluation of material integrity and structural reliability.

Almost all the literature on composites is less than three decades old. It appears in the form of reports, conference proceedings (ASTM, SEM, ICCM, ICM, SAMPE, AIAA, ASC, etc.), journals (*Journal of Composite Materials, Journal of Reinforced Plastics and Composites, Journal of Composites Technology and Research, Composites, Journal of Polymer Composites, Composites Science and Technology, Composites Engineering, Composite Structures*), and a few books and monographs, such as those by Jones [1], Christensen [2], Tsai and Hahn [3], Agarwal and Broutman [4], Hull [5], and Whitney, Daniel, and Pipes [6].

This chapter reviews briefly the state of mechanics of composites, with emphasis on experimental methods, and assesses the trends and prospects for the future of composites technology. It starts with the basic mechanics of composites, followed by a description of the experimental methodology applicable to composites, and applications to material characterization, biaxial testing, effects of stress concentrations, and nondestructive evaluation.

19-1 Mechanics of Composites

Mechanics of composites deals with the mechanical behavior of composite materials and structures. Depending on the scale of observation, this discipline can be classified into micromechanics, minimechanics, and macromechanics. In micromechanics the stress analysis studies the fiber–matrix interaction and failure analysis considers the various micromechanisms of failure, such as matrix, interface, and fiber failures. In minimechanics the smallest repeating element, such as a fiber and its surrounding matrix, is treated as a homogeneous but anisotropic element. Failure criteria are expressed as element or ply failure criteria. In macromechanics the entire laminate may be considered to be homogeneous, albeit anisotropic, and failure is described in terms of laminate failure criteria.

Micromechanics

The objective of micromechanics is to predict the behavior of the unidirectional lamina as a function of the fiber and matrix properties and the fiber volume fraction. These properties are referred to the three principal material axes of the lamina (Fig. 19-1). The properties of interest are mechanical properties (elastic constants and strengths) under longitudinal (x_1) tension and compression, transverse (x_2) tension and compression, and in-plane shear, as well as thermal expansion and moisture expansion coefficients, and transport properties.

COMPOSITE MATERIALS

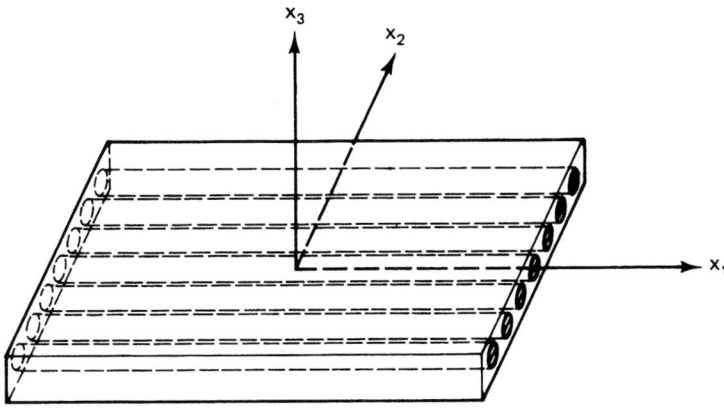

Figure 19-1 Principal material axes of a unidirectional lamina.

Under longitudinal tension, if we assume perfect bond between the matrix and the fibers, the longitudinal strains are uniform throughout and equal for the matrix and fibers.

$$\varepsilon_{1f} = \varepsilon_{1m} = \varepsilon_1 \tag{19-1}$$

This assumption leads to the simple "rule of mixtures" for the modulus

$$E_1 = E_f V_f + E_m(1 - V_f) \tag{19-2}$$

where E_1, E_f, E_m = moduli of composite, fiber, and matrix, respectively
V_f = fiber volume fraction

When the longitudinal stress is increased, the phase that has the lower ultimate strain, usually the fibers, fails first. The longitudinal tensile strength can be predicted by a rule of mixtures as

$$F_{1T} \simeq F_{fT} V_f + \sigma_m(1 - V_f) = F_{fT}\left(V_f + V_m \frac{E_m}{E_f}\right) \tag{19-3}$$

When the matrix limits the strength, as might be the case with the recently developed high-elongation fibers,

$$F_{1T} = F_{mT}\left(V_m + V_f \frac{E_f}{E_m}\right) \tag{19-4}$$

where F_{fT}, F_{mT} = fiber and matrix tensile strengths, respectively
V_f, V_m = fiber and matrix volume ratios, respectively
σ_m = matrix stress at a strain equal to fiber failure strain

The predictions above are only approximate, since the fiber strength varies from point to point and from fiber to fiber. To account for this statistical variability, Rosen [7] proposed the so-called cumulative weakening model. Fiber breaks occur at weak points. Following a fiber fracture, high shear stresses develop at the interface in the matrix, with loss of effective fiber length (Fig. 19-2). Possible failure mechanisms accompanying a fiber break are transverse matrix cracking, conical shear matrix fracture, and fiber debonding. As the load is increased further, the damage zone does not grow, but the damage sites increase in number and precipitate final failure. Failure

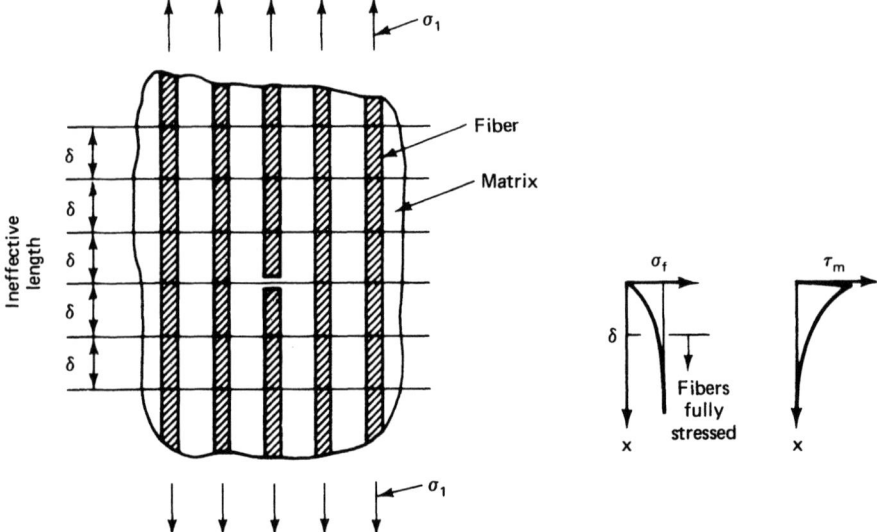

Figure 19-2 Failure model of unidirectional composite under longitudinal tensile loading. (From Ref. 7.)

modes vary with fiber volume ratio and interfacial bond strength relative to matrix strength. As the fiber volume fraction increases, the failure modes vary from brittle, to brittle with filament pull-out, to irregular (Fig. 19-3) [8].

In longitudinal compression, failure is assumed to be associated with microbuckling of the fibers within the matrix (Fig. 19-4). At low values of V_f the extensional or out-of-phase mode of microbuckling is predicted with a compressive strength of

$$F_{1c} \simeq 2V_f \left[\frac{V_f E_m E_f}{3(1 - V_f)} \right]^{1/2} \tag{19-5}$$

At higher values of V_f the shear or in-phase mode is predicted with a compressive strength of

$$F_{1c} \simeq \frac{G_m}{1 - V_f} \tag{19-6}$$

The most critical loading of a unidirectional composite is transverse tensile loading. In this case the matrix stresses and strains are the governing criteria of failure. The transverse modulus E_2 can be calculated by various methods, including energy methods based on variational principles giving upper and lower bonds [10]; elasticity methods employing finite elements, finite differences, and complex variables [11,12], and strength-of-materials methods [13].

The simplest computation of transverse modulus, based on the assumption of equal stresses in the matrix and fiber, gives

$$E_2 = \frac{E_f E_m}{E_f(1 - V_f) + E_m V_f} \tag{19-7}$$

The strength-of-materials approach by Greszczuk [13] yields

COMPOSITE MATERIALS

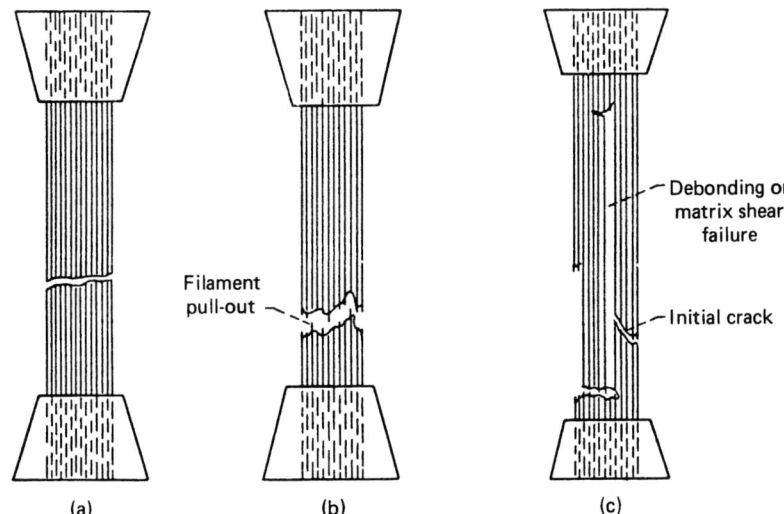

Figure 19-3 Longitudinal tensile failure modes for unidirectional composite: (a) brittle; (b) brittle with filament pull-out; (c) irregular. (From Ref. 8)

$$E_2 \simeq E_0\beta + E_m^*(1 - \beta) \tag{19-8}$$

where

$$E_0 = \frac{E_m^* + 2\beta(E_f - E_m^*)}{1 + \frac{2\beta(1 - 2\beta)}{E_m^* E_f}[(E_f - E_m^*)^2 - (\nu_m E_f - \nu_f E_m^*)^2]}$$

$$E_m^* = \frac{E_m}{1 - \nu_m^2}$$

$$\beta = \sqrt{\frac{V_f}{\pi}}$$

Halpin and Tsai [12] gave a simplified solution based on elasticity theory:

$$E_2 = \frac{E_m(1 + \xi\eta V_f)}{1 - \eta V_f} \tag{19-9}$$

where

$$\eta = \frac{E_f/E_m - 1}{E_f/E_m + \xi}$$

The parameter ξ depends on loading conditions and fiber packing and is normally obtained empirically by fitting curves to experimental results. The variation of the transverse modulus as a function of the fiber volume ratio is shown in Fig. 19-5 for some typical composite materials. Matrix stresses on a transverse plane of a unidirectional composite arise from matrix shrinkage during curing, differential thermal expansion, moisture absorption, and external loading.

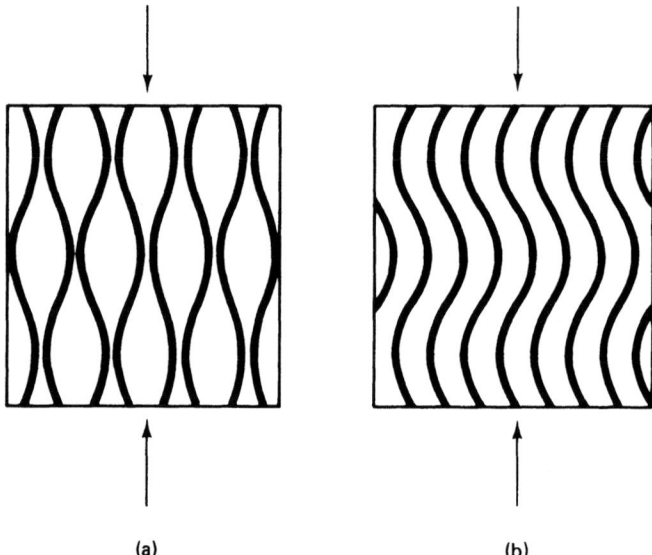

Figure 19-4 Schematic representation of microbuckling modes in a unidirectional composite under longitudinal compressive loading: (a) out-of-phase or extensional mode; (b) in-phase or shear mode. (From Ref. 9.)

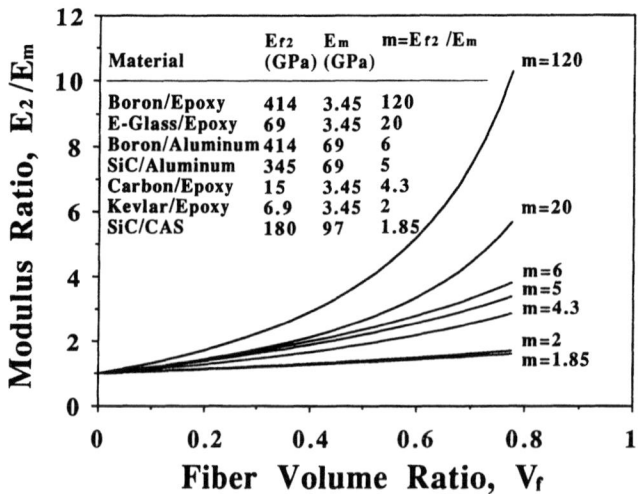

Figure 19-5 Transverse modulus of unidirectional composites as a function of fiber volume ratio. (See Halpin–Tsai equations, Ref. 12.)

COMPOSITE MATERIALS

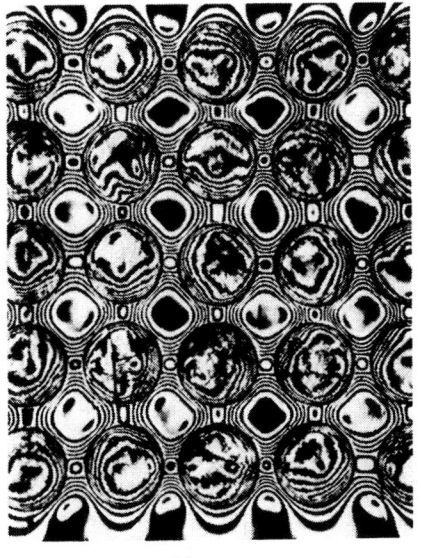 

(a) (b)

Figure 19-6 Isochromatic fringe patterns in slice from model subjected to (a) matrix shrinkage and (b) combined matrix shrinkage and external loading.

Transverse tensile loading of a unidirectional composite results in high strain concentrations in the matrix. In this case the matrix stresses and strains are the governing criteria of failure. Stress distributions in the matrix can be obtained analytically by finite-element methods and experimentally by means of two- and three-dimensional photoelastic models [11,14–19].

Marloff and Daniel [18] used stress-freezing and slicing techniques to determine stress distributions in the matrix of a unidirectional composite model subjected to matrix shrinkage and transverse loading. Isochromatic fringe patterns of slices taken from the models are shown in Fig. 19-6. The principle of superposition was used to separate the effects of shrinkage from those of combined shrinkage and loading. The distribution of shrinkage stresses along an axis of symmetry is shown in Fig. 19-7. Stress distributions due to external load along an axis of symmetry in the loading direction are shown in Fig. 19-8. These are presented in dimensionless form by dividing the stress components by the applied average stress. In this particular case the stress ratio in the direction of loading reaches a maximum value of 2.0 at the midpoint of the matrix section between the fibers.

The stress concentration factor, defined as the ratio of the maximum stress at the interface to the applied average stress, has been obtained analytically and by means of two- and three-dimensional photoelastic models. Its variation with fiber spacing is shown in Fig. 19-9. The quantity of importance is the strain concentration factor, which is related to the stress concentration factor as follows:

$$k_\varepsilon = \frac{\varepsilon_{\max}}{\varepsilon_0} = k_\sigma \frac{E_2}{E_m} \frac{(1 + \nu_m)(1 - 2\nu_m)}{1 - \nu_m} \tag{19-10}$$

where $\varepsilon_{\max}$ and ε_0 are the maximum and average strains, respectively. The variation of the strain concentration factor with fiber spacing for a composite model is illustrated in Fig. 19-10.

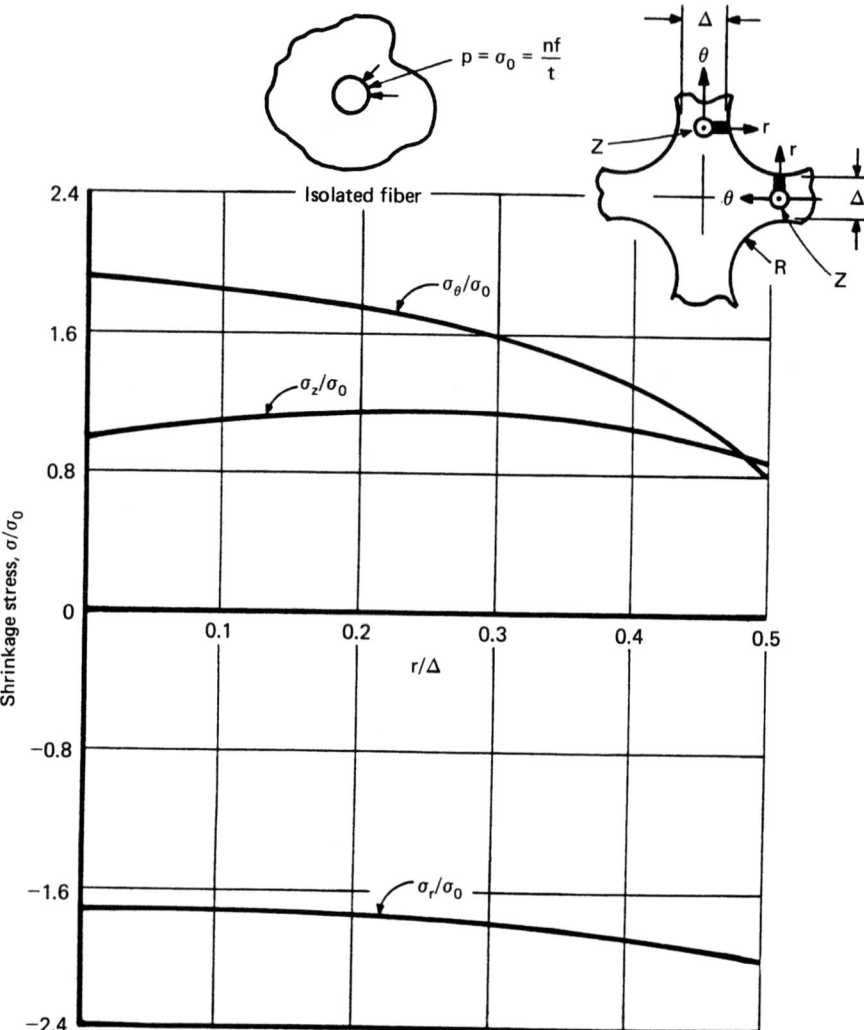

Figure 19-7 Distribution of shrinkage stresses across section between fibers, $\Delta/R = 0.50$.

The maximum strain at the interface due to combined transverse tensile loading and curing shrinkage is given by

$$\varepsilon_{max} = k_\varepsilon \varepsilon_0 + \varepsilon_R = k_\varepsilon \frac{\sigma_0}{E_2} + \varepsilon_R \tag{19-11}$$

where σ_0 = average applied stress
ε_R = radial shrinkage strain

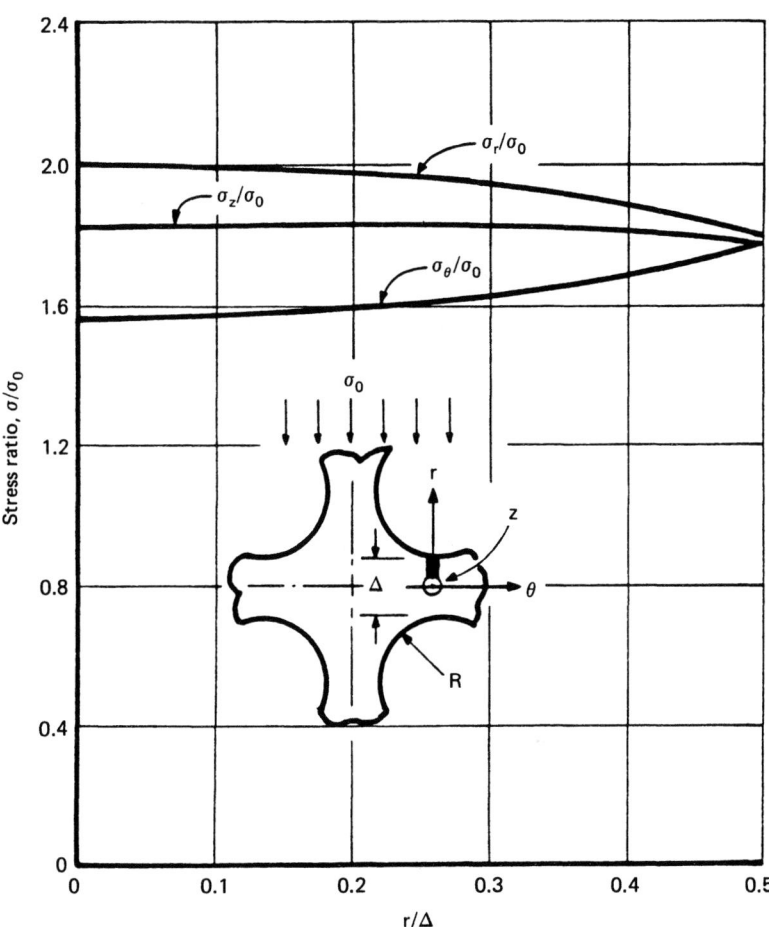

Figure 19-8 Stress distributions due to external load across section between fibers parallel to load direction, $\Delta/R = 0.50$.

Assuming a maximum tensile strain failure criterion and linear elastic behavior to failure for the matrix, we predict the following transverse tensile strength for a unidirectional composite [19]:

$$F_{2T} = \frac{1 - \nu_m}{k_\sigma(1 + \nu_m)(1 - 2\nu_m)}(F_{mT} - \varepsilon_R E_m) \tag{19-12}$$

For a typical glass–epoxy composite of 0.50 fiber volume ratio, the predicted strength would be $F_{2T} = 46.9$ MPa (6790 psi), whereas the actually measured strength is 47.3 MPa (6850 psi). This strength prediction is based on a local deterministic failure criterion at a point. Actual fracture should be based on global failure criteria, taking into account the statistical distribution of strengths and failure sites.

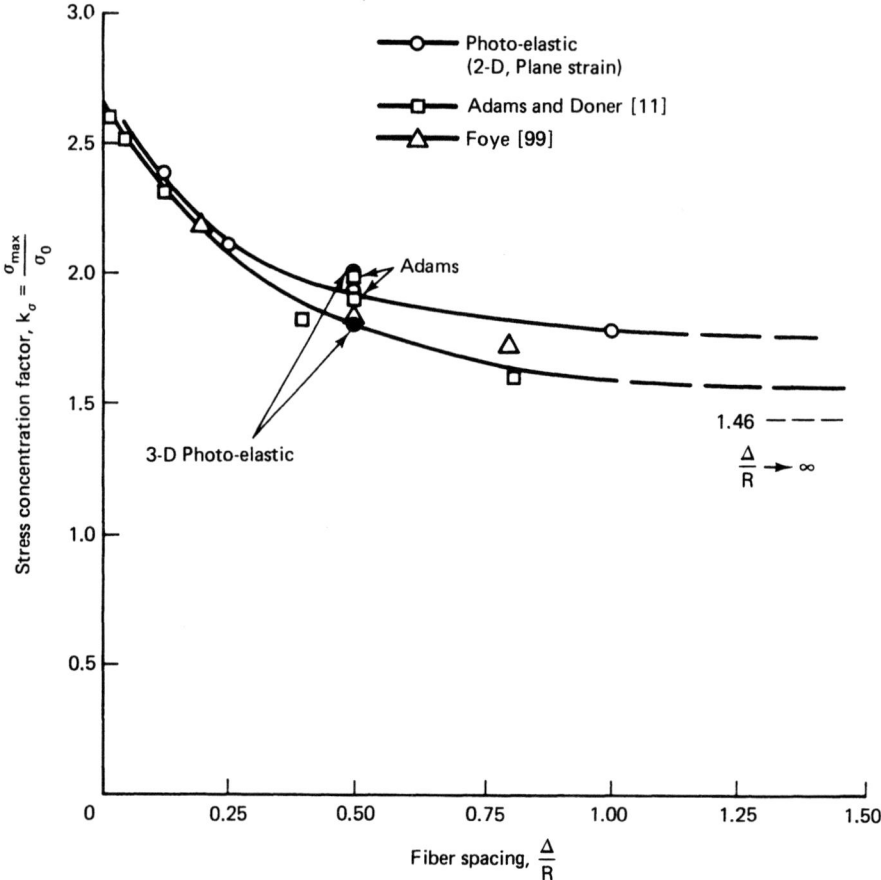

Figure 19-9 Stress concentration factor as a function of inclusion spacing for square array under transverse loading.

The behavior of a unidirectional composite under in-plane shear loading, like that under transverse tension, is matrix dominated. In-plane shear loading results in shear-stress concentration in the matrix, which can produce matrix shear failure, fiber–matrix debonding, or a combination of these. The shear-stress concentration factor k_τ can be calculated as follows:

$$k_\tau = \frac{1 - V_f/(1 - G_m/G_f)}{1 - (4V_f/\pi)^{1/2}(1 - G_m/G_f)} \tag{19-13}$$

where G_m and G_f are the matrix and fiber shear moduli, respectively.

A simple prediction for the in-plane shear strength is given by

$$F_{12} = \frac{F_{ms}}{k_\tau} \tag{19-14}$$

COMPOSITE MATERIALS

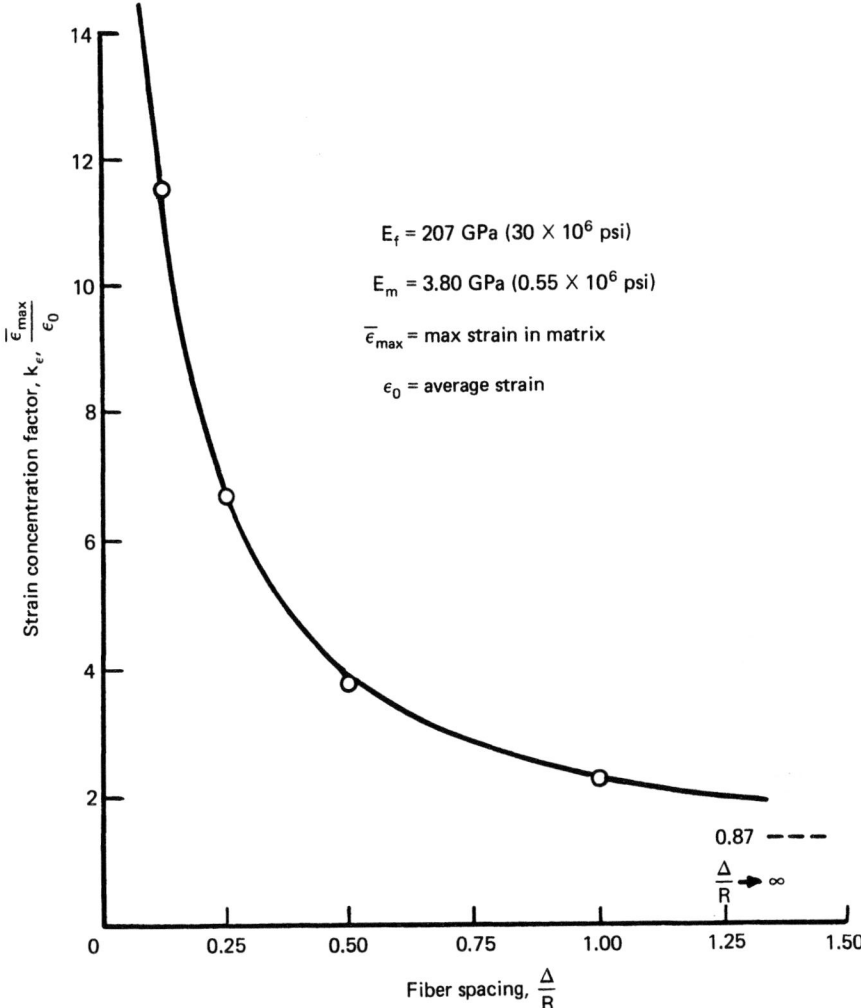

Figure 19-10 Strain concentration factor as a function of inclusion spacing for square array under transverse loading.

where F_{ms} is the matrix shear strength.

In addition to the mechanical properties discussed heretofore, the response of the composite lamina to hygrothermal variations is of great interest. The expansional hygrothermal strains are expressed as

$$\varepsilon_i^E = \alpha_i \Delta T + \beta_i \Delta C \qquad (19\text{-}15)$$

where $\varepsilon_i^E = \varepsilon_1^E, \varepsilon_2^E, \varepsilon_{12}^E$ hygrothermal strains
$\alpha_i = \alpha_1, \alpha_2, \alpha_{12}$ thermal expansion coefficients
$\beta_i = \beta_1, \beta_2, \beta_{12}$ moisture expansion coefficients
$\Delta T, \Delta C$ = changes in temperature and moisture concentration

and subscripts 1, 2, and 12 denote longitudinal, transverse, and in-plane shear response, respectively.

The coefficients of thermal expansion can be calculated in terms of the constituent properties [20]:

$$\alpha_1 = \frac{\overline{E\alpha}}{\overline{E}} \tag{19-16}$$

$$\alpha_2 = \alpha_f V_f (1 + \nu_f) + \alpha_m V_m (1 + \nu_m) - (\nu_f V_f + \nu_m V_m) \frac{\overline{E\alpha}}{\overline{E}} \tag{19-17}$$

where α_f, α_m = fiber and matrix thermal expansion coefficients
$\overline{E} = E_f V_f + E_m V_m$
$\overline{E\alpha} = E_f \alpha_f V_f + E_m \alpha_m V_m$

Expressions for β_i are exactly analogous; however, simplifications can be made when we assume (as is the case for boron and graphite fibers) that

$$\beta_f = 0$$

Elastic Behavior of Unidirectional Lamina

The usual analysis of composite structures is based on the macromechanical approach, where the lamina, as the basic building block, is treated as a homogeneous anisotropic layer. In the most general case an anisotropic body is characterized by 81 elastic constants, which are reduced to 36 by taking into account the symmetry of the stress and strain tensors, and to 21 by using energy considerations. The general orthotropic material, having three mutually perpendicular planes of elastic symmetry, is characterized by nine independent constants, which are reduced to five when the material has transverse isotropy. The unidirectional lamina is a transversely isotropic orthotropic material assumed to be under plane-stress conditions. In such a case the number of independent elastic constants is reduced to four.

Referring to the coordinate system in Fig. 19-1, the constitutive relations for the lamina are

$$\begin{bmatrix} \sigma_1 \\ \sigma_2 \\ \tau_{12} \end{bmatrix} = \begin{bmatrix} Q_{11} & Q_{12} & 0 \\ Q_{12} & Q_{22} & 0 \\ 0 & 0 & Q_{66} \end{bmatrix} \begin{bmatrix} \varepsilon_1 \\ \varepsilon_2 \\ \gamma_{12} \end{bmatrix} \tag{19-18}$$

where the components of the lamina stiffness matrix are

$$Q_{11} = E_1/(1 - \nu_{12}\nu_{21})$$

$$Q_{22} = E_2/(1 - \nu_{12}\nu_{21})$$

$$Q_{12} = \nu_{21}E_1/(1 - \nu_{12}\nu_{21}) = \nu_{12}E_2/(1 - \nu_{12}\nu_{21})$$

$$Q_{66} = G_{12}$$

E_1, E_2, G_{12} = longitudinal, transverse, and in-plane shear moduli, respectively

ν_{12}, ν_{21} = major and minor Poisson's ratios, corresponding to loading in the longitudinal and transverse directions, respectively

The two Poisson's ratios are related to the two principal Young's moduli as follows:

$$\nu_{12}E_2 = \nu_{21}E_1 \tag{19-19}$$

The inverse constitutive relations are

$$\begin{bmatrix} \varepsilon_1 \\ \varepsilon_2 \\ \gamma_{12} \end{bmatrix} = \begin{bmatrix} S_{11} & S_{12} & 0 \\ S_{12} & S_{22} & 0 \\ 0 & 0 & S_{66} \end{bmatrix} \begin{bmatrix} \sigma_1 \\ \sigma_2 \\ \tau_{12} \end{bmatrix} \tag{19-20}$$

where the components of the lamina compliance matrix are related to the engineering constants as follows:

$$S_{11} = 1/E_1$$

$$S_{22} = 1/E_2$$

$$S_{12} = -\nu_{12}/E_1 = -\nu_{21}/E_2$$

$$S_{66} = 1/G_{12}$$

Thus the unidirectional lamina can be completely characterized by four independent constants, such as E_1, E_2, G_{12}, and ν_{12}.

Normally, the lamina principal axes (1, 2) do not coincide with the loading or reference axes (x, y) (Fig. 19-11). Then the stress and strain components referred to the principal material axes (1, 2) can be expressed in terms of those referred to the loading axes (x, y) by the following transformation relations:

$$\begin{bmatrix} \sigma_1 \\ \sigma_2 \\ \tau_{12} \end{bmatrix} = [T] \begin{bmatrix} \sigma_x \\ \sigma_y \\ \tau_{xy} \end{bmatrix} \tag{19-21}$$

$$\begin{bmatrix} \varepsilon_1 \\ \varepsilon_2 \\ \tfrac{1}{2}\gamma_{12} \end{bmatrix} = [T] \begin{bmatrix} \varepsilon_x \\ \varepsilon_y \\ \tfrac{1}{2}\gamma_{xy} \end{bmatrix} \tag{19-22}$$

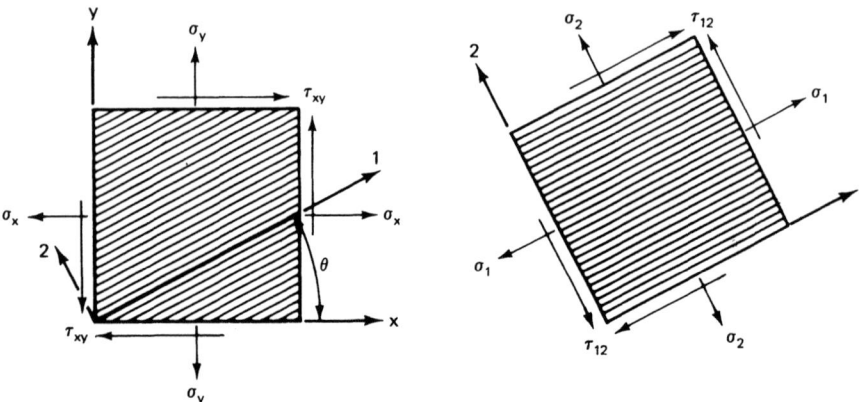

Figure 19-11 Stress components in unidirectional lamina referred to loading and material axes.

where the transformation matrix [T] is given by

$$[T] = \begin{bmatrix} m^2 & n^2 & 2mn \\ n^2 & m^2 & -2mn \\ -mn & mn & m^2 - n^2 \end{bmatrix}$$

where $m = \cos\theta$ and $n = \sin\theta$.

Equation (19-18) shows that when the lamina is loaded in tension or compression along the principal material directions, there is no shear strain with respect to these directions. Similarly, a shear stress τ_{12} produces only a shear strain γ_{12}. Thus there is no coupling between extensional and shear deformations. This is not the case when the lamina is loaded along arbitrary axes x and y. Then the stress–strain relations are expressed as

$$\begin{bmatrix} \sigma_x \\ \sigma_y \\ \tau_{xy} \end{bmatrix} = \begin{bmatrix} \overline{Q}_{11} & \overline{Q}_{12} & \overline{Q}_{16} \\ \overline{Q}_{12} & \overline{Q}_{22} & \overline{Q}_{26} \\ \overline{Q}_{16} & \overline{Q}_{26} & \overline{Q}_{66} \end{bmatrix} \begin{bmatrix} \varepsilon_x \\ \varepsilon_y \\ \gamma_{xy} \end{bmatrix} \qquad (19\text{-}23)$$

where the components of the transformed lamina stiffness matrix $[\overline{Q}]$ referred to an arbitrary system of axes, are given by

$$\bar{Q}_{11} = Q_{11}m^4 + 2(Q_{12} + 2Q_{66})m^2n^2 + Q_{22}n^2$$
$$\bar{Q}_{22} = Q_{11}n^4 + 2(Q_{12} + 2Q_{66})m^2n^2 + Q_{22}m^2$$
$$\bar{Q}_{12} = (Q_{11} + Q_{22} - 4Q_{66})m^2n^2 + Q_{12}(m^4 + n^4) \tag{19-24}$$
$$\bar{Q}_{66} = (Q_{11} + Q_{22} - 2Q_{12} - 2Q_{66})m^2n^2 + Q_{66}(m^4 + n^4)$$
$$\bar{Q}_{16} = (Q_{11} - Q_{12} - 2Q_{66})m^3n + (Q_{12} - Q_{22} + 2Q_{66})mn^3$$
$$\bar{Q}_{26} = (Q_{11} - Q_{12} - 2Q_{66})mn^3 + (Q_{12} - Q_{22} + 2Q_{66})m^3n$$

Similar transformation relations exist for the components of the compliance matrix in the strain–stress relations

$$\begin{bmatrix} \varepsilon_x \\ \varepsilon_y \\ \gamma_{xy} \end{bmatrix} = \begin{bmatrix} \bar{S}_{11} & \bar{S}_{12} & \bar{S}_{16} \\ \bar{S}_{12} & \bar{S}_{22} & \bar{S}_{26} \\ \bar{S}_{16} & \bar{S}_{26} & \bar{S}_{66} \end{bmatrix} \begin{bmatrix} \sigma_x \\ \sigma_y \\ \tau_{xy} \end{bmatrix} \tag{19-25}$$

This relationship can also be expressed in terms of engineering parameters as follows:

$$\begin{bmatrix} \varepsilon_x \\ \varepsilon_y \\ \gamma_{xy} \end{bmatrix} = \begin{bmatrix} \dfrac{1}{E_x} & -\dfrac{\nu_{yx}}{E_y} & \dfrac{\eta_{sx}}{G_{xy}} \\ -\dfrac{\nu_{xy}}{E_x} & \dfrac{1}{E_y} & \dfrac{\eta_{sy}}{G_{xy}} \\ \dfrac{\eta_{xs}}{E_x} & \dfrac{\eta_{ys}}{E_y} & \dfrac{1}{G_{xy}} \end{bmatrix} \begin{bmatrix} \sigma_x \\ \sigma_y \\ \tau_{xy} \end{bmatrix} \tag{19-26}$$

Here a quantitative measure of the coupling between extensional and shear deformations is given by the shear coupling coefficients defined as follows:

$$\eta_{sx} = \frac{\varepsilon_x}{\gamma_{xy}} \quad \text{when } \sigma_x = \sigma_y = 0$$

$$\eta_{sy} = \frac{\varepsilon_y}{\gamma_{xy}} \quad \text{when } \sigma_x = \sigma_y = 0 \tag{19-27}$$

$$\eta_{xs} = \frac{\gamma_{xy}}{\varepsilon_x} \quad \text{when } \sigma_y = \tau_{xy} = 0$$

$$\eta_{ys} = \frac{\gamma_{xy}}{\varepsilon_y} \quad \text{when } \sigma_x = \tau_{xy} = 0$$

These coefficients do not represent additional material constants, since they are related to the other constants. Elastic constants are interrelated as follows:

$$\frac{\nu_{12}}{\nu_{21}} = \frac{E_1}{E_2} \quad \frac{\nu_{xy}}{\nu_{yx}} = \frac{E_x}{E_y}$$
$$\frac{\eta_{xs}}{\eta_{sx}} = \frac{E_x}{G_{xy}} \quad \frac{\eta_{ys}}{\eta_{sy}} = \frac{E_y}{G_{xy}} \tag{19-28}$$

Using Eqs. (19-25), (19-26), and the transformation equations for the compliance matrix $[\overline{S}]$, we can obtain transformation relations in terms of the engineering constants. The results of such transformations are shown graphically in Fig. 19-12 for E_x, G_{xy}, ν_{xy}, and η_{sx} as a function of the angle between the reference (loading) axis and the fiber direction.

Strength of Unidirectional Lamina

The strength of a lamina, like the stiffness, is an anisotropic property (i.e., it varies with the reference direction). The principal stresses and strains depend on the loading and material anisotropy. The axes of principal stresses, principal strains, and material symmetry do not necessarily coincide. Since strength varies with direction, the maximum stress in general does not govern failure. In a manner analogous to stiffness transformation, it seems desirable to express strength along an arbitrary direction as a function of strength characteristics along the principal material directions.

A unidirectional lamina is characterized by the following strength properties:

F_{1T}: longitudinal tensile strength
F_{1C}: longitudinal compressive strength
F_{2T}: transverse tensile strength
F_{1C}: transverse compressive strength
F_{12} (or F_6): in-plane or intralaminar shear strength

The lamina strength can be described in the form of macromechanical failure theories. The best-known failure theories are the following:

1. Maximum stress theory
2. Maximum strain theory
3. Deviatoric strain-energy theory for anisotropic materials (Tsai–Hill)
4. Interaction tensor polynomial theory (Tsai–Wu)

According to the maximum stress theory, failure occurs when at least one stress component along one of the principal material directions exceeds the corresponding strength in that direction. The theory does not consider stress interaction. The failure criteria are expressed in the form of three subcriteria:

$$\sigma = \begin{cases} F_{1T} & \text{when } \sigma_1 > 0 \\ F_{1C} & \text{when } \sigma_1 < 0 \end{cases}$$

$$\sigma_2 = \begin{cases} F_{2T} & \text{when } \sigma_2 > 0 \\ F_{2C} & \text{when } \sigma_2 < 0 \end{cases} \qquad (19\text{-}29)$$

$$\tau_{12} = F_{12}$$

According to the maximum strain theory, the lamina fails when at least one of the strain components along one of the principal material axes exceeds the corresponding ultimate strain in that direction. This theory allows for some interaction of stress components due to Poisson's effect. It is expressed in the form of three subcriteria:

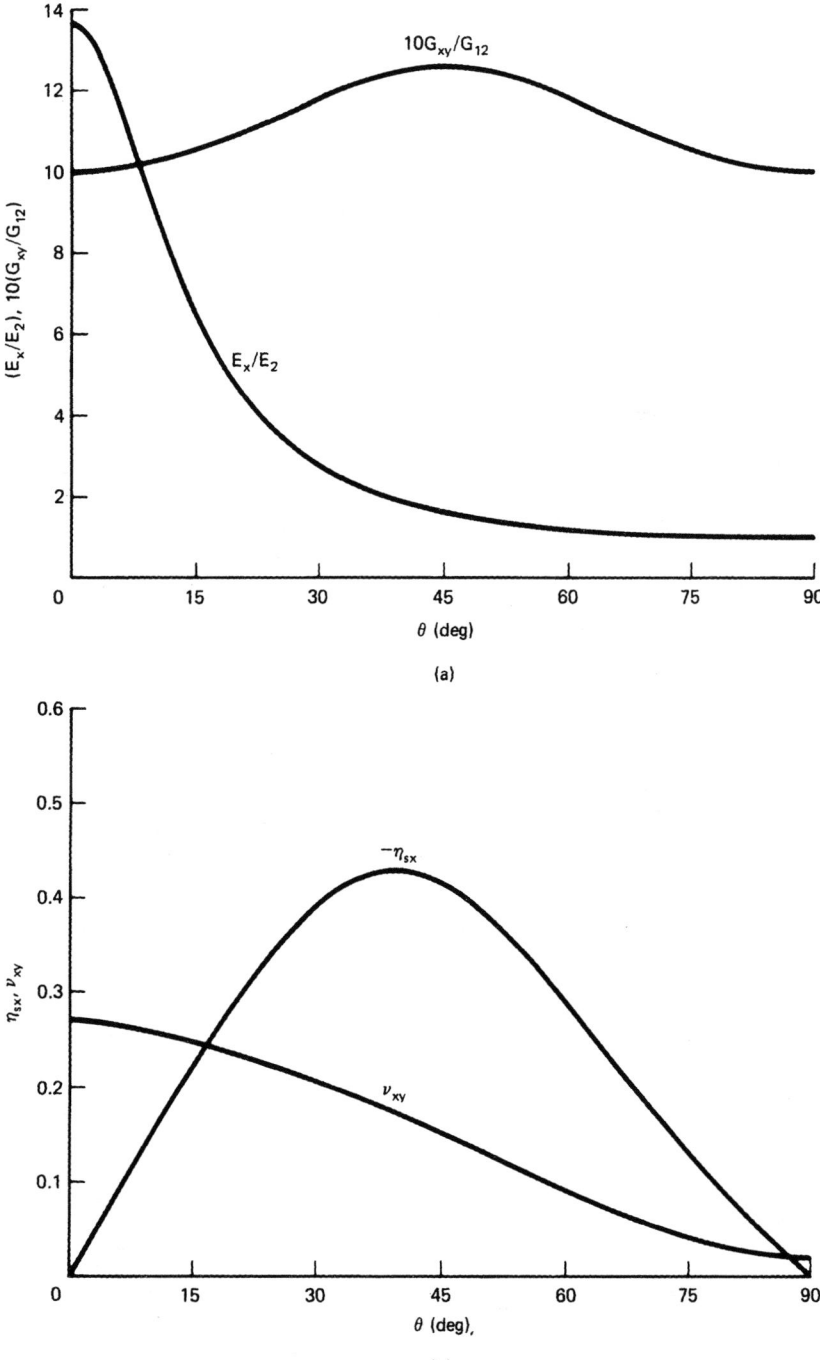

Figure 19-12 (a) Young's modulus and shear modulus of unidirectional composite as a function of fiber orientation (AS-4/3501-6 graphite–epoxy); (b) Poisson's ratio and shear coupling coefficient of unidirectional composite as a function of fiber orientation (AS-4/3501-6 graphite–epoxy).

$$\varepsilon_1 = \begin{cases} \varepsilon_{1T}^u & \text{when } \varepsilon_1 > 0 \\ \varepsilon_{1C}^u & \text{when } \varepsilon_1 < 0 \end{cases}$$

$$\varepsilon_2 = \begin{cases} \varepsilon_{2T}^u & \text{when } \varepsilon_2 > 0 \\ \varepsilon_{2C}^u & \text{when } \varepsilon_2 < 0 \end{cases} \tag{19-30}$$

$$\gamma_{12} = \gamma_{12}^u = 2\varepsilon_{12}^u$$

The deviatoric strain-energy theory is based on the von Mises failure criterion for isotropic materials, which was modified by Hill for anisotropic materials, and applied to composites by Tsai. The so-called Tsai–Hill criterion is expressed as

$$\frac{\sigma_1^2}{F_1^2} + \frac{\sigma_2^2}{F_2^2} + \frac{\tau_{12}^2}{F_{12}^2} - \frac{\sigma_1 \sigma_2}{F_1^2} = 1 \tag{19-31}$$

where

$$F_1 = \begin{cases} F_{1T} & \text{when } \sigma_1 > 0 \\ F_{1C} & \text{when } \sigma_1 < 0 \end{cases}$$

$$F_2 = \begin{cases} F_{2T} & \text{when } \sigma_2 > 0 \\ F_{2C} & \text{when } \sigma_2 < 0 \end{cases}$$

This criterion yields a failure envelope which is a closed surface in the σ_1, σ_2, τ_{12} space. The theory agrees with experimental results and allows for considerable interaction among σ_1, σ_2, τ_{12}. However, it does not distinguish between tensile and compressive strengths.

Improvements to the Tsai–Hill theory were introduced by Goldenblat and Kopnov [21] and Tsai and Wu [22]. The failure criterion is expressed in the form of a quadratic stress tensor polynomial

$$f_i \sigma_i + f_{ij} \sigma_i \sigma_j = 1 \tag{19-32}$$

where $i = 1, 2, 6$

f_i, f_{ij} = strength tensors of second and fourth order, respectively

For a lamina under a plane state of stress, Eq. (19-32) is expanded as [22]

$$f_1 \sigma_1 + f_2 \sigma_2 + f_6 \tau_{12} + f_{11} \sigma_1^2 + f_{22} \sigma_2^2 + f_{66} \tau_{12}^2$$
$$+ 2f_{12} \sigma_1 \sigma_2 + 2f_{16} \sigma_1 \tau_{12} + 2f_{26} \sigma_2 \tau_{12} = 1 \tag{19-33}$$

The linear terms in this equation allow the distinction between tensile and compressive strengths. The term f_{12} represents the interaction of σ_1 and σ_2. By applying elementary loading conditions to the lamina, we obtain values for the coefficients of the quadratic equation:

$$f_1 = \frac{1}{F_{1T}} + \frac{1}{F_{1C}} \qquad f_{11} = -\frac{1}{F_{1T} F_{1C}}$$

$$f_2 = \frac{1}{F_{2T}} + \frac{1}{F_{2C}} \qquad f_{22} = -\frac{1}{F_{2T} F_{2C}} \tag{19-34}$$

$$f_6 = f_{16} = f_{26} = 0$$

$$f_{66} = \frac{1}{F_{12}^2}$$

$$f_{12} \simeq -\tfrac{1}{2}(f_{11} f_{22})^{1/2}$$

One advantage of this theory is the invariance of the criterion with rotation of coordinates. In addition, it allows for transformation by means of normal transformation laws, and it has symmetry akin to stiffness and compliance matrices.

Elastic Behavior and Multidirectional Laminates

Most structural applications of composites involve multidirectional laminates, which are made up of two or more unidirectional laminae stacked together at various angles. Classical lamination theory predicts the behavior of the laminate from the properties of the individual laminae. The underlying assumptions of lamination theory are as follows:

1. The laminate is considered to be quasi-homogeneous and anisotropic.
2. Each layer of the laminate is orthotropic with transverse isotropy.
3. All layers have identical material properties (except for hybrid laminates).
4. The laminate and its layers are in a state of plane stress.
5. Displacements vary linearly through the thickness of the laminate and are continuous throughout.

The strains at any point in the laminate are related to the midplane strains and curvatures as

$$\begin{bmatrix} \varepsilon_x \\ \varepsilon_y \\ \gamma_{xy} \end{bmatrix} = \begin{bmatrix} \varepsilon_x^0 \\ \varepsilon_y^0 \\ \gamma_{xy}^0 \end{bmatrix} + z \begin{bmatrix} \kappa_x \\ \kappa_y \\ \kappa_{xy} \end{bmatrix} \quad (19\text{-}35)$$

The state of stress in the kth lamina can be expressed as

$$\begin{bmatrix} \sigma_x \\ \sigma_y \\ \tau_{xy} \end{bmatrix}_k = \begin{bmatrix} \overline{Q}_{11} & \overline{Q}_{12} & \overline{Q}_{16} \\ \overline{Q}_{12} & \overline{Q}_{22} & \overline{Q}_{26} \\ \overline{Q}_{16} & \overline{Q}_{26} & \overline{Q}_{66} \end{bmatrix}_k \begin{bmatrix} \varepsilon_x^0 + z\kappa_x \\ \varepsilon_y^0 + z\kappa_y \\ \gamma_{xy}^0 + z\kappa_{xy} \end{bmatrix} \quad (19\text{-}36)$$

It is usually convenient to deal with a system of forces and moments in the laminate rather than stresses in each lamina. These force and moment resultants for a laminate consisting of n plies are defined as (Fig. 19-13)

$$\begin{bmatrix} N_x \\ N_y \\ N_{xy} \end{bmatrix} = \sum_{k=1}^{n} \int_{h_{k-1}}^{h_k} \begin{bmatrix} \sigma_x \\ \sigma_y \\ \tau_{xy} \end{bmatrix} dz \quad (19\text{-}37)$$

$$\begin{bmatrix} M_x \\ M_y \\ M_{xy} \end{bmatrix} = \sum_{k=1}^{n} \int_{h_{k-1}}^{h_k} \begin{bmatrix} \sigma_x \\ \sigma_y \\ \tau_{xy} \end{bmatrix} z\, dz \quad (19\text{-}38)$$

Combining Eqs. (19-36) and (19-37), we obtain

$$\begin{bmatrix} N_x \\ N_y \\ N_{xy} \end{bmatrix} = \begin{bmatrix} A_{11} & A_{12} & A_{16} \\ A_{12} & A_{22} & A_{26} \\ A_{16} & A_{26} & A_{66} \end{bmatrix} \begin{bmatrix} \varepsilon_x^0 \\ \varepsilon_y^0 \\ \gamma_{xy}^0 \end{bmatrix} + \begin{bmatrix} B_{11} & B_{12} & B_{16} \\ B_{12} & B_{22} & B_{26} \\ B_{16} & B_{26} & B_{66} \end{bmatrix} \begin{bmatrix} \kappa_x \\ \kappa_y \\ \kappa_{xy} \end{bmatrix} \quad (19\text{-}39)$$

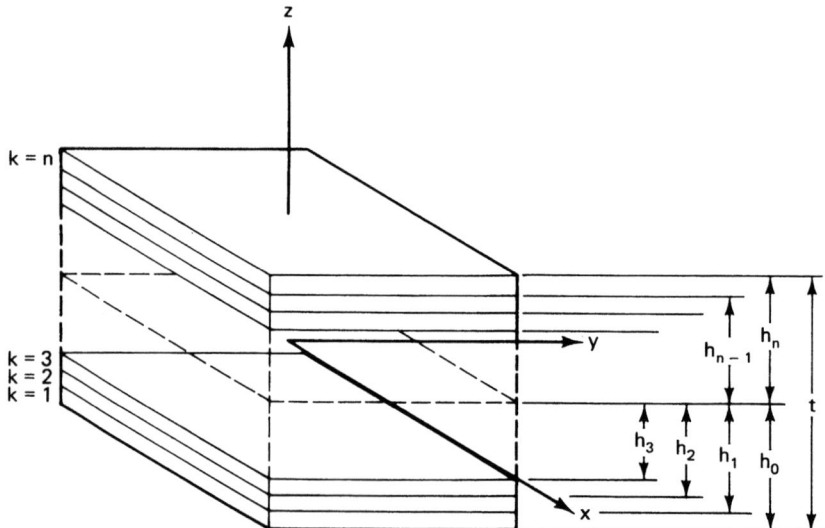

Figure 19-13 Notation for lamina coordinates within a laminate.

or

$$[N] = [A][\varepsilon^0] + [B][\kappa]$$

where

$$A_{ij} = \sum_{k=1}^{n} (\overline{Q}_{ij})_k (h_k - h_{k-1}) \tag{19-40}$$

$$B_{ij} = \frac{1}{2} \sum_{k=1}^{n} (\overline{Q}_{ij})_k (h_k^2 - h_{k-1}^2) \tag{19-41}$$

Combining Eqs. (19-36) and (19-38), we obtain

$$\begin{bmatrix} M_x \\ M_y \\ M_{xy} \end{bmatrix} = \begin{bmatrix} B_{11} & B_{12} & B_{16} \\ B_{12} & B_{22} & B_{26} \\ B_{16} & B_{26} & B_{66} \end{bmatrix} \begin{bmatrix} \varepsilon_x^0 \\ \varepsilon_y^0 \\ \gamma_{xy}^0 \end{bmatrix} + \begin{bmatrix} D_{11} & D_{12} & D_{16} \\ D_{12} & D_{22} & D_{26} \\ D_{16} & D_{26} & D_{66} \end{bmatrix} \begin{bmatrix} \kappa_x \\ \kappa_y \\ \kappa_{xy} \end{bmatrix} \tag{19-42}$$

or

$$[M] = [B][\varepsilon^0] + [D][\kappa]$$

where

$$D_{ij} = \frac{1}{3} \sum_{k=1}^{n} (\overline{Q}_{ij})_k (h_k^3 - h_{k-1}^3) \tag{19-43}$$

In the general case combining extension and curvature, the constitutive relations of the laminate are expressed as

$$\begin{bmatrix} N_x \\ N_y \\ N_{xy} \\ --- \\ M_x \\ M_y \\ M_{xy} \end{bmatrix} = \begin{bmatrix} A_{11} & A_{12} & A_{16} & | & B_{11} & B_{12} & B_{16} \\ A_{12} & A_{22} & A_{26} & | & B_{12} & B_{22} & B_{26} \\ A_{16} & A_{26} & A_{66} & | & B_{16} & B_{26} & B_{66} \\ --- & --- & --- & + & --- & --- & --- \\ B_{11} & B_{12} & B_{16} & | & D_{11} & D_{12} & D_{16} \\ B_{12} & B_{22} & B_{26} & | & D_{12} & D_{22} & D_{26} \\ B_{16} & B_{26} & B_{66} & | & D_{16} & D_{26} & D_{66} \end{bmatrix} \begin{bmatrix} \varepsilon_x^0 \\ \varepsilon_y^0 \\ \gamma_{xy}^0 \\ --- \\ \kappa_x \\ \kappa_y \\ \kappa_{xy} \end{bmatrix} \quad (19\text{-}44)$$

where A_{ij} = extensional stiffnesses

B_{ij} = coupling stiffness (coupling in-plane forces and curvatures and/or bending moments and extension)

D_{ij} = flexural stiffnesses

The general relations (19-44) are simplified for special types of laminates. For example, for midplane symmetric laminates the coupling stiffnesses become zero,

$$B_{ij} = 0$$

that is, there is no coupling between extension and flexure. Other simplifications result for other specific types of laminates.

Stresses and Strength of Multidirectional Laminates

The laminate constitutive relations discussed before allow the determination of the strains throughout the laminate, from which the strains in any individual lamina can be determined, first along the laminate axes (x, y) and then, by transformation, along the material axes of the lamina $(1, 2)$. Using the stress–strain equations for the lamina [Eq. (19-18)], we obtain the stress components along the principal material axes of that lamina. Thus each lamina can be fully assessed independently of its neighboring laminae.

To evaluate the strength of a laminate, new approaches are needed, since failure of one lamina does not imply failure of the laminate. Additional factors affecting the laminate strength analysis are (1) the coordinate mismatch between laminae, (2) varying lamina stiffnesses along laminate coordinate system, (3) the stacking sequence, which affects bending and coupling stiffnesses, and (4) curing conditions affecting residual stresses. Two general approaches to strength analysis have been followed.

1. *First-ply failure (FPF)*. The laminate is considered to have failed when the first layer fails.
2. *Ultimate laminate failure*. The laminate is considered to have failed when the maximum load level is exceeded, following multiple-ply failure (MPF).

The first approach (FPF) is conservative but requires low safety factors. The second approach (MPF) is advanced but requires high safety factors.

The procedure for the multiple-ply failure (MPF) approach consists of the following steps:

1. Lamina stresses are determined as a function of the unknown load(s), first in the laminate coordinate system (x, y), then transformed in the material coordinate system of each lamina.
2. By applying one of the failure criteria for the lamina (e.g., Tsai–Wu), the load at which the first ply fails (FPF) is determined.

3. The failed lamina is replaced with one having $E_2 = G_{12} = 0$, and new laminate [A], [B], [D] matrices are calculated.
4. Lamina stresses are recalculated, while verifying that remaining laminae do not fail immediately after failure of the first ply, due to stress redistribution.
5. The load is increased until another lamina fails, while verifying again that remaining plies do not fail immediately due to increased load sharing.
6. If all remaining plies cannot support the load, progressive failure occurs, leading to total failure.

19-2 Experimental Methods for Composite Materials

The design and analysis of composite structures relies heavily on experimental data. Testing of composites serves a variety of purposes. Specific types and applications of testing include the following:

1. Characterization of constituent materials (i.e., fiber and matrix) for use in micromechanics analyses. Knowing these properties, one can predict, in principle, the behavior of the lamina, hence of the laminates and structures.
2. Verification of micromechanics analyses, especially in the case of nonlinear, inelastic behavior, including effects of curing stresses, temperature, and moisture.
3. Qualification and characterization of the basic unidirectional lamina that forms the building block of all laminated structures. This characterization, especially the determination of irreversible and ultimate properties, is essential in providing data for subsequent stress analysis using lamination theory.
4. Experimental stress analysis by measuring strain distributions and using the appropriate constitutive relations.
5. Fracture characterization, including identification of fracture mechanisms and modes, initiation, and propagation.
6. Testing under special conditions of loading (e.g., fatigue, multiaxial, and high-rate testing).
7. Assessment of structural integrity and reliability using nondestructive evaluation methods.
8. Life prediction through accelerated testing.

A variety of experimental methods are used for the applications above. Generally, most experimental methods applicable to isotropic materials can be adapted to composite materials. Most of these deal with measurement of deformations or strains. Some employ modeling procedures and others are applied directly to the prototype. Some give point-per-point information and others give full-field information. The most commonly used methods are:

1. Photoelastic methods
2. Strain gages
3. Moiré methods
4. Holographic and speckle interferometric methods
5. Nondestructive evaluation methods, such as ultrasonics, x-radiography, acoustic emission, and thermography
6. Fractography

Reviews of various methods applicable to composites have been given in the literature [6,23,24].

Photoelastic Methods

Two-dimensional photoelastic scaled-up models of the longitudinal or transverse cross sections of a composite are used to determine stress distributions on the fiber scale [25,26,27]. Three-dimensional models, with appropriate selection of the fiber and matrix materials, have also been analyzed using standard stress-freezing and slicing techniques [18]. Stress distributions due to loading and curing shrinkage and stress and strain concentrations in the matrix can be obtained. Two-dimensional models have also been used in dynamic photoelastic studies of fracture mechanisms and crack propagation on the micromechanical scale.

To stimulate the prototype composite with regard to fiber–matrix bonding and fiber scale, two-dimensional microphotoelastic models have been used with actual fibers embedded in the matrix [7,16]. A more realistic modeling of fibrous composites is obtained by means of anisotropic photoelasticity using transparent birefringent composites [28–33]. The models are glass-fiber-reinforced plastics with the matrix and the fibers having the same index of refraction. Transparent fibrous composites can be made to simulate the anisotropy of opaque fibrous composites, such as boron–epoxy and graphite–epoxy. These transparent composites are treated as homogeneous materials with anisotropic elastic and optical properties. For orthotropic materials under plane-stress conditions, the stress-optic law takes the from

$$\begin{bmatrix} N_{11} \\ N_{22} \\ N_{12} \end{bmatrix} = \begin{bmatrix} B_{11} & B_{12} & 0 \\ B_{12} & B_{22} & 0 \\ 0 & 0 & B_{66} \end{bmatrix} \begin{bmatrix} \sigma_1 \\ \sigma_2 \\ \tau_{12} \end{bmatrix} \quad (19\text{-}45)$$

where N_{ij} = birefringence tensor
$B_{11}, B_{12}, B_{22}, B_{66}$ = components of fringe-value tensor

In terms of relative birefringence, the stress-optic law takes the form

$$N = \left[\left(\frac{\sigma_1}{2f_1} - \frac{\sigma_2}{2f_2} \right)^2 + \left(\frac{2\tau_{12}}{f_{12}} \right)^2 \right]^{1/2} \quad (19\text{-}46)$$

where N = relative birefringence per unit thickness
f_1, f_2, f_{12} = principal material fringe values, referred to principal material axes of composite

The latest advances in this area are the development of fabrication procedures for multidirectional laminates and digital image analysis procedures for analyzing low levels of birefringence [33,34].

The method of photoelastic coatings (see Chapter 5) can be applied directly to the usually opaque prototype composite. As in the case of isotropic substrates, the birefringence in the coating is related to the difference in principal strains by the strain-optic law [Eq. (5-52)]. The stresses in the composite substrate are computed from the strains using the appropriate anisotropic stress–strain relations. The analysis is simplified when the axes of material, geometric, and loading symmetry coincide. Then the coating birefringence is related to the principal stresses in the composite by

$$\frac{Nf_\varepsilon}{2h} = \sigma_1 \frac{1 + \nu_{12}}{E_1} - \sigma_2 \frac{1 + \nu_{21}}{E_2} \quad (19\text{-}47)$$

where f_ε is the strain fringe value of the coating material and h is the coating thickness. One of the major limitations of the method near free boundaries is the effect of Poisson's ratio mismatch. This problem, which exists also in applications to isotropic materials, can be further aggravated in applications to composites as a result of the wider range of Poisson's ratios of composite laminates. The Poisson's ratio mismatch introduces some uncertainty in the strain analysis near free boundaries over a distance of approximately four coating thicknesses from the boundary.

However, it has been shown experimentally that at a free boundary, the principal strain and the nonzero principal stress in the composite along the boundary are given by [35]

$$\varepsilon_1 = \frac{Nf_\varepsilon}{2h} \frac{1}{1 + \nu^c} \tag{19-48}$$

and

$$\sigma_1 = \frac{E_1}{1 + \nu^c} \frac{Nf_\varepsilon}{2h} \tag{19-49}$$

where ν^c is Poisson's ratio of the coating.

Electrical-Resistance Strain Gages

Commercially available foil gages are widely used in composites. The general principles of strain gages are discussed in Chapter 2. Of particular interest to composites is the response of the gages to temperature variations. A change of ΔT in ambient temperature produces the following resistance change in the gage:

$$\left(\frac{\Delta R}{R}\right)_{\Delta T} = (\alpha_\theta - \beta)S_g \Delta T + \gamma \Delta T \tag{19-50}$$

where α_θ is the coefficient of thermal expansion of the composite in the direction of the gage axis, β the thermal coefficient of expansion of the gage, and γ the temperature coefficient of resistivity of the gage alloy. Whereas in the case of isotropic materials it is possible to match the coefficients of thermal expansion of the gage and specimen, this is not feasible with composites, where the coefficient of thermal expansion is a highly anisotropic property.

A practical approach for temperature compensation of strain gages on composite materials relies on the use of a reference material of known, and preferably small, thermal expansion. For every type of gage applied to a composite material, a similar gage is applied to the reference material, which undergoes the same temperature history as the composite. Then the true thermal strain in the composite is given by

$$\varepsilon_t = \varepsilon_a - \varepsilon_r + \varepsilon_{tr} \tag{19-51}$$

where ε_t is the true strain, ε_a the apparent (uncorrected) strain in the composite, ε_r the apparent strain in the reference material, and ε_{tr} the known ($\alpha_r \Delta T$) thermal expansion of the reference material. A commonly used reference material is titanium silicate, with a coefficient of thermal expansion of 0.03×10^{-6} K^{-1} (0.017 μin./in./°F). To minimize the purely thermal response of the gage, it is preferable to use gages with zero expansion temperature compensation (i.e., $\beta \simeq 0$). The thermal response of such a gage mounted on a titanium silicate specimen is illustrated in Fig. 19-14.

Another important aspect of strain-gage applications to composites is the transverse sensitivity of the gage. This is expressed in terms of the cross-sensitivity factor K given by the gage manufacturer, which ranges between ± 0.01 and ± 0.02. The transverse sensitivity correction becomes appreciable whenever the strain to be measured is small compared to the strain in the direction perpendicular to the gage grid. This situation is encountered frequently in composites.

Composite laminates are amenable to embedment of strain gages between plies during the fabrication process. This allows for strain measurements in the interior of composite laminates, as well as monitoring of strains during curing and during subsequent service conditions. Techniques have been developed and applied to the measurement of subsurface strains in various composites

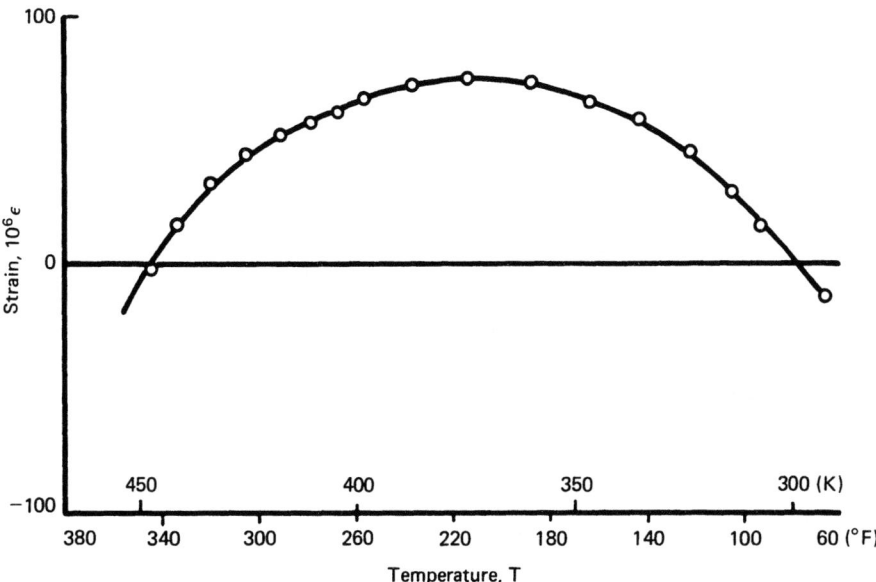

Figure 19-14 Apparent strain as a function of temperature of WK-00-125TM-350 gage bonded on titanium silicate.

under static and dynamic conditions [6,36–38]. It was demonstrated that conventional foil gages, when embedded, do not reduce the strength of the material and do not affect the average mechanical properties. Figure 19-15 shows a cross section through a boron–epoxy laminate with embedded strain gages.

Moiré Methods

The moiré method described in Chapter 6, being strictly geometric in nature, is independent of the anisotropy and inelasticity of the material. The sensitivity of the method depends on the density of the rulings used and on the magnitude of the deformation. Glass–epoxy composites in general, as well as graphite–epoxy and boron–epoxy laminates with a large number of ±45° plies, undergo sufficiently large deformations to yield analyzable moiré fringes using coarse model gratings of up to 40 lines/mm (1000 lines/in.). In many cases, however, the deformations to be measured are small. Therefore, the sensitivity of the method must be enhanced by other means, such as fringe interpolation and multiplication techniques [39].

In conventional practice, gratings of up to 40 lines/mm (1000 lines/in.) are applied to the specimen surface by photographic, etching, transfer, or bonding techniques. The specimen surface is normally prepared with a smooth and partially reflective substrate. Film gratings can be bonded to this reflective surface or photoprinted onto it by the photoresist process. During loading of the specimen, a transparent reference grating is placed near or in contact with the specimen grating, and the resulting fringe patterns are recorded photographically. Sometimes it is more suitable to photograph the specimen grating in the no-load and load conditions and form the moiré patterns on the bench subsequently. Moiré patterns are related to the displacement field, since they represent

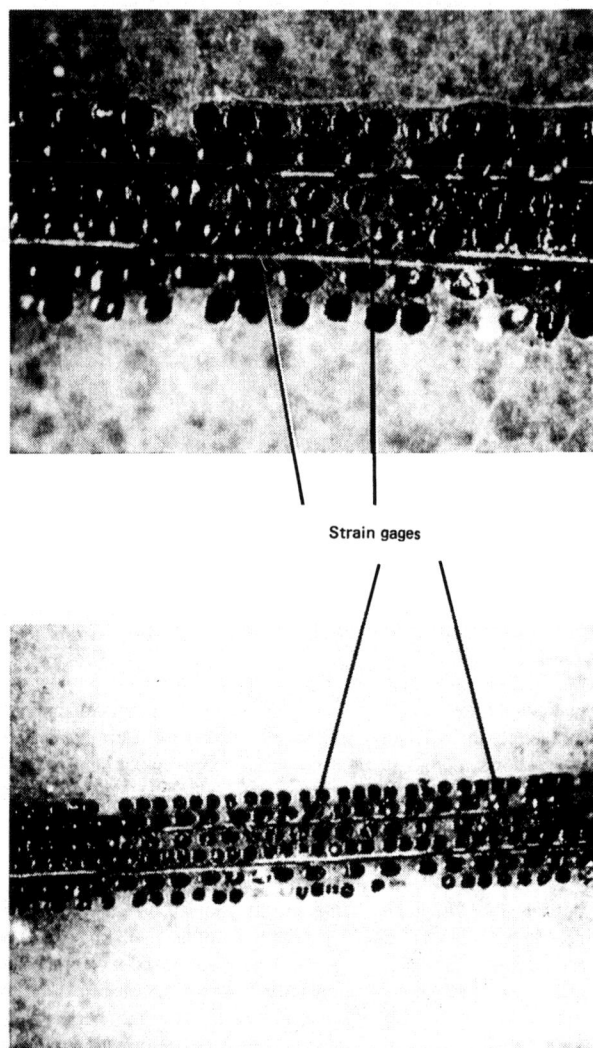

Figure 19-15 Section through a gage location of an instrumented [0/±45/0]$_s$ boron–epoxy specimen showing embedded three-gage rosettes on the second and fourth plies. (Note the separate sections of the gage grid exposed below the second ply.)

loci of points having the same component of displacement normal to the grating lines. Strains are obtained from the displacements by graphical, optical (mechanical), or numerical differentiation.

The greatest progress in sensitivity enhancement occurred with new developments in moiré interferometry using high-frequency gratings of up to the theoretical limit of approximately 4000 lines/mm (101,600 lines/in.) (Chapter 7). Moiré interferometric techniques are finding many applications in composites [40].

Holographic and Speckle Interferometric Methods

The basic principles of the holographic and speckle interferometric methods are discussed in Chapter 8. Since conventional holographic techniques are best suited for measuring out-of-plane displacements, many applications to composites deal with flexure and vibration of plates [41]. Another important application of holographic interferometry is the nondestructive evaluation (NDE) of composite material integrity, to be discussed later.

Speckle interferometry has some similarities to holographic interferometry, but it has some advantages over conventional interferometry. It has less stringent requirements regarding the setup and the light source. It allows the determination of in-plane and out-of-plane displacements directly. An important development in this area allows the simultaneous determination of derivatives of surface displacements along any direction and with variable sensitivity using a single photographic record (shearing specklegram) [42].

19-3 Composite Material Characterization

Characterization of Constituent Materials

Tensile properties of the fiber and matrix are of particular interest. Single fibers are tested in tension to determine modulus, strength, and ultimate strain. The method is discussed in detail in ASTM D3379-75 for fibers of modulus higher than 21 GPa. For matrix–resin systems that can be processed in thick sheets, a standard dogbone type of specimen is used, as described in ASTM Specification D638-72. Resin systems that cannot be fabricated in thick sheets are cast into thin sheet or film form. Thin strips can then be tested in tension, as described in ASTM D882-73.

Material Qualification

Composite materials are normally available from the manufacturer in the form of a tape consisting of unidirectional fibers impregnated with the matrix resin partially advanced (B-staged). The prepregs are supplied to a variety of specifications, such as fiber volume ratio, ply thickness, and degree of B-staging. Prior to fabrication of larger components and extensive material characterization, some preliminary quality-control tests are conducted. Flexural and interlaminar shear strengths are determined from unidirectional coupons and compared with published or expected values.

Qualification testing is done by means of beams subjected to three-point bending. The test specimens are normally 15-ply-thick unidirectional coupons. Flexural-strength coupons are 10.2 cm (4 in.) long and 1.27 cm (0.50 in.) wide, with a 6.3-cm (2.5-in.) span length. Shear-strength coupons are typically 1.5 cm (0.6 in.) long and 0.64 cm (0.25 in.) wide, with a 1-cm (0.4-in.) span length. The standard beam formulas below are used to determine strength values:

$$F_{1F} = \frac{3Pl}{2wt^2} \qquad (19\text{-}52)$$

$$F_{12} = \frac{3P}{4wt} \qquad (19\text{-}53)$$

where F_{1F}, F_{12} = flexural and shear strengths, respectively
P = load
w = beam width
t = thickness
l = length

Laminate fabrication and more extensive material characterization follow the satisfactory completion of the qualification tests.

Density

The density of a composite sample is usually determined by the displacement method (ASTM D792-66). The sample is first weighed in air and then suspended by a wire and weighed while totally immersed in a liquid of known density and good wettability, such as alcohol. Finally, the suspended wire is weighed immersed in the liquid to the same depth as during the sample weighing. Similar procedures are used for measuring the density of the polymeric resin matrix and the fiber separately.

Fiber Volume Ratio

The fiber volume ratio (FVR) can be determined by two basic methods. One method involves the dissolution of the matrix in a liquid solvent or, in the case of nonburning fibers, the burning of the matrix, with the weight of the remaining fibers and the fiber and composite densities used in the determination of the percentage volume of fiber. The second method, the gravimetric method, uses the densities of the constituent and composite materials as follows:

$$\text{FVR} = V_f = \frac{\rho_c - \rho_m}{\rho_f - \rho_m} \qquad (19\text{-}54)$$

where ρ_c, ρ_m, and ρ_f are the densities of composite, matrix, and fiber, respectively. The gravimetric method is valid if the composite is free of voids. In most cases the void content is less than 1%; thus the assumption of a void-free composite is reasonable. The void content can be checked in photomicrographs of composite cross sections.

Coefficients of Thermal Expansion

The coefficient of thermal expansion is determined by measuring thermal strains in a composite specimen as a function of temperature over a temperature range. Commercial strain gages have been found suitable for measuring small thermal strains in composites. In this case the strain-gage readings must be corrected for the purely thermal output of the gage [6]. Figure 19-16 shows thermal strains measured in unidirectional graphite–epoxy, Kevlar 49–epoxy, and S-glass–epoxy specimens. Both Kevlar 49–epoxy and graphite–epoxy exhibit negative thermal strains in the longitudinal (fiber) direction. The Kevlar 49–epoxy exhibits the largest positive transverse and negative longitudinal strains. The S-glass–epoxy undergoes the lowest thermal deformation in the transverse direction and the highest (positive) in the longitudinal direction. Coefficients of thermal

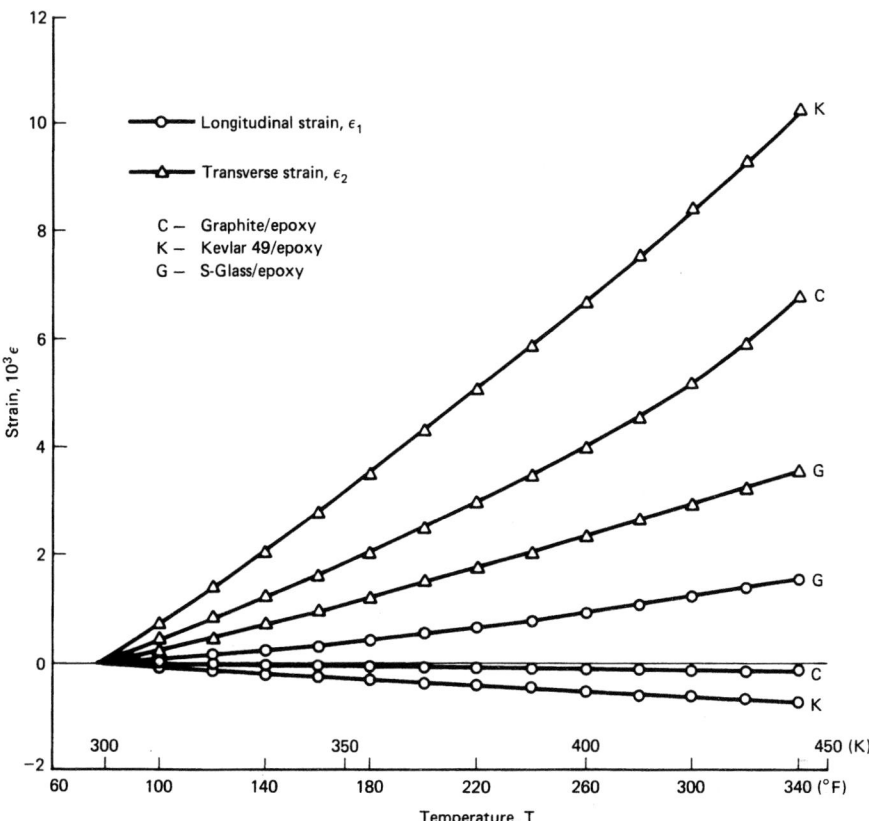

Figure 19-16 Thermal strains in unidirectional composites as a function of temperature.

expansion computed as the slopes of these curves are tabulated in Table 19-1 for eight composite materials. Coefficients are listed at room temperature [297 K (75°F)] and at the elevated temperature of 450 K (350°F). All graphite fiber composites exhibit negative thermal expansion in the fiber direction. The polyimide matrix composites do not show any variation of thermal coefficients with temperature. This is true at least up to the postcuring temperature of 589 K (600°F).

The expansion coefficients along an arbitrary coordinate system x–y rotated by an angle θ with respect to the fiber direction can be obtained by a linear coordinate transformation as

$$\alpha_x = \alpha_1 m^2 + \alpha_2 n^2$$
$$\alpha_y = \alpha_1 n^2 + \alpha_2 m^2 \qquad (19\text{-}55)$$
$$\alpha_{xy} = 2(\alpha_1 - \alpha_2)mn$$

where α_1, α_2 = longitudinal and transverse coefficients of thermal expansion
$m = \cos\theta$
$n = \sin\theta$

Table 19-1 Coefficients of Thermal Expansion of Unidirectional Composite Materials

Material	Longitudinal Coefficient of Thermal Expansion, α_1, 10^{-6} K^{-1} ($\mu\varepsilon$/°F)[a]		Transverse Coefficient of Thermal Expansion, α_2, 10^{-6} K^{-1} ($\mu\varepsilon$/°F)[a]	
	297 K (75°F)	450 K (350°F)	297 K (75°F)	450 K (350°F)
Boron–epoxy (Boron–AVCO 5505)	6.1 (3.4)	6.1 (3.4)	30.3 (16.9)	37.8 (21.0)
Boron–polyimide (Boron–WRD 9171)	4.9 (2.7)	4.9 (2.7)	28.4 (15.8)	28.4 (15.8)
Graphite–epoxy (Modmor I–ERLA 4289)	−1.1 (−0.6)	3.2 (1.3)	31.5 (17.5)	27.0 (15.0)
Graphite–epoxy (Modmor I–ERLA 4617)	−1.3 (−0.7)	−1.3 (−0.7)	33.9 (18.8)	83.7 (46.5)
Graphite–polymide (Modmor I–WRD 9371)	−0.4 (−0.2)	−0.4 (−0.2)	25.3 (14.1)	25.3 (14.1)
S-glass–epoxy (Scotchply–1009-26 5901)	3.8 (2.1)	3.8 (2.1)	16.7 (9.3)	54.9 (30.5)
S-glass–epoxy (S-glass–ERLA 4617)	6.6 (3.7)	14.1 (7.9)	19.7 (10.9)	26.5 (14.7)
Kevlar–epoxy (Kevlar 49–ERLA 4617)	−4.0 (−2.2)	−5.7 (−3.2)	57.6 (32.0)	82.8 (46.0)

[a] $\mu\varepsilon$, microstrain.

The effective coefficients for multidirectional laminates can be evaluated from the properties of the single lamina by means of lamination theory.

Coefficients of Moisture Expansion

The coefficients of moisture expansion are determined by measuring specimen deformations over a range of moisture weight gains [6]. This type of measurement is difficult because of moisture effects on conventional strain-gage adhesives. In the case of unidirectional composites, most of the swelling occurs in the transverse to the fiber direction, whereas in the case of multidirectional laminates, most of the swelling occurs in the thickness direction.

Static Tensile Properties of Unidirectional Composites

Uniaxial tensile tests are conducted on unidirectional laminae to determine the following properties

E_1 = longitudinal Young's modulus

E_2 = transverse Young's modulus

ν_{12} = major Poisson's ratio

ν_{21} = minor Poisson's ratio

F_{1T} = longitudinal tensile strength

F_{2T} = transverse tensile strength

ε_{1T}^u = ultimate longitudinal tensile strain

ε_{2T}^u = ultimate transverse tensile strain

Tensile specimens are straight-sided coupons of constant cross section with adhesively bonded, beveled glass–epoxy tabs. The longitudinal (0°) coupon is six plies thick and 1.27 cm (0.50 in.) wide, while the transverse (90°) coupon is eight plies thick and 2.54 cm (1.00 in.) wide. Both specimens have an overall length of 23 cm (9 in.) and a 15.2-cm (6-in.) gage length. The specimens are incrementally loaded to failure under uniaxial tensile loading with recording of strains at every step. Typical stress–strain curves for 0° and 90° graphite–epoxy specimens are shown in Figs. 19-17 and 19-18. The moduli and Poisson's ratios are determined from the initial slopes of the curves fitted through the data points. Results for tensile and other properties are summarized in Table 19-2. Typical fractures of unidirectional tensile coupons are shown in Figs. 19-19 and 19-20. In the case of 0° specimens, the failure mode consists of fiber breakage, matrix splitting, and fiber pull-out. The latter mechanism is much more pronounced in the brooming failure pattern of glass–epoxy. Transverse (90°) specimens fail in a brittle manner by matrix tensile failure between fibers.

Static Compressive Properties

Compression testing of composites is one of the most difficult types of testing because of the tendency for premature failure due to crushing or buckling. Over the years, many test methods have been developed and used, incorporating a variety of specimen designs and loading fixtures. A review of these methods is found in the SEM monograph [6].

Compression test methods can be classified into three broad categories. In the first (type I), specimens with a very short but unsupported gage length are used. One of these, the so-called Celanese test, makes use of coupon specimens 14.1 cm (5.5 in.) long, 15 to 20 plies thick, and 0.64 cm (0.25 in.) wide (ASTM D-3410-75). The coupons are tabbed with long tapered glass–epoxy tabs, leaving a gage section 1.27 cm (0.5 in.) long. Load is introduced through friction by means of split conical collet grips that fit into matching sleeves, which in turn fit into a snugly fitting cylindrical shell. One major disadvantage of this fixture is that it requires a perfect cone-to-cone contact. This contact is not normally achieved as a result of small variations in tab thickness. Instead, contact is limited to two lines on opposite sides of the specimen. This unstable condition causes a lateral shift in the grips, which then produces high frictional forces in the enveloping cylinder. This situation has resulted in erroneously high values for the stiffness and compressive strength.

The IITRI (Illinois Institute of Technology Research Institute) test method represents a modification of the method above [43]. The conical grips were replaced with trapezoidal wedges. This eliminates the problem of line contact, since surface-to-surface contact can be attained at all positions of the wedges. Furthermore, it permits precompression of the specimen tabs to prevent slippage in the early stages of loading. The lateral alignment of the fixture top and bottom halves is assured by a guidance system consisting of two parallel roller bushings. The specimen, grips, and fixture assembly are illustrated in Fig. 19-21. Typical results obtained with this fixture for a graphite–epoxy material are included in Table 19-2.

In the second category of test methods (type II), a relatively long, fully supported specimen is used. The test specimen is similar to the tensile coupon discussed before but slightly shorter and with longer tabs. The fixture provides contact support over the entire gage length of the specimen. The type II tests yield data similar to those obtained by the type I tests, although 0° specimens give consistently lower values. This may be due to some premature buckling despite the lateral support.

In the third category of compression test methods the composite laminate is bonded to a honeycomb core, which provides the required lateral support. In one case, two composite coupons are bonded to a honeycomb block and the sandwich specimen is loaded edgewise in direct compression. In the second case, a sandwich beam is made by bonding the composite laminate

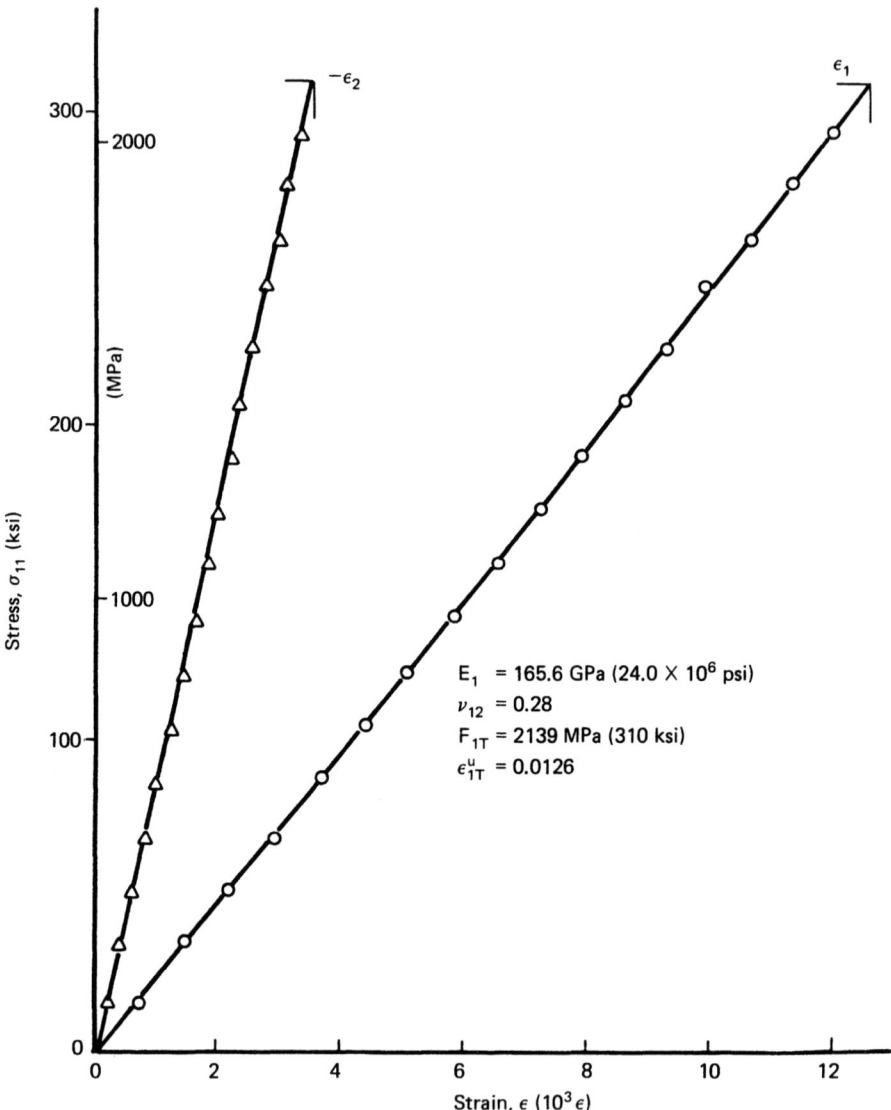

Figure 19-17 Stress–strain curves for [0₆] graphite–epoxy specimen under uniaxial tensile loading.

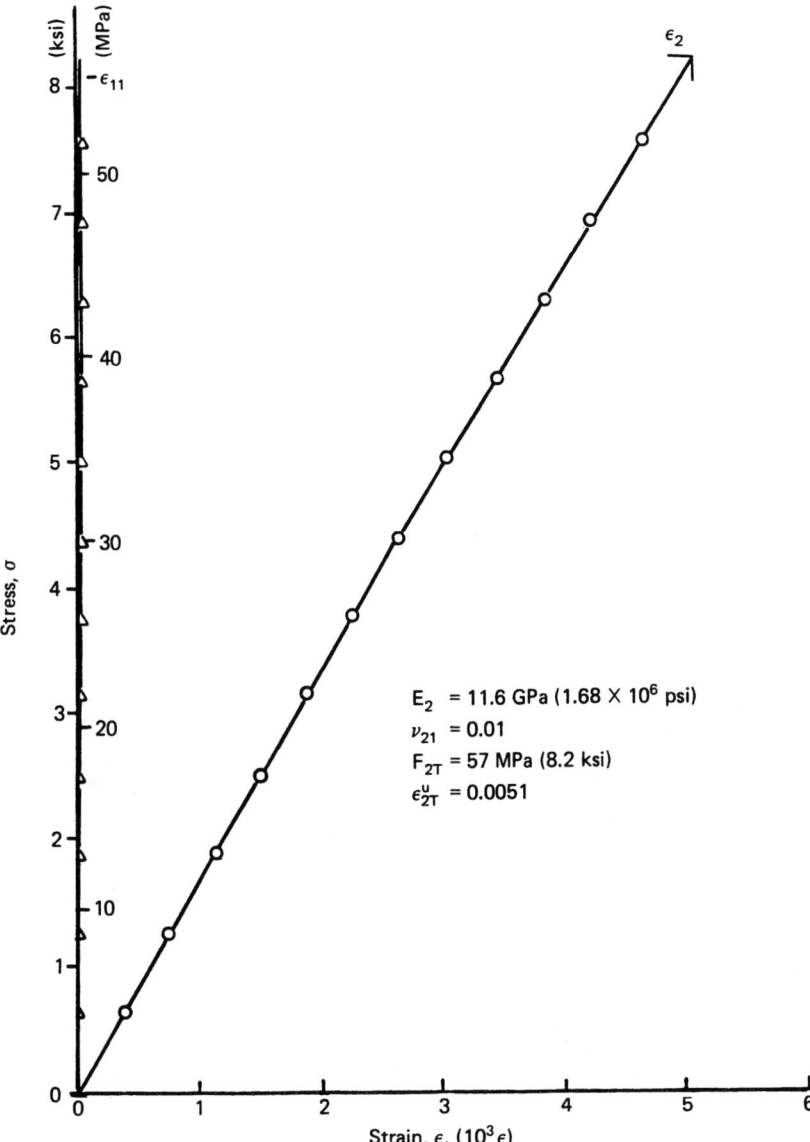

Figure 19-18 Stress–strain curves for [90₁₆] graphite–epoxy specimen under uniaxial tensile loading.

Table 19-2 Properties of Unidirectional Graphite–Epoxy (AS-4/3501-6)

Property	Value
Fiber volume ratio, V_f	0.65
Longitudinal modulus, E_1	145 GPa (21.0 × 10⁶ psi)
Transverse modulus, E_2	10.6 GPa (1.54 × 10⁶ psi)
Shear modulus, G_{12}	7.6 GPa (1.10 × 10⁶ psi)
Major Poisson's ratio, ν_{12}	0.27
Minor Poisson's ratio, ν_{21}	0.02
Longitudinal tensile strength, F_{1T}	2090 MPa (303 ksi)
Ultimate longitudinal tensile strain, ε^u_{1T}	0.0137
Longitudinal compressive strength, F_{1C}	1440 MPa (209 ksi)
Ultimate longitudinal compressive strain, ε^u_{1C}	0.0141
Transverse tensile strength, F_{2T}	64 MPa (9.3 ksi)
Ultimate transverse tensile strain, ε^u_{2T}	0.0060
Transverse compressive strength, F_{2C}	228 MPa (33 ksi)
Ultimate transverse compressive strain, ε^u_{2C}	0.0239
In-plane shear strength, F_{12}	71 MPa (10.3 ksi)
Ultimate shear strain, ε^u_{12}	0.0075

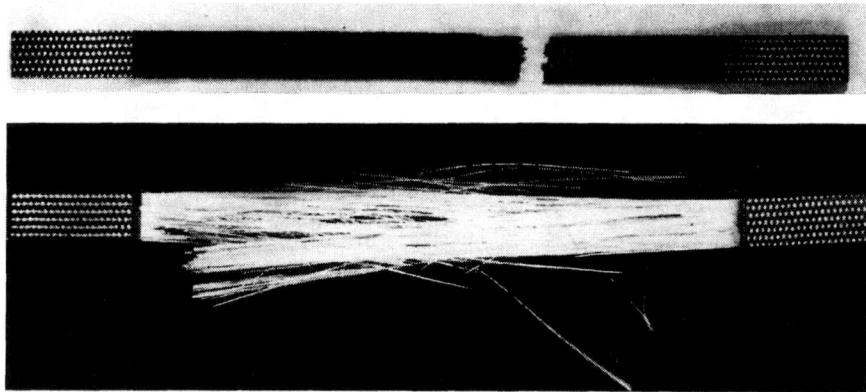

Figure 19-19 Typical fractures of unidirectional 0° tensile specimens: top, boron–epoxy; bottom, S-glass–epoxy.

Figure 19-20 Typical fractures of unidirectional 90° tensile specimens: top, graphite–high-modulus epoxy; bottom, graphic–polyimide.

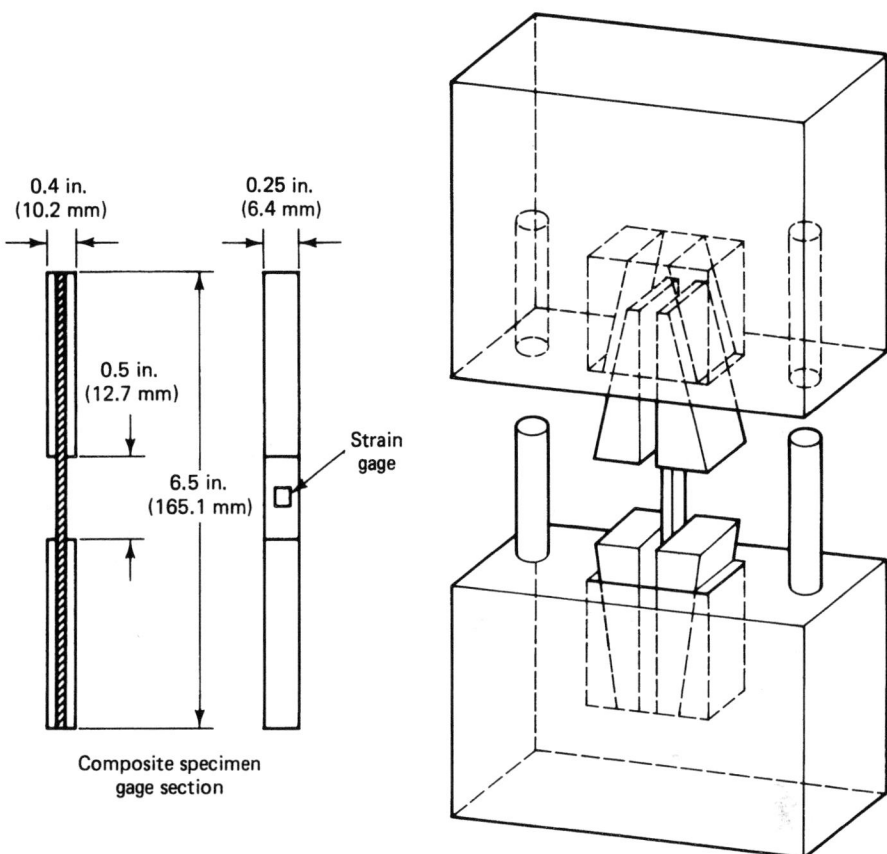

Figure 19-21 The IITRI compression test specimen and fixture.

to one side of the honeycomb core and a balancing metal face sheet on the other side. The beam is loaded in four-point bending to produce nearly uniform compression in the test section of the composite sheet. Results obtained from sandwich beam tests tend to be higher than those obtained by the other methods discussed, probably because of the restraint offered by the honeycomb and the biaxial state of stress induced in the composite face sheet.

In-Plane Shear Properties

Full characterization of a unidirectional composite requires the determination of lamina properties under in-plane shear parallel to the fibers (i.e., shear modulus G_{12}, shear strength F_{12}, and ultimate shear strain ε_{12}^u). There are four generally accepted test methods for determination of these properties: the $[\pm 45]_{ns}$ coupon test, the 10° off-axis test, the rail shear test, and the torsion test.

The first test method utilizes an eight-ply $[\pm 45]_{2s}$ coupon of the same dimensions as the 90° unidirectional tensile coupon discussed before [44]. When this coupon is subjected to a uniaxial tensile stress, the in-plane shear stress and in-plane shear strain referred to the fiber direction in each lamina are given by

$$\tau_{12} = \frac{\sigma_x}{2}$$
$$\gamma_{12} = 2\varepsilon_{12} = \varepsilon_x - \varepsilon_y$$
(19-56)

where σ_x is the applied uniaxial stress and ε_x and ε_y are the axial and transverse strains in the coupon, measured with two-gage rosettes. The in-plane (or intralaminar) shear modulus of the unidirectional lamina is obtained from the slope of the τ_{12} versus γ_{12} curve as

$$G_{12} = \frac{\sigma_x}{2(\varepsilon_x - \varepsilon_y)}$$
(19-57)

This method tends to overestimate the shear strength because of the constraint imposed on the lamina by the adjacent plies. The method does not take into account edge effects or the influence of the other stress components σ_1 and σ_2 on the lamina. A typical shear-stress versus shear-strain curve obtained from a $[\pm 45]_{2s}$ graphite–epoxy coupon is shown in Fig. 19-22.

The second test method is the 10° off-axis test [37,45]. The 10° angle is chosen to minimize the effects of longitudinal and transverse stress components σ_1 and σ_2 on the shear response. The specimen is a six-ply unidirectional coupon with the fibers oriented at 10° with the loading axis, 1.27 cm (0.5 in.) wide and 33 cm (13 in.) long. It is tabbed with tapered tabs and instrumented with three-gage rosettes on each side of the test section. The three-gage rosette normally has one gage element oriented axially (ε_x), one at 45° with the loading axis or 55° with the fiber direction (ε_{45}), and one transversely to the loading axis (ε_y). The specimen is subjected to a uniaxial tensile stress σ_x in increments up to failure. The intralaminar shear stress referred to the fiber coordinate system is given by

$$\tau_{12} = \sigma_x \sin \theta \cos \theta = 0.171 \sigma_x$$
(19-58)

The corresponding shear strain can be obtained in terms of the three strain components measured with the three-gage rosette as

$$\gamma_{12} = 2\varepsilon_{12} = (\varepsilon_x - \varepsilon_y) \sin 2\theta + [2\varepsilon_{45} - (\varepsilon_x + \varepsilon_y)] \cos 2\theta$$
(19-59)

where $\theta = 10°$. The in-plane shear modulus is obtained by plotting τ_{12} versus γ_{12} and taking the initial slope of the curve. The ultimate values of τ_{12} and γ_{12} determine the shear strength and ultimate shear strain. A typical shear-stress versus shear-strain curve obtained from a 10° off-axis graphite–epoxy specimen is shown in Fig. 19-23. Results obtained from such tests with graphite–epoxy are included in Table 19-2. This method tends to underestimate the ultimate properties as a result of interaction of the tensile stress across the fibers.

The third method of determining shear properties is the rail shear test, the two-rail or the three-rail test. In the three-rail test a rectangular coupon is clamped between three parallel pairs of rails (Fig. 19-24). The load is applied to one end of the middle rails and reacted at the opposite ends of the two outer pairs of rails. The average shear stress is obtained by dividing half the applied load by the thickness times the clamped length of the specimen. The shear strain is obtained from a single gage placed at the center of the exposed specimen at 45° with the rail axes. In this type of specimen, the state of stress near the ends is not pure shear and may thus result in premature failure due to the other stress components.

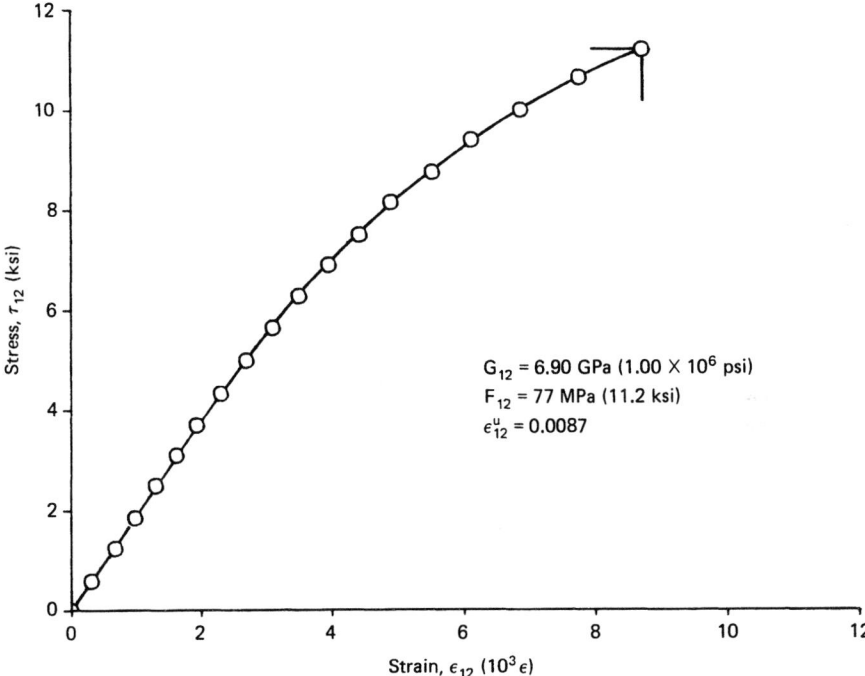

Figure 19-22 Shear stress versus shear strain for a $[\pm 45]_{2s}$ graphite-epoxy specimen under uniaxial tensile loading (AS-4/3501-6).

The fourth method is the torsion method, utilizing a solid rod or a hollow tubular specimen subjected to torque. A shear stress–shear strain curve can be obtained by plotting torque versus angle of twist or shear strain measured directly with strain gages. Such specimens are difficult to make and load. The solid rod is less desirable because of the shear-stress gradient across the section.

19-4 Biaxial Testing

The response of composite laminates to biaxial states of stress is difficult to predict analytically on the basis of lamina properties because of nonlinear behavior and failure-mode interactions. Several theories have been presented to date for predicting biaxial properties of composites, but they all rely to some extent on some limiting assumptions [22,46–48]. To check or verify some of these theories and to generate useful failure envelopes for design purposes, it is necessary to conduct extensive testing of composite laminates under biaxial states of stress. The application of a general in-plane biaxial state of stress, including tension, compression, and shear components, poses one of the most difficult problems in composite testing. Some of the basic requirements for a biaxial test specimen are:

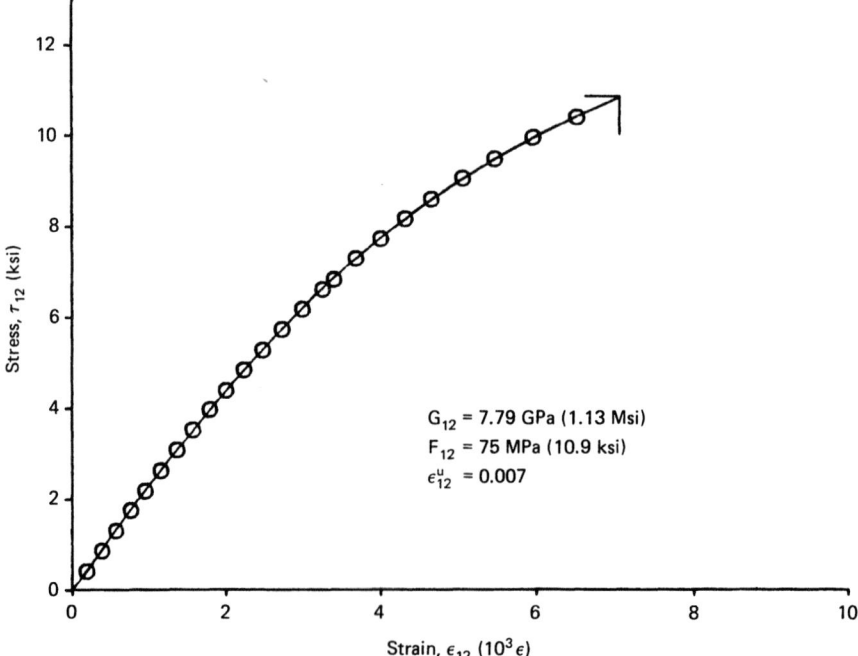

Figure 19-23 Stress–strain curve for a $[10]_6$ graphite–epoxy specimen under uniaxial tensile loading (AS-4/3501-6).

1. A significant volume of the material must be under a homogeneous state of stress.
2. Primary failure must occur in the test section.
3. The state of stress must be known without the need for secondary measurements or analysis.
4. It must be possible to vary the three in-plane stress components (σ_x, σ_y, τ_{xy}) independently.

A variety of specimen types and techniques have been proposed and used for biaxial laminate characterization. They include the off-axis coupon or ring, the cross-beam sandwich specimen, the bulge plate, the rectangular plate loaded in biaxial tension, and the thin-walled tubular specimen.

Off-Axis Specimen

Uniaxial loading along a direction other than one of the principal material axes produces a biaxial state of stress. This specimen has been used successfully in coupon and ring forms [49,50]. In the latter case, thin-walled rings with the principal material axes at an angle with the circumferential direction are subjected to internal pressure loading. Some of the limitations of this specimen are:

1. The biaxial normal stresses are always of the same sign.
2. There is no possibility for independent variation of the three stress components (nonproportional loading).

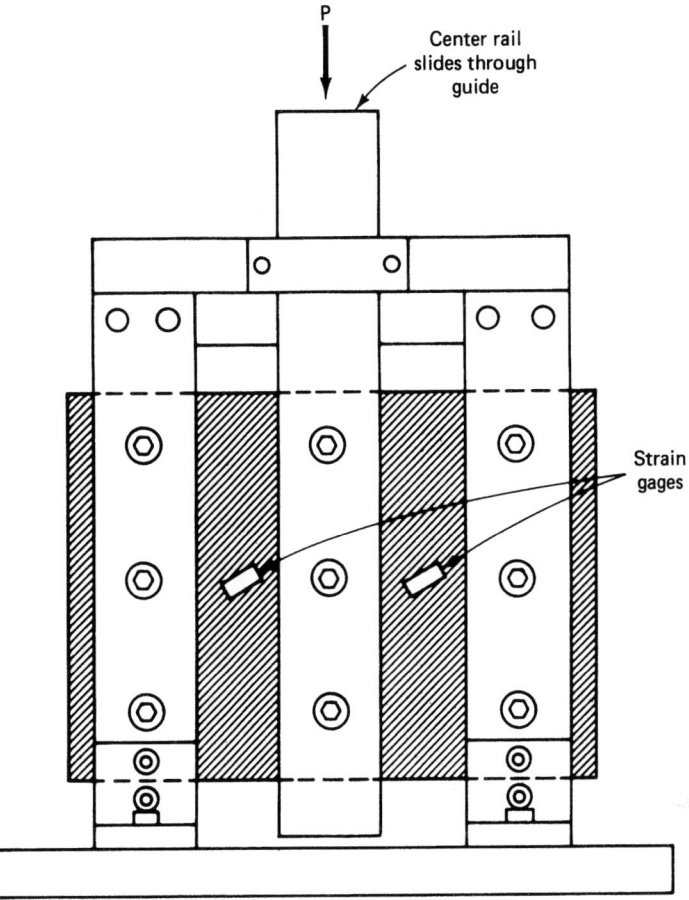

Figure 19-24 Three-rail shear fixture.

The off-axis laminate can also be tested in compression by using it as a skin in a honeycomb-core sandwich beam under pure bending. Within the limitations stated before, the specimen is simple to use and reliable.

Cross-Beam Sandwich Specimen

The sandwich specimen consists of two intersecting beams subjected to pure bending. In one of the many types discussed, the laminate is used as one skin of a honeycomb–core sandwich. This type of specimen allows, in principle, the application of tension–tension, tension–compression, and compression–compression loading in the central test section. One of the major limitations of this specimen is the disturbing influence of the corners on stress distributions in the test section and on fracture initiation at the points of stress concentration. The state of stress in the test section cannot be determined from the specimen geometry and applied loads without knowledge of the material properties.

Flat-Plate Specimen

The flat-plate specimen is usually a square plate subjected to tension–tension loading on its sides through fiberglass tabs. A variety of biaxial states of stress (in the tension–tension–shear space) can be achieved by rotating the principal material axes with respect to the loading directions. Nonproportional loading is possible to some degree. To ensure stress homogeneity in a reasonable test section and failure in this region, it is necessary to design the tabs and transition region very carefully.

This specimen is most suitable when the influence of biaxial stress on notches is investigated [51,52]. Tensile loading is introduced by means of four whiffle-tree grip linkages designed to ensure that four equal tensile loads are applied to each side of the specimen. A sketch of a biaxial plate specimen is shown in Fig. 19-25.

The major limitations of this type of test are that it is primarily limited to tension–tension loading, and in the case of unnotched specimens, it is very difficult to induce failure in the test section.

Thin-Walled Tubular Specimen

Of the various biaxial test specimens mentioned, the tubular specimen appears to be the most versatile and to offer the greatest potential. It offers the possibility of applying any desired biaxial state of stress with or without proportional loading. The state of generalized plane stress can be achieved by the independent application of axial loads, internal or external pressure, and torque.

Testing of composite tubular specimens has been discussed by several investigators who analyzed the various problems arising in this type of specimen [53–56]. One of the most frequent and most critical problems encountered is that of introducing and maintaining a uniform biaxial state of stress in the specimen test section and inducing failure in the test region. Some testing of tubular specimens has been done without any provisions for relieving end constraints [56,57]. To overcome or minimize end constraints and gripping problems, the concepts of tab and grip pressurization have been used [49,58]. These concepts, however, have not been implemented with full success because of the inherent difficulty of the problem.

A typical tubular specimen geometry is illustrated in Fig. 19-26. The specimen is designed to have end tabs for gripping and load introduction. These tabs are made of epoxy or glass–epoxy premachined and bonded to the specimen ends. The stiffness of these tabs in the axial and circumferential directions relative to those of the composite tube can be varied by varying the glass–epoxy layup and the tab thickness.

This specimen geometry has been analyzed extensively using finite-element methods [59]. The objective of these analyses was to minimize stress discontinuities in the transition between the test section and tabbed section of the specimen by varying the tab materials and geometry and the tab compensating pressure. The most critical loading from the point of view of stress discontinuity is pressure loading. The most critical stresses are the axial bending stresses in the transition region. The optimum transition between the tabbed and test sections was provided by tapered glass–epoxy tabs with tapered epoxy extensions. In the case of torsional loading, no shear–stress peaks are produced in the transition region. In the case of axial loading applied through grips at the end of the tab, high localized stress peaks are present at the outer surface of the tab. This localized stress can be alleviated somewhat by a slight rounding off of the grip ends, to prevent premature tab failure.

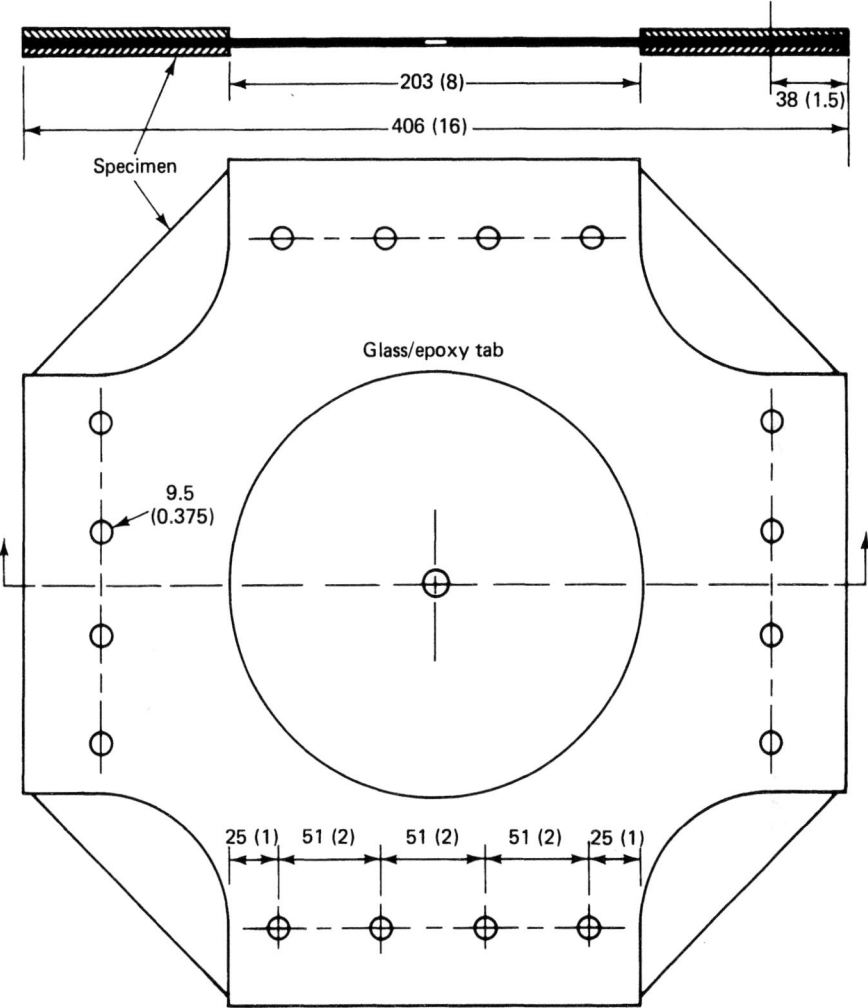

Figure 19-25 Sketch of biaxial composite specimen [dimensions in mm (in.)].

The preparation of tubular specimens must be done very carefully to ensure that the specimen is of a quality similar to that of flat laminates fabricated and cured in an autoclave. Composite tubes can be fabricated in any desired ply orientation and stacking sequence. Fabrication techniques have been developed and described in the technical literature [60–62]. The latter procedure consists of wrapping the prepreg composite tape around a cylindrical perforated mandrel and then, by means of internal pressure, expanding the prepreg tube against the wall of the cylindrical cavity mold tool. Figure 19-27 shows the consecutive layers of materials around the steel mandrel required in the fabrication process. The glass–epoxy tabs required for the specimen are fabricated as tubes in a similar manner. They are subsequently machined to size and bonded on the composite tube.

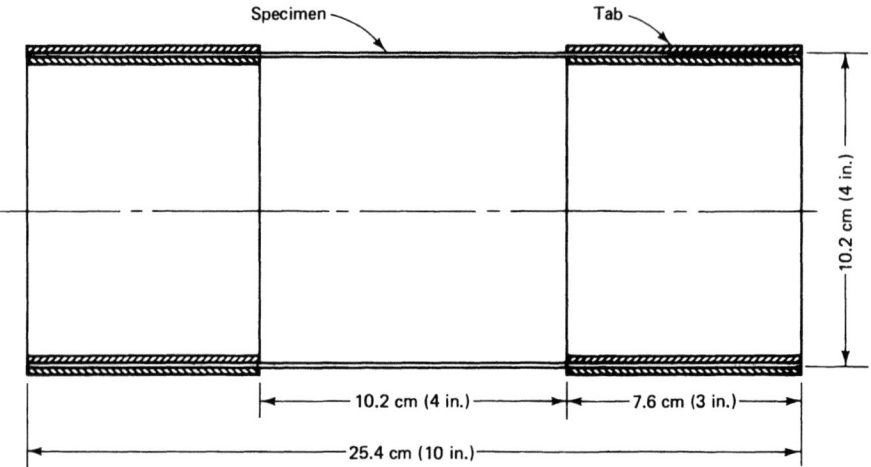

Figure 19-26 Thin-walled tubular specimen for biaxial testing of composites.

The complete system for biaxial testing of composite tubular specimens requires the introduction and control of internal and external pressures on the specimen, internal and external compensating pressures on the tabs, grip activation pressure, axial load, and torque. The axial loads and torques are introduced through segmented collet grips at the tabs by means of linear and torque actuators. All pressures applied directly to the specimen or to the actuators are applied and controlled independently of each other with an electrohydraulic system. A drawing of the test section of a recently designed biaxial testing system [63] is shown in Fig. 19-28.

The problem of biaxial testing of tubular specimens is inherently difficult and is not amenable to a complete solution. The analysis to date shows that it is possible to obtain valid tests in many cases for a variety of laminates over large portions of the failure envelope.

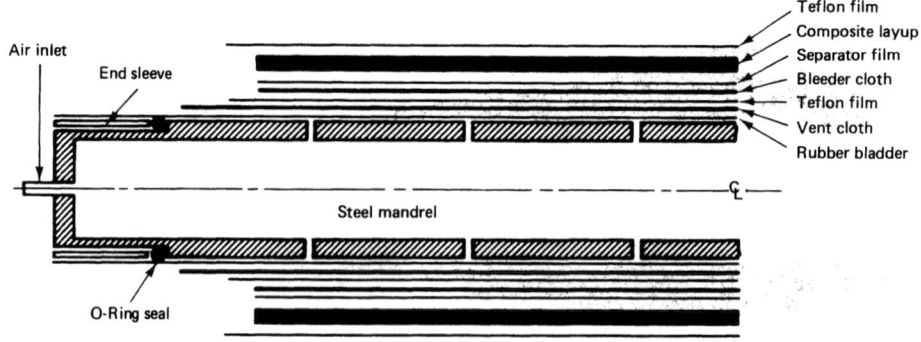

Figure 19-27 Center steel mandrel and sequence of material layers employed in composite laminate tube fabrication.

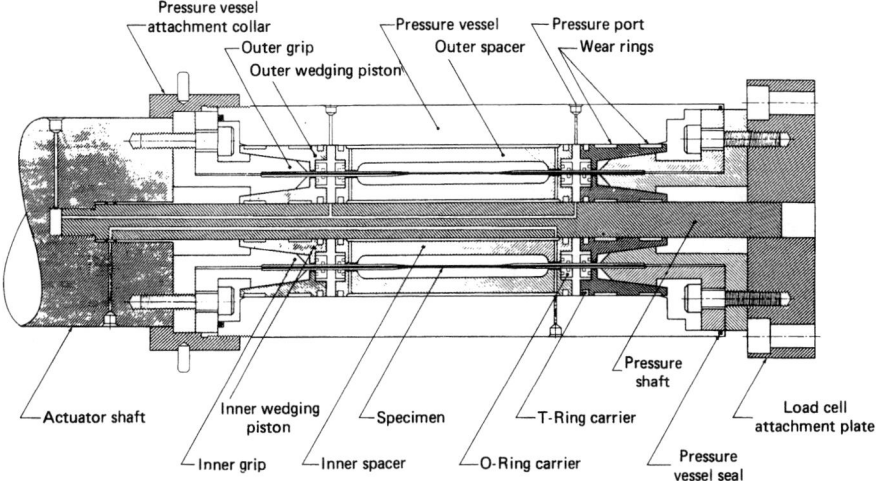

Figure 19-28 Cross section of test section assembly of biaxial testing system. (From Ref. 63.)

19-5 Effects of Stress Concentration

The behavior of composite laminates with stress concentrations is of great interest in design because of the resulting strength reduction and life reduction due to damage growth around these stress concentrations. Failure of laminated composites with stress concentrations has been dealt with using two major approaches. One approach is based on concepts of linear elastic fracture mechanics carried over from isotropic materials, while the other approach takes into consideration the actual stress distributions near the discontinuity. The first approach was used by Waddoups et al. [64], who assumed the existence of two fictitious Griffith-type cracks on the boundary of the hole, and by Cruse [65], who modeled the circular hole with a straight crack having an equivalent stress distribution near its tip. Following the second approach, Whitney and Nuismer [66] proposed simplified stress fracture criteria based on the actual stress distribution near the notch. According to the average stress criterion proposed, failure occurs when the average stress over an assumed characteristic dimension from the boundary of the notch equals the tensile strength of the unnotched material. Comparison with results from uniaxial tensile tests showed satisfactory agreement between predicted and measured strengths for a narrow range of values of the characteristic dimension.

In a similar approach proposed by Wu [48] and applied by Lo and Wu [67], lamina failure criteria are used and a characteristic dimension (volume) is postulated. Failure is said to occur if the average state of stress–strain on the boundary of this characteristic volume falls on the failure envelope of the lamina. To analyze an angle-ply laminate, this requires the study of progressive degradation of the various plies in the vicinity of the notch up to complete failure of the laminate. This approach is usually based on linear lamination theory and may not account for coupling of various failure modes of the different plies. Application of this analysis requires the definition of the characteristic volume, appropriate stress analysis for varying boundary conditions and material properties, and knowledge or determination of the failure criteria for the lamina.

Experimental methods using strain gages, photoelastic coatings, and moiré interferometry have proven very useful in studying the deformation and failure of composite laminates with circular holes and through-thickness cracks of various sizes. The effects on failure of laminate layup, staking sequence, notch size, and far-field stress biaxiality have been investigated. The approach used was to load composite plate specimens with holes or cracks under uniaxial and biaxial tension, measure deformations by means of experimental strain analysis techniques, and determine strain distributions, failure modes, and strength reduction ratios. Experimental results are compared with predictions based on linear elastic fracture mechanics, an average stress criterion, and a progressive degradation model.

Laminates with Holes

Experimental methods have been used to study the influence of material type, laminate layup, stacking sequence, notch size, and far-field stress biaxiality. The influence of the laminate layup has been studied for a variety of lamination geometries [68]. Laminates with a high percentage of 0° plies, but with sufficient 45° plies to mollify the stress concentration factor, are the strongest. The $[0/90]_{2s}$ construction with 50% 0° plies is not strong because of the high stress concentration factor. The $[\pm 45]_{2s}$ construction is the weakest because of the absence of 0° plies, although the stress concentration factor is the lowest. Stacking sequence was also found to have a noticeable influence on strength and failure patterns. Stacking sequences resulting in tensile interlaminar normal stresses near the boundary of the hole reduce the strength of the laminate. In some cases stacking sequence variations can cause drastic differences in strength related to changes in failure modes from catastrophic to noncatastrophic. The influence of hole diameter for uniaxially loaded plates has been described by using the average stress criterion [66]. The stress distribution along the horizontal (transverse) axis through the hole in an anisotropic plate can be approximated by the expression [69]

$$\frac{\sigma_y(x, 0)}{\sigma_0} \simeq 1 + \frac{1}{2}\rho^{-2} + \frac{3}{2}\rho^{-4} - \frac{k_\sigma - 3}{2}(5\rho^{-6} - 7\rho^{-8}) \qquad (19\text{-}60)$$

where $\sigma_y(x, 0)$ = vertical (axial) stress along horizontal axis
σ_0 = applied far-field axial stress
$\rho = x/a$
a = hole radius
k_σ = anisotropic stress concentration factor

According to the average stress criterion, failure occurs when the axial stress averaged over a characteristic distance a_0 from the hole boundary equals the strength F_0 of the unnotched laminate. The strength reduction ratio, the ratio of notched to unnotched strength, is expressed as

$$k_s = \frac{F_{yT}}{F_0} = \frac{2}{(1 + \xi)[2 + \xi^2 + (k_\sigma - 3)\xi^6]} \qquad (19\text{-}61)$$

where $\xi = a/(a + a_0)$.

Experimental results obtained for uniaxially loaded $[0_2/\pm 45]_s$ graphite–epoxy laminates with holes of various diameters are in good agreement with predictions based on the average stress criterion for a characteristic length $a_0 = 5$ mm, as shown in Fig. 19-29. One interesting result observed in this and similar cases is the existence of a threshold hole diameter, approximately 1.5 mm in this case, below which the laminate becomes notch insensitive.

The behavior of biaxially loaded composite plates with holes has been studied experimentally [51,70]. The stress distribution around a hole in a $[0/\pm 45/90]_s$ graphite–epoxy plate subjected to equal biaxial tension is illustrated by the isochromatic fringe pattern of the photoelastic coating (Fig. 19-30). Initially, the circumferential strain is uniform around the boundary of the hole. With increasing load, regions of high strain concentration with nonlinear response develop at eight

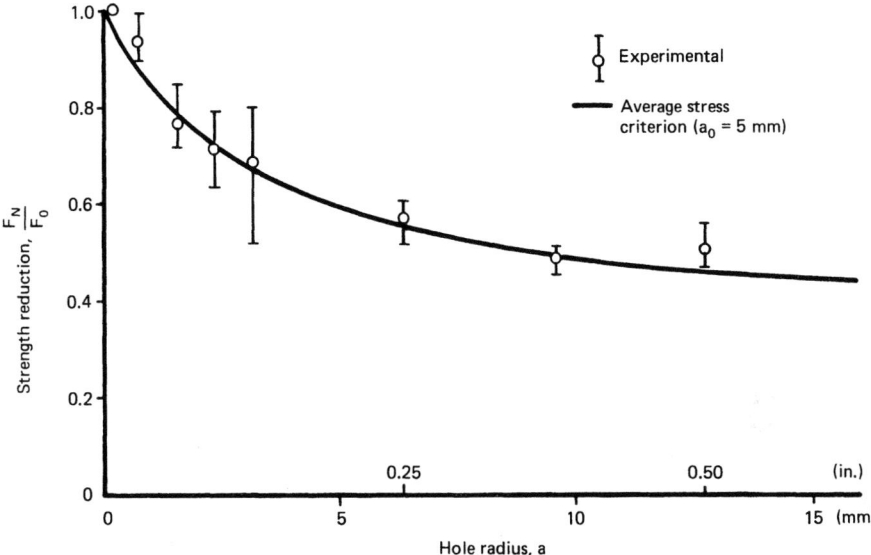

Figure 19-29 Strength reduction as a function of hole radius for $[0_2/\pm 45]_s$ graphite–epoxy plates with circular holes under uniaxial tensile loading.

characteristic locations 22.5° off the fiber axes. Failure is initiated at one or more points of high strain concentration (Fig. 19-31). Maximum strains at failure on the hole boundary reach values up to twice the ultimate strain of the unnotched laminate. The strength reduction ratio—the ratio of notched biaxial strength to unnotched uniaxial strength—is plotted versus hole radius in Fig. 19-32. These ratios are higher than corresponding values for uniaxial loading by approximately 30%. The variation of strength reduction ratio with hole diameter was satifactorily described by using an average biaxial stress criterion. Radial and circumferential stresses around the hole were averaged over an annulus 3 mm wide and compared with the biaxial strength envelope for the quasi-isotropic unnotched laminate. Results were also in good agreement with predictions based on a tensor polynomial failure criterion for the individual lamina and a progressive degradation model [67]. The strength reduction ratios for uniaxial and equal biaxial tensile loading represent lower and upper bounds for any biaxial tensile loading of this laminate. The strength reduction for any other tensile biaxiality ratio would fall between these two bounds and could be estimated approximately by interpolation.

Similar biaxial tests with $[0_2/\pm 45]_s$ laminates show more complex behavior not possible to describe with simplified failure criteria [70]. For the larger holes, characteristic points of fringe concentration were observed in the photoelastic coating at points 22.5° off the horizontal axis. The patterns change with hole diameter. For example, for a specimen with a 1.27-cm-diameter hole, in addition to the previously observed birefringence concentration at the 67° locations, there is pronounced concentration of equal or greater intensity at the 0° location (Fig. 19-33). The effect of hole diameter on the deformation pattern and failure modes is further illustrated by the isochromatic fringe patterns in the photoelastic coating around a 0.64-cm-diameter hole (Fig. 19-34). The fringe patterns at low loads indicated a nearly uniform strain distribution on the boundary of the hole. Surface cracking in the vertical direction occurs first. Birefringence (strain) concen-

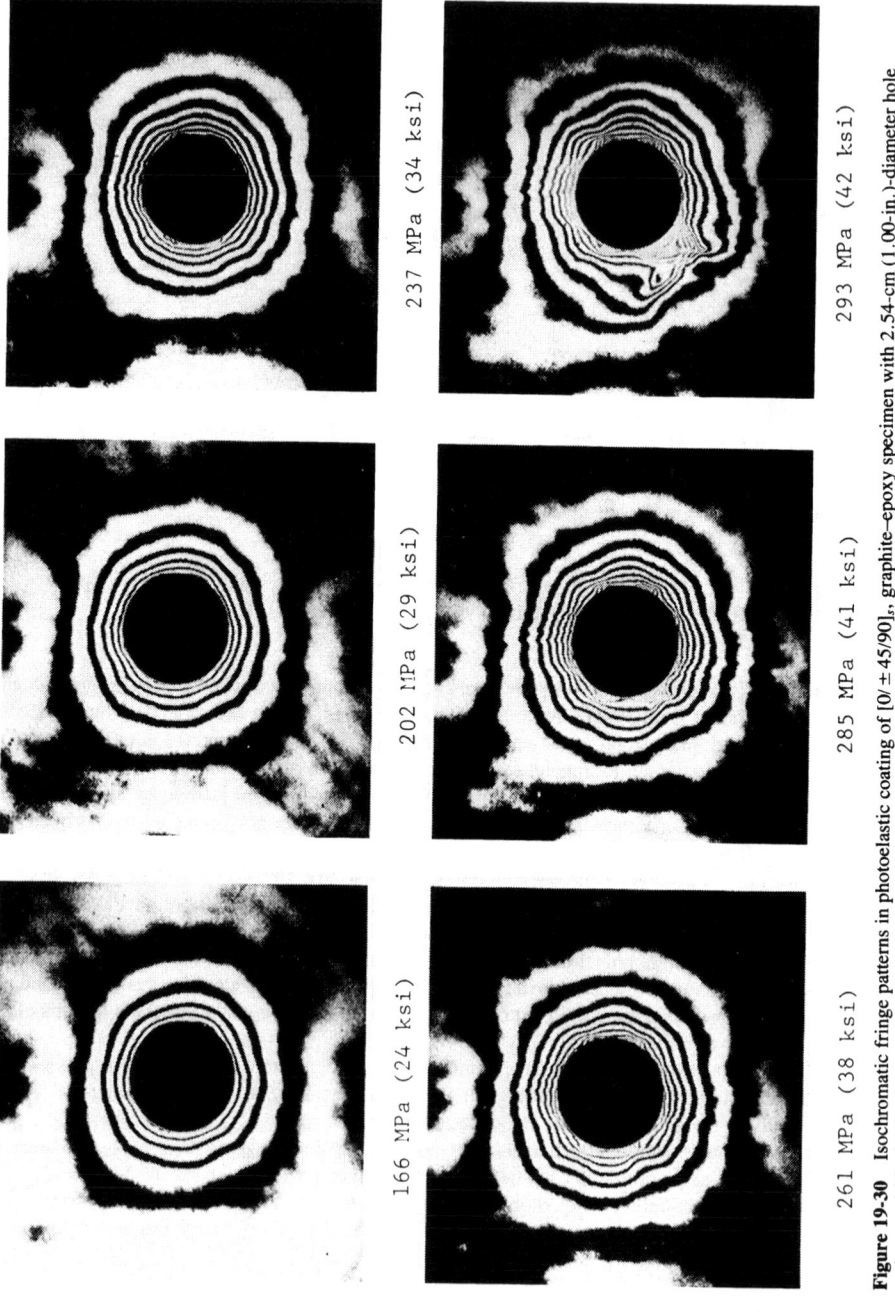

Figure 19-30 Isochromatic fringe patterns in photoelastic coating of [0/±45/90]$_s$ graphite–epoxy specimen with 2.54-cm (1.00-in.)-diameter hole under equal biaxial loading.

COMPOSITE MATERIALS

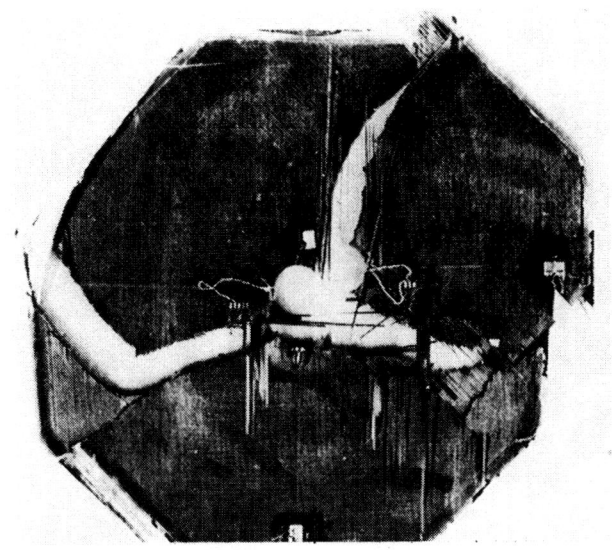

Figure 19-31 Failure pattern in $[0/\pm 45/90]_s$ graphite–epoxy specimen with 1.91-cm (0.75-in.)-diameter hole under equal biaxial loading.

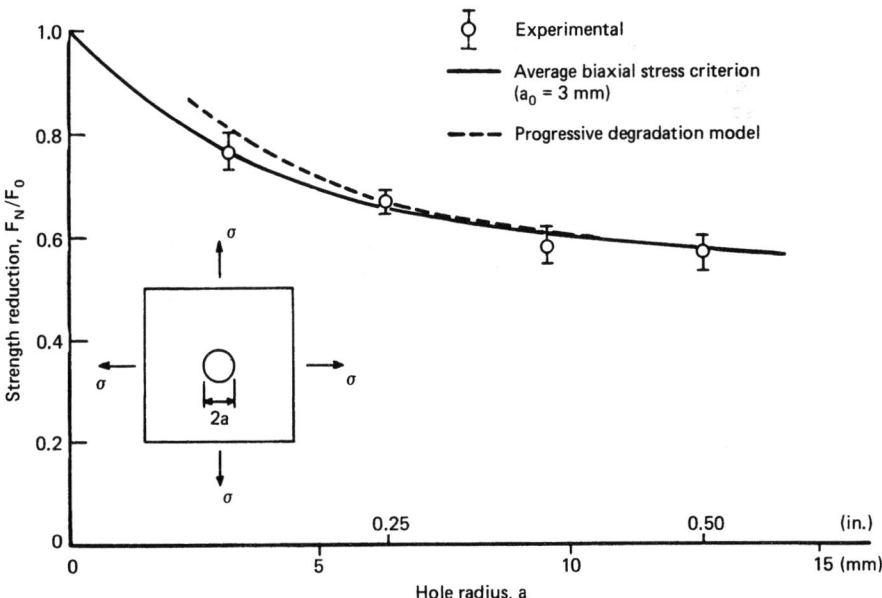

Figure 19-32 Strength reduction as a function of hole radius for $[0/\pm 45/90]_s$ graphite–epoxy plates with circular holes under 1:1 biaxial tensile loading.

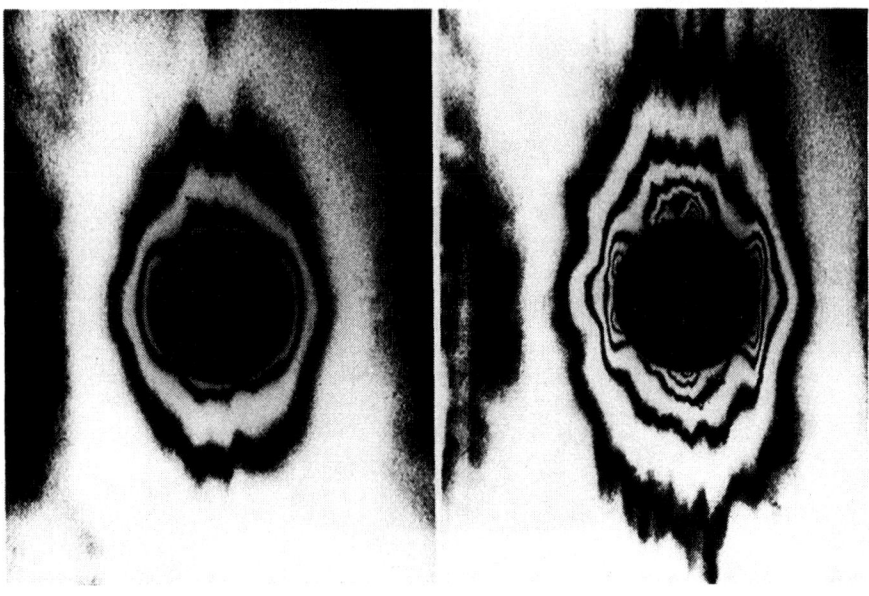

Figure 19-33 Isochromatic fringe patterns in photoelastic coating around 1.27-cm (0.50-in.)-diameter hole in $[0_2/\pm 45]_s$ graphite–epoxy specimen under biaxial loading $\sigma_{yy} = 2\sigma_{xx}$.

tration develops later at points 60° off the vertical axis. The vertical surface crack pattern gradually extends throughout the specimen, whereas the birefringence levels at the 60° and 90° locations remain unchanged. The stress level and location of failure initiation can be predicted approximately by comparing the circumferential stress and strength distributions around the hole boundary. Figure 19-35 shows the strength distribution obtained from tests of the $[0_2/\pm 45]_s$ unnotched laminate and the elastic stress distribution at the 152-MPa level of applied vertical stress. As can be seen, this stress level represents the limit beyond which nonlinear response and localized failure begin to take place at the 0° and 67.5° locations, respectively. As the applied stresses are increased, localized failure near the 67.5° location causes stress redistribution followed by local failure near the 0° location but away from the hole boundary. Two basic failure patterns were observed: (1) horizontal cracking initiating at points off the horizontal axis and accompanied by extensive delamination of the subsurface ±45° plies and (2) vertical cracking along vertical tangents to the hole, accompanied by delamination of the outer 0° plies. The first pattern is typical for the larger holes, whereas the second pattern becomes dominant as the hole diameter decreases. The complex interaction between the two sources and modes of localized failure, which depends on the hole diameter, makes a theoretical prediction of this type of failure very difficult.

Most of the work to date deals with static loading of laminates with holes. Some testing has been done on uniaxial cyclic loading. It is known, for example, that the residual strength of notched plates increases with cyclic loading. It is not known, however, how this behavior varies with the type of laminate and cyclic stress biaxiality, especially in the case of failure-mode interaction.

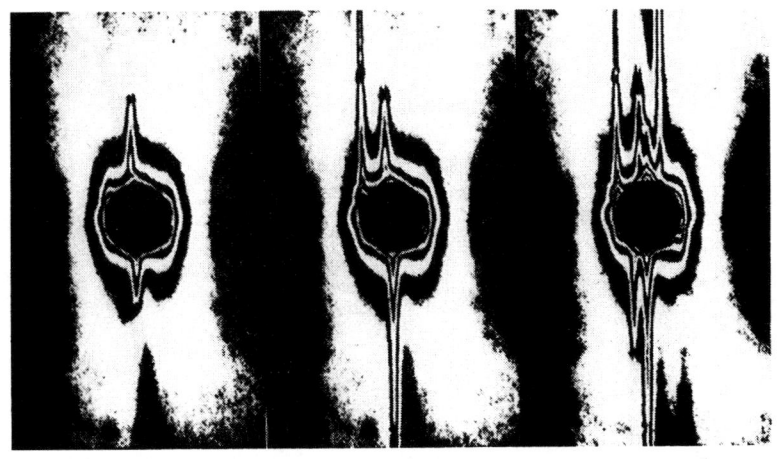

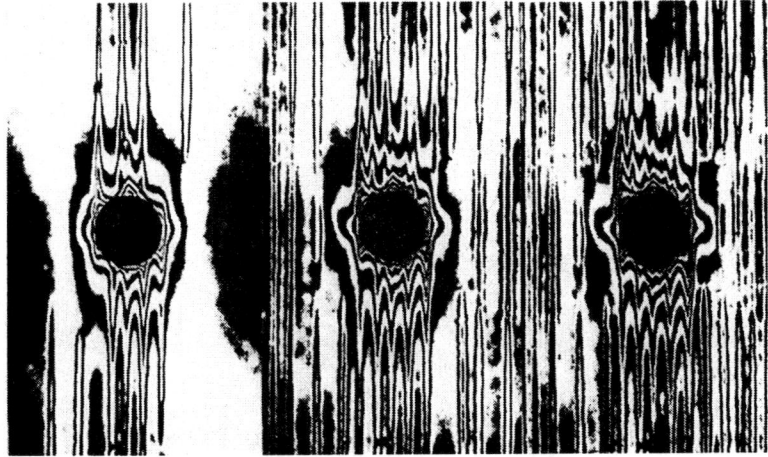

Figure 19-34 Isochromatic fringe patterns in photoelastic coating around 0.64-cm (0.25-in.)-diameter hole in $[0_2/\pm 45]_s$ graphite–epoxy specimen under biaxial loading $\sigma_y = 2\sigma_x$.

Laminates with Cracks

The deformation and failure of $[0/\pm 45/90]_s$ graphite–epoxy plates with cracks of different lengths have been investigated by experimental techniques [71,72]. In general, failure at the tip of the crack takes the form of a damage zone consisting of ply subcracking along fiber directions, local delaminations, and fiber breakage in adjacent plies along the initial subcracks. The strain distribution around the crack tip and the phenomenon of damage zone formation and growth for a uniaxially loaded plate are vividly illustrated by the isochromatic fringe patterns in the photoelastic coating (Fig. 19-36). A noticeable characteristic is the apparent extension of the damage zone at a 45° angle with the crack direction. The size of this zone increases with applied stress up to some

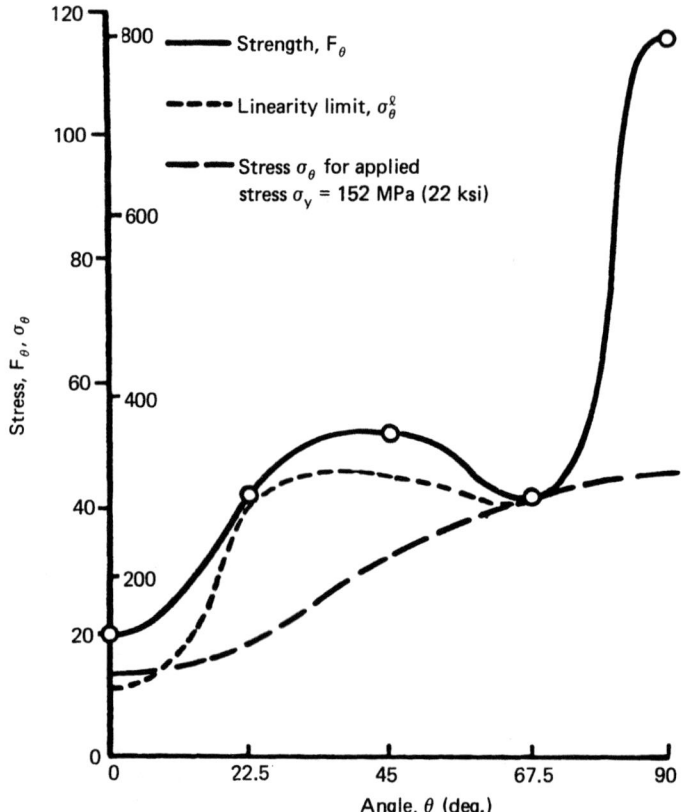

Figure 19-35 Distribution around boundary of hole of circumferential strength and stress for $[0_2/\pm 45]_s$ graphite–epoxy specimen under biaxial loading, $\sigma_y = 2\sigma_x$.

critical value at which point the specimen fails catastrophically. Far-field strains and the crack-opening displacement were obtained from moiré fringe patterns around the crack. The crack-opening displacement becomes nonlinear and increases at a faster rate at an applied stress corresponding to the rapid strain increase observed with strain gages.

The average stress criterion, defined in a similar manner as in the case of holes, predicts the following strength reduction ratio:

$$k_s = \frac{F_{yT}}{F_0} = \sqrt{\frac{a_0}{a_0 + 2a}} \qquad (19\text{-}62)$$

where a = half crack length
a_0 = characteristic distance from crack tip

Experimental results for the strength reduction ratio agree well with predictions based on the average stress criterion using a characteristic dimension $a_0 = 5$ mm (Fig. 19-37). Comparison of results with those from similar specimens with circular holes shows that strength reduction in this

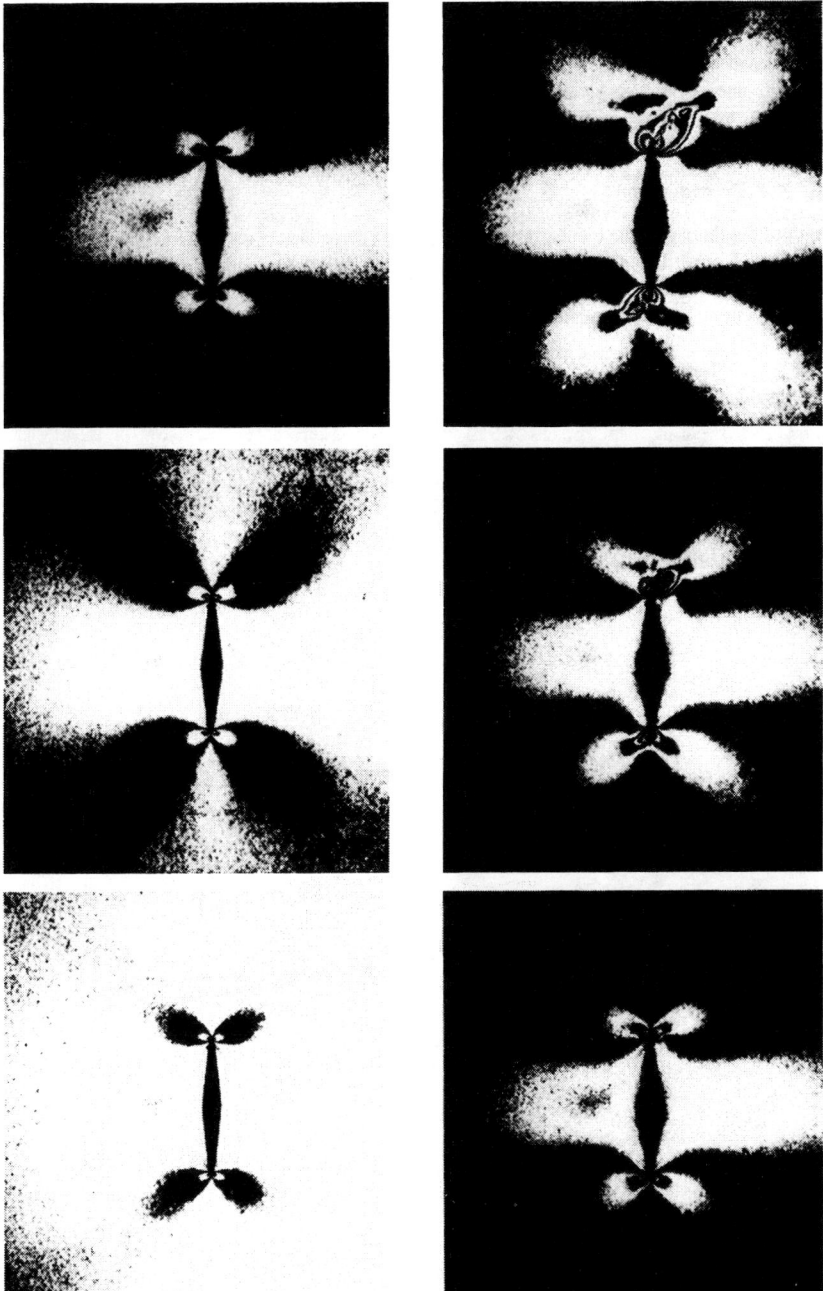

Figure 19-36 Isochromatic fringe patterns in photoelastic coating around 1.27-cm (0.50-in.) crack of $[0/\pm 45/90]_s$ graphite–epoxy specimen at various levels of applied stress.

case is nearly independent of notch geometry (i.e., specimens with holes and cracks of the same size have nearly the same strength). Experimental results for specimens with holes along with predictions based on an average stress criterion for holes are also shown in Fig. 19-37.

The results above can also be analyzed by linear elastic fracture mechanics. If the half crack length a is adjusted to include the length a_0 of the damage zone near the crack tip, the critical stress intensity factor is given by

$$K_Q = F_{yT}\sqrt{\pi(a + a_0)} \tag{19-63}$$

The length of the damage zone can be taken equal to the characteristic dimension a_0 in the average stress criterion (5 mm). It can also be approximated by the subcrack length near failure as detected by the photoelastic coating. It is seen in Fig. 19-38 that the critical stress-intensity factor, as modified to account for the critical damage zone at the crack tip, is nearly constant with crack length.

Similar studies have been conducted with quasi-isotropic plates under uniaxial loading having cracks of various lengths and at various angles with the loading direction [73] (Fig. 19-39). This type of loading produces a biaxial state of stress around the crack. The crack-opening displacement (COD) and crack-shearing displacement (CSD) were obtained directly from moiré fringe patterns (Fig. 19-40). For large angles of crack inclination, the COD is much larger than the CSD. The relationship is reversed for smaller angles (e.g., for $\theta = 15°$, the CSD is four times the COD). The deformation and damage around the crack tips were clearly illustrated in most cases by the fringe patterns of photoelastic coatings (Fig. 19-41). The maximum strain at the crack tip, as computed from the birefringence, is in excess of 0.03, or three times the ultimate strain of the

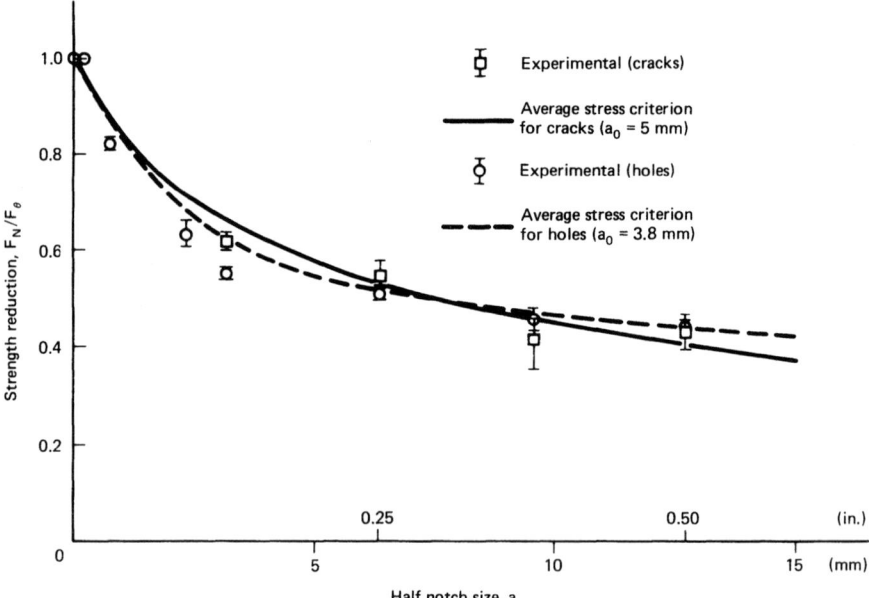

Figure 19-37 Strength reduction as a function of notch size for $[0/\pm 45/90]_s$ graphite–epoxy plates with circular holes and horizontal cracks under uniaxial tensile loading.

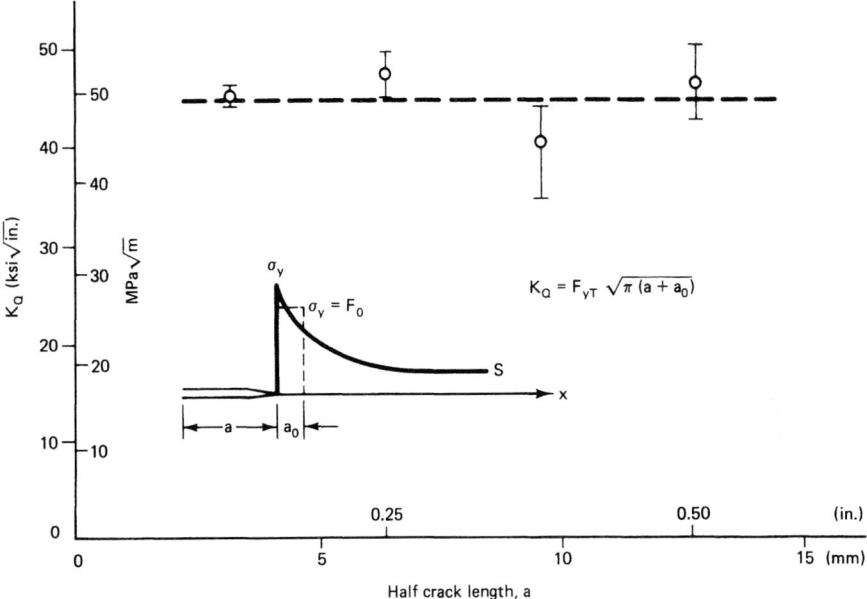

Figure 19-38 Critical stress-intensity factor as a function of crack length for $[0/\pm 45/90]_s$ graphite–epoxy plates with horizontal cracks under uniaxial tensile loading.

unnotched laminate. In most cases initial damage propagation occurs along subsurface fiber directions.

Results were analyzed by means of an average stress criterion, modified to include the projected crack length on the horizontal axis normal to the loading direction. The predicted strength reduction ratio is

$$k_s = \frac{F_{yT}}{F_0} = \sqrt{\frac{a_0}{a_0 + 2a \sin \theta}} \tag{19-64}$$

Experimental results are in agreement with this prediction when $a_0 = 4$ mm (0.16 in.), which is the maximum size of the damage zone just prior to failure as obtained from the isochromatic fringe patterns (Fig. 19-42). Results were also analyzed by means of linear elastic fracture mechanics, although the phenomena observed are clearly more complex. An empirical approach was used whereby a critical stress intensity factor was computed as follows:

$$K_Q = F_{yT}\sqrt{\pi(a \sin \theta + a_0)} \tag{19-65}$$

where a_0 represents a characteristic dimension. It was found that using a value for the parameter a_0 equal to 4 mm (0.16 in.), as before, the critical stress-intensity factor was nearly constant with angle of crack inclination as well as initial crack length (Fig. 19-43).

Similar quasi-isotropic graphite–epoxy plates with cracks were loaded under direct biaxial tension [52]. The specimens were square plates machined with the crack inclined at 30° and 60° with the sides of the plate (Fig. 19-44). The specimens were loaded under a 2:1 biaxial tensile loading as shown, with the stress at 30° to the crack twice as large as the stress at 60° to the

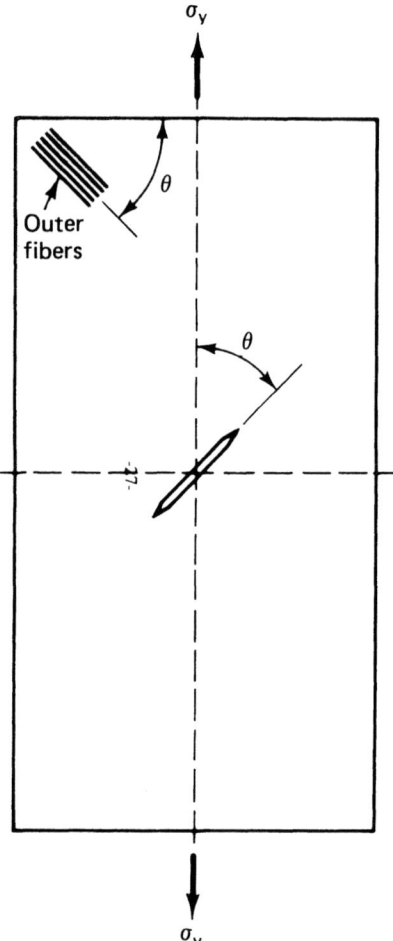

Figure 19-39 Uniaxially loaded [0/±45/90]$_s$ graphite–epoxy specimen with inclined crack.

crack. This biaxial loading produces a shear stress in the crack direction 0.35 times the stress normal to the crack and a normal stress in the crack direction 1.40 times the stress normal to the crack. Crack lengths between 0.64 and 2.54 cm were investigated. Isochromatic fringe patterns in the coating around a 1.27-cm-long crack are shown in Fig. 19-45. In addition to the primary propagation normal to the initial crack, there is crack extension along the original crack direction, probably along the fibers of the central plies of the laminate. There is also evidence of tertiary crack propagation normal to the initial crack direction but initiating at the tip of the subsurface extended crack. Experimental results agree satisfactorily with predictions obtained by using a maximum stress failure criterion for the individual lamina and a progressive degradation model

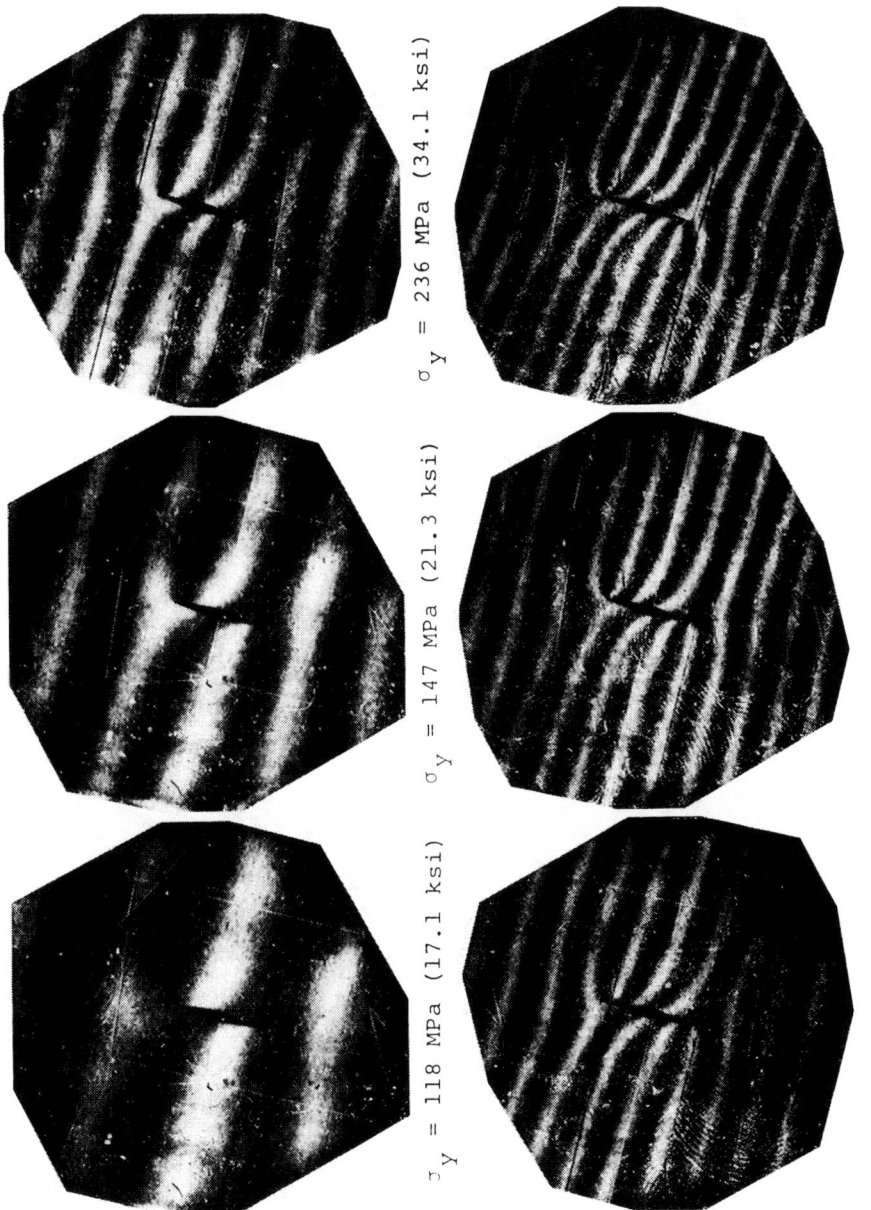

Figure 19-40 Moiré fringe patterns corresponding to displacements parallel to the crack in $[0/\pm 45/90]_s$ graphite–epoxy plate with crack at 15° with loading direction.

Figure 19-41 Isochromatic fringe patterns in photoelastic coating in $[0/\pm 45/90]_s$ graphite–epoxy plate with crack at 60° with loading direction.

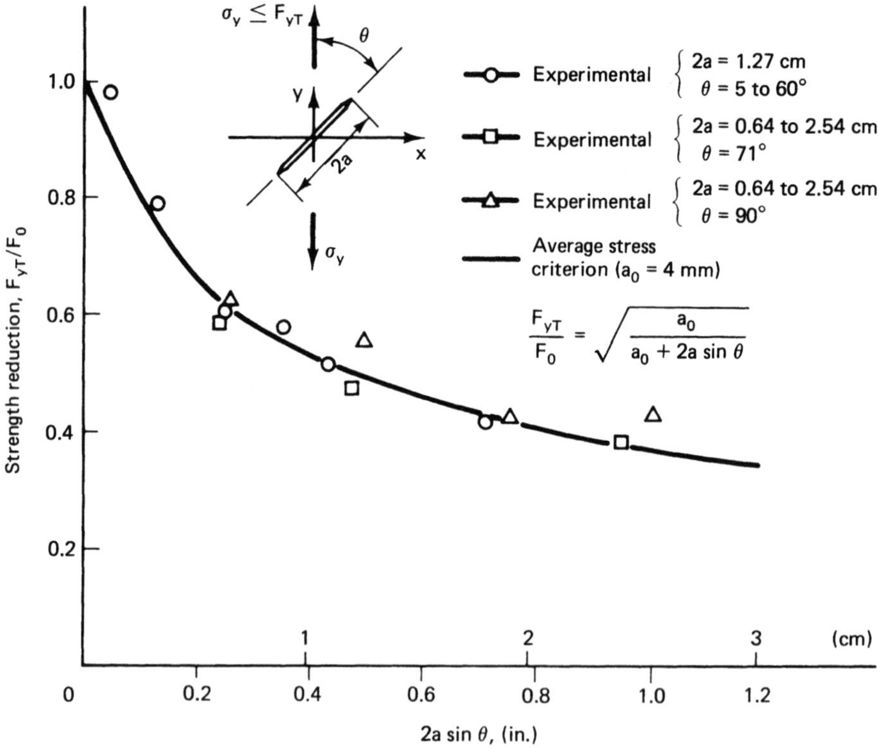

Figure 19-42 Strength reduction as a function of projected crack length for $[0/\pm 45/90]_s$ graphite–epoxy plates with cracks under uniaxial tensile loading.

$$\frac{F_{yT}}{F_0} = \sqrt{\frac{a_0}{a_0 + 2a \sin \theta}}$$

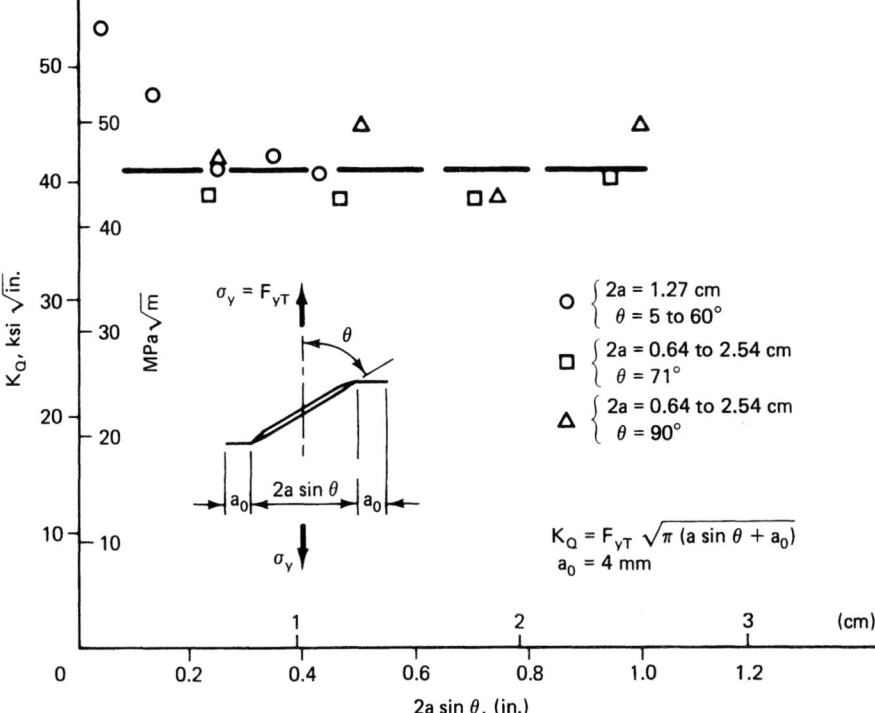

Figure 19-43 Critical stress intensity factor as a function of projected crack length for $[0/\pm 45/90]_s$ graphite-epoxy plates with cracks under uniaxial tensile loading.

(Fig. 19-46). The approach used in developing this model makes use only of the properties of a single lamina: strength, stiffness, and a characteristic dimension r_c in front of the crack tip [67]. Local failure of a lamina within the laminate is said to occur if the state of stress (strain) at a distance r_c from the crack tip at a critical angle θ_c falls on the failure envelope of the lamina.

19-6 Nondestructive Evaluation

The objectives of nondestructive evaluation of composite materials are to develop effective methods of flaw detection and characterization after fabrication and in service; to monitor flaw growth under loading and environmental variations; to assess flaw criticality (effects of defects); to establish accept/reject criteria; and to prescribe repair procedures where possible.

Flaws can be introduced in composite laminates during processing and fabrication. These include contaminants, porosity and voids, fiber–matrix debonding, unpolymerized resin, nonuniform fiber distribution, fiber misalignment, inclusions, fiber fractures, matrix crazing and fracture, ply gaps, and delaminations. Sources of stress concentration and damage growth, such as holes or cutouts

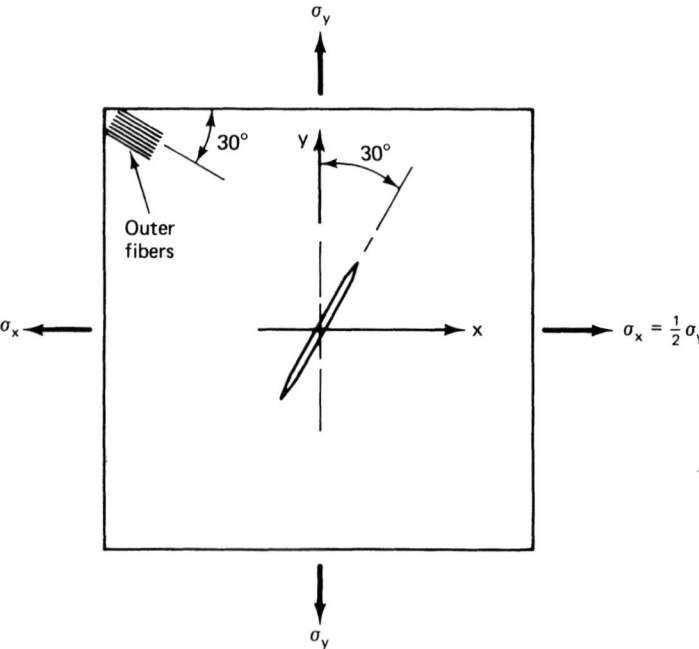

Figure 19-44 Biaxial loading of $[0/\pm45/90]_s$ graphite–epoxy specimens with cracks.

of various sizes and shapes, are introduced for design functions. Defects are also induced or enlarged in service with age, loading, and environmental conditions. These include matrix aging and degradation, moisture absorption, ultraviolet radiation, damage zones around initial stress concentrations, surface cracks and scratches, impact damage, and delamination.

A variety of nondestructive evaluation (NDE) techniques are used for evaluating the integrity of composite materials. They include radiographic (x-ray and neutron), optical (moiré, birefringent coatings, holography, and interferometry), thermographic, acoustic (acoustic wave, acoustic emission), embedded sensor, ultrasonic, and electromagnetic techniques. The various methods are best suited for detecting flaws of different types. Reviews of the state of the art of NDE of composites have been given by Daniel and Liber [74], Matzkanin et al. [75], and Henneke [76]. Specialized symposia have dealt with the subject of NDE of composites [77,78].

This section describes some of the NDE methods for composite materials, including their advantages and limitations.

Optical Methods

The optical category includes visual inspection, moiré, photoelastic coatings, holographic, and speckle interferometric methods.

Visual inspection is the most obvious method. The visibility of surface flaws, such as scratches and fiber breaks, can be enhanced with appropriate lighting and dye penetrants. Fluorescent penetrants combined with microscopic examination make a useful inspection tool. The process

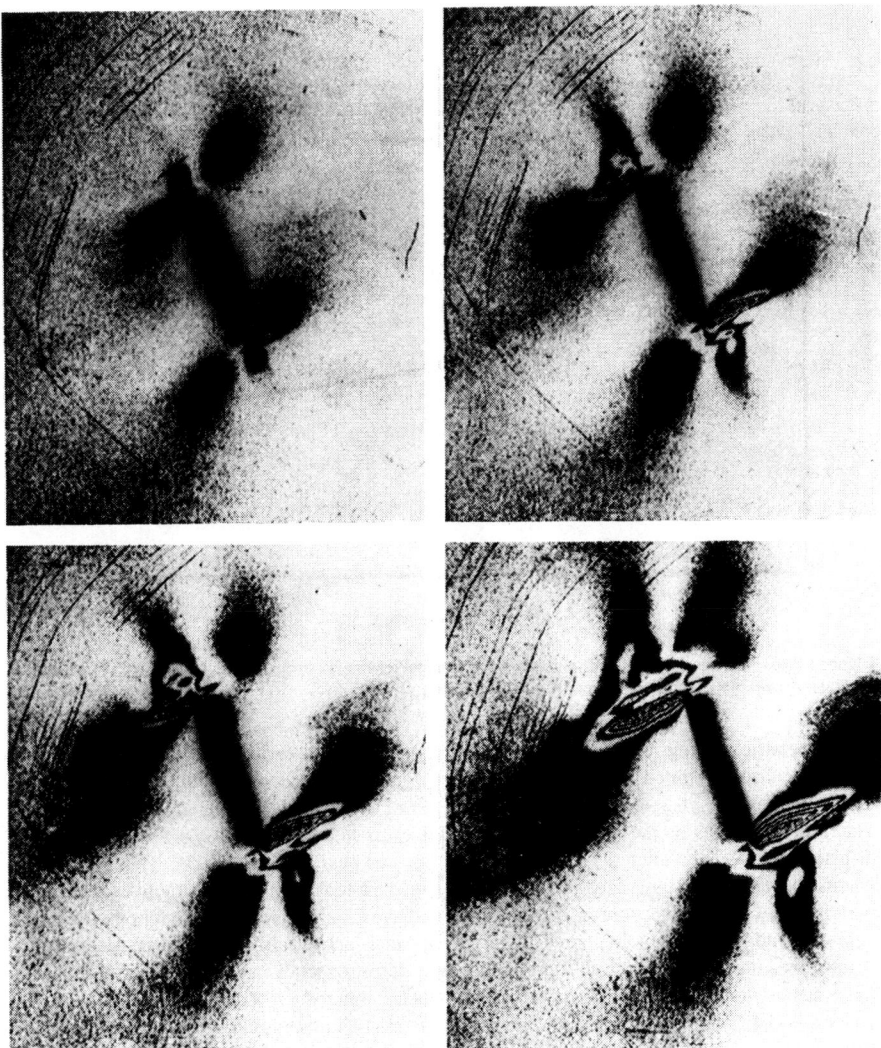

Figure 19-45 Isochromatic fringe patterns in photoelastic coating around 1.27-cm (0.5-in.) crack in $[0/\pm 45/90]_s$ graphite–epoxy specimen under biaxial loading $\sigma_{yy} = 2\sigma_{xx}$ at 30° with crack direction.

can be further improved with the use of television cameras for remote and systematic inspection. One of the most frequently used methods is the replication technique, where faithful replicas of the composite surface or edge are made and examined microscopically [79].

Moiré methods, discussed earlier, have been used successfully in measuring localized deformations and flaw growth in composites. The most recent developments make use of moiré interferometry for damage characterization of composites [40]. Moiré fringe patterns depicting damage in composite materials are shown in Fig. 19-47.

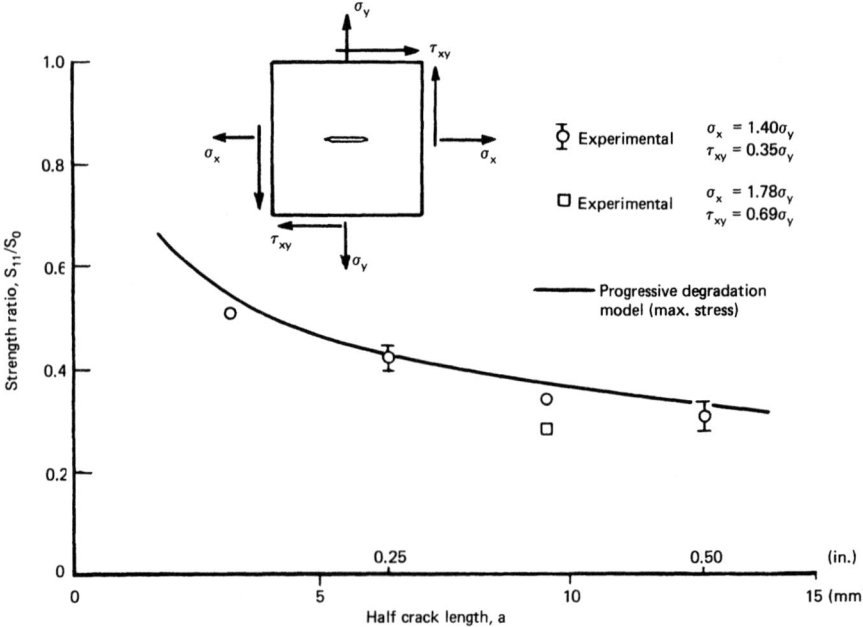

Figure 19-46 Comparison of experimental and theoretical results for strength reduction ratio for $[0/\pm 45/90]_s$ graphite–epoxy plates with cracks under biaxial loading.

Photoelastic coating techniques, as discussed and illustrated earlier, are very sensitive in detecting localized strain concentrations and damage growth in composites.

As mentioned earlier, one important application of holography is NDE of material integrity. The technique relies on the fact that if the state of stress in a component is changed, the surface displacements will be altered locally around surface or near surface defects. This alteration is manifested as a local anomaly in the overall fringe pattern. This nondestructive detection is performed by double-exposure or real-time holography. Cracking and delamination in composite laminates and bond defects in composite sandwich panels or adhesive joints can be detected. The type of loading applied to reveal the presence of a defect depends on the type of defect sought, the material properties, and the component geometry. Figure 19-48 shows an example of the application of holography [80] to the detection of fatigue-induced damage in a graphite–epoxy laminate. The fringe patterns were induced thermally by using a radiant heat source. Holography is an effective NDE technique capable of detecting both matrix cracks and fiber breaks in the surface plies. It can also give some information relative to the through-thickness distribution of delaminations. It does not, however, detect subsurface matrix cracking.

Another interferometric technique applicable to NDE of composites is speckle interferometry. The feasibility of the method as a nondustructive evaluation tool has been demonstrated, but its full potential has not been fully explored [81].

Thermographic Methods

Thermographic methods are based on the principle of thermal imaging of heat patterns. Two types of testing are used for generating a thermal image, active and passive testing. In active testing

Figure 19-47 Moiré fringe patterns in fatigue loaded [0/ ± 45/90]$_s$ graphite–epoxy specimen (U-displacement field; f = 2400 lines/mm = 60,960 lines/in.). (Courtesy of D. Post.)

heat is generated internally by cyclic loading of the specimen. The heat generated is a function of material properties, stress level, type of loading, and frequency. Stress concentrations around defects produce heat concentrations. In passive testing the specimen, heated by an external source and the heat conduction, and influenced by material and geometric discontinuities, produces thermal patterns to be recorded. There are two basic methods of thermal imaging, the chemical method employing liquid crystals and the electronic method employing an infrared camera.

The high sensitivity of liquid crystals to heat and their property of scattering visible radiation make them suitable for detecting flaws in composites. Local delaminations or inclusions resulting in nonuniform heat conduction during passive testing, and fiber fractures producing localized heat during active testing are easily manifested with liquid crystals [82,83]. These cholesteric liquid crystals, available as premixed liquids, pastes, or encapsulated sheets, selectively scatter certain wavelengths of light with increasing temperature in the following sequence: red, yellow, green, blue. The threshold temperature level for the first color to appear ranges from −7 to 70°C for different types of crystals. The sensitivity of the crystals varies between 3 and 15°C. Liquid crystals can be used in both active and passive testing to locate flaws, such as inclusions and internal fiber breaks, and to monitor damage development. They are inexpensive and simple to use, but they have limited temperature ranges and are susceptible to deterioration.

The electronic method of thermal imaging uses an infrared detector that scans the field, conditions the signals, and displays the output on a CRT and/or a video monitor. The technique of studying material integrity by means of heat patterns produced by vibratory excitation is called vibrothermography [84]. The vibrations are of low amplitude and high frequency (up to 18 kHz). Under this vibratory loading, heat patterns develop quickly and reach a steady state within seconds. The heat patterns produced are very sensitive to the excitation frequency. Typical thermographs

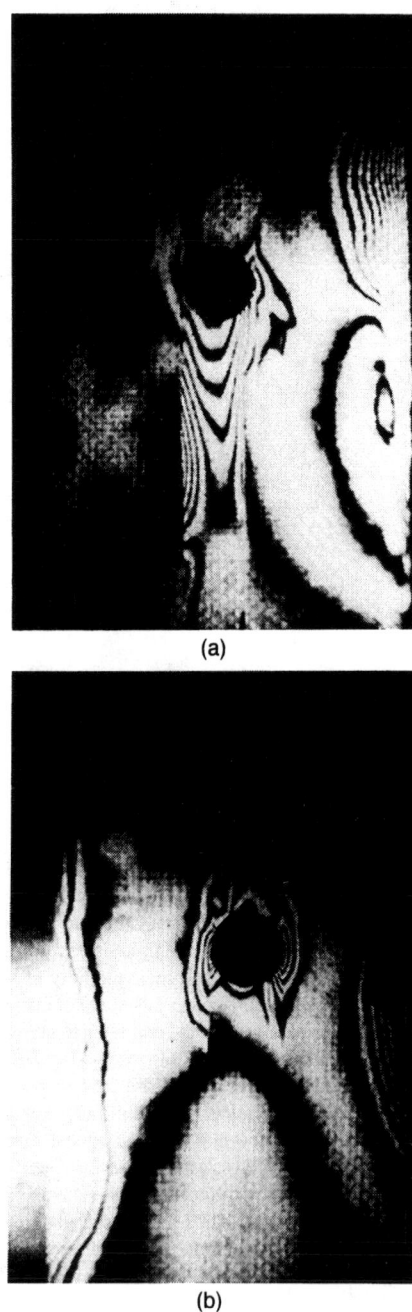

Figure 19-48 Thermally induced holographic fringe patterns in fatigue loaded $[(0/\pm 45/90)_2]_s$ graphite–epoxy specimen with circular hole: (a) front surface; (b) back surface (as viewed from the front through the specimen). (From Ref. 80.)

produced after the fatigue loading of a $[\pm 45/0]_s$ graphite–epoxy specimen with an initial internal flaw are shown in Fig. 19-49 [84]. The thermographs are compared with corresponding C-scans. The vibrothermographic technique is very sensitive in revealing the presence of delaminations, relatively large cracks, and matrix cracking. However, it does not provide as clear a definition of the flaw geometry as ultrasonic C-scanning (Fig. 19-49).

Radiographic Methods

Low-voltage radiography with soft x-rays has been found particularly suitable for NDE of composites [85]. The most effective means of detecting damage consisting of matrix cracks and delaminations is by enhancing the method with x-ray opaque penetrants. A commonly used penetrant is a solution of zinc iodide, Kodak Photoflow 200, isopropyl alcohol, and distilled water. The penetrant is applied to the damaged region and allowed to penetrate through the edges or surface cracks of the specimen. The specimen is then placed over an x-ray film and exposed to an x-ray source. Typical radiographs of a $[0/90_2]_s$ graphite–epoxy laminate loaded cyclically are shown in Fig. 19-50. Three types of damage mechanism can be identified: transverse matrix cracking, longitudinal matrix cracking, and localized delaminations. Figure 19-51 is a radiograph depicting damage in a quasi-isotropic graphite–epoxy plate subjected to the impact of a falling weight.

X-radiography is an effective and sensitive method of delineating damage in composites. However, it requires the use of penetrants that can be applied only if the defects sought communicate with the outer surfaces. Furthermore, ordinary radiographs give only a plan view of the damage without information on the through-thickness location of the defects. The latter limitation can be overcome in some cases of thin laminates by the use of stereoradiography [86]. In this approach, two radiographs are taken of the specimen at slightly different viewing angles. The pair of radiographs is examined by a stereo-optic viewer, revealing the three-dimensional distribution of the damage.

Ultrasonic Methods

The most widely used method of flaw detection and characterization is the ultrasonic one [74–76,87]. It is based on the attenuation of high-frequency sound passing through the specimen. The attenuation results from three sources: viscoelastic effects in the resin matrix, geometric dispersion caused by material heterogeneity, and geometric attenuation caused by internal defects such as delaminations and cracks. The effects of the latter are maximized by proper selection of the soundwave frequency.

The ultrasonic pulses are transmitted and received by piezoelectric transducers, usually in the range of 1 to 15 MHz. The transducers and the specimen are immersed in a tank of water to provide a uniform coupling medium for transmission of the ultrasonic waves. In some applications to large components in the aerospace industry, the parts are not immersed. Rather, a coupling medium is provided by a steady jet of water bridging the gap between the transducer and part.

Two modes of scanning are normally used. In the through-transmission mode two transducers are used: a transmitter in front of the specimen and a receiver in the back of the specimen. The presence of defects is related to the attenuation of the received signal. In the pulse-echo mode the ultrasound transmitted by the single transducer is received by the same transducer after reflection

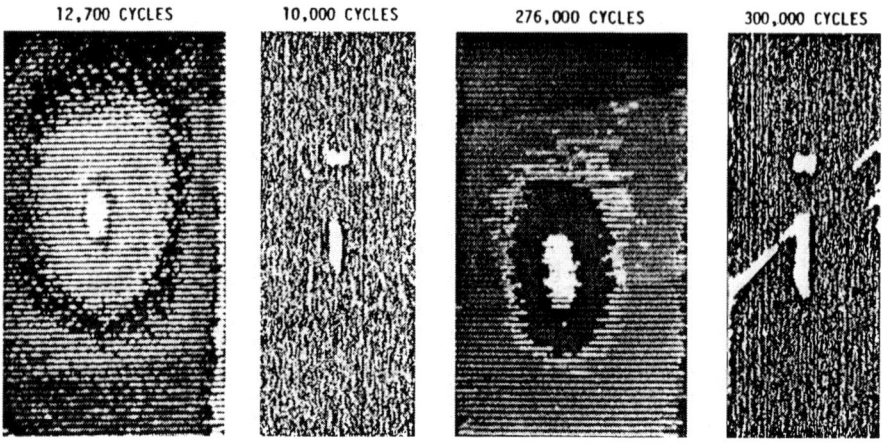

Figure 19-49 Thermographs and C-scans of a $[\pm 45/0]_s$ graphite–epoxy specimen with internal damage after fatigue loading. (From Ref. 84.)

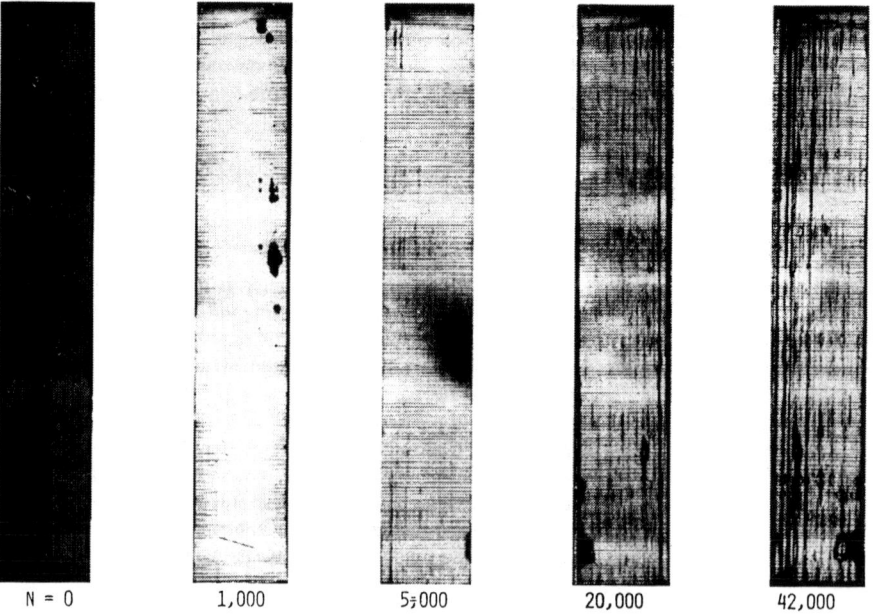

Figure 19-50 X-radiographs of $[0/90_2]_s$ graphite–epoxy specimen under fatigue loading with $R = 0.1$, $\sigma_{max} = 518$ MPa (75 ksi), for various numbers of cycles.

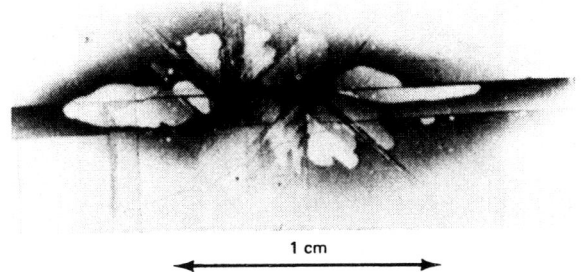

Figure 19-51 X-radiograph of $[0/-45/90/45]_s$, 12.7-cm (5-in.)-diameter clamped graphite–epoxy plate subjected to impact of a 294-g (0.648-lb) impactor from a height of 91.4 cm (36 in.).

from the back surface of the specimen (Fig. 19-52). The attenuation of the reflected pulse is related to internal defects in the laminate. A variation of the two techniques above is the pitch–catch method, which employs two transducers on the same side of the specimen. Pulses emitted by the transmitting transducer at an angle to the specimen surface are received by the receiving transducer after reflection from the back surface of the specimen.

Ultrasonic inspection records can be obtained in many forms, such as an amplitude–time display at a specific point on the specimen (A-scan), cross-sectional view of the specimen along a scan line (B-scan), and a series of scans covering the surface of the specimen and giving a plan view image (C-scan). Examples of A-scans and C-scans are shown in Figs. 19-53 and 19-54. Two modes of recording C-scans are shown. In the first mode, the pen-lift mode, an alarm circuit with a limit is used. The limit is adjusted to lift the pen of the x-y recorder whenever the gated peak voltage of the received pulse becomes smaller than a preset value corresponding to known flaws of a standard specimen (Fig. 19-54a). In the second mode, the analog mode, the gated peak

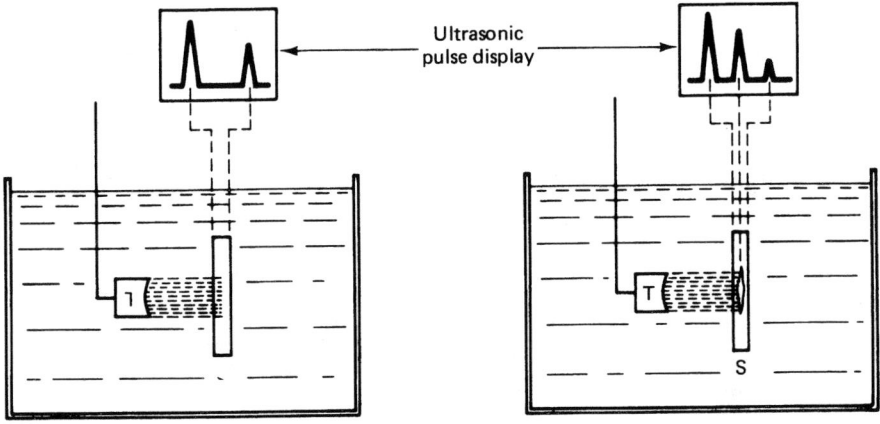

Figure 19-52 Ultrasonic pulse-echo method.

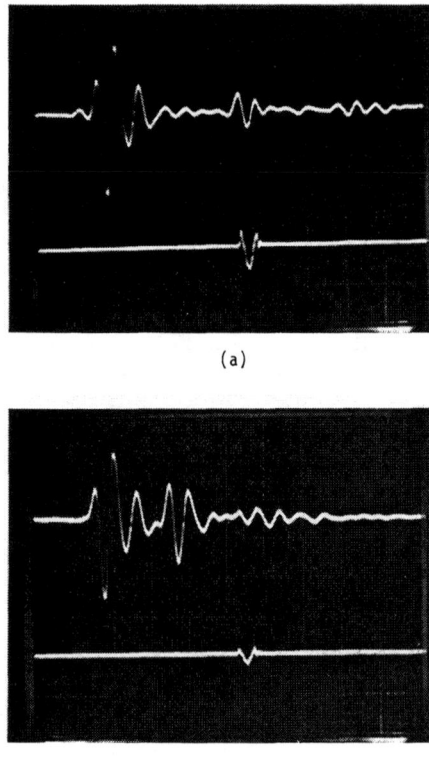

(a)

(b)

Figure 19-53 Amplitude–time records (A-scans) of ultrasonic pulse at two locations of FP/magnesium specimen: (a) unflawed location; (b) flawed location (0.5 V/div, 0.5 μs/div).

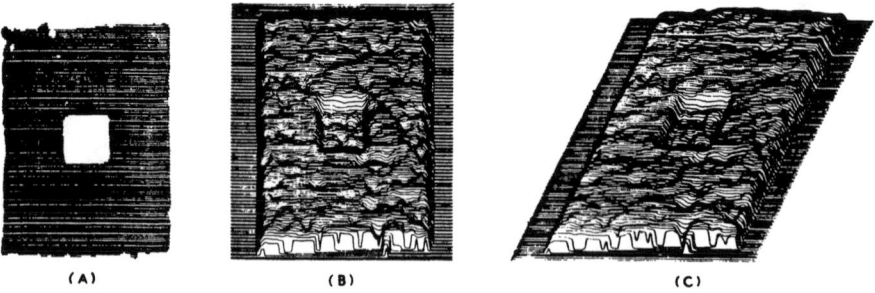

Figure 19-54 Ultrasonic C-scan of $[(0/\pm 45/90)_2]_s$ graphite–epoxy specimen with an embedded interply film patch: (a) pen-lift scan; (b) analog scan (normal); (c) offset analog scan (perspective).

COMPOSITE MATERIALS

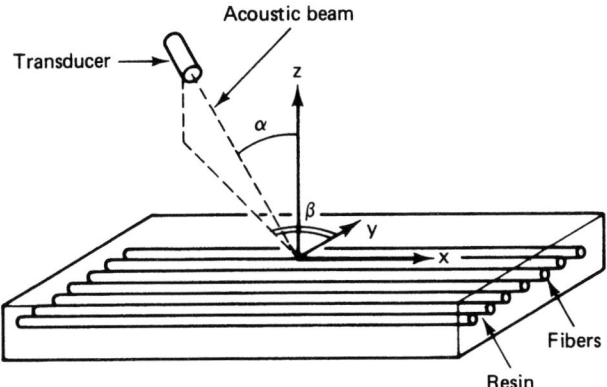

Figure 19-55 Detection of fiber orientation by ultrasonic back-scattering measurements; α, angle of incidence; β, angle between y axis and the transmitter beam trajectory on the layer plane.

voltage of the received pulse is recorded as a deflection of the pen normal to the scanning direction. The variations of this deflection are related to the presence of flaws. A typical analog record of the same specimen shown in Fig. 19-54a is shown in Fig. 19-54b. Figure 19-54c shows an offset analog scan, where a component of the x-axis signal is fed into the y-axis signal, resulting in a perspective view of the specimen.

Computer-aided procedures have been developed for recording and for more sophisticated processing of ultrasonic data. Plotting of the data in any desired form can be done at any time after inspection using different discriminator levels to detect smaller or larger flaws.

Conventional ultrasonics, with proper transducers, instrumentation, and settings, can detect large voids and delaminations with an accuracy of over 90%. In assessing the integrity of a composite laminate, one must also be able to determine nondestructively dispersed or bulk material characteristics, such as degree of cure, porosity, moisture content, layup, and fiber volume radio.

An ultrasonic technique suitable for measuring dispersed defects in the ultrasonic attenuation technique [88,89]. A comparison of the amplitudes of successive echoes from the specimen received by the transducer yields a value for the attenuation coefficient of the specimen material. Ultrasonic stress-wave techniques have been proposed for evaluating effects of microporosity, variations in fiber volume content, and for predicting tensile strength in composites [90].

One promising method is that employing ultrasonic backscattering measurements [91]. It was demonstrated that these techniques are capable of detecting matrix cracking, porosity, fiber orientation, and fiber misalignment in individual plies. The technique uses the setup shown in Fig. 19-55, which allows a transducer to be rotated about a normal to the surface axis at various angles of incidence. The maximum backscattered energy is received by the transmitting/receiving transducer when the transducer axis is normal to the defect. Thus the shape, location, and orientation of the defect can be determined approximately by angularly indexing the transducer. An example of detection of fiber orientations in a quasi-isotropic laminate is shown in Fig. 19-56 [91].

Acoustic Emission Method

The acoustic emission method is based on the phenomenon that the sudden release of energy inside a material results in emission of acoustic pulses. Energy release occurs as a result of

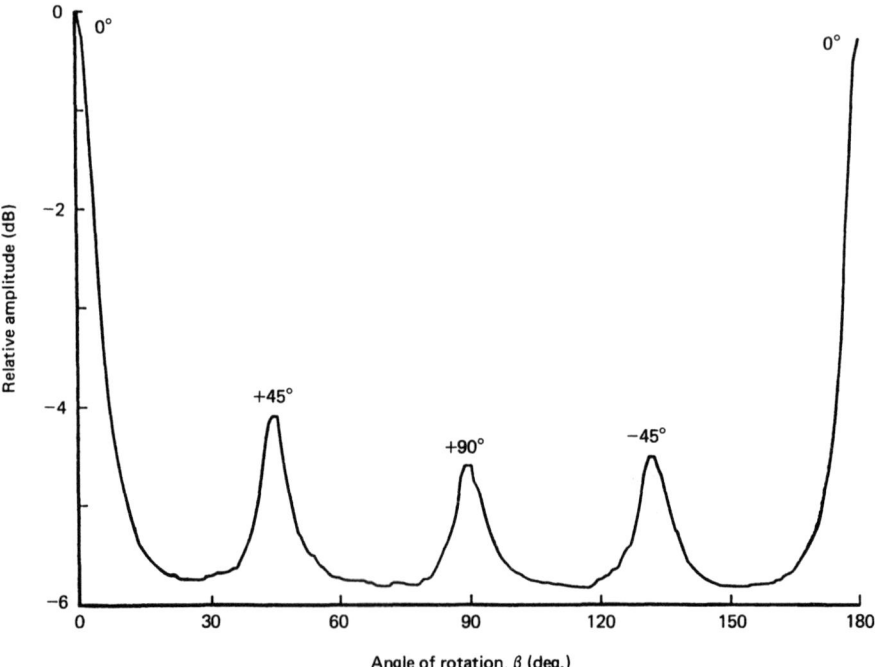

Figure 19-56 Backscattering from a quasi-isotropic $[0/\pm 45/90]_s$ graphite–epoxy specimen (angle of incidence, $\alpha = 30°$) (From Ref. [91]).

deformations or failure processes caused by loading. The primary failure mechanisms in fiber composites are matrix cracking, fiber debonding, and fiber breakage. The acoustic signals are detected by piezoelectric transducers in contact with the specimen through a coupling medium, electronically processed, and recorded. The usual procedure is to count the number of pulses above a preset amplitude threshold. The result can be recorded and presented in terms of cumulative number of counts, which indicates the extent of damage, or rate of counts, which is related to the rate of damage growth. The various failure mechanisms in composites produce signals of different amplitudes. Thus fiber breakage produces a higher acoustic emission activity than fiber debonding, which in turn produces more measurable counts than matrix cracking.

Acoustic emission is an integrating method, with the signals reflecting the integrated effects of specimen volume under load, source and transducer location, and type of damage. Thus, to be able to detect and discriminate defects and failure modes, it is not sufficient to count the number and rate of acoustic emission pulses. Attempts have been made to discriminate and identify the various microfracture modes by amplitude distribution and frequency spectral analysis of the acoustic pulses [92,93]. Computer pattern-recognition techniques have been used to analyze acoustic emission signals and to correlate frequency and amplitude distributions with physical and mechanical properties of the composite material, such as moisture content and strength.

Other Techniques

Internal deformation in composite laminates and structures during curing and throughout their service life can be monitored continuously with embedded strain gages supplemented with embedded thermocouples. Techniques for embedding such sensors in composite laminates and recording and interpreting the results have been developed and applied [36,37]. These techniques have been used to monitor the buildup of lamination residual stresses in angle-ply laminates during curing and the internal deformations during subsequent environmental and loading cycles. They can also be used for continual monitoring of composite structures in service for signs of structural degradation.

Electromagnetic techniques using microwaves have been used in glass–epoxy composites. Results to date have not been very conclusive. One of the major problems is that of composite material degradation due to exposure to high humidity and temperature. To date no suitable NDE techniques exist for measuring moisture absorption and matrix degradation.

Dispersed or bulk material characteristics, such as degree of cure, porosity, moisture content, and fiber volume fraction, are best correlated with bulk material properties such as stiffness and damping. Stiffness degradation in fatigue-loaded composites is an accepted measure of damage growth [95,96]. Such bulk properties can be measured using vibration or wave-propagation techniques. Most of the applications to date deal with stiffness measurements at various stages of the fatigue loading of the material. However, stiffness is not always a sensitive indicator of material damage, especially matrix cracking and delaminations. It is expected that damping would be a more sensitive indicator of matrix failures.

An effective technique for verifying results of nondestructive methods and for characterizing damage in composites has been described [97]. It is a destructive technique, called the "deply" technique, consisting of separating the laminate into its individual laminae without disturbing the fracture pattern. The laminate is penetrated with a gold chloride solution, which highlights the areas of delamination, matrix cracking, porosity, and fiber breakage.

19-7 Summary and Future Trends

An overview has been given of the status of mechanics of composites with emphasis on experimental aspects. Specific areas covered include the basic mechanics of composites, experimental methodology, material characterization and biaxial testing, effects of stress concentration, and nondestructive evaluation.

The section on mechanics of composites included a discussion of micromechanics and the behavior of the unidirectional lamina and the multidirectional laminate. Experimental work in micromechanics has been relatively limited since the late 1960s. However, the potential of experimental micromechanics has not been exhausted. This discipline can help in the evaluation of new matrix materials by studying the influence of matrix stiffness, ultimate strain, and fracture toughness. Experimental micromechanical models can be studied to observe localized fracture initiation and nucleation and to help develop local and global failure criteria for transversely loaded unidirectional composites. Further micromechanics research could be directed to the evaluation of residual, thermal, and moisture stresses on the behavior of composites. Recent analytical work has shown a relationship between residual stresses and cool down of epoxy–matrix composites [98]. Micromechanics could help evaluate such theoretical predictions.

Experimental methods discussed include strain gages, photoelastic methods, moiré methods, holographic, and speckle interferometric methods. The widespread use of computers in the laboratory is enabling the rapid acquisition and processing of large amounts of data. Strain gages are routinely used in material characterization with some applications unique to composites, such as embedded strain gages. Of the full-field optical methods, moiré interferometry seems to have great potential for composites. The introduction of digital image analysis procedures and the development of "half-fringe photoelasticity" enhances the use of transparent birefringent composites in anisotropic photoelastic studies. The application of embedded optical fibers for strain analysis is currently under investigation. A new full-field method based on the equations of thermoelasticity has been demonstrated for isotropic materials. Instrumentation utilizing this method (SPATE: see Chapter 14) is available and yields the sum of the principal stresses on the surface of the body. It is anticipated that the method will be extended to composite materials and structures.

The physical and mechanical characterization of the unidirectional lamina is a basic step for any experimental or theoretical analysis of composite structures. Not all methods discussed have been standardized, and there is considerable debate as to the best methods for determining the coefficients of moisture expansion, compressive, and in-plane shear properties. Testing techniques have been developed for high-rate testing including environmental effects. The subject of nonlinear characterization of composites, including temperature and moisture effects, has been studied by several investigators but is of fundamental importance and needs new approaches. Current research efforts are focusing on the behavior of the material under cyclic loading conditions. Damage mechanisms, damage accumulation, and the attendant degradation of stiffness and residual strength are being investigated. Future studies will be extended to include spectrum fatigue loading and environmental effects. Systematic experimental programs will be needed to evaluate and develop theories and models of damage accumulation. Future efforts also will be directed toward methods of accelerated testing for determining the reliability and life of structures.

The problem of biaxial testing is still not under control. However, in many specific cases valid data can be obtained. There is great need for valid biaxial data. Biaxial fatigue data are practically nonexistent.

Stress and failure analysis around stress concentrations was discussed at some length. However, most of the work to date deals with well-defined flaws, such as holes and through-the-thickness cracks. There is need to extend this work to study more realistic defects, such as partial delaminations, scratches, and impact-damaged areas. The fatigue behavior of laminates with notches, especially under biaxial loading, remains to be studied systematically.

In the area of composite joints there is great potential for application of strain analysis techniques. Studies similar to those conducted for open holes could be extended to bolted joints, to determine the deformation and failure initiation and propagation around loaded holes. Refined experimental techniques, such as moiré interferometry, are needed for determining in-situ properties of adhesives in bonded joints.

Nondestructive evaluation relies heavily on ultrasonics and x-radiography. There is need for establishing defect standards for adjusting and calibrating the nondestructive instrumentation. The most easily detectable flaws by ultrasonics are delaminations, inclusions, and surface scratches. However, flaws of other types, such as localized matrix cracking, fiber misalignment, and dispersed defects, are not easily detectable. Ultrasonic backscattering techniques have the potential of detecting fiber orientation, matrix cracking, and porosity [91]. Future efforts will be directed toward computer processing of data and image recognition methods. Ordinary x-radiography gives only a plan view of the damage; however, new stereo radiographic techniques can reveal the three-dimensional distribution of the damage. It is not certain that the latter method will develop into a practical NDE method. The method of real-time filmless radiography is currently under investigation and appears promising. The feasibility of computed tomography (CT) has been demonstrated, but the economic practicality of the method for composite materials is debatable. A novel technique for precise characterization of damage, albeit in a destructive manner, the "deply" technique has been developed [97]. Future efforts of NDE of composites will be directed to real-time methods and to applications such as the in-situ measurement of moisture and the evaluation of the strength of adhesive joints. New methods such as nuclear magnetic resonance spectrometry and acoustography will be investigated.

The ultimate goal of all nondestructive inspection is to assess the safety and remaining useful life of the composite structure. Following the detection and characterization of defects by NDE, it is essential to assess their criticality or effect on structural behavior and strength. This aspect of the problem remains the most important one and needs to be addressed by combined experimental and theoretical means.

The direction of future research will be greatly influenced by new developments in materials technology. For example, new types of graphite fiber are being produced with higher ultimate strains in the range 1.5 to 1.8%. The impact of such fibers on the behavior of composites has not yet been fully assessed. The moisture sensitivity of epoxy matrices continues to preoccupy many investigators. Thermoplastic matrices such as PEEK (polyether ether ketone) are being investigated and appear very attractive. They do not suffer from moisture absorption problems, and their composites are more easily amenable to repair. One new type of composite, developed at the Delft University of Technology, is fabricated by sandwiching layers of unidirectional Aramid (Kevlar) composite between thin (0.3 to 0.4 mm) sheets of aluminum alloy. This new material, called Arall, can be fabricated in a variety of sizes and shapes and can result in weight savings of 30 to 40% over conventional materials.

The sensitivity of conventional composites to delamination remains of concern. Techniques of three-dimensional reinforcement by means of graphite nails or stitching are being investigated. The use of woven-fabric and short-fiber reinforcement is receiving more attention.

The utilization of conventional and new materials is intimately related to the development of fabrication methods. The manufacturing process is one of the most important stages in assuring the quality of the finished product. Quality-control inspection and automation are being introduced in the manufacturing process.

Prospects for the future of composites technology are bright, as the cost of the basic materials and fabrication decreases and new high-volume applications appear. The initial driving force in the technology development was weight savings. Later, cost competitiveness with more conventional materials became equally important. To these two requirements today is added the need for assurance of quality, reproducibility, and predictability of behavior over the lifetime of the structure. Future research will be directed toward the measurement of durability, reliability, and life prediction of composite structures.

References

1. Robert M. Jones, *Mechanics of Composite Materials*, McGraw-Hill, New York, 1975.
2. Richard M. Christensen, *Mechanics of Composite Materials*, Wiley, New York, 1979.
3. Stephen W. Tsai and H. T. Hahn, *Introduction to Composite Materials*, Technomic, Lancaster, PA, 1980.
4. Bhagwan D. Agarwal and Lawrence J. Broutman, *Analysis and Performance of Fiber Composites*, Wiley, New York, 1980.
5. Derek Hull, *An Introduction to Composite Materials*, Cambridge University Press, Cambridge, 1981.
6. James M. Whitney, Isaac M. Daniel, and R. Byron Pipes, *Experimental Mechanics of Fiber Reinforced Composite Materials*, Society for Experimental Mechanics, Bethel, CT, 1984.
7. B. Walter Rosen, Tensile Failures of Fibrous Composites, *AIAA J.*, 2, no. 11 (1964), 1985.

8. Christos C. Chamis, Micromechanics Strength Theories, in *Fracture and Fatigue*, Vol. 5 (L. J. Broutman, Ed.), Academic Press, New York, 1974.

9. B. Walter Rosen, Mechanics of Composite Strengthening, Chap. 3 in *Fiber Composite Materials*, American Society for Metals, Metals Park, OH, 1965.

10. Zvi Hashin and B. Walter Rosen, The Elastic Moduli of Fiber-Reinforced Materials, *J. Appl. Mech., ASME, 31* (1964), 233.

11. Don Adams and D. R. Doner, Transverse Normal Loading of a Unidirectional Composite, *J. Compos. Mater., 1* (1967), 152–164.

12. John C. Halpin and Stephen W. Tsai, Environmental Factors in Composite Materials Design, Air Force Materials Lab Report AFML-TR-67-423, 1967.

13. Longin B. Greszczuk, Theoretical and Experimental Studies on Properties and Behavior of Filamentary Composites, *SPI, 21st Annu. Conf.*, Washington, DC, 1966.

14. D. P. Murphy and Don F. Adams, Energy Absorption Mechanisms During Crack Propagation in Metal Matrix Composites, Report UWME-DR-901-103-1, University of Wyoming, Laramie, 1979.

15. Isaac M. Daniel and August J. Durelli, Shrinkage Stresses Around Rigid Inclusions, *Exp. Mech., 2* (1962), 240–244.

16. Isaac M. Daniel, Micromechanics, in Structural Air-Frame Applications of Advanced Composite Materials, AFML-TR-69-101, II, 1969.

17. T. Koufopoulos and Pericles S. Theocaris, Shrinkage Stresses in Two-Phase Materials, *J. Compos. Mater., 3* (1969), 308–320.

18. Richard H. Marloff and Isaac M. Daniel, Three-Dimensional Photoelastic Analysis of a Fiber-Reinforced Composite Model, *Exp. Mech., 9* (1969), 156–162.

19. Isaac M. Daniel, Gerald M. Koller, and Tahei Niiro, Photoelastic Studies of Internal Stress Distributions of Unidirectional Composites, AMMRC-TR-80-56, 1980.

20. Richard A. Schapery, Thermal Expansion Coefficients of Composite Materials Based on Energy Principles, *J. Compos. Mater., 2* (1968), 380–404.

21. I. Goldenblat and V. A. Kopnov, Strength of Glass-Reinforced Plastics in the Complex Stress State, *Mekh. Polim., 1* (1965), 70–78; English translation: *Polym. Mech., 1* (1966), 54–60.

22. Stephen W. Tsai and Edward M. Wu, A General Theory of Strength for Anisotropic Materials, *J. Compos. Mater., 5* (January 1971), 58–80.

23. Isaac M. Daniel and Robert E. Rowlands, Experimental Stress Analysis of Composite Materials, ASME Paper 72-DE-6, 1972.

24. Charles W. Bert, Experimental Characterization of Composites, in *Structural Design and Analysis*, Part II, Vol. 8 (C. C. Chamis, Ed.), *Composite Materials* (L. J. Broutman and R. H. Krock, Eds.), Academic Press, New York, 1974.

25. Isaac M. Daniel, Photoelastic Investigation of Composites, in *Mechanics of Composites*, Vol. 2 (G. P. Sendeckyj, Ed.), *Composite Materials* (L. J. Broutman and R. H. Krock, Eds.), Academic Press, New York, 1974.

26. Isaac M. Daniel, Optical Methods for Testing Composite Materials, in *Failure Modes of Composite Materials with Organic Matrices and Their Consequences on Design*, AGARD-CP-163, 1975.

27. Isaac M. Daniel, Experimental Mechanics of Composite Materials, in *Mechanics of Composite Materials* (Z. Hashin and C. T. Herakovich, Eds.), Pergamon Press, Elmsford, NY, 1982, 473–496.

28. Hui Pih and C. E. Knight, Photoelastic Analysis of Anisotropic Fiber Composites, *J. Compos. Mater., 3* (1969), 94–107.

29. Robert C. Sampson, A Stress-Optic Law for Photoelastic Analysis of Orthotropic Composites, *Exp. Mech., 10* (1970), 210–215.

30. James W. Dally and R. Prabhakaran, Photo-Orthotropic-Elasticity, *Exp. Mech., 11* (1971), 346–356.

31. Charles W. Bert, Theory of Photoelasticity of Birefringent Filamentary Composites, *Fibre Sci. Technol.*, 5 (1972), 165–171.

32. C. E. Knight and Hui Pih, Orthotropic Stress-Optic Law for Plane Stress Photoelasticity of Composite Materials, *Fibre Sci. Technol.*, 9 (1976), 297–313.

33. Isaac M. Daniel, Gerald M. Koller, and Tahei Niiro, Development and Characterization of Orthotropic-Birefringent Materials, *Exp. Mech.*, 24 (1984), 135–143.

34. Arkady S. Voloshin and Christian P. Burger, Photo-Orthotropic Elasticity Through a Digital Image Analysis of Low Level Birefringence, *Proc. 7th Int. Conf. Exp. Stress Anal.*, Haifa, Israel, August 1982, 483–494.

35. James W. Dally and Ivo Alfirevich, Application of Birefringent Coatings to Glass-Fiber-Reinforced Plastics, *Exp. Mech.*, 9 (1969), 97–102.

36. Isaac M. Daniel, James L. Mullineaux, Franklin J. Ahimaz, and Theodore Liber, The Embedded Strain Gage Technique for Testing Boron/Epoxy Composites, in *Composite Materials: Testing and Design*, ASTM STP 497, American Society for Testing and Materials, Philadelphia, 1972, 257–272.

37. Isaac M. Daniel and Theodore Liber, Lamination Residual Stresses in Fiber Composites, NASA-CR-134826, 1975.

38. Isaac M. Daniel, Theodore Liber, and Ralph H. Labedz, Wave Propagation in Transversely Impacted Composite Laminates, *Exp. Mech.*, 19 (1979), 9–16.

39. Isaac M. Daniel, Robert E. Rowlands, and Daniel Post, Strain Analysis of Composites by Moiré Methods, *Exp. Mech.*, 13 (1973), 246–252.

40. Daniel Post, Moiré Interferometry for Damage Analysis of Composites, *Proc. SESA Spring Conf.*, Cleveland, 1983, 337–343.

41. Robert E. Rowlands, Theodore Liber, Isaac M. Daniel, and Peter G. Rose, Stress Analysis of Anisotropic Laminated Plates, *AIAA J.*, 12 (1974), 903–908.

42. Y. Y. Hung, Isaac M. Daniel, and Robert E. Rowlands, Full-Field Optical Strain Measurement Having Post-Recording Sensitivity and Direction Selectivity, *Exp. Mech.*, 18 (1978), 56–60.

43. Kenneth E. Hofer and P. N. Rao, A New Static Compression Fixture for Advanced Composite Materials, *J. Test. Eval.*, 5 (1977), 278–283.

44. B. Walter Rosen, A Simple Procedure for Experimental Determination of the Longitudinal Shear Modulus of Unidirectional Composites, *J. Compos. Mater.*, 6 (1972), 552–554.

45. Christos C. Chamis and John H. Sinclair, Ten-Degree Off-Axis Test for Shear Properties in Fiber Composites, *Exp. Mech.*, 17 (1977), 339–346.

46. P. H. Petit and Max E. Waddoups, A Method of Predicting the Nonlinear Behavior of Laminate Composites, *J. Compos. Mater.*, 3 (1969), 2–19.

47. R. S. Sandhu, Ultimate Strength Analysis of Symmetric Laminates, AFFDL-TR-73-137, 1974.

48. Edward M. Wu, Failure Criteria to Fracture Mode Analysis of Composite Laminates, AGARD-CP-163, 1975.

49. B. William Cole and R. Byron Pipes, Utilization of the Tubular and Off-Axis Specimens for Composite Biaxial Characterization, AFFDL-TR-72-130, 1972.

50. Robert E. Rowlands, Analytical–Experimental Correlation of Polyaxial States of Stress in Thornel–Epoxy Laminates, *Exp. Mech.*, 18 (1978), 253–260.

51. Isaac M. Daniel, Behavior of Graphite/Epoxy Plates with Holes Under Biaxial Loading, *Exp. Mech.*, 20 (1980), 1–8.

52. Isaac M. Daniel, Biaxial Testing of Graphite/Epoxy Laminates with Cracks, in *Test Methods and Design Allowables for Fibrous Composites* (C. C. Chamis, Ed.), ASTM STP 734. American Society for Testing and Materials, Philadelphia, 1981, 109–128.

53. Nicholas J. Pagano and James M. Whitney, Geometric Design of Composite Cylindrical Characterization Specimens, *J. Compos. Mater.*, 4 (1970), 360–378.

54. James M. Whitney, Glen C. Grimes, and Philip H. Francis, Effect of End Attachment on the Strength of Fiber-Reinforced Composite Cylinders, *Exp. Mech.*, 13 (1973), 185–192.

55. Timothy L. Sullivan and Christos C. Chamis, Some Important Aspects in Testing High Modulus Fiber Composite Tubes in Axial Tension, *Analysis of the Test Methods for High Modulus Fibers and Composites*, ASTM STP 521, American Society for Testing and Materials, Philadelphia, 1973.

56. Thomas R. Guess and Frank P. Gerstle, Jr., Deformation and Fracture of Resin Matrix Composites in Combined Stress States, *J. Compos. Mater.*, 7 (1977), 448–464.

57. U. Hütter, H. Schelling, and H. Krauss, An Experimental Study to Determine Failure Envelope of Composite Materials with Tubular Specimens Under Combined Loads and Comparison with Several Classical Criteria, AGARD-CP-163, 1975.

58. U. S. Lindholm, A. Nagy, L. M. Yeakley, and W. L. Ko, Design of a Test Machine for Biaxial Testing of Composite Laminate Cylinders, AFFDL-TR-75-83, 1975.

59. Isaac M. Daniel, Theodore Liber, Ray Vanderby, and Gerald M. Koller, Analysis of Tubular Specimen for Biaxial Testing of Composite Laminates, *Advances in Composite Materials, Proc. ICCM/3*, 1980, 840–855.

60. B. William Cole and R. Byron Pipes, Filamentary Composite Laminates Subjected to Biaxial Stress Fields, AFFDL-TR-73-115, 1974.

61. Donald N. Weed and Philip H. Francis, Process Development for the Fabrication of High Quality Composite Tubes, *Fibre Sci. Technol.*, 10 (1977), 89–100.

62. Theodore Liber, Isaac M. Daniel, Ralph H. Labedz, and Tahei Niiro, Fabrication and Testing of Composite Ring Specimens, *Proc. 34th Annu. SPI Tech. Conf., Reinf. Plast./Compos. Inst.*, Sect. 22-B, 1979.

63. Scott W. Shramm and Isaac M. Daniel, Test System for Conducting Biaxial Tests of Composite Laminates, Vol. II, *Test System Design*, IITRI Final Report on Air Force Contract F33615-77-C-3014, 1983.

64. Max E. Waddoups, J. R. Eisenmann, and B. E. Kaminski, Microscopic Fracture Mechanisms of Advanced Composite Materials, *J. Compos. Mater.*, 5 (1971), 446–454.

65. Thomas A. Cruse, Tensile Strength of Notched Laminates, *J. Compos. Mater.*, 7 (1973), 218–229.

66. James M. Whitney and Ralph J. Nuismer, Stress Fracture Criteria for Laminated Composites Containing Stress Concentrations, *J. Compos. Mater.*, 8 (1974), 253–265.

67. King Him Lo and Edward M. Wu, Failure Strength of Notched Composite Laminates, in *Advanced Composite Serviceability Program* (J. Altman, B. Burroughs, R. Hunziker, and D. Konishi, Eds.), AFWAL-TR-80-4092, 1980.

68. Isaac M. Daniel, Robert E. Rowlands, and James B. Whiteside, Effects of Material and Stacking Sequence on Behavior of Composite Plates with Holes, *Exp. Mech.*, 4 (1974), 1–9.

69. H. J. Konish and James M. Whitney, Approximate Stresses in an Orthotropic Plate Containing a Circular Hole, *J. Compos. Mater.*, 9 (1975), 157–166.

70. Isaac M. Daniel, Biaxial Testing of $[0_2/\pm 45]_s$ Graphite/Epoxy Plates with Holes, *Exp. Mech.*, 22 (1982), 188–195.

71. Isaac M. Daniel, Strain and Failure Analysis in Graphite/Epoxy Laminates with Cracks, *Exp. Mech.*, 18 (1978), 246–252.

72. Y. T. Yeow, Donald H. Morris, and Halbert F. Brinson, The Fracture Behavior of Graphite/Epoxy Laminates, *Exp. Mech.*, 19 (1979), 1–8.

73. Isaac M. Daniel, Mixed Mode Failure of Composite Laminates with Cracks, *Proc. 5th Int. Conf. Exp. Stress Anal.*, Montreal, Canada, June 1984, 320–327.

74. Isaac M. Daniel and Theodore Liber, Nondestructive Evaluation Techniques for Composite Materials, *Proc. 12th Symp. Nondestr. Eval.*, San Antonio, TX, 1979.

75. G. A. Matzkanin, G. L. Burkhardt, and C. M. Teller, Nondestructive Evaluation of Fiber Reinforced Epoxy Composites—A State of the Art Survey, Contract DLA 900-77-C-3733, Southwest Research Institute for U.S. Army Aviation Research and Development Command, 1979.

76. Edmund G. Henneke, Nondestructive Evaluation of Damage in Composite Materials, *Proc. 1983 Fall Meeting, SESA*, Salt Lake City, UT, 1983.

77. R. Byron Pipes (Ed.), *Nondestructive Evaluation and Flaw Criticality for Composite Materials*, ASTM STP 696, American Society for Testing and Materials, Philadelphia, 1979.

78. Wayne W. Stinchcomb (Ed.), *Mechanics of Nondestructive Testing*, Plenum Press, New York, 1980.

79. Kenneth L. Reifsnider, Some Fundamental Aspects of the Fatigue and Fracture Response of Composite Materials, *Proc. 14th Annu. Meeting Soc. Eng. Sci.*, Lehigh University, Bethlehem, PA, 1977.

80. George P. Sendeckyj, G. E. Maddux, and N. A. Tracy, Comparison of Holographic, Radiographic, and Ultrasonic Techniques for Damage Detection in Composite Materials, *Proc. ICCM/2*, Toronto, Canada, 1978, Metallurgical Society of AIME, 1037–1056.

81. Y. Y. Hung and Joseph D. Hovanesian, Nondestructive Testing by Speckle Shearing Interferometry, *Proc. 12th Symp. Nondestr. Eval.*, San Antonio, TX, 1979, 163–167.

82. Lawrence J. Broutman, T. Kobayashi, and D. Carrillo, Determination of Fracture Sites in Composite Materials with Liquid Crystals, *J. Compos. Mater.*, 3 (1969), 702–704.

83. J. A. Charles, Liquid Crystals for Flaw Detection in Composites, *Nondestructive Evaluation and Flaw Criticality for Composite Materials*, ASTM STP 696, American Society for Testing and Materials, Philadelphia, 1979, 72–82.

84. Kenneth L. Reifsnider, Edmund G. Henneke II, and Wayne W. Stinchcomb, The Mechanics of Vibrothermography, in *Mechanics of Nondestructive Testing* (W. W. Stinchcomb, Ed.), Plenum Press, New York, 1980, 249–276.

85. Francis Chang, J. C. Couchman, J. R. Eisenmann, and B. G. W. Yee, Application of a Special X-Ray Non-Destructive Testing Technique for Monitoring Damage Zone Growth in Composite Laminates, *Composite Reliability*, ASTM STP 580, American Society for Testing and Materials, Philadelphia, 1975, 176–190.

86. George P. Sendeckyj, G. E. Maddux, and E. Porter, Damage Documentation in Composites by Stereo Radiography, In *Damage in Composite Materials*, (K. L. Reifsnider, Ed.), ASTM STP 775, American Society for Testing and Materials, Philadelphia, 1982, 16–26.

87. Theodore Liber, Isaac M. Daniel, and Scott W. Schramm, Ultrasonic Techniques for Inspecting Flat and Cylindrical Composite Specimens, in *Nondestructive Evaluation and Flaw Criticality for Composite Materials*, ASTM STP 696 (R. B. Pipes, Ed.), American Society for Testing and Materials, Philadelphia, 1979, 5–25.

88. D. T. Hayford and Edmund G. Henneke, A Model for Correlating Damage and Ultrasonic Attenuation in Composites, *Composite Materials: Testing and Design*, ASTM STP 674, American Society for Testing and Materials, Philadelphia, 1979, 184–200.

89. Scott W. Schramm and Isaac M. Daniel, Nondestructive Evaluation of Metal-Matrix Composites, AF/DARPA Rev. Prog. Quantitative NDE, La Jolla, CA, 1982.

90. Alex Vary and K. J. Bowles, An Ultrasonic-Acoustic Technique for Nondestructive Evaluation of Fiber Composite Strength, *Proc. 33rd Annu. Tech. Conf., Reinf. Plast./Compos. Inst.*, Sec. 24-A, 1978.

91. Yoseph Bar-Cohen and Robert L. Crane, Nondestructive Evaluation of Fiber-Reinforced Composites with Backscattering Measurements, in *Composite Materials: Testing and Design* (I. M. Daniel, Ed.), ASTM STP 787, American Society for Testing and Materials, Philadelphia, 1982, 343–354.

92. Assa Rotem, The Discrimination of Micro-Fracture Mode of Fibrous Composite Material by Acoustic Emission Technique, *Fibre Sci. Technol.*, 10 (1977), 101–121.

93. Madhu Madhukar and Jonathan Awerbuch, Monitoring Damage Progression in Center-Notched Boron/Aluminum Laminates through Acoustic Emission, *Composite Materials: Testing and Design* (J. M. Whitney, Ed.), ASTM STP 893, American Society for Testing and Materials, Philadelphia, 1986, 337–367.

94. Lloyd J. Graham and R. K. Elsley, Characterization of Acoustic Emission Signals from Composites, *Proc. ARPA/AFML Rev. Prog. Quantitative NDE*, AFML-TR-78-55, 1978, 219–225.

95. Kevin T. O'Brien, Characterization of Delamination Onset and Growth in a Composite Laminate, *Damage in Composite Materials*, (K. L. Reifsnider, Ed.), ASTM STP 775, American Society for Testing and Materials, Philadelphia, 1982, 140–167.

96. E. T. Camponeschi and Wayne W. Stinchcomb, Stiffness Reduction as an Indicator of Damage in Graphite/Epoxy Laminates, *Composite Materials: Testing and Design* (I. M. Daniel, Ed.), ASTM STP 787, American Society for Testing and Materials, Philadelphia, 1982, 225–246.

97. Sam M. Freeman, Characterization of Lamina and Interlaminar Damage in Graphite–Epoxy Composites by the Deply

Technique *Composite Materials: Testing and Design* (Sixth Conf.), (I. M. Daniel, Ed.), ASTM STP 787, American Society for Testing and Materials, Philadelphia, 1982, 50–62.

98. Yechiel Weitsman, Residual Thermal Stresses Due to Cooldown of Epoxy–Resin Composites, *J. Appl. Mech., 46* (1979), 563–567.

99. Ray L. Foye, Structural Composites, *Quarterly Report 3,* Contract AF33(615)-5150, North American Aviation, Columbus Division, March 1967.

CHAPTER
20
Experimental Fracture Mechanics

C. W. Smith
*Virginia Polytechnic Institute
and State University
Blacksburg, Virginia*

Albert S. Kobayashi
*University of Washington
Seattle, Washington*

20-0 Introduction

Although basic concepts regarding the onset of unstable crack propagation were introduced by Griffith as early as 1919 in connection with his experiments on glass [1,2], it was not until the late 1940s that it was perceived both by Irwin [3] and by Orowan [4] that the Griffith concept could be carried over to the study of the quasi-brittle fracture of structural metals. This was accomplished by including in the Griffith energy balance equation a surface plastic work term (along with the elastic surface energy), which becomes dominant in normally ductile materials. Despite the usefulness of this development, it was not considered rigorous enough by some because of its energy-based concept. The energy method, however, became popular after the emergence of the "stress-intensity factor (SIF)," which was presented by Irwin in 1957 [5] and can be readily derived [6] from classical elasticity theory, as demonstrated by Westergaard [7], extensively by Muskhelishvili [8], Lekhnitskii [9], and others, and with earlier solutions by Sneddon [10] and Williams [11].

However, in addition to formalizing the role of the SIF in elastic stress analysis, it was simultaneously necessary to provide for the presence of a plastic zone within the zone dominated by the SIF in order to accommodate large amounts of plastic work. Irwin [12] recognized that this zone would increase the opening of the crack and modeled this effect with a small increase in the crack length to account for the influence of the plastic zone but not requiring a full elastoplastic analysis. Wells [13] introduced refinements based on crack-tip elastoplastic analysis which were subsequently developed and have formed a cornerstone for building fracture theory [14–16] based on small-scale yielding. The recent shift in analysis and design from ultrahigh- to medium-strength, high-toughness materials, however, has moved beyond small-scale yielding theory and brought about the development of elastoplastic fracture mechanisms. As expected, the crack-tip state of elastic–plastic fracture mechanisms is affected by the loading history and the size of the plastic region. Examples of the former can be seen in fatigue crack retardation or acceleration [17], which are attributed to variations in the crack-tip plastic zone size and accompanying residual stresses, under variable loading and unloading cycles. Large plastic regions also render small-scale yielding theory inaccurate for predicting the onset of fracture, since the state of stress cannot be approximated by the elastic stress with $1/\sqrt{r}$ singularity. Large-scale yielding also induces subcritical crack growth with increased loading, where the trailing wake of residual stresses offsets the elastic–plastic state of stress at the crack tip.

In contrast to the mature state of the science of linear elastic fracture mechanics with two decades of practical applications, concerted studies in dynamic fracture mechanics were not initiated until the mid-1970s. Prior to that time, dynamic crack propagation was modeled with a sequence of static analyses, and crack arrest was treated as the inverse of the onset of crack instability. Roberts and Wells [18] and Dulaney and Brace [19] used such static analysis together with Mott's dynamic extension [20] of Griffith's instability criterion to represent the crack velocity in terms of the modulus of elasticity, the density, and the critical crack length. Berry [21,22], on the other hand, was the first to realize that a crack could propagate under a stress larger than the critical stress of Griffith and introduced an empirical multiplying factor that was greater than unity and varied with the specimen boundary conditions. Since these energy theories were variations of Griffith's classical criterion, the resultant crack velocities were explicit functions of the initial crack length. The early but limited experimental investigations [23–25], which were conducted to establish design requirements for crack arresters in welded structures, unfortunately did not provide the information that was needed to assess the validity of the above-mentioned empirical relations on crack velocities.

Interest in dynamic fracture suddenly increased in the mid-1970s with the concerns for the safety of critical components in nuclear power plants subjected to seismic, projectile impact, and other accidental dynamic overloadings. The timely developments in new analytical and experimental techniques, coupled with the explosive advances in numerical techniques in fracture dynamics, were used effectively to study the basic mechanisms involved in rapid crack propagation. While controversies remain, plausible dynamic fracture criteria for constant-velocity crack propagation, dynamic crack arrest, dynamic crack curving, and dynamic crack branching have been proposed. A dynamic fracture criterion in the presence of large-scale yielding is also available. The obvious application of dynamic fracture criteria for predicting structural failure under impact loading is, however, hindered by the complexity of the dynamic state of stress in actual structural components.

20-1 Theoretical Background

Linear Elastic Fracture Mechanics (LEFM)

By treating the crack as a branch cut (Fig. 20-1a) and applying the principles of linear, small-deformation elasticity, Westergaard [7] and Muskhelishvili [8] obtained global field equations for the stress and displacement components for a straight-front crack under mode 1 conditions (Fig. 20-1b), which can be written in the following form[1,2]:

---

[1] See also Chapter 1.
[2] A list of the symbols and conventions used in this chapter precedes the References.

EXPERIMENTAL FRACTURE MECHANICS

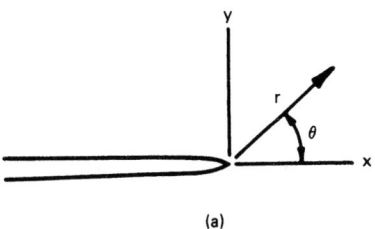

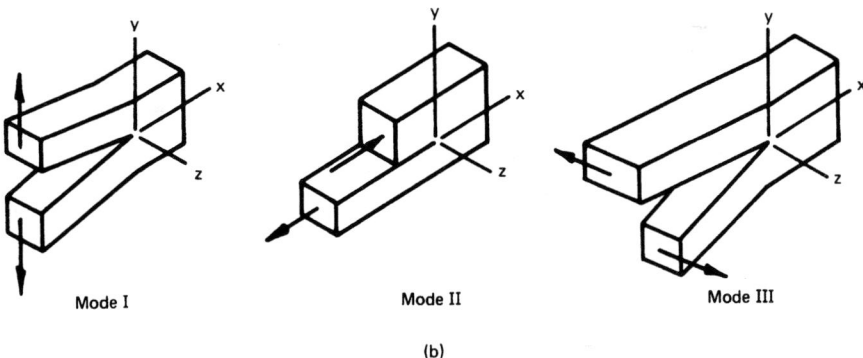

Figure 20-1 (a) Idealized crack profile; (b) local models of deformation.

$$\sigma_{xx} = \frac{K}{\sqrt{2\pi r}} \cos\frac{\theta}{2}\left(1 - \sin\frac{\theta}{2}\sin\frac{3\theta}{2}\right) + \sigma_{0x}$$
$$+ \sum_{n=3}^{\infty}\left(A_n \frac{n}{2}\right) r^{n/2-1}\left\{\left[2 + (-1)^n + \frac{n}{2}\right]\cos\left(\frac{n}{2} - 1\right)\theta\right.$$
$$\left. - \left(\frac{n}{2} - 1\right)\cos\left(\frac{n}{2} - 3\right)\theta\right\}$$

$$\sigma_{yy} = \frac{K}{\sqrt{2\pi r}} \cos\frac{\theta}{2}\left(1 + \sin\frac{\theta}{2}\sin\frac{3\theta}{2}\right)$$
$$+ \sum_{n=3}^{\infty}\left(A_n \frac{n}{2}\right) r^{n/2-1}\left\{\left[2 - (-1)^n - \frac{n}{2}\right]\cos\left(\frac{n}{2} - 1\right)\theta\right. \quad (20\text{-}1)$$
$$\left. - \left(\frac{n}{2} - 1\right)\cos\left(\frac{n}{2} - 3\right)\theta\right\}$$

$$\sigma_{xy} = \frac{K}{\sqrt{2\pi r}} \sin\frac{\theta}{2} \cos\frac{\theta}{2} \cos\frac{3\theta}{2}$$

$$- \sum_{n=3}^{\infty} \left(A_n \frac{n}{2}\right)^{n/2-1} \left\{\left[(-1)^n + \frac{n}{2}\right] \sin\left(\frac{n}{2} - 1\right)\theta\right.$$

$$\left. - \left(\frac{n}{2} - 1\right) \sin\left(\frac{n}{2} - 3\right)\theta\right\}$$

where σ_{0x} is the second-order nonsingular term, which appears only in the σ_{xx} component.

For the state of plane stress, the displacement components in the x and y directions, u_x and u_y, become

$$u_x = \frac{K}{\mu} \sqrt{\frac{r}{2\pi}} \cos\frac{\theta}{2} \left(\frac{1-\nu}{1+\nu} + \sin^2\frac{\theta}{2}\right) + \sigma_{0x} \frac{2}{(1+\nu)\mu} r \cos\theta$$

$$+ \sum_{n=3}^{\infty} \frac{A_n}{2\mu} r^{n/2} \left\{\frac{3-\nu}{1+\nu} \cos\frac{n\theta}{2} - \frac{n}{2} \cos\left(\frac{n}{2} - 2\right)\theta + \left[\frac{n}{2} + (-1)^n\right] \cos\frac{n\theta}{2}\right\}$$

$$u_y = \frac{K}{\mu} \sqrt{\frac{r}{2\pi}} \cos\frac{\theta}{2}\left(\frac{2}{1+\nu} - \cos^2\frac{\theta}{2}\right) - \sigma_{0x} \frac{2\nu}{\mu(1+\nu)} r \sin\theta$$

$$+ \sum_{n=3}^{\infty} \frac{A_n}{2\mu} r^{n/2} \left\{\frac{3-\nu}{1+\nu} \sin\frac{n\theta}{2} + \frac{n}{2} \sin\left(\frac{n}{2} - 2\right)\theta - \left[\frac{n}{2} + (-1)^n\right] \sin\frac{n\theta}{2}\right\}$$

(20-2a)

where μ = shear modulus
ν = Poisson's ratio

For the state of plane strain, the displacement components become

$$u_x = \frac{K_I}{\mu} \sqrt{\frac{r}{2\pi}} \cos\frac{\theta}{2}\left(1 - 2\nu + \sin^2\frac{\theta}{2}\right) + \sigma_{0x} \frac{2(1-\nu)}{\mu} r \cos\theta$$

$$+ \sum_{n=3}^{\infty} \frac{A_n}{2\mu} r^{n/2} \left\{(3 - 4\nu) \cos\frac{n\theta}{2} - \frac{n}{2} \cos\left(\frac{n}{2} - 2\right)\theta + \left[\frac{n}{2} + (-1)^n\right] \cos\frac{n\theta}{2}\right\}$$

$$u_y = \frac{K_I}{\mu} \sqrt{\frac{r}{2\pi}} \sin\frac{\theta}{2}\left(2 - 2\nu - \cos^2\frac{\theta}{2}\right) - \sigma_{0x} \frac{2\nu}{\mu} r \sin\theta$$

$$+ \sum_{n=3}^{\infty} \frac{A_n}{2\mu} r^{n/2} \left\{(3 - 4\nu) \sin\frac{n\theta}{2} + \frac{n}{2} \sin\left(\frac{n}{2} - 2\right)\theta - \left[\frac{n}{2} + (-1)^n\right] \sin\frac{n\theta}{2}\right\}$$

(20-2b)

The subscript I added to the opening mode of stress-intensity factor K denotes the state of plane strain when discussion is restricted to tensile fracture. No such designation exists for sliding mode or mode II (Fig. 20-1b) crack-tip deformation, where the plane-stress and plane-strain stress intensity factors are both referred to as K_{II}.

To extend the two-dimensional analysis to three dimensions, one must ascertain the crack-tip state of stress surrounding the curved crack front. Irwin [26] postulated the plane-strain state of stress and then derived the plane-strain stress-intensity K_I for an elliptical crack. Irwin's postulate was later verified [27,28]; specifically it was shown that the mode I local stresses can be represented in terms of a local coordinate system t-n-z, which is tangential and normal to the crack front as shown in Fig. 20-2a as

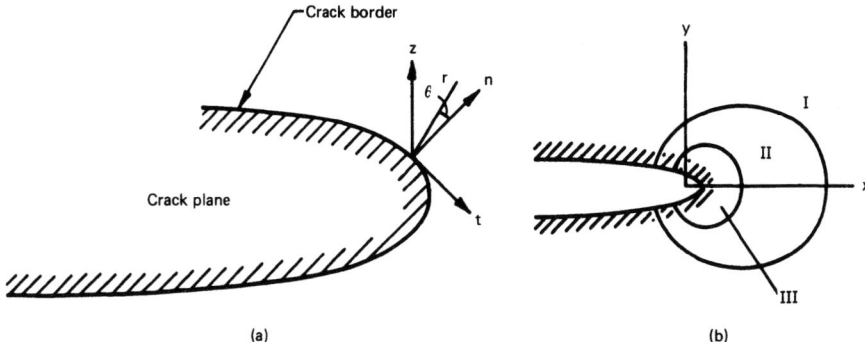

Figure 20-2 (a) Crack-tip coordinates along a curved crack front; (b) local zones around an opened crack profile.

$$\sigma_{tt} = 2\nu \left[\frac{K_I}{(2\pi r)^{1/2}} \cos \frac{\theta}{2} \right] + O(\sqrt{r})$$

$$\sigma_{nn} = \frac{K_I}{(2\pi r)^{1/2}} \cos \frac{\theta}{2} \left(1 - \sin \frac{\theta}{2} \sin \frac{3\theta}{2} \right) + O(\sqrt{r})$$

$$\sigma_{zz} = \frac{K_I}{(2\pi r)^{1/2}} \cos \frac{\theta}{2} \left(1 + \sin \frac{\theta}{2} \sin \frac{3\theta}{2} \right) + O(\sqrt{r})$$ (20-3)

$$\sigma_{tn} = O(\sqrt{r}) \quad \sigma_{tz} = O(\sqrt{r})$$

$$\sigma_{nz} = \frac{K_I}{(2\pi r)^{1/2}} \sin \frac{\theta}{2} \cos \frac{\theta}{2} \cos \frac{3\theta}{2} + O(\sqrt{r})$$

$$u_t = O(r)$$ (20-4)

$$u_n = \frac{K_I}{\mu} \left(\frac{r}{2\pi} \right)^{1/2} \cos \frac{\theta}{2} \left(1 - 2\nu + \sin^2 \frac{\theta}{2} \right) + O(r)$$

$$u_z = \frac{K_I}{\mu} \left(\frac{r}{2\pi} \right)^{1/2} \sin \frac{\theta}{2} \left(2 - 2\nu - \cos^2 \frac{\theta}{2} \right) + O(r)$$

Comparison of Eqs. (20-3) and (20-4) with Eqs. (20-1) and (20-2) shows that the states of stress and displacement in the vicinity of a curved crack front are similar to the two-dimensional plane-strain state discussed previously. Thus, knowing either σ_{zz} or u_z, the opening mode of stress-intensity factor K_I can be determined readily.

When a mode I load is applied to a branch cut (or crack), the crack surfaces separate and the crack front moves forward in the x direction, forming a profile that is approximately elliptical (Fig. 20-2b). The theoretically infinite state of stress at the crack tip imposes finite rotations and deformations very near the crack tip, along with other nonlinear effects (plasticity, viscoelasticity, etc.). To apply the principles of LEFM experimentally in measuring SIF values, we must make measurements outside this zone (zone III in Fig. 20-2b). On the other hand, the stress singularity at the crack tip, which is described by Eq. (20-1), will apply only near the crack tip, so our measurement zone (zone II) will also be bounded externally by zone I, which includes measurable nonsingular effects. Since measurements are not made at the crack tip itself, Eqs. (20-3) will need some modification to account for the influence of the nonsingular stress field in the measurement zone (zone II).

It is noted that since SIF values vary linearly with the elastic stress, they are proportional to the applied loads. When the SIF reaches a critical value, called K_{Ic} the crack is assumed to propagate unstably. In this chapter, however, we restrict ourselves to the experimental determination of subcritical SIF values (i.e., $K < K_{Ic}$).

Strain-Energy Release Rate and Stress-Intensity Factors

Griffith [1] based his analysis on the energy balance for a cracked elastic body. That is, he assumed that for a crack to grow, sufficient surplus strain energy would have to be stored in the body to accomplish the work of separating the material surfaces (i.e., surface energy). Thus, as a crack grows, the strain-energy release rate, given by

$$G = \frac{dU}{da} \tag{20-5}$$

where U is the strain energy and a is the crack length, will be in balance with the crack growth resistance R, given by

$$R = \frac{dW}{da} \tag{20-6}$$

where W is the surface work required to overcome the tractions holding a material body together along a surface and produce new fracture surface area. For stationary or stably growing cracks,

$$G = R$$

If $G > R$, unstable crack propagation occurs.

In a perfectly elastic brittle material, $W = 2A\gamma$, where γ is the specific surface energy and A the newly created fracture surface area. When a plastic zone exists, it serves as an energy sink, requiring much additional strain energy to produce crack growth. However, if the plastic zone is small and the plastic energy is assumed to lie in the fracture surface, it may be considered to be part of the surface work.

If one loads a cracked elastic body, fixes the load points, and allows an increment of crack growth to occur, the strain energy released during crack growth should just equal the work required to close the crack over the increment of growth. Thus, from Fig. 20-3 we may write the following [6]:

$$G = \frac{dU}{da} = -\lim_{\delta a \to 0} \frac{2}{\delta a} \int_0^{\delta a} \left[\frac{\sigma_{yy}(x, 0)u_y(\delta a - x, \pi)}{2} + \frac{\sigma_{xy}(x, 0)u_x(\delta a - x, \pi)}{2} \right. $$
$$\left. + \frac{\sigma_{yz}(x, 0)u_z(\delta a - x, \pi)}{2} \right] dx \tag{20-7}$$

If we now substitute the leading terms of Eqs. (20-1) and (20-2) (and also expressions for local forms of σ_{ij} and u_i for modes II and III) into Eq. (20-7) and integrate, we have

$$G = G_I + G_{II} + G_{III} \tag{20-8}$$

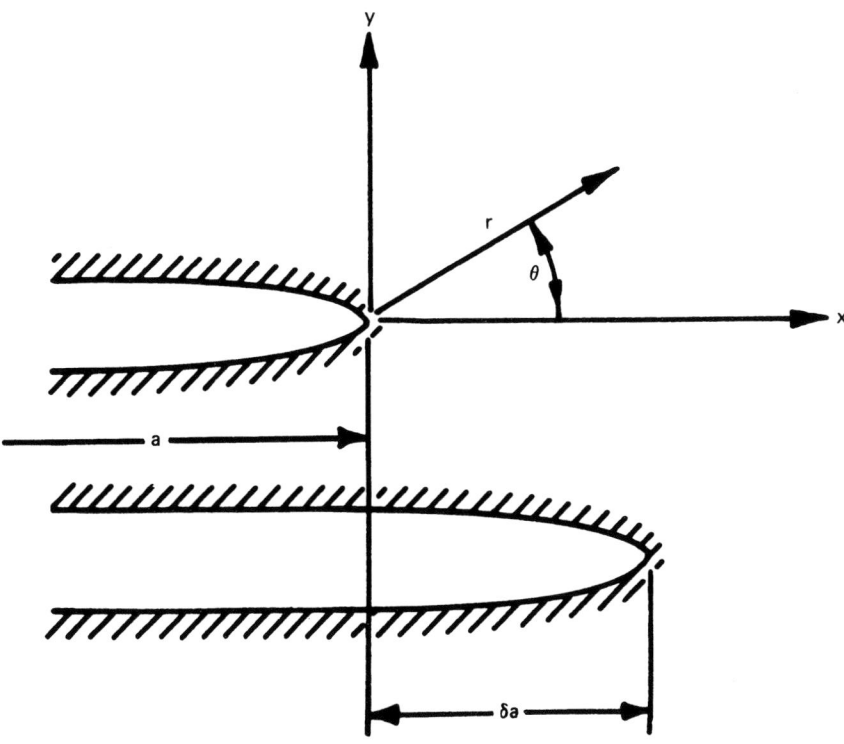

Figure 20-3 Crack extension in its plane.

where

$$G_I = \frac{1 - \nu^2}{E} K_I^2 \tag{20-9a}$$

$$G_{II} = \frac{1 - \nu^2}{E} K_{II}^2 \tag{20-9b}$$

$$G_{III} = \frac{1 + \nu}{E} K_{III}^2 \tag{20-9c}$$

which demonstrates the equivalence of the concepts of energy balance and stress intensity. Equations (20-9a) and (20-9b) are for plane strain. For plane stress the term $1 - \nu^2$ would be absent.

Strain-Energy Density Criterion

The strain-energy density criterion, $U_v = dU/dV$, developed by Sih [29], is based on the local density of the strain-energy field surrounding a crack. It assumes that fracture initiates when U_v reaches a critical value of U_c at a critical distance r_c from the crack tip.

The elastic strain energy U per unit volume V of the material is

$$U_v = \frac{1}{2E}(\sigma_{xx}^2 + \sigma_{yy}^2 + \sigma_{zz}^2) - \frac{\nu}{E}(\sigma_{xx}\sigma_{yy} + \sigma_{yy}\sigma_{zz} + \sigma_{zz}\sigma_{xx})$$
$$+ \frac{1+\nu}{E}(\sigma_{xy}^2 + \sigma_{zx}^2 + \sigma_{yz}^2) \qquad (20\text{-}10)$$

For a plane-strain condition,

$$\sigma_{zz} = \nu(\sigma_{xx} + \sigma_{yy}) \qquad (20\text{-}11)$$

and thus

$$U_v = \frac{dU}{dV} = \frac{1}{2E}[(\sigma_{xx}^2 + \sigma_{yy}^2) - 2\nu(\sigma_{xx}\sigma_{yy}) + 2(1+\nu)\sigma_{xy}^2] \qquad (20\text{-}12)$$

where E and ν are the modulus of elasticity and Poisson's ratio, respectively.

The energy density field U_v in the vicinity of the crack border may be expressed in the well-known form

$$U_v = \frac{S}{r} + \text{nonsingular terms} \qquad (20\text{-}13)$$

The coefficient of $1/r$ in Eq. (20-13) is referred to as the strain-energy density factor, S. Since the strain-energy density U_v approaches infinity as r tends to zero, a finite $U_v = U_c$ is evaluated at a critical distance r_c. The region inside r_c, where the material is expected to be nonlinear elastic or plastic, is called the core region.

The intensity of the strain energy density for the state of plane strain at a critical distance r_c can be written as

$$S_c = r_c(U_c) \qquad (20\text{-}14)$$

or

$$S_c = r_c\left\{\frac{1}{2E}[(\sigma_{xx}^2 + \sigma_{yy}^2) - 2\nu(\sigma_{xx}\sigma_{yy}) + 2(1+\nu)\sigma_{xy}^2]\right\} \qquad (20\text{-}15)$$

In all the foregoing, a straight crack front has been assumed. This allows treatment of modes I and II as either plane stress or plane strain, while mode III is separately treated as "antiplane" strain. When real cracks grow, however, even in "two-dimensional" problems, the crack fronts are not straight, and when boundaries and loadings are complex, the cracks may also be nonplanar. For the case of plane cracks in finite-thickness bodies, designers have often assumed the existence of plane strain in the central region and plane stress where the crack front intersects the boundary. Studies [30–33] have shown that the order of the stress singularity is different (where a crack intersects a free surface at right angles) from -0.5 and is a function of Poisson's ratio. For structural materials where $\nu = 0.3$, this effect is likely to be negligible. However, for rubberlike materials, rocket propellants, and so on, exhibiting higher values of Poisson's ratio, the effect may be significant. When a straight-front crack in a plate curves as it grows, one may conjecture that it tries to reestablish the inverse-square-root singularity along the crack front, but at the free surface, the new geometry is likely to be highly three-dimensional.

Thus it is well to remember that all crack problems are really three-dimensional, although our measurements, and often our algorithms for converting into SIF values, measurements in which three-dimensional effects are present, involve averaging through some thickness.

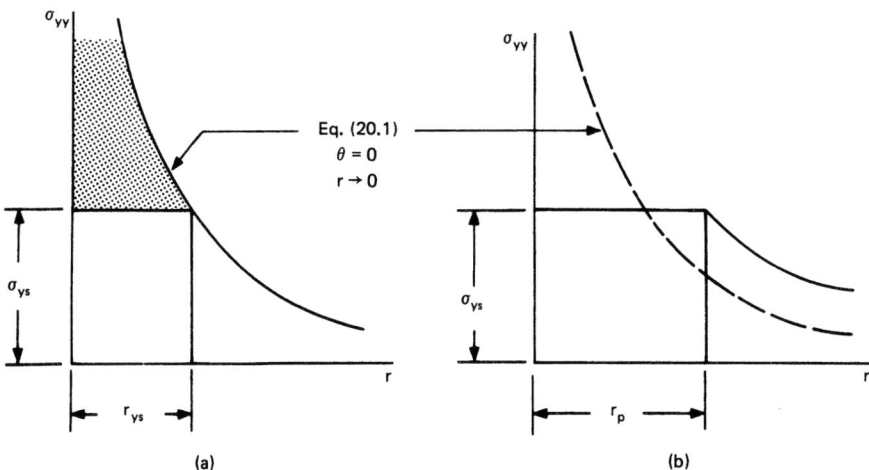

Figure 20-4 Redistribution of singular stresses due to the presence of the plastic zone (a) before and (b) after redistribution.

Elastoplastic Fracture Mechanics: The Irwin r_p Plasticity Adjustment Factor

As a natural extension of linear elastic fracture mechanics, Irwin [12,34] suggested that the effect of increased compliance due to the small-scale yielding at the crack tip could be simulated by a hypothetical extension r_p of the crack tip. Following Irwin and Paris [35] and Broek [36], if one assumes that the local mode I stress σ_{yy} [Eqs. (20-1)] along $y = 0$ reaches the yield strength at a distance r_{ys} from the physical crack tip, then (see Fig. 20-4a)

$$\sigma_{ys} = \frac{K_I}{(2\pi r_{ys})^{1/2}} \tag{20-16}$$

However, redistribution of the elastic stress represented by the shaded region of Fig. 20-4a produces further yielding, so that the elastic stress distribution must be moved to the right, modifying the stress distribution to that of Fig. 20-4b. This requires that

$$\int_0^{r_{ys}} \frac{K_I}{(2\pi r)^{1/2}} \, dr = r_p \sigma_{ys}$$

whence, noting that $r_{ys}^{1/2} = [1/(2\pi)^{1/2}] (K_I/\sigma_{ys})$, there results

$$r_p = \frac{1}{\pi} \left(\frac{K_I}{\sigma_{ys}} \right)^2 \tag{20-17}$$

Irwin suggested that if the physical crack length were increased by $r_p/2$ (Fig. 20-5), Eqs. (20-1) and (20-2) could still be used to calculate the near-tip stresses and displacements and the influence of the plastic zone would be approximately accounted for. The expression $K = Y\sigma a^{1/2}$ as adjusted by Irwin becomes

$$K = Y\sigma \left(a + \frac{r_p}{2} \right)^{1/2}$$

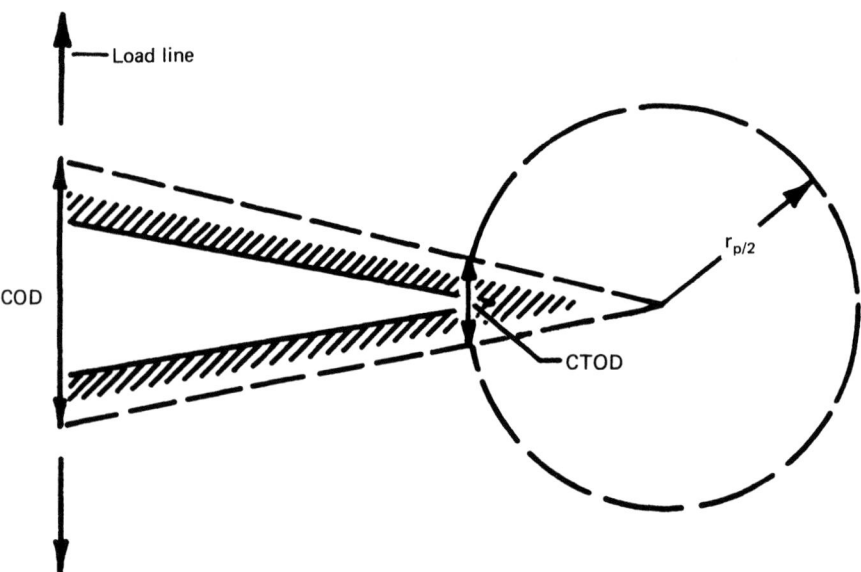

Figure 20-5 Crack-opening displacement (COD) at crack tip (CTOD) as influenced by the plastic zone.

where Y is the geometric correction factor, σ the applied stress, and a the characteristic crack length. For the state of plane strain, σ_{ys} is elevated to $\sqrt{3}\sigma_{ys}$ by the von Mises yield condition.

The plastic adjustment factor can also be used to estimate the increase in the crack-opening displacement (COD) due to the crack-tip stretch caused by local yielding. When the half-crack length a is replaced by the extended crack length of $a + r_p/2$ in the COD equation for an embedded through-crack in an infinite plate under uniform tension, the COD becomes

$$\text{COD} = 2u_y = \frac{4\sigma}{E}\left[\left(a + \frac{r_p}{2}\right)^2 - x^2\right]^{1/2} \tag{20-18}$$

where x is the distance along the crack from its center and E is the modulus of elasticity. The crack-opening stretch, which is the crack-tip opening displacement (CTOD) (Fig. 20-5), is obtained by

$$\text{CTOD} = 2u_y = \frac{4\sigma}{E}\left[\left(a + \frac{r_p}{2}\right)^2 - a^2\right]^{1/2}$$

$$\text{CTOD} \approx \frac{4\sigma}{E}(ar_p)^{1/2} = \frac{4K^2}{\pi E \sigma_{ys}} \tag{20-19}$$

Wells [13] derived the strain-energy release rate G using Eq. (20-18) and showed that the same CTOD of Eq. (20-19) is obtained for both plane-stress and plane-strain conditions.

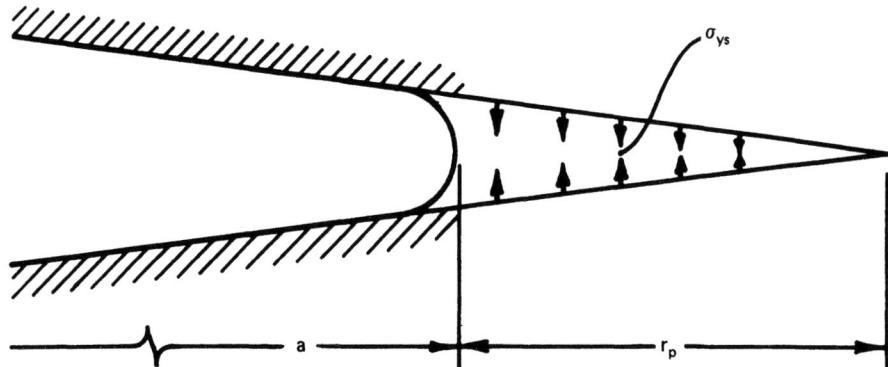

Figure 20-6 Dugdale strip yield zone at the crack tip of a semi-infinite crack.

Since its inception by Wells, the CTOD criterion for ductile fracture has been promoted by the British Welding Society, but difficulties [37] in computing as well as measuring the physical CTOD have hampered its wide acceptance. Recent developments in computational techniques, however, have reduced, if not eliminated, the former difficulties, and thus the CTOD criterion is receiving more attention [38,39] despite a decade of U.S. efforts in promoting the J-integral criterion.

Strip Yield Model

When thin fracture specimens of ductile metals are subjected to subcritical loading, a visible strip of necking often extends from the stationary or subcritical growing crack tip. Dugdale [40] modeled this neck region as an extension of the physical crack loaded with the yield strength as the prescribed tensile forces, which pull the two opposing hypothetical crack surfaces together, as shown in Fig. 20-6, by superimposing two elastic solutions. Similar modeling based on a continuous distribution of dislocations was advanced by Barenblatt [41] and Bilby et al. [42]. With the postulated surface tractions on the hypothetically extended crack tip, the elastoplastic problem in two dimensions reduces to a more manageable two-dimensional elastic problem.

The length of the Dugdale strip yield zone r_p is chosen to eliminate the singular state of stress at the physical crack tip. For a through-crack in an infinite solid subjected to uniform tension, the Dugdale strip yield zone length r_p can be computed from

$$\frac{a}{a + r_p} = \cos \frac{\pi \sigma}{2\sigma_{ys}} \qquad (20\text{-}20)$$

For $\sigma \ll \sigma_{ys}$, or in the presence of K_I-dominated small-scale yielding, r_p becomes

$$r_p = \frac{\pi K^2}{8\sigma_{ys}^2} \qquad (20\text{-}21)$$

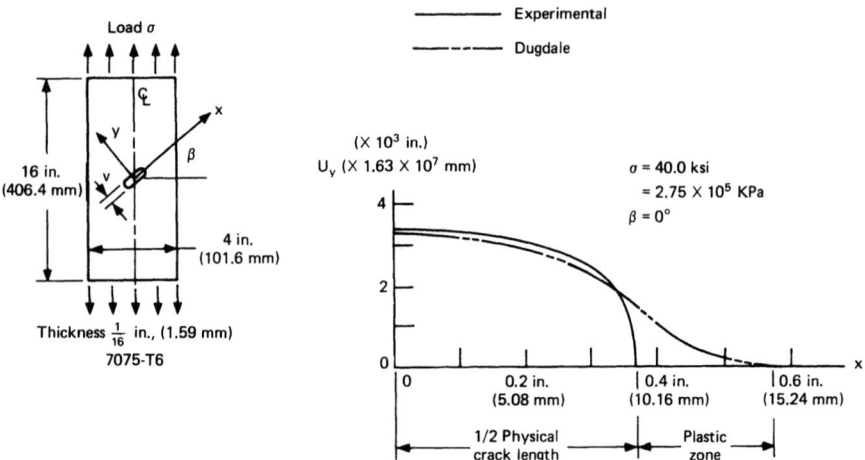

Figure 20-7 Experimental and theoretical crack-opening displacement in a centrally notched aluminum plate 7075-T6. (From Ref. 46.)

Furthermore, the CTOD at the physical crack tip for plane stress becomes

$$\text{CTOD} = 2u_y = \frac{8\sigma_{ys}a}{\pi E} \log \frac{a + r_p}{a} \tag{20-22}$$

The mathematical simplicity of the Dugdale strip yield model and its experimental support [43–45] have thus enabled this model to survive the onslaught of recent literature on ductile fracture. Figure 20-7 shows an experimental verification [46] of the CODs computed by using the Dugdale strip yield model. In particular, good agreement between the two CTODs is noted.

J Integral

The path-independent integrals in two-dimensional plastic solids, which were first presented by Eshelby [47] in 1956, gained sudden popularity with the introduction of Rice's J integral [48] in 1968. Using the deformation theory of plasticity, Rice showed that the J integral is equal to the

where μ = shear modulus
ν = Poisson's ratio

For the state of plane strain, the displacement components become

The J integral as defined by Rice [48] is

$$J = \int_s \left(U_v \, dx_2 - \sigma \frac{\partial u_i}{\partial x_1} \, ds \right) \tag{20-23}$$

where $U_v = U_v(\varepsilon_{ij})$, the strain-energy density
σ_i = components of stress vector on S (Fig. 20-8)
u_i = components of displacement on S (Fig. 20-8)

For a crack in a linear elastic solid, the J integral is equal to Irwin's strain energy release rate given by Eq. (20-9).

Other conservation integrals for elastostatics include those of Knowles and Sternberg [55], which was reinterpreted for fracture mechanics by Budiansky and Rice [56]. More recently, Atluri [57] has derived a path-independent integral incremental vector that can be used under finite-deformation, rate-independent dynamic elastoplastic conditions. This integral incremental vector could be particularly useful in characterizing crack extension under cyclic loading conditions in viscoplastic solids.

Elastodynamic Fracture Mechanics

Consider an arbitrary motion of a crack tip in a plane subjected to a general exterior stress field with a crack speed less than the characteristic Rayleigh-wave velocity. This dynamic state of stress is given by Freund and Clifton [58] in terms of the local rectangular (x, y) and polar (r, θ) coordinates, when the crack velocity vector is parallel to the x direction, as shown in Fig. 20-2b. When the second-order term of σ_{0x}, which is acting parallel to the direction of crack extension, is added to Freund's [58] near-field solution for a dynamic state of stress for a running crack, the following are obtained [59]:

$$\sigma_{xx} = \frac{K_I^{dyn}}{\sqrt{2\pi r}} F_{11}(c, c_1, c_2, \theta) + \frac{K_{II}^{dyn}}{\sqrt{2\pi r}} G_{11}(c, c_1, c_2, \theta) + \sigma_{0x}$$

$$\sigma_{yy} = \frac{K_I^{dyn}}{\sqrt{2\pi r}} F_{22}(c, c_1, c_2, \theta) + \frac{K_{II}^{dyn}}{\sqrt{2\pi r}} G_{22}(c, c_1, c_2, \theta) \qquad (20\text{-}24a)$$

$$\sigma_{xy} = \frac{K_I^{dyn}}{\sqrt{2\pi r}} F_{12}(c, c_1, c_2, \theta) + \frac{K_{II}^{dyn}}{\sqrt{2\pi r}} G_{12}(c, c_1, c_2, \theta)$$

where

$$F_{11}(c, c_1, c_2, \theta) = B_I(c)\left[-(1 + 2S_1^2 - S_2^2)f_{11} - \frac{4S_1S_2}{1 + S_2^2}f_{22}\right]$$

$$G_{11}(c, c_1, c_2, \theta) = B_{II}(c)\left[-(1 + 2S_1^2 - S_2^2)g_{11} - \frac{4S_1S_2}{1 + S_2^2}g_{22}\right]$$

$$F_{22}(c, c_1, c_2, \theta) = B_I(c)\left[-(1 + 2S_2^2)f_{11} + \frac{4S_1S_2}{1 + S_2^2}f_{22}\right] \qquad (20\text{-}24b)$$

$$G_{22}(c, c_1, c_2, \theta) = B_{II}(c)[(1 + 2S_2)g_{11} - (1 + S_2^2)g_{22}]$$

$$F_{12}(c, c_1, c_2, \theta) = B_I(c)[2S_1(g_{11} - g_{22})]$$

$$G_{12}(c, c_1, c_2, \theta) = B_{II}(c)\left[\frac{1}{2S_2}(4_1S_2)f_{11} - (1 + S_2^2)^2 f_{22}\right]$$

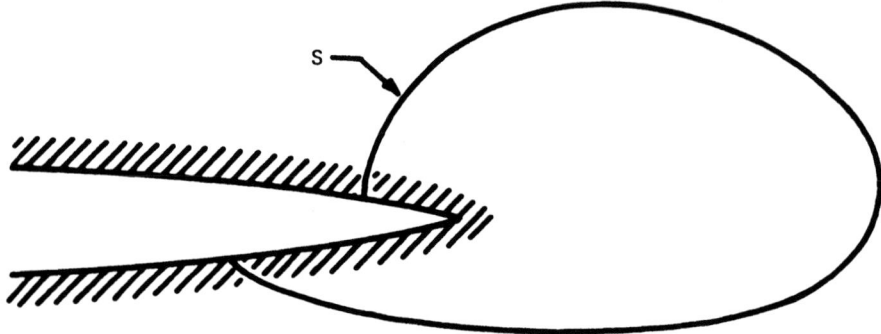

Figure 20-8 *J*-integral contour around a crack tip.

σ_{0x} is the remote stress or the nonsingular stress

$$f_{11} = [f(c_1) + g(c_1)]^{1/2}$$

$$g_{11} = [f(c_1) - g(c_1)]^{1/2}$$

$$f_{22} = [f(c_2) + g(c_2)]^{1/2}$$

$$g_{22} = [f(c_2) - g(c_2)]^{1/2} \qquad (20\text{-}24c)$$

$$f(c_1) = \frac{1}{[1 - (c^2/c_1^2)\sin^2\theta]^{1/2}} \qquad g(c_1) = \frac{\cos\theta}{1 - (c^2/c_1^2)\sin^2\theta}$$

$$f(c_2) = \frac{1}{[1 - (c^2/c_2^2)\sin^2\theta]^{1/2}} \qquad g(c_2) = \frac{\cos\theta}{1 - (c^2/c_2^2)\sin^2\theta}$$

$$B_I(c) = \frac{1 + S_2^2}{4S_1 S_2 - (1 + S_2^2)^2} \qquad B_{II}(c) = \frac{2S_2}{4S_1 S_2 - (1 + S_2^2)^2}$$

$$S_1^2 = 1 - \frac{c^2}{c_1^2} \qquad S_2^2 = 1 - \frac{c^2}{c_2^2}$$

where c, c_1, and c_2 are crack, dilatational-wave, and distortional-wave velocities, respectively.

The dynamic singular crack-tip stress field for small values of c/c_1 differs from the corresponding static stress field in that the largest principal singular tensile stress acts parallel to the y axis not only in mode I loading but also in mixed-mode loading for small values of K_{II}/K_I. Furthermore, the ratio of σ_{xx}/σ_{yy} is roughly equal to unity for both static and dynamic loads up to c/c_1 of 0.10. Situations arise however, for which a single-parameter characterization of K_I or K_{II}, in mode I or mode II, is not adequate because of spreading of the fracture process zone or reduction in size of the K-dominated singular zone. For example, even in a brittle solid, roughening of the fracture surface due to spreading of advanced cracking and incipient branching can subsequently enlarge the fracture process zone. As a second example, when dynamic photoelasticity is used to evaluate K_I and K_{II}, the isochromatic fringes very close to the crack tip normally cannot be used for practical reasons, such as light scattering from the caustic at the crack tip, crack-front curvature, and fringe clarity. The experimental stress analyst is thus frequently forced to take measurements from regions that border the valid region of the foregoing crack-tip field equations. In such cases, it is desirable to incorporate an additional stress field parameter, σ_{0x}. This inevitable involvement of the second-order term forms the basis of incorporating σ_{0x} in dynamic mixed-mode analysis of a moving crack.

20-2 Experimental Methods

It will be convenient to divide a study of experimental methods for determining fracture parameters into two parts: those involving measurements on

1. Opaque bodies
2. Transparent bodies

SIF Determination in Opaque Bodies

Four basic techniques have been employed to measure the SIF in opaque bodies, and the basis for each will be briefly discussed here. Specific details of the experimental techniques have been described earlier and are not repeated.

Compliance Method

When a cracked body is loaded with a force P to produce a load point displacement u (Fig. 20-9), we may define the compliance of the body by

$$C = \frac{u}{P} \tag{20-25}$$

or for an infinitesimal crack extension da,

$$C = \frac{du}{dP} \tag{20-26}$$

If the crack then extends an amount da under the load P, the strain-energy release rate may be expressed as

$$G = P\frac{du}{da} - \frac{dU}{da} \tag{20-27}$$

where

$$U = \tfrac{1}{2}pu = \tfrac{1}{2}P(CP) = \tfrac{1}{2}CP^2 \quad \text{and} \quad \frac{du}{da} = \frac{d(CP)}{da} = \frac{PdC}{da} + \frac{CdP}{da}$$

Thus, from Eq. (29-27),

$$G = \frac{P^2}{2}\frac{dC}{da} \tag{20-28}$$

per unit thickness of the body. Thus, from Eq. (20-9) for a body of thickness h, we have, for a mode I state of plane strain,

$$K_I^2 = \frac{E}{1-\nu^2}\frac{P^2}{2h}\frac{dC}{da} \tag{20-29}$$

Thus, by measuring the load and compliance of a cracked body for various crack lengths, K_I can

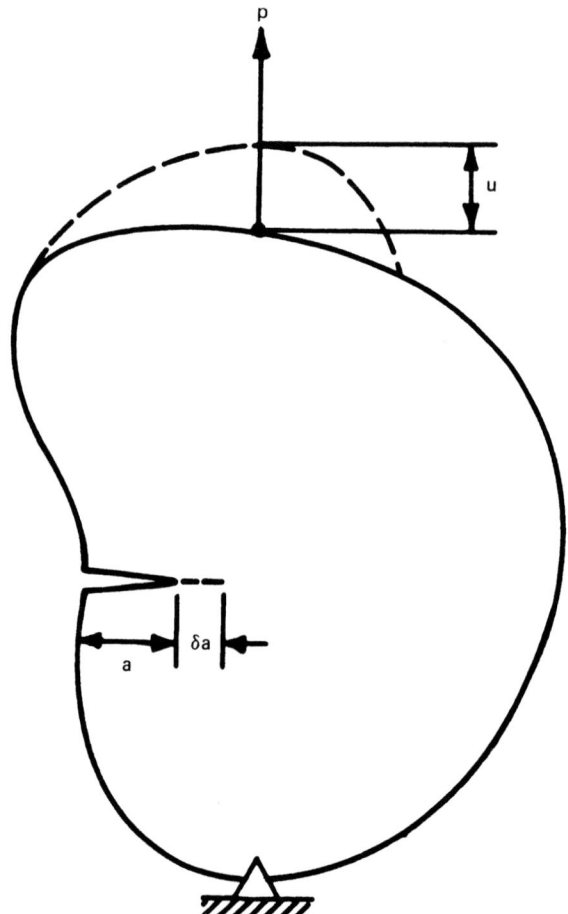

Figure 20-9 Cracked body for compliance calculation.

be determined. The procedure for generating data for use in Eq. (20-29) is described in Fig. 20-10. A detailed description of the compliance method is found in Ref. 60.

Although used primarily in dealing with cases of moderate crack-tip plasticity where LEFM may not apply, the J integral, which is described in the preceding section, can be used in LEFM, where

$$J = G \qquad (20\text{-}30)$$

Because K is expressible in terms of G (Eqs. 20-9), it follows that the experimental procedure for determining J is the same as for G for the LEFM case. The experiments may be carried out in one of two ways, as depicted in Fig. 20-11, where the shaded areas indicate the energy released due to an increment of crack growth, and can be shown to be equivalent, as follows.

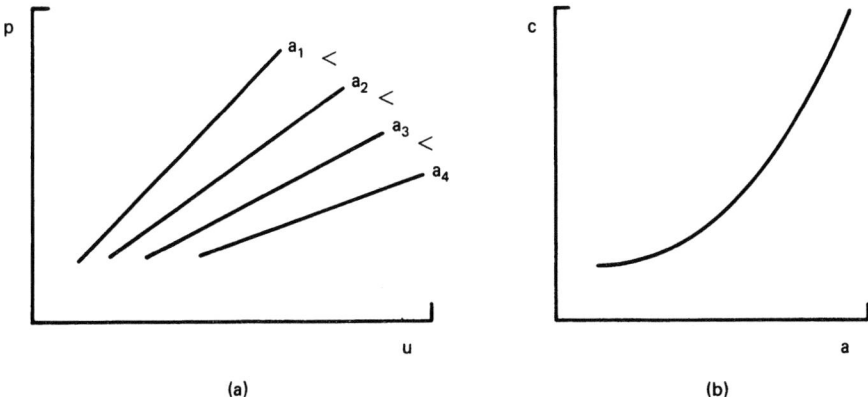

Figure 20-10 Determination of K_I from compliance: (a) determine slopes $C = du/dP$ for each crack length and (b) plot by measuring dC/da and computing K_I from Eq. (20-29).

From Fig. 20-11a:

$$dU = \frac{Pu}{2} + P\,du - \frac{P(u+du)}{2} = \frac{P\,du}{2} = \frac{PC\,dP}{2}$$

From Fig. 20-11b:

$$dU = \frac{Pu}{2} - \frac{(P-dP)u}{2} = \frac{u\,dP}{2} = \frac{CP\,dP}{2} \tag{20-31}$$

Thus G (or J) can be measured experimentally from a load displacement record as one of the shaded areas of Fig. 20-11 divided by the crack-growth increment. One may then determine K from Eq. (20-9).

The more important use of the J integral is the case in which a plastic zone is present. Assuming the material behavior to be adequately represented as nonlinearly elastic, one can still compute J, but $J \neq G$ here. However, J for this case is often used as a fracture parameter, which includes moderate plasticity.

As mentioned earlier, the J integral has been used together with the "L" integral [56] to determine the K_I and K_{II} stress intensity factors under mixed-mode loading conditions in LEFM [61]. For a through crack of length $2a$ in an infinite plate subjected to far-field loading of σ_{xx}, σ_{yy}, and σ_{xy},

$$L = \frac{-2aK_{II}(K_I + \sigma_{xx}\sqrt{\pi a})}{E} \tag{20-32}$$

$$J = \frac{K_I^2 + K_{II}^2}{E} \tag{20-33}$$

By evaluating experimentally the two contour integrals of Eqs. (20-32) and (20-33), K_I and K_{II} can be determined numerically. This determination of the integrands of J and L requires the knowledge of all stress components and rotation components along the contour. In a novel experimental approach, King and Herrmann [61] measured the variations in $\sigma_{xx} + \sigma_{yy}$ along two 12-mm-square regions surrounding the crack tip of a slanted crack in a 6061 T6 aluminum tension plate by acoustoelasticity.

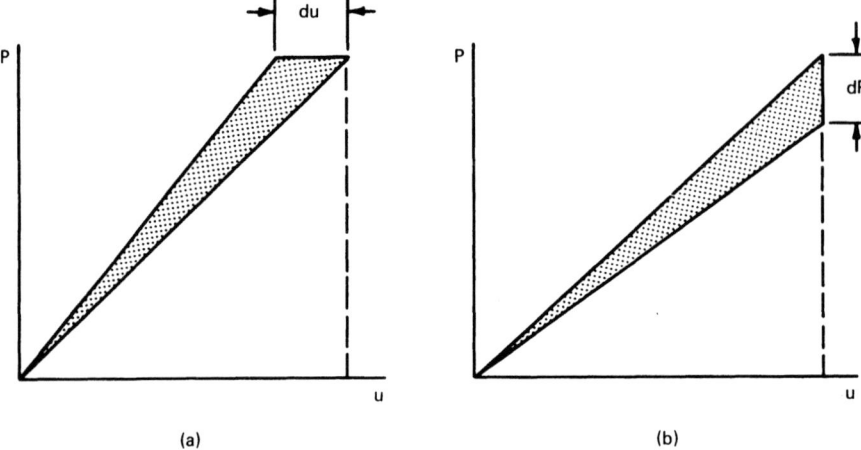

Figure 20-11 Conditions for calculation of strain-energy release rate: (a) crack extension under constant load; (b) crack extension under fixed displacement.

Acoustoelasticity, which was hailed as an analog to photoelasticity for opaque materials in 1959 [62], failed to achieve wide acceptance because of the unresolved transducer coupling effect and high sonic attenuation [63]. The resurgence of acoustoelasticity in the 1980s was due to improvements in instrumentation techniques but was mainly attributed to the ability to process large amounts of ultrasonic data by means of a computer-controlled scanning system [64–66].

Because the longitudinal ultrasonic waves provide the plane-stress isopachics, King and Herrmann used the ratios of the experimentally determined values versus corresponding theoretical values for an infinite plate to rescale the theoretical stress and rotation, which were then used to numerically determine the J and L values. While an error of 14% was reported, the procedure, with necessary improvements in the instrumentation and numerical techniques, shows potential for separation of K_I and K_{II} in mixed-mode loading conditions.

Clark et al. [67], on the other hand, used shear-wave measurements, which are referred to as acoustic birefringence, to determine point by point the acoustic birefringence in a 51 mm × 51 mm region of a 2024-T6 aluminum compact specimen. A 10-MHz ac-cut quartz shear-wave transducer of 1.8 mm diameter was used in a manual scanning process, and the estimated accuracy was approximately 5% and ±2° in birefringence and its direction measurements, respectively.

The acoustoelastic birefringence generated from 66 data points in the square region was reconstructed, and Sanford's procedure [68] was used to compute five coefficients in the LEFM crack-tip stress field by averaging the results of 100 computations using 20 randomly selected data points each time. Good agreement was noted between the corresponding coefficients, which were obtained from a similar photoelasticity experiment. The stress-intensity factor was computed from the coefficient of the first term, or the $1/\sqrt{r}$ term, in Eqs. (20-1).

The acoustoelastic technique is one of the few static stress analysis techniques available for opaque materials. As in two-dimensional photoelasticity, the thickness-averaged acoustic birefringence is not subject to the plane-stress constraint of the caustic method. Obvious improvement in the technique can be made by incorporating an automated scanning procedure with real-time data processing, which has been used by others [64]. Yet to be explored is the physical significance of acoustic birefringence associated with the crack-tip plastic region associated with ductile fracture.

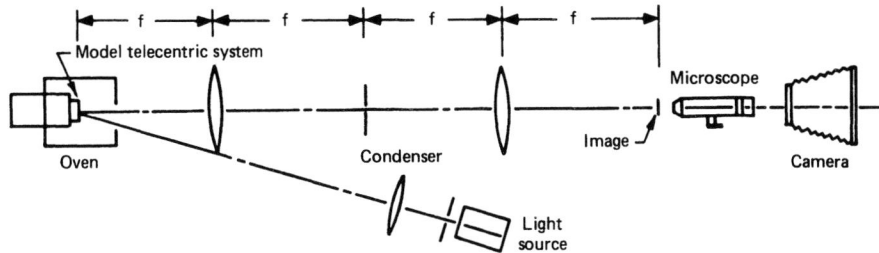

Figure 20-12 Optical setup for high-temperature studies. (From Ref. 71.)

Crack-Opening Displacement (COD) and Crack-Tip-Opening Displacement (CTOD)

As with the J integral, the COD method for obtaining the SIF is usually applied to problems involving small to moderate amounts of plasticity. However, it also provides a workable experimental method for determining the SIF for the case of LEFM where only small-scale yielding occurs. For example, for a through crack in an infinite flat plate under mode I loading, $K_I = \sigma(\pi a)^{1/2}$, so we use $K_I = \sigma[(\pi a + \pi r_p/2)]^{1/2}$ in Eq. (20-2b) for u_y; there results for the CTOD at the tip of the physical crack:

$$\text{CTOD} = 2u_y\left(\frac{r_p}{2}, \pi\right) = \frac{4(1-\nu^2)\sigma}{E}(ar_p)^{1/2} \tag{20-34}$$

or from Eq. (20-17),

$$\text{CTOD} = \frac{4(1-\nu^2)}{\pi E} \frac{K_I^2}{\sigma_{ys}} \tag{20-35}$$

Thus, by measuring the COD at the crack-tip (CTOD), K_I can be calculated.

Moiré Technique

The use of the moiré technique in elastoplastic fracture mechanics is not new [69,70]. Despite its obvious application to high-temperature nonlinear problems in fracture mechanics, the literature is relatively sparse in the fracture mechanics interpretation of the crack-tip displacement field determined by the moiré method. An exception is found in the analysis of externally notched rings sliced from a type 304 stainless tube, 7.1 mm OD and 0.38 mm thick, with electro-etched cross-line gratings of 40 lines/mm and subjected to a simulated internal pressure at 110°F [71]. Figure 20-12 shows the experimental setup for recording the distorted grating, which was analyzed by master gratings of 4 and 8 lines/mm. For the resultant u_x and u_y moiré fringe patterns the CTOD for slow-crack-growth initiation was found to be

$$\text{CTOD} = 0.976a\sigma^{5.78} \tag{20-36}$$

where the crack length a and the applied hoop stress are represented in millimeters and kilopascals, respectively.

Figure 20-13 shows that the initiation COD in this experiment remained relatively constant

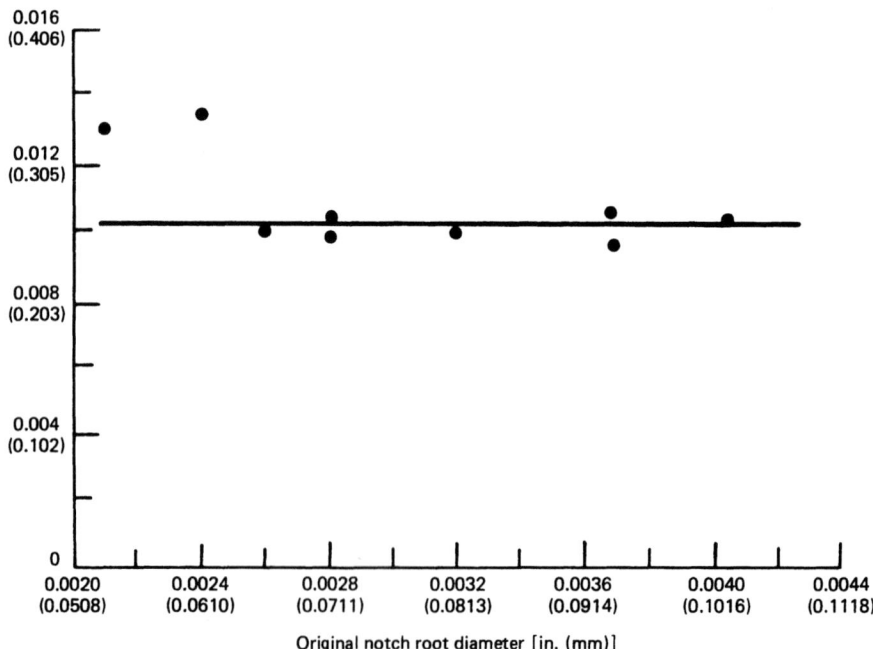

Figure 20-13 COD for initiation of slow crack growth as function of original root diameter. (From Ref. 71.)

despite the changes in the crack-tip bluntness. Sciammarella [71] then estimated the J integral for the initiation of slow crack growth by the following approximate formula, after Rice et al. [72]:

$$J_{in} = \frac{1}{bt}\left(2\int_0^{\delta_{cr}} P\, d\delta - P_T \delta_{cr}\right) \qquad (20\text{-}37)$$

where b is the ligament length, t the specimen thickness, δ_{cr} the displacement due to the presence of the crack between two reference sections for the load at the moment of crack initiation, and

$$P = \sigma A \qquad (20\text{-}38)$$

where σ is the hoop stress and A the specimen cross-sectional area. The values of δ_{cr} were obtained as

$$\delta_{cr} = \delta_{\text{total}} - \delta_{\text{no cr}} \qquad (20\text{-}39)$$

where δ_{total} is the displacement between two reference cross sections and

$$\delta_{\text{no cr}} = \delta_h \qquad (20\text{-}40)$$

where δ_h is the increment of hoop strain.

Interferometry

In general, interferometry involves the interference between two rays of light. The interference may be constructive or destructive. Controlled constructive interference can maximize the light intensity, while destructive interference can be controlled to eliminate light intensity. The former occurs when the light rays are separated by an even number of half-wavelengths, and the latter results when the separation consists of an odd number of half-wavelengths.

Interferometric methods have been employed extensively to study near-tip plastic strain and associated displacement fields in opaque cracked-body problems [73–75]. However, instances of the use of such methods to determine the SIF are much rarer in the literature. A survey of the use of interferometry in fracture mechanics has been provided by Packman [76].

For opaque bodies, data are normally obtained from the surface of the body surrounding the region where the crack intersects the free surface of the body. The optical data, in the form of fringes (which are usually proportional to the in-plane displacements), must then be analytically related to the SIF. If the measurements are made outside a near-tip nonlinear zone, Eqs. (20-2) for u_i may be used for that purpose. Recently, two successful applications of this approach have been made:

Moiré Interferometry[3]

Using a "virtual" grating created by splitting a laser beam and then recombining the components to set up a three-dimensional array of walls of constructive and destructive interference, the first author and his associates [77,78] have been able to generate near-tip moiré displacement fringes by viewing a deformed grating through the virtual grating and obtaining SIF estimates using the procedure outlined above. A schematic of the optical setup is shown in Fig. 20-14a together with a near-tip moiré pattern for u_x (Fig. 20-14b).

Dynamic Measurements

Sharpe and Sukere have successfully applied [79] an interferometric technique developed by Sharpe [80] to measure crack-opening displacements near the surface of opaque materials under dynamic load. The method involves making indentations on either side of the crack and illuminating the cracked region with a laser at normal incidence to the specimen surface (Fig. 20-15). By measuring the change in indentation spacing through the interferometric fringe patterns produced by overlap of the light rays diffracted from the two indentations, the crack-opening displacements may be measured and converted to an instantaneous SIF value through an algorithm derived from Eqs. (20-2).

Other Concepts

Gerberich [73] and Swedlow and Gerberich [81] utilized birefringent coatings on the surface of opaque specimens to produce residual photoelastic isochromatic patterns[4] in the reflected light field for the primary purpose of studying near-tip plastic strains. Conceptually, SIF values at the specimen surface also could be obtained from the isochromatics in the loaded state. However, the works cited, as well as other studies [82] suggest that the accuracy of the SIF measurement by this method may be limited.

SIF Determination in Transparent Bodies

Transparent models of opaque structures may be constructed to allow the use of transmission optical methods in the determination of SIF values. These methods may be divided into three main categories:

---

[3] See also Chapter 7.
[4] See the next subsection for a discussion of the determination of SIF values from photoelasticity.

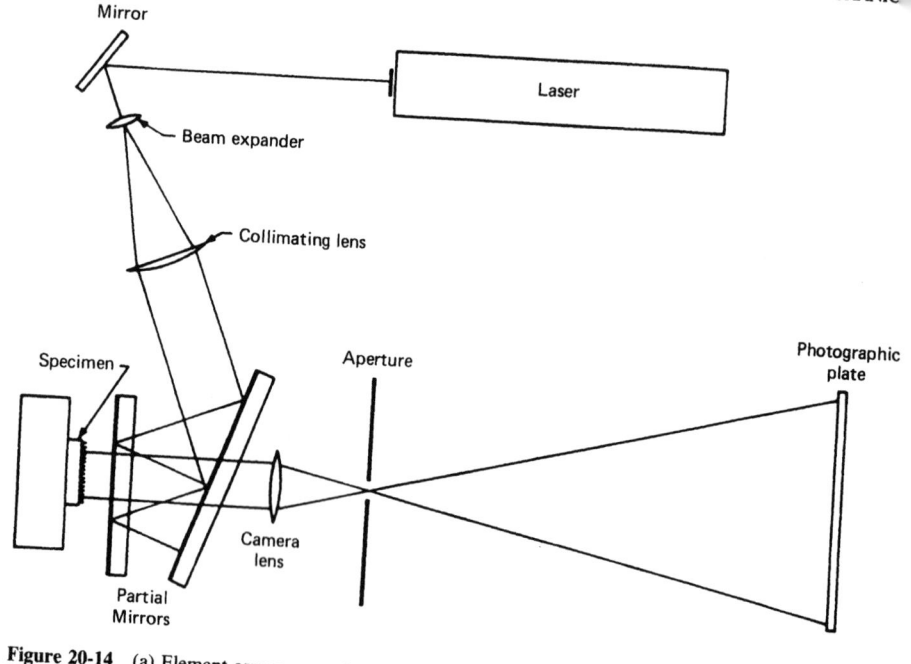

Figure 20-14 (a) Element arrangement for moiré analysis. (From Ref. 77.)

Figure 20-14 (b) A near-tip moiré fringe pattern. (From Ref. 77.)

EXPERIMENTAL FRACTURE MECHANICS

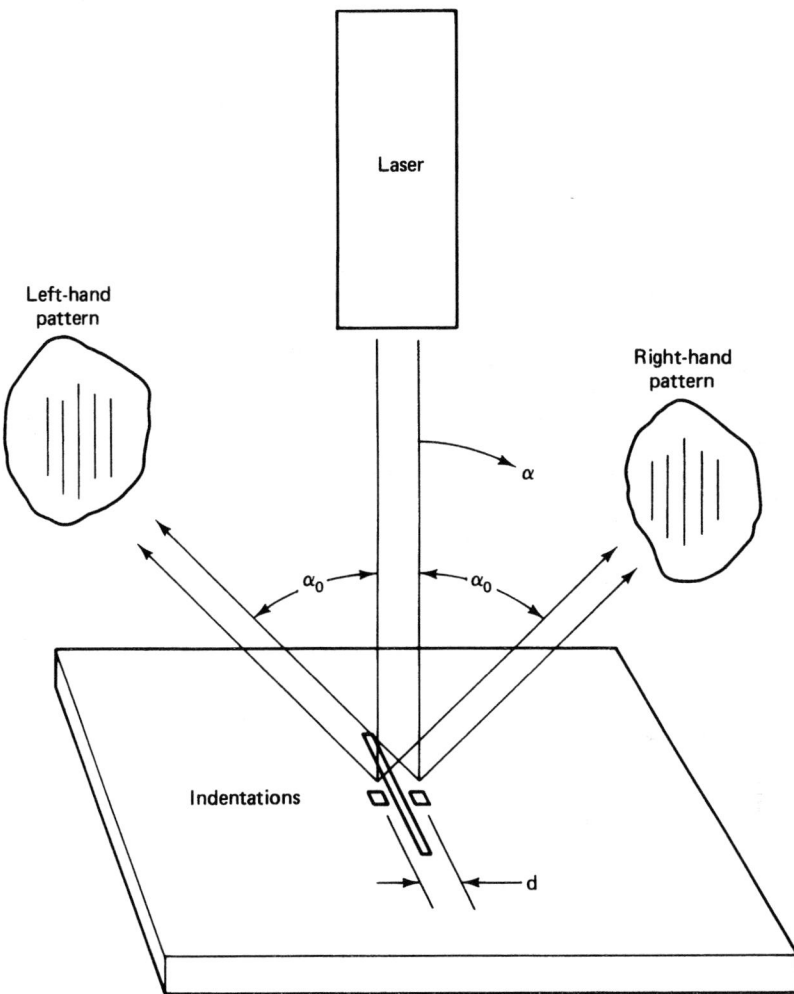

Figure 20-15 Displacement measurement by interferometric method. (From Ref. 79.)

1. Interferometric
2. Shadow or caustic method
3. Photoelasticity

Within these categories, several different methods have been developed for making optical measurements on loaded cracked bodies which can be, through suitable algorithms, converted into SIF values. Because the field theory of LEFM was initially developed for the plane case, we shall divide our discussion accordingly into:

Figure 20-16 Mode I stress fringe signature.

1. Two-dimensional methods, which include:
   a. Photoelasticity
   b. Crack-tip opening and holographic interferometry
   c. Shadow patterns (caustics)
2. Three-dimensional methods, which include:
   a. Frozen-stress photoelasticity
   b. Integrated frozen-stress moiré analysis
   c. Other concepts

Two-Dimensional Methods

Photoelastic Method

It was the shape or signature of the stress fringe pattern around a crack tip that led Irwin to the concept of the stress-intensity factor [83]. Because the stress fringe order is proportional to the magnitude of the maximum in-plane shearing stress, one may obtain a relation for τ_{max} from Eqs. (20-1) for mode I loading and then obtain a relation between the fringe order and the SIF in the following manner.

We first recognize that when the crack opens, we are prevented from taking data at the crack tip, as noted in Fig. 20-2b. Moreover, the mode I stress fringe signature (Fig. 20-16) suggests that the stress gradient near the crack tip is best defined along $\theta = \pi/2$. When we take data near but away from the crack tip, there will be some contribution to the stress fringe data from the nonsingular part of the stress field. To account for these effects, we may include in Eqs. (20-1), for stress, the components σ_{ij}^0 ($i, j = x, y$), where σ_{ij}^0 may be regarded as the Taylor-series expansion of the nonsingular stresses in the measurement zone. For a small zone sufficiently close to the crack tip, say within 1 mm of the crack tip, σ_{ij}^0 can be adequately approximated for the mode I case as being independent of r and replaces σ_{0x}. Thus within zone II of Fig. 20-2b,

$$\sigma_{ij} = \frac{K_I}{(2\pi r)^{1/2}} f_{ij}(\theta) - \sigma_{ij}^0(\theta) \qquad (i, j = x, y) \tag{20-41}$$

or along $\theta = \pi/2$,

$$\sigma_{ij} = \frac{K_I}{(2\pi r)^{1/2}} f_{ij}\left(\frac{\pi}{2}\right) - \sigma_{ij}^0\left(\frac{\pi}{2}\right) \qquad (i, j = x, y) \tag{20-42}$$

If we compute τ_{max} from Eqs. (20-42) and truncate to the same order as Eqs. (20-41), multiply by $(8\pi r)^{1/2}$, and normalize with respect to $\sigma(\pi a)^{1/2}$, where σ represents a remote unit stress, we obtain

$$\frac{K_{AP}}{\sigma(\pi a)^{1/2}} = \frac{K_I}{\sigma(\pi a)^{1/2}} + \frac{f(\sigma_{ij}^0)}{\sigma}\left(\frac{r}{a}\right)^{1/2} \tag{20-43}$$

where $f(\sigma_{ij}^0)$ are a set of constants and $K_{AP} = \tau_{max}(8\pi r)^{1/2}$, the "apparent" SIF. Equation (20-43) predicts that the normalized apparent SIF will vary linearly with the square root of the normalized distance of the measured data from the crack tip. By combining Eq. (20-43) with the photoelastic stress-optic law,

$$\tau_{max} = \frac{n'f'}{h} \tag{20-44}$$

where n' = photoelastic stress fringe order
h = model thickness
f' = material fringe value

one can compute values $K_{AP}/\sigma(\pi a)^{1/2}$ and $(r/a)^{1/2}$ and plot them. Figure 20-17 is an example of the result from which $K_I/\sigma(\pi a)^{1/2}$ is obtained.

When the algorithm above is applied to mixed-mode problems [84], it should be restricted to cases of $K_I > K_{II}$. However, this does not appear to be a serious restriction for growing cracks, for in the writer's experience, such cracks exhibit predominantly mode I fringe signatures (Fig. 20-18). However, Kobayashi et al. [85] have shown quite definitively that following branching of a rapidly propagating crack, a nearly pure shear mode may exist as a transient.

A number of other algorithms have been developed by other investigators [86–89], and these have been compared using near-tip data for both mode I and mixed-mode cases [90].

Interferometric Methods

When interferometric methods are employed with transparent materials, it is possible to illuminate the specimen with light normal or inclined to the crack plane and to retrieve optical information other than at the surface of the body. Sommer [91] focused attention on this approach in studying a center-cracked glass plate with a liquid-fluid crack loaded with oppositely directed crack-opening loads applied to the center of each crack face. He illuminated the crack surfaces with parallel monochromatic light. However, because of the loading arrangement, oblique incidence of the light to the specimen and crack surfaces was required. SIF values were computed from optical displacement data through a form of Eqs. (20-2). This approach was extended and summarized in Ref. 92. Crosley et al. [93], employing a double-cantilever-beam specimen, were able to simplify the optics by employing normal incidence, as illustrated in Fig. 20-19 for a specimen consisting

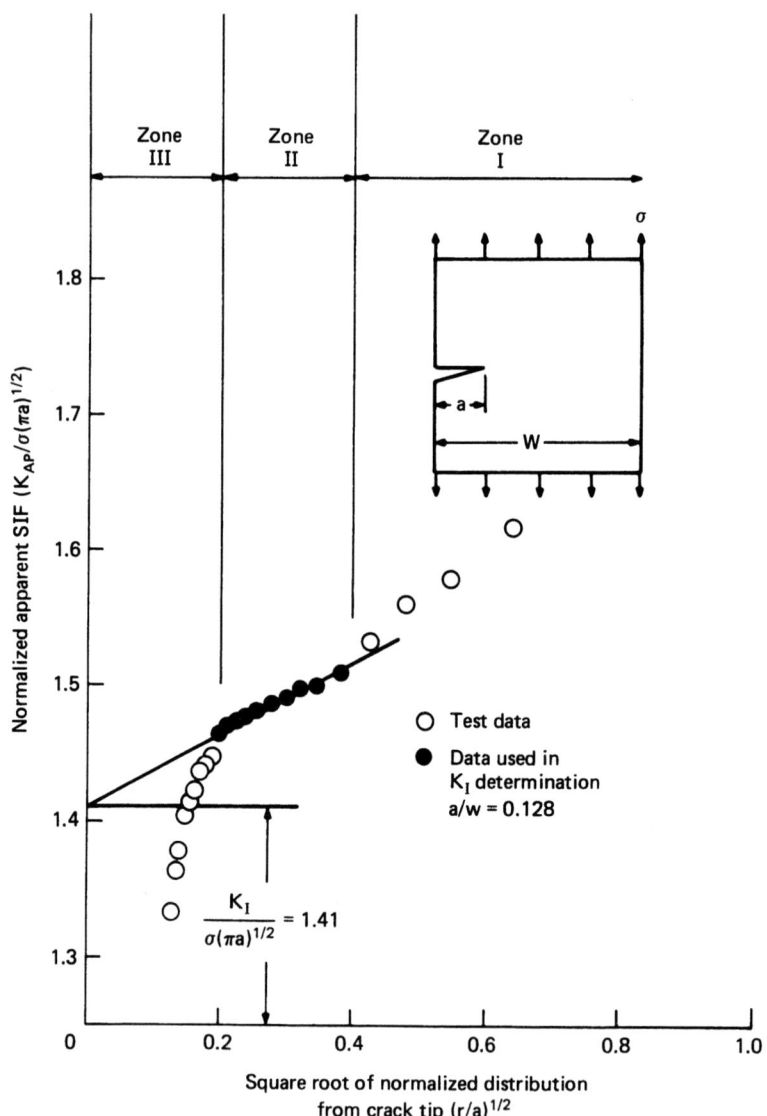

Figure 20-17 Estimation of K_I from a typical set of mode I photoelastic data. (From Ref. 102.)

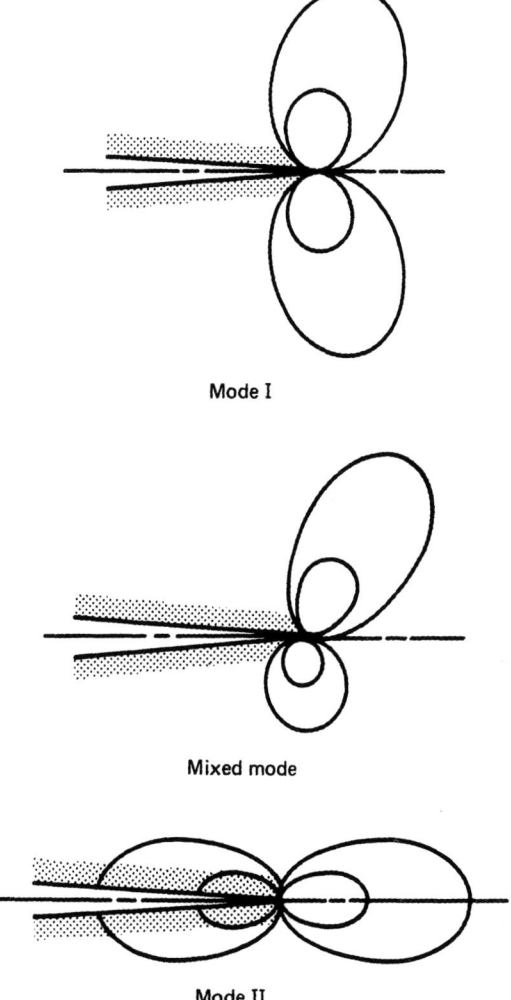

Figure 20-18 Three near-tip stress fringe signatures.

of two glass arms cemented together by a thin layer of epoxy. Since destructive interference occurred when

$$2u_y = \tfrac{1}{2}(n' - \tfrac{1}{2})\lambda' \tag{20-45}$$

where u_y = displacement of each cracked surface
n' = interferometric fringe order
λ' = wavelength of monochromatic light

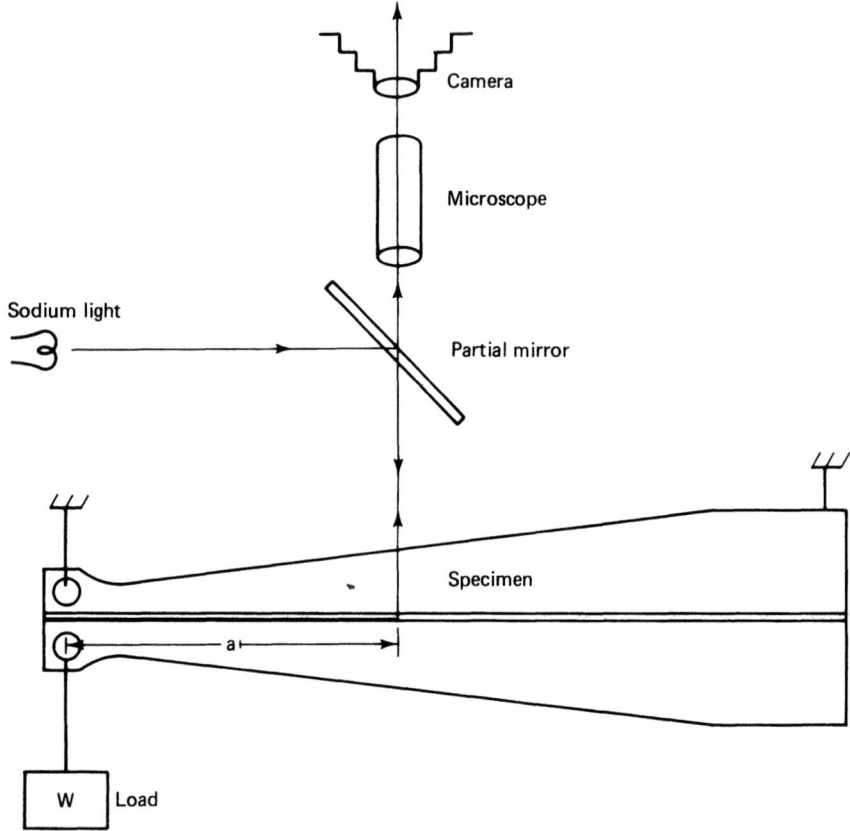

Figure 20-19 Schematic of test for interferometric crack-tip opening measurement. (From Ref. 93.)

then Eq. (20-2b) with u_y modified for plane stress could be used with Eq. (20-45) to compute the SIF. Comparisons were made with the empirical expression for K_I for the tapered double-cantilever-beam specimen, neglecting the influence of the adhesive bond layer. Equation (20-2) reduced to

$$u_y(r, \pi) = \left(\frac{4}{3\mu}\right) K_I \left(\frac{r}{2\pi}\right)^{1/2} \tag{20-46}$$

which predicts a parabolic crack profile. The fringe patterns, however, are available across the specimen thickness (Fig. 20-20), so three-dimensional variations in u_y could be obtained by this method.

A second interferometric method has been successfully used to estimate two-dimensional SIF values in transparent materials by Dudderar and O'Regan [94], who applied real-time transmission holographic interferometry to finite-width, edge-cracked Plexiglas specimens. In this method, a reference diffraction grating is formed by splitting and recombining two light beams. The grating will reconstruct one of the beams (object beam, Fig. 20-21) when replaced in its original position

EXPERIMENTAL FRACTURE MECHANICS

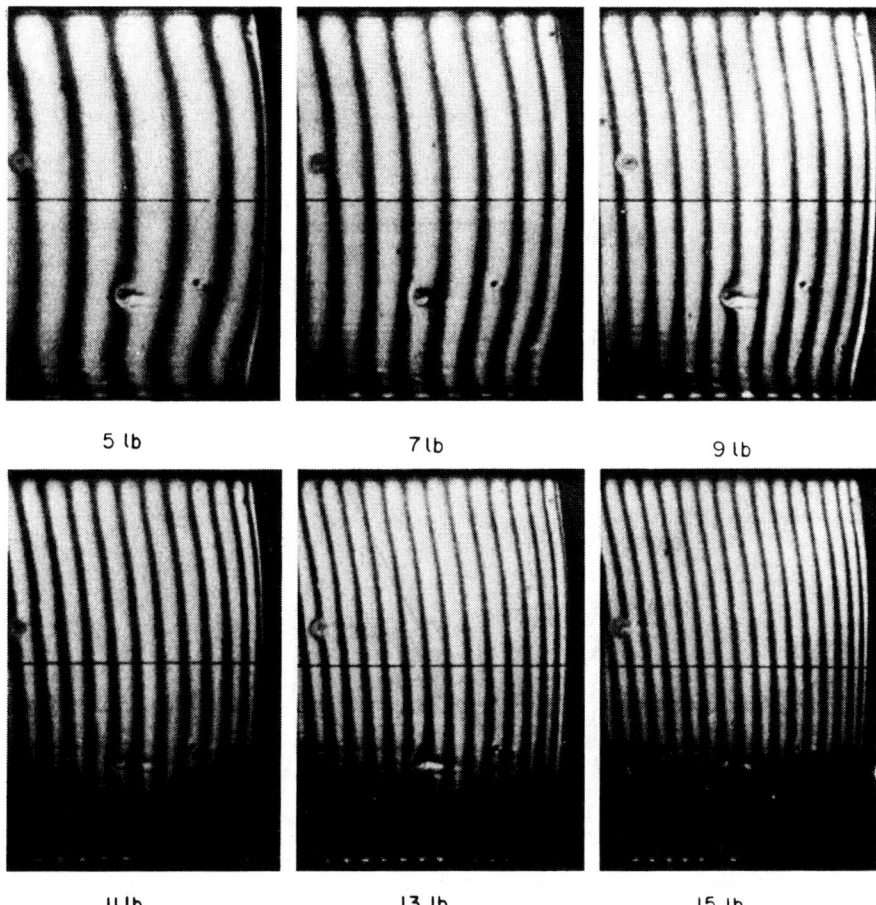

Figure 20-20 Interference fringes for crack-tip opening measurements. (From Ref. 93.)

and illuminated by the other (reference) beam. Holographic fringes thus obtained (Fig. 20-22) are proportional to the strain normal to the specimen surface. Assuming the existence of generalized plane stress, Eqs. (20-1) can be used to compute the mode I SIF.

Shadow Pattern (Caustic) Method[5]

Although most of the experimental methods discussed for obtaining SIFs were developed for other general-purpose uses and later applied to cracked bodies, the shadow pattern originated with Manogg [95] in an experimental study of cracked bodies and has been extensively developed and applied to a wide variety of static and dynamic problems. Excellent discussions of the method

---

[5] See also Chapter 9.

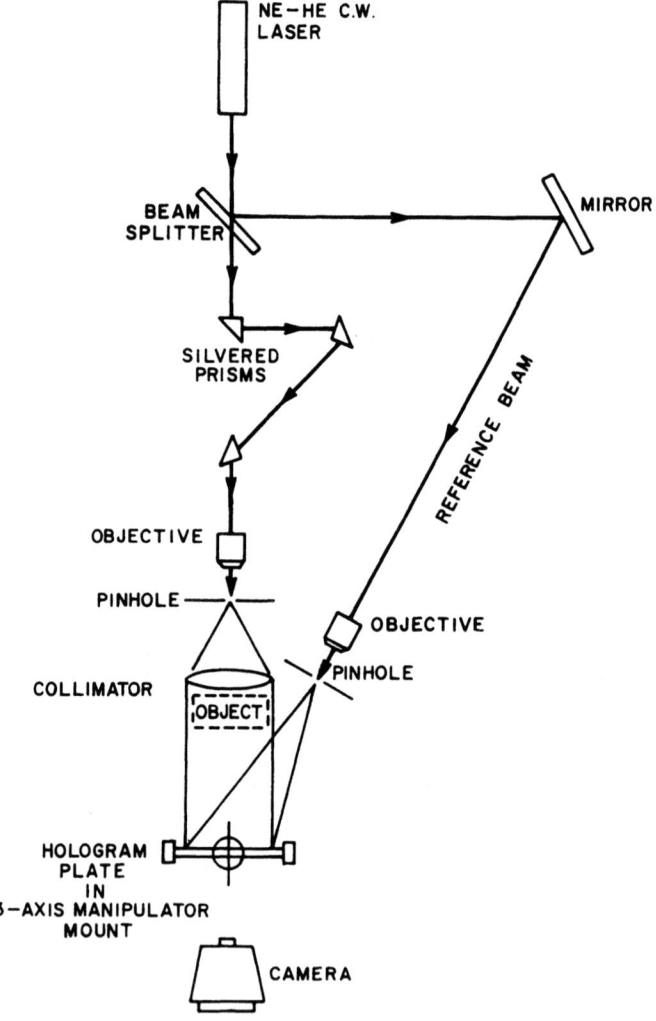

Figure 20-21 Optical arrangement for holographic fringes. (From Ref. 94.)

are provided by Theocaris [96] and Beinert and Kalthoff [97]. Moreover, its application has been developed for use on both transparent and opaque materials.

When a cracked specimen is loaded in mode I, the stress intensification around the crack tip causes a reduction of both specimen thickness and the refractive index of the specimen material. As a result, the material surrounding the crack tip acts as a divergent lens, diverting the light that originated from a point source away from the crack-tip region. This causes the image of the crack on a plane behind the specimen to exhibit a dark spot surrounded by a bright light, which is known as a caustic. The upper half of Fig. 20-23 shows this effect schematically for a transparent material, and the lower half shows that by using the light reflected from the mirror surface of an opaque

EXPERIMENTAL FRACTURE MECHANICS

Figure 20-22 Near-tip holographic interferometric fringes. (From Ref. 94.)

material, one obtains, on a virtual image plane behind the specimen, the same kind of pattern as for the transparent material. In the latter case, however, the light deflection results only from thickness change. Typical patterns produced in this way are shown in Fig. 20-24 for materials that are both optically isotropic and anisotropic. (Although two bright closed bands occur in the latter case, they are not markedly different in outline.)

By combining the geometric conditions of the method with a form of Eqs. (20-1), one can obtain the relation

$$K_I = C_0 D^{5/2} \tag{20-47}$$

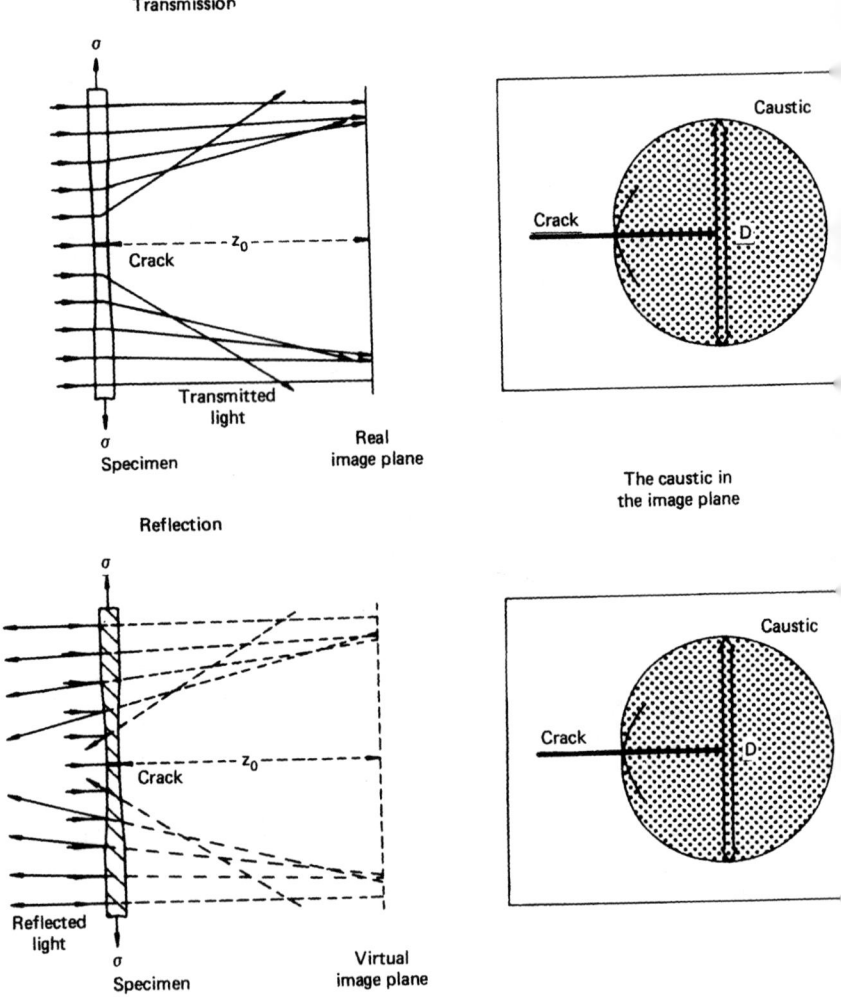

Figure 20-23 Schematic of shadow pattern development. (From Ref. 97.)

where C_0 is a coefficient containing the geometric conditions and D is shown in Figs. 20-23 a 20-24. Moreover, C_0 will be different for plane-stress and plane-strain conditions. Reference provides a detailed development of Eq. (20-47).

Caustics can also be generated by any deformed opaque specimen surface, including the obvic dimpling surrounding a ductile crack. Rosakis and Freund [98] used an asymptotic elastic–plas analysis to relate this dimpling to a plastic intensity factor. By postulating a Hutchinson–Ric Rosengren (HRR) singularity, J-deformation theory of plasticity, and the separation of θ and the plastic strain in the thickness direction is obtained as

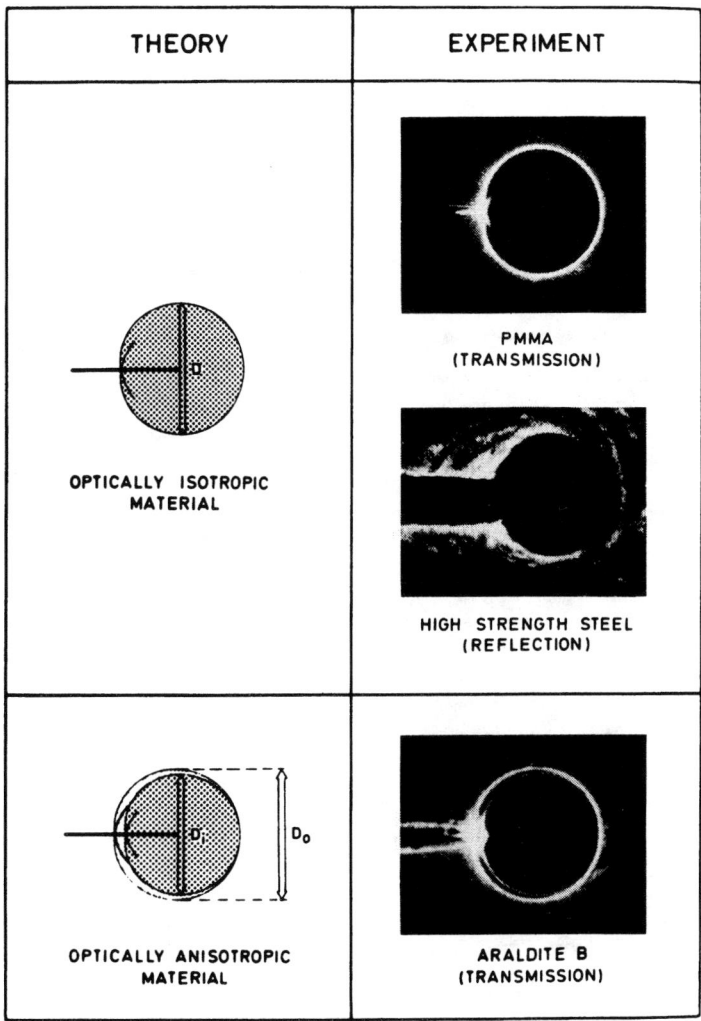

Figure 20-24 Geometric construction of the predicted initial curve (dashed) and caustic curve (solid). (From Ref. 97.)

$$\varepsilon_{zz}^P = -(\varepsilon_{rr}^P + \varepsilon_{\theta\theta}^P) \tag{20-48}$$

where the in-plane plastic strain components are given in terms of the stress components as

$$\varepsilon_{rr}^P = \frac{\alpha\sigma_{ys}}{E}\left(\frac{JE}{a\sigma_{ys}^2 I_n r}\right)^{n/(n+1)}\left(\frac{\sigma_e}{\sigma_{ys}}\right)^{n-1}\left(\sum_{rr} - \frac{1}{2}\sum_{\theta\theta}\right) \tag{20-49}$$

where a = material constant
$\sigma_e^2 = \frac{3}{2}S_{ij}S_{ij}$ and $S_{ij} = \sigma_{ij} - \frac{1}{3}\sigma_{kk}\sigma_{ij}$
n = hardening exponent
$\Sigma_{ij}(n, \theta) = \sigma_{ij}/\sigma_{ys}(JE/\alpha\sigma_{ys}^2 I_n r)^{-1/(n+1)}$
I_n = dimensionless quantity in the HRR relation

Although further verification study is necessary, the caustic method promises to provide an experimental procedure with which the J value can be determined directly using crack-tip measurements, in contrast to the ASTM designated far-field procedure, which is based on many simplifying assumptions.

Three-Dimensional Methods

Frozen-Stress Photoelastic Method

This method involves a technique that dates back to the studies of Oppel [99]. The term "frozen stress" is a misnomer in the sense that one does not obtain frozen-in macrostresses. Instead, the deformations and stress fringes are the frozen or fixed values. To accomplish stress freezing, one begins with a transparent material exhibiting certain diphase optical and mechanical properties. Readers are referred to Chapter 5 for details of the frozen-stress procedure.

There are two basic limitations to the frozen-stress method in general. The first is that it applies only to linear elastic behavior, and the second is that all stress-freezing materials are essentially incompressible above critical temperature (i.e., Poisson's ratio, $\nu \approx 0.5$).

The practical significance of the first limitation is that when one introduces a sharp crack into the body (theoretically a singularity), one expects a local region of nonlinearity very near the crack tip. Such a region is, in fact, observed photoelastically. This region may be due to nonlinear material behavior, crack-tip blunting, nonlinear strain-optic relations, or some combination of these effects. The important point is that this region must be identified and excluded from the SIF determination. The second limitation leads to an important consideration, especially when verifying the accuracy of the method by comparing with two-dimensional analytical solutions.

The effect of the elevated value of Poisson's ratio in cracked-body problems may be described as follows. When a through-thickness crack occurs in a body of finite thickness, a region of high constraint develops around the crack tip, characterized by the root radius of the crack tip and the plate thickness, the latter always being orders of magnitude larger than the former. This constraint dissipates rapidly with increasing distance from the crack tip and, away from the crack tip, generalized plane stress prevails in a plate-shaped body problem. The presence of the constraint variation, however, is a three-dimensional effect which is present in all real bodies, and when measurements are made near the crack tip, they will be associated with a state of nearly plane strain, while measurements away from the near-tip zone, such as compliance measurements, will be for a zone in which generalized plane stress exists. Srawley et al. [100] and Irwin [101] have suggested that near-tip measurements in such problems can be "converted" to two-dimensional results through the multiplying factor $(1 - \nu^2)^{1/2}$. This effect, while small for $\nu = 0.3$, is substantial when $\nu = 0.5$. The first author and his associates have extended these ideas and suggested the factor

$$\frac{[1 - (0.3)^2]^{1/2}}{[1 - (0.5)^2]^{1/2}} \tag{20-50}$$

for converting plate-type frozen-stress data for comparison with "two-dimensional" results where $\nu = 0.3$. This method has been found to be remarkably accurate for plate-shaped bodies. Details of this study are found in Ref. 102. When cracked bodies are not two-dimensional (as with surface flaws), this effect cannot be estimated in the same way, but estimates of one author and his associates based on an inverse-square-root algorithm indicate a maximum elevation in the SIF from the near-field measurements of the order of the experimental error (i.e., about 5%).

A complete analysis for all three local modes in a three-dimensional problem requires an extension of the algorithms used for two-dimensional problems. It has been shown [28, 103] that the stresses near the border of an elliptical crack, when expressed in terms of a set of local, moving, rectangular Cartesian coordinates in a plane perpendicular to the flaw border, have the same form as the stresses in a plane perpendicular to the border of a straight-front crack. Consider a half-space containing a part circular surface flaw in a plane at an angle β to the boundary with remote uniform tension parallel to the z' direction (Fig. 20-25). The local moving orthogonal coordinate system t-n-z is always oriented such that t is tangent to the flaw border and n is normal to the flaw border but both n and t are in the flaw plane [104]. The z axis is normal to the flaw plane. In such a problem, all three local modes of near-field deformation (i.e., modes I, II, and III) will be present as we move around the flaw border. We note that when $\alpha = 0$, mode III will be absent.

The stress distribution near the part-through crack in the data zone, corresponding to the opening mode of deformation, can be taken as [referring to Eqs. (20-3)]

$$\sigma_{nn} = \frac{K_I}{(2\pi r)^{1/2}} \cos\frac{\theta}{2}\left(1 - \sin\frac{\theta}{2}\sin\frac{3\theta}{2}\right) - \sigma^1_{nn} \qquad (20\text{-}51\text{a})$$

$$\sigma_{zz} = \frac{K_I}{(2\pi r)^{1/2}} \cos\frac{\theta}{2}\left(1 + \sin\frac{\theta}{2}\sin\frac{3\theta}{2}\right) - \sigma^1_{zz} \qquad (20\text{-}51\text{b})$$

$$\sigma_{nz} = \frac{K_I}{(2\pi r)^{1/2}} \sin\frac{\theta}{2}\left(\cos\frac{\theta}{2}\cos\frac{3\theta}{2}\right) - \sigma^1_{nz} \qquad (20\text{-}51\text{c})$$

where K_I is the mode I SIF and the coordinates r and θ are shown in Fig. 20-2. We use σ^1_{ij} to represent the contribution of the mode I regular stress field to the measurement zone, which is taken far enough from the crack tip to avoid the nonlinear zone very near the tip. Although they may generally be regarded as expressible in Taylor-series expansions, it turns out that only the leading terms of said series are normally necessary, so that σ^1_{ij} are constants for a given point and θ value on the flaw border but vary from point to point. It should be noted that in three-dimensional problems, the near-tip state of stress is generally neither plane stress nor plane strain in the measurement zone. Since this zone does not include the crack surfaces (e.g., for mode I, $\theta = \pi/2$), in deference to the possibly complex stress gradients, we allow all three components of σ^1_{ij} to be nonzero. Nonetheless, when a two-parameter (K_I, σ^1_{ij}) stress system is employed, the result is the same as if σ^1_{zz} and σ^1_{nz} were taken to be zero. For a larger number of parameters, however, the result would be different and more general. Stresses in the data zone corresponding to the mode II can be taken as

$$\sigma_{nn} = \frac{-K_{II}}{(2\pi r)^{1/2}} \sin\frac{\theta}{2}\left(2 + \cos\frac{\theta}{2}\cos\frac{3\theta}{2}\right) - \sigma^2_{nn} \qquad (20\text{-}52\text{a})$$

$$\sigma_{zz} = \frac{K_{II}}{(2\pi r)^{1/2}} \sin\frac{\theta}{2}\cos\frac{\theta}{2}\cos\frac{3\theta}{2} - \sigma^2_{zz} \qquad (20\text{-}52\text{b})$$

$$\sigma_{nz} = \frac{K_{II}}{(2\pi r)^{1/2}} \cos\frac{\theta}{2}\left(1 - \sin\frac{\theta}{2}\sin\frac{3\theta}{2}\right) - \sigma^2_{nz} \qquad (20\text{-}52\text{c})$$

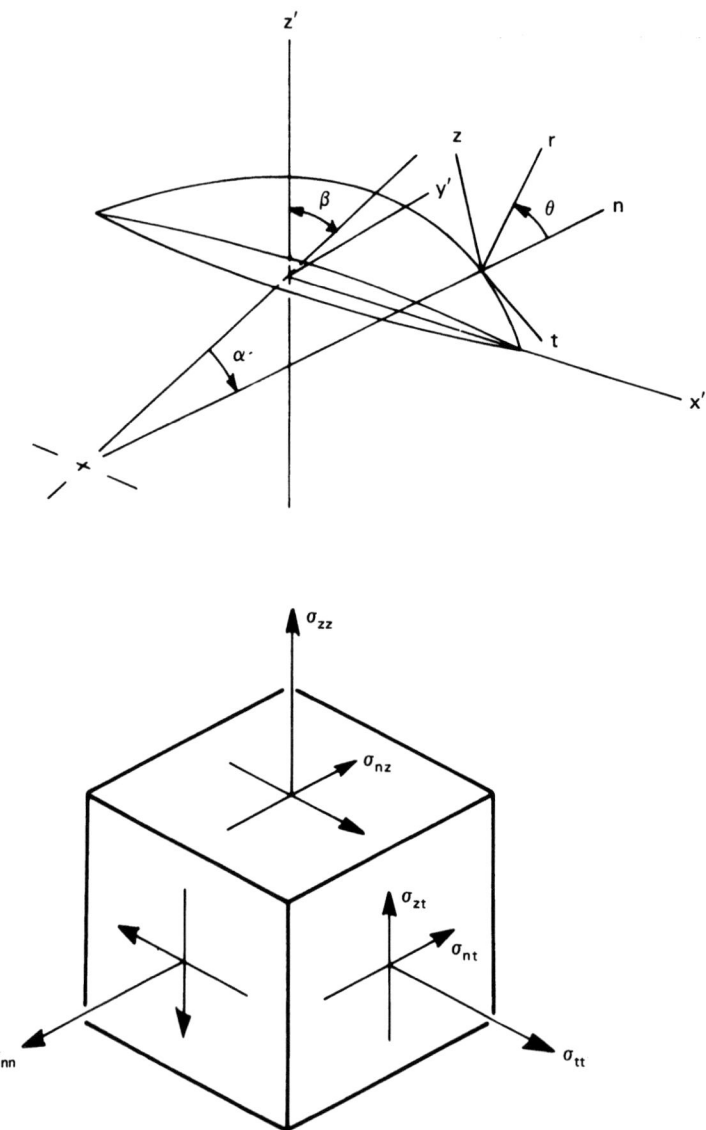

Figure 20-25 Mixed-mode flaw geometry and notation. (From Ref. 104.)

where K_{II} is analogous to K_I.

Finally, the stresses in the data zone corresponding to the mode III loading can be taken as

$$\sigma_{nt} = \frac{-K_{III}}{(2\pi r)^{1/2}} \sin\frac{\theta}{2} - \sigma_{nt}^3 \tag{20-53a}$$

$$\sigma_{zt} = \frac{K_{III}}{(2\pi r)^{1/2}} \cos\frac{\theta}{2} - \sigma_{zt}^3 \tag{20-53b}$$

Although σ_{ij}^{1+2} have no influence on the singular stress field itself, they do alter the isochromatic fringe pattern, which is proportional to the maximum in-plane shearing stress.

From the stress field given in Eqs. (20-51) and (20-52), the maximum shearing stress in the plane, n-z, perpendicular to the crack front can be obtained using

$$(\tau)_{\max}^{nz} = \left[\left(\frac{\sigma_{zz} - \sigma_{nn}}{2}\right)^2 + \sigma_{nz}^2\right]^{1/2} \tag{20-54}$$

Truncating to the same order as Eqs. (20-51) and (20-52), one gets

$$(\tau)_{\max}^{nz} = \frac{A}{r^{1/2}} + B \tag{20-55}$$

for fringe loops approaching the inner loops of Fig. 20-26, where

$$A = \frac{1}{\sqrt{8\pi}}[(K_I \sin\theta + 2K_{II} \cos\theta)^2 + (K_{II} \sin\theta)^2]^{1/2} \tag{20-56}$$

and

$$B = B(\sigma_{ij}^{1+2})$$

The maximum shearing stress in the n-z plane [the left side of Eqs. (20-54) and (20-55)] is determined photoelastically.

Now, in general, the effect of σ_{ij}^{1+2} involves both a folding and a change in eccentricity of the fringe loops (Fig. 20-26a). If folding occurs, θ_m, the angle along which the distance to a fringe from the crack tip is greatest, will vary with the fringe order n' and to obtain θ_m^0, the value of θ_m associated with K_I and K_{II}, one must plot θ_m versus r/a and extrapolate to the origin. In the experiments studied, θ_m was constant over the data range in the fashion indicated qualitatively by Fig. 20-26b. Data are always taken from the fringe loops leaning ahead of the crack tip, as indicated in Fig. 20-26a. Upon computing

$$\lim_{\substack{r_m \to 0 \\ \theta_m \to \theta_m^0}} (8\pi r_m)^{1/2} \frac{\partial(\tau)_{\max}^{nz}}{\partial\theta}(K_I, K_{II}, r_m, \theta_m, \sigma_{ij}^{1+2}) = 0 \tag{20-57}$$

one obtains

$$\left(\frac{K_{II}}{K_I}\right)^2 - \frac{4}{3}\left(\frac{K_{II}}{K_I}\right)\cot 2\theta_m^0 - \frac{1}{3} = 0 \tag{20-58}$$

Since θ_m^0 can be measured experimentally, K_{II}/K_I can be calculated from Eq. (20-58). Then by combining the stress-optic law with a modified form of Eq. (20-55), we obtain

$$(\tau)_{\max}^{nz} = \frac{fn'}{2h_t} = \frac{K_{AP}^*}{(8\pi r)^{1/2}} \tag{20-59}$$

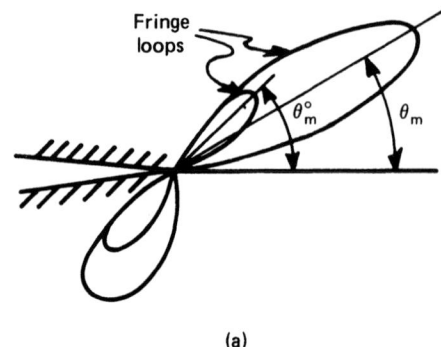

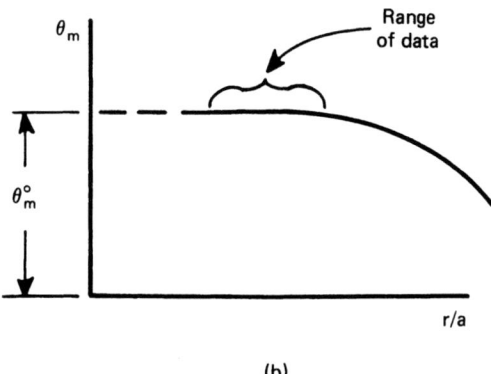

Figure 20-26 Determination of θ_m^0 for mixed-mode loading. (From ref. 104.)

where $K_{AP}^* = (\tau)_{max}^{nz} (8\pi r)^{1/2}$ is defined as the "apparent" mixed-mode SIF. Hence

$$K_{AP}^* = [(K_{IAP} \sin \theta_m + 2K_{IIAP} \cos \theta_m)^2 + (K_{IIAP} \sin \theta_m)^2]^{1/2} \quad (20\text{-}60)$$

and one can solve for the individual values of K_I and K_{II}

To do this, one must obtain

$$K^* = [(K_I \sin \theta_m^0 + 2K_{II} \cos \theta_m^0)^2 + (K_{II} \sin \theta_m^0)^2]^{1/2} \quad (20\text{-}61)$$

from K_{AP}^* by plotting $K_{AP}^* = (\tau)_{max}^{nz} (8\pi r)^{1/2}$ versus $(r/a)^{1/2}$, identifying a linear zone, and extrapolating to the origin. This will yield K^*. A typical set of fringe data illustrating such a determination is given in Fig. 20-27. Once K^*, K_{II}/K_I, and θ_m^0 are known, K_I and K_{II} can be calculated and then can be normalized using proper quantities.

Note that the approach above utilizes a two-parameter (A, B) model [Eq. (20-55)], since the linear zone can be located experimentally (Fig. 20-27). However, if one cannot locate such a zone experimentally, additional terms leading to an equation of the form

$$\tau_{max} = \frac{A}{r^{1/2}} + \sum_{m=0}^{\infty} B_m r^{m/2}$$

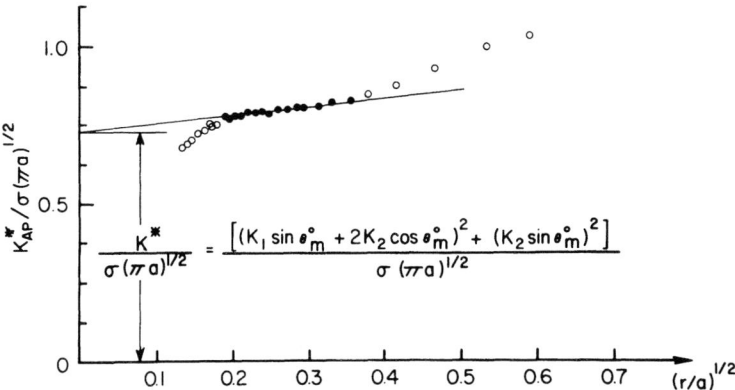

Figure 20-27 Estimating $K^*/\sigma(\pi a)^{1/2}$. (From Ref. 104.)

with suitable truncation criteria must be considered. Since such criteria are not yet established, the latter approach is avoided where possible and has not been found necessary in studies to date.

The stress distribution σ_{zz}, acting in a plane perpendicular to the crack surface and tangent to the crack front (z-t plane) or in a plane parallel to the z-t plane can be found from Eqs. (20-51) and (20-52).

$$\sigma_{zz} = \frac{K_I}{(2\pi r)^{1/2}} \cos\frac{\theta}{2}\left(1 + \sin\frac{\theta}{2}\sin\frac{3\theta}{2}\right)$$
$$+ \frac{K_{II}}{(2\pi r)^{1/2}} \sin\frac{\theta}{2}\left(\cos\frac{\theta}{2}\cos\frac{3\theta}{2}\right) - \sigma_{zz}^{1+2} \qquad (20\text{-}62)$$

To arrive at a value for σ_{zz}, experiments by one author indicate that the usual assumption of plane strain (for the plane problem) may not be valid here. However, if one assumes a state of nearly generalized plane strain such that the value of ε_{tt} can be considered to be constant over a portion of the length of the flaw border, the observed state of varying transverse constraint along the flaw border can apparently by approximated rather well. Thus we assume that

$$\varepsilon_{tt} = \frac{\sigma_{tt} - \nu(\sigma_{nn} + \sigma_{zz})}{E} = \bar{\varepsilon} \qquad (20\text{-}63)$$

where $\sigma_{tt} = E\bar{\varepsilon} + \nu(\sigma_{nn} + \sigma_{zz})$, where $\bar{\varepsilon}$ may be adjusted at intervals along the flaw border and where, again, from Eqs. (20-51) and (20-52),

$$\sigma_{nn} = \frac{K_I}{(2\pi r)^{1/2}} \cos\frac{\theta}{2}\left(1 - \sin\frac{\theta}{2}\sin\frac{3\theta}{2}\right)$$
$$- \frac{K_{II}}{(2\pi r)^{1/2}} \sin\frac{\theta}{2}\left(2 + \cos\frac{\theta}{2}\cos\frac{3\theta}{2}\right) - \sigma_{nn}^{1+2} \qquad (20\text{-}64)$$

for $\nu = 0.5$ (as in stress-freezing work), we then have

$$\sigma_{tt} = \frac{K_I}{(2\pi r)^{1/2}} \cos\frac{\theta}{2} - \frac{K_{II}}{(2\pi r)^{1/2}} \sin\frac{\theta}{2} - \sigma_{tt}^{1+2} + E\bar{\varepsilon} \qquad (20\text{-}65)$$

$$(\tau)_{\max}^{zt} = \left[\left(\frac{\sigma_{zz} - \sigma_{tt}}{2}\right)^2 + \sigma_{zt}^2\right]^{1/2} \qquad (20\text{-}66)$$

Consider the line normal to the crack surface that passes through the crack tip in the zt plane. For this case $\theta = \pi/2$, and when substituted in Eqs. (20-62), (20-65), (20-53b), and (20-66), there results

$$\sigma_{zz} = \frac{1}{4(\pi r)^{1/2}} (3K_I - K_{II}) - \sigma_{zz}^{1+2} \tag{20-67a}$$

$$\sigma_{tt} = \frac{1}{2(\pi r)^{1/2}} (K_I - K_{II}) - \sigma_{tt}^{1+2} + E\bar{\varepsilon} \tag{20-67b}$$

$$\sigma_{zt} = \frac{K_{III}}{2(\pi r)^{1/2}} - \sigma_{zt}^{1+2} \tag{20-67c}$$

Then for a two-parameter model, as before,

$$(\tau)_{max}^{tz} = \frac{C}{r^{1/2}} + D \tag{20-68}$$

where

$$C = \frac{1}{4(\pi r)^{1/2}} \left[\frac{1}{16}(K_I + K_{II})^2 + K_{III}^2 \right]^{1/2} \tag{20-69}$$

and

$$D = D(E\bar{\varepsilon}, \sigma_{ij}^{1+2}) \tag{20-70}$$

Now by combining the stress-optic law with a modified form of Eq. (20-68),

$$(\tau)_{max}^{zt} = \frac{fn'}{2h_n} = \frac{K_{AP}^{**}}{(8\pi r)^{1/2}} \tag{20-71}$$

where

$$K_{AP}^{**} = \sqrt{2} \left[\frac{1}{16}(K_{IAP} + K_{IIAP})^2 + K_{IIIAP}^2 \right]^{1/2} \tag{20-72}$$

the value of K_{III} can be obtained. To do this, one must obtain

$$K^{**} = \sqrt{2} \left[\frac{1}{16}(K_I + K_{II})^2 + K_{III}^2 \right]^{1/2} \tag{20-73}$$

from K_{AP}^{**} by plotting a $K_{AP}^{**} = (\tau)_{max}^{zt}(8\pi r)^{1/2}$ versus $(r/a)^{1/2}$ curve, identifying a linear zone, and extrapolating to the origin. This will yield K^{**}. At the points where the flaw border intersects the boundary of the plate, SIF values are uncertain and require a boundary layer analysis for an accurate evaluation.

As noted earlier, the photoelastic method may be applied directly to two-dimensional problems, but the frozen-stress method is required to obtain SIF distributions in three-dimensional problems. Two kinds of crack may be employed. When the shape of the crack front is prescribed, the crack can be produced (in limited configurations) by a machined slot. From studies by one author and others, it has been concluded that a narrow slot terminating in a 30° vee notch of root radius = 0.025 mm will yield photoelastically determined SIF values to within 2% of those in natural cracks of the same border configuration. We call such slots "artificial" cracks. Natural cracks are made by fixing a sharp blade normal to a surface and striking the blade with a hammer. A starter crack is initiated dynamically, then propagates into the material and arrests. In the frozen-stress process, these cracks may be extended by applying monotonically increasing loads above critical temperature

EXPERIMENTAL FRACTURE MECHANICS

until crack growth occurs. When the desired crack size is reached, loads are reduced, terminating flaw growth, and the stress-freezing cycle is completed by cooling under reduced load. The shapes of cracks produced in this manner are dictated by the loads and the boundary conditions applied to the model. Shapes that are quite complex may appear to be virtually identical to those produced by stable fatigue crack growth in metals when (1) only small-scale yielding occurs, (2) surface crack closure effects are negligible, and (3) stress ratios are within a limited range above or below unity. This is an important observation, for it implies that the frozen-stress method can open the way for estimating SIF distributions in three-dimensional cracked-body problems where neither the SIF distribution nor the flaw shape is known a priori.

The complete experimental procedure involves the following steps.

1. Construct the model from a stress-free diphase transparent photoelastic material either by inserting a machined artificial flaw to exact size or by tapping in a natural starter crack much smaller than the desired size.
2. For complex shapes, parts are then glued together with no crack tips near glue lines.
3. The model is then placed in a stress-freezing oven in an appropriate loading rig, heated above critical temperature, and loaded in the same fashion as the prototype. It is important that the test rig not restrain the model during the heating cycle, for the thermal coefficient of expansion of the model material is an order of magnitude higher than for structural metals. For natural cracks, sufficient load is applied to extend the starter crack to several times its initial size and then reduced to terminate flaw growth.
4. The model is cooled under load to room temperature and sliced in planes mutually orthogonal to the crack plane and the crack border (i.e., nz plane, Fig. 20-25). Slices should be stored in moisture-free air if the material is sensitive to time–edge effects.
5. Slices are coated with a liquid of the same refractive index as the model material and analyzed in a crossed circular polariscope using monochromatic light, which will produce isochromatics of dark and bright fringes, and utilizing the Tardy method for measuring fractional fringe orders.
6. Fringe-order data are then fitted to a simple least-squares computer program to establish a linear $K^*_{AP}/\sigma(\pi a)^{1/2}$ versus $(r/a)^{1/2}$ zone as in Fig. 20-27, and this line is used to obtain the normalized SIF* for a slice taken at a given location along the flaw border. The σ represents a load parameter such as remote stress (σ) or internal pressure (p).
7. Once $K^*/\sigma(\pi a)^{1/2}$ and θ_m^0 have been obtained for each slice, values of K_I and K_{II} may be computed from Eqs. (20-58) and (20-61).
8. Then, by removing a subslice from each primary slice and viewing along the n direction, we plot $K^{**}_{AP}/\sigma(\pi a)^{1/2}$ versus $(r/a)^{1/2}$, extrapolate as in Fig. 20-27 to obtain $K^{**}/\sigma(\pi a)^{1/2}$, and finally, compute K_{III} from Eq. (20-73).
9. Collecting results, we may finally plot values of K_I, K_{II}, and K_{III} versus a position along the crack front.

Examples of the application of this method to practical engineering problems are provided for both artificial [104] and natural [105] cracks in the open literature. These studies suggest an accuracy of $\pm 5\%$ for K_I estimates and about half of the mode I accuracy for the K_{II} and K_{III} estimates. Moreover, studies of stably growing natural cracks suggest that mode I always dominates the stress intensification, with the shear modes serving mainly to reorient the crack surfaces to new principal surfaces.

Integrated Frozen Stress—Moiré Interferometry Method

The foregoing discussion demonstrates that most of the experimental methods for obtaining SIF values are based on a form of Eq. (20-1) or (20-2). That is, the analytical foundations of LEFM

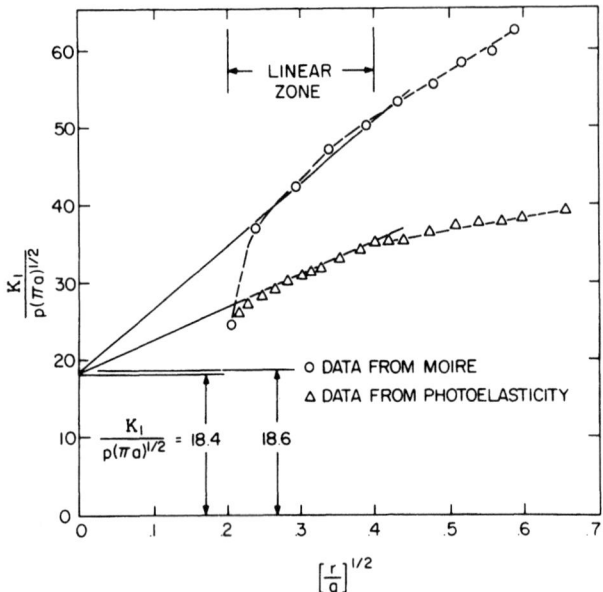

Figure 20-28 Linear zones in photoelastic and moiré data for the same photoelastic slice. (From Ref. 106.)

are essentially two-dimensional. Consequently, as noted earlier, it is not surprising that these foundations break down in the neighborhood of the region of intersection of a straight-front crack with a free surface. As noted earlier, prior to the studies described in Refs. 30–33, designers had employed plane-strain concepts in the interior of bodies and plane-stress concepts on the outer surface [i.e., Eqs. (20-1) and (20-2)]. To study these concepts experimentally, the first author and his associates have combined the experimental techniques for frozen-stress photoelasticity and high-density moiré interferometry. The combination involves the following experimental procedure:

1. Stress-freeze, slice, and determine K_I distributions using the procedure described earlier.
2. Apply linear gratings to the surface of the frozen slices.
3. Anneal slices and observe deformed grating through a virtual grating to obtain in-plane moiré displacements.
4. Convert moiré displacements to K_I values.

If step 4 is carried out using the plane-strain form of Eqs. (20-1) and (20-2) and the resulting $K_{AP}/p(\pi a)^{1/2}$ versus $(r/a)^{1/2}$ curves compared with those from step 1, the results shown in Fig. 20-28 will be typical for an interior slice. Although the extrapolated values of $K_I/p(\pi a)^{1/2}$ are in good agreement, the $K_{AP}/p(\pi a)^{1/2}$ curves diverge considerably. This divergence is interpreted to mean that the degree of constraint in the measurement zone is less than the assumed plane-strain conditions used in computing $K_{AP}/p(\pi a)^{1/2}$ values from the moiré data [106].

To study the variation in the exponent λ of r through a crack-front boundary layer adjacent to the free surface of a body, a four-point bending specimen under four-point loading was employed (Fig. 20-29). The foregoing experimental procedure was used, and the algorithm for converting the measured displacements into SIF values was taken as

EXPERIMENTAL FRACTURE MECHANICS

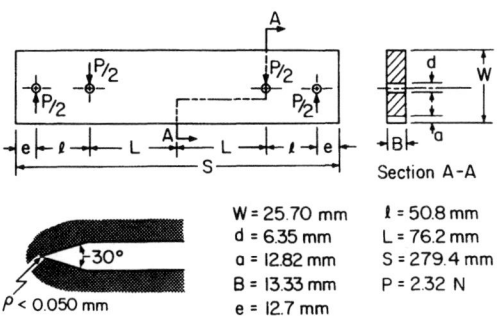

Figure 20-29 Four-point bending test geometry. (From Ref. 107.)

$$u_z = \overline{C}(K_\lambda)_{AP} r_u^{\lambda+} \tag{20-74}$$

where u_z = displacement component normal to crack surface
$\overline{C}$ = dimensional coefficient
$(K_\lambda)_{AP}$ = apparent stress eigenfactor
λ_u^+ = first eigenvalue of the three-dimensional solution [30] in the displacement field equation
$(\lambda_\sigma^+ = \lambda_u^+ - 1$ is first eigenvalue in the stress field equations)

Since these eigenvalues (λ_u^+ and λ_σ^+) are evaluated in a transition zone where the combined effect of the vertex singularity at the free surface and the LEFM singularity are believed to occur, we call these quantities pseudoeigenvalues.

Thus, by plotting $\log u_z^+$ versus $\log r$ across the specimen thickness, values of λ_u^+ could be obtained. Figure 20-30 [107] shows a typical result. From these values, a graph of λ_u versus specimen thickness was constructed. Very good agreement with Ref. 30 was obtained at the free

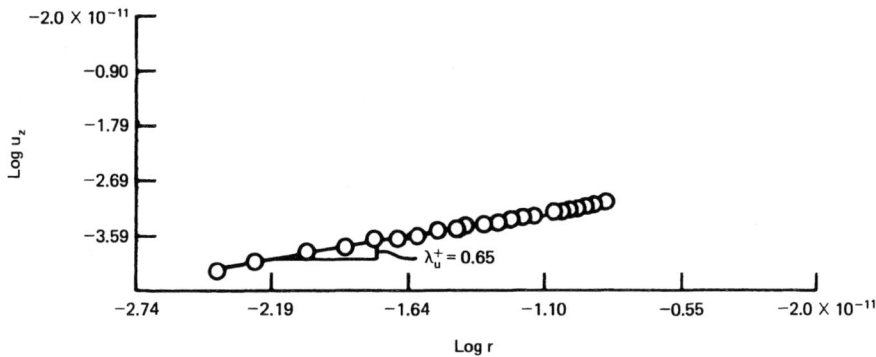

Figure 20-30 Variation of first near-tip displacement eigenvalue (λ_u) in the boundary layer. (From Ref. 107.)

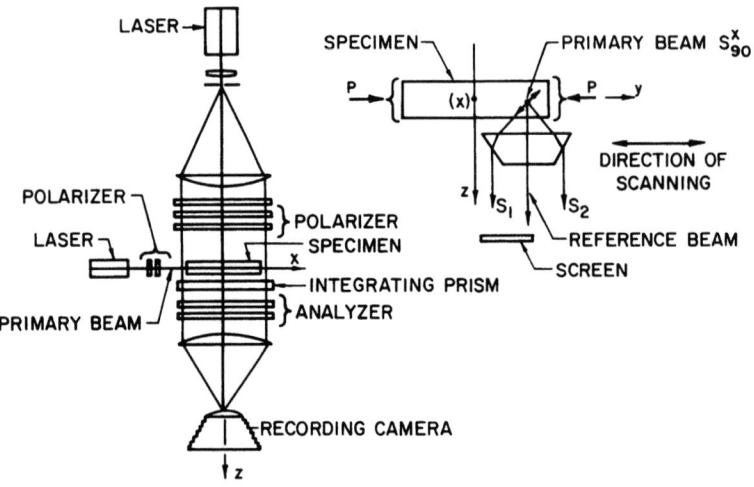

Figure 20-31 Integrated polariscope for isodyne photoelasticity. (From Ref. 111.)

surface. Results show a significant reduction in singularity order at the free boundary with a transition of λ_u^+ from 0.67 to 0.54 in a boundary layer thickness of about 10% of the specimen thickness for incompressible materials [107]. These results suggest that when making near-tip surface measurements on materials that are nearly incompressible and exhibit small-scale plasticity, classical LEFM breaks down and large errors in fracture parameters (of the order of 100%) can be introduced by using classical LEFM algorithms to interpret measured data. Investigations of this type are continuing [108].

This work illustrates that by combining experimental techniques, new information on the SIF and related concepts can be obtained.

Other Concepts

Conceptually, one might expect three-dimensional information to be extractable from cracked bodies through the use of scattered-light photoelasticity. In this approach, fringe patterns are obtained from observations made from directions other than along the direction of propagation of the incident light (often normal to the incident direction). The illumination of these patterns results from a scattering of the incident light by the material through which it is passing. It is not surprising, then, that the sharpness of the fringe pattern is degraded by this process. However, scattered light has been employed successfully to measure two-dimensional SIF values [109] and, quite recently, it has been adapted to the determination of SIF values in three-dimensional crack-body problems. Two different approaches have been used.

Isodyne Photoelasticity. Isodynes are curves of constant intensity of the normal forces acting on characteristic curves in a plane-stress field and are thus related to the first derivatives of the Airy stress function. Two isodyne fields related to two orthogonal characteristic curves completely define the elastic state of plane stress [110,111]. When modeled optically with the integrated polariscope, shown in Fig. 20-31, the photoelastic isodynes, which resemble the isochromatics obtained by scattered-light photoelasticity, are differentiated and combined with transmission-type

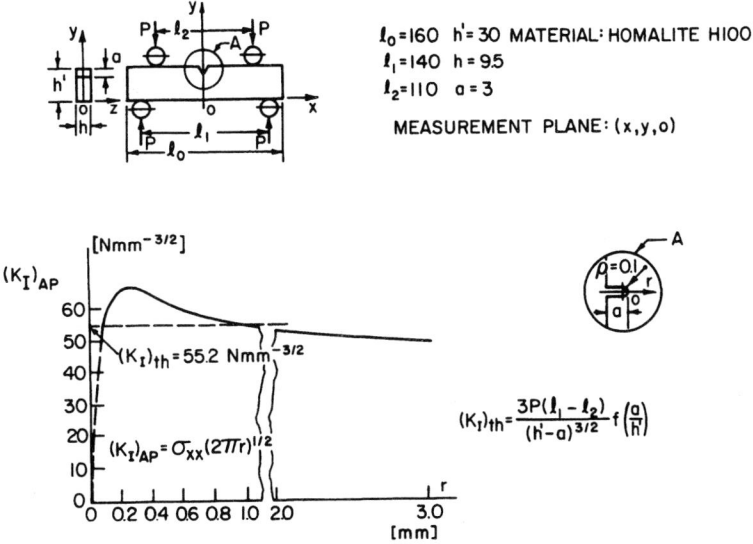

Figure 20-32 Four-point bend specimen with sharp notch: $(K_I)_{AP}$ for central plane was determined by isodyne technique. (From Ref. 113.)

isochromatic data. Pindera and his associates have adapted a two-dimensional isodyne photoelastic technique to the estimation of SIF values which may yield useful results in three-dimensional problems [112]. While the analytical foundations of the optical methods are extensive, the mode I SIF is estimated from the stress normal to the crack planes in the usual way [i.e., Eqs. (20-1) or (20-51)]. As with scattered-light photoelasticity, optical inhomogeneity generated by the high stress gradient in the vicinity of the crack tip may distort the photoelastic isodyne. The requirement for a plane-stress state, which is not a prerequisite in scattered-light photoelasticity, can be modulated by the "semiplane stress state" used by Pindera et al. [113], who then determined the stress-intensity factor at the midsection (plane of symmetry) of a four-point bend specimen, shown in Fig. 20-32. Compressive loading on the slot was used for spatial modulation in this analysis. Also shown in Fig. 20-32 is the variation in the apparent stress-intensity factor computed for various crack-tip distances; note the pronounced effect of the near-tip nonlinearity and crack-tip bluntness.

Assuming that the influence of optical inhomogeneity in the scattered light path can be quantified, the photoelastic isodyne technique shares the same advantage of three-dimensional scattered light photoelasticity which can be used to analyze the crack-tip state of stress under live load. The stress intensity factor can be computed more accurately if K is expressed directly in terms of the isodyne value, thus eliminating the extra numerical differentiation process in obtaining the stresses.

Optical Slicing. By passing two parallel sheets of polarized light through a loaded photoelastic model, an optical slice of material between the planes of light is isolated, and the light scattered from this slice is observed normal to the light planes. By locating such planes parallel to the nz plane (Fig. 20-25) and using equations of the form of Eqs. (20-51), deSailly and LaGarde have estimated SIF values for surface flaws [114].

Other Applications

Several experimental methods have been extended to a study of three-dimensional effects including plasticity. By combining the use of amplitude and shadow moiré on the outer surface with posttest metallographic sectioning along the crack front, information regarding the near-tip plastic zone around surface flaws was obtained [115]. In another study, by seeding with speckles only selected planes in a transparent model, variations in displacements through the thickness along the border of a surface flaw were measured by the random speckle method [116]. In another study [117], the SIF distribution through the thickness of an ASTM E–399 standard compact specimen with a straight-front crack was obtained by embedding gratings in parallel planes along the crack border and utilizing moiré techniques. These methods all focus on displacement measurements and include measurements in the plastic zone.

Hybrid Experimental-Numerical Analysis. One of the major obstacles that hinders the progress of experimental ductile fracture research is the undefined crack-tip states of stress and strain in the presence of large-scale yielding. Because the $1/\sqrt{r}$ singular state in linear elastic fracture mechanics which successfully models brittle fracture is a physical impossibility, other phenomenological models have been proposed. One such model is the Dugdale strip yield zone, which conveniently reduces the elastic–plastic crack-tip state to an elastic one. The Dugdale strip yield model used in a recent analysis [118] is a modification of the classical Dugdale model, where higher-order terms were added to increase the number of disposable parameters. Experimental data are then used to fit the disposable parameters associated with the Dugdale model, which is modified to fit the complex state associated with large-scale yielding, just as the stress-intensity factor is determined from photoelasticity and moiré fringe data. The adequacy of such models can be verified by matching other crack-tip data not used in the fitting process but generated numerically by the Dugdale model and independently by experiment. The extensive numerical experimentation necessary for this verification study in essence replaces the finite-element or boundary element method used in the traditional hybrid experimental-numerical stress analysis technique described in Chapter 17. The verified modified Dugdale model can then be used through the generation mode of hybrid experimental-numerical analysis to generate numerically various fracture parameters for evaluation.

The utility of the hybrid experimental-numerical analysis was demonstrated by applying it to stable crack growth under mixed-mode loading [118]. Isochromatics in a 1.6-mm-thick polycarbonate tensile specimen with a central slanted crack were recorded during a continuing stable crack growth period. The resultant Z-shaped crack was modeled by a straight Dugdale crack, which was modified to account for the residual stresses left behind in the wake of the rapidly extending crack, as shown in Fig. 20-33. The modification consisted of two unknown tangential forces acting at the physical crack tip. Lengths of the Dugdale strip yield zones ahead of the crack tip were measured from photoelastic records. These lengths coincided with the length of the theoretical values of the horizontal crack, thus justifying the use of the model of Fig. 20-33 to represent the Z-shaped cracks. The crack-tip stress field, which is represented by a polynomial stress function of the crack-tip coordinates, together with the two unknown tangential forces, was fitted to the recorded elastic isochromatics surrounding the plastic region using an overdeterministic fitting routine. Figure 20-34 shows the crack-tip opening angle (CTOA), which was computed by using the modified Dugdale model. For the two initial crack geometries it was found to be almost constant during the stable crack growth process.

While the hybrid experimental-numerical technique may not provide micromechanics insight to crack-tip mechanics, it can be used effectively to extract fracture parameters that otherwise cannot be measured directly.

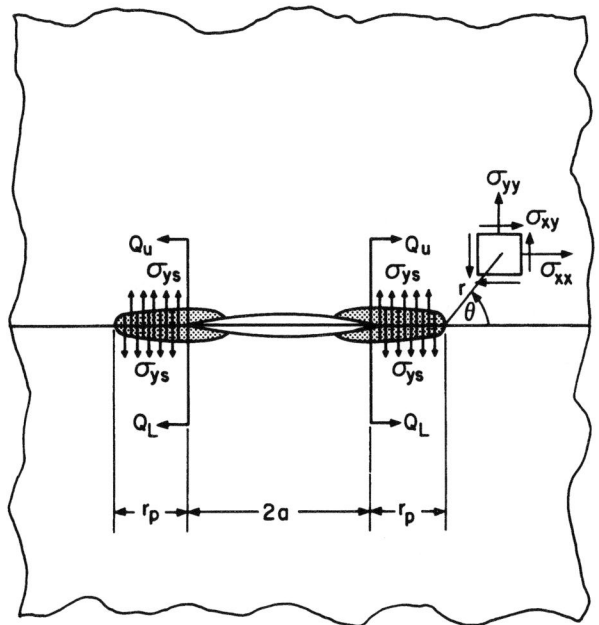

Figure 20-33 Modified Dugdale strip yield model. (From Ref. 118.)

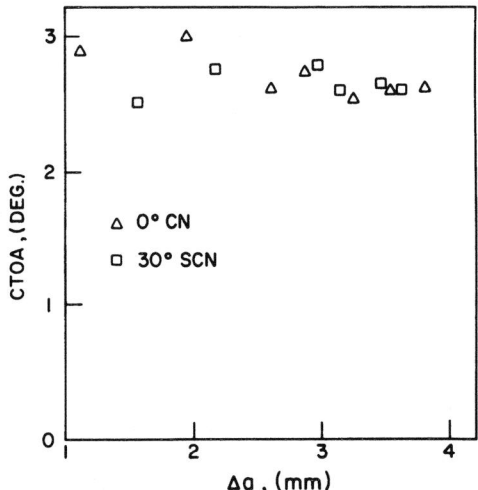

Figure 20-34 CTOA during stable crack growth of 0° CN and 30° SCN specimens. (From Ref. 118.)

Dynamic Photoelasticity and Caustics. Over the past two decades, optical techniques have been used extensively to analyze the dynamic state of stress surrounding a rapidly propagating crack. Dynamic stress-intensity factors were then computed by fitting a postulated static or dynamic crack-tip field to the whole-field experimental data, which were generated by an optical technique. Only limited data on fracturing brittle metal [119–121] were extracted using the two popular optical techniques (viz., dynamic caustics and dynamic photoelasticity). Due to the experimental difficulties involved, much of the published experimental data [122–133] thus relates to the dynamic response of photoelastic polymeric sheets.

Dynamic isochromatics and caustics surrounding a running crack often exhibit moderate asymmetry. Such photoelastic patterns were heretofore considered to be experimental abnormalities and were ignored by averaging the unsymmetric patterns during the data reduction process. Careful postmortem inspection of the photoelastic specimens, however, shows that higher σ_{0x} and slightly unsymmetric isochromatics are often associated with slightly curved crack patterns. With the development of a data reduction procedure [134] for evaluating dynamic K_{II} together with K_I and σ_{0x}, it has become possible to investigate the criteria above by extracting K_I, K_{II}, and σ_{0x} from the previously recorded dynamic isochromatics surrounding running crack-tips of curved cracks.

Figure 20-35 shows three frames from a 16-frame dynamic photoelastic record of a curving crack in a Homalite-100 dynamic tear test (DTT) specimen 9.5 mm ($\frac{3}{8}$ in.) thick and 88.9 mm by 400 mm (3.5 in. × 15 in.). This beam with a blunt initial crack of 6.4 mm (7/32 in.) length was impact loaded by a drop weight of 1.48 kg (3.25 lb). The crack emanated from the blunt sawcut crack and propagated through much of the height of the beam prior to curving near the region of impact loading. Further details on the experimental setup, crack-velocity measurements, and dynamic calibration of the Homalite-100 material used are given in Refs. 135 and 136. Figure 20-36 shows K_I, K_{II}, and σ_{0x}, which are computed by Eqs. (20-24a), obtained from the dynamic photoelastic pattern preceding and immediately after crack curving shown in Fig. 20-35. K_{II} is negligible at the point of instability and a pronounced fluctuation in σ_{0x} is noted.

Stable Crack Growth

In the early stages of the development of fracture mechanics technology, emphasis was placed on the prediction of the onset of crack instability in high-strength quasi-brittle materials and was strongly rooted in LEFM with the inclusion of small-scale yielding. However, with the development of tougher materials, the stable tearing of cracks into the plastic zone surrounding the crack tip began to occur more and more frequently. In focusing on stable crack growth, it is convenient to differentiate between an initial crack that has been loaded but has not yet grown (Fig. 20-37a) and one that has grown by tearing into the plastic zone. In the latter case (Fig. 20-37b) a plastic wake is present behind the crack tip. For purposes of discussion, the latter case can be divided into two categories:

1. Cases involving sufficient plasticity to invalidate LEFM
2. Cases in which LEFM and small-scale yielding with an HRR singularity still apply

Within the context of material response, case 1 is at the opposite extreme from classical quasi-brittle LEFM behavior. As noted in Section 20-1, the J-integral R curve has been used to analyze stable crack growth and instability associated with ductile tearing. Methodology involving the use of the Dugdale model to compute $(u_y)_R$ (called V_R) has been suggested by Newman [137] for including the displacement due to the plastic wake in stable crack growth under monotonic loading. This quantity is then used instead of J_R for studying stable tearing.

The material response associated with case 2 lies between quasi-brittle behavior, on the one

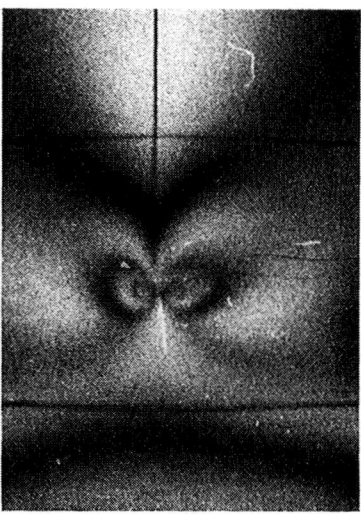

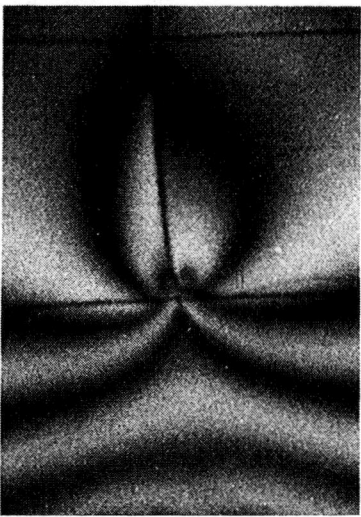

Figure 20-35 Dynamic photoelastic record of crack curving: (a) fifth frame, 100 μs, theoretical angle 0°, measured angle 0°; (b) eighth frame, 130 μs, theoretical angle 11°, measured angle 11°; (c) tenth frame, 160 μs, theoretical angle 23°, measured angle 26°. (From Ref. 135.)

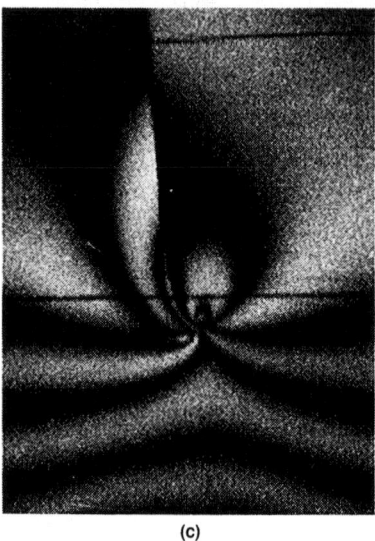

(c)

Figure 20-35 (*continued*)

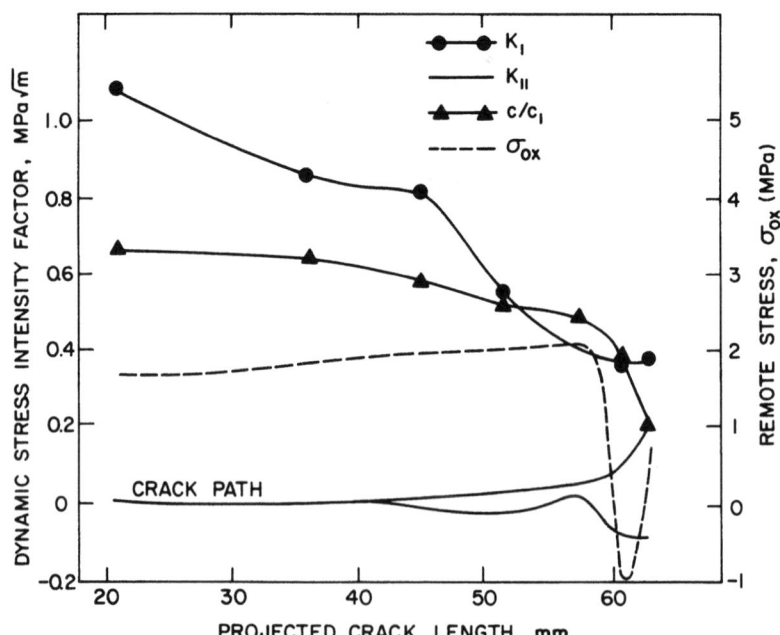

Figure 20-36 Values of K_I, K_{II}, and σ_{0x} from dynamic isochromatics. (From Ref. 136.)

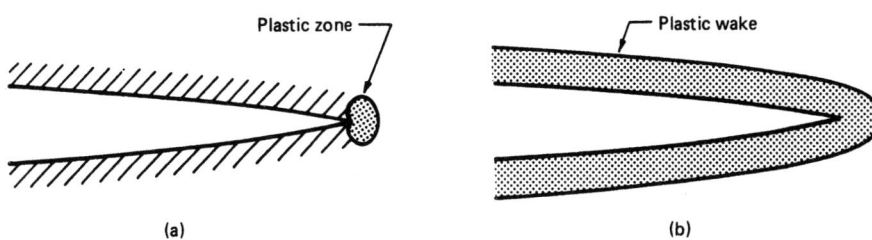

Figure 20-37 (a) Loaded crack that has not grown; (b) loaded crack that has grown.

hand, and ductile tearing on the other, and technology has developed for dealing with two kinds of problems in this class:

1. Fatigue crack growth
2. Time-dependent or creep crack growth

Fatigue Crack Growth

By far the largest number of in-service failures result from a scenario involving fluctuating, cyclic, or repeated loading, usually combined with a static load. Typically, a defect located near a stress raiser enlarges under repeated tensile stresses into a crack that continues to grow stably a small amount during each load cycle until the crack either penetrates the structure or attains critical length, after which unstable crack propagation occurs. Since SIF levels are normally below critical values, we may speak of K_{max} and K_{min}, which are proportional to σ_{max} and σ_{min} in any part of a load spectrum. Assuming that during a load cycle the crack length changes little, we may write

$$\Delta K(t) = K(t)_{max} - K(t)_{min} = Q\Delta\sigma(t)(\pi a)^{1/2} \tag{20-75}$$

where Q is the geometric factor appropriate to the cracked-body geometry. However, the values of K and ΔK are not the values normally associated with monotonically loaded cracked bodies. Instead, one uses "effective" values K_{eff} and ΔK_{eff}, which are introduced to account for the phenomenon of crack closure [138].

As noted above, when a cracked body is loaded in tension, a plastic zone forms around the crack tip (Fig. 20-37a). If the load is removed, an elastic compressive stress field forms around the plastic zone at the crack tip. The plastic zone keeps the crack propped open. If loading continues into compression, a compressive yield zone will form inside the plastic zone produced in tension. When a crack is grown by repeating this process, the crack surfaces develop plastic lips (Fig. 20-37b) so that when the crack is unloaded, it will be fully or partially closed. Thus some tensile stress level (σ_{op}) is required to open the crack. We define

$$\Delta K_{eff} = Q\Delta\sigma_{eff}(\pi a)^{1/2} \tag{20-76}$$

where $\Delta\sigma_{eff} = \sigma_{max} - \sigma_{op}$.

Equation (20-76) is generally greatly oversimplified, since fatigue crack fronts are not usually straight, and since less transverse constraint occurs near where the crack border intersects the body surface, the plastic zone will be larger here than in the interior of the body. Thus crack closure, while apparent on the surface of the body, may not occur on the interior of the body to the same extent as on the surface.

From the point of view of experimental mechanics, an important measurement in fatigue crack growth is the measurement of σ_{op}. This is usually accomplished by placing a gage across the

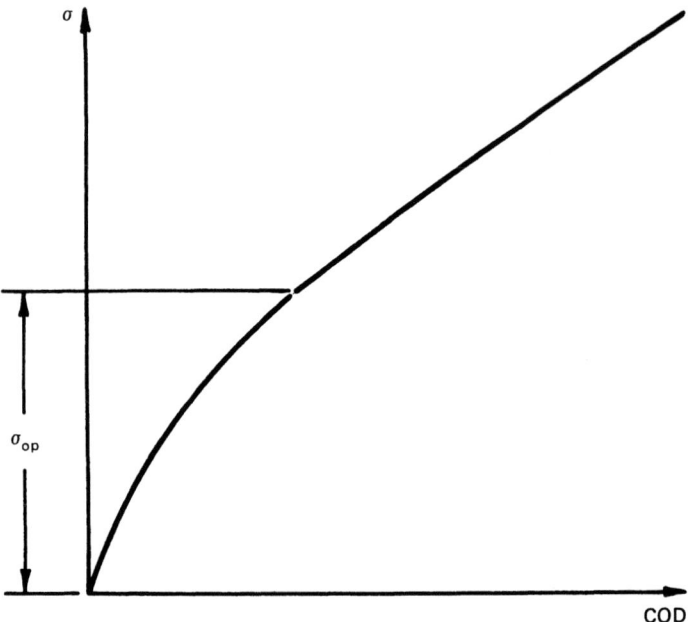

Figure 20-38 Determination of crack-opening stress.

crack and measuring the COD at the load line, or the CTOD. By plotting σ versus COD, a curve such as that shown in Fig. 20-38 is obtained. It is important to realize, however, that such a measurement is made on the surface of a body and, if the body is thick, it may not describe the true state of affairs inside the body. However, one normally expects closure effects to be greatest in thin bodies, and such measurements here would be expected to be more representative of the closure condition.

The approach to the measurement and prediction of fatigue crack growth is quite empirical. Experiments are conducted to measure the variation of crack length a with the number of applied stress cycles N. Data plots for various stress levels typically appear as shown in Fig. 20-39. Such curves collapse onto a single curve (Fig. 20-40) when values of crack growth rate da/dN and ΔK_{eff} are obtained.

Figure 20-40 contains at least three main regimes. A threshold region begins with a threshold value of ΔK_{eff} where no crack growth occurs and extends until the curve reaches a constant slope; an unstable region is characterized by a rapid increase in crack growth rate toward imminent fracture. The crack growth rate in the linear region is constant for most metals on a plot of log da/dN versus log ΔK_{eff} and leads to the empirical relation

$$\frac{da}{dN} = \tilde{C}(\Delta K_{\text{eff}})^m \qquad (20\text{-}77)$$

where $\tilde{C}$ and m are taken to be experimentally determined material constants. Values of m have been found to vary from 3 for steel to 4 for some aluminums.

In fatigue crack growth experiments one usually uses compact tension specimens prepared in accordance with ASTM E399-81, which gives the function form for $\Delta K_{\text{eff}} = \Delta K_{\text{eff}}(\Delta P_{\text{eff}}, a/w, h)$. The objective of such studies is to separate the variables a and N in Eq. (20-77) and integrate

EXPERIMENTAL FRACTURE MECHANICS

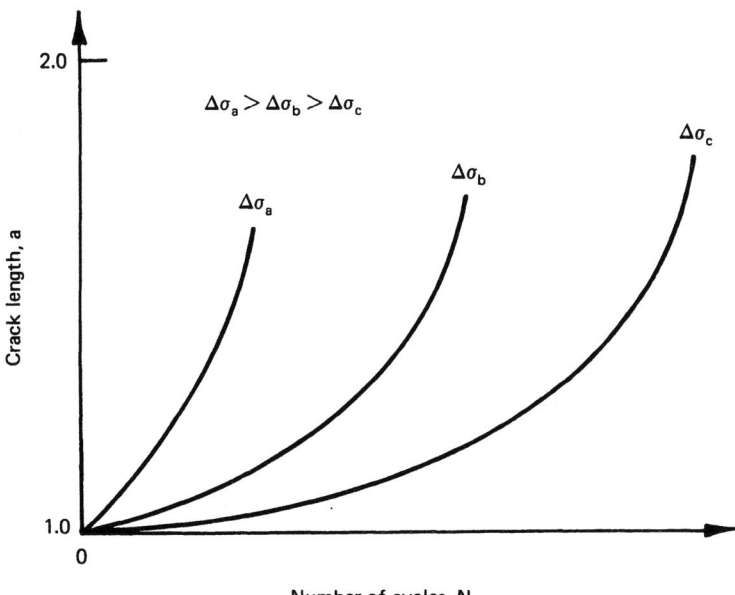

Figure 20-39 Variation of crack length with cycles under cyclic stress.

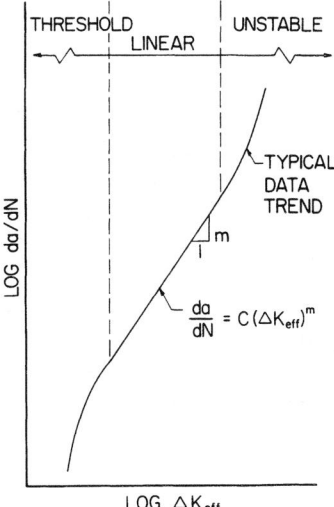

Figure 20-40 Typical crack growth rate curve.

the equation. Knowing the critical length of $a = a_c$, one can compute the number of cycles to failure (N_c) as

$$\int_{a0}^{a_c} \frac{da}{\tilde{C}(\Delta K_{\text{eff}})^m} = \int_0^{N_c} dN \tag{20-78}$$

If ΔK_{eff} is a complicated function of a, numerical integration may be necessary.

A number of factors need to be taken into account when determining cycles to failure:

1. If the load spectrum is not periodic, it should be duplicated if possible in the crack growth tests. This will provide a more accurate account of load sequence effects. In complex cases, special cycle counting methods may be useful [139].
2. If significant mean stress is present, it should be included in the load spectrum and da/dN can be expressed as a function of $R = K_{\min}/K_{\max}$ as well as the other variables.
3. Environmental effects (especially corrosion) may need to be accounted for, especially if they cause variation of $\tilde{C}$ and m.
4. Actually, Eq. (20-78) is one-dimensional. In some cases, complicated three-dimensional effects such as partial crack closure need to be accounted for.

Even when control is exercised over these factors. One may find that while little scatter appear on a graph of data such as Fig. 20-40, the percent error in computing N_c from a form of Eq. (20-78) may be 100%.

Fatigue crack growth studies involve extensive testing. Test methods usually follow ASTM E 647-81. Other pertinent test methods and practices for fracture mechanics testing are provided in the ASTM *Annual Book of Standards for Metals*. A series of papers extending fatigue experiments to surface cracks is found in Ref. 140. An excellent tutorial on this subject has been provided by Schijve [141].

Creep Crack Growth

The increased use of materials at very high temperatures has led to the necessity for measuring and analyzing time-dependent stable crack growth. The success experienced in correlating fatigue crack growth rate with the effective ΔK [Eq. (20-77)] led to studies which correlated data using

$$\dot{a} = \frac{da}{dt} = f(K) \tag{20-79}$$

Some success was found in establishing a linear relation between log $\dot{a}$ and log K for some metals [142]. However, as studies were extended to other materials, test geometries, and higher temperatures, it became clear that data scatter from such a simple correlation approach was too great.

Rice suggested that by introducing the rate terms $\dot{U}_v$ and $\dot{\mu}_i$ into the J-integral definition [Eq. (20-23)], where $\dot{U}_v = \int \sigma_{ij} \, d\dot{\varepsilon}_{ij}$, a steady-state, path-independent integral called C^* results. Landes and Begley [143] showed that use of C^* in place of K improved data correlation on a log $\dot{a}$ versus log C^* plot of test data (Fig. 20-41). Instead of load-displacement records, load-displacement rate records are used to evaluate C^*. However, Biner and Wilkinson [144] pointed out that in many practical situations, the time to establish steady-state conditions is comparable to the life of the structural element. They suggest that if one desires to correlate data for creep crack growth under non-steady-state conditions, a near-tip energy rate integral defined by

$$C_B(t) = \lim_{r \to 0} C(t) \tag{20-80}$$

be employed, where r is the distance to the path along which $C_B(t)$ is measured. A typical result

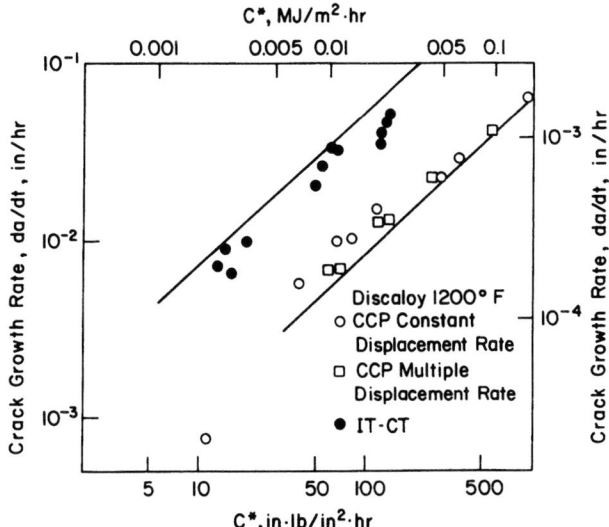

Figure 20-41 Correlation of creep crack growth test data with C^* for center-cracked panels (CCP) and compact tension (CT) specimens. (From Ref. 143.)

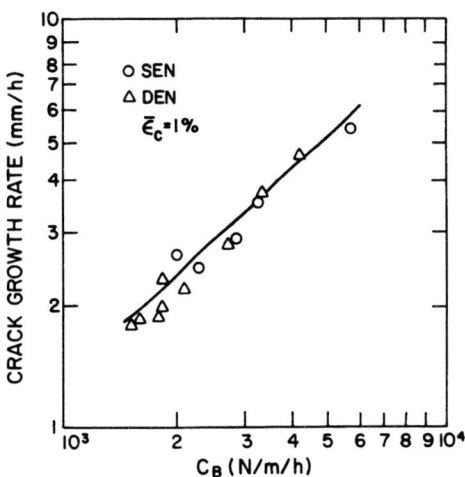

Figure 20-42 Use of C_B to eliminate geometry dependence for single-edge-notch (SEN) and double-edge-notch (DEN) plates. (From Ref. 144.)

for two test geometries (single and double edge-notched specimens) is shown in Fig. 20-42. Current studies [145] are directed toward connecting the continuum and microscopic concepts of time-dependent cracking.

20-3 Summary

The foregoing is intended as a state-of-the-art survey of experimental methods currently employed to estimate values of appropriate fracture parameters under various conditions of service. A conscious attempt was made to use linear elastic fracture mechanics as the common, unifying concept while identifying its limitations as extensions. Experimental details have been omitted in favor of presenting an overview of the concepts of the methods. These details can be obtained through key references offered.

Many important areas of fracture mechanics, such as analytical models, fracture toughness determination, and subcontinuum considerations, have been omitted to enable us to focus with brevity on the title subject.

List of Symbols

a	crack length (or half-crack length)
A_n	coefficients for σ_{ij}, u_i (n = integer)
c	crack velocity
c_1	dilatational-wave velocity
c_2	shear-wave velocity
C	compliance
$\bar{C}, \tilde{C}$	constants
C^*; $C(t)$, $C_B(t)$	steady state; non-steady-state energy rate integrals
C_0	geometric coefficient
D	dimension of caustic
E; μ	modulus of elasticity in tension; shear
f	photoelastic material fringe value
$F_{ij}, G_{ij}, i, j = 1, 2$ ($B_\mathrm{I}, B_\mathrm{II}, f, g$)	functions as indicated in text
$G_\mathrm{I}, G_\mathrm{II}, G_\mathrm{III}$	strain-energy release rates
h, h_z, h_n, w	body or slice thickness, body width
I_n	dimensionless quantity
J, J_in, L	contour integrals as described in text
J_2	strain invariant
$K, K_\mathrm{I}, K_\mathrm{II}, K_\mathrm{III}$	stress-intensity factors
$K_\mathrm{AP}, K_\mathrm{AP}^*, K_\mathrm{AP}^{**}$	"apparent" stress-intensity factors
$(K_\lambda)_\mathrm{AP}; K_\lambda$	apparent; first "eigenstress" factor
$K_\mathrm{I}^\mathrm{dyn}$	stress-intensity factor for a propagating crack
K^*, K^{**}	mixed mode, mode III stress-intensity factors
n'	photoelastic or interferometric fringe order
N, N_c	number of stress cycles of load, cycles of load to failure
p	crack face pressure
P	applied load
q, σ	remote unit stresses

	r_c	distance from crack tip to S_c
	r_m	distance from crack tip to furthest point on a given fringe
	r_p	distance from crack tip to boundary of elastic–plastic zone
	r_{ys}	distance from crack tip to point where elastic $\sigma_{yy} = \sigma_{ys}$
	R	crack growth resistance
	s	path of integration
	S, S_c	strain-energy density factor, critical of S
	S_{ij}	$\equiv \sigma_{ig} - \frac{1}{3}\sigma_{kk}\sigma_{ij}$
$u_i(i = x, y$ or $n, z)$		components of displacement
	$U; U_v$	total strain energy; strain-energy density
	V	material volume
	W	surface work to create crack
	Y, Q	geometric correction factors
	γ	specific surface energy
$\delta, \delta_{cr}, \delta_{total}, \delta_{no\ cr}, \delta_h$		displacements as described in text
	ΔK	$K_{max} - K_{min}$
	ΔK_{eff}	$K_{max} - K_{op}$
	$\Delta \sigma$	$\sigma_{max} - \sigma_{min}$
	$\Delta \sigma_{eff}$	$\sigma_{max} - \sigma_{op}$
	$\bar{\varepsilon}$	average value of ε_{tt}
	$\bar{\varepsilon}_c$	critical equivalent strain at failure
$\varepsilon_{ij}^P, i, j = 1, 2, 3,$ or r, θ, z		plastic strain components
	θ_m^0	value of θ_m as $r \to 0$
	λ'	wavelength of light
	λ_u^+	first eigenvalue in u_i
	λ_σ^+	first eigenvalue in σ_{ij}
	ν	Poisson's ratio
$\sigma_{ij}(i, j = x, y, z$ or $t, n, z)$		components of stress
$\sigma_{ij}^0, \sigma_{ij}^k(k = 1, 2, 3), \sigma_{0x}$		nonsingular stress components
	σ_e^2	$\equiv \frac{3}{2} S_{ij} S_{ij}$
	σ_{op}	crack-opening stress
	σ_{ys}	yield strength
	Σ_{ij}	as defined in text
	τ_{max}	maximum shear stress
	$(\tau)_{max}^{nz}, (\tau)_{max}^{tz}$	maximum shear stress in nz plane, tz plane
	dot ($\cdot$)	denotes d/dt

References

1. A. A. Griffith, The Phenomena of Flow and Rupture in Solids, *Phil. Trans. R. Soc. London, Ser. A, 221* (1921), 163–198.
2. A. A. Griffith, The Theory of Rupture, *Proc. 1st Int. Cong. Appl. Mech.*, 1924, 55–63.
3. G. R. Irwin, Fracture Dynamics, *Fract. Met.*, American Society of Metals, Metals Park, OH, 1948, 147–166.
4. E. Orowan, Energy Criteria of Fracture, *Weld. Res. Suppl., 20* (1955), 1575.

5. G. R. Irwin, Analysis of Stress and Strain Near the End of a Crack Transversing a Plate, *ASME Trans., J. Appl. Mech., 24*, no. 3 (September 1957), 361–364.
6. P. C. Paris and G. C. Sih, Stress Analysis of Cracks, *Fracture Toughness Testing and Its Applications*, ASTM-STP 381, American Society for Testing and Materials, Philadelphia, 1965, 30–83.
7. H. M. Westergaard, Bearing Pressures on Cracks, *ASME Trans., J. Appl. Mech., 6* (1939), A49–A53.
8. N. I. Muskhelishvili, *Some Basic Problems of the Theory of Elasticity,* 1956; transl. J. R. M. Radok, Noordhoff, Groningen, The Netherlands, 1963.
9. S. G. Lekhnitskii, *Theory of Elasticity of an Anisotropic Elastic Body,* Holden-Day, San Francisco, 1963.
10. I. Sneddon, The Distribution of Stress in the Neighborhood of a Crack in an Elastic Solid, *Proc. R. Soc. London, Ser. A, 87* (1946) 229–258.
11. M. L. Williams, On the Stress Distribution at the Base of a Stationary Crack, *ASME Trans., J. Appl. Mech., 24* (1957), 109–114.
12. G. R. Irwin, Plastic Zone Near a Crack and Fracture Toughness, *Proc. 7th Ordinance Mater. Res. Conf. (Sagamore)*, Syracuse University, Syracuse, NY, 1960, IV-63 to IV-78.
13. A. A. Wells, Application of Fracture Mechanics at and Beyond General Yielding, *Br. Weld. J.*, November 1963, 563–570.
14. F. A. McClintock and G. R. Irwin, Plasticity Aspects of Fracture Mechanics, *Fracture Toughness Testing and Its Applications,* ASTM-STP, American Society for Testing and Materials, Philadelphia, 1965, 84–113.
15. J. R. Rice and G. F. Rosengren, Plane Strain Deformation Near a Crack Tip in a Power Hardening Material, *J. Mech. Phys. Solids, 16* (1968), 1–12.
16. J. W. Hutchinson, Singular Behavior at the End of a Tensile Crack in a Hardening Material, *J. Mech. Phys. Solids, 16* (1968), 13–31.
17. D. Broek, Fatigue Crack Propagation, *Elementary Engineering Fracture Mechanics,* 3rd ed., Nijhoff, Hingham, MA, 1982, 250–287.
18. D. K. Roberts and A. A. Wells, The Velocity of Brittle Fracture, *Engineering,* December 1954, 820–821.
19. E. N. Dulaney and W. F. Brace, Velocity Behavior of a Growing Crack, *J. Appl. Phys., 31,* no. 12 (1960), 2233–2236.
20. N. R. Mott, Fracture of Metals, Some Theoretical Considerations, *Engineering,* January 1948, 16–18.
21. J. P. Berry, Some Kinetic Considerations of the Griffith Criterion for Fracture. I: Equations of Motion at Constant Force. *J. Mech. Phys. Solids, 8* (1960), 194–206.
22. J. P. Berry, Some Kinetic Considerations of the Griffith Criterion for Fracture. II: Equations of Motion at Constant Deformation, *J. Mech. Phys. Solids, 8* (1960), 207–216.
23. S. T. Rolfe and W. J. Hall, Strain Field Associated with Brittle Fracture Propagation in Wide Steel Plates, *Exp. Mech., 1* (1961), 113–119.
24. M. Yoshiki, T. Kanazawa, and H. Itagaki, An Analysis of the Propagation of Brittle Fracture, *Exp. Mech., 6* (1966), 458–462.
25. V. Schardin, Velocity Effects in Fracture, in *Fracture* (B. L. Averbach, D. K. Felbeck, G. T. Hahn, and D. A. Thomas, Eds.), Wiley, New York, 1959, 297–330.
26. G. R. Irwin, The Crack Extension Force for a Part-Through Crack in a Plate, *J. Appl. Mech. Trans. ASME, 29* (December 1962), 651–654.
27. F. W. Smith, A. F. Emery, and A. S. Kobayashi, Stress Intensity Factors for Cracks—I. Infinite Solid, *Trans. ASME, J. Appl. Mech., 34* (December 1967), 946–952.
28. M. K. Kassir and G. C. Sih, Three-Dimensional Stress Distribution Around Elliptical Crack Under Arbitrary Loadings, *Trans. ASME, J. Appl. Mech., 33* (September 1966), 601–611.
29. G. C. Sih, Some Basic Problems in Fracture Mechanics and New Concepts, *Eng. Fract. Mech., 5* (1973), 365–377.

30. J. P. Benthem, The Quarter Infinite Crack in a Half Space; Alternative and Additional Solutions, *Int. J. Solids Struct.*, 16 (1980), 119–130.

31. E. S. Folias, Method of Solution of a Class of Three-Dimensional Elastostatic Problems Under Mode I Loading, *Int. J. Fract.*, 16, no. 4 (1980), 335–348.

32. C. W. Smith and J. S. Epstein, Measurement of Near-Tip Fields Near the Right Angle Intersection of Straight Front Cracks, in *Time Dependent Fracture* (A. S. Krausz, Ed.), Nijhoff, Dordrecht, The Netherlands, 1985, 223–244.

33. C. W. Smith, J. S. Epstein, and O. Olaosebikan, Experimental Boundary Layer Studies in Three-Dimensional Fracture Problems, *Advances in Aerospace Structures, Materials and Dynamics*, ASME-AD-06, November 1983, 119–126.

34. G. R. Irwin, Relation of Crack Toughness Measurements to Practical Applications, *Weld. J., Res. Suppl.*, 41 (November 1962), 519s–528s.

35. G. R. Irwin and P. C. Paris, Fundamental Aspects of Crack Growth and Fracture, in *Fracture*, Vol. III, *Engineering Fundamentals and Environmental Effects* (H. Liebowitz, Ed.), Academic Press, New York, 1971, 1–46.

36. D. Broek, *Elementary Engineering Fracture Mechanics*, Nijhoff Kluver of Boston, Hingham, MA, 1982, 92–93.

37. R. W. Nichols and M. W. Dobson (Eds.), *Practical Fracture Mechanics for Structural Steel*, United Kingdom Atomic Energy Authority, 1969.

38. J. D. Landes, J. A. Begley, and G. A. Clarke (Eds.), *Elastic–Plastic Fracture*, ASTM-STP 668, American Society for Testing and Materials, Philadelphia, 1979.

39. C. F. Shih and J. P. Gudas (Eds.), *Elastic–Plastic Fracture, Second Symposium*, Vols. I and II, ASTM-STP 803, American Society for Testing and Materials, Philadelphia, 1983.

40. D. A. Dugdale, Yielding of Steel Sheet Containing Slits, *J. Mech. Phys. Solids*, 8 (1960), 100–104.

41. G. I. Barenblatt, The Mathematical Theory of Equilibrium of Cracks in Brittle Fracture, *Adv. Appl. Mech.*, 7 (1962), 59–129.

42. F. A. Bilby, A. H. Cottrell, and K. H. Swinden, The Spread of Plastic Yield from a Notch, *Proc. R. Soc. London, Ser. A*, 272 (1963), 304–310.

43. H. F. Brinson, The Ductile Fracture of Polycarbonate, *Exp. Mech.*, 10 (February 1970), 72–77.

44. H. F. Brinson and H. Gonzales, Jr., The Dugdale Model Adapted to the Transverse Bending of Plates, *Exp. Mech.*, 12 (March 1972), 130–135.

45. H. F. Brinson, J. H. Underwood, and A. R. Rosenfield, Yield Zones in Polymeric and Metallic Plates Under Pure Bending Containing a Through Crack, *Int. J. Fract.*, 9, no. 4 (December 1973), 405–420.

46. A. S. Kobayashi, W. L. Engstrom, and B. R. Simon, Crack Opening Displacement and Normal Strains in Centrally Notched Plates, *Exp. Mech.*, 9 (April 1969), 163–170.

47. J. D. Eshelby, The Continuum Theory of Lattice Defects, in *Solid State Physics* (F. Seitz and D. Turnall, Eds.), Academic Press, New York, 1956, 79–144.

48. J. R. Rice, A Path-Independent Integral and Approximate Analysis of Strain Concentration by Notches and Cracks, *Trans. ASME, J. Appl. Mech.*, 35 (June 1968), 379–386.

49. J. R. Rice, Mathematical Analysis in the Mechanics of Fracture, in *Fracture, An Advanced Treatise*, Vol. II (H. Liebowitz, Ed.), Academic Press, New York, 1968, 191–311.

50. J. D. Landes and J. A. Begley, Test Results from *J*-Integral Studies: An Attempt to Establish a J_{Ia} Testing Procedure, in *Fracture Analysis* (P. C. Paris and G. R. Irwin, Eds.), ASTM-STP 560, American Society for Testing and Materials, Philadelphia, 1974, 170–186.

51. J. D. Landes, H. Walker, and G. A. Clarke, Evaluation of Estimation Procedures Used in *J*-Integral Testing, in *Elastic–Plastic Fracture* (J. D. Landes, J. A. Begley, and G. A. Clarke, Eds.), ASTM-STP 668, American Society for Testing and Materials, Philadelphia, 1979, 266–287.

52. K. Ohji, A. Otsuka, and H. Kobayashi, Evaluation of Several J_{IC} Testing Procedures Recommended in Japan, in *Elastic–Plastic Fracture: Second Symposium*, Vol. II, *Fracture Resistance Curve and Engineering Applications*

(C. F. Shih and J. P. Gudas, Eds.), ASTM-STP 803, American Society for Testing and Materials, Philadelphia, 1983, II-398 to II-419.

53. P. C. Paris, H. Tada, A. Zahoor, and H. Ernst, The Theory of Instability of the Tearing Mode of Elastic–Plastic Crack Growth, in *Elastic–Plastic Fracture* (J. D. Landes, J. A. Begley, and G. A. Clarke, Eds.), ASTM-STP 668, American Society for Testing and Materials, Philadelphia, 1979, 5–36.

54. H. Ernst, Some Salient Features of the Tearing Instability Theory, in *Elastic–Plastic Fracture: Second Symposium, Vol. II, Fracture Resistance Curves and Engineering Applications* (C. F. Shih and J. P. Gudas, Eds.), ASTM-STP 803, American Society for Testing and Materials, Philadelphia, 1983, II-133 to II-155.

55. J. K. Knowles and E. Sternberg, On a Class of Conservation Laws in Linearized and Finite Elastostatics, *Arch. Ration. Mech. Anal.*, 44 (1972), 187–211.

56. B. Budiansky and J. R. Rice, Conservation Laws and Energy Release Rates, *Trans. ASME, J. Appl. Mech.*, 40 (March 1973), 201–203.

57. S. N. Atluri, Path-Independent Integrals in Finite Elasticity and Inelasticity, with Body Forces, Inertia, and Arbitrary Crack-Face Conditions, *Eng. Fract. Mech.*, 16, no. 3 (1982), 341–364.

58. L. B. Freund and R. J. Clifton, On the Uniqueness of Plane Elasto-Dynamic Solutions for Running Cracks, *J. Elasticity*, 4, no. 4 (1974), 293–299.

59. A. S. Kobayashi and M. Ramulu, Dynamic Stress Intensity Factors for Unsymmetric Dynamic Isochromatics, *Exp. Mech.*, 21 (1981), 41–49.

60. R. T. Busbey, D. M. Fisher, M. H. Jones, and J. E. Srawley, Compliance Measurements, in *Experimental Techniques in Fracture Mechanics* (A. S. Kobayashi, Ed.), Iowa State University Press and SESA, 1973, 76–95.

61. R. B. King and G. Herrmann, Acoustoelastic Determination of Forces on a Crack in Mixed-Mode Loading, *ASME J. Appl. Mech.*, 50 (June 1983), 379–382.

62. R. W. Benson and V. J. Raelson, Acoustoelasticity, *Prod. Eng.*, 30 (1959), 56–59.

63. J. Blinka and W. Sachse, Application of Ultrasonic-Pulse-Spectroscopy Measurements to Experimental Stress Analysis, *Exp. Mech.*, 16 (December 1976), 448–453.

64. G. S. Kino, J. B. Hunter, G. C. Johnson, A. R. Selfridge, D. M. Barnett, G. Herrmann, and C. R. Steele, Acoustoelastic Imaging of Stress Fields, *J. Appl. Phys.*, 50, no. 4 (1979), 530–535.

65. K. Okada, Stress-Acoustic Relation for Stress Measurements by Ultrasonic Technique, *J. Acoust. Soc. J. (E)*, 1, no. 3 (1980), 193–200.

66. K. Okada, Acoustoelastic Determination of Stress in Slightly Orthotropic Materials, *Exp. Mech.*, 21 (1981), 461–466.

67. A. V. Clark, R. B. Mignogna, and R. J. Sanford, Acousto-Elastic Measurement of Stress and Stress Intensity Factors Around Crack Tips, *Ultrasonics*, March 1983, pp. 57–64.

68. R. J. Sanford and R. Chona, An Analysis of Photoelastic Fracture Patterns with Sampled Least-Squares Methods, *Proc. 1981 SESA Spring Meeting*, June 1981, 273–276.

69. J. H. Underwood and D. P. Kendall, Measurement of the Strain Distribution in the Region of the Crack Tip, presented at ASTM E-24 Meeting, March 1967.

70. W. L. Hu and H. W. Liu, Crack Tip Strain—A Comparison of Finite Element Calculations and Moiré Measurements, in *Cracks and Fracture* (J. L. Swedlow and M. L. Williams, Eds.), ASTM-STP 601, American Society for Testing and Materials, Philadelphia, 1976, 622–634.

71. C. A. Sciammarella and M. P. K. Rao, Failure Analysis of Stainless Steel at Elevated Temperatures, *Exp. Mech.*, 19 (1979), 389–398.

72. J. R. Rice, P. C. Paris, and J. G. Merkle, Some Further Results on J-Integral Analysis and Estimates, in *Progress in Flaw Growth and Fracture Toughness Testing*, ASTM-STP 536, American Society for Testing and Materials, Philadelphia, 1973, 231–245.

73. W. W. Gerberich, Plastic Strains and Energy Density in Cracked Plates: I. Experimental Technique and Results, *Exp. Mech.*, 4, no. 11 (1964), 335–344.

74. J. H. Underwood, J. L. Swedlow, and D. P. Kendall, Experimental and Analytical Strains in an Edge Cracked Sheet, *J. Eng. Fract. Mech.*, 2, no. 3 (May 1971), 183–196.

75. H. W. Liu and J. S. Ke, Moiré Method, in *Experimental Techniques in Fracture Mechanics—2*, SESA Monograph 2 (A.S. Kobayashi, Ed.), 1975, 111–165.

76. P. F. Packman, The Role of Interferometry in Fracture Studies, in *Experimental Techniques in Fracture Mechanics—2*, SESA Monograph 2 (A. S. Kobayashi, Ed.), 1975, 59–87.

77. C. W. Smith, D. Post, and G. Nicoletto, Prediction of Sub-Critical Crack Growth Data from Model Experiments, *Devel. Theor. Appl. Mech.*, XI (1982), 167–179.

78. C. W. Smith, D. Post, G. Hiatt, and G. Nicoletto, Displacement Measurements Around Cracks in Three-Dimensional Problems by a Hybrid Experimental Technique, *Exp. Mech.*, 23, no. 1 (March 1983), 15–20.

79. W. N. Sharpe, Jr., and A. A. Sukere, Transient Response of a Central Crack to a Tensile Pulse, *Exp. Mech.*, 23, no. 1 (March 1983), 89–98.

80. W. N. Sharpe, Jr., Dynamic Strain Measurement with the Interferometric Strain Gage, *Exp. Mech.*, 10, no. 2 (1970), 89–92.

81. J. L. Swedlow and W. W. Gerberich, Plastic Strains and Energy Density in Cracked Plates, *Exp. Mech.*, 4, no. 12 (1964), 345–351.

82. J. Duffy, Accuracy of Birefringent Coating Method for Coatings of Arbitrary Thickness, *Exp. Mech.*, 1, no. 1 (1961), 21–32.

83. G. R. Irwin, private communication, 1982.

84. D. G. Smith and C. W. Smith, Photoelastic Determination of Mixed-Mode Stress Intensity Factors, *Eng. Fract. Mech.*, 4 (1972), 357–366.

85. A. S. Kobayashi, M. Ramulu, and B. S. J. Kang, Dynamic Crack Branching—A Photoelastic Evaluation, *Fract. Mech.*, 15 (1984).

86. W. B. Bradley and A. S. Kobayashi, An Investigation of Propagating Cracks by Dynamic Photoelasticity, *Exp. Mech.*, 10, no. 3 (1970), 106–113.

87. P. S. Theocaris and E. E. Gdoutos, A Photoelastic Determination of K_I Stress Intensity Factors, *Eng. Fract. Mech.*, 7, no. 2 (1975), 331–339.

88. J. W. Dally and R. J. Sanford, A General Method for Determining Mixed-Mode Stress Intensity Factors from Isochromatic Fringe Patterns, *Eng. Fract. Mech.*, 11, no. 4 (1979), 621–634.

89. H. P. Rossmanith and G. R. Irwin, Analysis of Dynamic Isochromatic Crack Tip Stress Patterns, Research Report, Department of Mechanical Engineering, University of Maryland, Baltimore, 1979, 443 pp.

90. C. W. Smith and O. Olaosebikan, On the Extraction of Mixed-Mode Stress Intensity Factors from Near-Tip Photoelastic Data in Three-Dimensional Problems, *Proc. 5th Int. Cong. Exp. Mech.*, June 1984, 511–519.

91. E. Sommer, An Optical Method for Determining the Crack-Tip Stress Intensity Factor, *Eng. Fract. Mech.*, 1, no. 4 (1970), 705–718.

92. E. Sommer, Experimental Determination of Stress Intensity Factor by COD Measurements, in *Mechanics of Fracture*, Vol. 7 (G. C. Sih, Ed.), Noordhoff Int. Pub., Leyden, The Netherlands, 1981, 331–347.

93. P. B. Crosley, S. Mostovoy, and E. J. Ripling, An Optical Interference Method for Experimental Stress Analysis of Cracked Structures, *Eng. Fract. Mech.*, 3, no. 4 (1971), 421–433.

94. T. D. Dudderar and R. O'Regan, Measurement of the Strain Field Near a Crack Tip in Polymethylmethacrylate by Holographic Interferometry, *Exp. Mech.*, 11, no. 2 (1971), 49–56.

95. P. Manogg, Investigation of the Rupture of a Plexiglas Plate by Means of an Optical Method Involving High Speed Filming of the Shadows Originating Around Holes Drilled in the Plate, *Int. J. Frac. Mech.*, 2 (1966), 604–613.

96. P. S. Theocaris, Elastic Stress Intensity Factors Evaluated by Caustics, in *Experimental Evaluation of Stress Concentration and Stress Intensity Factors—Mechanics of Fracture*, Vol. 7 (G. C. Sih, Ed.), Noordhoff Int. Pub., Leyden, The Netherlands, 1981, 189–252.

97. J. Beinert and J. F. Kalthoff, Experimental Determination of Dynamic Stress Intensity Factors by Shadow Patterns,

in *Experimental Evaluation of Stress Concentration and Stress Intensity Factors—Mechanics of Fracture*, Vol. 7 (G. C. Sih, Ed.), Noordhoff Int. Pub., Leyden, The Netherlands, 1981, 281–330.

98. A. J. Rosakis and L. B. Freund, Optical Measurement of Plastic Strain Concentration at a Crack Tip in a Ductile Steel Plate, *J. Eng. Mater. Tech.*, *104* (April 1982), 115–120.

99. G. Oppel, Photoelastic Investigation of Three-Dimensional Stress and Strain Conditions, NACA TM 824, transl. J. Vanier, 1937.

100. J. R. Srawley, M. W. Jones, and B. Gross, Experimental Determination of the Dependence of Crack Extension Force on Crack Length for a Single Edge Notch Tensile Specimen, NASA-TN-D2396, May 1964.

101. G. R. Irwin, Measurement Challenges in Fracture Mechanics, William Murray Lecture, Society of Experimental Stress Analysis Fall Meeting, October 1973.

102. C. W. Smith, J. J. McGowan, and M. Jolles, Effects of Artificial Cracks and Poisson's Ratio upon Photoelastic Stress Intensity Determination, *Exp Mech.*, *16*, no. 5 (1976), 188–193.

103. G. C. Sih and H. Liebowitz, Mathematical Theories of Brittle Fracture, Chap. 2 of *Mathematical Fundamentals*, Vol. II of *Fracture*, Academic Press, New York, 1968, 150.

104. C. W. Smith, W. H. Peters, and A. T. Andonian, Mixed-Mode Stress Intensity Distributions for Part Circular Surface Flaws, *Eng. Fract. Mech.*, *13* (1979), 615–629.

105. C. W. Smith, W. H. Peters, W. T. Hardrath, and T. S. Fleischman, Stress Intensity Distributions in Nozzle Corner Cracks of Complex Geometry, *Trans. 5th Int. Conf. Struct. Mech. React. Technol.*, Vol. G, Paper G4/4, 1979, 8 pp.

106. C. W. Smith, D. Post, and G. Nicoletto, Experimental Stress Intensity Distributions in Three Dimensional Cracked Body Problems, *Proc. J. Conf. Exp. Mech.*, SESA-JSME, May 1982, 196–200.

107. C. W. Smith, O. Olaosebikan, and J. S. Epstein, A Proposed Rationale for Accounting for Boundary Layer Effects in Designing against Fracture in Three Dimensional Problems, *Proc. 11th Can. Fract. Conf.*, June 1984, 209–220.

108. C. W. Smith and D. Constantinescu, On the Intersection of Crack Borders with Free Surfaces; An Engineering Interpretation of Optical Experimental Results, *Proc. 2nd Int. Conf. Photomechanics and Speckle Metrology*, 1991 Vol. 1554, 102–115.

109. E. L. Ross, G. Kaminsky, and J. C. Conway, Jr., Measurement of Mode I Stress Intensity Factors by Scattered Light Photoelasticity, *Proc. Soc. Expr. Stress Analysis*, Vol. 39, 1982, 117–120.

110. J. T. Pindera, Analytical Foundations of the Isodyne Photoelasticity, *Mech. Res. Commun.*, *8* (1981), 391–397.

111. S. B. Mazurkiewics and J. T. Pindera, Integrated Plane Photoelastic Method—Application of Photoelastic Isodynes, *Exp. Mech.*, *19* (July 1979), 225–234.

112. J. T. Pindera and B. R. Krasnowski, Determination of Stress Intensity Factors in Thin and Thick Plates Using Isodyne Photoelasticity, *Fract. Probl. Solutions Energy Ind.*, 1981, 147–156.

113. J. T. Pindera, B. R. Krasnowski, and M. J. Pindera, An Analysis of Semi-Plane Stress States in Fracture Mechanics and Composite Structures Using Isodyne Photoelasticity, *Proc. 1982 J. Conf. Exp. Mech.*, SESA, May 1982, 417–421.

114. R. de Sailly and A. LaGarde, Surface Crack Analysis by a Optical Slicing Method of Three-Dimensional Photoelasticity, (in French), *Proc. 7th Int. Conf. Exp. Stress Anal.*, August 1982, 315–329.

115. J. S. Epstein, W. R. Lloyd, and W. G. Reuter, Three-Dimensional Displacement Measurements of Elastic–Plastic Surface Flaws, *Analytical, Numerical and Experimental Aspects of Three-Dimensional Fracture Processes*, ASME AMD Vol. 91, 1988, 33–50.

116. F. P. Chiang and H. Lu, Interior Crack-Tip Elastic–Plastic Field Measurement by Embedded Random Speckle Method, *Analytical, Numerical and Experimental Aspect of Three-Dimensional Fracture Processes*, ASME AMD Vol. 91, 1988, 205–214.

117. S. Dhar, G. L. Cloud, and S. Palebut, Measurement of Three-Dimensional Mode I Stress Intensity Factors Using Multiple Embedded Grid Metals Technique, *J. Theor. Appl. Fract. Mech.*, *12* (1989), 141–147.

118. Y. J. Sun, O. S. Lee, and A. S. Kobayashi, Crack Tip Plasticity Under Mixed-Mode Loading, *Proc. ICF Symp. Fract. Mech.*, Beijing, China, November 1983.

119. J. F. Kalthoff, J. Beinert, S. Winkler, and W. Klemm, Experimental Analysis of Dynamic Effects in Different

Crack Arrest Test Specimens, in *Crack Arrest Methodology and Applications* (G. T. Hahn and M. F. Kanninen, Eds.), ASTM-STP 711, American Society for Testing and Materials, Philadelphia, 1980, 109–128.

120. T. Kobayashi and J. W. Dally, Dynamic Photoelastic Determination of the à-K Relation for 4340 Alloy Steel, in *Crack Arrest Methodology and Applications* (G. T. Hahn and M. F. Kanninen, Eds.), ASTM-STP 71, American Society for Testing and Materials, Philadelphia, 1980, 189–239.

121. J. F. Kalthoff, W. Boehme, S. Winkler, and W. Klemm, Measurements of Dynamic Stress Intensity Factors in Impacted Bend Specimens, in *C.S.N.I. Specialist Meeting on Instrumented Precracked Charpy Testing* (R. A. Wullaert, Ed.), EPRI NP-2102-LD, C.S.N.I. No. 67, November 1981, 2–19.

122. A. A. Betser, A. S. Kobayashi, O. S. Lee, and B. S.-J. Kang, Crack Tip Dynamic Isochromatics in the Presence of Small Scale Yielding, *Exp. Mech., 22* (1982), 132–138.

123. A. A. Wells and D. Post, The Dynamic Stress Distribution Surrounding a Running Crack—A Photoelastic Analysis, *Proc. Soc. Exp. Stress Anal., 16* (1958), 69–92.

124. A. J. Rosakis and L. B. Freund, The Effect of Crack-Tip Plasticity on the Determination of Dynamic Stress Intensity Factors by the Optical Method of Caustics, *J. Appl. Mech., 48* (1981), 302–308.

125. W. B. Bradley and A. S. Kobayashi, Fracture Dynamics—A Photoelastic Investigation, *Eng. Fract. Mech., 3* (1971), 317–332.

126. J. W. Dally, Dynamic Photoelastic Studies of Fracture, *Exp. Mech., 19* (1979), 349–367.

127. A. S. Kobayashi and S. Mall, Dynamic Fracture Toughness of Homalite-100, *Exp. Mech., 18* (1978), 11–18.

128. H. P. Rossmanith and A. Shukla, Dynamic Photoelastic Investigation of Interaction of Stress Waves with Running Cracks, *Exp. Mech., 21* (1981), 415–422.

129. F. Katsamanis, D. Raftopulos, and P. S. Theocaris, Static Dynamic Stress Intensity Factors by the Method of Transmitted Caustics, *J. Eng. Mater. Technol., 99* (1977), 105–109.

130. W. Goldsmith and F. Katsamanis, Fracture of Notched Polymeric Beams Due to Central Impact, *Exp. Mech., 19* (1979), 235–244.

131. A. Rosakis, Analysis of the Optical Method of Caustics for Dynamic Crack Propagation, *Eng. Fract. Mech., 13* (1980), 331–347.

132. K. Ravi-Chandar and W. G. Knauss, Dynamic Crack-Tip Stresses Under Stress Wave Loading—A Comparison of Theory and Experiment, *Int. J. Fract., 20* (1982), 204–222.

133. K. Ravi-Chandar, An Experimental Investigation into the Mechanics of Dynamic Fracture, Ph.D. thesis, California Institute of Technology, Pasadena, 1982.

134. A. S. Kobayashi and M. Ramulu, Dynamic Stress Intensity Factors for Unsymmetric Dynamic Isochromatics, *Exp. Mech., 21* (January 1981), 41–48.

135. A. S. Kobayashi and C. F. Chan, A Dynamic Photoelastic Analysis of Dynamic-Tear-Test Specimen, *Exp. Mech., 16*, no. 5 (1976), 176–181.

136. M. Ramulu and A. S. Kobayashi, Dynamic Crack Curving—A Photoelastic Evaluation, *Proc. J. Conf. Exp. Mech.,* May 1982, 258–262.

137. J. C. Newman, Jr., Prediction of Stable Crack Growth and Instability Using the V_R Curve Method, *5th Int. Cong. Exp. Mech.,* Montreal, June 1984.

138. W. Elber, The Significance of Crack Closure, *Damage Tolerance in Aircraft Structures*, ASTM-STP 486, American Society for Testing and Materials, Philadelphia, 1971, 230–242.

139. N. E. Dowling, Fatigue Life Prediction for Complex Load versus Time Histories, *Pressure Vessels & Piping: Design Technology: A Decade of Progress*, ASME Book No. G00213, 1982, 487–498.

140. W. G. Reuter, J. H. Underwood, and J. C. Newman, Jr. (Eds.) *Surface-Crack Growth: Models, Experiments and Structures*, ASTM-STP 1060, American Society for Testing and Materials, Philadelphia, 1990, 215–414.

141. J. Schijve, Four Lectures on Fatigue Crack Growth, *J. Eng. Fract. Mech., 11* (1979), 167–221.

142. J. G. Kaufman, K. O. Bogardus, D. A. Maurey, and R. C. Malcolm, Creep Cracking in 2219-T851 Plate at Elevated

Temperatures, *Mechanics of Crack Growth,* ASTM-STP 590, American Society for Testing and Materials, Philadelphia, 1976, 149–168.
143. J. D. Landes and J. A. Begley, A Fracture Mechanics Approach to Creep Crack Growth, *Crack Growth,* ASTM-STP 590, American Society for Testing and Materials, Philadelphia, 1976, 128–148.
144. S. B. Biner and D. S. Wilkinson, Numerical Modelling of Creep Crack Growth, *Proc. 11th Can. Fract. Conf.,* June 1984, 3–15.
145. A. Argon, The Mechanisms of Growth of Cracks in Creeping Alloys, *11th Can. Fract. Conf.,* June 1984.

CHAPTER
21

Digital Image Processing

Yoshiharu Morimoto

Osaka University
Osaka, Japan

21-0 Introduction

Purpose of Image Processing

Optical methods such as photoelasticity, grid method, geometric moiré, moiré topography, moiré interferometry, and holographic interferometry, are very useful whole-field methods for measuring stress, strain, deformation, and shape in contrast to point-wise methods, such as the strain-gage method. Manual analysis of the very large amount of two-dimensional data obtained by these whole-field methods is time-consuming and tedious. Moreover, a subjective process is normally used in extracting useful information from these data, which usually are scattered, and thus the results may depend on the analyst. If the data are analyzed with the aid of computers, however, processing is very fast and the subjective element is almost eliminated. The processing of images aided by a computer is called "digital image processing," "digital picture processing," or "image processing" [1–10].

Image processing not only provides labor-saving, high-speed processing and accurate analysis, but also can extract useful information that might be suppressed when conventional methods are used. It is particularly useful in the analysis of very large amounts of data that would be cumbersome to process manually. Additionally, image processing generates new analytical methods.

The computer analysis of images entails sampling and digitization as discussed in the next section. For example, the most popular sampled image size is 512×480 pixels with 256 gray levels. The amount of data for this image is huge so that computers with very large memory and high speed processing are required for image processing. Formerly these computers were very expensive, but the availability of personal computers and workstations has eliminated this restriction, and high-speed image processing can be performed at little expense with high accuracy, using only personal computers [4,7,8].

This chapter describes procedures for developing image-processing systems and for analyzing images. The use of these systems allows the introduction of image-processing software for measuring stress, strain, deformation, and shape, and especially fringe pattern analyses, as well as new analysis methods [10].

Table 21-1 Examples Applying Image Processing for Measurements of Stress, Strain, Deformation, and Shape

Measurement Method	Analyzed Object	Measured Value	Ref.
Photoelasticity	Fringe	Stress	11–17
Moiré method	Fringe	Displacement strain	
Geometric moiré	Fringe	Displacement	18–26
Scanning moiré	Fringe	Displacement	27–34
Shift (shearing)	Fringe	Strain	35–36
Moiré interferometry	Fringe	Displacement	37
Grid method	Grid	Displacement, shape	38–42
Triangulation	Grid, dots	Shape	43–46
Holographic interferometry	Fringe	Displacement	47–53
Speckle interferometry	Speckle	Displacement	54–58
Shearography	Speckle	Strain	59
General fringe	Fringe	Displacement, etc.	60–62
Correlation	Random pattern	Displacement	63–65
Fracture	Facet, crack	Shape	66–67
Thermoelasticity	Temperature	Stress	68–69
Copper electroplating	Particle shape	Stress	70
Particle analysis	Particle shape	Shape	71

History of Image Processing for Experimental Mechanics

Image processing is widely used to analyze whole-field images. Table 21-1 shows examples applying image processing to analyses of stress, strain, deformation, and shape [11–71].

In 1967, Sciammarella and Sturgion [18] introduced a computer to analyze a fringe pattern by using a one-dimensional digital filter and the phases of a fringe pattern. For two-dimensional, whole-field analysis, in 1977 Idesawa et al. [27] proposed a scanning moiré method for measuring a three-dimensional shape by means of a fringe pattern obtained by sampling a deformed grating on virtual grating lines, which is a reference grating. In 1978 Ueda et al. [28] proposed an electronic grating moiré method, recording a deformed grating with a TV camera, thus producing a reference grating electronically and superimposing it on the deformed grating image to obtain a contour map of an object. In 1979 Müller et al. [11] and Seguchi et al. [12] introduced independently an image-processing method for analyzing photoelastic fringe patterns. Patterson [15] described the history of automated photoelastic analysis. In 1983 Hariharan et al. [47] proposed the phase-shift method for fringe patterns, which provided high accuracy. In 1983 Takeda and Mutoh [43] proposed Fourier transform profilometry in the field of moiré topography. In 1984 Morimoto et al. [29,30] developed the scanning moiré method using the thinning out of the scanning lines of a TV camera or sampling points of an image processor [8]. Morimoto et al. [23–26] also developed Fourier transform moiré and grid method using the Fourier spectra of deformed grating images. The details of these methods are discussed in Section 21-7.

21-1 Analog and Digital Images [1–8]

Analog and Digital Systems

There are two groups of image-processing systems. One group consists of analog systems, which use optics, photography, and video techniques. The other is comprised of digital systems using

DIGITAL IMAGE PROCESSING

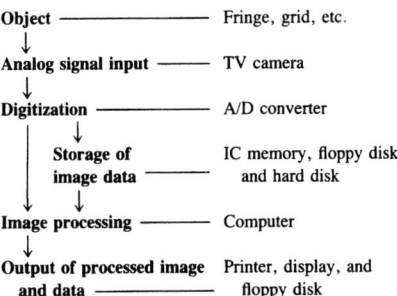

Figure 21-1 Flow of image-processing steps and devices.

electronic digital computers. The use of analog systems including lenses, diffraction gratings, cameras, and TV devices provide very fast and low-cost processing but lacks flexibility. The digital systems are flexible, accurate, and reproducible, although they are expensive and slow. These disadvantages, however, can be overcome with the development of electric parts, such as integrated circuits (ICs).

In this chapter, which treats digital systems, "image processing" means digital image processing using computers.

Procedures of Image Processing

To analyze images by computer, analog data must be transformed into digital images, which consist of digitized brightness data at discrete sampling points, by using an analog-to-digital (A/D) converter. The digitized data are stored in IC memory, processed to perform the desired image-processing operations, and then output to printers and display devices. The flow of these procedures is shown in Fig. 21-1.

Sampling and Quantizing

To digitize an image, its brightness is sampled at intersections called sampling points (Fig. 21-2a). The sampled brightness is digitized as shown in Fig. 21-2b, with the result appearing in Fig. 21-2c. The most popular method for sampling an image is the use of a TV camera and an A/D converter. The image signal from a TV camera for the NTSC system used in the United States and Japan is shown schematically in Fig. 21-3. An image is constructed of 525 scanning lines which consists of two fields. Every second scanning line makes one field. After one-sixtieth second from the start of scanning the first field, the scanning of the second field starts.

The signal intensity shows the brightness on the scanning lines. The signal is output as the one continuous wave of voltage (Fig. 21-4a). This voltage signal is quantized by an A/D converter and stored in the IC memory as digital values as shown in Fig. 21-4b. The digitized value is called the "gray value," "gray level," or "gray scale." The voltage is high when the scanning line is on white, and it is low when the scanning line is on black. The voltage is proportional to the brightness of the sampling point in an ideal system. In practice, the digitized gray value is not proportional to the brightness because of the nonlinearity of the image sensor and the A/D converter. Figure 21-5 shows a schematic example of the relationship between brightness value of an object and the digitized gray value. When the gray value instead of the brightness value is quantitatively used in image analysis, this relationship (i.e., of brightness value to gray value) must be checked.

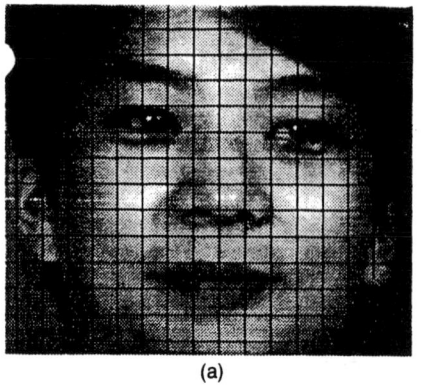

(a)

```
101  74  34  27  23  43  75  69  32  27  19  33  29  32  30 101
 95  30  23  24  62  87 108 105  82  37  21  28  18  28  27  34
 14  23  21  29  44 132 176 157 145 104  32  26  24  27  19  12
 18  20  19  76 119 159 188 203 190 137  97  28  30  21  29  19
 16  21  22  94 129 167 220 230 206 204 203  80  27  23  14  16
 15  22  44 107 122 149 132 169 218 203 149  89  98 100  17  19
 13  17  93 119 120 130 175 182 224 182 177 182 170 106  29  22
     20 109 120 130 141 141 142 185 166  70  23  42 138  56  17
 96  31 108 144 181 202 200 159 196 145 199 165 175 141 100  22
119  52 103 140 202 215 221 130 194 163 213 218 221 178 103  40
113 119  98 136 187 212 138  38 163  71 117 231 226 162  76  94
109 148  94 139 165 193 175 168 177 181 192 193 178 134 106  94
101  92  86 142 148 149 150 111 168 158 148 166 154 117 154  69
103 106  83 133 157 160 136 114 126 107 152 167 159 102  68  69
103 106  24  93 135 150 164 153 154 179 165 144 130  72  69  68
101 107  98 105 106 115 149 161 167 173 158 123  92  67  65  66
```

(b)

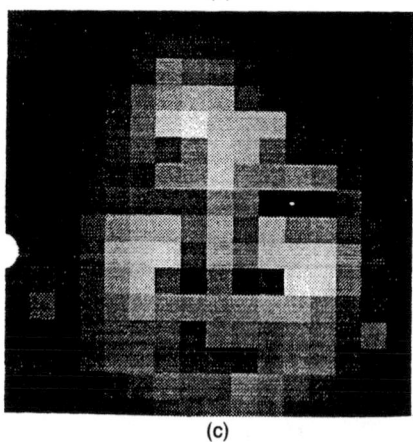

(c)

Figure 21-2 Sampling and quantizing of image: (a) Original image and sampling points; (b) digitized data; (c) digitized image.

DIGITAL IMAGE PROCESSING

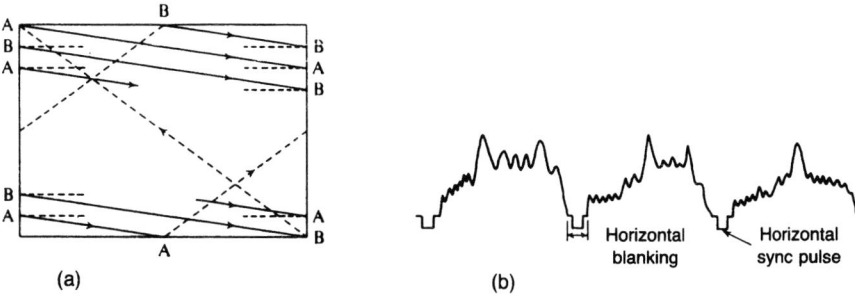

Figure 21-3 Image signal of the NTSC system: (a) scanning lines (A, first field; B, second field); (b) voltage signal on output line.

Some image-processing hardware systems (image processors) are adjustable and can linearize this curve. Most image processors can adjust the maximum and minimum voltages to the maximum and minimum gray levels of the processors, respectively. Alternatively, adjustment of the aperture of the camera lens improves the relationship.

By sampling and quantizing, the object is expressed as an image consisting of digital values $f(x,y)$, at the sampling point: that is, at the two-dimensional coordinates (x,y). Each sampled point is called a pixel; the digital value at any point is termed "gray value," "gray level," or "gray scale" as mentioned above. The pixel number is usually taken from 64 (6 bits) to 4096 (12 bits) in the horizontal and vertical directions. If the pixel sizes (the distance between the neighboring pixels) in the horizontal and vertical directions are different, the correction in rotation of an image must be calculated. A square pixel (i.e., horizontal pixel equal in size to the vertical) is therefore good for measurement. A higher number of pixels results in a finer position resolution. A larger number of pixels and gray levels provides a more realistic and more beautiful image. However, when these numbers are large, a large amount of IC memory is required and the calculation speed for one image becomes slow. Therefore suitable sampling point and gray level numbers must be selected according to the purpose for which the image is being processed.

The gray-level number is usually taken as an integer value from 2 to 256 levels. More than 64 gray levels are required to obtain a natural image for the human eye. Many analyses in engineering are performed with gray levels below 64, and occasionally two levels of white and black provide

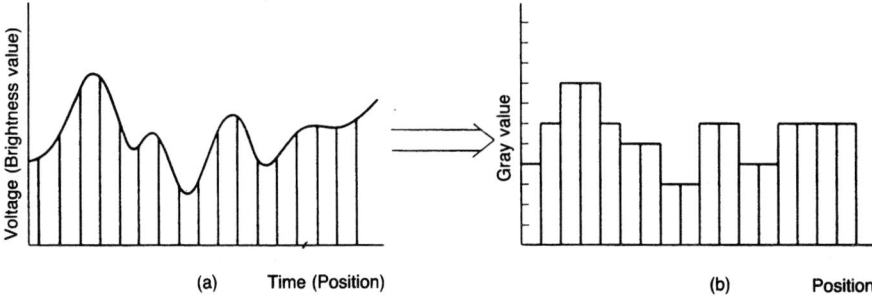

Figure 21-4 Schematic explanation of sampling and quantizing: (a) analog signal; (b) digital signal.

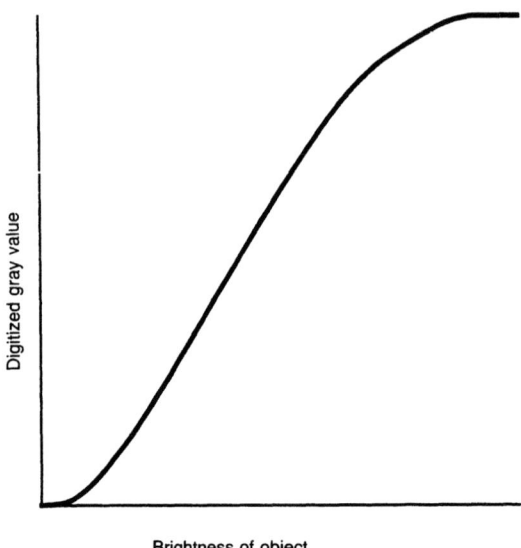

Figure 21-5 Schematic relationship between brightness of object and digitized gray value.

sufficient information. The most popular image-processing system using a TV camera treats 512 pixels in horizontal, 480 pixels in vertical, and 256 gray levels.

While a monochromatic image is sufficient for almost all cases discussed in this chapter, color image processing may become popular in the near future. Color image processing calls for gray levels (brightness levels) for each of the three primary colors (red, green, and blue), hence a larger memory size and longer calculation time are needed. For a monochromatic image, each gray level is related to a color, and the pseudocolor results in a powerful image for presentation.

The stored information of an image is obtained by multiplying the pixel number by the bit number of the gray level and by the number of primary colors. For example, when an image has a pixel size of 512 × 480 and there are 256 (= 8 bits) gray levels, a monochromatic output will require a total of 512 × 480 × 8 × 1 = 1,966,080 bits. The image can express 2 to the 1,966,080th power combinations.

Figure 21-6 shows the differences obtained by changing the pixel number for one image. Figure 21-7 shows what happens when the gray-level number is changed. Although the printer used to output these images does not have good resolution, the readers can see the differences between these images.

Sampling Theorem and Aliasing

When an object is sampled by a TV camera or an image processor, the resultant image sometimes shows an image that is not similar to the original object. For example, if an object with a stripe pattern is recorded by a TV camera, a moiré pattern appears (Fig. 21-8). This phenomenon is called aliasing in signal-processing theory. When the sampling frequency (the reciprocal of the sampling pitch, i.e., pixel size) is smaller than two times of the highest spatial frequency of the object, aliasing appears.

DIGITAL IMAGE PROCESSING

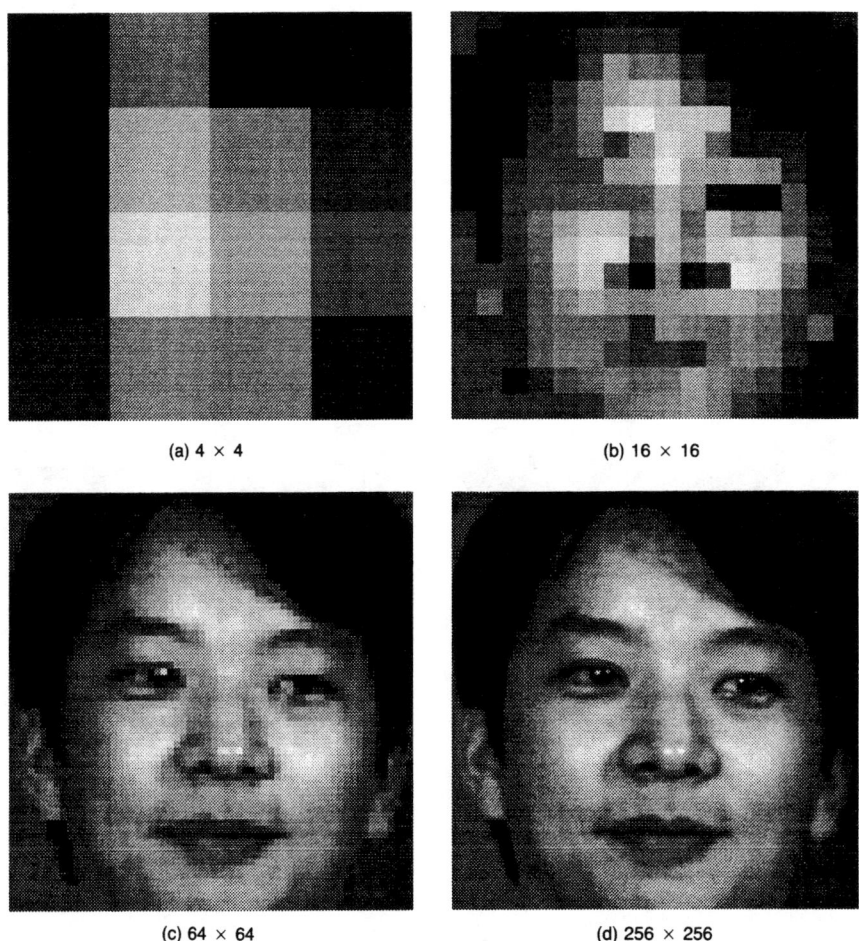

Figure 21-6 Differences in output obtained by changing the number of pixels: (a) 4 × 4 pixels; (b) 16 × 16 pixels; (c) 64 × 64 pixels; (d) 256 × 256 pixels.

This is the sampling (Whittaker–Kotelnikov–Shannon) theorem [3]. The sampling pitch should therefore be smaller than a half of the smallest pitch of the object. Indeed, aliasing can be used to advantage in the moiré method, since moiré is a type of aliasing.

21-2 Image-Processing Systems (Hardware)

Total Systems

A typical digital image-processing system is usually composed of a TV camera for image input, an A/D converter for digitization of images, IC image memory for image storage, a digital-to-

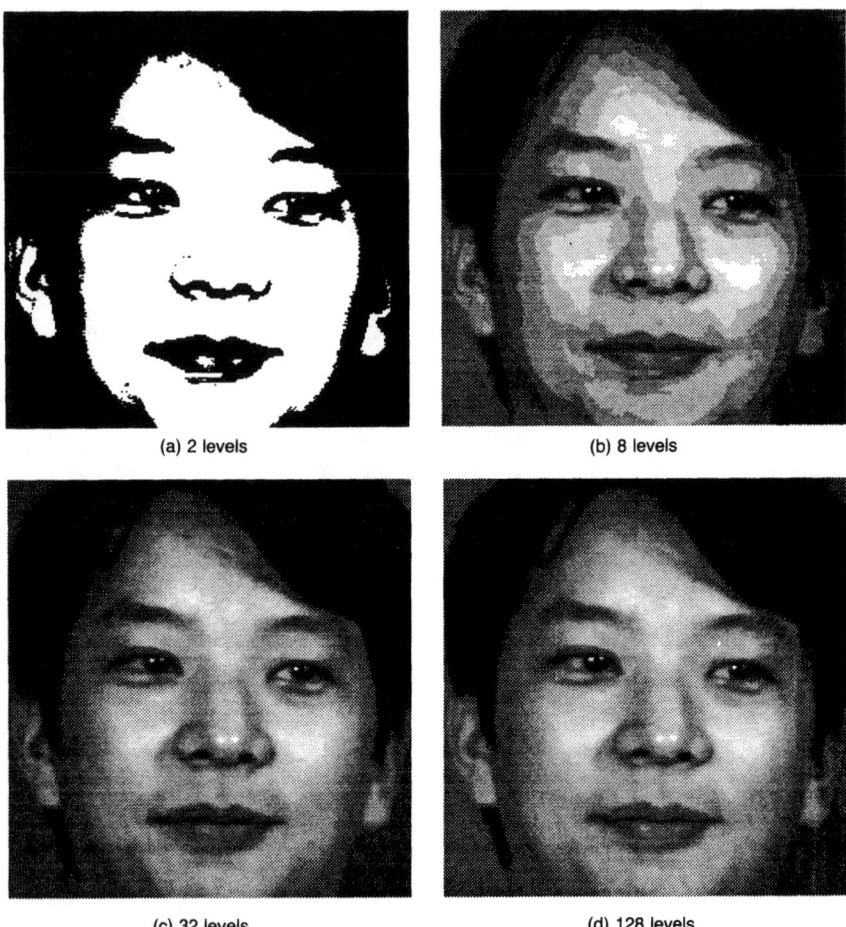

Figure 21-7 Differences in output obtained by changing the number of gray levels: (a) 2 levels; (b) 8 levels; (c) 32 levels; (d) 128 levels.

analog (D/A) converter for image output to a monitor display, and a computer for control of the system and for the image-processing calculations. The systems are roughly divided into three groups according to hardware components:

1. Special-use systems with special image processors
2. Combination systems (image-processing unit plus a wide-use computer)
3. System using only a wide-use computer

System 1 usually has image-processing chips or boards (digital signal processors) with fast processing speed. Real-time processing is usually performed: processing time is one-thirtieth second for one image in the NTSC system. A general-purpose system is expensive; a special-use system

DIGITAL IMAGE PROCESSING

Figure 21-8 Example of moiré patterns, which appear on the ventilation slots of a monitor in a picture taken by a TV camera.

sometimes is cheaper. For example, a real-time system for measuring displacement and strain distributions using the thinning-out theory of the scanning moiré method and the shift method of the geometric moiré approach [34] (see Section 21-8) is very cheap.

System 2 has an image-processing unit (sometimes an image-processing board) with an A/D converter, IC memory for image data storage, and a D/A converter to show the processed image on a monitor. The unit is controlled by a general-purpose computer such as a personal computer or a workstation. Calculations for image processing are also performed by the computer.

System 3 uses only a general-purpose computer such as a personal computer or a workstation, which does not have special image input and output devices for a TV camera and special IC image memory for image storage. This system uses only a standard computer system. If image data are recorded by another image-processing system and stored in the computer memory (RAM) through a floppy disk or a computer network, image processing is performed by using an image-processing program alone. The processed image is output on the monitor display of the computer. It is also possible to input images from inexpensive peripheral devices of the personal computer, such as image scanners.

System 1 is suitable for inspection, which requires high-speed processing, on factory mass-production lines. For research in which objects and procedures are not fixed, and for small frequency of image input and large time of analysis, system 2 is recommended. If an image input device is purchased cooperatively and the images stored on floppy disk, system 3 is very cheap because it is now possible for all researchers to have access to personal computers.

Sample Image-Processing System

A typical image-processing system using a personal computer, corresponding to system 2, consists of a TV camera and an image-processing unit—an A/D converter, a D/A converter and digital memory. It sometimes involves a monitor display or output ports for a monitor display or a video cassette recorder. This unit is also called an image grabber. If an image-processing board having the same ability as the image-processing unit is used, it is stored in the computer. The computer (3 in Fig. 21-9), which performs image-processing calculations and controls the peripheral devices,

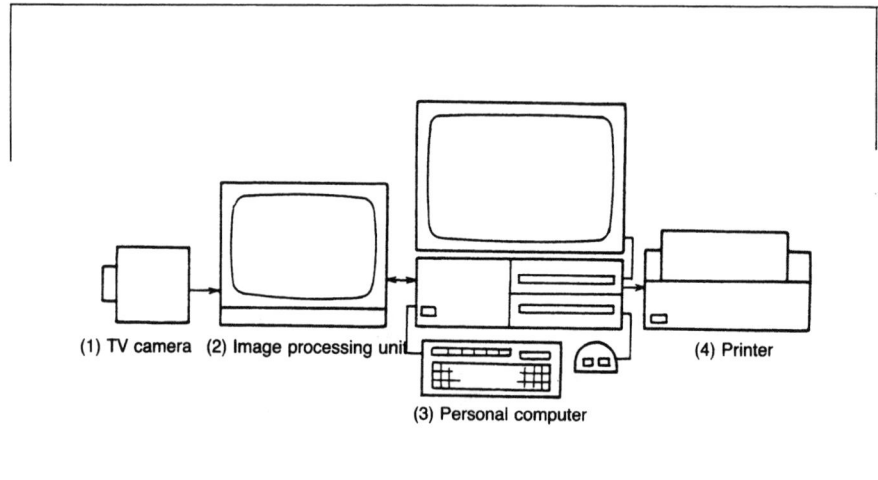

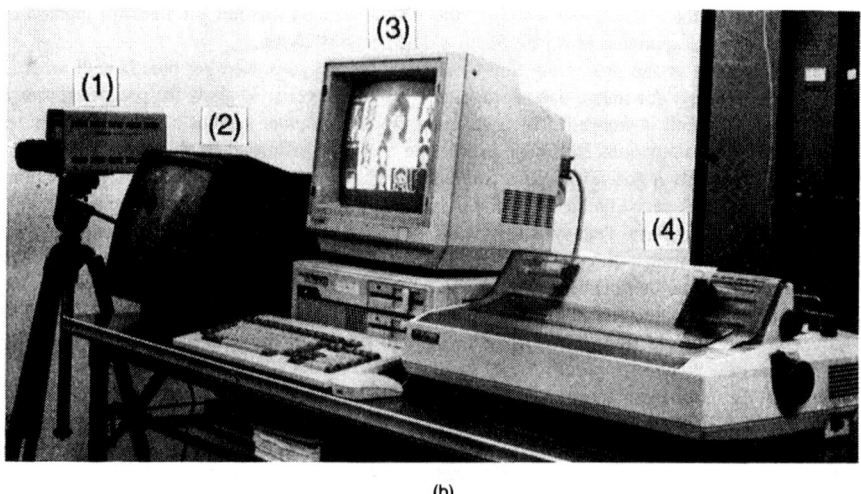

Figure 21-9 Typical image-processing system: (a) schematic concept; (b) working setup.

includes or connects floppy disk units, a hard disk unit, a keyboard, and a mouse. A printer (4) serves for outputting hard copies of images. The image processing unit, including a TV camera, can rapidly input images from a TV camera or a video cassette recorder. If an image scanner is used instead of the image-processing unit including the TV camera, image input is inexpensive although the input speed is slow. The devices illustrated in Fig. 21-9 are discussed further in the subsections that follow.

Computers

Initially, minicomputers were used to perform image processing because of their large memory size and high-speed processing ability. Since the cost of the minicomputer was relatively high, only few people had access to this approach. Now, personal computers are popular. However, until a few years ago, the memory capacity of personal computers was too small to handle the calculations needed for image processing, especially complex Fourier transformations of images, and a PC operating system suitable for image processing was lacking.

With recent rapid advances in personal computers, computers with high-speed central processing units and large memory capacity can be used for image processing. For using computers as system 3 mentioned above, the following points may be considered.

1. High resolution display (more than 512 × 400 pixels)
2. High gray levels (more than 16 gray levels; 256 gray levels is better)
3. Display for many colors with a look-up table
4. Support for Assembler and C languages
5. Support for many software tools
6. Support for many peripheral devices

Input Devices (TV Camera and Image Scanner)

A TV camera system is the most popular input device. It performs high-speed inputting. It is easy to combine a zoom lens or a microscope and a video cassette recorder. Since the clock frequency of the image processor for the TV system is higher than the clock frequency for the present High-generation of computers, it is difficult to store image data directly in computer memories. As a result, dedicated high-speed IC memory usually accompanies the image-processing unit.

As for the sensors of TV cameras for image processing, solid-state sensors such as the charge-coupled device (CCD) type and the complementary metal oxide–semiconductor (C-MOS) type are better than vacuum-tube sensors. Although the interference between the pixels of the sensor and the pixels of the image processor sometime creates a moiré pattern, dimensional accuracy is very good, sensor size is small, and variations due to temperature and time are very small. Such sensors are rugged, have long life, are operated with low power, and have high sensitivity at low light levels. To obtain high-resolution images of 512 to 8192 pixels per line, linear CCD array sensors are used. An area image is obtained by moving the linear array.

A camera using a linear image sensor can be used much like a TV camera, although the time needed for image input is long. An image scanner usually has a linear CCD image sensor and is easily used as a facsimile machine or a copier, although the image size is fixed. The resolution is high and the input clock can be adjusted to the computer clock; that is, output data are directly stored in the computer memory.

Output Devices (Printer, Display, and Memory)

The images captured and processed in a computer are output to a printer or a monitor display, and the image data are stored on a floppy disk or a hard disk.

21-3 Image-Processing Programs (Software)

An image has a large amount of information with much redundancy, and thus there are many methods of extracting the desired information. Even if one image is similar to another image, the best processing procedure with the highest processing speed and minimum error will not be the same for the two images. If optimal processing is not required, a general-purpose image-processing program is useful. In practice, however, it is difficult to process all images by a general-purpose program. Some calculations (analysis from the coordinates of measured points to the strain, etc.) should be programmed by the user. When the process is limited and specified, as in the inspection of a mass-production conveyor line in a factory, it is better to make a special program for high-speed processing.

Some general-purpose image-processing program packages that work on minicomputers, personal computers, and workstations are available to the public. For examples, SPIDER [2] (Subroutine Package for Image Data Enhancement and Recognition) works on minicomputers and workstations. PIMPOM [7,8] (Program of IMage Processing On Microcomputer) works on personal computers and is described in Section 21-5. These programs are useful for large volumes of image processing. To apply these programs to an individual analysis, however, users must customize their software by revising these programs or adding new ones.

Languages for Image Processing

Assembler, C, FORTRAN, and BASIC (listed in order of programming difficulty and faster processing speed) are the main languages used for image processing. Assembler is very fast but has poor portability to other computers, since it is computer dependent. Recently, C language has become the most popular for image processing because of its high-speed calculation, ease of programming, and portability to other computers. Combinations of two languages such as Assembler and BASIC provide fast calculations and easy programming.

Commands and Menus

To obtain the desired results from an original image taken by a TV camera, a series of image-processing operations is accomplished by invoking commands or making menu selections step by step. Commands and menus sometimes request parameters that are difficult for beginners to remember. It is more convenient to select parameters by responding to questions posed by the computer or by automatically remembering the parameters in the former processing. That is, analysis is performed by conversationally selecting menu items to analyze an image step by step or through automated processing.

Remarks on Programming for Image Processing

Although the techniques of programming for image processing depend on the purpose and the hardware, the following tips are useful.

1. *High-Speed Processing.* The numbers for image data are usually confined 0 to 255 on gray values and 0 to 511 on position. It is usually useless to use real numbers for image data and calculations. If integer numbers are used instead of real numbers, the speed of calculation is improved, and the memory requirements are greatly reduced. Since the combination in calculations between a small number of image data is not large, prepared look-up tables are useful for high-speed processing. The calculation for all pixels takes more time than the calculation for all columns of the table. If the table is used, the results of calculation are obtained faster because one looks at the table only according to the image data.

   It is an advantage if area size can be selected, because processing an area that is small in relation to the desired area is faster than processing the whole area. If only the program for the calculation repeated on all pixels is written in Assembler, the calculation proceeds faster

than it would if the program were written in a language other than Assembler, and the programming is easier than the programming of a whole program written in Assembler.
2. *Memory Economy.* Image data use a large amount of memory. When the program takes up a great deal of memory, the area remaining for data may be insufficient. Therefore if the image size can be selected freely in programming, the image can be analyzed by selecting a small area. Image data composed of integer numbers also save memory.
3. *Use of Computer Functions.* In image processing, display of images is important. The display of an image on a computer monitor depends on the hardware–software system of the computer. BASIC and C have some commands for accessing the video memory for showing an image on the computer monitor. By using these commands, the display of images becomes easier and faster. If the gray levels of an image for image processing are adjusted to the gray levels of the video memory, the video memory can be used as an image memory for image processing. This saves the memory for image data.
4. *Operational ease.* The "undo" function, to return to an earlier image from the current image, is useful for evaluating the computer's processing ability and finding the best combination. If a series of image-processing commands can be automatically remembered and repeated, processing for similar images can be performed automatically.

21-4 Algorithms for Image Processing

Since the amount of information associated with an image is immense, there are many methods of extracting the desired information from the image. These methods use algorithms in each step of the process; that is, complex procedures of image processing are composed of a series of many simple algorithms. The small number of popular general-purpose algorithms are described in many references [1–10].

This section explains some popular general-purpose algorithms and presents sample applications.

Image Coordinates

The coordinates of an image to be subjected to digital processing are usually different from ordinary coordinates. The origin of the coordinate is taken at the upper left corner of the image. Although the x axis is to the right of the origin, the y axis is usually downward from the origin as shown in Fig. 21-10. Because the TV camera scanning lines and the address of computer memory run along this direction, the pixel in the upper left corner has coordinates (0,0) and the coordinate increases with pixel unit.

Histograms

To check whether a recorded image is appropriate, the original brightness distribution of the real original object is compared to the brightness distribution (gray-value distribution) of the recorded image. It is, however, difficult to check the brightness distribution of the real original object. In such cases, a gray-value histogram of the recorded image provides the necessary information, as shown in the following examples.

If the gray values are almost equal in frequency, digitization may be appropriate, although it depends on the brightness distribution of the object. When the frequency of the maximum and the minimum gray values is very large or very small, the illumination or the aperture of the camera may be changed, or the level of the A/D converter may be changed. Figure 21-11 shows gray-value histograms. In Fig. 21-11a, almost all gray values of the histogram have nonzero frequency;

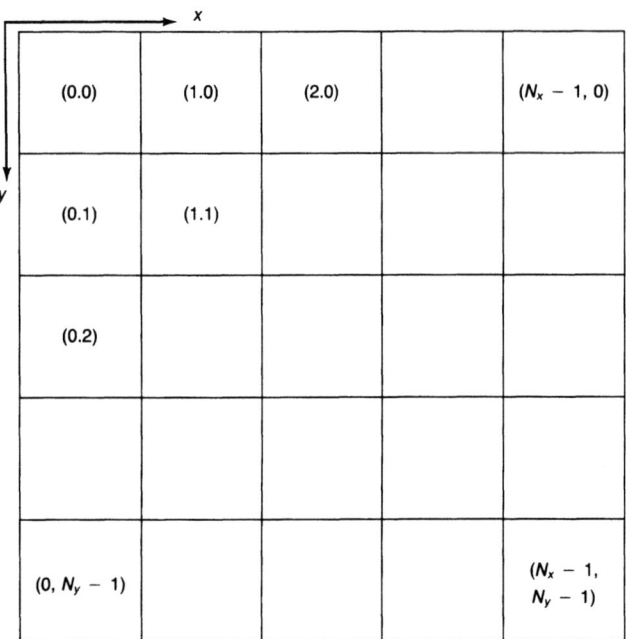

Figure 21-10 Image-processing coordinates.

this image has appropriate exposure. Figure 21-11b is dark, and the histogram distribution is clustered on the low values.

Gray-Value Transformations

When the original gray value at position (x,y) is $f(x,y)$, the processed gray value $g(x,y)$ by operation of gray-value transform F_A is expressed as

$$g(x,y) = F_A(f(x,y)) \qquad (21\text{-}1)$$

For example, when the gray-value histogram is clustered around low values, a high-contrast image is obtained by changing each gray value to another value. That is, each gray value is changed so that the new histogram has a distribution in the entire region of the gray values. Figure 21-12 was obtained by changing the gray values of Fig. 21-11b, to improve the *contrast*.

The gray-value transform is sometimes used to extract the characteristics of an image. For example, in recognition of the size of a white object, the object is put on the black background and taken in an image processor. The histogram shown in Fig. 21-13 has two separated groups, around 10 and 170. By *thresholding* the image at the value in the valley of the histogram, the image is separated into two parts, white and black. The white part is the object and the black part is the background. The area of the object is calculated by counting the pixels in the white part.

In contrast, by counting the pixels in the area of a black object on a white background, we can measure the black area of the thresholded image. Alternatively, when the original image is

DIGITAL IMAGE PROCESSING

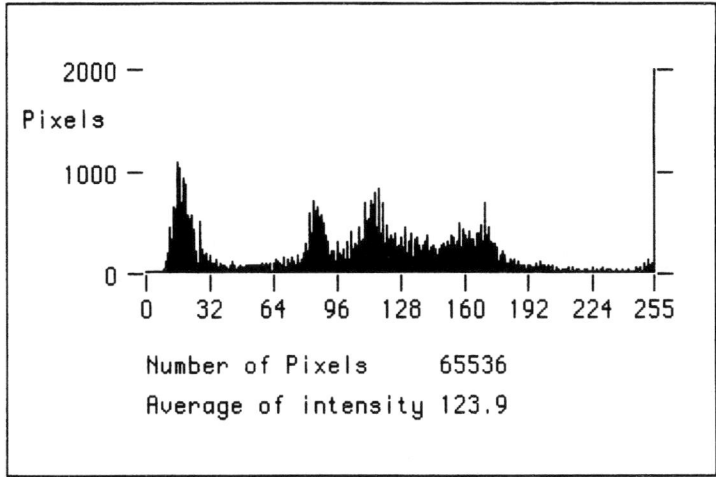

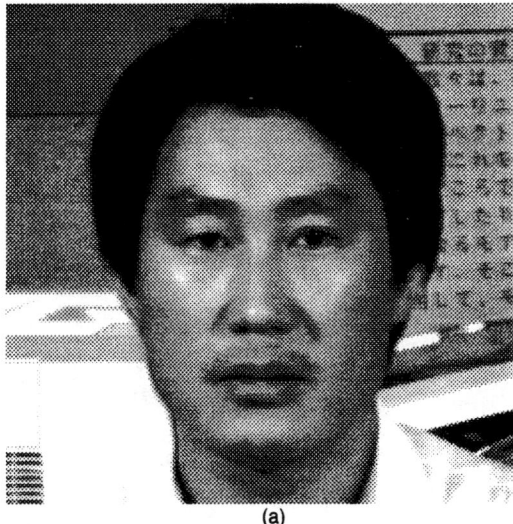

(a)

Figure 21-11 Comparison of histograms between proper and dark images. (a) image of proper brightness and its histogram; (b) dark image and its histogram.

reversed (the negative image of the original), we can use the same program mentioned for the black background. The *reverse* is obtained by subtracting the original gray value from the maximum gray value of the image processor. Such thresholding and reverse processes also belong to the gray-value transform.

In the display of an image processor, the gray value is expressed in the brightness of the point. In a monochromatic image, the gray value is proportional to the brightness value of the point. In a color image, the gray value (brightness) of each primary color is proportional to the brightness value of the respective primary color. However, for a monochromatic image, if each gray value

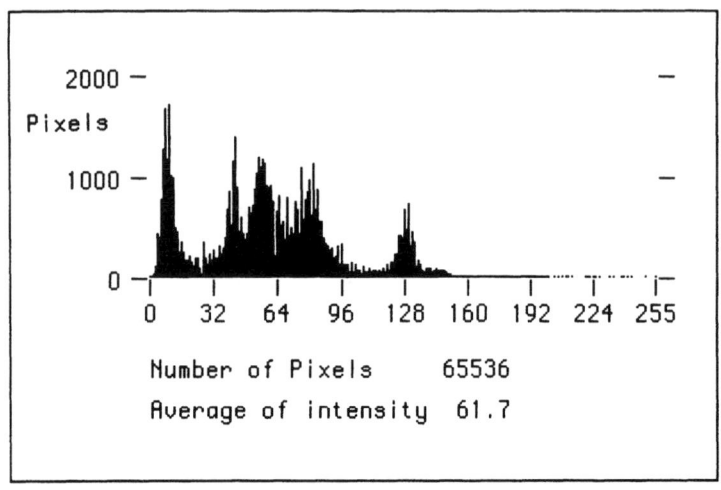

(b)

Figure 21-11 (continued)

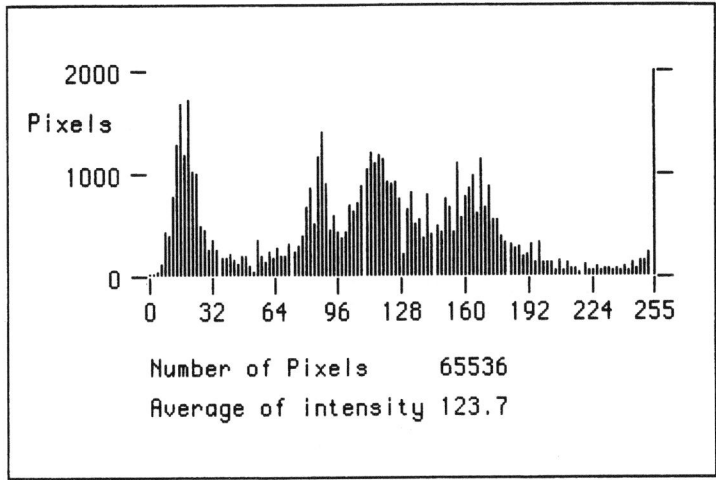

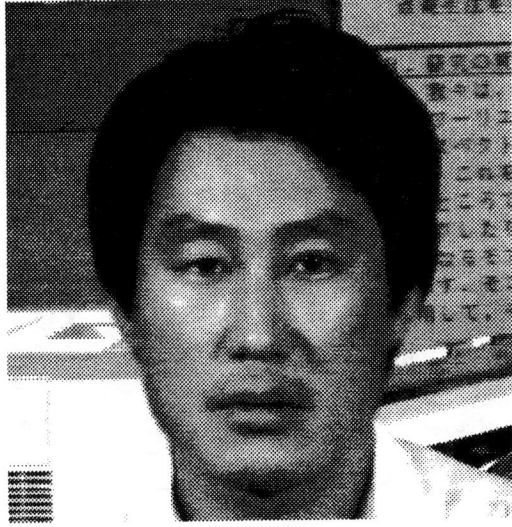

Figure 21-12 Example of gray-value transform (from overly dark image of Fig. 21-11b).

corresponds to a color, the color image has an artificial color called a *pseudocolor*. Images using pseudocolor project a strong emphasis. For example, by changing the color of the look-up table, we can show the regions with tensile strain in green and the regions with compressive strain in blue. The regions with higher strain value are expressed in brighter colors.

The table that shows the colors corresponding to gray values is called the look-up table. Since the total number of the values in the look-up table is less than the total number of pixels, it takes

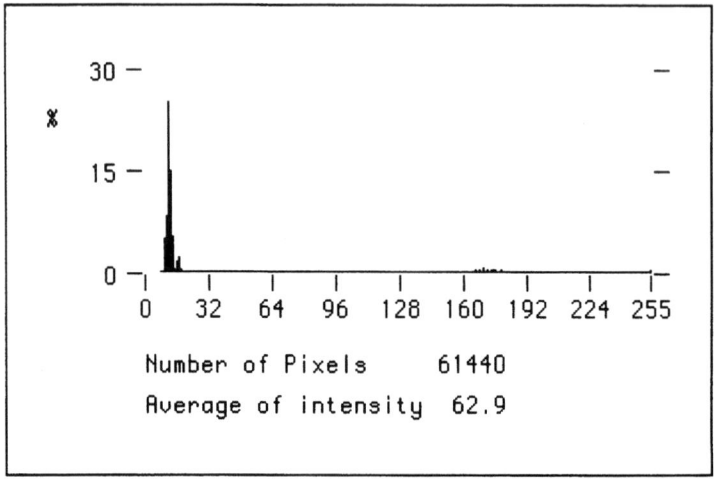

(a)

Figure 21-13 Example of thresholding: (a) original image and its histogram; (b) thresholded image and its histogram.

less time to change the look-up table in the computer than to change the gray values of all the pixels in an image. Therefore, the calculation time for changing the look-up table is shorter. Since changing the look-up table does not change the original data of the image, it is useful to check the gray-value transform by changing the look-up table. For example, we can try to change the threshold value for obtaining the best bilevel (two-value) image.

Figure 21-14 shows some examples of the relationship between original gray values and transformed gray values.

DIGITAL IMAGE PROCESSING

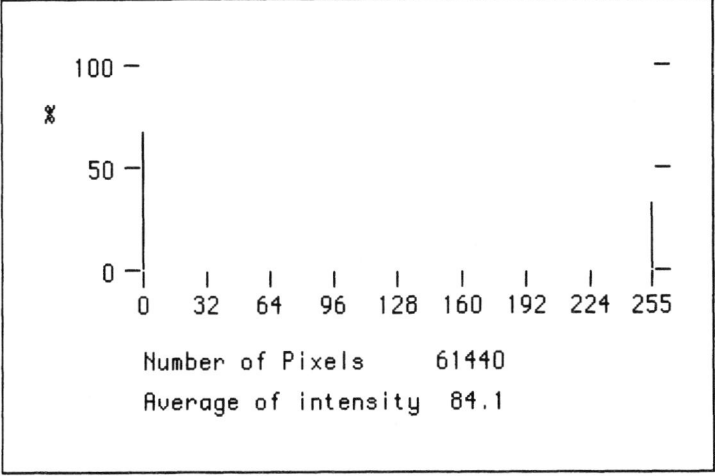

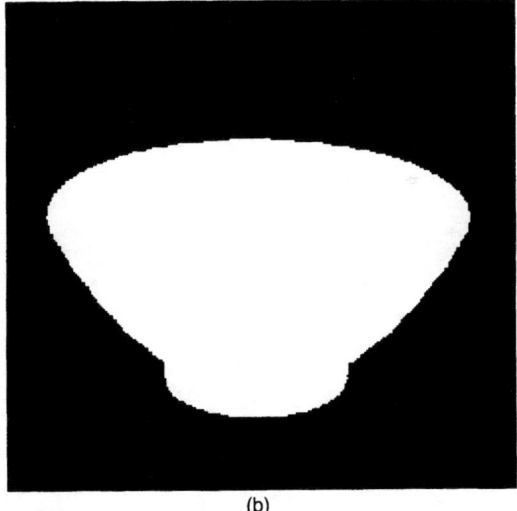

(b)

Figure 21-13 *(continued)*

Montage

"Montage" provides a new image $g(x,y)$ by an operation F_B between two original images $f_1(x,y)$ and $f_2(x,y)$.

$$g(x,y) = F_B(f_1(x,y), f_2(x,y)) \tag{21-2}$$

By using this operation, it is possible to investigate the difference between two images, as in dynamic image analysis, or to insert a different image into the background of the image of a

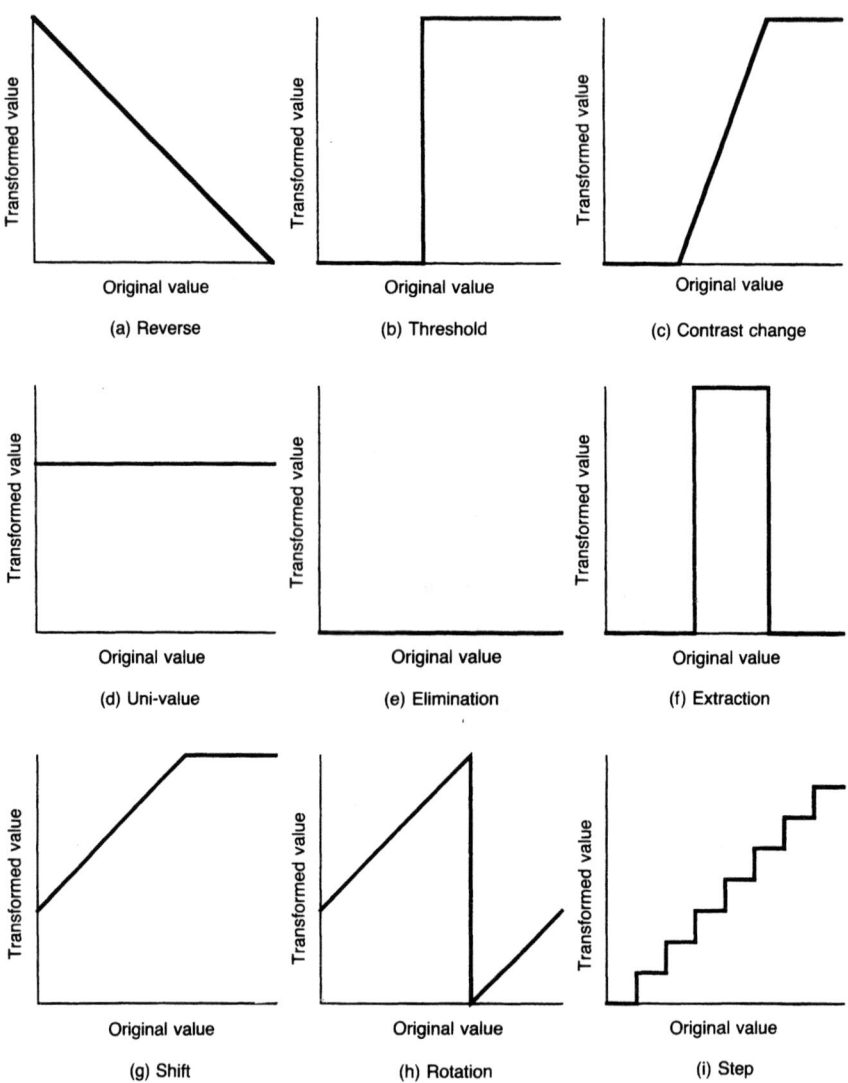

Figure 21-14 Some examples of the relationship between original and transformed gray values.

person. Figure 21-15 shows an example of detecting moving objects by calculating the *difference* of two images taken at different times.

Masking (Filtering)

For noise reduction and enhancement processing (e.g., filtering, averaging, edge detecting), the relationship between the gray values at several neighboring pixels is considered. In one-dimensional

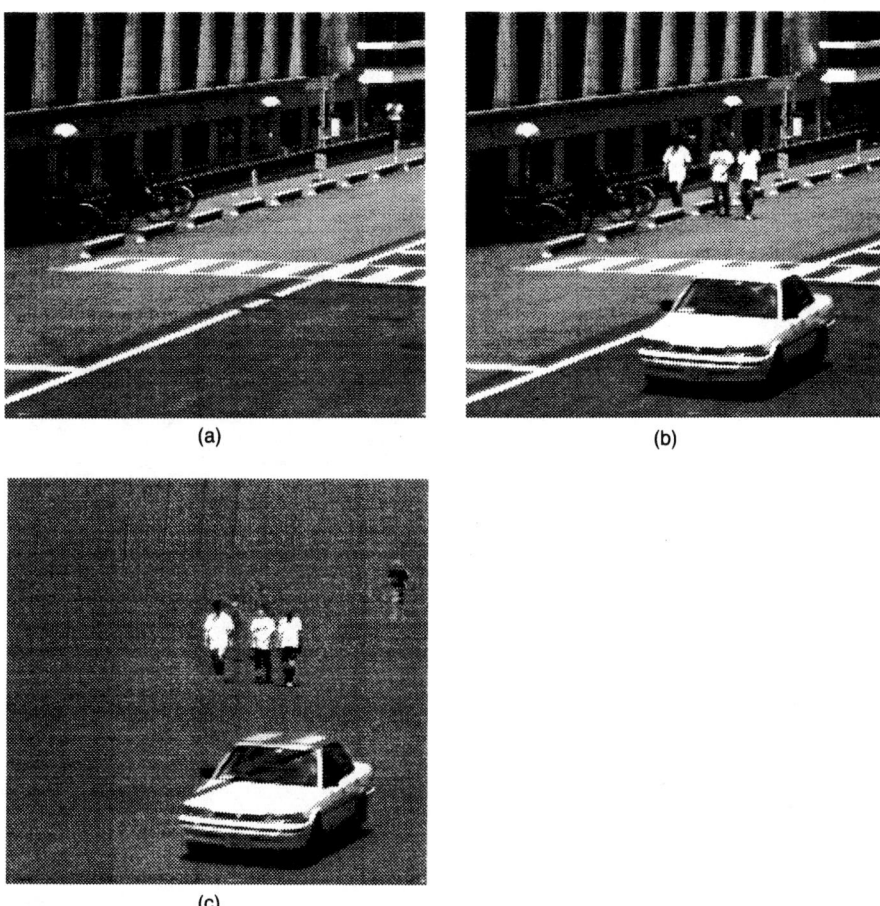

Figure 21-15 Example of montage (detecting a moving object): (a) image at time t_0; (b) image at time t_1; (c) image obtained by subtraction of two images plus constant.

analysis, a mask consisting of several continuous pixels on a line is used. In two-dimensional analysis, square masks of 3 × 3 pixels, 5 × 5 pixels, and so on are used.

One-Dimensional Masks

For x-directional one-dimensional analysis, the transform F_{c1} is expressed as

$$g_a(x,y) = F_{c1}\{f(x-i,y), f(x-i+1,y), \ldots, f(x,y), \ldots, \\ f(x+i-1, y), f(x+i, y)\} \tag{21-3}$$

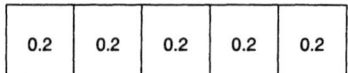

(a) Averaging (Low-pass filter)

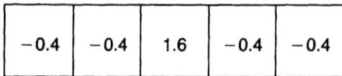

(b) High-pass filter

Figure 21-16 One-dimensional masks for (a) averaging (low-pass filter) and (b) high-pass filter.

In this case, the mask size is $2i + 1$. For example, when $i = 2$—that is, mask size is 5—averaging (i.e., *low-pass filter*) is expressed as

$$g_{a1}(x,y) = \frac{\{f(x - 2, y) + f(x - 1, y) + f(x,y) + f(x + 1, y) + f(x + 2, y)\}}{5} \quad (21\text{-}4)$$

The multiplication factor for each pixel is called the weight of the mask. In this case each pixel weight is 0.2. This mask is expressed as shown in Fig. 21-16a.

To eliminate the uneven illumination that accompanies low frequency, we use a *high-pass filter*:

$$\begin{aligned}
g_{a2}(x,y) &= c_1\{f(x,y) - g_{a1}(x,y)\} + c_2 \\
&= c_1\{-0.2f(x - 2, y) - 0.2f(x - 1, y) + 0.8f(x,y) - 0.2f(x + 1, y) \\
&\quad - 0.2f(x + 2, y)\} + c_2
\end{aligned}$$

When $c_1 = 2$, this mask is expressed as shown in Fig. 21-16b. Figure 21-17 is the example using the masks from Fig. 21-16.

Two-Dimensional Masks

For two-dimensional analysis, the x- and y-directional one-dimensional masks are combined. The most popular mask size is 3×3 (Fig. 21-18).

The general form of a 3×3 mask is expressed as follows:

$$\begin{aligned}
g_b(x,y) = F_{c2}\{&f(x - 1, y - 1), f(x, y - 1), f(x + 1, y - 1), \\
&f(x - 1, y), f(x,y), f(x + 1, y), \\
&f(x - 1, y + 1), f(x, y + 1), f(x + 1, y + 1)\}
\end{aligned} \quad (21\text{-}6)$$

Averaging is expressed as

$$\begin{aligned}
g_{b1}(x,y) = \{&f(x - 1, y - 1) + f(x, y - 1) + f(x + 1, y - 1) \\
&+ f(x - 1, y) + f(x,y) + f(x + 1, y) \\
&+ f(x - 1, y + 1) + f(x, y + 1) + f(x + 1, y + 1)\}/9
\end{aligned} \quad (21\text{-}7)$$

The weight in each pixel of the mask is shown in Fig. 21-19a. Figure 21-19b is another *averaging* mask with double weight at the center pixel.

The *x-directional differential* is expressed as

$$\begin{aligned}
g_{b2}(x,y) &= \frac{\partial f(x,y)}{\partial x} \\
&= \{f(x + 1, y - 1) - f(x - 1, y - 1) \\
&\quad + f(x + 1, y) - f(x - 1, y) \\
&\quad + f(x + 1, y + 1) - f(x - 1, y + 1)\}/6
\end{aligned} \quad (21\text{-}8)$$

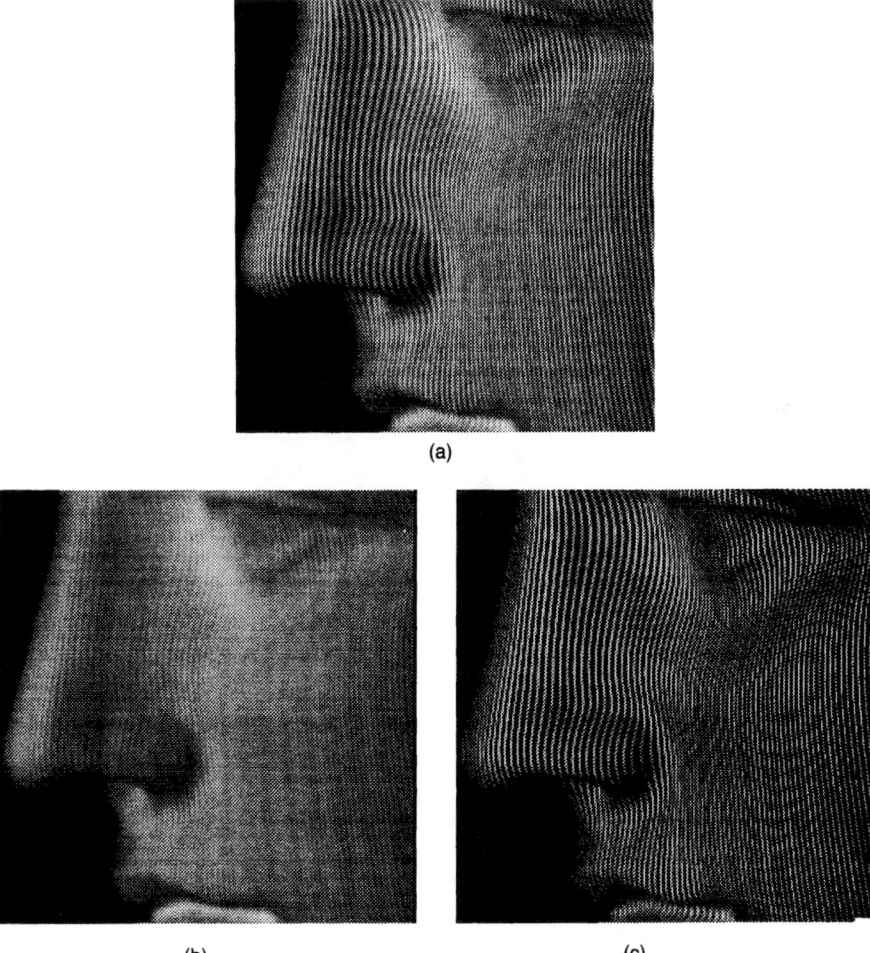

Figure 21-17 Example of the use of one-dimensional mask: (a) original image; (b) averaging by 5-pixel mask; (c) high-pass filter expressed in Eq. (21-5) with $c_1 = 2$ and $c_2 = 0$.

In image processing, the *differential* is usually expressed as the sum of the absolutes of the x- and y-directional differentials. This differential is called the Sobel filter, which has double weights for the center line and row. The weights are shown in Fig. 21-19. This filter is useful to detect the edge lines of an object.

The *Laplacian* is

$$g_{b3}(x,y) = \frac{\partial^2 f(x,y)}{\partial x^2} + \frac{\partial^2 f(x,y)}{\partial y^2}$$

$$= \{f(x-1, y-1) + f(x, y-1) + f(x+1, y-1) \quad (21\text{-}9)$$

$(x-1, y-1)$	$(x, y-1)$	$(x+1, y-1)$
$(x-1, y)$	(x, y)	$(x+1, y)$
$(x-1, y+1)$	$(x, y+1)$	$(x+1, y+1)$

Figure 21-18 A two-dimensional mask (3 × 3).

$$+ f(x-1, y) - 8f(x,y) + f(x+1, y)$$
$$+ f(x-1, y+1) + f(x, y+1) + f(x+1, y+1)\}/3$$

where the center line and row have four times the weight and the neighbor lines and rows have one time the weight, that is

$$\frac{\partial^2 f(x,y)}{\partial x^2} = \frac{\left\{\frac{\partial^2 f(x, y-1)}{\partial x^2} + 4\frac{\partial^2 f(x, y)}{\partial x^2} + \frac{\partial^2 f(x, y+1)}{\partial x^2}\right\}}{6} \quad (21\text{-}10)$$

The terms in the curly braces are

$$\frac{\partial^2 f(x,y)}{\partial x^2} = \frac{\partial f(x+\frac{1}{2}, y)}{\partial x} - \frac{\partial f(x-\frac{1}{2}, y)}{\partial x} \quad (21\text{-}11)$$

$$\frac{\partial f(x+\frac{1}{2}, y)}{\partial x} = f(x+1, y) - f(x,y) \quad (21\text{-}12)$$

Figure 21-19d shows the weight of each pixel of the mask. Figure 21-9e, also a Laplacian mask, was obtained by setting one times weight at the center line and row and zero times weight at the neighbor lines and rows. The Laplacian is also useful for detecting the edge lines of an object.

Figure 21-20 shows examples of averaging, differential, and Laplacian mask processing.

Next, let us consider bilevel white and black images. *Thinning* is an operation for obtaining the centers of thick lines such as fringe lines or grating lines. Theoretically, when the pixels at the edge of a thick line that do not cut the line and are not terminal points of the line are removed in iterative fashion, the pixels left show the centerline of the thick line. Figure 21-21 illustrates thinning with respect to the removable edges and the unremovable terminal points.

Examples of detecting centerlines by thinning and boundary lines by boundary detection are shown in Fig. 21-22.

In thinning and boundary detection, if the line may be connected to one of 8 pixels surrounding the pixel, the connection is called an 8-connection. If it may be connected to one of 4 pixels (upper, lower, left, and right of the pixel), the connection is called a 4-connection. That is, there are no diagonal connections in a 4-connection. Examples of the 8- and 4-connections are shown in Fig. 21-23.

DIGITAL IMAGE PROCESSING

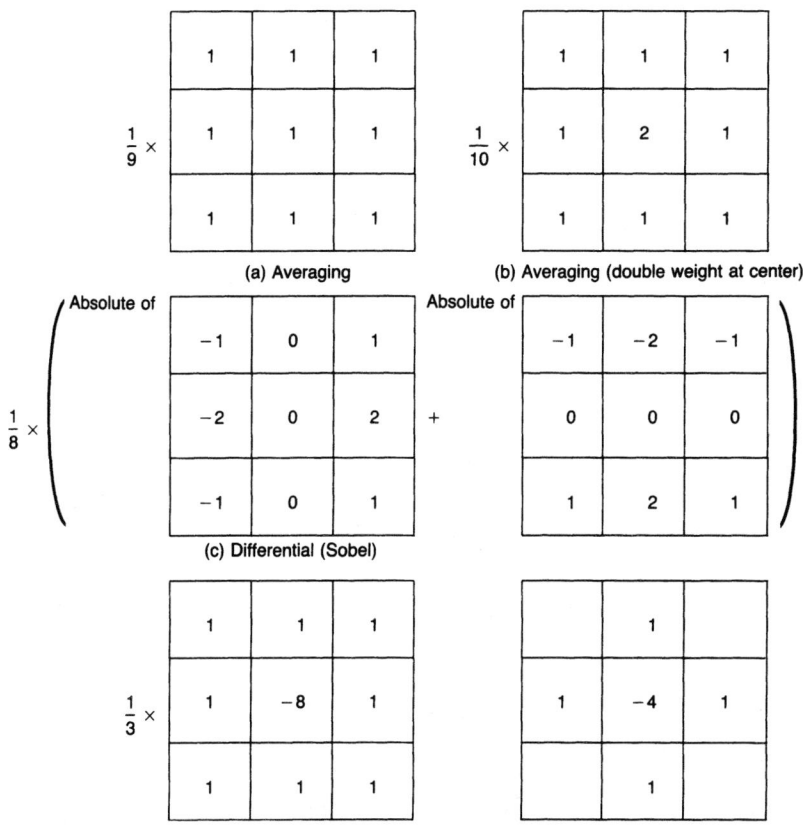

Figure 21-19 Two-dimensional masks: (a) averaging; (b) averaging (double weight at center); (c) differential (Sobel); (d) Laplacian (with weights in neighbor pixels); (e) Laplacian (without weights in neighbor pixels).

Geometric Transformations

Enlargement, shrinkage, rotation, and parallel movement are expressed in *affine transforms*. When an original point (x_0, y_0) with gray value $f_1(x_0, y_0)$ moves to another point (x_1, y_1) after geometric transformation, the gray values $g(x_1, y_1)$ are not changed. Then

$$g(x_1, y_1) = f(x_0, y_0) \tag{21-13}$$

where

$$\begin{bmatrix} x_1 \\ y_1 \end{bmatrix} = \begin{bmatrix} a_{11} & a_{12} \\ a_{21} & a_{22} \end{bmatrix} \begin{bmatrix} x_0 \\ y_0 \end{bmatrix} + \begin{bmatrix} a_{13} \\ a_{23} \end{bmatrix} \tag{21-14}$$

Equation (21-14) is an affine transform. The factors $a_{11}, \ldots, a_{23}$ are constant depending on the transform. *Enlargement, shrinkage, rotation,* and *parallel movement* are given by different values for these factors. For example, enlargement or shrinkage of b_x and b_y in the x- and y-directions respectively, yields

$$a_{11} = b_x, \ a_{12} = 0, \ a_{13} = 0, \ a_{21} = 0, \ a_{22} = b_y, \ a_{23} = 0 \tag{21-15}$$

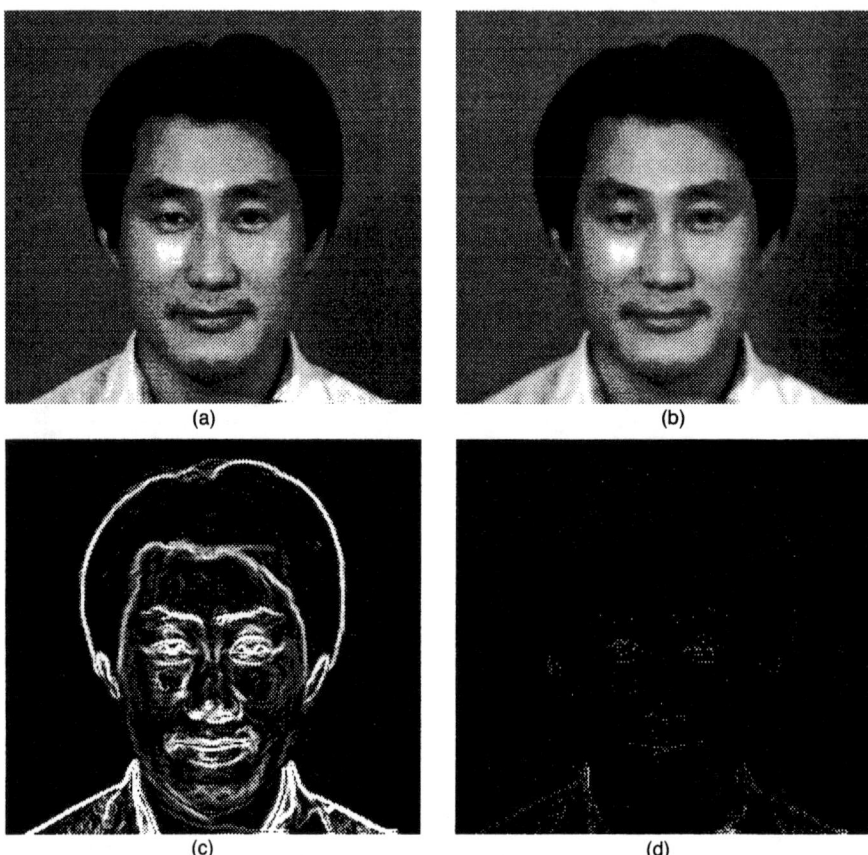

Figure 21-20 Examples of mask processing: (a) original image; (b) by averaging; (c) by differentiation; (d) by means of Laplacian.

For rotation by angle θ,

$$a_{11} = \cos\theta, a_{12} = -\sin\theta, a_{13} = 0, a_{21} = \sin\theta, a_{22} = \cos\theta, a_{23} = 0 \qquad (21\text{-}16)$$

For parallel movement, the amounts of shifts in the x and y directions are substituted in a_{13} and a_{23}, respectively. Figure 21-24 presents examples of the types of geometric transform mentioned.

Fourier Transformations

Fourier transformations are useful in mechanical engineering because it allows the analyst to express a given function as a combination of cosinusoidal waves. In image processing, the brightness function of an image can be expressed by a Fourier transform. By computing the Fourier transform of an image, the frequency components of the image can be detected. Proper Fourier filtering allows noise to be eliminated and image data to be compressed.

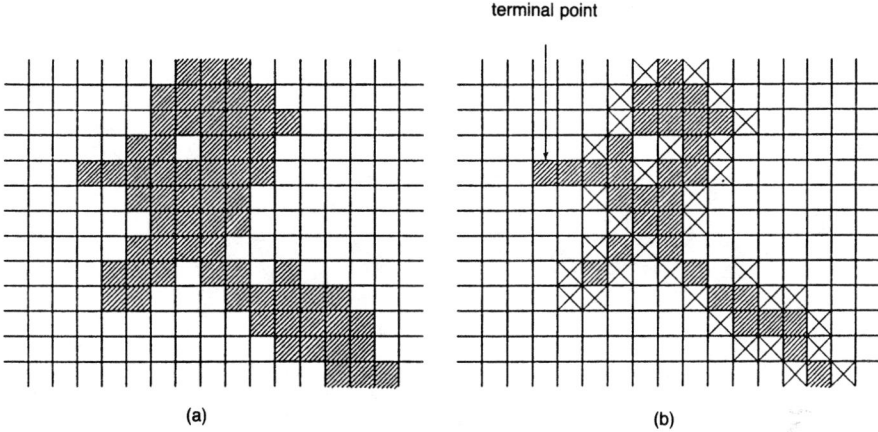

Figure 21-21 Removable edge and unremovable terminals in thinning: (a) original image; (b) after 1 time thinning (points belonging to boundary that were removed indicated by ×).

Compression of image data, or image data reduction serves to decrease the amount of data in an image without loss of useful information. For an image size of 256 × 256, and a gray level of 256 (8 bits), the total bit number is very large: 256 × 256 × 8 = 524,288 bits. Such enormous amounts of data are not necessary in extracting certain characteristics of an image. Redundancies can be removed, and characteristics can be clarified. Decreasing the amount of data in an image permits the image data to be sent more quickly via communication lines and stored more easily in memory.

Noise usually has special regions of the frequency plane depending on their source. For example, uneven illumination has low frequencies, and electric noise, which causes one-dot noise, has high frequencies. By using Fourier filtering, these noisy parts of an image can be eliminated; however, the original image may be distorted at such parts as edges.

As an example of the Fourier transform, let us consider an image composed of only three cosinusoidal waves, although this is an extreme case.

$$f(x, y) = \sum_{i=1}^{3} c_i \{\cos 2\pi(u_i x + v_i y)\} + c_0 \tag{21-17}$$

The Fourier transform of this image is

$$F(u, v) = \sum_{i=1}^{3} \frac{c_i}{2} \{\delta(u - u_i, v - v_i) + \delta(u + u_i, v + v_i)\} + c_0\delta(0, 0) \tag{21-18}$$

where c_0 is a constant value of the average brightness of the image. The number of data is 12 (c_i, u_i, and v_i, $i = 0, 1, 2$, and 3). If the 12 data are stored in memory, because the number of data in the image can be reconstructed. The amount of memory needed is greatly decreased because the number of data in the original image (256 × 256 = 65536) is now 12, after data compression. Although this is an extreme example, the data in an ordinary image can be compressed by Fourier transformation. Figure 21-25 shows the example with 12 data. Figure 21-25b is the Fourier transformation of the original image (Fig. 21-25a).

When the coefficients of the highest and the lowest frequencies have been eliminated from this

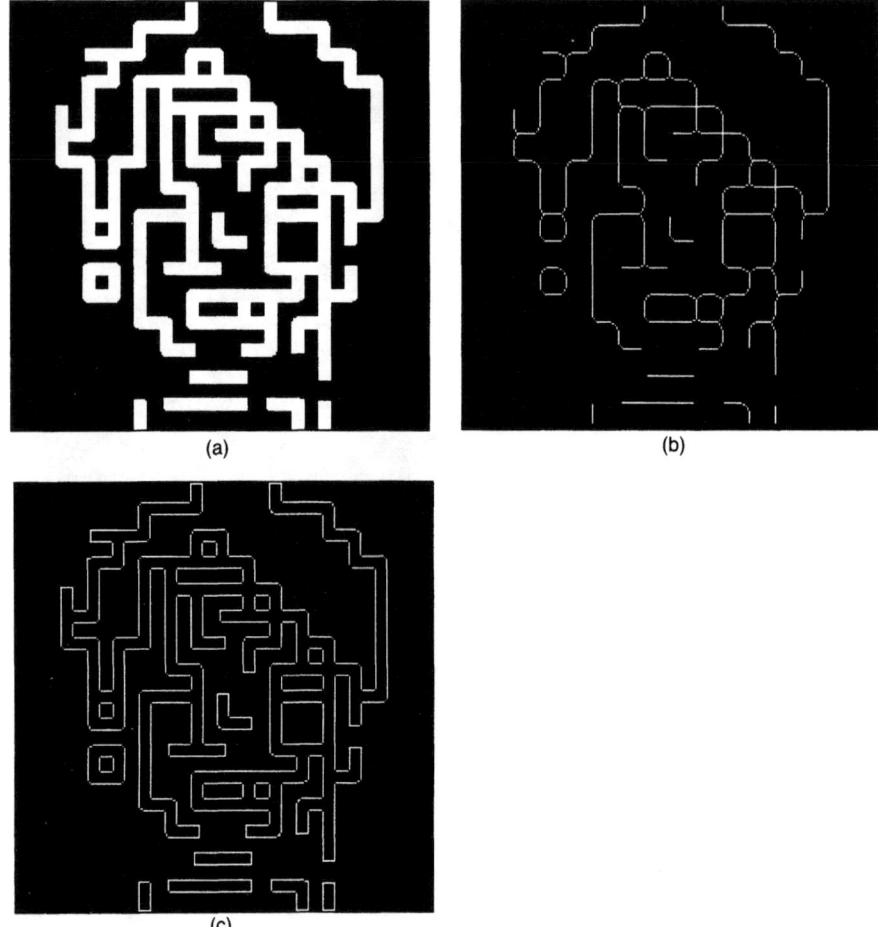

Figure 21-22 Examples of thinning and boundary detection: (a) original image; (b) result after thinning; (c) result after boundary detection.

spectrum, and the coefficient of the medium frequency multiplied by a constant value, the image reconstructed by inverse Fourier transformation of the spectrum is

$$f_2(x, y) = c'_2 \cos \{2\pi(u_2 x + v_2 y)\} \tag{21-19}$$

as shown in Fig. 21-25c.

This Fourier transform is a real number calculation and the calculation time is large. The fast Fourier transform (FFT) is usually used in such calculations. The calculation speed is high, although the image size (length of data) is limited in 2 to an integer (2^n) pixels. The mixed-radix fast Fourier transform (MRFFT) [72] is fast for many-integer numbers, although it some limitations.

An image can be composed of many rectangular brightness distributions instead of cosinusoidal

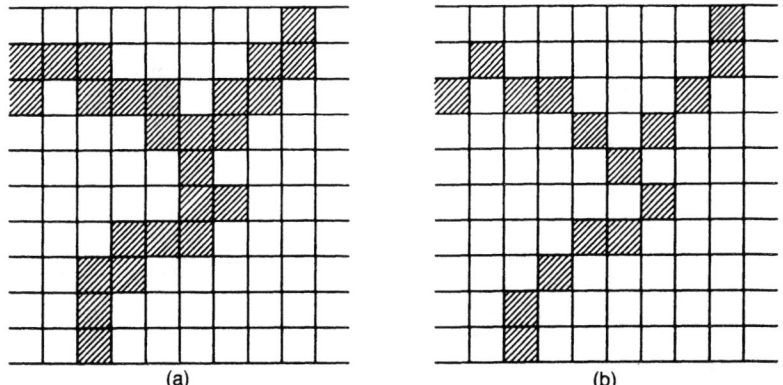

Figure 21-23 Examples of (a) 4-connection and (b) 8-connection thinning.

brightness distributions. The Hadamard transform can be used to compute the frequency of the rectangular information. Calculation speed is very high but the phase information is not present.

21-5 Examples of Image Processing

Image-Processing Program: PIMPOM

The image-processing program called PIMPOM [4,7,8] (Program of IMage Processing On Microcomputer) was developed to analyze images on personal computers. If image data are stored on a floppy disk, the program can be run on an ordinary personal computer (NEC PC-9801 with Japanese MS-DOS) without special devices for image processing. This program treats an image of 640 × 400 pixels with 16 gray levels, which can change to pseudocolors from 4096 colors. The program is written in BASIC and Assembler. Each process is performed by selecting one of the menus and supplying the parameters asked by the computer. This menu selection is useful for trial-and-error work. Batch processing is easy because the steps can be repeated by remembering the order and the parameters of each one. Although this program is widely used and the calculation speed is very high, it cannot treat complex data, as are used in calculating the Fourier transform. Another PIMPOM program running on the Macintosh II was developed using the C language. This program treats 256 gray levels and image size is free, although it depends on how much memory the computer has. The menus are shown in Fig. 21-26. An image display obtained by this program is shown in Fig. 21-27. The images shown in Section 21-4 also were processed by PIMPOM.

21-6 Fringe Pattern Analysis

This section discusses fringe pattern analysis, a popular approach for investigating whole-field distribution of displacement and strain. Kujawinska reported the details of the automatic fringe pattern analysis [62].

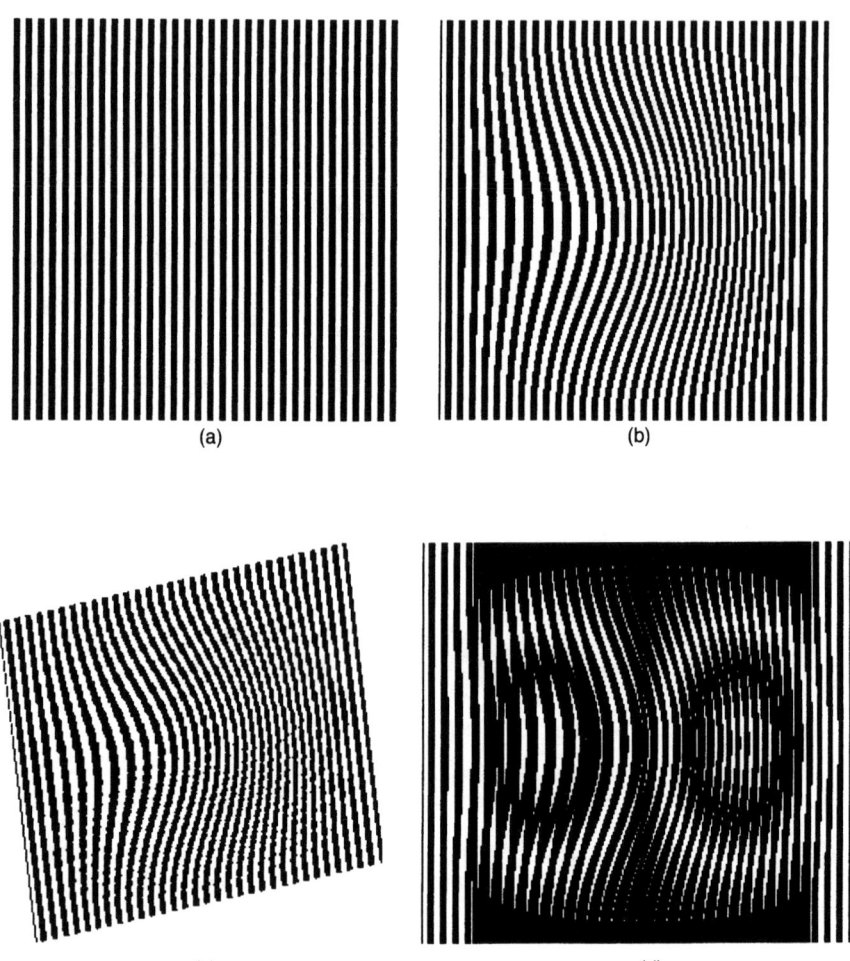

Figure 21-24 Examples of geometric transformation: (a) original grating undeformed; (b) original grating deformed; (c) affine transform of Fig. (b) ($a_{11} = 0.9$, $a_{12} = 0.1$, $a_{13} = 0$, $a_{21} = -0.2$, $a_{22} = 0.8$, $a_{23} = 0$); (d) parallel movement of deformed grating and superposing on original one (shift method); (e) shrinkage of undeformed grating and superposition on original deformed grating (mismatched moiré method); (f) rotation of undeformed grating and superposition on original deformed grating (misalignment moiré method).

Basic Discussion of Fringe Pattern Analysis

The brightness distribution of a fringe pattern obtained by coherent optical methods is approximately cosinusoidal as follows:

$$f = f_0\{1 + \gamma \cos(\phi - \alpha)\} \tag{21-20}$$

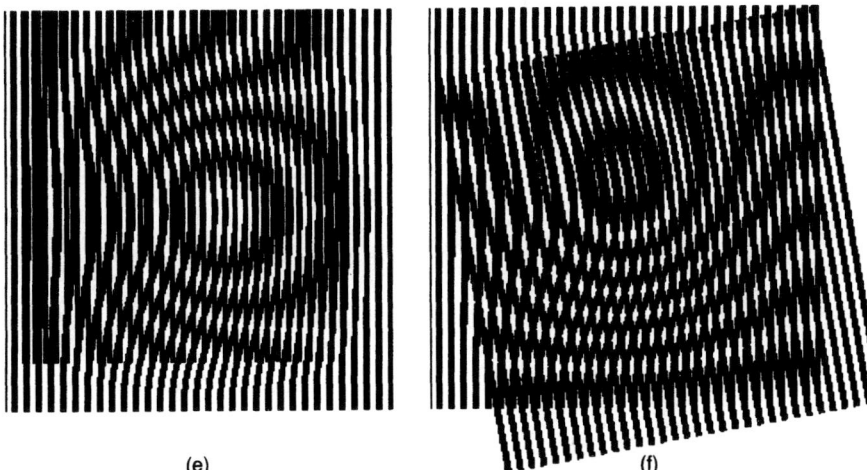

(e)   (f)

Figure 21-24 (*continued*)

where f_0 is the average brightness, γ is the amplitude of the brightness, ϕ is the optical phase of the fringes (which is a function of position X), and α is the phase-shift amount (which is given by changing an optical path). For example, in geometric moiré, when the pitch of the grating before deformation is p_0 and the reference grating is shifted by x_0, and a point x before deformation moves to point X after displacement u, the brightness distribution of the fringe pattern is approximately expressed by Eq. (21-20), where $\phi = 2\pi(X - x)/p_0 = 2\pi u/p_0$ is the phase of the fringe, $m(= \phi/2\pi)$ is the fringe order, $\alpha = 2\pi x_0/p_0$ is the phase-shift amount of the reference grating (i.e., the phase of the origin is $-\alpha$) and $(\phi - \alpha)$ is the phase of the shifted fringe.

If f_0 and γ are constants, the brightness of the fringe at each point corresponds accurately to the phase, that is, the fringe order in a decimal number, and the displacement data are easily obtained in the whole field. It is, however, difficult to measure accurately the brightness of fringes because of the nonlinearity of the image sensor and noise such as uneven illumination and stains on the surface of the object. If the brightness distribution is not basically cosinusoidal, analysis is still possible.

Let us discuss analysis methods when f_0 and γ are not constants but functions of position depending on optical noise. At the places where the phases of fringes are $n\pi$ (n is an integer), the points are the lightest or the darkest points locally. These points are the fringe centers, and it is easier to obtain the lightest or the darkest points than to obtain the positions with other degrees of brightness. When $\alpha = 0$, the fringe order at the lightest point in a local area is an integer. The fringe order at the darkest point in a local area is an integer and a half. Measurement of the positions of the fringe center points was very popular method in early analysis.

When the brightness is measured accurately, the phases can be analyzed directly by the phase-shift method, which analyzes a few images obtained by different phase shifts.

When the brightness distribution is not cosinusoidal, the phase can be analyzed if the first harmonic of the Fourier spectrum of the image can be extracted.

Let us discuss image-processing techniques for the above-mentioned cases: the fringe center position analysis method, the phase-shift method, and the Fourier transform method.

1000 HANDBOOK ON EXPERIMENTAL MECHANICS

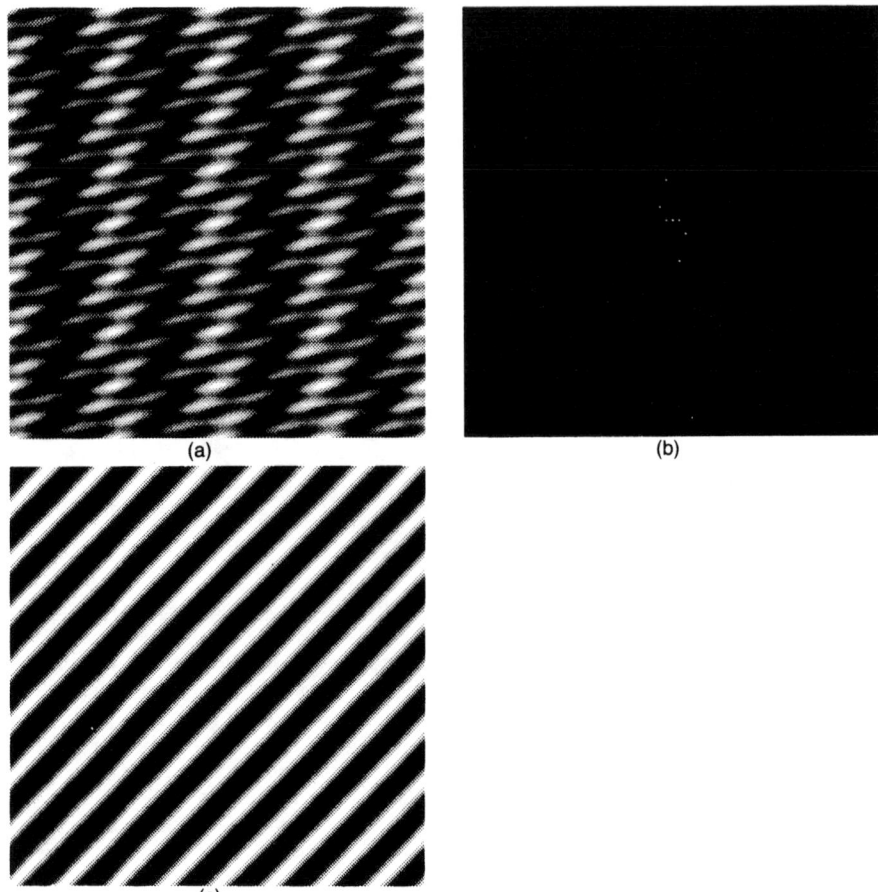

Figure 21-25 (a) Original image composed of three cosinusoidal waves; (b) Fourier transform of this image; (c) the same image after Fourier filtering.

Fringe Center Position Analysis Method [30,32,33,49,61]

Methods for obtaining the positions of fringe line centers include tracking the ridge or the valley of fringe brightness distribution and finding the location at which the slope of the brightness is zero. However, these methods are not successful in practical cases because of the noise in the images. The most popular method is the method of using thinning of threshold fringes. Let us explain the process by using an analysis obtained by holographic interferometry to measure strain distribution as shown in Fig. 21-28 [49].

Figure 21-28a shows a specimen composed of aluminum, epoxy resin, and aluminum layers. The left edge is hit by a steel plate. The stress wave is recorded on a holographic plate by a double-pulsed ruby laser. The recorded holographic plate (hologram) is observed from left and right sides in front of the hologram to obtain the two-dimensional displacement information. Figure

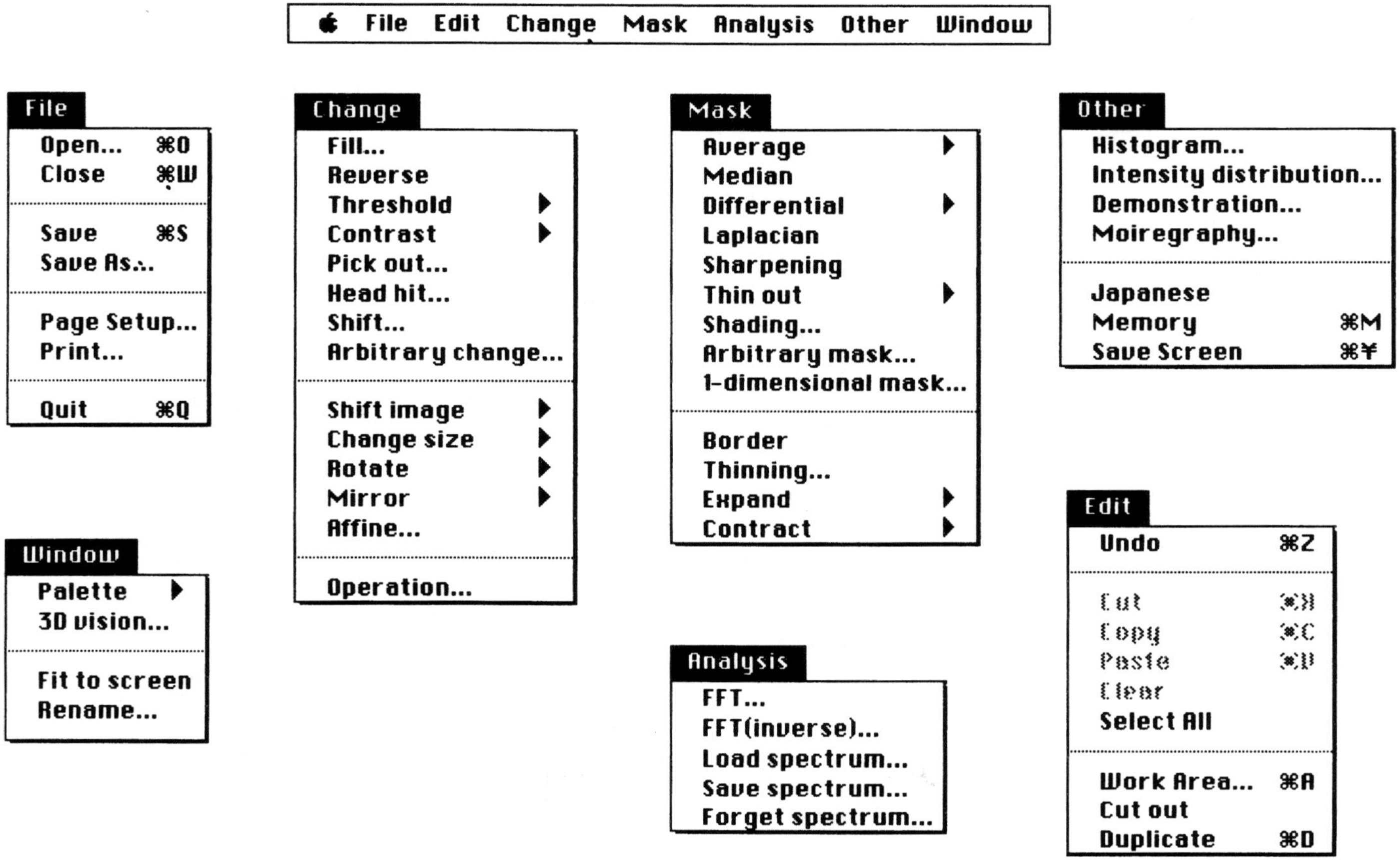

Figure 21-26 Menus of PIMPOM, an image-processing program for the Macintosh II.

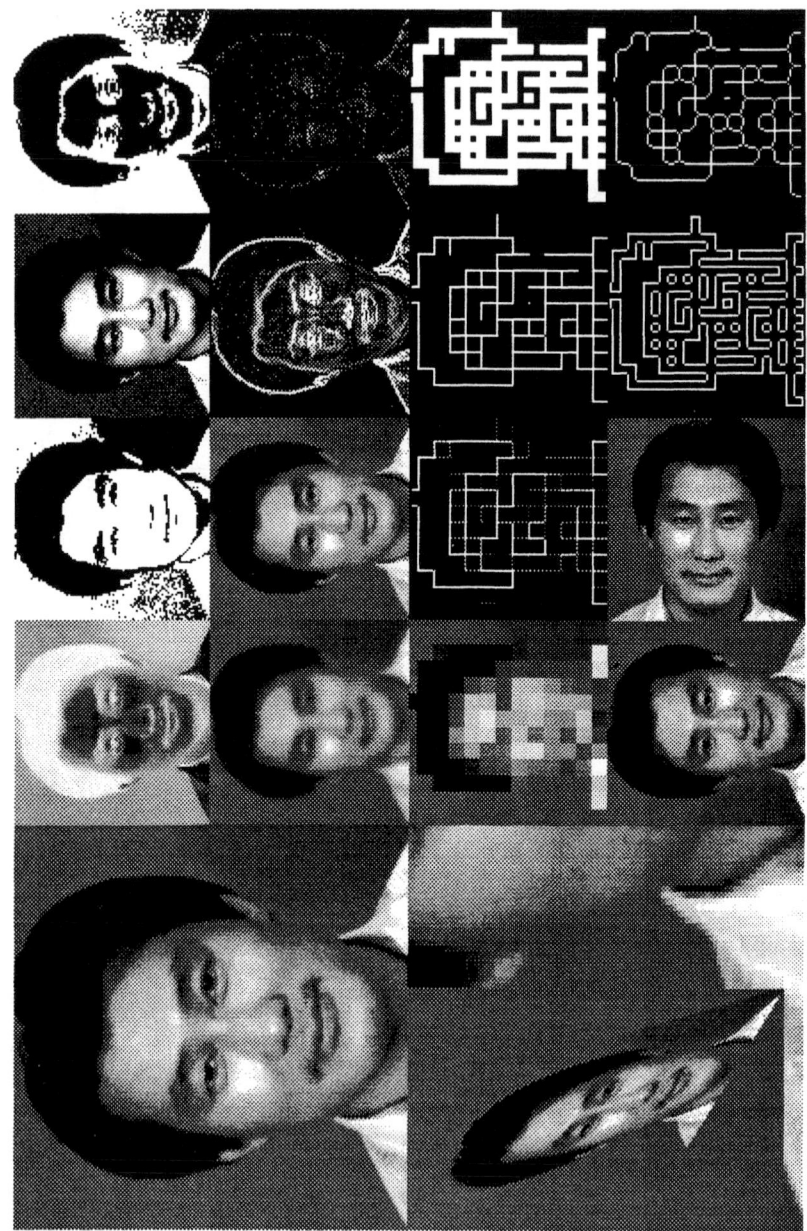

Figure 21-27 Image processing by PIMPOM on a Macintosh II.

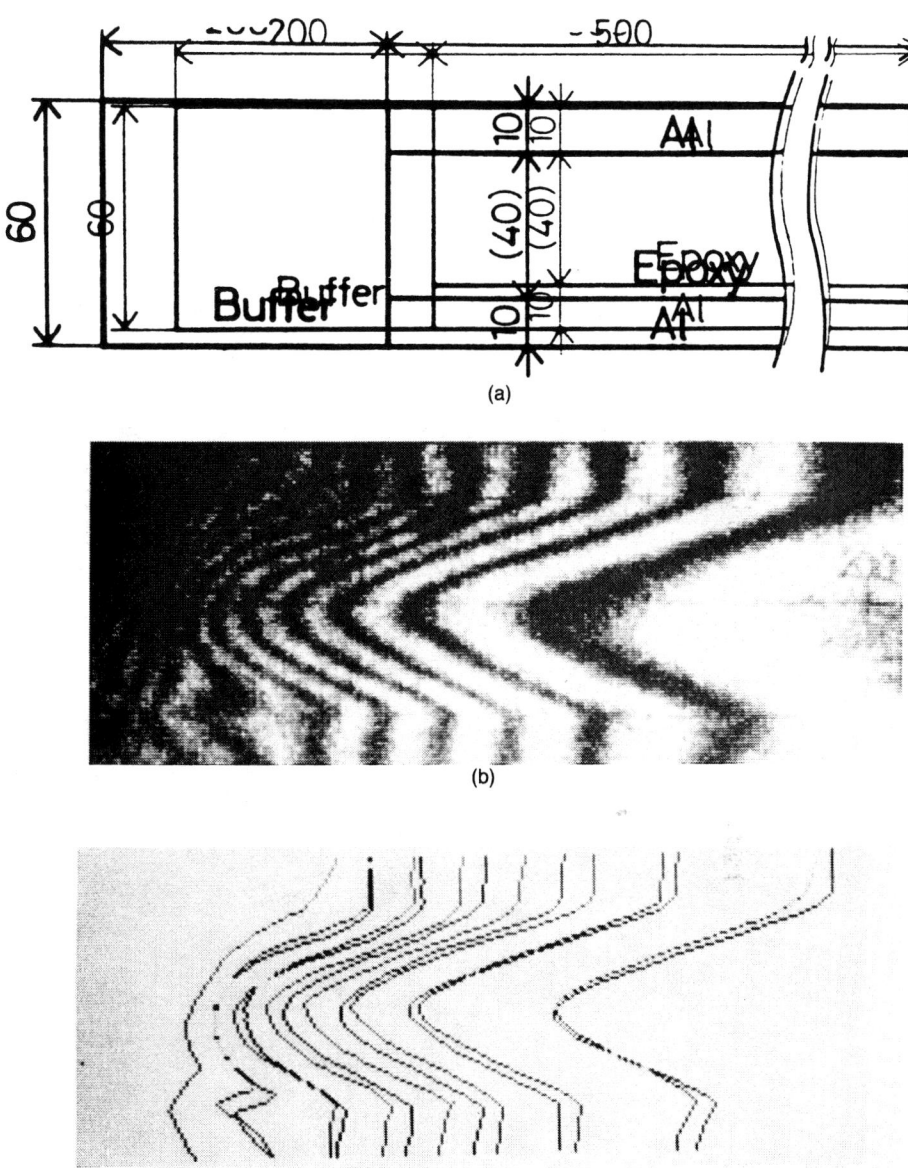

Figure 21-28 Strain distribution analysis by method of thinning fringes obtained by holographic interferometry: (a) schematic of specimen; (b) fringe pattern observed from right side; (c) thinning of fringe lines; (d) displacement distribution; (e) shear-stress distribution.

1004　HANDBOOK ON EXPERIMENTAL MECHANICS

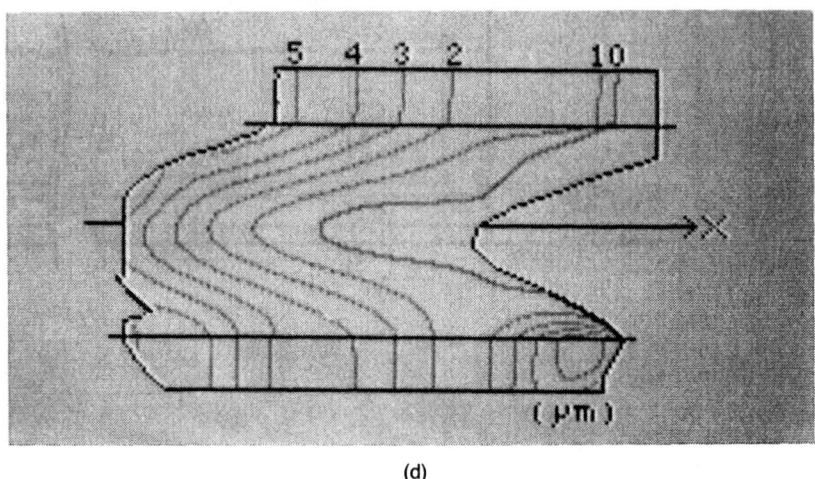

(d)

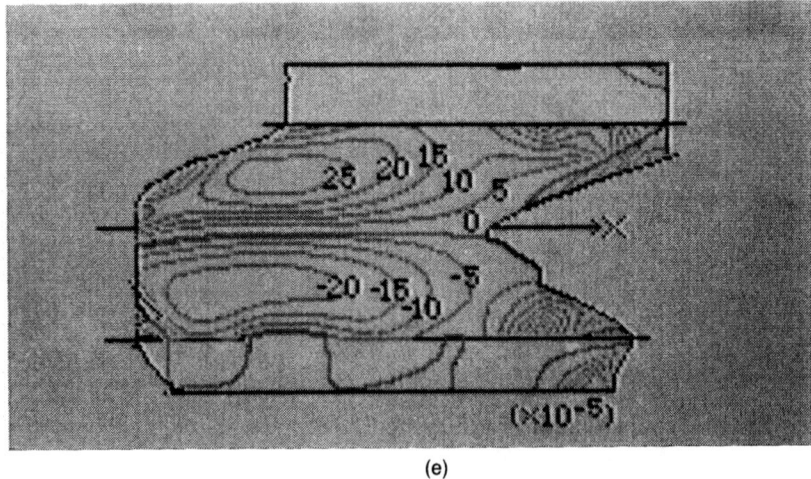

(e)

Figure 21-28 (*continued*)

DIGITAL IMAGE PROCESSING

21-28b is the image of interference fringes observed from right side. The fringe centerlines are obtained by thresholding and thinning the fringes. Figure 21-28c shows the centerlines of fringes of Fig. 21-28b and fringes observed from left side. By analyzing the fringe orders at the positions of the corresponding points between the two images, the displacement of the point is determined, as shown in Fig. 21-28d. By differentiating the displacement, the strain distribution is obtained. The stress is calculated from the constitutive equation, as shown in Fig. 21-28e.

The procedure is basically as follows:

1. Recording two images from different directions
2. Filtering to reduce noise
3. Thresholding, to obtain bilevel images
4. Noise erasing and connecting of cut lines
5. Thinning to obtain the centerlines of the fringe lines
6. Reading the coordinates of the fringe centerlines
7. Calculating the displacements from the data of the coordinates of the fringe centerlines and the fringe orders of the corresponding points of two images
8. Interpolating the displacements by a spline function
9. Calculating the strain distribution obtained by differentiating the spline function
10. Calculating the stress distribution by using the constitutive equation
11. Displaying the results on a TV monitor or outputting them as hard copy from a printer

The histogram of gray values is useful to check whether the image input is good. For a fringe pattern, the best histogram should have all the gray values from the maximum to the minimum value of the image processor, and the histogram distribution with respect to the gray levels may be cosinusoidal. If the histogram does not have the maximum or minimum values, or is not cosinusoidal, the input devices including the camera aperture and the A/D converter may not be adjusted in the gray value width or may not have linearity. However, the fringe center position analysis method can accept the nonlinearity.

Filtering (item 2) is not necessary for an image of good quality. If the image has some point white dots (or black dots) in the black fringes (or white fringes), however, averaging filtering or median filtering (low-pass filter) before thresholding may eliminate such dot noise.

Expanding and shrinking operations after thresholding can cut flaws or connect the lines with cut parts. For an image with uneven, low-frequency brightness, it is difficult to obtain the centerlines of the fringes from one thresholded image. The centerline pattern in the whole field is obtained by combining some images obtained by changing the threshold levels or by using a high-pass filter. Fixed noise sources such as stains on the camera lens may be erased by calculating the difference from the image before deformation without fringes. Whether the fringe order difference between the neighboring fringes is $+1$, -1, or 0 can be known by moving the view point (i.e., camera position) and checking the moving direction of the fringe lines. In the case of moiré fringes, the difference may be found by checking the moving direction of the fringe lines when the reference grating is shifted, that is, when α in Eq. (21-20) (i.e., x_0) is changed.

Fringe orders are obtained only at the positions of the fringe centers. An interpolation such as a spline function is used to obtain continuous data. The distribution is usually displayed as the monochromatic brightness distribution, with pseudocolors corresponding to the values or a three-dimensional wire frame mesh on a TV monitor.

Phase-Shift Method [47,51,53,55]

Unlike the fringe center position analysis method, the phase-shift method does not require any interpolation when the brightness distribution of a fringe pattern is cosinusoidal. The displacement

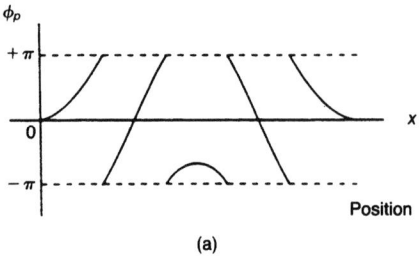

(a)

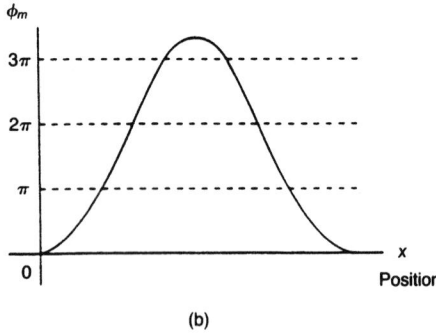

(b)

Figure 21-29 Connection of phases: (a) band-limited computer phase output; (b) connected phases.

at each point of the whole field of an image can be obtained by measuring the phase of the cosinusoidal wave at that point. Let us consider the case in which the brightness distribution of the fringe is expressed as shown in Eq. (21-20). If the phase ϕ at each point is determined, the displacement can be analyzed. The phase function is determined when fringe patterns are recorded three times by changing the shift α in Eq. (21-20) by known amounts. For example, when the shift amounts $\alpha_i (i = 1, 2,$ and $3)$ are $\pi/4$, $3\pi/4$, and $5\pi/4$, and the brightness value at each point is f_i, the phase can be calculated by the following equation.

$$\phi = \tan^{-1}\left[\frac{f_3 - f_2}{f_1 - f_2}\right] \quad (21\text{-}21)$$

If the values of $\cos(\phi - \alpha)$ in Eq. (21-20) at some points on different fringe lines are equal, the brightness values are also equal. Moreover, the arctangent function is confined to the range of $-\pi$ to π even when the actual phase is even out of this range. Therefore, sudden phase jumping often occurs (Fig. 21-29a). The phase values should be smoothly connected by adding or subtracting 2π to the phase value of the neighboring point as shown in Fig. 21-29b.

The phase-shift method requires neither the determination of the fringe center nor the consideration of increasing or decreasing direction of the fringe order. Since brightness or phase information at all points is used, the spatial resolution is very high if the brightness distribution is cosinusoidal.

Since the ratio of brightness differences is used in Eq. (21-21), the errors, which depend on uneven light or dirt on a camera lens, are decreased, the calculation is simple, and therefore the calculation speed is high. It is necessary to record a few images with accurately shifted phase since the brightness distribution must be cosinusoidal. The image input device should be of good quality.

To shift the phase in geometric moiré, the reference grating may be shifted by using a micrometer. For two beam interference fringes, one of the interfering wavefronts may be moved slightly by moving a mirror or a glass wedge. It is impossible to apply this method to fringe or grating patterns without a phase shift such as the grid method.

Fourier Transform Method [18,43,23–26]

To interpolate the data between fringes, Sciammarella and Sturgion [18] analyzed the phases of mismatched fringes by using the Fourier spectrum of the fringe image. More recently, Takeda and Mutoh [43] introduced Fourier transform profilometry (FTP), which is a new moiré topography method. In this method, the three-dimensional shape of an object is measured by calculating the phase of the Fourier spectrum of the image of a grating projected on the object. Morimoto et al. [29–33] proposed the scanning moiré method for measuring strain distribution. In this method, the moiré pattern appears when the scanning lines of the TV image of the deformed grating are thinned out. The appearance is explained as aliasing in the "sampling theory" of signal processing, that is, shifting of the Fourier spectrum of the deformed grating toward the origin of the frequency.

In addition, Morimoto et al. [23–26] presented new methods analyzing one-dimensional strain distribution by using the Fourier transform of a grating image. One method is a type of moiré method that uses the first harmonic [23] or one of the higher harmonics [24] to show contour lines of displacement. Another is a mismatched moiré method, which is extended to a grid method [26]. These methods are obtained by combining and extending Sciammarella's idea, Takeda's method, and Morimoto's scanning moiré method. These methods are extended to two-dimensional strain analysis [25]. A two-dimensional cross grating is used to measure two directional components of the displacement. Chiang [73] also separated the two components from the images of the moiré fringes or gratings by optical filtering.

The moiré method developed by Morimoto et al. is called the Fourier transform moiré method (FTMM). The grid method is called the Fourier transform grid method (FTGM). These two methods combined to make the Fourier transform moiré and grid method (FTMGM). The theory and applications of the FTMGM are detailed in the next section.

Let us show the basic idea of the Fourier transform moiré method. The behavior of a moiré fringe pattern is similar to the beating of sound, which occurs when two waves of sound are superimposed. If the frequencies of the waves are f_A and f_B, the frequency of the beating is $f_A - f_B$ (Fig. 21-30a). On the other hand, moiré fringe patterns appear when two gratings are superimposed. If the spatial frequencies of the gratings are f'_A and f'_B, the spatial frequency of the moiré fringe is also $f'_A - f'_B$ (Fig. 21-30b). The FTMM basically use this principle.

The Fourier transform of the grating image is used to measure the frequency and the phase of the model grating. The spatial frequencies of the reference grating are constant, and the frequency of the deformed model grating is analyzed by using the Fourier spectrum of the deformed grating image. Then, by subtracting the frequency of the reference grating from the frequency of the deformed model grating, the moiré fringe frequency is obtained. The moiré fringes become cosinusoidal by extracting the first harmonic of the spectrum. The phase of the cosinusoidal brightness corresponds to the displacement.

If the extracted first harmonic is not shifted and the inverse Fourier transform of the first harmonic is calculated, the brightness distribution of the original grating changes from rectangular to cosinusoidal as shown in Fig. 21-31. The same phase of the cosinusoidal functions before and after deformation corresponds to the same point on the object. The position difference of the two points gives the displacement.

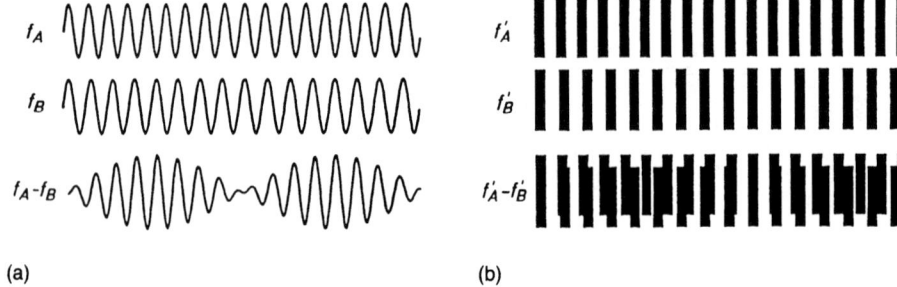

Figure 21-30 (a) Beating of sound waves; (b) moiré pattern of gratings.

This method is accurate because the extraction of the first harmonic automatically eliminates low-frequency and high-frequency noise. However, the discrete Fourier transform automatically connects the data at both ends of the region. If the data are different at both ends of the region, the discrete Fourier transform produces virtual high-frequency components, which induce error. When the first harmonic cannot be extracted alone, as happens when the strain change is very large, the high-frequency components are filtered by the process. This lead to errors.

21-7 Fourier Transform Moiré and Grid Method

In the Fourier transform moiré and grid method (FTMGM) [23–26], interpolation between fringes or grating lines is performed naturally by using the phases of the cosinusoidal brightness distribution which are obtained from the first harmonic of the Fourier spectrum of the image of the fringes or grating lines. This process is simple and, because it lacks subjective elements, human errors can be excluded, leading to accurate measurements of displacement, strain, strain rate, and height.

The theory of the FTMGM has been detailed in Refs. 23 through 26. First, the FTMGM using the first harmonic of the Fourier spectrum of a deformed grating pattern [23] was proposed. Second, a higher harmonic of the spectrum [24] was used. Next, a mismatched moiré method and a grid method [26] were developed. Furthermore, the FTMGM was extended to encompass two-dimensional analysis [25]. This section outlines the theoretical treatment of the one-dimensional analysis only.

Displacement and Strain Analysis

Figure 21-32 shows the brightness intensity distribution of a model grating before deformation. It consists of alternate dark and bright stripes and is utilized for the displacement measurement of an object in the direction normal to the grating lines.

The expansion of the grating brightness function $f(x)$ in a Fourier series is

$$f(x) = \sum_{n=-\infty}^{\infty} C_n \exp\{jn\omega_0(x - x_0)\} \qquad (21\text{-}22)$$

where

$$C_n = E\frac{\sin(n\pi b_0/p_0)}{n\pi} \qquad \omega_0 = \frac{2\pi}{p_0} \qquad (21\text{-}23)$$

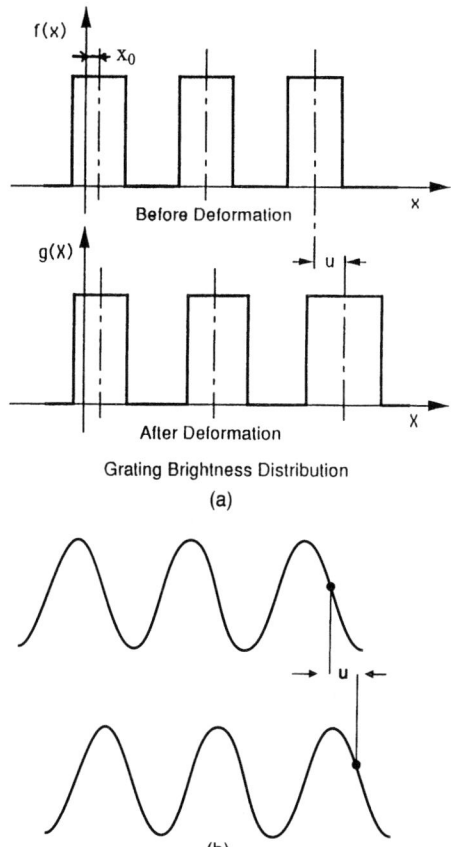

Figure 21-31 Change of brightness from rectangular to cosinusoidal by extracting first harmonic of grating image: (a) original brightness distribution; (b) brightness distribution obtained from first harmonic.

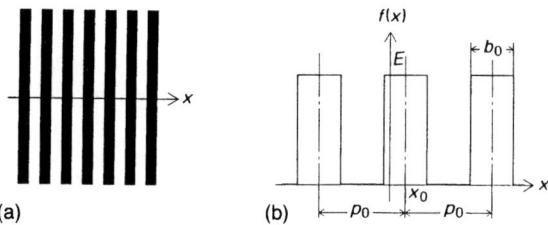

Figure 21-32 Brightness intensity distribution of a model grating before deformation: (a) grating; (b) brightness distribution.

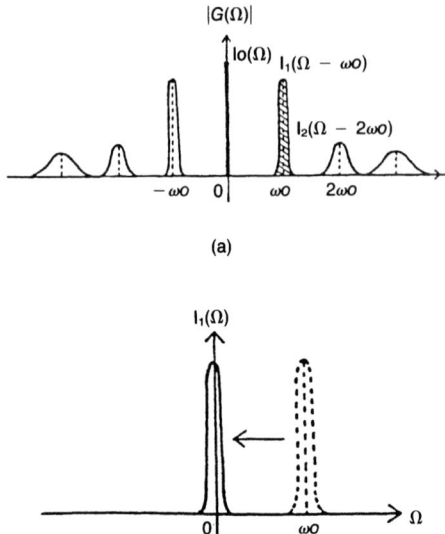

Figure 21-33 (a) Fourier spectrum of deformed grating; (b) shifted first harmonic.

and j is the imaginary unit, n is an integer, p_0 is the grating pitch, and b_0 is the width of the bright part of the grating. Now, let us consider the deformation of this grating which is printed on an object before deformation. When the displacement at any point X after deformation is $u(X)$, the relationship between point X and the initial position x of this point before deformation is expressed as

$$X = x + u(X) \tag{21-24}$$

The grating brightness function $g(X)$ at point X after deformation corresponds to the grating brightness $f(x)$ in the initial position x. Thus,

$$g(X) = f(x) = f\{X - u(X)\} = \sum_{n=-\infty}^{\infty} C_n \exp[jn\omega_0\{X - u(X) - x_0\}]$$
$$= \sum_{n=-\infty}^{\infty} i_n(X)\exp(jn\omega_0 X) \tag{21-25}$$

where

$$i_n(X) = C_n \exp[-jn\omega_0\{u(X) + x_0\}] \tag{21-26}$$

The Fourier transform of the grating pattern expressed in Eq. (21-25) is

$$G(\Omega) = \int_{-\infty}^{\infty} g(X)\exp(-j\Omega X)dX = \sum_{n=-\infty}^{\infty} I_n(\Omega - n\omega_0) \tag{21-27}$$

where $I_n(\Omega)$ is the Fourier transform of $i_n(X)$. Figure 21-33a shows schematically the power spectrum obtained by the transform.

Now from Fig. 21-33, the first harmonic $I_1(\Omega - \omega_0)$ indicated with diagonal lines extracted by filtering. By shifting both the real and imaginary parts of the first harmonic by ω_0 toward the origin of the frequency axis Ω, we obtain $I_1(\Omega)$ as shown in Fig. 21-33b. The computation of the inverse Fourier transform of $I_1(\Omega)$ gives

$$I_1(X) = \frac{1}{2\pi} \int_{-\infty}^{\infty} I_1(\Omega)\exp(j\Omega X)d\Omega = C_1\exp\{j\theta_1(X)\} \tag{21-28}$$

where

$$\theta_1(X) = -\omega_0\{u(X) + x_0\} \tag{21-29}$$

This image shows a complex moiré pattern with a real part $\text{Re}\{i_1(X)\}$ and an imaginary part $\text{Im}\{i_1(X)\}$. The phase $\theta_1(X)$ of the complex moiré fringes is computed from the complex moiré pattern as

$$\theta_1(X) = \tan^{-1}\frac{\text{Im}\{i_1(X)\}}{\text{Re}\{i_1(X)\}} \tag{21-30}$$

From Eqs. (21-29) and (21-30), the displacement $u(X)$ is obtained.

$$u(X) = -\frac{1}{\omega_0}\tan^{-1}\frac{\text{Im}\{i_1(X)\}}{\text{Re}\{i_1(X)\}} - x_0 \tag{21-31}$$

The Eulerian strain at a point X is given as a derivative of the displacement $u(X)$.

$$\varepsilon_X = \frac{\partial u(X)}{\partial X} = -\frac{1}{\omega_0}\frac{\partial \theta_1}{\partial X} \tag{21-32}$$

If a higher harmonic [24] is selected instead of the first harmonic, a greater number of fringes is obtained. If the first harmonic is shifted by any amount ω_0' instead of ω_0, the result corresponds to a mismatched moiré method [26]. If ω_0' is zero, it corresponds to the grid method [26]. By using a two-dimensional grating and two-dimensional Fourier transformation, two-dimensional displacement components and strain distributions can be analyzed [25].

Now we give some examples of the measurement of displacement, deflection, and shape.
Figure 21-34 shows the flow diagram of a general FTMGM process.

Strain Measurement of Cracked Rubber Plate

Figure 21-35 shows strain analysis for a rubber plate with a crack at the center. The digitized original image appears in Fig. 21-35a. The image size is 240 × 240 pixels. The one-dimensional spectrum is calculated on each y-directional (vertical) line. The Fourier spectrum of the image is shown in Fig. 21-35b. The first harmonic is extracted from the spectrum by a filter and shifted toward the origin by the frequency of the undeformed grating. The inverse Fourier transform of the shifted first harmonic shows a clear moiré pattern (Fig. 21-35c). By differentiating the phases of the fringes, the strain distribution can be obtained, as shown in Fig. 21-35d.

Two-Dimensional Strain Analysis of a Rubber Fender [25]

Figure 21-36 presents the analysis of two-dimensional strain distribution in a rubber fender. The original image, with two-dimensional cross grating, and the Fourier power spectrum of the

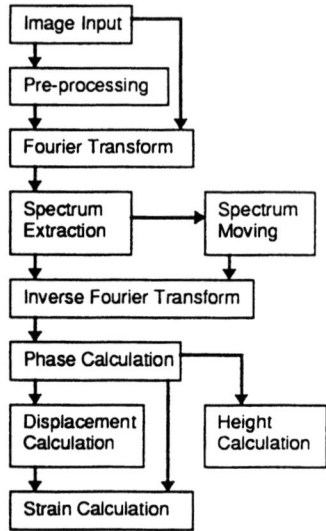

Figure 21-34 Flow diagram of the Fourier transform moiré and grid method.

deformed grating, are shown (Fig. 21-36a,b), indicating the horizontal and vertical directions show the X- and Y-directional frequencies Ω_X and Ω_Y, respectively. The origin of the frequency is centered. The image size is 480 × 480 pixels and the grating pitch before deformation is 4 pixels, so that the first harmonic frequency is 120. The first-harmonic frequency of the Y-directional grating before deformation appears at the point of 120 pixels in the Ω_Y direction and 0 pixels in the Ω_X direction from the frequency origin. The first Ω_Y directional harmonic frequency of the deformed grating exists about the point (0, 120).

The first harmonic is extracted and shifted toward the origin by 120 pixels as shown in Fig. 21-36c. From the inverse discrete Fourier transform of the shifted first harmonic, complex moiré patterns appear (Fig. 21-36d). These fringes show the Y-directional equal displacement lines. By calculating the phases of the fringes and differentiating the phase, the strain distribution shown in Fig. 21-36e is obtained. The inverse Fourier transform of the unshifted first harmonic extracts only the Y-directional grating lines from Fig. 21-36a, as shown in Fig. 21-36f. The brightness distribution of the grating is changed from rectangular to cosinusoidal, and the grating becomes clearer because the first harmonic extraction eliminate dot noise, shading, and X-directional grating lines automatically. The displacement is also analyzed from the phase of the cosinusoidal grating lines. This corresponds to a grid method.

Measurement of Small Deflections by Holographic Interferometry

The FTMGM can be used in the analysis of large deformations as shown above because the number of pixels on a line in an image is not so many. However, if the method is combined with a carrier fringe pattern produced by a laser method such as moiré interferometry, holographic interferometry, or speckle interferometry, small displacements can be measured. It is possible to measure small (submicrometer) displacements by analyzing the fringe phases. Let us show an example of the analysis of three-point bending of a beam by holographic interferometry.

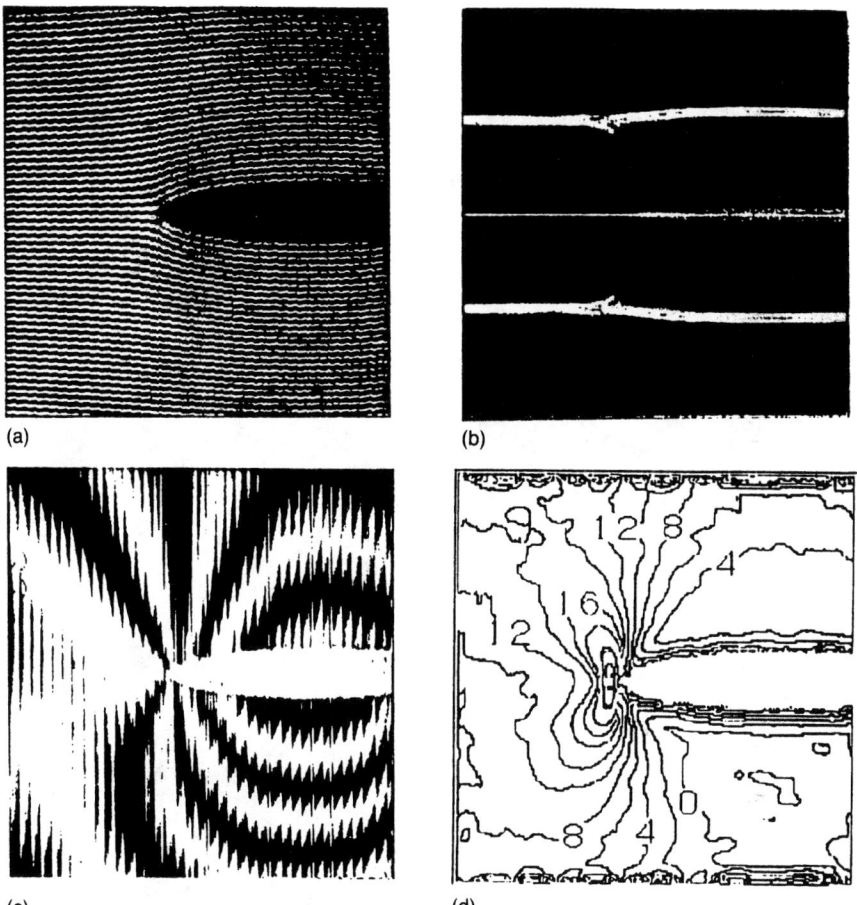

Figure 21-35 Strain analysis of a cracked rubber plate: (a) original image; (b) Fourier spectrum; (c) moiré pattern obtained by inverse Fourier transform of shifted first harmonic (real part); (d) strain distribution in the Y direction (%).

We apply a maximum deflection of 1.5 μm to the center of the beam. The carrier pattern is produced by moving the light path in the reconstruction. The holographic fringe pattern with carrier is shown in Fig. 21-37a. The deflection distribution obtained by the analysis is shown in Fig. 21-37b. The theoretical distribution along the centerline compares with the experimental results in Fig. 21-37c.

Shape Measurement

By applying FTMGM to a stereoscopic measurement method using a projected grating, the shape of a three-dimensional object can be measured. A grating is projected onto an object and the image is recorded by an image processor. The grating lines of the image are deformed according

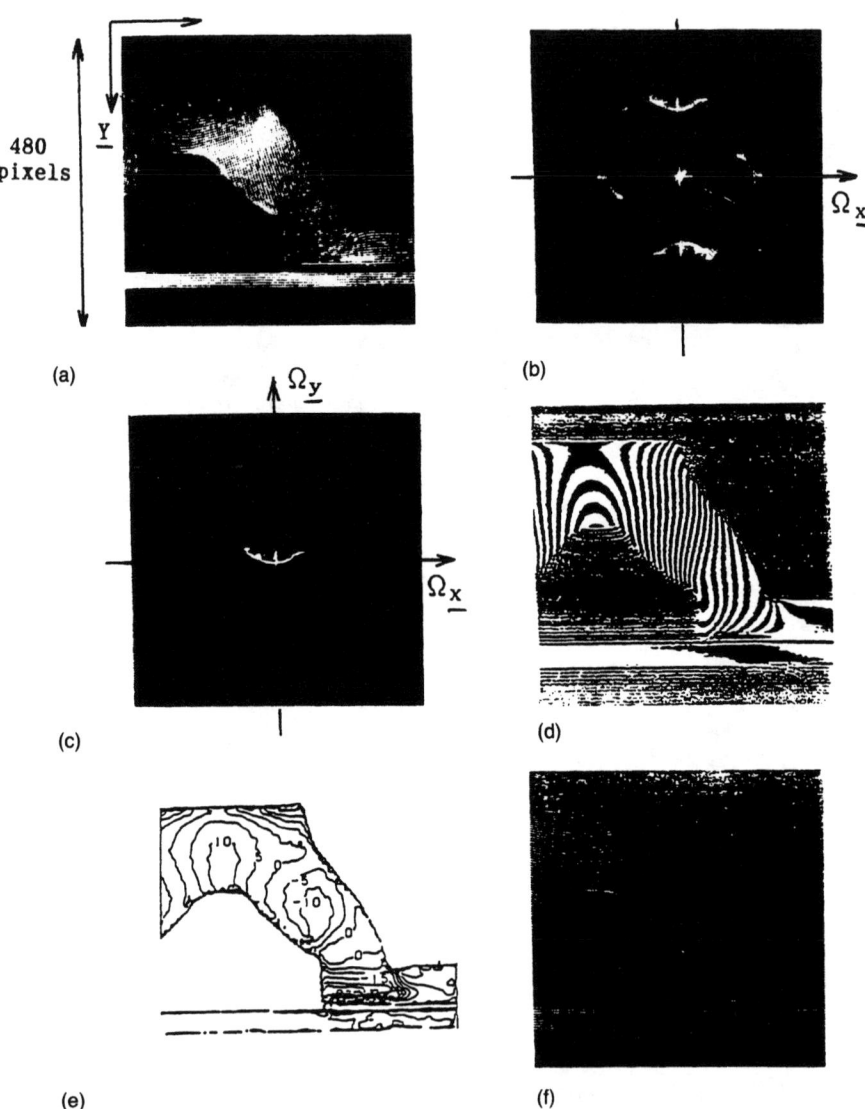

Figure 21-36 Two-dimensional strain analysis of a rubber fender: (a) original image; (b) Fourier spectrum; (c) extracted and shifted (120-pixel shift, real part) Y-directional first harmonic of (b); (d) complex moiré pattern; (e) strain distribution in the Y direction (%); (f) complex grating obtained from unshifted Y-directional first harmonic.

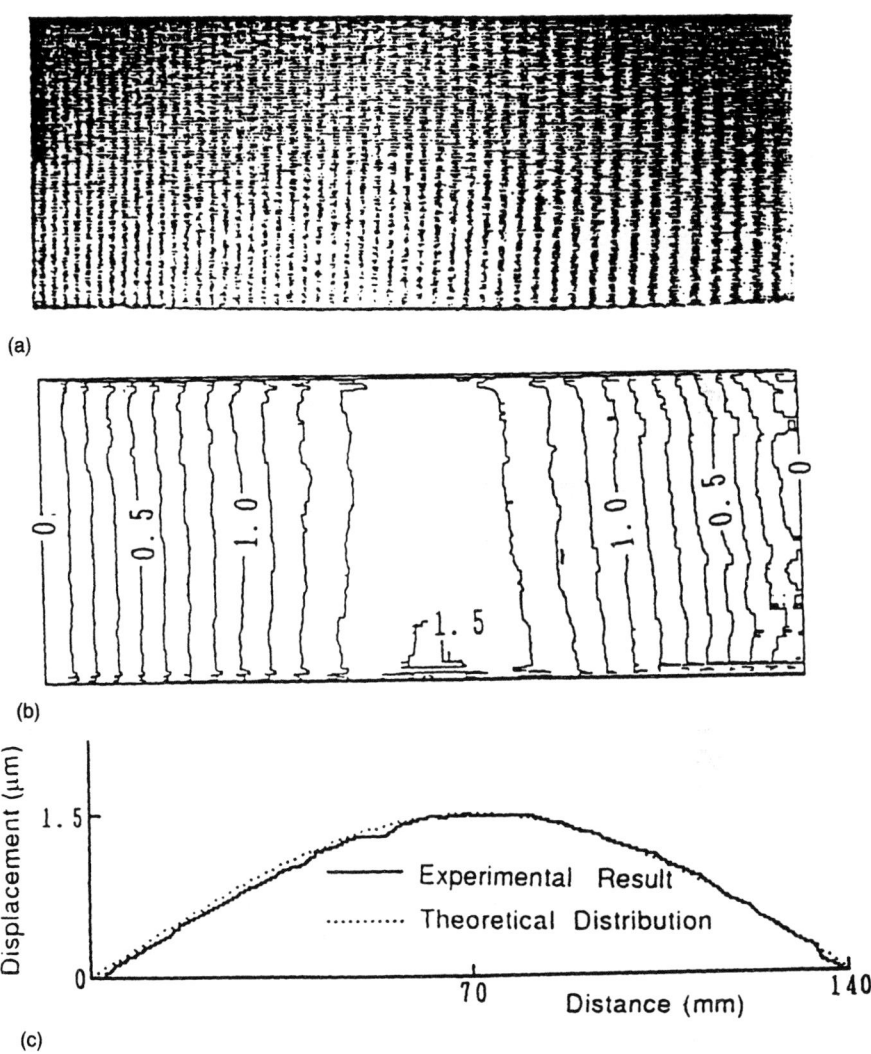

Figure 21-37 FTMGM analysis of a fringe pattern obtained by holographic interferometry: (a) holographic fringe pattern with carrier; (b) deflection distribution analyzed by FTMGM; (c) comparison of theoretical and experimental results.

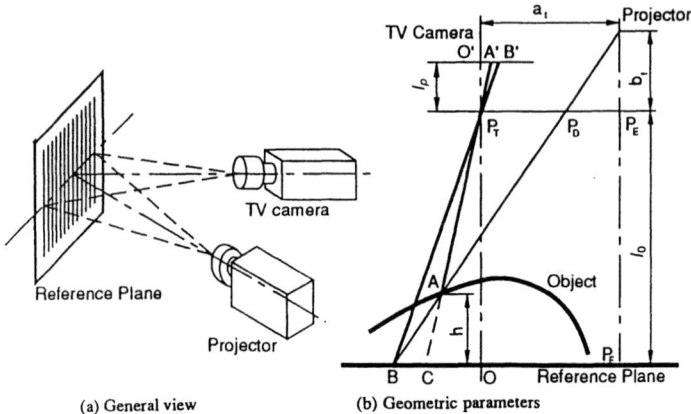

(a) General view (b) Geometric parameters

Figure 21-38 (a) Schematic setup for the stereoscopic method; (b) geometric parameters.

to the shape of the object. By applying the FTMGM to the deformed grating image, the displacements of the grating lines can be analyzed. These displacements correspond to the heights, which show the shape of the object.

Figure 21-38 shows a schematic of the setup for this method. A grating is projected onto an object and the deformed grating is recorded by a TV camera. A point light projected at point B on a reference flat plane moves to point C on the object when the object is displaced out of the reference plane. The relationship between the movement $\overline{BC}$ (displacement $u(X)$) and the height $h(X)$ is expressed as follows:

$$h(X) = \frac{l_0 \overline{BC}}{\overline{BC} + \overline{P_T P_D}} \tag{21-33}$$

where $\overline{P_T P_D}$ is obtained from

$$\overline{P_T P_D} = \frac{a_t l_0 - b_1(\overline{BC} + \overline{CO})}{l_0 + b_1} \tag{21-34}$$

Measurement of the Shape of an Artificial Leg by Slit Light Projection Method

Three-dimensional shape can be measured by the FTGM. Figure 21-39 shows how the dimensions of a mold for an artificial leg may be determined with sufficient accuracy to minimize the patient's discomfort. Noncontact measurement of the shape of the leg in the region of the site of amputation is desirable, and a stereoscopic measurement method was developed for this purpose. As shown schematically in Fig. 21-29a, a slit light beam is projected on the cut end of the leg of a patient. In this case, a mold model of the leg is used. The deformed slit light on the model is recorded by a TV camera and digitized. This operation is repeated 60 times by rotating the optical system around the leg. The images obtained after each 6° rotation are superposed, after shifting by constant pixels. The resultant image (Fig. 21-39b) is analyzed by the FTGM to obtain the displacements

DIGITAL IMAGE PROCESSING

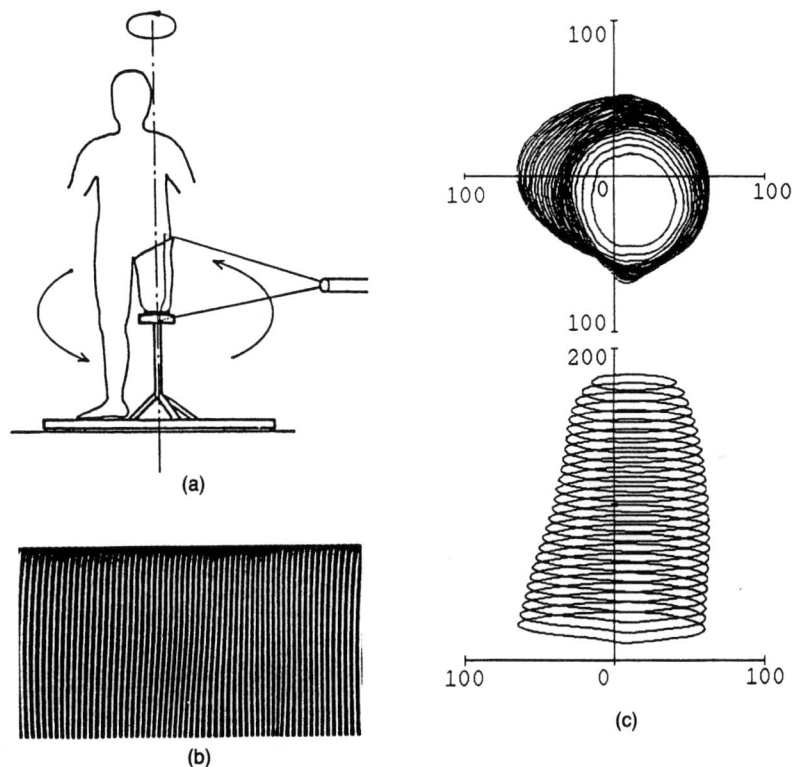

Figure 21-39 Shape analysis by FTMGM of socket for an amputated leg: (a) setup for measurement; (b) image constructed by superposing; (c) shape of object.

of the deformed slit lines. The shape is obtained from the displacements of the slit lines by using Eq. (21-33). The shape of the cross section of the leg obtained by this analysis is shown in Fig. 21-39c.

21-8 Special Image-Processing Systems (Hardware)

Real-Time Analyzer of Displacement and Strain Distributions [34]

An image-processing system to measure displacement and strain distributions on the basis of the geometric moiré, the scanning moiré method, and the shift method (also called the shearing method or the mechanical differentiation) was developed [34]. The analysis is a real-time method that does not require photographic processing, and it is carried out without contacting the specimen. This section begins by describing the phenomenon of moiré appearance by the scanning moiré method. Then the hardware configuration is explained followed by an application of the system to strain measurement.

Scanning Moiré Method

In the conventional geometric moiré method, a moiré pattern appears as the interference between a model grating and a master grating. However, a moiré pattern also appears when the model grating lines are sampled by the scanning lines of a TV camera or by the pixels of an image processor [29–34]. If the sampling points of the image processor are regarded as the master grating with X- and Y-directional crossed lines, the geometric relationship between the model grating lines, the sampling points, and the fringe lines is obtained in the same way as in conventional geometric moiré. The moiré fringe appearance by the scanning moiré method can be explained by using sampling theory and the Fourier transform of the image [31], or by using the geometric relationships [33].

Figure 21-40 illustrates the appearance of moiré fringes as visualized by the scanning moiré method. Figure 21-40a shows the positions of the sampling points of an image processor as black points. Figure 21-40b shows the undeformed grating lines drawn or projected on a specimen. Before deformation, the TV camera is adjusted so that the grating lines are perpendicular to the horizontal scanning lines. Each line of the model grating coincides with each corresponding horizontal sampling point of the image processor. The model grating is sampled at the points of Fig. 21-40a by the image processor. If a sampling point is on a black line of the model grating, a black-point image is obtained. If it is on a white line, a white-point image is obtained. Figure 21-40c shows the images obtained by these processes. In this case, all point images are black; there is no moiré pattern.

The image resulting when the strain of a deformed specimen is small (Fig. 21-40d) appears in Fig. 21-40e. The pitch of the model grating is nearly equal to the pitch of the sampling points so that the white-point images and the black-point images appear in groups, which produce a moiré fringe pattern.

In a large-deformation case (Fig. 21-40f), the pitch of the model grating is about 2.6 times larger than that of the scanning lines; these sampled images are shown in Fig. 21-40g, in which no moiré pattern can be discerned. However, if the point images from alternate horizontal sampling points are thinned out, as in Fig. 21-40h, the images are separated into white and black groups showing moiré fringes. Conversely, Fig. 21-40i shows an image that consists of the thinned-out line images of Fig. 21-40h but differing in phase by π from Fig. 21-40h in term of both thinning-out and moiré fringes. Moreover, if every third horizontal sampling point is picked up from Fig. 21-40g, the pattern shown in Fig. 21-40j is obtained, and this too is a moiré pattern different from Fig. 21-40h. In Fig. 21-40j, the first sampling point of the three continuous points of Fig. 21-40g is selected. If, instead, the second sampling point is selected, the image in Fig. 21-40k is obtained. Figure 21-40l shows the image obtained by selecting the third point.

The phase of the moiré fringes corresponds to the thinning-out phase. The direction that increases the fringe order can be determined by increasing the thinning-out phase. The direction of fringe movement when the thinning-out phase is increased indicates the direction of increasing fringe order. If the center positions of the fringes in Fig. 21-40j are regarded as having continuous integer fringe order, the center positions of the fringes in Fig. 21-40k have the order: integer fringe order $+ \frac{1}{3}$. In this case, the fringes move to the left and the fringe order increases toward the negative direction. Therefore, if the thinning-out phase is changed, fractional fringe orders are obtained.

Finally, if all the sampling point images that were thinned out in Fig. 21-40l are replaced by the preceding point images, which are picked up in Fig. 21-40l, the moiré fringes become clearer, as shown in Fig. 21-40m. In this process, however, the center positions of the resultant fringes are shifted by a half of N pixels.

These thinning-out processes correspond to a mismatch in the conventional geometric moiré method. Although Fig. 21-40 shows only two brightness values, white and black, a gray halftone

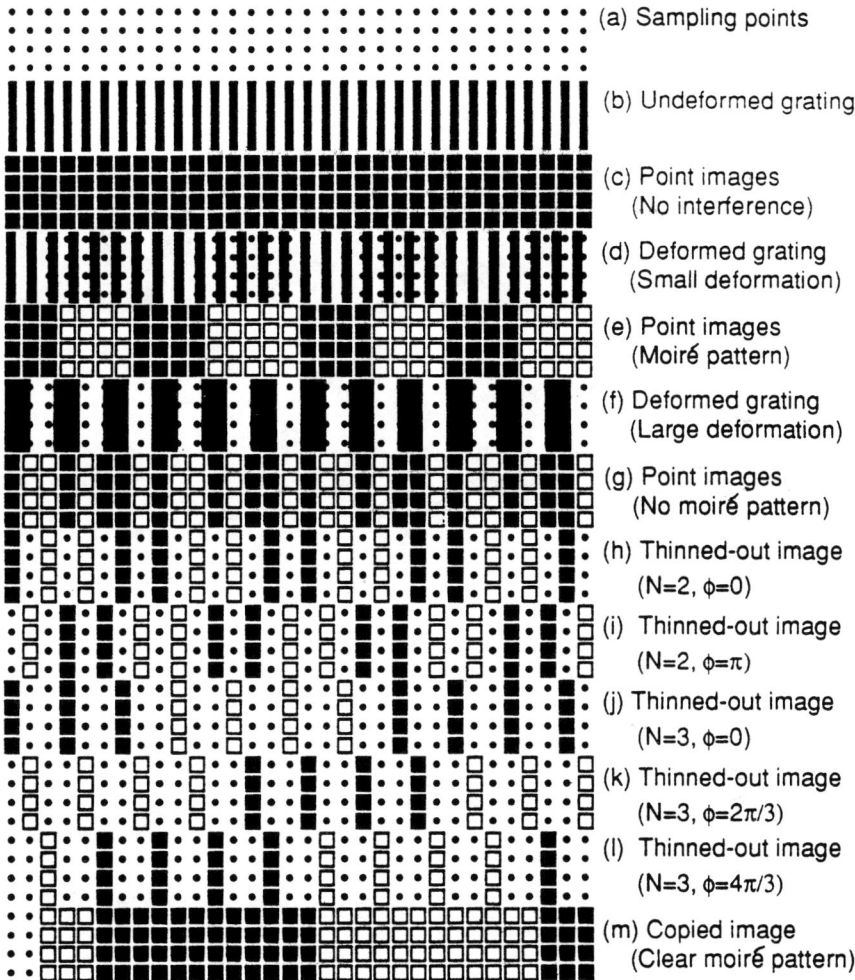

Figure 21-40 Appearance of moiré fringes generated by scanning moiré method: the images sampled by sampling points show moiré patterns; thinned-out images show different patterns according to the thinning-out index N and the phase ϕ.

image is usually obtained. The contrast of the moiré fringes depends on the size of the sampling point and the brightness distribution of the model grating.

When the pitch of the sampling points of an image processor is considerably smaller than that of the model grating lines, the image shows only the grating lines and no clearly discernible moiré pattern. In such a case, if the sampling points are periodically picked up, the image shows a moiré pattern. By changing the pickup pitch (i.e., the thinning-out index N), different moiré patterns

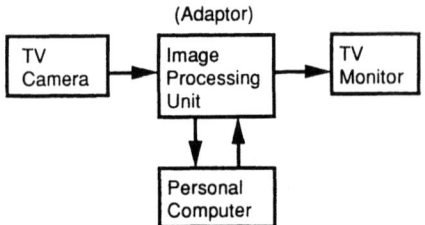

Figure 21-41 Flow diagram of image-processing system for real-time displacement and strain analyzer.

can be obtained. This corresponds to the mismatched moiré method. It is easy to change the pitch using image processing.

Flow Diagram of System

Figure 21-41 shows the flow diagram of this image-processing system. The main part of this system is an adapter for processing an input image to obtain displacement or strain distributions and to output the processed image to a TV monitor display. The TV camera, the TV monitor, and the personal computer are commonly used hardware. The image recorded by the TV camera is digitized by the image-processing unit and stored in the frame memory in the unit. The digital image is processed to obtain displacement or strain distributions in real-time. That is, calculation time for one image is 1/30 second and the frame speed is 30 frames/s. The processed image is output to a TV monitor. The operation is controlled by a personal computer. Figure 21-42 shows the flow diagram of the image-processing unit.

Application

An explanation of the functions is explained using a simulation model (Fig. 21-43). Figure 21-43a shows the deformed grating. Before deformation, these lines are straight and equally spaced. The size is a 200-mm square, and the grating pitch is 4 mm, corresponding to 8 pixels.

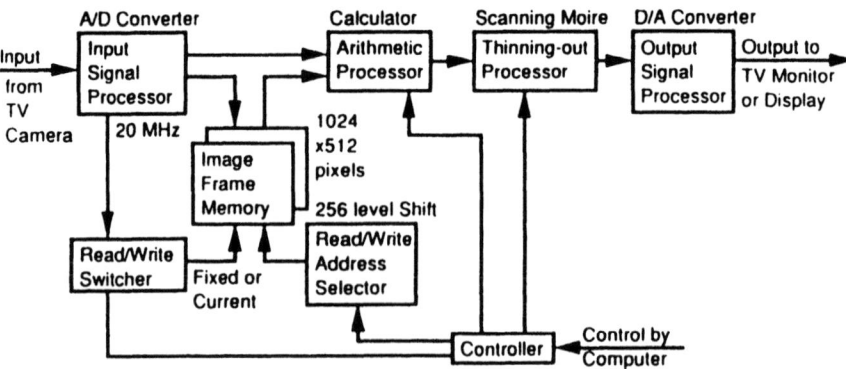

Figure 22-42 Flow diagram of image-processing unit.

DIGITAL IMAGE PROCESSING

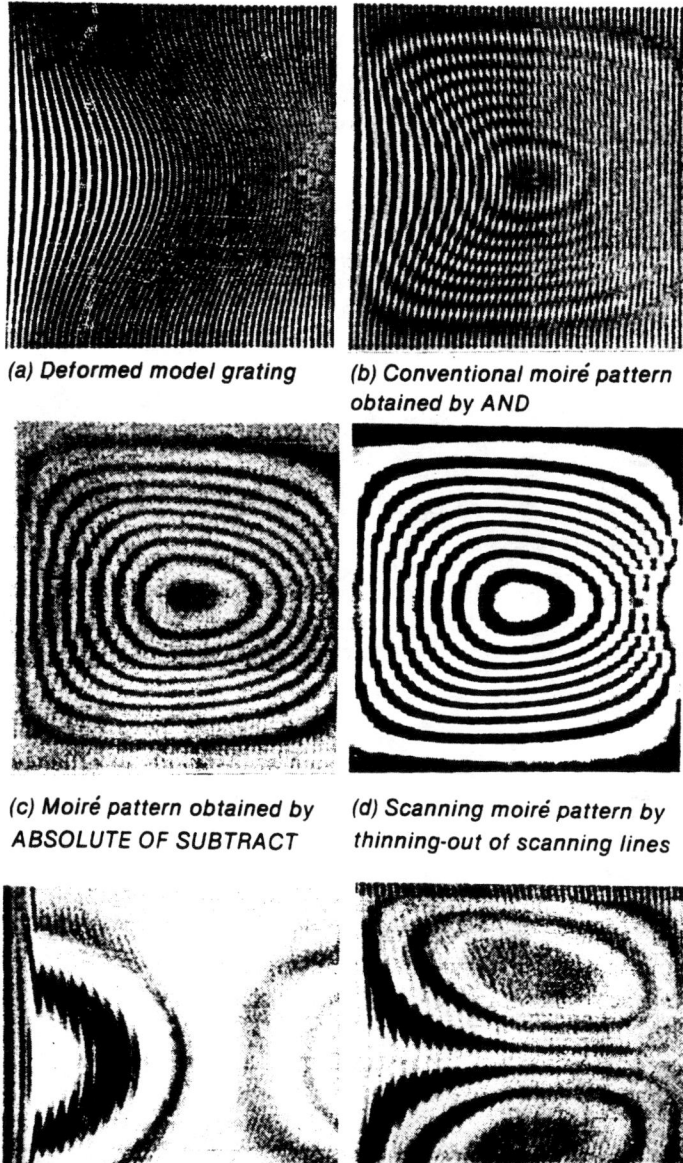

(a) Deformed model grating
(b) Conventional moiré pattern obtained by AND
(c) Moiré pattern obtained by ABSOLUTE OF SUBTRACT
(d) Scanning moiré pattern by thinning-out of scanning lines
(e) Moiré pattern showing normal strain by horizontal shift
(f) Moiré pattern showing shear strain by vertical shift

Figure 21-43 Application of system to simulate model gratings.

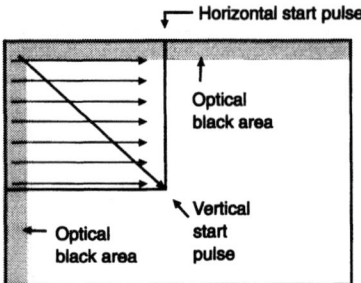

Figure 21-44 Scanning area of high-speed video camera sensor.

One-dimensional deformation is considered where the grating lines deform horizontally. The fringe lines obtained by superposing the deformed grating lines on the image of the master grating appear in Fig. 21-43b and c. The fringe pattern of Fig. 21-43b is obtained by selecting the function AND. The fringe pattern in Fig. 21-43c, obtained by ABSOLUTE OF SUBTRACT, is clearer than that of Fig. 21-43b because the original grating lines in Fig. 21-43c are almost eliminated. Figure 21-43d shows the fringes obtained by horizontal thinning out of the scanning lines of the deformed grating (Fig. 21-43a). In this case, the thinning-out index N is 8. The fringe lines in Fig. 21-43e, obtained by horizontal shift, show normal strain contours in 20% increments. Figure 21-43f, obtained by vertical shift, shows shear-strain contours in 50% increments. Displacement and strain distributions are easily obtained in real-time using this system.

High-Speed Video Camera [41]

Because the frame speed of a high-speed camera is naturally large, an enormous amount of recording material such as film or tape must be used when it is difficult to predict the start time of a phenomenon, such as lightning or crack propagation. Waste of recording materials can be avoided if the phenomenon is cyclically recorded on rewritable IC memory instead of film or tape, if the recording is stopped when the phenomenon is completed, and if the images are stored in IC memory. Based on this principle, a system of a high-speed video camera with IC image memory was made.

Hardware

A new commercial MOS-type solid-state image sensor serves as the image sensor of the camera. It is possible to change the area size for the output into any size by an suitable choice of software. The scanning area is changed as shown in Fig. 21-44. The scanning speed for one frame can be changed: the smaller, the area selected, the higher the frame speed. In other words, the scanning area of the MOS-type image sensor is changed by selecting horizontal and vertical switches by sending a horizontal start pulse and a vertical start pulse.

Therefore, by combining the changes of the intervals of the horizontal and vertical start pulses, the maximum frame speed is about 300,000 frames/s. In this case, the frame size is only one pixel. If the optical black reference pixels are erased and the clock speed is doubled, the theoretical maximum frame speed is more than 10,000,000 frames/s, though the image size is one pixel. The frame speed or image size of this video camera is controlled by a personal computer.

DIGITAL IMAGE PROCESSING 1023

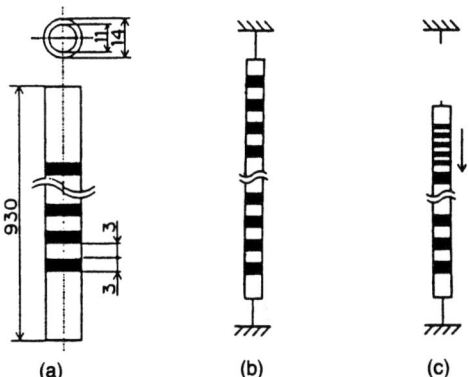

Figure 21-45 Schematics of rubber tube: (a) before stretching; (b) before unloading (stretched); (c) unloading.

The signal from the vertical output line is stored in IC memory by using an A/D converter or is stored on a video tape recorder. If the image signal is sent to the same IC memory by cyclically rewriting, the desired images are stored by sending a trigger when the phenomenon is over. This corresponds to a three-dimensional (x, y, t) transient memory.

Application to an Unloading Wave Propagating in a Rubber Tube

Stress-wave propagation in a rubber tube is analyzed by using a high-speed video camera and the Fourier transform grid method. A rubber tube with a grating painted on its surface is shown in Fig. 21-45a. The grating pitch is 6 mm. The rubber tube is stretched as shown in Fig. 21-45b. The grating pitch is stretched 6.77 mm and the camera position is adjusted to coincide with a pitch of 16 pixels length of the frame memory. When the upper end is cut, the unloading stress wave propagates in the tube as shown in Fig. 21-45c. The propagation of the unloading wave is imaged by the high-speed video camera. The images are analyzed to obtain displacement, velocity, strain, and strain-rate distributions by using the FTGM.

In this experiment, the scanning area size is set so that the horizontal length is 512 pixels and the vertical length is 2 pixels. This deformation is considered to be one dimensional, so that the vertical 2 pixels are separated into two different time data. Thus, the frame rate in this case corresponds to 15,730 frames/s. The images obtained by the high-speed video camera are recorded on a video tape recorder. The analog signals of the images on the video tape recorder are converted to digital images by an A/D converter and stored in the image memory. In this system, of course, it is possible to connect the high-speed video camera directly to the image memory.

The original image obtained by this camera is shown in Fig. 21-46a. Because the camera is rotated 90°, the left side of the image corresponds to the upper part of the rubber tube. The right corresponds to the lower part. The downward direction shows the time coordinate. Figure 21-46a shows the grating lines moving from the left to the right when an unloading wave is propagating. The x-directional Fourier transform is computed on each time line. In Figure 21-46b, which shows the Fourier power spectrum of Fig. 21-46a, the horizontal direction shows the frequency. The origin of the frequency is the vertical line at the center of the figure. The right domain is the positive frequency, which is proportional to the distance from the origin,

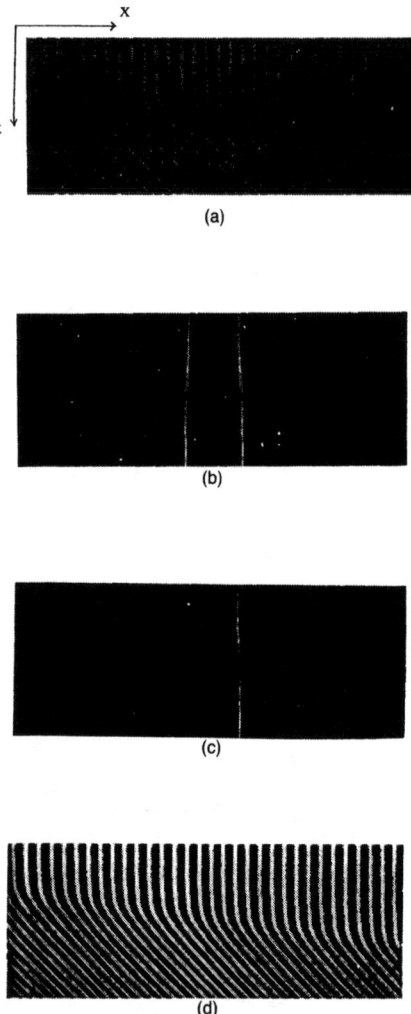

Figure 21-46 Analysis of propagation in a rubber tube of an unloading wave: (a) original image; (b) Fourier spectrum; (c) extracted first harmonic; (d) complex grating; (e) displacement distribution (mm); (f) strain distribution (m/s); (g) velocity distribution (%); (h) strain-rate distribution (s^{-1}).

and the left domain is the negative frequency. The brightness intensity of this figure corresponds to the power intensity of the frequency. Figure 21-46c shows the extracted first harmonic from Fig. 21-46b. Figure 21-46d shows the real part of the complex grating pattern obtained from the inverse Fourier transform of the extracted harmonic. Figure 21-46e shows the displacement distribution obtained by calculating the phases of the complex grating. By differentiating the displacement, the distributions of the strain, velocity, and strain rate are obtained as shown in Fig. 21-46f through h.

DIGITAL IMAGE PROCESSING   1025

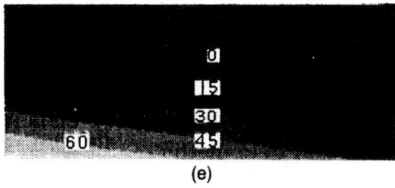

(e)

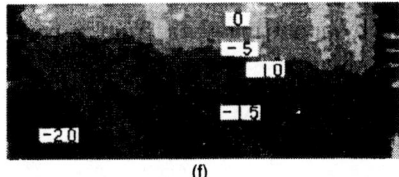

(f)

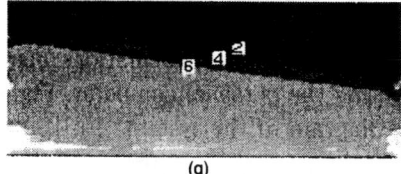

(g)

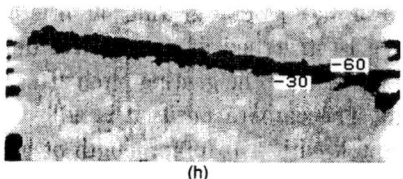

(h)

Figure 21-46 (*continued*)

21-9 Summary and Future Trends

Image processing is commonly used in the whole-field analysis of an image. It saves time and labor, and provides accurate results. Moreover, it creates new methods for analysis.

Video cameras and still cameras with built-in computers, or computers with built-in cameras, will be popular in the future. These devices will be as small as a present compact camera. Everybody will be able to process images easily using artificial intelligence. Every engineer will be able to check the strain distribution of objects in situ by using such equipment.

The integer gray values of 0 to 255 are treated in ordinary image processors at present. To calculate accurately higher-order differentials of the data obtained by image processing in experimental mechanics, image processing dealing with more highly resolved gray levels is required. Real-number gray values and complex-number gray values will be popular in experimental mechanics, although highly precise image input devices are necessary.

Although the resolution of an image taken by a TV camera at the present time is not very high, the performance of image-processing systems is improving as advances are made with computers and devices such as high-definition TV (HDTV) systems and high-resolution digital still cameras. In the future, the accuracy will be better than that of the strain-gage method.

Image analysis of multi-dimensional phenomena, which are functions of space, time, phase, and so on, will be popular according as the increase of memory capacity in computers. However, present computation speed of computers is too slow for the multi-dimensional image processing. Optical computers and neural computers will permit high-speed analysis for the processing and they will perform real-time analysis for a video movie or high-speed video images.

Experimental results do not always coincide with the results obtained by the theoretical methods such as the finite-element method (FEM) and the boundary-element method (BEM). Reasons for these differences may include inaccurate boundary conditions and constitutive equations. A system combining accurate experimental methods using image processing, such as the FTMGM and FEM and/or BEM, will be constructed in the future to produce consistent results by reestimating the boundary conditions and constitutive equations as an inverse problem. Highly accurate analysis of experimental data in a whole field will be performed, and internal stress and strain will be accurately analyzed by combining the FEM and the BEM.

References

Image Processing

1. W. K. Pratt, *Digital Image Processing*, Wiley-Interscience, New York, 1978.
2. Densoken (Electrotechnical Laboratory) (Ed.), *SPIDER User's Manual* (in Japanese), Tsukuba, Japan, 1980.
3. A. Rosenfeld and A. C. Kak, *Digital Picture Processing* 2nd ed., Vols. 1 and 2, Academic Press, New York, 1982.
4. Y. Morimoto, *Gazoushori (Image Processing)* (in Japanese), Baifukan, Tokyo, 1984.
5. M. P. Ekstrom (Ed.), *Digital Image Processing Techniques*, Academic Press, Orlando, 1984.
6. R. C. Gonzalez and P. Wintz, *Digital Image Processing* 2nd ed., Addison-Wesley, Reading, MA, 1987.
7. Y. Morimoto (Ed.), *Handbook on Image Processing by Personal Computer*, Vol. 3, *Program of Image Processing on Personal Computer (PIMPOM)* (in Japanese), Kyoritsu-Shuppan, Tokyo, 1988.
8. Y. Morimoto (Ed.), *Handbook on Image Processing by Personal Computer*, Vol. 2, *Image Processing by BASIC* (in Japanese), Kyoritsu-Shuppan, Tokyo, 1989.
9. Y. Seguchi, Image Processing Technique for Research of Strength of Materials and Its Applications, *J. Soc. Mater. Sci. Jpn.* (in Japanese), *35*, no. 389 (1986), 95–105.
10. Y. Morimoto, Image Processing Aided Analyses of Stress, Strain, Deformation and Shape, *Trans. JSME* (in Japanese), *55*, no. 511 (1989), 365–372.

Photoelasticity

11. R. K. Müller and L. R. Saackel, Complete Automatic Analysis of Photoelastic Fringes, *Exp. Mech.*, *19*, no. 7 (1979), 245–251.
12. Y. Seguchi, Y. Tomita, and M. Watanabe, Computer-Aided Fringe-Pattern Analyzer—A Case of Photoelastic Fringe, *Exp. Mech.*, *19*, no. 10 (1979), 362–370.
13. A. S. Voloshin and C. P. Burger, Half-Fringe Photoelasticity: A New Approach to Whole-Field Stress Analysis, *Exp. Mech.*, *23*, no. 3 (1983), 304–313.

14. E. Umezaki, T. Tamaki, and S. Takahashi, Image Analysis of Photoelastic Fringes (1st Report, A Method for Determination of Fringe Orders and Directions of Principal Stress in the Whole Region of a 2-Dimensional Model by Image Processing), *Trans. JSME* (in Japanese), 52, no. 474A (1986), 561–566.

15. E. A. Patterson, Automated Photoelastic Analysis, *Strain*, 24, no. 1 (1988), 15–20.

16. W. C. Wang, C. Y. Chen, and W. H. Chen, Half-Fringe Photoelastic Analysis of Rectangular Plates Subjected to Transverse Loading, *Opt. Eng.*, 22, no. 8 (1988), 636–640.

17. S. Mawatari, M. Takashi, and Y. Toyoda, Whole-Area Photoelastic Analysis by Image Processing on the Principal Stress Direction and Separation of Isochromatics from Isoclinics, *Trans. JSME* (in Japanese), 55, no. 514A (1989), 1423–1428.

Geometric Moiré

18. C. A. Sciammarella and D. L. Sturgion, Digital-Filtering Techniques Applied to the Interpolation of Moiré-Fringes Data, *Exp. Mech.*, 7, no. 11 (1967), 468–475.

19. W. R. J. Funnell, Image Processing Applied to the Interactive Analysis of Interferometric Fringes, *Appl. Opt.*, 20, no. 18 (1981), 3245–3250.

20. K. Gasvik, Moiré Technique by Means of Digital Image Processing, *Appl. Opt.*, 22, no. 23, (1983), 3543–3548.

21. Y. Morimoto, Strain Measurement Technique by Moiré Method, *Kikai no Kenkyu* (in Japanese), 37, no. 10 (1985), 1137–1144.

22. A. S. Voloshin, C. P. Burger, R. E. Rowlands, and T. G. Richard, Fractional Moiré Strain Analysis Using Digital Imaging Techniques, *Exp. Mech.*, 26, no. 3 (1986), 254–258.

23. Y. Morimoto, Y. Seguchi, and T. Higashi, Moiré Analysis of Strain by Fourier Transform, *Trans. JSME* (in Japanese), 54, no. 504A (1988), 1546–1552: *JSME Int. J.* (English trans.), Ser. I, 32, no. 4 (1989), 540–546.

24. Y. Morimoto, Y. Seguchi, and T. Higashi, Application of Moiré Analysis of Strain Using Fourier Transform, *Opt. Eng.*, 27, no. 8 (1988), 650–656.

25. Y. Morimoto, Y. Seguchi, and T. Higashi, Two-Dimensional Moiré Method and Grid Method Using Fourier Transform, *Exp. Mech.*, 29, no. 4 (1989), 399–404.

26. Y. Morimoto, Y. Seguchi, and T. Higashi, Strain Analysis by Mismatch Moiré Method and Grid Method Using Fourier Transform, *Comput. Mech.*, 6, no. 1 (1990), 1–10.

Scanning Moiré Method

27. M. Idesawa, T. Yatagai, and T. Soma, Scanning Moiré Method and Automatic Measurements of 3-D Shapes, *Appl. Opt.*, 16, no. 8 (1977), 2152–2162.

28. T. Ueda, Real Time Moiré Topography Equipment Controlled by Microcomputer, *Trans. Inst. Electro. Commun. Eng. Jpn.* (in Japanese), J61-D, no. 5 (1978), 299–306.

29. Y. Morimoto and T. Shiraishi, Fringe Pattern Analysis by Image Processing Using Personal Computer, *J. Soc. Mater. Sci. Jpn.* (in Japanese) 33, no. 367 (1984), 495–500.

30. Y. Morimoto and T. Hayashi, Deformation Measurement During Powder Compaction by a Scanning Moiré Method, *Exp. Mech.*, 24, no. 2 (1984), 112–116.

31. Y. Morimoto, T. Hayashi, and N. Yamaguchi, Strain Measurement by Scanning-Moiré Method, *Bull. JSME*, 27, no. 233 (1984), 2347–2352.

32. Y. Morimoto and T. Hayashi, Scanning-Moiré Method, in *Photoelasticity* (M. Nishida, Ed.), Springer-Verlag, Tokyo, 1986, 47–52.

33. Y. Morimoto, Strain Measurement by Scanning Moiré Method and Image Processing, *J. JSNDI* (in Japanese) 35, no. 2 (1986), 62–70.

34. Y. Morimoto, Y. Seguchi, and N. Suese, Real-Time Analyzer of Displacement and Strain Distributions Using Moiré Method, *Exp. Tech.*, 15, no. 1, (1991), 36–39.

Shift (Shearing) Method

35. Y. Morimoto, T. Hayashi, and T. Inatani, Strain Measurement Using Moiré by Shifting Model Grating and Its Image Processing, *J. Soc. Mater. Sci. Jpn.* (in Japanese) *35*, no. 396 (1986), 1077–1082.
36. J. J. Wasowski and L. M. Wasowski, Computer-Based Optical Differentiation of Fringe Patterns, *Exp. Tech., 11*, no. 3 (1987), 16–18.

Moiré Interferometry

37. M. E. Tuttle and D. L. Graesser, Compression Creep of Graphite/Epoxy Laminates Monitored Using Moiré Interferometry, *Opt. Lasers Eng., 12*, no. 2 & 3 (1990), 151–171.

Grid Method

38. N. Soneda, S. Yoshimura, and G. Yagawa, Measurement of the Dynamic Strain Field Near a Crack Tip Using Computer Picture Processing, *Trans. JSME* (in Japanese) *52*, no. 476A (1986), 1105–1109.
39. A. Kawasaki, M. Obata, and H. Shimada, Computer-Aided Fine-Grid Method for Local-Strain Measurement Near a Crack-Tip in $2\frac{1}{4}$ Cr-1Mo Steel at 773 K, *Exp. Tech., 11*, no. 8 (1987), 24–27.
40. Y. Morimoto, Y. Seguchi, and T. Inatani, Strain Analysis by Grid Method and Its Automatization Using Image Processing, *J. JSNDI* (in Japanese) *36*, no. 8 (1987), 546–551.
41. Y. Morimoto, Y. Seguchi, and M. Yamashita, Development of a High-Speed Video Camera and Measurement of Strain Rate Distribution Under Propagation of a Stress Wave, *Trans. JSME* (in Japanese) *55*, no. 517 (1989), 2027–2032; *JSME Int. J.* (English trans.), Ser. I, *34*, no. 1 (1991), 37–43.
42. J. S. Sirkis, System Response to Automated Grid Methods, *Opt. Eng., 29*, no. 12 (1990), 1485–1493.

Triangulation

43. M. Takeda and K. Mutoh, Fourier Transform Profilometry for the Automatic Measurement of 3-D Object Shapes, *Appl. Opt., 22*, no. 24 (1983), 3977–3982.
44. M. Halioura, R. S. Krishnamurthy, H. C. Liu, and F. P. Chiang, Automated 360° Profilometry of 3-D Diffuse Objects, *Appl. Opt., 24*, no. 14 (1985), 2193–2196.
45. Y. Shirai, *Three Dimensional Computer Vision*, Springer-Verlag, New York, Berlin, Heidelberg, 1987.
46. G. R. Fernie, G. Griggs, S. Baltlett, and K. Lunau, Shape Sensing for Computer Aided Below-Knee Prosthetic Socket Design, *Prosthet. Orthotics Int., 9* (1985), 12–16.

Holographic Interferometry

47. P. Hariharan, B. F. Oreb, and N. Brown, A Digital Phase-Measurement System for Real-Time Holographic Interferometry, *Opt. Commun., 41*, no. 6 (1982), 393–396.
48. P. Hariharan, B. F. Oreb, and N. Brown, Real-Time Holographic Interferometry: A Microcomputer System for the Measurement of Vector Displacements, *Appl. Opt., 22*, no. 6 (1983), 876–880.
49. T. Hayashi, R. Ugo, and Y. Morimoto, Experimental Observation of Stress Waves Propagating in Laminated Composites, *Exp. Mech., 26*, no. 2 (1986), 169–174.
50. D. R. Matthys, T. D. Dudderar, and J. A. Gilbert, Automated Analysis of Holointerferograms for the Determination of Surface Displacement. *Exp. Mech., 28*, no. 1 (1988), 86–91.
51. M. Kujawinska and D. W. Robinson, Multichannel Phase-Stepped Holographic Interferometry, *Appl. Opt., 27*, no. 2 (1988), 312–319.
52. Y. Morimoto, Y. Seguchi, and H. Nakao, Real-Time Holographic Interferometry by Superimposing Carrier Patterns Using Image Processing Technique, *J. Jpn. Soc. Nondestr. Insp.* (in Japanese), *37*, no. 8 (1988), 630–635.
53. P. K. Rastogi, M. Barillot, and G. H. Kaufmann, Comparative Phase Shifting Holographic Interferometry, *Appl. Opt., 30*, no. 7 (1991), 772–728.

Speckle Interferometry

54. S. Nakadate, T. Yatagai, and H. Saito, Electric Speckle Pattern Interferometry Using Digital Image Processing Techniques, *Appl. Opt., 19*, no. 11 (1980), 1879–1883.

55. K. Creath, Phase-Shifting Speckle Interferometry, *Appl. Opt., 24*, no. 18 (1985), 3053–3058.

56. E. Diez, D. Chambless, W. Swinson, and J. Turner, Image Processing Techniques in Laser-Speckle Photography with Application to Hybrid Stress Analysis, *Exp. Mech., 26*, no. 3 (1986), 230–237.

57. H. D. Navone and G. H. Kaufmann, Two-Dimensional Digital Processing of Speckle Photography Fringes. 3: Accuracy in Angular Determination, *Appl. Opt., 26*, no. 1 (1987), 154–156.

58. I. Yamaguchi, Advances in the Laser Speckle Strain Gauge, *Opt. Eng., 27*, no. 3 (1988), 214–218.

Shearography

59. D. W. Templeton, Computerization of Carrier Fringe Data Acquisition, Reduction, and Display, *Exp. Tech., 11*, no. 11 (1987), 26–30.

General Fringe Analysis

60. S. Toyooka and M. Tominaga, Spatial Fringe Scanning for Optical Phase Measurement, *Opt. Commun., 51*, no. 2 (1984), 68–71.

61. T. Y. Chen and C. E. Taylar, Computerized Fringe Analysis in Photomechanics, *Exp. Mech., 29*, no. 3 (1989), 323–329.

62. M. Kujawinska, Automatic Fringe Pattern Analysis in Optical Methods of Testing, *Warsaw Univ. Technol. Sci. Rep., 138* (1990), 1–80.

Correlation

63. T. C. Chu, W. F. Ranson, M. A. Sutton, and W. H. Peters, Applications of Digital-Image-Correlation Techniques to Experimental Mechanics, *Exp. Mech., 25*, no. 3 (1985), 232–244.

64. T. Arai, M. Obata, K. Nishimura, and H. Shimada, Non-Contact Two-Dimensional Displacement Measurement by Digital Image Correlation, *J. JSNDI* (in Japanese), *37*, no. 8 (1988), 643–648.

65. M. A. Hamed, Object-Motion Measurements Using Pulse-Echo Acoustical Speckle and Two-Dimensional Correlation, *Exp. Mech., 27*, no. 3 (1987), 250–254.

Fracture Analysis

66. S. K. Sasaki, A. Ishii, and Y. Ochi, An Application of Image Processing to Measuring Surface Fatigue Crack Propagation Path (Some Metallurgical Factors Affecting the Scatter of Fatigue Crack Initiation and Propagation), *Trans. of JSME* (in Japanese), *53*, no. 493 (1987), 1780–1785.

67. K. Minoshima, T. Nagasaki and K. Komai, Quantitative Analysis of Facet Sizes of Cleavage Failures in Sharpy Impact Tests Using a Computer Image Processing Technique, *Trans. JSME* (in Japanese), *54*, no. 504, (1988), 1559–1565.

Thermoelasticity

68. M. Shiratori, T. Miyoshi, A. Maruyama, and T. Nakanishi, Real-Time Stress Analysis by a Scanning Infrared Camera, *Trans. JSME* (in Japanese), *52*, no. 478 (1986), 1553–1558.

69. Y. M. Huang, H. H. AbdelMohsen, and R. E. Rowlands, Determination of Individual Stresses Thermoelastically, *Exp. Mech., 30*, no. 1 (1990), 88–94.

Copper Electroplating

70. A. Kato, Stress Management by Cooper Electroplating Aided by a Personal Computer, *Exp. Mech., 27*, no. 2 (1987), 132–137.

Particle Analysis

71. Y. Morimoto, T. Hayashi, and A. Nakanishi, Mechanical Process of Dynamic Powder Compaction, *3rd Conf. Mech. Prop. High Rates of Strain, Oxford, Inst. Phys. Conf. Ser. 70* (1984), 427–434.

Non-Image Processing

72. R. C. Singleton, An Algorithm for Computing the Mixed Radix Fast Fourier Transform, *IEEE Trans. Audio Electroacoust. AU-17*, no. 2 (1969), 93–103.

73. F. P. Chiang, Techniques of Optical Spatial Filtering Applied to the Processing of Moiré-Fringe Patterns, *Exp. Mech., 9*, no. 11 (1969), 523–526.

CHAPTER
22

Statistical Analysis of Experimental Data

James W. Dally
University of Maryland
College Park, Maryland

22-0 Introduction

Experimental measurements inherently exhibit variation. If an experiment is repeated many times using precise instruments, the observation or measurement will vary. This variability, which is fundamental to all measuring systems, is due to two factors. First, the quantity being measured may exhibit significant variation. For example, in a materials study to determine fatigue life at a specified stress level, large differences in the number of cycles to failure are noted when several specimens are tested. This is a variation inherent in the fatigue process, and it is observed in the life measurements. Second, the measuring system, including the transducer, signal-conditioning equipment, readout device, and operator, introduce error in the determination. This error may be systematic or random depending on its source. An instrument operated out of calibration produces a systematic error, whereas reading errors due to interpolation on a chart are random. The accumulation of random errors in the measuring system produces a variation that must be examined in relation to the magnitude of the quantity being measured.

The data obtained from repeated measurements represent an array of readings that does not give an exact result. To extract the maximum information from these data, statistical methods are employed. The first step in the statistical treatment of data is to establish the distribution, often, by graphical representation. Next, this statistical distribution is characterized with a measure of its central value, such as mean, median, or mode. Finally, the spread or dispersion of the distribution is determined in terms of the variance or the standard deviation.

With elementary statistical methods, the experimentalist can reduce a large amount of data by defining the type of distribution, the most-expected value (mean), and the expected variation from the mean value (standard deviation). Summarizing data in this manner is the most meaningful form for application to design problems or for communication to other engineers who need the results of the experiments.

Table 22-1 Listing of Data in Order of Increasing Magnitude for the Fatigue Life N of an Aluminum Alloy

Sample Number	$\log_{10} N$	Sample Number	$\log_{10} N$
1	5.00	21	5.32
2	5.00	22	5.32
3	5.03	23	5.32
4	5.05	24	5.33
5	5.07	25	5.34
6	5.10	26	5.36
7	5.10	27	5.36
8	5.12	28	5.40
9	5.12	29	5.40
10	5.14	30	5.43
11	5.17	31	5.44
12	5.18	32	5.48
13	5.21	33	5.55
14	5.21	34	5.60
15	5.24	35	5.71
16	5.24	36	5.76
17	5.27	37	5.78
18	5.28	38	5.87
19	5.30	39	5.89
20	5.31	40	6.00

22-1 Characterizing Statistical Distributions

Representing Data

Consider an experiment that has been conducted n times, where a single variable, say strength, has been measured. The data obtained represent a sample of size n from an infinite population of all possible measurements that could have been made. The simplest way to present the data is to list the measurements in order of increasing magnitude, as shown in Table 22-1. These data can be rearranged into eight groups to give a frequency distribution, as shown in Table 22-2. The advantage of representing data in a frequency distribution is that the central tendency is more clearly illustrated.

The shape of the distribution function, which represents the $\log_{10}$ of the fatigue life, is indicated by the data groupings presented in Table 22-2. However, a graphical presentation of the group

Table 22-2 Frequency Distribution for Fatigue Life ($\log_{10} N$)

Group Interval	Number of Observations in Group	Relative Frequency	Cumulative Frequency
4.90–5.04	3	0.075	0.075
5.05–5.19	9	0.225	0.300
5.20–5.34	13	0.325	0.625
5.35–5.49	7	0.175	0.800
5.50–5.64	2	0.050	0.850
5.65–5.79	3	0.075	0.925
5.80–5.94	2	0.050	0.975
5.95–6.09	1	0.025	1.000

STATISTICAL ANALYSIS OF EXPERIMENTAL DATA 1033

data, known as a histogram (Fig. 22-1a), shows the distribution with its central tendency and variability more clearly. Superimposed on the histogram is the curve showing the relative frequency of the occurrence of a group of measurements. Note that the points for the relative frequency are plotted at the midpoint of the group interval. Figure 22-1b shows the cumulative-frequency diagram that represents these data. The quantity shown as the cumulative frequency is the number of readings/total measurements that have a value less than a specified value of life. It is evident from Fig. 22-1 that the cumulative frequencies represent the running sum of the relative frequencies and that in plotting Fig. 22-1 the end value for the group interval is used for the abscissa.

Measures of Central Tendency

While the histogram or frequency distributions graphically define the statistical distribution function, it is often preferable to present numerical data to define the characteristics of the distribution. One characteristic is the central tendency of the data. The most commonly employed measure of the central tendency of data is the sample mean $\bar{x}$, which is defined as

$$\bar{x} = \sum_{i=1}^{n} \frac{x_i}{n} \tag{22-1}$$

where x_i = ith value of the quantity
n = total number of measurements

Because of time and costs involved in conducting tests, the number of measurements is limited and the sample mean $\bar{x}$ is only an estimate of the true arithmetic mean of the population. It will be shown later that $\bar{x}$ approaches μ as the number of measurements increases.

The mean value of the data presented in Table 22-1 is $\bar{x} = 5.34$, which indicates an average fatigue life of 221,300 cycles. The median and the mode are also measures of central tendency. The median is the central value in a group of ordered data. Reference to Table 22-1 shows the 20th or the 21st listing as the central value (since the sample size n was an even number). Taking the 21st listing as the central value gives the median as 5.31, indicating a median fatigue life of 204,200 cycles. The mode is the most frequent value that occurs in a data set which corresponds to the peak of the relative frequency curve. Examination of Fig. 22-1 shows that the mode is 5.27, which corresponds to a fatigue life of 186,200 cycles.

It is evident that a typical set of data gives different values for the three measures of central tendency. There are two reasons for this difference. First, the population from which the data set was drawn may not be Gaussian, where the three measures are expected to coincide. Second, even if the population is Gaussian, the number of measurements n is usually small and deviations due to a small sample size are to be expected.

Measures of Dispersion

It is possible that two distributions of data will have the same mean but different dispersions, as shown in the relative frequency diagrams given in Fig. 22-2. There are several different measures of dispersion, including range, mean deviation, variance and standard deviation. Standard deviation, which is the most common, is defined as

$$S_i = \left[\frac{\sum_{i=1}^{n} (x_i - \bar{x})^2}{n - 1} \right]^{1/2} \tag{22-2}$$

Since the sample size n is small, S_x represents an estimate of the true standard deviation σ of

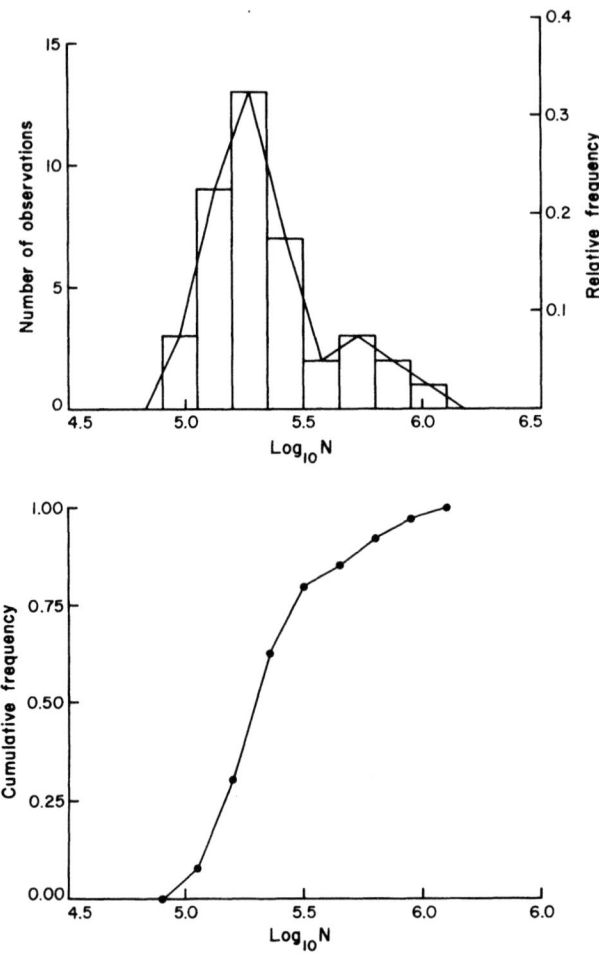

Figure 22-1 (a) Histogram with a superimposed relative frequency diagram; (b) cumulative frequency diagram.

the population. With preprogrammed electronic computers, computation of the mean and standard deviation is easily and quickly performed.

It is often important to compare the dispersion of the data relative to its mean. The comparison is known in statistics as the coefficient of variation cv given by

$$cv = \left(\frac{S_x}{\bar{x}}\right) 100 \tag{22-3}$$

The coefficient of variation represents a normalized parameter that indicates the variability of the data in relation to their mean.

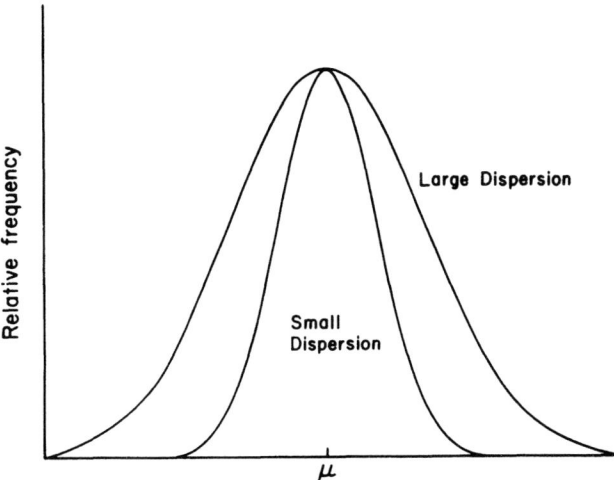

Figure 22-2 Relative frequency diagram showing data with small and large dispersion.

22-2 Statistical Distribution Functions

Gaussian Distribution

As the sample size is increased and the group width is decreased, the relative frequency diagram illustrated in Fig. 22-1 will approach a smooth curve (a theoretical distribution curve) known as a distribution function. Many different distributions have been developed in statistics;[1] however, the distribution that is most frequently used in experimental mechanics is the normal or Gaussian distribution, defined in Fig. 22-3. This distribution is extremely important because it describes random errors in measurements and variations observed in many strength determinations.

The normal distribution is completely defined by two parameters, the mean μ and standard deviation σ, or according to the relation for the relative frequency f given by

$$f(z) = \frac{1}{\sqrt{2\pi}} e^{-(z^2/2)} \tag{22-4}$$

where

$$z = \frac{x - \mu}{\sigma} \tag{22-5}$$

Experimental data with finite sample sizes are analyzed to obtain $\bar{x}$ as an estimate of μ and S, as an estimate of σ. This procedure permits the experimentalist to use data drawn from small samples to represent the entire population.

The primary advantage of the Gaussian distribution function given in Eq. (22-4) is its use in predicting the probability P of observing a measurement within an interval between z_1 and z_2.

---

[1] Binomial, Poisson, exponential, hypergeometric, Weibull, chi-square, Student's t, F, and Gumbel are other useful statistical distribution functions.

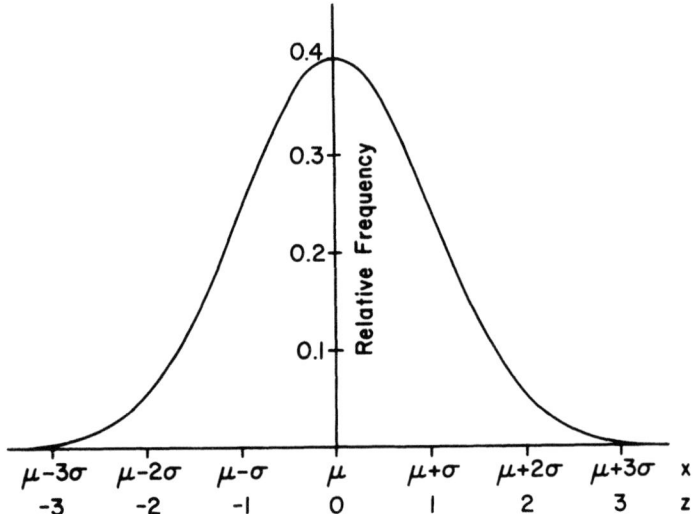

Figure 22-3 The normal or Gaussian distribution function.

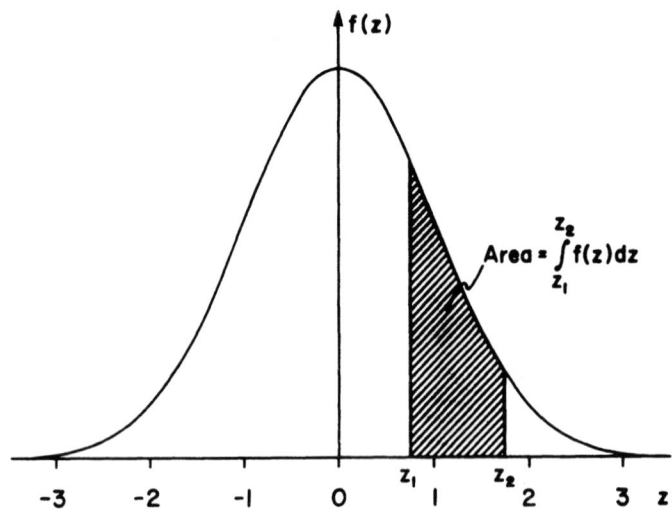

Figure 22-4 Probability of a measurement of x between limits of z_1 and z_2. Note that the total area under the $f(z) - z$ curve is 1.

Figure 22-4 is a graphical representation of the probability: hatched area under the curve gives the probability that a measurement will occur within the interval between z_1 and z_2. This probability is given by

$$P(z_1, z_2) = \int_{z_1}^{z_2} f(z)\, dz = \frac{1}{\sqrt{2\pi}} \int_{z_1}^{z_2} e^{-(z^2/2)}\, dz \qquad (22\text{-}6)$$

The results for the integral shown in Eq. (22-6) are obtained from standard tables found in any text on statistics. Reference to these tables shows that the probability of a measurement or observation between the limits z_1 and z_2 are

z_1	z_2	$P(z_1, z_2)$
$\bar{x} - S_x$	$\bar{x} + S_x$	0.683
$\bar{x} - 2S_x$	$\bar{x} + 2S_x$	0.954
$\bar{x} - 3S_x$	$\bar{x} + 3S_x$	0.997

If it is assumed that the data presented in Table 22-1 are represented by a normal distribution, it may be shown from Eqs. (22-1) and (22-4) that $\bar{x} = 5.34$ and that $S_x = 0.255$. Also, it may be stated that the probability of a fatigue life less than 100,000 cycles (i.e., $\log_{10} N < 5$) is 9.18%. This result is obtained directly from Eq. (22-6) and tables that give $P(-\infty, -1.333) = 0.0918$.

Weibull Distribution

In investigation of the loss of strength of materials due to brittle fracture, crack initiation toughness, or fatigue life, researchers often find that the Weibull distribution provides a more suitable approach to the statistical analysis of the available data. The Weibull distribution function $P(x)$ is defined as

$$\begin{aligned} P(x) &= 1 - \exp -\left(\frac{x - x_0}{b}\right)^m & \text{for } x > x_0 \\ P_x &= 0 & \text{for } x < x_0 \end{aligned} \qquad (22\text{-}7)$$

where x_0, b, and m are the three parameters that define this distribution function. In studies of strength, $P(x)$ is taken as the probability of failure when a stress of x is placed on a specimen. The parameter x_0 is the zero strength, since $P(x) = 0$ for $x < x_0$. The constants b and m are known as the scale parameter and the Weibull slope parameter (modulus), respectively.

Four Weibull distribution curves are presented in Fig. 22-5 for the cases of $m = 2, 5, 10$, and 20, with $x_0 = 3$ and $b = 10$. These curves illustrate the two important features of the Weibull distribution. First, there is a threshold strength x_0, and if the applied stress is less than x_0, the probability of failure is zero. Second, the Weibull distribution curves are not symmetric, and the distortion in the S-shaped curves is controlled by the Weibull slope parameter. The nonsymmetry results in a long, low tail on the classical S curve, which represents the cumulative probability of failure and indicates a very small but finite probability of failure at relatively low stress levels, provided, of course, $x > x_0$. The application of the Weibull distribution to predict failure rates of 1% or less of the population is particularly important in engineering projects that call for reliabilities of 99% or greater.

To utilize the Weibull distribution requires knowledge of the Weibull parameters x_0, b, and m. In experimental investigations, it is necessary to conduct experiments and obtain a relatively large data set for the determination of x_0, b, and m. Consider as an illustration Weibull's own work in statistically characterizing the fiber strength of Indian cotton. In this example an unusually large sample ($n = 3000$) was studied by measuring the load to fracture in grams for each fiber. The

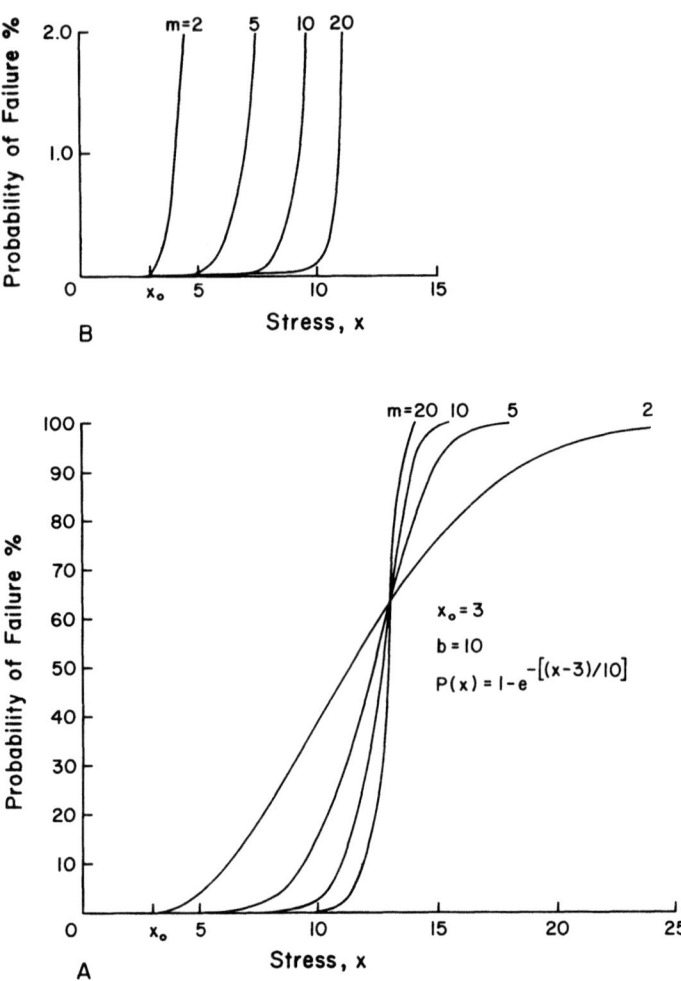

Figure 22-5 (a) The Weibull distribution function; (b) expanded scale $P(x)$ emphasizing the ability of the Weibull distribution to be employed to predict failures at low stress levels.

strength data obtained are placed in sequential order, with the lowest value (corresponding to $k = 1$) first and the largest value (corresponding to $k = 3000$) last. The probability of failure $P(x)$ at a load x is then determined from

$$P = \frac{k}{n + 1} \tag{22-8}$$

where k = order number of the sequenced data
n = total sample size

STATISTICAL ANALYSIS OF EXPERIMENTAL DATA 1039

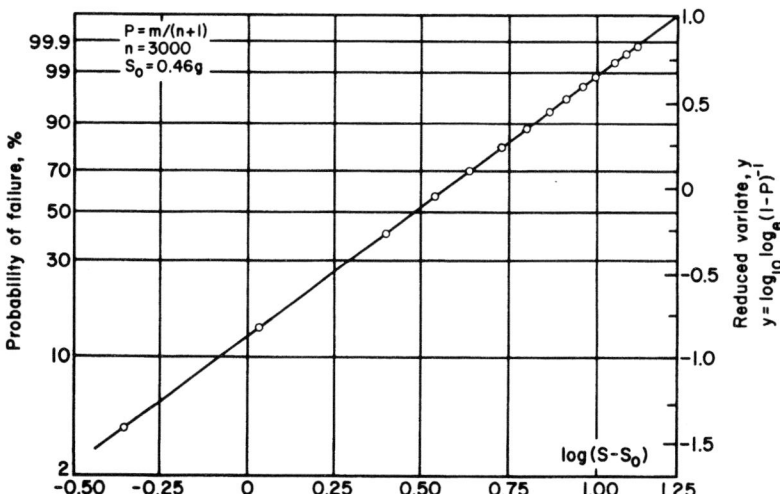

Figure 22-6 Fiber strength of Indian cotton shown as a graph with Weibull's reduced variate. (From Ref. 9.)

At this stage it is possible to prepare a graph of probability of failure $P(x)$ as a function of strength x to obtain a curve similar to that shown in Fig. 22-5. However, to determine the Weibull parameters x_0, b, and m requires additional conditioning of the data. From Eq. (22-7) it is evident that

$$\exp\left(\frac{x - x_0}{b}\right)^m = [1 - P(x)]^{-1} \tag{22-9}$$

Taking the natural log of both sides of Eq. (22-9) yields

$$\left(\frac{x - x_0}{b}\right)^m = \ln[1 - P(x)]^{-1} \tag{22-10}$$

Taking the $\log_{10}$ of both sides of Eq. (22-10) gives a relation for the slope parameter m.

$$m = \frac{\log_{10} \ln[1 - P(x)]^{-1}}{\log_{10}(x - x_0) - \log_{10} b} \tag{22-11}$$

The numerator of Eq. (22-11) is the reduced variate $y = \log_{10} \ln[1 - P(x)]^{-1}$ used for the ordinate in preparing a graph of the conditioned data as indicated in Fig. 22-6. Note that y is a function of P alone, and for this reason both the P and the y scales can be displayed on the ordinates (see Fig. 22-6). The lead term in the denominator of Eq. (22-11) is the reduced variate $x = \log_{10}(x - x_0)$ used for the abscissa in Fig. 22-6.

In the Weibull example, x_0 was adjusted to 0.46 g so that the data would fall on a straight line when plotted against the reduced x and y variates.

The constant b is determined from the condition that

$$\log_{10} b = \log_{10}(x - x_0) \quad \text{when } y = 0 \tag{22-12}$$

Note from Fig. 22-6 that $y = 0$ when $\log_{10}(x - x_0) = 0.54$, which gives $b = 0.54$.

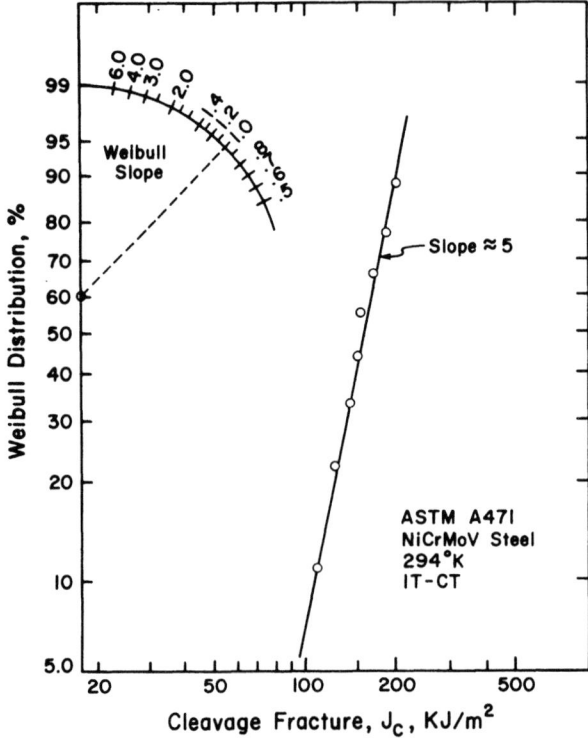

Figure 22-7 Probability of crack initiation as a function of toughness J_c showing a Weibull distribution. (From Ref. 13.)

Finally, m is given by the slope of the straight line when the data are plotted in terms of the reduced variates x and y. In this example problem $m = 1.48$.

Another example of the use of the Weibull distribution is illustrated in Fig. 22-7, where Landes and Shaffer show the toughness parameter J_c for ASTM A 471 steel in terms of a Weibull distribution. In this example Weibull paper was employed, which permits the reduced variates y and x to be displayed directly in terms of the measured quantities P and J_c. The inset, a Weibull slope scale, provides a graphical estimate of m, which in this case was about 5.

22-3 Confidence Intervals for Predictions

Having represented experimental data with a normal distribution by using estimates of the mean $\bar{x}$ and standard deviation S, and having made predictions about the occurrence of measurements, a question arises regarding the confidence that can be placed in either the estimates or the predictions. One cannot be totally confident in the predictions or estimates because of sampling error.

To illustrate sampling error, consider a series of samples each containing n measurements taken from the same population to determine $\bar{x}_1$, $\bar{x}_2$, $\bar{x}_3$, and so on. The distribution of the $\bar{x}$'s is normal,

STATISTICAL ANALYSIS OF EXPERIMENTAL DATA

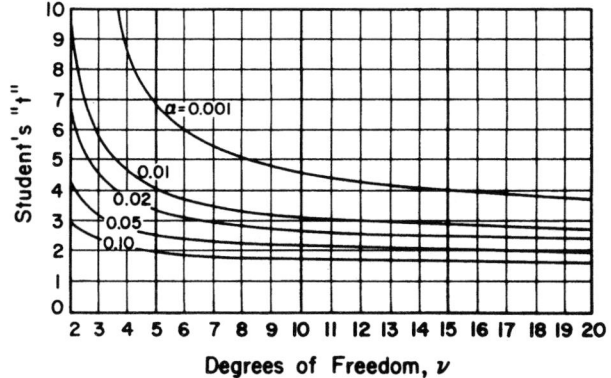

Figure 22-8 Student's t statistic as a function of degrees of freedom ν, with α the probability of exceeding t as a parameter.

with the same mean value as the distribution of the individual measurement x. However, the standard deviation of $\bar{x}$ may be represented by

$$S_{\bar{x}} = \frac{S_x}{\sqrt{n}} \tag{22-13}$$

Knowing the standard deviation of $\bar{x}$ permits confidence intervals to be established for the mean. The interval within which the true population mean μ is located, is given by

$$(\bar{x} - zS_{\bar{x}}) < \mu < (\bar{x} + zS_{\bar{x}}) \tag{22-14}$$

where $\bar{x} - zS_{\bar{x}}$ = lower confidence limit
$\bar{x} + zS_{\bar{x}}$ = upper confidence limit

and where z is defined in Eq. (22-5).

If z is taken as 3, the confidence that μ will be located within the confidence interval defined in Eq. (22-14) is 99.7%. For confidence levels of 99.9, 99.0, 95.0 and 90.0, the corresponding z values are 3.30, 2.57, 1.96, and 1.65, respectively.

When the sample size is very small ($n < 20$), the standard deviation S_x does not provide a reliable estimate of the standard deviation σ of the population, and Eq. (22-14) should not be employed. The bias introduced by the very small samples is corrected for by modifying Eq. (22-14) to read as

$$(\bar{x} - t(\alpha)S_{\bar{x}}) < \mu < (\bar{x} + t(\alpha)S_{\bar{x}}) \tag{22-15}$$

where $t(\alpha)$ is the statistic known as Student's t, which is presented in Fig. 22-8. The parameter α is the level of significance, which is the probability of exceeding a given value of t. The value of t depends on the number of degrees of freedom ν. The degrees of freedom equal the number of independent measurements employed in the determination. In this instance, one degree of freedom was lost in determining x, and thus $\nu = n - 1$.

The term $t(\alpha)S_{\bar{x}}$ in Eq. (22-15) represents the measure from the estimated mean $\bar{x}$ to one or the other of the confidence limits. This term may be used to estimate the sample size required to produce an estimate of the mean $\bar{x}$ with a specified reliability. Noting that half the bandwidth of the confidence interval is $\delta = t(\alpha)S_{\bar{x}}$ and using Eq. (22-13), it is apparent that the sample size is given by

$$n = \left[\frac{t(\alpha)S_x}{\delta}\right]^2 \tag{21-16}$$

Consider the data in Table 22-1, where $\bar{x} = 5.34$ and $S_x = 0.255$. If this estimate of μ is to be accurate to $\pm 3\%$, $\delta = (0.03)(5.34) = 0.1602$. Moreover, if the estimate is to be made with a confidence level of 99.9%, $z = t = 3.30$ for $\alpha = 0.001$. Then a required sample size of $n = 27.6$ is obtained from Eq. (22-16). Thus the sample size of $n = 40$ for the data set presented in Table 22-1 is more than adequate to make an estimate of the mean of log N with a 99.9% confidence of accuracy within $\pm 3\%$.

22-4 Comparison of Means

Since the Student's t distribution compensates for the effect of small-sample bias and converges to the normal distribution in large samples, it is a very useful statistic[2] in engineering application. One important application is to test the difference between two means to determine whether the difference is significant or due to random variation. Consider as an example two sets of yield-strength data where $n_1 = 20$, $\bar{x}_1 = 541$ MPa (78.4 ksi), and $S_{x1} = 51.7$ MPa (6.04 ksi); $n_2 = 25$, $\bar{x}_2 = 563$ MPa (81.6 ksi); and $S_{x2} = 38.3$ MPa (5.56 ksi).

We want to determine whether the difference between the yield strengths is due to a difference in the steel or to a random statistical difference. The first step is to compute the standard deviation in the difference of the means, $S(\bar{x}_1 - \bar{x}_2)$, by

$$S^2_{\bar{x}_2 - \bar{x}_1} = S^2_P\left(\frac{1}{n_1} + \frac{1}{n_2}\right) = \frac{S^2_P(n_1 + n_2)}{n_1 n_2} \tag{22-17}$$

where S^2_P is the pooled variance given by

$$S^2_P = \frac{(n_1 - 1)S^2_{x1} + (n_2 - 1)S^2_{x2}}{n_1 + n_2 - 2} \tag{22-18}$$

Next compute the statistic, t, according to

$$t = \frac{|\bar{x}_2 - \bar{x}_1|}{S_{\bar{x}_2 - \bar{x}_1}} \tag{22-19}$$

and compare this value of t with $t(\alpha)$ to determine whether the difference in the means is significant. The value of $t(\alpha)$ depends on the degrees of freedom $\nu = n_1 + n_2 - 2$ and the level of significance of the comparison. The levels of significance commonly employed are 5% and 1%. The 5% level of significance means that the probability of a random variation being taken for a real difference is only 5%. Comparisons at 1% levels of significance are 99% certain; however, in such a strong test, real differences in the means may be attributed to random error.

Returning now to the example, note that Eq. (22-18) can be used to determine $S^2_P = 230$ MPa (33.37 ksi) and Eq. (22-17) to determine $S^2_{\bar{x}_2 - \bar{x}_1} = 20.7$ MPa (3.00 ksi). The statistic t, by Eq.

---

[2] The statistic $t = (\bar{x} - \mu)\sqrt{n}\, S_x$.

STATISTICAL ANALYSIS OF EXPERIMENTAL DATA

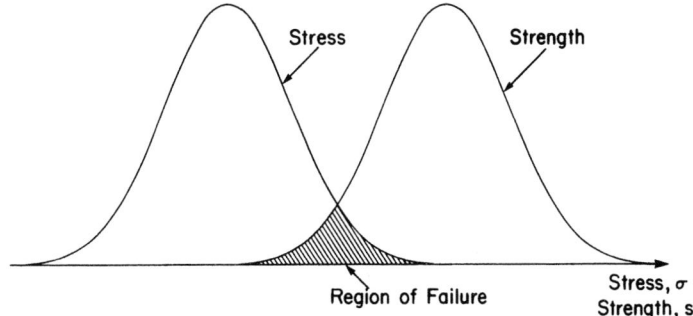

Figure 22-9 Superimposed normal distribution curves for strength and stress showing region of failure.

(22-19), is $t = 1.848$. Referring to Fig. 22-8 or a table for ν, using $t = 43$ and $\alpha = 0.10$, one finds that $t(\alpha) = 1.68$. The value of $\alpha = 0.10$ is used for a 5% level-of-significance test because the comparison is one-sided (i.e., $\bar{x}_2 > \bar{x}_1$). The comparison shows that $t = 1.848 > t(\alpha) = 1.68$; thus it can be concluded with a 95% level of confidence that the yield strength of the second shipment of steel was higher than that of the first shipment.

22-5 Statistical Safety Factor

In experimental mechanics it is often necessary to determine the stresses acting on a component and its strength in order to predict whether failure will occur or if the component is safe. This can be a difficult prediction if both the stress σ_{ij} and the strength S_y are variables, since failure will occur only in the region of overlap of the two distribution functions, as shown in Fig. 22-9.

To determine the probability of failure or, conversely, the reliability, the statistic z_R is computed from

$$z_R = \frac{(\bar{x}_s - \bar{x}_\sigma)}{S_{(s-\sigma)}} \tag{22-20}$$

where

$$S_{(s-\sigma)} = (S_s^2 + S_\sigma^2)^{1/2} \tag{22-21}$$

and the subscripts s and σ refer to strength and stress, respectively.

The reliability associated with the value of z_R determined from Eq. (22-20) may be determined from a table showing the area $A(z_R)$ under a standard normal distribution curve by using

$$R = 0.5 + A(z_R) \tag{22-22}$$

Typical value of R as a function of the statistic z_R are given in Table 22-3.

The reliability determined in this manner incorporates a safety factor of 1. If a safety factor of N is to be specified together with a reliability, Eqs. (22-20) and (22-21) are rewritten to give a modified relation for z_R as

$$Z_R = \frac{\bar{x}_s - N\bar{x}_\sigma}{(S_s^2 + S_\sigma^2)^{1/2}} \tag{22-23}$$

Table 22-3 Reliability R as a Function of the Statistic z_R

R (%)	z_R
50	1
90	1.288
95	1.645
99	2.300
99.9	3.095
99.99	3.700

which can be rearranged to give the safety factor N as

$$N = \frac{1}{\bar{x}_\sigma}(\bar{x}_s - z_R \sqrt{S_s^2 + S_\sigma^2}) \tag{22-24}$$

22-6 Statistical Conditioning of Data

In discussing the normal distribution function, it was observed that measurement error could be described with this distribution. It was further observed that the standard deviation of the estimate mean $S_{\bar{x}}$ could be reduced by increasing the number of readings n insofar as costs of the measurements permit.

Two exceptions should be carefully noted in this statistical approach to improving data. First, systematic error is not a random variable, and statistical procedures will not help to minimize these errors. There is no substitute for using accurately calibrated and properly zeroed instruments in the measurements.

The second exception is the result of a mistake due to an erroneously recorded data point. Often such a data point is out of place in comparison with the other data collected and the experimentalist must decide whether it is due to a mistake (hence to be rejected) or to some unusual but real condition (hence to be retained). A statistical procedure known as Chauvenet's criterion provides a consistent basis for the decision to reject or retain data.

Chauvenet's criterion involves computation of a deviation ratio DR defined as

$$\text{DR} = \frac{x - \bar{x}}{S_x} \tag{22-25}$$

for each data point x. The data point is rejected when $\text{DR} > \text{DR}_0$ and retained otherwise. The value of DR_0 depends on the number of measurements n as shown in Table 22-4.

When a data point is rejected, the mean $\bar{x}$ and standard deviation S_x are recalculated; however, Chauvenet's criterion is applied only once.

22-7 Regression Analysis

Many experiments involve the measurement of one dependent variable, which may depend on one or more independent variables, $x_1, x_2, \ldots, x_k$. Regression analysis provides the statistical

Table 22-4 Chauvenet's Deviation Ratio DR_0

Number of Measurements, n	Deviation Ratio, DR_0
2	1.15
3	1.38
4	1.54
5	1.65
7	1.80
10	1.96
15	2.13
25	2.33
50	2.57
100	2.81
300	3.14
500	3.29
1000	3.48

approach for conditioning the data obtained in experiments in which two or more related quantities are measured.

Linear Regression

Frequently, measurements of two parameters are made in experimental mechanics to empirically relate one dependent variable, say y, to an independent variable x. Because both x and y are variables, the data points scatter on a graph of y versus x as shown in Fig. 22-10, and the placement of a line defining the mathematical relation

$$y = mx + b \tag{22-26}$$

through these scattered data points may be difficult. The statistical procedure used is called the least-squares method, which determines the slope m and the intercept b in Eq. (22-26) in a way designed to minimize the sum of the squared deviations of the data points from the line.

Applying the least-squares method to minimize $\Sigma(y_i - y)^2$ gives the slope m and the intercept b as

$$m = \frac{\Sigma x \, \Sigma y - n \, \Sigma xy}{(\Sigma x)^2 - N \, \Sigma x^2} \tag{22-27}$$

and

$$b = \frac{\Sigma y - m \, \Sigma x}{n} \tag{22-28}$$

which define the best straight line through the data points.

It is possible to determine a coefficient of correlation R^2 to indicate the relative amount of the variation in y which is due to x. For the linear relation between y and x,

$$R^2 = 1 - \frac{n-1}{n-2} \frac{\{y^2\} - m\{xy\}}{\{y^2\}} \tag{22-29}$$

where $\{y^2\} = \Sigma y^2 - (\Sigma y)^2/n$
$\{xy\} = \Sigma xy - (\Sigma x)(\Sigma y)/n$

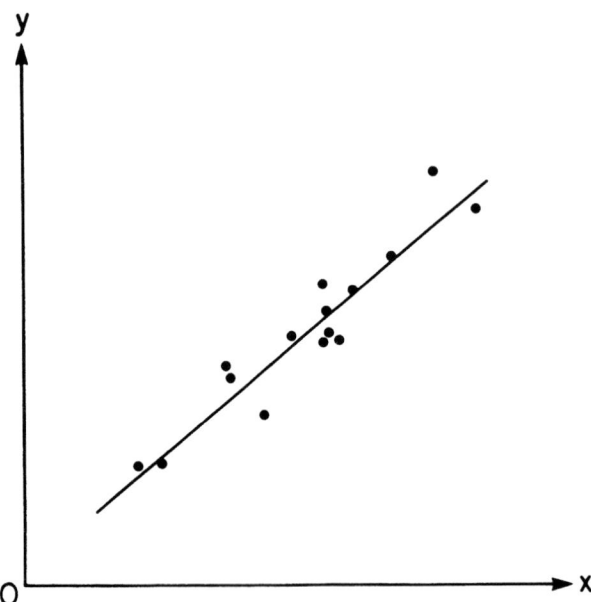

Figure 22-10 Linear regression analysis is used to fit a least-squares line through scattered data points.

When the value of the coefficient of correlation is relatively large (0.8 to 1.0), most of the variation in the dependent variable y has been accounted for in terms of the independent variable x, and the linear regression relation provided in Eq. (22-26) is reliable. However, when the coefficient of correlation is low, Eq. (22-26) is inadequate. Perhaps the relationship between x and y is not linear or other independent variables not accounted for in Eq. (22-26) may be influencing y.

The method of linear regression can be extended to nonlinear relations provided the data can be transformed by using log-log or linear-log representation. When the data are of the form

$$y = ax^m \qquad (22\text{-}30)$$

taking logarithms of both sides gives

$$\log y = m \log x + \log a \qquad (22\text{-}31)$$

Equation (22-31) can be transformed into

$$y' = mx' = b \qquad (22\text{-}32)$$

which is identical to Eq. (22-26) if the substitutions $y' = \log y$, $x' = \log x$, and $b = \log a$ are made.

Similarly, when the relation is exponential with

$$y = ae^{mx} \qquad (22\text{-}33)$$

taking logarithms gives

STATISTICAL ANALYSIS OF EXPERIMENTAL DATA

$$\ln y = mx + \ln a \tag{22-34}$$

Equation (22-34) can be transformed into

$$y' = mx + b \tag{22-35}$$

which is identical to Eq. (22-26) if the substitutions $y' = \ln y$ and $b = \ln a$ are made.

These techniques for applying the least-squares method to nonlinear data minimize the sum of the squared deviations in logarithms of the data rather than the data proper. For at least some cases, which include the finite-life region of the stress-cycle diagram for ferrous materials, this procedure is recommended, since the logarithms of fatigue life for many materials loaded at a constant stress amplitude are normally distributed.

Multivariate Regression

When the dependent variable y is a function of several independent variables $x_1, x_2, \ldots, x_k$, the approach is to write the multivariate regression equation, given by

$$y = a + b_1 x_1 + b_2 x_2 + \cdots + b_k x_k \tag{22-36}$$

where $a, b_1, b_2, \ldots, b_k$ are regression coefficients.

The regression coefficients are determined by using the method of least squares where the $\Sigma(y_i - y)^2$ is minimized. To show this process, note from Eq. (22-36) that

$$\Sigma(y_i - y)^2 = \Sigma(y_i - a - b_1 x_1 - b_2 x_2 - \ldots - b_k x_k)^2 = \Delta^2 \tag{22-37}$$

For $\Sigma(y_i - y)^2$ to be a minimum implies that

$$\frac{\partial \Delta^2}{\partial a} = 2[\Sigma(y_i - a - b_1 x_1 - b_2 x_2 - \cdots - b_k x_k)(-1)] = 0$$

$$\frac{\partial \Delta^2}{\partial b_1} = 2[\Sigma(y_i - a - b_1 x_1 - b_2 x_2 - \cdots - b_k x_k)(-x_1)] = 0$$

.
.
.
$$\tag{22-38}$$

$$\frac{\partial \Delta^2}{\partial b_k} = 2[\Sigma(y_i - a - b_1 x_1 - b_2 x_2 - \cdots - b_k x_k)(-x_k)] = 0$$

These relations lead to a set of $k + 1$ equations, which may be solved for the unknown regression coefficients $a, b_1, b_2, \ldots, b_k$.

$$an + b_1 \sum x_2 + \cdots + b_k \sum x_k = \sum y_i$$
$$a \sum x_1 + b_1 \sum x_1^2 + b_2 \sum x_1 x_2 + \cdots + b_k \sum x_1 x_k = \sum y_i x_i \tag{22-39}$$
$$a \sum x_k + b_1 \sum x_1 x_k + b_2 \sum x_2 x_k + \cdots + b_k \sum x_k^2 = \sum y_i x_k$$

The correlation coefficient R^2 is again used to determine the degree of association between the dependent and independent variables. For multiple regression equations, the correlation coefficient is given by

$$R^2 = 1 - \frac{n-1}{n-k} \frac{\{y^2\} - b_1\{yx_1\} - b_2\{yx_2\} - \cdots - b_k\{yx_k\}}{\{y^2\}} \qquad (22\text{-}40)$$

where $\{yx_k\} = \Sigma yx_k - (\Sigma y)(\Sigma x_k)/n$.

This analysis is for linear, noninteracting, independent variables; however, the analysis can be extended to include cases in which the regression equations would have higher-order and cross-products terms. The nonlinear terms can enter the regression equation in an additive manner and are treated as an extra variable. With well-established computer routines for regression analysis, the set of $(k + 1)$ simultaneous equations given by Eq. (22-39) can be solved quickly and inexpensively and no difficulties are encountered in adding extra terms to account for nonlinearities and interactions.

22-8 Field Applications of Least-Squares Methods

The least-squares method is an important mathematical process used in regression analysis to obtain the regression coefficients. In Ref. 15 Sanford showed that the least-squares methods could be extended to field analysis of data obtained with optical techniques (photoelasticity, moiré, holography, etc.). With these optical methods a fringe order N related to a field quantity such as stress, strain, or displacement can be measured at a large number of points over a field (x, y). The applications require an analytical representation of the field quantities as a function of position (x, y) over the field. Several important problems, including calibration, fracture mechanics, and contact stresses, have analytical solutions in which select terms require experimental data for complete evaluation. Two examples will be described, which introduce both the linear and nonlinear least-squares methods applied over a field (x, y).

Linear Least-Squares Methods

Consider a calibration model in photoelasticity as the first example and write the equation for the fringe order $N(x, y)$ as

$$N(x, y) = \frac{h}{f_\sigma} G(x, y) + E(x, y) \qquad (22\text{-}41)$$

where $G(x, y)$ = analytical representation of the difference of the principal stresses in the calibration model
h = model thickness
f_σ = material fringe value
$E(x, y)$ = residual birefringence

Take a linear distribution for $E(x, y)$ as

$$E(x, y) = Ax + By + C \qquad (22\text{-}42)$$

Now, for any selected position (x_k, y_k) in the field where N_k is determined, write

$$N_k(x_k, y_k) = \frac{h}{f_\sigma} G_k(x_k, y_k) + Ax_k + By_k + C \qquad (22\text{-}43)$$

Note that Eq. (22-43) is linear in terms of the unknowns (h/f_σ), A, B, and C. For M selected data points with $M > 4$, an overdeterministic system of linear equations results from Eq. (22-43). This system of equations can be expressed in matrix form as

STATISTICAL ANALYSIS OF EXPERIMENTAL DATA

$$[N] = [a][w] \tag{22-44}$$

where

$$[N] = \begin{bmatrix} N_1 \\ N_2 \\ \cdot \\ \cdot \\ \cdot \\ N_M \end{bmatrix} \quad [a] = \begin{bmatrix} G_1 & x_1 & y_1 & 1 \\ G_2 & x_2 & y_2 & 1 \\ \cdot & & & \cdot \\ \cdot & & & \cdot \\ \cdot & & & \cdot \\ G_M & x_M & y_M & 1 \end{bmatrix} \quad [w] = \begin{bmatrix} h/f_\sigma \\ A \\ B \\ C \end{bmatrix}$$

The solution set of M equations for the unknowns h/f_σ, A, B, and C can be achieved in a least-squares sense through the use of matrix methods. Note that

$$[a]^T[N] = [c][w] \tag{22-45}$$

where $[c] = [a]^T[a]$, and that

$$[w] = [c]^{-1}[a]^T[N] \tag{22-46}$$

where $[c]^{-1}$ is the inverse of $[c]$. Solution of the matrix $[w]$ gives the column elements, which are the unknowns. This form of solution is easy to accomplish on a minicomputer, which can be programmed in BASIC, and the matrix manipulations can be written in a few lines.

The matrix algebra outlined above is equivalent to minimizing the cumulative error ε, which is

$$\varepsilon = \sum_{k=1}^{M} \left[\frac{h}{f_\sigma} G(x_k, y_k) + Ax_k + By_k + C - N_k \right]^2 \tag{22-47}$$

The matrix operations apply the least-squares criteria, which require that

$$\frac{\partial \varepsilon}{\partial (h/f_\sigma)} = \frac{\partial \varepsilon}{\partial A} = \frac{\partial \varepsilon}{\partial B} = \frac{\partial \varepsilon}{\partial C} = 0 \tag{22-48}$$

The advantage of this statistical approach to the calibration of model materials in optical arrangements is the use of full-field data to reduce errors due to discrepancies in either the model materials or the optical systems.

Nonlinear Least-Squares Methods

In the preceding section a linear least-squares method provided a direct approach to improving the accuracy of calibration with a single-step computation of an overdeterministic set of linear equations. In other photoelastic experiments involving either the determination of unknowns arising in stresses near a crack tip or contact stresses near a concentrated load, the governing equations are nonlinear in terms of the unknown quantities. In these cases the procedure to be followed involves linearizing the governing equations, applying the least-squares criteria to the linearized equations, and iterating to converge to an accurate solution for the unknowns.

To illustrate this statistical approach, consider a photoelastic experiment that yields an isochromatic fringe pattern near the tip of a crack in a specimen subjected to mixed-mode loading. In this example, there are three unknowns, K_I, K_{II}, and σ_{0x}, which are related to the experimentally determined fringe orders N_k at positions (r_k, θ_k). The governing equation for this mixed-mode fracture problem is

$$\left(\frac{Nf_\sigma}{h}\right)^2 = \frac{1}{2\pi r}[(K_I \sin\theta + 2K_{II}\cos\theta)^2 + (K_{II}\sin\theta)^2]$$

$$+ \frac{2\sigma_{0x}}{\sqrt{2\pi r}} \sin\frac{\theta}{2}[K_I \sin\theta(1 + 2\cos\theta) \quad (22\text{-}49)$$

$$+ K_{II}(1 + 2\cos^2\theta + \cos\theta)] + \sigma_{0x}^2$$

To solve Eq. (22-49) in an overdeterministic sense, form the function $f(K_I, K_{II}, \sigma_{0x})$ as

$$f_k(K_I, K_{II}, \sigma_{0x}) = \frac{1}{2\pi r}[(K_I \sin\theta_k + 2K_{II}\cos\theta_k)^2 + (K_{II}\sin\theta_k)^2]$$

$$+ \frac{2\sigma_{0x}}{\sqrt{2\pi r_k}} \sin\frac{\theta_k}{2} [K_I \sin\theta_k(1 + 2\cos\theta_k)$$

$$+ K_{II}(1 + 2\cos^2\theta_k + \cos\theta_k)] \quad (22\text{-}50)$$

$$+ \sigma_{0x}^2 - \left(\frac{N_k f_\sigma}{h}\right)^2 = 0$$

where $k = 1, 2,$ or 3 and r_k, θ_k are coordinates defining a point on an isochromatic fringe of order N_k.

Next, perform a Taylor-series expansion of Eq. (22-50) to obtain

$$(f_k)_{i+1} = (f_k)_i + \left(\frac{\partial f_k}{\partial K_I}\right)_i \Delta K_I + \left(\frac{\partial f_k}{\partial K_{II}}\right)_i \Delta K_{II} + \left(\frac{\partial f_k}{\partial \sigma_{0x}}\right)_i \Delta \sigma_{0x} \quad (22\text{-}51)$$

where i refers to the ith iteration step and ΔK_I, ΔK_{II}, and $\Delta \sigma_{0x}$ are corrections to the earlier estimate of K_I, K_{II}, and σ_{0x}. It is evident from Eq. (22-50) that corrections should be made given $f_k(K_I, K_{II}, \sigma_{0x}) = 0$. This fact leads to the iterative equation

$$\left(\frac{\partial f_k}{\partial K_I}\right)_i \Delta K_I + \left(\frac{\partial f_k}{\partial K_{II}}\right)_i \Delta K_{II} + \left(\frac{\partial f_k}{\partial \sigma_0}\right)_i \Delta \sigma_{0x} = -(f_k)_i \quad (22\text{-}52)$$

In matrix form the set of m equations represented by Eq. (22-52) can be written as

$$[\mathbf{f}] = [\mathbf{a}][\mathbf{\Delta K}]$$

where the matrices are defined as

$$[\mathbf{f}] = \begin{bmatrix} f_1 \\ \cdot \\ \cdot \\ \cdot \\ f_m \end{bmatrix} \quad [\mathbf{a}] = -\begin{bmatrix} \frac{\partial f_1}{\partial K_I} & \frac{\partial f_1}{\partial K_{II}} & \frac{\partial f_1}{\partial \sigma_{0x}} \\ \cdot & \cdot & \cdot \\ \frac{\partial f_m}{\partial K_I} & \frac{\partial f_m}{\partial K_{II}} & \frac{\partial f_m}{\partial \sigma_{0x}} \end{bmatrix} \quad [\mathbf{\Delta K}] = \begin{bmatrix} \Delta K_I \\ \Delta K_{II} \\ \Delta \sigma_{0x} \end{bmatrix} \quad (22\text{-}53)$$

The least-squares minimization process is accomplished by multiplying from the left both sides of Eq. (22-53) by the transpose of matrix $[\mathbf{a}]$, to give

$$[\mathbf{a}]^T[\mathbf{f}] = [\mathbf{a}]^T[\mathbf{a}][\mathbf{\Delta K}] \quad (22\text{-}54)$$

or

$$[d] = [c][\Delta K]$$

where $[d] = [a]^T[f]$
$[c] = [a^T][a]$

Finally, the correction terms are given by

$$[\Delta K] = [c]^{-1}[d] \tag{22-55}$$

The solution of Eq. (22-55) gives ΔK_I, ΔK_{II}, and $\Delta \sigma_{0x}$, which are used to correct initial estimates of K_I, ΔK_{II}, and σ_{0x}, and obtain a better fit of the function f to m data points.

The procedure is executed on a minicomputer programmed in BASIC. One starts by assuming initial values for K_I, ΔK_{II}, and σ_{0x}. Then the elements of the matrices [f] and [a] are computed for each of the m data points. The correction matrix $[\Delta K]$ is computed from Eq. (22-55), and finally the estimates of the unknowns are corrected by noting that

$$(K_I)_{i+1} = (K_I)_i + \Delta K_I$$

$$(K_{II})_{i+1} = (K_{II})_i + \Delta K_{II} \tag{22-56}$$

$$(\sigma_{0x})_{i+1} = (\sigma_{0x})_i + \Delta \sigma_{0x}$$

This is repeated until each element in $[\Delta K]$ becomes acceptably small. Since convergence is quite rapid, the number of iterations required for accurate estimates of the unknowns is usually small.

22-9 Chi-Square Analysis

The chi-square test is used in statistics to verify the use of a specific distribution function to represent the population from which a set of data has been obtained. The chi-square statistic χ^2 is defined as

$$\chi^2 = \sum_{i=1}^{k} \left[\frac{(n_o - n_e)^2}{n_e} \right]_i \tag{22-57}$$

where n_o = actual number of observations in the ith group interval
n_e = expected number of observations in the ith group interval based on the specified distribution
k = total number of group intervals

The value of χ^2 is computed to determine how well the data fit the assumed statistical distribution. If $\chi^2 = 0$, the match is perfect. Values of $\chi^2 > 0$ indicate the possibility that the data are not represented by the specified distribution. The probability that the value of χ^2 occurred because of random variation is given in Fig. 22-11. Note that the degree of freedom is given by

$$\nu = n - k \tag{22-58}$$

where n = number of group intervals
k = number of imposed conditions on the distribution

As an example of the application of the χ^2 test, consider the data for fatigue life presented in Table 22-2. The test is to determine whether this population can be represented by a Gaussian distribution with $\bar{x} = 5.34$ and $S_x = 0.255$, where $x = \log_{10} N$. The χ^2 computations are summarized in Table 22-5, where the observed number of specimens within 10 different group

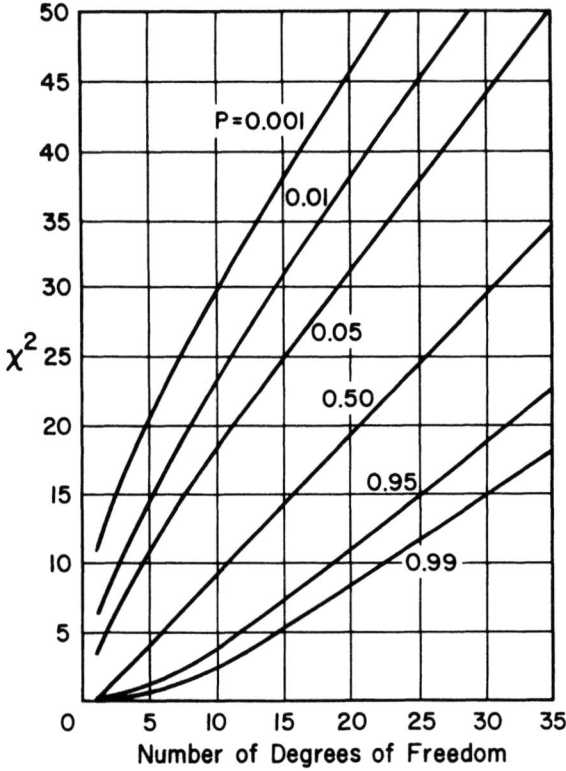

Figure 22-11 Expected values of χ^2 as a function of the number of degrees of freedom with probability of valid representation as a parameter.

Table 22-5 χ^2 Computation for Grouped Fatigue Life ($LOG_{10}N$) Data

Group Interval	n_o	n_e	$(n_o - n_e)^2/n_e$
0–4.89	0	1.55	1.55
4.90–5.04	3	3.24	0.02
5.50–5.19	9	6.34	1.17
5.20–5.34	13	8.87	1.92
5.35–5.49	7	8.87	0.39
5.50–5.64	2	6.34	2.97
5.65–5.79	3	3.24	0.02
5.80–5.94	2	1.14	0.65
5.95–6.09	1	0.28	1.85
6.10–∞	0	0.06	0.06
			$\chi^2 = 10.6$

STATISTICAL ANALYSIS OF EXPERIMENTAL DATA

Table 22-6 Observed (O) and Expected (E) Inspection Results

	Number Observed	Number Expected	$(O - E)^2/E$
Passed	910	900	0.111
Failed	90	100	1.000
			$1.111 = \chi^2$

intervals are listed. The expected number of observations were computed by using the areas under the Gaussian distribution function associated with the widths of the group interval. The values of χ^2 for the 10 different group intervals were added to give the total $\chi^2 = 10.6$ for the entire set of 40 observations.

The number of degrees of freedom was $10 - 2 = 8$, since only two constraints are imposed on the Gaussian distribution function (i.e., $\bar{x}$ and S_x). Referring to Fig. 22-11, one can state with a 23% probability level that the data can be represented by a Gaussian distribution. This result indicates that there is not a strong reason to reject the Gaussian distribution to represent the population. If χ^2 had been larger and the probability associated with the higher χ^2 value between 5 and 10%, the use of the Gaussian distribution would be questionable. For probabilities less than 5%, the Gaussian distribution would be rejected. Note that for probabilities greater than 90%, one should question the authenticity of the data instead of the goodness of the fit.

The χ^2 statistic also can be used for contingency testing, in which the sample is classified according to one of two criteria—pass or fail. Consider, as an example, an inspection procedure with a particular type of strain gage where 10% of the gages are rejected due to etching imperfections in the grids. In an effort to reduce this rejection rate, the manufacturer has introduced several new clean-room techniques, which are expected to improve the quality of the grid. In the first lot of 1000 gages, the failure rate was reduced by 9%. Is this reduced failure rate due to chance variation, or have the new clean-room techniques improved the manufacturing process? The computation of χ^2 is illustrated in Table 22-6.

Examining the results of Fig. 22-11 with $\nu = 1$ gives a probability in which χ^2 exceeds 1.111 of about 31%; thus the test shows a strong indication that the clean-room improvements have reduced rejection rates. Stronger statistical statements could be made after the testing of the second lot of strain gages has been completed. At this point, 2000 gages would have been inspected with $\chi^2 = 2.222$, and the probability of χ^2 exceeding this value due to random variation is only 15%.

22-10 Summary

These notes provide a summary of some of the more useful methods of statistics that can be applied to experimental mechanics. The treatment of these methods has been limited by constraints on the length of this chapter, and the developments are often brief and cursory. The reader is urged to explore the subject matter in much more detail by reading several of the books and reports listed below.

Readings on Statistics

Introductory Books
1. D. Blackwell, *Basic Statistics*, McGraw-Hill, New York, 1969.

2. G. W. Snedecor and W. G. Cochran, *Statistical Methods*, 6th ed., Iowa State University Press, Ames, 1967.

3. P. W. Zehna, *Introductory Statistics*, Prindle, Weber & Schmidt, Boston, 1974.

4. Y. Chou, *Probability and Statistics for Decision Making*, Holt, Rinehart and Winston, New York, 1972.

Books for Engineering, Production, and Research

5. R. M. Bethea, B. S. Duran, and T. L. Bouillion, *Statistical Methods for Engineers and Scientists*, Dekker, New York, 1975.

6. O. L. Davies and P. L. Goldsmith (Eds.), *Statistical Methods in Research and Production*, Hafner, New York, 1972.

Books with More Direct Application of Statistics to Experimental Mechanics

7. H. D. Young, *Statistical Treatment of Experimental Data*, McGraw-Hill, New York, 1962.

8. G. M. Bragg, *Principles of Experimentation and Measurement*, Prentice-Hall, Englewood Cliffs, NJ, 1974.

9. W. Weibull, *Fatigue Testing and Analysis of Results*, Pergamon Press, Elmsford, NY, 1961.

10. J. P. Holman, *Experimental Methods for Engineers*, 2nd ed., McGraw-Hill, New York, 1966.

11. A. J. Durelli, E. A. Phillips, and C. H. Tsao, *Introduction to the Theoretical and Experimental Analysis of Stress and Strain*, McGraw-Hill, New York, 1958.

12. ASTM, *A Tentative Guide for Fatigue Testing and the Statistical Analysis of Fatigue Data*, STP-91-A, American Society for Testing and Materials, Philadelphia, 1958.

Selected Articles

13. J. D. Landes and D. H. Shaffer, Statistical Characterization of Fracture in the Transition Region, Scientific Paper 79-ID3-JINTF-P4, Westinghouse R&D Center, Pittsburgh, PA, 1979.

14. W. R. Andrews, Small Specimen Brittle Fracture Toughness Testing, Report DF 79MPL250, Materials and Processing Laboratory, General Electric Co., Schenectady, NY, 1979.

15. R. J. Sanford, Application of the Least Squares Method to Photoelastic Analysis, *Exp. Mech., 20* (1980), 192–197.

16. R. J. Sanford and J. W. Dally, A General Method for Determining Mixed-Mode Stress Intensity Factors from Isochromatic Fringe Patterns, *Eng. Fract. Mech., 11* (1979), 621–633.

Author Index

Abbis, J. B., 362
AbelMoshen, H. H., 1029
Aben, H., 261, 265, 275
Adams, D., 900
Adams, F. D., 386
Adams, L. H., 262
Adams, P. H., 577
Adelmohsen, M. H., 598
Adkins, J. E., 36
Agarwal, B. D., 261, 265, 830, 899
Ahimaz, F. J., 901
Ahmad, S., 37
Ahmed, M., 555
Aiken, R., 599
Ajovalasit, A., 801, 825
Aleksandrov, E. B., 404
Alferness, R., 364
Alfirevich, I., 901
Allemang, R. J., 635, 745, 747, 749
Allison, I. M., 248, 262
Almen, J. O., 823
Alpsten, G. A., 823
Anastasi, R., 578
Anderson, G. P., 826
Andonian, A. T., 966
Andrews, D. R., 292, 296
Andrews, W. R., 782, 1054
Ansevin, R. W., 827
Appl, G. J., 824
Aprahamian, R., 366, 404

Arai, T., 1029
Arcan, M., 261, 266
Archbold, E., 366, 404, 405
Argon, A., 968
Argyris, J. H., 13, 35
Aritome, H., 364
Asher, G. W., 745
Askegaard, V., 826
Asundi, A., 364
Atluri, S., 1, 34, 35, 36, 782, 917, 964
Auld, B. A., 555
Awerbuch, J., 903
Ayre, R. S., 745

Babcock, S. G., 578
Bach, F., 826
Baek, T. K., 263
Baker, W. E., 634
Bakis, C. E., 596, 598
Balas, J., 757, 782
Baltelett, S., 1028
Banerjee, P. K., 37
Bar-Cohen, Y., 77, 903
Baracat, W. A., 323, 360, 362
Barenblatt, G. I., 915, 963
Barillot, M., 1028
Barishpolsky, B. M., 754, 781
Barker, D. M., 261, 386, 405
Barnett, D. M., 826, 964
Barrett, C. S., 823

Bartolotta, P. A., 577
Barton, J. R., 555
Basehore, M. L., 360, 362
Bathe, K. J., 36, 37
Bathgate, R. G., 796, 824
Bayer, M., 578
Bazergui, A., 828
Beaney, E. M., 825
Beatty, J., 578
Beatty, M. F., 76
Begley, J. A., 963, 968
Beinert, J., 474, 475, 783, 934, 965, 966
Beissner, R. E., 555, 828
Belgen, M. H., 582, 597
Benson, R. W., 812, 825, 826, 964
Benthem, J. P., 963
Bergman, M., 748
Bergman, R. H., 827
Berman, A., 749
Berry, J. P., 906, 962
Bert, C. W., 798, 824
Bethea, R. M., 1054
Betser, A. A., 967
Bever, M. B., 582, 597
Bhat, G., 578
Bilby, F. A., 915, 963
Billaud, J. F., 747
Billing, B., 362
Billington, E. W., 37
Biner, S. B., 968
Bingham, E. C., 37
Biot, M. A., 36, 597
Bishop, R. E. D., 746
Bishop, S. M., 362
Bizon, P. T., 573, 578
Black, P. H., 823
Blackwell, D., 1054
Blake, J. N., 77
Bland, D. R., 36
Blinka, J., 814, 826, 964
Blomquist, D. S., 163
Bloss, R. L., 566, 577
Blount, E. I., 823
Boag, J. M., 823
Bobardus, K. O., 967
Böhme, W., 475, 967
Boiten, R. G., 791, 824
Bonch-Bruvich, A. M., 365, 372, 404
Boone, P. M., 295
Boresi, A. P., 39
Born, M., 475
Bossaert, W., 295
Bouillion, T. L., 1054

Boussinesq, J., 35
Bover, M. B., 582, 597
Bowie, G. E., 786, 801, 825
Bowles, D. F., 360, 362
Bowles, K. J., 903
Boyce, B. R., 597
Boyle, J. T., 590, 598
Brace, W. F., 906, 962
Bradley, W. B., 953, 965, 967
Bragg, G. M., 1054
Brasfield, R. G., 577
Bream, R. G., 598
Brenden, B. B., 825
Brew, D., 578
Bridgman, P. W., 634
Brill, W. A., 264
Brinson, H. F., 265, 889, 951
Briones, A. R., 404
Brittain, J. O., 577
Broadway, J. A., 366, 393, 405
Broch, J. T., 162
Broek, D., 782, 913, 962, 963
Broek, S., 475
Brooks, E. W., 360
Brooks, R. E., 365, 404
Broutman, L. J., 830, 899, 903
Brown, C. A., 782
Brown, D. L., 635, 745, 746, 747, 749
Brown, G. M., 404
Brown, N., 1028
Brull, M. S., 266
Bryngdahl, O., 365, 404
Bucaro, J. A., 68, 72, 76, 77
Buckens, F., 826
Budiansky, B., 917, 964
Burch, J. M., 295, 404
Burger, C. P., 165, 248, 261, 263, 264, 887, 1026, 1027
Burkhardt, G. L., 902
Burton, D., 555
Busbey, R. T., 964
Bush, A. J., 800, 824
Butler, C. D., 68, 76
Butters, J. N., 405
Bynum, J. E., 824
Byron, R., 902

Calfo, F. D., 573, 578
Callabresi, M. L., 578
Camponeschi, E. T., 903
Campus, F., 787, 824
Canistraro, H., 578
Capecchi, D., 749

AUTHOR INDEX

Carbon, G. D., 747
Carome, E., 76
Cate, W. T., 824
Cedrone, N. P., 826
Cernosek, J., 233, 236, 262
Chambless, D., 1029
Chambless, D. A., 366, 393, 405
Chamis, C. C., 900, 901, 902
Chan, C. F., 967
Chan, C. T., 571, 578
Chan, W. K., 599
Chandrashekhara, K., 781
Chang, B. J., 364
Chang, F., 903
Chang, F. H., 555
Change, N. D., 163
Chapman, J. R., 826
Charles, J. A., 903
Chaturvedi, S. K., 261, 262, 265
Chen, C. Y., 1027
Chen, T. L., 295
Chen, T. Y. 1029
Chen, W. H., 1027
Cheng, Y. F., 264
Chermahinmi, R. G., 265
Chevalier, Y., 749
Chewning, S. W., 76
Chiang, F. P., 295, 296, 386, 387, 405, 578, 966, 1007, 1028, 1029
Chick, B. B., 812, 826
Childs, W. H. J., 262
Choi, H. Y., 634
Chona, R., 964
Chou, Y., 1054
Chowley, J., 749
Christensen, R. M., 536, 830, 899
Christner, B. K., 554
Chu, T. C., 390, 366, 405, 1029
Clark, A. B. J., 263
Clark, A. V., 909, 922, 962
Clark, G. A., 964
Claus, R. O., 76, 77
Clifton, R. J., 917, 964
Cloud, G., 386, 405, 573, 578
Cloud, G. L., 966
Cobb, R. E., 747
Cochran, W. G., 1054
Cohen, J. B., 822, 827
Coker, E. G., 262, 781
Cole, B. W., 901, 902
Cole, J. H., 68, 76
Collatz, L., 36
Collins, R., 782

Compton, K. T., 582, 597
Constantinescu, D., 966
Conway, J. C., Jr., 966
Cook, R. D., 37
Cooper, G. P., 599
Cordiano, H. V., 824
Corke, M., 77
Cormeau, I. C., 35
Couchman, J. C., 903
Cowles, B. A., 571, 578
Cox, L. J., 599
Cox, P. M., 599
Craig, R. R., Jr., 745
Crane, R. L., 903
Creath, K., 1029
Crecraft, D. I., 826
Crites, N., 577
Crosley, P. B., 476, 965
Cross, R. W., 578
Crosswy, F. L., 163
Crowley, J., 747
Cruse, T. A., 36, 871, 902
Cullity, B. D., 827
Culshaw, B., 76
Cummings, W. M., 586, 588, 589, 597, 598
Curran, D. R., 826
Czarnek, R., 360, 363

Dally, J. W., 39, 76, 117, 163, 262, 263, 264, 265, 294, 295, 361, 525, 781, 782, 965, 967, 900, 901
Dandliker, R., 490
Dandridge, G. H., 76, 77
Daniel, I. M., 265, 295, 362, 830, 835, 900, 901, 902
Dankin, J., 76, 77
Dantu, M., 295
Dardy, H. D., 76
Davies, O. L., 1054
Day, M. F., 577
de Sailly, R., 949, 966
de Veubeke, B. F., 664, 745
de Wolf, T. N., 525
Deblauewe, F., 746
Dechaene, R., 295
Deeds, W. E., 578
del Rio, C. J., 295
Delameter, W. R., 825
Deleu, E. N., 295
deLorenzi, H. G., 782
Denys, R. M., 295
DePaula, R., 76
Dhar, S., 966
Dhir, R. K., 825

Diez, E., 1029
DiMarscio, P. S., 827, 828
Dimonds, R. A., 598
Dittbenner, G. R., 76
Dobson, M. W., 963
Dodd, C. V., 578
Dodge, F. T., 634
Doherty, J. E., 527
Dölle, H., 827
Doner, D. R., 900
Dowling, N. E., 967
Downs, B., 750
Drenner, D. C., 561, 577
Driscoll, W. G., 261
Drucker, D. C., 8, 35, 262, 782
Druez, J., 828
Drzik, M., 782
Dudderar, T. D., 932, 965, 1028
Duffy, D. E., 386
Duffy, J., 965
Dugdale, D. A., 915, 916, 950, 951, 963
Duggan, M. F., 571, 578
Dulaney, E. N., 906, 962
Duncan, J. L., 598
Duncan, W. J., 634
Dundurs, J., 634
Dunn, S., 598
Dunphy, J. R., 77
Duran, B. S., 1054
Durelli, A. J., 262, 294, 295, 361, 362, 491, 525, 781, 900, 1054
Duvall, W. I., 825
Dyson, B. F., 577

Edelman, S., 163
Eisenmann, J. R., 902
Ekstrom, M. P., 1026
El-Hout, J. N., 264
Elber, W., 967
Elliott, K. B., 747
Ellis, G., 496, 525
Elsley, R. K., 903
Emery, A. F., 782, 783, 962
Englund, D. R., 565, 577
Engstrom, W. L., 963
Enke, N. F., 583, 597, 599
Ennos, A. E., 366, 404, 406
Enoch, J. M., 782
Epstein, J. S., 363, 963, 966
Erf, R. K., 490
Eringen, A. C., 34, 36, 40
Erisman, E. R., 262
Ernst, H., 964

Eshelby, J. D., 17, 35, 963
Evensen, D. A., 404
Ewins, D. J., 162

Fairhurst, C., 825
Feix, M., 745
Feng, Z., 591, 598
Fernie, G. R., 1028
Fidler, R., 577
Fillod, R., 746
Filon, L. N. G., 217, 262
Finnie, I., 35
Fisher, D. M., 964
Fisher, J. L., 555
Fitzgerald, J. E., 36
Flaman, M. T., 800, 802, 825
Flanagan, J. H., 262
Flannelly, W. G., 749
Fleischman, T. S., 966
Flugge, S., 34
Folias, E. S., 963
Forgacs, R. L., 812, 826
Forlifer, W. R., 746
Formenti, D., 749
Forno, C., 295
Foster, J. H., 561, 576
Fourney, M. E., 386
Fouth, D. A., 748
Foye, R. L., 904
Francis, P. H., 902
Freeman, S. M., 903
Freire, J. L. F., 264
French, D. N., 827
Freund, L. B., 474, 475, 917, 936, 964, 966, 967
Freynik, H. S., 76
Frick, R. P., 821, 827
Friedenwald, J. S., 782
Frocht, M. M., 217, 264, 781
Fuchs, E. A., 579
Fukuoka, H., 827
Fukuzono, K., 748
Fumagalli, E., 634
Fung, Y. C., 36
Funnell, W. R. J., 1027

Gaal, P. S., 572, 578
Gabor, D., 365, 404
Gadd, C. W., 828
Gallagher, R. H., 36
Gardner, C. G., 555
Gaspar, B. C., 598
Gasvik, K., 1027
Gaukroger, D. R., 745

Gazzara, C. P., 827
Gdoutos, E. E., 265, 965
Gehlen, P. C., 782
Gerberich, W. W., 925, 964, 965
Gersch, W., 670, 748
Gerstle, F. P., Jr., 902
Giallorenzi, T. J., 76
Gilbert, J. A., 1028
Gillette, O. L., 577
Gladwell, G. M. L., 746
Glatz, J., 555
Gloster, J., 782
Glover, R. J., 823
Goldenblat, I., 900
Goldsmith, P. L., 1054
Gomide, H. A., 264
Gonzalez, R. C., 1026
Gottenburg, W. G., 366, 404
Graesser, D. L., 1028
Graham, L. J., 903
Grannell, S. N., 36
Grant, R. M., 404
Gravitz, S. I., 749
Greaves, L. J., 582, 597
Green, A. E., 36
Green, R. L., 260, 265
Greszczuk, L. B., 900
Griffith, A. A., 905, 910, 961
Griggs, G., 1028
Grimes, G. C., 901
Groom, J. J., 828
Gross, B., 966
Gruber, G. J., 555
Gryagoridis, J., 264
Gudas, J. P., 963
Guess, T. R., 902
Guild, J., 297, 314, 361
Gunther, M., 76
Gupta, B., 824
Gurtman, G. A., 827
Gutta, R. F., 598, 597
Guyan, R. J., 749

Hachimine, K., 404
Hahn, G. T., 499, 782
Hahn, T. H., 830, 889
Haigren, J. A., 823
Haines, K. A., 365, 404
Hales, R., 571, 578
Halioua, M., 276, 292
Halioura, M., 1028
Hall, W. J., 962
Halpin, J. C., 900

Hamed, M. A., 1029
Hamilton, R., 590, 598
Hans, S., 161, 162
Harding, K. G., 296
Hardrath, W. T., 966
Hardwood, N., 597, 598
Hariharan, P., 970, 1028
Harris, F. C., 262, 296
Harris, H. G., 634
Harting, D. R., 577
Hartman, G. A., 577
Harwood, N., 586, 588, 589, 598
Hashin, Z., 900
Haslach, W., 76
Hasumura, T., 265
Hawkes, I., 825
Hawkins, F. J., 745
Hayashi, T., 295
Hayashi, T., 1027, 1028
Hayes, D. J., 825
Hayes, J. K., 564, 577
Heflinger, L. O., 404
Heller, M., 598
Heller, W. R., 35
Henneke, E. G., 902
Herakovich, C. T., 363
Hercher, M., 578
Herrmann, G., 921, 964
Hetényi, M., 217
Hiatt, G., 363, 965
Higashi, T., 1027
Highsmith, A. L., 360
Hildebrand, B. P., 365, 404, 825
Hill, R., 34
Hirschberg, M. H., 577
Hirsh, A. E. 260, 265
Ho, B. L., 748
Hoare, F. E., 826
Hochstein, P. A., 572, 578
Hocker, G. B., 68, 76
Hodges, P. G., 36
Hodulak, L., 782
Hofer, K. E., 525, 901
Hogg, D., 77
Hohenemser, K., 35
Holister, A. S., 262
Holloway, D. C., 263, 405
Holman, J. P., 1054
Hondo, M., 265
Honeycutt, E. H., 265
Horger, O. J., 823
Horiguchi, T., 77
Houi, A., 749

Hovanesian, J. D., 264
Hsu, D. K., 828
Hsu, N. N., 828
Hu, C. P., 405
Hu, W. L., 964
Hu, X., 162
Huang, T. C., 823
Huang, Y. M., 590, 598, 1029
Hudson, L. D., 559, 576
Huetter, U., 902
Hughes, W. F., 262
Hull, S., 815, 889
Hulse, C. O., 576
Hung, Y. Y., 366, 405
Hunter, A. R., 264
Hunter, J. B., 964
Huntley, H. E., 634
Hutchinson, J. W., 475, 936, 962
Hutley, M. C., 321, 364
Hyer, M. W., 363

Ibanez, P., 745
Ibrahim, S. R., 748
Idesawa, M., 970, 1027
Ilic, D. B., 826
Ipsen, D. C., 634
Ireland, D. R., 475
Irons, B., 750
Irons, B. M., 37
Irwin, G. R., 905, 913, 928, 938, 962, 963, 965
Ishil, A., 1029
Issacson, E. de St. Q., 634
Issacson, M. de St. Q., 634
Itagaki, H., 962
Ito, K., 264, 265
Iwashimizu, Y., 826

Jackson, C. M., 577
Jackson, D. A., 77
Jackson, H. S., 555
Jacobs, K. A., 754, 781
Jacobson, A. D., 404, 525
Jacobson, R. H., 525
Jaisingh, G., 405
James, M. R., 827
Javornicky, J., 264
Jessop, H. T., 262
Joakimides, N., 474
Joh, D., 360, 363
Johnson, G. C., 826, 964
Johnson, R. L., 264
Johnson, W., 36
Jolles, M., 966

Jolly, D., 555
Jolly, W. D., 555
Jones, E., 163
Jones, M. H., 964
Jones, M. W., 966
Jones, R., 598, 599, 830, 899
Jordan, E. H., 571, 572, 578
Jorgenson, W. E., 578
Jou, J. Y., 783
Juang, J. N., 748

Kabe, A. M., 746
Kachanov, A., 37
Kageyama, K., 598
Kak, A. C., 1026
Kalb, H. T., 163
Kalman, R. E., 748
Kalthoff, J. F., 407, 474, 475, 476, 767, 934, 965, 966
Kaminski, B. E., 966
Kanazawa, T., 962
Kanda, H., 747, 749
Kang, B. S. J., 266, 965
Kanninen, M. F., 766, 782
Kao, T. Y., 295
Kapp, D. A., 77
Kardestuncer, H., 34
Kassir, M. K., 962
Kata, A., 1029
Katsamanis, F., 967
Katz, J., 490
Kaufman, J. G., 967
Kaufmann, G. H. 1028, 1029
Kawasaki, A., 1028
Ke, J. S., 965
Kelly, L. G., 749
Kelsey, R. A., 771, 792, 795, 796, 824
Kendall, D. P., 964
Kersey, A. D., 77
Keusseyan, R. L., 571, 578
Keys, R. J., 597
Khetan, R. P., 405
Kikuchi, M., 598
Kikuta, T., 555
Kim, B. Y., 77
King, R. B., 921, 964
Kino, G. S., 827, 952
Kirby, G. C., 265
Kirsch, G., 787, 788, 790, 791, 807, 824
Kistler, W. P., 163
Kleiber, M., 35
Klemm, W., 475, 967
Kline, S. J., 634

Klosterman, A., 746
Knight, C. E., 900
Knowles, J. K., 917, 964
Kobayashi, A. S., 1, 35, 36, 261, 266, 295, 525, 579, 751, 782, 783, 905, 929, 963, 964, 965, 967
Kobayashi, H., 963
Kobayashi, T., 783, 903, 967
Koenig, L. N., 264
Koistinen, P., 827
Koller, G. M., 265, 900, 901
Kolsky, H., 825
Komai, K., 1029
Konish, H. J., 902
Koo, K. P., 77
Kopnov, V. A., 900
Koufopoulos, T., 900
Kowalski, H. C., 264
Krasnowski, B. R., 966
Krauss, H., 902
Krautkrammer, J., 555
Krawitz, A. D., 827
Krenevez, J. P., 747
Krishnamurthy, H., 296
Krishnamurthy, R. S., 1028
Kromer, F. J., 800, 824
Kubomura, K., 826
Kugawinska, M., 1029
Kujawinsak, 997, 1028
Kundrat, J., 747
Kurishima, T., 77
Kuske, A., 276

Labedz, R. H., 901
Laerman, K. H., 753, 781
LaGarde, A., 245, 262, 263, 966
Lake, B. R., 824
Lallement, G., 746
Lally, M., 746
Lally, N., 746
Landes, J. D., 963, 968, 1054
Langhaar, H. L., 602, 634
Langman, R. W., 748
Lanius, S. J., 565, 577
Lant, C. T., 575, 579
Lawrence, C., 782
Layton, M. R., 72, 77
Leaity, G. P., 599
Lee, C. E., 70, 77
Lee, F., 825
Lee, G. H., 781
Lee, G. M., 749
Lee, H., 555
Lee, O. S., 966

Leeman, E. R., 825
Leendertz, J. A., 366, 404, 490
Lei, J. F., 561, 577
Leisure, R. G., 828
Leith, E. N., 364, 365, 367, 404
Lekhnitskii, S. G., 905, 962
Leleux, G., 747
Lembregts, F., 746, 747, 749
Lemcoe, M. M., 561, 576, 578
Lenk J. D., 117
Lentner, H. R., 828
Leon, S., 364
Leon, Z., 787, 824
Lesniak, J. R., 590, 598
Letner, H. R., 827
Leuridan, J., 749
Leuridan, R. J., 747, 749
Leven, M. M., 234, 236, 262
Lewis, R. C., 745
Li, C. Y., 571, 578, 602
Li, Q., 77
Liaw, B. M., 783
Libby, H. L., 555
Liber, T., 312, 901, 902
Liebowitz, H., 35, 966
Lindholm, U. S., 571, 578
Link, M., 749
Litenberg, F. K., 288, 296
Liu, H. C., 1028
Liu, H. W., 964, 965
Liu, K., 76
Liu, K. C., 579
Livnat, A., 364
Lloyd, B. E., 598
Lloyd, W. R., 966
Lo, K. H., 871, 902
Loader, A. J., 598
Lohr, D., 598
Loreck, R., 264
Loss, F. J., 783
Love, A. E., 35
Loveday, M. S., 577
Lu, H., 966
Lu, W. Y., 827
Lubowinski, S., 360, 363
Lunau, K., 1028
Luo, S., 748
Lynch, J. J., 824
Lynn, P. P., 36

Ma, C. C., 474
Ma, L. C., 576
Macagno, E. O., 602, 634

Machida, S., 783
Mack, R., 77
MacKinnon, L., 826
Maddux, G. E., 386
Madhuka, M., 903
Magrab, E. B., 67, 163
Malcolm, R. C., 967
Mall, S., 767, 967
Mallik, D., 263, 265
Malmo, J. T., 578
Malvern, L. E., 36
Mamaros, T. C., 825
Manning, B. W., 802, 803, 825
Manogg, P., 407, 474, 933, 965
Marburger, R. E., 827
Marchant, A. C., 323, 362
Marchant, M. J., 362
Marion, R. H., 573, 578
Mark, R., 261, 266
Markus, A. M., 77
Marloff, R. H., 900
Marom, E., 490
Marschall, C. W., 782
Maruyama, A., 1029
Mason, W. P., 40, 64, 76
Mathar, J., 824
Matsui, S., 321, 341, 364
Matsumoto, E., 263, 264
Matthys, D. R., 1028
Matzkanin, G. A., 828
Maurey, D. A., 967
Mawatari, S., 1027
Mazurkiewics, S. B., 262, 966
McCalvey, L. F., 76
McClintock, F. A., 962
McClung, F. J., 366, 404
McConnell, K. G., 79, 119, 163
McDonach, A., 362, 363, 828
McDonald, B. A., 827
McEwen, W. K., 750, 761, 782
McGaw, M. A., 577
McGowan, J. J., 782, 966
McGrew, J., 749
McKannan, E. C., 825
McKeever, B., 746
McKelvie, J., 322, 362, 363, 598, 828
McKenzie, A. K., 598
McKenzie, P., 363, 828
McLachlan, K., 824
McLaughlin, T. F., 362
McMaster, R. C., 554, 555, 586
McSkimin, H. J., 812, 825
Measures, R. M., 76, 77

Medlock, R. S., 76
Melcher, D. M., 827
Mellor, P. B., 36
Meltz, G., 77
Mendelson, A., 37
Merhib, C. P., 555
Meriwether, H. D., 827
Merkle, J. G., 964
Merrill, R. H., 825
Mesmer, G., 771, 824
Metherell, A. F., 555
Meyer, R., 577
Migliore, H., 828
Mignogna, R. B., 964
Mikullcik, E. C., 748
Miller, J. A., 825
Miller, M. S., 77
Miner, R. V., Jr., 578
Minoshima, K., 1029
Mirimand, N., 747
Mirza, M. S., 634
Mishler, H. W., 828
Miskioglu, R. G., 263, 264
Mitchell, L. D., 745, 747
Miyoshi, T., 1029
Monch, E., 264
Mooney, M., 34
Mordakacherry, J. M., 599
Morey, W. W., 77
Morgan, T. A., 825
Morimoto, Y., 295, 969, 970, 1007, 1026, 1027
Moriwaki, K., 364
Morosow, G., 745
Morris, D. H., 264, 902
Morse, S., 295
Moslehy, F. A., 756, 781
Mostovoy, S., 965
Mott, N. R., 906, 962
Mountain, D. S., 582, 597, 599
Mroz, Z., 35
Mueller, R. K., 248, 262, 970
Mulc, A., 264
Mullen, S. J., 554
Muller, R. K., 1026
Mullineaux, J. L., 901
Mulot, M. M., 296
Murakawa, H., 36
Murphy, D. P., 900
Murphy, G., 634
Murphy, K. A., 76, 77
Murphy, M., 599
Murty, M. V. R. K., 364
Mushkelishvili, N. I., 17, 35, 905, 962

AUTHOR INDEX

Muthart, R. E., 554
Mutoh, K., 1007, 1028
Myers, A., 826

Nagasaki, T., 1029
Nagy, A., 902
Naka, H., 827
Nakadate, S., 1028
Nakanishi, A., 1029
Nakanishi, T., 1029
Nakao, H., 1028
Nakazawa, M., 827
Namba, S., 364
Natke, H. G., 745
Navone, H. D., 1029
Nemat-Nasser, S., 35
Nemeth, M. P., 363
Newman, J. C., Jr., 952, 967
Newwar, A. M., 803, 824, 825
Nichola, W. E., 827
Nichols, R. W., 963
Nicoletto, G., 360, 362, 363, 965, 966
Nightingale, C. M., 564, 577
Niiro, T., 265, 900, 901
Nishimura, K., 1029
Nishioka, T., 35, 36, 782
Nisida, M., 265, 824, 828
Noltingk, B. E., 577
Noor, A. K., 37
Noronda, P. J., 826
Norrie, D., 34
Norris, E. B., 577
Nuismer, R. J., 871, 902
Nurse, P., 248, 262
Nyquist, G. N., 782

O'Brien, K. T., 903
O'Regan, R., 932, 965
Obata, M., 1028, 1029
Obert, L., 825
Ochi, Y., 1029
Ohashi, Y., 264
Ohji, K., 963
Okada, K., 826, 964
Okimura, H., 577
Okubo, S., 525
Olaosebikan, O., 265, 963, 965, 966
Oliver, D. E., 597
Oplinger, D. W., 295
Oppel, G., 938, 966
Oreb, B. F., 1028
Orowan, E., 905, 961
Osias, J. R., 782

Oster, G., 295
Otsuka, A., 963

Packman, P. F., 965
Pagano, N. J., 901
Page, S. W. J., 598
Palebut, S., 966
Pandit, S. M., 748
Pappa, R. S., 745, 748
Paris, P. C., 475, 913, 962, 964
Park, Y. S., 163
Parks, J. S., 261, 265
Parks, V. J., 267, 295, 360, 361, 386, 405, 525
Patterson, E. A., 1027
Pease, D. M., 578
Peebles, F. N., 260, 265
Peiffer, J., 578
Pendered, J. W., 746
Pennington, D., 163
Perkins, E. S., 782
Perl, M., 782
Perry, C. C., 39, 76, 825
Perzyna, P., 12, 35
Peters, W. H., 102, 265, 365, 405, 782, 966, 1029
Peterson, E. L., 747
Petit, P. H., 901
Phillips, E. A., 262, 525, 1054
Pierce, E. T., 163
Pih, H., 265, 579, 901
Pilkey, W. D., 34, 37
Pindera, J. T., 245, 262, 966
Pipes, R. B., 899, 901, 902
Piranda, J. L., 746
Pirodda, L., 292, 296
Podleschny, R., 475
Polzin, M. H., 828
Pommiesr, M., 782
Popelar, C. H., 782
Porter, E., 903
Post, D., 262, 295, 297, 323, 362, 363, 901, 965, 967
Potter, R. T., 582, 597
Potter, R. W., 746
Powell, R. L., 366, 404
Prabhakaran, R., 265, 900
Prados, J. W., 265
Prager, W., 34, 36, 37
Pratt, W. K., 1026
Prevey, P. S., 827
Priest, R. G., 76
Prince, J. M., 555
Pritty, D., 362
Procter, E., 825

Prony, R., 747
Prosser, J. C., 561, 562, 576

Quinlan, P. M., 36
Quinn, R. A., 555
Quinney, H., 34

Rades, M., 749
Radke, R., 578
Raelson, V. J., 826, 964
Raftopulos, D., 967
Ramalu, M., 266, 783, 964, 965, 967
Ranganayakamma, B., 296
Ranson, W. F., 262, 263, 365, 405, 756, 782, 1029
Rao, G. V., 754, 781
Rao, M. P. K., 574, 578, 964
Rao, P. N., 901
Rao, S. S., 37
Rashleigh, S. C., 76, 77
Rastogi, P. K., 1028
Ravi-Chandar, K., 967
Rayleigh, L., 602, 626, 634
Raymondo, P., 561, 576
Redner, S., 248, 262, 822, 825, 827
Reed, K. W., 36
Reemsnyder, H. S., 823
Reifsnider, K. L., 596, 598, 903
Rendler, N. J., 795, 796, 798, 802, 824
Reuter, W. G., 966
Rice, J. R., 475, 917, 936, 962, 963, 964
Richard, T. G., 1027
Rider, W., 825
Riley, W. F., 39, 76, 79, 117, 254, 262, 263, 264, 294, 295, 297, 361, 781
Riparbelli, C., 775, 790, 791, 824
Ripling, E. J., 965
Ritter, R., 296
Robert, A., 825
Roberts, D. K., 906, 962
Robertson, G., 262
Robinson, D. W., 1028
Rocca, R., 582, 597
Rocklin, T., 749
Rodgerd, E. H., 555
Rogers, G. L., 262
Rogers, S. A., 599
Rohsenow, W. M., 634
Rolfe, S. T., 962
Rollins, F. R., 826
Rosakis, A. J., 407, 474, 936, 966, 967
Rose, P. G., 295, 901
Rosen, B. W., 831, 900, 901
Rosenfeld, A., 1026

Rosenfield, A. R., 782
Rosengren, G. F., 475, 936, 962
Ross, F. B., 554
Rossmanith, H. P., 263, 965, 967
Rost, R. W., 747
Rost, R., 746, 747, 749
Rotem, A., 903
Rowlands, R. E., 295, 322, 362, 590, 598, 785, 825, 826, 889, 900, 901, 902, 1027, 1029
Rowlands, S. N., 555
Ruiz, C., 363
Rummel, W. D., 554
Russ, S. M., 577
Russell, A. D., 36
Russell, S. S., 579
Ruud, C. O., 822, 827, 828
Ryall, T. G., 593, 599
Rybicki, E. F., 782, 828

Saache, L., 248, 262
Saackel, L. R., 1026
Sablik, M. J., 555
Sachse, W., 814, 826, 964
Saito, H., 365, 404, 1028
Sakurai, T., 35
Salerno, V. L., 824
Sampson, R. C., 900
Sandhu, R. S., 901
Sandifer, J. P. 801, 825
Sandor, B. I., 583, 597, 599
Sanford, R. J. 261, 263, 265, 965, 1054
Sas, D., 746
Sasaki, S. K., 1029
Savin G. N., 17, 35
Schajer, G. S., 802, 803, 825
Schardin, H., 902
Schijve, J., 958, 967
Schlammarella, C. A., 970, 1007, 1027
Schlegal, W. A., 782
Schmale, D. T., 570, 578
Schmank, M. J., 827
Schmueser, D. W., 828
Schramm, S. W., 902, 903
Schueler, C. F., 555
Schuring, D. J., 634
Sciammarella, C. A., 295, 361, 578, 964
Scott, M. P., 826
Sedov, L. I., 37, 634
Seguchi, Y., 970, 1026, 1027
Seidelmann, W., 475
Selfridge, A. R., 964
Sendeckyj, G. P., 903
Seo, L., 783

AUTHOR INDEX

Shaffer, D. H., 1054
Shahbender, R. A., 827
Shao, Y., 578
Shapery, R. A., 900
Sharp, J. J., 634
Sharpe, D. R., 748
Sharpe, W. N., Jr., 557, 576, 579, 965
Shewchuk, J., 824
Shih, C. F., 766, 782, 963
Shih, C. Y., 746
Shimada, H., 1028, 1029
Shin, S., 364
Shinohara, A., 827
Shirai, Y., 1028
Shiraishi, T., 1027
Shiratori, M., 1029
Shukla, A., 362, 385, 783, 967
Shumely, M., 782
Sigel, G. H., Jr., 76
Sigl, C. C., 555
Sih, G. C., 35, 475, 962, 966
Sih, G. C., 599
Simmons, E. E., Jr., 43, 76
Simon, B., 963
Sims, D. L., 578
Sinclair, J. H., 901
Singh, A., 263
Singleton, R. C., 1029
Sirkis, J. S., 76, 1028
Sizemore, K. S., 163
Skingle, C. W., 745
Sladek, J., 782
Sloane, E., 746
Smith, C. S., 40, 64, 76
Smith, C. W., 264, 265, 905, 963, 965
Smith, D. G., 965
Smith, E. R., 163
Smith, F. W., 962
Smith, J. E., 564, 569, 577
Smith, R. A., 599
Smith, R. T., 826
Sneddon, I., 905, 962
Snedecor, G. W., 1054
Snitzer, E., 77
Snoeys, R., 746, 747
Soete, W., 787, 788, 796, 824
Solid, J. E., 365, 371, 404, 427
Soltész, 475
Soma, T., 1027
Sommargren, G. E., 489
Sommer, E., 965
Soneda, N., 1028
Sorenson, J. E., 577

Sowers, J. D., 750
Sparrow, J. G., 599
Spencer, G. H., 364
Srawley, J. R., 938, 954, 964, 966
St. Helen, R., 782
Stahle, C. V., Jr., 734, 746, 747
Stanely, P., 590, 598, 599
Stange, W. A., 576
Stanley, P., 598, 599
Stehlin, P., 76
Steindler, R., 577
Sternberg, E., 917, 964
Stetson, K. A., 362, 366, 386, 404, 405, 477, 490
Stinchcomb, W. W., 598, 903
Stokey, W. F., 262, 525
Stonesifer, R. W., 782, 828
Stowell, W. R., 577
Stroke, G. W., 365, 404
Sturgion, D. L., 970, 1007, 1027
Suese, N., 1027
Suga, S., 364
Sukere, A. A., 925, 965
Sullivan, T. L., 902
Sumner, G., 571, 578
Sun, Y. J., 966
Sutera, S. P., 265
Sutton, M. A., 365, 405, 579, 1029
Suzuki, H., 748
Swanson, G., 364
Swanson, W. M., 265
Swedlow, J. L., 965
Sweryn, A. J., 746
Swinson, W. F., 262
Swinson, W., 1029
Sys, W. M., 295
Szafranski, F. J., 565, 577
Szucs, G. M., 634

Tada, H., 77, 964
Takabayashi, H., 824
Takahasi, S., 1027
Takasaki, H., 296
Takashi, M., 1027
Takeda, M., 1007, 1028
Tall, L., 823
Tamaki, T., 1027
Tapanes, E., 76
Targoff, W. P., 749
Tateda, M., 77
Taylar, C. E., 1029
Taylor, C. E., 76, 405
Taylor, E. S., 634
Taylor, G, I., 34

Taylor, G. A., 745
Taylor, H. F., 77
Tebedge, N., 823
Teller, C. M., 828
Templeton, D. W., 1029
Ten Gate, W., 791, 824
Tenney, D. R., 362
Teramoto, T., 783
Thalmann, R., 490
Theiault, J. P., 77
Theocaris, P. S., 265, 295, 361, 407, 475, 900, 934, 965, 967
Thomas, T. Y., 37
Thompson, G. L., 824
Thomson, R. A., 264
Thomson, W., (Lord Kelvin), 35, 39, 76, 581, 597
Throop, J. F., 823
Thurston, R. N., 40, 64, 76
Toda, H., 827
Tokarcik, A. G., 828
Tominaga, M., 1029
Tomita, Y., 1026
Tomlinson, G. R., 749
Toyoda, Y., 1027
Toyooka, S., 1029
Trail-Nash, R. W., 664, 745, 903
Tretheway, M. W., 749
Truell, R., 826
Truesdell, C., 34
Tsai, S. W., 830, 900
Tsao, C. H., 525, 1054
Turner, J. L., 579
Turner, J., 1029
Turner, R. D., 76
Turner, W. B., 598
Tuttle, M. E., 1028
Tventen, A. B., 77
Twa, G. J., 564, 577

Udd, E., 77
Ueda, T., 970, 1027
Ueki, K., 598
Ugo, R., 1028
Umeagukwu, I., 766
Umezaki, E., 1027
Underwood, J. H., 964
Upatnieks, J., 365, 367, 404
Urabe, Y., 783

Valanis, K. C., 35
Valis, T., 76, 77
Vallem, J. H., 322, 362
Van der Auweraer, H., 746, 749

Van Karsen, C., 747
Van Loon, P., 746
van der Werft, R., 782
Vancrombrugge, R., 787, 791, 796, 824
Vanderby, R., 889
Vanherck, P., 746
Vary, A., 903
Vaughan, W., 261
Vensarkar, A., 76
Vick, L. W., 598
Vigness, I., 795, 796, 798, 802, 824
Vijayakumar, K., 35
Vinckier, A. G., 295
Vinh, T., 749
Vold, H., 745, 746, 747, 749
Volian, A., 749
Voloshin, A. S., 263, 265, 266, 901, 1026, 1027
Voorhes, D., 578

Wada, B. K., 746
Waddoups, M. E., 871, 901, 902
Wade, C. A., 77
Wade, G., 555
Wadsworth, N., 362
Waidelich, D. L., 555
Walker, C. A., 322
Walker, H., 362, 363, 828, 963
Walle, L. J. V., 824
Walters, D. J., 578
Wang, W. C., 1027
Want, K. Y., 555
Ward, M., 579
Waslowski, L. M., 1028
Wasowski, J. J., 1028
Wasserman, M., 295
Watanabe, M., 1026
Watanabe, O., 35
Watson, R. B., 362
Wayland, H., 265
Weber, W., 581, 597
Weed, D. N., 902
Weibull, W., 1054
Weise, R. A., 561, 562, 576
Weissman, E. M., 362, 363
Weitsman, Y., 904
Wells, A. A., 905, 906, 915
Weng, A. S., 598, 599
Wert, J. J., 826
Westergaard, H. M., 905, 962
Westline, P. S., 634, 934
White, R. N., 76, 634
Whitefield, J. K., 264
Whiteside, J. B., 902

AUTHOR INDEX

Whitney, J. M., 830, 871
Wilkinson, D. S., 968
Willemin, J. F., 490
Willheim, A., 824
Williams, D. R., 579
Williams, F. T., 825
Williams, J. R., 598
Williams, M. L., 475, 905, 962
Williams, R., 746, 749
Wilson, W. K., 782
Winkler, S., 474, 475, 476
Wintz, P., 1026
Witt, F., 825
Wnuk, S. P., 577
Wolters, W. J., 405
Woo, S., 761, 763, 782
Wood, J. D., 363
Wrisley, D. L., 663, 745
Wu, E. M., 871, 901, 902
Wu, T. T., 561, 576
Wuerker, R. F., 404

Yagawa, G., 1028
Yalin, M. S., 634
Yamada, Y., 35
Yamaguchi, I., 404, 575, 579, 1029
Yamaguchi, N., 1027
Yamane, T., 826
Yamashita, M., 1028
Yang, K. H., 579
Yatagal, T., 1027, 1028
Yavelak, J. J., 828

Yazdi, A. R., 571, 578
Yeakley, L. M., 571, 578, 903
Yee, B. G. W., 888
Yeh, C., 77
Yeow, Y. T., 902
Yih, H. R., 598
Yoshikawa, N., 77
Yoshiki, M., 962
Yoshimura, N., 35
Yoshimura, S., 1028
Yoshinari, H., 783
Young, D. F., 601
Young, H. D., 1054

Zachary, L. W., 264
Zaehna, P. W., 1054
Zakhoor, A., 964
Zandeman, F., 262
Zawada, L. P., 577
Zener, C., 597
Zhang, D., 598, 599
Zhang, L., 747
Zhang, P., 263, 264
Zhao, L., 576
Zhao, L. B., 576
Zheng, H., 579
Ziegler, H., 35
Zienkiewick, O. C., 36, 37
Zimmerman, B. D., 77
Zimmerman, R., 746, 747
Zwerling, C., 295

Subject Index

Acoustic impedance, 575
Acoustoelasticity, 797
Amplifier, 79
  built-in, 144
  charge, 142
  differential, 84
  differentiating, 85
  integrating, 85
  inverting, 83
  noninverting, 83
  operational, 81
  summing, 85
Amplitude modulation, 90
Araldite B, 417

Bauschinger effect, 7
Birefringence, 177
  flow, 258
Birefringent coating, 245
Boundary element method, 25, 756
Brewster's angle, 172
Brittle coating:
  analysis, 496
  ceramic, 518
  creep correction, 519
  failure chart, 500
  law of failure, 494
  principal stress, 493, 509
  refrigeration, 509
  relaxation cracking, 499

  resin-based, 516
  sensitivity, 517
  state of stress, 491

Calcite, 179
Calibration:
  charge amplifier system, 158
  high-input-impedance amplifier system, 158
  photoelastic model material, 218
  strain gage, 48
  transducer, 152, 688
Capacitance gage, 565
Caustics:
  constants, 417
  double, 423
  dynamic, 434
  elastic-plastic, 441
  higher-order effects, 438
  mapping equations, 416, 427
  model materials, 451
  physical principle, 407
  reflection, 409
  single, 423
  stress concentration, 410, 420
  stress intensity factor determination, 428
  transmission, 409
Celluloid, 218
Chain gage, 44
Columbia resin, CR39, 216, 440
Compatibility equations, 754

Compensator, 211
  Soleil-Babinet, 190, 211
  Tardy, 208
Compliance method, 919
Composites:
  acoustic emission, 895
  biaxial testing, 865
  fiber volume ratio, 856
  mechanical properties, 862
  mechanics, 830
  micromechanics, 830
  multidirectional, 847
  unidirectional, 831
Constant-current excitation, 96, 102
Constant-current source, 92
Constant-voltage excitation, 94
Constant-voltage source, 91
Converter, 89, 90
Corneoscleral envelope, 757
Crack:
  Dugdale strip yield zone, 915, 951
  elliptical, 940
  part circular, 940
Crack arrest, 460
Crack propagation:
  dynamics, 460, 463, 952
Crack-tip deformation:
  mixed-mode, 432, 931, 942
  mode I (opening mode), 425, 907
  mode II (sliding mode), 425, 907
  mode III (tearing mode), 425, 907
Cranz-Schardin camera, 452
Creep, 10
  compliance, 6
  function, 31

Deformation:
  infinitesimal, 1, 616
Digital image processing:
  algorithm, 981
  aliasing, 971
  filtering, 988
  Fourier transform, 1007
  geometric transform, 993
  gray value, 982
  holographic interferometry, 1012
  moiré and grid method, 1008, 1018
  sampling and quantitizing, 971
  shape measurement, 1013
Digital signal processing, 672
  digitization equations, 674
Displacement:
  crack-tip, 18, 425

Displacement transducer:
  seismic, 135
  variable-inductance, 128
Double refraction, 177

Eddy-current:
  physical principle of operation, 533
  skin depth, 536
Elasticity, 2
Epoxy, 216
Excitation signals, 709
Extensometer:
  electromechanical, 570
  electro-optical, 572

Filter, 86
  high-pass RC, 86
  low-pass RC, 87
  second-order, 89, 123
Finite-difference method, 21, 754, 758
Finite element method:
  dynamic, 766
  elastic-plastic, 766
  theoretical background, 22
  three-dimensional, 765
  two-dimensional, 768
Flying spot scanner, 763
Fourier transform:
  discrete, 676
  fast, 393
  image processing, 1007
Fracture mechanics:
  COA, 771
  COD, 914
  CTOA, 951
  CTOD, 914
  elastic-plastic, 763, 913
  linear elastic, 906
  stable crack growth, 952
  strip yield model, 915
Fracture specimen:
  compact, 768
  double-cantilever beam specimen, 462, 778
  single-edge-notch, 440
Frequency-response function, 643, 665, 687
Fresnel's ellipsoid, 182
Fringe-multiplication, 204

Geometric moiré method:
  calculus approach, 276
  geometric approach, 274
  high-temperature, 573
  in-plane, 267

mismatch, 281
out-of-plane, 286
shadow, 289
super-moiré analysis, 284
Glan Taylor prism, 181
Glass, 217, 440, 778
Graphite-epoxy, unidirectional composite, 862
Grating:
 auxiliary specimen, 326
 diffraction, 309
 geometric moiré, 269
 specimen, 320
 virtual, 312

Hardening:
 isotropic, 8
 kinematic, 8
Hologram:
 double-exposed, 367
 dynamic, 385
 heterodyne, 479
Holographic interferometry:
 double-exposure, 379
 fringe processing, 1008
Holography, time-average, 383
Homalite, 216, 440, 772, 952
HRR field, 936
Hydrostatic pressure, 5

Ibrahim time-domain approach, 668
Image processing:
 analog, 970
 digital, 970
Impact loading, 456, 467, 619
Impulse-response function, 641
Index:
 ellipsoid, 182
 refraction, 180
Interference:
 fringe order, 306
 impure two-beam, 302
 pure two-beam, 501
Interferometry:
 Mach-Zender, 477
 strain-displacement gage, 574
 stress intensity factor determination, 927, 933
Isochromatics, 195
 color matching, 213
 crack-tip, 472
 matrix shrinkage, 835
 multiplication, 205
Isoclinics, 197
Isoelastic alloy, 43

Isostatics, 221
Isothetics, 285

J integral, 916, 937

Karma alloy, 42
KRAK-GAGE, 771

Lame's parameter, 4
Laplace transform, 644
Least square method, 693
Light source wavelength, 204
Liquid penetrant inspection, 530

Methods for magnetizing materials, 532
Modal analysis:
 experimental, 635, 659
 parameter estimation, 711
 models, 715
 theory, 638
Moiré fringe:
 carrier pattern, 340
 counting, 329
 image processing, 1008
 multiplication, 286, 358
 optical filtering, 340
 recording, 270
 reflection, 286
 rigid-body rotation, 326
Moiré interferometry:
 achromatic system, 335
 fundamentals, 298, 315
 mold, 320
 optical systems, 323, 333
 out-of-plane motion, 338
 replication, 322
 stress intensity factor determination, 924
Moiré method, composite, 883
Multiplexing:
 frequency-division, 111
 time-division, 111

Noise, electric, 111
Nondestructive evaluation:
 eddy-current inspection, 533
 holography, 380
 magnetic inspection, 532
 penetrant inspection, 527
 radiographic inspection, 539
 ultrasonic inspection, 544
Normal-mode forced excitation, 663

Oblique-incidence method, 224

Optical-fiber strain sensor, 57
Optical heterodyning:
  basic technology, 478
  concomitant fringe pattern, 482
  phase meter, 478
  scintillation noise, 479
  stored fringe pattern, 479

Photoelastic coating, 247
  composite, 850
Photoelasticity:
  composite materials, 850
  computer-aided, 248
  dynamic, 252, 766
  isodyne, 948
  model materials, 216
  orthotropic, 260
  scattered light, 239
  stress freezing, 334, 754, 938
  stress intensity factor determination, 928
  theory, 165
  three-dimensional, 333, 938
  two-dimensional, 219, 928
Photography, 209
Photoplasticity, 258
Photothermoelasticity, 256
Plane strain, 5
Plane stress, 5
Plasticity, 7
Plates:
  half-wave, 187
  quarter-wave, 187
  retardation, 183
  wave, 183
Point load solution, 15
Poisson's ratio, 4, 618
Polariscope:
  circular, 194
  collimated-beam, 203
  color matching, 213
  diffused light, 202
  errors, 200
  first invariant, 206
  light source, 204
  plane, 193
  scattered-light, 244
  white light, 215
Polarization:
  fine grids, 175
  reflection, 169
  scattering, 173
Polarized light:
  circularly, 167
  elliptically, 166
  linearly (plane), 166
Polarizer:
  crossed, 176
  dichoric, 176
  parallel, 176
  wire-grid, 176
Polaroid sheet, 176
Polycarbonate, 216
Polymethyl methacrylate, 217, 440
Polyurethane, 218
Potentiometer:
  circuit, 94
  variable resistance, 126
Pressure transducer, 138

Radiography:
  composite, 891
  exposure equivalent factor, 541
  half-life, 542
RC differentiator, 122
RC integrator, 121
Refraction index, 177
Relaxation:
  function, 31
  modulus, 6
  tensor, 6
Residual stress, 785
Residual stress measurement:
  Bathgate's method, 796
  Bert's method, 798
  Boiten and Ten Gate's method, 791
  Cordiano and Salerno's method, 796
  hole drilling method, 786
  Kelsey's method, 791
  Mather's method, 787
  Nishida and Takabayashi's method, 793
  numerical analysis, 801
  Rendler and Vigness' method, 795
  Riparbelli's method, 790
  Shewchuk's method, 800
  Soete and Vancrombrugge's method, 787
  Rock mechanics, hole boring, 806

Sensor:
  heterodyne optical strain, 488
  microswitch, 130
  photovoltaic, 130
  piezoelectric, 140
  velocity, 131
Shadow optical method:
  arrangement, 448
  image, 412

SUBJECT INDEX

instrumentation, 450
light distribution, 421
Similarity relation, 257
Similitude:
  differential equations, 629
  dimensions, 603
  fluid-flow variables, 622
  heat transfer, 625
  models, 611, 627
  pi term, 610
  theory, 624
Signal conditioning, 79
Singular point, 220
Slip ring, 110
Snell's law, 169
SPATE, 581
  theory, 582
Speckle interferometry, 756
  heterodyne, 484
  high-temperature, 605
  shearing, 375, 388
  speckle metrology, 392
  speckle pattern digital image, 394
Speckle photography:
  heterodyning, 482
  image processing, 393
  laser, 390
  single-beam, 372
Stable crack growth, 952
Statistical analysis, 1031
  chi-square analysis, 1051
  conditioning of data, 1044
  least-square method, 1048
  multivariate regression, 1047
  regression analysis, 1044
  safety factor, 1043
Statistical distribution:
  central tendency, 1033
  confidence intervals, 1040
  dispersion, 1033
  Gaussian, 1035
  means, 1042
  Weibull, 1037
Strain:
  Green-Lagrange, 1, 16
  mechanical differentiation, 341
  sensitivity, 40
Strain energy density, 5
Strain energy density criterion, 911
Strain-energy release rate, 910
Strain gage:
  adhesive, 46
  Boeing, 566

Bragg grating, 74
bonded resistance, 43
capacitance, 565
carrier, 44
CERL-Planner, 566
composite, 852
electric resistance, 40
foil, 40, 45
high-temperature, 557
optical fiber, 67
polarimetric, 71
resistance, 48, 559
semiconductor, 64
transverse sensitivity, 50
twin-core optical fiber, 73
wire, 560
weldable, 46
Strain-gage instrumentation, 79
Strain measurement:
  dynamic, 561
  static, 563
Stress:
  Cauchy, 2
  crack-tip, 17, 424, 925
  deviator, 7
  gage, 47
  Kirchhoff, 2
  Piola-Kirchhoff, 2
  principal, 221
  trajectory, 222
  yield, 7
Stress analysis, hybrid
  experimental-numerical, 751
Stress concentration factor, composite, 87
Stress intensity factor:
  critical, 910
  dynamic, 19, 430, 772
  static, 18, 420, 893
Stress-optic law, 190
Switching:
  three-pole, 108
  single-pole, 108

Telemetry, 111
Thermoelasticity, 581
Thermography:
  composite, 886
  SPATE, 581
Three-point bend specimen, 457
Transducer mounting method, 687

Ultrasonic inspection:
  composite, 891

surface-wave technique, 809
Ultrasonics:
  acoustic impedance, 545
  basic principle, 544
  critical angle, 546
  new technology, 552

Voltage follower, 82, 140
Viscoelasticity, 6
Viscoplasticity, 10

Wavefront, warped, 324
Wheatstone bridge, 97

X-ray technique, 540, 891

Yield condition:
  Tresca, 7
  von Mises, 7
Young's modulus, 4